Near-Capacity Multi-Functional MIMO Systems

Near-Capacity Multi-Functional MIMO Systems

Sphere-Packing, Iterative Detection and Cooperation

Lajos Hanzo, Osamah Alamri, Mohammed El-Hajjar and Nan Wu

All of
University of Southampton, UK

IEEE

WILEY

A John Wiley and Sons, Ltd, Publication

This edition first published 2009

Registered office
John Wiley & Sons Ltd, The Atrium, Southern Gate, Chichester, West Sussex, PO19 8SQ, United Kingdom

For details of our global editorial offices, for customer services and for information about how to apply for permission to reuse the copyright material in this book please see our website at www.wiley.com.

Library of Congress Cataloging-in-Publication Data

Hanzo, Lajos.
Near-capacity multi functional MIMO systems : sphere-packing, iterative detection, and cooperation / Lajos Hanzo ... [et al.].
p. cm.
Includes bibliographical references and index.
ISBN 978-0-470-77965-1 (cloth)
1. MIMO systems. I. Hanzo, Lajos, 1952-
TK5103.2.N43 2009
621.384–dc22

2008049883

A catalogue record for this book is available from the British Library.

ISBN 9780470779651 (H/B)

Set in 10/12pt Times by Sunrise Setting Ltd, Torquay, UK.
Printed in Singapore by Markono Print Media Pte Ltd.

We dedicate this monograph to the numerous contributors of this field, many of whom are listed in the author index.

The classic Shannon–Hartley law suggests that the achievable channel capacity increases logarithmically with the transmit power. By contrast, the MIMO capacity increases linearly with the number of transmit antennas, provided that the number of receive antennas is equal to the number of transmit antennas. With the further proviso that the total transmit power is increased proportionately to the number of transmit antennas, a linear capacity increase is achieved upon increasing the transmit power, which justifies the spectacular success of MIMOs . . .

Contents

About the Authors

Lajos Hanzo FREng, FIEEE, FIET, DSc received his degree in electronics in 1976 and his doctorate in 1983. During his 31-year career in telecommunications he has held various research and academic posts in Hungary, Germany and the UK. Since 1986 he has been with the School of Electronics and Computer Science, University of Southampton, UK, where he holds the chair in telecommunications. He has co-authored 17 books on mobile radio communications totaling in excess of 10 000 pages, published in excess of 800 research papers, acted as TPC Chair of several IEEE conferences, presented keynote lectures and been awarded a number of distinctions. Currently he is directing an academic research team, working on a range of research projects in the field of wireless multimedia communications sponsored by industry, the Engineering and Physical Sciences Research Council (EPSRC) UK, the European IST Programme and the Mobile Virtual Centre of Excellence (VCE), UK. He is an enthusiastic supporter of industrial and academic liaison and he offers a range of industrial courses. He is also an IEEE Distinguished Lecturer as well as a Governor of both the IEEE ComSoc and the VTS. He is the acting Editor-in-Chief of the IEEE Press. For further information on research in progress and associated publications please refer to http://www-mobile.ecs.soton.ac.uk.

Osamah Rashed Alamri received his BS degree with first class honours in electrical engineering from King Fahd University of Petroleum and Minerals (KFUPM), Dhahran, Saudi Arabia, in 1997, where he was ranked first with a 4.0 GPA. In 2002, he received his MS degree in electrical engineering from Stanford University, California, USA. He submitted his PhD thesis in October 2006 and published in excess of 20 research papers while working towards his PhD degree with the Communications Group, School of Electronics and Computer Science, University of Southampton, UK. His research interests include sphere packing modulation, space-time coding, turbo coding and detection, multi-dimensional mapping and MIMO systems. At the time of writing he is continuing his investigations as a post-doctoral researcher.

Mohammed El-Hajjar received his BEng degree (with distinction) in electrical engineering from the American University of Beirut (AUB), Lebanon, and his MSc degree (with distinction) in radio frequency communication systems from the University of Southampton, UK. Since October 2005, he has been working towards his PhD degree with the Communications Group, School of Electronics and Computer Science, University of Southampton, UK. He is the recipient of several academic awards from the AUB as well as the University of Southampton. His research interests include sphere packing modulation, space-time coding, differential space-time spreading, adaptive transceiver design and cooperative communications. In 2008 he completed his PhD thesis and joined Ensigma in Chepstow, Wales, UK, as a wireless system architect.

Nan Wu received his BEng in electronics engineering in 2003 from Dalian University of Technology, China. He then moved to the UK and received his MSc degree (with distinction) and PhD from the University of Southampton, UK in 2004 and 2008, respectively. His research interests are in the area of wireless communications, including space-time coding, channel coding and cooperative MIMO systems. In September 2008 he joined the National Institute of Standards and Technology (NIST) in the USA as a guest researcher working on cross-layer designs.

Other Wiley–IEEE Press Books on Related Topics

For detailed contents and sample chapters please refer to www.wiley.com and www-mobile.ecs.soton.ac.uk

- L. Hanzo, J. Blogh, S. Ni: *3G, HSPA and FDD versus TDD Networking: Smart Antennas and Adaptive Modulation, 2nd Edition*, ISBN: 978-0-470-75420-7, 596 pages, February 2008
- L. Hanzo, P. Cherriman, J. Streit: *Video Compression and Communications: From Basics to H.261, H.263, H.264, MPEG4 for DVB and HSDPA-Style Adaptive Turbo-Transceivers, 2nd Edition*, ISBN: 978-0-470-51849-6, 702 pages, September 2007
- L. Hanzo, C. Somerville, J. Woodard: *Voice and Audio Compression for Wireless Communications, 2nd Edition*, ISBN: 978-0-470-51581-5, 880 pages, August 2007
- L. Hanzo, T. Keller: *OFDM and MC-CDMA: A Primer*, ISBN: 978-0-470-03007-3, 430 pages, April 2006
- A. Molisch: *Wireless Communications*, ISBN: 978-0-470-84887-6 (HB) / 978-0-470-84888-3 (PB), 668 pages, September 2005, £85.00 (HB) / £45 (PB)
- L. Hanzo, S. X. Ng, T. Keller, W. Webb: *Quadrature Amplitude Modulation: From Basics to Adaptive Trellis-Coded, Turbo-Equalised and Space-Time Coded OFDM, CDMA and MC-CDMA Systems, 2nd Edition*, ISBN: 978-0-470-09468-6, 1036 pages, September 2004
- L. Hanzo, L.-L. Yang, E.-L. Kuan, K. Yen: *Single and Multi-Carrier DS-CDMA: Multi-User Detection, Space-Time Spreading, Synchronisation, Networking and Standards*, ISBN: 978-0-470-86309-1; 1104 pages, August 2003
- L. Hanzo, M. Münster, B. J. Choi, T. Keller: *OFDM and MC-CDMA for Broadband Multi-User Communications, WLANs and Broadcasting*, ISBN: 978-0-470-85879-0, 1014 pages, July 2003
- L. Hanzo, T. H. Liew, B. L. Yeap: *Turbo Coding, Turbo Equalisation and Space-Time Coding for Transmission over Fading Channels*, ISBN: 978-0-470-84726-8, 766 pages, July 2002
- L. Hanzo, C. H. Wong, M. S. Yee: *Adaptive Wireless Transceivers: Turbo-Coded, Turbo-Equalized and Space-Time Coded TDMA, CDMA, and OFDM Systems*, ISBN: 978-0-470-84689-6, 752 pages, February 2002

Preface

The family of recent wireless standards included the optional employment of Multiple-Input Multiple-Output (MIMO) techniques. This was motivated by the observation according to the classic Shannon–Hartley law that the achievable channel capacity increases logarithmically with the transmit power. In contrast, the MIMO capacity increases linearly with the number of transmit antennas, provided that the number of receive antennas is equal to the number of transmit antennas. With the further proviso that the total transmit power is increased in proportion to the number of transmit antennas, a linear capacity increase is achieved upon increasing the transmit power, which justifies the spectacular success of MIMO systems.

Hence, this volume explores recent research advances in MIMO techniques as well as their limitations. The basic types of multiple-antenna-aided wireless systems are classified and their benefits are characterized. We also argue that under realistic propagation conditions, when for example the signals associated with the MIMO elements become correlated owing to shadow fading, the predicted performance gains may erode substantially. Furthermore, owing to the limited dimensions of pocket-sized handsets, the employment of multiple antenna elements at the mobile station is impractical. In this scenario only the family of distributed MIMO elements relying on the cooperation of potentially single-element mobile stations is capable of eliminating the correlation of the signals impinging on the MIMO elements, as is discussed in this book. We also report on a variety of avant-garde hybrid MIMO designs to set out promising future research directions.

Our intentions with this book are as follows.

1. First, to pay tribute to all researchers, colleagues and valued friends who have contributed to the field. Hence, this book is dedicated to them, since without their quest for better MIMO solutions for wireless communications this monograph could not have been conceived. They are too numerous to name here, hence they appear in the author index of the book. Our hope is that the conception of this monograph on the topic will provide an adequate portrayal of the community's research and will further fuel this innovation process.

2. We expect to stimulate further research by exposing open research problems and by collating a range of practical problems and design issues for the practitioners. The coherent further efforts of the wireless research community are expected to lead to the solution of the range of outstanding problems, ultimately providing us with flexible MIMO-aided wireless transceivers exhibiting a performance close to information-theoretical limits.

Acknowledgements

We are indebted to our many colleagues who have enhanced our understanding of the subject. These colleagues and valued friends, too numerous to be mentioned individually, have influenced our views concerning the subject of the book. We thank them for the enlightenment gained from our collaborations on various projects, papers and books. We are particularly grateful to our academic colleagues Professor Sheng Chen, Dr Soon-Xin Ng, Dr Rob Maunder and Dr Lie-Liang Yang. We would also like to express our appreciation to Sohail Ahmed, Andreas Ahrens, Jos Akhtman, Jan Brecht, Jon Blogh, Nicholas Bonello, Marco Breiling, Marco del Buono, Sheng Chen, Peter Cherriman, Stanley Chia, Byoung Jo Choi, Joseph Cheung, Jin-Yi Chung, Peter Fortune, Thanh Nguyen Dang, Sheyam Lal Dhomeja, Lim Dongmin, Dirk Didascalou, Stephan Ernst, Eddie Green, David Greenwood, Chen Hong, Hee Thong How, Bin Hu, Ming Jiang, Thomas Keller, Lingkun Kong, Choo Leng Koh, Ee Lin Kuan, W. H. Lam, Wei Liu, Kyungchun Lee, Xiang Liu, Fasih Muhammad, Matthias Münster, Song Ni, C. C. Lee, M. A. Nofal, Xiao Lin, Chee Siong Lee, Tong-Hooi Liew, Noor Shamsiah Othman,Raja Ali Riaz, Vincent Roger-Marchart, Redwan Salami, Professor Raymond Steele, Shinya Sugiura, David Stewart, Clare Sommerville, Tim Stevens, Shuang Tan, Ronal Tee, Jeff Torrance, Spyros Vlahoyiannatos, Jin Wang, Li Wang, William Webb, Chun-Yi Wei, Hua Wei, Stefan Weiss, John Williams, Seung-Hwang Won, Jason Woodard, Choong Hin Wong, Henry Wong, James Wong, Andy Wolfgang, Lei Xu, Chong Xu, Du Yang, Wang Yao, Bee-Leong Yeap, Mong-Suan Yee, Kai Yen, Andy Yuen, Jiayi Zhang, Rong Zhang and many others with whom we enjoyed an association.

We also acknowledge our valuable associations with the Virtual Centre of Excellence in Mobile Communications, in particular with its chief executive, Dr Walter Tuttlebee, and other members of its Executive Committee, namely Dr Keith Baughan, Professor Hamid Aghvami, Professor Mark Beach, Professor John Dunlop, Professor Barry Evans, Professor Peter Grant, Dr Dean Kitchener, Professor Steve MacLaughlin, Professor Joseph McGeehan, Dr Tim Moulsley, Professor Rahim Tafazolli, Professor Mike Walker and many other valued colleagues. Our sincere thanks are also due to John Hand and Nafeesa Simjee, EPSRC, UK, for supporting our research. We would also like to thank Dr Joao Da Silva, Dr Jorge Pereira, Bartholome Arroyo, Bernard Barani, Demosthenes Ikonomou and other valued colleagues from the Commission of the European Communities, Brussels, Belgium.

Similarly, our sincere thanks are due to Katharine Unwin, Mark Hammond, Sarah Hinton and their colleagues at John Wiley & Sons, Ltd, in Chichester, UK. Finally, our sincere gratitude is due to the numerous authors listed in the author index, as well as to those whose work was not cited owing to space limitations, for their contributions to the state of the art, without whom this book would not have materialized.

Lajos Hanzo, Osamah Alamri, Mohammed El-Hajjar and Nan Wu
School of Electronics and Computer Science,
University of Southampton, UK

Chapter 1

Problem Formulation, Objectives and Benefits

The objective of this light-hearted introductory chapter is to provide a brief rudimentary exposure of the pivotal aspects of the book. Our treatment in this chapter is conceptual, rather than mathematically motivated, with the objective of characterizing the attainable diversity gains, multiplexing gains and beamforming gains. All issues touched upon in this chapter are revisited in a more rigorous mathematical approach in the remaining chapters.

Digital communication exploiting Multiple-Input Multiple-Output (MIMO) wireless channels has recently attracted considerable attention as one of the most significant technical breakthroughs in modern communications. Soon after its invention, the technology seems to have the potential to be part of large-scale standards-driven commercial wireless products and networks such as broadband wireless access systems, Wireless Local Area Networks (WLANs), third-generation (3G) networks and beyond [1]. The 3G systems are expected to have the capability to support circuit and packet data at high bit rates. Rates of 144 kbit s^{-1} or higher in high mobility (vehicular) traffic, 384 kbit s^{-1} for pedestrian traffic and 2 Mbit s^{-1} or higher for indoor traffic are targeted [2]. Wireless systems that employ multiple antennas provide a promising platform for achieving such high rates because of the improved bit/symbol capacity compared with the Single-Input Single-Output (SISO) systems [3].

As shown in Figure 1.1, MIMO systems can be defined as wireless communication systems for which the transmitting end as well as the receiving end is equipped with multiple antenna elements. The basic concept of MIMO is that the transmitted signals from all transmit antennas are combined at each receive antenna element in such a way as to improve the Bit Error Rate (BER) performance or the data rate (bit s^{-1}) of the transmission. Both the network's Quality of Service (QoS) and the operator's revenues can be increased significantly because of this advantage of MIMO systems. One can think of MIMO systems as an extension to smart antennas. However, the idea of using antenna arrays for improving the wireless transmission was introduced several decades ago.

Space-Time Processing (STP) is the core concept of MIMO systems. Time is the natural dimension of digital communication data. Space refers to the spatial dimension inherent in the use of multiple spatially distributed antennas. Most of the current interest in Space-Time Coding (STC) is driven by discoveries in the late 1980s and early 1990s that multiple antennas can exploit a rich wireless scattering environment and benefit from the multipath

Near-Capacity Multi-Functional MIMO Systems L. Hanzo, O. R. Alamri, M. El-Hajjar and N. Wu

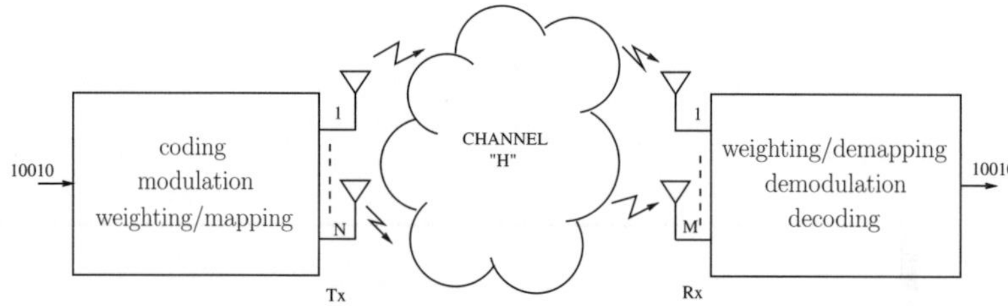

Figure 1.1: MIMO wireless transmission system.

fading nature of the wireless channel. Current research mostly focuses on channel modeling and measurement, and on the design of modulation and coding techniques that take into consideration the two-dimensional nature of STP (i.e. the space and time dimensions) [4].

1.1 The Wireless Channel and the Concept of Diversity

The key characteristics of the mobile radio channel in contrast to the Gaussian channel are small-scale fading and multipath propagation [5]. Small-scale fading, which is usually simply called fading, refers to the rapid fluctuation of signal strength over a short travel distance or period of time. Fading is primarily caused by multipath propagation of the transmitted signal, which creates replicas of the transmitted signal that arrive at the receiver with different delays. These versions of the transmitted signal combine either constructively or destructively at the receiver resulting in fluctuations in the amplitude and phase of the resultant signal. Other factors that have an influence on the small-scale fading include velocity of the mobile station, speed of the surrounding objects and the transmission bandwidth of the signal [5]. During severe fading, the transmitted signal cannot be determined by the receiver unless some less-attenuated version of it is available. This usually can be achieved by introducing some sort of diversity in the transmitted signal. The three most common diversity techniques are as follows [6].

- **Temporal diversity**. An example of temporal diversity is channel coding with time interleaving. The receiver is provided with several versions of the transmitted signal as redundancy in the temporal domain.

- **Frequency diversity**. This type of diversity is based on the phenomenon that the structure of multipath propagation depends on the frequency of the transmitted wave. Thus, redundancy in the frequency domain provides the receiver with several replicas of the transmitted signal that experience different fading at any particular time instant.

- **Antenna or space diversity**. In order to create space diversity, several spatially separated or differentially polarized antennas are employed. This generates redundancy of the transmitted signal in the spatial domain, where each replica undergoes a different propagation path. In this context, diversity order refers to the number of decorrelated spatial branches available at the transmitter or receiver, where the probability of losing a signal decreases exponentially with increasing diversity order.

It is always desirable to employ all forms of diversity in order to combat the adverse effects of the wireless channel [7]. However, it is sometimes impractical to employ a particular type of diversity in a specific situation [8]. For example, temporal diversity is ineffective in slow fading channels especially for delay-sensitive applications. In addition, antenna diversity at the mobile unit induces design impracticality. The most common systems that employ different types of diversity techniques for the sake of improving the performance of wireless transmission/reception are STC and MIMO schemes. Next, a brief historical overview on STC and MIMO systems is presented summarizing the main contributions in this field.

1.2 Diversity and Multiplexing Trade-offs in Multi-functional MIMO Systems

1.2.1 Classification of MIMO Systems

Again, our objective in this light-hearted section is to provide a brief conceptual overview of the material discussed in significantly more detail in Parts I and III of the book. More specifically, we briefly consider the design alternatives of different MIMO schemes, while considering the attainable diversity gains, multiplexing gains and beamforming gains. Our easy-reading conceptual treatment in this section aims to avoid the rigor of mathematics, which is left for the detailed approach of the remaining chapters.

Here we would like to commence with a brief classification of different MIMO schemes, which are categorized as diversity techniques, multiplexing schemes, multiple access arrangements and beamforming techniques. We then introduce two multi-functional MIMO families. These multi-functional MIMOs are capable of combining the benefits of several MIMO schemes and hence they attain an improved performance in terms of both their BER and their throughput. The first multi-functional MIMO family represents the recently proposed Layered Steered Space-Time Codes (LSSTCs), which combine the triple benefits of Space-Time Block Codes (STBCs), Vertical Bell Labs Layered Space-Time (V-BLAST) schemes and beamforming. The other multi-functional MIMO scheme is referred to as Layered Steered Space-Time Spreading (LSSTS) and combines the benefits of Space-Time Spreading (STS), V-BLAST and beamforming with those of the generalized Multicarrier Direct Sequence Code Division Multiple Access (MC DS-CDMA). We also compare the attainable diversity, multiplexing and beamforming gains of the different MIMO schemes in order to document the advantages of the multi-functional MIMOs over conventional MIMO schemes.

Recently, there has been a growing demand for flexible and bandwidth-efficient transceivers capable of supporting the explosive expansion of the Internet and the continued dramatic increase in demand for high-speed multimedia wireless services. Advances in channel coding made it feasible to approach Shannon's capacity limit in systems equipped with a single antenna [9], but fortunately these capacity limits can be further extended with the aid of multiple antennas. Recently, MIMO systems have attracted considerable research attention and are considered as one of the most significant technical breakthroughs in contemporary communications.

Explicitly, the MIMO schemes can be categorized as diversity techniques, multiplexing schemes, multiple access methods, beamforming as well as multi-functional MIMO arrangements, as shown in Figure 1.2. Spatial diversity can be attained by employing multiple antennas at the transmitter or the receiver. Multiple antennas can be used to transmit and receive appropriately encoded replicas of the same information sequence in order to achieve diversity and hence to obtain an improved BER performance. In the context of diversity techniques, the antennas are spaced as far apart as possible, so that the signals transmitted

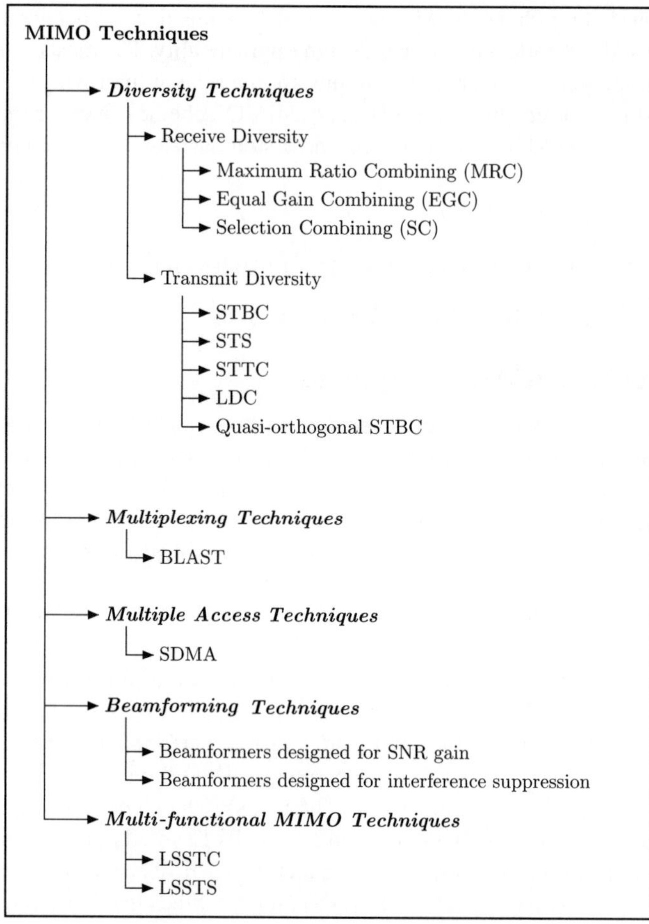

Figure 1.2: Classification of MIMO techniques.

to or received by the different antennas experience independent fading and hence we attain the highest possible diversity gain.

A simple spatial diversity technique, which does not involve any loss of bandwidth, is constituted by the employment of multiple antennas at the receiver, where several techniques can be employed for combining the independently fading signal replicas, including Maximum Ratio Combining (MRC), Equal Gain Combining (EGC) and Selection Combining (SC), as shown in Figure 1.2. Several transmit, rather than receive, diversity techniques have also been proposed in the literature [8, 10–12], as shown in Figure 1.2. In [11], Alamouti proposed a witty transmit diversity technique using two transmit antennas, the key advantage of which was the employment of low-complexity single-receive-antenna-based detection, which avoids the more complex joint detection of multiple symbols. The decoding algorithm proposed in [11] can be generalized to an arbitrary number of receive antennas using MRC, EGC or SC. Alamouti's achievement inspired Tarokh *et al.* [12] to generalize the concept of transmit diversity schemes to more than two transmit antennas, contriving the generalized concept of STBCs. The family of STBCs is capable of attaining the same diversity gain as Space-Time Trellis Codes (STTCs) [8] at a lower decoding complexity, when employing the same number of transmit antennas. However, a disadvantage of STBCs when compared with STTCs is that they employ unsophisticated repetition-coding and hence provide no coding gain. Furthermore, inspired by the philosophy of STBCs, Hochwald *et al.* [13] proposed the transmit diversity concept known as STS for the downlink of Wideband Code Division Multiple Access (WCDMA) that is capable of achieving the highest possible transmit diversity gain.

Regretfully, the STBC and STS designs of [12, 13] contrived for more than two transmit antennas result in a reduction of the achievable throughput per channel use. An alternative idea invoked for constructing full-rate STBCs for complex-valued modulation schemes and more than two antennas was suggested in [14]. Here the strict constraint of perfect orthogonality was relaxed in favor of achieving a higher data rate. The resultant STBCs were referred to as quasi-orthogonal STBCs [14].

The STBC and STS designs offer, at best, the same data rate as an uncoded single-antenna system, but they provide an improved BER performance compared with the family of single-antenna-aided systems by providing diversity gains. In contrast to this, several high-rate space-time transmission schemes having a normalized rate higher than unity have been proposed in the literature. For example, high-rate space-time codes that are linear both in space and time, namely the family of the so-called Linear Dispersion Codes (LDCs), were proposed in [10]. LDCs provide a flexible trade-off between emulating STC and/or spatial multiplexing.

STBCs and STTCs are capable of providing diversity gains for the sake of improving the achievable system performance. However, this BER performance improvement is often achieved at the expense of a rate loss, since STBCs and STTCs may result in a throughput loss compared with single-antenna-aided systems. As a design alternative, a specific class of MIMO systems was designed for improving the attainable spectral efficiency of the system by transmitting different signal streams independently over each of the transmit antennas, hence resulting in a multiplexing gain. This class of MIMOs subsumes the Bell Labs Layered Space-Time (BLAST) scheme and its relatives [15]. The BLAST scheme aims to increase the system throughput in terms of the number of bits per symbol that can be transmitted in a given bandwidth at a given integrity.

In contrast to the family of BLAST schemes, where multiple antennas are activated by a single user to increase the user's throughput, Space Division Multiple Access (SDMA) [16] employs multiple antennas for the sake of supporting multiple users. SDMA exploits the unique user-specific Channel Impulse Response (CIR) of the different users for separating their received signals. On the other hand, in beamforming arrangements [16], typically $\lambda/2$-spaced antenna elements are used for the sake of creating a spatially selective transmitter/receiver beam, where λ represents the carrier's wavelength. Beamforming is employed for providing a beamforming gain by mitigating the effects of various interfering signals, provided that they arrive from sufficiently different directions. In addition, beamforming is capable of suppressing the effects of co-channel interference, hence allowing the system to support multiple users by angularly separating them. Again, this angular separation becomes feasible only on condition that the corresponding users are separable in terms of the angle of arrival of their beams.

Finally, multi-functional MIMOs, as the terminology suggests, combine the benefits of several MIMO schemes including diversity gains and multiplexing gains as well as beamforming gains. As mentioned earlier, V-BLAST is capable of achieving the maximum attainable multiplexing gain, while STBC can achieve the full achievable antenna diversity gain facilitated by the number of independently fading diversity channels. Hence, it was proposed in [17] to combine these two techniques in order to provide both antenna diversity and spectral efficiency gains. Furthermore, the combined array processing proposed in [17] was improved in [18] by optimizing the decoding order of the different antenna layers. An iterative decoding algorithm was proposed in [18] that results in achieving the full receive diversity gain for the combined V-BLAST STBC system facilitated by the number of independently fading diversity channels. On the other hand, in [19] the authors presented a transmission scheme referred to as Double Space-Time Transmit Diversity (D-STTD), which consists of two STBC layers at the transmitter, which is equipped with four transmit antennas, while the receiver is equipped with two antennas. Furthermore, in order to achieve additional performance gains, beamforming has been combined both with spatial diversity and with spatial multiplexing techniques. STBC has been combined with beamforming in order to attain an improved Signal-to-Noise Ratio (SNR) gain in addition to the diversity gain [20].

This contribution provides a light-hearted perspective on further research advances in the field of multi-functional MIMO systems and demonstrates how diversity, multiplexing and beamforming gains are achieved by multi-functional MIMOs. More explicitly, in Section 1.2.2 we elaborate on the design of two novel multi-functional MIMOs, which are characterized by diversity gain and multiplexing gain as well as beamforming gain. In Section 1.2.3 we quantify the achievable performance of the different MIMO schemes. A comparison of the different MIMO schemes expressed in terms of their diversity, multiplexing and beamforming gains is presented in Section 1.2.4, followed by our brief conclusions.

1.2.2 Multi-functional MIMO Systems

Space-time codes have been designed for the sake of attaining the highest possible diversity gain, where the diversity order of STBC schemes is $(N_t \times N_r)$, where N_t is the number of transmit antennas and N_r represents the number of receive antennas. However, the STBC schemes were not designed for attaining any multiplexing gain; quite the contrary, in some STBC designs there is a rate loss, which results in a reduced throughput in comparison to Single-Input Single-Output (SISO) systems. On the other hand, the V-BLAST scheme was designed for attaining the maximum achievable multiplexing gain equal to the number of transmit antennas, although it does not attain a high diversity gain. Therefore, the appealing concept of multi-functional MIMO schemes designed for combining the benefits of STBC and BLAST schemes arises, in order to provide both diversity and multiplexing gains. In the following we describe two novel multi-functional MIMO schemes that can attain diversity gain, multiplexing gain and beamforming gain, while employing low-complexity linear receivers.

1.2.2.1 Layered Steered Space-Time Codes

The first multi-functional MIMO scheme combines the benefits of the V-BLAST scheme and of STBCs as well as of beamforming. Thus, the proposed system benefits from the multiplexing gain of V-BLAST, from the diversity gain of STBCs and from the SNR gain of the beamformer. This multi-functional MIMO scheme was referred to as a Layered Steered Space-Time Code (LSSTC) [21].

A block diagram of the proposed LSSTC scheme is given in Figure 1.3. The system's architecture in Figure 1.3 has N_t transmit Antenna Arrays (AAs) spaced sufficiently far apart in order to experience independent fading and hence to achieve transmit diversity. The L_{AA} elements of each of the AAs are spaced at a distance of $\lambda/2$ for the sake of achieving a

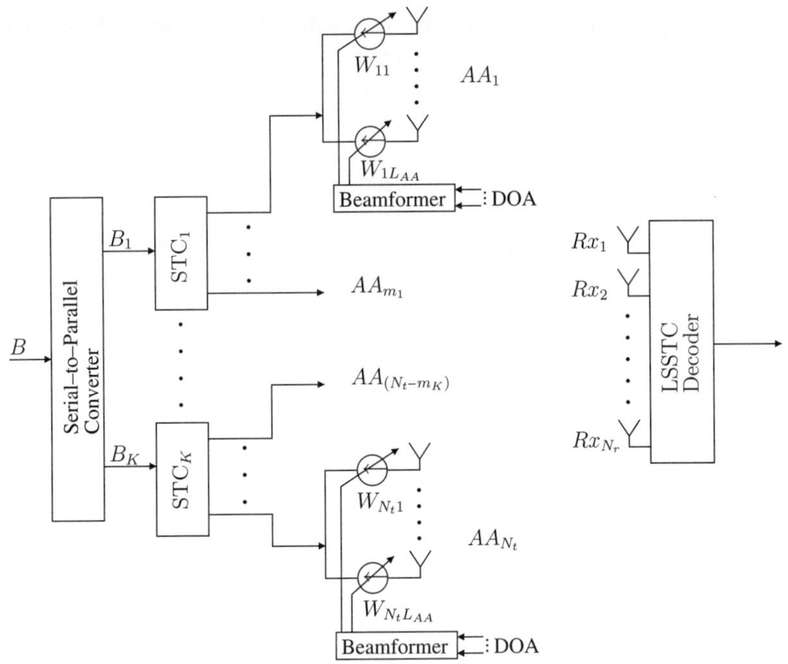

Figure 1.3: LSSTC system block diagram.

beamforming gain. Furthermore, the receiver is equipped with $N_r \geqslant N_t$ antennas. According to Figure 1.3, a block of B input information symbols is serial-to-parallel converted to K groups of symbol streams of length $B_1, B_2, \ldots, B_K$, where $B_1 + B_2 + \cdots + B_K = B$. Each group of B_k symbols, $k \in [1, K]$, is then encoded by a component space-time code STC_k associated with m_k transmit AAs, where $m_1 + m_2 + \cdots + m_K = N_t$.

The L_{AA}-dimensional spatio-temporal CIR vector spanning the mth transmitter AA, $m \in [1, \ldots, N_t]$, and the nth receiver antenna, $n \in [1, \ldots, N_r]$, can be expressed as $\mathbf{h}_{nm}(t) = \mathbf{a}_{nm}(t)\delta(t - \tau_k)$, where τ_k is the signal's delay and $\mathbf{a}_{nm}(t)$ is the CIR of the mnth link between the mth AA and the nth receive antenna. Based on the assumption that the array elements are separated by half a wavelength, we have $\mathbf{a}_{nm}(t) = \alpha_{nm}(t) \cdot \mathbf{d}_{nm}$, where $\alpha_{nm}(t)$ is a Rayleigh faded envelope and $\mathbf{d}_{nm}$ is an L_{AA}-dimensional vector, whose elements are based on the Direction of Arrival (DOA) of the signal to the receiver. As for the AA-specific DOA, we consider a scenario where the distance between the transmitter and the receiver is significantly higher than that between the AAs and thus we can assume that the signals arrive at the different AAs in parallel, i.e. the DOA at the different AAs is the same. In this scenario, the MRC-criterion based transmit beamformer, which constitutes an effective solution to maximizing the antenna gain, is the optimum beamformer.

The decoder applies Group Successive Interference Cancelation (GSIC) based on the Zero Forcing (ZF) algorithm [17] for decoding the received signal. The most beneficial decoding order of the STC layers is determined on the basis of detecting the highest-power layer first for the sake of a high correct detection probability. For simplicity, let us consider the case of $K = 2$ STBC layers, where layer 1 is detected first, which allows us to eliminate the interference caused by the signal of layer 2. However, the proposed concept is applicable to arbitrary STCs and to an arbitrary number of layers K. For this reason, the decoder of layer 1 has to compute a matrix $\mathbf{Q}$, so that we have $\mathbf{Q} \cdot \widehat{\mathbf{H}}_2 = 0$, where $\widehat{\mathbf{H}}_2$ represents the channel matrix of the second STBC layer whose nmth element is α_{nm}. Therefore, the decoder computes an orthonormal basis for the left null space of $\widehat{\mathbf{H}}_2$ and assigns the vectors of the basis to the rows of $\mathbf{Q}$. Multiplying $\mathbf{Q}$ by the received signal matrix $\mathbf{Y}$ suppresses the interference of layer 2 originally imposed on layer 1 and generates a signal which can be decoded using Maximum Likelihood (ML) STBC detection. Then, the decoder subtracts the remodulated contribution of the decoded symbols of layer 1 from the composite twin-layer received signal $\mathbf{Y}$. Finally, the decoder applies direct STBC decoding to the second layer, since the interference imposed by the first layer has been eliminated. This group-interference cancelation procedure can be generalized to arbitrary N_t and K values.

1.2.2.2 LSSTS

The LSSTC scheme of Section 1.2.2.1 combines the benefits of V-BLAST, STBC and beamforming and hence is characterized by a diversity gain and a multiplexing gain as well as a beamforming gain. However, a drawback of the LSSTC scheme is the fact that the number of receive antennas N_r should be at least equal to the number of transmit antennas N_t. This condition is not very practical when employing pocket-sized Mobile Stations (MS) that are limited in size and complexity. The LSSTC scheme can be applied in a scenario where two Base Stations (BS) cooperate or a BS is communicating with a MIMO-aided laptop. Therefore, in order to allow communication between a BS and a MS accommodating fewer antennas than the transmitting BS while employing simple linear receivers, the LSSTS scheme described in the following can be employed.

A block diagram of the LSSTS scheme is shown in Figure 1.4. The LSSTS scheme combines the benefits of V-BLAST, STS and beamforming with generalized MC DS-CDMA [22] for the sake of achieving a multiplexing gain and a spatial and frequency diversity gain as well as a beamforming gain. The LSSTS scheme described in this section employs $N_t = 4$ transmit antennas and $N_r = 2$ receive antennas and employs a linear receiver to decode the received signal.

The system architecture employed in Figure 1.4 for the proposed scheme is equipped with $N_t = 4$ transmit AAs spaced sufficiently far apart in order to experience independent fading. The L_{AA} elements of each of the AAs are spaced at a distance of $\lambda/2$ for the sake of achieving beamforming. The system can support K users transmitting at the same time and using the same carrier frequencies, while they can be differentiated by the user-specific spreading code $\bar{\mathbf{c}}_k$, where $k \in [1, K]$. In addition, in the generalized MC DS-CDMA system considered, the subcarrier frequencies are arranged in a way that guarantees that the same STS signal is spread to and hence transmitted by the specific V subcarriers having the maximum possible frequency separation, so that they experience independent fading and achieve the maximum attainable frequency diversity.

The system considered employs the generalized MC DS-CDMA scheme of [22] using UV subcarriers. The transmitter schematic of the kth user is shown in Figure 1.4, where a block of UN_t data symbols $\mathbf{x}$ is Serial-to-Parallel (S/P) converted to U parallel subblocks. Afterwards, each set of N_t symbols is S/P converted to $G = 2$ groups, where each group is encoded using the $N_{tg} = 2$ antenna-aided STS procedure of [13], where the transmitted signal is spread to N_{tg} transmit antennas with the aid of the orthogonal spreading codes of $\{\bar{\mathbf{c}}_{k,1}, \bar{\mathbf{c}}_{k,2}, \ldots, \bar{\mathbf{c}}_{k,N_{tg}}\}$, $k = 1, 2, \ldots, K$. The spreading codes $\bar{\mathbf{c}}_{k,1}$ and $\bar{\mathbf{c}}_{k,2}$ are generated from the same user-specific spreading code $\bar{\mathbf{c}}_k$ as in [13]. The discrete symbol duration of the orthogonal STS codes is $N_{tg}N_e$, where N_e represents the kth user's Time Domain (TD) Spreading Factor (SF).

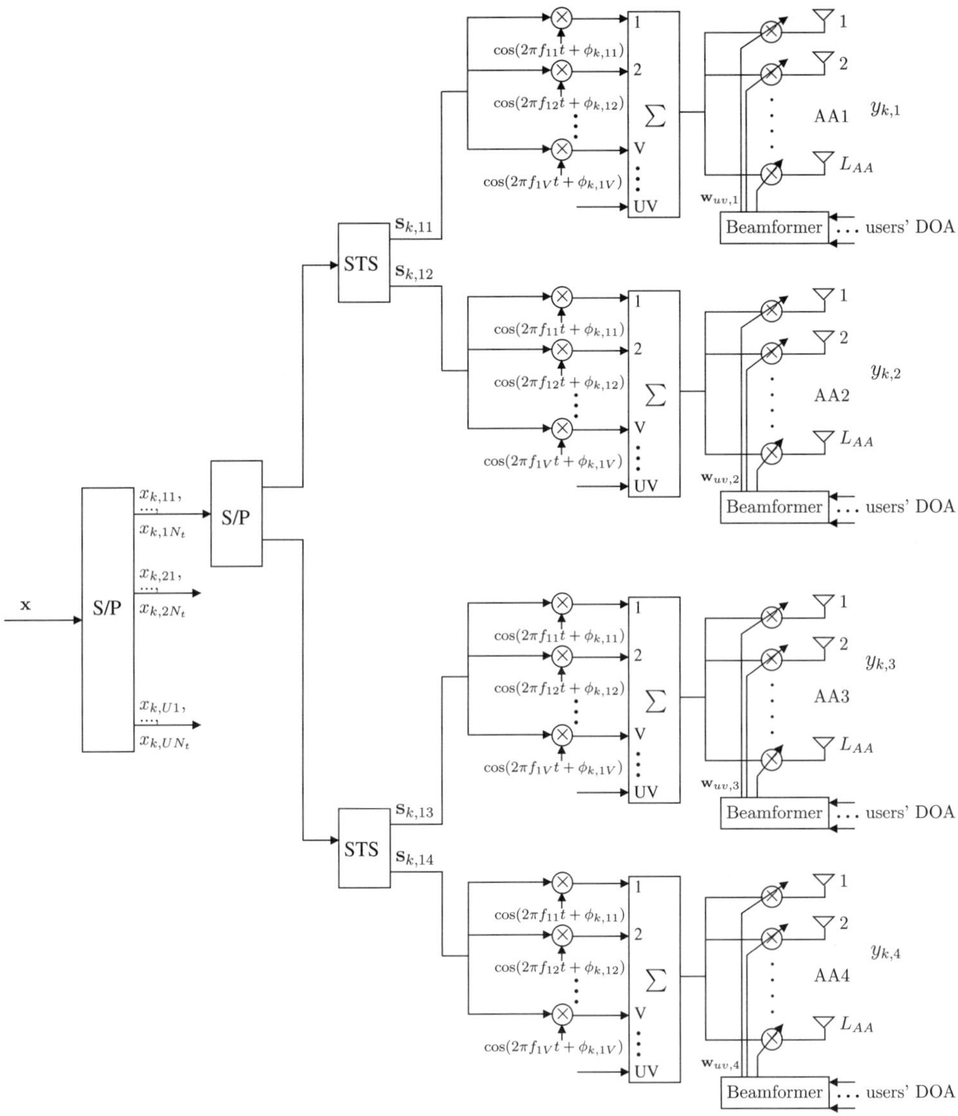

Figure 1.4: The kth user's LSSTS-aided generalized MC DS-CDMA transmitter model.

The UN_t outputs of the UG STS blocks modulate a group of subcarrier frequencies $\{f_{u,1}, f_{u,2}, \ldots, f_{u,V}\}$. Since each of the U subblocks is spread to and hence conveyed with the aid of V subcarriers, a total of UV subcarriers are required in the MC DS-CDMA system considered. The UV subcarrier signals are superimposed on each other in order to form the complex-valued modulated signal for transmission. Finally, according to the kth user's channel information, the UVN_t signals of the kth user are weighted by the transmit weight vector $\mathbf{w}_{uv,n}^{(k)}$ determined for the uvth subcarrier of the kth user, which is generated for the nth AA. Assuming that the system employs a modulation scheme transmitting D bits-per-symbol, then the bandwidth efficiency of the LSSTS-aided generalized MC DS-CDMA system is given by $2UD$ bits-per-channel-use.

The uvth CIR considered in the case of LSSTS is the same as that considered in the previous section for LSSTC. Assuming that the K users' data are transmitted synchronously over a dispersive Rayleigh fading channel, decoding is carried out in two steps: first Successive Interface Cancellation (SIC) is performed according to [19], followed by the STS decoding procedure of [13].

Finally, after combining the $k = 1$st user's identical replicas of the same signal transmitted by spreading over V subcarriers, the decision variables corresponding to the symbols transmitted in the uth subblock can be expressed as $\widetilde{x}_{1,u} = \sum_{v=1}^{V} \widetilde{x}_{1,uv}$. Therefore, the decoded signal has a diversity order of $2V$. More explicitly, second-order spatial diversity is attained from the STS operation and a diversity order of V is achieved as a benefit of spreading by the generalized MC DS-CDMA scheme, where the subcarrier frequencies are arranged in a way that guarantees that the same STS signal is spread to and hence transmitted by the specific V subcarriers having the maximum possible frequency separation, so that they experience as independent fading as possible.

1.2.3 Expected Performance and Discussions

In this section we compare the BER performance of the different MIMO schemes to that of the SISO system. We compare BPSK modulated systems, while considering transmissions over correlated Rayleigh fading channels associated with a normalized Doppler frequency of 0.01.

According to Figure 1.5, the V-BLAST system employing $(N_t, N_r) = (4, 4)$ antennas has a slightly better BER performance than the SISO system, despite its quadrupled throughput. Also observe in Figure 1.5 that the slope of the BER curves of both the V-BLAST and of the SISO system is similar, which suggests that V-BLAST does not attain a high diversity gain, but it is capable of attaining a high multiplexing gain. In addition, Figure 1.5 shows that the STBC system employing $(N_t, N_r) = (2, 1)$ attains a better BER performance than the SISO and V-BLAST schemes due to the diversity gain attained by the STBC. Further diversity gain can be attained by the four-antenna-aided STBC employing four receive antennas, which results in a diversity order of 16. As shown in Figure 1.5 the four-antenna-aided STBC scheme employing four receive antennas is capable of attaining around 15 dB gain at a BER of 10^{-5} over the twin-transmit-antenna-aided STBC system using $N_r = 1$. However, a drawback of the four-antenna-aided system is that it results in a throughput loss, where four symbols are transmitted in eight time slots, resulting in a rate of $1/2$.

Observe in Figure 1.5 that the LSSTS scheme employing $(N_t, N_r) = (4, 2)$ and $V = 1$ attains an identical BER performance to that of the twin-transmit-antenna-aided STBC system. This means that the LSSTS scheme employing $V = 1$ has a diversity order of two, similar to the twin-antenna-aided STBC. On the other hand, the LSSTS scheme attains twice the throughput of the twin-transmit-antenna-aided STBC scheme. In addition, when V is increased from one to four, the achievable BER performance improves due to the additional frequency diversity gain attained.

A further performance improvement is attained by the LSSTC scheme in conjunction with $(N_t, N_r) = (4, 4)$ compared to the LSSTS scheme. The LSSTC scheme employs more antennas than the LSSTS scheme and hence attains both a higher diversity order and a better BER performance. Furthermore, Figure 1.5 shows the performance improvements attained by beamforming, where the LSSTC scheme employing $L_{AA} = 4$ attains around 6 dB performance improvement at a BER of 10^{-5} over its counterpart employing $L_{AA} = 1$,

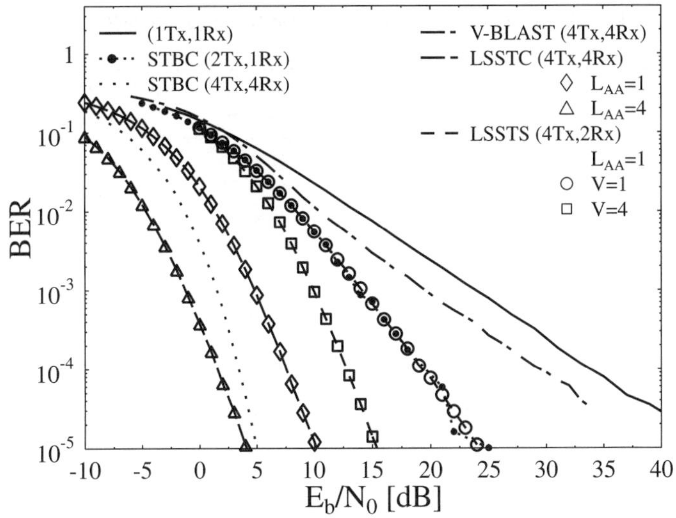

Figure 1.5: BER performance comparison of the SISO, STBC, V-BLAST, LSSTC and LSSTS schemes, while communicating over a correlated Rayleigh fading channel associated with a normalized Doppler frequency of $f_d = 0.01$.

provided that the DOA is perfectly known. Finally, a comparison between the STBC and LSSTC schemes using $(N_t, N_r) = (4, 4)$ reveals that the STBC arrangement attains a better performance than the LSSTC scheme employing $L_{AA} = 1$. This is due to the fact that the STBC scheme has a higher diversity gain, while the LSSTC scheme attains a throughput that is four times that of its STBC counterpart.

1.2.4 Diversity versus Multiplexing Trade-offs in MIMO Systems

According to our previous discussions, different MIMO schemes have different structures and hence a different BER as well as throughput performance. Explicitly, the STBC scheme is capable of attaining the highest possible spatial diversity gain, while having no multiplexing gain; in fact, some STBC structures result in a throughput loss. On the other hand, the V-BLAST scheme is capable of achieving the maximum possible multiplexing gain, while attaining a low diversity gain, depending on the choice of the V-BLAST decoder employed. Furthermore, we have introduced the LSSTC and LSSTS multi-functional MIMO designs that are capable of attaining diversity and multiplexing as well as beamforming gains.

Table 1.1 compares the diversity, multiplexing and beamforming gains of the different MIMO schemes for different configurations. In Table 1.1, N_t and N_r stand for the number of transmit and receive antennas, respectively, while L_{AA} represents the number of elements per transmit AA and V denotes the number of subcarriers employed by the generalized MC DS-CDMA system. In addition, the number of layers represents the number of antenna layers that are used for transmitting different data symbols at the same time, for the sake of attaining a multiplexing gain.

As shown in Table 1.1, the STBC schemes are capable of attaining a full diversity order of $(N_t \times N_r)$, while achieving no multiplexing or beamforming gain. In contrast, in the case of four-antenna- and eight-antenna-aided STBC schemes, the multiplexing gain is $1/2$, resulting in half the throughput of the SISO scheme. For example, in the four-antenna-aided

Table 1.1: Comparison of the gains achieved by various MIMO schemes.

	N_t	N_r	L_{AA}	V	Number of layers	Diversity order	Multiplexing order	Beamforming order
STBC	2	N_r	1	1	1	$2 \times N_r$	1	1
	4	N_r	1	1	1	$4 \times N_r$	$1/2$	1
	8	N_r	1	1	1	$8 \times N_r$	$1/2$	1
V-BLAST	2	2	1	1	2	1	2	1
ZF-SIC	4	4	1	1	4	1	4	1
	8	8	1	1	8	1	8	1
LSSTC	4	4	L_{AA}	1	2	4	2	L_{AA}
	8	8	L_{AA}	1	2	16	1	L_{AA}
	8	8	L_{AA}	1	4	4	4	L_{AA}
LSSTS	4	2	L_{AA}	V	2	$2 \times V$	2	L_{AA}

STBC scheme, four symbols are transmitted in eight time slots and similarly for the eight-antenna-aided STBC scheme, eight complex-valued symbols are transmitted in 16 time slots. On the other hand, as shown in Table 1.1, the V-BLAST scheme can attain a multiplexing gain of N_t, since the different antennas transmit different symbols in the same time slot. For example, for the V-BLAST scheme employing $(N_t, N_r) = (4, 4)$, the transmitter transmits four different symbols from the four different antennas in the same time slot, which results in a quadrupled multiplexing gain in comparison to that of the SISO scheme. Observe in Table 1.1 that the diversity order of V-BLAST employing the ZF-SIC is one for different (N_t, N_r) configurations. The diversity order of the V-BLAST scheme employing ZF-SIC is $(N_r - N_t + 1)$.

The LSSTC scheme combines the benefits of STBC and V-BLAST as well as of beamforming, as discussed earlier. This becomes clear in Table 1.1, where it is shown that the LSSTC scheme attains a diversity gain and a multiplexing gain as well as a beamforming gain. In the case of the $(N_t, N_r) = (4, 4)$ configuration, two twin-antenna STBC layers are implemented, which results in a diversity order of four and a multiplexing order of two. This is due to the fact that four symbols are transmitted from the four transmit antennas in two time slots. In addition, when L_{AA} elements are used per AA, then a beamforming gain can be attained. In the $(N_t, N_r) = (8, 8)$ configuration, two different schemes can be implemented. The first scheme is a two-layer configuration with each layer constituted by a four-antenna STBC scheme. The other configuration employs four layers of the twin-antenna STBC scheme. The two configurations result in the different diversity and multiplexing gains shown in Table 1.1.

Finally, in the LSSTS scheme four transmit and two receive antennas are employed, where the transmit antennas are separated into two STS layers. The diversity order achieved by the LSSTS scheme is $(2 \times V)$ as discussed in Section 1.2.2.2. The multiplexing order of the LSSTS scheme is two, since four symbols are transmitted in two time slots. Moreover, the LSSTS scheme is capable of attaining a beamforming gain, when $L_{AA} > 1$ elements per AA are used.

In this section a brief classification of MIMO schemes was presented based on their attainable diversity and multiplexing or beamforming gains. We also investigated the design of multi-functional MIMO schemes that are capable of combining the benefits of several MIMO schemes and hence attaining diversity and multiplexing as well as beamforming gains. More explicitly, we introduced two multi-functional MIMO schemes: LSSTC and LSSTS. The LSSTC combines the benefits of STBC and V-BLAST as well as beamforming, while the LSSTS combines the advantages of STS, V-BLAST and beamforming with those of generalized MC DS-CDMA, while supporting multiple users. Finally, a comparison between the BER performance as well as the diversity, multiplexing and beamforming gains of the different MIMO schemes reveals that multi-functional MIMOs are capable of attaining an improved performance over STBC and V-BLAST schemes.

1.3 Coherent versus Non-coherent Detection for STBCs Using Co-located and Cooperative Antenna Elements

1.3.1 Motivation

Again, our objective in this conceptually motivated section is to provide a brief overview of the material discussed in intricate detail in Part II of the monograph. More specifically, we briefly consider the design alternatives of various STBCs documented in the open literature, focusing our attention on the so-called orthogonal design approach and on the layered method. We introduce the generalized concept of the linear dispersion STC architecture, which allows us to strike a compromise between the above-mentioned two approaches. We also demonstrate that the powerful linear dispersion structure is capable of unifying the entire suite of existing schemes. As a further benefit, it offers extra design flexibility so that diverse system requirements can be satisfied. Furthermore, after characterizing the fundamental relationship between STBCs and Differential STBCs (DSTBCs), we highlight the benefits of non-coherently detected schemes, which are capable of exploiting the advantages of multiple antennas, while circumventing the potentially excessive burden of multi-antenna channel estimation. In addition, we demonstrate that the linear dispersion structure can also be applied in systems where the multiple-antenna array is formed in a distributed fashion by multiple single-antenna-aided cooperating MSs. Hence, the design of co-located and cooperative MIMO systems aiming for achieving diversity is linked from a linear dispersion perspective.

As argued above, apart from employing multiple antennas at the transmitter and receiver in a 'co-located' fashion, a Virtual Antenna Array (VAA) may also be formed by a group of cooperating single-antenna-aided MSs. The resultant cooperative MIMO system is capable of offering similar degrees of freedom to those of a co-located MIMO system having independently fading signal impinging on the antenna elements. In other words, the distributed MIMO elements are capable of mimicking the functionality of the co-located MIMO elements.

The advantage of a MIMO system can be exploited in two ways: to increase the reliability of the system by providing a diversity gain [23] and/or to increase the data rate by providing multiplexing gain [23]. In fact, it has been shown that there is a fundamental trade-off between the achievable diversity gain and the attainable multiplexing gain for a given MIMO system [23]. The term spatial multiplexing gain refers to the fact that one can use multiple antennas to achieve a higher throughput at the cost of an increased SNR requirement. On the other hand, the concept of spatial diversity is to provide multiple independently fading replicas of the transmitted signal for the receiver with the aid of the MIMO channel. If indeed these replicas are faded independently, it is unlikely that all copies of the transmitted signal are in a deep fade simultaneously. Therefore, the receiver is expected to reliably decode the transmitted signal using these independently faded received signals. Finally, apart from the 'spatial' dimension, diversity can also be achieved in both the temporal and the frequency domains.

In many practical scenarios, reliable wireless communications may not be guaranteed, even when multiple antennas have been employed. For example, when large-scale shadow fading contaminates the wireless links, all of the channels tend to fade together rather than independently, hence eroding the achievable diversity gain. Therefore, the concept of 'cooperative diversity' [24] has been proposed in the literature; this is a technique designed for providing diversity using the single antennas of other nodes in the cellular network as 'virtual' antennas.

In this section the design philosophies of spatial-diversity-oriented STBCs designed for open-loop MIMO systems are presented from a unique linear dispersion perspective. More explicitly, Section 1.3.2 demonstrates various design guidelines of STBCs and presents a general framework to unify all of the existing coherently detected STBCs found in the open literature. Section 1.3.3 exploits the linkage between coherently detected STBCs and non-coherently detected STBCs having no Channel State Information (CSI), which enables our general framework to incorporate the subclass of differential encoding/decoding techniques. Similarly, Section 1.3.4 examines the fundamental linkage between co-located and cooperative MIMO systems. Hence, cooperative STBCs can be created by applying the general linear dispersion framework. The expected performance is illustrated in Section 1.3.5, along with the related brief conclusions.

1.3.2 Evolution of STBCs

In order to exploit both the spatial and the temporal domains offered by a MIMO system, STBCs transmit a signal matrix $\mathbf{S}$ conveying the source information. For a MIMO system having M transmit and N receive antennas, a STBC scheme can be designed to transmit Q symbols using T channel slots. The STBC scheme may be described by the parameter combination $(MNTQ)$ having the normalized throughput of $R = Q/T$. In other words, the concept of STBCs is to design a set of matrices $\mathbf{S}$ satisfying both the throughput and the diversity order requirements under certain complexity constraints. However, the number of matrices to be designed may become excessive, when the system is operating at a high normalized throughput facilitated by a high number of antennas. This challenge was mainly addressed from two perspectives in the open literature, namely from the orthogonal and the layered approaches, both of which are highlighted in the following.

1.3.2.1 Orthogonal Approach

The 'orthogonal' approach was first proposed in [11], which was later generalized in [25]. The philosophy behind Orthogonal STBCs (OSTBCs) is that the space-time signal matrix $\mathbf{S}$ has to be an orthogonal/unitary matrix, where the orthogonality embedded in $\mathbf{S}$ is capable of decoupling the transmitted multi-antenna-coded symbol streams into independent single-antenna symbols. For example, Alamouti's scheme [11] can be characterized as follows:

$$\mathbf{G}_2 = \begin{pmatrix} s_1 & s_2 \\ -s_2^* & s_1^* \end{pmatrix}. \tag{1.1}$$

Although the associated single-stream decoding procedure is appealing, the orthogonality of the multi-antenna streams limits the choice of modulation schemes and restricts the antenna configurations supported. Hence, the OSTBCs are unable to reach a high normalized throughput. On the other hand, when relaxing the above-mentioned orthogonality of the OSTBCs, a potentially higher normalized throughput can be achieved, as exemplified by the family of Quasi-Orthogonal STBCs (QOSTBCs) [14].

Furthermore, the design of OSTBCs involves the inevitable throughput versus diversity gain trade-off characterized in [26]. This is because each spatial and temporal slot is used to convey either a new symbol to increase the throughput or a redundant symbol to attain diversity. However, with the aid of the recent advances in high-throughput full-diversity STBCs [27], it has been shown that it is not necessary to sacrifice the throughput in favor of achieving diversity or vice versa. (Using the average pairwise symbol error probability analysis technique [25], it follows that the maximum attainable diversity order of a STBC scheme designed for an $(M \times N)$-element MIMO system is $D_{\text{full}} = MN$.) The philosophy of these schemes is that they impose redundancy on the space-time codeword, but additional diversity gain can be achieved by spreading the information across both the spatial and temporal domains. We will elaborate on this idea in more detail using LDCs [10, 27] in Section 1.3.2.3.

1.3.2.2 Layered Approach

The motivation for introducing a 'layered' structure into the STBCs is that of supporting high-throughput communications, as exemplified by schemes such as the BLAST architecture [15]. Although conventional BLAST-type schemes were not designed for achieving diversity, they provide a new insightful angle for STBC designs. For example, the $\mathbf{G}_2$ STBC of Equation (1.1) can be considered as a scheme consisting of two layers. The first layer only conveys the information symbol s_1, which is 'repetition-coded' and mapped to one of the diagonals of Equation (1.1). The second layer contains the symbol s_2, which is mapped to the other diagonal of Equation (1.1). Each layer occupies half of the four spatial and temporal slots and the inherent orthogonality enables the receiver to separate the two layers to facilitate simple single-layer decoding. Since having more layers has the promise of an increased normalized throughput, the question of how many layers a STBC codeword can accommodate arises. In the literature, the authors of [28] proposed a class of STBCs employing a unitary Time Variant Linear Transformation (TVLT), which is capable of transmitting using T layers, while maintaining a normalized throughput of $R = M$. Furthermore, high-throughput Threaded Algebraic STBCs (TASTBCs) [29] were proposed in order to support up to M adaptively reconfigurable layers. One of the intriguing features of both TVLTs and TASTBCs is that both of them are capable of achieving the full attainable spatial diversity order of MN, while maintaining a high throughput. This is guaranteed by mapping the signals of each layer across all of the antennas, in order to achieve spatial diversity, as exemplified in Equation (1.1). Again, this philosophy will be further augmented using the linear dispersion structure below.

1.3.2.3 Linear Dispersion Codes

At this stage of our discussions, the challenge of STBC design becomes that of finding an appropriate way of relaxing the orthogonality, while maintaining the maximum diversity order of MN. Alternatively, we may cast this design dilemma as contriving a 'layered' structure capable of exploiting all of the spatial and temporal diversity resources available

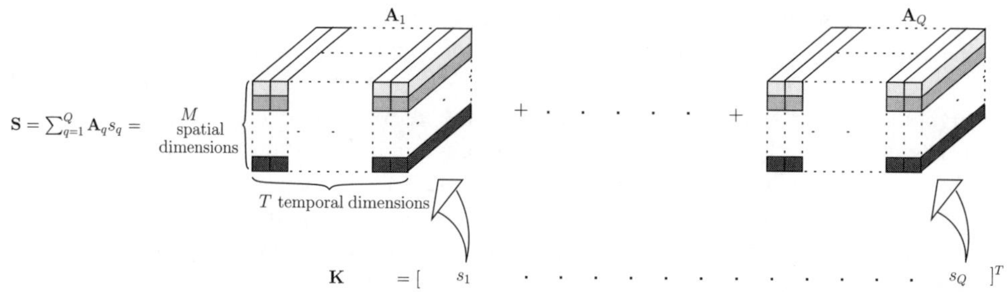

Figure 1.6: The space-time codeword $\mathbf{S}$ employing the linear dispersion structure.

for a single layer, while supporting multiple layers. Remarkably, these design objectives may be satisfied by the family of LDCs proposed by Hassibi and Hochwald [10]. The revolutionary concept of LDCs [10, 27] invokes a matrix-based linear modulation framework, where each space-time transmission matrix $\mathbf{S}$ is generated by a linear combination of so-called dispersion matrices and the weights of the components are determined by the associated transmitted symbol vector. More explicitly, given an information symbol vector $\mathbf{K} = [s_1, s_2, \ldots, s_Q]^{\mathrm{T}}$ constituted by symbols of an arbitrary modulation constellation, the transmitted space-time matrix $\mathbf{S}$ may be defined as (see [27]) $\mathbf{S} = \sum_{q=1}^{Q} \mathbf{A}_q s_q$, where each symbol s_q is dispersed to the M spatial and T temporal dimensions using a specific dispersion matrix $\mathbf{A}_q \in \zeta^{M \times T}$ and $\mathbf{S}$ is attained by the linear combination of all of the weighted dispersion matrices. A schematic of the LDCs is specifically visualized in Figure 1.6.

Hence, it is beneficial to revisit the OSTBCs of Section 1.3.2.1 using the linear dispersion structure of Figure 1.6. More explicitly, if each individual dispersion matrix $\mathbf{A}_q$ is an orthogonal matrix and the set of dispersion matrices are orthogonal to each other, the resultant space-time codeword $\mathbf{S}$ has to be an orthogonal/unitary matrix achieving full diversity [30]. This design philosophy is that of the family of Linear STBCs (LSTBCs) [30]. Furthermore, by appropriately choosing the set of dispersion matrices depicted in Figure 1.6, the degree of orthogonality can be adjusted, leading to an increased design flexibility.

When compared with the layered STBCs discussed in Section 1.3.2.2, the LDCs of Figure 1.6 consist of Q layers corresponding to the number of symbols transmitted per space-time block. Since the parameter Q is unrestricted, LDCs are capable of supporting an arbitrary number of layers. Furthermore, rather than using only some of the $(M \times T)$ slots available, each layer of the LDCs spans all of the dimensions available, resulting in a high number of legitimate dispersion matrices. However, it may not be feasible to perfectly separate the high number of superimposed layers, unless sophisticated multi-stream receivers are used.

In summary, the LDCs of Figure 1.6 subsume all of the above-mentioned STBCs exhibiting diverse characteristics by simply employing different sets of dispersion matrices. Hence, the LDCs provide a natural framework for satisfying diverse design criteria. In the following, we offer a range of further remarks concerning LDCs.

- LDCs are suitable for arbitrary transmit and receive antenna configurations, combined with arbitrary modulation schemes.

- The maximum achievable diversity order of a LDC scheme is $N \cdot \min(M, T)$. This implies that increasing T beyond M does not provide any further advantage in terms of an increased diversity, whereas having $T < M$ could decrease the maximum achievable spatial diversity order. As expected, the receive diversity order is determined by the number of receive antennas N alone. For the proof of this please refer to [27].

- Every transmitted signal launched from each antenna is the linear combination of all of the information symbols weighted by a set of dispersion matrices of Figure 1.6, which ensures that different replicas of the same information symbol are attainable. In other words, LDCs demonstrate that spatial diversity can also be achieved without transmitting redundant information at the cost of reducing the normalized throughput R.

- All of the dispersion matrices $\mathbf{A}_q$ of Figure 1.6 can be described with the aid of a single Dispersion Character Matrix (DCM) χ specified by $\chi = [vec(\mathbf{A}_1), vec(\mathbf{A}_2), \ldots, vec(\mathbf{A}_Q)]$, where the $vec(\cdot)$ operation represents the vertical stacking of the columns of an arbitrary matrix. The benefit of using a single DCM is that there is no need to design Q separate dispersion matrices.

- The linear nature of Figure 1.6 enables the receiver to recover the source information, provided that the CSI is perfectly known at the receiver.

As far as the design of the set of dispersion matrices or the DCM is concerned, in their original form, the LDCs [10] were optimized to maximize the ergodic capacity. In practice, the channel's input is constituted by non-Gaussian symbols, such as discrete-amplitude Phase Shift Keying (PSK) and Quadrature Amplitude Modulation (QAM) signals, where a Discrete-input Continuous-output Memoryless Channel (DCMC) is encountered. Therefore, the more pertinent DCMC capacity is employed to optimize the LDCs [31]. Naturally, other optimization criteria can also be imposed on the DCM. For example, the authors of [27] specifically designed the LDCs to maximize the ergodic capacity, while maintaining a low BER. On the other hand, LDCs can also be optimized using the so-called determinant criterion [25], where having a non-vanishing determinant is guaranteed.

1.3.3 Differential STBCs using Co-located Antenna Elements

The primary focus of the codes discussed in Section 1.3.2 has been cases where the receiver has knowledge of the CSI. In practice, knowledge of the CSI is typically acquired using a channel sounding sequence, which has an exponentially increasing complexity as a function of the number of antennas. Furthermore, the relative frequency of estimating the channel has to be increased in proportion to the Doppler frequency. Finally, an excessive number of training symbols may be required. Hence, precious transmit power as well as valuable bandwidth are wasted. Therefore, differentially encoded low-complexity schemes, dispensing with pilot-based channel estimation and invoking non-coherent detection became attractive.

The conventional Differential PSK (DPSK) designed for single-antenna-aided systems differentially encodes the information between successive transmission symbols; thus the information can be recovered without the knowledge of the CSI, provided that the CSI does not change substantially between them. A DPSK scheme is constituted by the serial concatenation of a PSK modulator and a differential encoder. Here, we extend this philosophy to multiple-antenna-aided systems and the schematic of the resultant DSTBC system is portrayed in Figure 1.7.

More explicitly, the 'space-time mapper' of Figure 1.7 maps the nth differentially encoded transmission matrix $\mathbf{S}_n$ to all of the spatial and temporal slots, whereas the 'differential encoder' of Figure 1.7 correlates the consecutive transmission matrices by

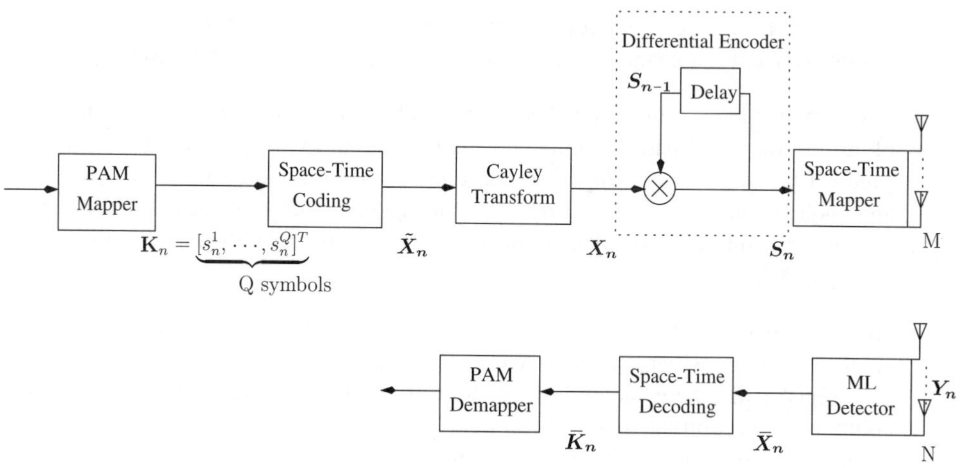

Figure 1.7: Schematic of a DLDC $(MNTQ)$ scheme equipped with M transmit and N receive antennas and employing the Cayley transform, while transmitting Q symbols over T time slots using $\mathbf{S}_n$. PAM, Pulse Amplitude Modulation.

$\mathbf{S}_n = \mathbf{S}_{n-1} \cdot \mathbf{X}_n$, where $\mathbf{X}_n$ is the space-time coded information matrix. Furthermore, this differential encoding process restricts the set of matrices $\mathbf{X}_n$ to be unitary, otherwise the product $\mathbf{S}_n = \mathbf{X}_n \mathbf{X}_{n-1} \cdots \mathbf{X}_1$ may become zero, infinity, or both in different spatial and temporal directions. In other words, the challenge of designing DSTBCs can be described as that of designing a family of STBCs where all the space-time matrices $\mathbf{X}_n$ are unitary.

Recall that the OSTBCs detailed in Section 1.3.2.1 generate orthogonal/unitary matrices by default; hence they become natural candidates for employment in DSTBC designs. Since the LDCs exhibit a high design flexibility as demonstrated in Section 1.3.2.3, it is desirable to retain the linear dispersion architecture in the differential design. Hence, the 'space-time encoding' block of Figure 1.7 is configured to generate space-time codewords $\tilde{\mathbf{X}}_n$ obeying the linear dispersion structure of Figure 1.6. However, even if each individual dispersion matrix is a unitary matrix, there is no guarantee that their weighted sum $\tilde{\mathbf{X}}_n$ will automatically become a unitary matrix. Hence, the 'Cayley transform' [10] block of Figure 1.7 is introduced in order to provide an efficient way of projecting the linearly structured matrix $\tilde{\mathbf{X}}_n$ into a unique unitary matrix $\mathbf{X}_n$, which potentially facilitates the differential encoding of Figure 1.7. Apart from its additional computational complexity, the Cayley transform requires the employment of real-valued modulation schemes in order to generate the unitary matrices. Also note that the DSTBCs typically suffer from a 3 dB penalty in comparison to the STBCs having perfect CSI, owing to the doubled equivalent channel noise encountered during the detection.

In summary, the design of DSTBCs based on coherently detected STBCs is facilitated by the 'unitary' constraint. Furthermore, the philosophy of LDCs may be extended to the differential encoding domain and the resultant Differential LDCs (DLDCs) based on the Cayley transform [32] provide a general framework for DSTBCs. Consequently, DLDCs are capable of supporting arbitrary antenna configurations as well as a dynamically reconfigurable throughput. Similarly to LDCs, all of the dispersion matrices of a DLDC scheme can be characterized by a single DCM. On the other hand, the performance of DLDCs is affected by the rate of Doppler-induced channel fluctuations.

1.3.4 Cooperative STBCs using Distributed Antenna Elements

The STBC techniques detailed in Sections 1.3.2 and 1.3.3 provide promising solutions in the context of co-located MIMO systems requiring reliable wireless communications at high rates. However, it may not always be practical to accommodate multiple antennas at the MSs, owing to cost, size and other hardware limitations. A further limitation of having co-located MIMO elements is that even at relatively large element separations their elements may not benefit from independent fading, when subjected to shadow-fading imposed, for example, by large-bodied vehicles or other shadowing local paraphernalia. As a remedy, the concept of cooperative MIMOs has been proposed for cellular systems as an attempt to attain a better communication efficiency beyond that permitted by a single node's resources. More specifically, a group of mobile nodes, known as relays, 'share' their antennas with other users to create a VAA to provide spatial diversity gain.

Owing to the philosophical similarities between the cooperative MIMO and the co-located MIMO systems, numerous space-time block coding techniques have been 'transplanted' into relay-aided schemes in order to achieve cooperative diversity, based on either Amplify-and-Forward (AF) strategies or Decode-and-Forward (DF) arrangements. It was Laneman and Wornell [24] who first proposed to employ OSTBCs for cooperative MIMO systems, where each relay transmits according to a different column of the orthogonal STBC matrix. In this section, we focus our attention on the employment of the LDC structure in the context of cooperative MIMO systems, namely in twin-layer Cooperative LDCs (CLDCs).

Figure 1.8 portrays the schematic of the cooperation-aided Uplink (UL) system using the above-mentioned twin-layer CLDC. As seen in Figure 1.8, each transmission block consists of two intervals, namely the broadcast interval of duration T_1 and the cooperation interval of length T_2. This scheme supports the cooperation of M relays transmitting Q information symbols per block to the base station equipped with N receive antennas, provided that the total number of channel uses T obeys $T = T_1 + T_2$.

The assumptions and the rationale of this model are summarized as follows.

- All of the relays of Figure 1.8 are assumed to transmit synchronously. Quasi-synchronous transmissions can be accomplished, when the relative delays between the relays are significantly shorter than the symbol duration.
- All of the nodes of Figure 1.8 are assumed to have a single antenna and hence operate in half-duplex mode, i.e. at any point of time, a node can either transmit or receive. This constraint is imposed to prevent the high-power transmit signal from contaminating the low-power received signal, for example, by the non-linear distortion-induced out-of-bound emissions routinely encountered at the transmitter.
- All of the relays of Figure 1.8 transmit and receive on the same frequency as the source node, in order to avoid wasting or occupying additional bandwidth.
- No communication is permitted between the relays, in an effort to minimize the total network traffic. The relays may use the same previously unallocated time slot for their reception and transmission.
- Since the simple AF strategy is adopted, only linear combination operations are performed at the relays before retransmitting the signals dispersed to the cooperating MIMO elements to the BS.
- We confine the total number of channel uses of the twin-layer CLDC scheme of Figure 1.8 to T. Hence, by appropriately adjusting the parameters T_1 or T_2, different

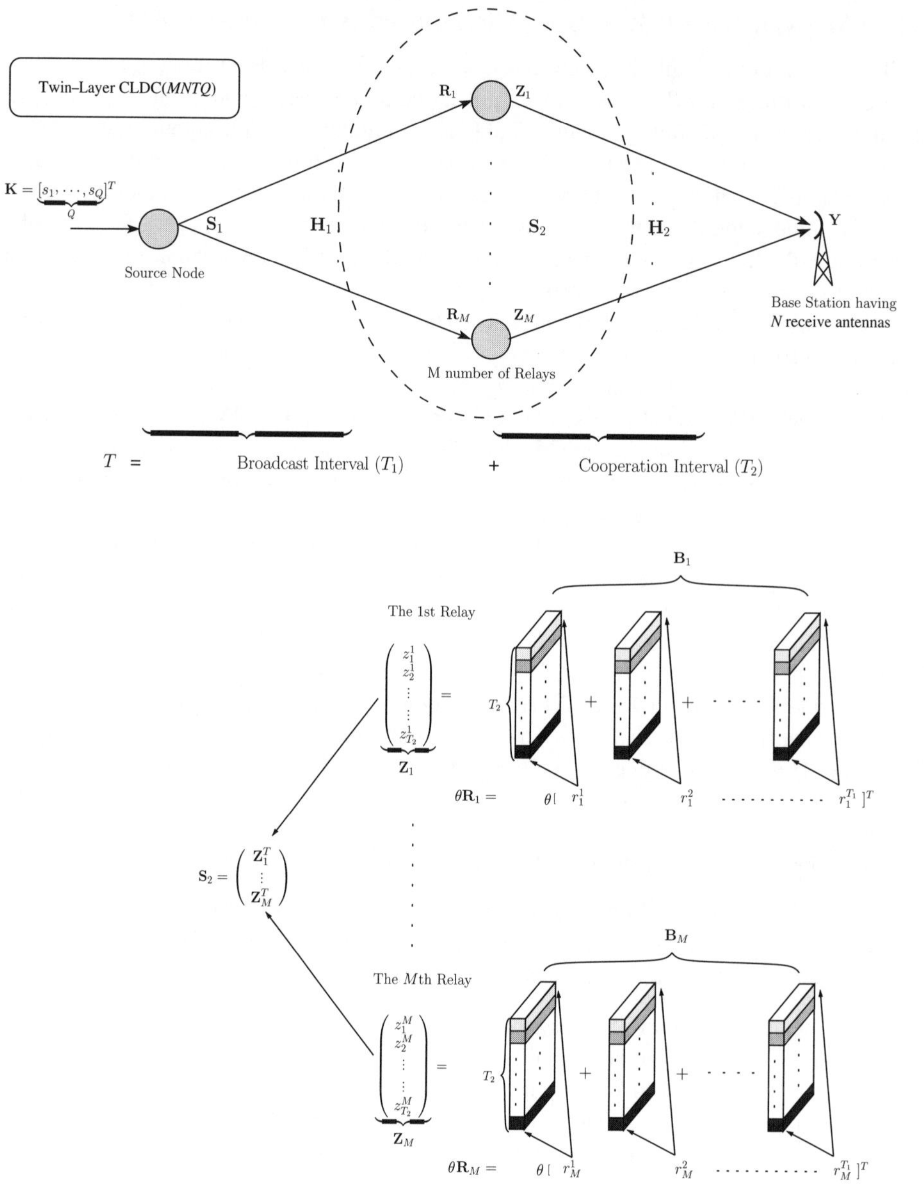

Figure 1.8: Schematic of the cooperation-aided uplink system employing twin-layer CLDCs.

degrees of freedom can be provided for the broadcast interval as well as for the cooperation interval.

- At any given time, the total transmit power of the twin-layer CLDC scheme of Figure 1.8 is normalized to unity.

During the broadcast interval T_1, the first-layer dispersion matrix χ_1 is responsible for dispersing the source information vector $\mathbf{K} = [s_1, \ldots, s_Q]^{\mathrm{T}}$ to all of the T_1 temporal slots.

The resultant space-time codeword $\mathbf{S}_1$ of Figure 1.8 is transmitted through independent source-to-relay Rayleigh fading channels. The received signal vector of the mth relay is $\mathbf{R}_m$. During the cooperation interval T_2, the relays form a VAA and cooperatively transmit the space-time codeword $\mathbf{S}_2$ of Figure 1.8 to the BS based on CLDC's second-layer dispersion matrix. More explicitly, the process of forming the cooperative space-time codeword $\mathbf{S}_2$ for the cooperation interval is also visualized in Figure 1.8. Each relay contributes one row of the cooperative space-time codeword $\mathbf{S}_2$ by dispersing the corresponding normalized received signal vector $\theta\mathbf{R}_m = \theta[r_m^1, \ldots, r_m^{T_1}]^{\mathrm{T}}$ to the available T_2 temporal slots using the pre-assigned dispersion matrix. Again, the dispersion matrices from all of the relays can be characterized by a single DCM χ_2. Hence, the receiver can recover the source information by exploiting the linearity of the CLDCs, provided that the CSI as well as the dispersion matrices χ_1 and χ_2 are known at the receiver. Since only T_2 time slots are used for achieving cooperative diversity and the relays only have access to the noisy version of the transmitted information, the maximum achievable diversity order of the CLDCs becomes $D \approx N \cdot \min(M, T_2)$.

Hence, we conclude that the fundamental difference between cooperative and co-located MIMO systems is the existence of the broadcast interval, which is used by the relays to attain preferably perfect but typically imperfect source information, depending on the specific cooperation strategy employed. As a result, instead of employing a single DCM as in the LDC scheme, the CLDC scheme requires a pair of DCMs (χ_1, χ_2) in order to characterize the transmission regime of both the broadcast interval and the cooperation interval, respectively.

1.3.5 Performance for Imperfect Channel Estimates and Shadow-fading

In this section, we provide a set of comparisons between the family of LDCs, DLDCs and CLDCs in order to evaluate their advantages as well as limitations, when communicating in small-scale or large-scale fading scenarios as well as when having perfect or imperfect CSI at the receiver. All of the simulation parameters are listed in Table 1.2. Observe in Table 1.2 that we set $M = T$ and assume that the channels were subjected to Rayleigh fading having $f_d = 10^{-2}$ in order to enable the adequate operation of the DLDCs based on the Cayley transform. Hence, the group of LDCs and DLDCs have the potential to achieve the full attainable diversity order of $D = N \cdot \min(M, T)$ (see [27]) in comparison to the reduced maximum diversity order $D \approx N \cdot \min(M, T_2)$ of the CLDCs.

For Comparison A of Table 1.2, Figure 1.9 characterizes the achievable throughput of the group of LDCs, DLDCs and CLDCs recorded at $\mathrm{BER} = 10^{-4}$, when the wireless channels were subjected to small-scale Rayleigh fading and perfect CSI was available at the receiver. Observe in Figure 1.9 that the class of LDCs is capable of operating at the lowest SNR at a certain throughput. The group of DLDCs suffers from the usual 3 dB SNR penalty in comparison to that of the LDCs, since no CSI was exploited. The family of CLDCs operates at SNRs further away from that of the LDCs, owing to their reduced achievable diversity order as well as due to having the noisy rather than perfect version of the source information.

For Comparison B of Table 1.2, Figure 1.10 characterizes the effective throughput of the LDCs, the DLDCs and the CLDCs of Table 1.2 recorded at $\mathrm{BER} = 10^{-4}$, when the wireless channels were subjected to small-scale Rayleigh fading and the receiver has imperfect CSI. We assume that the channel estimation errors obey the Gaussian distribution and the degree of the CSI estimation errors is governed by the ratio ω (dB) with respect to the received signal power. Hence, the perfect CSI scenario corresponds to $\omega = -\infty$. Observe in Figure 1.10 that the family of DLDCs demonstrated a significant advantage over the LDCs at a high

Table 1.2: Comparison of the LDCs of Section 1.3.2.3, the DLDCs of Section 1.3.3 and the CLDCs of Section 1.3.4, when communicating over small-scale/large-scale fading channels and having perfect/imperfect CSI at the receiver.

	LDC	DLDC	CLDC
M	3	3	3
N	2	2	2
T	3	3	$T_1 = 1, T_2 = 2$
Q	1, 2, 3	1, 2, 3	1, 2, 3
Modulation	BPSK	BPSK	BPSK
Mapping	Gray mapping	Gray mapping	Gray mapping
Detector	ML	ML	ML
Doppler frequency	$f_d = 10^{-2}$	$f_d = 10^{-2}$	$f_d = 10^{-2}$
Diversity	$D = 6$	$D = 6$	$D \approx 4$
Comparison A	Small-scale Rayleigh fading, perfect CSI in Figure 1.9		
Comparison B	Small-scale Rayleigh fading, imperfect CSI in Figure 1.10		
Comparison C	Large-scale shadowing, perfect CSI in Figure 1.11		

throughput, even when the channel estimation errors were as low as $\omega = -10$ dB. Since the group of CLDCs has an error floor higher than $\text{BER} = 10^{-4}$, the associated throughput curve was omitted from Figure 1.10. This phenomenon suggested that the CLDCs are more sensitive to the channel estimation errors than the LDCs. This is because the CLDCs require the CSI knowledge of both the source-to-relay and the relay-to-destination channels in comparison to the classic single-phase direct transmission regime of co-located MIMO systems.

Finally, in Comparison C of Table 1.2, Figure 1.11 characterizes the throughput of the LDCs, the DLDCs and the CLDCs of Table 1.2 recorded at $\text{BER} = 10^{-4}$, when the communication channels were subjected to shadowing and the receiver had access to perfect CSI. The shadow-fading effect was assumed to have log-normal distribution and was governed by a random Gaussian variable having zero mean and a standard deviation of Ω (dB). Observe in Figure 1.11 that the family of CLDCs designed for the cooperative MIMO systems has the best ability to combat the effect of large-scale shadowing with the aid of relays. Compared with the small-scale Rayleigh fading performance curves of Figure 1.9, the SNR required for the group of LDCs in Table 1.2 to maintain a BER of 10^{-4} increased by about 17 dB, even though the receiver had access to perfect CSI.

Again, our investigations indicate that the LDCs obeying the structure of Figure 1.6 are ideal for small-scale fading environments, when near-perfect CSI is available. On the other hand, the DLDCs having the structure of Figure 1.7 constitute the most appropriate solution, when the CSI is unavailable or the channel estimation would impose severe errors. Finally, when large-scale shadowing dominates the achievable performance, the family of CLDCs obeying the structure of Figure 1.8 remains capable of maintaining reliable wireless communications.

In conclusion of this brief section, the design guidelines of diverse STBCs schemes found in the open literature were considered. More explicitly, we demonstrated that the linear dispersion structure unifies the orthogonal approach as well as the layered architecture. The flexibility of the LDCs allows them to be designed according to diverse constraints and to be adapted to various MIMO scenarios. Furthermore, the linkage between STBCs and

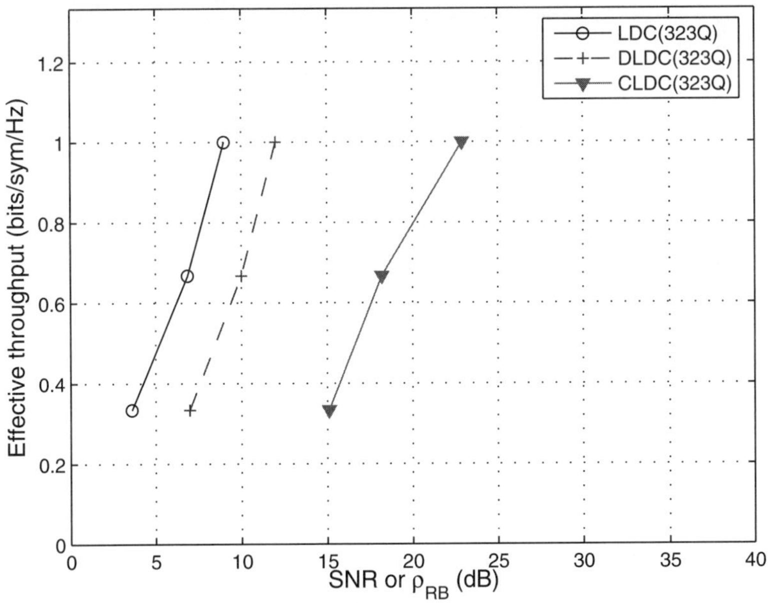

Figure 1.9: Throughput comparison for the LDCs of Section 1.3.2.3, the DLDCs of Section 1.3.3 and the CLDCs of Section 1.3.4 recorded at $\text{BER} = 10^{-4}$, when communicating over small-scale Rayleigh fading channels and assuming that the perfect CSI was known by the receiver. All of the system parameters are summarized in Table 1.2.

DSTBCs may be established with the aid of the 'unitary' constraint, which enables us to invoke the LDC structure also in the differential encoding domain. In the case of cooperative MIMO systems, the flexible linear structure remains applicable, but an additional dispersion character matrix is required to characterize the transmissions during the broadcast interval. Finally, the rudimentary performance results provided explicitly demonstrated the suitable application scenarios for the various LDCs, DLDCs and CLDCs considered.

1.4 Historical Perspective and State-of-the-Art Contributions

1.4.1 Co-located MIMO Techniques

MIMO systems exhibit higher capacity than single-antenna-aided systems. Multiple antennas can be used to provide diversity gains and hence a better BER performance or multiplexing gains, in order to attain a higher throughput. In addition, multiple antennas can be used at the transmitter or receiver in order to attain a beamforming gain. On the other hand, multiple antennas can be employed in order to attain diversity gains and multiplexing gains as well as beamforming gains as shown in Figure 1.2. The terminology of co-located MIMOs refers to the systems where the multiple antennas are located at the same transmitter or receiver station. In the following, we give an overview of the family of multiple antennas, when used for achieving diversity, multiplexing or beamforming gains.

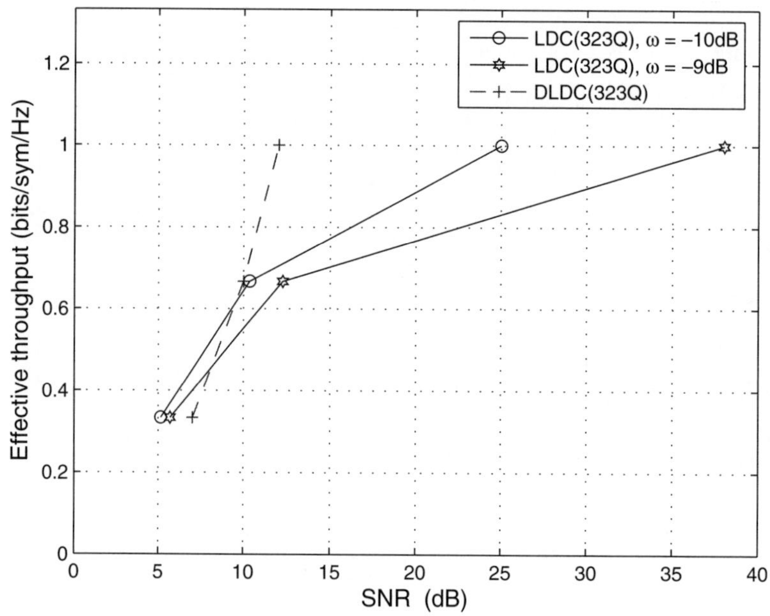

Figure 1.10: Throughput comparison for the LDCs of Section 1.3.2.3, the DLDCs of Section 1.3.3 and the CLDCs of Section 1.3.4 recorded at $\text{BER} = 10^{-4}$, when communicating over small-scale Rayleigh fading channels and having imperfect CSI governed by ω (dB). All of the system parameters are summarized in Table 1.2.

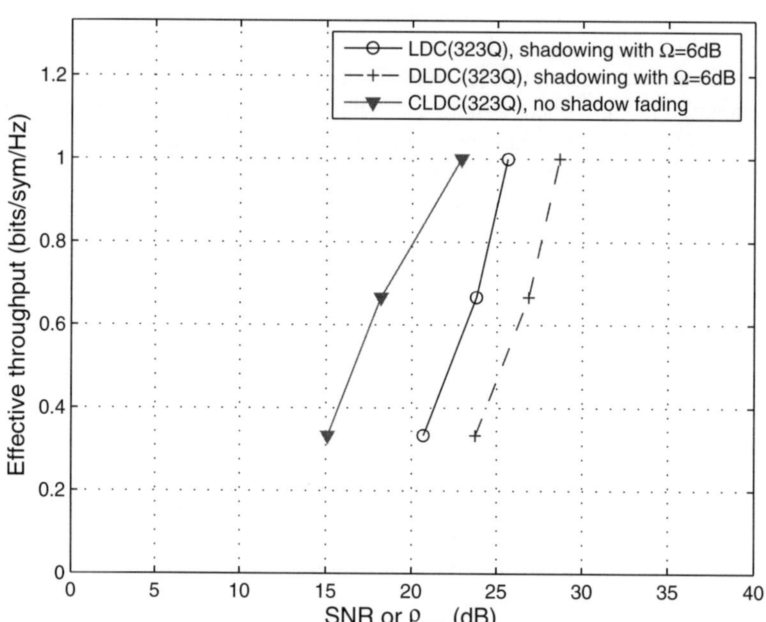

Figure 1.11: Throughput comparison for the LDCs of Section 1.3.2.3, the DLDCs of Section 1.3.3 and the CLDCs of Section 1.3.4 recorded at $\text{BER} = 10^{-4}$, when the channels were subjected to large-scale shadowing and assuming that the perfect CSI was known at the receiver. All of the system parameters are summarized in Table 1.2.

1.4.1.1 Diversity Techniques

Communication in the presence of channel fading has been one of the grand research challenges in recent times. In a fading channel, the associated severe attenuation often results in decoding errors. A natural way of overcoming this problem is to allow the receiver to have several replicas of the same transmitted signal, while assuming that at least some of them are not severely attenuated. This technique is referred to as diversity, where it is possible to attain diversity gains by creating independently fading signal replicas in the time, frequency or spatial domain.

The major coherent spatial diversity techniques are summarized in Tables 1.3 and 1.4 (also see Section 1.2.1).

A common feature of all the above-mentioned schemes is that they use coherent detection, which assumes the availability of accurate CSI at the receiver. In practice, the CSI of each link between each transmit and each receive antenna pair has to be estimated at the receiver either blindly or using training symbols. However, channel estimation invoked for all of the antennas substantially increases both the cost and the complexity of the receiver. Furthermore, when the CSI fluctuates dramatically from burst to burst, an increased number of training symbols has to be transmitted, potentially resulting in an undesirably high transmission overhead and wastage of transmission power. Therefore, it is beneficial to develop low-complexity techniques that do not require any channel information and thus are capable of mitigating the complexity of MIMO-channel estimation.

A detection algorithm designed for Alamouti's scheme [11] was proposed in [52], where the channel encountered at time instant t was estimated using the pair of symbols detected at time instant $t-1$. The algorithm, nonetheless, has to estimate the channel during the very first time instant using training symbols and hence is not truly differential. Tarokh and Jafarkhani [53, 62] proposed a differential encoding and decoding algorithm for Alamouti's scheme [11] using real-valued phasor constellations in which the transmitted signal can be demodulated either with or without CSI at the receiver. The resultant differential decoding aided non-coherent receiver performs within 3 dB from the coherent receiver assuming perfect channel knowledge at the receiver. The differential scheme of [53] was restricted to complex-valued PSK modulation. The twin-antenna-aided differential STBC scheme of [53] was extended to QAM constellations in [58].

DSTBC schemes designed for multiple antennas were proposed in [56] for real-valued constellations. Afterwards, the authors of [58, 59] developed a DSTBC scheme that supports non-constant modulus constellations combined with four transmit antennas. This extension, however, requires knowledge of the received power in order to appropriately normalize the received signal. The received power was estimated blindly using the received differentially encoded signals without invoking any channel estimation techniques or transmitting any pilot symbols. In [54], a differential modulation scheme was proposed for the sake of attaining transmit diversity based on unitary space-time codes [63]. The proposed scheme can be employed in conjunction with an arbitrary number of transmit antennas. Around the same time, a similar differential scheme was also proposed in [55] based on the employment of group codes.

Zhu and Jafarkhani [60] proposed a differential modulation scheme based on QOSTBCs, which were compared with that of [56] and resulted in a reduced BER as a benefit of providing full diversity. In addition, a new class of QOSTBCs was proposed in [61], which presented a simple differential decoding scheme that avoids signal constellation expansion. The major contributions on differential spatial diversity techniques are summarized in Table 1.5.

Table 1.3: Major coherent spatial diversity techniques (Part 1).

Year	Author(s)	Contribution
1959	Brennan [33]	Introduced and provided analysis for the three combining techniques: SC, MRC and EGC.
1991	Wittneben [34]	Proposed a bandwidth-efficient transmit diversity technique, where different BSs transmit the same signal.
1993	Wittneben [35]	Proposed a modulation diversity scheme in a system equipped with multiple transmit antennas.
	Seshadri and Winters [36]	Proposed a transmit diversity scheme that was inspired by the delay diversity design of Wittneben [35].
1994	Winters [37]	Proved that the diversity advantage of the scheme proposed in [34] is equal to the number of transmit antennas.
1996	Eng *et al.* [38]	Compared several diversity combining techniques in a Rayleigh fading transmission with coherent detection and proposed a new second-order SC technique.
1998	Alamouti [11]	Discovered a transmit diversity scheme using two transmit antennas with simple linear processing at the receiver.
	Tarokh *et al.* [8]	Proposed a complete study of design criteria for maximum diversity and coding gains in addition to the design of STTCs.
1999	Tarokh *et al.* [12, 25]	Generalized Alamouti's diversity scheme [11] to more than two transmit antennas.
	Guey *et al.* [39]	Derived the criterion for designing the maximum transmit diversity gain.
2001	Hochwald *et al.* [13]	Proposed the twin-antenna-aided STS scheme.
	Jafarkhani [14]	Designed rate-one STBC, which are quasi-orthogonal and provide partial diversity gain.
2002	Hassibi and Hachwald [10]	Proposed the LDCs that provide a flexible trade-off between space-time coding and spatial multiplexing.
	Stoica and Ganesan [40]	Compared the performance of STBCs when employing different estimation/detection techniques and proposed a blind detection scheme dispensing with the pilot symbols transmission for channel estimation.

Table 1.4: Major coherent spatial diversity techniques (Part 2).

Year	Author(s)	Contribution
2003	Wang and Xia [42]	Derived upper bounds for the rates of complex orthogonal STBCs.
	Su *et al.* [43]	Introduced the concept of combining OSTBC designs with the principle of sphere packing.
2005	Zhang and Gulliver [44]	Derived the capacity and probability of error expressions for PSK/PAM/QAM modulation with STBC for transmission over Rayleigh, Ricean and Nakagami fading channels.
2006	Liew and Hanzo [45]	Studied the performance of STTC and STBC in the context of wideband channels using adaptive orthogonal frequency division multiplex modulation.
2007	Alamri *et al.* [46]	Modified the SP demapper of [43] for the sake of accepting the *a priori* information passed to it from the channel decoder as extrinsic information.
2008	Luo and Leib [47]	Combined OSTBCs with delay diversity and designed special symbol mappings for maximizing the coding advantage.

1.4.1.2 Multiplexing Techniques

STBC and STTC are capable of providing diversity gains for the sake of improving the achievable system performance. However, this BER performance improvement is often achieved at the expense of a rate loss since the STBC and STTC may result in a throughput loss compared with single-antenna-aided systems. As a design alternative, a specific class of MIMO systems was designed for improving the attainable spectral efficiency of the system by transmitting the signals independently from each of the transmit antennas, hence resulting in a multiplexing gain.

The basic principle of spatial multiplexing can be summarized as follows. The source bit sequence at the transmitter side is split into N_t sequences, which are modulated and then transmitted simultaneously from the N_t transmit antennas using the same carrier frequency. At the receiver side, interference cancelation is employed in order to separate the different transmitted signals. In the case of narrowband frequency flat fading, there are several decoding algorithms designed for interference cancelation at the receiver side of the spatial multiplexing systems. The different receivers can be characterized by a trade-off between the achievable performance and the complexity imposed. A low-complexity receiver is constituted by the ZF or the Minimum Mean Square Error (MMSE) technique [64, 65]. However, when we employ the ZF receiver, the attainable BER performance is typically poor in addition to imposing the condition that the number of receive antennas should be at least equal to the number of transmit antennas. The optimum receiver is the ML receiver [6], which is capable of achieving full diversity gain, i.e. the same diversity order, as the number of receive antennas. However, a major drawback of the ML receiver is its complexity, which grows exponentially with the number of transmit antennas and the number of bits per symbol

Table 1.5: Major differential spatial diversity techniques.

Year	Author(s)	Contribution
1998	Tarokh *et al.* [52]	Proposed a detection algorithm for the Alamouti scheme [11] dispensing with channel estimation.
1999	Tarokh and Jafarkhani [53]	Proposed a differential encoding/decoding of Alamouti's scheme [11] with PSK constellations.
2000	Hochwald and Swelders [54]	Proposed a differential modulation scheme for transmit diversity based on unitary space-time codes.
	Hughes [55]	Proposed a differential modulation scheme that is based on group codes.
2001	Jafarkhani and Tarokh [56]	Proposed a differential detection scheme for the multiple-antenna STBC [12].
2002	Schober and Lampe [57]	Proposed non-coherent receivers for DSTM that can provide satisfactory performance in fast fading, unlike the conventional differential schemes that perform poorly in fast fading.
2003	Hwang *et al.* [58, 58]	Extended the scheme of [56] to QAM constellations.
2004	Nam *et al.* [59]	Extended the scheme of [58, 58] to four transmit antennas and QAM constellations.
2005	Zhu and Jafarkhani [60]	Proposed a differential modulation scheme based on quasi-orthogonal STBCs, which when compared with that of [56] results in a lower BER and provides full diversity.
2007	Song and Burr [61]	Proposed a new class of quasi-orthogonal STBCs and presented a simple differential decoding scheme for the proposed structures that avoids signal constellation expansion.

employed by the modulation scheme. Fortunately, the complexity of the ML decoders can be reduced by employing sphere decoders [66–68] that are capable of achieving a similar performance to the ML decoders at a fraction of their complexity.

In [69] Foschini proposed a multi-layer MIMO structure, known as the Diagonal Bell Labs' Layered Space-Time (D-BLAST) scheme, which is in principle capable of approaching the substantial capacity of MIMO systems. (The diagonal approach implies that the signal mapped to the consecutive antenna elements is delayed in time, which has the potential of subjecting the delayed signal components of a space-time symbol to more independent fading, hence leading to a potential diversity gain.) The D-BLAST signal may be subjected to low-complexity linear processing for decoding the received signals. However, the diagonal approach suffers from a potentially high implementation complexity that led Wolniansky *et al.* to propose another version of BLAST, which is known as V-BLAST [15]. In V-BLAST, each transmit antenna simultaneously transmits independent data over the same carrier frequency band. At the receiver side, provided that the number of receive antennas is higher than or equal to the number of transmit antennas, a low complexity *serial* decoding algorithm may be applied to detect the transmitted data. The V-BLAST transceiver is capable of providing a substantial increase of a specific user's effective bit rate without the need for any increase in the transmitted power or the system's bandwidth. However, its impediment is that it was not designed for exploiting transmit diversity. Furthermore, the decision errors of a particular antenna's detector propagate to other bits of the multi-antenna symbol, when erroneously canceling the effects of the sliced bits from the composite signal. The V-BLAST detector first selects the layer (in the case of V-BLAST, this corresponds to each of the transmit antennas) with the largest SNR and estimates the transmitted bits of that layer, while treating the other layers as interference. The detected symbol is then subtracted from the received signal and then the layer with the second highest SNR is selected for decoding. The procedure is repeated for all of the layers. The BER performance of each layer is different and it depends on the received SNR of each layer. The first decoded layer has the highest SNR, while the layers detected later have a higher diversity order, since they suffer from less interference.

The BLAST detection algorithm is based on SIC, which was originally proposed for multi-user detection in Code Division Multiple Access (CDMA) systems [76]. Several BLAST detectors have been proposed in the literature for either reducing the complexity [74, 77–81] or for improving the attainable BER performance [73, 82–87]. An alternative design approach contrived for spatial multiplexing using fewer receive antennas than transmit antennas was proposed in [88] based on group Maximum *A Posteriori* (MAP) detection. In [72, 89] a spatial multiplexing scheme referred to as Turbo-BLAST was proposed, which uses quasi-random interleaving in conjunction with an iterative receiver structure, in order to separate the individual layers. The major spatial multiplexing techniques are summarized in Table 1.6.

1.4.1.3 Beamforming Techniques

According to Sections 1.4.1.1 and 1.4.1.2, it becomes clear that multiple antennas can be used for the sake of attaining either spatial diversity or spatial multiplexing gains. However, multiple antennas can also be used in order to improve the SNR at the receiver or the Signal-to-Interference-plus-Noise Ratio (SINR) in a multi-user scenario. This can be achieved by employing beamforming techniques [16, 90]. Beamforming constitutes an effective technique of reducing the multiple access interference, where the antenna gain is increased in the direction of the desired user, whilst reducing the gain towards the interfering users.

In a wireless communications scenario the transmitted signals propagate via several paths and hence are received from different directions/phases at the receiver. If the directions of the different propagation paths are known at the transmitter or the receiver, then beamforming techniques can be employed in order to direct the received beam pattern in the direction of the specified antenna or user [91, 92]. Hence, significant SNR gains can be achieved in comparison to a single antenna system. On the transmitter side, when the DOA of the dominant paths at the receiver is known for the transmitter, then the transmit power is concentrated in the direction of the target user, and less power is wasted in the other directions.

On the other hand, beamforming can be used in order to reduce the co-channel interference or multi-user interference. When using beamforming, each user adjusts their beam pattern to ensure that there are nulls in the directions of the other users, while there is a high directivity in the direction of the desired receiver [16, 93]. Hence, the system attains a SINR gain.

Table 1.6: Major spatial multiplexing techniques.

Year	Author(s)	Contribution
1996	Foschini [69]	Studied the encoding and decoding of the diagonal BLAST structure.
1998	Wolniansky *et al.* [15]	Introduced the V-BLAST architecture for reducing the implementation complexity of the diagonal approach.
1999	Golden *et al.* [70]	Provided the first real-time BLAST demonstrations.
2001	Benjebbour *et al.* [71]	Introduced the MMSE receiver for V-BLAST and introduced an ordering scheme for improving the attainable performance.
2002	Sellathurai and Haykin [72]	Studied the combination of BLAST architecture with that of a turbo code to improve its performance.
2003	Wubben *et al.* [73]	Proposed a detector for improving the attainable performance of V-BLAST.
2004	Zhu *et al.* [74]	Proposed a complexity-reduction algorithm for BLAST detectors.
2005	Huang *et al.* [75]	Proposed a new detection algorithm for BLAST based on the concept of particle filtering and provided a near ML performance at a reasonable complexity.

1.4.1.4 Multi-functional MIMO Techniques

V-BLAST is capable of achieving full multiplexing gain, while STBC can achieve full antenna diversity gain. Hence, it was proposed in [17] to combine the two techniques to provide both antenna diversity and spectral efficiency gains. More specifically, it was proposed that the antennas at the transmitter be partitioned into layers, where each layer uses STBC. At the receiver side, successive group interference cancelation can be applied to each layer before decoding the signals using ML STBC decoding. Therefore, by combining V-BLAST and STBC, an improved transmit diversity gain can be achieved as compared with pure V-BLAST, while ensuring that the overall bandwidth efficiency is higher than that of pure STBC owing to the independence of the signals transmitted by different STBC layers. Furthermore, the combined array processing proposed in [17] was improved in [18] by optimizing the decoding order of the different antenna layers. An iterative decoding algorithm was proposed in [18] that results in a full receive diversity gain for the combined V-BLAST STBC system.

In [19] the authors presented a transmission scheme referred to as Double Space-Time Transmit Diversity (D-STTD), which consists of two STBC layers at the transmitter, which is equipped with four transmit antennas, while the receiver is equipped with two antennas. The decoding of D-STTD presented in [19] is based on a linear decoding scheme presented in [105], where the authors provided a broad overview of STC and signal processing designed for high data rate wireless communications. A two-user scheme was presented in [105], where each user is equipped with a twin-antenna-aided STBC scheme transmitting at the same carrier frequency and in the same time slot. A two-antenna-aided receiver was implemented for the sake of decoding the two users' data, while eliminating the interference imposed by the users on each other's data. An extension to the idea of combining interference cancelation with STBC techniques was presented in [95, 97], where the STBC and interference cancelation arrangements were combined with CDMA for the sake of increasing the number of users supported by the system. A ZF decoder designed for the D-STTD was presented in [102] for the sake of reducing the decoding complexity. Finally, the authors of [96, 106] presented further results that compare the performance of STBC versus D-STTD and extended the applicability of the D-STTD scheme to more than two STBC layers.

Furthermore, in order to achieve additional performance gains, beamforming has been combined with spatial diversity as well as spatial multiplexing techniques. STBC has been combined with beamforming in order to attain a higher SNR gain in addition to the diversity gain [20, 99, 100, 107–109]. In [20], the authors combined conventional transmit beamforming with STBC, assuming that the transmitter has partial knowledge of the channel and derived a performance criterion for a frequency-flat fading channel. In addition, a particularly efficient solution was developed in [20] for the specific case of independently fading channel coefficients. More explicitly, the transmission scheme of [20] combines the benefits of conventional beamforming with those of orthogonal STBC. Furthermore, in [100] the performance of combined beamforming and STBC has been analyzed as a function of the number of antenna array groups. Explicitly, Zhu and Lim [100] compared the performance of the system combining beamforming with STBC, while using either a single or two antenna arrays, and studied the effect of the DOA on the attainable system performance. Finally, multiplexing techniques have been combined with beamforming techniques in [110–112]. The major multi-functional MIMO techniques are summarized in Tables 1.7 and 1.8.

1.4.2 Distributed MIMO Techniques

Wireless channels suffer from multipath propagation of the signals that results in channel fading. Employing multiple transmit antennas is a beneficial method that can be used to counteract the effects of the channel fading by providing diversity gains. Transmit diversity results in a significantly improved BER performance, when the different transmit antennas are spatially located so that the paths arriving from each transmit antenna to the destination experience independent fading, which can be achieved by having a distance between the different antennas that is significantly higher than the carrier's wavelength. However, considering a handheld mobile phone, it is not a feasible option to position the transmit antennas sufficiently far apart in order to achieve independent fading. On the other hand, the spatial fading correlation caused by insufficiently high antenna spacing at the transmitter or receiver of a MIMO system results in a degradation of both the achievable capacity and the BER performance of MIMO systems. The problem of correlation of the transmit signals can be circumvented by introducing a new class of MIMOs also referred to as distributed MIMOs or cooperative communications [113, 114].

The basic idea behind cooperative communications can be traced back to the idea of the relay channel, which was introduced in 1971 by Van der Meulen [115]. Cover and El Gamal [116] characterized the relay channel from an information theoretic point of view. In [118] Sendonaris *et al.* generalized the conventional relay model, where there is one source, one relay and one destination, to multiple nodes that transmit their own data as well as serving as relays for each other. The scheme of [118] was referred to as 'user cooperation diversity'. Sendonaris *et al.* presented in [113, 114] a simple user-cooperation methodology based on a DF signaling scheme using CDMA. In [119] the authors reported

Table 1.7: Major multi-functional MIMO techniques (Part 1).

Year	Author(s)	Contribution
1998	Naguib *et al.* [94]	Presented a multi-user scenario where each user employs STBC and the receiver applies interference cancelation for eliminating the co-channel interference and then uses ML decoding for the STBC of each user.
1999	Tarokh *et al.* [17]	Proposed to combine STBC with V-BLAST in order to provide both antenna diversity and spectral efficiency gains.
2000	Huang and Viswanathan [95]	Extended the idea of combining interference cancelation with STBC to multi-user scenarios using CDMA.
2001	Stamoulis *et al.* [96]	Proposed a simple decoder for the two-user system where each user employs STBC, and showed how the decoder can be extended to more users and then extended the results for frequency-selective channels.
2002	Onggosanusi *et al.* [19]	Presented the D-STTD scheme, which consists of two STBC blocks at the transmitter, which is equipped with four antennas, while the receiver is equipped with two antennas.
	Jongren *et al.* [20]	Combined conventional transmit beamforming with STBC assuming that the transmitter has partial knowledge of the channel and derived criteria for improving the system performance.
	Huang *et al.* [97]	Introduced a transmission scheme that can achieve transmit diversity and spatial separation and proposed a generalisation of the V-BLAST detector for CDMA signals.
	Soni *et al.* [98]	Designed a hybrid downlink technique for achieving both transmit diversity and transmit beamforming combined with DS-CDMA.
2003	Liu and Gunawan [99]	Combined the twin-antenna-aided Alamouti STBC with ideal beamforming in order to show that the system can attain a better performance while keeping full diversity and unity rate.
2004	Tao and Cheng [18]	Improved the design of [17] by optimizing the decoding order of the different antenna layers. Also proposed an iterative decoder that can achieve full diversity.
	Zhu and Lim [100]	Compared the performance of two systems combining beamforming with STBC, while using a single or two antenna arrays, and studied the effect of the DOA on the performance of the two schemes.

Table 1.8: Major multi-functional MIMO techniques (Part 2).

Year	Author(s)	Contribution
2005	Zhao and Dubey [101]	Compared the performance of the combined diversity and multiplexing systems while employing ZF, QR and MMSE group interference cancelation techniques.
	Lee *et al.* [102]	Proposed a computationally efficient ZF decoder for the D-STTD scheme [19] that achieves similar performance to the conventional ZF decoder but with less complexity.
2007	Sellathurai *et al.* [103]	Investigated the performance of multi-rate layered space-time coded MIMO systems and proposed a framework where each of the layers is encoded independently with different rates subject to equal per-layer outage probabilities.
2008	Ekbatani and Jafarkhani [104]	Combined STBC and transmit beamforming while using limited-rate channel state information at the transmitter. Also proposed a combined coding, beamforming and spatial multiplexing scheme over multiple-antenna multi-user channels that enables a low-complexity joint interference cancelation.
	Luo and Leib [47]	Considered a new class of full-diversity STCs that consist of a combination of delay transmit diversity with orthogonal STBCs and specially designed symbol mappings.

data rate gains and a decreased sensitivity to channel variations, where it was concluded that cooperation effectively mimics the multi-antenna scenario with the aid of single-antenna terminals. Dohler *et al.* [121] introduced the concept of VAAs that emulates Alamouti's STBC for single-antenna-aided cooperating users. Space-time coded cooperative diversity protocols for exploiting spatial diversity in a cooperative scenario were proposed in [24].

Cooperative communications has been shown to offer significant performance gains in terms of various performance metrics, including diversity gains [24, 124, 144] as well as multiplexing gains [128]. Hunter and Nosratinia [120] proposed the novel philosophy of coded cooperation schemes, which combine the idea of cooperation with the classic channel coding methods. Extension to the framework of coded cooperation was presented in [126], where the diversity gain of coded cooperation was increased with the aid of ideas borrowed from the area of space-time codes. In addition, a turbo-coded scheme was proposed in [126] in the framework of cooperative communications. Furthermore, the analysis of the performance benefits of channel codes in a coded-cooperation-aided scenario was performed in [127]. Laneman *et al.* [124] developed and analyzed cooperative diversity protocols and compared the DF, AF, selection relaying and incremental relaying signaling strategies.

Recently, there has been substantial research interest in the idea of soft relaying, where the relay passes soft information to the destination. In [129], it was argued that the DF signaling loses soft information and, hence, it was proposed to use soft DF signaling, where all operations are performed using the Log-Likelihood Ratio (LLR)-based representation of soft information. It was shown in [129] that the soft DF philosophy outperforms the DF and the AF signaling strategies. In [137] soft DF was also used, where the soft information was quantized, encoded and superimposed before transmission to the destination. In [134] soft-information-based relaying was employed in a turbo-coding scheme, where the relay derives parity – checking Binary PSK (BPSK) symbol estimates for the received source information and forwards the symbols to the destination. In [129, 134, 137] soft information relaying has been used, where it was shown that soft DF attains a better performance than hard DF. Furthermore, in [130, 135] distributed source coding techniques have been adopted for employment in wireless cooperative communications in order to improve the attainable performance. The major distributed MIMO techniques are summarized in Tables 1.9, 1.10 and 1.11.

1.5 Iterative Detection Schemes and their Convergence Analysis

The concept of concatenated codes was proposed in [145]. However, at the time of its conception it was deemed to have an excessive complexity and hence it failed to initiate immediate research interest. It was not until the discovery of turbo codes [146] that efficient iterative decoding of concatenated codes became a reality at a low complexity by employing simple constituent codes. Since then, the appealing iterative decoding of concatenated codes has inspired numerous researchers to extend the technique to other transmission schemes consisting of a concatenation of two or more constituent decoding stages [147–163].

For example, in [154] iterative decoding was invoked for exchanging extrinsic information between a soft-output symbol detector and an outer channel decoder in order to combat the effect of Inter-Symbol Interference (ISI). In [155] iterative decoding was carried out by exchanging information between an outer convolutional decoder and an inner Trellis Coded Modulation (TCM) decoder. The authors of [156, 157] presented a unified theory of Bit-Interleaved Coded Modulation (BICM). On the other hand, the employment of the iterative detection principle in [158] was considered for iterative soft demapping in the context of BICM, where a soft demapper was used between the multilevel demodulator and the channel decoder. In addition, iterative multi-user detection and channel decoding was proposed in [162] for CDMA schemes. Finally, in [163] an iteratively detected scheme was proposed for the Rayleigh fading MIMO channel, where an OSTBC scheme was considered as the inner code combined with an additional block code as the outer channel code.

It was shown in [166] that a recursive inner code is needed in order to maximise the interleaver gain and to avoid the average BER floor, when employing iterative decoding. This principle has been adopted by several authors designing serially concatenated schemes, where unity-rate inner codes were employed for designing low-complexity iterative-detection-aided schemes suitable for bandwidth- and power-limited systems having stringent BER requirements [167, 168, 170, 177, 180].

Semi-analytical tools devised for analysing the convergence behavior of iteratively decoded systems have attracted considerable research attention [167, 169, 172–175, 178, 181, 182]. In [169], ten Brink proposed the employment of the so-called EXtrinsic Information Transfer (EXIT) characteristics for describing the flow of extrinsic information between the soft-in soft-out constituent decoders. The computation of EXIT charts was further simplified in [174] to a time averaging, when the Probability Density Functions (PDFs) of

Table 1.9: Major distributed MIMO techniques (Part 1).

Year	Author(s)	Contribution
1971	Van der Meulen [115]	Investigated a simple three-node relay channel incorporating a transmitter, a relay and a receiver using a time-sharing approach.
1979	Cover and El Gamal [116]	Characterized the relay channel from an information theoretic point of view.
1983	Willems [117]	Introduced a partially cooperative communications scenario where the encoders are connected by communication links with finite capacities, which permit both encoders to communicate with each other. The paper also established the capacity region of the multiple access channel with partially cooperating encoders.
1998	Sendonaris *et al.* [118]	Generalized the relay model to multiple nodes that transmit their own data as well as serving as relays for each other.
2001	Laneman *et al.* [119]	Built upon the classical relay channel and exploited space diversity available at distributed antennas through coordinated transmission and processing by cooperating radios.
2002	Hunter and Nosratinia [120]	Proposed a user cooperation scheme for wireless communications in which the idea of cooperation was combined with the existing channel coding methods.
	Dohler *et al.* [121]	Introduced the concept of VAAs that emulates Alamouti's STBC for single-antenna-aided cooperating users.
2003	Sendonaris *et al.* [113, 114]	Presented a simple user-cooperation methodology based on a DF signaling scheme using CDMA.
	Laneman and Warnell [24]	Developed space-time coded cooperative diversity protocols for exploiting spatial diversity in a cooperation scenario, which can also be used for higher spectral efficiencies than repetition-based schemes.
	Valenti and Zhao [122, 123]	Proposed a turbo-coding scheme in a relay network.
2004	Laneman *et al.* [124]	Developed and analyzed cooperative diversity protocols and compared the DF, AF, selection relaying and incremental relaying.
	Nabar *et al.* [125]	Analyzed the spatial diversity performance of various signaling protocols.

Table 1.10: Major distributed MIMO techniques (Part 2).

Year	Author(s)	Contribution
2004	Janani *et al.* [126]	Presented two extensions to the coded cooperation framework [120]: increased the diversity of coded cooperation via ideas borrowed from space-time codes and applied turbo codes in the proposed relay framework.
	Stefanov and Erkip [127]	Analyzed the performance of channel codes that are capable of achieving the full diversity provided by user cooperation in the presence of noisy interuser channels.
2005	Azarian *et al.* [128]	Proposed cooperative signaling protocols that can achieve the diversity–multiplexing trade-off.
	Sneessens and Vandendorpe [129]	Proposed a soft DF signaling strategy that can outperform the conventional DF and AF.
	Hu and Li [130]	Proposed Slepian–Wolf cooperation that exploits distributed source coding technologies in wireless cooperative communication.
	Yu and Li [131]	Compared the AF and DF signaling schemes in practical scenarios.
2006	Hunter *et al.* [132, 133]	Developed the idea of coded cooperation [120] by computing BER and Frame Error Rate (FER) bounds as well as the outage probability of coded cooperation.
	Li *et al.* [134]	Employed soft information relaying in a BPSK modulated relay system employing turbo coding.
	Hu and Li [135]	Proposed Wyner–Ziv cooperation as a generalization of the Slepian–Wolf cooperation [130] with a compress-and-forward signaling strategy.
	Host-Madsen [136]	Derived upper and lower bounds for the capacity of four-node *ad hoc* networks with two transmitters and two receivers using cooperative diversity.
2007	Bui and Yuan [137]	Proposed soft information relaying where the relay LLR values are quantized, encoded and superimposedly modulated before being forwarded to the destination.
	Khormuji and Larsson [138]	Improved the performance of the conventional DF strategy by employing constellation rearrangement in the source and the relay.
	Bao and Li [139]	Combined the benefits of AF and DF and proposed a new signaling strategy referred to as decode–amplify–forward.
	Xiao *et al.* [140]	Introduced the concept of network coding in cooperative communications.

Table 1.11: Major distributed MIMO techniques (Part 3).

Year	Author(s)	Contribution
2008	Yue *et al.* [141]	Compared the multiplexed coding and superposition coding in the coded cooperation system.
	Zhang *et al.* [142]	Proposed a distributed space-frequency coded cooperation scheme for communication over frequency-selective channels.
	Wang and Giannakis [143]	Introduced the complex field network coding approach that can mitigate the throughput loss in the conventional signaling schemes and attain full diversity gain.

the information communicated between the input and output of the constituent decoders are both symmetric and consistent. A tutorial introduction to EXIT charts can be found in [181]. The concept of EXIT chart analysis has been extended to three-stage concatenated systems in [171, 174, 178]. The major contributions on iterative detection and its convergence analysis are summarized in Tables 1.12 and 1.13.

1.6 Outline and Novel Aspects of the Monograph

1.6.1 Outline of the Book

Having briefly reviewed the literature of STC, concatenated schemes and iterative decoding, and having studied the convergence behavior of iterative schemes, let us now outline the organization of the book.

- **Chapter 2: Space-Time Block Code Design using Sphere Packing**

  In this chapter, we consider the theory and design of STBCs using SP modulation, referred to here as STBC-SP. We first summarize the design criteria of space-time coded communication systems in Section 2.2. In Section 2.3, we emphasize the design criteria relevant for time-correlated fading channels, where both the pairwise error probability and the corresponding design criterion are presented in Section 2.3.2. In Section 2.4, orthogonal space-time designs combined with SP modulation are considered for space-time signals, where the motivation behind the adoption of SP modulation in conjunction with orthogonal design is discussed in Section 2.4.2. Section 2.4.4 discusses the problem of constructing a SP constellation having a particular size L. The capacity of STBC-SP schemes employing $N_t = 2$ transmit antennas is derived in Section 2.4.5, demonstrating that STBC-SP schemes exhibit a higher capacity than conventionally modulated STBC schemes. Finally, the performance of STBC-SP schemes is presented in Section 2.5, demonstrating that STBC-SP schemes are capable of outperforming STBC schemes that employ conventional modulation (i.e. PSK, QAM).

- **Chapter 3: Turbo Detection of Channel-coded STBC-SP Schemes**

  In this chapter, we demonstrate that the performance of STBC-SP systems can be further improved by concatenating SP-aided modulation with channel coding and

Table 1.12: Major concatenated schemes and iterative detection (Part 1).

Year	Author(s)	Contribution
1966	Forney [145]	Promoted concatenated codes.
1974	Bahl *et al.* [164]	Invented the MAP algorithm.
1993	Berrou *et al.* [146]	Invented the turbo codes and showed that the iterative decoding is an efficient way of improving the attainable performance.
1995	Robertson *et al.* [165]	Proposed the log-MAP algorithm that results in similar performance to the MAP algorithm but with significantly lower complexity.
	Divsalar and Pollara [147]	Extended the turbo principle to multiple parallel concatenated codes.
1996	Benedetto and Montorsi [148]	Extended the turbo principle to serially concatenated block and convolutional codes.
1997	Benedetto *et al.* [155]	Proposed an iterative detection scheme where iterations were carried out between the outer convolutional code and an inner TCM decoder.
	Caire *et al.* [156, 157]	Presented the BICM concept with its design rules.
	Li and Ritcey [159–161]	Presented the BICM with iterative detection scheme.
1998	Benedetto *et al.* [149, 166]	Studied the design of multiple serially concatenated codes with interleavers.
	ten Brink *et al.* [158]	Introduced a soft demapper between the multilevel demodulator and the channel decoder in an iteratively detected coded system.
1999	Wang and Poor [162]	Proposed iterative multi-user detection and channel decoding for coded CDMA systems.
2000	Divsalar *et al.* [167, 168]	Employed unity-rate inner codes for designing low-complexity iterative detection schemes suitable for bandwidth- and power-limited systems having stringent BER requirements.
	ten Brink [169]	Proposed the employment of EXIT charts for analyzing the convergence behavior of iteratively detected systems.
2001	Lee [170]	Studied the effect of precoding on serially concatenated systems with ISI channels.
	ten Brink [171, 172]	Extended the employment of EXIT charts to three-stage parallel concatenated codes.
	El Gamal and Hammons [173]	Used SNR measures for studying the convergence behavior of iterative decoding.

Table 1.13: Major concatenated schemes and iterative detection (Part 2).

Year	Author(s)	Contribution
2002	Tüchler [174]	Simplified the computation of EXIT charts.
		Extended the EXIT chart analysis to three-stage serially concatenated systems.
	Tüchler *et al.* [175]	Compared several algorithms predicting the decoding convergence of iterative decoding schemes.
2003	Sezgin *et al.* [163]	Proposed an iterative detection scheme where a block code was used as an outer code and STBC as an inner code.
2004	Tüchler [176]	Proposed a design procedure for creating systems exhibiting beneficial decoding convergence depending on the block length.
2005	Lifang *et al.* [177]	Showed that non-square QAM can be decomposed into parity-check block encoder having a recursive nature and a memoryless modulator. Iterative decoding was implemented with an outer code for improving the system performance.
	Brännström *et al.* [178]	Considered EXIT chart analysis for multiple concatenated codes using three-dimensional charts and proposed a way for finding the optimal activation order.
2008	Maunder *et al.* [179]	Designed irregular variable length codes for the near-capacity design of joint source and channel coding aided systems.

performing demapping as well as channel decoding iteratively. The SP demapper of [43] is further developed for the sake of accepting the *a priori* information passed to it from the channel decoder as extrinsic information. Two realizations of a novel bit-based iterative-detection-aided STBC-SP scheme are presented, namely a Recursive Systematic Convolutional (RSC) coded turbo-detected STBC-SP scheme and a binary Low-Density Parity Check (LDPC)-coded turbo-detected STBC-SP arrangement. Our system overview is provided in Section 3.2. In Section 3.3, we show how the STBC-SP demapper is modified for exploiting the *a priori* knowledge provided by the channel decoder, which is essential for the employment of iterative demapping and decoding. EXIT chart analysis is invoked in Section 3.4 in order to study and design the turbo-detected schemes proposed in Section 3.2. We propose ten different Anti-Gray Mapping (AGM) schemes that are specifically selected from all of the possible mapping schemes for $L = 16$ in order to demonstrate the different extrinsic information transfer characteristics associated with different bit-to-symbol mapping schemes. The slope of the EXIT curves corresponding to the different AGM schemes increases gradually in fine steps. This attractive characteristic is a result of the multi-dimensional constellation mapping and having this property is essential for designing near-capacity turbo-detected systems. The performance of the turbo-detected bit-based STBC-SP schemes is presented in Section 3.5, where we investigate the relation between the achievable BER and the mutual information at the input as well as at the output of the outer decoders. In addition, the predictions of our EXIT chart analysis are verified by generating the actual decoding trajectories and BER curves. The effect of interleaver depth is also addressed, since matching the predictions of the EXIT chart analysis is only guaranteed when employing large interleaver depths. The BER performance of the proposed channel-coded STBC-SP scheme is compared with that of an uncoded STBC-SP scheme [43] and to that of a channel-coded conventionally modulated STBC scheme.

- **Chapter 4: Turbo Detection of Channel-coded DSTBC-SP Schemes**

  In Chapters 2 and 3, we assume that the channel state information is perfectly known at the receiver. This, however, requires sophisticated channel estimation techniques, which imposes excess cost and complexity. In Chapter 4, we consider the design of novel SP modulated differential STBC schemes, referred to here as DSTBC-SP, that require no channel estimation. We describe in Section 4.2.1 how DSTBC schemes are constructed using SP modulation. The performance of uncoded DSTBC-SP schemes is considered in Section 4.2.2, where we compare the performance of different DSTBC-SP schemes against equivalent conventional DSTBC schemes under various channel conditions. Simulation results are provided for systems having different BPS rates in conjunction with appropriate conventional and SP modulation schemes. In Section 4.3, we propose novel bit-based RSC-coded turbo-detected DSTBC-SP schemes. The system's architecture is outlined in Section 4.3.1. The EXIT chart analysis of Section 3.4 is employed in Section 4.3.2 in order to design and analyze the convergence behavior of the proposed turbo-detected RSC-coded DSTBC-SP schemes. In Section 4.3.3, we investigate the performance of the proposed RSC-coded DSTBC-SP schemes. The actual decoding trajectories and BER performance curves are also provided, when using various interleaver depths.

- **Chapter 5: Three-stage Turbo-detected STBC-SP Schemes**

  The conventional two-stage turbo-detected schemes introduced in Chapter 3 suffer from a BER floor, preventing them from achieving infinitesimally low BER values, since the inner coding stage is of non-recursive nature. In Chapter 5, we circumvent this deficiency by proposing a three-stage turbo-detected STBC-SP scheme, where a rate-one recursive inner precoder is employed to avoid having a BER floor. Section 5.2 provides a brief description of the proposed three-stage system. We consider three different types of channel codes for the outer encoder, namely a repeater, an RSC code and an Irregular Convolutional Code (IRCC). Our three-dimensional EXIT chart analysis is presented in Section 5.3.2, where its simplified two-dimensional projections are provided in Section 5.3.3. In Section 5.3.4, we employ the powerful technique of EXIT tunnel-area minimization for near-capacity operation. More specifically, we exploit the well-understood properties of conventional 2D EXIT charts that a narrow but nonetheless open EXIT-tunnel represents a near-capacity performance. Consequently, we invoke IRCCs for the sake of appropriately shaping the EXIT curves by minimizing the area within the EXIT-tunnel using the procedure of [175, 176]. In Section 5.4, an upper bound on the maximum achievable rate is calculated based on the EXIT chart analysis. More explicitly, a procedure is proposed for calculating a tighter upper bound of the maximum achievable bandwidth efficiency of STBC-SP schemes based on the area property of the EXIT charts discussed in Section 5.3.4. The design procedure is summarized in Algorithm 5.1. The performance of the three-stage turbo-detected STBC-SP schemes is demonstrated and characterized in Section 5.5, where we discuss the actual decoding trajectories, BER performance and the effect of interleaver depth on the achievable performance. We also investigate the E_b/N_0 distance to capacity for the three-stage RSC-coded as well as for the IRCC-coded STBC-SP schemes, when employing various interleaver depths and using different numbers of three-stage iterations. Finally, in Section 5.5.5, the performance of both the three-stage RSC-coded and the IRCC-coded STBC-SP schemes are compared, when employing various interleaver depths, while using different numbers of three-stage iterations.

- **Chapter 6: Symbol-based Channel-coded STBC-SP Schemes**

  In all previous chapters, iterative decoding is employed at the bit level. In contrast, in this chapter, we explore a range of further design options and propose a purely symbol-based scheme, where symbol-based turbo detection is carried out by exchanging extrinsic information between an outer non-binary LDPC code and a rate-one non-binary inner precoder. The motivation behind the development of this symbol-based scheme is that a reduced transmit power may be required, when symbol-based rather than bit-based iterative decoding is employed [183]. The systems architectures of the proposed symbol-based and turbo-detected scheme and its equivalent bit-based scheme are presented in Section 6.2. Symbol-based iterative decoding is discussed in Section 6.3, where it is demonstrated how the *a priori* information is removed from the decoded *a posteriori* probability with the aid of symbol-based element-wise division for the sake of generating the extrinsic probability. Section 6.4 provides our non-binary EXIT chart analysis. More specifically, in Section 6.4.1 we demonstrate how non-binary EXIT charts can be generated without generating an L-dimensional histogram [184] since the complexity of this operation may become higher than conducting full-scale BER or Symbol Error Rate (SER) simulations, when the number of bits per symbol is high. In Section 6.4.2, we address the problem of generating the *a priori* symbol probabilities, when the binary bits within each non-binary symbol are assumed be either independent or not. Accordingly, a detailed procedure is described in Section 6.4.2.2 for creating the *a priori* symbol probabilities, when the binary bits of each non-binary symbol may no longer be assumed to be independent. The results of our non-binary EXIT chart analysis are provided in Section 6.4.3, where the EXIT charts of both the symbol-based and bit-based schemes are compared, demonstrating that the symbol-based schemes require a lower transmit power and a lower number of decoding iterations for achieving a performance comparable to that of their bit-based counterparts. The performance of the symbol-based and bit-based LDPC-coded STBC-SP schemes is investigated in Section 6.5, in terms of the actual decoding trajectories and the attainable BER performance. The effect of employing various interleaver depths or, equivalently, LDPC output block lengths on the achievable performance is also considered in Section 6.5.

- **Chapter 7: Linear Dispersion Codes: An EXIT Chart Perspective**

  The theory and design of LDCs designed for co-located MIMO systems is investigated and a novel irregular near-capacity scheme using LDCs as the inner constituent code is proposed. Section 7.2 presents LDC models suitable for describing OSTBCs and for non-orthogonal STBCs. Furthermore, a novel method of optimizing the LDCs according to their DCMC capacities is proposed. In Section 7.3, the relationship between various STBC designs and LDCs is exploited in detail, demonstrating the flexibility of the linear dispersion framework. Table 7.3 specifically characterizes the evolution of STBCs in terms of their rate, diversity and flexibility. As far as channel-coded schemes are concerned, Section 7.4 investigates various design issues related to two-stage concatenated convolutional coded LDCs as well as to three-stage precoder-assisted LDCs with the aid of EXIT chart, including their maximum achievable rates and their iteration parameters. Section 7.5 investigates the irregular code design principle originally derived for IRCCs and employs a similar concept to design a family of Irregular Precoded Linear Dispersion Codes (IR-PLDCs) as the inner constituent code of a Serial Concatenated Code (SCC). In Sections 7.5.1–7.5.3, different degrees of irregularity are imposed on both the inner IR-PLDCs and the outer IRCCs.

- **Chapter 8: Differential Space-Time Block Codes: A Universal Approach**

  In Chapter 8, we exploit the linear dispersion structure in the context of non-coherently detected MIMO systems as well as characterizing the effective throughput achieved by an irregular SCC scheme. In Section 8.2, the multi-antenna-aided DSTBC's system architecture is derived from the conventional single-antenna-aided DPSK scheme, followed by the characterization of the fundamental relationship between STBCs and DSTBCs in Section 8.2.3. After characterizing the performance of DSTBCs based on various orthogonal constraints in Section 8.3, Section 8.4 proposes the family of DLDCs based on the Cayley transform. In Section 8.5, we introduce the concept of SP modulation [185] and jointly design the SP modulation and DSTBCs. The convolutional-coded SP-aided DSTBC scheme of Figure 8.18 is capable of approaching the capacity at a specific SNR. Finally, in Section 8.6 the irregular design philosophy is imposed on both the inner and outer codes. Again, the resultant IRCC-coded Irregular Precoded Differential Linear Dispersion Codes (IR-PDLDCs) of Figure 8.25 has the potential of operating near the attainable capacity across a wide range of SNRs.

- **Chapter 9: Cooperative Space-Time Block Codes**

  In Chapter 9, we apply the linear dispersion structure to the family of relay-aided cooperative schemes and characterize the maximum achievable throughput achieved by the irregular system. More explicitly, Section 9.1 justifies the need for cooperation and portrays the system architecture in Figure 9.2. The mathematical model of the proposed twin-layer CLDCs as well as the rationale of our assumptions are discussed in Sections 9.2.1–9.2.3. The fundamental link between LDCs and CLDCs is exploited in Section 9.2.4, followed by their achievable performance recorded in Section 9.2.5. Similarly, we impose the irregular design philosophy in the context of cooperative MIMO systems in Section 9.3. The resultant IRCC-coded Irregular Precoded Cooperative Linear Dispersion Codes (IR-PCLDCs) of Figure 9.16 become capable of achieving a flexible effective throughput according to the SNR encountered, while maintaining an infinitesimally low BER.

- **Chapter 10: Differential Space-Time Spreading**

  This chapter introduces the idea of differential STS and its combination with SP modulation. The chapter first reviews the concept of differential encoding in Section 10.2. It is shown that differential encoding requires no CSI at the receiver and thus eliminates the complexity of channel estimation at the expense of a 3 dB performance loss compared with the coherently detected system assuming perfect CIR recovery at the receiver. In Section 10.3, we outline the encoding and decoding processes of the

Differential Space-Time Spreading (DSTS) scheme, when combined with conventional modulation schemes such as PSK and QAM. In Section 10.3.3, the philosophy of DSTS using SP modulation is introduced based on the fact that the diversity product of the DSTS design is improved by maximizing the Minimum Euclidean Distance (MED) of the DSTS symbols, which is motivated by the fact that SP has the best known MED in the real-valued space. Section 10.3.4 discusses the problem of constructing a SP constellation having a particular size L. The capacity of DSTS-SP schemes employing $N_t = 2$ transmit antennas is derived in Section 10.3.6, followed by the performance characterization of a twin-antenna-aided DSTS scheme in Section 10.3.7, demonstrating that the DSTS scheme is capable of providing full diversity. Our results demonstrate that DSTS-SP schemes are capable of outperforming DSTS schemes dispensing with SP.

The four-antenna-aided DSTS design is characterized in Section 10.4, where it is demonstrated that the DSTS scheme can be combined with conventional real- and complex-valued constellations as well as with SP modulation. It is also shown that the four-dimensional SP modulation scheme is constructed differently in the case of two transmit antennas than when employing four transmit antennas. The capacity of the four-antenna-aided DSTS-SP scheme is also derived for different spectral efficiency systems, while employing a variable number of receive antennas in Section 10.4.5. Finally, Section 10.4.6 presents the simulation results obtained for the four-antenna-aided DSTS scheme, when combined with conventional as well as SP modulations.

► **Chapter 11: Iterative Detection of Channel-coded DSTS Schemes**

In this chapter, two realizations of a novel iterative-detection-aided DSTS-SP scheme are presented, namely an iteratively detected RSC-coded DSTS-SP scheme as well as an iteratively detected RSC-coded and Unity Rate Code (URC) precoded DSTS-SP arrangement. The iteratively detected RSC-coded DSTS-SP scheme is described in detail in Section 11.2. In Section 11.2.1, we show how the DSTS-SP demapper was modified for exploiting the *a priori* knowledge provided by the channel decoder, which is essential for the employment of iterative detection.

The concept of EXIT charts is introduced in Section 11.2.2 as a tool designed for studying iterative-detection-aided schemes. Then, we propose a novel technique for computing the maximum achievable bandwidth efficiency of the system based on EXIT charts in Section 11.2.3, followed by a discussion of the system's performance. Section 11.2.5 presents an application of the iteratively detected RSC-coded DSTS-SP system, where an Adaptive Multi-rate Wideband (AMR-WB) source codec was employed by the system in order to demonstrate the attainable performance improvements.

In addition, in Section 11.3 we propose an iteratively detected RSC-coded and URC-precoded DSTS-SP scheme that is capable of eliminating the error floor exhibited by the system of Section 11.2, which hence performed closer to the system's achievable capacity. In Section 11.3.1 we present an overview of the system operation, followed by a discussion of the results in Section 11.3.2. In Section 11.3.3 we present an application of the proposed system, while employing Irregular Variable Length Codes (IRVLCs) as our outer code for the sake of achieving a near-capacity performance.

- **Chapter 12: Adaptive DSTS-assisted Iteratively Detected SP Modulation**

  In this chapter we propose a novel adaptive DSTS-aided system that exploits the advantages of differential encoding and iterative decoding, as well as SP modulation, while adapting the system parameters for the sake of achieving the highest possible bandwidth efficiency, as well as maintaining a given target BER. The proposed adaptive DSTS-SP scheme benefits from a substantial diversity gain, while using four transmit antennas without the need for pilot-assisted channel envelope estimation and coherent detection. The proposed scheme reaches the target BER of 10^{-3} at a SNR of about 5 dB and maintains it for SNRs in excess of this value, while increasing the effective throughput. The system's bandwidth efficiency varies from 0.25 bit s^{-1} to $16\ \text{Hz}^{-1}$.

- **Chapter 13: Layered Steered Space-Time Codes**

  In this chapter, we propose a multi-functional MIMO scheme, that combines the benefits of V-BLAST codes and of space-time codes as well as of beamforming. Thus, the proposed system benefits from the multiplexing gain of the V-BLAST, from the diversity gain of the space-time codes and from the SNR gain of the beamformer. The MIMO scheme is referred to as LSSTC. To further enhance the attainable system performance and to maximize the coding advantage of the proposed transmission scheme, the system is also combined with multi-dimensional SP modulation.

  In Section 13.2 we outline the encoding and decoding processes of this multi-functional MIMO scheme when combined with conventional as well as SP modulation schemes. Then, in Section 13.3 we quantify the capacity of the proposed multi-functional MIMO and present the capacity limits for a system employing $N_t = 4$ transmit antennas, $N_r = 4$ receive antennas and a variable number L_{AA} of elements per AA. Furthermore, in Section 13.4.3 we quantify the upper bound of the achievable bandwidth efficiency of the system based on the EXIT charts obtained for the iteratively detected system.

  To further enhance the attainable system performance, the proposed MIMO scheme is serially concatenated with both an outer code and a URC, where three different receiver structures are presented by varying the iterative detection configuration of the constituent decoders/demapper. In Section 13.4.1 we provide a brief description of the iteratively detected two-stage RSC-coded LSSTC-SP scheme, where extrinsic information is exchanged between the outer RSC decoder and the inner URC decoder, while no iterations are carried out between the URC decoder and the SP demapper. The convergence behavior of the iterative-detection-aided system is analyzed using EXIT charts in Section 13.4.1.1. In Section 13.4.1.2, we employ the powerful technique of EXIT tunnel-area minimization, for the sake of achieving a near-capacity operation. Consequently, we invoke IRCCs for the sake of appropriately shaping the EXIT curves by minimizing the area within the EXIT-tunnel using the procedure of [174, 176].

  In Section 13.4.2 we present an iteratively detected three-stage RSC-coded LSSTC scheme, where extrinsic information is exchanged between the three constituent decoders, namely the outer RSC decoder, the inner URC decoder and the demapper. Three-dimensional EXIT charts are presented in Section 13.4.2.1, followed by Section 13.4.2.2, where the simplified two-dimensional projections of the three-dimensional EXIT charts are provided. Finally, in Section 13.5 we discuss our performance results and characterize the three iteratively detected LSSTC schemes proposed. Explicitly, the SP-aided system is capable of operating within 0.9, 0.6 and 0.4 dB from the maximum achievable rate limit. However, to operate within 0.6 dB from the maximum achievable rate limit, the system imposes twice the complexity compared with a system operating within 0.9 dB from this limit. On the other hand, to operate as close as 0.4 dB from the maximum achievable rate limit, the system imposes 20 times higher complexity in comparison to that operating within 0.9 dB from the maximum achievable rate limit. The proposed design principles are applicable to an arbitrary number of antennas and diverse antenna configurations as well as modem schemes. In contrast, the Quadrature Phase Shift Keying (QPSK) modulated three-stage iteratively detected system is capable of operating within 1.54 dB from the maximum achievable rate limit and thus the SP modulated system outperforms its QPSK-aided counterpart by about 1 dB at a BER of 10^{-6}.

► Chapter 14: DL LSSTS-aided Generalized MC DS-CDMA

A multi-functional multi-user MIMO scheme that combines the benefits of V-BLAST, of STS and of generalized MC DS-CDMA as well as of beamforming is presented in this chapter. The proposed system is referred to as LSSTS-aided generalized MC DS-CDMA, and it benefits from a multiplexing gain, a spatial diversity gain, a frequency diversity gain and a beamforming gain.

In Section 14.2 the proposed LSSTS scheme's transmitter structure is characterized and then the decoding process is illustrated. Afterwards, in order to increase the number of users supported by the system, Frequency Domain (FD) spreading is applied in the generalized MC DS-CDMA in addition to the TD spreading action of the STS. A user-grouping technique is employed that minimizes the FD interference coefficient for the users in the same TD group.

To further enhance the achievable system's performance, the proposed MIMO scheme is serially concatenated with an outer code combined with a URC, where three different iteratively detected systems are presented in Section 14.4. EXIT charts are used to study the convergence behavior of the proposed systems and in Section 14.4.1 we propose a LLR post-processing technique for the soft output of the QPSK demapper in order to improve the achievable system performance. In Section 14.5 we discuss our performance results and characterize the three proposed iteratively detected schemes, while employing $N_t = 4$ transmit AAs, $N_r = 2$ receive antennas, L_{AA} elements per AA and V subcarriers supporting K users.

► Chapter 15: Distributed Turbo Coding

A cooperative communication scheme referred to as Distributed Turbo Coding (DTC) is presented in Chapter 15. In the proposed scheme, two users are cooperating, where each user's transmitter is constituted by an RSC code and an interleaver followed by a SP mapper. In Section 15.2 we provide an overview of cooperative communications and the background of the major cooperative signaling strategies including AF, DF and coded cooperation. In Section 15.3 the DTC scheme is presented, where a two-phase cooperation scheme is proposed. In the first phase, the two users exchange their data, while in the second phase the two users simultaneously transmit their data to the BS. In Section 15.4 we characterize the attainable system performance and study the effects of varying the inter-user channel characteristics on the performance of the uplink DTC scheme.

► **Chapter 16: Conclusions and Future Research**

This chapter summarizes the main findings of the book, giving cognizance to the current trends in the research community, and outlines a range of suggestions for future research.

1.6.2 Novel Aspects of the Book

The book is based on a diverse range of publications, which are listed at the back and covers the following novel research aspects.

- The achievable performance of several STBC-SP schemes employing various SP constellation sizes L is investigated, where the constellation points are first chosen based on a minimum energy criterion. Then, an exhaustive computer search is conducted in order to find the set of L points having the highest MED from the entire set of constellation points satisfying the minimum energy criterion.

- A turbo-detected SP modulation aided STBC scheme is proposed, where the SP demapper is further developed for the sake of accepting the *a priori* information passed to it from the channel decoder as extrinsic information [186, 187].

- In order to portray the different EXIT characteristics associated with different bit-to-SP-symbol mapping schemes, ten different AGM schemes are developed that are selected specifically from all of the possible mapping schemes available for $L = 16$. The slope of the EXIT curves corresponding to the different AGM schemes increases gradually in fine steps demonstrating the advantages of multi-dimensional constellation mapping. Exhibiting a gradually increasing EXIT characteristic is essential for the sake of designing near-capacity turbo-detected systems. The proposed turbo-detected STBC-SP scheme is optimized using EXIT charts [188].

- A differential turbo-detected SP modulation aided STBC scheme that requires no CSI is proposed and its performance is optimized using EXIT charts [189].

- A three-stage serially concatenated turbo-detected STBC-SP scheme is proposed that is capable of achieving infinitesimally low BER values, where the performance is no longer limited by a BER floor. The convergence behavior of the three-stage system is analyzed and designed with the aid of three-dimensional EXIT charts and their two-dimensional projections, resulting in a near-capacity performance [190, 191].

- A procedure is proposed for calculating a tighter upper bound on the maximum achievable bandwidth efficiency of concatenated schemes based on the so-called 'area property' of the EXIT charts [190, 191].

- A purely symbol-based scheme is proposed, where symbol-based turbo detection is carried out by exchanging extrinsic information between an outer non-binary LDPC code and a rate-one non-binary inner precoder. The convergence behavior of the proposed symbol-based scheme is analyzed using novel symbol-based EXIT charts [46, 192–194].

- LDCs are used in order to unify the family of orthogonal and non-orthogonal STBC designs. This unified structure enables us to examine existing STBCs from a novel perspective. More explicitly, we characterize the linkage between existing

STBCs found in the open literature and LDCs in terms of both their mathematical representations and their design philosophies. Furthermore, we propose to optimize the LDCs from a capacity maximization perspective, namely by maximizing the DCMC capacity of the LDCs.

- We demonstrate the fundamental relationship between STBCs and DSTBCs, which enables us to extend the STBC design philosophy to DSTBCs. Furthermore, the Cayley transform [10] is introduced as an efficient way of constructing unitary matrices for their description. The resultant DLDCs based on the Cayley transform exhibit similar characteristics to those of their coherently detected LDC counterparts.
- The fundamental relationship between co-located and cooperative MIMO systems is investigated, which is facilitated by the establishment of the broadcast interval. In other words, CSTBCs are designed to provide spatial diversity with the aid of a two-phase transmission regime. Hence, we propose the family of twin-layer CLDCs, which inherit the flexible linear dispersion structure and are specifically designed to exploit the above-mentioned two-phase transmission regime.
- A quantitative comparative study of LDCs, DLDCs and CLDCs is conducted, since all of them are based on the linear dispersion structure. Our investigations suggest that the family of LDCs is suitable for co-located MIMO systems employing coherent detection. The class of DLDCs is more applicable, when no CSI is available at the receiver. When relay-aided cooperative transmission is necessary to avoid the performance erosion imposed by shadow fading, our specifically designed twin-layer CLDCs are more beneficial.
- The concept of irregular coding [195] is documented and extended to a broad range of systems. More explicitly, the irregular design principle is applied in the context of the inner code of a serial concatenated coding scheme. The resultant inner IR-PLDC scheme facilitates the system's near-capacity operation across a wide range of SNRs. Similarly, we propose the IR-DLDCs for non-coherent MIMO systems and IR-CLDCs for cooperative MIMO systems.
- A DSTS scheme is proposed, which is advocated for the sake of achieving a high transmit diversity gain in a multi-user system, while eliminating the complexity of MIMO channel estimation. In addition, the system is combined with multi-dimensional SP modulation, which is capable of maximizing the coding advantage of the transmission scheme by jointly designing and detecting the sphere-packed DSTS symbols. The capacity of the DSTS-SP scheme is quantified analytically, where it is shown that the DSTS-SP system attains a higher capacity than its counterpart dispensing with SP [196, 197].
- Iteratively detected DSTS-SP schemes are designed for near-capacity operation, where EXIT charts are used for analyzing the convergence behavior of the iterative detection. The outer code used in the iterative-detection-aided systems is a RSC code, while the inner code is SP mapper in the first system and a URC in the second system, where the URC is capable of eliminating the error floor present in the BER performance of the system dispensing with URC [196, 198–200].
- An algorithm is devised for computing the maximum achievable rate of the DSTS system using EXIT charts, where the maximum achievable rate obtained using EXIT charts matches closely with the analytically computed capacity [196].

- An adaptive DSTS-SP scheme is proposed in order to maximize the system's throughput. The adaptive scheme exploits the advantages of differential encoding and iterative decoding as well as SP modulation. The achievable integrity and bit rate enhancements of the system are determined by the following factors: the specific transmission configuration used for transmitting data from the four antennas, the SF used and the RSC encoder's code rate [201].

- The merits of V-BLAST, STC and beamforming are amalgamated in a LSSTC multi-functional MIMO scheme for the sake of achieving a multiplexing gain and a diversity gain as well as a beamforming gain. In addition, the capacity of the LSSTC-SP scheme is quantified analytically [202].

- Furthermore, in order to characterize the LSSTC scheme, three iteratively detected LSSTC-SP receiver structures are proposed, where iterative detection is carried out between the outer code's decoder, the intermediate code's decoder and the LSSTC-SP demapper. The three systems are capable of operating within 0.9, 0.4 and 0.6 dB from the maximum achievable rate limit of the system. A comparison between the three iteratively detected schemes reveals that a carefully designed two-stage iterative-detection-aided scheme is capable of operating sufficiently close to capacity at a lower complexity, when compared to a three-stage system employing RSC or a two-stage system employing an IRCC as the outer code [21,203].

- A multi-functional MIMO combining STS, V-BLAST and beamforming with generalized MC DS-CDMA is proposed and referred to as LSSTS. The LSSTS scheme is capable of achieving a spatial diversity gain, frequency diversity gain and a multiplexing gain as well as beamforming gain.The number of users supported can be extended by employing combined TD and FD spreading [204].

- A novel LLR post-processing technique is devised for improving the iteratively detected LSSTS system's performance [204].

- Finally, ideas from cooperative communications and turbo coding are combined to form a DTC, where turbo coding is employed by exchanging extrinsic information between the outer codes' decoders in the two cooperating users' handsets.

Part I

Coherent Versus Differential Turbo Detection of Sphere-packing-aided Single-user MIMO Systems

List of Symbols in Part I

General Notation

- The superscript * is used to indicate complex conjugation. Therefore, a^* represents the complex conjugate of the variable a.
- The superscript ^{T} is used to indicate the matrix transpose operation. Therefore, $\mathbf{a}^{\mathrm{T}}$ represents the transpose of the matrix $\mathbf{a}$.
- The superscript $^{\mathcal{H}}$ is used to indicate the complex conjugate transpose operation. Therefore, $\mathbf{a}^{\mathcal{H}}$ represents the complex conjugate transpose of the matrix $\mathbf{a}$.
- The notation $\hat{x}$ represents the estimate of x.

Special Symbols

$a_{l,i}$ The ith coordinate of the lth SP symbol $\mathbf{s}^l$.

$\mathcal{A}$ The area under a curve.

$\mathbf{A}$ The *a priori* probability matrix of a non-binary decoder.

$\mathbf{A}_{ldpc}$ The *a priori* probability matrix of the non-binary LDPC decoder.

$\mathbf{A}_{urc}$ The *a priori* probability matrix of the symbol-based unity-rate decoder.

B The number of binary bits corresponding to a constellation symbol.

$\mathbf{b}$ A block of B binary bits.

b_i The binary bit at position i in $\mathbf{b}$.

c Outer channel coded bits.

$\mathbf{C}$ The space-time signal matrix.

c_t^i The complex symbol transmitted by transmit antenna i at time slot t.

$\mathbb{C}^n$ The n-dimensional complex space.

$C_{\text{DCMC}}^{\text{STBC-SP}}$	The DCMC capacity of STBC-SP schemes.
D	The dimension of a D-dimensional signal set.
$\mathcal{D}$	The depth of the random interleaver.
$\mathbf{D}_{ldpc}$	The *a posteriori* probability matrix of the non-binary LDPC decoder.
$\mathbf{D}_{urc}$	The *a posteriori* probability matrix of the symbol-based unity-rate decoder.
$E[k]$	The expected value of k.
E_b	Bit energy.
E_s	Symbol energy.
E_{total}	The total energy of a constellation set.
f_D	The normalized Doppler frequency.
G	The feedforward generator polynomial of RSC codes.
G_r	The feedback generator polynomial of RSC codes.
$\mathbf{G}_{2^k}$	An orthogonal design of size $(2^k \times 2^k)$.
h_i	The channel impulse response from transmit antenna i for single-receive antenna systems.
$h_{i,j}$	The channel impulse response from transmit antenna i to receive antenna j.
I_0	The bit-wise unconditional mutual information.
I_{3S}	The number of three-stage iterations.
I_A	The mutual information associated with the *a priori* information.
I_{A_D}	The mutual information associated with the *a priori* LLR values $L_{D,a}$ of the outer channel decoder.
I_{A_M}	The mutual information associated with the *a priori* LLR values $L_{M,a}$ of the SP demapper.
I_E	The mutual information associated with the extrinsic information.
I_{E_D}	The mutual information associated with the extrinsic LLR values $L_{D,e}$ of the outer channel decoder.
I_{E_M}	The mutual information associated with the extrinsic LLR values $L_{M,e}$ of the SP demapper.
I_{ext}	The number of external joint iterations.
I_{int}	The number of LDPC internal iterations.
$\mathbf{I}_n$	The identity matrix of size $(n \times n)$.
$\mathcal{K}$	The constraint length of RSC codes.

K The rank of a matrix.

K_{ldpc} LDPC output block length.

L The size of the legitimate modulation constellation S.

$L_{D,a}$ The *a priori* LLR values of the outer channel decoder.

$L_{D,e}$ The extrinsic LLR values of the outer channel decoder.

$L_{D,i,p}$ The LLR values of the original uncoded systematic information bits.

$L_{D,p}$ The *a posteriori* LLR values of the outer channel decoder.

$L_{M,a}$ The *a priori* LLR values of the SP demapper.

$L_{M,e}$ The extrinsic LLR values of the SP demapper.

$L_{M,p}$ The *a posteriori* LLR values of the SP demapper.

N_t Number of transmit antennas.

N_r Number of receive antennas.

n_A The zero-mean Gaussian random variable used for modeling the *a priori* information input.

n_f A normalization factor.

P Number of subcodes in a family of subcodes (e.g. IRCCs).

$\mathbf{Q}$ The soft-metric probability matrix produced by the symbol-based SP demodulator.

R Coding rate.

$\mathbf{R}$ The channel correlation matrix.

r_i The received sphere packing symbol at time instant t.

$\mathbb{R}^n$ The n-dimensional real-valued Euclidean space.

$\mathbf{s}$ A SP symbol.

S The legitimate constellation set.

S_0^k The subset of the legitimate constellation set S that contains all symbols having $b_k = 0$.

S_1^k The subset of the legitimate constellation set S that contains all symbols having $b_k = 1$.

$\mathbf{s}^l$ The lth SP legitimate symbol.

T The number of time slots needed for transmitting a specific number of symbols.

T_s Signaling period.

T_{sp} The transfer function from SP to complex signals.

T_{sym} Symbol period.

$\mathbf{v}$ Non-binary LDPC encoded integer symbols.

$\mathbf{w}$ A four-dimensional Gaussian random variable.

W Bandwidth.

y_t^j The received complex signal at receive antenna j at time slot t.

z_t^j The complex Additive White Gaussian Noise (AWGN) noise at receive antenna j at time instant t.

α_i Weight coefficient of the ith subcode.

ζ_R The diversity product for time-correlated fading channels having a correlation matrix $\mathbf{R}$.

ζ_{rapid} The diversity product for rapid fading channels.

ζ_{static} The diversity product for quasi-static fading channels.

η Bandwidth efficiency.

θ_i The phase shift of the channel impulse response h_i.

π Interleaver.

π^{-1} Deinterleaver.

ρ The SNR.

σ_n^2 The complex noise's variance.

Chapter 2

Space-Time Block Code Design using Sphere Packing

2.1 Introduction

Most of the schemes outlined in Chapter 1 assumed encountering one of two different idealized channel conditions, namely either quasi-static or rapid fading [8, 11, 12, 39, 50, 54, 205–216]. However, practical wireless communication channels exhibit both spatial and temporal correlation. The space-time codes that have been specifically designed for quasi-static or rapid fading channels may not be optimum when they are employed in practice. Therefore, the specific design criteria adopted should take into account the typical fading rates encountered, which determine the amount of correlation. In this chapter, a general design criterion is considered that takes into account the amount of correlation encountered.

In [43], the design of robust space-time modulation schemes designed for time-correlated Rayleigh fading channels was considered. It was assumed that the wireless channel is only time-correlated. Specific design criteria adopted for the time-correlated Rayleigh fading channel were derived. The design concept of maximizing the diversity product [54, 208] was generalized in [43] in order to take the effects of the temporal correlation into account. The lower and upper bounds of the general diversity product were also established in order to assist in the development of systematic space-time signal design procedures. In order to maximize the achievable coding advantage for space-time signals that achieve full diversity, a class of space-time block coded signals was constructed by combining orthogonal space-time code design with SP modulation, which is referred to here as STBC-SP.

The rest of the chapter is organized as follows. First, the design criteria devised for space-time signals are presented in Section 2.2, where the channel model is discussed along with the design criteria invoked for quasi-static and rapid fading channels. Then, the design of space-time signals recommended for time-correlated fading channels is reviewed in Section 2.3, where the concept of a general diversity product is presented. In Section 2.4, the motivation behind the adoption of SP modulation combined with orthogonal design is discussed, where the specific signal design devised for two transmit antennas is considered in more detail. Finally, the performance of STBC-SP schemes is presented in Section 2.5,

Near-Capacity Multi-Functional MIMO Systems L. Hanzo, O. R. Alamri, M. El-Hajjar and N. Wu

demonstrating that the amalgamated STBC-SP schemes outperform STBC arrangements that employ conventional modulation (i.e. PSK, QAM).

2.2 Design Criteria for Space-Time Signals

2.2.1 Channel Model

A wireless communication system having N_t transmit and N_r receive antennas is considered. The space-time modulator first divides the input information bits into blocks of B bits and maps each block to the appropriate space-time signal selected from the signal set of size $L = 2^B$. The space-time signal is then transmitted over the N_t transmit antennas using T time slots as detailed in [8, 43]. In general terms, each space-time signal can be expressed using the following $(T \times N_t)$-dimensional matrix

$$\mathbf{C} = \begin{bmatrix} c_1^1 & c_1^2 & \cdots & c_1^{N_t} \\ c_2^1 & c_2^2 & \cdots & c_2^{N_t} \\ \vdots & \vdots & \ddots & \vdots \\ c_T^1 & c_T^2 & \cdots & c_T^{N_t} \end{bmatrix}_{T\times N_t}, \tag{2.1}$$

where c_t^i denotes the symbol transmitted by transmit antenna i, for $i = 1, \ldots, N_t$, in time slot t, for $t = 1, \ldots, T$. The above space-time signal satisfies the following energy constraint

$$E[\|\mathbf{C}\|_F^2] = N_t T, \tag{2.2}$$

where $E[\cdot]$ refers to the expected value and $\|\mathbf{C}\|_F$ is the Frobenius norm of $\mathbf{C}$, defined as [217]

$$\|\mathbf{C}\|_F^2 = \mathrm{tr}(\mathbf{C}^{\mathcal{H}}\mathbf{C}) = \mathrm{tr}(\mathbf{C}\mathbf{C}^{\mathcal{H}}) = \sum_{t=1}^{T}\sum_{i=1}^{N_t} |c_t^i|^2, \tag{2.3}$$

where tr$(\cdot)$ denotes the trace of a matrix and $(\cdot)^{\mathcal{H}}$ denotes the complex conjugate transpose of a matrix. The signal y_t^j received by antenna j at time t is given by

$$y_t^j = \sqrt{\frac{\rho}{N_t}} \sum_{i=1}^{N_t} c_t^i h_{i,j}(t) + z_t^j, \quad t = 1, \ldots, T, \tag{2.4}$$

where z_t^j is the complex AWGN encountered at receive antenna j at time t, which has zero mean and unit variance, while $h_{i,j}(t)$ is the channel coefficient characterizing the non-dispersive link between transmit antenna i and receive antenna j at time t. The channel coefficients are modeled as zero-mean complex-valued Gaussian random variables having a unit variance and they are assumed to be known at the receiver, but unknown at the transmitter. It is also assumed that the channel has temporal correlation but no spatial correlation. In other words, the channel coefficients $h_{i,j}(t)$ are independent for different indices (i, j), but correlated in the time domain. Furthermore, ρ is defined as the SNR per space-time signal at each receive antenna, and it is independent of the number of transmit antennas.

In [218, 219], the received signal of Equation (2.4) was rewritten in a vectorial form as

$$\mathbf{Y} = \sqrt{\frac{\rho}{N_t}}\mathbf{D}\mathbf{H} + \mathbf{Z}, \tag{2.5}$$

where $\mathbf{D}$ is an $(N_rT \times N_tN_rT)$-dimensional matrix constructed from the space-time signal matrix $\mathbf{C}$ as follows

$$\mathbf{D} = \begin{bmatrix} \mathbf{D}_1 & \mathbf{D}_2 & \cdots & \mathbf{D}_{N_t} & 0 & 0 & \cdots & 0 & \cdots & 0 & 0 & \cdots & 0 \\ 0 & 0 & \cdots & 0 & \mathbf{D}_1 & \mathbf{D}_2 & \cdots & \mathbf{D}_{N_t} & \cdots & 0 & 0 & \cdots & 0 \\ & & \vdots & & & & \vdots & & \ddots & & & \vdots & \\ 0 & 0 & \cdots & 0 & 0 & 0 & \cdots & 0 & \cdots & \mathbf{D}_1 & \mathbf{D}_2 & \cdots & \mathbf{D}_{N_t} \end{bmatrix}_{N_rT \times N_tN_rT}, \tag{2.6}$$

where

$$\mathbf{D}_i = \mathrm{diag}(c_1^i, c_2^i, \ldots, c_T^i), \quad i = 1, \ldots, N_t, \tag{2.7}$$

which corresponds to the ith column of the space-time signal matrix $\mathbf{C}$. The (N_tN_rT)-dimensional non-dispersive channel coefficient vector $\mathbf{H}$ can be written as

$$\mathbf{H} = \begin{bmatrix} \mathbf{h}_{1,1}^T & \cdots & \mathbf{h}_{N_t,1}^T & \mathbf{h}_{1,2}^T & \cdots & \mathbf{h}_{N_t,2}^T & \cdots & \mathbf{h}_{1,N_r}^T & \cdots & \mathbf{h}_{N_t,N_r}^T \end{bmatrix}_{N_tN_rT\times 1}^{\mathrm{T}}, \tag{2.8}$$

where $(\cdot)^{\mathrm{T}}$ denotes the transpose of a matrix and finally we have

$$\mathbf{h}_{i,j} = \begin{bmatrix} h_{i,j}(1) & h_{i,j}(2) & \cdots & h_{i,j}(T) \end{bmatrix}_{T\times 1}^{\mathrm{T}}. \tag{2.9}$$

The received signal vector $\mathbf{Y}$ is written as

$$\mathbf{Y} = \begin{bmatrix} y_1^1 & \cdots & y_T^1 & y_1^2 & \cdots & y_T^2 & \cdots & y_1^{N_r} & \cdots & y_T^{N_r} \end{bmatrix}_{N_rT\times 1}^{\mathrm{T}}. \tag{2.10}$$

Finally, the noise vector $\mathbf{Z}$ has the form

$$\mathbf{Z} = \begin{bmatrix} z_1^1 & \cdots & z_T^1 & z_1^2 & \cdots & z_T^2 & \cdots & z_1^{N_r} & \cdots & z_T^{N_r} \end{bmatrix}_{N_rT\times 1}^{\mathrm{T}}. \tag{2.11}$$

Example 2.2.1 ((2 × 2) systems). In order to illustrate the employment of the vectorial representation of Equation (2.5), we consider the simple scenario, when we have $N_t = N_r = T = 2$. The matrix $\mathbf{D}$ can be written as follows:

$$\mathbf{D} = \begin{bmatrix} \mathbf{D}_1 & \mathbf{D}_2 & 0 & 0 \\ 0 & 0 & \mathbf{D}_1 & \mathbf{D}_2 \end{bmatrix}_{4\times 8},$$

where

$$\mathbf{D}_i = \begin{bmatrix} c_1^i & 0 \\ 0 & c_2^i \end{bmatrix}_{2\times 2},$$

which leads to

$$\mathbf{D} = \begin{bmatrix} c_1^1 & 0 & c_1^2 & 0 & 0 & 0 & 0 & 0 \\ 0 & c_2^1 & 0 & c_2^2 & 0 & 0 & 0 & 0 \\ 0 & 0 & 0 & 0 & c_1^1 & 0 & c_1^2 & 0 \\ 0 & 0 & 0 & 0 & 0 & c_2^1 & 0 & c_2^2 \end{bmatrix}_{4\times 8}.$$

The non-dispersive CIR vector $\mathbf{H}$ is written as

$$\mathbf{H} = \begin{bmatrix} h_{1,1}(1) & h_{1,1}(2) & h_{2,1}(1) & h_{2,1}(2) & h_{1,2}(1) & h_{1,2}(2) & h_{2,2}(1) & h_{2,2}(2) \end{bmatrix}_{8\times 1}^{\mathrm{T}}.$$

Accordingly, the received signal $\mathbf{Y}$ expressed in Equations (2.5) and (2.10) for the (2×2) system under consideration is written as

$$\mathbf{Y} = \begin{bmatrix} y_1^1 \\ y_2^1 \\ y_1^2 \\ y_2^2 \end{bmatrix} = \begin{bmatrix} c_1^1 \cdot h_{1,1}(1) + c_1^2 \cdot h_{2,1}(1) + z_1^1 \\ c_2^1 \cdot h_{1,1}(2) + c_2^2 \cdot h_{2,1}(2) + z_2^1 \\ c_1^1 \cdot h_{1,2}(1) + c_1^2 \cdot h_{2,2}(1) + z_1^2 \\ c_2^1 \cdot h_{1,2}(2) + c_2^2 \cdot h_{2,2}(2) + z_2^2 \end{bmatrix}.$$

2.2.2 Pairwise Error Probability and Design Criterion

Consider Equation (2.10) and let $\mathbf{D}$ as well as $\tilde{\mathbf{D}}$ be two different matrices corresponding to two different space-time signals $\mathbf{C}$ and $\tilde{\mathbf{C}}$, respectively. According to [218,219], the pairwise error probability between $\mathbf{D}$ and $\tilde{\mathbf{D}}$ can be upper bounded as

$$P(\mathbf{D} \rightarrow \tilde{\mathbf{D}}) \leq \binom{2K-1}{K} \left(\prod_{i=1}^{K} \gamma_i \right)^{-1} \left(\frac{\rho}{N_t} \right)^{-K}, \qquad (2.12)$$

where K is the rank of $(\mathbf{D} - \tilde{\mathbf{D}})\mathbf{R}(\mathbf{D} - \tilde{\mathbf{D}})^{\mathcal{H}}$, $\gamma_1, \gamma_2, \ldots, \gamma_K$ are the non-zero eigenvalues of $(\mathbf{D} - \tilde{\mathbf{D}})\mathbf{R}(\mathbf{D} - \tilde{\mathbf{D}})^{\mathcal{H}}$ and $\mathbf{R} = E[\mathbf{H}\mathbf{H}^{\mathcal{H}}]$ is the correlation matrix of $\mathbf{H}$. Maximizing the pairwise error probability of Equation (2.12), rather than that between the modulated symbols of the individual time slots, implies that the individual space-time signals have to have the maximum possible Euclidean distance.

The authors of [218, 219] have also proposed a general design criterion based on the upper bound of the pairwise error probability expressed in Equation (2.12) that consists of two parts:

(i) The minimum rank of $(\mathbf{D} - \tilde{\mathbf{D}})\mathbf{R}(\mathbf{D} - \tilde{\mathbf{D}})^{\mathcal{H}}$ should be maximized.

(ii) The minimum value of the product $\prod_{i=1}^{K} \gamma_i$ should be maximized.

This criterion is consistent with the well-known space-time signal design criteria outlined in [8,39] for the extreme cases of the quasi-static and the rapid fading channel models, which are reproduced here for completeness.

- **Quasi-static fading channels.** The minimum rank of

$$\boldsymbol{\Delta} \triangleq (\mathbf{C} - \tilde{\mathbf{C}})(\mathbf{C} - \tilde{\mathbf{C}})^{\mathcal{H}}, \qquad (2.13)$$

over all pairs of distinct space-time signals $\mathbf{C}$ and $\tilde{\mathbf{C}}$ should be maximized. According to [54,208], if $\boldsymbol{\Delta}$ is of full rank for any pair of distinct signals $\mathbf{C}$ and $\tilde{\mathbf{C}}$, in other words, if columns are independent of each other, then the so-called diversity product is given by [54, 208]

$$\zeta_{\text{static}} = \frac{1}{2\sqrt{N_t}} \min_{\mathbf{C} \neq \tilde{\mathbf{C}}} |\det(\boldsymbol{\Delta})|^{1/2T}, \qquad (2.14)$$

where the diversity product or coding advantage is defined as the estimated SNR gain over an uncoded system having the same diversity order as the coded system [8].

- **Rapid fading channels.** The minimum number of non-zero rows of $(\mathbf{C} - \tilde{\mathbf{C}})$ should be maximized for any pair of distinct signals $\mathbf{C}$ and $\tilde{\mathbf{C}}$. If there is no zero row in $(\mathbf{C} - \tilde{\mathbf{C}})$, then the diversity product is given by [8, 39]

$$\zeta_{\text{rapid}} = \frac{1}{2\sqrt{N_t}} \min_{\mathbf{C} \neq \tilde{\mathbf{C}}} \left(\prod_{t=1}^{T} \|\mathbf{c}_t - \tilde{\mathbf{c}}_t\|_F^2 \right)^{1/2T}, \qquad (2.15)$$

where $\mathbf{c}_t$ and $\tilde{\mathbf{c}}_t$ are the tth rows of $\mathbf{C}$ and $\tilde{\mathbf{C}}$, respectively.

2.3 Design Criteria for Time-correlated Fading Channels

In this section, the design criteria devised for time-correlated fading channels in [43] are considered. The pairwise error probability of the space-time signals and related design criteria are presented in Section 2.3.2. The generalized diversity product along with its lower and upper bounds are also discussed in Sections 2.3.3.1 and 2.3.3.2, respectively.

2.3.1 Preliminaries

The channel's correlation matrix $\mathbf{R}$ can be written as

$$\begin{aligned}\mathbf{R} &= E[\mathbf{H}\mathbf{H}^{\mathcal{H}}] \\ &= \text{diag}(\mathbf{R}_{1,1}, \ldots, \mathbf{R}_{N_t,1}, \mathbf{R}_{1,2}, \ldots, \mathbf{R}_{N_t,2}, \ldots\ldots, \mathbf{R}_{1,N_r}, \ldots, \mathbf{R}_{N_t,N_r}), \end{aligned} \tag{2.16}$$

where

$$\mathbf{R}_{i,j} = E[\mathbf{h}_{i,j}\mathbf{h}_{i,j}^{\mathcal{H}}] \tag{2.17}$$

corresponds to the time-domain correlation matrix of the channel coefficients describing the link between transmit antenna i and receive antenna j. When using Jakes' fading model [220], all of the time-domain correlation matrices $\mathbf{R}_{i,j}$ are the same. In other words, the temporal correlation of the signal transmitted from transmit antenna i to receive antenna j is the same. Hence, the correlation matrix of Equation (2.16) can be written as

$$\mathbf{R} = \mathbf{I}_{N_tN_r} \otimes \breve{\mathbf{R}}, \tag{2.18}$$

where $\otimes$ denotes the tensor product [217], $\mathbf{I}_{N_tN_r}$ is the identity matrix of size $(N_tN_r \times N_tN_r)$ and R is the time correlation matrix of the signal transmitted between antenna i and antenna j, defined as

$$\breve{\mathbf{R}} \triangleq \mathbf{R}_{i,j} = \begin{bmatrix} r_{1,1} & \cdots & r_{1,T} \\ \vdots & \ddots & \vdots \\ r_{T,1} & \cdots & r_{T,T} \end{bmatrix}_{T\times T}. \tag{2.19}$$

From Equations (2.6), (2.7) and (2.18), one can write [43]

$$\begin{aligned}&(\mathbf{D} - \tilde{\mathbf{D}})\mathbf{R}(\mathbf{D} - \tilde{\mathbf{D}})^{\mathcal{H}} \\ &= \mathbf{I}_{N_r} \otimes \left[\sum_{i=1}^{N_t}(\mathbf{D}_i - \tilde{\mathbf{D}}_i)\breve{\mathbf{R}}(\mathbf{D}_i - \tilde{\mathbf{D}}_i)^{\mathcal{H}}\right] \\ &= \mathbf{I}_{N_r} \\ &\quad \otimes \begin{bmatrix} \sum_{i=1}^{N_t} |c_1^i - \tilde{c}_1^i|^2 r_{1,1} & \sum_{i=1}^{N_t}(c_1^i - \tilde{c}_1^i)(c_2^i - \tilde{c}_2^i)^* r_{1,2} & \cdots & \sum_{i=1}^{N_t}(c_1^i - \tilde{c}_1^i)(c_T^i - \tilde{c}_T^i)^* r_{1,T} \\ \sum_{i=1}^{N_t}(c_2^i - \tilde{c}_2^i)(c_1^i - \tilde{c}_1^i)^* r_{2,1} & \sum_{i=1}^{N_t} |c_2^i - \tilde{c}_2^i|^2 r_{2,2} & \cdots & \sum_{i=1}^{N_t}(c_2^i - \tilde{c}_2^i)(c_T^i - \tilde{c}_T^i)^* r_{2,T} \\ \vdots & \vdots & \ddots & \vdots \\ \sum_{i=1}^{N_t}(c_T^i - \tilde{c}_T^i)(c_1^i - \tilde{c}_1^i)^* r_{T,1} & \sum_{i=1}^{N_t}(c_T^i - \tilde{c}_T^i)(c_2^i - \tilde{c}_2^i)^* r_{T,2} & \cdots & \sum_{i=1}^{N_t} |c_T^i - \tilde{c}_T^i|^2 r_{T,T} \end{bmatrix} \\ &= \mathbf{I}_{N_r} \otimes \{[(\mathbf{C} - \tilde{\mathbf{C}})(\mathbf{C} - \tilde{\mathbf{C}})^{\mathcal{H}}] \circ \breve{\mathbf{R}}\} \\ &= \mathbf{I}_{N_r} \otimes \{\mathbf{\Delta} \circ \breve{\mathbf{R}}\}, \end{aligned} \tag{2.20}$$

where $\mathbf{\Delta}$ is defined in Equation (2.13), and $\circ$ denotes the Hadamard product [217], which is defined as follows.

Definition 2.3.1. Let $\mathbf{A} = \{a_{i,j}\}$ and $\mathbf{B} = \{b_{i,j}\}$ be two matrices of dimension $(m \times n)$. The Hadamard product of $\mathbf{A}$ and $\mathbf{B}$ is defined as

$$\mathbf{A} \circ \mathbf{B} = \begin{bmatrix} a_{1,1}b_{1,1} & \cdots & a_{1,n}b_{1,n} \\ \vdots & \ddots & \vdots \\ a_{m,1}b_{m,1} & \cdots & a_{m,n}b_{m,n} \end{bmatrix}. \tag{2.21}$$

2.3.2 Pairwise Error Probability and Design Criterion

Upon combining Equation (2.20) with Equation (2.12), the upper bound of the pairwise space-time symbol error probability between $\mathbf{C}$ and $\tilde{\mathbf{C}}$ can be written as [43]

$$P(\mathbf{C} \rightarrow \tilde{\mathbf{C}}) \leq \binom{2rN_r - 1}{rN_r} \left(\prod_{i=1}^{r} \lambda_i \right)^{-N_r} \left(\frac{\rho}{N_t} \right)^{-rN_r}, \tag{2.22}$$

where r is the rank of $\mathbf{\Delta} \circ \breve{\mathbf{R}}$, and $\lambda_1, \lambda_2, \ldots, \lambda_r$ are the non-zero eigenvalues of $\mathbf{\Delta} \circ \breve{\mathbf{R}}$. According to Equation (2.22), the space-time code design criteria devised for time-correlated fading channels can be formulated as follows [43].

(a) **Design for diversity advantage.** The minimum rank of $\mathbf{\Delta} \circ \breve{\mathbf{R}}$ should be maximized over all pairs of distinct space-time signals $\mathbf{C}$ and $\tilde{\mathbf{C}}$.

(b) **Design for coding advantage.** The minimum value of the product $\prod_{i=1}^{r} \lambda_i$ over all pairs of distinct space-time signals $\mathbf{C}$ and $\tilde{\mathbf{C}}$ should be maximized.

2.3.3 Coding Advantage

The above-mentioned coding advantage or the diversity product was used in [54,208] in order to characterize the performance of different STC schemes that achieve an identical diversity advantage. The concept of the diversity product was generalized in [43] to time-correlated fading channels. Undoubtedly, this generalization renders the evaluation and comparison of different space-time coding schemes more straightforward. In addition, upper and lower bounds on the generalized diversity product were developed in [43].

2.3.3.1 Generalized Diversity Product

According to the upper bound of the pairwise space-time symbol error probability expressed in Equation (2.22), and assuming that $\mathbf{\Delta} \circ \breve{\mathbf{R}}$ is of full rank, the generalized diversity product can be written as [43]:

$$\zeta_R = \frac{1}{2\sqrt{N_t}} \min_{\mathbf{C} \neq \tilde{\mathbf{C}}} |\det(\mathbf{\Delta} \circ \breve{\mathbf{R}})|^{1/2T}. \tag{2.23}$$

It may be directly observed that for quasi-static fading channels (i.e. for $\mathbf{\Delta} \circ \breve{\mathbf{R}} = \mathbf{\Delta}$), Equation (2.23) reduces to Equation (2.14), namely to

$$\zeta_{\text{static}} = \frac{1}{2\sqrt{N_t}} \min_{\mathbf{C} \neq \tilde{\mathbf{C}}} |\det(\mathbf{\Delta})|^{1/2T}.$$

For rapid fading channels the correlation matrix $\breve{\mathbf{R}}$ indicates that unless the time-lag considered is zero, the corresponding correlation coefficient is also zero, which is expressed as

$$\breve{\mathbf{R}} = \mathbf{I}_T = \begin{bmatrix} 1 & 0 & \cdots & 0 \\ 0 & 1 & \cdots & 0 \\ \vdots & \vdots & \ddots & \vdots \\ 0 & 0 & \cdots & 1 \end{bmatrix}_{T\times T}.$$

In addition, it may be directly observed that for this case $\mathbf{\Delta} \circ \breve{\mathbf{R}}$ is a diagonal matrix whose entries are the diagonal entries of $\mathbf{\Delta}$, namely

$$(\mathbf{\Delta} \circ \breve{\mathbf{R}})_{\mathrm{rapid}} = \begin{bmatrix} \|\mathbf{c}_1 - \tilde{\mathbf{c}}_1\|_F^2 & 0 & \cdots & 0 \\ 0 & \|\mathbf{c}_2 - \tilde{\mathbf{c}}_2\|_F^2 & \cdots & 0 \\ \vdots & \vdots & \ddots & \vdots \\ 0 & 0 & \cdots & \|\mathbf{c}_T - \tilde{\mathbf{c}}_T\|_F^2 \end{bmatrix}_{T\times T}, \tag{2.24}$$

where $\mathbf{c}_t$ and $\tilde{\mathbf{c}}_t$, $t = 1, \ldots, T$, are the tth rows of $\mathbf{C}$ and $\tilde{\mathbf{C}}$, respectively. Consequently, Equation (2.23) reduces to Equation (2.15), namely to

$$\zeta_{\mathrm{rapid}} = \frac{1}{2\sqrt{N_t}} \min_{\mathbf{C}\neq\tilde{\mathbf{C}}} \left(\prod_{t=1}^{T} \|\mathbf{c}_t - \tilde{\mathbf{c}}_t\|_F^2 \right)^{1/2T}.$$

2.3.3.2 Upper and Lower Bounds on the Generalized Diversity Product

It was shown in [43] that the generalized diversity product, ζ_R of Equation (2.23) is lower bounded by ζ_{static} of Equation (2.14) and upper bounded by ζ_{rapid} of Equation (2.15). More specifically, if a set of space-time signals characterized in terms of its dimension $(T \times N_t)$ has L elements and $\mathbf{\Delta} \circ \breve{\mathbf{R}}$ is of full rank for any pair of distinct space-time signals $\mathbf{C}$ and $\tilde{\mathbf{C}}$, then the diversity product ζ_R of these signals, devised for a fading channel and defined by the time-correlation matrix $\breve{\mathbf{R}}$, satisfies [43]

$$\max\{\zeta_{\mathrm{static}}, |\det(\breve{\mathbf{R}})|^{1/2T}\zeta_{\mathrm{rapid}}\} \leq \zeta_R \leq \zeta_{\mathrm{rapid}} \leq \sqrt{\frac{L}{2(L-1)}}. \tag{2.25}$$

According to Equation (2.25), the diversity product ζ_R is determined by ζ_{rapid} if the time correlation matrix $\breve{\mathbf{R}}$ is of full rank, since $|\det(\breve{\mathbf{R}})|^{1/2T}\zeta_{\mathrm{rapid}} \leq \zeta_R \leq \zeta_{\mathrm{rapid}}$. In contrast, if $\breve{\mathbf{R}}$ does not have full rank, the diversity product ζ_R is no longer determined by ζ_{rapid}, since we have $\det(\breve{\mathbf{R}}) = 0$, which leads to $0 \leq \zeta_R \leq \zeta_{\mathrm{rapid}}$. Another observation inferred from Equation (2.25) is that we have $\zeta_{\mathrm{static}} \leq \zeta_R \leq \zeta_{\mathrm{rapid}}$, which suggests that the diversity product ζ_R will be independent of the channel's correlation matrix $\breve{\mathbf{R}}$ for all space-time codes having $\zeta_{\mathrm{static}} = \zeta_{\mathrm{rapid}}$. Furthermore, the problem of designing robust space-time signals for time-correlated fading channels can be reduced to that of designing space-time signals for quasi-static fading channels, if ζ_{static} achieves the upper bound of $\sqrt{L/2(L-1)}$ quantified in Equation (2.25) or at least is close to it, since ζ_R will also achieve or will at least be close to this upper bound for any time-correlated fading channel. Based on these results, all space-time signals designed for quasi-static fading channels, such as cyclic space-time codes [54], space-time codes derived from orthogonal design [11, 12, 205, 206], parametric space-time codes [210], Cayley space-time codes [209], etc., can also be used for communication over time-correlated fading channels.

2.4 Orthogonal Space-Time Code Design using SP

In this section, orthogonal space-time code design using SP modulation is considered. First, a rudimentary introduction to SP as a geometric problem is provided in Section 2.4.1, followed by the general concept and the motivation of combining orthogonal space-time code design using SP are discussed in Section 2.4.2. Then, the specific signal design proposed for $N_t = 2$ transmit antennas is provided in Section 2.4.3 in order to shed further light on the concept of combining orthogonal design with SP. The SP constellation construction is described in Section 2.4.4. Finally, the capacity of STBC-SP schemes is derived in Section 2.4.5

2.4.1 General Concept of SP

In this section we present a brief overview of SP, addressing some of the fundamental issues, such as the problem of packing spheres in a given n-dimensional Euclidean space $\mathbb{R}^n$ and the problem of determining how many spheres can just touch another sphere of the same size in this space. The latter problem is referred to as the 'kissing number problem'. More detailed discussions on these issues may be found in [221].

2.4.1.1 The SP Problem

The classic SP problem is to find out how densely identical non-overlapping spheres can be packed in a specific space. For the traditional three-dimensional space, the SP problem may be exemplified by considering a large storage area, and asking what is the highest number of identical balls that can be packed into this space. The packing problem becomes trivial if we consider wooden cubes, for example, instead of the balls. The number of wooden cubes that we can pack would be equal to the volume of the storage area divided by the volume of one wooden cube, assuming that no space would be left over. In other words, we can ideally fill 100% of the storage space since the wooden cubes fit together with no space wasted in between. In contrast, there is always some wasted space in between packed spheres.

Generally, the density Δ of a SP is defined as the proportion of a volume occupied by the spheres, which may be defined as [221]

$$\begin{aligned}\Delta &= \text{proportion of the space that is occupied by the spheres,}\\ &= \frac{\text{volume of one sphere}}{\text{volume of total space}}.\end{aligned} \tag{2.26}$$

The SP problem is concerned with finding the densest packing of equal spheres in a specific n-dimensional space. This geometric problem has numerous practical applications.

2.4.1.2 Representation of n-dimensional Real Euclidean Space $\mathbb{R}^n$

A phasor point in the n-dimensional real-valued Euclidean space $\mathbb{R}^n$ may be represented by a string of n real numbers [221]

$$x = (x_1, x_2, x_3, \ldots, x_n).$$

Accordingly, a sphere in $\mathbb{R}^n$ with a center $u = (u_1, \ldots, u_n)$ and radius ρ consists of all points $x = (x_1, x_2, x_3, \ldots, x_n)$ that satisfy the following constraint

$$(x_1 - u_1)^2 + (x_2 - u_2)^2 + \cdots + (x_n - u_n)^2 = \rho^2.$$

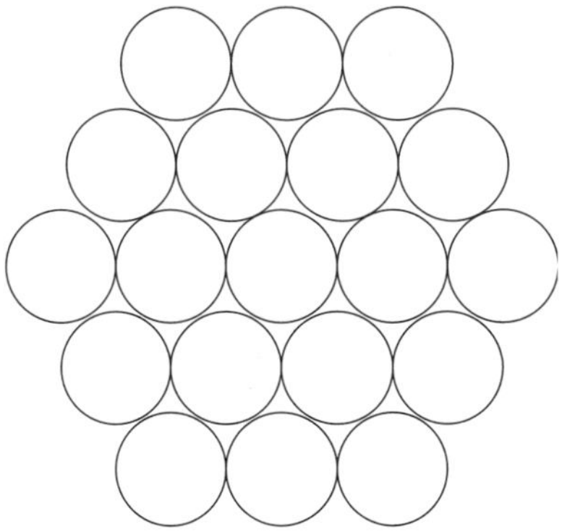

Figure 2.1: SP (or circle-packing) in the two-dimensional Euclidean space.

The graphical representation of the SP problem is only feasible in two or three dimensions. Generally, a SP in $\mathbb{R}^n$ is unambiguously described by specifying the centers u and the radius ρ, where all manipulations are carried out using coordinates. For example, the four-dimensional real-valued Euclidean space $\mathbb{R}^4$ consists of phasor points having four coordinates instead of the conventional three coordinates that represent the three-dimensional real Euclidean space $\mathbb{R}^3$. Example phasor points in the four-dimensional real Euclidean space $\mathbb{R}^4$ include $(1.0, -2.3, 3.4, 0.7)$, $(5.6, 8.0, 14.2, -7.2), \ldots$.

2.4.1.3 Kepler Conjecture

In the three-dimensional Euclidean space $\mathbb{R}^3$, the best packing is that where the centers of the spheres form a so-called face-centered cubic lattice [221], where the spheres only occupy $\pi/\sqrt{18} \approx 0.7405$ of the total space. Consequently, the density Δ of the face-centered cubic lattice packing is about 0.7405. Johannes Kepler conjectured in 1611 that this is the maximum attainable density in the three-dimensional Euclidean space $\mathbb{R}^3$. This was known as the Kepler conjecture. In 1998, Thomas Hales announced the proof of the Kepler conjecture [222, 223], where he based his proof on the approach suggested by Laszló Fejes Tóth in 1953 and employed complex computer calculations in order to check many individual cases. The Kepler conjecture has been extended to dimensions higher than three, where it is still an open mathematical problem.

2.4.1.4 Kissing Numbers

Another important issue, closely related to the SP problem, is known as the kissing number problem. With respect to our three-dimensional bowling balls example mentioned in Section 2.4.1.1, the kissing number problem asks what is the highest number of balls that can be arranged so that they all touch, or 'kiss', another ball of the same size [221]. Generally, the kissing number problem asks for the highest number of n-dimensional spheres that can kiss another sphere of a similar size.

In the one-dimensional Euclidean space, the answer is two, while in the two-dimensional Euclidean space, it is six, as seen in Figure 2.1. In the three-dimensional Euclidean space, the answer is now known to be 12. The answer is also known in 8 and 24 dimensions to be 240 and 196 560, respectively. However, in four dimensions the highest kissing number is only known to be either 24 or 25.

Table 2.1: Sphere packing and kissing number in several dimensions. The bold-faced values refer to those that were proven optimal. Copyright © Conway and Sloane, [221], 1999.

	Dimension										
	1	2	3	4	5	6	7	8	12	16	24
Densest packing	$\mathbf{Z}$	$\mathbf{A_2}$	$\mathbf{A_3}$	D_4	D_5	E_6	E_7	E_8	K_{12}	Λ_{16}	Λ_{24}
Highest kissing number	**2**	**6**	**12**	24	40	72	126	**240**	7 564	320	**196 560**

2.4.1.5 *n*-dimensional Packings

The one-dimensional lattice Z, representing the integers, is the densest packing in one dimension with a density of $\Delta = 1$. In two dimensions, the densest packing is also known, with a density of $\Delta = \pi/\sqrt{12} \approx 0.9069$. Figure 2.1 shows the hexagonal lattice packing, which is the densest in two dimensions. In three dimensions, the face-centered cubic lattice is the densest packing with a density of $\Delta = \pi/\sqrt{18} \approx 0.7405$.

In the four-dimensional real Euclidean space $\mathbb{R}^4$, the lattice D_4 is deemed to be the densest SP arrangement, although this fact is not yet proven. More specifically, D_4 may be defined as a lattice that consists of all phasor points having integer coordinates (a_1, a_2, a_3, a_4), which are subjected to the SP constraint of $a_1 + a_2 + a_3 + a_4 = k$, where k is an even integer. The sphere centered at (0, 0, 0, 0) has 24 spheres around it, centered at the points $(\pm 1, \pm 1, 0, 0)$, where any choice of signs and any ordering of the coordinates is legitimate [221].

The densest packing in the eight-dimensional real-valued Euclidean space $\mathbb{R}^8$ is the E_8 lattice, which is a discrete subfield of $\mathbb{R}^8$ that consists of all phasor points having natural or half-natural[1] coordinates $(a_1, \ldots, a_8)$ that are subjected to the SP constraint of $a_1 + \cdots + a_8 = k$, where k is an even integer. The existence of E_8 was proven in 1867 by Smith [224], but its first quadratic form expression was given in 1873 by Korkine and Zolotare [225]. It was proven to be the densest packing in eight dimensions in 1979 [226, 227]. Table 2.1 summarizes the SP schemes and the highest kissing numbers in several dimensions.

2.4.1.6 Applications of SP

In this section, some example applications are briefly presented in order to justify our interest in finding dense packings of n-dimensional spheres. These applications include pure geometry, number theory, coding theory and code search as well as approximation problems.

The SP problem originally appeared as an interesting problem in pure geometry, which was mentioned by Hilbert in his list of open problems in 1900 [228, 229]. Several authors demonstrated great interest in the SP problem from the perspective of geometry [230–235]. Moreover, applications of SP in number theory include, for example, solving the so-called Diophantine equations and the geometry of numbers [236–238]. The sphere packing problem also has a direct connection with the construction of error correction codes. More specifically, the construction of an (n, M, d)-code[2] can be viewed as finding the densest packing of spheres having a radius of $\rho = (d-1)/2$ in n dimensions [221]. Another application of SP is in solving n-dimensional search or approximation problems [239–242]. A recently proposed

[1] If a is a natural number then $b = a + \frac{1}{2}$ is a half-natural number.

[2] A code of length n, containing M codewords and with minimum distance d is referred to as an (n, M, d)-code that can detect and correct up to $(d-1)/2$ errors [6].

search method that has attracted considerable attention is sphere decoding [67, 243], which provides an accurate estimation similar to the optimum ML solution but with a far lower complexity.

2.4.2 SP-aided STBC Concept

The concept of orthogonal space-time signal design has become synonymous with STC, following the remarkable discovery of Alamouti [11]. Since then, it has attracted considerable further attention [244]. The orthogonal space-time code design can be described with the aid of the following recursive procedure [207] as follows. Let $\mathbf{G}_1(x_1) = x_1\mathbf{I}_1$, and

$$\mathbf{G}_{2^k}(x_1, x_2, \ldots, x_{k+1}) = \begin{bmatrix} \mathbf{G}_{2^{k-1}}(x_1, x_2, \ldots, x_k) & x_{k+1}\mathbf{I}_{2^{k-1}} \\ -x_{k+1}^*\mathbf{I}_{2^{k-1}} & \mathbf{G}_{2^{k-1}}^{\mathcal{H}}(x_1, x_2, \ldots, x_k) \end{bmatrix}, \tag{2.27}$$

for $k = 1, 2, \ldots,$ where x_{k+1}^* is the complex conjugate of x_{k+1}. Then, $\mathbf{G}_{2^k}(x_1, x_2, \ldots, x_{k+1})$ constitutes an orthogonal space-time code generator matrix design of size $(2^k \times 2^k)$, which maps the complex variables representing $(x_1, x_2, \ldots, x_{k+1})$ to $N_t = 2^k$ transmit antennas. In other words, $x_1, x_2, \ldots, x_{k+1}$ represent $(k+1)$ complex modulated symbols to be transmitted from $N_t = 2^k$ transmit antennas in $T = 2^k$ time slots. The effective throughput of $\mathbf{G}_{2^k}$ is $(k+1)/2^k$ complex symbols per time slot, which is the maximum throughput that may be achieved by an orthogonal space-time code design having a square-shaped generator matrix [207].

Space-time signals can be constructed directly from the orthogonal space-time code design $\mathbf{G}_{2^k}$ of Equation (2.27) for $N_t = 2^k$, $k = 1, 2, \ldots,$ transmit antennas as [207]

$$\mathbf{C} = \sqrt{\frac{2^k}{k+1}}\mathbf{G}_{2^k}(x_1, x_2, \ldots, x_{k+1}), \tag{2.28}$$

where $x_1, x_2, \ldots, x_{k+1}$ are specifically chosen from a constellation space satisfying the following condition:

$$E[|x_1|^2 + |x_2|^2 + \cdots + |x_{k+1}|^2] = k+1.$$

The multiplicative term $\sqrt{2^k/(k+1)}$ in Equation (2.28) is a normalisation factor used for ensuring that the space-time signal $\mathbf{C}$ of Equation (2.28) satisfies the energy constraint of Equation (2.2), which dictates that

$$E[\|\mathbf{C}\|_F^2] = N_t T = 2^k \cdot 2^k = 2^{2k}, \tag{2.29}$$

since the space-time signal $\mathbf{C}$ of Equation (2.28) is of size $(2^k \times 2^k)$. The energy of the signal $\mathbf{C}$ in Equation (2.28) is computed as follows

$$\begin{aligned} E\|\mathbf{C}\|_F^2 &= \left(\frac{2^k}{k+1}\right) \cdot E[2^k|x_1|^2 + 2^k|x_2|^2 + \cdots + 2^k|x_{k+1}|^2] \\ &= \left(\frac{2^k}{k+1}\right) \cdot 2^k \cdot E[|x_1|^2 + |x_2|^2 + \cdots + |x_{k+1}|^2] \\ &= \left(\frac{2^k}{k+1}\right) \cdot 2^k \cdot (k+1) \\ &= 2^{2k}, \end{aligned}$$

which satisfies the energy constraint of Equation (2.29).

Let $\mathbf{C}$ and $\tilde{\mathbf{C}}$ be two distinct space-time signals constructed directly from the orthogonal space-time code design $\mathbf{G}_{2^k}$ of Equation (2.27)

$$\mathbf{C} = \sqrt{\frac{2^k}{k+1}}\mathbf{G}_{2^k}(x_1, x_2, \ldots, x_{k+1}) \quad \text{and} \quad \tilde{\mathbf{C}} = \sqrt{\frac{2^k}{k+1}}\mathbf{G}_{2^k}(\tilde{x}_1, \tilde{x}_2, \ldots, \tilde{x}_{k+1}),$$

then we have

$$\mathbf{C} - \tilde{\mathbf{C}} = \sqrt{\frac{2^k}{k+1}}\mathbf{G}_{2^k}(x_1 - \tilde{x}_1, x_2 - \tilde{x}_2, \ldots, x_{k+1} - \tilde{x}_{k+1}),$$

and

$$\boldsymbol{\Delta} = (\mathbf{C} - \tilde{\mathbf{C}})(\mathbf{C} - \tilde{\mathbf{C}})^{\mathcal{H}} = (\mathbf{C} - \tilde{\mathbf{C}})^{\mathcal{H}}(\mathbf{C} - \tilde{\mathbf{C}}) = \frac{2^k}{k+1}\sum_{i=1}^{k+1}|x_i - \tilde{x}_i|^2\mathbf{I}_{2^k}, \tag{2.30}$$

implying that $\boldsymbol{\Delta}$ is a diagonal matrix of size $(2^k \times 2^k)$. Observe that all of the diagonal entries in $\breve{\mathbf{R}}$ of Equation (2.19) are unity, yielding

$$\boldsymbol{\Delta} \circ \breve{\mathbf{R}} = \begin{bmatrix} \frac{2^k}{k+1}\sum_{i=1}^{k+1}|x_i - \tilde{x}_i|^2 & 0 & \cdots & 0 \\ 0 & \frac{2^k}{k+1}\sum_{i=1}^{k+1}|x_i - \tilde{x}_i|^2 & \cdots & 0 \\ \vdots & \vdots & \ddots & \vdots \\ 0 & 0 & \cdots & \frac{2^k}{k+1}\sum_{i=1}^{k+1}|x_i - \tilde{x}_i|^2 \end{bmatrix}_{2^k \times 2^k}, \tag{2.31}$$

and, therefore,

$$\det(\boldsymbol{\Delta} \circ \breve{\mathbf{R}}) = \left(\frac{2^k}{k+1}\sum_{i=1}^{k+1}|x_i - \tilde{x}_i|^2\right)^{2^k}. \tag{2.32}$$

Now, with the aid of Equations (2.23) and (2.32), as well as by observing that $T = N_t = 2^k$, the diversity product evaluated for space-time signals constructed from the orthogonal space-time code design $\mathbf{G}_{2^k}$ of Equation (2.27) can written as [43]

$$\begin{aligned} \zeta_R &= \frac{1}{2\sqrt{2^k}} \min_{\mathbf{C} \neq \tilde{\mathbf{C}}} |\det(\boldsymbol{\Delta} \circ \breve{\mathbf{R}})|^{1/2 \cdot 2^k} \\ &= \frac{1}{2\sqrt{2^k}} \min_{\mathbf{C} \neq \tilde{\mathbf{C}}} \left(\frac{2^k}{k+1}\sum_{i=1}^{k+1}|x_i - \tilde{x}_i|^2\right)^{2^k/2 \cdot 2^k} \\ &= \frac{1}{2\sqrt{k+1}} \min_{\mathbf{C} \neq \tilde{\mathbf{C}}} \left(\sum_{i=1}^{k+1}|x_i - \tilde{x}_i|^2\right)^{1/2} \\ &= \frac{1}{2\sqrt{k+1}} \min_{(x_1, x_2, \ldots, x_{k+1}) \neq (\tilde{x}_1, \ldots, \tilde{x}_{k+1})} \left(\sum_{i=1}^{k+1}|x_i - \tilde{x}_i|^2\right)^{1/2}, \end{aligned} \tag{2.33}$$

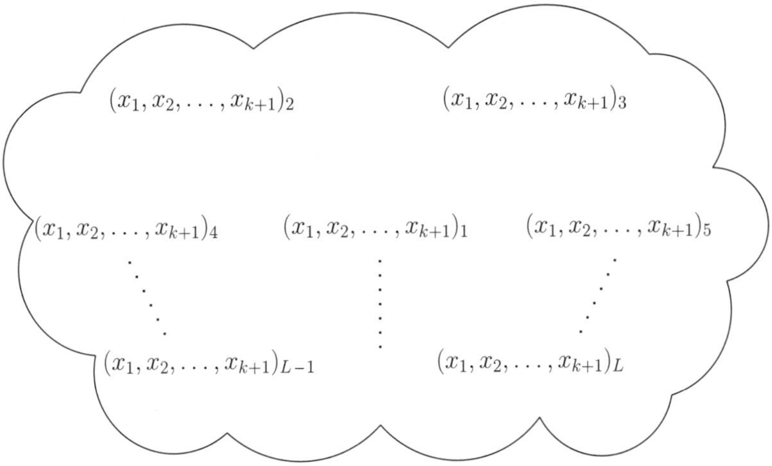

Figure 2.2: The L legitimate $(k+1)$-dimensional complex vectors.

which shows that the diversity product of space-time signals constructed from the orthogonal design $\mathbf{G}_{2^k}$ of Equation (2.27) is independent of the correlation matrix $\breve{\mathbf{R}}$, provided that the space-time signals achieve full diversity (i.e. that $\mathbf{\Delta} \circ \breve{\mathbf{R}}$ is of full rank).

Assuming that there are L legitimate space-time signals, which the encoder can choose from, Equation (2.33) dictates that the diversity product is determined by the MED of the L number of $(k+1)$-dimensional complex vectors $(x_1, x_2, \ldots, x_{k+1})_l \in \mathbb{C}^{k+1}$, $l = 0, \ldots, L-1$. Therefore, in order to maximize the diversity product, the L $(k+1)$-dimensional complex vectors must be designed by ensuring that they have the best MED in the $(k+1)$-dimensional complex-valued space $\mathbb{C}^{k+1}$, as illustrated in Figure 2.2. If each of the L $(k+1)$-dimensional complex vectors is expressed using its real and imaginary components, so that we have

$$(x_1, \ldots, x_{k+1})_l \quad \Longleftrightarrow \quad (a_1 + ja_2, \ldots, a_{2(k+1)-1} + ja_{2(k+1)})_l, \tag{2.34}$$

for $l = 0, \ldots, L-1$, then each of these complex vectors can be represented in the stylised space shown in Figure 2.3. It may be directly observed from Figure 2.3 that the design problem can be readily transformed from the $(k+1)$-dimensional complex-valued space $\mathbb{C}^{k+1}$ to the $2(k+1)$-dimensional real-valued Euclidean space $\mathbb{R}^{2(k+1)}$, as portrayed in Figure 2.4. It was proposed in [43] to use SP for combining the individual antennas' signals into a joint space-time design, since they have the best known MED in the $2(k+1)$-dimensional real-valued Euclidean space $\mathbb{R}^{2(k+1)}$(see [221]) which directly maximizes the diversity product expression of Equation (2.33).

To summarize the general concept, space-time signals having the maximum Euclidean distance may be constructed from an Alamouti-style orthogonal design using SP modulation for $N_t = 2^k, k = 1, 2, 3, \ldots,$ transmit antennas. When conventional modulation methods (i.e. QAM, PSK) are used in the design of space-time signals, the symbols $x_1, x_2, \ldots, x_{k+1}$ are chosen independently of the modulation constellation. In contrast, in the case of SP, these symbols are designed jointly in order to further increase the coding advantage, as suggested above. Assuming that there are L space-time signals that the encoder can choose from and each space-time signal is transmitted over $T = N_t$ consecutive time slots, the effective throughput of the space-time modulated scheme is $(\log_2 L)/N_t$ bits per space-time channel access, where a space-time channel access is again defined as the transmission of one of

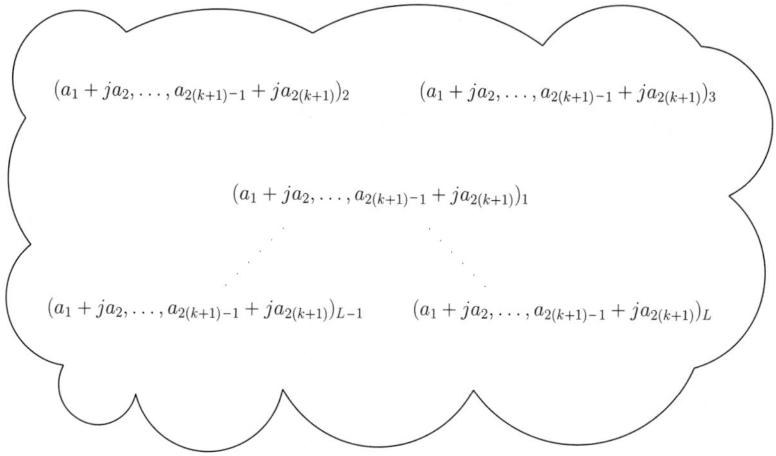

Figure 2.3: The L legitimate $(k+1)$-dimensional complex vectors represented by their real and imaginary components.

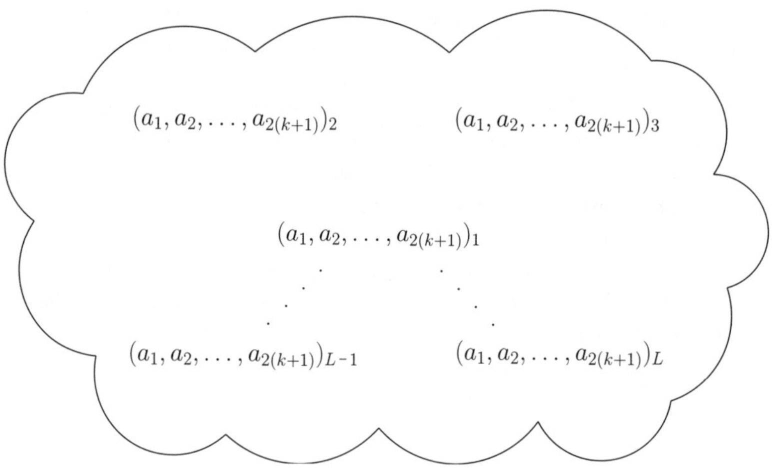

Figure 2.4: The L legitimate $2(k+1)$-dimensional real-valued vectors.

the L legitimate space-time signals over the N_t transmit antennas within the corresponding N_t time slots. When assuming rectangular Nyquist filtering, the resultant spectral efficiency becomes $(\log_2 L)/N_t$ bit s^{-1} Hz^{-1}. Without loss of generality, $\mathbf{G}_2(x_1, x_2)$ is considered in Section 2.4.3, in order to illustrate the concept of orthogonal space-time code design using sphere packing for $N_t = 2$ transmit antennas.

2.4.3 Signal Design for Two Transmit Antennas

The orthogonal space-time signal design for $N_t = 2$ transmit antennas is characterized as

$$\mathbf{G}_2(x_1, x_2) = \begin{bmatrix} x_1 & x_2 \\ -x_2^* & x_1^* \end{bmatrix}, \tag{2.35}$$

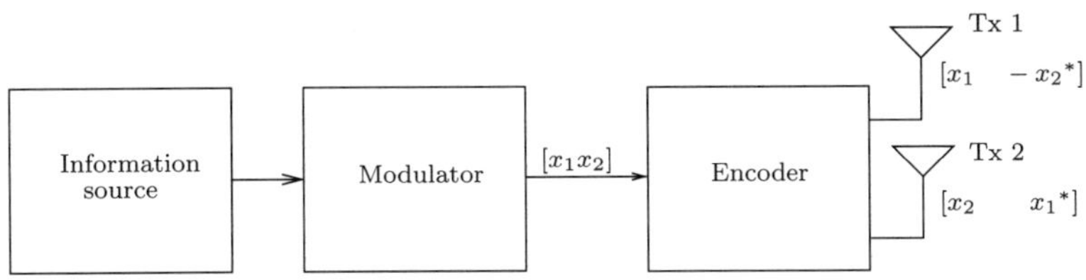

Figure 2.5: Block diagram of the two-transmitter STBC scheme.

Table 2.2: Encoding and transmission sequence for the two-transmitter STBC.

	Antenna 1	Antenna 2
Time slot t	x_1	x_2
Time slot $t+1$	$-x_2^*$	x_1^*

which was introduced by Alamouti [11], where the rows and columns of Equation (2.35) represent the temporal and spatial dimensions, corresponding to two consecutive time slots and two transmit antennas, respectively. In this section, space-time encoding designed for systems employing $N_t = 2$ transmit antennas is presented first in Section 2.4.3.1, while in Section 2.4.3.2 the details of the receiver and ML detection algorithm are given. Then, in Section 2.4.3.3 a range of systems using $N_t = 2$ transmit antennas and multiple receive antennas are discussed. Finally, in Section 2.4.3.4 we show how SP is combined with orthogonal space-time signal design employing N_t transmit antennas.

2.4.3.1 G_2 Space-Time Encoding

As shown in Figure 2.5, the space-time encoder is preceded by a modulator. Each group of B information bits is first modulated using an L-ary modulation scheme, where we have $B = \log_2 L$. Then, the encoder operates on a block of two consecutive modulated symbols x_1 and x_2, mapping them to the transmit antennas according to the generator matrix seen in Equation (2.35).

Each activation of the encoder considers two consecutive time slots. During the first time slot, x_1 is transmitted from the first antenna and simultaneously x_2 is transmitted from the second antenna. During the second time slot, $-x_2^*$ is transmitted from the first antenna and simultaneously x_1^* is transmitted from the second antenna. The columns of the generator matrix represent antennas and the rows represent time slots. Therefore, encoding is carried out in both the space and time domains. Table 2.2 shows the encoding and transmission sequence for the two-transmitter STBC scheme.

2.4.3.2 Receiver and ML Decoding

Figure 2.6 shows the block diagram for the receiver, when a single receive antenna is employed. Let $h_1(t)$ and $h_2(t)$ denote the complex-valued fading channel coefficients of the first and second antennas with respect to the receive antenna at time instant t, respectively. It is assumed that the channel coefficients are constant over two consecutive time slots. Hence,

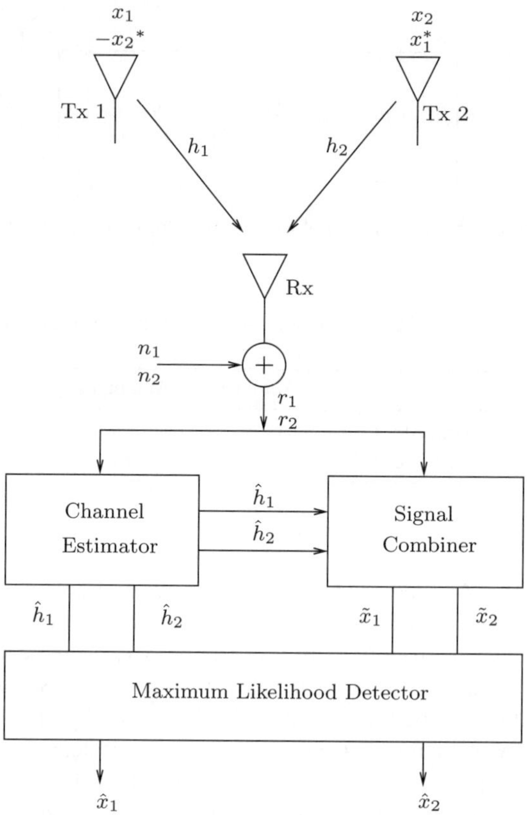

Figure 2.6: Receiver for the two-transmitter STBC.

they can be written as follows

$$h_1(t) = h_1(t+1) = h_1 = |h_1|e^{j\theta_1}, \tag{2.36}$$

$$h_2(t) = h_2(t+1) = h_2 = |h_2|e^{j\theta_2}, \tag{2.37}$$

where $|h_i|$ and θ_i, $i = 1, 2$, are the amplitude and phase shift of the complex-valued channel coefficient for the path spanning from transmit antenna i to the receive antenna.

As shown in Figure 2.6, r_1 and r_2 represent the signals received over the two consecutive time slots t and $t + 1$, respectively, which can be written as

$$r_1 = h_1 x_1 + h_2 x_2 + n_1, \tag{2.38}$$

$$r_2 = -h_1 x_2^* + h_2 x_1^* + n_2, \tag{2.39}$$

where n_1 and n_2 are independent complex-valued AWGN samples at time t and $t + 1$, respectively, with zero mean and a power spectral density of $N_0/2$ per dimension.

Assuming perfect channel estimation, the channel's fading coefficients, h_1 and h_2, are known at the receiver. The availability of the CSI at the receiver will improve the performance of the decoder. The fading coefficient estimates h_1 and h_2 are used by the decoder to find the most likely transmitted signals, as we will show in Equations (2.43) and (2.44). The ML decoder decides on the specific pair of signals $(\hat{x}_1, \hat{x}_2)$ that minimizes the following distance metric:

$$d^2(r_1, h_1\hat{x}_1 + h_2\hat{x}_2) + d^2(r_2, -h_1\hat{x}_2^* + h_2\hat{x}_1^*), \tag{2.40}$$

over all possible values of $\hat{x}_1$ and $\hat{x}_2$, while assuming that all of the signals in the modulation constellation are equiprobable. The squared Euclidean distance between two complex-valued signals, say x and y, can be expressed as

$$d^2(x, y) = (x - y)(x^* - y^*). \tag{2.41}$$

Upon using Equation (2.41), the detection criterion of Equation (2.40) can be further manipulated as follows:

$$\begin{aligned} &d^2(r_1, h_1\hat{x}_1 + h_2\hat{x}_2) + d^2(r_2, -h_1\hat{x}_2^* + h_2\hat{x}_1^*) \\ &\quad = (r_1 - h_1\hat{x}_1 - h_2\hat{x}_2)(r_1^* - h_1^*\hat{x}_1^* - h_2^*\hat{x}_2^*) + (r_2 + h_1\hat{x}_2^* - h_2\hat{x}_1^*)(r_2^* + h_1^*\hat{x}_2 - h_2^*\hat{x}_1) \\ &\quad = (|h_1|^2 + |h_2|^2)(|\hat{x}_1|^2 + |\hat{x}_2|^2) - h_1^* r_1 \hat{x}_1^* - h_2 r_2^* \hat{x}_1^* - h_1 r_1^* \hat{x}_1 - h_2^* r_2 \hat{x}_1 \\ &\qquad - h_2^* r_1 \hat{x}_2^* + h_1 r_2^* \hat{x}_2^* - h_2 r_1^* \hat{x}_2 + h_1^* r_2 \hat{x}_2 + |r_1|^2 + |r_2|^2. \end{aligned} \tag{2.42}$$

Let us define $\tilde{x}_1$ and $\tilde{x}_2$ as two decision statistics constructed by combining the received signals r_1 and r_2 with the perfectly estimated CSI as follows:

$$\tilde{x}_1 = h_1^* r_1 + h_2 r_2^*, \tag{2.43}$$

$$\tilde{x}_2 = h_2^* r_1 - h_1 r_2^*. \tag{2.44}$$

Moreover, $d^2(\tilde{x}_1, \hat{x}_1)$ and $d^2(\tilde{x}_2, \hat{x}_2)$ can be expressed using Equations (2.43) and (2.44) along with Equation (2.41) as follows:

$$\begin{aligned} d^2(\tilde{x}_1, \hat{x}_1) &= (\tilde{x}_1 - \hat{x}_1)(\tilde{x}_1^* - \hat{x}_1^*) \\ &= |\hat{x}_1|^2 + |\tilde{x}_1|^2 - h_1^* r_1 \hat{x}_1^* - h_2 r_2^* \hat{x}_1^* - h_1 r_1^* \hat{x}_1 - h_2^* r_2 \hat{x}_1, \end{aligned} \tag{2.45}$$

and

$$\begin{aligned} d^2(\tilde{x}_2, \hat{x}_2) &= (\tilde{x}_2 - \hat{x}_2)(\tilde{x}_2^* - \hat{x}_2^*) \\ &= |\hat{x}_2|^2 + |\tilde{x}_2|^2 - h_2^* r_1 \hat{x}_2^* + h_1 r_2^* \hat{x}_2^* - h_2 r_1^* \hat{x}_2 + h_1^* r_2 \hat{x}_2. \end{aligned} \tag{2.46}$$

Upon substituting Equations (2.45) and (2.46) into Equation (2.42) and omitting all terms that are not functions of $\hat{x}_1$ or $\hat{x}_2$, the resultant decoding criterion can be expressed as

$$(\hat{x}_1, \hat{x}_2) = \arg \min_{(\hat{x}_1, \hat{x}_2) \in C} (|h_1|^2 + |h_1|^2 - 1)(|\hat{x}_1|^2 + |\hat{x}_2|^2) + d^2(\tilde{x}_1, \hat{x}_1) + d^2(\tilde{x}_2, \hat{x}_2), \tag{2.47}$$

where C is the set of all possible modulated symbol pairs $(\hat{x}_1, \hat{x}_2)$. Furthermore, expressing r_1 and r_2 in Equations (2.43) and (2.44), respectively, using Equations (2.38) and (2.39) yields

$$\tilde{x}_1 = (|h_1|^2 + |h_2|^2)x_1 + h_1^* n_1 + h_2 n_2^*, \tag{2.48}$$

$$\tilde{x}_2 = (|h_1|^2 + |h_2|^2)x_2 - h_1 n_2^* + h_2^* n_1. \tag{2.49}$$

The fact that the decision statistics $\tilde{x}_i$ only depend on x_i, $i = 1, 2$, enables us to separate the ML decoding rule of Equation (2.47) into two independent decoding criteria for x_1 and x_2 as follows:

$$\hat{x}_1 = \arg \min_{\hat{x}_1 \in S} (|h_1|^2 + |h_2|^2 - 1)|\hat{x}_1|^2 + d^2(\tilde{x}_1, \hat{x}_1), \tag{2.50}$$

$$\hat{x}_2 = \arg \min_{\hat{x}_2 \in S} (|h_1|^2 + |h_2|^2 - 1)|\hat{x}_2|^2 + d^2(\tilde{x}_2, \hat{x}_2), \tag{2.51}$$

where S represents the legitimate constellation set. Observe furthermore in Equations (2.48) and (2.49) that even if one of the fading coefficients is small owing to a deep magnitude fade, the chances are that the other coefficient is simultaneously high owing to their independent fading. For the special case of M-PSK signal constellations, the amplitude of all possible modulated symbols, $\hat{x}_i$, $i = 1, 2$, are constant leading to the following further simplified decoding criteria:

$$\hat{x}_1 = \arg \min_{\hat{x}_1 \in S} d^2(\tilde{x}_1, \hat{x}_1), \tag{2.52}$$

$$\hat{x}_2 = \arg \min_{\hat{x}_2 \in S} d^2(\tilde{x}_2, \hat{x}_2). \tag{2.53}$$

2.4.3.3 G$_2$ STC using Multiple Receive Antennas

As motivation for this section, it is worth noting at this early stage that the employment of multiple transmit antennas requires the proportionate reduction of the individual antennas' transmit power for the sake of their fair comparison. This requires for example, for $N_t = 2$, that transmit diversity provides more than 3 dB gain in order to compensate for halving the transmit power of both antennas, which is typically only realistic with the aid of multiple receive antennas. Motivated by this, the two-transmitter STC scheme of Section 2.4.3.2 can be extended to systems employing an arbitrary number of receive antennas, N_r. The decoding criteria must be modified to include the effect of the multiple receive antennas. However, the encoding and transmission is identical to the case of a single receive antenna. Let r_1^j and r_2^j denote the received signals by the jth receive antenna at time t and $t+1$, respectively

$$\begin{aligned} r_1^j &= h_{1,j}x_1 + h_{2,j}x_2 + n_1^j, \\ r_2^j &= -h_{1,j}x_2^* + h_{2,j}x_1^* + n_2^j, \end{aligned} \tag{2.54}$$

where $h_{i,j}$, $i = 1, 2$, $j = 1, \ldots, N_r$, represents the complex-valued fading coefficient for the path spanning from transmit antenna i to receive antenna j, and n_1^j as well as n_2^j are the AWGN samples at time instants t and $t+1$, respectively, at receive antenna j.

The decision statistics $\tilde{x}_1$ and $\tilde{x}_2$ which are constructed by the receiver based on the linear combination of the received signals are expressed as [9, 11, 245]

$$\begin{aligned} \tilde{x}_1 &= \sum_{j=1}^{N_r} h_{1,j}^* r_1^j + h_{2,j}(r_2^j)^* \\ &= \sum_{j=1}^{N_r} [(|h_{1,j}|^2 + |h_{2,j}|^2)x_1 + h_{1,j}^* n_1^j + h_{2,j}(n_2^j)^*], \end{aligned} \tag{2.55}$$

$$\begin{aligned} \tilde{x}_2 &= \sum_{j=1}^{N_r} h_{2,j}^* r_1^j - h_{2,j}(r_2^j)^* \\ &= \sum_{j=1}^{N_r} [(|h_{1,j}|^2 + |h_{2,j}|^2)x_2 + h_{2,j}^* n_1^j - h_{1,j}(n_2^j)^*]. \end{aligned} \tag{2.56}$$

Based on Equations (2.43), (2.44), (2.55) and (2.56), the following rule can be deduced for the linear combination of the received signals [9]:

For $\tilde{x}_i$, $i=1,2$, multiply r_j, $j=1,2$, with the conjugate of the corresponding complex-valued fading coefficient h if x_i is present in r_j as suggested by Equations (2.43), (2.44), (2.55) and (2.56). Otherwise, if x_i^* is present in r_j, multiply r_j^* with the corresponding complex-valued fading coefficient h. Then, the resultant products are added or subtracted from the rest, depending on the sign of the term in the expression of the received signal r_j.

It is plausible from Equations (2.55) and (2.56) that the number of independent propagation paths is proportional to the number of receivers N_r, namely $2 \cdot N_r$. Therefore, as argued at the beginning of this section, if one path is in a deep fade, other paths are still likely to provide a high-reliability link for the transmitted signal, provided that the paths are spatially uncorrelated. This advantage is valid even if a single receive antenna is used, because in that case there are two independent paths corresponding to the first and second transmit antennas.

The ML decoding criteria used by the receiver for the two independent signals x_1 and x_2 are given by [11, 245]

$$\hat{x}_1 = \arg \min_{\hat{x}_1 \in S} \left[\left(\sum_{j=1}^{N_r} (|h_{1,j}|^2 + |h_{2,j}|^2) - 1 \right) |\hat{x}_1|^2 + d^2(\tilde{x}_1, \hat{x}_1) \right], \tag{2.57}$$

$$\hat{x}_2 = \arg \min_{\hat{x}_2 \in S} \left[\left(\sum_{j=1}^{N_r} (|h_{1,j}|^2 + |h_{2,j}|^2) - 1 \right) |\hat{x}_2|^2 + d^2(\tilde{x}_2, \hat{x}_2) \right]. \tag{2.58}$$

2.4.3.4 G$_2$ Orthogonal Design using SP

The $\mathbf{G}_2$ generator matrix of Equation (2.35) is reproduced here for convenience:

$$\mathbf{G}_2(x_1, x_2) = \begin{bmatrix} x_1 & x_2 \\ -x_2^* & x_1^* \end{bmatrix}.$$

According to Alamouti's design [11] for example, x_1 and x_2 represent conventional BPSK modulated symbols transmitted in the first and second time slots and no effort is made to jointly design a signal constellation for the various combinations of x_1 and x_2. For the sake of generalizing our treatment, let us assume that there are L legitimate space-time signals $\mathbf{G}_2(x_{l,1}, x_{l,2})$, $l=0,1,\ldots,L-1$, where L represents the number of sphere-packed modulated symbols. The transmitter has to choose the modulated signal from these L legitimate symbols, which have to be transmitted over the two antennas in two consecutive time slots, where the throughput of the system is given by $(\log_2 L)/2$ bits per channel access. In contrast to Alamouti's independent design of the two time slots' signals [11], our aim is to design $x_{l,1}$ and $x_{l,2}$ jointly, so that they have the best MED from all other $(L-1)$ legitimate transmitted space-time signals [43], since this minimizes the system's space-time symbol error probability. Let $(a_{l,1}, a_{l,2}, a_{l,3}, a_{l,4})$, $l=0,\ldots,L-1$, be phasor points belonging to the four-dimensional real-valued Euclidean space $\mathbb{R}^4$, where each of the four elements $a_{l,d}$, $d=1,\ldots,4$, gives one coordinate of the two time slots' complex-valued phasor points. Hence, $x_{l,1}$ and $x_{l,2}$ may be written as

$$\begin{aligned} \{x_{l,1}, x_{l,2}\} &= T_{sp}(a_{l,1}, a_{l,2}, a_{l,3}, a_{l,4}) \\ &= \{a_{l,1} + ja_{l,2}, a_{l,3} + ja_{l,4}\}, \end{aligned} \tag{2.59}$$

where the SP function T_{sp} represents the mapping of the SP symbols $(a_{l,1}, a_{l,2}, a_{l,3}, a_{l,4})$ to the complex-valued symbols $x_{l,1}$ and $x_{l,2}$, $l = 0, \ldots, L-1$.

In the four-dimensional real-valued Euclidean space $\mathbb{R}^4$, the lattice D_4 is defined as a SP having the best MED from all other $(L-1)$ legitimate constellation points in $\mathbb{R}^4$ (see [221]). More specifically, D_4 may be defined as a lattice that consists of all legitimate sphere-packed constellation points having integer coordinates $(a_{l,1}, a_{l,2}, a_{l,3}, a_{l,4})$, $l = 0, \ldots, L-1$, uniquely and unambiguously describing the L legitimate combinations of the two time slots' modulated symbols in Alamouti's scheme, but subjected to the SP constraint of [221]

$$a_{l,1} + a_{l,2} + a_{l,3} + a_{l,4} = k_l, \quad l = 0, \ldots, L-1, \tag{2.60}$$

where k_l may assume any even integer value. Alternatively, D_4 may be defined as the integer span of the vectors $\mathbf{v}_1$, $\mathbf{v}_2$, $\mathbf{v}_3$ and $\mathbf{v}_4$ that form the rows of the following generator matrix [221]:

$$\begin{bmatrix} \mathbf{v}_1 \\ \mathbf{v}_2 \\ \mathbf{v}_3 \\ \mathbf{v}_4 \end{bmatrix} \triangleq \begin{bmatrix} 2 & 0 & 0 & 0 \\ 1 & 1 & 0 & 0 \\ 1 & 0 & 1 & 0 \\ 1 & 0 & 0 & 1 \end{bmatrix}. \tag{2.61}$$

We may infer from the above definition in Equation (2.61) that D_4 contains the centers $(2, 0, 0, 0)$, $(1, 1, 0, 0)$, $(1, 0, 1, 0)$ and $(1, 0, 0, 1)$. It also contains all linear combinations of these points. For example, $\mathbf{v}_1 - 3 \cdot \mathbf{v}_3 = (2, 0, 0, 0) - 3 \cdot (1, 0, 1, 0) = (-1, 0, -3, 0)$ is a legitimate phasor point in D_4.

Assuming that $S = \{\mathbf{s}^l = [a_{l,1}\ a_{l,2}\ a_{l,3}\ a_{l,4}] \in \mathbb{R}^4 : 0 \leq l \leq L-1\}$ constitutes a set of L legitimate constellation points from the lattice D_4 having a total energy of

$$E_{\text{total}} \triangleq \sum_{l=0}^{L-1} (|a_{l,1}|^2 + |a_{l,2}|^2 + |a_{l,3}|^2 + |a_{l,4}|^2), \tag{2.62}$$

and upon introducing the notation

$$\begin{aligned} \mathbf{C}_l &= \sqrt{\frac{2L}{E_{\text{total}}}} \mathbf{G}_2(x_{l,1}, x_{l,2}), \\ &= \sqrt{\frac{2L}{E_{\text{total}}}} \mathbf{G}_2(a_{l,1} + ja_{l,2}, a_{l,3} + ja_{l,4}), \quad l = 0, \ldots, L-1, \end{aligned} \tag{2.63}$$

we have a set of space-time signals, $\{\mathbf{C}_l : 0 \leq l \leq L-1\}$, whose diversity product is determined by the MED of the set of L legitimate constellation points in S. The transformation of the L legitimate two-dimensional complex vectors $(x_{l,1}, x_{l,2})$, $l = 0, \ldots, L-1$, to the L legitimate four-dimensional real-valued vectors $(a_{l,1}, a_{l,2}, a_{l,3}, a_{l,4})$, $l = 0, \ldots, L-1$, is depicted in Figures 2.7 and 2.8.

The normalization factor $\sqrt{2L/E_{\text{total}}}$ in Equation (2.63) is used for ensuring that the space-time signal of Equation (2.63) satisfies the energy constraint of Equation (2.2), which can be derived as follows.

Let us define n_f as a normalization factor to be used in Equation (2.63) for ensuring that the space-time signal satisfies the energy constraint of Equation (2.2). Then the space-time signal of Equation (2.63) can be written as

$$\mathbf{C}_l = n_f \begin{bmatrix} x_{l,1} & x_{l,2} \\ -x_{l,2}^* & x_{l,1}^* \end{bmatrix}. \tag{2.64}$$

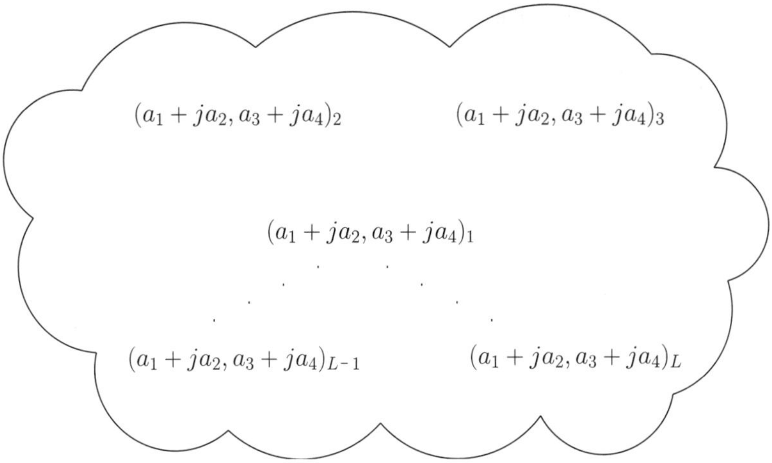

Figure 2.7: The L legitimate two-dimensional complex vectors for $\mathbf{G}_2$ space-time signals.

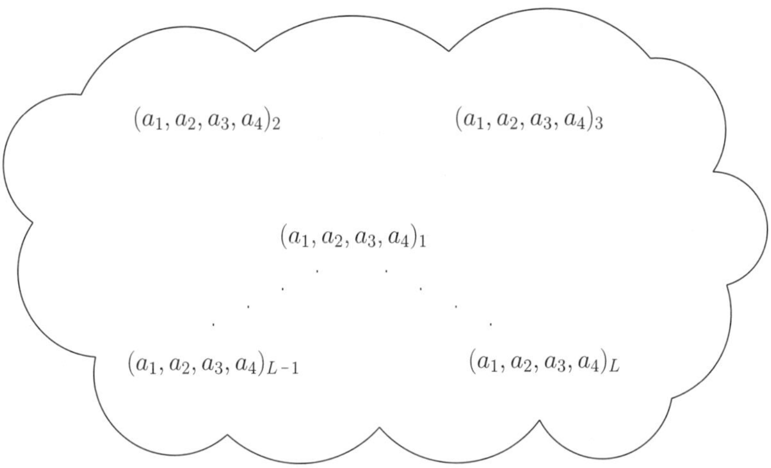

Figure 2.8: The L legitimate four-dimensional real-valued vectors for $\mathbf{G}_2$ space-time signals.

The energy of the space-time signal formulated in Equation (2.64) is given by

$$\begin{aligned} E[\|\mathbf{C}_l\|_F^2] &= E[n_f^2 \cdot (|x_{l,1}|^2 + |x_{l,2}|^2 + |x_{l,2}^*|^2 + |x_{l,1}^*|^2)] \\ &= 2 \cdot n_f^2 \cdot E[|x_{l,1}|^2 + |x_{l,2}|^2]. \end{aligned} \tag{2.65}$$

Expressing Equation (2.65) with the aid of Equation (2.59) yields

$$E[\|\mathbf{C}_l\|_F^2] = 2 \cdot n_f^2 \cdot E[|a_{l,1}|^2 + |a_{l,2}|^2 + |a_{l,3}|^2 + |a_{l,4}|^2]. \tag{2.66}$$

Observe from Equation (2.62) that the average energy of a single SP symbol is given by

$$E[|a_{l,1}|^2 + |a_{l,2}|^2 + |a_{l,3}|^2 + |a_{l,4}|^2] = \frac{E_{\text{total}}}{L}. \tag{2.67}$$

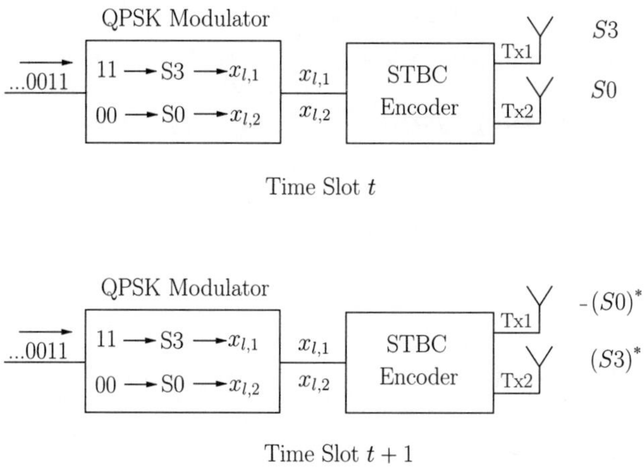

Figure 2.9: Transmission of two QPSK symbols during two consecutive time slots.

Then, Equation (2.66) becomes

$$\begin{aligned} E[\|\mathbf{C}_l\|_F^2] &= 2 \cdot n_f^2 \cdot \frac{E_{\text{total}}}{L} \\ &= T \cdot N_t \quad \text{(from Equation (2.2))} \\ &= 2 \cdot 2 = 4. \end{aligned} \tag{2.68}$$

Hence, with the aid of Equation (2.68) we arrive at

$$n_f = \sqrt{\frac{2L}{E_{\text{total}}}}. \tag{2.69}$$

The following example illustrates how SP modulation may be implemented in combination with $\mathbf{G}_2$ space-time coded systems.

Example 2.4.1. Assume that there are $L = 16$ different legitimate space-time signals, $\mathbf{G}_2(x_{l,1}, x_{l,2})$, $l = 0, \ldots, 15$, that the encoder can choose from. We consider two optional modulation schemes, namely conventional QPSK modulation and SP modulation.

- Conventional QPSK modulation. There are four legitimate two-bit QPSK symbols, $S0$, $S1$, $S2$, and $S3$, that can be used for representing any of the symbols $x_{l,1}$ and $x_{l,2}$, $l = 0, \ldots, 15$. The transmission regime of the two consecutive time slots is demonstrated in Figure 2.9.

- SP modulation. We need $L = 16$ phasor points selected from the lattice D_4, $(a_{l,1}, a_{l,2}, a_{l,3}, a_{l,4})$, $l = 0, \ldots, 15$, in order to jointly represent each space-time signal $(x_{l,1}, x_{l,2})$, $l = 0, \ldots, 15$, according to Equation (2.59), as depicted in Figure 2.10. Table 2.3 shows the effective throughput and the associated transmission block sizes for different values of L.

2.4.4 SP Constellation Construction

Since the orthogonal $\mathbf{G}_2$ space-time signal, which is constructed from the SP scheme of Equation (2.63), is multiplied by a factor that is inversely proportional to $\sqrt{E_{\text{total}}}$, namely

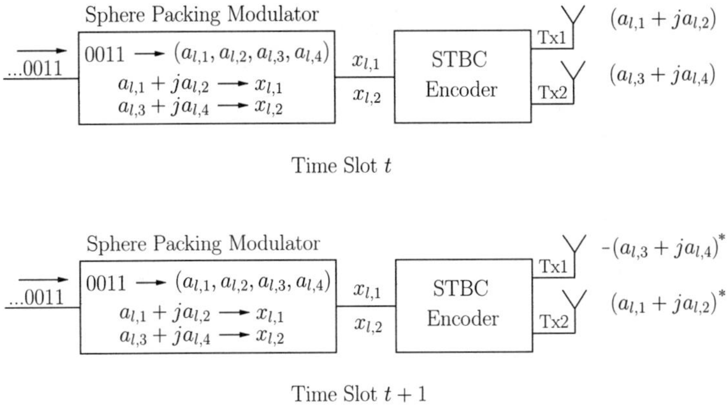

Figure 2.10: Transmission of a single sphere-packed symbol during two consecutive time slots.

Table 2.3: Throughput of SP-aided $\mathbf{G}_2$ systems for different SP signal set sizes L.

L	Block size (bits)	Throughput (bit s^{-1} Hz^{-1})
4	2	1
8	3	1.5
16	4	2
32	5	2.5
64	6	3
128	7	3.5
256	8	4
512	9	4.5
1024	10	5
2048	11	5.5
4096	12	6

by $\sqrt{2L/E_{\text{total}}}$, it is desirable to choose a specific subset of L points from the entire set of legitimate constellation points hosted by D_4, which results in the minimum total energy E_{total} while maintaining a certain minimum distance amongst the SP symbols. Viewing this design trade-off from a different perspective, if more than L points satisfy the minimum total energy constraint, an exhaustive computer search is carried out to determine the optimum choice of the L points out of all possible points, which possess the highest MED, hence minimizing the SP symbol error probability. For this purpose, the legitimate constellation points hosted by D_4 are categorized into layers or shells based on their norms or energy (i.e. distance from the origin) as seen in Table 2.4. For example, it was shown in [221] that the first layer consists of 24 legitimate constellation points hosted by D_4 having an identical minimum energy of $E = 2$. In simple terms, the SP symbol centered at $(0, 0, 0, 0)$ has 24 minimum-distance or closest-neighbor SP symbols around it, centered at the points $(\pm 1, \pm 1, 0, 0)$, where any choice of signs and any ordering of the coordinates is legitimate.

Example 2.4.2. Assume that there are $L = 16$ different legitimate space-time signals, $\mathbf{G}_2(x_{l,1}, x_{l,2})$, $l = 0, \ldots, 15$, which the encoder can choose from. Then, we need the

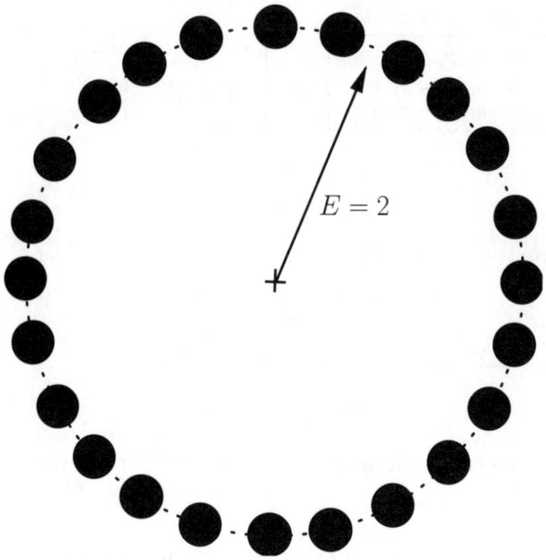

Figure 2.11: The 24 first-layer SP constellation points hosted by D_4 having the minimum energy of $E = 2$.

most 'meritorious' $L = 16$ phasor points from the lattice D_4 for representing the space-time signals. According to the above-mentioned minimum energy constraint, only the 24 legitimate constellation points hosted by the first SP layer of Table 2.4 are considered, as shown in Figure 2.11. Then, an exhaustive computer search is employed to determine the optimum choice of the $L = 16$ points out of the 24 possible first-layer points, which possess the highest MED.

Table 2.4 provides a summary of the constellation points hosted by the first ten layers in the four-dimensional lattice D_4. In order to generate the full list of SP regimes for a specific layer, we have to apply all legitimate permutations and signs for the corresponding constellation points given in Table 2.4.

2.4.5 Capacity of STBC-SP Schemes

In this section, the capacity of STBC-SP schemes having $N_t = 2$ transmit and N_r receive antennas is derived based on the results found in [246]. Assuming perfect channel estimation, the complex-valued channel output symbols received during two consecutive STBC time slots at receiver j are first diversity-combined in order to extract the estimates $\tilde{x}_{1,j}$ and $\tilde{x}_{2,j}$ of the most likely transmitted symbols $x_{l,1}$ and $x_{l,2}$, as seen in Equations (2.48) and (2.49), resulting in [247]

$$\tilde{x}_{1,j} = \sum_{i=1}^{N_t} |h_{i,j}|^2 \cdot x_{l,1} + \acute{n}_{1,j} = \chi^2_{2N_t,j} \cdot x_{l,1} + \acute{n}_{1,j}, \tag{2.70}$$

$$\tilde{x}_{2,j} = \sum_{i=1}^{N_t} |h_{i,j}|^2 \cdot x_{l,2} + \acute{n}_{2,j} = \chi^2_{2N_t,j} \cdot x_{l,2} + \acute{n}_{2,j}, \tag{2.71}$$

for $j = 1, \ldots, N_r$, where again, $h_{i,j}$ represents the complex-valued Rayleigh channel coefficient between the ith transmit antenna and the jth receive antenna; $\chi^2_{2N_t,j} = \sum_{i=1}^{N_t} |h_{i,j}|^2$

Table 2.4: The first ten layers of D_4.

Layer	Constellation points				Norm	Number of combinations
0	0	0	0	0	0	1
1	±1	±1	0	0	2	24
2	±2	0	0	0	4	8
	±1	±1	±1	±1	4	16
3	±2	±1	±1	0	6	96
4	±2	±2	0	0	8	24
5	±2	±2	±1	±1	10	96
	±3	±1	0	0	10	48
6	±3	±1	±1	±1	12	64
	±2	±2	±2	0	12	32
7	±3	±2	±1	0	14	192
8	±2	±2	±2	±2	16	16
	±4	0	0	0	16	8
9	±4	±1	±1	0	18	96
	±3	±2	±2	±1	18	192
	±3	±3	0	0	18	24
10	±4	±2	0	0	20	48
	±3	±3	±1	±1	20	96

represents the chi-squared distributed random variable having $2N_t$ degrees of freedom at receiver j; $\acute{n}_{1,j}$ as well as $\acute{n}_{2,j}$ are the zero-mean complex Gaussian random variables at receiver j during the first and second time slots, respectively, and having a variance of $\sigma^2_{\acute{n}_j} = \chi^2_{2N_t,j} \cdot \sigma^2_n$, where σ^2_n is the original noise variance per dimension. The received sphere-packed symbol $\mathbf{r}_j$ of receiver j is then constructed from the estimates $\tilde{x}_{1,j}$ and $\tilde{x}_{2,j}$ using the inverse function of T_{sp} introduced in Equation (2.59) as

$$\mathbf{r}_j = T^{-1}_{sp}(\tilde{x}_{1,j}, \tilde{x}_{2,j}), \tag{2.72}$$

where we have $\mathbf{r}_j = [\tilde{a}_{j,1}\ \tilde{a}_{j,2}\ \tilde{a}_{j,3}\ \tilde{a}_{j,4}] \in \mathbb{R}^4$. The received sphere-packed symbol $\mathbf{r}_j$ can be written as

$$\mathbf{r}_j = \chi^2_{2N_t,j} \cdot \sqrt{\frac{2L}{E_{\text{total}}}} \cdot \mathbf{s}^l + \mathbf{w}_j, \tag{2.73}$$

where we have $\mathbf{s}^l = [a_{l,1}\ a_{l,2}\ a_{l,3}\ a_{l,4}] \in S$, $0 \le l \le L-1$. Furthermore, $\mathbf{w}_j = [w_{j,1}\ w_{j,2}\ w_{j,3}\ w_{j,4}] \in \mathbb{R}^4$ is a four-dimensional real-valued Gaussian random variable having a covariance matrix of $\sigma^2_{w_j} \cdot \mathbf{I}_{N_D} = \sigma^2_{\acute{n}_j} \cdot \mathbf{I}_{N_D} = \chi^2_{2N_t,j} \cdot \sigma^2_n \cdot \mathbf{I}_{N_D}$, where we have $N_D = 4$, since the SP symbol constellation S is four-dimensional. According to Equation (2.73), the conditional probability $p(\mathbf{r}_j|\mathbf{s}^l)$ of receiving a four-dimensional signal $\mathbf{r}_j$, given that a four-dimensional L-ary SP symbol $\mathbf{s}^l \in S$, $0 \le l \le L-1$, was transmitted over the Rayleigh channel of Equation (2.73), is given by the following PDF:

$$\begin{aligned} p(\mathbf{r}_j|\mathbf{s}^l) &= \frac{1}{(2\pi\sigma^2_{w_j})^{N_D/2}} \exp\left(-\frac{1}{2\sigma^2_{w_j}}(\mathbf{r}_j - \beta_j \cdot \mathbf{s}^l)(\mathbf{r}_j - \beta_j \cdot \mathbf{s}^l)^T\right) \\ &= \frac{1}{(2\pi\sigma^2_{w_j})^{N_D/2}} \exp\left(-\frac{1}{2\sigma^2_{w_j}}\left(\sum_{d=1}^{N_D}(\tilde{a}_{j,d} - \beta_j \cdot a_{l,d})^2\right)\right), \end{aligned} \tag{2.74}$$

where

$$\beta_j = \chi^2_{2N_t,j} \cdot \sqrt{\frac{2L}{E_{\text{total}}}}.$$

Let us define $\vec{\mathbf{r}} = (\mathbf{r}_1, \ldots, \mathbf{r}_{N_r})^{\mathrm{T}}$ as the N_r-element real-valued four-dimensional received signal vector. Using Equation (2.74), the conditional probability $p(\vec{\mathbf{r}}|\mathbf{s}^l)$ is given by [6]

$$\begin{aligned} p(\vec{\mathbf{r}}|\mathbf{s}^l) &= \prod_{j=1}^{N_r} p(\mathbf{r}_j|\mathbf{s}^l), \\ &= \frac{1}{\prod_{j=1}^{N_r}(2\pi\sigma^2_{w_j})^{N_D/2}} \exp\left(- \sum_{d=1}^{N_D}\sum_{j=1}^{N_r} \frac{(\tilde{a}_{j,d} - \beta_j \cdot a_{l,d})^2}{2\sigma^2_{w_j}} \right). \end{aligned} \tag{2.75}$$

The channel capacity valid for STBC schemes using N_D-dimensional L-ary signaling [248] over the DCMC [6] was derived in [246] from that of the DMC [249]. Accordingly, the channel capacity derived for STBC-SP schemes using N_D-dimensional L-ary signaling may be written as [246, 250]

$$\begin{aligned} C^{\text{STBC-SP}}_{\text{DCMC}} = \max_{p(\mathbf{s}^0),\ldots,p(\mathbf{s}^{L-1})} \sum_{l=0}^{L-1} \underbrace{\int_{-\infty}^{\infty} \cdots \int_{-\infty}^{\infty}}_{N_D\text{-fold}} p(\vec{\mathbf{r}}|\mathbf{s}^l) p(\mathbf{s}^l) \\ \cdot \log_2 \left(\frac{p(\vec{\mathbf{r}}|\mathbf{s}^l)}{\sum_{k=0}^{L-1} p(\vec{\mathbf{r}}|\mathbf{s}^k) p(\mathbf{s}^k)} \right) d\vec{\mathbf{r}} \quad \text{(bits per symbol)}, \end{aligned} \tag{2.76}$$

where $p(\mathbf{s}^l)$ is the probability of occurrence for the transmitted sphere packing symbol $\mathbf{s}^l$ and $p(\vec{\mathbf{r}}|\mathbf{s}^l)$ is expressed in Equation (2.75). The right-hand side of Equation (2.76) is maximized when the transmitted SP symbols are equiprobably distributed, i.e. when we have $p(\mathbf{s}^l) = 1/L, l = 0, \ldots, L-1$, which leads to achieving the full capacity [249]. The right-hand side of Equation (2.76) may be further simplified as follows [246]:

$$\begin{aligned} \log_2 \left(\frac{p(\vec{\mathbf{r}}|\mathbf{s}^l)}{\sum_{k=0}^{L-1} p(\vec{\mathbf{r}}|\mathbf{s}^k) p(\mathbf{s}^k)} \right) &= -\log_2 \left(\frac{1}{L} \sum_{k=0}^{L-1} \frac{p(\vec{\mathbf{r}}|\mathbf{s}^k)}{p(\vec{\mathbf{r}}|\mathbf{s}^l)} \right) \\ &= \log_2(L) - \log_2 \sum_{k=0}^{L-1} \exp(\Psi_{l,k}), \end{aligned} \tag{2.77}$$

where $\Psi_{l,k}$ is expressed as [246]

$$\begin{aligned} \Psi_{l,k} &= \sum_{d=1}^{N_D}\sum_{j=1}^{N_r} \left(\frac{-(\tilde{a}_{j,d} - \beta_j \cdot a_{k,d})^2 + (\tilde{a}_{j,d} - \beta_j \cdot a_{l,d})^2}{2\sigma^2_{w_j}} \right) \\ &= \sum_{d=1}^{N_D}\sum_{j=1}^{N_r} \left(\frac{(\beta_j \cdot (a_{l,d} - a_{k,d}) + w_{j,d})^2 + (w_{j,d})^2}{2\sigma^2_{w_j}} \right). \end{aligned} \tag{2.78}$$

Now, substituting Equation (2.77) into Equation (2.76) and observing that we have $p(\mathbf{s}^l) = 1/L$, yields [246]

$$C_{\text{DCMC}}^{\text{STBC-SP}} = \frac{\log_2(L)}{L} \sum_{l=0}^{L-1} \underbrace{\int_{-\infty}^{\infty} \cdots \int_{-\infty}^{\infty}}_{N_D\text{-fold}} p(\vec{\mathbf{r}}|\mathbf{s}^l)\, d\vec{\mathbf{r}}$$
$$- \frac{1}{L} \sum_{l=0}^{L-1} \underbrace{\int_{-\infty}^{\infty} \cdots \int_{-\infty}^{\infty}}_{N_D\text{-fold}} p(\vec{\mathbf{r}}|\mathbf{s}^l) \log_2 \sum_{k=0}^{L-1} \exp(\Psi_{l,k})\, d\vec{\mathbf{r}}$$
$$= \log_2(L) - \frac{1}{L} \sum_{l=0}^{L-1} E\bigg[\log_2 \sum_{k=0}^{L-1} \exp(\Psi_{l,k}) \bigg| \mathbf{s}^l\bigg] \quad \text{(bits per symbol)}, \qquad (2.79)$$

where $E[\xi|\mathbf{s}^l]$ is the expectation of ξ conditioned on $\mathbf{s}^l$. The expectation in Equation (2.79) can be estimated using a sufficiently high number of $\chi^2_{2N_t,j}$ and $\mathbf{w}_j$ realizations with the aid of Monte Carlo simulations for $j = 1, \ldots, N_r$.

The resultant bandwidth efficiency as defined in [250] is computed by normalizing the channel capacity with respect to the bandwidth W and the signaling period T_s, resulting in [246, 250]

$$\eta_{\text{DCMC}}^{\text{STBC-SP}}(E_b/N_0) = \frac{C_{\text{DCMC}}^{\text{STBC-SP}}}{N_D/2} \quad (\text{bit s}^{-1}\ \text{Hz}^{-1}). \qquad (2.80)$$

Figure 2.12 shows the DCMC capacity evaluated from Equation (2.79) for the four-dimensional SP-modulation-assisted STBC scheme for $L = 4$, 16, 64 and 256, when employing $N_t = 2$ transmit antennas as well as $N_r = 1$, 2 and 6 receive antennas, respectively. The Continuous-input Continuous-output Memoryless Channel (CCMC) [6] capacity of the MIMO scheme was also plotted for comparison in Figure 2.12 based on [251]. Figure 2.13 demonstrates and compares the achievable bandwidth efficiency of various SP-modulated STBC schemes and identical-throughput conventionally modulated STBC schemes. The specific modulation type employed for the various schemes is outlined in Table 2.5. Figure 2.13 explicitly illustrates that a higher bandwidth efficiency may be attained when employing SP modulation in conjunction with STBC schemes having $N_t = 2$ transmit antennas.

2.5 STBC-SP Performance

In this section, the two-transmit-antenna-based scheme of Section 2.4.3 is considered. Simulation results are provided for systems having different Bits Per Symbol (BPS) throughputs in conjunction with the appropriate conventional and SP modulation schemes, as outlined in Table 2.5. Observe that $T = 2$ time slots are required for transmitting a single sphere-packed symbol, when using the $N_t = 2$ transmit-antenna-based scheme of Section 2.4.3. In contrast, two conventionally modulated symbols are transmitted during the same time period. Therefore, the throughput of the SP modulation scheme has to be twice that of the conventional modulation scheme in order to create systems having an identical overall BPS throughput. This explains the specific choices of L in Table 2.5. Results are also shown for systems employing $N_r = 1, 2, 3, 4, 5,$ and 6 receive antennas, when communicating over a correlated Rayleigh fading channel having a normalized Doppler frequency of $f_D = 0.1$. The channel's complex fading envelope is assumed to be constant over the transmission

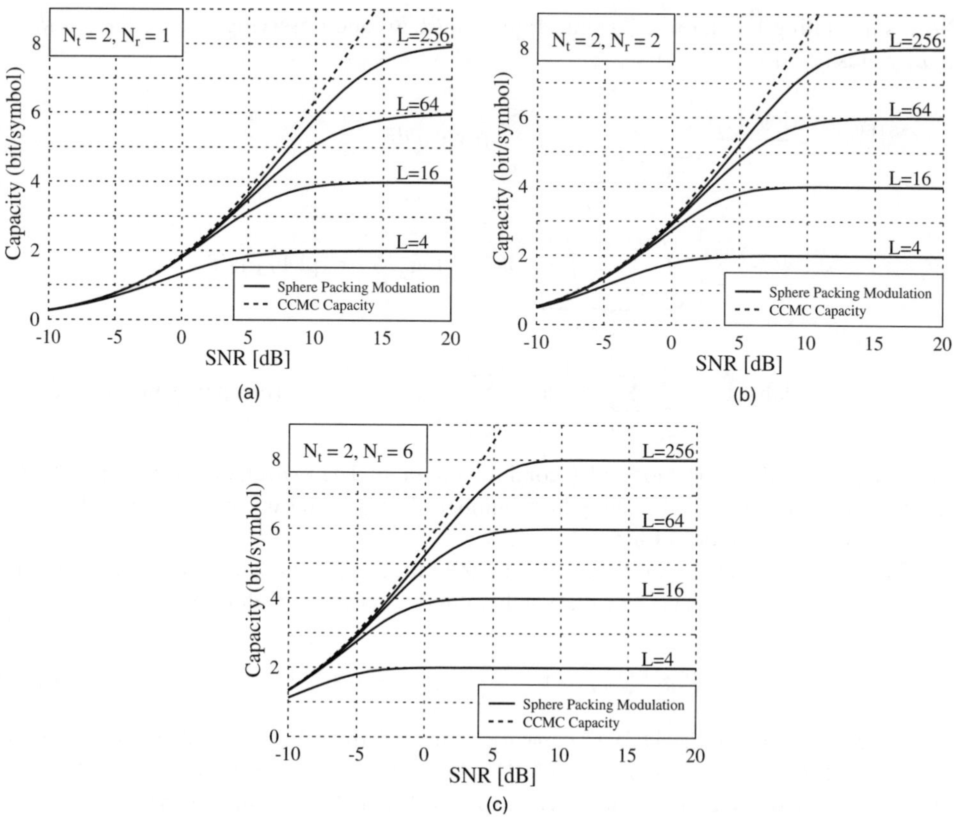

Figure 2.12: Capacity of STBC-SP-based schemes evaluated from Equation (2.79) and using $L =$ 4, 16, 64 and 256, when employing $N_t = 2$ transmit and N_r receive antennas for communicating over a correlated SPSI Rayleigh fading channel having a normalized Doppler frequency of $f_D = 0.1$: (a) $N_r = 1$; (b) $N_r = 2$; (c) $N_r = 6$.

Table 2.5: Conventional and SP modulation employed for different BPS throughputs.

Throughput (BPS)	Conventional modulation	SP modulation
1	BPSK	$L = 4$
2	QPSK	$L = 16$
3	8-PSK	$L = 64$
4	16-QAM	$L = 256$

period of $T = 2$ time slots of a single space-time signal or, equivalently, the transmission period of a sphere-packed symbol. This type of channel will be referred to here as a Sphere Packing Symbol Invariant (SPSI) channel. In this section, both the achievable BER and the SP Symbol Error Rate (SP-SER) are considered. The SP-SER represents the block error rate, where the block size is $B = \log_2 L$ bits, which is also synonymous to the Space-Time Symbol Error Rate (ST-SER).

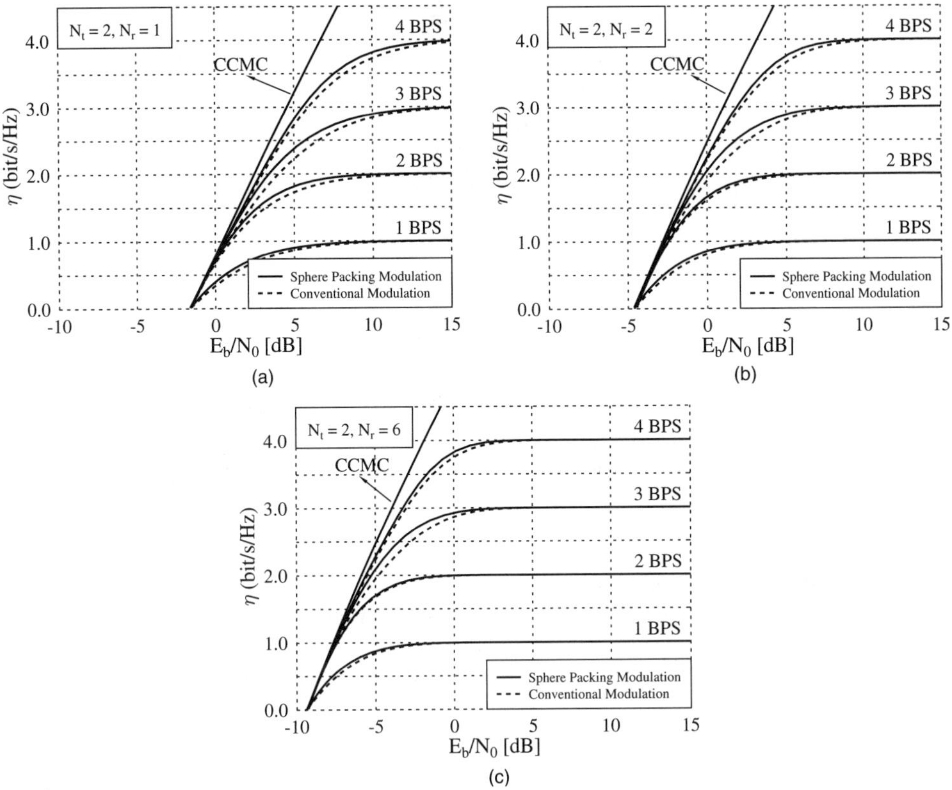

Figure 2.13: Bandwidth efficiency of STBC-SP-based schemes evaluated from Equation (2.80) for $L = 4$, 16, 64 and 256, when employing $N_t = 2$ transmit and N_r receive antennas for communicating over a correlated SPSI Rayleigh fading channel having a normalized Doppler frequency of $f_D = 0.1$: (a) $N_r = 1$; (b) $N_r = 2$; (c) $N_r = 6$.

SP-aided orthogonal design schemes promise to provide improved SP-SER (or ST-SER), when compared with the same metric of conventionally modulated orthogonal STBC-based schemes. This promise is based on the fact that SP modulation improves the diversity product (or coding advantage) of Equation (2.33), which is derived from the upper bound outlined in Equation (2.22) for the pairwise error probability between any two distinct space-time signals. On the other hand, the BER performance of SP-aided orthogonal space-time code design schemes is not always guaranteed to be better than that of conventionally modulated space-time signals, since SP modulation specifically optimizes the MED between two distinct space-time signals constructed using Equation (2.28), but not between the individual constituent symbols, $x_1, x_2, \ldots, x_{k+1}$, which conventional-modulation-based STBC optimizes. However, it is demonstrated later in this section that upon increasing the number of receive antennas, the achievable SP-SER performance improvement increases, which in turn leads to further BER performance improvements for the SP-aided orthogonal STBC schemes, as compared with conventionally modulated orthogonal STBC schemes.

Figure 2.14 shows the SP-SER performance curves of different orthogonal STBC schemes in combination with both conventional and SP modulation for the different BPS throughputs, as outlined in Table 2.5. The systems employ $N_t = 2$ transmit antennas and $N_r = 1$ receive antenna for communicating over a correlated SPSI Rayleigh fading channel

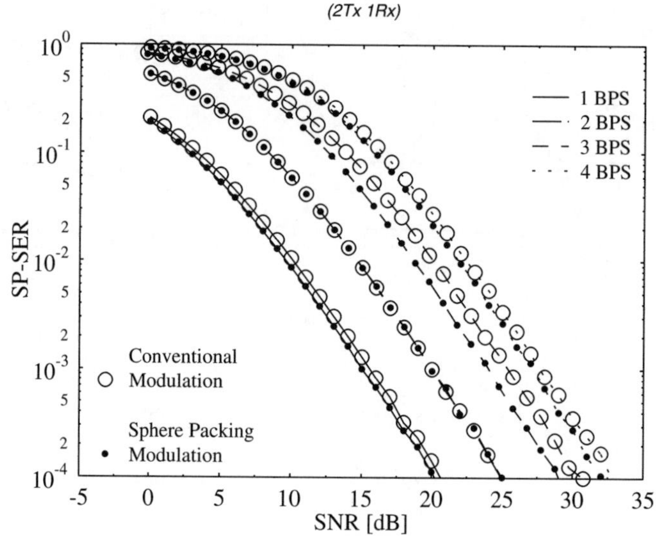

Figure 2.14: SP-SER of different orthogonal STBC schemes in combination with both conventional and SP modulation for the different BPS throughputs of Table 2.5, when employing $N_t = 2$ transmit and $N_r = 1$ receive antennas for communicating over a correlated SPSI Rayleigh fading channel having a normalized Doppler frequency of $f_D = 0.1$.

having a normalized Doppler frequency of $f_D = 0.1$. It is seen from Figure 2.14 that for a particular BPS throughput, the two curves corresponding to the conventional scheme and to the SP modulation scheme have the same asymptotic slope (i.e. diversity level). This agrees with the observation stated in [11], namely that $\mathbf{G}_2$-based space-time schemes dispensing with SP modulation also achieve full diversity. Therefore, SP is not expected to improve the asymptotic slope of the performance curves. Nonetheless, Figure 2.14 shows that orthogonal STBC using SP offers a coding advantage over the conventionally modulated orthogonal STBC design. For example, SP modulation having a throughput of 3 BPS and $L = 64$ achieves a coding gain of about 1.2 dB over classic 8-PSK modulated STBC at a SP-SER of 10^{-4}. The corresponding BER performance curves are shown in Figure 2.15. The BER performances of SP modulation and conventional modulation are identical for systems having rates of 1 and 2 BPS since it can be shown that QPSK, for example, constitutes a SP scheme. However, Figure 2.15 also shows that conventional-modulation-based STBC outperforms SP modulation when higher BPS throughputs are considered if we employ $N_r = 1$ receive antenna.

Figures 2.16–2.25 illustrate the effect of increasing the number of receive antennas from $N_r = 2$ to 6, respectively. The figures demonstrate that upon increasing the number of receive antennas, the SP-SER performance advantage of SP modulation over conventional modulation increases. However, this increase becomes negligible when employing more than $N_r = 3$ antennas. It is also illustrated in Figures 2.17, 2.19, 2.21, 2.23 and 2.25 that the BER performance of SP modulation is improved compared with that of conventional modulation when increasing the number of receive antennas.

For example, as seen in Figure 2.15, the BER performance of 8-PSK is better than that of SP modulation having $L = 64$ when employing $N_r = 1$ receive antenna. However, the BER performance of SP improves when increasing the number of receive antennas,

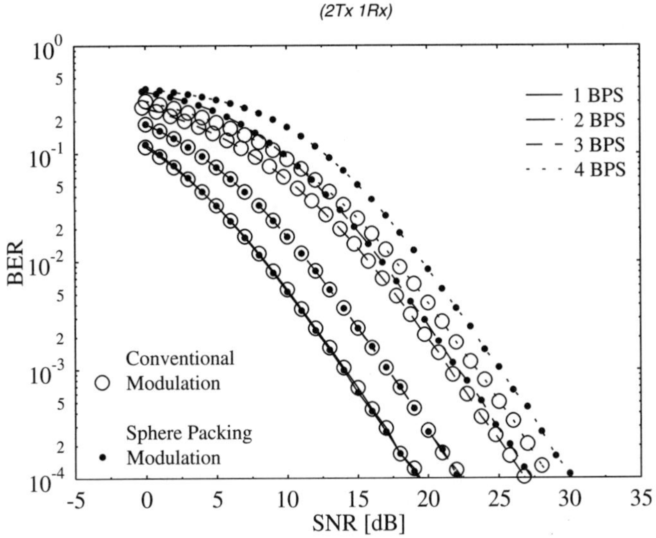

Figure 2.15: BER of different orthogonal STBC schemes in combination with both conventional and SP modulation for the different BPS throughputs of Table 2.5, when employing $N_t = 2$ transmit and $N_r = 1$ receive antennas for communicating over a correlated SPSI Rayleigh fading channel having a normalized Doppler frequency of $f_D = 0.1$.

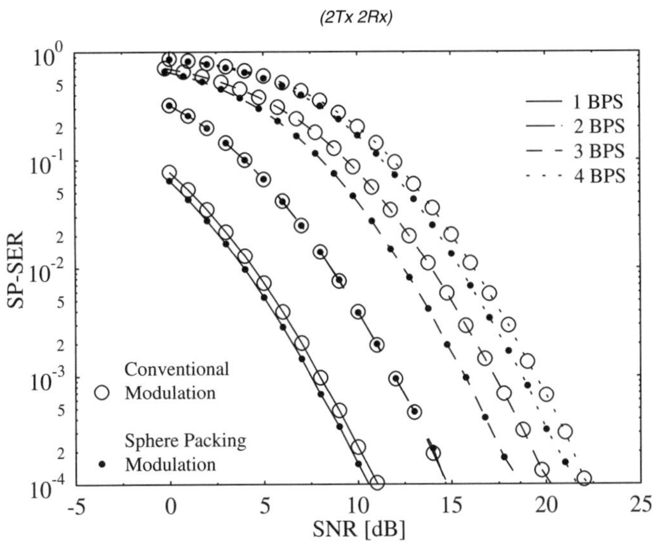

Figure 2.16: SP-SER of different orthogonal STBC schemes in combination with both conventional and SP modulation for the different BPS throughputs of Table 2.5, when employing $N_t = 2$ transmit and $N_r = 2$ receive antennas for communicating over a correlated SPSI Rayleigh fading channel having a normalized Doppler frequency of $f_D = 0.1$.

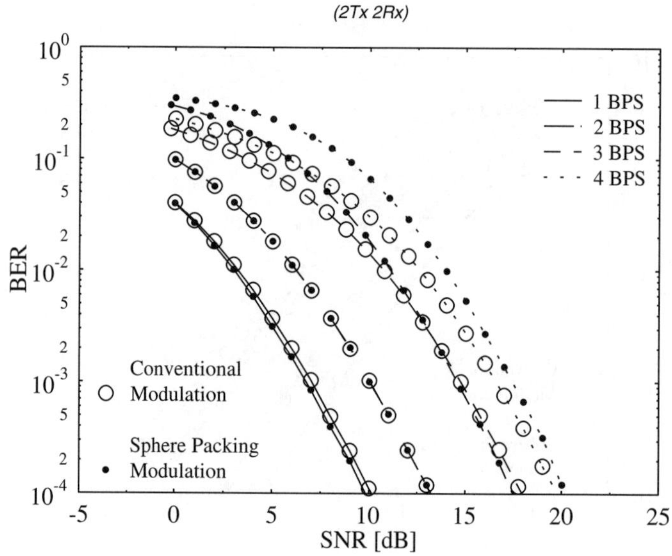

Figure 2.17: BER of different orthogonal STBC schemes in combination with both conventional and SP modulation for the different BPS throughputs of Table 2.5, when employing $N_t = 2$ transmit and $N_r = 2$ receive antennas for communicating over a correlated SPSI Rayleigh fading channel having a normalized Doppler frequency of $f_D = 0.1$.

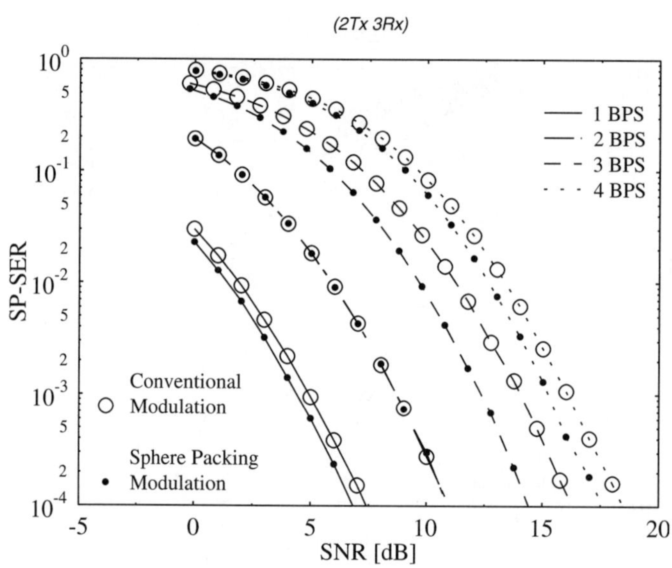

Figure 2.18: SP-SER of different orthogonal STBC schemes in combination with both conventional and SP modulation for the different BPS throughputs of Table 2.5, when employing $N_t = 2$ transmit and $N_r = 3$ receive antennas for communicating over a correlated SPSI Rayleigh fading channel having a normalized Doppler frequency of $f_D = 0.1$.

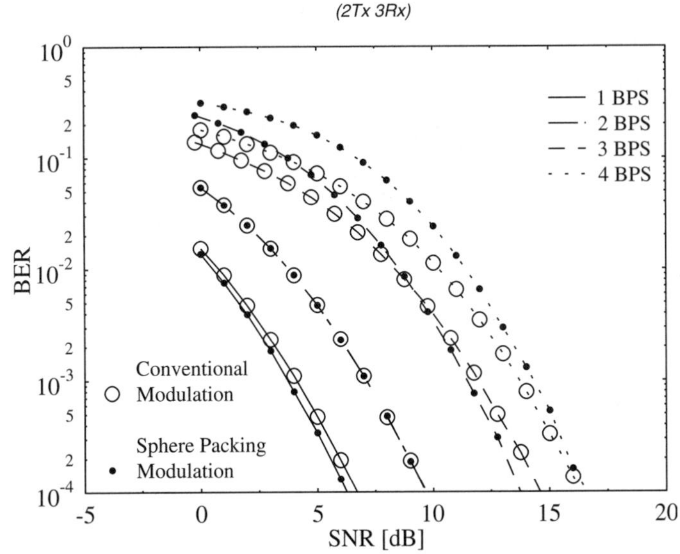

Figure 2.19: BER of different orthogonal STBC schemes in combination with both conventional and SP modulation for the different BPS throughputs of Table 2.5, when employing $N_t = 2$ transmit and $N_r = 3$ receive antennas for communicating over a correlated SPSI Rayleigh fading channel having a normalized Doppler frequency of $f_D = 0.1$.

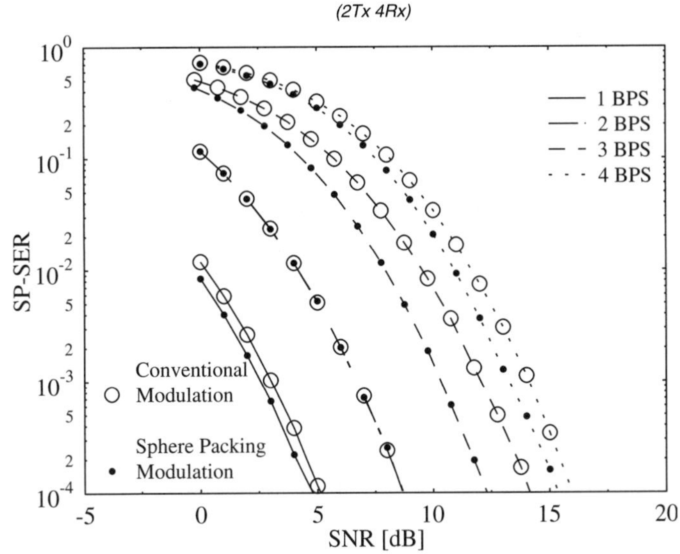

Figure 2.20: SP-SER of different orthogonal STBC schemes in combination with both conventional and SP modulation for the different BPS throughputs of Table 2.5, when employing $N_t = 2$ transmit and $N_r = 4$ receive antennas for communicating over a correlated SPSI Rayleigh fading channel having a normalized Doppler frequency of $f_D = 0.1$.

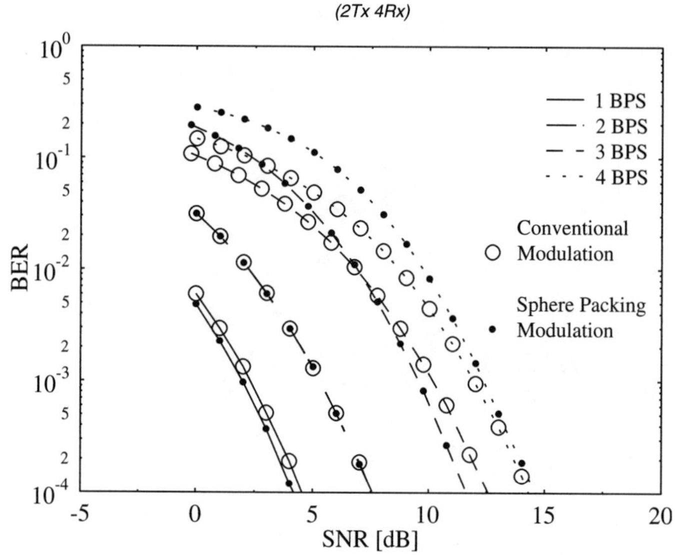

Figure 2.21: BER of different orthogonal STBC schemes in combination with both conventional and SP modulation for the different BPS throughputs of Table 2.5, when employing $N_t = 2$ transmit and $N_r = 4$ receive antennas for communicating over a correlated SPSI Rayleigh fading channel having a normalized Doppler frequency of $f_D = 0.1$.

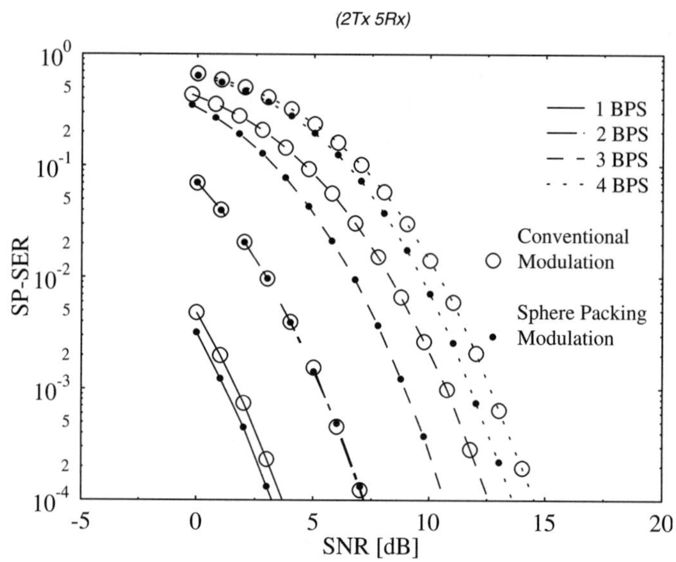

Figure 2.22: SP-SER of different orthogonal STBC schemes in combination with both conventional and SP modulation for the different BPS throughputs of Table 2.5, when employing $N_t = 2$ transmit and $N_r = 5$ receive antennas for communicating over a correlated SPSI Rayleigh fading channel having a normalized Doppler frequency of $f_D = 0.1$.

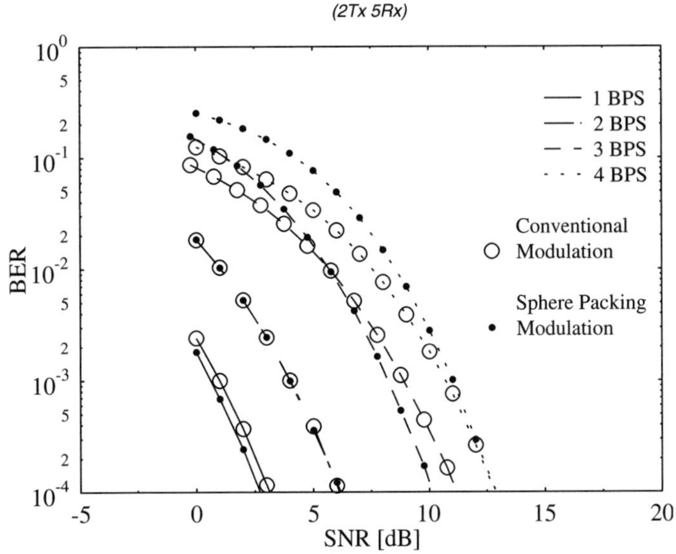

Figure 2.23: BER of different orthogonal STBC schemes in combination with both conventional and SP modulation for the different BPS throughputs of Table 2.5, when employing $N_t = 2$ transmit and $N_r = 5$ receive antennas for communicating over a correlated SPSI Rayleigh fading channel having a normalized Doppler frequency of $f_D = 0.1$.

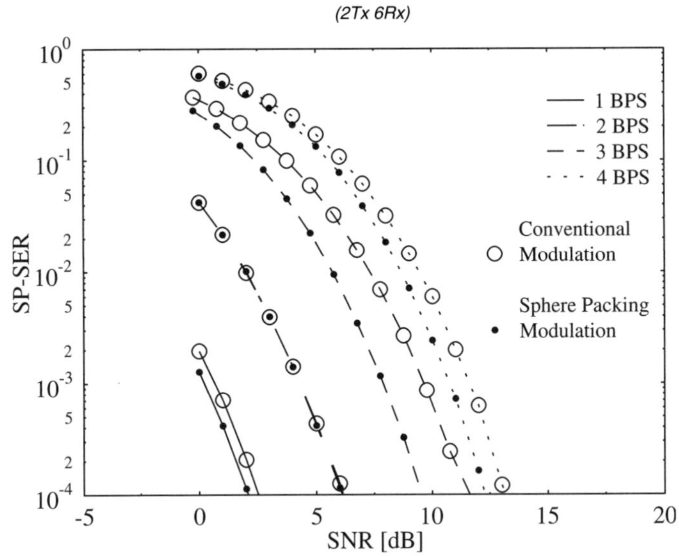

Figure 2.24: SP-SER of different orthogonal STBC schemes in combination with both conventional and SP modulation for the different BPS throughputs of Table 2.5, when employing $N_t = 2$ transmit and $N_r = 6$ receive antennas for communicating over a correlated SPSI Rayleigh fading channel having a normalized Doppler frequency of $f_D = 0.1$.

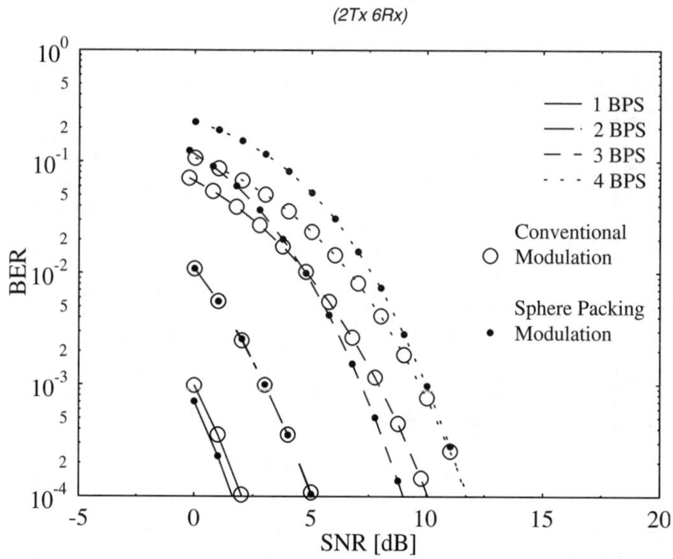

Figure 2.25: BER of different orthogonal STBC schemes in combination with both conventional and SP modulation for the different BPS throughputs of Table 2.5, when employing $N_t = 2$ transmit and $N_r = 6$ receive antennas for communicating over a correlated SPSI Rayleigh fading channel having a normalized Doppler frequency of $f_D = 0.1$.

Table 2.6: Coding gains of SP-modulation-aided STBC over conventional STBC at a SP-SER of 10^{-4} for the schemes characterized in Figures 2.14–2.25.

	Coding gains (dB)			
	1 BPS	2 BPS	3 BPS	4 BPS
$N_r = 1$	0.4	0.0	1.2	0.6
$N_r = 2$	0.5	0.0	1.6	0.6
$N_r = 3$	0.6	0.0	2.0	0.9
$N_r = 4$	0.4	0.0	1.9	0.7
$N_r = 5$	0.5	0.0	1.9	0.9
$N_r = 6$	0.4	0.0	2.1	0.9

leading to a coding advantage of about 1 dB over 8-PSK when employing $N_r = 6$ receive antennas. The coding gains achieved by SP modulation over conventional modulation schemes at an SP-SER and a BER of 10^{-4} for the schemes characterized in Figures 2.14–2.25 are summarized in Tables 2.6 and 2.7. The negative coding gain values indicate where conventional modulation outperforms SP modulation. Figures 2.26–2.28 illustrate the SNR values required to achieve a SP-SER of 10^{-4} using the different orthogonal STBC schemes in combination with both conventional and SP modulation for the different BPS throughputs of Table 2.5, when employing $N_t = 2$ transmit and $N_r = 1$, 3 and 6 receive antennas, respectively, for communicating over a correlated SPSI Rayleigh fading channel with a normalized Doppler frequency of $f_D = 0.1$.

Table 2.7: Coding gains of SP-modulation-aided STBC over conventional STBC at a BER of 10^{-4} for the schemes characterized in Figures 2.14–2.25.

	Coding gains (dB)			
	1 BPS	2 BPS	3 BPS	4 BPS
$N_r = 1$	0.0	0.0	−0.5	−1.0
$N_r = 2$	0.3	0.0	0.5	−0.5
$N_r = 3$	0.5	0.0	0.9	−0.3
$N_r = 4$	0.4	0.0	1.0	−0.3
$N_r = 5$	0.4	0.0	1.0	0.0
$N_r = 6$	0.4	0.0	1.0	0.1

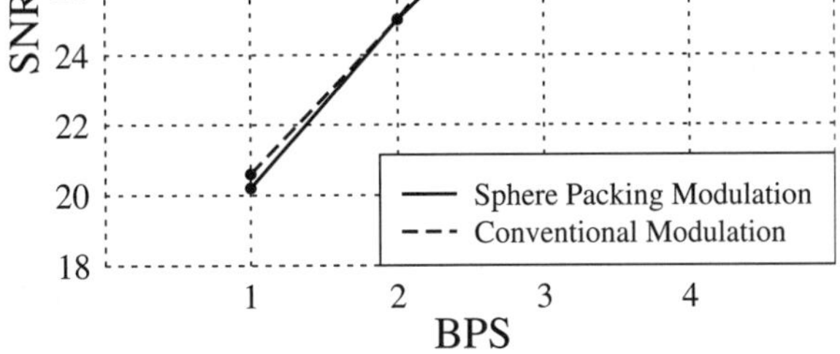

Figure 2.26: SNR required to achieve a SP-SER of 10^{-4} by different orthogonal STBC schemes in combination with both conventional and SP modulation for the different BPS throughputs of Table 2.5, when employing $N_t = 2$ transmit and $N_r = 1$ receive antennas for communicating over a correlated SPSI Rayleigh fading channel having a normalized Doppler frequency of $f_D = 0.1$.

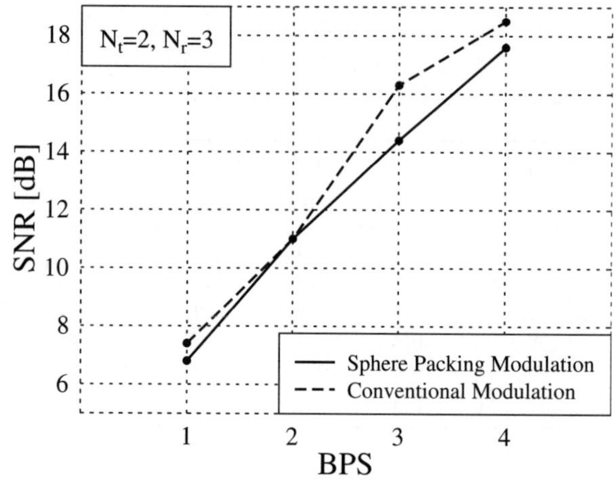

Figure 2.27: SNR required to achieve a SP-SER of 10^{-4} by different orthogonal STBC schemes in combination with both conventional and SP modulation for the different BPS throughputs of Table 2.5, when employing $N_t = 2$ transmit and $N_r = 3$ receive antennas for communicating over a correlated SPSI Rayleigh fading channel having a normalized Doppler frequency of $f_D = 0.1$.

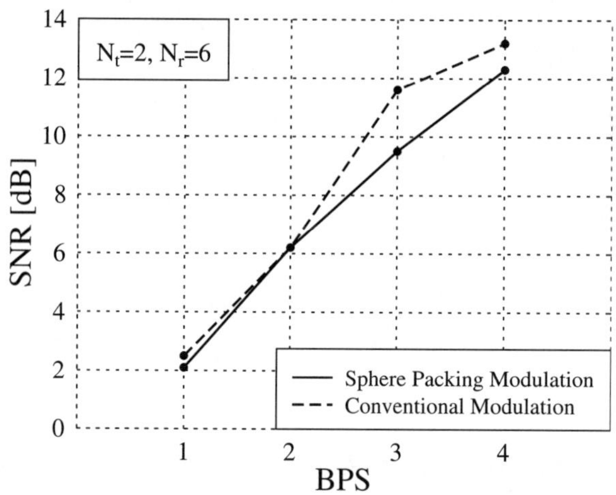

Figure 2.28: SNR required to achieve a SP-SER of 10^{-4} by different orthogonal STBC schemes in combination with both conventional and SP modulation for the different BPS throughputs of Table 2.5, when employing $N_t = 2$ transmit and $N_r = 6$ receive antennas for communicating over a correlated SPSI Rayleigh fading channel having a normalized Doppler frequency of $f_D = 0.1$.

Observe in Figures 2.14–2.25 that, as alluded to before, the performance curves of QPSK modulation and SP modulation having $L = 16$ (i.e. 2 BPS schemes) are identical. This phenomenon is due to the fact that QPSK modulation constitutes a special case of SP modulation, when it is combined with $\mathbf{G}_2$ space-time signals. More specifically, consider the $\mathbf{G}_2$ space-time signal defined as $\mathbf{G}_2(x_{l,1}, x_{l,2})$, $l = 0, \ldots, 15$. If $x_{l,1}$ and $x_{l,2}$ are chosen independently from the QPSK modulation constellation, then the 16 legitimate space-time signals produced will be identical to the 16 legitimate space-time signals constructed using Equation (2.63), where $(a_{l,1}, a_{l,2}, a_{l,3}, a_{l,4})$, $l = 0, \ldots, 15$, correspond to the 16 SP constellation points hosted by D_4 that are centered at all possible permutations of $(\pm 1, \pm 1, \pm 1, \pm 1)$ and have a normalization factor of $1/\sqrt{2}$. These SP constellation points belong to the second layer of D_4 seen in Table 2.4.

2.6 Chapter Conclusions

It was shown in this chapter that the diversity product of orthogonal space-time signals is determined by the MED of the $(k+1)$-dimensional complex vectors $(x_1, x_2, \ldots, x_{k+1})$. In order to maximize the diversity product, it was proposed in [43] to use SP having the best known MED in the $2(k+1)$-dimensional real-valued Euclidean space $\mathbb{R}^{2(k+1)}$ (see [221]). The capacity analysis provided in Section 2.4.5 demonstrated that SP-aided orthogonal STBC design has potential performance improvements over conventionally modulated orthogonal STBC schemes. Furthermore, our simulation results presented in Section 2.5 showed that SP-aided orthogonal STBC provides some coding gain over conventionally modulated orthogonal STBC schemes. Tables 2.6 and 2.7 summarize the coding gains achieved by SP modulation over conventional modulation at a SP-SER and a BER of 10^{-4} for the schemes characterized in Figures 2.14–2.25.

2.7 Chapter Summary

In this chapter we first summarized the design criteria of space-time coded communication systems in Section 2.2. Specifically, our quasi-static and rapidly fading channels were described in Sections 2.2.1 and 2.2.2, respectively. In Section 2.3, we outlined the design criteria used for time-correlated fading channels, where both the pairwise SP symbol error probability as well as the corresponding design criteria were presented in Section 2.3.2. The concept of diversity product, which was introduced in [54, 208] and generalized in [43] in order to account for the effects of temporal correlation, was discussed in Section 2.3.3.1. Furthermore, both lower and upper bounds on the generalized diversity product discussed in Section 2.3.3.1 were provided in Section 2.3.3.2.

In Section 2.4, the philosophy of orthogonal STBC design using SP modulation was considered for space-time signals, where the motivation behind the adoption of SP modulation was discussed in Section 2.4.2. The SP signal design derived for $N_t = 2$ transmit antennas was provided in Section 2.4.3, further illustrating the concept of combining an orthogonal STBC design with SP. Section 2.4.4 discussed the problem of constructing a SP constellation having a particular size L. The constellation points were first chosen based on the minimum energy criterion. Then, an exhaustive computer search was conducted for all of the SP symbols having the lowest possible energy, in order to find the specific set of L points having the best MED from all of the other constellation points satisfying the minimum energy criterion. The capacity of STBC-SP schemes employing $N_t = 2$ transmit antennas was derived in Section 2.4.5. Finally, the performance of STBC-SP schemes was characterized in Section 2.5, demonstrating that STBC-SP schemes are capable of outperforming STBC schemes that employ conventional modulation (i.e. PSK, QAM). The coding gains achieved by the SP-assisted STBC over conventional STBC at an SP-SER and a BER of 10^{-4} are summarised in Tables 2.6 and 2.7 for the schemes characterized in Figures 2.14–2.25.

In the next chapter, we demonstrate that the performance of STBC-SP systems can be further improved by concatenating SP-aided modulation with channel coding and performing demapping as well as channel decoding iteratively. The SP demapper of [43] is further developed for the sake of accepting the *a priori* information passed to it from the channel decoder as extrinsic information. The convergence behavior of this concatenated scheme is investigated with the aid of EXIT charts.

Chapter 3

Turbo Detection of Channel-coded STBC-SP Schemes*

3.1 Introduction

In Chapter 2, a recently proposed space-time signal construction method that combines orthogonal design with SP (STBC-SP) has been introduced. STBC-SP schemes have shown useful performance improvements over Alamouti's conventional orthogonal design. In this chapter, the performance of STBC-SP systems is improved by developing novel bit-based turbo-detected schemes.

Iterative decoding of spectrally efficient modulation schemes was considered by several authors [9, 252–254]. In [158], the employment of the turbo principle was considered for iterative soft demapping in the context of BICM, where a soft demapper was used between the multilevel demodulator and the channel decoder. In [163], a turbo-coding scheme was proposed for the MIMO Rayleigh fading channel, where a block code was employed as an outer channel code, while an orthogonal STBC scheme was considered as the inner code.

Recently, the convergence behavior of iterative decoding has attracted considerable attention [167, 169, 172, 173, 255–259]. In order to determine the E_b/N_0 convergence threshold of randomly constructed irregular LDPC codes transmitted over the AWGN channel, the authors of [255] proposed the employment of a density evolution algorithm, which was also invoked in [256, 257] for the sake of constructing LDPC codes capable of operating at low E_b/N_0 values. SNR-based measures were used in [167, 173] to study the convergence of iterative decoders, while the authors of [258] investigated the convergence behavior of unity-rate inner codes based on a combination of SNR measures and mutual information. In [169, 172], ten Brink proposed the employment of the EXIT characteristics between a concatenated decoder's output and input for describing the flow of extrinsic information through the soft-in/soft-out constituent decoders. A tutorial introduction to EXIT charts can be found in [181]. In addition, several algorithms predicting the convergence of iterative decoding schemes were compared in [259].

Motivated by the performance improvements reported in [43] and [158], we propose a novel bit-based system that exploits the advantages of both the iterative demapping and

*Parts of this chapter are based on the collaborative research outlined in [188].

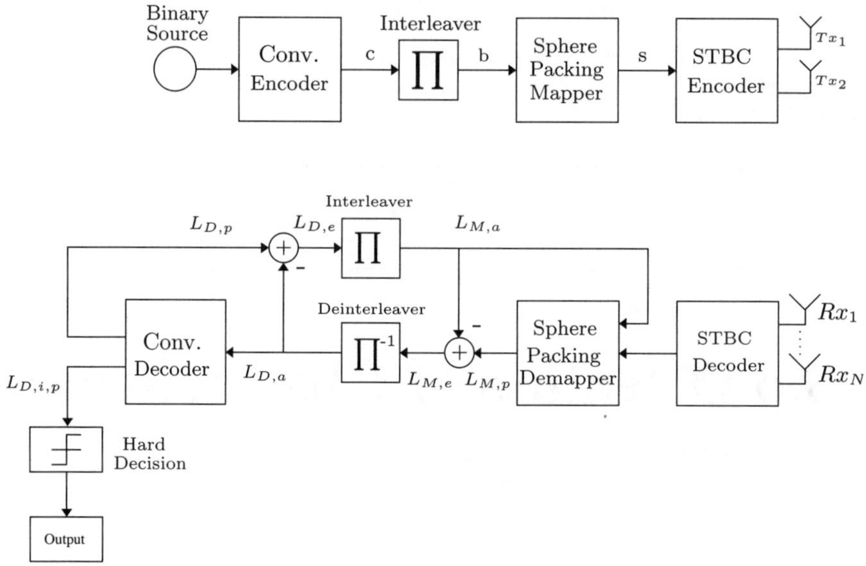

Figure 3.1: Bit-based turbo detection RSC-coded STBC-SP system.

decoding techniques of [158] and those of the STBC-SP scheme of [43]. The STBC-SP demapper of [43] was further developed for the sake of accepting the *a priori* information passed to it from the channel decoder as extrinsic information. As a benefit of the proposed solution, we demonstrate in Section 3.5 that the proposed turbo-detection-aided STBC-SP scheme is capable of providing an E_b/N_0 gain of 20.2 dB at a BER of 10^{-5} over the STBC-SP scheme of [43].

In this chapter, two realizations of a novel bit-based iterative-detection-aided STBC-SP scheme are presented, namely a RSC-coded turbo-detected STBC-SP scheme and a binary LDPC-coded turbo-detected STBC-SP arrangement. Our system overview is provided in Section 3.2. Section 3.3 shows how the STBC-SP demapper is modified to exploit the *a priori* knowledge provided by the channel decoder. Section 3.4 provides the EXIT chart analysis of the turbo-detected bit-based scheme, while our simulation results and discussions are provided in Section 3.5. Finally, the chapter is concluded in Section 3.6.

3.2 System Overview

3.2.1 RSC-coded Turbo-detected STBC-SP scheme

The schematic of the entire RSC-coded turbo-detected STBC-SP scheme is shown in Figure 3.1, where the transmitted source bits are first convolutionally encoded and then interleaved by a random bit interleaver. A rate $R = \frac{1}{2}$ RSC code is considered in this chapter. After channel interleaving, the SP mapper first maps B channel-coded binary bits $\mathbf{b} = (b_0, \ldots, b_{B-1})$ to a SP-modulated symbol $\mathrm{s} \in S$ such that we have $\mathrm{s} = map_{sp}(\mathbf{b})$, where $B = \log_2 L$. The STBC-SP encoder then maps the SP-modulated symbol s to a space-time signal $\mathbf{C}_l = \sqrt{2L/E_{\text{total}}}\mathbf{G}_2(x_{l,1}, x_{l,2})$, $0 \leq l \leq L-1$, using Equation (2.59). Subsequently, each space-time signal is transmitted over $T = 2$ time slots using two transmit antennas, as shown in Equation (2.35) and Table 2.2.

In this chapter, we consider an SPSI correlated narrowband Rayleigh fading channel, based on the Jakes fading model [220] and associated with a normalized Doppler frequency of $f_D = f_d T_{sym} = 0.1$, where f_d is the Doppler frequency and T_{sym} is the symbol period. The complex fading envelope is thus assumed to be constant across the transmission period of a space-time coded symbol spanning $T = 2$ time slots. The complex AWGN of $n = n_I + jn_Q$ is also added to the received signal, where n_I and n_Q are two independent zero-mean Gaussian random variables having a variance of $\sigma_n^2 = \sigma_{n_I}^2 = \sigma_{n_Q}^2 = N_0/2$ per dimension, where $N_0/2$ represents the double-sided noise power spectral density expressed in W/Hz.

As shown in Figure 3.1, the received complex-valued symbols are demapped to their LLR representation for each of the B coded bits per STBC-SP symbol. The *a priori* LLR values of the demodulator are subtracted from the *a posteriori* LLR values for the sake of generating the extrinsic LLR values $L_{M,e}$ and then the LLRs $L_{M,e}$ are deinterleaved by a soft-bit deinterleaver, as seen in Figure 3.1. Next, the soft bits $L_{D,a}$ are passed to the convolutional decoder in order to compute the *a posteriori* LLR values $L_{D,p}$ provided by the Log-MAP algorithm [165] for all of the channel-coded bits. During the last iteration, only the LLR values $L_{D,i,p}$ of the original uncoded systematic information bits are required, which are passed to a hard decision decoder in order to determine the estimated transmitted source bits. The extrinsic information $L_{D,e}$ is generated by subtracting the *a priori* information from the *a posteriori* information according to $L_{D,p} - L_{D,a}$, which is then fed back to the STBC-SP demapper as the *a priori* information $L_{M,a}$ after appropriately reordering them using the interleaver of Figure 3.1. The STBC-SP demapper exploits the *a priori* information for the sake of providing improved *a posteriori* LLR values, which are then passed to the channel decoder and in turn back to the STBC-SP demodulator for further iterations.

3.2.2 Binary LDPC-coded Turbo-detected STBC-SP scheme

The schematic of the entire binary LDPC-coded turbo-detected STBC-SP scheme is shown in Figure 3.2, where a rate $R = \frac{1}{2}$ binary LDPC code is employed. Observe that channel interleaving is not required between the binary LDPC encoder and the SP mapper, since the LDPC parity check matrix is randomly constructed, where each of the parity check equations is checking several random bit positions in a codeword, which has a similar effect to that of the channel interleaver. A sophisticated encoding as well as turbo-detection procedure similar to that described in Section 3.2.1 is outlined in Figure 3.2. However, based on the *a posteriori* LLR values $L_{D,p}$ recorded at the output of the LDPC decoder, a tentative hard decision will be made during each turbo-detection iteration and the resultant codeword will be checked by the LDPC code's parity check matrix. If the resultant vector is an all-zero sequence, then a legitimate codeword has been found, and the hard-decision-based sequence will be output. Otherwise, if the maximum affordable number of iterations has not been reached, the *a priori* information, $L_{D,a}$, is subtracted from the *a posteriori* LLR values, $L_{D,p}$, for the sake of generating the extrinsic information, $L_{D,e}$, which is appropriately interleaved and fed back to the demodulator for the next iteration, as seen in Figure 3.2. The process continues, until the affordable maximum number of iterations has been encountered or a legitimate codeword has been found.

3.3 Iterative Demapping

For the sake of simplicity, a system having a single receive antenna is considered, although its extension to systems having more than one receive antenna is straightforward. Assuming

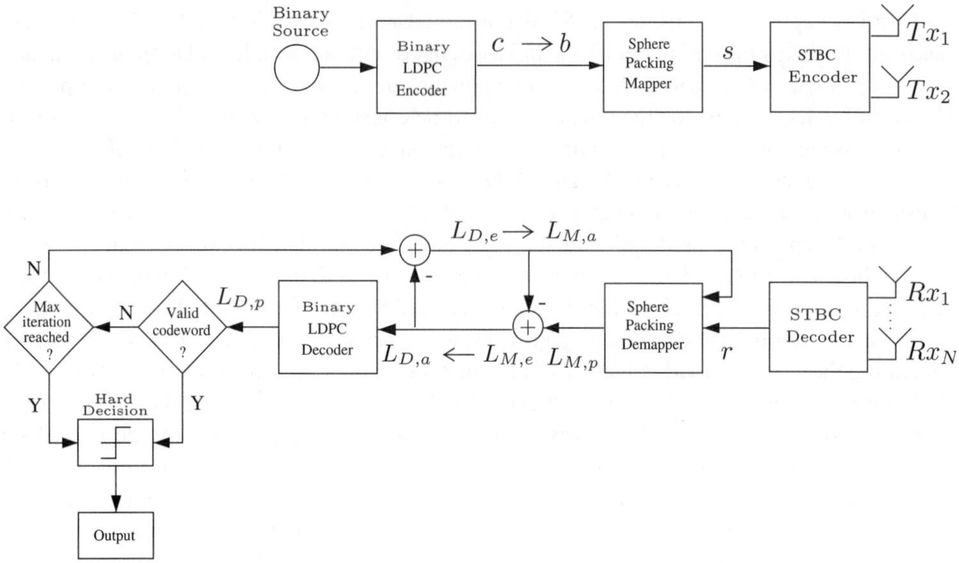

Figure 3.2: Bit-based turbo-detection LDPC-coded STBC-SP system.

perfect channel estimation, the complex-valued channel output symbols received during two consecutive time slots are first diversity-combined in order to extract the estimates $\tilde{x}_1$ and $\tilde{x}_2$ of the most likely transmitted symbols $x_{l,1}$ and $x_{l,2}$ as was described in Section 2.4.3.2, resulting in

$$\tilde{x}_1 = (|h_1|^2 + |h_2|^2) \cdot x_{l,1} + \acute{n}_1, \tag{3.1}$$

$$\tilde{x}_2 = (|h_1|^2 + |h_2|^2) \cdot x_{l,2} + \acute{n}_2, \tag{3.2}$$

where h_1 and h_2 represent the complex-valued channel coefficients corresponding to the first and second transmit antenna, respectively, and $\acute{n}_1$ as well as $\acute{n}_2$ are zero-mean complex Gaussian random variables with variance $\sigma_{\acute{n}}^2 = (|h_1|^2 + |h_2|^2) \cdot \sigma_n^2$. A received sphere-packed symbol $\mathbf{r}$ is then constructed from the estimates $\tilde{x}_1$ and $\tilde{x}_2$ using Equation (2.59) as

$$\mathbf{r} = T_{sp}^{-1}(\tilde{x}_1, \tilde{x}_2), \tag{3.3}$$

where $\mathbf{r} = \{[\tilde{a}_1\ \tilde{a}_2\ \tilde{a}_3\ \tilde{a}_4] \in \mathbb{R}^4\}$. The received sphere-packed symbol $\mathbf{r}$ can be written as

$$\mathbf{r} = h \cdot \sqrt{\frac{2L}{E_{\text{total}}}} \cdot \mathbf{s}^l + \mathbf{w}, \tag{3.4}$$

where $h = (|h_1|^2 + |h_2|^2)$, $\mathbf{s}^l \in S$, $0 \le l \le L-1$, and $\mathbf{w}$ is a four-dimensional real-valued Gaussian random variable having a covariance matrix of $\sigma_w^2 \cdot \mathbf{I}_{N_D} = \sigma_{\acute{n}}^2 \cdot \mathbf{I}_{N_D} = h \cdot \sigma_n^2 \cdot \mathbf{I}_{N_D}$, where $N_D = 4$, since the symbol constellation S is four-dimensional. According to Equation (3.4), the conditional PDF $p(\mathbf{r}|\mathbf{s}^l)$ is given by

$$\begin{aligned} p(\mathbf{r}|\mathbf{s}^l) &= \frac{1}{(2\pi\sigma_w^2)^{N_D/2}} \exp\left(-\frac{1}{2\sigma_w^2}(\mathbf{r} - \alpha \cdot \mathbf{s}^l)(\mathbf{r} - \alpha \cdot \mathbf{s}^l)^T\right), \\ &= \frac{1}{(2\pi\sigma_w^2)^{N_D/2}} \exp\left(-\frac{1}{2\sigma_w^2}\left(\sum_{i=1}^{N_D}(\tilde{a}_i - \alpha \cdot a_{l,i})^2\right)\right), \end{aligned} \tag{3.5}$$

where we have $\alpha = h \cdot \sqrt{2L/E_{\text{total}}}$.

The SP symbol $\mathbf{r}$ carries B channel-coded binary bits $\mathbf{b} = (b_0, \ldots, b_{B-1})$. The LLR-value of bit b_k for $k = 0, \ldots, B-1$ can be written as [158]

$$L(b_k|\mathbf{r}) = L_a(b_k) + \ln \frac{\sum_{\mathbf{s}^l \in S_1^k} p(\mathbf{r}|\mathbf{s}^l) \cdot \exp\Big(\sum_{j=0, j\neq k}^{B-1} b_j L_a(b_j)\Big)}{\sum_{\mathbf{s}^l \in S_0^k} p(\mathbf{r}|\mathbf{s}^l) \cdot \exp\Big(\sum_{j=0, j\neq k}^{B-1} b_j L_a(b_j)\Big)}, \tag{3.6}$$

where S_1^k and S_0^k are subsets of the symbol constellation S such that $S_1^k \triangleq \{\mathbf{s}^l \in S : b_k = 1\}$ and, likewise, $S_0^k \triangleq \{\mathbf{s}^l \in S : b_k = 0\}$. In other words, S_i^k represents all symbols of the set S, where we have $b_k = i \in \{0, 1\}$, $k = 0, \ldots, B-1$. Using Equation (3.5), we can write Equation (3.6) as

$$\begin{aligned} L(b_k|\mathbf{r}) &= L_a(b_k) \\ &\quad + \ln \frac{\sum_{\mathbf{s}^l \in S_1^k} \exp\Big(-(1/2\sigma_w^2)(\mathbf{r} - \alpha \cdot \mathbf{s}^l)(\mathbf{r} - \alpha \cdot \mathbf{s}^l)^T + \sum_{j=0, j\neq k}^{B-1} b_j L_a(b_j)\Big)}{\sum_{\mathbf{s}^l \in S_0^k} \exp\Big(-(1/2\sigma_w^2)(\mathbf{r} - \alpha \cdot \mathbf{s}^l)(\mathbf{r} - \alpha \cdot \mathbf{s}^l)^T + \sum_{j=0, j\neq k}^{B-1} b_j L_a(b_j)\Big)} \\ &= L_{M,a} + L_{M,e}. \end{aligned} \tag{3.7}$$

Finally, the max-log approximation of Equation (3.7) is as follows:

$$\begin{aligned} L(b_k|\mathbf{r}) = L_a(b_k) &+ \max_{\mathbf{s}^l \in S_1^k} \left[-\frac{1}{2\sigma_w^2}(\mathbf{r} - \alpha \cdot \mathbf{s}^l)(\mathbf{r} - \alpha \cdot \mathbf{s}^l)^T + \sum_{j=0, j\neq k}^{B-1} b_j L_a(b_j) \right] \\ &- \max_{\mathbf{s}^l \in S_0^k} \left[-\frac{1}{2\sigma_w^2}(\mathbf{r} - \alpha \cdot \mathbf{s}^l)(\mathbf{r} - \alpha \cdot \mathbf{s}^l)^T + \sum_{j=0, j\neq k}^{B-1} b_j L_a(b_j) \right]. \end{aligned} \tag{3.8}$$

Example 3.3.1 (Iterative demapping for $L = 4$). In order to explain the theory behind Equation (3.6), let us consider a SP modulation scheme associated with $L = 4$. The binary bits b_k, $k = 0, 1$, corresponding to each SP symbol $\mathbf{s}^l$, $l = 0, \ldots, 3$, are outlined in Table 3.1. The LLR-value of b_0, for example, may be written as [158]

$$\begin{aligned} L(b_0|\mathbf{r}) &= \ln \frac{p(b_0 = 1|\mathbf{r})}{p(b_0 = 0|\mathbf{r})} \\ &= \ln \frac{p(b_0 = 1, b_1 = 0|\mathbf{r}) + p(b_0 = 1, b_1 = 1|\mathbf{r})}{p(b_0 = 0, b_1 = 0|\mathbf{r}) + p(b_0 = 0, b_1 = 1|\mathbf{r})}. \end{aligned} \tag{3.9}$$

Using Bayes' rule[1], Equation (3.9) may be expressed as

$$\begin{aligned} &L(b_0|\mathbf{r}) \\ &\quad = \ln \frac{p(\mathbf{r}|b_0 = 1, b_1 = 0) \cdot p(b_0 = 1, b_1 = 0) + p(\mathbf{r}|b_0 = 1, b_1 = 1) \cdot p(b_0 = 1, b_1 = 1)}{p(\mathbf{r}|b_0 = 0, b_1 = 0) \cdot p(b_0 = 0, b_1 = 0) + p(\mathbf{r}|b_0 = 0, b_1 = 1) \cdot p(b_0 = 0, b_1 = 1)}. \end{aligned} \tag{3.10}$$

[1]Let X and Y denote two random variables. According to Bayes' rule, the conditional probability $p(X|Y)$ may be written as [260]

$$p(X|Y) = \frac{p(Y|X) \cdot p(X)}{p(Y)}.$$

Table 3.1: Binary bits corresponding to SP symbols, when $L = 4$.

	Binary bits	
SP symbol	b_0	b_1
$\mathbf{s}^0$	0	0
$\mathbf{s}^1$	1	0
$\mathbf{s}^2$	0	1
$\mathbf{s}^3$	1	1

Since bit-interleaving is employed, it is reasonable to assume that b_0 and b_1 are independent of each other. Hence, Equation (3.10) may be manipulated as follows:

$$\begin{aligned}
&L(b_0|\mathbf{r})\\
&= \ln \frac{p(\mathbf{r}|b_0=1,b_1=0)\cdot p(b_0=1)\cdot p(b_1=0)+p(\mathbf{r}|b_0=1,b_1=1)\cdot p(b_0=1)\cdot p(b_1=1)}{p(\mathbf{r}|b_0=0,b_1=0)\cdot p(b_0=0)\cdot p(b_1=0)+p(\mathbf{r}|b_0=0,b_1=1)\cdot p(b_0=0)\cdot p(b_1=1)}\\
&= \ln \frac{p(b_0=1)}{p(b_0=0)} + \ln \frac{p(\mathbf{r}|b_0=1,b_1=0)\cdot p(b_1=0)+p(\mathbf{r}|b_0=1,b_1=1)\cdot p(b_1=1)}{p(\mathbf{r}|b_0=0,b_1=0)\cdot p(b_1=0)+p(\mathbf{r}|b_0=0,b_1=1)\cdot p(b_1=1)}\\
&= L_a(b_0) + \ln \frac{p(\mathbf{r}|b_0=1,b_1=0)+p(\mathbf{r}|b_0=1,b_1=1)\cdot (p(b_1=1)/p(b_1=0))}{p(\mathbf{r}|b_0=0,b_1=0)+p(\mathbf{r}|b_0=0,b_1=1)\cdot (p(b_1=1)/p(b_1=0))}\\
&= L_a(b_0) + \ln \frac{p(\mathbf{r}|b_0=1,b_1=0)+p(\mathbf{r}|b_0=1,b_1=1)\cdot \exp(\ln(p(b_1=1)/p(b_1=0)))}{p(\mathbf{r}|b_0=0,b_1=0)+p(\mathbf{r}|b_0=0,b_1=1)\cdot \exp(\ln(p(b_1=1)/p(b_1=0)))}\\
&= L_a(b_0) + \ln \frac{p(\mathbf{r}|b_0=1,b_1=0)+p(\mathbf{r}|b_0=1,b_1=1)\cdot \exp(L_a(b_1))}{p(\mathbf{r}|b_0=0,b_1=0)+p(\mathbf{r}|b_0=0,b_1=1)\cdot \exp(L_a(b_1))}\\
&= L_a(b_0) + \ln \frac{p(\mathbf{r}|\mathbf{s}^1)+p(\mathbf{r}|\mathbf{s}^3)\cdot \exp(L_a(b_1))}{p(\mathbf{r}|\mathbf{s}^0)+p(\mathbf{r}|\mathbf{s}^2)\cdot \exp(L_a(b_1))},
\end{aligned} \tag{3.11}$$

which is generalized in Equation (3.6).

3.4 Binary EXIT Chart Analysis

The main objective of employing EXIT charts, proposed by ten Brink [169, 172], is to predict the convergence behavior of the iterative decoder by examining the evolution of the input/output mutual information exchange between the inner and outer decoders in consecutive iterations. The application of EXIT charts is based on two assumptions, namely that upon assuming large interleaver lengths:

- the *a priori* LLR values are fairly uncorrelated;
- the probability density function of the *a priori* LLR values is Gaussian.

3.4.1 Transfer Characteristics of the Demapper

As seen in Figures 3.1 and 3.2, the inputs of the sphere-packing demapper are the noise-contaminated channel observations and the *a priori* information $L_{M,a}$ generated by the outer channel decoder. The demapper outputs the *a posteriori* information $L_{M,p}$, subtracts the *a priori* and hence produces the extrinsic information $L_{M,e}$ as shown in Section 3.3.

Based on the above-mentioned two assumptions, the *a priori* input $L_{M,a}$ can be modeled by applying an independent zero-mean Gaussian random variable n_A having a variance of σ_A^2. In conjunction with the outer channel coded and interleaved bits $b \in \{0, 1\}$ of Figures 3.1 and 3.2 or equivalently $x \in \{-1, +1\}$, the *a priori* input $L_{M,a}$ can be written as [169]

$$L_{M,a} = \mu_A \cdot x + n_A, \tag{3.12}$$

where $\mu_A = \sigma_A^2/2$ since $L_{M,a}$ is a LLR value obeying the Gaussian distribution [261]. Accordingly, the conditional PDF of the *a priori* input $L_{M,a}$ is

$$p_A(\zeta|X = x) = \frac{1}{\sqrt{2\pi}\sigma_A} \exp\left(-\frac{(\zeta - (\sigma_A^2/2)\cdot x)^2}{2\sigma_A^2}\right). \tag{3.13}$$

The mutual information of $I_{A_M} = I(X; L_{M,a})$, $0 \le I_{A_M} \le 1$, between the outer coded and interleaved bits x and the LLR values $L_{M,a}$ is used to quantify the information content of the *a priori* knowledge [262]:

$$I_{A_M} = \frac{1}{2}\cdot \sum_{x=-1,+1} \int_{-\infty}^{+\infty} p_A(\zeta|X = x)\cdot \log_2 \frac{2\cdot p_A(\zeta|X = x)}{p_A(\zeta|X = -1) + p_A(\zeta|X = +1)}\, \mathrm{d}\zeta. \tag{3.14}$$

Using Equation (3.13), Equation (3.14) can be expressed as

$$I_{A_M}(\sigma_A) = 1 - \frac{1}{\sqrt{2\pi}\sigma_A}\int_{-\infty}^{+\infty} \exp\left(-\frac{(\zeta - \sigma_A^2/2)^2}{2\sigma_A^2}\right)\cdot \log_2[1 + e^{-\zeta}]\, \mathrm{d}\zeta. \tag{3.15}$$

For notational simplicity and in order to highlight the dependence of I_{A_M} on σ_A, the following abbreviation is introduced [169, 172]:

$$J(\sigma) := I_{A_M}(\sigma_A = \sigma), \tag{3.16}$$

$$\lim_{\sigma \leftarrow 0} J(\sigma) = 0, \quad \lim_{\sigma \leftarrow \infty} J(\sigma) = 1, \qquad \sigma > 0. \tag{3.17}$$

The function $J(\sigma)$ is monotonically increasing and therefore its inverse exists. Figure 3.3 shows a plot of $J(\sigma)$ as a function of σ. It was shown in [175] that the mutual information between the equiprobable bits X and their respective LLRs L for symmetric and consistent[2] L-values always simplifies to

$$I(X; L) = 1 - \int_{-\infty}^{+\infty} p(L|X = +1)\cdot \log_2[1 + e^{-L}]\, \mathrm{d}L,$$

$$I(X; L) = 1 - E_{X=+1}\{\log_2[1 + e^{-L}]\}. \tag{3.18}$$

In order to quantify the information content of the extrinsic LLR values $L_{M,e}$ at the output of the demapper, the mutual information $I_{E_M} = I(X; L_{M,e})$ can be used, which is computed as in Equation (3.14) using the PDF p_E of the extrinsic output expressed as

$$I_{E_M} = \frac{1}{2}\cdot \sum_{x=-1,+1} \int_{-\infty}^{+\infty} p_E(\zeta|X = x)\cdot \log_2 \frac{2\cdot p_E(\zeta|X = x)}{p_E(\zeta|X = -1) + p_E(\zeta|X = +1)}\, \mathrm{d}\zeta. \tag{3.19}$$

[2]The LLR values are symmetric if their PDF is symmetric: $p(-\zeta|X = +1) = p(\zeta|X = -1)$. In addition, all LLR values with symmetric distributions satisfy the consistency condition [175]:

$$p(-\zeta|X = x) = e^{-x\zeta} p(\zeta|X = x).$$

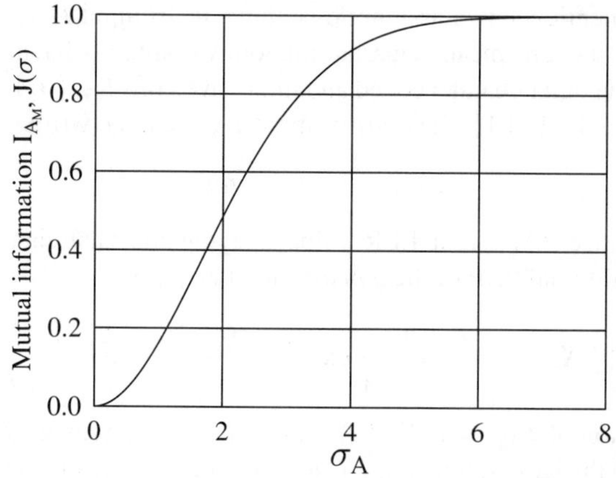

Figure 3.3: Mutual information I_{A_M} as a function of σ_A and evaluated from Equation (3.15).

Considering I_{E_M} as a function of both I_{A_M} and the E_b/N_0 value encountered, the demapper's EXIT characteristic is defined as [169, 172]

$$I_{E_M} = T_M(I_{A_M}, E_b/N_0). \tag{3.20}$$

Figure 3.4 illustrates how the EXIT characteristic $T_M(I_{A_M}, E_b/N_0)$ is calculated for a specific $(I_{A_M}, E_b/N_0)$-input combination. First the noise variance σ_n of the wireless channel is set according to the E_b/N_0 value considered. Then, σ_A is calculated based on the specific value of I_{A_M} where the EXIT curve has to be evaluated using $\sigma_A = J^{-1}(I_{A_M})$ expressed from Equation (3.16) and plotted in Figure 3.3. Next, σ_A is used for creating $L_{M,a}$ according to Equation (3.12), which is applied as the *a priori* input of the demapper. Finally, the mutual information of $I_{E_M} = I(X; L_{M,e})$, $0 \leq I_{E_M} \leq 1$, between the outer coded and interleaved bits x and the LLR values $L_{M,e}$ is calculated using Equation (3.19) with the aid of the PDF p_E of the extrinsic output $L_{M,e}$. This requires the determination of the distribution p_E by means of Monte Carlo simulations. However, according to [175], by invoking the ergodicity theorem in Equation (3.18), namely by replacing the expected value by the time average, the mutual information can be estimated using a sufficiently large number of samples even for non-Gaussian or unknown distributions, which may be expressed as [175]

$$I(X; L_{M,e}) = 1 - E_{X=+1}\{\log_2[1 + e^{-L_{M,e}}]\} \approx 1 - \frac{1}{N}\sum_{n=1}^{N} \log_2[1 + e^{-x(n)\cdot L_{M,e}(n)}]. \tag{3.21}$$

Figure 3.5 shows the extrinsic information transfer characteristics of the SP demapper in conjunction with $L = 16$ and different mapping schemes between the interleaver's output and the STBC encoder. As expected, Gray mapping does not provide any iteration gain upon increasing the mutual information at the input of the demapper. However, using a variety of different AGM schemes [158] results in different extrinsic information transfer characteristics, as illustrated by the different slopes seen in Figure 3.5. The ten different AGM mapping schemes shown in Figure 3.5 are specifically selected from all of the possible mapping schemes for $L = 16$ in order to demonstrate the different extrinsic information transfer characteristics associated with different bit-to-symbol mapping schemes. There are a total of 16! different mapping schemes. The details of the Gray mapping as well as the various AGM schemes considered in this chapter are given in Appendix A.

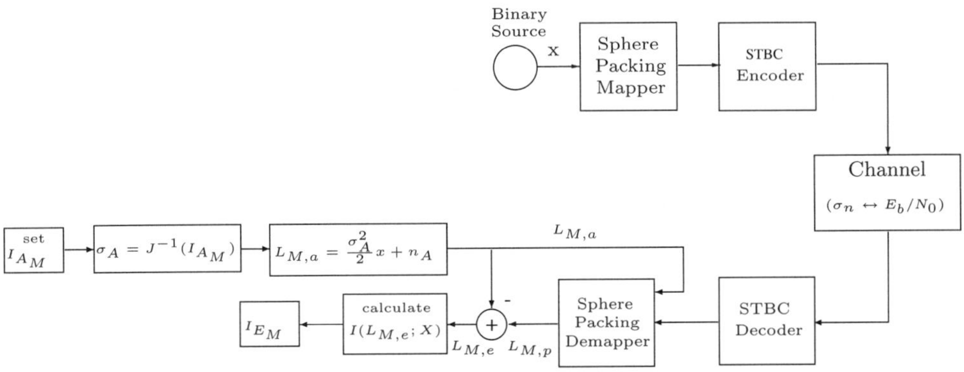

Figure 3.4: Evaluation of the demapper transfer characteristic.

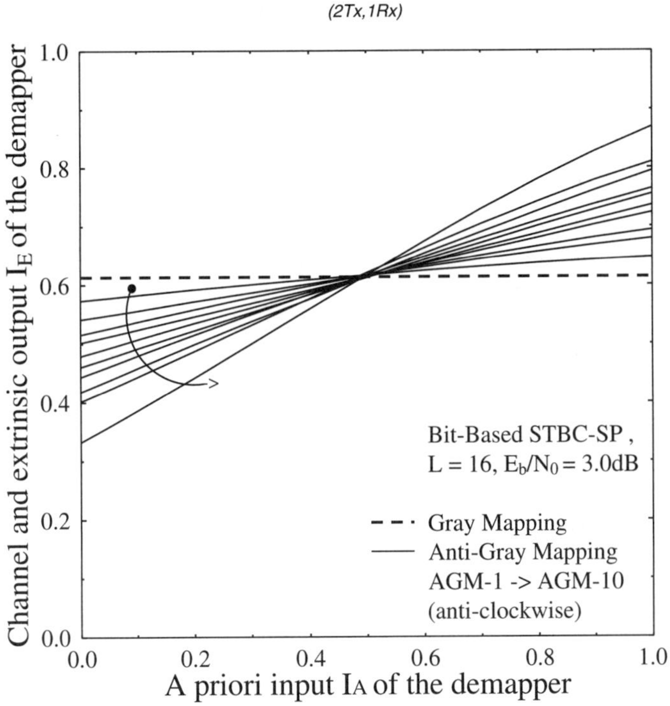

Figure 3.5: SP demapper EXIT characteristics for different bits-to-SP-symbol mapping schemes at $E_b/N_0 = 3.0$ dB for $L = 16$.

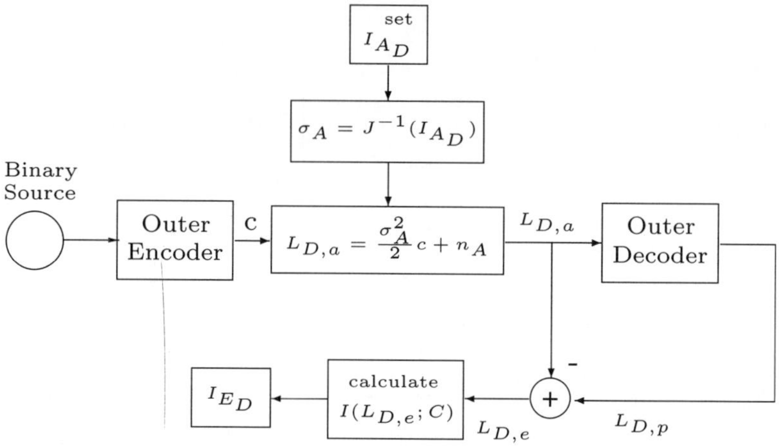

Figure 3.6: Evaluation of the outer channel decoder transfer characteristic.

3.4.2 Transfer Characteristics of the Outer Decoder

The EXIT characteristic of the outer channel decoder describes the relationship between the outer channel coded input $L_{D,a}$ and the outer channel decoded extrinsic output $L_{D,e}$. The input of the outer channel decoder consists only of the *a priori* input $L_{D,a}$ provided by the SP demapper. Therefore, the EXIT characteristic of the outer channel decoder is independent of the E_b/N_0-value and, hence, may be written as

$$I_{E_D} = T_D(I_{A_D}), \tag{3.22}$$

where $I_{A_D} = I(C; L_{D,a})$, $0 \leq I_{A_D} \leq 1$, is the mutual information between the outer channel coded bits c and the LLR values $L_{D,a}$ and similarly $I_{E_D} = I(C; L_{D,e})$, $0 \leq I_{E_D} \leq 1$, is the mutual information between the outer channel coded bits c and the LLR values $L_{D,e}$. The computational model of evaluating the EXIT characteristic of the outer channel decoder is shown in Figure 3.6. As can be seen from the figure, the procedure is similar to that of the SP demapper shown in Figure 3.4, except that its value is independent of the SNR. Again, $I_{E_D} = I(C; L_{D,e})$ can be computed either by evaluating the histogram approximation of p_E (see [169, 172]) and then applying Equation (3.19) or, more conveniently, by the time-averaging method [175] of Equation (3.21) as

$$I(C; L_{D,e}) = 1 - E\{\log_2[1 + e^{-L_{D,e}}]\} \approx 1 - \frac{1}{N}\sum_{n=1}^{N} \log_2[1 + e^{-c(n)\cdot L_{D,e}(n)}]. \tag{3.23}$$

The EXIT characteristics of several half-rate RSC codes having different constraint lengths are shown in Figure 3.7. The generator polynomials employed are given in Table 3.2 in their octal representation, where G_r is the feedback polynomial and G is the feedforward polynomial. Figure 3.7 demonstrates that for $I_{A_D} > 0.5$, the set of RSC codes having higher constraint lengths converge faster upon increasing I_{A_D} than the RSC codes having smaller constraint lengths. This behavior is due to the fact that the higher constraint-length RSC codes exhibit a better minimum free distance than shorter constraint-length RSC codes [6]. Similarly, Figure 3.8 illustrates the EXIT characteristics of a half-rate binary LDPC code having an average column weight of 2.5 and using different numbers of internal LDPC iterations. As intuitively expected, the figure confirms that the EXIT characteristics of the LDPC decoder improve as the number of internal LDPC iterations is increased.

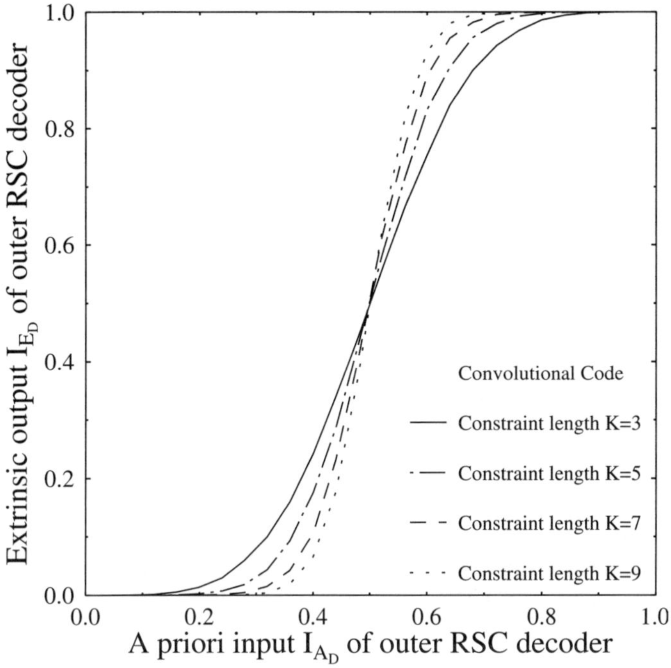

Figure 3.7: EXIT characteristics of several half-rate RSC codes having different constraint lengths.

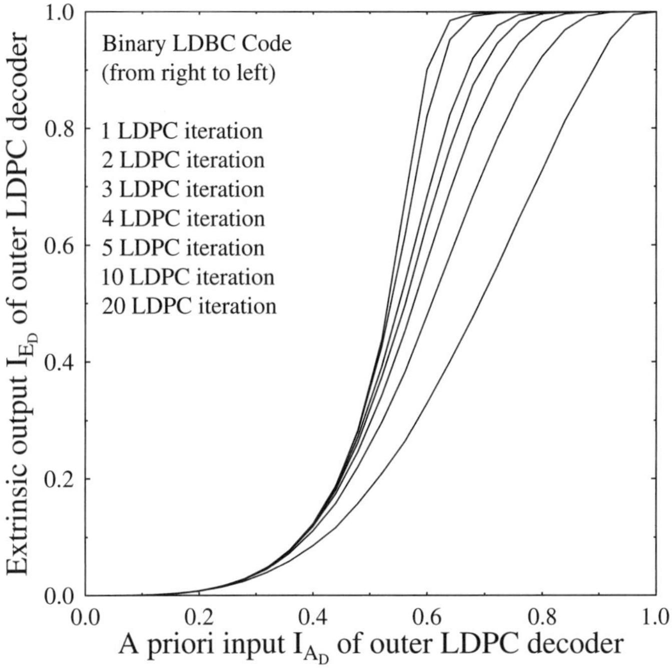

Figure 3.8: EXIT characteristics of a half-rate binary LDPC code having an average column weight of 2.5 and using different numbers of LDPC internal iterations.

Table 3.2: Half-rate RSC codes parameters.

Constraint length	Generator polynomials in octals	
$\mathcal{K}$	G_r	G
3	5	7
4	15	17
5	35	23
6	53	75
7	133	171
8	247	371
9	561	753

3.4.3 Extrinsic Information Transfer Chart

The exchange of extrinsic information in the decoder schematics of Figures 3.1 and 3.2 is visualized by plotting the EXIT characteristics of the SP demapper and outer channel decoder in a joint diagram. This diagram is known as the EXIT chart [169, 172]. The outer channel decoder's extrinsic output I_{E_D} becomes the SP demapper's *a priori* input I_{A_M}, which is represented on the x-axis. Similarly, on the y-axis, the SP demapper's extrinsic output I_{E_M} becomes the outer channel decoder's *a priori* input I_{A_D}. Accordingly, the axes of Figures 3.7 and 3.8 are swapped intentionally for the sake of creating the EXIT chart as seen in Figures 3.9 and 3.10.

Figure 3.9 shows the EXIT chart of a turbo-detection channel-coded STBC-SP scheme employing AGM (AGM-9) of Figure 3.5 in combination with outer RSC code having constraint length $\mathcal{K} = 5$ when communicating over a correlated Rayleigh fading channel having $f_D = 0.1$. Ideally, in order for the exchange of extrinsic information between the SP demapper and the outer RSC decoder to converge at a specific E_b/N_0 value, the EXIT characteristic curve of the SP demapper at the E_b/N_0 value of interest and the EXIT characteristic curve of the outer RSC decoder should only intersect at the $(1.0, 1.0)$ point. If this condition is satisfied, then a so-called convergence tunnel [169, 172] appears on the EXIT chart. The narrower the tunnel, the more iterations are required to reach the $(1.0, 1.0)$ point. If the two EXIT characteristic curves, however, intersect at a point infinitesimally close to the $I_{E_D} = 1.0$ line rather than at the $(1.0, 1.0)$ point, then a moderately low BER could still be achieved, but the BER will not become as low as in the schemes when the intersection is at the $(1.0, 1.0)$ point. These types of tunnels are referred to here as semi-convergent tunnels. Observe in Figure 3.9 that a semi-convergent tunnel exists at $E_b/N_0 = 2.5$ dB. This implies that according to the predictions of the EXIT chart seen in Figure 3.9, the iterative decoding process is expected to converge to a moderately low BER at $E_b/N_0 = 2.5$ dB. The validity of this prediction is, however, dependent on how accurately the two EXIT chart assumptions outlined at the beginning of Section 3.4 are satisfied. These EXIT-chart-based convergence predictions are usually verified by the actual iterative decoding trajectory, as discussed in Section 3.5. Similarly, the EXIT chart of the turbo-detected channel-coded STBC-SP scheme of Figure 3.2 employing the AGM (AGM-6) scheme of Figure 3.5 in combination with a binary outer LDPC code having an average column weight of 2.5 and using ten internal LDPC iterations is portrayed in Figure 3.10 when communicating over a SPSI-correlated Rayleigh

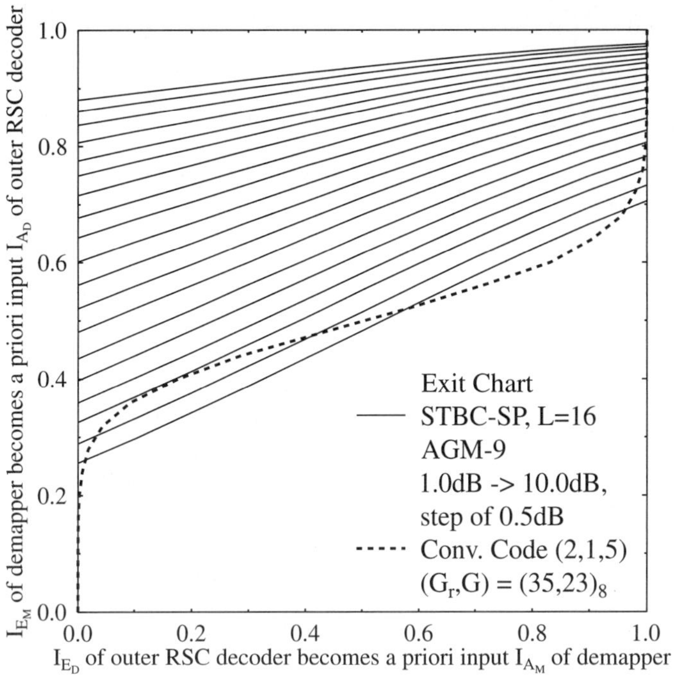

Figure 3.9: EXIT chart of a turbo-detected RSC channel-coded STBC-SP scheme employing AGM (AGM-9) of Figure 3.5 in combination with outer RSC code having constraint length $\mathcal{K} = 5$ when communicating over a correlated Rayleigh fading channel having $f_D = 0.1$.

fading channel having $f_D = 0.1$. According to the figure, iterative decoding convergence tending to moderately low BER values becomes possible for $E_b/N_0 > 2.0$ dB for this particular system arrangement, as a semi-convergent tunnel is beginning to take shape upon increasing the SNR beyond $E_b/N_0 = 2.0$ dB.

Appendix B provides the complete list of EXIT charts for the bit-based turbo-detected STBC-SP schemes of Figures 3.1 and 3.2, when employing the mapping schemes of Figure 3.5 in combination with outer RSC and binary LDPC codes.

3.5 Performance of Turbo-detected Bit-based STBC-SP Schemes

Without loss of generality, we considered a SP modulation scheme associated with $L = 16$ using two transmit and a single receiver antenna in order to demonstrate the performance improvements achieved by the proposed system. All simulation parameters are listed in Table 3.3.

3.5.1 Performance of RSC-coded STBC-SP Scheme

3.5.1.1 Mutual Information and Achievable BER

Observe in the EXIT charts provided in Appendix B.1 that once a semi-convergent tunnel is formed, the intersection point of the EXIT characteristic curves of the SP demapper and the

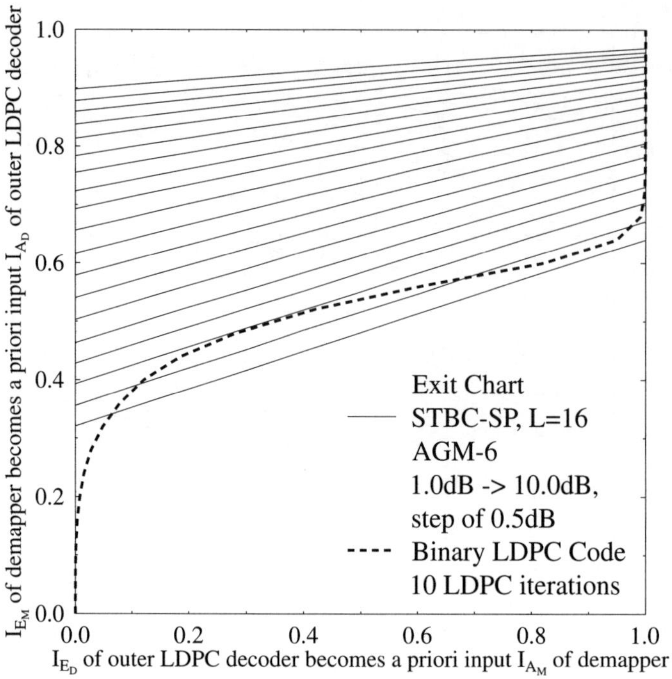

Figure 3.10: EXIT chart of a turbo-detected binary LDPC channel-coded STBC-SP scheme employing AGM (AGM-6) of Figure 3.5 in combination with outer binary LDPC code having an average column weight of 2.5 and using ten internal LDPC iterations when communicating over a correlated Rayleigh fading channel having $f_D = 0.1$.

Table 3.3: Bit-based system parameters.

Modulation	SP with $L = 16$
Number of transmitters	2
Number of receivers	1
Channel type	SPSI-correlated Rayleigh fading
Normalized Doppler frequency	0.1
RSC code rate	0.5
Average LDPC column weight	2.5
LDPC code rate	0.5
LDPC decoding field	$GF(2)$
System throughput	1 BPS

outer RSC decoder slides gradually towards the $(1.0, 1.0)$ point upon increasing the SNR. In order to investigate how the position of the intersection point affects the BER performance, Figure 3.11 shows the achievable BER as a function of the mutual information I_{E_D} at the output of the RSC decoder for different constraint lengths. According to Figure 3.11, the intersection point should be at least at $I_{E_D} = 0.985$ in order to achieve a BER of 10^{-3}, which is independent of the RSC code's constraint length. This is true because Figure 3.11 relates the mutual information at the output of the RSC decoder to the achievable BER. Figure 3.12,

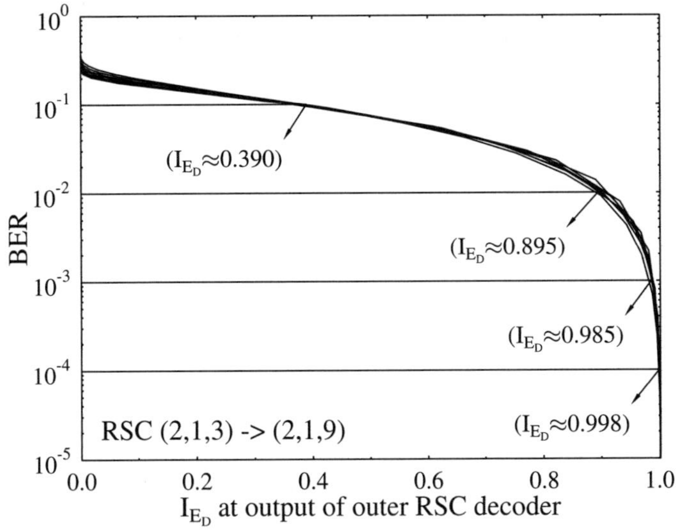

Figure 3.11: BER of different half-rate RSC decoders versus their extrinsic output I_{E_D}, when increasing the code's constraint length from $\mathcal{K} = 3$ to 9.

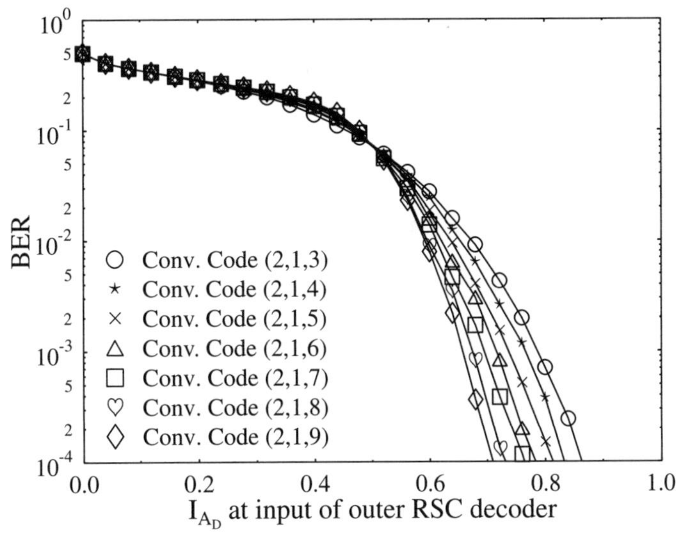

Figure 3.12: BER of different half-rate RSC decoders versus their *a priori* input I_{A_D}.

however, relates the mutual information at the input of the RSC decoder to the achievable BER. The effect of the code's constraint length becomes evident in Figure 3.12, since RSC codes having higher constraint lengths require lower I_{A_D} values in order to achieve a similar BER. Table 3.4 summarizes the I_{A_D} values required for achieving a BER of 10^{-4} at the input of the RSC decoders of Table 3.2.

It was reported in [158] that there is a strong correlation between the average bit-wise capacity computed using no *a priori* information, which corresponds to the unconditional average bit-wise mutual information I_0 of the symbol constellation and the achievable BER

Table 3.4: Required I_{A_D} at the input of the RSC decoders of Table 3.2 for achieving a BER of 10^{-4}.

Constraint length	Required I_{A_D}
3	0.863
4	0.832
5	0.813
6	0.783
7	0.764
8	0.731
9	0.710

performance when iterative demapping and turbo detection are employed. The achievable performance depends on the specific assignment of the bits to each symbol in the constellation. In other words, the achievable performance depends on the mapping scheme. Different STBC-SP mapping schemes spanning a wide range of different I_0 values were investigated in Figure 3.13 for the sake of demonstrating this phenomenon. Figure 3.13 characterizes the achievable BER performance against the unconditional bit-wise mutual information I_0 for different STBC-SP mappings at $E_b/N_0 = 4.0$ dB in conjunction with $L = 16$ and an outer RSC code having a constraint length of $\mathcal{K} = 5$, when employing the system parameters outlined in Table 3.3 and using an interleaver depth of $\mathcal{D} = 4000$ bits. The unconditional average bit-wise mutual information I_0 of a specific mapping scheme at a particular channel condition can be computed using the model in Figure 3.4 when $I_{A_M} = 0$. Figure 3.13 also confirms the interesting fact stated in [158] that the choice of the optimum bits-to-symbol mapping scheme is dependent on the number of iterations used. For example, at $E_b/N_0 = 4.0$ dB Gray mapping is the optimum scheme when no iteration is used at $I_0 = 0.68$. In addition, as shown in Figure 3.13, the bits-to-symbol mapping scheme associated with $I_0 = 0.54$ at $E_b/N_0 = 4.0$ dB is the optimum mapping when three or more iterations are employed. These observations are only valid for the specific RSC code under consideration. Other RSC codes are usually associated with different optimum mapping schemes, as illustrated by the EXIT charts seen in Appendix B.1.

3.5.1.2 Decoding Trajectory and the Effect of the Interleaver Depth

Figure 3.14 illustrates the actual decoding trajectory of the turbo-detected RSC channel-coded STBC-SP scheme of Figure 3.9 at $E_b/N_0 = 2.5$ dB and using an interleaver depth of $\mathcal{D} = 10^6$ bits. The zigzag path in Figure 3.14 represents the actual EXIT between the SP demapper and the outer RSC channel decoder. Since a long interleaver is employed, the assumptions outlined at the beginning of Section 3.4 are justified and hence the EXIT chart prediction has been attained. However, the decoding trajectories shown in Figures 3.15 and 3.16 are different from the EXIT chart prediction because shorter interleaver lengths are used. The achievable I_{E_D} when employing different interleaver lengths is demonstrated in Figure 3.17. The BER of the schemes of Figure 3.17 is shown in Figure 3.18 when using ten external joint iterations where the difference in the achievable BER at $E_b/N_0 = 2.5$ dB is evident when increasing the interleaver length to $\mathcal{D} = 10^6$ bits. In addition, observe the turbo cliff at $E_b/N_0 = 2.5$ dB upon increasing the interleaver length.

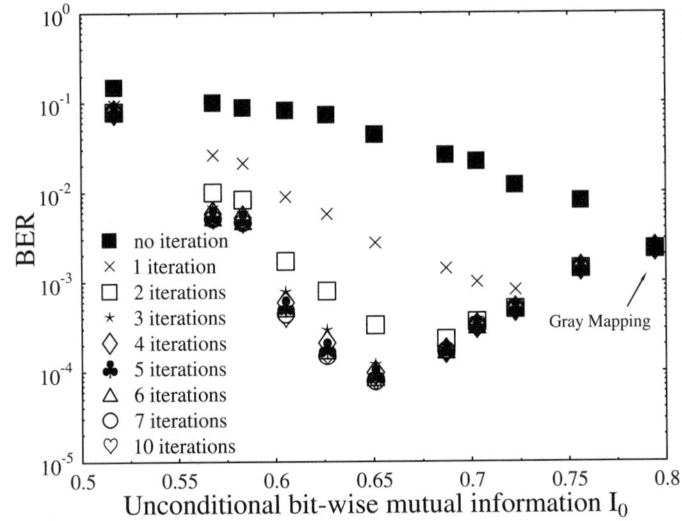

Figure 3.13: BER versus unconditional bit-wise mutual information I_0 between 0.4205 and 0.6785 for different STBC-SP mappings at $E_b/N_0 = 4.0$ dB in conjunction with $L = 16$, when employing the system parameters outlined in Table 3.3 and using an interleaver depth of $\mathcal{D} = 4000$ bits.

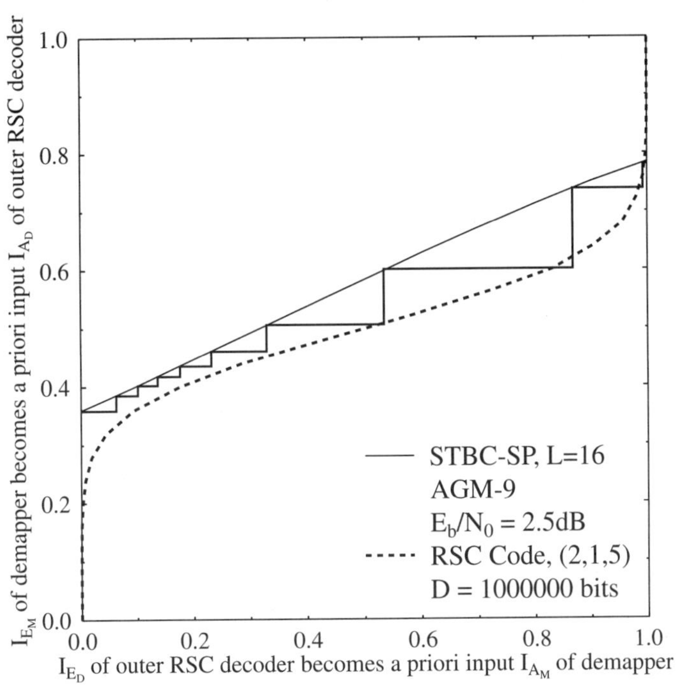

Figure 3.14: Decoding trajectory of a turbo-detected RSC channel-coded STBC-SP scheme employing AGM (AGM-9) in combination with the outer RSC code (2,1,5) and the system parameters outlined in Table 3.3 and operating at $E_b/N_0 = 2.5$ dB with an interleaver depth of $\mathcal{D} = 10^6$ bits.

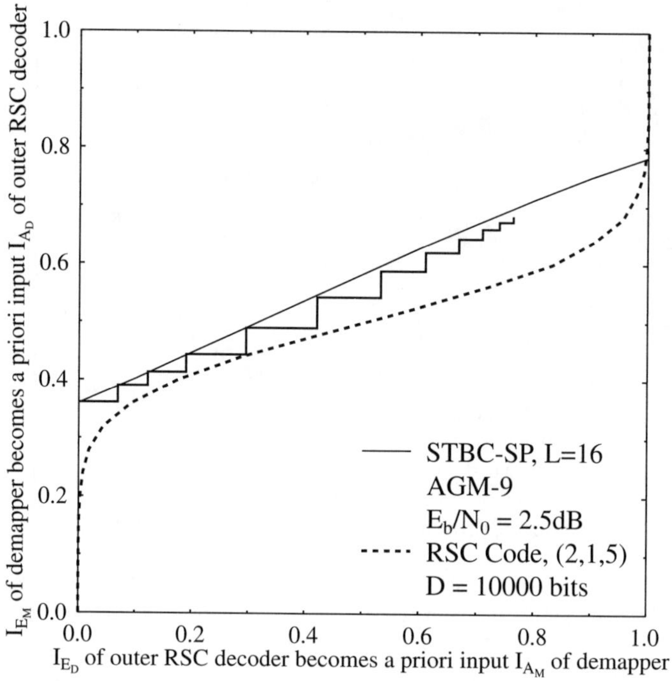

Figure 3.15: Decoding trajectory of turbo-detected RSC channel-coded STBC-SP scheme employing AGM (AGM-9) in combination with the outer RSC code (2,1,5) and the system parameters outlined in Table 3.3 and operating at $E_b/N_0 = 2.5$ dB with an interleaver depth of $\mathcal{D} = 10^4$ bits.

3.5.1.3 BER Performance

Figure 3.19 compares the performance of the proposed convolutional-coded STBC-SP scheme employing AGM (AGM-9) and Gray mapping against that of an identical-throughput 1 BPS uncoded STBC-SP scheme and a conventional orthogonal STBC design as well as against a RSC-coded QPSK-modulated STBC scheme, when employing the system parameters outlined in Table 3.3 and using an interleaver depth of $\mathcal{D} = 10^6$ bits. The QPSK-modulated STBC system employs a set-partitioning mapping scheme reminiscent of TCM [263]. Observe in Figure 3.19 by comparing the two Gray mapping STBC-SP curves that no BER improvement was obtained when ten turbo-detection iterations were employed in conjunction with Gray mapping, which was reported also in [158] and evident from the flat curve of the Gray mapping in Figure 3.5. In contrast, AGM (AGM-9) achieved a useful performance improvement in conjunction with iterative demapping and decoding. Explicitly, Figure 3.19 demonstrates that a coding advantage of about 20.2 dB was achieved at a BER of 10^{-5} after ten iterations by the convolutional-coded AGM-9 STBC-SP system over both the uncoded STBC-SP [43] and the conventional orthogonal STBC-design-based [11, 12] schemes for transmission over the correlated Rayleigh fading channel considered. In addition, coding advantages of approximately 3.2 and 2.0 dB were attained over the 1 BPS throughput RSC-coded Gray mapping STBC-SP and the RSC-coded QPSK-modulated STBC schemes, respectively.

Figure 3.20 demonstrates the attainable performance of various AGM-based RSC-coded STBC-SP schemes in conjunction with $L = 16$, when employing the system parameters

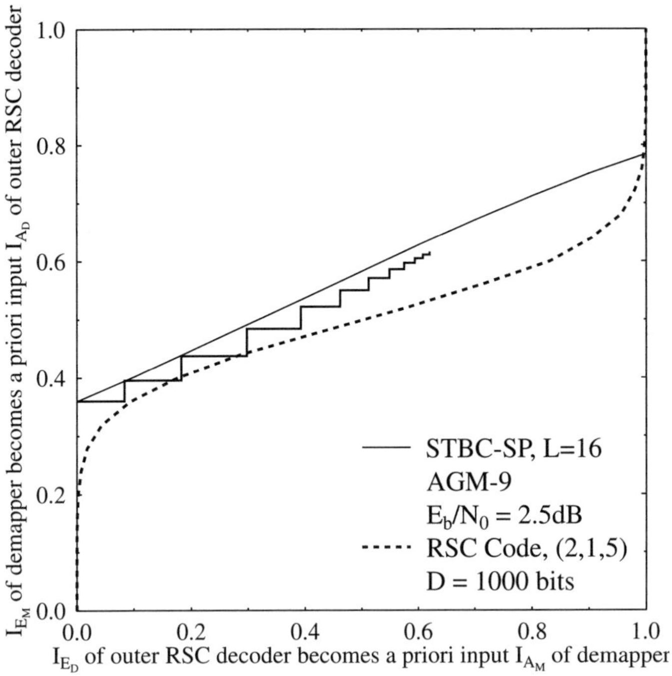

Figure 3.16: Decoding trajectory of turbo-detected RSC channel-coded STBC-SP scheme employing AGM (AGM-9) in combination with the outer RSC code (2,1,5) and the system parameters outlined in Table 3.3 and operating at $E_b/N_0 = 2.5$ dB with an interleaver depth of $\mathcal{D} = 10^3$ bits.

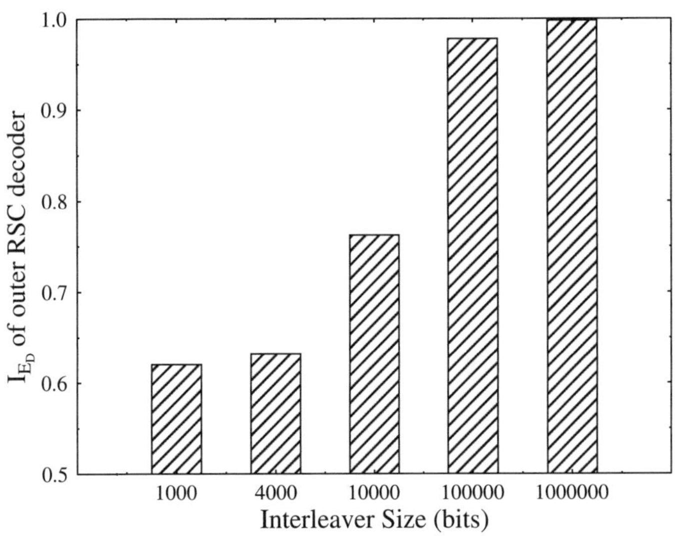

Figure 3.17: Achievable extrinsic information of turbo-detected RSC channel-coded STBC-SP scheme employing AGM (AGM-9) in combination with the outer RSC code (2,1,5) and the system parameters outlined in Table 3.3 and operating at $E_b/N_0 = 2.5$ dB with different interleaver depths.

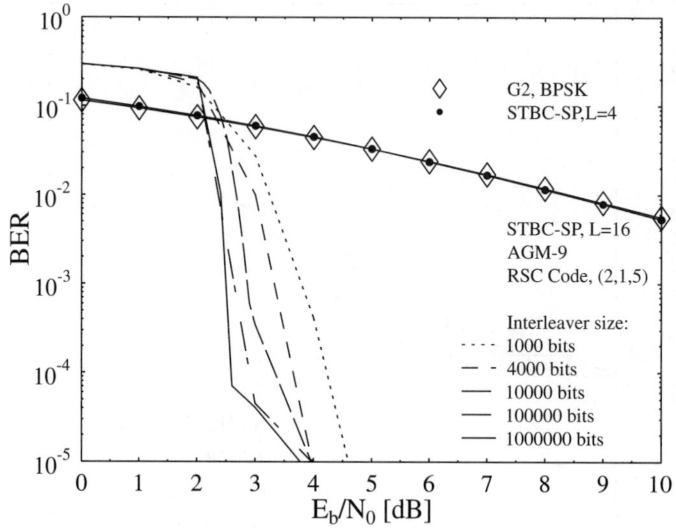

Figure 3.18: Performance comparison of AGM (AGM-9)-based RSC-coded STBC-SP schemes in conjunction with $L = 16$ against an identical-throughput 1 BPS uncoded STBC-SP scheme using $L = 4$ and against Alamouti's conventional G_2-BPSK scheme, when employing the system parameters outlined in Table 3.3 and using different interleaver depths after ten external joint iterations.

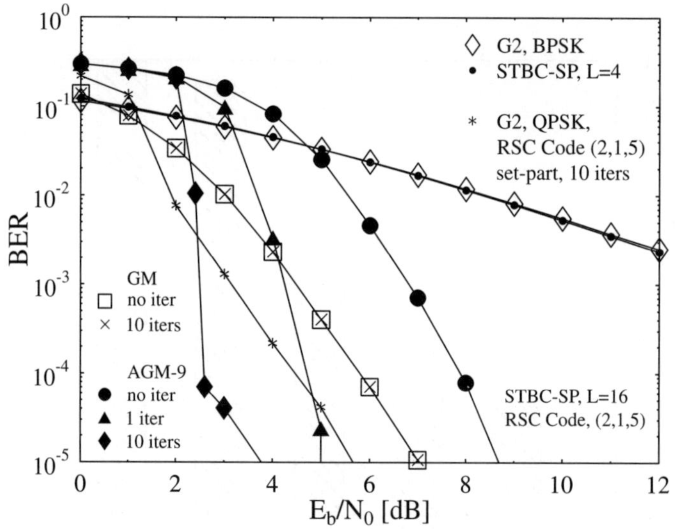

Figure 3.19: Performance comparison of AGM (AGM-9) and Gray mapping (GM)-based RSC-coded STBC-SP schemes in conjunction with $L = 16$ against an identical-throughput 1 BPS uncoded STBC-SP scheme using $L = 4$ and against Alamouti's conventional G_2-BPSK scheme as well as against a RSC-coded QPSK-modulated STBC scheme, when employing the system parameters outlined in Table 3.3 and using an interleaver depth of $\mathcal{D} = 10^6$ bits.

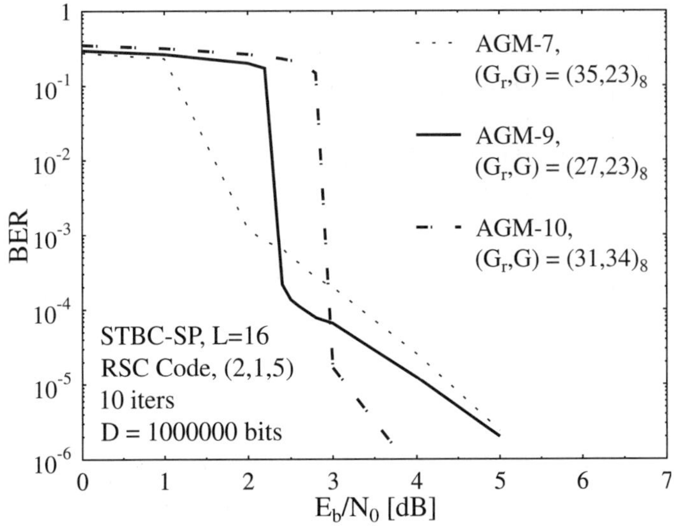

Figure 3.20: Performance comparison of various AGM-based RSC-coded STBC-SP schemes in conjunction with $L = 16$, when employing the system parameters outlined in Table 3.3 and using an interleaver depth of $\mathcal{D} = 10^6$ bits.

outlined in Table 3.3 and using an interleaver depth of $\mathcal{D} = 10^6$ bits after ten iterations. Observe in Figure 3.20 that the three schemes perform differently. For example, the AGM-7-based scheme exhibits a turbo cliff at a lower E_b/N_0 value than the others, when combined with the specific RSC code having the octal generator polynomials of $(G_r, G) = (35, 23)_8$. However, the AGM-7-based scheme has the highest BER floor. Therefore, different AGM schemes may be combined with specific RSC codes for the sake of designing systems satisfying specific criteria, such as for example attaining an early convergence or a lower BER floor. The EXIT charts of the three different schemes seen in Figure 3.20 are illustrated in Figure 3.21.

3.5.2 Performance of Binary LDPC-coded STBC-SP Scheme

3.5.2.1 Mutual Information and Achievable BER

Figure 3.22 shows the achievable BER as a function of the mutual information I_{A_D} at the input of the binary LDPC decoder, when performing different numbers of internal LDPC iterations. According to Figure 3.22, a mutual information of about $I_{A_D} = 0.97$ has to be forwarded to the binary LDPC decoder in order to achieve a BER of 10^{-4}, when performing one internal LDPC iteration. However, this requirement drops to about $I_{A_D} = 0.8$ and $I_{A_D} = 0.7$, when performing 5 and 20 internal LDPC iterations, respectively. Table 3.5 summarises the I_{A_D} values at the input of the binary LDPC decoder required for achieving a BER of 10^{-4}, when performing different numbers of internal LDPC iterations.

3.5.2.2 Decoding Trajectory and Effect of Interleaver Depth

Figure 3.23 illustrates the decoding trajectory of the turbo-detected binary LDPC channel-coded STBC-SP scheme employing AGM (AGM-6) in combination with a half-rate outer binary LDPC code having an average column weight of 2.5 and using ten internal LDPC

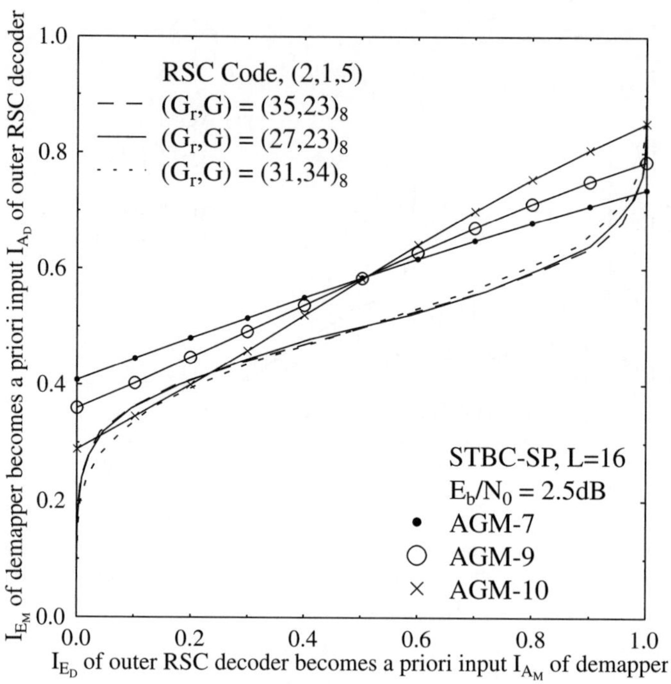

Figure 3.21: EXIT chart of turbo-detected RSC-coded STBC-SP systems employing various AGM schemes from Figure 3.5 in combination with outer RSC codes having constraint length $\mathcal{K} = 5$ and different generator polynomials, when employing the system parameters outlined in Table 3.3.

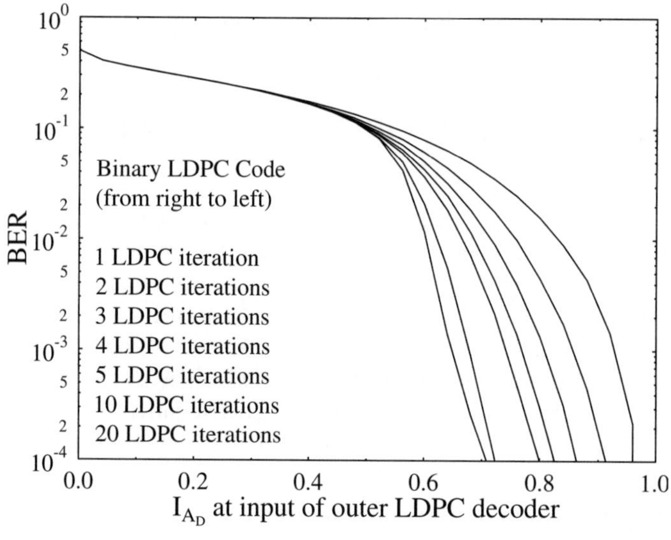

Figure 3.22: BER of a half-rate binary LDPC decoder versus *a priori* input I_{A_D} when employing different numbers of internal LDPC iterations.

Table 3.5: Required I_{A_D} for achieving a BER of 10^{-4} when performing different numbers of internal LDPC iterations.

Number of internal LDPC iterations	Required I_{A_D}
1	0.970
2	0.913
3	0.863
4	0.826
5	0.800
10	0.723
20	0.708

iterations, when communicating over the channel outlined in Table 3.3 and operating at $E_b/N_0 = 2.5$ dB after ten joint external iterations. The actual iterative performance is different from the prediction of the EXIT chart, since a short interleaver depth of $\mathcal{D} = 1500$ bits is employed, which does not guarantee that the LLR values have a Gaussian distribution. Figure 3.24 shows our performance comparison of the AGM-(AGM-6) and Gray-mapping-based LDPC-coded STBC-SP schemes in conjunction with $L = 16$ against an identical-throughput 1 BPS uncoded STBC-SP scheme using $L = 4$ and against Alamouti's conventional G_2-BPSK scheme, when communicating over the channel outlined in Table 3.3 and using an outer half-rate binary LDPC code having an average column weight of 2.5 as well as ten internal LDPC iterations. Figure 3.24 demonstrates that a coding gain of about 19.3 dB is attained by the LDPC-coded STBC-SP scheme after ten joint external iterations against both the identical-throughput uncoded STBC-SP scheme and Alamouti's conventional G_2-BPSK scheme, when using an output block length of $\mathcal{D} = 1500$ bits. However, a better performance is predicted by the EXIT chart seen in Figure 3.10, where a semi-convergent tunnel exits at $E_b/N_0 = 2.5$ dB and, hence, a lower BER could be attained upon increasing the output block length or, equivalently, the interleaver length. Figure 3.25 shows the decoding trajectory of the turbo-detected binary LDPC channel-coded STBC-SP scheme of Figures 3.23 and 3.24, when using an increased output block length of $K_{ldpc} = 10\,000$ bits. It is evident from Figure 3.25 that an improved performance is achieved when employing a larger interleaver length, which is also demonstrated by the BER curves in Figure 3.26. According to Figure 3.26, a coding gain of about 21.2 dB is attained by the LDPC-coded STBC-SP scheme after ten joint external iterations against both the identical-throughput uncoded STBC-SP scheme and Alamouti's conventional G_2-BPSK scheme, when using an output block length of $K_{ldpc} = 10\,000$ bits. In addition, a coding advantage of approximately 1.8 dB was attained over the 1 BPS throughput LDPC-coded Gray mapping STBC-SP scheme.

3.5.2.3 Effect of Internal LDPC Iterations and Joint External Iterations

In order to investigate the effects of performing different numbers of internal LDPC iterations and joint external iterations on the performance of the binary LDPC-coded STBC-SP scheme of Figure 3.2, Figure 3.27 demonstrates the achievable coding gain for different combinations of internal and external iterations. More specifically, Figure 3.27 shows the achievable coding gain of a bit-based binary LDPC-coded STBC-SP scheme employing AGM-6 in conjunction

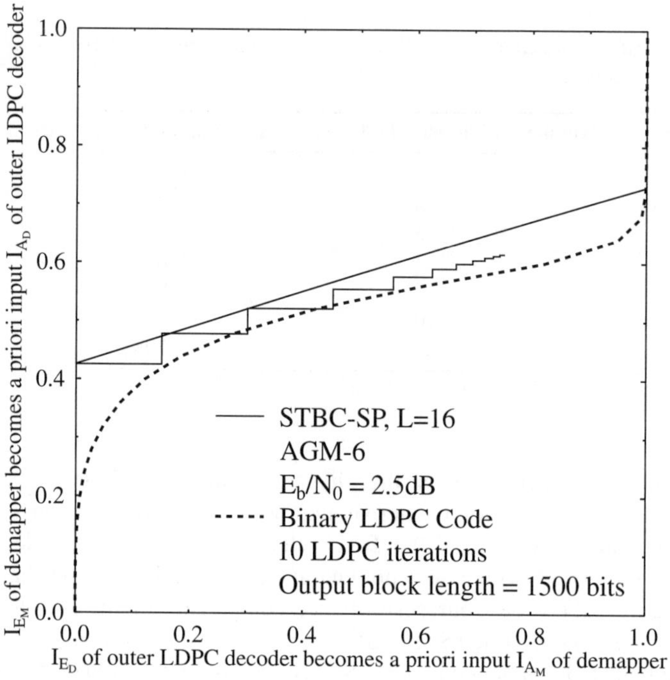

Figure 3.23: Decoding trajectory of turbo-detected binary LDPC channel-coded STBC-SP scheme employing AGM (AGM-6) in combination with outer half-rate binary LDPC code and system parameters of Table 3.3 when using ten internal LDPC iterations and operating at $E_b/N_0 = 2.5$ dB after ten joint external iterations.

with different combinations of joint and LDPC iterations as compared with the identical-throughput 1 BPS uncoded STBC-SP scheme of [43] and Alamouti's conventional G_2-BPSK scheme [11] at a BER of 10^{-5}, when communicating over the channel outlined in Table 3.3 and using a half-rate outer binary LDPC code having an output block length of 1500 bits. The terms '0 joint iteration' and 'no joint iterations', used in Figure 3.27 and subsequent figures, refer to the open-loop scenario, where no extrinsic information is fed back from the outer decoder to the SP demapper. Observe in Figure 3.27 that for a specific fixed number of internal LDPC iterations, the attainable coding gain improvement becomes negligible after carrying out two joint external iterations. In other words, most of the BER improvements are achieved during the first two iterations. This observation is illustrated in both Figures 3.28 and 3.29, which show the performance improvement attained upon increasing the number of joint external iterations, while fixing the number of internal LDPC iterations to one and five iterations, respectively. The effect of increasing the number of internal LDPC iterations while fixing the number of joint external iterations can also be observed from Figure 3.27, which demonstrates that the coding gain improvements become negligible after encountering five internal LDPC iterations. Figures 3.30 and 3.31 illustrate the effect of increasing the number of internal LDPC iterations on the performance of the bit-based binary LDPC-coded STBC-SP scheme, while fixing the number of joint external iterations to one and five iterations, respectively.

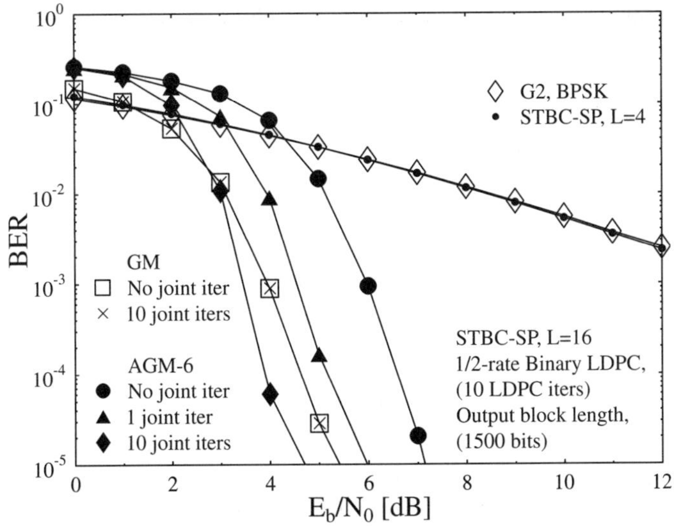

Figure 3.24: Performance comparison of AGM (AGM-6) and Gray mapping (GM)-based LDPC-coded STBC-SP schemes in conjunction with $L = 16$ against an identical-throughput 1 BPS uncoded STBC-SP scheme using $L = 4$ and against Alamouti's conventional G_2-BPSK scheme, when employing system parameters of Table 3.3 and using an outer half-rate binary LDPC code having an average column weight of 2.5 and ten internal LDPC iterations.

3.6 Chapter Conclusions

In this chapter, we have proposed a novel system that exploits the advantages of both iterative demapping and turbo detection [158] as well as those of the STBC-SP scheme of [43]. Our investigations demonstrated that significant performance improvements may be achieved when the AGM STBC-SP is combined with outer channel decoding and iterative demapping as compared with the Gray-mapping-based systems. Subsequently, the EXIT chart was used to search for bits-to-symbol mapping schemes that converge at lower E_b/N_0 values. Several STBC-SP mapping schemes covering a wide range of EXIT characteristics were investigated. When using an appropriate bits-to-symbol mapping scheme and ten turbo-detection iterations, gains of about 20.4 and 21.2 dB at a BER of 10^{-5} were obtained by the convolutional-coded and LDPC-coded STBC-SP schemes, respectively, over the identical-throughput 1 BPS uncoded STBC-SP benchmarker scheme [43]. In addition, coding advantages of approximately 3.2 and 2.0 dB at a BER of 10^{-5} were attained over the 1 BPS throughput RSC-coded GM STBC-SP and the RSC-coded QPSK-modulated STBC schemes, respectively.

3.7 Chapter Summary

In this chapter, two realizations of a novel bit-based iterative-detection aided STBC-SP scheme have been presented, namely a RSC-coded turbo-detected STBC-SP scheme and a binary LDPC-coded turbo-detected STBC-SP arrangement. Our system overview was provided in Section 3.2. The LDPC-coded scheme of Figure 3.2 did not require channel interleaving, since the LDPC parity check matrix is randomly constructed, where each of the

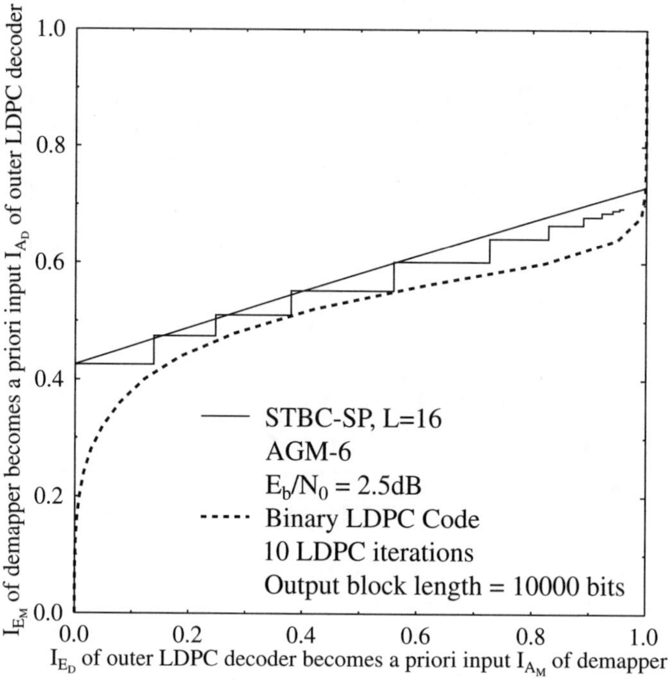

Figure 3.25: Decoding trajectory of turbo-detected binary LDPC channel-coded STBC-SP scheme employing AGM (AGM-6) in combination with the outer half-rate binary LDPC code and system parameters of Table 3.3 when using ten internal LDPC iterations and operating at $E_b/N_0 = 2.5$ dB after ten joint external iterations.

parity check equations is checking several random bit positions in a codeword, which has a similar effect to that of the channel interleaver. In Section 3.3, we showed how the STBC-SP demapper was modified for exploiting the *a priori* knowledge provided by the channel decoder, which is essential for the employment of iterative demapping and decoding.

EXIT chart analysis was invoked in Section 3.4 in order to study and design the turbo-detected schemes proposed in Section 3.2. Measuring the demapper's EXIT characteristics was explained in Section 3.4.1 and in Figure 3.4. We proposed ten different AGM schemes in Figure 3.5 that are specifically selected from all of the possible mapping schemes for $L = 16$ in order to demonstrate the different EXIT characteristics associated with different bit-to-symbol mapping schemes. Both the Gray mapping and the various AGM mapping schemes considered in this chapter are detailed in Appendix A. In Section 3.4.2, we discussed how the EXIT characteristics of an outer decoder in a serially concatenated scheme may be calculated. Figure 3.6 summarizes the calculation process.

The performance of the turbo-detected bit-based STBC-SP schemes was presented in Section 3.5. First, we considered the performance of the RSC-coded turbo-detected STBC-SP scheme in Section 3.5.1. The relation between the achievable BER and the mutual information at the input as well as at the output of the outer RSC decoder was discussed in Section 3.5.1.1. The predictions of our EXIT chart analysis outlined in Section 3.4.3 were verified by generating the actual decoding trajectories in Section 3.5.1.2. The effect of interleaver depth was also addressed in Section 3.5.1.2, since matching the predictions of the EXIT chart analysis is only guaranteed when employing large interleaver depths.

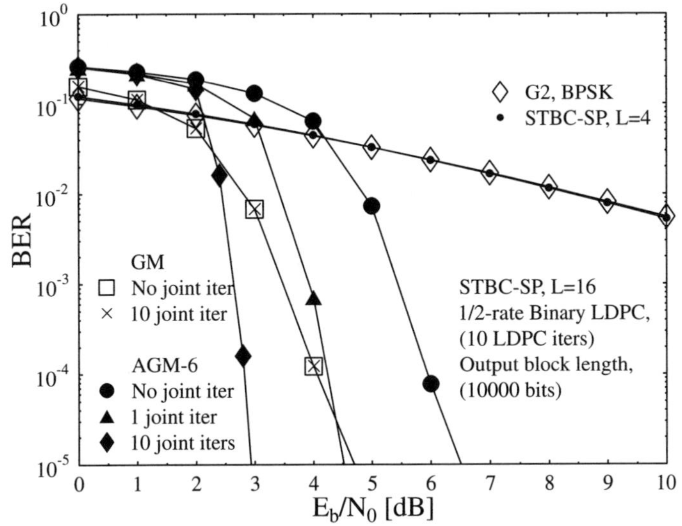

Figure 3.26: Performance comparison of AGM (AGM-6) and Gray mapping (GM)-based LDPC-coded STBC-SP schemes in conjunction with $L = 16$ against an identical-throughput 1 BPS uncoded STBC-SP scheme using $L = 4$ and against Alamouti's conventional G_2-BPSK scheme, when employing the system parameters of Table 3.3 and using an outer half-rate binary LDPC code having an average column weight of 2.5 and ten internal LDPC iterations.

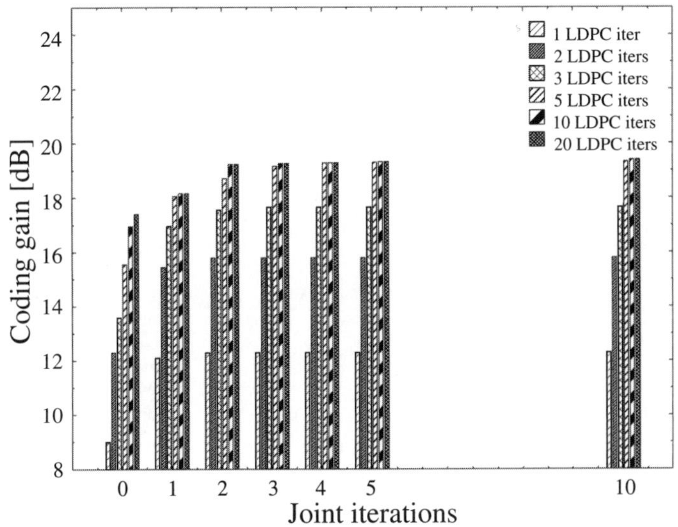

Figure 3.27: Coding gain of a bit-based binary LDPC-coded STBC-SP scheme employing AGM-6 in conjunction with different combinations of joint and LDPC iterations as compared with the identical-throughput 1 BPS uncoded STBC-SP scheme of [43] and Alamouti's conventional G_2-BPSK scheme [11] at a BER of 10^{-5}, when communicating over the channel outlined in Table 3.3 and using an outer half-rate binary LDPC code having an output block length of 1500 bits.

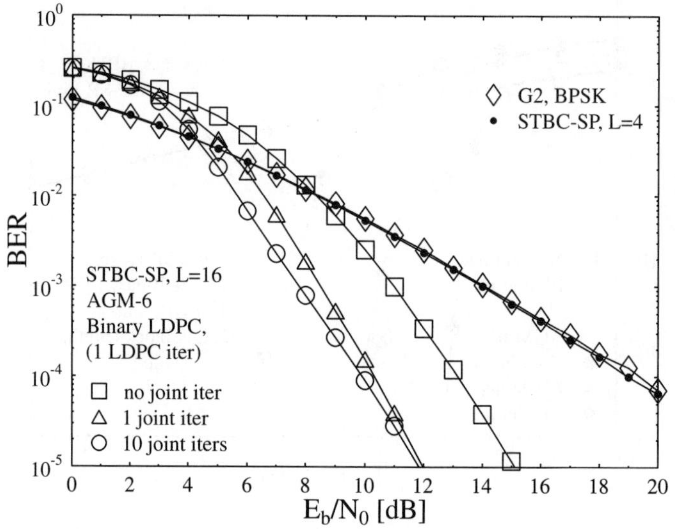

Figure 3.28: Performance of AGM (AGM-6) based LDPC-coded STBC-SP schemes in conjunction with $L = 16$, when employing the system parameters of Table 3.3 and using an outer half-rate binary LDPC code having an output block length of 1500 bits and one internal LDPC iteration.

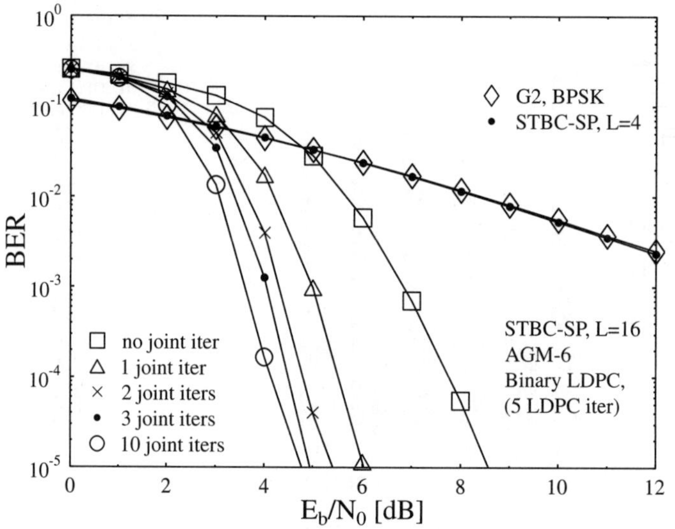

Figure 3.29: Performance of AGM (AGM-6) based LDPC-coded STBC-SP schemes in conjunction with $L = 16$, when employing the system parameters of Table 3.3 and using an outer half-rate binary LDPC code having an output block length of 1500 bits and five internal LDPC iterations.

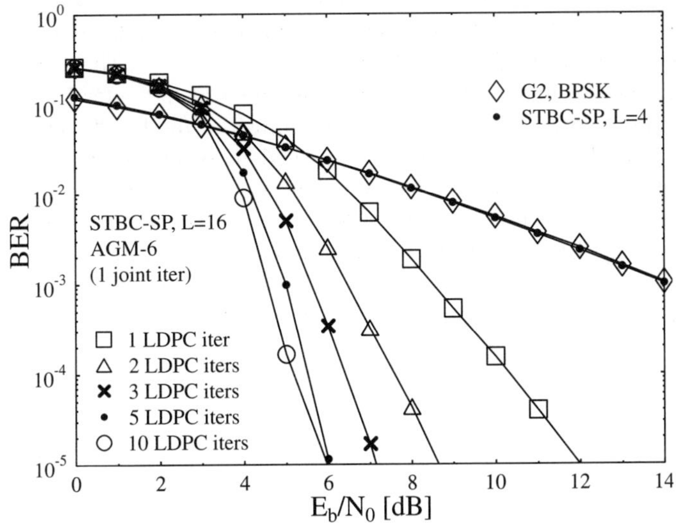

Figure 3.30: Performance of AGM (AGM-6) based LDPC-coded STBC-SP schemes in conjunction with $L = 16$, when employing the system parameters of Table 3.3 and using an outer half-rate binary LDPC code having an output block length of 1500 bits and one joint external iteration.

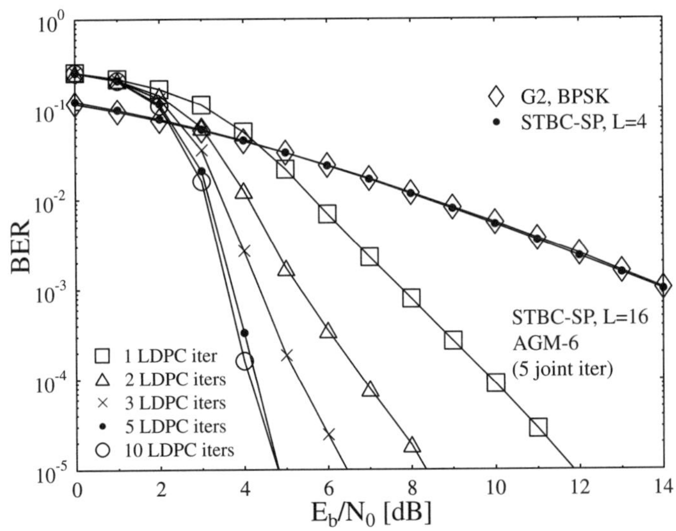

Figure 3.31: Performance of AGM (AGM-6) based LDPC-coded STBC-SP schemes in conjunction with $L = 16$, when employing the system parameters of Table 3.3 and using an outer half-rate binary LDPC code having an output block length of 1500 bits and five joint external LDPC iterations.

The BER performance of the proposed RSC-coded STBC-SP scheme was compared in Section 3.5.1.3 with that of an uncoded STBC-SP scheme [43] and with that of an RSC-coded conventionally modulated STBC scheme. Second, we considered the performance of the LDPC-coded turbo-detected STBC-SP scheme in Section 3.5.2. The relation between the achievable BER and the mutual information at the input of the outer LDPC decoder was discussed in Section 3.5.2.1 and Figure 3.22. The effect of the LDPC output block length K_{ldpc} on the achievable performance was investigated in Section 3.5.2.2, while the effect of internal LDPC iterations and joint external iterations was studied in Section 3.5.2.3.

In this chapter, we assumed that the CSI is perfectly known at the receiver. This, however, requires sophisticated channel estimation techniques, which imposes excess cost and complexity. In the next chapter, we consider the design of various SP-modulated differential STBC schemes that require no channel estimation.

Chapter 4

Turbo Detection of Channel-coded DSTBC-SP Schemes*

4.1 Introduction

In Chapter 2, the concept of a novel STBC design using SP modulation (STBC-SP) was introduced and it was demonstrated that STBC-SP schemes outperform STBC schemes that employ conventional modulation schemes, such as PSK and QAM. In Chapter 3, turbo detection of channel-coded STBC-SP schemes was considered, where several novel bit-based turbo-detected STBC-SP schemes were proposed. All of the schemes considered in Chapters 2 and 3 assumed perfect channel knowledge at the receiver. In practice the CSI of the link spanning from each transmit antenna to each receive antenna has to be estimated at the receiver using, for example, training or pilot symbols. However, channel estimation increases the cost and complexity of the receiver. Furthermore, when the channel varies dramatically from transmission burst to transmission burst, a high channel sounding overhead is required for every transmission burst.

Alternatively, schemes that require no CSI knowledge may be developed. A detection algorithm designed for Alamouti's scheme [11] was proposed in [52], where the channel encountered at time instant t was estimated using the pair of symbols detected at time instant $t-1$. The algorithm, nonetheless, has to estimate the channel during the very first time instance using training symbols. Tarokh and Jafarkhani [53, 62] proposed a differential encoding and decoding technique for Alamouti's scheme [11] and hence the transmitted signal can be demodulated both with and without CSI at the receiver. The resultant differential-decoding-aided non-coherent receiver performs within 3 dB of the coherent receiver. However, this scheme expands the modulated signal constellation and its applications are limited to systems having two transmit antennas when using a complex-valued modulated constellation and to systems with eight or fewer transmit antennas when using a real-valued phasor constellation. The complex constellation was also restricted to PSK schemes, and was extended to QAM constellations in [58, 264]. This extension, however, requires knowledge of the received power in order to appropriately normalize the received signal. In 2000, Hochwald and Sweldens [54] proposed a differential modulation scheme for

*Parts of this chapter are based on the collaborative research outlined in [189].

Near-Capacity Multi-Functional MIMO Systems L. Hanzo, O. R. Alamri, M. El-Hajjar and N. Wu

the sake of attaining transmit diversity based on unitary space-time codes [63]. The proposed scheme can be employed in conjunction with an arbitrary number of transmit antennas. Around the same time, a similar differential scheme was also proposed by Hughes [55], which is based on the employment of group codes.

In this chapter, we combine the DSTBC schemes of [62] and [264] with SP modulation. The chapter is organized as follows. In Section 4.2, we describe how the SP-aided DSTBC schemes, referred to here as the DSTBC-SP schemes, are constructed. In Section 4.3, we propose a novel bit-based iterative-detection-aided DSTBC-SP scheme. Finally, the chapter is concluded in Section 4.4.

4.2 Differential STBC using SP Modulation

This section describes how DSTBC schemes are constructed using SP modulation. More specifically, a DSTBC signal design employing SP modulation is considered in Section 4.2.1. Section 4.2.2 compares the performance of different DSTBC-SP schemes against equivalent conventional DSTBC schemes under various channel conditions.

4.2.1 DSTBC Signal Design using SP Modulation

Some of the material summarized in this section was discussed in more detail in Section 2.4. Nonetheless, a brief overview is provided here for the sake of offering a self-contained treatment of the subject. Orthogonal transmit diversity designs can be described recursively [207] as follows. Let $\mathbf{G}_1(g_1) = g_1\mathbf{I}_1$, and

$$\mathbf{G}_{2^k}(g_1, \ldots, g_{k+1}) = \begin{bmatrix} \mathbf{G}_{2^{k-1}}(g_1, \ldots, g_k) & g_{k+1}\mathbf{I}_{2^{k-1}} \\ -g_{k+1}^*\mathbf{I}_{2^{k-1}} & \mathbf{G}_{2^{k-1}}^{\mathcal{H}}(g_1, \ldots, g_k) \end{bmatrix},$$

for $k = 1, 2, 3, \ldots,$ where g_{k+1}^* is the complex conjugate of g_{k+1}, $\mathbf{G}_{2^{k-1}}^{\mathcal{H}}(g_1, \ldots, g_k)$ is the Hermitian of $\mathbf{G}_{2^{k-1}}(g_1, \ldots, g_k)$ and $\mathbf{I}_{2^{k-1}}$ is a $(2^{k-1} \times 2^{k-1})$ identity matrix. Then, $\mathbf{G}_{2^k}(g_1, g_2, \ldots, g_{k+1})$ constitutes an orthogonal design of size $(2^k \times 2^k)$, which maps the complex variables representing $(g_1, g_2, \ldots, g_{k+1})$ to $N_t = 2^k$ transmit antennas. In other words, $g_1, g_2, \ldots, g_{k+1}$ represent $k + 1$ complex modulated symbols to be transmitted from 2^k transmit antennas in $T = 2^k$ time slots. It was shown in [43] that the diversity product quantifying coding advantage of an orthogonal transmit diversity scheme is determined by the MED of the vectors $(g_1, g_2, \ldots, g_{k+1})$. Therefore, in order to maximize the achievable coding advantage, it was proposed in [43] to use SP schemes that have the best known MED in the $2(k + 1)$-dimensional real-valued Euclidean space $\mathbb{R}^{2(k+1)}$ (see [221]).

In this chapter, differential space-time systems [58, 62] employing $N_t = 2$ transmit antennas are considered, which are characterized by the generator matrix of [11]

$$\mathbf{G}_2(g_1, g_2) = \begin{bmatrix} g_1 & g_2 \\ -g_2^* & g_1^* \end{bmatrix}, \tag{4.1}$$

and the rows and columns of Equation (4.1) represent the temporal and spatial dimensions, corresponding to two consecutive time slots and two transmit antennas, respectively. The transmission is initialized by sending arbitrary symbols $g_1(1)$ and $g_2(1)$ using Equation (4.1) during the first and second time slots from the first and second transmit antennas. At time $2t + 1$, $t = 1, 2, \ldots,$ a block of B bits arrives at the encoder, where each $B/2$ bits are independently modulated using a $2^{B/2}$-ary modulation constellation producing $x_1(2t + 1)$

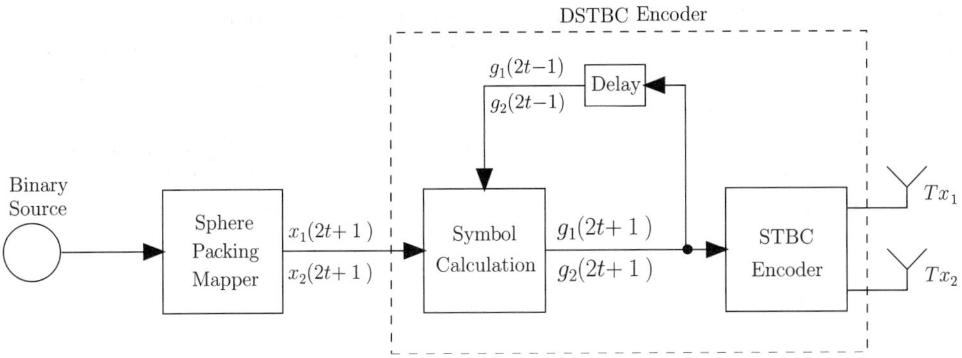

Figure 4.1: A DSTBC-SP encoder.

and $x_2(2t+1)$. Now, for $t \geq 1$ the transmission symbols $g_1(2t+1)$ and $g_2(2t+1)$ are calculated as follows [58]:

$$g_1(2t+1) = \frac{1}{n_f} \cdot [x_1(2t+1)g_1(2t-1) - x_2(2t+1)g_2^*(2t-1)],$$
$$g_2(2t+1) = \frac{1}{n_f} \cdot [x_1(2t+1)g_2(2t-1) + x_2(2t+1)g_1^*(2t-1)], \quad (4.2)$$

where

$$n_f = \sqrt{|g_1(2t-1)|^2 + |g_2(2t-1)|^2}.$$

More specifically, $g_1(2t+1)$ and $g_2(2t+1)$ are transmitted from the first and second transmit antennas, respectively, at time $2t+1$. In contrast, $-g_2^*(2t+1)$ and $g_1^*(2t+1)$ are transmitted from the first and second transmit antennas, respectively, at time $2t+2$. The DSTBC-SP encoder is illustrated in Figure 4.1.

After differentially decoding the received signals during four time slots and assuming that a single receive antenna is employed, the following estimates on $x_1(2t+1)$ and $x_2(2t+1)$ are produced [58]

$$\tilde{x}_1(2t+1) = (|h_1|^2 + |h_2|^2) \cdot n_f \cdot x_1(2t+1) + \acute{n}_1,$$
$$\tilde{x}_2(2t+1) = (|h_1|^2 + |h_2|^2) \cdot n_f \cdot x_2(2t+1) + \acute{n}_2, \quad (4.3)$$

where h_1 and h_2 represent the complex-valued channel coefficients corresponding to the first and second transmit antenna, respectively, and $\acute{n}_1$ as well as $\acute{n}_2$ are zero-mean complex Gaussian random variables with variance $\sigma_{\acute{n}}^2 = 2 \cdot (|h_1|^2 + |h_2|^2) \cdot n_f \cdot \sigma_n^2$, while σ_n^2 is the original noise variance. Observe from Equation (4.3) that when the received signals are differentially decoded, the resultant signals will be scaled versions of $x_1(2t+1)$ and $x_2(2t+1)$, which are corrupted by complex AWGN similar to the $\mathbf{G}_2$ STBC of [11, 12]. This observation implies that the diversity product of differential space-time systems [58, 62] is determined by the MED of all legitimate vectors (x_1, x_2), where the time index is removed for notational simplicity. According to [58, 62] for example, x_1 and x_2 represent independent conventional BPSK modulated symbols and no effort is made to jointly design a symbol constellation for the various combinations of x_1 and x_2. For the sake of generalizing our treatment, let us assume that there are L legitimate vectors $(x_{l,1}, x_{l,2})$, $l = 0, \ldots, L-1$, where L represents the number of sphere-packed modulated symbols. The encoder, then, has to choose the modulated symbol associated with each block of B bits from these L legitimate symbols, which determines the signals to be transmitted over the two antennas in two consecutive time slots using Equation (4.2), where the throughput of the system is given by $(\log_2 L)/2$ bits per channel use. In contrast to the independent design of $x_{l,1}$ and $x_{l,2}$ (see [58, 62]), our aim is to design $x_{l,1}$ and $x_{l,2}$ jointly, such that they have the best MED from all other $(L-1)$ legitimate symbols, since this minimizes the system's error probability. Let $(a_{l,1}, a_{l,2}, a_{l,3}, a_{l,4})$, $l = 0, 1, \ldots, L-1$, be phasor points selected from the four-dimensional real-valued Euclidean space $\mathbb{R}^4$, where each of the four elements $a_{l,1}, a_{l,2}, a_{l,3}, a_{l,4}$ gives one coordinate of the complex-valued phasor points. Hence, $x_{l,1}$ and $x_{l,2}$ may be written as

$$\begin{aligned}\{x_{l,1}, x_{l,2}\} &= T_{\text{sp}}(a_{l,1}, a_{l,2}, a_{l,3}, a_{l,4}) \\ &= \{a_{l,1} + ja_{l,2}, a_{l,3} + ja_{l,4}\}.\end{aligned} \tag{4.4}$$

In the four-dimensional real-valued Euclidean space $\mathbb{R}^4$, the lattice D_4 is defined as a SP having the best MED from all other $(L-1)$ legitimate constellation points in $\mathbb{R}^4$ (see [221]), as discussed in Section 2.4.3.4. More specifically, D_4 may be defined as a lattice that consists of all legitimate sphere-packed constellation points having integer coordinates $[a_1\ a_2\ a_3\ a_4]$ uniquely and unambiguously describing the legitimate combinations of the modulated symbols $x_{l,1}$ and $x_{l,2}$, but subjected to the SP constraint of

$$a_1 + a_2 + a_3 + a_4 = k,$$

where k is an even integer. Assuming that

$$S = \{\mathbf{s}^l = [a_{l,1}, a_{l,2}, a_{l,3}, a_{l,4}] \in \mathbb{R}^4 : 0 \le l \le L-1\}$$

constitutes a set of L legitimate constellation points from the lattice D_4 having a total energy of

$$E_{\text{total}} \triangleq \sum_{l=0}^{L-1} (|a_{l,1}|^2 + |a_{l,2}|^2 + |a_{l,3}|^2 + |a_{l,4}|^2),$$

and upon introducing the notation

$$\mathbf{C}_l = \sqrt{\frac{2L}{E_{\text{total}}}}(x_{l,1}, x_{l,2}), \quad l = 0, \ldots, L-1, \tag{4.5}$$

we have a set of complex constellation symbols, $\{\mathbf{C}_l : 0 \le l \le L-1\}$, whose diversity product is determined by the MED of the set of L legitimate constellation points in S.

4.2.2 Performance of DSTBC-SP Schemes

In this section, the two-transmit-antenna-aided scheme of Section 4.2.1 is considered. Simulation results are provided for systems having different numbers of BPS in conjunction with appropriate conventional and SP modulation schemes, as outlined in Table 4.1. Observe that two consecutive time slots are required for transmitting a single SP symbol when using the two-transmit-antenna-based scheme of Section 4.2.1. In contrast, two conventionally modulated symbols are transmitted during the same time period. Therefore, the throughput of the SP modulation scheme has to be twice as high as that of the conventional modulation

Table 4.1: Conventional and SP modulation employed for different BPS rates.

Rate (BPS)	Conventional modulation	SP modulation
1	BPSK	$L = 4$
2	QPSK	$L = 16$
3	8-PSK	$L = 64$

scheme in order to compensate for the potential rate loss and to produce systems having an identical overall BPS throughput. This explains the specific choices of L in Table 4.1. Our results are also presented in terms of BER and SP-SER performance curves for various systems employing $N_r = 1,\ 2,\ 3$, and 4 receive antennas for communicating over two types of Rayleigh fading channels, namely block and SPSI Rayleigh fading channels, where SP-SER and SPSI were defined in Section 2.5.

4.2.2.1 Block Rayleigh Fading Channels

The channel is assumed to be constant over the transmission period of one frame. SP-aided DSTBC schemes promise to provide improved SP-SER, when compared with conventionally modulated DSTBC schemes. This promise is based on the fact that SP modulation optimizes the MED of the set of complex constellation symbols constructed using Equation (4.5). On the other hand, the BER performance of SP-aided DSTBC schemes is not necessarily guaranteed to be better than that of conventionally modulated DSTBC schemes, since SP modulation does not optimize the MED of the individual constituent symbols $x_{l,1}$, $x_{l,2}$, which is the objective of conventional modulation schemes. However, it is demonstrated below that increasing the number of receive antennas will lead to further BER improvements for the SP-aided DSTBC schemes in comparison with conventionally modulated DSTBC schemes.

Figure 4.2 shows the SP-SER performance curves of different DSTBC schemes in conjunction with different conventional as well as SP modulation schemes at various BPS throughput values as outlined in Table 4.1. All systems employ two transmit and one receive antennas for communicating over the block Rayleigh fading channel under consideration. It is evident from Figure 4.2 that for a particular BPS throughput, the two curves corresponding to the conventional modulation scheme and to the SP modulation scheme have the same asymptotic slope (i.e. diversity order). This observation is similar to those stated in [11, 58, 62], namely that $\mathbf{G}_2$-based space-time coded systems are capable of achieving full diversity. Accordingly, it is not expected that the asymptotic slope of the performance curves would improve by merely employing new modulation schemes without introducing another level of concatenating coding, invoking outer channel codes. However, Figure 4.2 shows that the SP-SER performance of DSTBC schemes may be improved by employing SP modulation. For example, SP modulation having $L = 64$ achieves a coding gain of about 1.2 dB over 8-PSK modulation at an SP-SER of 10^{-4}. The resultant BER performance curves are shown in Figure 4.3. The BER performance of SP modulation and conventional modulation are identical for systems having rates of 1 and 2 BPS. However, the figure shows that the advantage of conventional modulation over SP modulation diminishes upon increasing the SNR.

Figures 4.4–4.9 illustrate the beneficial effect of increasing the number of receive antennas from $N_r = 2$ to 4 antennas, respectively. Observe in Figures 4.5, 4.7 and 4.9 that

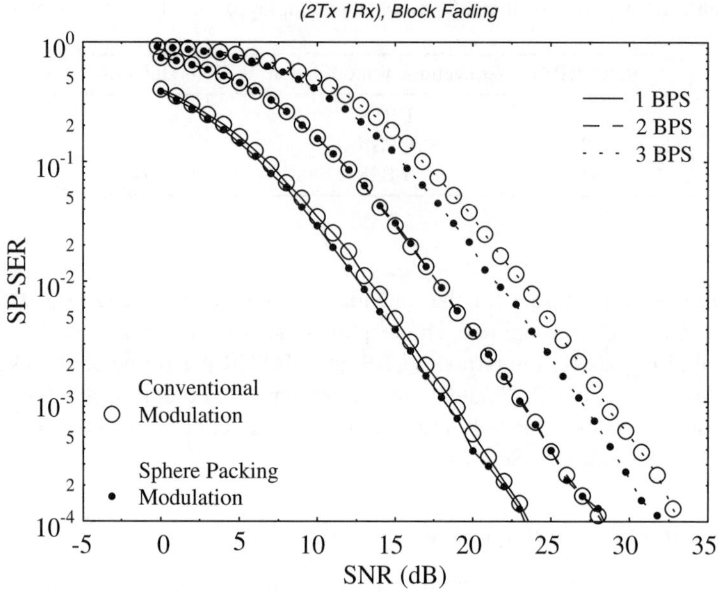

Figure 4.2: SP-SER of different DSTBC schemes in combination with conventional modulation and SP modulation for different BPS rates as outlined in Table 4.1 and employing $N_t = 2$ transmit antennas and $N_r = 1$ receive antenna, when communicating over a block Rayleigh fading channel.

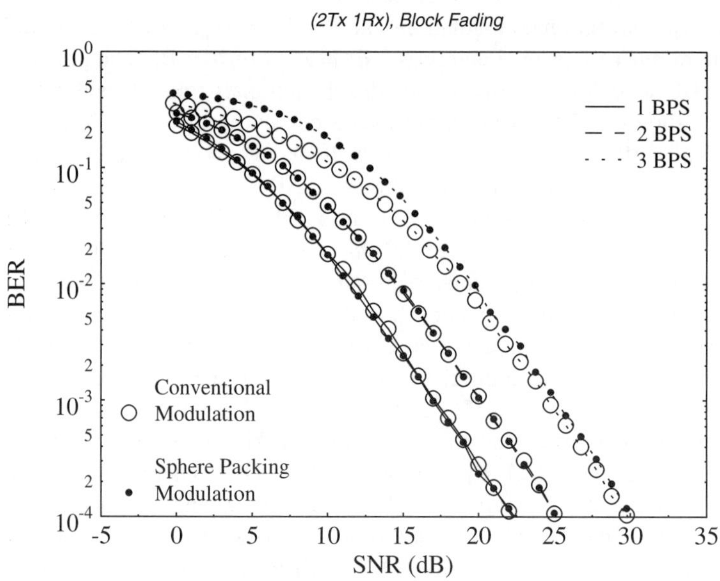

Figure 4.3: BER of different DSTBC schemes in combination with conventional modulation and SP modulation for different BPS rates as outlined in Table 4.1 and employing $N_t = 2$ transmit antennas and $N_r = 1$ receive antenna, when communicating over a block Rayleigh fading channel.

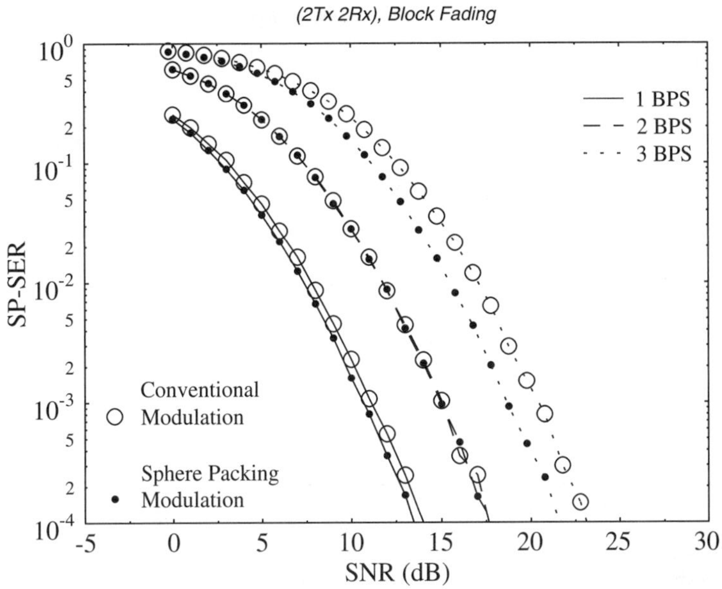

Figure 4.4: SP-SER of different DSTBC schemes in combination with conventional modulation and SP modulation for different BPS rates as outlined in Table 4.1 and employing $N_t = 2$ transmit antennas and $N_r = 2$ receive antennas, when communicating over a block Rayleigh fading channel.

the BER performance of SP modulation improves in comparison to that of conventional modulation when increasing the number of receive antennas, especially for schemes having throughput of 1 and 3 BPS. Observe, however, in Figures 4.2 to 4.9 that the BER and SP-SER performance curves of 2 BPS throughput QPSK modulation and those of the identical-throughput SP modulation having $L = 16$ are identical. This phenomenon is due to the fact that QPSK modulation is a special case of the SP modulation constellation constructed using Equation (4.5); this was discussed in detail in Section 2.5.

The attainable coding gains of SP modulation over conventional modulation are summarized in Table 4.2 for the schemes characterized in Figures 4.2, 4.4, 4.6 and 4.8 at a SP-SER of 10^{-4}, when communicating over the block Rayleigh fading channel under consideration.

4.2.2.2 SPSI Rayleigh Fading Channels

In this section, the channel is assumed to be constant over the transmission period of one SP symbol (i.e. two consecutive time slots). This type of channel was referred to in Section 2.5 as a SPSI channel. The channel is also assumed to be correlated and has a normalized Doppler frequency of $f_D = 0.01$. Figures 4.10–4.17 portray the SP-SER and BER performance curves of different DSTBC schemes in conjunction with different conventional as well as SP modulation schemes at various BPS throughput rates as outlined in Table 4.1. The key points discussed in Section 4.2.2.1 also apply to the results seen in Figures 4.10–4.17. The coding gains of SP modulation over conventional modulation are summarized in Table 4.3 for the schemes characterized in Figures 4.10, 4.12, 4.14 and 4.16, at a SP-SER of 10^{-4}, when communicating over the SPSI correlated Rayleigh fading channel under consideration.

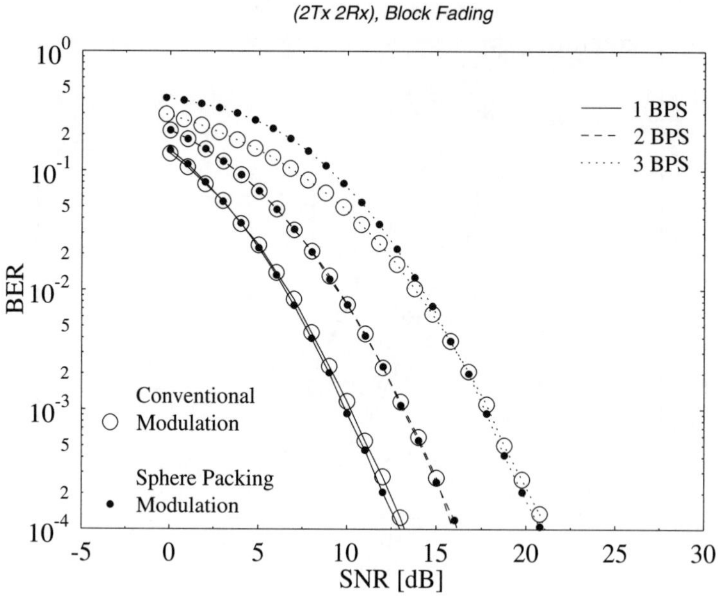

Figure 4.5: BER of different DSTBC schemes in combination with conventional modulation and SP modulation for different BPS rates as outlined in Table 4.1 and employing $N_t = 2$ transmit antennas and $N_r = 2$ receive antennas, when communicating over a block Rayleigh fading channel.

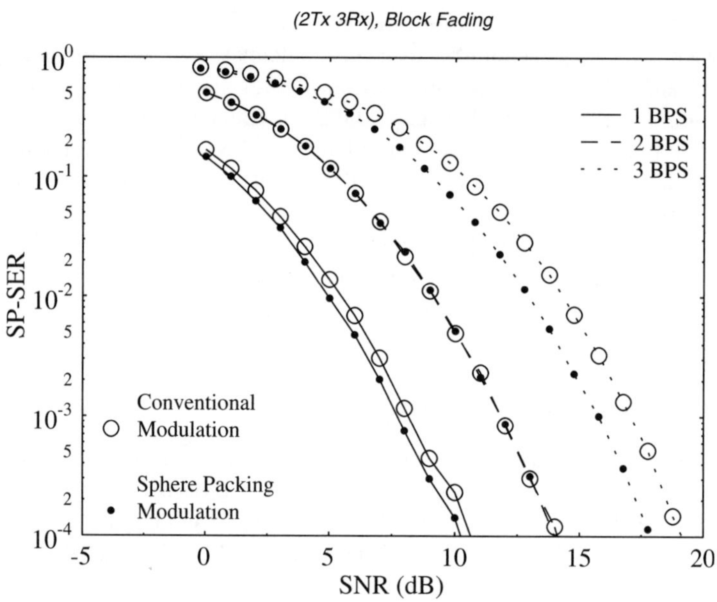

Figure 4.6: SP-SER of different DSTBC schemes in combination with conventional modulation and SP modulation for different BPS rates as outlined in Table 4.1 and employing $N_t = 2$ transmit antennas and $N_r = 3$ receive antennas, when communicating over a block Rayleigh fading channel.

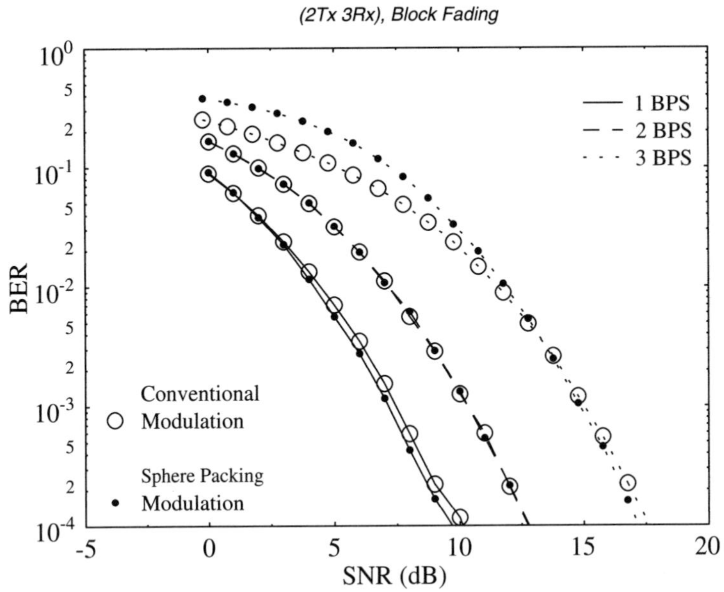

Figure 4.7: BER of different DSTBC schemes in combination with conventional modulation and SP modulation for different BPS rates as outlined in Table 4.1 and employing $N_t = 2$ transmit antennas and $N_r = 3$ receive antennas, when communicating over a block Rayleigh fading channel.

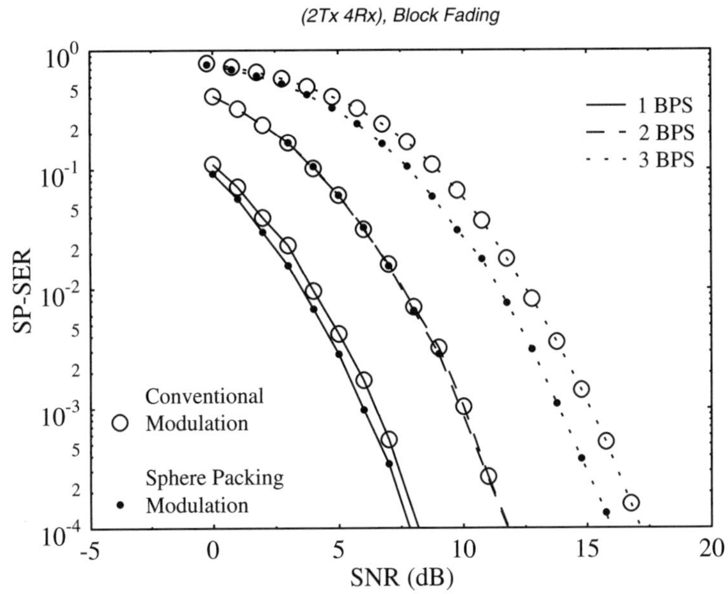

Figure 4.8: SP-SER of different DSTBC schemes in combination with conventional modulation and SP modulation for different BPS rates as outlined in Table 4.1 and employing $N_t = 2$ transmit antennas and $N_r = 4$ receive antennas, when communicating over a block Rayleigh fading channel.

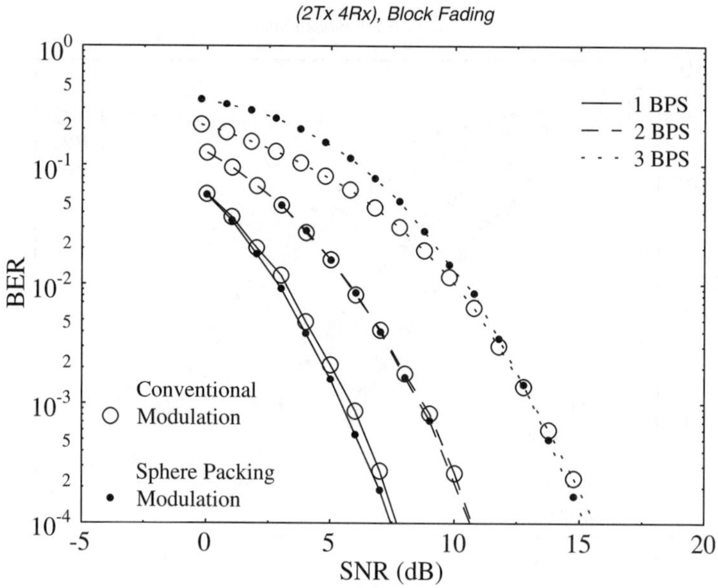

Figure 4.9: BER of different DSTBC schemes in combination with conventional modulation and SP modulation for different BPS rates as outlined in Table 4.1 and employing $N_t = 2$ transmit antennas and $N_r = 4$ receive antennas, when communicating over a block Rayleigh fading channel.

Table 4.2: Coding gains of SP modulation over conventional modulation at a SP-SER of 10^{-4} for the schemes of Figures 4.2, 4.4, 4.6 and 4.8, when communicating over a block Rayleigh fading channel.

	Coding gains (dB)		
	1 BPS	2 BPS	3 BPS
$N_r = 1$	0.3	0.0	1.2
$N_r = 2$	0.5	0.0	1.4
$N_r = 3$	0.4	0.0	1.3
$N_r = 4$	0.4	0.0	1.3

Table 4.3: Coding gains of SP modulation over conventional modulation at a SP-SER of 10^{-4} for the schemes of Figures 4.10, 4.12, 4.14 and 4.16, when communicating over a SPSI correlated Rayleigh fading channel having a normalized Doppler frequency of $f_D = 0.01$.

	Coding gains (dB)		
	1 BPS	2 BPS	3 BPS
$N_r = 1$	0.5	0.0	—
$N_r = 2$	0.5	0.0	2.1
$N_r = 3$	0.5	0.0	1.6
$N_r = 4$	0.4	0.0	1.4

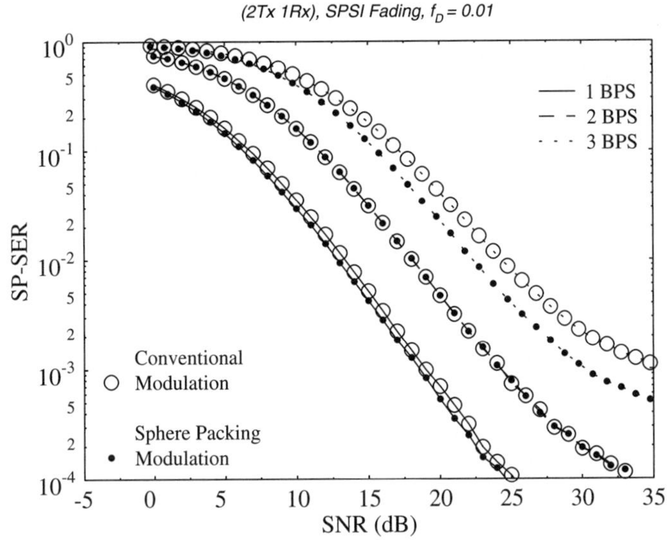

Figure 4.10: SP-SER of different DSTBC schemes in combination with conventional modulation and SP modulation for different BPS rates as outlined in Table 4.1 and employing $N_t = 2$ transmit antennas and $N_r = 1$ receive antenna, when communicating over a SPSI correlated Rayleigh fading channel having a normalized Doppler frequency of $f_D = 0.01$.

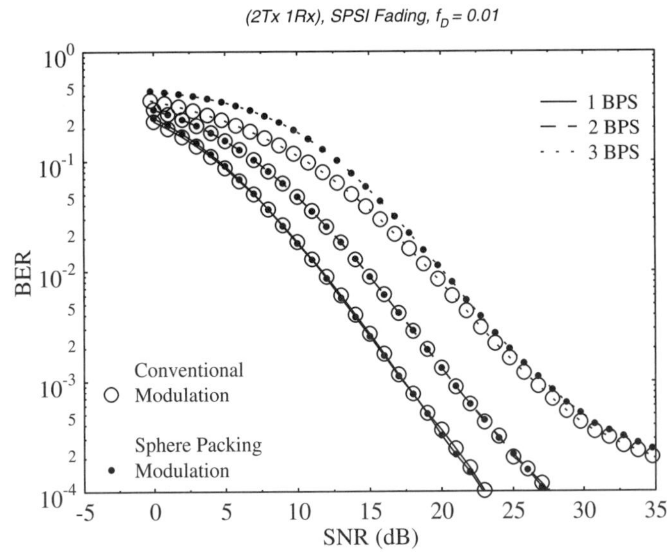

Figure 4.11: BER of different DSTBC schemes in combination with conventional modulation and SP modulation for different BPS rates as outlined in Table 4.1 and employing $N_t = 2$ transmit antennas and $N_r = 1$ receive antenna, when communicating over a SPSI correlated Rayleigh fading channel having a normalized Doppler frequency of $f_D = 0.01$.

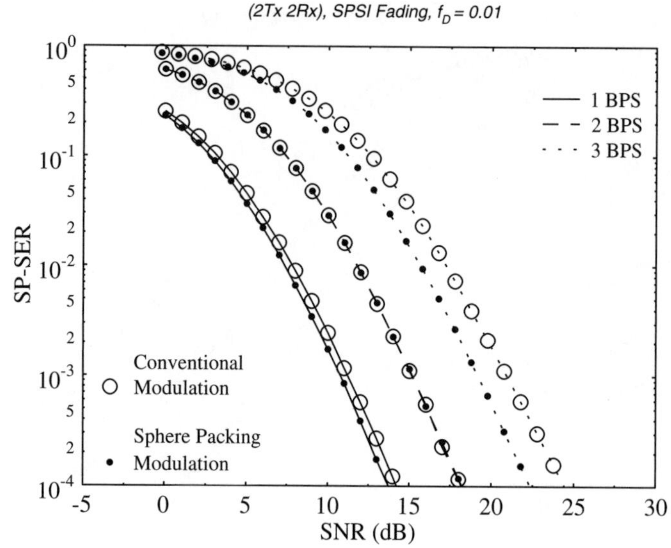

Figure 4.12: SP-SER of different DSTBC schemes in combination with conventional modulation and SP modulation for different BPS rates as outlined in Table 4.1 and employing $N_t = 2$ transmit antennas and $N_r = 2$ receive antennas, when communicating over a SPSI correlated Rayleigh fading channel having a normalized Doppler frequency of $f_D = 0.01$.

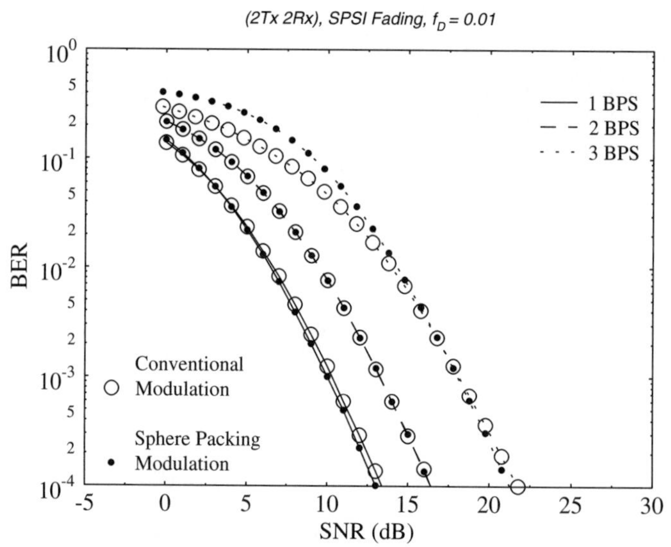

Figure 4.13: BER of different DSTBC schemes in combination with conventional modulation and SP modulation for different BPS rates as outlined in Table 4.1 and employing $N_t = 2$ transmit antennas and $N_r = 2$ receive antennas, when communicating over a SPSI correlated Rayleigh fading channel having a normalized Doppler frequency of $f_D = 0.01$.

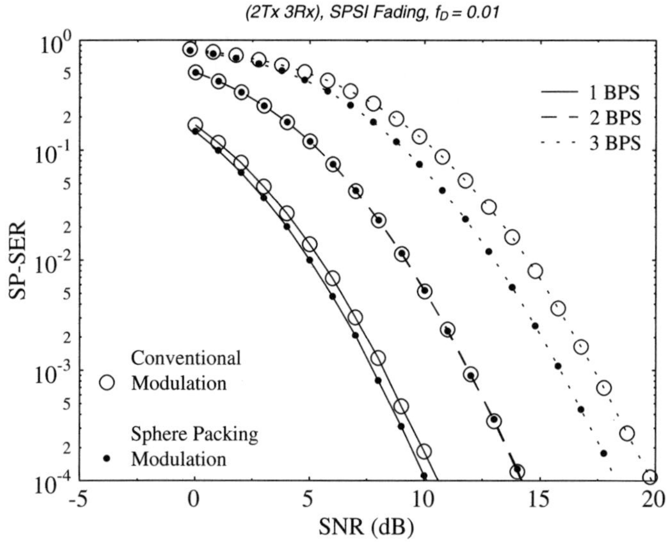

Figure 4.14: SP-SER of different DSTBC schemes in combination with conventional modulation and SP modulation for different BPS rates as outlined in Table 4.1 and employing $N_t = 2$ transmit antennas and $N_r = 3$ receive antennas, when communicating over a SPSI correlated Rayleigh fading channel having a normalized Doppler frequency of $f_D = 0.01$.

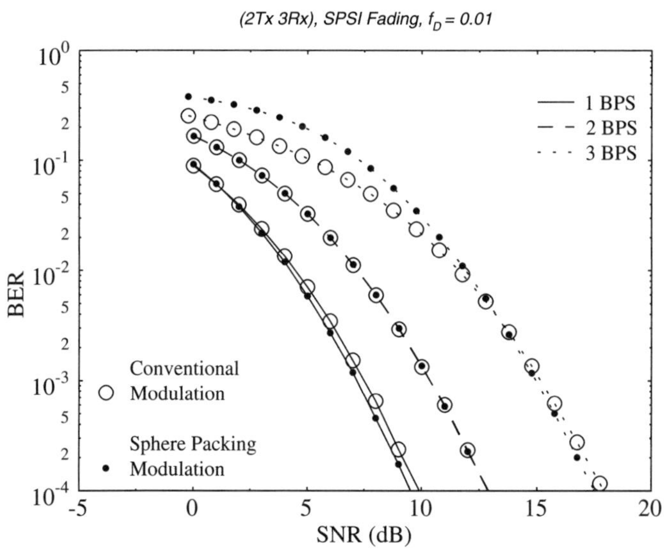

Figure 4.15: BER of different DSTBC schemes in combination with conventional modulation and SP modulation for different BPS rates as outlined in Table 4.1 and employing $N_t = 2$ transmit antennas and $N_r = 3$ receive antennas, when communicating over a SPSI correlated Rayleigh fading channel having a normalized Doppler frequency of $f_D = 0.01$.

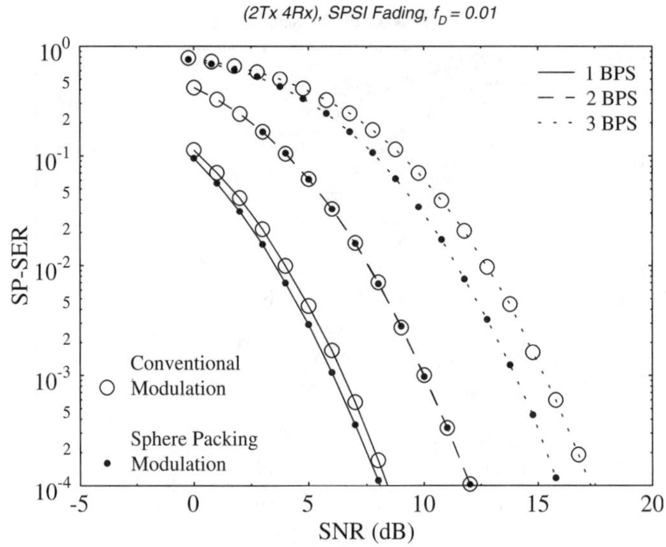

Figure 4.16: SP-SER of different DSTBC schemes in combination with conventional modulation and SP modulation for different BPS rates as outlined in Table 4.1 and employing $N_t = 2$ transmit antennas and $N_r = 4$ receive antennas, when communicating over a SPSI correlated Rayleigh fading channel having a normalized Doppler frequency of $f_D = 0.01$.

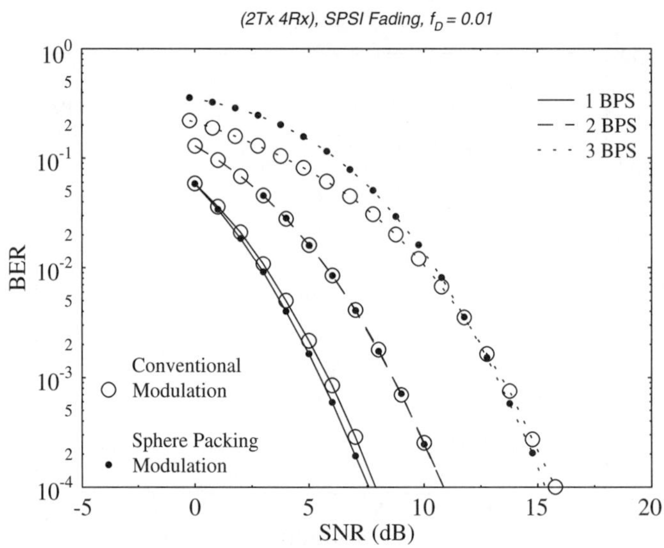

Figure 4.17: BER of different DSTBC schemes in combination with conventional modulation and SP modulation for different BPS rates as outlined in Table 4.1 and employing $N_t = 2$ transmit antennas and $N_r = 4$ receive antennas, when communicating over a SPSI correlated Rayleigh fading channel having a normalized Doppler frequency of $f_D = 0.01$.

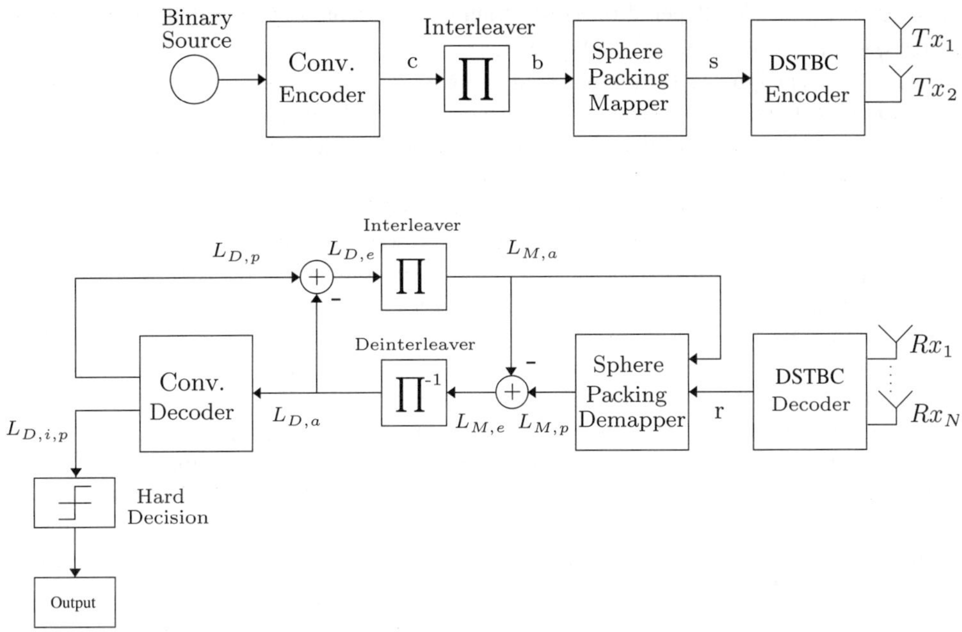

Figure 4.18: RSC-coded turbo-detected DSTBC-SP system.

4.3 Bit-based RSC-coded Turbo-detected DSTBC-SP Scheme

In this section, a novel bit-based RSC-coded turbo-detected DSTBC-SP scheme is discussed in detail. Our system overview is outlined in Section 4.3.1. Section 4.3.2 provides the EXIT chart analysis of the turbo-detected RSC-coded DSTBC-SP scheme, while our simulation results and discussions are offered in Section 4.3.3.

4.3.1 System Overview

The schematic of the entire system is shown in Figure 4.18, where the transmitted source bits are convolutionally encoded and then interleaved by a random bit interleaver. A rate $R = \frac{1}{2}$ RSC code was employed. After channel interleaving, the SP mapper first maps B channel-coded bits $\mathbf{b} = (b_0, \ldots, b_{B-1})$ to a legitimate constellation point $\mathbf{s}^l \in S$ from the lattice D_4, where we have $B = \log_2 L$. The mapper then maps the constellation point $\mathbf{s}^l$ to complex symbols $x_{l,1}$ and $x_{l,2}$ using Equations (4.4) and (4.5). Subsequently, the DSTBC encoder calculates the symbols to be transmitted according to Equation (4.2) over $T = 2$ consecutive time slots using two transmit antennas, as shown in Equation (4.1).

In this chapter, we consider a correlated narrowband Rayleigh fading channel, associated with a normalized Doppler frequency of $f_D = f_d T_{sym} = 0.01$, where f_d is the Doppler frequency and T_{sym} is the symbol duration. The complex fading envelope is assumed to be constant across the transmission period of two SP symbols spanning $T = 4$ time slots. The complex AWGN of $n = n_I + jn_Q$ is also added to the received signal, where n_I and n_Q are two independent zero-mean Gaussian random variables having a variance of

$\sigma_n^2 = \sigma_{n_I}^2 = \sigma_{n_Q}^2 = N_0/2$ per dimension, where $N_0/2$ represents the double-sided noise power spectral density expressed in W Hz^{-1}.

As shown in Figure 4.18, the received complex-valued symbols are first differentially decoded by the DSTBC decoder. Then, the decoded symbols are passed to the SP demapper, where they are demapped to their LLR representation for each of the B coded bits per SP symbol. The *a priori* LLR values of the demodulator are subtracted from the *a posteriori* LLR values for the sake of generating the extrinsic LLR values $L_{M,e}$, and then the LLRs $L_{M,e}$ are deinterleaved by a soft-bit deinterleaver, as seen in Figure 4.18. Next, the soft bits $L_{D,a}$ are passed to the convolutional decoder in order to compute the *a posteriori* LLR values $L_{D,p}$ provided by the Log-MAP algorithm [165] for all the channel-coded bits. During the last iteration, only the LLR values $L_{D,i,p}$ of the original uncoded systematic information bits are required, which are passed to a hard decision decoder in order to determine the estimated transmitted source bits. The extrinsic information $L_{D,e}$ is generated by subtracting the *a priori* information from the *a posteriori* information according to $L_{D,p} - L_{D,a}$, which is then fed back to the DSTBC-SP demapper as the *a priori* information $L_{M,a}$ after appropriately reordering them using the interleaver of Figure 4.18. The SP demapper exploits the *a priori* information for the sake of providing improved *a posteriori* LLR values, which are then passed to the channel decoder and, in turn, back to the SP demodulator for further iterations. More detailed discussions on the iterative demapping process and how the SP demapper is modified for exploiting the *a priori* knowledge provided by the channel decoder were provided in Section 3.3.

4.3.2 EXIT Chart Analysis

The bit-based EXIT chart theory was introduced in Section 3.4. EXIT charts will be employed here in order to predict the convergence behavior of the iterative decoder seen in Figure 4.18 by examining the evolution of the input/output mutual information exchange between the inner SP demapper and the outer RSC decoder in consecutive iterations. Without loss of generality, we consider a SP modulation scheme associated with $L = 16$ using two transmit and a single receiver antenna.

Figure 4.19 shows the EXIT characteristics of the SP symbol-to-bit demapper in conjunction with $L = 16$ and different mapping schemes between the interleaver's output and the SP mapper. Observe that Gray mapping does not provide any iteration gain upon increasing the mutual information at the input of the demapper, which was also reported in [158]. The reason for this observation is that the adjacent Gray-coded symbols differ from that considered in a single bit-position and hence no extrinsic information is gleamed from the remaining identical bits. This situation is reversed when using different AGM schemes [158], resulting in different EXIT characteristics, as illustrated by the different slopes seen in Figure 4.19. The ten different AGM mapping schemes shown in Figure 4.19 are specifically selected from all of the possible mapping schemes for $L = 16$ in order to demonstrate the different EXIT characteristics associated with different bit-to-symbol mapping schemes. There are a total of 16! different mapping schemes.

Figure 4.20 shows the EXIT chart of a turbo-detection-aided, channel-coded DSTBC-SP scheme employing the AGM (AGM-10) of Figure 4.19 in conjunction with the system parameters outlined in Table 4.5 and the outer RSC code having a constraint length $\mathcal{K} = 3$ and the generator polynomials defined in Table 4.4. Similarly, Figure 4.21 shows the EXIT chart of a turbo-detection aided, channel-coded DSTBC-SP scheme employing the AGM (AGM-9) of Figure 4.19 in conjunction with the system parameters outlined in Table 4.5 and the outer RSC code having a constraint length $\mathcal{K} = 5$ and the generator polynomials

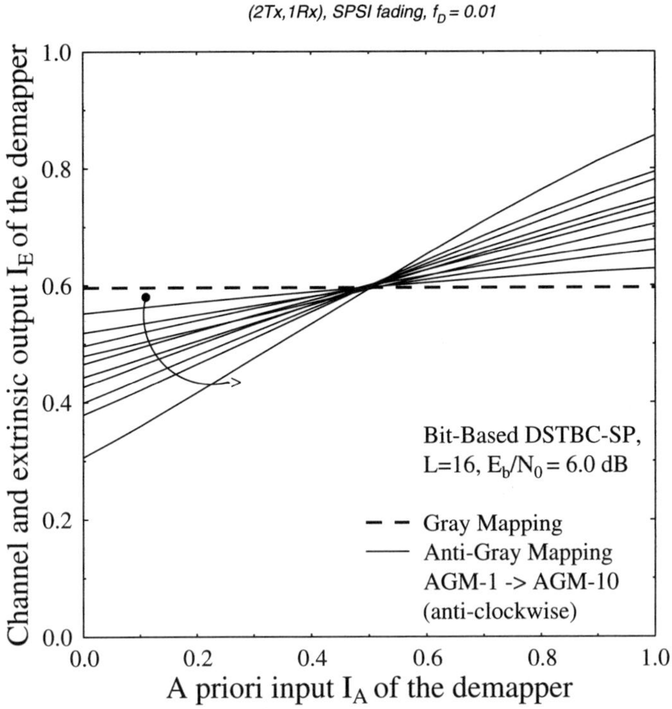

Figure 4.19: SP demapper EXIT characteristics for different bits to SP symbol mapping schemes at $E_b/N_0 = 6.0$ dB for $L = 16$.

defined in Table 4.4. Ideally, in order for the exchange of extrinsic information between the SP demapper and the outer RSC decoder to converge at a specific E_b/N_0 value, the EXIT curve of the SP demapper recorded at the E_b/N_0 value of interest and the EXIT characteristic curve of the outer RSC decoder should only intersect at the $(I_{A_D}, I_{E_D}) = (1.0, 1.0)$ point. If this condition is satisfied, then a so-called convergence tunnel [169] appears in the EXIT chart. Even if there is no open tunnel in the EXIT chart, but the two EXIT curves intersect at a point infinitesimally close to the $I_{E_D} = 1.0$ line rather than at the $(1.0, 1.0)$ point, then a sufficiently low BER may still be achievable. These types of tunnels were referred to in Section 3.4.3 as semi-convergent tunnels. The narrower the tunnel, the closer the system operates to the Shannon limit and hence a high number of iterations are required for reaching the intersection point. Observe in Figures 4.20 and 4.21 that semi-convergent tunnels exist at $E_b/N_0 = 6.0$ dB. This implies that according to the predictions of the EXIT charts seen in Figures 4.20 and 4.21, the iterative decoding process is expected to converge and hence a low BER may be attained at $E_b/N_0 = 6.0$ dB. The validity of this prediction is, however, dependent on how accurately the two EXIT chart assumptions outlined at the beginning of Section 3.4 are satisfied. These EXIT-chart-based convergence predictions are verified by the actual iterative decoding trajectory in Section 4.3.3.

Appendix C provides the complete list of EXIT charts for the RSC-coded turbo-detected DSTBC-SP scheme of Figure 4.18, when employing the mapping schemes of Figure 4.19 in combination with various outer RSC codes.

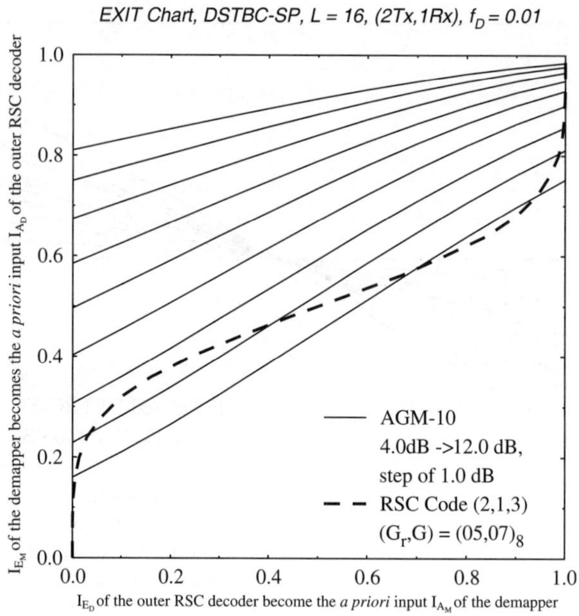

Figure 4.20: EXIT chart of a turbo-detected RSC channel-coded DSTBC-SP scheme employing AGM (AGM-10) of Figure 4.19 in combination with outer RSC code having constraint length $\mathcal{K} = 3$, when communicating over a SPSI Rayleigh fading channel having $f_D = 0.01$.

Table 4.4: Half-rate RSC codes parameters.

Constraint length	Generator polynomials in octals	
$\mathcal{K}$	G_r	G
3	05	07
5	35	23

Table 4.5: RSC-coded turbo-detected DSTBC-SP system parameters.

Modulation	SP with $L = 16$
Number of transmitters	2
Number of receivers	1
Channel type	SPSI correlated Rayleigh fading
Normalized Doppler frequency	0.01
RSC code rate	0.5
System throughput	1 BPS

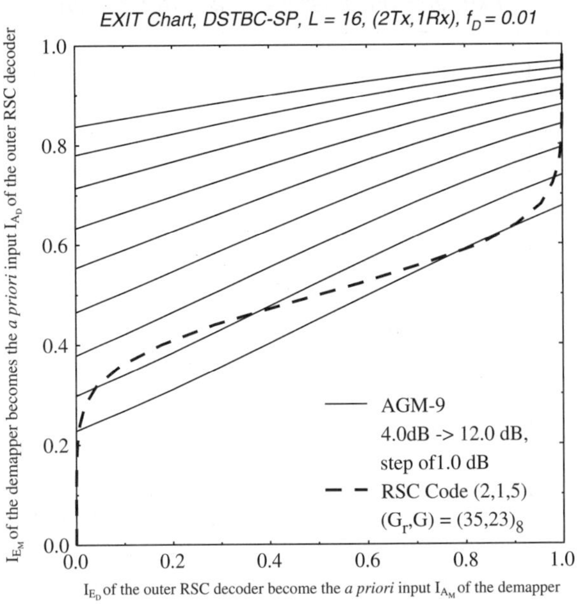

Figure 4.21: EXIT chart of a turbo-detected RSC channel-coded DSTBC-SP scheme employing AGM (AGM-9) of Figure 4.19 in combination with outer RSC code having constraint length $\mathcal{K} = 5$, when communicating over a SPSI Rayleigh fading channel having $f_D = 0.01$.

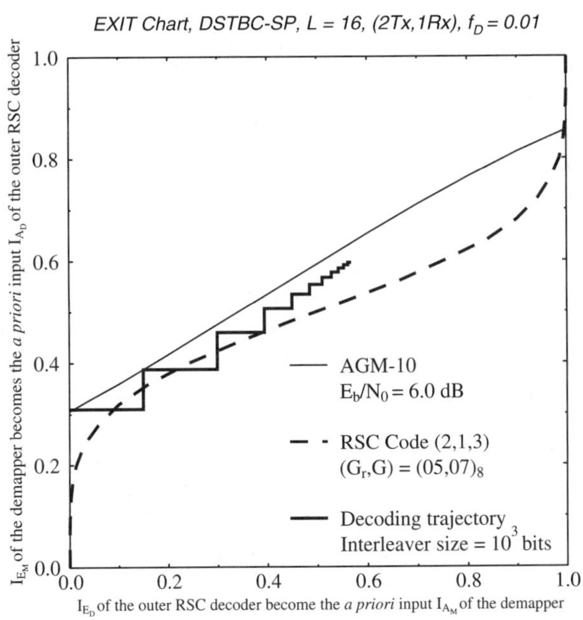

Figure 4.22: Decoding trajectory of a turbo-detected RSC channel-coded DSTBC-SP scheme employing AGM (AGM-10) in combination with the outer RSC code (2,1,3) and the system parameters outlined in Table 4.5 and operating at $E_b/N_0 = 6.0$ dB with an interleaver depth of $\mathcal{D} = 10^3$ bits.

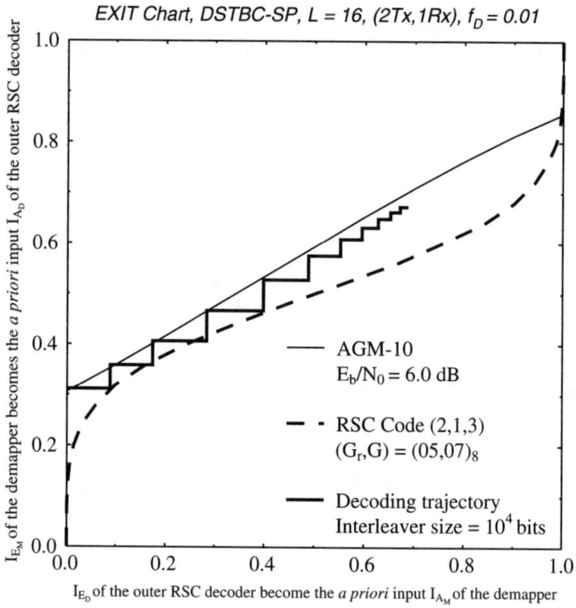

Figure 4.23: Decoding trajectory of a turbo-detected RSC channel-coded DSTBC-SP scheme employing AGM (AGM-10) in combination with the outer RSC code (2,1,3) and the system parameters outlined in Table 4.5 and operating at $E_b/N_0 = 6.0$ dB with an interleaver depth of $\mathcal{D} = 10^4$ bits.

4.3.3 Performance of the RSC-coded DSTBC-SP scheme

All simulation parameters associated with the RSC-coded system of Figure 4.18 are listed in Table 4.5. Figures 4.22–4.25 illustrate the actual decoding trajectories of the turbo-detected RSC-coded DSTBC-SP scheme of Figure 4.20 at $E_b/N_0 = 6.0$ dB, when using different interleaver depths. The zigzag paths seen in Figures 4.22–4.25 represent the actual EXIT between the sphere-packing demapper and the outer RSC channel decoder. Observe in Figure 4.25 that since a long interleaver is employed, the assumptions outlined at the beginning of Section 3.4 are justified and hence the EXIT-chart-based convergence prediction of the step-wise linear actual decoding trajectory is quite accurate. In contrast, the decoding trajectories shown in Figures 4.22–4.24 deviate more substantially from the EXIT chart prediction, because shorter interleaver lengths are used. Figures 4.26 and 4.27 highlight the influence of interleaver depth on the system's attainable performance. More specifically, Figure 4.26 demonstrates the achievable extrinsic information at the output of the RSC channel decoder of the turbo-detected RSC-coded DSTBC-SP scheme of Figure 4.20 at $E_b/N_0 = 6.0$ dB, when using different interleaver depths and after ten external joint iterations. Furthermore, Figure 4.27 shows a performance comparison of the AGM (AGM-10) based RSC-coded DSTBC-SP scheme, when using different interleaver depths and after ten external joint iterations.

Figure 4.28 compares the attainable performance of the proposed RSC-coded DSTBC-SP scheme employing both AGM (AGM-10) and Gray mapping against that of an identical-throughput 1 1BPS uncoded DSTBC-SP scheme using $L = 4$ and against a RSC-coded QPSK-modulated DSTBC scheme, when employing the system parameters outlined in Table 4.5 and using an interleaver depth of $\mathcal{D} = 10^6$ bits. The QPSK-modulated DSTBC

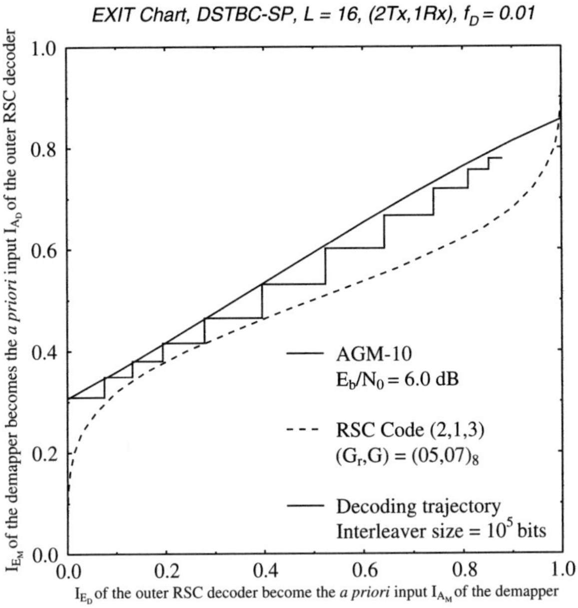

Figure 4.24: Decoding trajectory of a turbo-detected RSC channel-coded DSTBC-SP scheme employing AGM (AGM-10) in combination with the outer RSC code (2,1,3) and the system parameters outlined in Table 4.5 and operating at $E_b/N_0 = 6.0$ dB with an interleaver depth of $\mathcal{D} = 10^5$ bits.

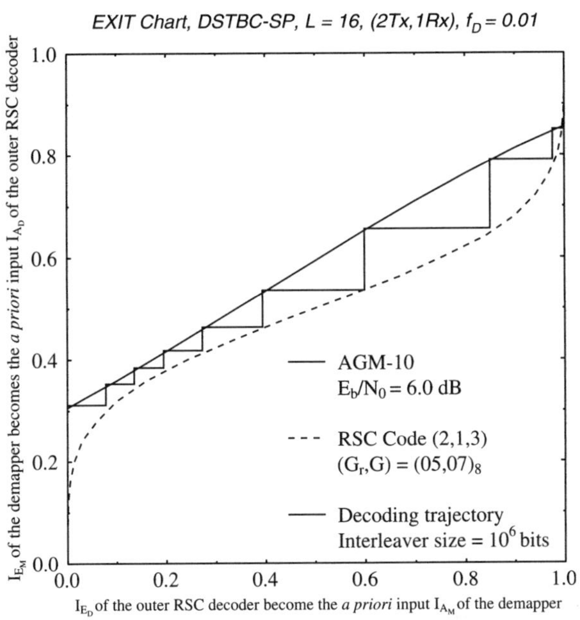

Figure 4.25: Decoding trajectory of a turbo-detected RSC channel-coded DSTBC-SP scheme employing AGM (AGM-10) in combination with the outer RSC code (2,1,3) and the system parameters outlined in Table 4.5 and operating at $E_b/N_0 = 6.0$ dB with an interleaver depth of $\mathcal{D} = 10^6$ bits.

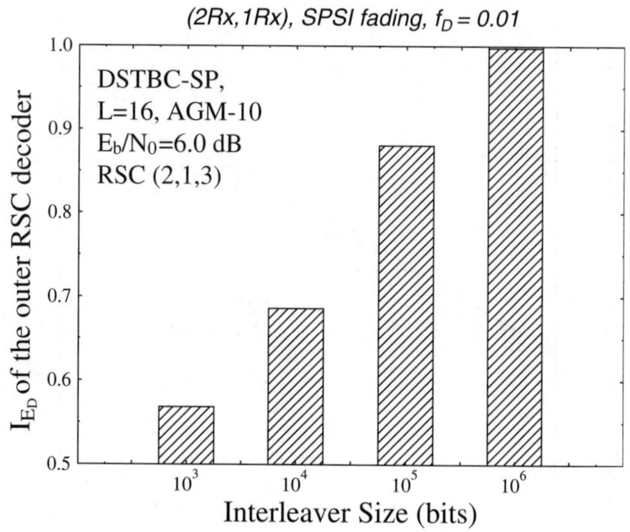

Figure 4.26: Achievable extrinsic information of a turbo-detected RSC channel-coded DSTBC-SP scheme employing AGM (AGM-10) in combination with the outer RSC code (2,1,3) and the system parameters outlined in Table 4.5 and operating at $E_b/N_0 = 6.0$ dB with different interleaver depths.

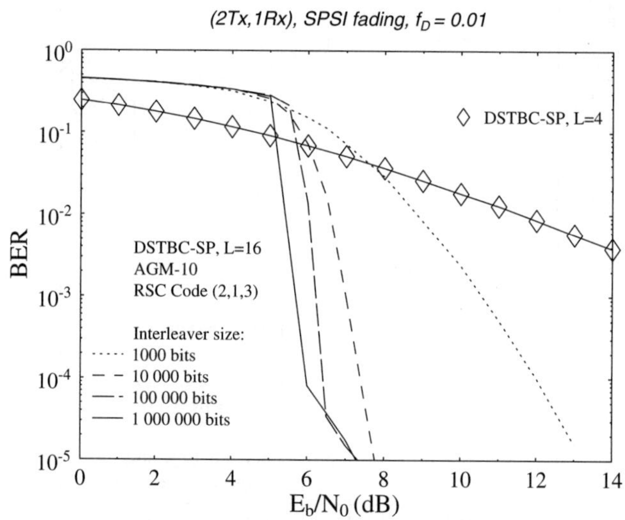

Figure 4.27: Performance comparison of AGM (AGM-10) based RSC-coded DSTBC-SP schemes in conjunction with $L = 16$ against an identical-throughput 1 BPS uncoded DSTBC-SP scheme using $L = 4$, when employing the system parameters outlined in Table 4.5 and using different interleaver depths after ten external joint iterations.

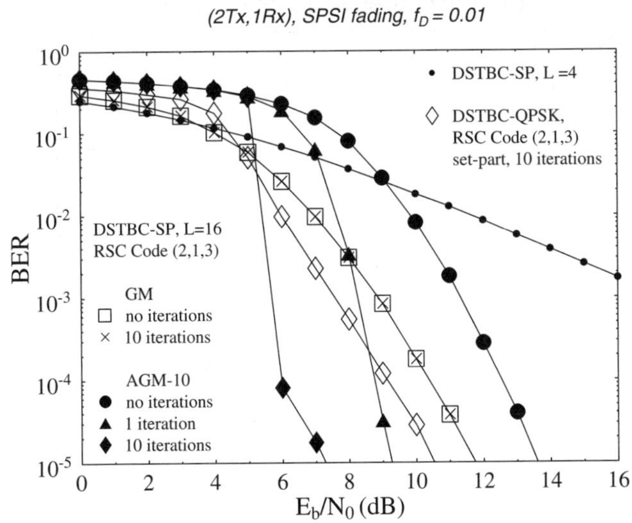

Figure 4.28: Performance comparison of AGM (AGM-10) and Gray mapping (GM)-based RSC-coded DSTBC-SP schemes in conjunction with $L = 16$ against an identical-throughput 1 BPS uncoded DSTBC-SP scheme using $L = 4$ and against a RSC-code QPSK-modulated DSTBC scheme, when employing the system parameters outlined in Table 4.5 and using an interleaver depth of $\mathcal{D} = 10^6$ bits.

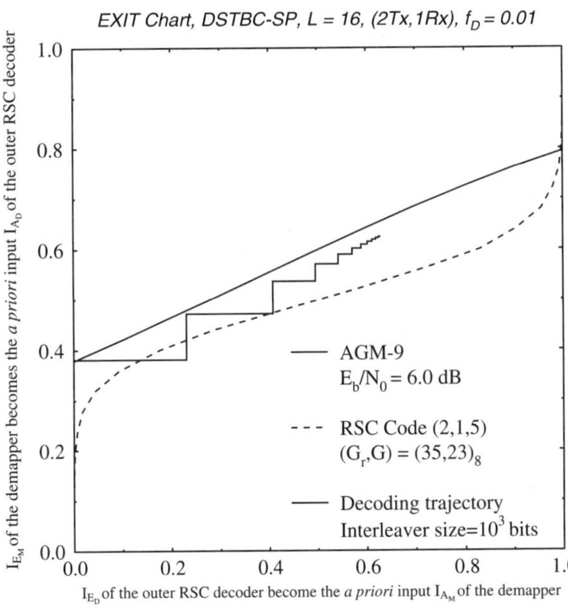

Figure 4.29: Decoding trajectory of a turbo-detected RSC channel-coded DSTBC-SP scheme employing AGM (AGM-9) in combination with the outer RSC code (2,1,5) and the system parameters outlined in Table 4.5 and operating at $E_b/N_0 = 6.0$ dB with an interleaver depth of $\mathcal{D} = 10^3$ bits.

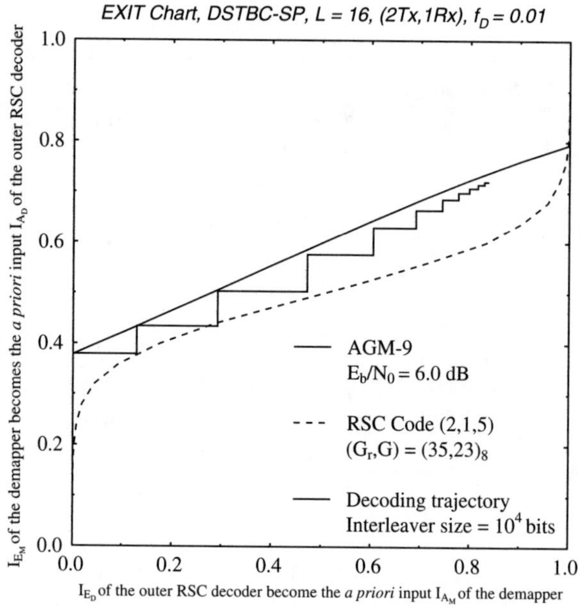

Figure 4.30: Decoding trajectory of a turbo-detected RSC channel-coded DSTBC-SP scheme employing AGM (AGM-9) in combination with the outer RSC code (2,1,5) and the system parameters outlined in Table 4.5 and operating at $E_b/N_0 = 6.0$ dB with an interleaver depth of $\mathcal{D} = 10^4$ bits.

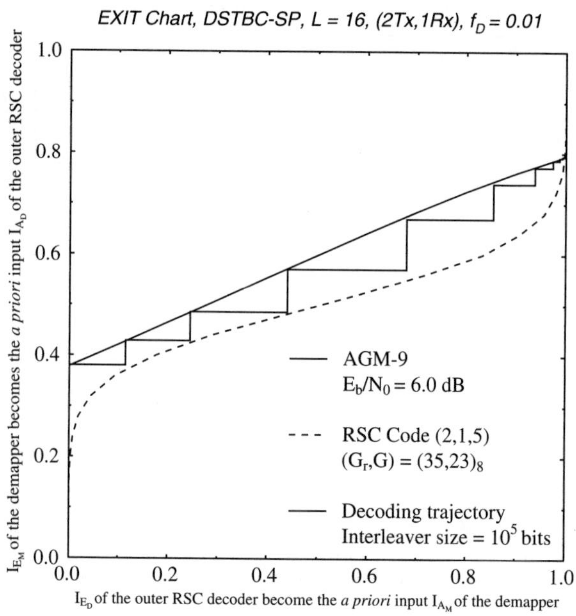

Figure 4.31: Decoding trajectory of a turbo-detected RSC channel-coded DSTBC-SP scheme employing AGM (AGM-9) in combination with the outer RSC code (2,1,5) and the system parameters outlined in Table 4.5 and operating at $E_b/N_0 = 6.0$ dB with an interleaver depth of $\mathcal{D} = 10^5$ bits.

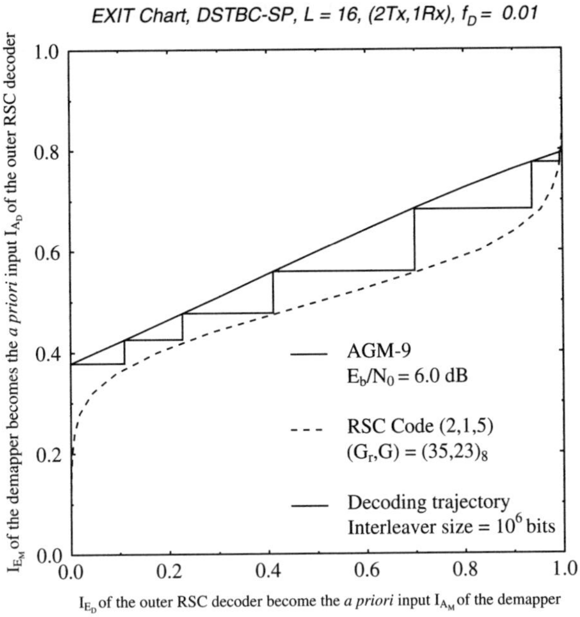

Figure 4.32: Decoding trajectory of a turbo-detected RSC channel-coded DSTBC-SP scheme employing AGM (AGM-9) in combination with the outer RSC code (2,1,5) and the system parameters outlined in Table 4.5 and operating at $E_b/N_0 = 6.0$ dB with an interleaver depth of $\mathcal{D} = 10^6$ bits.

system employs a set-partitioning mapping scheme reminiscent of TCM [263]. Observe in Figure 4.28 by comparing the two Gray mapping DSTBC-SP curves that no BER improvement was obtained when ten turbo-detection iterations were employed in conjunction with Gray mapping. This phenomenon was also reported in [158] and it becomes evident from the horizontal curve characterizing Gray mapping in Figure 4.19. In contrast, AGM (AGM-10) of Figure 4.19 achieved a useful performance improvement in conjunction with iterative demapping and decoding. Explicitly, Figure 4.28 demonstrates that a coding advantage of about 23.7 dB was achieved at a BER of 10^{-5} after ten iterations by the RSC-coded AGM-10 DSTBC-SP system over the uncoded DSTBC-SP for transmission over the correlated Rayleigh fading channel considered. In addition, coding advantages of approximately 4.5 and 3.3 dB were attained over the 1 BPS throughput RSC-coded Gray mapping DSTBC-SP scheme and the RSC-coded QPSK-modulated DSTBC scheme, respectively.

Figures 4.29–4.32 illustrate the actual decoding trajectories of the turbo-detected RSC-coded DSTBC-SP scheme of Figure 4.21 at $E_b/N_0 = 6.0$ dB, when using different interleaver depths. Observe in Figures 4.31 and 4.32 that again, since long interleavers are employed, the assumptions outlined at the beginning of Section 3.4 are justified and hence the EXIT-chart-based convergence prediction of the step-wise linear actual decoding trajectory is quite accurate. In contrast, the decoding trajectories shown in Figures 4.29 and 4.30 deviate from the EXIT chart prediction, because shorter interleaver lengths are used. Observe also the difference between Figures 4.31 and 4.32, where more iterations are required for approaching the intersection point when the interleaver depth drops from 10^6 to 10^5 bits. The influence of the interleaver depth on the system's attainable performance is further highlighted in

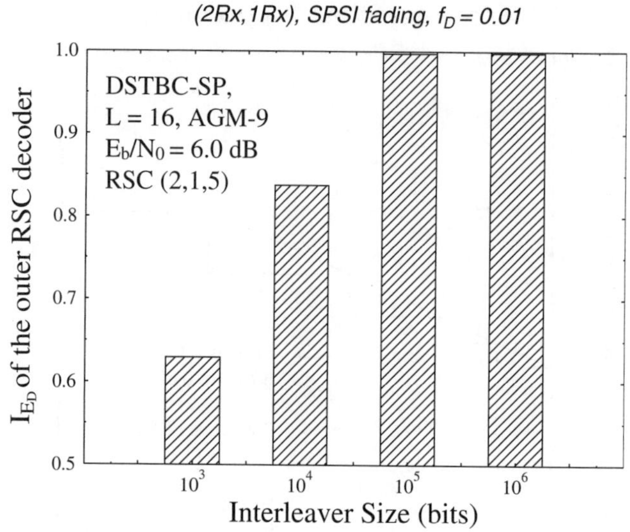

Figure 4.33: Achievable extrinsic information of a turbo-detected RSC channel-coded DSTBC-SP scheme employing AGM (AGM-9) in combination with the outer RSC code (2,1,5) and the system parameters outlined in Table 4.5 and operating at $E_b/N_0 = 6.0$ dB with different interleaver depths.

Figures 4.33–4.35. Figure 4.33 demonstrates the achievable extrinsic information at the output of the RSC channel decoder of the turbo-detected RSC-coded DSTBC-SP scheme of Figure 4.21 at $E_b/N_0 = 6.0$ dB, when using different interleaver depths and after ten external joint iterations. In addition, Figure 4.34 shows a performance comparison of the AGM-9-based RSC-coded DSTBC-SP scheme, when using different interleaver depths and after ten external joint iterations. Observe in Figures 4.33 and 4.34 that the system's attainable performance is almost identical after ten external joint iterations, when using an interleaver depth of 10^5 or 10^6. However, Figure 4.35 illustrates the achievable BER of the turbo-detected RSC channel-coded DSTBC-SP scheme of Figure 4.21, when operating at $E_b/N_0 = 6.0$ dB and using different interleaver depths as well as different numbers of iterations. According to Figure 4.35, four more iterations are necessitated by the system employing an interleaver depth of $\mathcal{D} = 10^5$ bits in order to achieve a BER comparable with that of the system employing an interleaver depth of $\mathcal{D} = 10^6$ bits, when operating at $E_b/N_0 = 6.0$ dB.

Figure 4.36 compares the attainable performance of the proposed RSC-coded DSTBC-SP scheme employing both AGM (AGM-9) and Gray mapping against that of an identical-throughput 1 BPS uncoded DSTBC-SP scheme using $L = 4$ and against a RSC-coded QPSK-modulated DSTBC scheme, when employing the system parameters outlined in Table 4.5 and using an interleaver depth of $\mathcal{D} = 10^6$ bits. The QPSK-modulated DSTBC system employs a set-partitioning mapping scheme. Observe in Figure 4.36 by comparing the two Gray mapping (GM) DSTBC-SP curves that no BER improvement was obtained when ten turbo-detection iterations were employed in conjunction with Gray mapping as expected. In contrast, AGM-9 of Figure 4.19 achieved a useful performance improvement in conjunction with iterative demapping and decoding. Explicitly, Figure 4.36 demonstrates that a coding advantage of about 23.7 dB was achieved at a BER of 10^{-5} after ten iterations by the RSC-coded AGM-9 DSTBC-SP system over the uncoded DSTBC-SP for

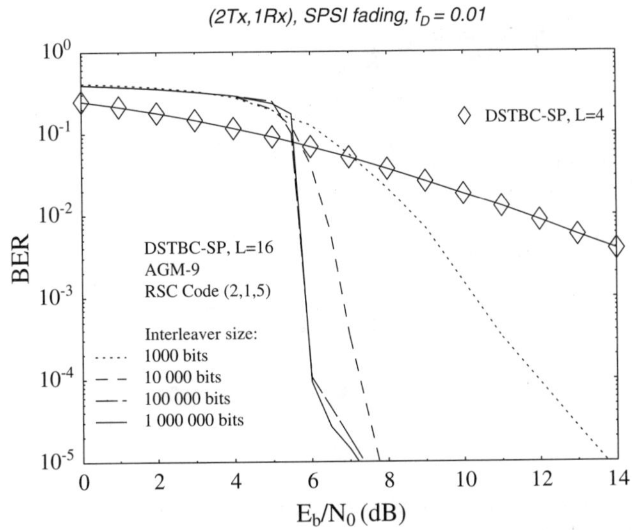

Figure 4.34: Performance comparison of AGM (AGM-9) based RSC-coded DSTBC-SP schemes in conjunction with $L = 16$ against an identical-throughput 1 BPS uncoded DSTBC-SP scheme using $L = 4$, when employing the system parameters outlined in Table 4.5 and using different interleaver depths after ten external joint iterations.

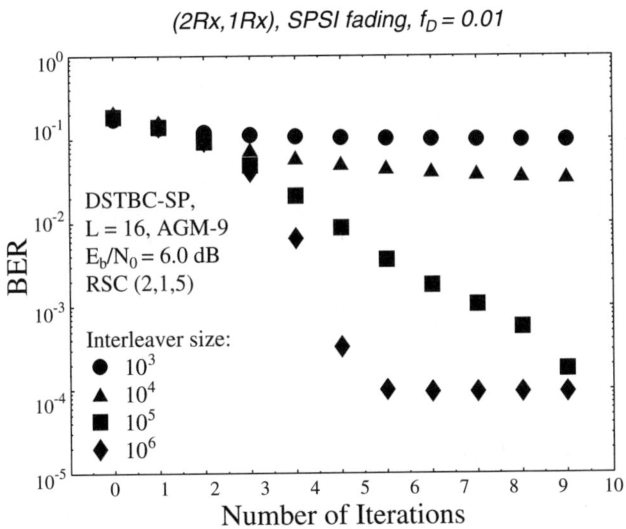

Figure 4.35: Achievable BER of a turbo-detected RSC channel-coded DSTBC-SP scheme employing AGM (AGM-9) in combination with the outer RSC code (2,1,5) and the system parameters outlined in Table 4.5 and operating at $E_b/N_0 = 6.0$ dB with different interleaver depths and numbers of iterations.

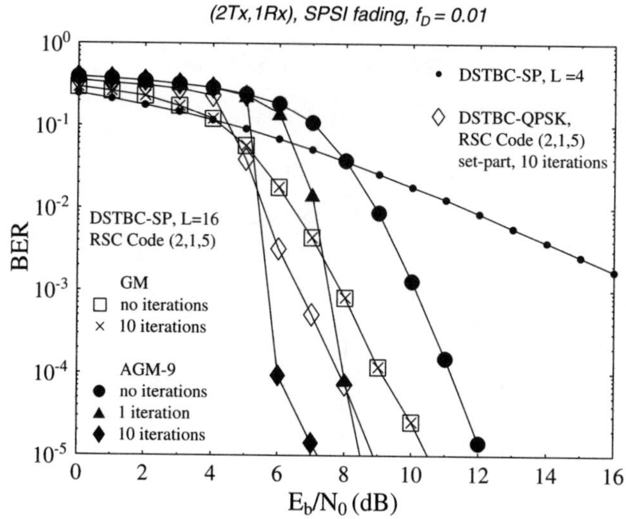

Figure 4.36: Performance comparison of AGM (AGM-9) and Gray mapping (GM)-based RSC-coded DSTBC-SP schemes in conjunction with $L = 16$ against an identical-throughput 1 BPS uncoded DSTBC-SP scheme using $L = 4$ and against a RSC-coded QPSK-modulated DSTBC scheme, when employing the system parameters outlined in Table 4.5 and using an interleaver depth of $\mathcal{D} = 10^6$ bits.

transmission over the correlated Rayleigh fading channel considered. In addition, coding advantages of approximately 3.3 and 1.7 dB were attained over the 1BPS-throughput RSC-coded GM DSTBC-SP scheme and the RSC-coded QPSK-modulated DSTBC scheme, respectively.

Figure 4.37 compares the achievable performance of the AGM-10-based RSC-coded DSTBC-SP scheme of Figure 4.20 against that of the AGM-9-based RSC-coded DSTBC-SP scheme of Figure 4.21, when employing the system parameters outlined in Table 4.5 and using an interleaver depth of $\mathcal{D} = 10^6$ bits. Observe that both schemes achieve a similar BER performance after ten turbo iterations. However, the AGM-9-based scheme of Figure 4.21 outperforms the AGM-10-based scheme of Figure 4.20, when performing a lower number of turbo iterations. This phenomenon is attributed to the fact that the AGM-9-based RSC-coded DSTBC-SP scheme of Figure 4.21 invokes an outer RSC code having a higher constraint length than that of the AGM-10 based scheme of Figure 4.20. The phenomenon may also be explained by observing Figure 3.12, which relates the mutual information at the input of the RSC decoder to the achievable BER.

4.4 Chapter Conclusions

In this chapter, we have extended the concept of SP modulation [43] to DSTBC schemes. The simulation results presented in Section 4.2.2 demonstrated that SP-aided DSTBC schemes provide some coding gain over conventionally modulated DSTBC schemes. Table 4.2 summarizes the coding gains of SP-modulation-aided DSTBC schemes over conventionally modulated DSTBC schemes at a SP-SER of 10^{-4}, when communicating over a block Rayleigh fading channel. Table 4.3 summarizes the coding gains of SP-modulation-aided DSTBC schemes over conventionally modulated DSTBC schemes at a SP-SER of 10^{-4},

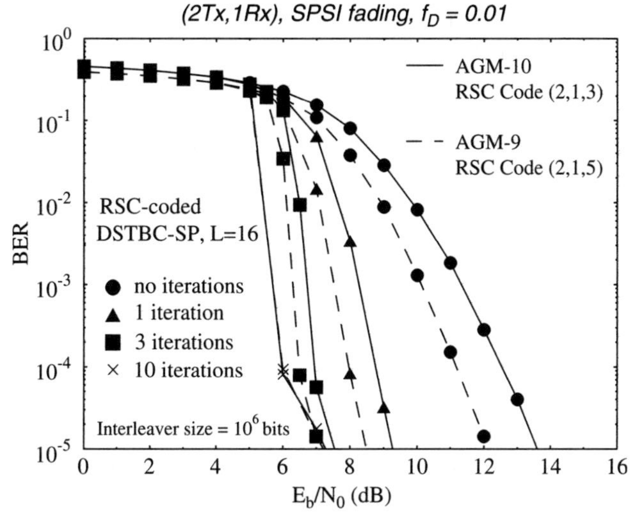

Figure 4.37: Performance comparison of AGM-10- and AGM-9-based RSC-coded DSTBC-SP schemes in conjunction with outer RSC codes having constraint lengths of $\mathcal{K} = 3$ and 5, respectively, when employing the system parameters outlined in Table 4.5 and using an interleaver depth of $\mathcal{D} = 10^6$ bits.

when communicating over a SPSI correlated Rayleigh fading channel having a normalized Doppler frequency of $f_D = 0.01$.

We also proposed in Section 4.3 a novel bit-based system that exploits the advantages of both iterative demapping and turbo detection [158], as well as those of the SP modulation proposed in [43]. Our investigations demonstrated that significant performance improvements may be achieved when the AGM DSTBC-SP scheme is combined with outer channel decoding and iterative demapping, as compared with the Gray-mapping-based systems. Subsequently, EXIT charts were used to search for the optimum bit-to-symbol mapping schemes that converge at the lowest possible E_b/N_0 values. Several DSTBC-SP mapping schemes covering a wide range of EXIT characteristics were investigated. When using an appropriate bit-to-symbol mapping scheme and ten turbo detection iterations, E_b/N_0 gains of about 23.7 and 3.3 dB at a BER of 10^{-5} were obtained by the RSC-coded DSTBC-SP scheme over the identical-throughput 1 BPS uncoded DSTBC-SP benchmarker scheme and over a turbo-detected system based on the DSTBC scheme of [58, 62].

4.5 Chapter Summary

In this chapter, we have proposed a novel DSTBC-SP system that exploits the advantages of the differential STBC schemes of [62] and [264] as well as those of the SP modulation of [43]. We described in Section 4.2.1 how DSTBC schemes are constructed using SP modulation. The performance of uncoded DSTBC-SP schemes was considered in Section 4.2.2, where we compared the performance of different DSTBC-SP schemes against equivalent conventional DSTBC schemes under various channel conditions. Simulation results were provided for systems having different BPS rates in conjunction with appropriate conventional and SP modulation schemes, as outlined in Table 4.1. In Section 4.2.2.1, the channel was assumed to be constant over the transmission period of one frame; this was referred to as a block-fading Rayleigh channel. The attainable coding gains of SP modulation over conventional modulation at a SP-SER of 10^{-4} were summarized in Table 4.2, while the BER performance was illustrated in Figures 4.2–4.9. In Section 4.2.2.2, the channel was assumed to be constant over the transmission period of one SP symbol (i.e. two consecutive time slots). This type of channel was referred to in Section 2.5 as a SPSI channel. The channel was also assumed to be correlated and had a normalized Doppler frequency of $f_D = 0.01$. Table 4.3 summarized the coding gains of SP modulation over conventional modulation at a SP-SER of 10^{-4}. The BER performance of DSTBC-SP schemes, when communicating over the SPSI correlated Rayleigh fading channel considered, was characterized in Figures 4.10–4.17.

In Section 4.3, we proposed novel bit-based RSC-coded turbo-detected DSTBC-SP schemes. The system's architecture was outlined in Section 4.3.1, where the schematic of the proposed arrangement was provided in Figure 4.18. The EXIT chart analysis of Section 3.4 was employed in Section 4.3.2 in order to design and analyze the convergence behavior of the proposed turbo-detected RSC-coded DSTBC-SP schemes. Figures 4.20 and 4.21 illustrated two AGM-based DSTBC-SP schemes in conjunction with outer RSC codes having constraint lengths of $\mathcal{K} = 3$ and 5, which are specifically designed for low BER floors. Appendix C provides the complete list of EXIT charts for the RSC-coded turbo-detected DSTBC-SP scheme of Figure 4.18, when employing the mapping schemes of Figure 4.19 in combination with various outer RSC codes.

In Section 4.3.3, we investigated the performance of the proposed RSC-coded DSTBC-SP schemes, when employing the simulation parameters listed in Table 4.5. The actual decoding trajectories of the proposed AGM-10-based scheme of Figure 4.20 were provided in Figures 4.22–4.25, when using various interleaver depths. Similarly, the actual decoding trajectories of the proposed AGM-9-based scheme of Figure 4.21 were provided in Figures 4.29–4.32, when using various interleaver depths. Finally, the BER performance of both schemes was demonstrated in Figures 4.27, 4.28 and 4.34–4.37.

In the next chapter, we propose a three-stage turbo-detected STBC-SP scheme, where a rate-one recursive inner precoder is employed. The objective of employing the rate-one recursive inner precoder is to avoid the BER floor often experienced by two-stage turbo-detected schemes, especially when the inner code is not recursive, such as in the turbo-detected schemes introduced in this chapter and in Chapter 3.

Chapter 5

Three-stage Turbo-detected STBC-SP Schemes*

5.1 Introduction

Conventional two-stage turbo-detected schemes as introduced in Chapters 3 and 4 typically suffer from a BER floor, preventing them from achieving infinitesimally low BER values, especially when the inner coding stage is of a non-recursive nature. In this chapter we circumvent this deficiency by proposing a three-stage turbo-detected STBC-SP scheme, where a rate-one recursive inner precoder is employed to avoid having a BER floor. The ultimate rationale of this chapter is to use a novel three-dimensional EXIT-chart-based technique to jointly design the two time slots' STBC signal by near-optimally combining them into an iteratively detected SP symbol.

The turbo principle of [146] was extended to multiple parallel concatenated codes in 1995 [147], to serially concatenated codes in 1996 [148] and to multiple serially concatenated codes in 1998 [149]. The appeal of concatenated coding is that low-complexity iterative detection replaces the potentially more complex optimum decoder, such as that of [265]. It was shown in [166] that a recursive inner code is needed in order to maximize the interleaver gain and to avoid the formation of a BER floor, when employing iterative decoding. This principle has been adopted by several authors designing serially concatenated schemes, where rate-one inner codes were employed for designing low-complexity turbo codes suitable for bandwidth- and power-limited systems having stringent BER requirements [170, 177, 180, 266, 267].

The convergence behavior of iterative decoding was discussed in Chapter 3. The employment of the EXIT characteristics between a concatenated decoder's output and input for describing the flow of extrinsic information through the soft-in/soft-out constituent decoders was pioneered by ten Brink [169]. Since then, EXIT chart analysis has attracted considerable attention [169, 174, 175, 178, 268, 269]. The computation of EXIT charts was further simplified in [175] to a time average, for scenarios when the PDFs of the communicated information at the input and output of the constituent decoders are both

*Parts of this chapter are based on the collaborative research outlined in [190].

symmetric and consistent. Furthermore, the concept of EXIT chart analysis has been extended to three-stage concatenated systems in [174, 178, 268, 269].

In this chapter, we propose a capacity-approaching three-stage turbo-detected STBC-SP scheme, where iterative decoding is carried out between three constituent decoders, namely a STBC-SP demapper, an inner rate-one recursive *A Posteriori* Probability (APP)-based decoder and an outer APP-based decoder. An upper bound on the maximum achievable rate is calculated based on the EXIT charts of the STBC-SP demapper. At a spectral efficiency of $\eta = 1$ bit s^{-1} Hz^{-1}, the upper bound of the maximum achievable rate is within 0.5 dB of the capacity, and our proposed three-stage scheme operates within 1.0 dB of the capacity. The rationale of the proposed architecture is explicit: (1) SP modulation maximizes the coding advantage of the transmission scheme by jointly designing and detecting the SP symbols hosting the two time slots' STBC symbols; (2) the inner rate-one recursive decoder maximizes the interleaver gain and hence avoids having a BER floor; and (3) the outer IRCCs [175, 176] minimize the area of the EXIT chart's convergence tunnel and hence facilitate near-capacity operation [270].

This chapter is organized as follows. In Section 5.2, a brief description of our three-stage system is presented. Section 5.3 provides our three-dimensional EXIT chart analysis along with its simplified two-dimensional projections. In Section 5.4 an upper bound on the maximum achievable rate is also calculated based on the EXIT chart analysis. The performance of the three-stage turbo-detected STBC-SP scheme is demonstrated and provided in Section 5.5. Finally, we conclude in Section 5.6.

5.2 System Overview

5.2.1 Encoder

The schematic of the entire system is shown in Figure 5.1, where the transmitted source bits $\mathbf{u}_1$ are encoded by the outer channel Encoder I having a rate of R_I. The outer channel encoded bits $\mathbf{c}_1$ are then interleaved by the first random bit interleaver, where the randomly permuted bits $\mathbf{u}_2$ are fed through the rate-one Encoder II. More specifically, Encoder II is the simple rate-one accumulator shown in Figure 5.2, which is described by the pair of octal generator polynomials $(G/G_r) = (2/3)_8$, where G_r is the feedback polynomial. The concatenated coded bits $\mathbf{c}_2$ at the output of the rate-one encoder are interleaved by the second random bit interleaver, producing the permuted bits $\mathbf{u}_3$. After channel interleaving, the SP mapper first maps blocks of B channel-coded bits $\mathbf{b} = b_{0,\ldots,B-1} \in \{0, 1\}$ to the $L = 2^B$ legitimate four-dimensional SP modulated symbols $\mathbf{s}^l \in S$, where $S = \{\mathbf{s}^l = [a_{l,1}\ a_{l,2}\ a_{l,3}\ a_{l,4}] \in \mathbb{R}^4 : 0 \leq l \leq L-1\}$ constitutes a set of L legitimate constellation points selected from the lattice D_4 (see [221]) having a total energy of $E_{\text{total}} \triangleq \sum_{l=0}^{L-1}(|a_{l,1}|^2 + |a_{l,2}|^2 + |a_{l,3}|^2 + |a_{l,4}|^2)$, as detailed in Section 2.4.3.4. The STBC encoder then maps each SP modulated symbol $\mathbf{s}^l$ to the space-time signal $\mathbf{C}_l$ of Equation (2.63) as [43]:

$$\mathbf{C}_l = \sqrt{\frac{2L}{E_{\text{total}}}}\mathbf{G}_2(x_{l,1}, x_{l,2}), \quad 0 \leq l \leq L-1,$$

where $x_{l,1}$ and $x_{l,2}$ are complex-valued symbols constructed from the four-dimensional real-valued coordinates of the SP symbol $\mathbf{s}^l$ in order to maximize the coding advantage of the space-time signal $\mathbf{C}_l$ [43], since the lattice D_4 has the best MED in the four-dimensional real-valued Euclidean space $\mathbb{R}^4$ (see [221]). Specifically, $x_{l,1}$ and $x_{l,2}$ may be written using

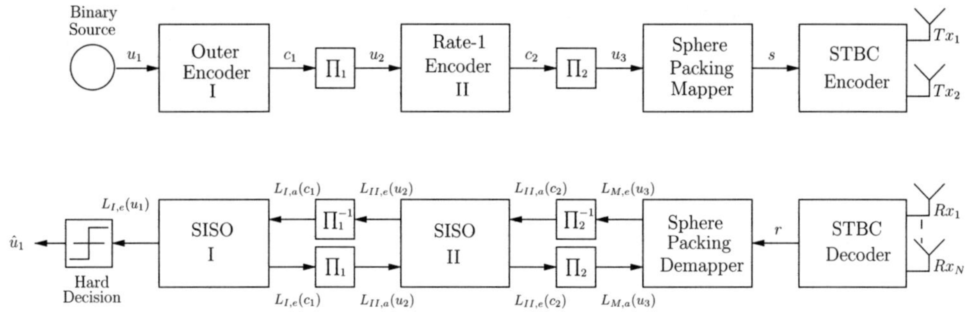

Figure 5.1: Three-stage serially concatenated system.

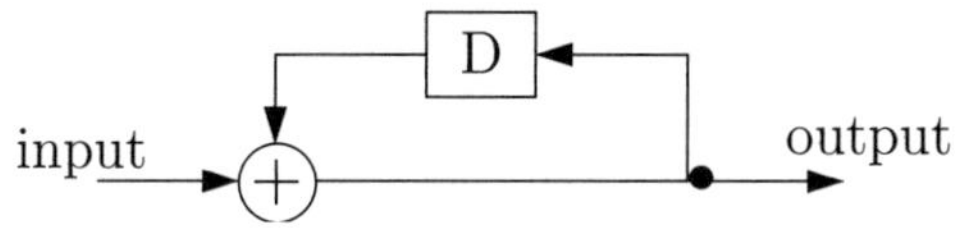

Figure 5.2: Rate-one accumulator.

Table 5.1: Encoder I types employed in various three-stage schemes.

System	Encoder I
RA-coded scheme	Half-rate repeater code
RSC-coded scheme	Half-rate recursive systematic convolutional code
IRCC-coded scheme	Half-rate irregular convolutional code

Equation (8.63), which is reproduced here for convenience:

$$\{x_{l,1}, x_{l,2}\} = T_{\text{sp}}(a_{l,1}, a_{l,2}, a_{l,3}, a_{l,4}) = \{a_{l,1} + ja_{l,2}, a_{l,3} + ja_{l,4}\}.$$

Furthermore, $\mathbf{G}_2(x_{l,1}, x_{l,2})$ is the space-time transmission matrix of Equation (2.35), which is reproduced here for convenience:

$$\mathbf{G}_2(x_1, x_2) = \begin{bmatrix} x_1 & x_2 \\ -x_2^* & x_1^* \end{bmatrix},$$

where the rows and columns represent the temporal and spatial dimensions, corresponding to two consecutive time slots and two transmit antennas, respectively.

In this chapter, we consider three different types of codes for Encoder I, namely a repeater, a RSC code and a IRCC. The resultant schemes are outlined in Table 5.1, where RA refers to a 'repeat-accumulate' [271] configuration.

5.2.2 Channel Model

In this chapter, we continue our discourse by considering a SPSI-correlated narrowband Rayleigh fading channel, based on Jakes' fading model [220] associated with a normalized Doppler frequency of $f_D = f_d T_s = 0.1$, where f_d is the Doppler frequency and T_s is the symbol period. The complex-valued fading envelope is assumed to be constant across the transmission period of a space-time coded symbol spanning $T = 2$ time slots. The complex AWGN of $n = n_I + jn_Q$ is also added to the received signal, where n_I and n_Q are two independent zero-mean Gaussian random variables having a variance of $\sigma_n^2 = \sigma_{n_I}^2 = \sigma_{n_Q}^2 = N_0/2$ per dimension, where $N_0/2$ represents the double-sided noise power spectral density expressed in W Hz^{-1}.

5.2.3 Decoder

As shown in Figure 5.1, the received complex-valued symbols are first decoded by the STBC decoder in order to produce the received SP soft symbols $\mathbf{r}$, where each SP symbol represents a block of B coded bits as described in Section 3.3. Then, iterative demapping/decoding is carried out between the SP demapper, APP-based soft-in/soft-out module II and APP-based soft-in/soft-out module I, where extrinsic information is exchanged between the three constituent demapper/decoder modules. More specifically, $L_{\cdot,a}(\cdot)$ in Figure 5.1 represents the *a priori* information, expressed in terms of the LLRs of the corresponding bits, whereas $L_{\cdot,e}(\cdot)$ represents the extrinsic LLRs of the corresponding bits. The iterative process is performed for a number of consecutive iterations. During the last iteration, only the LLR values $L_{I,e}(\mathbf{u_1})$ of the original uncoded systematic information bits $\mathbf{u}_1$ are required, which are passed to a hard decision decoder in order to determine the estimated transmitted source bits $\hat{\mathbf{u}}_1$ as shown in Figure 5.1.

5.3 EXIT Chart Analysis

5.3.1 Preliminaries

It was shown in Section 3.4 that the main objective of employing EXIT charts [169] is to predict the convergence behavior of the iterative decoder by examining the evolution of the input/output mutual information exchange between the inner and outer decoders in consecutive iterations. The application of EXIT charts is based on the two assumptions that upon assuming large interleaver lengths: (1) the *a priori* LLR values are fairly uncorrelated; and (2) the *a priori* LLR values exhibit a Gaussian PDF. In this section, the approach presented in [178] is adopted in order to provide the EXIT chart analysis of the proposed three-stage system of Figure 5.1.

Let $I_{\cdot,a}(x)$, $0 \leq I_{\cdot,a}(x) \leq 1$, denote the mutual information (MI) [272] between the *a priori* LLRs $L_{\cdot,a}(x)$ as well as the corresponding bits x and let $I_{\cdot,e}(x)$, $0 \leq I_{\cdot,e}(x) \leq 1$, denote the MI between the extrinsic LLRs $L_{\cdot,e}(x)$ and the corresponding bits x, where the subscript $(\cdot)$ is used to distinguish the different constituent decoders, i.e. Decoder I, Decoder II and the SP demapper.

5.3.2 Three-dimensional EXIT Charts

As seen from Figure 5.1, the input of Decoder II is constituted by the *a priori* input $L_{II,a}(c_2)$ and the *a priori* input $L_{II,a}(u_2)$ provided after bit-deinterleaving by the SP demapper and Decoder I, respectively, which are illustrated in Figures 5.3 and 5.4. Therefore, the EXIT characteristic of Decoder II can be described by the following two EXIT functions [169, 178]:

$$I_{II,e}(c_2) = T_{II,c_2}(I_{II,a}(u_2), I_{II,a}(c_2)), \quad (5.1)$$

$$I_{II,e}(u_2) = T_{II,u_2}(I_{II,a}(u_2), I_{II,a}(c_2)), \quad (5.2)$$

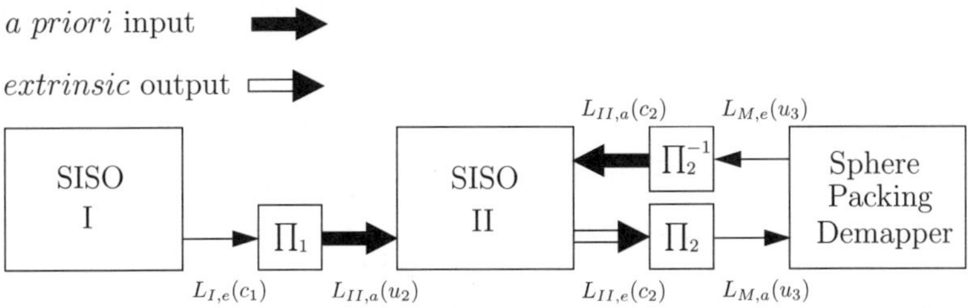

Figure 5.3: *A priori* input and extrinsic output corresponding to Equation (5.1).

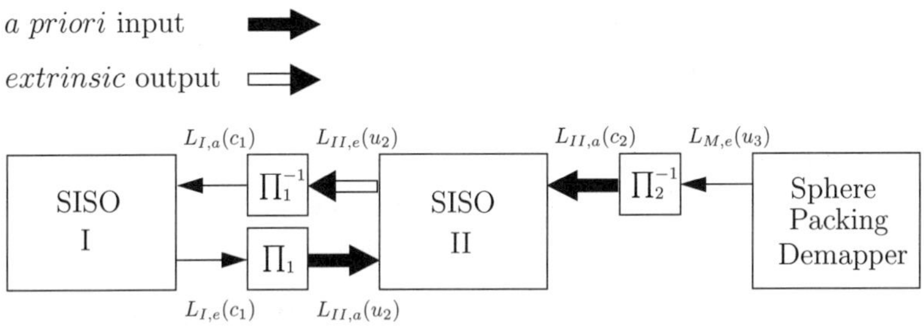

Figure 5.4: *A priori* input and extrinsic output corresponding to Equation (5.2).

which are illustrated by the three-dimensional surfaces drawn in dotted lines in Figures 5.5 and 5.6, respectively. On the other hand, the EXIT characteristic of the SP demapper as well as that of Decoder I are each dependent on a single *a priori* input, namely on $L_{M,a}(u_3)$ and $L_{I,a}(c_1)$, respectively, both of which are provided by the rate-one Decoder II after appropriately ordering the bits, as seen in Figure 5.1. The EXIT characteristic of the SP demapper is also dependent on the E_b/N_0 value. Consequently, the corresponding EXIT functions for the SP demapper and Decoder I, respectively, may be written as

$$I_{M,e}(u_3) = T_{M,u_3}(I_{M,a}(u_3), E_b/N_0), \tag{5.3}$$

$$I_{I,e}(c_1) = T_{I,c_1}(I_{I,a}(c_1)), \tag{5.4}$$

which are illustrated by the three-dimensional surfaces drawn in solid lines in Figures 5.5 and 5.6, respectively.

Equations (5.2) to (5.4) may be represented with the aid of two three-dimensional EXIT charts. More specifically, the three-dimensional EXIT chart of Figure 5.5 can be used to plot Equations (5.1) and (5.3), which describe the EXIT relation between the SP demapper and Decoder II. Similarly, the three-dimensional EXIT chart of Figure 5.6 can be used to describe the EXIT relation between Decoder II and Decoder I by plotting Equations (5.2) and (5.4). Figures 5.5 and 5.6 show an example of these three-dimensional EXIT charts, when Encoder I is a half-rate memory-one RSC code having octally represented generator polynomials of $(G_r, G) = (3, 2)_8$, while Encoder II is the simple rate-one accumulator described in Section 5.2.1 and Figure 5.2.

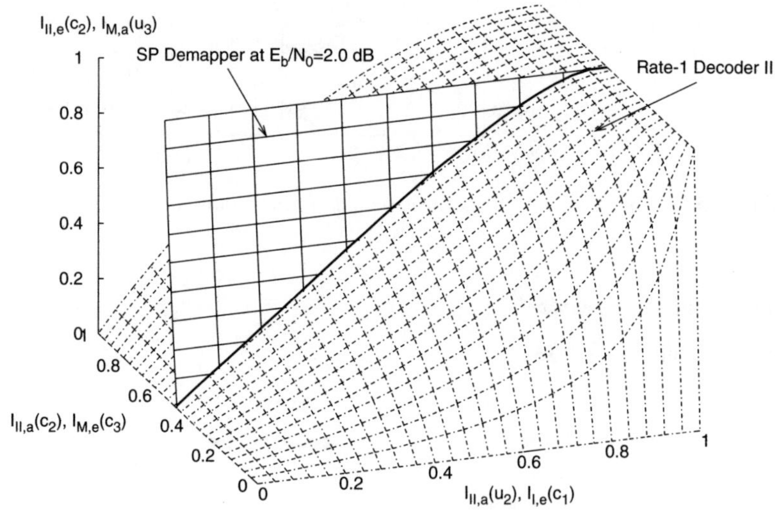

Figure 5.5: Three-dimensional EXIT chart of Decoder II and the SP demapper at $E_b/N_0 = 2.0$ dB.

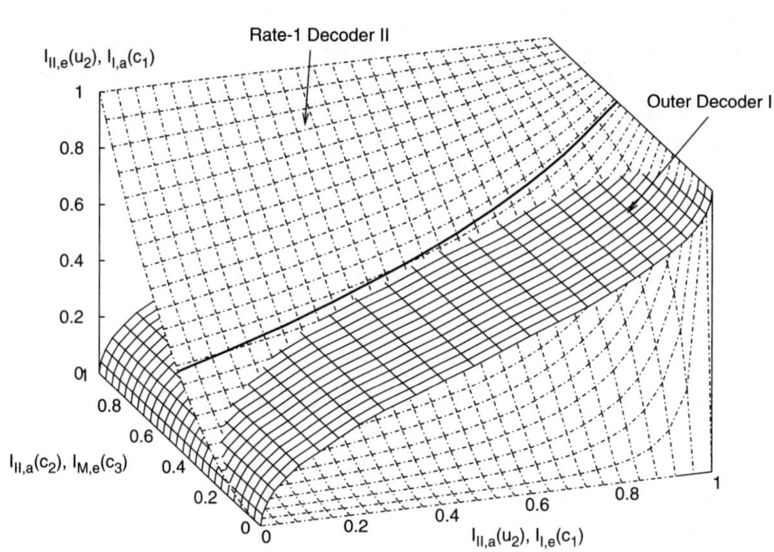

Figure 5.6: Three-dimensional EXIT chart of Decoder II and Decoder I with projection from Figure 5.5.

5.3.3 Two-dimensional EXIT Chart Projections

The three-dimensional EXIT charts of Figures 5.5 and 5.6 are somewhat cumbersome to interpret as well as to plot. Hence, in this section we derive their unique and unambiguous two-dimensional representations, which can be interpreted in the usual way.

The intersection of the surfaces in Figure 5.5, shown as a thick solid line, portrays the best achievable performance when exchanging mutual information between the SP demapper and the rate-one Decoder II for different fixed values of $I_{II,a}(u_2)$ spanning the range of $[0, 1]$. Each $(I_{II,a}(u_2), I_{II,a}(c_2), I_{II,e}(c_2))$ point belonging to the intersection line in Figure 5.5 uniquely specifies a three-dimensional point $(I_{II,a}(u_2), I_{II,a}(c_2), I_{II,e}(u_2))$ in Figure 5.6, according to the EXIT function of Equation (5.2). Therefore, the line corresponding to the $(I_{II,a}(u_2), I_{II,a}(c_2), I_{II,e}(c_2))$ points along the thick line of Figure 5.5 is projected to the solid line shown in Figure 5.6, while the two-dimensional projection of the solid line in Figure 5.6 at $I_{II,a}(c_2) = 0$ onto the plane spanned by the lines $(I_{II,a}(u_2), I_{II,e}(u_2))$ and $(I_{I,e}(c_1), I_{I,a}(c_1))$ is shown in Figure 5.7, represented by the dotted line at $E_b/N_0 = 2.0$ dB. This projected EXIT curve may be written as

$$I_{II,e}(u_2) = T^{p}_{II,u_2}(I_{II,a}(u_2), E_b/N_0). \tag{5.5}$$

Projected two-dimensional EXIT charts of similar nature are used throughout the rest of the book for the sake of describing the convergence behavior of the three-stage turbo-detected STBC-SP scheme. More details on the related three-dimensional to two-dimensional EXIT chart projection are provided in [178].

Figure 5.7 shows the EXIT curve of the SP demapper, when operating at $E_b/N_0 = 2.0$ dB and employing an AGM (AGM-1) scheme, which is described in Appendix A and in Table A.2. Figure 5.7 also shows the EXIT curve of the outer RSC Decoder I and the two-dimensional-projected EXIT curves of the combined SP demapper and the rate-one Decoder II at different E_b/N_0 values, when employing AGM-1 of Table A.2. Observe in Figure 5.7 that an open convergence tunnel is taking shape for the three-stage RSC-coded scheme upon increasing the SNR beyond $E_b/N_0 = 2.0$ dB. This implies that according to the predictions of the two-dimensional EXIT chart seen in Figure 5.7, the iterative decoding process is expected to converge to the $(1.0, 1.0)$ point and hence an infinitesimally low BER may be attained beyond $E_b/N_0 = 2.0$ dB. In contrast, for the traditional two-stage turbo-detected STBC-SP scheme, there would be a BER floor preventing it from achieving an infinitesimally low BER owing to the non-recursive nature of the SP demapper, which also prevents the intersection of the EXIT curves of the SP demapper and the outer RSC Decoder I from reaching the $(1.0, 1.0)$ point of convergence, despite increasing the SNR or the number of iterations. In contrast to this, the three-stage scheme of Figure 5.1 becomes capable of achieving an infinitesimally low BER, as suggested by the EXIT-chart predictions of Figure 5.7.

Figure 5.8 shows the EXIT curve of the SP demapper, when operating at $E_b/N_0 = 1.8$ dB and employing the AGM-7 scheme described in Appendix A and in Table A.8. The figure also shows the EXIT curve of the outer half-rate Decoder I of the precoder and the two-dimensional-projected EXIT curves of the combined SP demapper and the rate-one Decoder II at different E_b/N_0 values, when employing the AGM-7 scheme of Table A.8. Despite the lower complexity of the repeater as compared with that of the RSC code, Figure 5.8 demonstrates that an open convergence tunnel is taking shape for the three-stage RA-coded scheme upon increasing the SNR beyond $E_b/N_0 = 1.8$ dB. However, the convergence tunnel shown in Figure 5.8 becomes narrower as the system approaches the $(1.0, 1.0)$ point, creating a convergence 'bottleneck', which decreases the achievable iteration

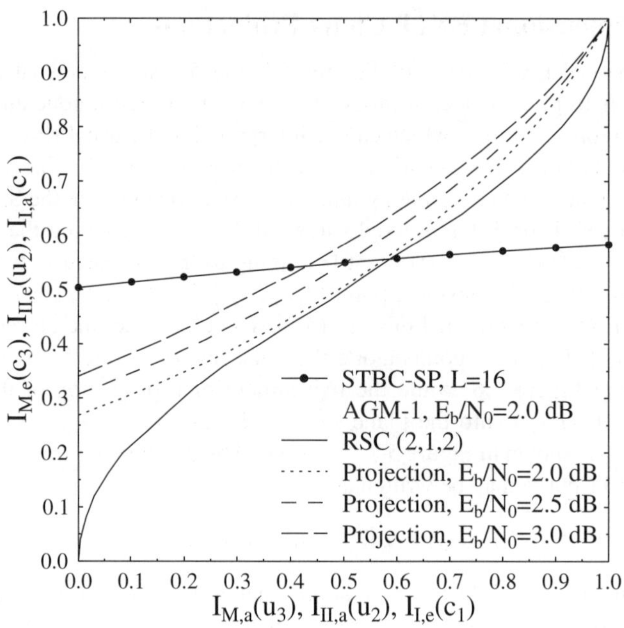

Figure 5.7: Two-dimensional projection of the EXIT charts of the three-stage RSC-coded STBC-SP scheme, when employing AGM-1 of Table A.2.

gain, despite increasing the system's decoding complexity. Furthermore, observe in Figure 5.8 that increasing the SNR value does not avoid the convergence bottleneck, which is expected to result in a residual BER floor when a finite number of decoding iterations is employed. The predictions of the EXIT charts seen in Figures 5.7 and 5.8 are verified in Section 5.5 using actual decoding trajectories and BER performance curves.

5.3.4 EXIT Tunnel-area Minimization for Near-capacity Operation

In this section we exploit the well-understood properties of conventional two-dimensional EXIT charts that a narrow and open EXIT tunnel represents a near-capacity performance. Therefore, we invoke IRCCs for the sake of appropriately shaping the EXIT curves by minimizing the area within the EXIT tunnel using the procedure of [175, 176].

Let $\mathcal{A}_I$ and $\bar{\mathcal{A}}_I$ be the area under the EXIT curve $T_{I,c_1}(i)$ of Equation (5.4) and its inverse $T_{I,c_1}^{-1}(i)$, $i \in [0, 1]$, respectively, which are expressed as

$$\mathcal{A}_I = \int_0^1 T_{I,c_1}(i)\, \mathrm{d}i, \quad \bar{\mathcal{A}}_I = \int_0^1 T_{I,c_1}^{-1}(i)\, \mathrm{d}i = 1 - \mathcal{A}_I. \tag{5.6}$$

Similarly, the area $\mathcal{A}_{II}^p$ is defined under the EXIT curve $T_{II,u_2}^p(i)$ of Equation (5.5). It was observed in [175, 273] that for the APP-based outer Decoder I, the area $\bar{\mathcal{A}}_I$ may be approximated by $\bar{\mathcal{A}}_I \approx R_I$, where the equality $\bar{\mathcal{A}}_I = R_I$ was later shown in [270] for the family of Binary Erasure Channels (BECs). The area property of $\bar{\mathcal{A}}_I \approx R_I$ implies that the lowest SNR convergence threshold occurs when we have $\mathcal{A}_{II}^p = R_I + \epsilon$, where ϵ is an infinitesimally small number, provided that the following convergence constraints hold [176]:

$$T_{II,u_2}^p(0) > 0, \quad T_{II,u_2}^p(1) = 1, \quad T_{II,u_2}^p(i) > T_{I,c_1}^{-1}(i), \quad \text{for all } i \in [0, 1). \tag{5.7}$$

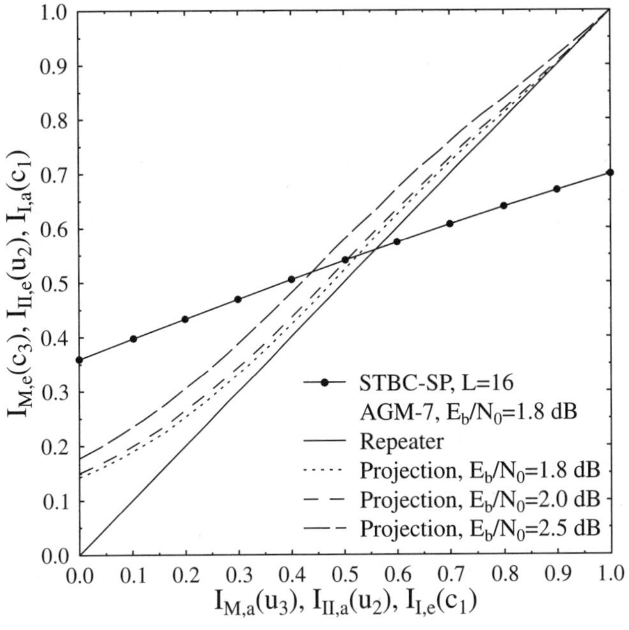

Figure 5.8: Two-dimensional projection of the EXIT charts of the three-stage RA-coded STBC-SP scheme, when employing the AGM-7 of Table A.8.

Observe in Figure 5.7, however, that there is a 'larger-than-necessary' tunnel area between the projected EXIT curve $T^p_{II,u_2}(i)$ and the EXIT curve $T^{-1}_{I,c_1}(i)$ of the outer half-rate RSC code at $E_b/N_0 = 2.0$ dB. This implies that the BER curve is farther from the achievable capacity than necessary, despite the fact that the specific bit-to-SP symbol mapping scheme of AGM-1 and the half-rate RSC code employed in Figure 5.7 were specifically optimized for convergence at a low E_b/N_0 value. More quantitatively, the area under the projected EXIT curve $T^p_{II,u_2}(i)$ is $\mathcal{A}^p_{II} \approx 0.55$ at $E_b/N_0 = 2.0$ dB, which is larger than the outer code rate of $R_I = 0.50$. Therefore, according to Figure 5.7 and to the area property of $\bar{\mathcal{A}}_I \approx R_I$, a lower E_b/N_0 convergence threshold may be attained, provided that the constraints outlined in Equation (5.7) are satisfied. In other words, the EXIT curve $T^{-1}_{I,c_1}(i)$ of the outer code should match the two-dimensional projected EXIT curve $T^p_{II,u_2}(i)$ of Figure 5.7 more closely. Hence we invoke IRCCs [175, 176] as outer codes that exhibit flexible EXIT characteristics, which can be optimized to more closely match the two-dimensional-projected EXIT curve $T^p_{II,u_2}(i)$ of Figure 5.7, rendering the near-capacity code optimization a simple curve-fitting process.

A rate $R_I = 1/2$ IRCC scheme constituted by a set of $P = 17$ subcodes having different code rates R^i_I, $i = 1, \ldots, P$, was constructed in [176]. The IRCC was designed from a systematic half-rate memory-four mother code defined by the octally represented generator polynomials $(G_r, G) = (31, 27)_8$, where puncturing was employed to obtain the code rates of $R^i_I > R_I$, while the code rates of $R^i_I < R_I$ were created by adding more generators and by puncturing. The two additional generators employed in [176] are defined by the octally represented polynomials of $G_1 = (35)_8$ and $G_1 = (35)_8$, where the resultant $P = 17$ subcodes have coding rates spanning the range of $[0.1, 0.9]$. Each of the $P = 17$ subcodes encodes a specific fraction of the uncoded bits determined by the weighting coefficient, α_i, $i = 1, \ldots, P$. Assuming an overall average code rate of R_I, the following conditions must be satisfied:

$$\sum_{i=1}^{P} \alpha_i = 1, \quad R_I = \sum_{i=1}^{P} \alpha_i R_I^i, \quad \text{and} \quad \alpha_i \in [0, 1], \quad \text{for all } i. \tag{5.8}$$

The EXIT function $T_{I,c_1}(I_{I,a}(c_1))$ of Equation (5.4) corresponding to the IRCC may be constructed from the EXIT functions of the $P = 17$ subcodes, $T_{I,c_1}^i(I_{I,a}(c_1))$, $i = 1, \ldots, P$. More specifically, the EXIT function $T_{I,c_1}(I_{I,a}(c_1))$ of the IRCC is the weighted superposition of the $P = 17$ EXIT functions $T_{I,c_1}^i(I_{I,a}(c_1))$, $i = 1, \ldots, P$, as follows [176]:

$$T_{I,c_1}(I_{I,a}(c_1)) = \sum_{i=1}^{P} \alpha_i T_{I,c_1}^i(I_{I,a}(c_1)). \tag{5.9}$$

Figure 5.9 shows the EXIT functions of the $P = 17$ subcodes used in [176]. Now the coefficients α_i are optimized with the aid of the iterative algorithm of [175], so that the EXIT curve of the resultant IRCC closely matches the two-dimensional projected EXIT curve $T_{II,u_2}^p(i)$ at the specific E_b/N_0 value, where we have $\mathcal{A}_{II}^p \approx 0.50$. Observe in Figure 5.10 that we have $\mathcal{A}_{II}^p \approx 0.51$ at $E_b/N_0 = 1.5$ dB, indicating that this E_b/N_0 value is close to the lowest attainable convergence threshold when employing a half-rate outer code. Figure 5.10 also shows the two-dimensional-projected EXIT curve of the resultant IRCC, where the optimized weighting coefficients are as follows:

$$\begin{aligned}[\alpha_1, \ldots, \alpha_{17}] = [&0,\ 0.0559066,\ 0.236757, 0,\ 0,\ 0,\ 0.23844,\ 0,\ 0,\\ &0.0306247,\ 0,\ 0.205574,\ 0, 0,\ 0.110076,\ 0,\ 0.122621].\end{aligned} \tag{5.10}$$

5.4 Maximum Achievable Bandwidth Efficiency

Both the capacity and the bandwidth efficiency of the STBC-SP schemes considered were derived in Section 2.4.5. In this section, a procedure is proposed for calculating a tighter upper bound of the maximum achievable bandwidth efficiency of STBC-SP schemes based on the area property of $\bar{\mathcal{A}}_I \approx R_I$ of the EXIT charts discussed in Section 5.3.4. The proposed procedure is applied in this section for calculating the maximum achievable bandwidth efficiency of the three-stage turbo-coded STBC-SP scheme associated with an SP signal constellation size of $L = 16$ considered in this chapter. More explicitly, it was shown in Section 5.3.4 that the outer Decoder I may have a maximum rate of $R_I^{\max} \approx \mathcal{A}_{II}^p$ at a specific E_b/N_0 value, where $\mathcal{A}_{II}^p$ is the area under the projected EXIT curve of the SP demapper and the rate-one Decoder II of Equation (5.5). Therefore, if $\mathcal{A}_{II}^p$ is calculated for different E_b/N_0 values, the maximum achievable bandwidth efficiency may be formulated as a function of the $\underline{E_b/N_0}$ value as follows:

$$\begin{aligned}\eta_{\max}(\underline{E_b/N_0}) &= B \cdot R_{\text{STBC-SP}} \cdot R_I^{\max}\\ &\approx B \cdot R_{\text{STBC-SP}} \cdot \mathcal{A}_{II}^p(E_b/N_0) \quad (\text{bit s}^{-1}\ \text{Hz}^{-1}),\end{aligned} \tag{5.11}$$

where $B = \log_2(L)$ is the number of bits per SP symbol and $R_{\text{STBC-SP}} = \frac{1}{2}$, since $T = 2$ time slots are needed to transmit one SP symbol according to Equations (2.35) and (2.63). In addition, $\underline{E_b/N_0}$ and E_b/N_0 are related as follows:

$$\underline{E_b/N_0} = E_b/N_0 + 10 \log\left(\frac{R_o}{\mathcal{A}_{II}^p(E_b/N_0)}\right) \quad (\text{dB}), \tag{5.12}$$

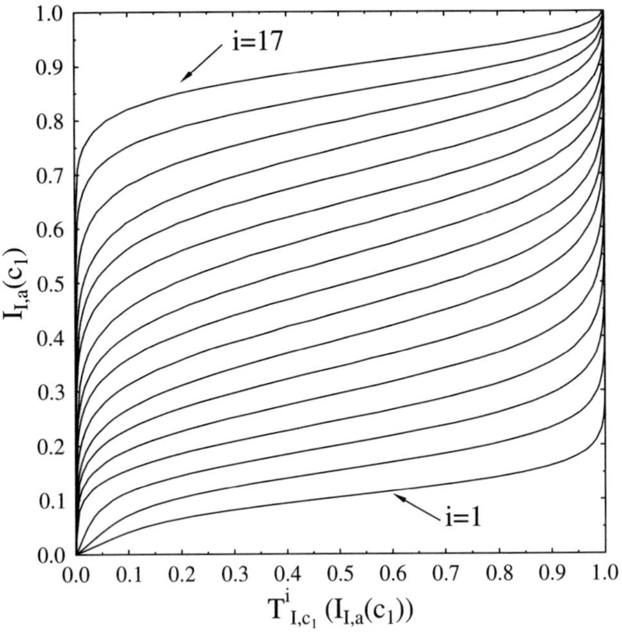

Figure 5.9: EXIT functions of the $P = 17$ subcodes of [176].

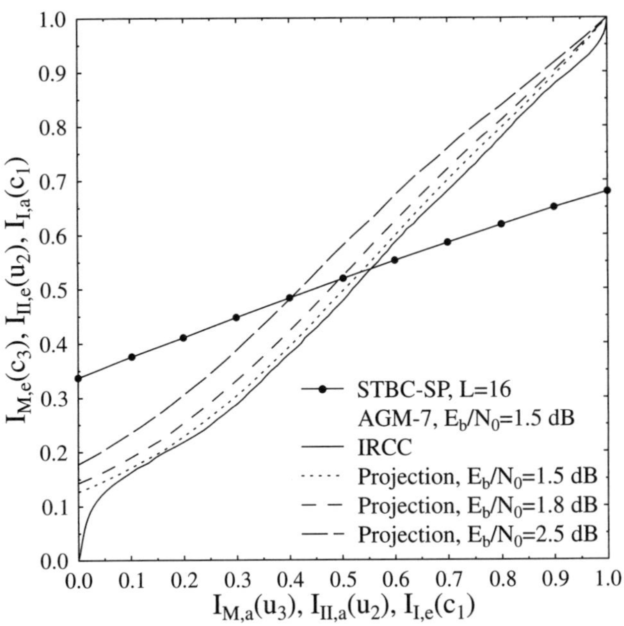

Figure 5.10: Two-dimensional projection of the EXIT charts of the three-stage IRCC-coded STBC-SP scheme, when employing AGM-7 of Table A.8.

Step 1: Let $R_I = R_o$.

Step 2: Let $E_b/N_0 = \rho_{\min}$ dB.

Step 3: Calculate N_0.

Step 4: Let $I_{M,a}(u_3) = 0$.

Step 5: Activate the SP demapper.

Step 6: Save $I_{M,e}(u_3) = T_{M,u_3}(I_{M,a}(u_3), E_b/N_0)$.

Step 7: Let $I_{M,a}(u_3) = I_{M,a}(u_3) + \epsilon$. If $I_{M,a}(u_3) \leq 1.0$, go to **Step 5**.

Step 8: Calculate $\mathcal{A}_M(E_b/N_0) = \int_0^1 T_{M,u_3}(i, E_b/N_0)\, di$.

Step 9: Calculate $\underline{E_b/N_0}$ using Equation (5.12).

Step 10: Save $\eta_{\max}(\underline{E_b/N_0})$ of Equation (5.11).

Step 11: Let $E_b/N_0 = E_b/N_0 + \epsilon$. If $E_b/N_0 \leq \rho_{\max}$ (dB), go to **Step 3**.

Step 12: Output $\eta_{\max}(\underline{E_b/N_0})$ from **Step 10**.

Algorithm 5.1: Maximum achievable bandwidth efficiency using EXIT charts.

where R_o is the original outer code rate used when generating the two-dimensional projected EXIT curves of the SP demapper and the rate-one Decoder II of Equation (5.5) corresponding to the different $\mathcal{A}^p_{II}$ values. A simple procedure may be used to calculate the maximum achievable bandwidth efficiency of Equation (5.11). For computational simplicity, the area $\mathcal{A}_M$ under the EXIT curve T_{M,u_3} of Equation (5.3) may be used instead of the area $\mathcal{A}^p_{II}$ under the two-dimensional projected EXIT curve T^p_{II,u_2} of Equation (5.5), since $\mathcal{A}_M = \mathcal{A}^p_{II}$, when $R_{II} = 1$. More specifically, the maximum achievable bandwidth efficiency of Equation (5.11) can be calculated using the procedure in Algorithm 5.1 for $E_b/N_0 \in [\rho_{\min}, \rho_{\max}]$, assuming that R_o is an arbitrary rate and ϵ is a small constant.

Observe that $\rho_{\min}$ and $\rho_{\max}$ are adjusted accordingly in order to produce the desired range of the resultant $\underline{E_b/N_0}$ values. Furthermore, the output of Algorithm 5.1 is independent of the specific choice of R_o, since Equation (5.12) would always adjust the E_b/N_0 values, regardless of R_o. For example, R_o may be set to the desired final R_I to be employed in the three-stage system.

The resultant maximum achievable bandwidth efficiency is demonstrated in Figure 5.11, which is slightly lower than the bandwidth efficiency of Equation (2.80), i.e. we have $\eta_{\max} < \eta^{\text{STBC-SP}}_{\text{DCMC}}$. Observe that the bandwidth efficiency calculated using Equation (2.80) and using the EXIT charts as well as Equation (5.11) were only proven to be equal for the family of BECs [270]. Nonetheless, similar trends have been observed for both AWGN and ISI channels [174, 176], when APP-based decoders are used for all decoder blocks [270]. However, the discrepancy between the two bandwidth efficiency curves shown in Figure 5.11 that are calculated using Equations (2.80) and (5.11) is due to the fact that the SP demapper is not an APP-based decoder. Nevertheless, the bandwidth efficiency calculated based on

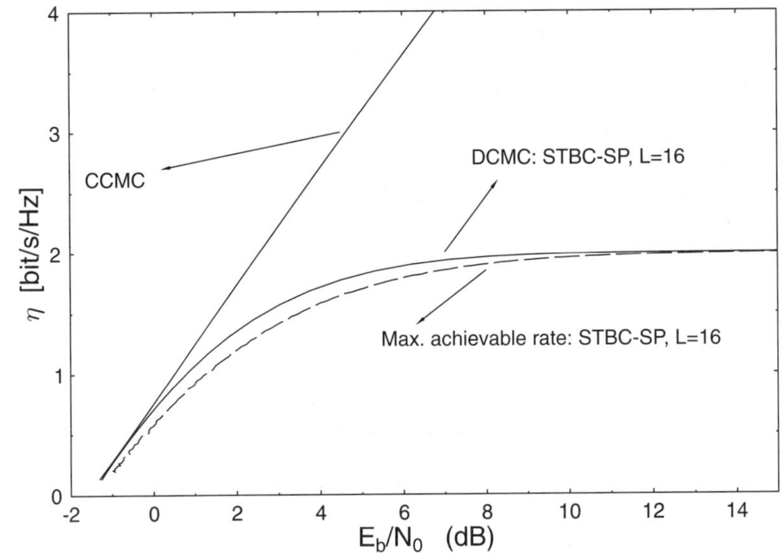

Figure 5.11: Bandwidth efficiency of the STBC-SP-based system with $L = 16$, when employing $N_t = 2$ transmit antennas and $N_r = 1$ receive antenna.

the EXIT charts using Equation (5.11) and Algorithm 5.1 constitutes a tighter bound on the maximum achievable bandwidth efficiency of the system.

Figure 5.11 shows that at a bandwidth efficiency of $\eta = 1$ bit s^{-1} Hz^{-1}, the capacity limit for the STBC-SP scheme is about $E_b/N_0 = 1.3$ dB, which is within 0.2 dB from the prediction of our EXIT chart analysis seen in Figure 5.10, where convergence is predicted at $E_b/N_0 = 1.5$ dB.

5.5 Performance of Three-stage Turbo-detected STBC-SP Schemes

5.5.1 System Parameters

In this chapter, we have considered a SP modulation scheme associated with $L = 16$ using two transmit and a single receiver antenna in order to demonstrate the performance improvements achieved by the proposed system. The communication channel is a SPSI-correlated Rayleigh fading channel, as described in Section 5.2.2. Three different outer encoder types were considered for Encoder I, namely a half-rate repeater, a half-rate memory-one RSC code and a half-rate memory-four IRCC constructed using the $P = 17$ subcodes combined according to the weighting coefficients of Equation (5.10). Encoder II is a simple rate-one accumulator, described by the pair of octal generator polynomials $(G/G_r) = (2/3)_8$. A three-stage iteration involves the following decoder activation sequence: (SP demapper–Decoder II–SP demapper–Decoder II–Decoder I–Decoder II). The overall system throughput is 1 BPS. All simulation parameters are listed in Table 5.2.

Table 5.2: System parameters.

Modulation	SP with $L = 16$
Number of transmitters	2
Number of receivers	1
Channel	Correlated Rayleigh fading
Normalized Doppler frequency	0.1
Outer Encoder I	**(1)** Half-rate repeater **(2)** Half-rate memory-one RSC code, $(G_r, G) = (3, 2)_8$ **(3)** Half-rate memory-four IRCC, $P = 17$, weighting coefficients of Equation (5.10)
Rate-one Encoder II	Rate-one memory-one RSC code, $(G/G_r) = (2/3)_8$
A system iteration	SP demapper $\rightarrow$ Decoder II $\rightarrow$ SP demapper $\rightarrow$ Decoder II $\rightarrow$ Decoder I $\rightarrow$ Decoder II
System throughput	1 BPS

5.5.2 Three-stage RA-coded STBC-SP Scheme

5.5.2.1 Decoding Trajectory

EXIT-chart-based convergence predictions are usually verified by the actual iterative decoding trajectory. Figure 5.8 shows that the three-stage turbo-detected RA-coded STBC-SP scheme is expected to converge at $E_b/N_0 = 1.8$ dB, where convergence to the $(1.0, 1.0)$ point requires an excessive number of three-stage iterations due to the convergence tunnel's bottleneck. Figure 5.12 illustrates the actual decoding trajectory of the three-stage RA-coded STBC-SP scheme of Figure 5.1 at $E_b/N_0 = 2.0$ dB, when using an interleaver depth of $\mathcal{D} = 10^6$ bits and $I_{3S} = 40$ three-stage iterations. The zigzag path seen in Figure 5.12 represents the actual extrinsic information transfer between the SP demapper and the rate-one Decoder II on one hand and the outer repeater Decoder I on the other.

5.5.2.2 BER Performance

Figure 5.13 compares the performance of the three-stage RA-coded STBC-SP scheme employing AGM-7 of Table A.8 against that of an identical-throughput 1 BPS uncoded STBC-SP scheme [43] using $L = 4$ as well as in comparison to Alamouti's conventional G_2-BPSK scheme [11]. The system is also benchmarked against the two-stage RSC-coded STBC-SP scheme of Figure 3.19 detailed in Section 3.5.1, when employing the system parameters outlined in Table 5.2 and using an interleaver depth of $\mathcal{D} = 10^6$ bits. Figure 5.13 demonstrates that the three-stage RA-coded scheme is prevented from achieving infinitesimally low BER values owing to the convergence tunnel's bottleneck seen in Figure 5.8, which is reflected as a residual BER floor in Figure 5.13. The two-stage turbo-detected STBC-SP scheme shown in Figure 5.13 also suffers from a residual BER floor, which is due to the absence of an open convergence tunnel to the $(1.0, 1.0)$ point, as seen in Figure 3.9. Observe that the two-stage turbo-detected STBC-SP scheme uses only $I_{2S} = 10$ two-stage iterations, since the advantage of employing any further iterations

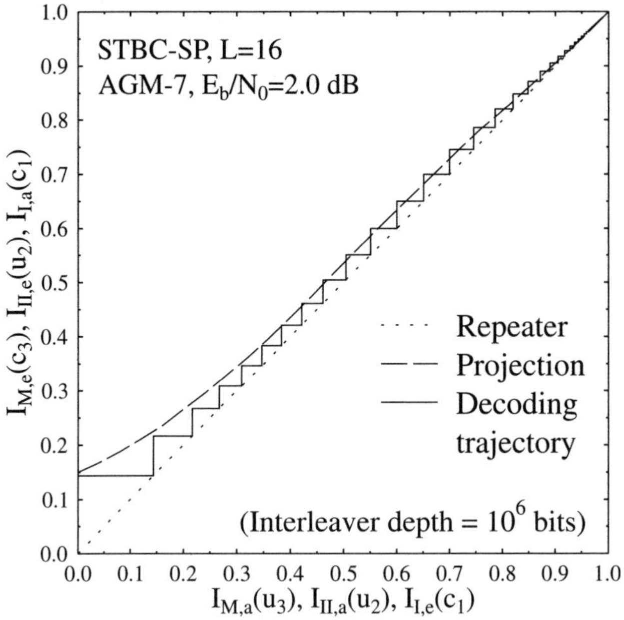

Figure 5.12: Decoding trajectory of the three-stage RA-coded STBC-SP scheme, when employing AGM-7 of Table A.8 in combination with the system parameters outlined in Table 5.2 and operating at $E_b/N_0 = 2.0$ dB with an interleaver depth of $\mathcal{D} = 10^6$ bits after $I_{3S} = 40$ three-stage iterations.

diminishes owing to the presence of a BER floor. The maximum rate limit seen in Figure 5.13 corresponds to the maximum achievable bandwidth efficiency calculated in Section 5.4.

5.5.2.3 Effect of Interleaver Depth

Figure 5.14 illustrates the effect of employing a shorter interleaver depth on the attainable performance of the three-stage RA-coded scheme. More specifically, Figure 5.14 characterizes the performance of the AGM-7-based RA-coded three-stage STBC-SP scheme in conjunction with $L = 16$, when employing the system parameters outlined in Table 5.2 and using an interleaver depth of $\mathcal{D} = 10^4$ bits. As expected, employing a shorter interleaver depth would render the convergence tunnel even narrower, which in turn is reflected as a higher BER floor, as shown in Figure 5.14.

5.5.3 Three-stage RSC-coded STBC-SP Scheme

5.5.3.1 Decoding Trajectory

Figure 5.7 shows that the three-stage RSC-coded STBC-SP scheme is expected to converge at $E_b/N_0 = 2.0$ dB. However, convergence to the $(1.0, 1.0)$ point requires an excessive number of three-stage iterations, since the convergence tunnel becomes extremely narrow at its central section. Nevertheless, convergence to the $(1.0, 1.0)$ point becomes more practical upon increasing the SNR beyond $E_b/N_0 > 2.0$ dB. Figure 5.15 illustrates the actual decoding trajectory of the three-stage RSC-coded STBC-SP scheme of Figure 5.1 at $E_b/N_0 = 2.2$ dB, when using an interleaver depth of $\mathcal{D} = 10^6$ bits and $I_{3S} = 28$ three-stage iterations.

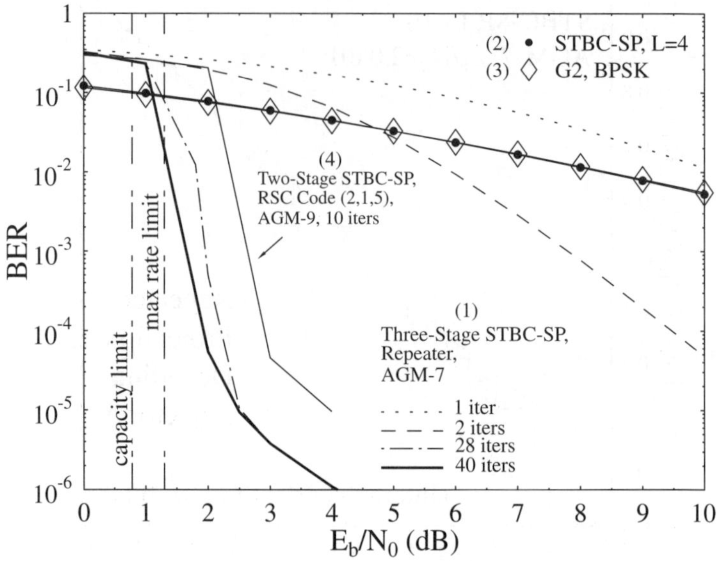

Figure 5.13: Performance comparison of the AGM-7-based (1) RA-coded three-stage STBC-SP scheme in conjunction with $L = 16$ against an identical-throughput 1 BPS uncoded STBC-SP scheme (2) using $L = 4$ and against Alamouti's conventional $\mathbf{G}_2$-BPSK scheme (3) as well as against a two-stage RSC-coded STBC-SP scheme (4) when employing the system parameters outlined in Table 5.2 and using an interleaver depth of $\mathcal{D} = 10^6$ bits.

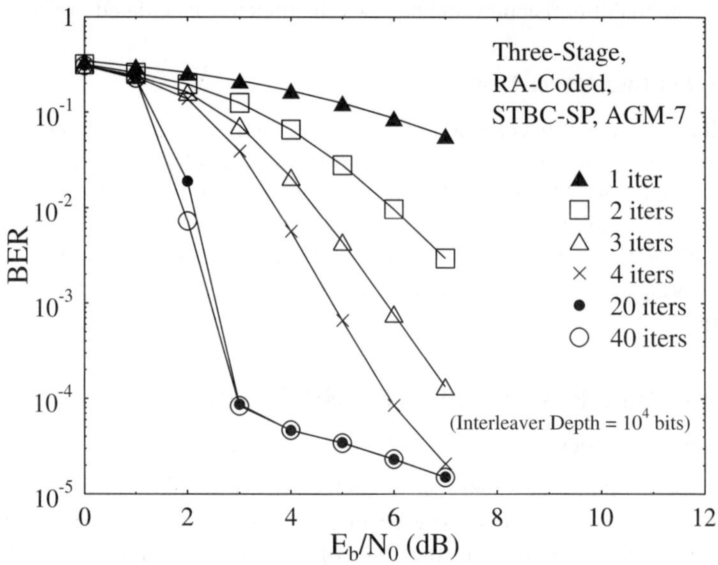

Figure 5.14: Performance of the AGM-7-based RA-coded three-stage STBC-SP scheme in conjunction with $L = 16$, when employing the system parameters outlined in Table 5.2 and using an interleaver depth of $\mathcal{D} = 10^4$ bits.

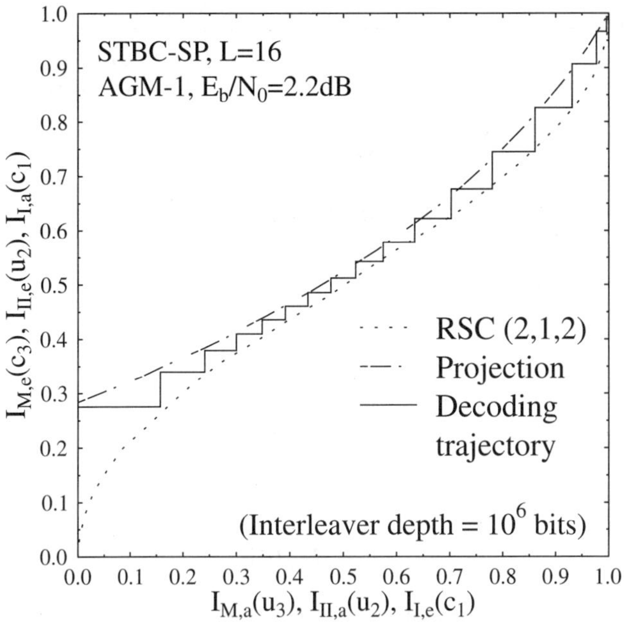

Figure 5.15: Decoding trajectory of the three-stage RSC-coded STBC-SP scheme, when employing the AGM-1 of Table A.2 in combination with the system parameters outlined in Table 5.2 and operating at $E_b/N_0 = 2.2$ dB with an interleaver depth of $\mathcal{D} = 10^6$ bits after $I_{3S} = 28$ three-stage iterations.

5.5.3.2 BER Performance

Figure 5.16 compares the performance of the three-stage RSC-coded STBC-SP scheme employing the AGM-1 scheme of Table A.2 against that of an identical-throughput 1 BPS uncoded STBC-SP scheme [43] using $L = 4$ as well as in comparison to Alamouti's conventional $\mathbf{G}_2$-BPSK scheme [11]. Again, the system is benchmarked against the two-stage RSC-coded STBC-SP scheme of Figure 3.19 detailed in Section 3.5.1, when employing the system parameters outlined in Table 5.2 and using an interleaver depth of $\mathcal{D} = 10^6$ bits. Figure 5.16 demonstrates that the three-stage RSC-coded scheme is capable of achieving infinitesimally low BER values owing to the absence of any BER floor, which is in contrast to the three-stage RA-coded scheme of Figure 5.13 and to the two-stage turbo-detected STBC-SP scheme of Figure 3.19 outlined in Section 3.5.1. Again, the two-stage turbo-detected STBC-SP scheme uses only $I_{2S} = 10$ two-stage iterations, since the advantage of employing any further iterations diminishes owing to the presence of a BER floor. Explicitly, Figure 5.16 demonstrates that a coding advantage of about 22.1 dB was achieved at a BER of 10^{-5} after $I_{3S} = 28$ iterations by the three-stage RSC-coded STBC-SP system over both the uncoded STBC-SP [43] and the conventional orthogonal STBC-design-based [11, 12] schemes for transmission over the correlated Rayleigh fading channel considered. In addition, a coding advantage of approximately 1.7 dB was attained over the 1 BPS throughput RSC-coded AGM-9 STBC-SP scheme of Figure 3.19 in Section 3.5.1 at the expense of an increased decoding complexity due to the employment of the rate-one decoder and the additional three-stage iterations.

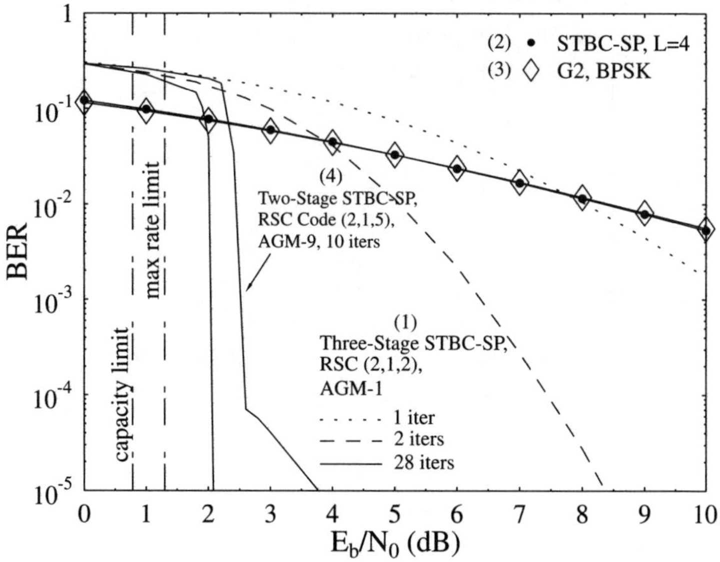

Figure 5.16: Performance comparison of the AGM-1-based (1) RSC-coded three-stage STBC-SP scheme in conjunction with $L = 16$ against an identical-throughput 1 BPS uncoded STBC-SP scheme (2) using $L = 4$ and against Alamouti's conventional $\mathbf{G}_2$-BPSK scheme (3) as well as against a two-stage RSC-coded STBC-SP scheme (4) when employing the system parameters outlined in Table 5.2 and using an interleaver depth of $\mathcal{D} = 10^6$ bits.

5.5.3.3 Effect of Interleaver Depth

The EXIT chart predictions are typically closely met when sufficiently high interleaver depths are employed, as mentioned in Section 5.3.1. Moreover, from a practical perspective it is always informative to investigate the achievable performance when employing shorter interleaver depths, while using different numbers of three-stage iterations. Figures 5.17 and 5.18 show the achievable coding gain of the three-stage STBC-SP scheme against Alamouti's conventional identical-throughput 1 BPS $\mathbf{G}_2$-BPSK scheme at BERs of 10^{-3} and 10^{-5}, respectively, when employing various interleaver depths and different numbers of three-stage iterations. In addition, Figure 5.19 investigates the effects of various interleaver depths on how closely the system approaches the capacity limit, when increasing the number of three-stage iterations. The capacity limit and maximum achievable limit seen in Figures 5.17 and 5.18 refer to the achievable coding gains of systems achieving capacity and achieving the maximum achievable bandwidth efficiency calculated in Section 5.4, respectively.

5.5.4 Three-stage IRCC-coded STBC-SP Scheme

5.5.4.1 Decoding Trajectory

Figure 5.10 shows that the three-stage IRCC-coded STBC-SP scheme is expected to converge at $E_b/N_0 = 1.5$ dB, where convergence to the $(1.0, 1.0)$ point requires an excessive number of three-stage iterations. However, convergence to the $(1.0, 1.0)$ point becomes more feasible for $E_b/N_0 > 1.5$ dB. Figure 5.20 illustrates the actual decoding trajectory of the three-stage

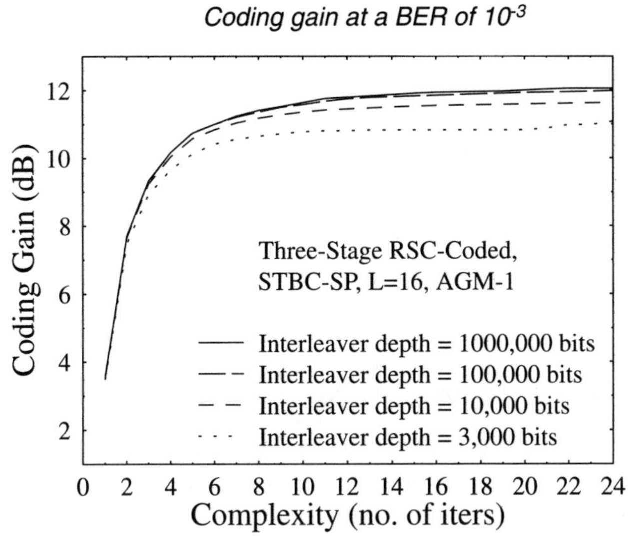

Figure 5.17: Achievable coding gain of the three-stage RSC-coded STBC-SP scheme against Alamouti's identical-throughput (1 BPS) conventional $\mathbf{G}_2$-BPSK scheme at a BER of 10^{-3} in combination with the system parameters outlined in Table 5.2, when employing various interleaver depths and differents numbers of three-stage iterations.

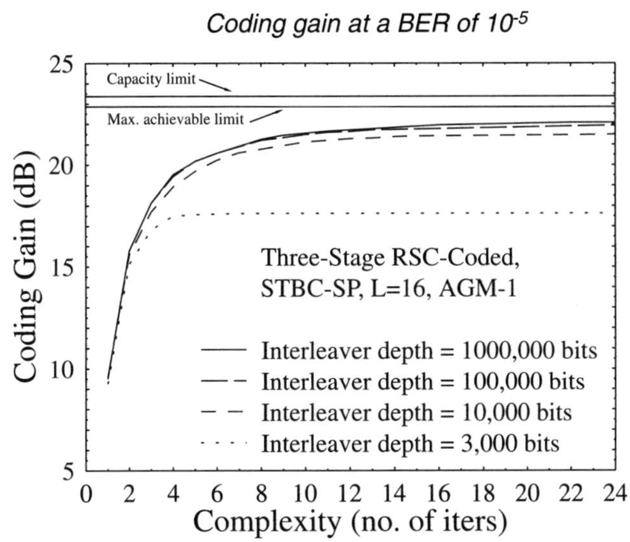

Figure 5.18: Achievable coding gain of the three-stage RSC-coded STBC-SP scheme against Alamouti's identical-throughput (1 BPS) conventional $\mathbf{G}_2$-BPSK scheme at a BER of 10^{-5} in combination with the system parameters outlined in Table 5.2, when employing various interleaver depths and different numbers of three-stage iterations.

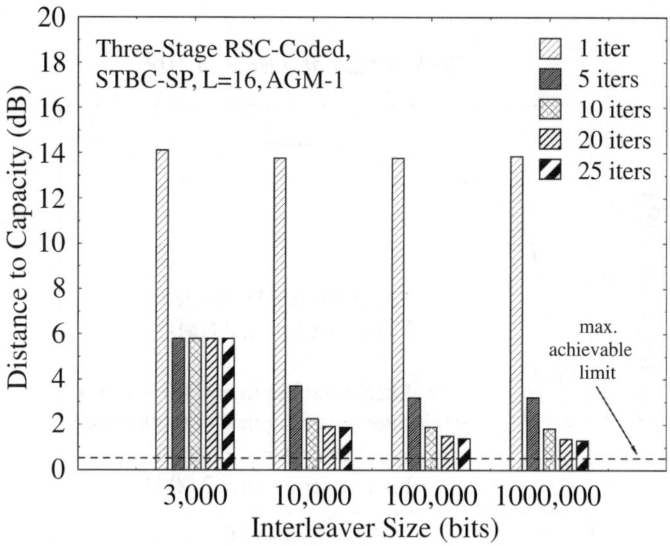

Figure 5.19: Achievable distance to capacity of the three-stage RSC-coded STBC-SP scheme in combination with the system parameters outlined in Table 5.2, when employing various interleaver depths and different numbers of three-stage iterations.

IRCC-coded STBC-SP scheme of Figure 5.1 at $E_b/N_0 = 1.8$ dB, when using an interleaver depth of $\mathcal{D} = 10^6$ bits and $I_{3S} = 33$ three-stage iterations. The zigzag path seen in Figure 5.20 represents the actual extrinsic information transfer between the SP demapper and the rate-one Decoder II on one hand and the outer IRCC Decoder I on the other.

5.5.4.2 BER Performance

Figure 5.21 compares the performance of the proposed three-stage IRCC-coded STBC-SP scheme employing AGM-7 against that of an identical-throughput 1 BPS uncoded STBC-SP scheme [43] using $L = 4$ and against Alamouti's conventional $\mathbf{G}_2$-BPSK scheme [11]. The system is again benchmarked against the two-stage RSC-coded STBC-SP scheme of Figure 3.19 in Section 3.5.1, when employing the system parameters outlined in Table 5.2 and using an interleaver depth of $\mathcal{D} = 10^6$ bits. Similar to the three-stage RSC-coded scheme of Figure 5.16, Figure 5.21 demonstrates that the proposed IRCC-coded turbo-detected scheme is capable of achieving infinitesimally low BER values. In other words, its performance is not limited by a BER floor. Explicitly, Figure 5.21 demonstrates that a coding advantage of about 22.3 dB was achieved at a BER of 10^{-5} after $I_{3S} = 28$ iterations by the three-stage turbo-detected IRCC-coded STBC-SP system over both the uncoded STBC-SP [43] and the conventional orthogonal STBC-design-based [11, 12] schemes for transmission over the correlated Rayleigh fading channel considered. In addition, a coding advantage of approximately 2.0 dB was attained over the 1 BPS throughput RSC-coded AGM-9 STBC-SP scheme of Figure 3.19 in Section 3.5.1 at the expense of an increased decoding complexity owing to the employment of the rate-one decoder and the additional three-stage iterations. According to Figure 5.21, the three-stage turbo-detected STBC-SP scheme operates within approximately 1.0 dB from the capacity limit of Equation (2.80) and 0.5 dB from the maximum achievable bandwidth efficiency limit of Equation (5.11).

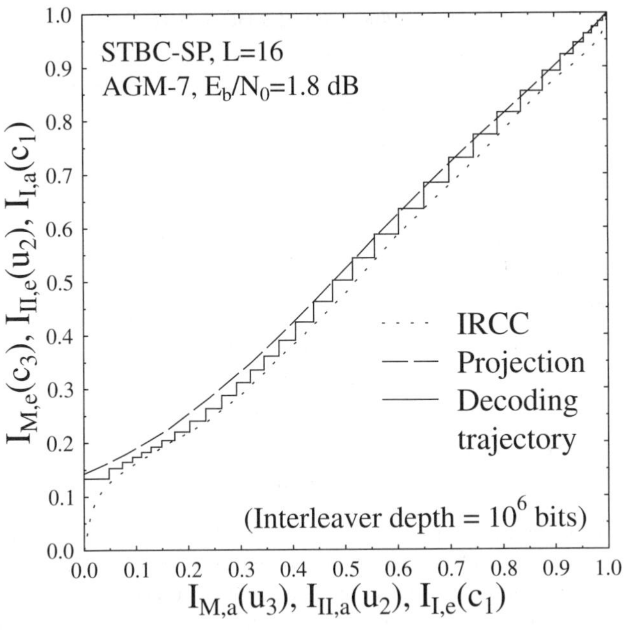

Figure 5.20: Decoding trajectory of the three-stage IRCC-coded STBC-SP scheme, when employing AGM-7 of Table A.8 in combination with the system parameters outlined in Table 5.2 and operating at $E_b/N_0 = 1.8$ dB with an interleaver depth of $\mathcal{D} = 10^6$ bits after $I_{3S} = 33$ three-stage iterations.

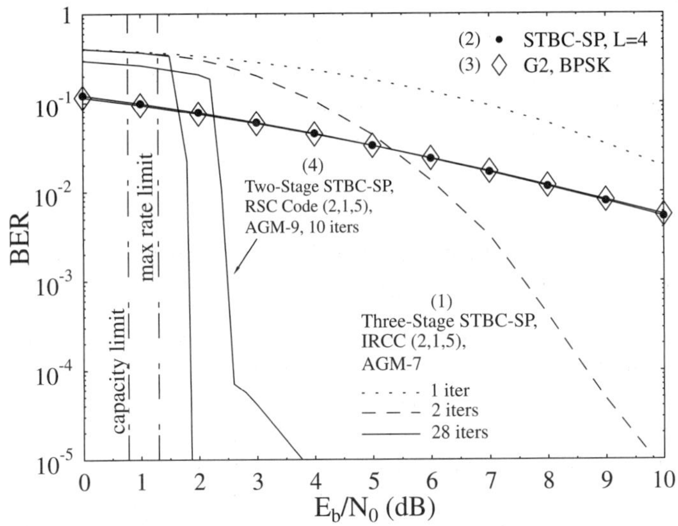

Figure 5.21: Performance comparison of the AGM-7-based (1) IRCC-coded three-stage STBC-SP scheme in conjunction with $L = 16$ against an identical-throughput 1 BPS uncoded STBC-SP scheme (2) using $L = 4$ and against Alamouti's conventional G_2-BPSK scheme (3) as well as against a two-stage RSC-coded STBC-SP scheme (4), when employing the system parameters outlined in Table 5.2 and using an interleaver depth of $\mathcal{D} = 10^6$ bits.

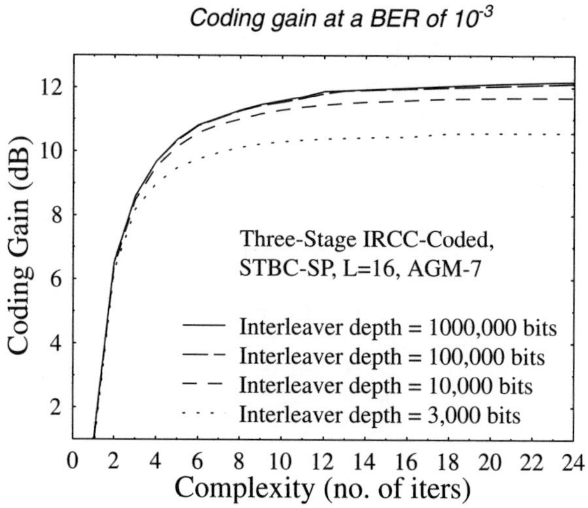

Figure 5.22: Achievable coding gain of the three-stage IRCC-coded STBC-SP scheme against Alamouti's identical-throughput (1 BPS) conventional $\mathbf{G}_2$-BPSK scheme at a BER of 10^{-3} in combination with the system parameters outlined in Table 5.2, when employing various interleaver depths and different numbers of three-stage iterations.

5.5.4.3 Effect of Interleaver Depth

The achievable performance was also investigated, when employing various interleaver depths, while using different number of three-stage iterations. Figures 5.22 and 5.23 show the achievable coding gain of the three-stage IRCC-coded STBC-SP scheme against Alamouti's conventional identical-throughput 1 BPS $\mathbf{G}_2$-BPSK scheme at BERs of 10^{-3} and 10^{-5}, respectively, when employing various interleaver depths and different numbers of three-stage iterations. In addition, Figure 5.24 investigates the effects of the interleaver depths on how closely the system approaches the capacity limit, when increasing the number of three-stage iterations.

5.5.5 Performance Comparison of RSC-coded and IRCC-coded Three-stage STBC-SP Schemes

In this section, the performance of the three-stage RSC-coded and IRCC-coded STBC-SP schemes is compared, when employing various interleaver depths, while using different numbers of three-stage iterations. Figures 5.25–5.27 compare the achievable coding gain of the three-stage RSC-coded and IRCC-coded STBC-SP schemes against Alamouti's identical-throughput 1 BPS conventional $\mathbf{G}_2$-BPSK scheme at a BER of 10^{-3} and 10^{-5} in combination with the system parameters outlined in Table 5.2, when employing different numbers of three-stage iterations and the interleaver depths of $\mathcal{D} = 10^4$, 10^5 and 10^6 bits, respectively. Figures 5.25–5.27 illustrate that the three-stage RSC-coded STBC-SP scheme always outperforms its IRCC-coded counterpart when employing a low number of three-stage iterations, i.e. less than $I_{3S} = 12$ iterations. This phenomenon is observed because the convergence tunnel of the three-stage RSC-coded scheme of Figure 5.7 exhibits a wider opening than that of the three-stage IRCC-coded scheme characterized in Figure 5.10, since the IRCC was specially designed for minimizing the EXIT tunnel's area and hence

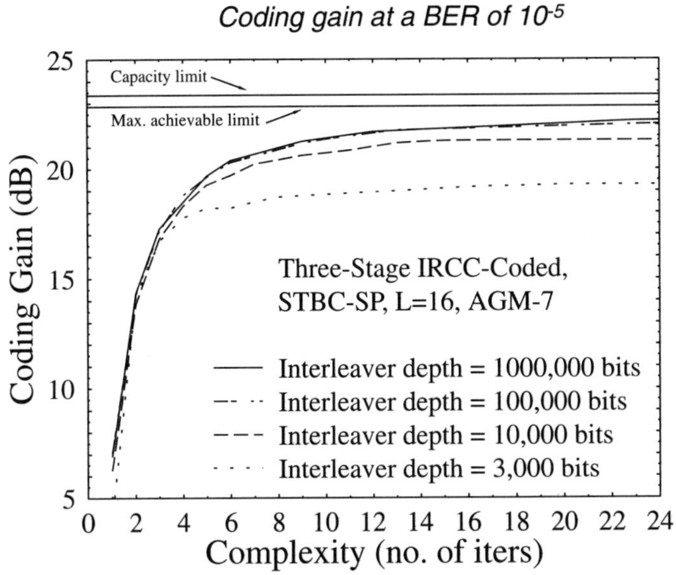

Figure 5.23: Achievable coding gain of the three-stage IRCC-coded STBC-SP scheme against Alamouti's identical-throughput (1 BPS) conventional $\mathbf{G}_2$-BPSK scheme at a BER of 10^{-5} in combination with the system parameters outlined in Table 5.2, when employing various interleaver depths and different numbers of three-stage iterations.

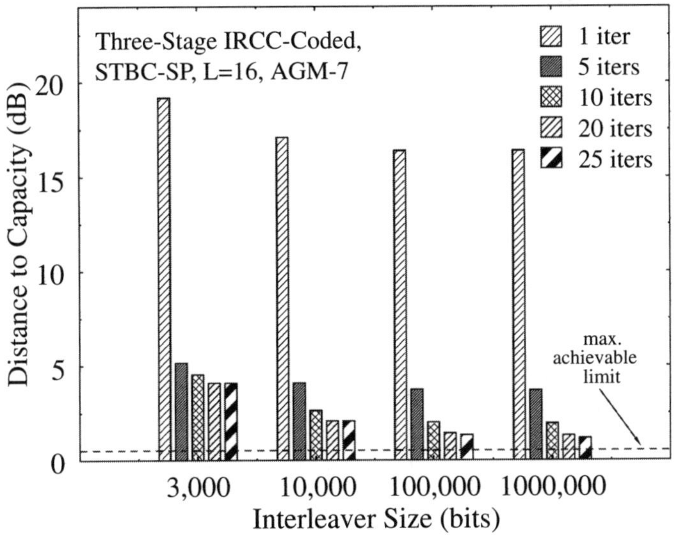

Figure 5.24: Achievable distance to capacity of the three-stage IRCC-coded STBC-SP scheme in combination with the system parameters outlined in Table 5.2, when employing various interleaver depths and different numbers of three-stage iterations.

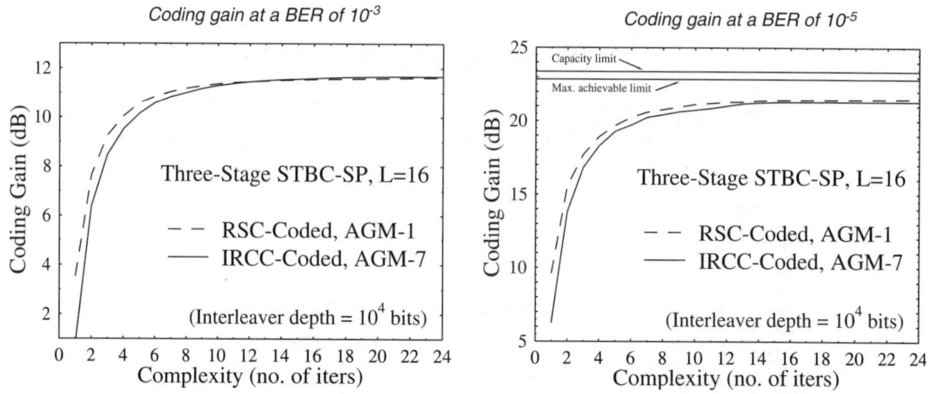

Figure 5.25: Comparison of achievable coding gain of the three-stage RSC and IRCC-coded STBC-SP schemes against Alamouti's identical-throughput (1 BPS) conventional $\mathbf{G}_2$-BPSK scheme at BER of 10^{-3} and 10^{-5} in combination with the system parameters outlined in Table 5.2, when employing different numbers of three-stage iterations and an interleaver depth of $\mathcal{D} = 10^4$ bits.

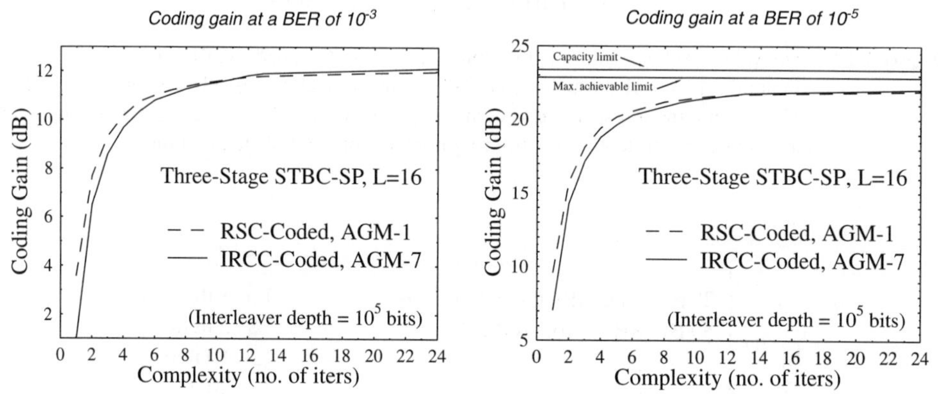

Figure 5.26: Comparison of achievable coding gain of the three-stage RSC and IRCC-coded STBC-SP schemes against Alamouti's identical-throughput (1 BPS) conventional $\mathbf{G}_2$-BPSK scheme at BER of 10^{-3} and 10^{-5} in combination with the system parameters outlined in Table 5.2, when employing different numbers of three-stage iterations and an interleaver depth of $\mathcal{D} = 10^5$ bits.

to facilitate convergence at low SNR values. However, the performance of the three-stage IRCC-coded STBC-SP scheme improves in comparison to its RSC-coded counterpart, when employing further iterations. In addition, the performance of the three-stage IRCC-coded STBC-SP scheme generally improves when increasing the interleaver depth, since the system performance approaches the predictions of the EXIT charts more closely.

5.6 Chapter Conclusions

We have proposed a three-stage serial concatenated turbo-detected STBC-SP scheme that is capable of achieving infinitesimally low BER values, where the performance is not limited

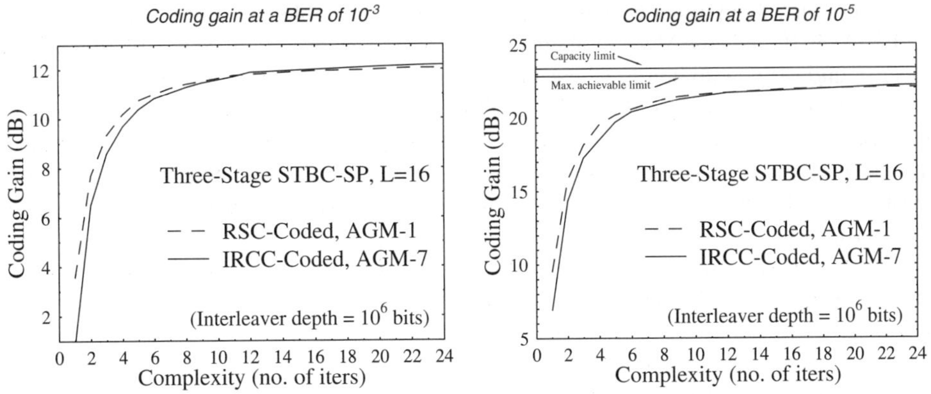

Figure 5.27: Comparison of achievable coding gain of the three-stage RSC and IRCC-coded STBC-SP schemes against Alamouti's identical-throughput (1 BPS) conventional $\mathbf{G}_2$-BPSK scheme at BER of 10^{-3} and 10^{-5} in combination with the system parameters outlined in Table 5.2, when employing different numbers of three-stage iterations and an interleaver depth of $\mathcal{D} = 10^6$ bits.

by a BER floor, which is routinely encountered in conventional two-stage systems. The convergence behavior of the three-stage system was analyzed with the aid of novel three-dimensional EXIT charts and their two-dimensional projections [174, 178]. With the advent of two-dimensional projections, an IRCC [175,176] was constructed for the sake of matching the projected EXIT curve of the SP demapper and the rate-one inner decoder leading to a near-capacity performance. The capacity of the STBC-SP scheme was calculated in Section 2.4.5, and a procedure was proposed in this chapter for calculating a tighter upper bound on the maximum achievable bandwidth efficiency of the three-stage system using EXIT chart analysis. Our proposed three-stage scheme operated within about 1.0 dB from the capacity limit and within 0.5 dB of the maximum achievable bandwidth efficiency limit. The performance of the three-stage schemes was investigated, when using various interleaver depths and employing different numbers of three-stage iterations.

5.7 Chapter Summary

In this chapter, we have proposed a three-stage serial concatenated turbo-detected STBC-SP scheme, which is capable of avoiding the formation of a pronounced BER floor. Section 5.2 provided a brief description of the proposed three-stage system, where the schematic of the entire system was shown in Figure 5.1. We considered three different types of channel codes for Encoder I, namely a repeater, a RSC code and an IRCC. The resultant schemes are outlined in Table 5.1. Our three-dimensional EXIT chart analysis was presented in Section 5.3.2, where its simplified two-dimensional projections were provided in Section 5.3.3. In Section 5.3.4, we employed the powerful technique of EXIT tunnel-area minimization for near-capacity operation. More specifically, we exploited the well-understood properties of conventional two-dimensional EXIT charts that a narrow but nonetheless open EXIT tunnel represents a near-capacity performance. Consequently, we invoked IRCCs for the sake of appropriately shaping the EXIT curves by minimizing the area within the EXIT-tunnel using the procedure of [175, 176].

In Section 5.4, an upper bound on the maximum achievable rate was calculated based on the EXIT chart analysis. More explicitly, a procedure was proposed for calculating a tighter upper bound of the maximum achievable bandwidth efficiency of STBC-SP schemes based on the area property of $\bar{\mathcal{A}}_I \approx R_I$ of the EXIT charts discussed in Section 5.3.4. The proposed procedure was applied in Section 5.4 for calculating the maximum achievable bandwidth efficiency of the three-stage turbo-coded STBC-SP scheme associated with the SP signal constellation size of $L = 16$ considered in this chapter. The design procedure was summarized in Algorithm 5.1.

The performance of the three-stage turbo-detected STBC-SP schemes was demonstrated and characterized in Section 5.5, where all simulation parameters were outlined in Section 5.5.1 and Table 5.2. In Section 5.5.2, we considered the performance of the three-stage RA-coded STBC-SP scheme, where the actual decoding trajectories, BER performance and the effect of interleaver depth on the achievable performance were discussed in Sections 5.5.2.1, 5.5.2.2 and 5.5.2.3, respectively. We observed from the BER curves seen in Figure 5.13 that the performance of the RA-coded STBC-SP scheme was limited by a BER floor, despite the employment of a recursive inner precoder. This observation is attributed to the convergence tunnel's bottleneck seen in Figure 5.8. The performance of the three-stage RSC-coded as well as IRCC-coded STBC-SP schemes was characterized in Sections 5.5.3 and 5.5.4, respectively. The effect of interleaver depth on the attainable performance of both the RSC-coded and the IRCC-coded schemes was investigated in Sections 5.5.2.3 and 5.5.4.3, respectively. The E_b/N_0 distance to capacity was summarized in Figures 5.19 and 5.24 for the three-stage RSC-coded as well as for the IRCC-coded STBC-SP schemes, respectively, when using the system parameters outlined in Table 5.2 and employing various interleaver depths as well as different numbers of three-stage iterations. Finally, in Section 5.5.5, the performance of the three-stage RSC-coded and IRCC-coded STBC-SP schemes was compared, when employing various interleaver depths, while using different numbers of three-stage iterations. More specifically, Figures 5.25–5.27 compared the achievable coding gain of the three-stage RSC-coded and IRCC-coded STBC-SP schemes against Alamouti's identical-throughput 1 BPS conventional $\mathbf{G}_2$-BPSK scheme at a BER of 10^{-3} and 10^{-5} in combination with the system parameters outlined in Table 5.2, when employing different numbers of three-stage iterations and the interleaver depths of $\mathcal{D} = 10^4$, 10^5 and 10^6 bits, respectively.

Both in this chapter and in all previous chapters, iterative decoding was employed at the bit level. In contrast, in the next chapter, we explore a range of further design options and propose a purely symbol-based scheme, where symbol-based turbo detection is carried out by exchanging extrinsic information between an outer non-binary LDPC code and a rate-one non-binary inner precoder. The motivation behind the development of this symbol-based scheme is that a reduced transmit power may be required when symbol-based rather than bit-based iterative decoding is employed [183].

Chapter 6

Symbol-based Channel-coded STBC-SP Schemes*

6.1 Introduction

In the previous chapters, iterative decoding was employed at the bit level. However, it was shown in [183] that a reduced transmit power may be required when symbol-based rather than bit-based iterative decoding is employed. Motivated by these observations, in this chapter we propose a purely symbol-based scheme, where symbol-based turbo detection is carried out by exchanging extrinsic information between an outer non-binary LDPC code and a rate-one non-binary inner precoder. The rationale of using a rate-one inner precoder is that it provides valuable extrinsic information at a low complexity and without reduction of the effective throughput.

Surprisingly, the family of LDPC codes originally devised by Gallager as early as 1962 [274] remained more or less unexploited until after the discovery of turbo codes in 1993 [146]. Since then, however, LDPC codes have experienced a renaissance [275] and attracted substantial research interest. MacKay and Neal demonstrated in [276] that despite their simple decoding structure, LDPC codes are also capable of operating near the channel capacity. Richardson and Urbanke [255] suggested the employment of a differential belief propagation decoding algorithm for binary LDPC codes using the fast Fourier transform (FFT) for reducing the decoding complexity imposed. In 1998, Davey and MacKay proposed a non-binary version of LDPC codes [277], which was potentially capable of outperforming binary LDPC codes. When using Richardson and Urbanke's FFT-based decoding algorithm [255], the complexity of non-binary LDPCs increases only linearly with respect to the size of the associated Galois field.

The convergence behavior of bit-based iterative decoding was studied in Chapter 3. More specifically, bit-based EXIT chart techniques were detailed in Section 3.4. Recently, EXIT charts have been extended to iterative non-binary coding schemes in [278], where a histogram-based approximation of the extrinsic information's PDF was required in order to compute the mutual information. Moreover, the evaluation of the extrinsic information's histogram becomes computationally demanding for a large number of bits per symbol. Hence,

*Parts of this chapter are based on the collaborative research outlined in [193].

a reduced-complexity method of computing non-binary EXIT charts was proposed in [184], which dispenses with the above-mentioned histogram computation and may be considered as a generalization of the approach presented in [279].

Motivated by the performance improvements reported in [43, 177, 183, 266, 277], we propose a novel symbol-based iterative scheme. We demonstrate that the proposed non-binary turbo-detection-aided STBC-SP scheme is capable of providing further performance improvements over both the STBC-SP scheme of [43] and a bit-based LDPC-coded turbo-detected STBC-SP scheme. The novel non-binary EXIT charts of [184, 278] are employed for designing our non-binary scheme. The rationale of the proposed architecture is explicit: (1) SP modulation maximizes the coding advantage of the transmission scheme by jointly designing and detecting the SP symbols hosting the two time slots' STBC symbols[1]; (2) the inner rate-one encoder and its low-complexity recursive decoder beneficially distributes the extrinsic information without reducing the effective throughput, maximizes the interleaver gain at a given length and hence avoids having a BER floor; and (3) symbol-based iterative decoding outperforms its bit-based counterpart.

This chapter is organized as follows. The system's architecture is presented in Section 6.2, where we describe the proposed symbol-based and turbo-detected scheme as well as its bit-based counterpart. Symbol-based iterative decoding is discussed in Section 6.3. Section 6.4 provides our EXIT chart analysis, while our simulation results are discussed in Section 6.5. Finally, we conclude in Section 6.6.

6.2 System Overview

6.2.1 Symbol-based LDPC-coded STBC-SP Scheme

The schematic of the non-binary arrangement is shown in Figure 6.1. The source bits are encoded by a rate $R = \frac{1}{2}$ non-binary LDPC encoder [277], to generate the LDPC encoded symbols $\mathbf{v} = (v_0, v_1, \ldots, v_{K_{ldpc}-1})$, $v_k \in GF(q)$, where K_{ldpc} is the LDPC output block length and q is the size of the LDPC decoding field. The LDPC encoded symbols are then precoded by a non-binary rate-one encoder, before each of them is mapped to the corresponding SP modulated symbol $\mathbf{s}^l \in S$, $0 \le l \le L-1$. There is a natural one-to-one mapping between l and the elements of the non-binary LDPC code defined over $GF(q)$, where we have $L = q$, allowing us to create a purely symbol-based system. Again, the rate-one precoder shown in Figure 6.1 is also a non-binary encoder, defined by the pair of octal generator polynomials $(G/G_r) = (2/3)_8$, where G denotes the feedforward output and G_r is the feedback to the input using a modulo q addition. Observe that channel interleaving is not required between the non-binary LDPC encoder and the rate-one encoder, since the LDPC parity check matrix is randomly constructed, where each of the parity check equations is checking several random $GF(q)$ symbol positions in a codeword, which has a similar effect to that of the channel interleaver. The SP mapper then maps the precoded symbols to legitimate four-dimensional SP modulated symbols $\mathbf{s}^l \in S$, where $S = \{\mathbf{s}^l = [a_{l,1}\ a_{l,2}\ a_{l,3}\ a_{l,4}] \in \mathbb{R}^4 : 0 \le l \le L-1\}$ constitutes a set of L legitimate constellation points selected from the lattice D_4 (see [221]) having a total energy of $E_{\text{total}} \triangleq \sum_{l=0}^{L-1}(|a_{l,1}|^2 + |a_{l,2}|^2 + |a_{l,3}|^2 + |a_{l,4}|^2)$, as detailed in Section 2.4.3.4. The STBC encoder then maps each SP modulated symbol $\mathbf{s}^l$ to

[1] In contrast, Alamouti detected two seemingly independent QPSK space-time symbols, although their amalgam constitutes a combined symbol.

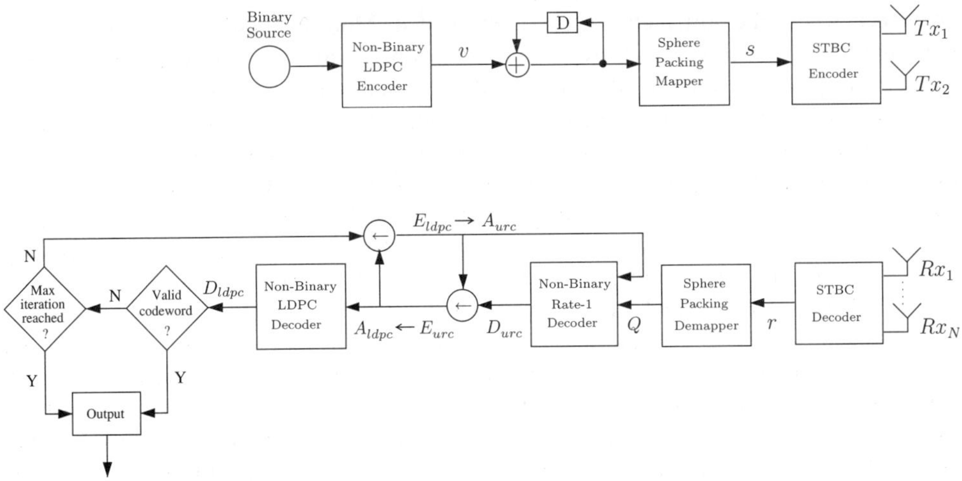

Figure 6.1: Symbol-based turbo-detection STBC-SP system.

the space-time signal $\mathbf{C}_l$ of Equation (2.63) as [43]

$$\mathbf{C}_l = \sqrt{\frac{2L}{E_{\text{total}}}}\mathbf{G}_2(x_{l,1}, x_{l,2}), \quad 0 \le l \le L-1,$$

where $x_{l,1}$ and $x_{l,2}$ are complex-valued symbols constructed from the four-dimensional real-valued coordinates of the SP symbol $\mathbf{s}^l$ in order to maximize the coding advantage of the space-time signal $\mathbf{C}_l$ [43], since the lattice D_4 has the best MED in the four-dimensional real-valued Euclidean space $\mathbb{R}^4$ (see [221]). Specifically, $x_{l,1}$ and $x_{l,2}$ may be written using Equation (8.63), which is reproduced here for convenience:

$$\{x_{l,1}, x_{l,2}\} = T_{\text{sp}}(a_{l,1}, a_{l,2}, a_{l,3}, a_{l,4}) = \{a_{l,1} + ja_{l,2}, a_{l,3} + ja_{l,4}\}.$$

Furthermore, $\mathbf{G}_2(x_{l,1}, x_{l,2})$ is the space-time transmission matrix of Equation (2.35), which is reproduced here for convenience:

$$\mathbf{G}_2(x_1, x_2) = \begin{bmatrix} x_1 & x_2 \\ -x_2^* & x_1^* \end{bmatrix},$$

where the rows and columns represent the temporal and spatial dimensions, corresponding to two consecutive time slots and two transmit antennas, respectively.

In this chapter, we consider a SPSI-correlated narrowband Rayleigh fading channel, based on the Jakes fading model [220] and associated with a normalized Doppler frequency of $f_D = f_d T_s = 0.1$, where f_d is the Doppler frequency and T_s is the symbol period. The complex fading envelope is assumed to be constant across the transmission period of a space-time coded symbol spanning $T = 2$ time slots. The complex AWGN of $n = n_I + jn_Q$ is also added to the received signal, where n_I and n_Q are two independent zero-mean Gaussian random variables having a variance of $\sigma_n^2 = \sigma_{n_I}^2 = \sigma_{n_Q}^2 = N_0/2$ per dimension, where $N_0/2$ represents the double-sided noise power spectral density expressed in W Hz^{-1}.

As shown in Figure 6.1, the received complex-valued symbols are first decoded by the STBC decoder to produce a received sphere-packed symbol $\mathbf{r}$, which is fed into the SP demapper, where the soft metric $\mathbf{Q}(k)$ is calculated. More explicitly, the notation $\mathbf{Q}(k)$ represents the soft metric passed from the SP demapper to the non-binary rate-one decoder based on the probability of the kth symbol of the encoded codeword by the rate-one encoder, as shown in Equation (6.4) of Section 6.3. As seen in Figure 6.1, the rate-one decoder processes these soft metrics in conjunction with the *a priori* information, $\mathbf{A}_{urc}$, in order to generate the APP, $\mathbf{D}_{urc}$, where the subscript *urc* refers to the URC. More specifically, the *a priori* information, $\mathbf{A}_{urc}$, is provided by the LDPC decoder as the soft metric for the LDPC encoded symbols. After removing the *a priori* information, $\mathbf{A}_{urc}$, from the APP denoted by $\mathbf{D}_{urc}$ using symbol-based element-wise division, as shown in Section 6.3, $\mathbf{A}_{ldpc}$ is passed as *a priori* information to the LDPC decoder, which carries out a specified number of LDPC iterations and produces the decoded APP $\mathbf{D}_{ldpc}$. Based on the APP, a tentative hard decision will be made and the resultant codeword will be checked by the LDPC code's parity check matrix. If the resultant vector is an all-zero sequence, then a legitimate codeword has been found, and the hard-decision-based sequence will be output. Otherwise, if the maximum affordable number of iterations has not been reached, the *a priori* information, $\mathbf{A}_{ldpc}$, is removed from the APP denoted by $\mathbf{D}_{ldpc}$ using symbol-based element-wise division and fed back to the non-binary rate-one decoder for the next iteration after appropriately reordering them using the interleaver, as seen in Figure 6.1. This process continues, until the affordable maximum number of iterations has been encountered or a legitimate codeword has been found.

6.2.2 Bit-based LDPC-coded STBC-SP Scheme

The structure of the bit-based scheme is shown in Figure 6.2, which is identical to its symbol-based counterpart seen in Figure 6.1, except that a binary, rather than non-binary, LDPC code is invoked in order to investigate the employment of bit interleaving and bit-based iterative decoding. Bit-to-symbol probability conversion is required, when passing extrinsic information from the binary LDPC decoder to the rate-one decoder. On the other hand, symbol-to-bit probability conversion is required, when passing extrinsic information from the rate-one decoder to the binary LDPC decoder. The bit probabilities at the input as well as at the output of the binary LDPC decoder are represented by their LLRs, where $L_{ldpc,a}$ refers to the *a priori* LLR values of the binary LDPC decoder, while $L_{ldpc,p}$ refers to the *a posteriori* LLR values.

6.3 Symbol-based Iterative Decoding

For the sake of simplicity, a system having a single receive antenna is considered, although its extension to systems having more than one receive antenna is straightforward. Assuming perfect channel estimation, the complex-valued channel output symbols received during two consecutive time slots are first diversity-combined in order to extract the estimates $\tilde{x}_1$ and $\tilde{x}_2$ of the most likely transmitted symbols $x_{l,1}$ and $x_{l,2}$ as was described in Section 2.4.3.2, resulting in

$$\tilde{x}_1 = (|h_1|^2 + |h_2|^2) \cdot x_{l,1} + \acute{n}_1,$$
$$\tilde{x}_2 = (|h_1|^2 + |h_2|^2) \cdot x_{l,2} + \acute{n}_2,$$

where h_1 and h_2 represent the complex-valued channel coefficients corresponding to the first and second transmit antenna, respectively, and $\acute{n}_1$ as well as $\acute{n}_2$ are zero-mean complex

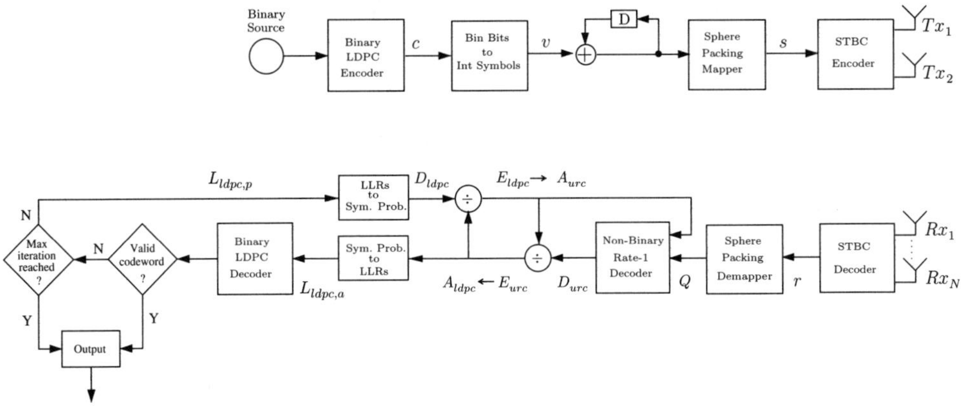

Figure 6.2: Bit-based turbo-detection STBC-SP system.

Gaussian random variables with variance $\sigma_{\tilde{n}}^2 = (|h_1|^2 + |h_2|^2) \cdot \sigma_n^2$. A received sphere-packed symbol $\mathbf{r}$ is then constructed from the estimates $\tilde{x}_1$ and $\tilde{x}_2$ using Equation (8.63) as

$$\mathbf{r} = T_{\text{sp}}^{-1}(\tilde{x}_1, \tilde{x}_2),$$

where $\mathbf{r} = \{[\tilde{a}_1\ \tilde{a}_2\ \tilde{a}_3\ \tilde{a}_4] \in \mathbb{R}^4\}$. The received sphere-packed symbol $\mathbf{r}$ can be written as shown in Equation (3.4), which is reproduced here for convenience:

$$\mathbf{r} = h \cdot \sqrt{\frac{2L}{E_{\text{total}}}} \cdot \mathbf{s}^l + \mathbf{w},$$

where $h = (|h_1|^2 + |h_2|^2)$, $\mathbf{s}^l \in S$, $0 \leq l \leq L-1$, and $\mathbf{w}$ is a four-dimensional real-valued Gaussian random variable having a covariance matrix of $\sigma_w^2 \cdot \mathbf{I}_{N_D} = \sigma_{\tilde{n}}^2 \cdot \mathbf{I}_{N_D} = h \cdot \sigma_n^2 \cdot \mathbf{I}_{N_D}$, where $N_D = 4$, since the symbol constellation S is four-dimensional. According to Equation (3.4), the conditional PDF $p(\mathbf{r}|\mathbf{s}^l)$ is given by

$$\begin{aligned} p(\mathbf{r}|\mathbf{s}^l) &= \frac{1}{(2\pi\sigma_w^2)^{N_D/2}} \exp\left[-\frac{1}{2\sigma_w^2}(\mathbf{r} - \alpha \cdot \mathbf{s}^l)(\mathbf{r} - \alpha \cdot \mathbf{s}^l)^{\mathrm{T}}\right], \\ &= \frac{1}{(2\pi\sigma_w^2)^{N_D/2}} \exp\left[-\frac{1}{2\sigma_w^2}\left(\sum_{i=1}^{4}(\tilde{a}_i - \alpha \cdot a_{l,i})2\right)\right], \end{aligned} \tag{6.1}$$

where we have $\alpha = h \cdot \sqrt{2L/E}$ and $(\cdot)^{\mathrm{T}}$ represents the transpose of a vector.

Similarly, the conditional PDF $p(\mathbf{s}^l|\mathbf{r})$ is given by

$$\begin{aligned} p(\mathbf{s}^l|\mathbf{r}) &= \frac{p(\mathbf{r}|\mathbf{s}^l) \cdot p(\mathbf{s}^l)}{p(\mathbf{r})}, \\ p(\mathbf{s}^l|\mathbf{r}) &= \frac{p(\mathbf{r}|\mathbf{s}^l) \cdot p(\mathbf{s}^l)}{\sum_{l=0}^{L-1}[p(\mathbf{r}|\mathbf{s}^l) \cdot p(\mathbf{s}^l)]}. \end{aligned} \tag{6.2}$$

Since the LDPC codeword consists of K_{ldpc} $GF(q)$ symbols, the SP demapper of Figure 6.1 will process K_{ldpc} received sphere-packed symbols, $(\mathbf{r}_0, \mathbf{r}_1, \ldots, \mathbf{r}_{K_{ldpc}-1})$, at a time to produce the following $(K_{ldpc} \times L)$ soft-metric matrix using Equation (6.2):

$$\mathbf{Q} = \left[\mathbf{Q}(0) \quad \mathbf{Q}(1) \quad \cdots \quad \mathbf{Q}(K_{ldpc}-1)\right]^{\mathrm{T}}, \tag{6.3}$$

where

$$\mathbf{Q}(k) = \left[p(\mathbf{s}_k = \mathbf{s}^0|\mathbf{r}_k) \quad p(\mathbf{s}_k = \mathbf{s}^1|\mathbf{r}_k) \quad \cdots \quad p(\mathbf{s}_k = \mathbf{s}^{L-1}|\mathbf{r}_k)\right], \tag{6.4}$$

for $k = 0, 1, \ldots, K_{ldpc} - 1$. All of the probabilities corresponding to a specific row in $\mathbf{Q}$, which correspond to a specific received symbol, should be normalized so that they sum up to unity.

The non-binary rate-one decoder of Figure 6.1 then processes the soft-metric matrix $\mathbf{Q}$ of Equation (6.3) in conjunction with the *a priori* information, $\mathbf{A}_{urc}$, in order to produce a decoded APP matrix, $\mathbf{D}_{urc}$, of size $(K_{ldpc} \times L)$ using a standard implementation of the forward–backward recursion based APP algorithm[2] [164]. During the first iteration, $p(\mathbf{s}^l)$, $0 \leq l \leq L-1$, seen in Equation (6.2) has to be set to $1/q$, since no *a priori* information is yet available from the LDPC decoder. The *a priori* knowledge fed into the rate-one decoder of Figure 6.1 is removed from the decoded APP matrix, $\mathbf{D}_{urc}$, using symbol-based element-wise division [278] for the sake of generating the extrinsic probability matrix, $\mathbf{E}_{urc}$, which is then fed into the LDPC decoder as the *a priori* knowledge, $\mathbf{A}_{ldpc}$, as alluded to before. More specifically, the following $(K_{ldpc} \times L)$ *a priori* information matrix is constructed:

$$\mathbf{A}_{ldpc} = \left[\mathbf{A}_{ldpc}(0) \quad \mathbf{A}_{ldpc}(1) \quad \cdots \quad \mathbf{A}_{ldpc}(K_{ldpc}-1)\right]^{\mathrm{T}}, \tag{6.5}$$

where

$$\mathbf{A}_{ldpc}(k) = \left[p(v_k = v^0) \quad p(v_k = v^1) \quad \cdots \quad p(v_k = v^{L-1})\right],$$

and we have

$$p(v_k = v^l) = \frac{(d_{urc})_k^l}{(a_{urc})_k^l}, \quad 0 \leq l \leq L-1 \text{ and } 0 \leq k \leq K_{ldpc} - 1, \tag{6.6}$$

while $(d_{urc})_k^l$ as well as $(a_{urc})_k^l$ refer to the elements at the cross-over point of the kth row and lth column of the matrices $\mathbf{D}_{urc}$ and $\mathbf{A}_{urc}$, respectively. In addition, v^l, $l = 0, \ldots, L-1$, represent all legitimate non-binary symbols at the input of the non-binary rate-one code and at the output of the non-binary LDPC code of Figure 6.1. Again, the probabilities corresponding to a specific row of the matrix $\mathbf{A}_{ldpc}$ should be normalized, so that the values add up to unity. The LDPC decoder exploits the *a priori* information, $\mathbf{A}_{ldpc}$, for the sake of producing a decoded soft-metric, $\mathbf{D}_{ldpc}$. Again, the *a priori* information, $\mathbf{A}_{ldpc}$, is removed from the decoded APP matrix, $\mathbf{D}_{ldpc}$, by symbol-based element-wise division for the sake of generating $\mathbf{E}_{ldpc}$, which is passed to the rate-one decoder of Figure 6.1 as the *a priori* knowledge, $\mathbf{A}_{urc}$, for further iterations, until a legitimate codeword is found or the affordable maximum number of iterations has been exhausted.

6.4 Non-binary EXIT Chart Analysis

6.4.1 Calculation of Non-binary EXIT Charts

The main objective of employing EXIT charts [169, 172] is to predict the convergence behavior of the iterative decoder by examining the evolution of the input/output mutual information exchange between the inner and outer decoders in consecutive iterations. Denoting the mutual information between two random variables X and Y as $I(X;Y)$, the

[2] The APP terminology is used here rather than the Maximum *A posteriori* Probability (MAP), since the soft-input soft-output rate-one decoder is computing probabilities rather than their maximums [155].

average *a priori* information, $I_{A_{urc}}$, at the input of the inner non-binary rate-one decoder and the average extrinsic information, $I_{E_{urc}}$, at the output of the inner non-binary rate-one decoder can be defined as [279]

$$\begin{aligned} I_{A_{urc}} &:= \frac{1}{M}\sum_{i=0}^{M-1} I(V_i; \mathbf{A}_{urc}(i)), \\ I_{E_{urc}} &:= \frac{1}{M}\sum_{i=0}^{M-1} I(V_i; \mathbf{E}_{urc}(i)), \end{aligned} \tag{6.7}$$

where V_i is an L-ary random variable representing the ith integer symbol, v_i, at the input of the rate-one encoder of Figure 6.1 and M is the total number of legitimate symbols v_i. Note that $\mathbf{A}_{urc}(i)$ and $\mathbf{E}_{urc}(i)$ are vectors of random variables corresponding to the ith row of the matrices $\mathbf{A}_{urc}$ and $\mathbf{E}_{urc}$, respectively. The transfer characteristic T_{urc} of the inner rate-1 decoder is a function of $I_{A_{urc}}$ and E_b/N_0 expressed as $I_{E_{urc}} = T_{urc}(I_{A_{urc}}, E_b/N_0)$. Similarly, the average *a priori* information, $I_{A_{ldpc}}$, at the input of the outer non-binary LDPC decoder and the average extrinsic information, $I_{E_{ldpc}}$, at the output of the outer non-binary LDPC decoder can be defined as

$$\begin{aligned} I_{A_{ldpc}} &:= \frac{1}{M}\sum_{i=0}^{M-1} I(V_i; \mathbf{A}_{ldpc}(i)), \\ I_{E_{ldpc}} &:= \frac{1}{M}\sum_{i=0}^{M-1} I(V_i; \mathbf{E}_{ldpc}(i)), \end{aligned} \tag{6.8}$$

where the transfer characteristic T_{ldpc} of the outer non-binary LDPC decoder is given by $I_{E_{ldpc}} = T_{ldpc}(I_{A_{ldpc}})$, which does not depend on the E_b/N_0 values. The exchange of extrinsic information in the system schematic of Figure 6.1 is visualized by plotting the extrinsic information transfer characteristics of the inner non-binary rate-one decoder and the outer non-binary LDPC decoder in a joint diagram. This diagram is known as the EXIT chart [169, 172]. As seen in Figure 6.1, the outer LDPC decoder's extrinsic output $I_{E_{ldpc}}$ becomes the inner rate-one decoder's *a priori* input $I_{A_{urc}}$, which is usually represented on the x-axis of the EXIT chart. Similarly, on the y-axis of the EXIT chart, the inner rate-one decoder's extrinsic output $I_{E_{urc}}$ becomes the outer LDPC decoder's *a priori* input $I_{A_{ldpc}}$, as seen in Figure 6.1.

The MI $I(V_i; \mathbf{A}_{urc}(i))$ in Equation (6.7) can be expressed as [278, 279]

$$I(V_i; \mathbf{A}_{urc}(i)) = \sum_{v_i=0}^{L-1} \int_{\mathrm{a}_{urc}(i)} p(\mathrm{a}_{urc}(i)|v_i)p(v_i)\log_2\left(\frac{p(\mathrm{a}_{urc}(i)|v_i)}{p(\mathrm{a}_{urc}(i))}\right) d\mathrm{a}_{urc}(i), \tag{6.9}$$

with

$$p(\mathrm{a}_{urc}(i)) = \sum_{v_i=0}^{L-1} p(\mathrm{a}_{urc}(i)|v_i)p(v_i), \tag{6.10}$$

and the *a priori* probabilities $p(v_i)$ for the indices v_i. The L-dimensional integration in (6.9) can be evaluated numerically, where the PDF $p(\mathrm{a}_{urc}(i)|v_i)$ may be obtained analytically by generating $\mathbf{A}_{urc}(i)$ according to the procedures described in Sections 6.4.2.1 and 6.4.2.2, depending on whether the $\log_2(L)$ bits corresponding to V_i are assumed to be independent or not. The term $I(V_i; \mathbf{E}_{urc}(i))$ can also be expressed using Equations (6.9) and (6.10), where $a_{urc}(i)$ is replaced with $e_{urc}(i)$. This requires the determination of the distribution of $p(e_{urc}(i)|v_i)$ by means of Monte Carlo simulations and computing an L-dimensional histogram [169, 278, 280]. However, a more efficient computation of non-binary EXIT functions was proposed in [184] that requires neither the computation of the L-dimensional histogram nor the evaluation of the L-dimensional integration in Equation (6.9). It was shown in [184] that by averaging over a sufficiently large number of blocks of length K_{ldpc}, the MI I_E can be estimated as

$$I_E = -H(V_1) + E\left\{ \frac{1}{K_{ldpc}} \sum_{i=0}^{K_{ldpc}-1} \sum_{l=0}^{L-1} e_k^l \right\}, \tag{6.11}$$

where e_k^l refers to the element at the cross-over point of the kth row and lth column of the matrices $\mathbf{E}_{urc}$ or $\mathbf{E}_{ldpc}$ and the entropy $H(V_1)$ can be readily determined from the *a priori* L-ary symbol distributions $p(v_i)$. For example, if we have $p(v_i = l) = 1/L$, for $l = 0, 1, \ldots, L-1$ (i.e. equiprobable L-ary symbols), then we arrive at $H(V_1) = -\log_2(L)$.

6.4.2 Generating the *A Priori* Symbol Probabilities

6.4.2.1 Case I: The Binary Bits of a Non-binary Symbol are Independent

Let $\mathbf{b}^{v_i} = (b_0^{v_i}, \ldots, b_{B-1}^{v_i})$ denote the binary vector corresponding to the natural mapping of the non-binary L-ary symbol v_i, where we have $L = 2^B$. The symbol probabilities corresponding to a non-binary L-ary symbol v_i, $i = 0, \ldots, K_{ldpc} - 1$, at the input of the non-binary rate-one decoder of Figure 6.2 are constructed from the LLRs of the randomly permuted bits $\mathbf{b}^{v_i} = (b_0^{v_i}, \ldots, b_{B-1}^{v_i})$, since a binary LDPC code is employed, which is equivalent to a bit interleaver, as alluded to in Section 6.2. Hence, for the bit-based turbo-detected scheme of Figure 6.2, it may be reasonably assumed that the binary bits within each non-binary symbol are independent of each other. In [278], a specific procedure was proposed for generating the *a priori* symbol probabilities for this case by observing the L-ary received symbol v_i arriving over a B-dimensional Gaussian channel, where again we have $B = \log_2(L)$. More specifically, the binary vector $\mathbf{b}^{v_i} = (b_0^{v_i}, \ldots, b_{B-1}^{v_i})$ is BPSK modulated to the following vector $\mathbf{x}^{v_i} = (x_0^{v_i}, \ldots, x_{B-1}^{v_i})$, where we have $x_k^{v_i} \in \{-1, +1\}$, $k = 0, \ldots, B-1$. Then, each BPSK modulated symbol $x_k^{v_i}$ is transmitted across an independent AWGN channel. This is equivalent to generating the *a priori* LLRs corresponding to the binary bits as described in Section 3.4. Hence, a simple LLRs to symbol probability conversion is carried out for the sake of generating the *a priori* matrix $\mathbf{A}$.

6.4.2.2 Case II: The Binary Bits of a Non-binary Symbol are not Independent

In this section, we consider the scenario when the binary bits within a non-binary symbol may no longer be assumed to be independent, because we employ symbol interleaving, as in the symbol-based scheme of Figure 6.1. Therefore, the procedure described in Section 6.4.2.1 may not be adopted to create *a priori* symbol probabilities that lead to sufficiently accurate EXIT charts. However, a beneficial procedure was proposed in [281, 282] that models the *a priori* symbol probabilities with sufficient accuracy. Let v denote an input symbol of the non-binary rate-one code, where the corresponding SP modulated symbol is s. In addition, let $f_{\mathrm{sp}}(\cdot)$ denote the SP mapping function, so that we have $\mathrm{s} = f_{\mathrm{sp}}(v)$. Assume that the SP modulated symbol s is transmitted across an AWGN channel, resulting in the following received signal:

$$\mathbf{r} = \mathbf{s} + \mathbf{n}, \tag{6.12}$$

where $\mathbf{n}$ is a four-dimensional real-valued Gaussian random variable having a covariance matrix of $\bar{\sigma}_n^2 \cdot \mathbf{I}_4$ and $\mathbf{I}_4$ is the identity matrix of size (4×4). The probability of occurrence for $\mathbf{s}$ is equal to that of v, since the SP mapping constitutes a memoryless operation, i.e. we have $p(\mathbf{s} = \mathbf{s}^i) = p(v = v^i)$, where $\mathbf{s}^i$, $i = 0, \ldots, L-1$, represent all legitimate SP modulated symbols, while v^i, $i = 0, \ldots, L-1$, denote all legitimate non-binary symbols at the input of the non-binary rate-one code, so that we have $\mathbf{s}^i = f_{\text{sp}}(v^i)$. The MI between $\mathbf{s}$ and $\mathbf{r}$, at a given probability of occurrence for $\mathbf{s}$, is given by [249]

$$\begin{aligned} I(\mathbf{s}, \mathbf{r}) &= \sum_{i=0}^{L-1} \int_{\mathbf{r}} p(\mathbf{s}^i, \mathbf{r}) \log_2 \left(\frac{p(\mathbf{s}^i, \mathbf{r})}{p(\mathbf{s}^i) p(\mathbf{r})} \right) d\mathbf{r}, \\ &= H(\mathbf{s}) - H(\mathbf{s}|\mathbf{r}), \end{aligned} \tag{6.13}$$

where $H(\mathbf{s})$ is the entropy of $\mathbf{s}$, expressed as

$$H(\mathbf{s}) = -\sum_{i=0}^{L-1} p(\mathbf{s}^i) \log_2(p(\mathbf{s}^i)), \tag{6.14}$$

and $H(\mathbf{s}|\mathbf{r})$ is the conditional entropy of $\mathbf{s}$ given $\mathbf{r}$, which is expressed as [246, 249]

$$H(\mathbf{s}|\mathbf{r}) = \sum_{i=0}^{L-1} p(\mathbf{s}^i) E\left[\log_2 \left(\sum_{j=0}^{L-1} \frac{p(\mathbf{s}^j)}{p(\mathbf{s}^i)} \exp(\mathbf{\Psi}_{i,j}) \right) \bigg| \mathbf{s}^i \right], \tag{6.15}$$

where $\exp(\mathbf{\Psi}_{i,j}) = p(\mathbf{r}|\mathbf{s}^j)/p(\mathbf{r}|\mathbf{s}^i)$ and $p(\mathbf{r}|\mathbf{s})$ is the conditional Gaussian PDF, while the exponent $\mathbf{\Psi}_{i,j}$ is written as [246]

$$\mathbf{\Psi}_{i,j} = \frac{-(\mathbf{s}^i - \mathbf{s}^j + \mathbf{n})(\mathbf{s}^i - \mathbf{s}^j + \mathbf{n})^{\mathrm{T}} + \mathbf{n} \cdot \mathbf{n}^{\mathrm{T}}}{2\bar{\sigma}_n^2}, \tag{6.16}$$

where $\mathbf{s}^i$, $\mathbf{s}^j$ and $\mathbf{n}$ are vectors of dimension (1×4) and $(\cdot)^{\mathrm{T}}$ denotes the transpose of a vector. Equation (6.13) may be further simplified, since the non-binary input symbols v are assumed to be equiprobable, which also means that we have $p(\mathbf{s}^i) = 1/L$, $i = 0, \ldots, L-1$. Accordingly, Equation (6.13) may be written as [246]

$$I(\mathbf{s}, \mathbf{r}) = \log_2(L) - \frac{1}{L} \sum_{i=0}^{L-1} E\left[\log_2 \left(\sum_{j=0}^{L-1} \exp(\mathbf{\Psi}_{i,j}) \right) \bigg| \mathbf{s}^i \right]. \tag{6.17}$$

The expectation $E[\cdot]$ in Equation (6.17) is calculated using a sufficiently high number of AWGN samples $\mathbf{n}$ with the aid of Monte Carlo simulations. Equation (6.17) is the maximum MI between $\mathbf{s}$ and $\mathbf{r}$, since $\mathbf{s}^i$, $i = 0, \ldots, L-1$, are equiprobably distributed. Therefore, the MI in Equation (6.17) represents the AWGN channel's capacity for the STBC MIMO system using the four-dimensional L-ary SP signal constellation for transmissions over the DCMC when employing $N_r = 1$ receive and $N_t = 2$ transmit antennas [246]. Figure 6.3 shows the capacity of the STBC-SP-based scheme for $L = 4$, 8 and 16, when employing $N_r = 1$ receive as well as $N_t = 2$ transmit antennas and communicating over an AWGN channel. Observe also from Equation (6.17) that $I(\mathbf{s}, \mathbf{r})$ may be expressed as a function of $\bar{\sigma}_n^2$ as follows [281, 282]:

$$I(\mathbf{s}, \mathbf{r}) = J(\bar{\sigma}_n^2), \quad J(\bar{\sigma}_n^2) \in [0, \log_2(L)], \tag{6.18}$$

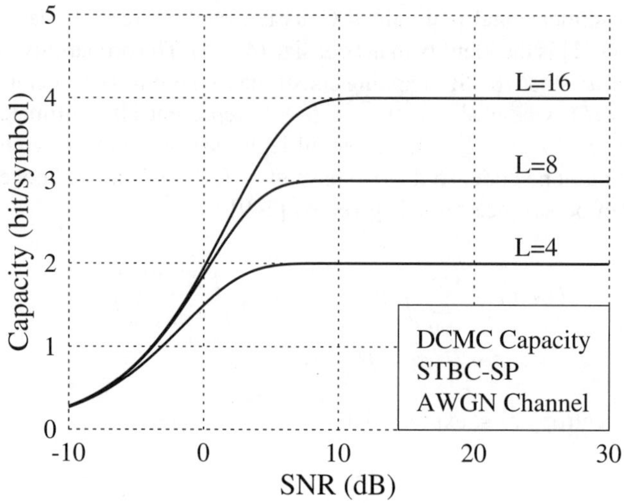

Figure 6.3: Capacity of the STBC-SP-based scheme with $L = 4$, 8 and 16, when employing $N_t = 2$ transmit antennas and $N_r = 1$ receive antenna and communicating over an AWGN channel.

where $J(\bar{\sigma}_n^2)$ is monotonically decreasing with respect to $\bar{\sigma}_n^2$. Figure 6.4 illustrates $J(\bar{\sigma}_n^2)$ as a function of $\bar{\sigma}_n^2$ for different SP constellation sizes L.

The $J(\bar{\sigma}_n^2)$ function of Figure 6.4 may now be used to generate a received signal $\bar{\mathbf{r}} = \mathbf{s} + \bar{\mathbf{n}}$, where the MI between s and $\bar{\mathbf{r}}$ is of a specific value I_A. More explicitly, let the *a priori* information of s be denoted as [281, 282]

$$I_A(\mathbf{s}) = I(\mathbf{s}, \mathbf{r}) = J(\bar{\sigma}_n^2). \tag{6.19}$$

Now, at a given $I_A(\mathbf{s})$ value, the corresponding noise variance $\bar{\sigma}_n^2$ may be found using the inverse of the $J(\bar{\sigma}_n^2)$ function of Figure 6.4, namely

$$\bar{\sigma}_n^2 = J^{-1}(I_A(\mathbf{s})). \tag{6.20}$$

Then, the following received signal is created:

$$\bar{\mathbf{r}} = \mathbf{s} + \bar{\mathbf{n}}, \tag{6.21}$$

where $\bar{\mathbf{n}}$ is an AWGN sample having a variance of $\bar{\sigma}_n^2$. Finally, the *a priori* symbol probabilities for $\mathbf{s}^i$, $i = 0, \ldots, L-1$, at the specific $I_A(\mathbf{s})$ value of interest may calculated as follows:

$$p(\mathbf{s} = \mathbf{s}^i) = \frac{1}{2\pi\bar{\sigma}_n^2} \exp\left(\frac{-(\bar{\mathbf{r}} - \mathbf{s}^i)(\bar{\mathbf{r}} - \mathbf{s}^i)^{\mathrm{T}}}{2\bar{\sigma}_n^2}\right), \tag{6.22}$$

which are used to create the matrices $\mathbf{A}_{urc}$ and $\mathbf{A}_{ldpc}$ containing our *a priori* knowledge, when generating the corresponding non-binary EXIT charts, where we have $p(\mathbf{s} = \mathbf{s}^i) = p(v = v^i)$, $i = 0, \ldots, L-1$. The procedure in Algorithm 6.1 summarizes the steps required for generating the matrix $\mathbf{A}$ of *a priori* knowledge corresponding to a vector of non-binary symbols $\mathbf{v} = (v_0, v_1, \ldots, v_{K_{ldpc}-1})$, where we have $v_k \in GF(L)$. Assume that ϵ is a small constant.

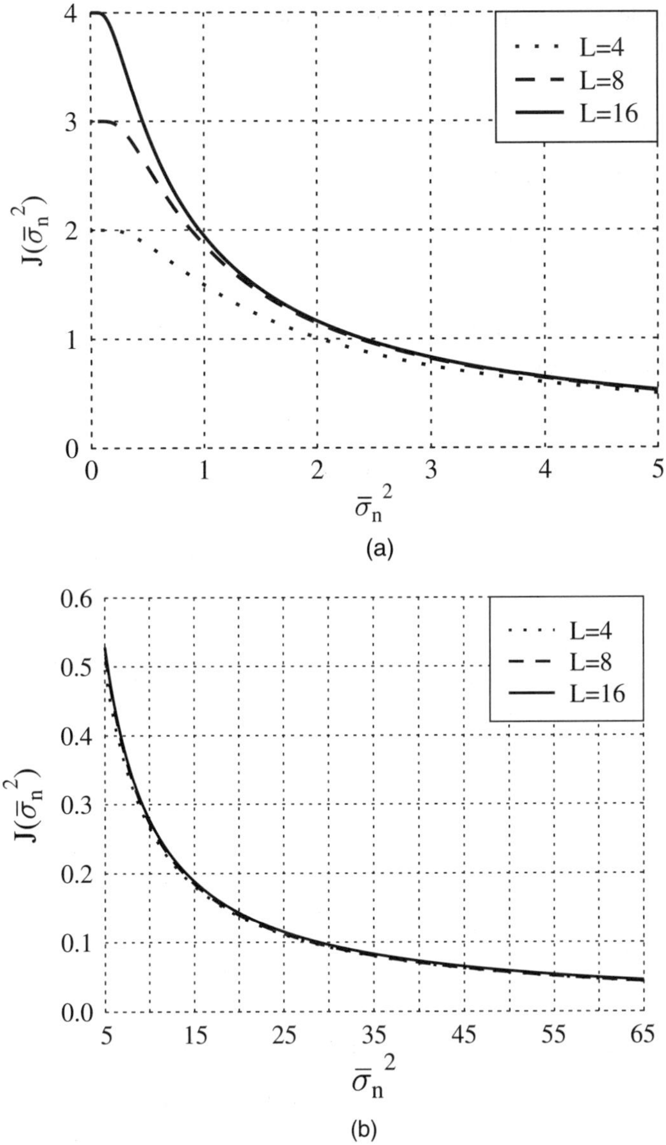

Figure 6.4: Plots of $J(\bar{\sigma}_n^2)$ as a function of $\bar{\sigma}_n^2$ for different SP constellation sizes L: (a) $\bar{\sigma}_n^2 \in [0, 5]$; (b) $\bar{\sigma}_n^2 \in [5, \ldots, 65]$.

6.4.3 EXIT Chart Results

6.4.3.1 EXIT Charts of Symbol-based Schemes

The EXIT charts of all symbol-based schemes were created assuming that the binary bits corresponding to a non-binary symbol are not independent, hence the procedure described in Section 6.4.2.2 was employed when creating the *a priori* information. Figure 6.5 shows the EXIT chart of the symbol-based LDPC-coded STBC-SP scheme of Figure 6.1 in combination with $L = 4$ and the half-rate outer LDPC code [277] defined over $GF(4)$, when operating

Step 1: Compute $J(\bar{\sigma}_n^2)$ using Equations (6.17) and (6.18).

Step 2: Let $I_A(\mathbf{s}) = 0$.

Step 3: Calculate $\bar{\sigma}_n^2 = J^{-1}(I_A(\mathbf{s}))$.

Step 4: Create a noise sample $\bar{\mathbf{n}}$ having a variance of $\bar{\sigma}_n^2$.

Step 5: Create a received signal $\bar{\mathbf{r}} = \mathbf{s} + \bar{\mathbf{n}}$, where $\mathbf{s} = f_{\text{sp}}(v_k)$.

Step 6: Calculate $p(\mathbf{s} = \mathbf{s}^i)$, $i = 0, \ldots, L-1$, from Equation (6.22).

Step 7: Normalize the L probabilities from **Step 6** and save them in the kth row of the *a priori* matrix $\mathbf{A}$.

Step 8: Let $I_A(\mathbf{s}) = I_A(\mathbf{s}) + \epsilon$. If $I_A(\mathbf{s}) \leq \log_2(L)$, go to **Step 3**.

Step 9: Output the *a priori* matrix $\mathbf{A}$ from **Step 7**.

Algorithm 6.1: Generating the *a priori* symbol probabilities, when the binary bits within a non-binary symbol are not independent [281, 282].

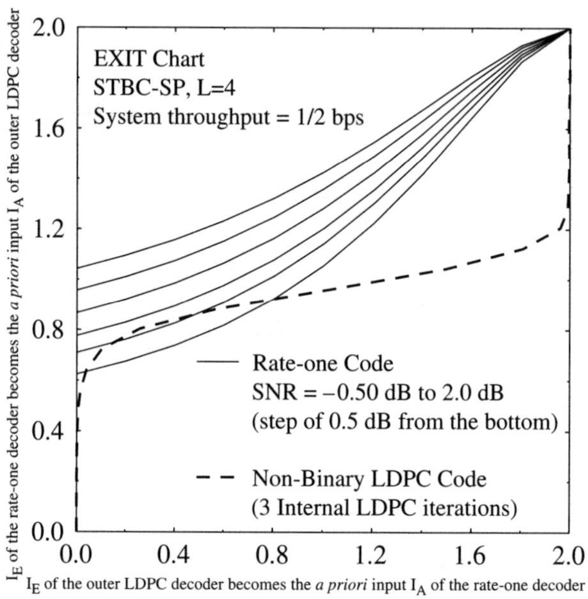

Figure 6.5: EXIT chart of a symbol-based LDPC-coded STBC-SP scheme in combination with the half-rate outer LDPC code defined over $GF(4)$, using three internal LDPC iterations and the system parameters outlined in Table 6.1.

at different SNR values and using $I_{\text{int}} = 3$ internal LDPC iterations as well as the system parameters outlined in Table 6.1. This is referred to as Scheme 1. Ideally, in order for the exchange of extrinsic information between the rate-one decoder and the outer LDPC decoder of Figure 6.1 to converge at a specific SNR value, the EXIT curve of the rate-one decoder recorded at the SNR value of interest and that of the outer LDPC decoder should only

Table 6.1: System parameters.

	Scheme 1	Scheme 2	Scheme 3
SP constellation size	$L = 4$	$L = 8$	$L = 16$
Number of transmitters		2	
Number of receivers		1	
Channel		Correlated Rayleigh fading	
Normalized Doppler frequency		0.1	
Average LDPC column weight		2.5	
LDPC coding rate		$\frac{1}{2}$	
Non-binary LDPC decoding field	$GF(4)$	$GF(8)$	$GF(16)$
Rate-one decoding field	$GF(4)$	$GF(8)$	$GF(16)$
System throughput (BPS)	$1/2$	$3/4$	1
LDPC-coded blocklength		1488 to 12 000 bits	

Table 6.2: The approximate SNR values where a turbo cliff is expected to occur, based on the EXIT chart curves for Schemes 1, 2 and 3 outlined in Table 6.1.

	Approximate SNR values (dB)	
	Symbol-based scheme	Bit-based scheme
Scheme 1	0.2	0.6
Scheme 2	2.0	2.3
Scheme 3	4.2	4.2

intersect at the point of $(I_A, I_E) = (2.0, 2.0)$. If this condition is satisfied, then a so-called open convergence tunnel [169, 172] appears in the EXIT chart. The narrower the tunnel, the closer the system's performance to the channel capacity [270] and hence in the spirit of Shannon's information theory more iterations are required for reaching the $(2.0, 2.0)$ point. Observe in Figure 6.5 that an open convergence tunnel is starting to take shape for SNR values higher than 0.0 dB. This implies that according to the predictions of the EXIT chart seen in Figure 6.5, the symbol-based Scheme 1 of Table 6.1 is expected to exhibit a BER turbo cliff at a SNR slightly higher than 0.0 dB.

Figures 6.6 and 6.7 show the EXIT chart of the symbol-based LDPC-coded STBC-SP scheme of Figure 6.1 in combination with $L = 8$ and 16, while employing the half-rate outer LDPC code [277] defined over $GF(8)$ and $GF(16)$, respectively, when operating at different SNR values and using $I_{\text{int}} = 3$ internal LDPC iterations as well as the system parameters outlined in Table 6.1. These two schemes are referred to as Schemes 2 and 3 in Table 6.1, respectively. The EXIT chart of Figure 6.6 shows that the symbol-based Scheme 2 is expected to exhibit an open convergence tunnel and hence a BER turbo cliff at SNR values in the proximity of 2.0 dB. Similarly, the EXIT chart of Figure 6.7 demonstrates that the symbol-based Scheme 3 of Table 6.1 is expected to exhibit a BER turbo cliff at SNR values in the proximity of 4.0 dB.

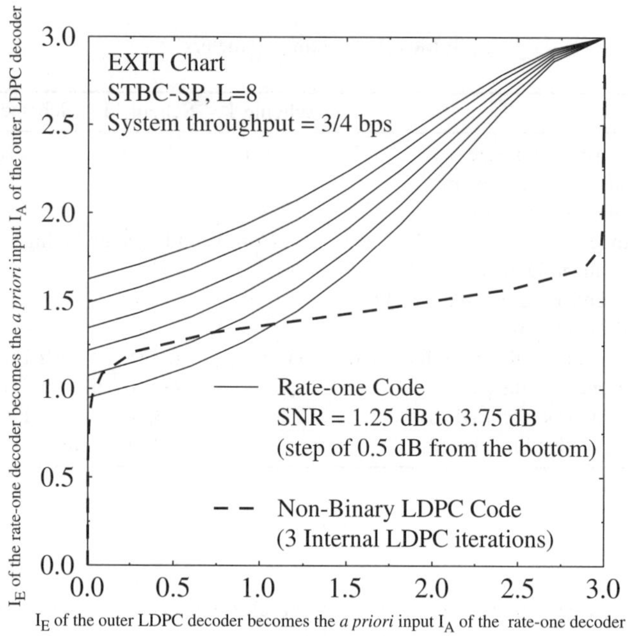

Figure 6.6: EXIT chart of a symbol-based LDPC-coded STBC-SP scheme in combination with the half-rate outer LDPC code defined over $GF(8)$, using three internal LDPC iterations and the system parameters outlined in Table 6.1.

6.4.3.2 EXIT Charts of Bit-based Schemes

The EXIT charts of all bit-based schemes were created assuming that the binary bits corresponding to a non-binary symbol are independent, hence the procedure described in Section 6.4.2.1 was employed when creating the *a priori* information. The EXIT charts of the bit-based schemes may also be created using binary EXIT charts [169, 172] by modeling the LLRs shown in Figure 6.2, as described in Section 3.4. Figures 6.8–6.10 illustrate the EXIT chart of the bit-based LDPC-coded STBC-SP scheme of Figure 6.2 in combination with $L = 4$, 8 and 16, respectively, while employing the half-rate outer binary LDPC code [274] defined over $GF(2)$, when operating at different SNR values and using $I_{\text{int}} = 3$ internal LDPC iterations as well as the system parameters outlined in Table 6.1. According to the EXIT charts seen in Figures 6.8–6.10, the bit-based schemes associated with $L = 4$, 8 and 16 are expected to converge at SNR values of approximately 0.5, 2.25 and 4.0 dB, respectively.

6.4.3.3 Comparison of the EXIT Charts of Symbol-based and Bit-based Schemes

In this section, we shed further light on the convergence behavior of both the symbol-based and the bit-based schemes by comparing their EXIT curves. Figure 6.11 shows the EXIT chart of the symbol-based LDPC-coded STBC-SP scheme of Figure 6.1 in combination with $L = 4$ and the half-rate outer LDPC code [277] defined over $GF(4)$, when operating at a SNR of 0.5 dB and using $I_{\text{int}} = 3$ internal LDPC iterations as well as the system parameters outlined in Table 6.1. Figure 6.11 also shows the EXIT chart of an identical-throughput $\frac{1}{2}$ BPS bit-based LDPC-coded STBC-SP scheme in combination with the half-rate outer LDPC code defined over $GF(2)$. Observe in Figure 6.11 that the symbol-based Scheme 1 of Table 6.1 exhibits an open convergence tunnel at $\text{SNR} = 0.5$ dB, while the equivalent bit-based scheme

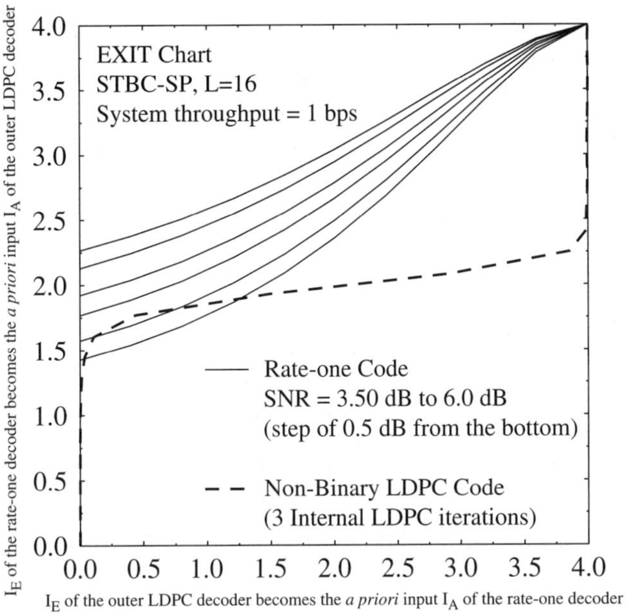

Figure 6.7: EXIT chart of a symbol-based LDPC-coded STBC-SP scheme in combination with the half-rate outer LDPC code defined over $GF(16)$, using three internal LDPC iterations and the system parameters outlined in Table 6.1.

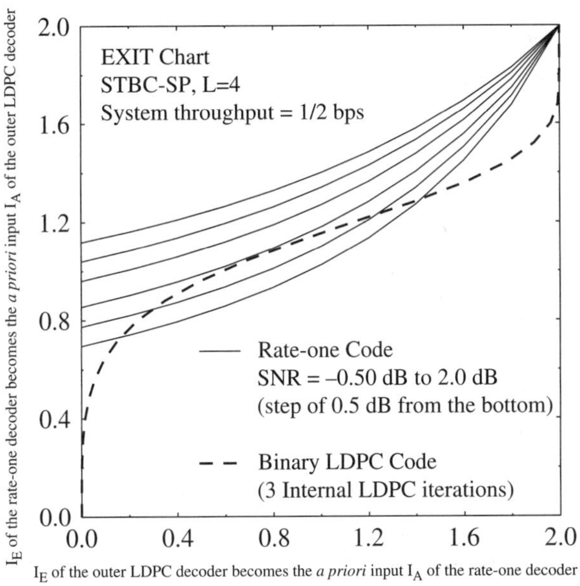

Figure 6.8: EXIT chart of a bit-based LDPC-coded STBC-SP scheme in combination with the half-rate outer LDPC code defined over $GF(2)$, using three internal LDPC iterations and the system parameters outlined in Table 6.1.

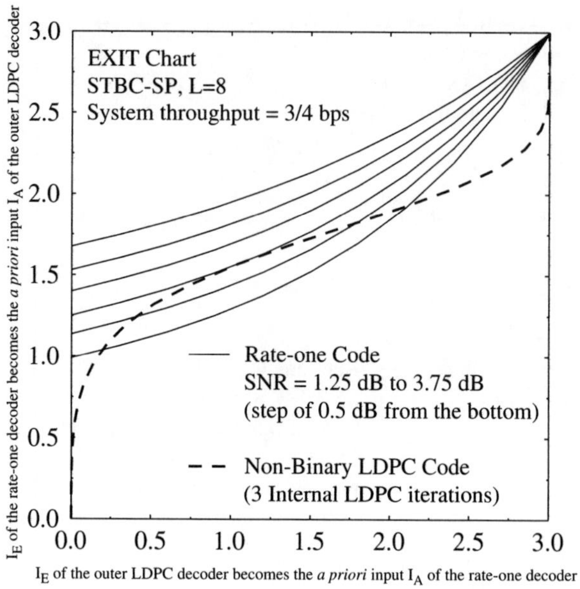

Figure 6.9: EXIT chart of a bit-based LDPC-coded STBC-SP scheme in combination with the half-rate outer LDPC code defined over $GF(2)$, using three internal LDPC iterations and the system parameters outlined in Table 6.1.

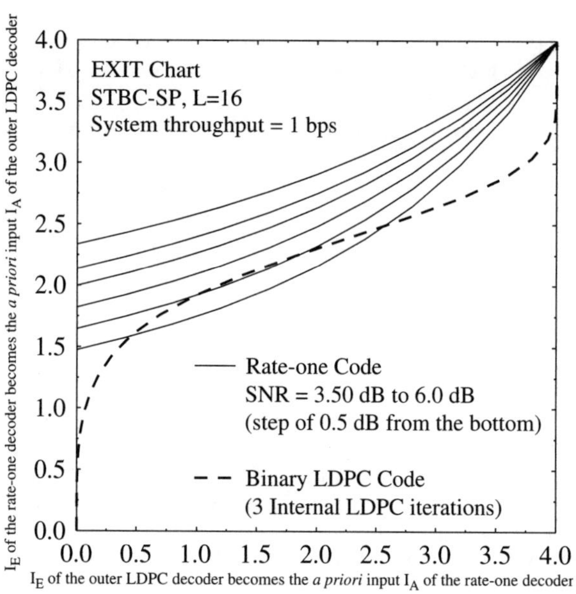

Figure 6.10: EXIT chart of a bit-based LDPC-coded STBC-SP scheme in combination with the half-rate outer LDPC code defined over $GF(2)$, using three internal LDPC iterations and the system parameters outlined in Table 6.1.

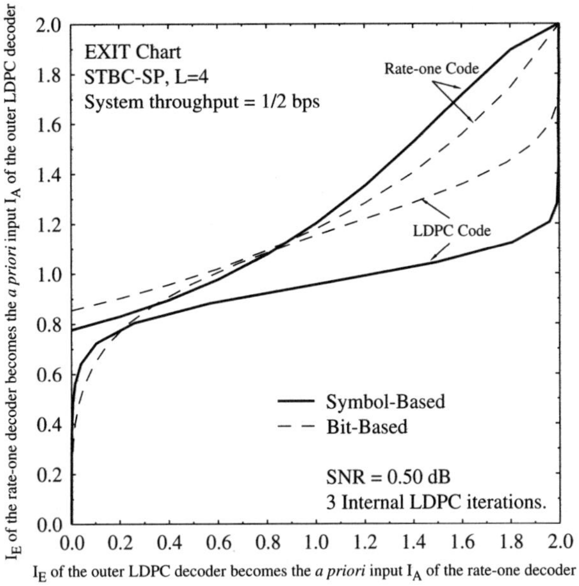

Figure 6.11: EXIT chart of symbol-based and bit-based LDPC-coded STBC-SP schemes in combination with the half-rate outer LDPC code defined over $GF(4)$ and $GF(2)$, respectively, using three internal LDPC iterations and the system parameters outlined in Table 6.1.

requires higher SNR values before an open convergence tunnel can be formed. This implies that according to the predictions of the EXIT chart seen in Figure 6.11, the symbol-based Scheme 1 of Table 6.1 is expected to have a lower convergence threshold than its bit-based counterpart and hence the former will exhibit a BER turbo cliff at a lower SNR value.

Figures 6.12 and 6.13 show the EXIT charts of the symbol-based Schemes 2 and 3 of Table 6.1, when using $I_{\text{int}} = 3$ internal LDPC iterations and operating at SNRs of 2.25 and 4.50 dB, respectively. Figures 6.12 and 6.13 also show the EXIT charts of the equivalent-throughput bit-based schemes. Observe in Figure 6.13 that although both the symbol-based and bit-based schemes require similar SNR values in order to exhibit an open convergence tunnel, the symbol-based scheme exhibits a wider tunnel. Hence, a lower number of iterations is needed to reach the convergence point of $(I_A, I_E) = (4.0, 4.0)$. These EXIT-tunnel-based convergence predictions are usually verified by the actual EXIT trajectory of iterative decoding as well as by the BER curves, as is discussed in Section 6.5. In general, once an open convergence tunnel is formed, the symbol-based schemes always exhibit a wider tunnel than the bit-based schemes, which leads to a lower number of required iterations. The SNR values predicted by the EXIT charts where a turbo cliff is expected to occur are listed in Table 6.2 for both the symbol-based and the bit-based LDPC-coded STBC-SP schemes.

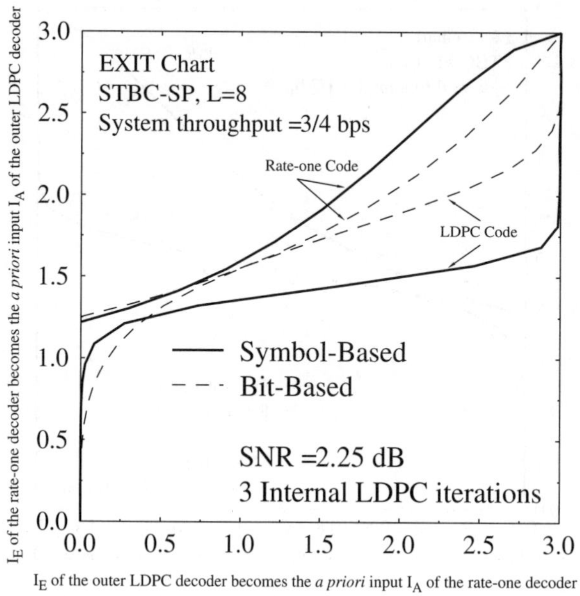

Figure 6.12: EXIT chart of symbol-based and bit-based LDPC-coded STBC-SP schemes in combination with the half-rate outer LDPC code defined over $GF(8)$ and $GF(2)$, respectively, using three internal LDPC iterations and the system parameters outlined in Table 6.1.

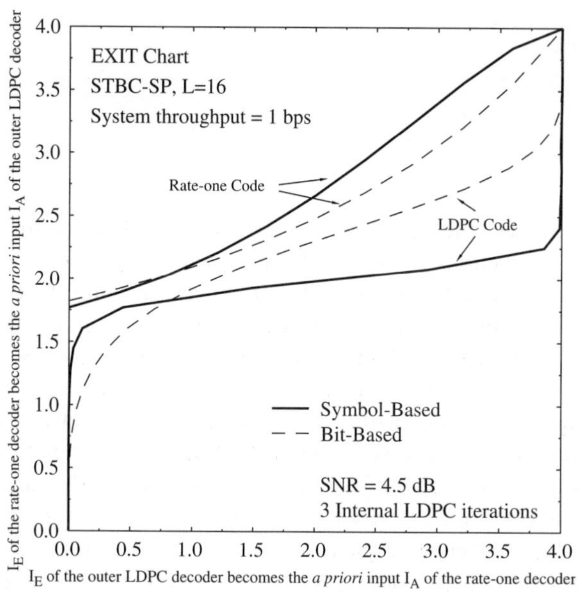

Figure 6.13: EXIT chart of symbol-based and bit-based LDPC-coded STBC-SP schemes in combination with the half-rate outer LDPC code defined over $GF(16)$ and $GF(2)$, respectively, using three internal LDPC iterations and the system parameters outlined in Table 6.1.

6.5 Performance of Bit-based and Symbol-based LDPC-coded STBC-SP Schemes

6.5.1 System Parameters

We considered a SP modulation scheme associated with $L = 4$, 8 and 16 using two transmit and a single receiver antenna to demonstrate the performance improvements achieved by the proposed system. All simulation parameters are listed in Table 6.1. The bit-based schemes corresponding to the symbol-based Schemes 1, 2 and 3 use the same system parameters as outlined in Table 6.1, except that a binary LDPC code defined over $GF(2)$ is employed instead of the non-binary LDPC codes. Recall from Section 2.4.4 that there are more than L legitimate SP symbols in the lattice D_4 and hence the required L SP symbols were chosen according to the minimum energy and highest MED criterion of Section 2.4.4.

6.5.2 Decoding Trajectory

Figure 6.14 illustrates the actual decoding trajectory of the turbo-detected symbol-based non-binary LDPC-coded STBC-SP scheme of Figure 6.13, when operating at $\text{SNR} = 5.0$ dB after $I_{\text{ext}} = 5$ joint external iterations and $I_{\text{int}} = 3$ internal LDPC iterations. The 'zigzag path' seen in Figure 6.14 represents the actual EXIT between the rate-one inner decoder and the outer non-binary LDPC decoder at $\text{SNR} = 5.0$ dB. The deviation of the decoding trajectory from the prediction of the EXIT chart is due to the fact that a finite LDPC output blocklength of $K_{ldpc} = 6000$ bits is employed, rendering the assumption of having Gaussian distributed symbol probabilities only approximately valid. This assumption was exploited when creating $\mathbf{A}_{urc}$ and $\mathbf{A}_{ldpc}$ for the sake of generating the appropriate *a priori* information I_A according to the procedure outlined in Section 6.4.2.2, in order to characterize the EXIT curves of the constituent decoders. Figure 6.15 illustrates the decoding trajectory of the turbo-detected bit-based binary LDPC-coded STBC-SP scheme of Figure 6.13, when operating at $\text{SNR} = 5.0$ dB after $I_{\text{ext}} = 15$ joint external iterations and $I_{\text{int}} = 3$ internal LDPC iterations. Observe in Figures 6.14 and 6.15 that more joint external iterations are required by the bit-based scheme in order to converge than by the symbol-based scheme.

6.5.3 BER Performance

Figure 6.16 compares the attainable performance of the symbol-based non-binary LDPC [277] and the bit-based binary LDPC-coded [274] STBC-SP schemes using the system parameters of Table 6.1 after $I_{\text{ext}} = 5$ joint external iterations and $I_{\text{int}} = 3$ internal LDPC iterations, when using a LDPC output blocklength of $K_{ldpc} = 12\,000$ bits. Observe that the turbo cliffs seen in Figure 6.16 occur at SNR values that are slightly different from that predicted by the EXIT charts of Section 6.4.3 and outlined in Table 6.2. Again, the discrepancy observed is due to the fact that a finite interleaver depth of $K_{ldpc} = 12\,000$ bits is employed. The approximate SNR values where a turbo cliff occurs were extracted from the BER curves seen in Figure 6.16 and are listed in Table 6.3. More explicitly, the turbo cliff was deemed to be reached when the achievable BER dips below 10^{-3}.

6.5.4 Effect of Interleaver Depth

The effect of employing different interleaver sizes or, equivalently, LDPC output block lengths on the achievable performance of both the symbol-based and bit-based schemes is illustrated in Figures 6.17–6.20. More specifically, Figures 6.17–6.19 compare the attainable

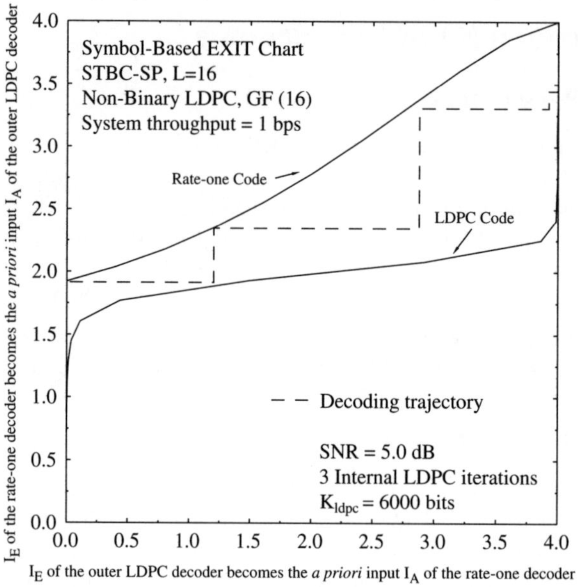

Figure 6.14: Decoding trajectory of the symbol-based half-rate non-binary LDPC-coded [277] STBC-SP scheme defined over $GF(16)$ in combination with the system parameters outlined in Table 6.1 and operating at $E_b/N_0 = 5.0$ dB after five joint external iterations and three internal LDPC iterations.

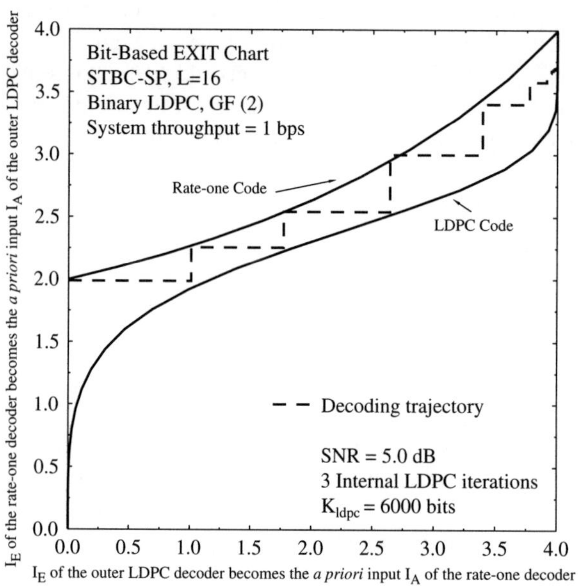

Figure 6.15: Decoding trajectory of the bit-based half-rate binary LDPC-coded [274] STBC-SP scheme in combination with the system parameters outlined in Table 6.1 and operating at $E_b/N_0 = 5.0$ dB after 15 joint external iterations and three internal LDPC iterations.

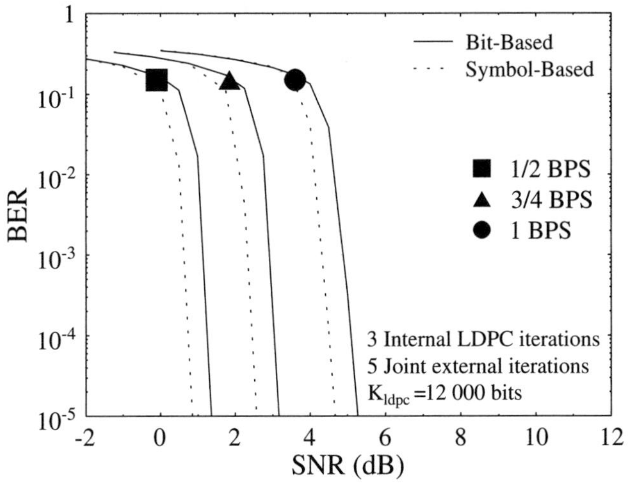

Figure 6.16: Performance of symbol-based and bit-based LDPC-coded STBC-SP schemes in combination with the system parameters outlined in Table 6.1 after five joint external iterations and three internal LDPC iterations, when using a LDPC output blocklength of $K_{ldpc} = 12\,000$.

Table 6.3: The approximate SNR values where a turbo cliff occurs, based on the BER curves seen in Figure 6.16 for Schemes 1, 2 and 3 outlined in Table 6.1.

	Approximate SNR Values (dB)	
	Symbol-based scheme	Bit-based scheme
Scheme 1	0.6	1.1
Scheme 2	2.3	2.9
Scheme 3	4.3	4.9

performance of the symbol-based non-binary LDPC- [277] and the bit-based binary LDPC-coded [274] STBC-SP schemes using the system parameters of Table 6.1 after $I_{\text{ext}} = 5$ joint external iterations and $I_{\text{int}} = 3$ internal LDPC iterations, when using LDPC output blocklengths of $K_{ldpc} = 1488$, 3000 and 6000 bits, respectively. In addition, Figure 6.20 characterizes the performance of the symbol-based half-rate LDPC-coded [277] STBC-SP schemes using the system parameters of Table 6.1 after $I_{\text{ext}} = 5$ joint external iterations and $I_{\text{int}} = 3$ internal LDPC iterations for various LDPC output block lengths. The SNR values required for achieving a BER of 10^{-5} are highlighted in Figure 6.21 versus the system throughput for both the bit-based and symbol-based LDPC-coded STBC-SP schemes using the system parameters of Table 6.1, when using three internal LDPC iterations and a LDPC output blocklength of $K_{ldpc} = 12\,000$ bits.

6.6 Chapter Conclusions

In this chapter we have proposed a novel symbol-based iterative scheme that exploits the advantages of non-binary LDPC codes [277], those of the rate-one inner codes of [266] as

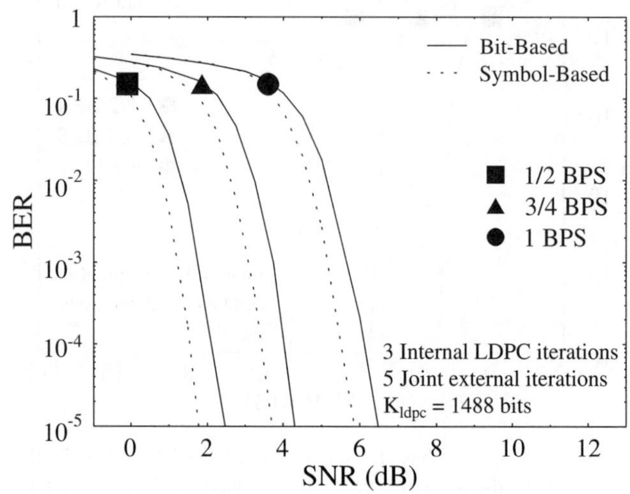

Figure 6.17: Performance of symbol-based and bit-based LDPC-coded STBC-SP schemes in combination with the system parameters outlined in Table 6.1 after five joint external iterations and three internal LDPC iterations, when using a LDPC output blocklength of $K_{ldpc} = 1488$.

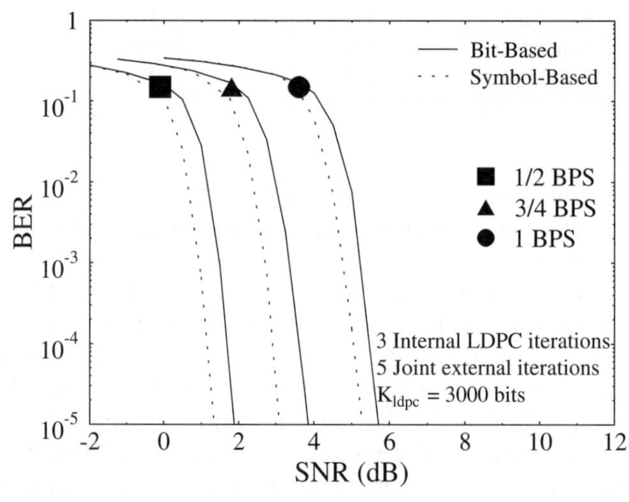

Figure 6.18: Performance of symbol-based and bit-based LDPC-coded STBC-SP schemes in combination with the system parameters outlined in Table 6.1 after five joint external iterations and three internal LDPC iterations, when using a LDPC output blocklength of $K_{ldpc} = 3000$.

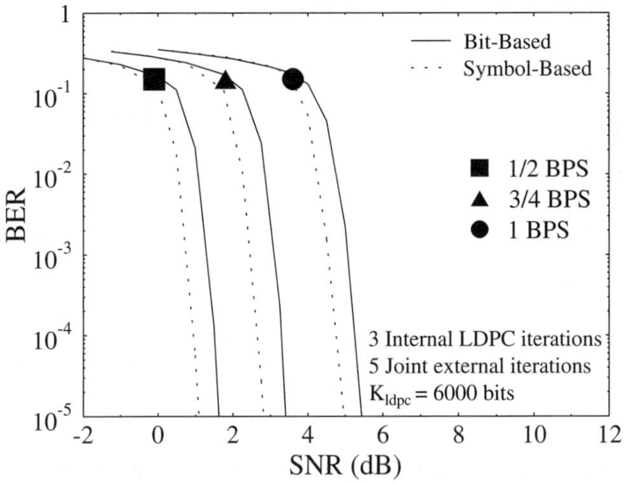

Figure 6.19: Performance of symbol-based and bit-based LDPC-coded STBC-SP schemes in combination with the system parameters outlined in Table 6.1 after five joint external iterations and three internal LDPC iterations, when using a LDPC output blocklength of $K_{ldpc} = 6000$.

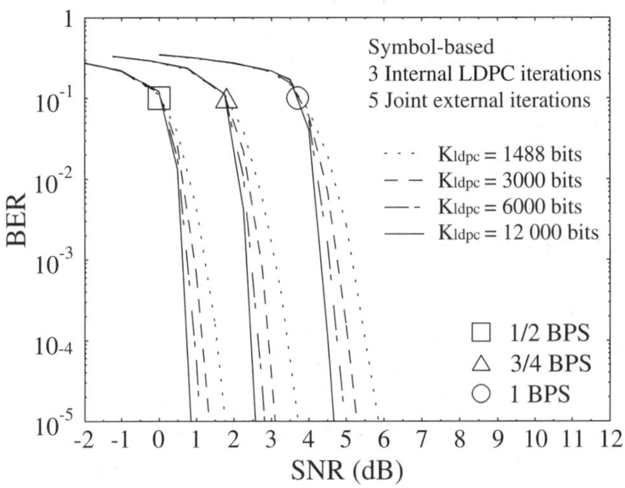

Figure 6.20: Performance of symbol-based half-rate LDPC-coded [277] STBC-SP schemes in combination with the system parameters outlined in Table 6.1 after five joint external iterations and three internal LDPC iterations while using different LDPC output block lengths.

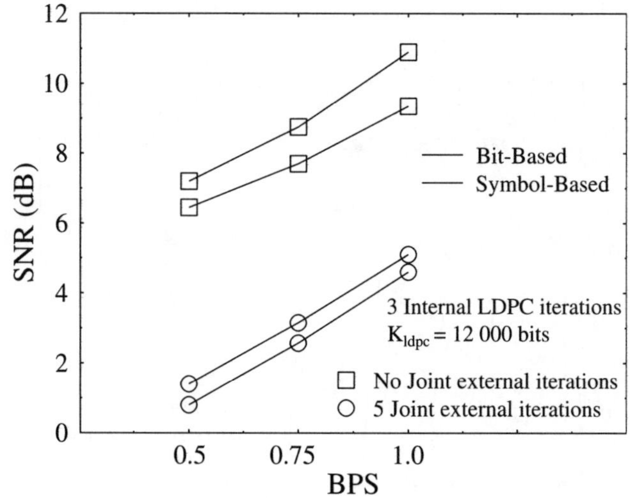

Figure 6.21: SNR values required to achieve a BER of 10^{-5} versus different system throughput for both the bit-based and symbol-based LDPC-coded STBC-SP schemes in combination with the system parameters outlined in Table 6.1, when using three internal LDPC iterations and a LDPC output blocklength of $K_{ldpc} = 12\ 000$ bits.

well as those of the STBC-SP scheme of [43]. Our investigations in Section 6.5 demonstrated that attractive performance improvements may be achieved by the proposed scheme over the equivalent-throughput bit-based schemes. Again, it was demonstrated in Figures 6.11–6.13 by EXIT chart analysis and in Figures 6.16–6.19 by the corresponding BER performance curves that the symbol-based scheme is capable of outperforming its bit-based counterpart and that both designs had an edge over Alamouti's now classic STBC scheme, which dispensed with the SP-based joint design of the QPSK space-time symbols. The performance of the symbol-based schemes was investigated in Figure 6.20, when employing various interleaver depths in order to illustrate the effect of the LDPC output block length on the attainable performance.

6.7 Chapter Summary

In this chapter, we proposed a novel symbol-based iterative scheme, where the system's architecture was presented in Section 6.2. Section 6.2.1 provided a detailed description of the proposed symbol-based and turbo-detected scheme, where a non-binary LDPC code was combined with a symbol-based interleaving. The equivalent bit-based scheme was described in Section 6.2.2, where a binary LDPC code was combined with a bit-based interleaver. Symbol-based iterative decoding was discussed in Section 6.3, where it was demonstrated how the *a priori* information $\mathbf{A}$ is removed from the decoded APP matrix $\mathbf{D}$ with the aid of symbol-based element-wise division for the sake of generating the extrinsic probability matrix $\mathbf{E}$.

Section 6.4 provided our non-binary EXIT chart analysis. More specifically, in Section 6.4.1 we demonstrated how non-binary EXIT charts can be generated without generating an L-dimensional histogram [184] since the complexity of this operation may become higher than conducting full-scale BER or SER simulations, when the number of bits per symbol is high. In Section 6.4.2, we addressed the problem of generating the *a priori* symbol probabilities when the binary bits within each non-binary symbol are assumed be either independent or not. More specifically, the binary bits within each non-binary symbol were assumed to be independent of each other when a binary LDPC code was employed, which is equivalent to bit interleaving, as alluded to in Section 6.2. Generating the *a priori* symbol probabilities for this particular assumption was presented in Section 6.4.2.1. In contrast, the binary bits of each non-binary symbol are no longer independent when employing a non-binary LDPC code and performing symbol-based decoding. Accordingly, a detailed procedure was described in Section 6.4.2.2 for creating the *a priori* symbol probabilities when the binary bits of each non-binary symbol may no longer be assumed to be independent. The results of our non-binary EXIT chart analysis were provided in Section 6.4.3, where the novel non-binary EXIT charts were used to study the convergence of the proposed symbol-based schemes in Section 6.4.3.1. On the other hand, in Section 6.4.3.2 non-binary EXIT charts were used to explore the convergence of the bit-based LDPC-coded STBC-SP schemes. The EXIT charts of the symbol-based and bit-based schemes were compared in Section 6.4.3.3. It was explicitly demonstrated in Figures 6.11–6.13 and Table 6.2 that the symbol-based schemes required a lower transmit power and a lower number of decoding iterations to achieve a performance comparable to that of their bit-based counterparts.

The performance of the symbol-based and bit-based LDPC-coded STBC-SP schemes was investigated in Section 6.5 when employing the system parameters outlined in Section 6.5.1 and Table 6.1. First, the actual decoding trajectories were presented in Section 6.5.2, where the mismatch seen in Figures 6.14 and 6.15 between the actual trajectories and the EXIT curves was a consequence of employing a finite interleaver depth of $K_{ldpc} = 6000$ bits. The attainable BER performance of both the symbol-based and the bit-based schemes was demonstrated in Section 6.5.3, where Figure 6.16 compares the achievable performance of all symbol-based schemes outlined in Table 6.1 against that of their bit-based counterparts. Furthermore, Table 6.3 lists the approximate SNR values where a turbo cliff occurs, based on the BER curves seen in Figure 6.16. The effect of employing various interleaver depths or, equivalently, LDPC output block lengths on the achievable performance was considered in Section 6.5.4 and Figures 6.17–6.20.

Part II

Coherent Versus Differential Turbo Detection of Single-user and Cooperative MIMOs

List of Symbols in Part II

Matrices and Vectors

$\mathbf{A}_n$	The nth dispersion matrix
$\mathbf{B}_n$	The nth dispersion matrix
$\mathbf{H}$	The CSI matrix
$\mathbf{I}$	The identity matrix
$\mathbf{K}$	A symbol vector containing Q symbols
$\mathbf{K}_n$	The nth symbol vector containing Q symbols
$\mathbf{K}_f$	All of the possible combinations from Q symbols
$\mathbf{R}_m$	The received signal vector at the mth relay
$\mathbf{S}$	The space-time transmission matrix
$\mathbf{S}_n$	The nth space-time transmission matrix
$\mathbf{S}_\triangle$	The difference matrix between two distinctive space-time codewords
$\mathbf{U}$	The reference matrix
$\mathbf{V}$	The complex-valued AWGN signal matrix
$\mathbf{V}_n$	The nth complex-valued AWGN signal matrix
$\mathbf{X}_n$	The nth space-time coded matrix
$\mathbf{Y}$	The received signal matrix at the BS
$\mathbf{Y}_n$	The nth received signal matrix at the BS
$\mathbf{Z}_m$	The signal vector transmitted from the mth relay
$\boldsymbol{\chi}$	The Dispersion Character Matrix (DCM) of the co-located MIMO system
$\boldsymbol{\chi}_1$	The DCM characterizing the broadcast interval
$\boldsymbol{\chi}_2$	The DCM characterizing the cooperation interval
$\boldsymbol{\lambda}$	The weighting coefficient vector of the inner code of an irregular scheme
γ	The weighting coefficient vector of the outer code of an irregular scheme

Variables

C_{MIMO}	The coherent MIMO channel's CCMC capacity
$C_{\text{MIMO}}^{\text{non}}$	The non-coherent MIMO channel's CCMC capacity
C_{LDC}	The CCMC capacity of a LDC scheme

$C_{\text{LDC}}^{\text{DCMC}}$	The DCMC capacity of a LDC scheme
D	The total spatial diversity gain
D_{tx}	The transmit spatial diversity gain
F	The number of legitimate combinations of the vector $\mathbf{K}$
I_A	The *a priori* information input of the inner code
$\bar{I}_A$	The *a priori* information input of the outer code
I_E	The extrinsic information output of the inner code
$\bar{I}_E$	The extrinsic information output of the outer code
L	The number of constellation points
M	The number of transmit antennas (relays)
N	The number of receive antennas
T	The number of time slots used per block during the transmission
T_1	The number of time slots used per block during the broadcast interval
T_2	The number of time slots used per block during the cooperation interval
Q	The number of symbols transmitted per space-time block
k	The number of outer iterations of a Serial Concatenated Code (SCC) scheme
j	The number of inner iterations of a SCC scheme
P	The total number of component codes of an irregular SCC system
P_{out}	The number of outer component codes of an irregular SCC system
P_{in}	The number of inner component codes of an irregular SCC system
s_n	The nth L-PSK modulated symbol
R_{STBC}	The symbol rate of a STBC scheme
$\bar{R}$	The code rate of a STBC scheme in bits per channel use
α_n	The real part of symbol s_n
β_n	The imaginary part of symbol s_n
$\acute{L}$	The number of layers within a STBC scheme
θ	The normalization factor
Ξ	The spatial multiplexing gain
ρ	The SNR
ρ_{SR}	The SNR at the relays
ρ_{RB}	The SNR at the BS
Ω	The standard deviation of a Gaussian distribution in decibels
σ_0	The complex Gaussian noise variance at the receiver
σ_{SR}	The complex Gaussian noise variance at the relays
σ_{RB}	The complex Gaussian noise variance at the BS
$\varpi(\rho)$	The decoding complexity of an irregular scheme at SNR ρ
η	The minimum rank of $\mathbf{S}_{\triangle}$
∇_n	The nth non-zero eigenvalue of $\mathbf{S}_{\triangle}$
Θ_{2n}^2	The chi-square random variable with dimension $2n$

Mathematical Operations

$\sum$	Sum operation
$\otimes$	Kronecker product
$\log[\cdot]$	Logarithm operation
$\max(\cdot)$	The maximum value of a matrix/vector
$\min(\cdot)$	The minimum value of a matrix/vector

$vec(\cdot)$	Vertical stacking of the columns of a matrix
$row(\cdot)$	Vertical stacking of the rows of a matrix
$\mathrm{tr}(\cdot)$	Trace operation of a matrix
$QR(\cdot)$	The QR-decomposition of a square matrix
$\det(\cdot)$	The determinant operation
$p(\cdot)$	The probability density function
$\Vert \cdot \Vert^2$	The second-order norm
$Real\{\cdot\}$	Real part of a complex value
$E\{\cdot\}$	Expectation of a random variable
$\mathbf{A}^H$	Matrix/vector Hermitian adjoint
$\mathbf{A}^{-1}$	Matrix inverse
$\mathbf{A}^{\mathrm{T}}$	Matrix/vector transpose
$\mathbf{A}^*$	Matrix/vector/variable complex conjugate

Symbols

$\Re^n$	n-dimensional real-valued Euclidean space
Π	Interleaver
Π^{-1}	Deinterleaver
Γ_n	The EXIT function of the nth irregular code
$\bar{E}$	The total power of a SP symbol
t^{i_1,i_2}	The (i_1, i_2)th entry of the matrix $\mathbf{S}$

Channel Parameters

f_d	The normalized Doppler frequency
δ	The spatial correlation coefficient
ω	The degree of channel estimation error in decibels
$\Im$	The channel coherence time

Chapter 7

Linear Dispersion Codes: An EXIT Chart Perspective

7.1 Introduction and Outline

The design of coding schemes for MIMO systems illustrated in Figure 7.1 operating at high SNRs involves a trade-off between the achievable rate at which the system's capacity increases and the rate at which the error probability decays [23]. This inherent trade-off provides a distinction between transmit diversity schemes such as the family of OSTBCs [25] that sacrifice the achievable rate in exchange for maximum reliability and the class of Spatial Division Multiplexing (SDM) arrangements, such as those belonging to the BLAST architecture [15]. SDM schemes are capable of supporting transmission rates close to the MIMO channel's capacity, but do so without fully benefiting from the diversity potential of the channel. Since OSTBCs and BLAST schemes were designed to achieve the two extremes of the trade-off scale, there is considerable interest in developing design methods for schemes that provide different trade-offs in terms of the achievable rate and the error probability, which are applicable for employment in a broad range of antenna configurations. In other words, the space-time transmission matrix $\mathbf{S}$ in Figure 7.1 provides a total of $(M \times T)$ transmission slots, when considering both the spatial and temporal domains. The design of STBCs aims to answer the question of how to use the available resources most efficiently.

Unitary Space-Time Modulation (USTM) was proposed independently by Hughes [55] and Hochwald *et al.* [54, 63] in an early attempt to design flexible STBCs that are capable of achieving the highest attainable diversity gain. Instead of specifically designing either the modulation schemes used or the inner structure of the space-time transmission matrix $\mathbf{S}$ of Figure 7.1, the philosophy of USTM [54, 55] is to directly maximize the mutual information between $\mathbf{S}$ and the received signal matrix $\mathbf{Y}$. However, the complexity of the optimization increases exponentially with both the number of antennas and the number of BPS. Hence, the optimization of high-rate USTM becomes infeasible. Hence, the problem of designing simple STBCs that exhibit both high rate and full diversity remains to be tackled.

The set of LDCs, first proposed by Hassibi and Hochwald [10], constitutes a wide-ranging class of STBCs exhibiting diverse characteristics. Hence, this family encompasses numerous existing schemes, providing a natural framework in which such design problems can be

Near-Capacity Multi-Functional MIMO Systems L. Hanzo, O. R. Alamri, M. El-Hajjar and N. Wu

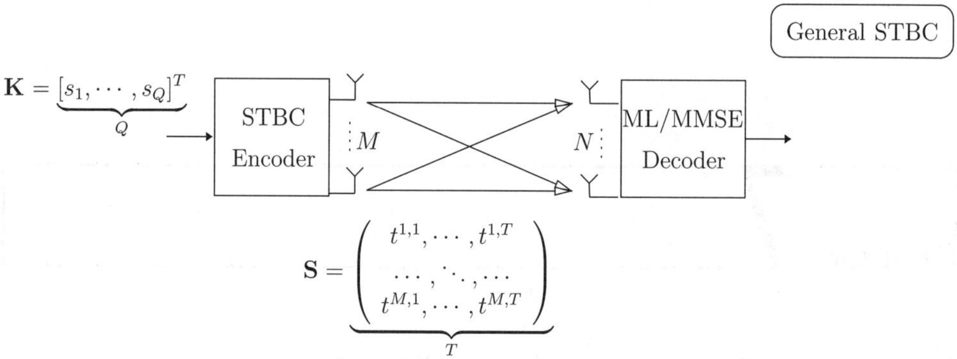

Figure 7.1: Schematic of MIMO system equipped with M transmit and N receive antennas, when transmitting Q symbols over T time slots using the space-time matrix $\mathbf{S}$.

posed. While some recently developed codes, such as the TASTBCs [283] as well as TVLT codes [28] and those proposed in [284], possess many desirable features, they remain a subset of the LDC framework. Given the generality of the LDC framework and the high degrees of design freedom, the focus of this chapter is on the design and further development based on LDCs.

The revolutionary concept of LDCs [10] invokes a matrix-based linear modulation framework seen in Figures 7.2 and 7.3, where each space-time transmission matrix $\mathbf{S}$ is generated by a linear combination of so-called dispersion matrices and the weights of the components are determined by the transmitted symbol vector $\mathbf{K}$ of Figure 7.1. These figures will be revisited in more depth during our further discourse. The structure of the dispersion matrices is governed by the DCM χ to be outlined in detail in Equation (7.16) and Appendix D. The dispersion matrices can be designed to maximize the ergodic MIMO capacity [3] as is further highlighted in Section 7.2. However, the LDCs proposed in [10] only optimized the ergodic capacity, which did not necessarily guarantee that a low error probability was achievable [27, 285]. The fact that the probability of a codeword error can be upper bounded by the largest Pairwise Symbol Error Probability (PSEP) [286] is often exploited. Therefore, LDCs minimizing the maximum PSEP based on frame theory [287] were proposed in [27]. On the other hand, LDCs can be optimized using the determinant criterion [52] using the same technique of the Golden code [288, 289]. However, in this chapter we propose a novel method to optimize the dispersion matrices, namely maximizing the DCMC capacity.

The achievable ergodic capacity of the LDCs [10,27,285] is also referred to as the CCMC capacity [286], where the channel's input is assumed to be a continuous-amplitude discrete-time Gaussian-distributed signal and the capacity is restricted only by either the signaling energy or the bandwidth. Therefore, we refer to the CCMC capacity as the unrestricted bound. Naturally, in practice the channel's input is constituted by non-Gaussian symbols, although in certain circumstances, when for example high-order QAM-aided symbols are transmitted over multiple Orthogonal Frequency-Division Multiplexing (OFDM) carriers this assumption becomes valid owing to the central limit theorem. In the more realistic context of discrete-amplitude PSK and QAM signals, we encounter a DCMC. Therefore, the DCMC capacity is more pertinent in the design of practical channel-coded modulation schemes. We demonstrate in Section 7.2 that LDCs achieving the same CCMC capacity may attain a different DCMC

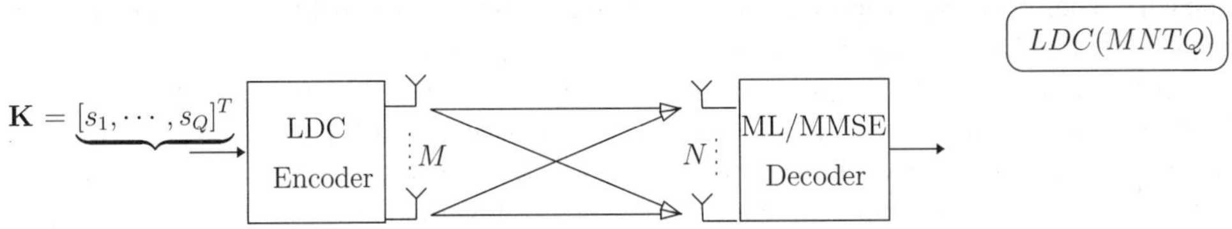

$$\mathbf{S} = \underbrace{\begin{pmatrix} t^{1,1}, \cdots, t^{1,T} \\ \cdots, \ddots, \cdots \\ t^{M,1}, \cdots, t^{M,T} \end{pmatrix}}_{T}$$

$$= \begin{pmatrix} a_1^{1,1}, \cdots, a_q^{1,T} \\ \cdots, \ddots, \cdots \\ a_1^{M,1}, \cdots, a_1^{M,T} \end{pmatrix} \alpha_1 \cdots + \begin{pmatrix} a_Q^{1,1}, \cdots, a_Q^{1,T} \\ \cdots, \ddots, \cdots \\ a_Q^{M,1}, \cdots, a_Q^{M,T} \end{pmatrix} \alpha_Q$$

$$+ j \begin{pmatrix} b_1^{1,1}, \cdots, b_1^{1,T} \\ \cdots, \ddots, \cdots \\ b_1^{M,1}, \cdots, b_1^{M,T} \end{pmatrix} \beta_1 \cdots + j \begin{pmatrix} b_Q^{1,1}, \cdots, b_Q^{1,T} \\ \cdots, \ddots, \cdots \\ b_Q^{M,1}, \cdots, b_Q^{M,T} \end{pmatrix} \beta_Q$$

$$= \mathbf{A}_1\alpha_1 + \cdots + \mathbf{A}_Q\alpha_Q + j\mathbf{B}_1\beta_1 + \cdots + j\mathbf{B}_Q\beta_Q$$

Figure 7.2: Schematic of a MIMO system equipped with M transmit and N receive antennas employing the LDC($MNTQ$) structure of [10], while transmitting Q symbols over T time slots using the space-time matrix $\mathbf{S}$.

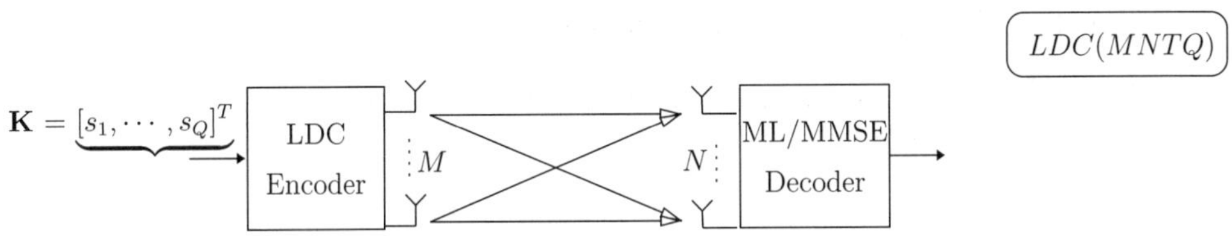

$$\mathbf{S} = \underbrace{\begin{pmatrix} t^{1,1}, \cdots, t^{1,T} \\ \cdots, \ddots, \cdots \\ t^{M,1}, \cdots, t^{M,T} \end{pmatrix}}_{T}$$

$$= \begin{pmatrix} a_1^{1,1}, \cdots, a_q^{1,T} \\ \cdots, \ddots, \cdots \\ a_1^{M,1}, \cdots, a_1^{M,T} \end{pmatrix} s_1 \cdots + \begin{pmatrix} a_Q^{1,1}, \cdots, a_Q^{1,T} \\ \cdots, \ddots, \cdots \\ a_Q^{M,1}, \cdots, a_Q^{M,T} \end{pmatrix} s_Q$$

$$= \mathbf{A}_1 s_1 + \cdots + \mathbf{A}_Q s_Q$$

Figure 7.3: Schematic of a MIMO system equipped with M transmit and N receive antennas employing the LDC($MNTQ$) structure of [27], while transmitting Q symbols over T time slots using the space-time matrix $\mathbf{S}$.

capacity. Therefore, we propose an optimization method that is capable of maximizing both the CCMC and DCMC capacity of the LDCs at the same time.

SCCs are capable of attaining an infinitesimally low BER, while maintaining a manageable decoding complexity. The discovery of 'turbo codes' [9, 146] considerably improved the attainable performance gains of concatenated codes by exchanging extrinsic information between concatenated component codes. It has been shown in [149] that a high coding gain can be achieved using SCCs with the aid of iterative decoding. Since LDCs have the ability to approach the potential capacity of MIMO systems, it is natural to serially concatenate, for example, a simple convolutional channel code as the outer code and employ LDCs as the inner code in order to approach the MIMO capacity, while having a near error-free BER performance.

The concept of 'irregularity' was proposed in [176, 195] for SCCs, when IRCCs were adopted as the outer channel code. Since IRCCs exhibited flexible EXIT characteristics [195], shaping the outer code's EXIT curve in order to match the inner code's EXIT curve becomes possible. However, near-capacity IRCC schemes may require an excessive number of iterations at the receiver to achieve an infinitesimally low BER, which may exceed the affordable complexity budget of mobile handsets.

Motivated by the above-mentioned flexibility of the irregular outer code design philosophy, in this chapter we circumvent the IRCC-related outer code limitations by proposing IR-PLDCs as the inner rather than the outer code, and serially concatenate the resultant IR-PLDCs with regular/irregular outer channel codes in order to achieve an infinitesimally low BER when employing iterative decoders. The novel contributions of this chapter are listed as follows [31, 290].

- LDCs are optimized by maximizing both the CCMC and DCMC capacity.
- The links between existing STBCs and LDCs are explored, and the necessary condition for STBCs to achieve both high rate and full diversity is investigated.
- We investigate the maximum achievable rate of precoded LDCs with the aid of EXIT charts, when using both ML and MMSE detectors.
- We propose IR-PLDCs as the inner code of SCCs and employ regular/irregular convolutional codes as the outer code. Hence, the resultant SCCs become capable of operating near the attainable MIMO channel capacity for a wide range of SNRs.

The outline of this chapter is as follows. In Section 7.2 we present the system models and the optimization of the LDCs, followed by the achievable performance in terms of their capacity and attainable BER. In Section 7.3, we characterize the link between existing STBCs found in the open literature and LDCs in terms of both mathematical representations and design philosophy. In Section 7.4, the EXIT chart analysis of LDCs is provided in order to characterize the convergence behavior of the serial concatenated scheme with/without unity rate precoders. Section 7.5 presents the detailed design of the proposed IR-PLDCs with the aid of EXIT charts, when a simple outer code is employed. Numerical results are also presented for characterizing the IR-PLDCs operating at different SNRs, when the decoding complexity is also taken into account. Furthermore, in Section 7.5.2, we compare the IRCCs and the proposed IR-PLDCs, where the irregularity is imposed in the outer and inner codes, respectively. In order to further illustrate the irregular concept, in Section 7.5.3 we propose an IRCC-coded IR-PLDC scheme, where the irregularity is evenly distributed between the inner and outer codes. Finally, our concluding remarks are provided in Section 7.6.

7.2 Linear Dispersion Codes

7.2.1 Channel Model

A MIMO communication system equipped with M transmit antennas and N receive antennas is illustrated in Figure 7.1. The space-time encoder takes the input symbol vector $\mathbf{K} = [s_1, s_2, \ldots, s_Q]^{\mathrm{T}}$ and generates a space-time transmission matrix $\mathbf{S}$. Then $\mathbf{S}$ is transmitted synchronously through the propagation environment using M transmit antennas. The signal received at each antenna is therefore given by a superposition of M transmitted signals corrupted both by AWGN and multiplicative fading. When the space-time transmission matrix $\mathbf{S}$ spans T symbol intervals, the received signal matrix $\mathbf{Y} \in \zeta^{N \times T}$ can be written as follows:

$$\mathbf{Y} = \mathbf{HS} + \mathbf{V}, \tag{7.1}$$

where $\mathbf{V}$ is a $(N \times T)$-element matrix whose columns represent realizations of an independent and identically distributed (i.i.d.) complex AWGN process with zero-mean and variance σ_0 determined by SNR ρ. The CIR matrix $\mathbf{H} \in \zeta^{N \times M}$ is assumed to be constant during a space-time codeword of T symbol periods and change independently from one space-time matrix to the next, which corresponds to experiencing quasi-static fading. Each entry of $\mathbf{H}$ represents the fading coefficient between a transmit–receive antenna pair. We assume that the scattering imposed by the propagation environment is sufficiently rich for the channel coefficients to be modelled as i.i.d. zero-mean complex-valued Gaussian random variables having a common variance of 0.5 per real dimension. The entries of the channel matrix are assumed to be known to the receiver, but not to the transmitter. The space-time transmission matrix $\mathbf{S} \in \zeta^{M \times T}$ is given by

$$\mathbf{S} = \begin{pmatrix} t^{1,1} & \ldots & t^{1,T} \\ t^{2,1} & \ldots & t^{2,T} \\ \vdots & \ddots & \vdots \\ t^{M,1} & \ldots & t^{M,T} \end{pmatrix}, \tag{7.2}$$

where t^{i_1,i_2} denotes the signal transmitted from the i_1th antenna in the i_2th time slot. We are concerned with designing the signal matrix $\mathbf{S}$ obeying the power constraint $E\{\mathrm{tr}(\mathbf{SS}^H)\} = T$, to ensure that the total transmission power is constant.

Given an arbitrary signal constellation, we follow [52] and define the symbol rate of a STBC having the structure of Figure 7.1 by

$$R_{\mathrm{STBC}} = \frac{Q}{T}. \tag{7.3}$$

Under this definition, a rate-two code corresponds to a STBC that transmits on average two symbols per channel use. For example, Alamouti's twin-antenna $\mathbf{G}_2$ code [11] has a rate of $R_{\mathbf{G}_2} = Q/T = 1$.

7.2.2 LDC Model of [10]

Let each space-time transmission matrix $\mathbf{S}$ formulated in Equation (7.2) represent the linear combination of Q space-time symbols, such as L-PSK or L-QAM symbols, which are dispersed over both space and time with the motivation of exploiting both the spatial and temporal diversity, as seen in Figure 7.2. We simply refer to this structure as a LDC. Hence,

$\mathbf{S}$ obeys

$$\mathbf{S} = \sum_{q=1}^{Q} (\mathbf{A}_q \alpha_q + j\mathbf{B}_q \beta_q), \tag{7.4}$$

where α_q and β_q are the real and imaginary parts of the qth transmitted symbol having a variance of 0.5, while $\mathbf{A}_q$ and $\mathbf{B}_q$ are real-valued or complex-valued dispersion matrices. We normalize $\mathbf{S}$ to $E\{\mathrm{tr}(\mathbf{S}^H\mathbf{S})\} = T$, which limits the total transmit power of the space-time codeword. It may be readily shown that this normalization imposes the following normalization on the matrices $\mathbf{A}_q$ and $\mathbf{B}_q$:

$$\sum_{q=1}^{Q} (\mathrm{tr}(\mathbf{A}_q^H \mathbf{A}_q) + \mathrm{tr}(\mathbf{B}_q^H \mathbf{B}_q)) = 2T. \tag{7.5}$$

The idea behind the decoding of LDCs is to exploit the linearity of Equation (7.4) with respect to the variables, leading to efficient V-BLAST-like decoding schemes [15]. We denote the real part of the received signal matrix $\mathbf{Y}$ by $\mathbf{Y}_R$ and the imaginary part by $\mathbf{Y}_I$. Similarly, we introduce the notation $\mathbf{H}_R$, $\mathbf{H}_I$, $\mathbf{V}_R$ and $\mathbf{V}_I$, and denote the ith columns of these matrices by $y_{R,i}$, $y_{I,i}$, $h_{R,i}$, $h_{I,i}$, $v_{R,i}$ and $v_{I,i}$. Let us also define

$$\bar{\mathbf{A}}_q = \begin{bmatrix} \mathbf{A}_{R,q} & -\mathbf{A}_{I,q} \\ \mathbf{A}_{I,q} & \mathbf{A}_{R,q} \end{bmatrix}, \quad \bar{\mathbf{B}}_q = \begin{bmatrix} -\mathbf{B}_{I,q} & -\mathbf{B}_{R,q} \\ \mathbf{B}_{R,q} & -\mathbf{B}_{I,q} \end{bmatrix}, \quad \underline{\mathbf{h}}_i = \begin{bmatrix} h_{R,i} \\ h_{I,i} \end{bmatrix}, \tag{7.6}$$

where we have $\bar{\mathbf{A}}_q \in \zeta^{2T\times 2M}$, $\bar{\mathbf{B}}_q \in \zeta^{2T\times 2M}$ and $\underline{\mathbf{h}}_n \in \zeta^{2M\times 1}$. Furthermore, define

$$\bar{\mathbf{H}} = \begin{bmatrix} \bar{\mathbf{A}}_1\underline{\mathbf{h}}_1 & \bar{\mathbf{B}}_1\underline{\mathbf{h}}_1 & \cdots & \bar{\mathbf{A}}_Q\underline{\mathbf{h}}_1 & \bar{\mathbf{B}}_Q\underline{\mathbf{h}}_1 \\ \vdots & \vdots & \ddots & \vdots & \vdots \\ \bar{\mathbf{A}}_1\underline{\mathbf{h}}_N & \bar{\mathbf{B}}_1\underline{\mathbf{h}}_N & \cdots & \bar{\mathbf{A}}_Q\underline{\mathbf{h}}_N & \bar{\mathbf{B}}_Q\underline{\mathbf{h}}_N \end{bmatrix}, \tag{7.7}$$

where we have $\bar{\mathbf{H}} \in \zeta^{2NT\times 2Q}$ and each element of $\bar{\mathbf{H}}$ is hence an element of the $\zeta^{2T\times 1}$ matrix. For example, we have $\bar{\mathbf{A}}_1\underline{\mathbf{h}}_1 \in \zeta^{2T\times 1}$.

Then the whole system can be represented by [10]

$$\begin{bmatrix} y_{R,1} \\ y_{I,1} \\ \vdots \\ y_{R,N} \\ y_{I,N} \end{bmatrix} = \bar{\mathbf{H}} \begin{bmatrix} \alpha_1 \\ \beta_1 \\ \vdots \\ \alpha_Q \\ \beta_Q \end{bmatrix} + \begin{bmatrix} v_{R,1} \\ v_{I,1} \\ \vdots \\ v_{R,N} \\ v_{I,N} \end{bmatrix}, \tag{7.8}$$

where the matrix denoting the received signal has a size of $(2NT \times 1)$, while the matrix hosting the transmitted symbols is of size $(2Q \times 1)$. The resultant equivalent channel $\bar{\mathbf{H}}$ is known to the receiver, because the original channel $\mathbf{H}$ and the dispersion matrices $\mathbf{A}_q$ and $\mathbf{B}_q$ are all known to the receiver. Equation (7.8) explicitly portrays the linear relation between the equivalent channel's input and output vectors. Therefore, ML and MMSE detectors as well as sphere decoders [291] can be readily employed to recover the Q transmitted symbols.

The dispersion matrices $\mathbf{A}_q$ and $\mathbf{B}_q$ of Figure 7.2 are specifically designed to maximize the mutual information between the equivalent transmit and receive vectors in Equation (7.8). This guarantees that the resultant LDCs minimize the potential mutual information penalty. This design procedure can be formalized as follows [10].

Given the equivalent MIMO channel matrix $\bar{\mathbf{H}}$ of Equation (7.7), the achievable capacity C_{LDC} of this MIMO system employing LDCs at SNR ρ is given by [10]

$$C_{\text{LDC}}(\rho, M, N, T, Q) = \max\left(\frac{1}{2T}E\left\{\log_2\left[\det\left(\mathbf{I} + \frac{\rho}{M}\bar{\mathbf{H}}\bar{\mathbf{H}}^H\right)\right]\right\}\right), \quad (7.9)$$

where $\mathbf{I} \in \zeta^{2NT\times 2NT}$ denotes an identity matrix. We choose $\mathbf{A}_q$ and $\mathbf{B}_q$ in order to maximize C_{LDC} for a given Q, subject to one of the following constraints:

(i) $\sum_{q=1}^{Q}(\text{tr}(\mathbf{A}_q^H\mathbf{A}_q) + \text{tr}(\mathbf{B}_q^H\mathbf{B}_q)) = 2T$, which represents our total transmission power constraint to be satisfied during T time slots;

(ii) $\text{tr}(\mathbf{A}_q^H\mathbf{A}_q) = \text{tr}(\mathbf{B}_q^H\mathbf{B}_q) = T/Q$, $q = 1, \ldots, Q$, implying that each space-time signal component is transmitted with the same overall power from the M antennas during the T consecutive time slots, which corresponds to T channel activations;

(iii) $\mathbf{A}_q^H\mathbf{A}_q = \mathbf{B}_q^H\mathbf{B}_q = (T/MQ)\mathbf{I}$, $q = 1, \ldots, Q$, indicating that α_q and β_q are dispersed with equal energy in all spatial and temporal dimensions.

We now continue by offering a few remarks concerning this model.

1. Clearly, the mutual information achieved by the LDC obeying Equation (7.9) is less than or equal to the MIMO channel capacity C_{MIMO} of [3]

$$C_{\text{MIMO}}(\rho, M, N) = E\left\{\log_2\left[\det\left(\mathbf{I} + \frac{\rho}{M}\mathbf{H}\mathbf{H}^H\right)\right]\right\}, \quad (7.10)$$

   where we have $\mathbf{I} \in \zeta^{N\times N}$. The corresponding lower bound can be derived [3]:

$$C_{\text{MIMO}}(\rho, M, N) \geq \min(M, N)\log_2\left[\frac{\rho}{M}\right] + \sum_{i=\max(M,N)-\min(M,N)+1}^{\max(M,N)} E\{\log_2[\Theta_{2i}^2]\}, \quad (7.11)$$

   where Θ_{2i}^2 is a chi-squared distributed random variable with dimension $2i$. Moreover, this lower bound is asymptotically tight at high SNRs. We observe that this is equivalent to the capacity of $\min(M, N)$ subchannels. In other words, the multiple-antenna-aided channel has $\min(M, N)$ degrees of freedom to communicate.

2. The solution of Equation (7.9) subject to any of the above-mentioned constraints (i)–(iii) is non-unique. The judicious choice of the dispersion matrices allows LDCs to satisfy other criteria, such as the rank criterion [52] detailed in Section 7.3.1, without sacrificing mutual information.

3. The optimization stipulated in Equation (7.9) is carried out for a specific SNR ρ, although the authors of [10] argue that the optimization process insensitive to the specific value of ρ in the range of $\rho \geq 20$ and the resultant LDCs generally perform well over a wide SNR range.

4. Observe that if Q is increased, the mutual information between the transmitted signal and the received signal is expected to increase. In contrast, when Q is decreased, a higher coding gain may be obtained. This implies a trade-off between the achievable capacity C_{LDC} and the decoding complexity imposed, and we give further details in Section 7.5.1.3.

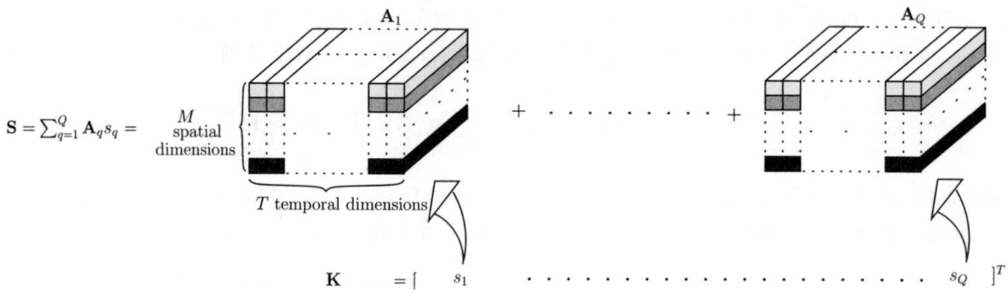

Figure 7.4: The space-time codeword $\mathbf{S}$ formulated based on Equation (7.12).

5. This design criterion is not linked directly with the diversity design criterion given in [52]; therefore, it does not guarantee good BER performance.

The general LDC model of Figure 7.2 is suitable for us to demonstrate that LDCs subsume the existing class of STBCs based on orthogonal designs. Since the conjugate operations seen in Equation (7.30) constitute the key feature of the Alamouti scheme [11], as well as of other STBCs based on orthogonal constraints, it requires that the real and imaginary parts of the transmitted symbols be modulated separately. However, for non-orthogonal STBCs, we do not have significant performance differences between separately modulated and in-phase/quadrature-phase combined modulated codes. Therefore, in order to simplify the related discussions as well as to reduce the size of the equivalent channel matrix $\bar{\mathbf{H}}$ of Equation (7.7), another LDC model is presented in the next section.

7.2.3 LDC Model of [27]

The schematic of the LDCs using the model of [27] is illustrated in Figure 7.3. Given the vector $\mathbf{K} = [s_1, s_2, \ldots, s_Q]^{\mathrm{T}}$ of L-PSK or L-QAM modulated transmission symbols seen in Figure 7.3, the transmitted space-time matrix $\mathbf{S}$ may be defined as [27]

$$\mathbf{S} = \sum_{q=1}^{Q} \mathbf{A}_q s_q, \tag{7.12}$$

which is visualized in Figure 7.4. More explicitly, each symbol s_q is dispersed to the M spatial and T temporal dimensions using a specific dispersion matrix $\mathbf{A}_q$ and the transmission space-time codeword $\mathbf{S}$ is attained by the linear combination of all of the weighted dispersion matrices, as seen in Figure 7.4.

Therefore, the codeword is uniquely and unambiguously determined by the set of dispersion matrices $\mathbf{A}_q$ that are known to both the transmitter and the receiver, which are arranged to be linked by $(M \times N)$ independent CIRs. Note that in contrast to Equation (7.4), this model modulates the real and imaginary parts of the symbols using the same dispersion matrix $\mathbf{A}_q$, rather than using another dispersion matrix $\mathbf{B}_q$. The transmitted codewords should satisfy the power constraint given by

$$\mathrm{tr}\left(\sum_{q=1}^{Q} \mathbf{A}_q^H \mathbf{A}_q\right) = T. \tag{7.13}$$

More strictly, each spatial and temporal slot should be allocated the same transmission power of

$$\mathrm{tr}(\mathbf{A}_q^H \mathbf{A}_q) = \frac{T}{Q}. \tag{7.14}$$

Similarly to Equation (7.8), it is desirable to rewrite the input–output matrix relationship of Equation (7.1) in an equivalent vectorial form. Define the $vec()$ operation as the vertical stacking of the columns of an arbitrary matrix. Subjecting both sides of Equation (7.1) to the $vec()$ operation gives the equivalent system matrix:

$$\bar{\mathbf{Y}} = \bar{\mathbf{H}}\chi\mathbf{K} + \bar{\mathbf{V}}, \tag{7.15}$$

where $\bar{\mathbf{Y}} \in \zeta^{NT\times 1}$, $\bar{\mathbf{H}} \in \zeta^{NT\times MT}$, $\chi \in \zeta^{MT\times Q}$, $\mathbf{K} \in \zeta^{Q\times 1}$ and $\bar{\mathbf{V}} \in \zeta^{NT\times 1}$. More explicitly, χ is referred to as the DCM, which is defined as

$$\chi = [vec(\mathbf{A}_1), vec(\mathbf{A}_2), \ldots, vec(\mathbf{A}_Q)], \tag{7.16}$$

while $\bar{\mathbf{H}}$ in Equation (7.15) is given by

$$\bar{\mathbf{H}} = \mathbf{I} \otimes \mathbf{H}, \tag{7.17}$$

where $\otimes$ denotes the Kronecker product and $\mathbf{I}$ is the identity matrix having a size of $(T \times T)$. The ML estimation of the transmitted signal vector $\mathbf{K}$ is formulated as

$$\bar{\mathbf{K}} = \arg\{\min(\|\bar{\mathbf{Y}} - \bar{\mathbf{H}}\chi\mathbf{K}_f\|^2)\}, \tag{7.18}$$

where $\mathbf{K}_f$ denotes all of the possible combinations of the Q transmitted symbols.

Using the equivalent input–output relationship of Equation (7.15) and applying the results of [251], the ergodic capacity of the LDCs subjected to quasi-static Rayleigh fading is given by

$$C_{\mathrm{LDC}}(\rho, M, N, T, Q) = \max_{\mathrm{tr}(\chi\chi^H)} \frac{1}{T} E\{\log[\det(\mathbf{I} + \rho\bar{\mathbf{H}}\chi\chi^H\bar{\mathbf{H}}^H)]\}, \tag{7.19}$$

where we have $\mathbf{I} \in \zeta^{NT\times NT}$.

The equivalent system model of Equation (7.15) is important, because it provides another view of the LDC structure. More explicitly, when transmitting space-time matrix $\mathbf{S} \in \zeta^{M\times T}$ over a MIMO channel $\mathbf{H} \in \zeta^{N\times M}$, it is equivalent to the transmission of the corresponding symbol vector $\mathbf{K}$ over an equivalent MIMO channel $\bar{\mathbf{H}}\chi \in \zeta^{NT\times Q}$. Applying the lower bound of Equation (7.11) to the equivalent MIMO channel $\bar{\mathbf{H}}\chi$, it can be shown that the capacity achieved by LDCs spanning T time slots is determined by $\min(NT, Q)$. Therefore, when we have $Q \geq MT$, there will be no further improvement of the LDCs' CCMC capacity of Equation (7.19).

Recall that the optimization procedure of Equation (7.9) may be used to find near-capacity LDCs by exhaustively searching the entire design space of LDCs. The resultant solution is not guaranteed to find the global maximum of the cost function. However, by maximizing the ergodic capacity of Equation (7.19) it is possible to obtain a near-capacity LDC by choosing Q according to M and T. More explicitly, upon substituting $Q \geq MT$ into Equation (7.19) and ensuring that

$$\chi\chi^H = \frac{1}{M}\mathbf{I}, \tag{7.20}$$

which satisfies the power constraint of $\mathrm{tr}(\chi\chi^H) = T$, we arrive at an equivalent channel in the form of the $(M \times N)$-element MIMO channel of Equation (7.10). Since the Q transmitted symbols hosted by the vector $\mathbf{K}$ of Figure 7.3 are jointly decoded as in Equation (7.18), the decoding complexity increases exponentially with Q. Hence, it is sufficient to choose $Q = MT$ rather than $Q > MT$ to achieve the full CCMC capacity of Equation (7.10).

In many practical scenarios, it may be desirable to maintain $Q < MT$ in order to reduce the decoding complexity, memory and latency or to satisfy various rate constraints. To accommodate these constraints, χ can be designed by removing the appropriate number of columns from a scaled unitary matrix to arrive at a DCM χ that satisfies

$$\chi^H \chi = \frac{T}{Q}\mathbf{I}. \tag{7.21}$$

Typically, when we have $Q < MT$, there is a loss of ergodic capacity, since it is no longer possible to exploit all of the degrees of freedom $\min(NT, Q)$ of Equation (7.11) provided by the equivalent MIMO channel $\bar{\mathbf{H}}\chi$.

This structure of LDCs guarantees to approach the MIMO channel's capacity quantified in Equation (7.10), depending on the specific choice of Q and T. It does not, however, guarantee a good performance in terms of error probability. By minimizing the maximum PSEP [52] of Equation (7.26), to be detailed in Section 7.3.1, the authors of [27] optimized LDCs that achieve both high rates and full diversity.

We now make a range of further remarks concerning this model.

- LDCs are suitable for arbitrary transmit and receive antenna configurations, combined with arbitrary modulation schemes.

- LDCs are capable of transforming an $(M \times N)$-antenna MIMO system into a $(Q \times NT)$-element equivalent MIMO system by exploiting their inherent linearity.

- Since the real and imaginary parts of the transmitted symbols are dispersed using different dispersion matrices obeying the LDC model of Figure 7.2, the size of the equivalent MIMO system is $(2Q \times 2NT)$, as explicitly shown in Equation (7.8). The non-orthogonal model of Equation (7.15) reduces the equivalent model to a size of $(Q \times NT)$.

- All of the dispersion matrices $\mathbf{A}_q$ of Equation (7.12) can be described with the aid of the single DCM χ characterized in Equation (7.16).

- The LDC's equivalent CIR of Equation (7.15) can be appropriately adjusted by employing a different DCM χ.

Theorem 7.1. *When we have $Q \geq MT$, any DCM χ satisfying Equation (7.20) is an optimal LDC which is capable of approaching the MIMO channel's capacity.*

Theorem 7.2. *When we have $Q < MT$, the ergodic capacity is approached within a margin which is proportional to Q/T.*

Theorem 7.3. *The diversity order of the LDC scheme is less than or equal to $N \cdot \min(M, T)$. This implies that increasing T beyond M does not provide any further advantage in terms of an increased spatial transmit diversity, where the receive diversity is determined by N alone.*

For the proofs of Theorems 7.1–7.3, please refer to [27].

7.2.4 Maximizing the Discrete LDC Capacity

In order to generate a DCM χ defined in Equation (7.16), a random search algorithm was adopted [27]. To elaborate a little further, the random search algorithm of [27] randomly generates a matrix χ from some specific distribution, for example the Gaussian distribution, to satisfy the capacity constraint formulated in Theorems 7.1 and 7.2. Then, the corresponding diversity order and the coding gain is maximized by checking the rank and determinant criteria [52] detailed in Section 7.3.1, when performing an exhaustive search through the entire set of legitimate dispersion matrices.

The random search algorithm has the advantage of providing a wide variety of legitimate LDCs and typically provides a good performance [27]. In fact, codes obtained using a random DCM often perform well without checking the rank and determinant criteria. Since the search is random, they are unable to guarantee finding the maximum of the determinant function of Equation (7.27). Other efforts found in the literature of optimizing LDCs include schemes designed for minimizing the Block Error Ratio (BLER) [292].

In this section, we adopt the aforementioned random search algorithm to optimize LDCs. However, we propose to optimize the LDCs by maximizing the corresponding DCMC capacity. The rationale and advantages of optimizing the DCMC capacity are listed as follows.

- According to the equivalent system matrix of Equation (7.15), transmitting the signal matrix $\mathbf{S}$ through the MIMO channel $\mathbf{H}$ employing the LDC structure of Figure 7.3 can be viewed as transmitting a symbol vector $\mathbf{K}$ through an equivalent MIMO channel given by $\bar{\mathbf{H}}\chi$ of Equation (7.15). Therefore, we propose to maximize the equivalent MIMO channel's capacity directly.

- The LDCs that achieve a higher DCMC capacity at a certain SNR typically exhibit a good BER performance in the high-SNR region, since achieving a higher DCMC capacity implies that they are capable of providing a higher integrity as a result of their higher diversity gain and coding gain.

- The proposed optimization procedure is capable of taking into account the modulated signal constellations, such the specific L-PSK or L-QAM constellations. Hence, the total number of possible combinations for the legitimate space-time symbol vector $\mathbf{K}$ is $F = L^Q$.

The conditional probability of receiving the space-time signal vector $\bar{\mathbf{Y}}$ of Equation (7.15), given a signal vector K_f for $f \in (1, \ldots, F)$ transmitted over a slowly Rayleigh fading channel is determined by the PDF of the noise, as seen in the following equation:

$$p(\bar{\mathbf{Y}}|K_f) = \frac{1}{\pi} \cdot \exp(-\|\bar{\mathbf{Y}} - \bar{\mathbf{H}}\chi K_f\|^2). \tag{7.22}$$

The DCMC capacity of the ML-detected MIMO system using L-QAM or L-PSK signaling, when we have the equivalent CIR $\bar{\mathbf{H}}\chi$ of Equation (7.15), is given by

$$C_{\text{LDC}}^{\text{DCMC}} = \frac{1}{T} \max_{p(K_1),\ldots,p(K_F)} \sum_{f=1}^{F} \int_{-\infty}^{\infty} \cdots \int_{-\infty}^{\infty} p(\bar{\mathbf{Y}}|K_f)p(K_f) \cdot \log_2 \left[\frac{p(\bar{\mathbf{Y}}|K_f)}{\sum_{g=1}^{F} p(\bar{\mathbf{Y}}|K_g)p(K_g)} \right] d\bar{\mathbf{Y}} \quad (\text{BPS Hz}^{-1}), \tag{7.23}$$

where the right-hand side of Equation (7.23) is maximized, when we have equiprobable space-time symbols obeying $p(K_f) = 1/F$ for $f = 1, \ldots, F$. Hence, (7.23) can be simplified as [293]

$$C_{\rm LDC}^{\rm DCMC} = \frac{1}{T}\left(\log_2[F] - \frac{1}{F}\sum_{f=1}^{F} E\left\{\log_2\left[\sum_{g=1}^{F}\exp(\Psi_{f,g})\middle|K_f\right]\right\}\right). \tag{7.24}$$

where $\Psi_{f,g}$ within the expectation value in curly braces in Equation (7.24) is given by

$$\Psi_{f,g} = -\|\bar{\mathbf{H}}\chi(K_f - K_g) + \bar{\mathbf{V}}\|^2 + \|\bar{\mathbf{V}}\|^2. \tag{7.25}$$

Since the equivalent system model represents the transmission regime of T time slots, the DCMC capacity of LDCs has to be divided by T, as seen in Equations (7.23) and (7.24).

Given the DCMC capacity of the LDC$(MNTQ)$ family, the following random search algorithm can be derived.

1. Randomly generate the complex-valued matrix $\bar{\chi} \in \zeta^{MT\times MT}$ using the Gaussian distribution.

2. If we arrange the system to satisfy $Q \geq MT$, the candidate DCM has to be a unitary matrix according to Theorem 7.1. It has been shown in [294] that a complex-valued matrix can be factored into the product of a unitary matrix and an upper triangular matrix using the QR decomposition [294]. Thus, a random DCM can be obtained by $\chi = (1/\sqrt{M})QR(\bar{\chi})$, which satisfies Equation (7.20).

3. In contrast, if we confine the LDC schemes to $Q < MT$, the DCM χ has to satisfy Equation (7.21) and it can be generated by retaining the first Q columns of the unitary matrix obtained using the QR decomposition of $\sqrt{T/Q}QR(\bar{\chi})$.

4. Having searched through the entire set of legitimate dispersion character matrices, we choose that particular matrix that maximizes the DCMC capacity of Equation (7.24).

The LDCs generated by the above algorithm are sufficiently diverse to represent the entire legitimate space and additionally they are capable of maximizing the DCMC capacity. We found that 100 000 random searches are typically sufficient for generating LDCs exhibiting a good BER performance. In Appendix D, we summarize all of the DCMs χ derived for all of the LDCs used in this chapter.

7.2.5 Performance Results

We have argued above that LDCs described by the parameters $(MNTQ)$ are suitable for arbitrary transmit and receive antenna configurations, when transmitting the Q symbols of the vector $\mathbf{K}$ during T time periods as seen in Figure 7.3. In this section, we present our simulation results for characterizing LDCs having different parameters and quantify their capacity with the aid of Theorems 7.1–7.3. For the moment, we concentrate on the achievable LDC performance in terms of the attainable ergodic capacity and BER. Gray labeling was assumed for the bit-to-symbol mapping and a ML detector using Equation (7.18) was employed. QPSK modulation was employed in all of the simulations. We adopted the non-orthogonal LDC structure of Figure 7.3 and the DCM χ was generated using the method presented in Section 7.2.4.

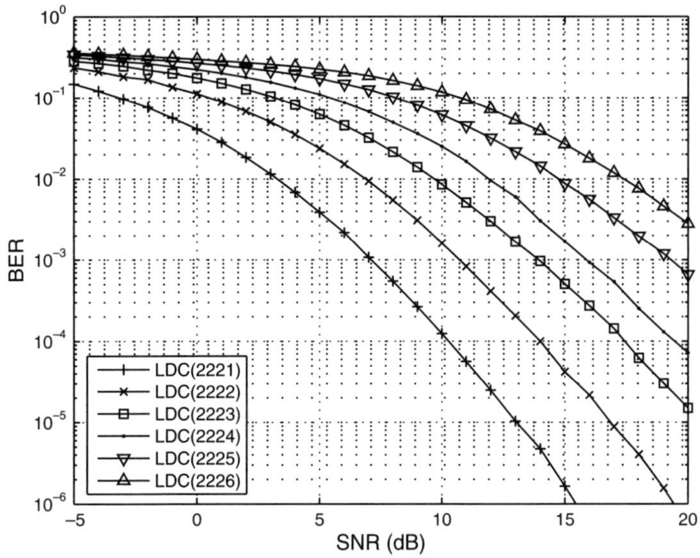

Figure 7.5: BER of a family of QPSK-modulated LDCs obeying the structure of Figure 7.3 having $M = 2$, $N = 2$, $T = 2$ and $Q = 1, 2, 3, 4, 5, 6$ using a ML detector, when transmitting over i.i.d. Rayleigh-fading channels.

Figure 7.5 portrays the BER performance of LDCs having rates of $R_{\mathrm{LDC}} = 0.5$, 1, 1.5, 2, 2.5 and 3, which were adjusted by fixing (MNT) while gradually increasing Q. Observe from the shape of the curves in Figure 7.5 that all of the codes do achieve the maximum diversity order of $D = 4$ that a (2×2)-element MIMO system is capable of providing. Owing to the power constraint of Equation (7.13), the average transmit energy of each symbol was decreased when increasing Q, which results in a BER degradation. The ML detector jointly and simultaneously decodes the Q symbols of the vector $\mathbf{K}$ seen in Figure 7.3. When we have $Q = 1$, the transmitted signal is a single element in a two-dimensional space obeying Equation (7.15). Since Gray labeling is used, the MED is maximized. However, when we have $Q > 1$, since each transmitted symbol is expected to fade independently, as seen in Equations (7.8) and (7.15), there is no guarantee that Gray mapping will ensure that the MED is maximized within the $2Q$-dimensional space.

The corresponding achievable CCMC capacity of LDCs having $M = 2$, $N = 2$, $T = 2$ and $Q = 1, 2, 3, 4, 5, 6$ using a ML detector is shown in Figure 7.6. The MIMO channel capacity of Equation (7.10) is also shown as an upper bound. Observe that LDCs having $Q \geq MT = 4$ have already achieved the maximum attainable diversity order of $D = 4$ and the MIMO channel capacity of Equation (7.10). Therefore, there is no point in selecting $Q > MT$ in terms of the achievable CCMC capacity of Equation (7.19), especially since this will increase the decoding complexity. In case of $Q < MT$, observe in Figure 7.6 that the CCMC capacity increases in proportion to Q/T, as stated in Theorem 7.2.

The achievable DCMC capacity of LDCs having $M = 2$, $N = 2$, $T = 2$ and $Q = 1, 2, 3, 4, 5, 6$ using a ML detector is shown in Figure 7.7 as plotted according to Equation (7.23). Recall that when we have $Q \geq MT$, the CCMC MIMO channel capacity has already been achieved as seen in Figure 7.6. In contrast, Figure 7.7 demonstrates that the DCMC capacity calculated using Equation (7.23) increases with Q, even when we have

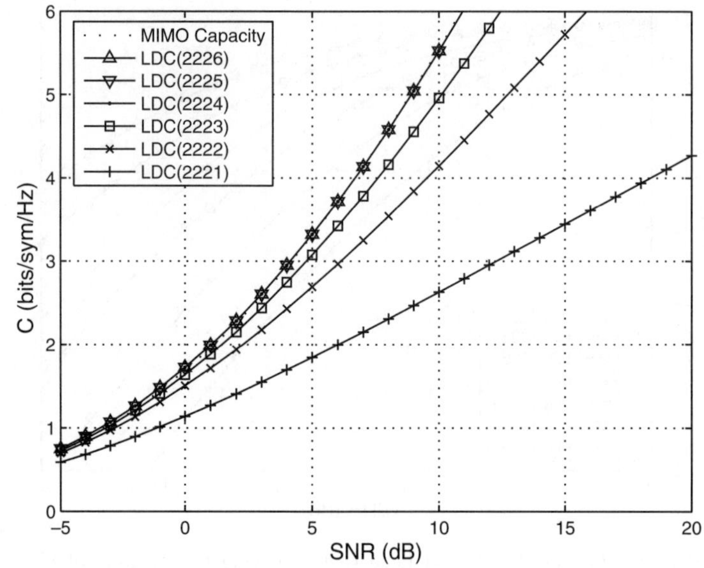

Figure 7.6: CCMC capacity comparison of a family of LDCs obeying the structure of Figure 7.3 having $M = 2$, $N = 2$, $T = 2$ and $Q = 1, 2, 3, 4, 5, 6$ using a ML detector, as plotted from Equation (7.9). The MIMO capacity curve is coincident with the LDC(2226), LDC(2225) and LDC(2224) curves.

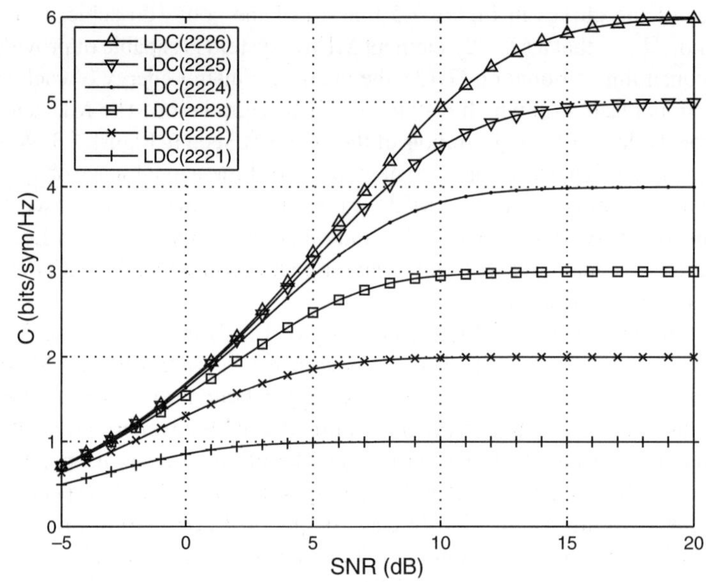

Figure 7.7: DCMC capacity comparison of LDCs obeying the structure of Figure 7.3 having $M = 2$, $N = 2$, $T = 2$ and $Q = 1, 2, 3, 4, 5, 6$ using a ML detector, as plotted from Equation (7.23), where QPSK modulation was employed.

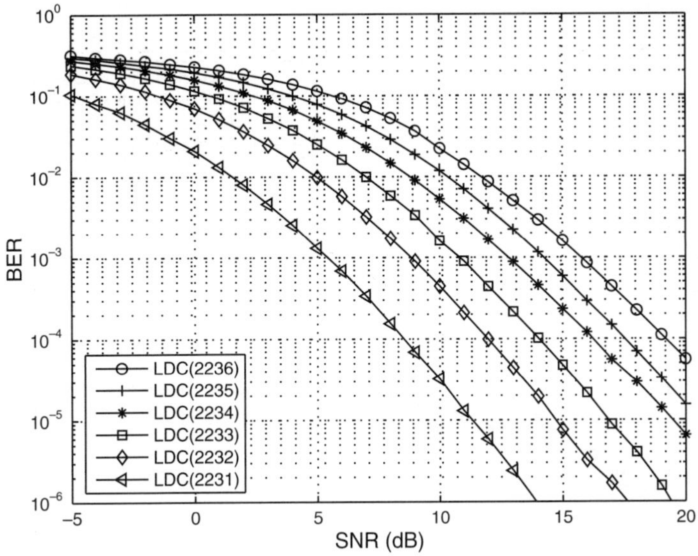

Figure 7.8: BER of a family of QPSK-modulated LDCs obeying the structure of Figure 7.3 having $M = 2$, $N = 2$, $T = 3$ and $Q = 1, 2, 3, 4, 5, 6$ using an ML detector, when transmitting over i.i.d. Rayleigh-fading channels.

$Q > MT$. Naturally, it is possible to further improve the throughput of the system by transmitting more BPS, i.e. by using higher-order modulation schemes.

Figure 7.8 shows the BER performance of a group of QPSK-modulated LDCs having $M = 2, N = 2, T = 3$ and $Q = 1, 2, 3, 4, 5, 6$ using a ML detector. Again, a BER degradation is observed upon increasing the number of symbols transmitted per space-time block. Compared with the family of LDCs having $T = 2$ and characterized in Figure 7.5, the group of LDCs characterized in Figure 7.8 has the same spatial diversity order of $D = 4$. This observation is supported by Theorem 7.3, where the transmit diversity order is determined by $\min(M, T)$. However, it is still interesting to observe that the identical-rate LDC pairs of [LDC(2222), LDC(2233)] and [LDC(2224), LDC(2236)] characterized in Figures 7.5 and 7.8 exhibit the same BER performance.

Furthermore, Figure 7.9 characterizes the DCMC capacity achieved by the equal-rate LDC pairs of [LDC(2222), LDC(2233)] and [LDC(2224), LDC(2236)] using a ML detector in conjunction with QPSK modulation. Not surprisingly, according to Equation (7.23), these equal-rate equal-diversity-order LDCs achieve the same DCMC capacity. This implies that increasing T and maintaining $T > M$, while fixing the rate to $R_{\text{LDC}} = Q/T$, do not benefit the system in terms of an improved BER performance and increased achievable capacity. In fact, maintaining the minimum value of T is desirable owing to the associated complexity considerations.

Figure 7.10 quantifies the achievable BER performance of the QPSK-modulated equal-rate LDC pairs [LDC(2212), LDC(2224)] obeying the structure of Figure 7.3 using a ML detector, when transmitting over i.i.d. Rayleigh-fading channels. According to Theorem 7.3, the LDC(2212) scheme achieves $D = 2$, whereas $D = 4$ diversity gain is achievable using the LDC(2224) scheme. The BER performance recorded in Figure 7.10 demonstrates a clear gap between these schemes in the high-SNR region.

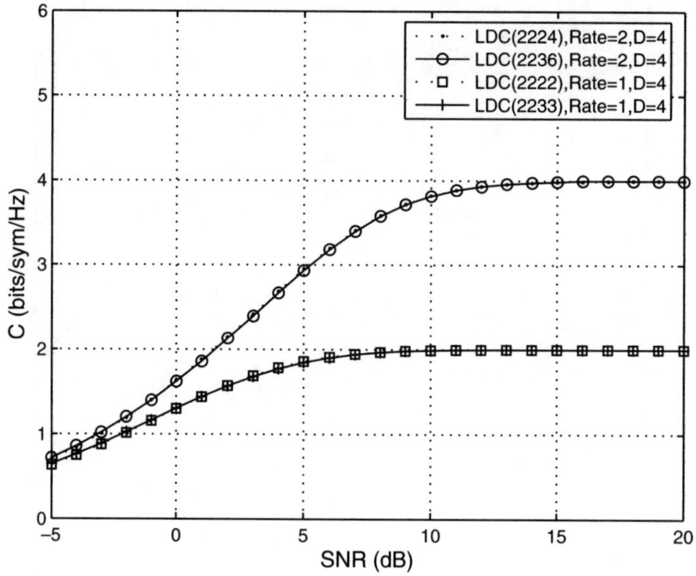

Figure 7.9: DCMC capacity comparison of QPSK-modulated equal-rate LDC pairs [LDC(2222), LDC(2233)] and [LDC(2224), LDC(2236)] obeying the structure of Figure 7.3 using an ML detector, as evaluated from Equation (7.23).

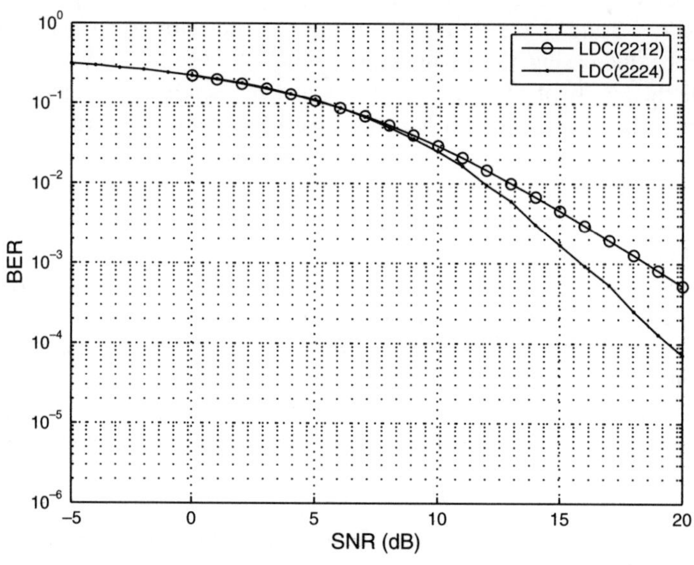

Figure 7.10: BER of QPSK-modulated equal-rate LDC pairs [LDC(2212), LDC(2224)] obeying the structure of Figure 7.3 using a ML detector, when transmitting over i.i.d. Rayleigh-fading channels.

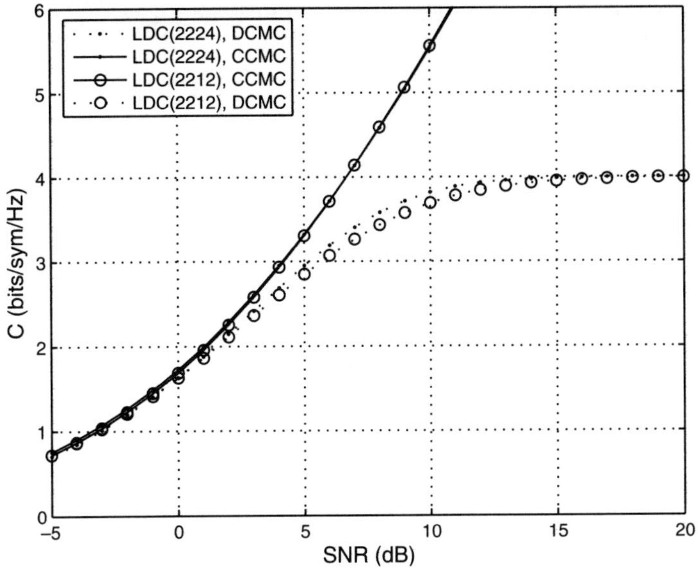

Figure 7.11: The CCMC and DCMC capacities of equal-rate LDC pairs [LDC(2212), LDC(2224)] obeying the structure of Figure 7.3 using a ML detector, as plotted using Equations (7.9) and (7.23). For plotting the DCMC capacity, QPSK modulation was employed.

The CCMC and DCMC capacities achieved by the LDC(2212) and LDC(2224) schemes plotted using Equations (7.9) and (7.23) are shown in Figure 7.11. According to Theorem 7.1, the only requirement for LDCs to achieve the MIMO channel's full capacity is that the DCM χ has to be unitary. Therefore, despite different diversity orders being achieved by the LDC(2212) and LDC(2224) schemes, they attain the same CCMC capacity. However, the LDC(2224) arrangement has a higher DCMC capacity characterized in Equation (7.23) in the SNR region of 5 dB $< \rho <$ 15 dB, since having a higher spatial diversity order provides a higher degree of protection, as shown in Figure 7.10.

Based on the discussions above as well as on Theorems 7.1–7.3 our observations may be summarized as follows.

Corollary 7.1. *A group of LDCs $(MNTQ)$ having M transmit and N receive antennas, as well as a fixed rate of $Q/T(T > M)$, exhibits the same diversity order and the same capacity. Consequently, they have an identical rate and an identical BER performance.*

In Figure 7.12 we plot the BER performance of a group of QPSK-modulated LDCs having the same fixed rate of $R_{\text{LDC}} = 2$, while using different MIMO antenna structures. Here, we used the LDC(2224) scheme as our benchmark. When increasing the number of transmit antennas to $M = 3$, according to Theorem 7.3, the LDC(3224) scheme exhibits the same diversity order of $D = 4$. However, the increased value of M enables the LDC(3224) arrangement to achieve a better coding gain, as shown in Figure 7.12. On the other hand, when increasing N to 3, the LDC(2324) arrangement achieved a diversity order of $D = 6$, which resulted in a significant SNR gain in Figure 7.12. Again, when an additional transmit antenna is employed compared with the LDC(2324) scheme, the resultant LDC(3324) arrangement demonstrated a better coding gain.

The capacity associated with changing the number of transmit and receive antennas is portrayed in Figure 7.13. Observe that the LDCs associated with $N = 3$, which have a higher

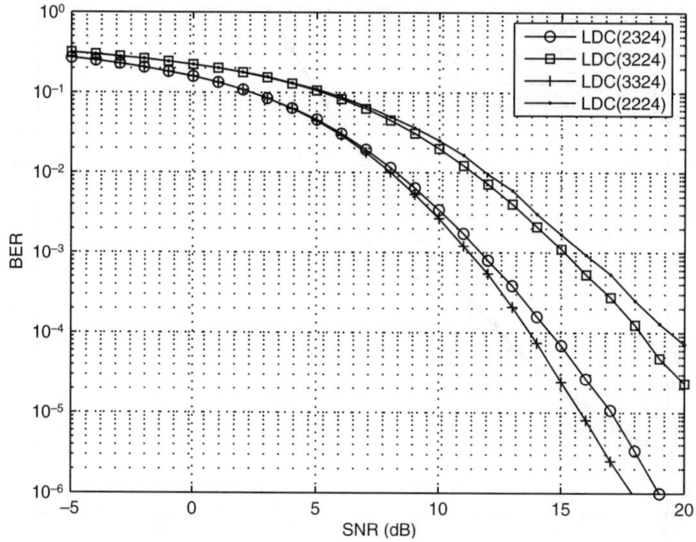

Figure 7.12: BER of a family of QPSK-modulated LDCs obeying the structure of Figure 7.3 having $M = 2, 3, N = 2, 3, T = 2$ and $Q = 4$ using a ML detector, when transmitting over i.i.d. Rayleigh-fading channels.

diversity order, also achieved a substantially higher CCMC/DCMC capacity compared with the LDCs having $N = 2$. On the other hand, the effect of employing more transmit antennas ($M \geq T$), which results in an increased coding gain in Figure 7.12 is insignificant in terms of the increased achievable capacity, as quantified in Figure 7.13.

7.2.6 Summary

In this section, we characterize the performance of the LDCs presented in Section 7.2.5 in terms of the effective throughput and the coding gains, where the latter is defined as the SNR difference, expressed in decibels, at a BER of 10^{-4} between various LDCs and the identical throughput single-antenna-aided systems.

In Figure 7.14, we plot the effective throughput against the SNR required to achieve $\text{BER} = 10^{-4}$ for a group of LDCs having $N = 1, 2, 3, 4$ receive and $M = 2$ transmit antennas using $T = 2$ time slots to transmit $Q = 1$ symbols per space-time block. An increased effective throughput was achieved by employing high-order modulation schemes, rather than by increasing the value of Q. For this particular group of $\text{LDC}(2N21)$, 9.5 dB is required for increasing the effective throughput from 0.5 to 2 (BPS Hz^{-1}).

Figure 7.15 plots the SNR requirement against the number of receive antennas N for a family of LDCs having $T = 2$, $Q = 4$ to achieve $\text{BER} = 10^{-4}$. For a given number of receive antennas N, the advantage of employing $M = 2, 3, 4$ transmit antennas is up to 6.5 dB compared with that of the single-antenna scheme, which increases gradually, as the total spatial diversity order is increased from $D = 2$ to 4. Also observe in Figure 7.15 that increasing N significantly reduces the SNR required to achieve $\text{BER} = 10^{-4}$, when we fix the number of transmit antennas M. In fact, the resultant SNR advantage of increasing N from 1 to 5 is about 25 dB.

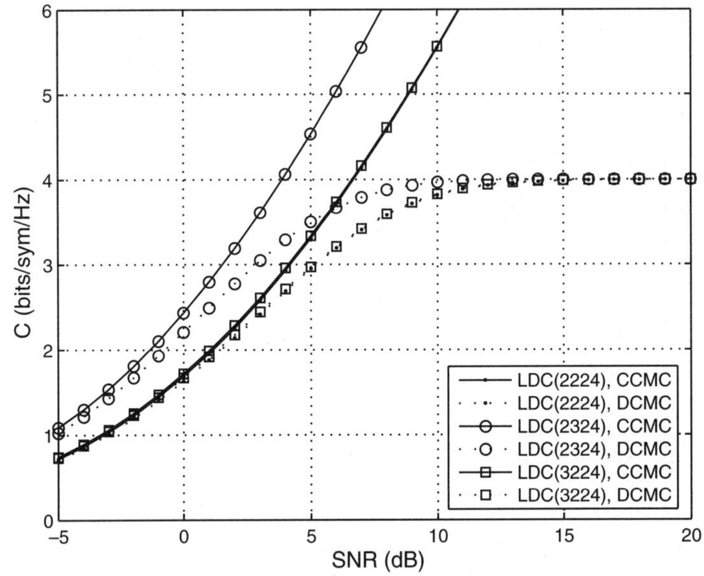

Figure 7.13: The CCMC and DCMC capacity of LDCs having $M = 2, 3$, $N = 2, 3$, $T = 2$ and $Q = 4$ obeying the structure of Figure 7.3 using a ML detector, as plotted using Equations (7.9) and (7.23). For plotting the DCMC capacity, QPSK modulation was employed.

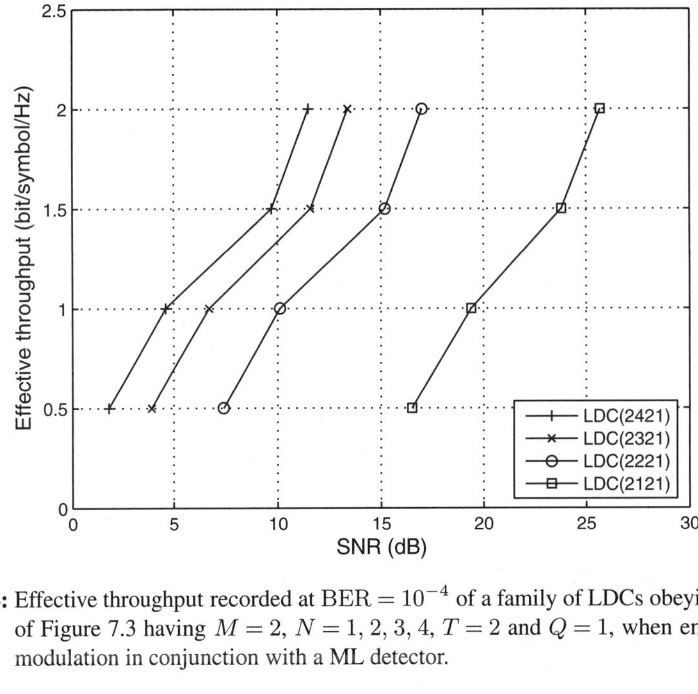

Figure 7.14: Effective throughput recorded at $\text{BER} = 10^{-4}$ of a family of LDCs obeying the structure of Figure 7.3 having $M = 2$, $N = 1, 2, 3, 4$, $T = 2$ and $Q = 1$, when employing QPSK modulation in conjunction with a ML detector.

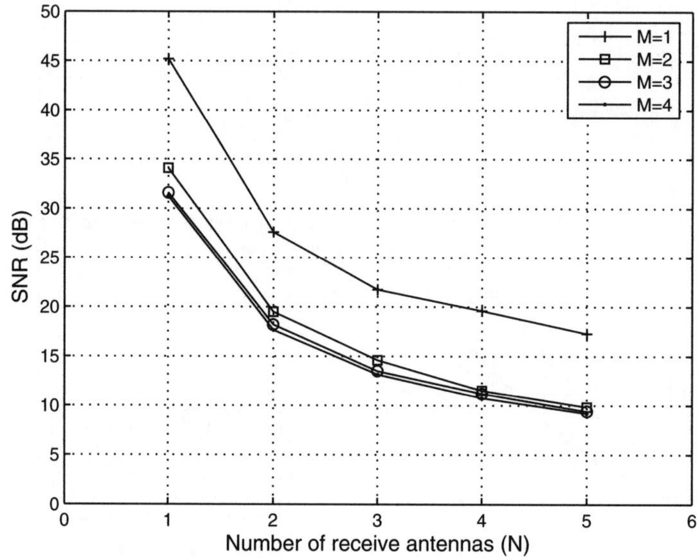

Figure 7.15: SNRs for a family of LDC schemes obeying the structure of Figure 7.3 having $M = 1, 2, 3, 4$, $N = 1, 2, 3, 4, 5$, $T = 2$ and $Q = 4$ to achieve BER $= 10^{-4}$, when employing QPSK modulation in conjunction with a ML detector.

Table 7.1: Coding gains of a family of LDCs have an effective throughput of 1 (BPS Hz^{-1}).

LDC	Diversity order	Modulation	Coding gain (dB)
LDC(2211)	2	BPSK	17.5
LDC(2221)	4	QPSK	23.9
LDC(2222)	4	BPSK	23.65
LDC(2231)	4	8PSK	20.8
LDC(2233)	4	BPSK	23.7

Furthermore, Tables 7.1 and 7.2 characterize the coding gains of a family of LDCs having an effective throughput of 1 and 2 (BPS Hz^{-1}), respectively, compared with the single-antenna system having the identical effective throughput. The corresponding modulation schemes employed are also listed in Tables 7.1 and 7.2. More particularly, for schemes having an effective throughput of 1 (BPS Hz^{-1}), the largest coding gain is observed when LDC(2233) was employed in conjunction with BPSK modulation, while the BPSK-modulated LDC(2224) arrangement achieved the highest coding gain for an effective throughput of 2 (BPS Hz^{-1}).

7.3 Link Between STBCs and LDCs

In this section, we first continue our discourse by reviewing the main results in the state-of-art literature related to STBCs, including the so-called rank criterion [52], the determinant criterion [52], the diversity versus multiplexing gain trade-off [23] and the diversity versus

Table 7.2: Coding gains of a family of LDCs have an effective throughput of 2 (BPS Hz^{-1}).

LDC	Diversity order	Modulation	Coding gain (dB)
LDC(2211)	2	QPSK	17.7
LDC(2212)	2	BPSK	17.4
LDC(2221)	4	16QAM	19.9
LDC(2222)	4	QPSK	23.0
LDC(2224)	4	BPSK	23.3
LDC(2231)	4	64QAM	16.0
LDC(2232)	4	8PSK	20.7
LDC(2233)	4	QPSK	23.0

rate trade-off [26]. Second, we investigate the relationship between all of the major representatives of STBCs reported in the open literature and LDCs from both mathematical and design concept perspectives.

7.3.1 Review of Existing STBC Knowledge

The PSEP of mistaking a space-time transmission matrix $\mathbf{S}$ of Equation (7.2) for another matrix $\bar{\mathbf{S}}$, denoted as $p(\bar{\mathbf{S}} \rightarrow \mathbf{S})$, depends only on the distance between the two matrices after transmission through the channel and the noise power, which is upper bounded by [52]:

$$p(\bar{\mathbf{S}} \rightarrow \mathbf{S}) \leq \frac{1}{(\nabla_1 \cdots \nabla_\eta)^N} \cdot \left(\frac{\rho}{4M}\right)^{\eta N}. \tag{7.26}$$

- **Rank criterion [52].** Observe in Equation (7.26) that $(\rho/4M)^{\eta N}$ dominates how fast the error decays with the SNR and the total diversity order is determined by ηN. More explicitly, the transmit diversity order of a STBC scheme is equal to the minimum rank η of the difference matrix $\mathbf{S}_\triangle = (\mathbf{S} - \bar{\mathbf{S}})$. Accordingly, η has to be found by searching through all distinct codeword pairs $\mathbf{S}$ and $\bar{\mathbf{S}}$, where $\bar{\mathbf{S}} \neq \mathbf{S}$.

- **Determinant criterion [52].** Also observe in Equation (7.26) that $1/(\nabla_1 \cdots \nabla_\eta)^N$ has to be minimized, which determines the coding gain of a STBC scheme. Furthermore, the determinant criterion states the minimum of

$$(\nabla_1 \cdots \nabla_\eta) \tag{7.27}$$

  evaluated over all distinct space-time codeword pairs determines the achievable coding gain and must be maximized, where the coefficients ∇_i are the non-zero eigenvalues of $\mathbf{S}_\triangle \mathbf{S}_\triangle^H$.

- **Full diversity.** Using the average PSEP analysis technique of Equation (7.26), it follows that the maximum attainable diversity order of a STBC scheme designed for an $(M \times N)$-element MIMO system is [12, 52]

$$D_{\text{full}} = MN, \tag{7.28}$$

  which implies that the space-time difference matrix $\mathbf{S}_\triangle$ should have full rank.

- **Full rate.** Since it is possible to transmit up to one 'independent' symbol per antenna per time slot, the symbol rate could be

$$R_{\text{full}} = M. \tag{7.29}$$

Note that LDCs obeying Equation (7.12) are capable of achieving a rate $R_{\text{LDC}} > M$, since the transmission matrix is the weighted sum of all of the symbols as exemplified in Figures 7.5, 7.6 and 7.8. However, the terminology 'full rate' remains useful in the discussion of a STBC scheme's CCMC capacity. For example, we have demonstrated in Equation (7.19) that when the LDC's rate $R_{\text{LDC}} \geq M$, there is no further improvement in terms of the achievable CCMC capacity. On the other hand, using the term 'full rate' is inappropriate for quantifying the DCMC capacity, since the corresponding capacity increases upon increasing the rate, as seen in Figure 7.7.

At this stage it is important to contrast the above-mentioned full-rate, full-diversity schemes, which were primarily conceived for providing transmit and receive diversity against the family of SDM MIMOs that were contrived for attaining a multiplexing gain, although they may also provide some diversity gain. The terminology of full-rate and full-diversity schemes is typically invoked in the context of STCs, although quantifying the throughput and diversity gain of SDM schemes would also be beneficial. Nonetheless, to the best of our knowledge, no parallel terminology has been used in the context of SDM schemes.

- **Diversity Multiplexing Gain Trade-off.** The authors of [23] showed that for a MIMO channel, there is a fundamental trade-off between the achievable diversity gain and the attainable multiplexing gain. More explicitly, achieving a higher spatial multiplexing gain comes at the price of sacrificing diversity gain, when employing a continuous-valued signal alphabet that grows linearly with the logarithm of the SNR. In other words, the diversity and multiple gain trade-off quantifies how rapidly the throughput of a STBC scheme can increase with the SNR, while having a certain diversity order. Any transmission arrangement employing a fixed-throughput modulation scheme fails to approach the achievable capacity at high SNRs, because the capacity increases with the SNR, while the throughput of this fixed modem arrangement does not.

- **Rate Diversity Trade-off.** In [26, 295], the authors argued that in MIMO systems, one can also increase the attainable transmission rate at the expense of a certain loss in the diversity gain, which reflects the associated rate versus diversity trade-off. This trade-off is characterized by $R_{\text{STBC}} \leq M - D_{tx} + 1$ (see [26]), where D_{tx} denotes the spatial transmit diversity order. With the aid of recent advances in high-rate full-diversity STBCs [27, 284, 296], it has been shown that it is not necessary to sacrifice rate in order to achieve diversity and vice versa. However, considering the rate versus diversity trade-off is still valuable for the analysis of STBCs obeying a certain structure, which is characterized by the so-called transmit symbol separability defined as follows.

Definition 7.1 (Transmit Symbol Separability (TSS)).

- If **all** of the entries of the transmitted space-time signal matrix $\mathbf{S}$ constitute a transformed version[1] of a single rather than several symbols from a specific modulated signal constellation, then this STBC is said to obey the property of full TSS.

[1]Transformation includes scalar multiplication, Hermitian transpose and conjugate operation.

- If **some** of the entries in the transmitted space-time signal matrix $\mathbf{S}$ are constituted by a combination of several symbols from a specific modulated signal constellation, then the STBC is said to obey the property of partial TSS.
- If **all** of the entries in the transmitted space-time matrix $\mathbf{S}$ are constituted by a combination of several symbols from a specific modulated constellation, then the STBC is deemed to be non-separable.

For example, the STBC scheme transmitting the signal $((1/\sqrt{2})s_1 + (1/\sqrt{2})s_2)$ is non-separable, while Alamouti's scheme [11] transmits either $\pm s_1$ or $\pm s_1^*$ and hence exhibits full TSS. Let us now consider the family of STBCs from a TSS perspective in more detail.

Property 7.1. *If a STBC scheme exhibits full TSS, then there exists a trade-off between the maximum rate and the maximum achievable spatial diversity. If a STBC scheme has the property of partial TSS or it is non-separable, it has the potential to achieve both high rate and full diversity at the same time.*

Discussion. When full TSS is maintained, each transmitted symbol only contains its own information. Therefore, there is a trade-off between increasing the achievable rate by sending more independent symbols and increasing the diversity by transmitting redundant information. For example, Alamouti's STBC and the classic V-BLAST scheme achieve the two extremes, respectively. On the other hand, even when partial TSS is maintained, some of the transmitted symbols may carry information related to multiple symbols, and then it is possible to achieve some grade of diversity, despite operating at high rate. Furthermore, the family of STBCs does not maintain TSS, but nonetheless has the potential of achieving high rate and full diversity.

7.3.2 Orthogonal STBCs

STBCs based on orthogonal designs in order to achieve full spatial diversity were first proposed in [11], and later were generalized in [25]. The philosophy behind OSTBCs is that each transmission space-time signal matrix $\mathbf{S}$ satisfies an orthogonality constraint. The orthogonality embedded in $\mathbf{S}$ enables the receiver to decouple the transmitted multi-antenna-coded symbol streams into independent symbols. Thus, simple ML detection can be carried out. Unfortunately, STBCs satisfying the orthogonality constraint exist for only a few specific choices of the parameters $(MNTQ)$ and they do not achieve the ergodic capacity of Equation (7.10), especially not when multiple receive antennas are employed.

For example, given a symbol vector $\mathbf{K} = [s_1, s_2]^{\mathrm{T}}$, the simple $\mathbf{G}_2$ space-time code of [25] can be written as

$$\begin{aligned}
\mathbf{G}_2 &= \begin{pmatrix} s_1 & s_2 \\ -s_2^* & s_1^* \end{pmatrix} \\
&= \begin{pmatrix} \alpha_1 & 0 \\ 0 & \alpha_1 \end{pmatrix} + j \begin{pmatrix} \beta_1 & 0 \\ 0 & -\beta_1 \end{pmatrix} + \begin{pmatrix} 0 & \alpha_2 \\ -\alpha_2 & 0 \end{pmatrix} + j \begin{pmatrix} 0 & \beta_2 \\ \beta_2 & 0 \end{pmatrix} \\
&= \begin{pmatrix} 1 & 0 \\ 0 & 1 \end{pmatrix} \alpha_1 + j \begin{pmatrix} 1 & 0 \\ 0 & -1 \end{pmatrix} \beta_1 + \begin{pmatrix} 0 & 1 \\ -1 & 0 \end{pmatrix} \alpha_2 + j \begin{pmatrix} 0 & 1 \\ 1 & 0 \end{pmatrix} \beta_2 \\
&= \mathbf{A}_1\alpha_1 + j\mathbf{B}_1\beta_1 + \mathbf{A}_2\alpha_2 + j\mathbf{B}_2\beta_2,
\end{aligned} \tag{7.30}$$

where $s_1 = \alpha_1 + j\beta_1$ and $s_2 = \alpha_2 + j\beta_2$. Observe that Equation (7.30) is fully characterized by the LDC structure of Figure 7.2. Following a similar decomposition process applied to

Equation (7.30), it may be readily shown that other OSTBCs such as the $\mathbf{G}_3$, $\mathbf{G}_4$, $\mathbf{H}_3$ and $\mathbf{H}_4$ schemes of [25] can also be fully specified by the dispersion structure of Figure 7.2.

From the perspective of LDCs, the family of OSTBCs disperses each multi-antenna space-time symbol to specific spatial and temporal slots according to a certain pattern, as seen in Equation (7.30). The pattern is fully characterized by the orthogonality of the dispersion matrices. Hence, both full TSS and simple ML detection are guaranteed.

It is worth mentioning that the OSTBCs $\mathbf{H}_3$ and $\mathbf{H}_4$ [25] exhibit partial TSS, whereas their counterparts $\mathbf{G}_3$ and $\mathbf{G}_4$ [25] possess the property of full TSS. Therefore, $\mathbf{H}_3$ and $\mathbf{H}_4$ achieve a higher rate than the $\mathbf{G}_3$ and $\mathbf{G}_4$ codes, although none of them are capable of achieving a symbol rate higher than one.

7.3.3 QOSTBCs

The main benefits of an orthogonal design are its simple decoding and full transmit diversity potential. When relaxing the simple separate decoding property of the multi-antenna streams, a potentially higher rate can be achieved. In [14], the family of QOSTBCs was proposed for the sake of pursuing high-rate transmission, while maintaining a certain diversity order. The class of QOSTBCs is capable of decoupling the symbol streams into groups, where each decoding group contains two symbols rather than a single symbol. Hence, a higher ML decoding complexity is imposed compared with that of the OSTBCs.

The construction of QOSTBCs can be derived directly from that of OSTBCs. For example, a (4×4)-antenna QOSTBC codeword matrix is given by [14]:

$$\mathbf{S} = \begin{pmatrix} \mathbf{G}_2(s_1, s_2) & \mathbf{G}_2(s_3, s_4) \\ -\mathbf{G}_2(s_3, s_4)^* & \mathbf{G}_2(s_1, s_2)^* \end{pmatrix}. \tag{7.31}$$

Therefore, it is straightforward to rewrite the QOSTBC transmission matrix of Equation (7.31) using the LDC structure of Figure 7.2 following the same procedure as in the context of Equation (7.30). Compared with the $\mathbf{G}_4$ OSTBC code of [25], the QOSTBC of Equation (7.31) achieves twice the symbol rate at the cost of sacrificing half of the transmit diversity order, which constitutes a manifestation of the rate versus diversity trade-off formulated in the context of Property 7.1.

Clearly, the LDC structure of Figure 7.2 subsumes the family of QOSTBCs. Similarly to OSTBCs, QOSTBCs disperse each multi-antenna space-time symbol to specific space-time slots obeying the relaxed orthogonality constraint. Observe from Equation (7.31) that full TSS is also ensured, which implies that the class of QOSTBCs obeys the rate versus diversity trade-off.

In later works [297, 298], the idea of constellation rotation was introduced in order to overcome the potential diversity loss. However, the associated diversity gain improvement accrued from modulation diversity [299], rather than from the spatial diversity addressed in this chapter.

7.3.4 LSTBCs based on Amicable Orthogonal Designs

Although QOSTBCs exhibit a higher design flexibility than OSTBCs, they are still not suitable for MIMO systems having flexible $(MNTQ)$ parameter combinations and they are unable to achieve either full diversity or a symbol rate higher than one. The further pursuit of high-diversity STBCs leads to the design of LSTBCs [30, 300–302], which are defined as

$$\mathbf{S} = \sum_{q=1}^{Q} (\alpha_q \mathbf{A}_q + j\beta_q \mathbf{B}_q). \tag{7.32}$$

The philosophy behind LSTBCs is to find specific orthogonal dispersion matrices $\mathbf{A}_q$ and $\mathbf{B}_q$, which are capable of separating Q transmitted symbols at the receiver. In other words, Q symbols are mapped to M transmit antennas with the aid of a set of orthogonal matrices.

To accomplish this design goal, the dispersion matrices designed for real-valued symbols have to satisfy the following requirement [300]:

$$\mathbf{A}_{i_1}\mathbf{A}_i^H = \mathbf{I} \quad (i = i_1), \tag{7.33}$$

$$\mathbf{A}_{i_1}\mathbf{A}_i^H = -\mathbf{A}_i\mathbf{A}_{i_1}^H \quad (i \neq i_1), \tag{7.34}$$

where $i, i_1 = 1, 2, \ldots, Q$. Full spatial diversity order is guaranteed by Equation (7.33), while Equation (7.34) ensures that the set of dispersion matrices are orthogonal to each other. For complex-valued modulated symbols, the dispersion matrices $\mathbf{A}_q$ and $\mathbf{B}_q$ should satisfy [300]

$$\mathbf{A}_{i_1}\mathbf{A}_i^H = \mathbf{I}, \quad \mathbf{B}_{i_1}\mathbf{B}_i^H = \mathbf{I} \quad (i = i_1), \tag{7.35}$$

$$\mathbf{A}_{i_1}\mathbf{A}_i^H = -\mathbf{A}_i\mathbf{A}_{i_1}^H, \quad \mathbf{B}_{i_1}\mathbf{B}_i^H = -\mathbf{B}_i\mathbf{B}_{i_1}^H \quad (i \neq i_1), \tag{7.36}$$

$$\mathbf{A}_{i_1}\mathbf{B}_i^H = \mathbf{B}_i\mathbf{A}_{i_1}^H, \quad (1 < i, i_1 < Q). \tag{7.37}$$

Equation (7.35) ensures that full spatial transmit diversity order can be achieved and Equation (7.36) ensures the orthogonality within the dispersion matrices $\mathbf{A}_i$ and $\mathbf{B}_i$. The orthogonality between the dispersion matrices $\mathbf{A}_i$ and $\mathbf{B}_i$ is guaranteed by Equation (7.37).

Recall that the family of OSTBCs discussed in Section 7.3.2 obey the orthogonal constraint of

$$\mathbf{S}\mathbf{S}^H = \mathbf{I}. \tag{7.38}$$

If the LSTBCs obeying the structure of Equation (7.32) satisfy the orthogonal constraint of Equation (7.38), it can be shown that the set dispersion matrices of $\mathbf{A}_q$ and $\mathbf{B}_q$ have to satisfy Equations (7.35), (7.36) and (7.37). The design of such a set of matrices is referred to as *amicable orthogonal* design and more details can be found in [303]. In other words, LSTBCs constitute a family of LDCs obeying the structure of Figure 7.2 that satisfy the orthogonality constraint of Equation (7.38). Thus, it becomes clear that the LSTBCs have to obey the rate versus diversity trade-off, owing to their full TSS property, as stated in Property 7.1.

7.3.5 Single-symbol-decodable STBCs based on QOSTBCs

Observe in Equations (7.35), (7.36) and (7.37) that the orthogonality imposed enables the transmitted space-time symbol streams to be separated by a set of dispersion matrices $\mathbf{A}_q$ and $\mathbf{B}_q$. However, ensuring the orthogonality for each of the Q transmitted symbols according to Equation (7.37) may not be necessary, if the real and imaginary parts of a transmitted space-time symbol are jointly detected. If we eliminate this constraint, the requirement for the set of dispersion matrices becomes [304]

$$\mathbf{A}_{i_1}\mathbf{A}_i^H = -\mathbf{A}_i\mathbf{A}_{i_1}^H, \quad \mathbf{B}_{i_1}\mathbf{B}_i^H = -\mathbf{B}_i\mathbf{B}_{i_1}^H \quad (i \neq i_1), \tag{7.39}$$

$$\mathbf{A}_{i_1}\mathbf{B}_i^H = \mathbf{B}_i\mathbf{A}_{i_1}^H \quad (i \neq i_1). \tag{7.40}$$

Compared with Equation (7.37), Equation (7.40) excludes the scenarios where $i = i_1$, which eliminates the orthogonality within each transmitted symbol of Figure 7.2. In the open literature, STBCs obeying Equations (7.39) and (7.40) are referred as Single-Symbol-Decodable STBCs (SSD-STBCs) [303, 304].

From another point of view, SSD-STBCs constitute the class of LDCs having the structure of Figure 7.2 that obeys the constraints of Equations (7.39) and (7.40). Consequently, the full TSS property is maintained, which inevitably imposes the rate versus diversity trade-off.

Note that the STBCs presented in Sections 7.3.2 to 7.3.5 can be characterized using the LDC structure of Figure 7.2, since they all possess some degrees of orthogonality. However, the orthogonality imposed not only restricts the design of STBCs, but imposes the rate versus diversity trade-off [295] characterized by having the property of full TSS.

7.3.6 Space-Time Codes using TVLT

For the sake of achieving high rate and full diversity in the context of STBCs, the authors of [28] proposed a class of STBCs employing unitary TVLT.

A TVLT code can be represented using the notation of TVLT(M, N, T, MT), where the MT symbols mapped to the M antennas are transmitted using T time slots and received by N antennas, hence maintaining a symbol rate of $R = M$. More explicitly, a TVLT scheme firstly separates MT symbols into T substreams or layers, which contain M symbols. Then, each substream is separately modulated by a specific vector for transmission in a single time slot. For example, a TVLT(2224) scheme disperses a substream containing symbols s_1 and s_2 during the first time slot and disperses the other substream containing symbols s_3 and s_4 during the second time slot. This process is expressed as follows:

$$\mathbf{S} = \begin{pmatrix} a_1^{11} & 0 \\ a_1^{21} & 0 \end{pmatrix} s_1 + \begin{pmatrix} a_2^{11} & 0 \\ a_2^{21} & 0 \end{pmatrix} s_2 + \begin{pmatrix} 0 & a_3^{12} \\ 0 & a_3^{32} \end{pmatrix} s_3 + \begin{pmatrix} 0 & a_4^{12} \\ 0 & a_4^{22} \end{pmatrix} s_4, \tag{7.41}$$

where $a_q^{i,j}$ denotes the entries of the dispersion matrices $\mathbf{A}_q$ of Equation (7.12). Furthermore, the rank criterion [52] discussed in Section 7.3.1 states that the minimum rank of the difference matrix $\mathbf{S}_\triangle = (\mathbf{S} - \bar{\mathbf{S}})$ is equal to the transmit diversity order of a STBC scheme. For the TVLT(2224) example of Equation (7.41), the difference transmission matrix is given by

$$\mathbf{S}_\triangle = \begin{pmatrix} a_1^{11} & 0 \\ a_1^{21} & 0 \end{pmatrix} s_{1\triangle} + \begin{pmatrix} a_2^{11} & 0 \\ a_2^{21} & 0 \end{pmatrix} s_{2\triangle} + \begin{pmatrix} 0 & a_3^{12} \\ 0 & a_3^{32} \end{pmatrix} s_{3\triangle} + \begin{pmatrix} 0 & a_4^{12} \\ 0 & a_4^{22} \end{pmatrix} s_{4\triangle}. \tag{7.42}$$

Observe in Equation (7.42) that maintaining full transmit diversity is guaranteed, as long as the non-zero columns of the dispersion matrices are independent of each other.

Clearly, the class of TVLT codes obeying similar structure to Equation (7.41) constitute a subset of LDCs following the schematic of Figure 7.3. The TVLT codes can be obtained using the LDC model of Equation (7.12), under the constraint of dispersing each substream to a single time slot. In other words, the different substreams are separated using time-division.

7.3.7 Threaded Algebraic Space-Time Codes

Recently, another family of STBCs has been proposed in order to achieve both high rate and full diversity, which is referred as to TASTBCs [29, 173, 283]. The main rationale behind this framework is to partition the Q symbols into $\acute{L}$ ($\acute{L} \leq M$) substreams and disperse each substream to a $(M \times 1)$-element vector with the aid of a set of dispersion matrices. The resultant vector is arranged diagonally into the space-time transmission matrix $\mathbf{S}$ of Figure 7.1 in order to span across the entire spatial and temporal domain encompassed by the TASTBC design. The $(M \times 1)$-element dispersed vector is referred as a 'thread' in [29]. The $\acute{L}$ dispersed vectors are designed to be 'orthogonal' to each other by employing an appropriately chosen scaling factor [29].

In order to expound a little further, we use the well-known $\mathbf{G}_2$ space-time matrix to demonstrate the philosophy of TASTBCs, where $\mathbf{G}_2$ is given by

$$\mathbf{G}_2 = \begin{pmatrix} s_1 & s_2 \\ -s_2^* & s_1^* \end{pmatrix}. \tag{7.43}$$

With a small modification, Alamouti's scheme can be written as follows:

$$\mathbf{G}_2 = \begin{pmatrix} s_1 & \sqrt{-1} \cdot s_2 \\ \sqrt{-1} \cdot s_2^* & s_1^* \end{pmatrix}. \tag{7.44}$$

It is straightforward to verify that the modified representation seen in Equation (7.44) has the same properties as the original Alamouti scheme of Equation (7.43). However, the modified representation portrayed in Equation (7.44) falls within the scope of TASTBCs. More particularly, the first substream containing the information symbol s_1 is dispersed by

$$\acute{L}_1 = \begin{pmatrix} s_1 \\ s_1^* \end{pmatrix} = \begin{pmatrix} 1 \\ 1 \end{pmatrix} \alpha_1 + j \begin{pmatrix} 1 \\ -1 \end{pmatrix} \beta_1. \tag{7.45}$$

Similarly, the second substream containing symbol s_2 is dispersed by

$$\acute{L}_2 = \sqrt{-1} \cdot \begin{pmatrix} s_2 \\ s_2^* \end{pmatrix} = \sqrt{-1} \cdot \begin{pmatrix} 1 \\ 1 \end{pmatrix} \alpha_2 + j\sqrt{-1} \cdot \begin{pmatrix} 1 \\ -1 \end{pmatrix} \beta_2. \tag{7.46}$$

Note that in this case $\sqrt{-1}$ is the aforementioned scaling factor, which guarantees the orthogonality between two threads.

The example of Equations (7.45) and (7.46) demonstrates that TASTBCs can be fully characterized by the general LDC structure of Figure 7.2. More explicitly, the orthogonality between the threads imposed on the space-time codeword $\mathbf{S}$ of Figure 7.2 restricts each substream of TASTBC to only partially explore the space-time recourses available. In contrast, LDC schemes disperse each information symbol to all of the space-time dimensions available.

It also interesting to compare the TVLT and TASTBC schemes. They share the same concept of dividing the Q symbols into substreams and then disperse each substream independently. The separation of substreams is achieved by time division for TVLT schemes, whereas the TASTBC arrangements employ a unique scaling factor for differentiating and separating the substreams, which was $\sqrt{-1}$ in Equations (7.45) and (7.46).

7.3.8 Summary

We have argued in Sections 7.3.2 to 7.3.7 that the family of OSTBCs, QOSTBCs, LSTBCs, SSD-STBCs, TVLTs and TASTBCs can be described using the unified structure of LDCs provided in Figures 7.2 and 7.3, where LDC$(MNTQ)$ represents a MIMO system employing M transmit and N receive antennas, transmitting Q symbols using T time slots. All of the representatives of this LDC family imposed different degrees of 'orthogonality' on the

Table 7.3: Diversity, rate, complexity and design flexibility comparison for various STBCs.

STBC	Rate	Diversity	TSS	Complexity	Orthogonality	Flexibility
OSTBC [52]	≤ 1	MN	Full	1	Full	Minimum
LSTBC [300]	≤ 1	MN	Full	1	Full	$\downarrow$
SSD-STBC [303]	1	$MN/2$	Full	1	$\downarrow$	$\downarrow$
QOSTBC [14]	1	$MN/2$	Full	2	$\downarrow$	$\downarrow$
TVLT [28]	M	MN	Partial	M	$\downarrow$	$\downarrow$
TASTBC [29]	M	MN	Partial	M	$\downarrow$	$\downarrow$
LDC [10, 27]	M	MN	None	Q	None	Maximum

general LDC framework of Figures 7.2 and 7.3. In this section, we compare the above-mentioned STBCs in terms of their diversity order and rate as well as complexity, and characterize the evolution of STBCs.

Assuming that symbol-based ML decoding is used and the same modulation scheme is employed by all of the STBCs considered, the decoding complexity ϖ imposed is associated with the number of symbols decoded at a time. The associated design flexibility can be quantified in terms of the number of practical MIMO solutions for a specific parameter combination of (M, N, T, Q).

In Table 7.3, we listed the rate, diversity and estimated decoding complexity of the diverse STBCs that have been discussed in Sections 7.3.2 to 7.3.7. Let us now continue with their brief characterization.

- Maintaining at least partial TSS constitutes a necessary condition for the design of high-rate, full-diversity STBCs. In simple physical terms, the TSS characterizes the degree of interdependence among the symbols within the transmitted space-time matrix $\mathbf{S}$. If each transmitted signal encapsulates the information of more than one information symbol, then high rate and full diversity can potentially be achieved at the same time. In contrast, if each transmitted signal encapsulates a single original input symbol's information, then the corresponding STBCs can only achieve either full diversity or a rate $R_{\text{STBC}} \leq 1$, as shown in Table 7.3.

- When decreasing the degree of orthogonality embedded in the space-time matrix $\mathbf{S}$ of Figure 7.1, typically the degree of design flexibility is increased. For example, the OSTBC design of Section 7.3.2 enjoys simple single-symbol ML decoding as a result of its full orthogonality. However, it has a very limited choice of dispersion matrices and the number of transmit antennas supported is limited. In contrast, LDCs' non-orthogonal structure of Figure 7.3 potentially facilitates an unlimited number of dispersion matrices for arbitrary MIMO antenna configurations, as shown in Equation (7.19).

- Observe from Table 7.3 that the degree of TSS is useful for characterizing the decoding complexity imposed. For example, high-rate full-diversity LDCs of [27] have to jointly decode Q symbols per space-time block. On the other hand, for example, the QOSTBCs of Section 7.3.3 are capable of separating the symbols into two-symbol decoding pairs, owing to their orthogonality.

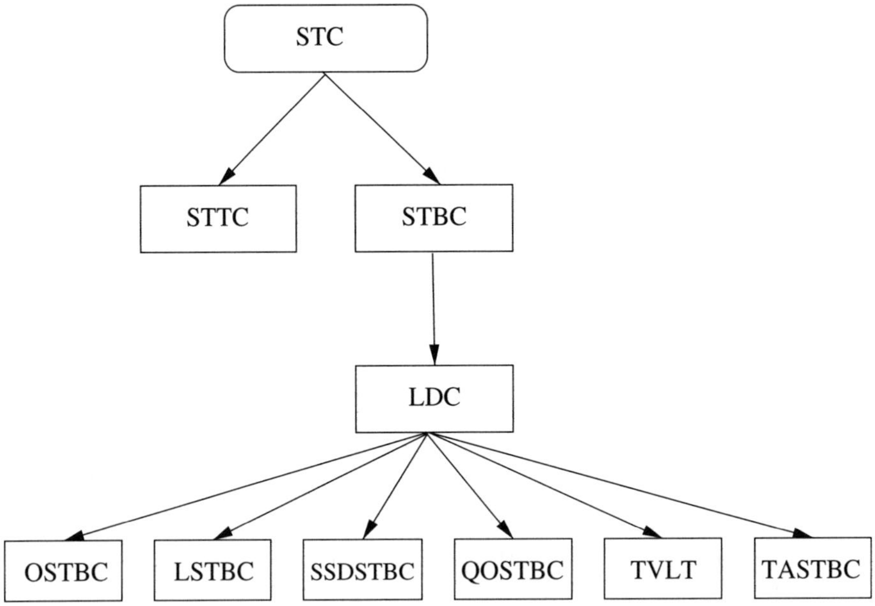

Figure 7.16: Classification of STC techniques.

- The flexibility of LDCs is also related to the number of space-time slots used by each symbol. For example, Equation (7.41) demonstrates the structure of TVLT [28] codes, which disperses each symbol merely to M out of the total MT space-time slots, whereas LDCs ensure that each space-time slot contains information related to all of the information symbols, as characterized in Equation (7.12).

Finally, as illustrated in Figure 7.16, we portray the family of LDCs as a prominent class of STP techniques, uniting the class STBCs [9]. More explicitly, the family of STC may be classified in two major categories, namely STBCs and STTCs [9, 50, 52]. The general LDC structure of Figures 7.2 and 7.3 is capable of providing diverse solutions to meet the challenge of achieving both full diversity and high rate. Hence, LDCs subsume many existing space-time block coding schemes, as seen in Figure 7.16.

7.4 EXIT-chart-based Design of LDCs

7.4.1 Analyzing Iteratively Detected LDCs

In this section, we analyze a serial concatenated channel-coded LDC scheme using iterative decoding. The system design is approached from a capacity maximization perspective with the aid of EXIT charts. More explicitly, the EXIT chart analysis enables us to estimate the maximum achievable rate of the proposed channel-coded LDC scheme using both ML and MMSE detectors, when employing the LDCs optimized in Section 7.2.4. Provided the capacity results illustrated in Section 7.2.5, we demonstrate how far the serial concatenated system operates from the LDCs DCMC capacity.

Figure 7.17 plots the schematic of a serial concatenated RSC-encoded LDC system employing iterative decoding. The information bits are first encoded by a convolution code. Here a simple half-rate $\mathrm{RSC}(215)$ code was used. Then, the interleaved bits are fed into the

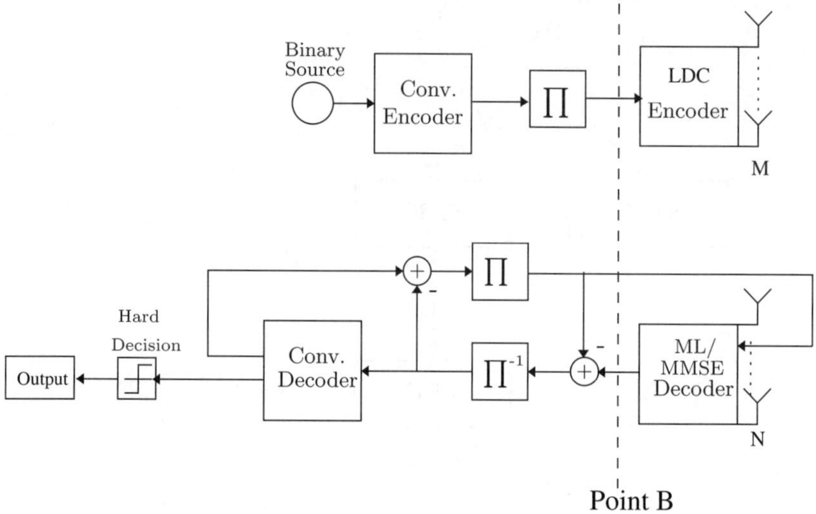

Figure 7.17: Schematic of a serial concatenated RSC-coded LDC system employing iterative decoding.

LDC encoder of Figure 7.17. Here, the LDC block also incorporates bit-to-symbol mapping using Gray labeling. The dispersion operation maps each symbol vector $\mathbf{K}$ containing Q symbols to the space-time transmission matrix $\mathbf{S}$ defined in Equation (7.12). At the receiver, extrinsic information is exchanged between the soft-in soft-out ML/MMSE and RSC detectors. More details about how the extrinsic information is calculated can be found in [305] and the references therein. This simple SCC scheme employing iterative decoders has the advantage of exploring both the spatial diversity provided by the LDCs and the temporal diversity offered by the RSC code. To show the exchange of extrinsic information, the right-hand side of Point B seen in Figure 7.17 is considered as the inner code and the left-hand side is the outer code.

The EXIT chart of the half-rate RSC-coded LDC(2224) scheme of Figure 7.17 having a diversity order of $D = 4$ and using the ML decoder is shown in Figure 7.18. In this chapter, I_A denotes the *a priori* information available for the inner code, which is provided by the extrinsic output of the outer code. In contrast, I_E denotes the extrinsic output of the inner code, which also contributes to the *a priori* input for the outer code. Observe that the intersection points of the inner and outer EXIT curves approach $I_A = 1.0$ as the SNR increases, where an infinitesimally low BER is expected.

Figure 7.19 shows the EXIT chart of the RSC(215)-coded LDC(2224) scheme having a diversity order of $D = 4$ and using a MMSE detector. Note that the area under the EXIT curves using the MMSE detector is smaller than that of its ML detection counterpart of Figure 7.18 for any given SNR.

The so-called 'area property' [270, 273] of EXIT charts may be formulated by stating that the area under the outer RSC code's EXIT curve is approximately equal to its code rate R_{out}. Thus, if we assume that the area under the EXIT curve of an outer code can be perfectly matched to the area under the inner code's EXIT curve at any SNR ρ, then it is possible to approximate the maximum achievable rate of a serial concatenated scheme by evaluating the area under the EXIT curves, given the rate of the inner block R_{in}, which is expressed as

$$C(\rho) = \log_2(L) \cdot R_{\text{in}} \cdot R_{\text{out}}, \tag{7.47}$$

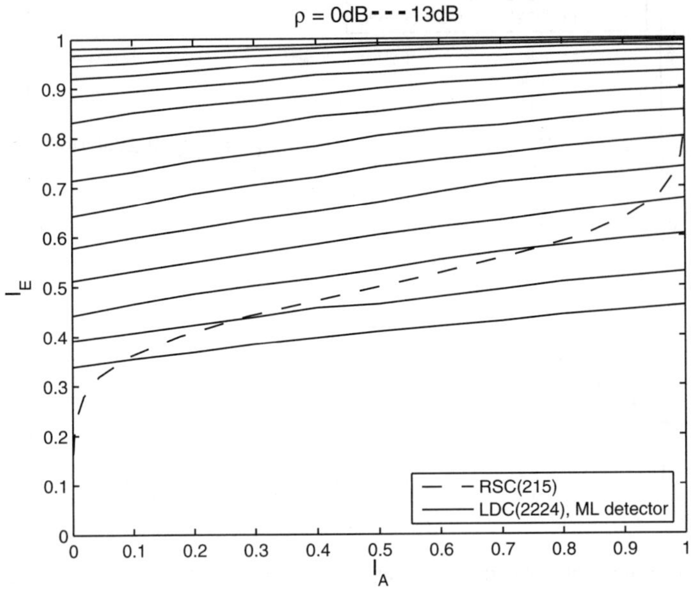

Figure 7.18: EXIT chart of the RSC(215)-coded LDC(2224) scheme of Figure 7.17 having a diversity order of $D = 4$, when employing a ML detector as well as using QPSK modulation.

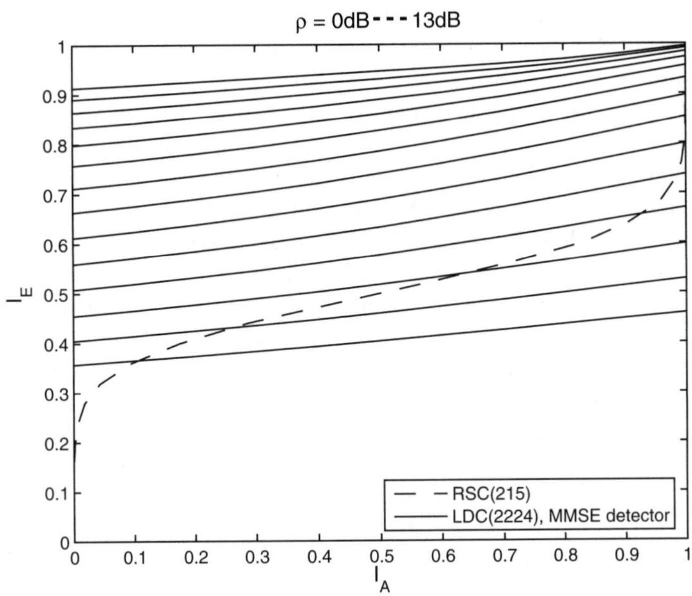

Figure 7.19: EXIT chart of the RSC(215)-coded LDC(2224) scheme of Figure 7.17 having a diversity order of $D = 4$, when employing a MMSE detector as well as using QPSK modulation.

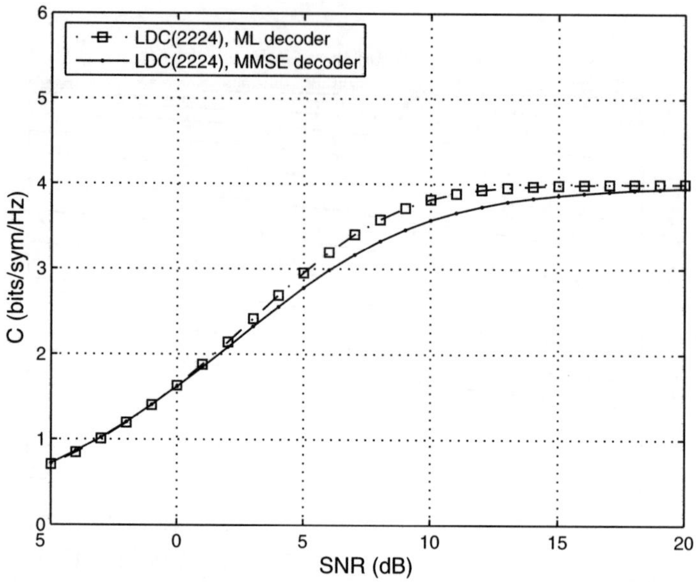

Figure 7.20: Maximum achievable rates comparison of half-rate RSC-coded LDC(2224) scheme of Figure 7.17 using QPSK modulation, when employing both ML and MMSE detectors.

where R_{out} is approximated by the area under the inner code's EXIT curve and L-PSK or L-QAM modulation is used.

Using Equation (7.47), we are able to plot the maximum achievable rates of the proposed serial concatenated RSC-coded LDC(2224) scheme using both ML and MMSE detectors in Figure 7.20. Observe that the system employing ML detection achieved a higher rate across the entire SNR region, since more decoding complexity was invested. Note that the achieved rate loss associated with employing a MMSE detector compared with the ML detector is not constant. The maximum rate loss was recorded at about $\rho = 7$ dB. It is interesting to compare the LDC(2224) scheme's maximum achievable rates employing the ML detector with its DCMC capacity curve plotted in Figure 7.7, where we observe that the two curves appear to be identical. This observation implies that it is feasible to achieve the MIMO channel's capacity using a SCC scheme under the assumption of using a variable-rate outer code.

To illustrate this a little further, Figure 7.21 quantifies the MMSE decoder's maximum achievable rate loss compared with that of the ML detector using the above-mentioned EXIT-aided method as a function of the SNR. For the rate-two LDC(2224) scheme, the peak of the bell-shaped capacity loss curve appears at $\rho = 7$ dB, where approximately 6.8% throughput is lost. When lower rate LDCs were employed, the rate loss was decreased and the peak of the curves gradually shifted to lower SNRs, as seen in Figure 7.21.

Ideally, in order to achieve an infinitesimally low BER, the inner and outer EXIT curves should only intersect at the $(I_A, I_E) = (1.0, 1.0)$ point. If this condition is satisfied, then a so-called convergence tunnel [169] appears in the EXIT chart. Even if there is no open tunnel in the EXIT chart leading to the $(1.0, 1.0)$ point, but the two curves intersect at a point close to $I_A = 1.0$, then a sufficiently low BER may still be achievable. Observe in Figure 7.19 that the intersection of inner and outer EXIT curves takes place before reaching the $I_A = 1.0$ point, unless the SNR is sufficiently high.

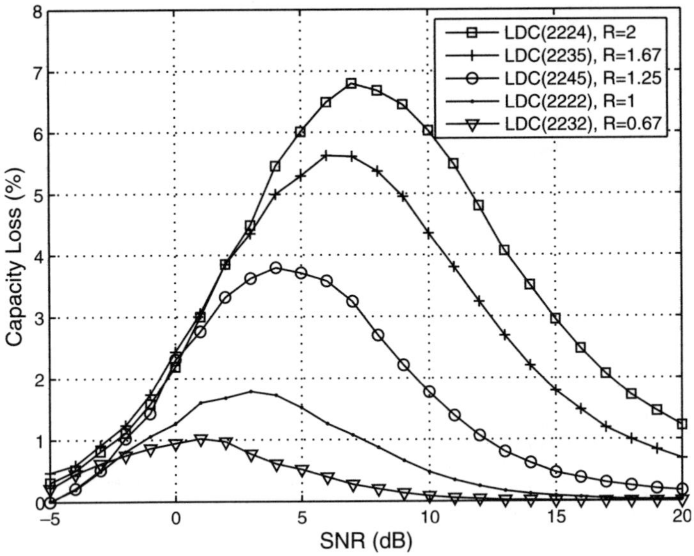

Figure 7.21: MMSE detector's maximum achievable rate loss for a group of RSC-coded LDC schemes of Figure 7.17 using QPSK modulation, when recorded at Point B of Figure 7.17.

In Figure 7.22, we characterize the BER performance of the RSC(215)-coded LDC(2224) having a diversity order of $D = 4$, while using QPSK modulation in conjunction with a MMSE decoder. Since the slope of the EXIT curves shown in Figure 7.19 is relatively low, the SCC scheme of Figure 7.17 reaches the best attainable performance after $k = 4$ iterations. The BER illustrated in Figure 7.22 gradually decreases upon increasing the SNR, which corresponds to the fact that the inner and outer EXIT curves' intersection point shifts to $(1.0, 1.0)$ point on the EXIT chart with the SNRs, as seen in Figure 7.19.

Figure 7.23 compares the EXIT curves of RSC(215)-coded LDC schemes of Figure 7.17 having $M = 2$, $N = 2$, $T = 2$ and $Q = 1, 2, 3, 4$, when employing QPSK modulation in conjunction with a MMSE decoder. Observe that when Q increases, the area under the EXIT chart decreases. It is also interesting to observe in Figure 7.23 that the curves become increasingly horizontal when Q is decreased. In fact, when we have $Q = 1$, the EXIT curve becomes a horizontal line. Recall that Gray mapping is employed for the information symbols; hence for the case of $Q = 1$, there is no extrinsic-information-induced improvement, as the *a priori* information increases. However, when we have $Q > 1$, each transmitted symbol is subjected to independent fading, and hence Gray mapping may no longer guarantee the largest Euclidean distance in a $2Q$-dimensional space. Therefore, the resultant EXIT curves' slope is increased, as Q increases.

Despite its good BER performance, the simple convolutional-coded LDC scheme of Figure 7.17 has a deficiency. More explicitly, the flatness of the EXIT curves seen in Figure 7.23 prevents the intersection from reaching the $(I_A, I_E) = (1.0, 1.0)$ point. Therefore, an infinitesimally low BER can only be achieved at high SNRs. Driven by the desire of designing iterative-decoding-aided LDCs having an infinitesimally low BER, the family of convolutional-coded precoder-aided LDCs is proposed in the next section.

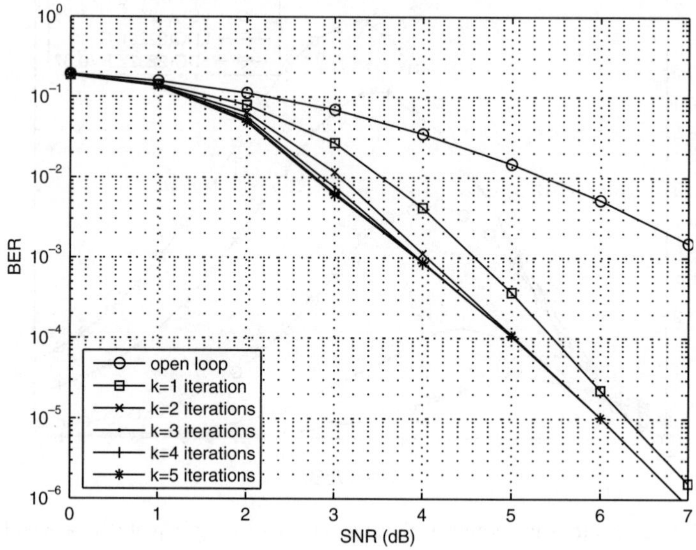

Figure 7.22: BER performance of the RSC(215)-coded LDC(2224) scheme of Figure 7.17 having a diversity order of $D = 4$ using QPSK modulation in conjunction with a MMSE detector.

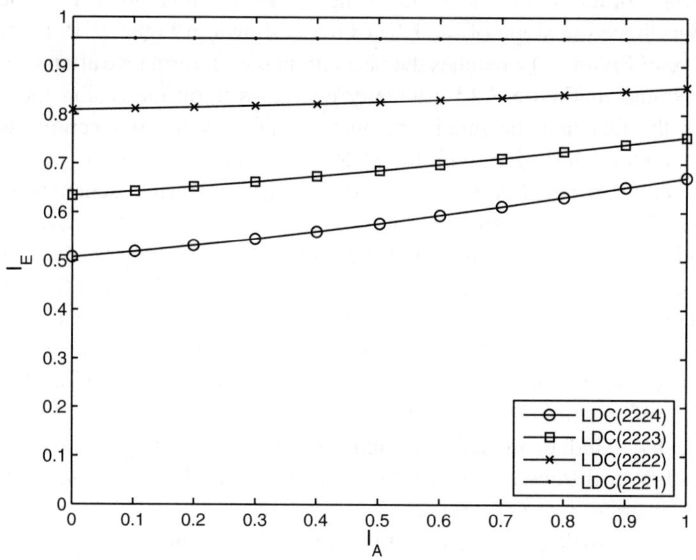

Figure 7.23: EXIT chart comparison for the QPSK-modulated RSC(215)-coded LDC(222Q) schemes of Figure 7.17 at $\rho = 3$ dB, where we have $Q = 1, 2, 3, 4$ and use a MMSE detector.

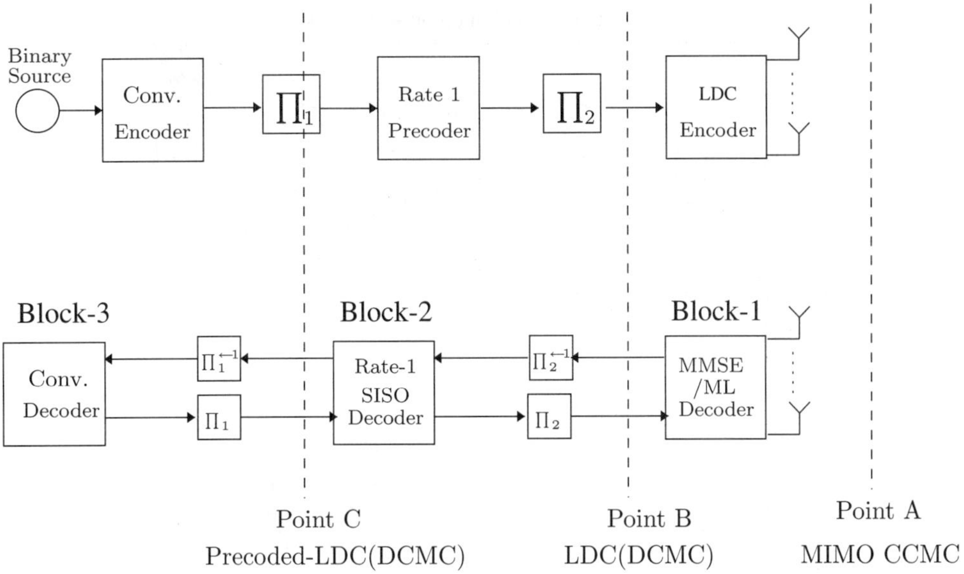

Figure 7.24: Schematic of a three-stage RSC-coded and precoder-aided LDC using iterative decoding.

7.4.2 Analyzing Iteratively Detected Precoded LDCs

The schematic of a RSC-coded and precoder-aided LDC system is shown in Figure 7.24. Compared with the non-precoded scheme of Figure 7.17, a rate-one precoder is placed between the convolutional encoding block and the LDC encoder, complemented by a second interleaver. The resultant schematic constitutes a three-stage system, where the extrinsic information is passed through the three decoder blocks according to a predefined activation order. Block-1 of Figure 7.24, namely the MMSE/ML decoder, receives its input information from the MIMO channel and its feedback information from the precoder's decoder. The intermediate Block-2 of Figure 7.24 benefits from the information provided by both the convolutional decoder and the LDC decoder. Its two outputs are forwarded to the surrounding blocks at both its sides. An extrinsic information exchange cycle between Block-1 and Block-2 of Figure 7.24 is defined as an inner iteration. The third decoding block, namely the convolutional decoder, exchanges information with the precoder. Hence, a single associated extrinsic information exchange cycle between the precoder's decoder and the convolutional decoder is referred to as an outer iteration.

Let us briefly discuss the features of the system structure of Figure 7.24. For example, if we split the system into two constituent parts at Point B of Figure 7.24, Block-1 has the same EXIT characteristics as the two-stage system of Figure 7.17, where the $(I_A, I_E) = (1.0, 1.0)$ point cannot be reached at all, unless the associated SNR is sufficiently high. However, if we split the system at Point C of Figure 7.24, we demonstrate at a later stage that the inner Precoded Linear Dispersion Codes (PLDCs) become capable of reaching the $(I_A, I_E) = (1.0, 1.0)$ point. Observe furthermore in Figure 7.24 that the equivalent channel at Point C is recursive, i.e. has an Infinite Impulse Response (IIR) as a benefit of using the precoder, whereas the equivalent channel at Point B is non-recursive.

It was argued, for example, in [306] that the activation order of the decoding blocks of a three-stage system substantially affects both the achievable performance and the detection

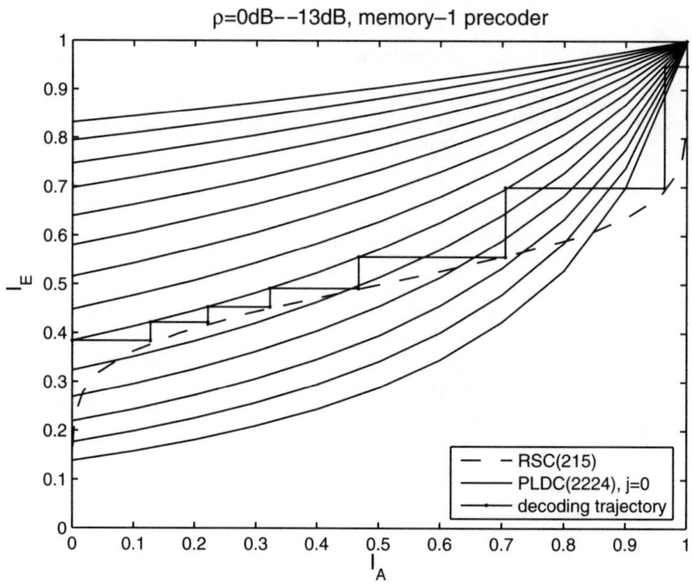

Figure 7.25: EXIT chart of the RSC(215)-coded memory-one precoder-aided LDC(2224) scheme of Figure 7.24 using QPSK modulation in conjunction with a MMSE detector when using $j = 0$ inner iterations and the decoding trajectory recorded at $\rho = 5$ dB.

complexity imposed at the receiver. The decoding activation order of the proposed three-stage system of Figure 7.24 is set to [(Block-1, Block-2)$^{j+1}$, Block-3]$^{k+1}$, where j and k are the number of inner and outer iterations, respectively. More explicitly, $(j + 1)$ iterations are invoked between Block-1 and Block-2 of Figure 7.24, followed by exchanging their joint extrinsic information with Block-3 $(k + 1)$ times. We propose to answer the question of how many inner iterations per outer iteration are necessary, from the capacity-approaching perspective. Ideally, a small value of j is desirable in the interest of minimizing the overall decoding complexity. In order to carry out a comparison between the non-precoded scheme of Figure 7.17 and the precoded schemes of Figure 7.24, in our forthcoming investigations a half-rate RSC(215) code is employed as the outer code and the length of the interleaver is set to 10^6 bits.

Figure 7.25 plots the EXIT chart of the three-stage RSC(215)-coded PLDC(2224) scheme of Figure 7.24 when using $j = 0$ inner iterations and the associated decoding trajectory recorded at $\rho = 5$ dB. Observe that the employment of the precoder facilitates the convergence of all of the inner EXIT curves to the $(I_A, I_E) = (1.0, 1.0)$ point, since the precoder is capable of gleaning extrinsic information from all of the bits within a frame. In contrast, for the non-precoded scheme of Figure 7.17, the extrinsic information can only be extracted from the bits within one information symbol. The decoding trajectory shown at $\rho = 5$ dB required $k = 6$ outer iterations to achieve an infinitesimally low BER.

When $j = 1$ inner iteration is employed in the three-stage RSC(215)-coded PLDC (2224) scheme of Figure 7.24, the corresponding EXIT chart is plotted in Figure 7.26. We observe that the area under the EXIT curves is higher than that of the system employing $j = 0$ inner iterations, as characterized previously in Figure 7.25, which implies having potentially increased maximum achievable rates. Note that the increased-area open EXIT tunnel between

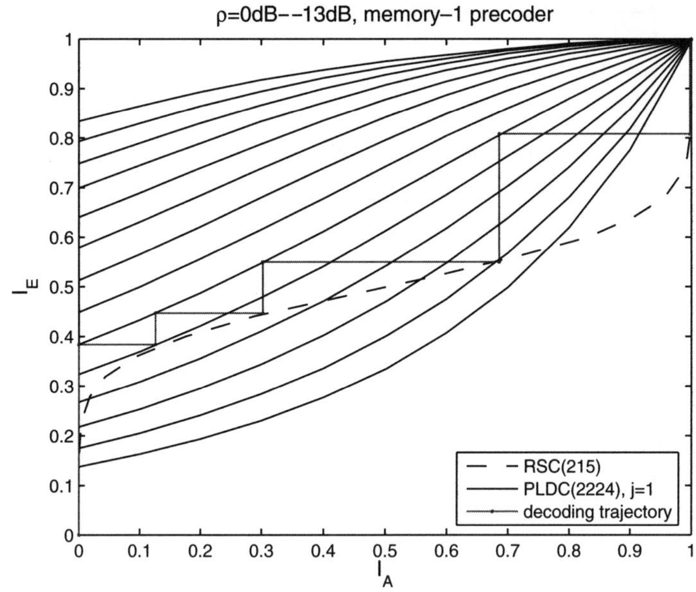

Figure 7.26: EXIT chart of the RSC(215)-coded memory-one precoder-aided LDC(2224) scheme of Figure 7.24 using QPSK modulation in conjunction with a MMSE detector when using $j = 1$ inner iteration and the decoding trajectory recorded at $\rho = 5$ dB.

the inner and outer code's EXIT curves benefits the decoding trajectory recorded at $\rho = 5$ dB, since only $k = 4$ outer iterations are necessary to converge to the $(I_A, I_E) = (1.0, 1.0)$ point.

Figure 7.27 quantifies the maximum achievable rates for the three-stage memory-one PLDC schemes of Figure 7.24 having different rates in conjunction with different numbers of inner iterations j. For each set of comparisons, the LDC's capacity measured at Point B of Figure 7.24 is plotted as the benchmark. For the rate $R_{2224} = 2$ scheme, a substantial maximum achievable rate gap is observed between the LDC scheme measured at Point B of Figure 7.24 and the corresponding PLDC scheme recorded at Point C of Figure 7.24, when the number of inner iterations is $j = 0$. However, when we have $j = 1$, the aforementioned rate loss is eliminated and a further increase of j has only a modest additional rate improvement. In fact, the maximum achievable rate loss is less than 1%, when we have $j = 1$. For the PLDC(2222) scheme having a rate of $R_{2222} = 1$, we observe in Figure 7.27 that although the aforementioned maximum achievable rate loss compared with the associated LDC's achievable capacity is still present, when employing $j = 0$ inner iterations, the associated discrepancy is narrower than that seen for the PLDC(2224) scheme. Observe in Figure 7.27 for the PLDC(2241) scheme having a rate of $R_{2241} = 0.25$ that there is no maximum achievable rate loss even in the absence of inner iterations.

The above observations are related to the EXIT characteristics of the LDC's MMSE decoding block, which is shown in Figure 7.23. When a single symbol is transmitted, i.e. we have $Q = 1$, the EXIT curve is a horizontal line. Therefore, regardless of the number of inner iterations employed, Block-1 of Figure 7.24 always outputs the identical extrinsic information. When Q is increased, the EXIT curves of Block-1 seen in Figure 7.23 become steeper, therefore higher extrinsic information can be obtained upon increasing the *a priori* information by using a higher number of inner iterations. Therefore, the resultant maximum achievable rate observed in Figure 7.27 at Point C of Figure 7.24 has an increasing gap with

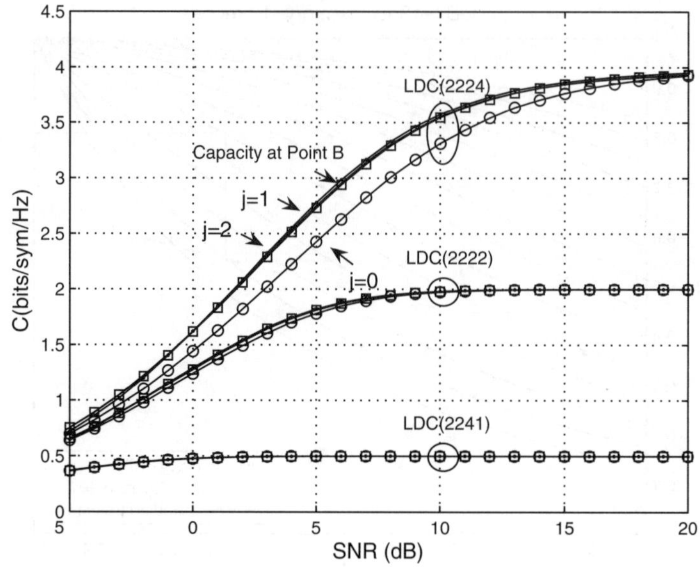

Figure 7.27: Comparison of the maximum achievable rates recorded at Point C of Figure 7.24 for the various PLDC schemes having $j = 0, 1, 2$ inner iterations when using QPSK modulation in conjunction with a MMSE detector.

respect to the one observed at Point B, when a higher number of symbols Q is transmitted by each LDC block.

In Figure 7.28, we characterize the achievable BER performance of the three-stage RSC(215)-coded memory-one precoder-aided LDC(2224) scheme using $j = 0$ inner iterations. The EXIT chart of Figure 7.25 predicted that an open convergence tunnel will be formed at $\rho = 5$ dB, which is evidenced by the BER performance of Figure 7.28, where $k = 6$ outer iterations were required to achieve an infinitesimally low BER. The comparison of Figures 7.22 and 7.28 reveals that the precoded scheme is capable of achieving a lower BER in the high-SNR region, compared with its non-precoded counterpart. When employing $j = 1$ inner iterations, the corresponding BER performance is plotted in Figure 7.29. This scheme reached the so-called turbo-cliff associated with an infinitesimally low BER at about $\rho = 4.2$ dB using $k = 10$ iterations, where a 0.8 dB SNR gain is achieved compared with the scheme employing $j = 0$ inner iterations.

Another advantage of employing a precoder is that by changing the generator polynomial and/or the precoder's memory size, it is possible to shape the EXIT curves without changing the area below them. The schematic of the unity-rate precoder having different memories is portrayed in Figure 7.30. The generator polynomial characterized in octal form determines the portion of the connections in Figure 7.30.

Figure 7.31 portrays the EXIT curves of the three-stage unity-rate precoder-aided LDC(2221) scheme having various precoder memories and polynomials at $\rho = -1$ dB. Observe that the area under the EXIT curve is constant to 0.79 for all of the curves, although their shapes are different. Note that the memory-one precoder has the highest I_E starting point at the left on the ordinate axis of Figure 7.31, when we have $I_A = 0$. Higher-memory precoded systems have an EXIT curve starting at a significantly lower I_E value, although

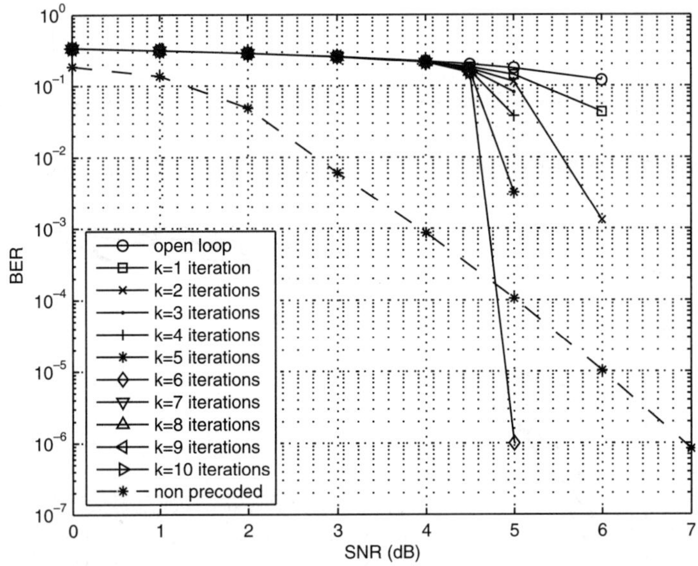

Figure 7.28: BER performance of the three-stage RSC(215)-coded PLDC(2224) scheme of Figure 7.24, employing $j = 0$ inner iterations using MMSE detection in conjunction with QPSK modulation.

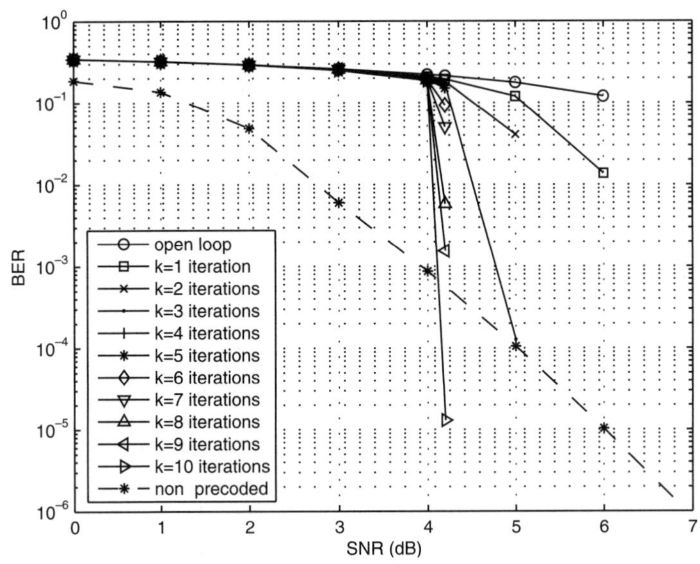

Figure 7.29: BER performance of the three-stage RSC(215)-coded PLDC(2224) scheme of Figure 7.24, employing $j = 1$ inner iterations using MMSE detection in conjunction with QPSK modulation.

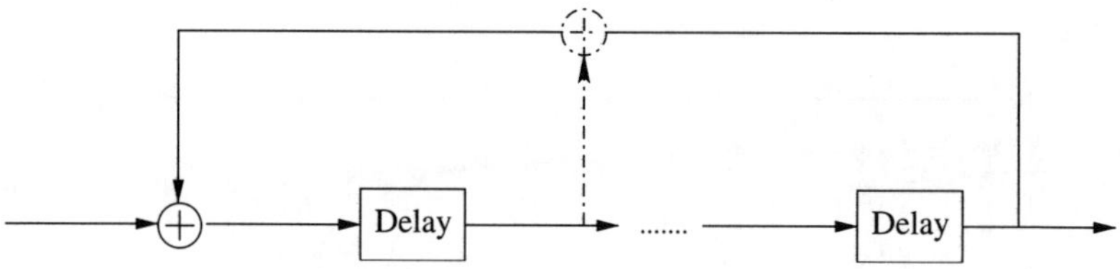

Figure 7.30: Schematic of the unity-rate precoder having different memories.

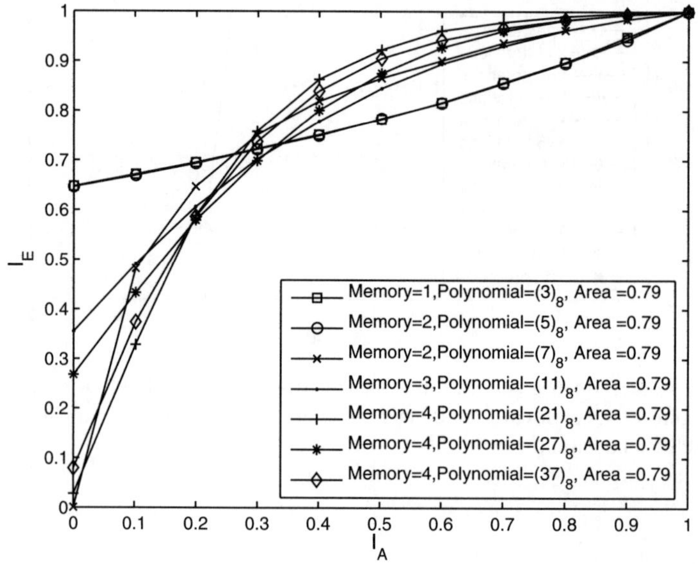

Figure 7.31: EXIT chart comparison of the three-stage RSC(215)-coded PLDC(2221) scheme of Figure 7.24, having different precoder memories and polynomials of Figure 7.30 at $\rho = -1$ dB, when using a MMSE detector in conjunction with QPSK modulation.

they are capable of benefiting from the extrinsic information more substantially in the middle range of I_A, as seen in Figure 7.31.

7.4.3 Summary

In this section, we summarize the coding advantage of the proposed serial concatenated schemes operating with/without the memory-one precoders, as portrayed in Figures 7.17 and 7.24, respectively. For the half-rate RSC(215)-coded schemes of Figures 7.17 and 7.24, the LDC(2224) scheme using QPSK modulation in conjunction with a MMSE detector was employed. Therefore, the resultant effective throughput is 1 (BPS Hz^{-1}). The uncoded LDC(2224) scheme using BPSK modulation having the identical throughput was used as a benchmark. The coding gain was recorded between the uncoded scheme and the RSC(215)-coded LDC scheme at both BER $= 10^{-4}$ and 10^{-5} from Figures 7.22, 7.28 and 7.29.

Table 7.4 summarizes the coding gains of the two-stage non-precoded system of Figure 7.17 and three-stage precoded LDC(2224) scheme of Figure 7.24 as well as the number of inner and outer iterations required to achieve the target BER. At a BER of 10^{-4}, an increased coding gain is observed, when a higher decoding complexity was invested in terms of employing precoders and increasing the number of inner iterations. Note that the advantage

Table 7.4: Coding gains of the RSC(215)-coded LDC(2224) scheme with/without a memory-one precoder having an effective throughput of 1 (BPS Hz^{-1}).

	Number of inner iterations (j)	Number of outer iterations (k)	BER of 10^{-4} (dB)	BER of 10^{-5} (dB)
Two-stage system of Figure 7.17	0	5	9.3	11.8
Three-stage system of Figure 7.24	0	10	9.5	12.9
Three-stage system of Figure 7.24	1	10	10.1	13.5

of employing a memory-one precoder having a single inner iteration over the non-precoded scheme is 0.8 dB. In contrast, the precoding advantage increased to 1.7 dB when the coding gain was quantified at a BER of 10^{-5}. In conclusion, the RSC-coded precoder-aided LDC scheme of Figure 7.24 outperforms the non-precoded scheme of Figure 7.17 in the low-BER region.

7.5 EXIT-chart-based Design of IR-PLDCs

In this section, we propose the novel IR-PLDCs as the inner code of a SCC scheme. The throughput of the SCC system is determined by the rate R_{out} of the outer channel code, the rate R_{in} of the IR-PLDC and the modulation scheme employed. Since unity-rate precoders are used, R_{in} is purely determined by the rate of the LDCs employed. The SCC scheme employing the IR-PLDC as the inner code is designed under the following assumptions.

- QPSK modulation is employed by all of the component codes of the proposed IR-PLDC scheme, although employing different modulation schemes is equally feasible.
- Rate-one precoders are employed by all of the component codes of the IR-PLDC, since lower-rate precoders introduce a capacity loss.
- The receiver has knowledge of the DCMs as well as the weighting coefficients $\boldsymbol{\lambda}$ and knows when to activate a specific dispersion matrix.

Before proceeding to the detailed description of the proposed serial concatenated IR-PLDC scheme, we summarize the salient results of the previous sections. First, the LDCs optimized according to Equation (7.24) achieve the maximum possible rate. Second, the number of inner iterations j required to attain the maximum achievable rate is $j = 1$ for $Q > 1$ or $j = 0$ for $Q = 1$, for the precoder-aided LDCs.

7.5.1 RSC-coded IR-PLDC Scheme

Figure 7.32 portrays the system model of the proposed serially concatenated RSC-coded IR-PLDC scheme. At the transmitter, a frame of information bits u_1 is encoded by a simple RSC encoder. Then the encoded bits c_1 are interleaved by a random interleaver, yielding the outer encoded bits u_2. Then the 'irregular partitioner' of Figure 7.32 feeds the appropriately selected fraction of u_2 into the various PLDC component codes according to a predefined weighting coefficient vector $\boldsymbol{\lambda}$. The inner IR-PLDC contains P_{in} component codes. The unity-rate precoders may exhibit different EXIT characteristics by employing different memories and generator polynomials, as seen in Figure 7.30. However, for the proposed system of

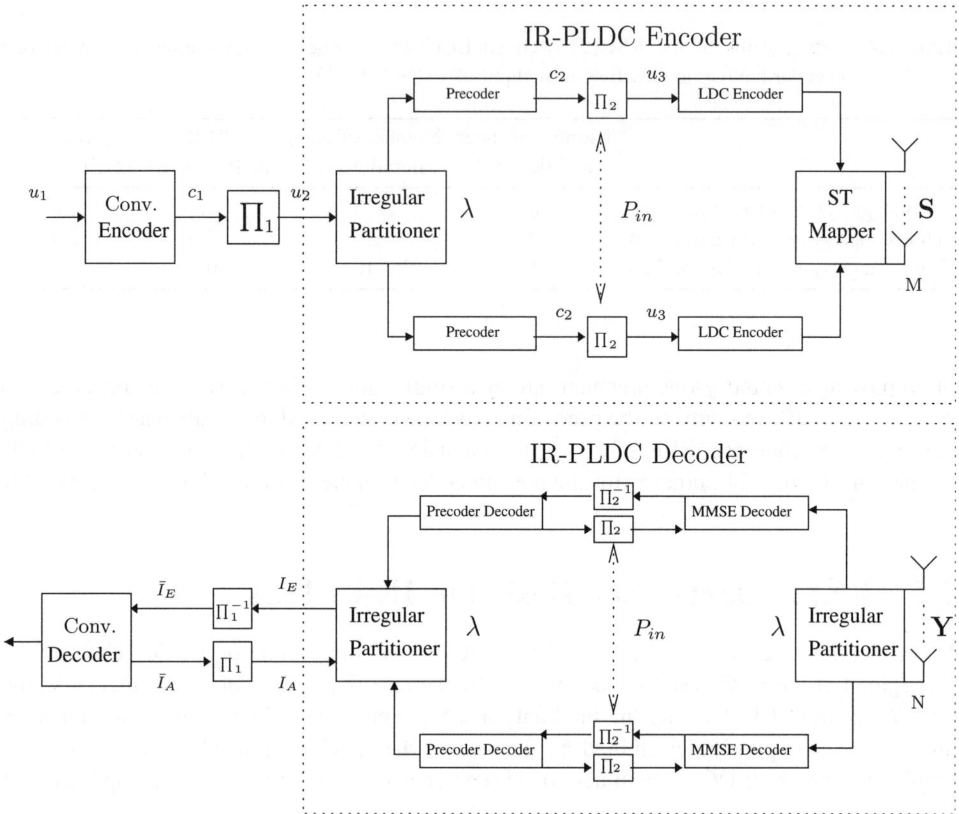

Figure 7.32: Schematic of the serial concatenated RSC-coded IR-PLDC system using iterative decoding.

Figure 7.32, memory-one precoders are employed for all of the PLDC components. More explicitly, within each PLDC encoder, the resultant precoded bits c_2 are interleaved by a second interleaver, yielding the interleaved bits u_3, which are fed to the bit-to-symbol mapper inside the LDC block of Figure 7.32. After QPSK modulation and space-time dispersion, the space-time transmission matrix $\mathbf{S}$ of Equation (7.12) is mapped to the $(M \times T)$ spatial and temporal dimensions by the 'space-time mapper' and transmitted over the uncorrelated Rayleigh fading channel contaminated by AWGN at each receive antenna.

At the receiver of Figure 7.32, again the 'irregular partitioner' determines the portion of the received signal matrices $\mathbf{Y}$ and the *a priori* information to each PLDC decoder, according to the weighting coefficient vector $\boldsymbol{\lambda}$. Then, an iterative decoding structure is employed, where extrinsic information is exchanged between three soft-in soft-out modules, namely the MMSE detector, the precoder and the outer RSC decoder in a number of consecutive iterations. To be specific, in Figure 7.32, I_A denotes the *a priori* information for the IR-PLDC represented in terms of LLRs, where I_E denotes the extrinsic information output of the inner code also expressed in terms of LLRs. We use $\bar{I}_A$ and $\bar{I}_E$ to represent the *a priori* input and extrinsic output of the outer RSC decoder, respectively. Note that the intermediate rate-one precoder processes two *a priori* inputs arriving from the MMSE detector and the outer decoder and generates two extrinsic outputs as well. Since the activation of different

PLDC components is implemented by employing different dispersion matrices, the hardware cost is modestly low.

In our forthcoming EXIT chart analysis, the precoders' decoders and the MMSE decoders are considered as a single inner decoding block, namely the IR-PLDC's decoder of Figure 7.32. The advantage of this representation is that the IR-PLDC's extrinsic information output I_E is only determined by the received signal matrix $\mathbf{Y}$ and the *a priori* input I_A, but remains unaffected by the extrinsic information exchange between the precoder's decoder and the MMSE detector. Thus, we can project the three-stage system into a two-stage system and hence the conventional two-dimensional EXIT chart analysis [169, 270] is applicable.

Following the approach of [178], we now carry out the EXIT chart analysis of the proposed RSC-coded IR-PLDC system. The inner IR-PLDC block has an *a priori* input given by I_A and the channel output $\mathbf{Y}$ of Figure 7.32. Thus, the corresponding EXIT functions are

$$I_E = \Gamma_{u_2}[I_A, \rho], \tag{7.48}$$

and for the outer RSC code the EXIT function is

$$\bar{I}_E = \Gamma_{c_1}[\bar{I}_A]. \tag{7.49}$$

According to the area properties [273] of EXIT charts introduced in Section 7.4.1, the area under the EXIT curve of an outer code is approximately equal to $(1 - R_{\text{out}})$. More explicitly, we have

$$R_{\text{out}} \approx 1 - \int_0^1 \Gamma_{c_1}(i)\, di, \tag{7.50}$$

suggesting that a higher rate code has a lower area under its EXIT curve. This observation is the rationale behind the design of IRCCs proposed in [195], where the aim is to minimize the area of the open EXIT tunnel, because this facilitates a near-capacity operation. However, we demonstrate in Section 7.5.1.1 that in contrast to IRCCs, the area under the inner IR-PLDC's EXIT curve does not have a linear relationship with its code rate. This property results in a different design procedure and objective of the IR-PLDCs in comparison with the IRCCs.

7.5.1.1 Generating Component Codes for IR-PLDCs

The employment of irregular codes was proposed by Tüchler and Hagenauer [176, 195], where IRCCs were used as an outer channel code. IRCCs are constituted by a family of convolutional codes having different code rates. They were specifically designed with the aid of EXIT charts [172], for the sake of improving the convergence behavior of iteratively decoded systems. In [195], the authors have proved that the aggregate EXIT function of an irregular code containing P component codes can be obtained from the linear combination of that of its component codes, under the assumption that the PDF of the LLRs is symmetric and continuous. More explicitly, the EXIT function of an irregular code is given by

$$\Gamma_{ir} = \sum_{i=1}^{P} \lambda_i \Gamma_i(I_{\text{input}}), \tag{7.51}$$

where λ_i represents the weighting coefficients of the ith component having transfer function Γ_i.

Inspired by the above-mentioned beneficial properties of an irregular system, in this section we propose an IR-PLDC scheme for employment as our inner rather than outer code for the SCC system. The EXIT function of the proposed IR-PLDC scheme is constituted

Table 7.5: The $P_{\text{in}} = 11$ LDC component codes of the IR-PLDC scheme of Figure 7.32 generated for a MIMO system having $M = 2$ and $N = 2$ antennas and employing QPSK modulation and a MMSE detector.

Index	M	N	T	Q	R_{LDC}	Inner iteration	Complexity
0	2	2	2	1	0.5	0	1571
1	⋮	⋮	⋮	2	1	1	4086
2	⋮	⋮	⋮	3	1.5	1	5030
3	⋮	⋮	⋮	4	2	1	5974
4	⋮	⋮	3	1	0.33	0	3285
5	⋮	⋮	⋮	2	0.67	1	8562
6	⋮	⋮	⋮	4	1.33	1	12 546
7	⋮	⋮	⋮	5	1.67	1	14 538
8	⋮	⋮	4	1	0.25	0	5639
9	⋮	⋮	⋮	3	0.75	1	18 126
10	⋮	⋮	⋮	5	1.25	1	24 974

by the superimposed combination of its component codes' EXIT functions determined by Equation (7.51). Clearly, each PLDC component's EXIT curve as well as the corresponding weighting coefficient vector $\boldsymbol{\lambda}$ play a crucial role in shaping the resultant aggregate EXIT function. Each component PLDC of the proposed IR-PLDC scheme of Figure 7.32 is constituted by an independent LDC combined with a unity-rate precoder. Different PLDC rates can be obtained by varying the number of transmitted symbols Q and the number of time slots T used per space-time block. Naturally, maintaining low values of Q and T is desirable for the sake of maintaining a low encoding/decoding complexity.

For example, in order to design an inner IR-PLDC coding scheme containing $P_{\text{in}} = 11$ components for a MIMO configuration having $M = 2$ transmit and $N = 2$ receive antennas, we commence by setting $T = 2$. Hence, all of the components have the potential of achieving the maximum diversity order of $D = 4$ according to Theorem 7.3. By setting $Q = 1$, we are able to optimize the DCMC capacity of LDC(2221) using Equation (7.24). Consequently, we can obtain more components by gradually increasing the Q value to increase the rate. We impose the limit of $Q \leq MT$ for the sake of maintaining a low complexity, although employing a higher value of Q is equally feasible. Hence, by increasing the value of T and maximizing the corresponding DCMC capacity of each LDC, we can generate a set of beneficial LDCs. Again, low Q and T values are desirable for the sake of maintaining a low complexity. The resultant $P_{\text{in}} = 11$ component codes designed for our IR-PLDC scheme are listed in Table 7.5.

Hence, for a MIMO system associated with M transmit and N receive antennas, the universal algorithm of generating P_{in} component codes for an IR-PLDC scheme can be formulated as follows.

- **Step 1.** Set the number of time slots to $T = M$ in order to ensure that all of the resultant component LDCs have the same maximum achievable diversity order of $N \cdot \min(M, T)$, as argued in Theorem 7.3.

- **Step 2.** Set $Q = 1$ and find the specific LDCs by searching through the entire set of legitimate codes, which maximize the DCMC capacity of Equation (7.24).

- **Step 3.** Set $Q := Q + 1$ for the sake of increasing the attainable throughput and repeat the previous optimization procedure using Equation (7.24), until the maximum number of transmitted symbols reaches $Q = MT$.

- **Step 4.** If the LDC having a rate of $R_{\text{LDC}} = Q/T$ exists, discard the current code having the same rate but a larger value of Q and T, for the sake of minimizing the complexity according to Corollary 7.1.

- **Step 5.** If the number of component codes found is less than P_{in}, then set $T := T + 1$ and proceed to Step 2. Otherwise, terminate the search process.

Naturally, the resultant code rates may not be expected to be evenly distributed, owing to having a limited number of legitimate combinations of T and Q.

Since unity-rate precoders are employed, the rate of each PLDC block is equal to the component LDC's rate. The number of inner iterations listed in Table 7.5 quantifies the number of iterations carried out between the precoder and the MMSE detector, which was optimized from the capacity approaching perspective, as explained in detail in Figure 7.27. The component codes generated from the above-mentioned algorithm ensure that the resultant P_{in} component codes have the lowest possible complexity. More explicitly, the complexity of each PLDC component code is jointly determined by the precoder's complexity, the MMSE detector's complexity and the number of inner iterations. In order to quantify the complexity in a unified manner, we count the number of addition and multiplication operations required to calculate a single LLR value in the logarithmic domain. Since the number of addition and multiplication operations can be quantified in terms of the so-called Add–Compare–Select (ACS) arithmetic operations, the complexity of each PLDC component is quantified by the ACS operations per LLR computation. Observe in Table 7.5 that when the value of T is fixed, the complexity is increased by increasing the value of Q. Furthermore, increasing the value of T typically resulted in a substantially increased complexity.

Figure 7.33 shows the EXIT charts of the above-mentioned $P_{\text{in}} = 11$ component PLDCs of Table 7.5 at $\rho = 0$ dB using memory-one precoders, which are used as component codes for the proposed IR-PLDC scheme of Figure 7.32. The shape of all of the EXIT curves is similar, since they are all combined with the memory-one precoders. Furthermore, Figure 7.34 portrays the EXIT curves of the same set of LDC component codes of Table 7.5 in conjunction with memory-three precoders. Clearly, although changing the size of the precoder's memory does not affect the area under the curves, it has a substantial impact on the shape of the curves. It is worth mentioning at this stage that a sufficiently diverse set of curves is necessary for employment in a flexible IR-PLDC scheme for the sake of maximizing the achievable rates as well as minimizing the detector's complexity.

Figure 7.35 quantifies the maximum achievable rates of the inner IR-PLDC scheme of Figure 7.32 using the EXIT charts of the $P_{\text{in}} = 11$ memory-one precoder-aided LDC component codes listed in Table 7.5. Recall from Section 7.4.1 that the maximum achievable rate is attained using Equation (7.47) under the assumption of having perfectly matched inner and outer EXIT curves, which results in an infinitesimally low EXIT tunnel area.

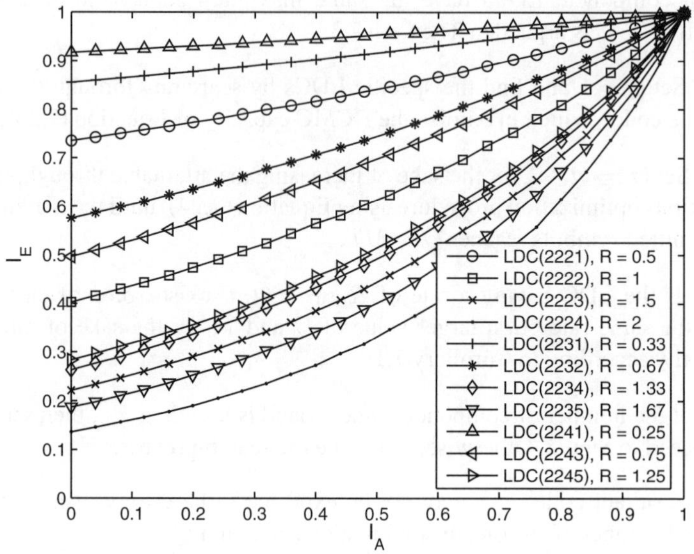

Figure 7.33: EXIT chart of the $P_{\text{in}} = 11$ PLDCs of Table 7.5 at $\rho = 0$ dB using memory-one precoders, which are used as component codes for the proposed IR-PLDC scheme of Figure 7.32.

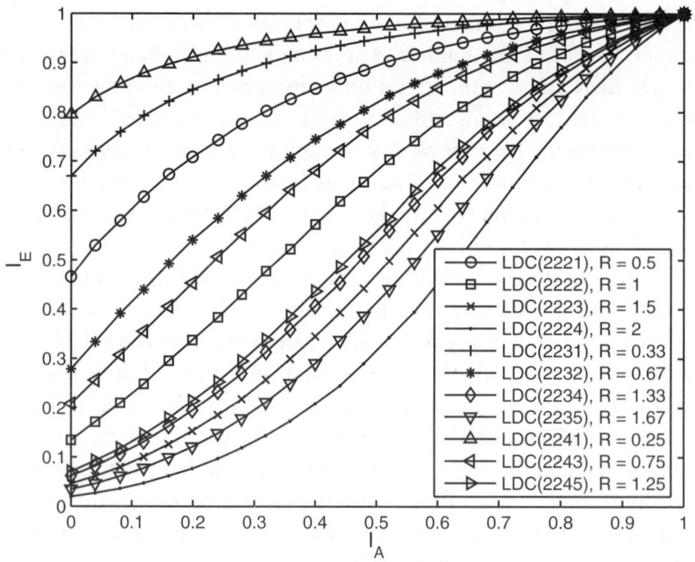

Figure 7.34: EXIT chart of the $P_{\text{in}} = 11$ PLDCs of Table 7.5 at $\rho = 0$ dB using memory-three precoders, which are used as component codes for the proposed IR-PLDC scheme of Figure 7.32.

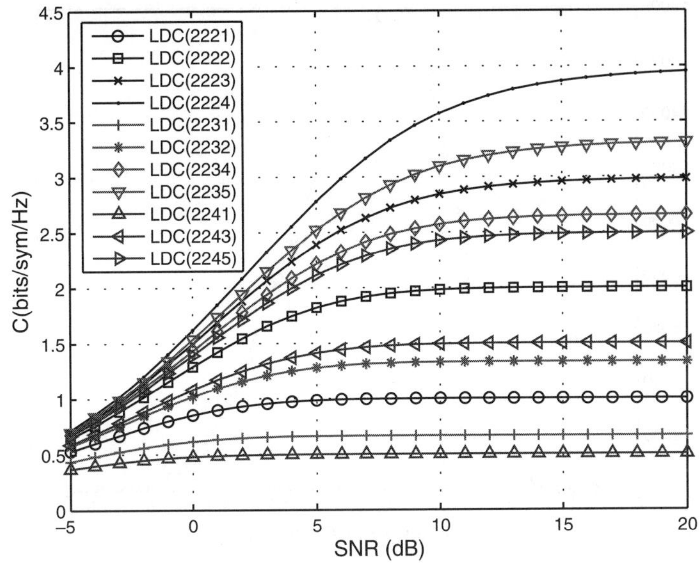

Figure 7.35: Maximum achievable rates of the $P_{\text{in}} = 11$ PLDCs listed in Table 7.5 as the component codes for the proposed IR-PLDC scheme of Figure 7.32, when employing QPSK modulation as well as a MMSE detector.

Observe in Figure 7.35 that a high-rate PLDC component code is capable of attaining a high maximum achievable rate. However, this does not necessarily imply imposing a higher complexity. For example, Table 7.5 shows that the high-rate PLDC(2224) scheme imposes a substantially lower complexity than that of the PLDC(2235) arrangement, since the former scheme maintains lower T and Q values.

Let us use λ_i, $i = (1, 2, \ldots, P_{\text{in}})$, to represent the specific fraction of I_E conveying the corresponding extrinsic information, which are fed into the IR-PLDC encoders/decoders of Figure 7.32. Therefore, the weighting coefficients $\boldsymbol{\lambda} = [\lambda_1, \ldots, \lambda_{P_{\text{in}}}]$ have to satisfy

$$\sum_{i=1}^{P_{\text{in}}}, \lambda_i = 1, \quad \lambda_i \in [0, 1], \tag{7.52}$$

and the aggregate rate R_{in} of the inner IR-PLDC scheme is given by

$$\frac{1}{R_{\text{in}}} = \sum_{i=1}^{P_{\text{in}}} \lambda_i \frac{1}{R_{(i,\text{LDC})}}. \tag{7.53}$$

In order to further illustrate the flexibility provided by the IR-PLDCs using the weighting coefficient vector $\boldsymbol{\lambda}$, Figure 7.36 plots the possible number of combinations against the inner IR-PLDC scheme's rate R_{in} given in Equation (7.53), under the assumption that we have $\lambda_i \in [0, 0.1, 0.2, \ldots, 0.9, 1]$. The bell-shaped distribution of Figure 7.36 exhibits a peak at approximately $R_{\text{in}} = 0.8$. At extremely low and high rates in the vicinity of $R_{\text{in}} = 0.25$ and $R_{\text{in}} = 2$, respectively, the number of legitimate combinations gradually reduced, owing to the lack of irregularity. Again, Figure 7.36 was generated using an exhaustive search for the weighting coefficient vector $\boldsymbol{\lambda}$ having a step size of 0.1 under the constraint of Equation (7.52).

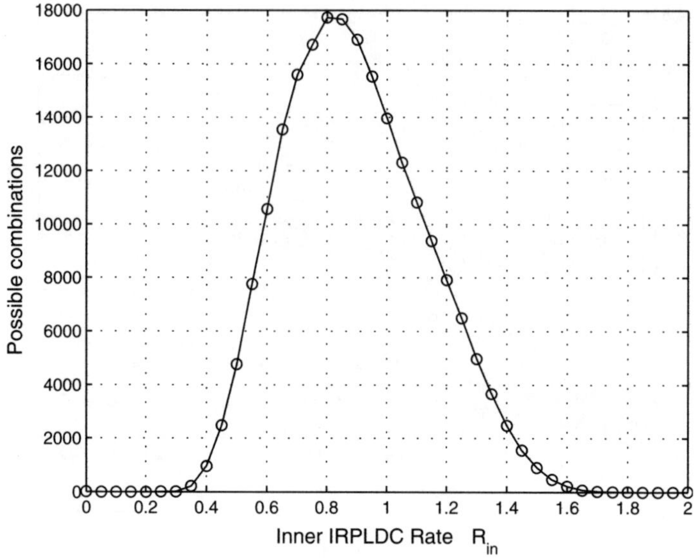

Figure 7.36: The number of possible combinations for the IR-PLDC scheme of Figure 7.32 against its rate using the $P_{\text{in}} = 11$ component codes listed in Table 7.5, where $\lambda_i \in [0, 0.1, 0.2, \ldots, 1]$.

Figure 7.37 attempts to characterize the flexibility of the IR-PLDC scheme of Figure 7.32 in terms of the associated area under the EXIT curves at $\rho = -1$ dB by plotting a group of IR-PLDCs' EXIT charts having different weighting coefficient vectors. Note that all of the IR-PLDCs' EXIT curves plotted in Figure 7.37 maintain an identical aggregate rate of $R_{\text{in}} = 0.85$. Figure 7.36 has revealed that there are around 17 500 possibilities, given the weighting coefficients $\lambda_i \in [0, 0.1, 0.2, \ldots, 0.9, 1]$. Figure 7.37 merely illustrates 12 EXIT curves out of the total 17 500 possible combinations. Observe in Figure 7.37 that the area under the curves exhibits a significant variation, which implies that the MIMO channel's capacity achieved by the IR-PLDC using different weighting coefficient vectors $\boldsymbol{\lambda}$ is different, despite the fact that they share the same aggregate rate $R_{\text{in}} = 0.85$.

In conclusion, the outer code's 'area property' of Equation (7.50) reflects a linear relationship between its rate R_{out} and the area under the EXIT curve, hence the area under the EXIT curves of the IRCC scheme [195] is equal to the aggregate rate, regardless of the shape of the curves. In contrast, the proposed inner IR-PLDC scheme of Figure 7.32 does not obey the linear property, since the area under the EXIT curve quantifies the MIMO channel's capacity achieved using an IR-PLDC scheme and its rate is given by Equation (7.53). Our arguments are further justified by Figure 7.37.

7.5.1.2 Maximum-rate RSC-coded IR-PLDCs

The channel-coded IR-PLDC system of Figure 7.32 employs a simple RSC code ($P_{\text{out}} = 1$) as the outer code, while using the $P_{\text{in}} = 11$ component codes characterized in Table 7.5 as the inner code. Hence, we have a total number of $P = P_{\text{in}} + P_{\text{out}} = 12$ component codes. Each PLDC component processes a fraction of the input information according to its weighting coefficient λ_i, which has to satisfy Equation (7.52), and the resultant aggregate inner code rate R_{in} is given by Equation (7.53).

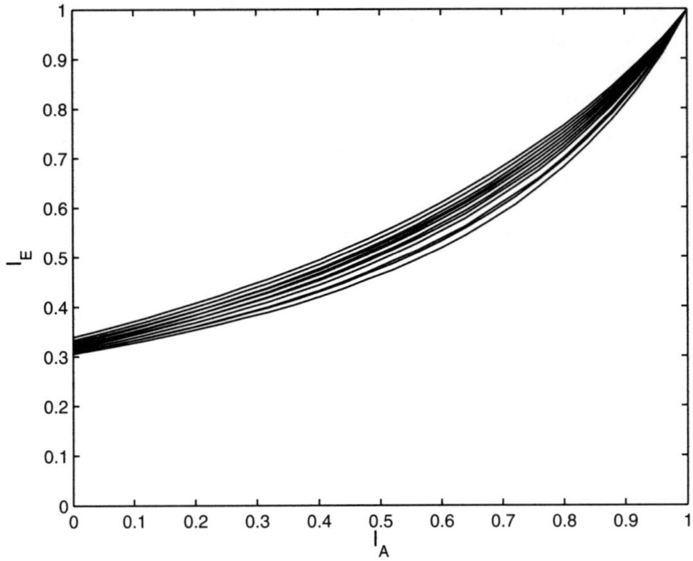

Figure 7.37: A group of EXIT charts of the IR-PLDC scheme of Figure 7.32 using the component codes of Table 7.5 at $\rho = -1$ dB, while having a fixed rate $R_{\rm in} = 0.85$ achieved by employing different weighting coefficient vectors $\boldsymbol{\lambda}$.

In order to achieve an infinitesimally low BER at a specific SNR, an open EXIT tunnel should exist in the EXIT chart. Assuming that each component code's EXIT curve is represented by l points, the EXIT function $\Gamma_{\rm irr}$ at SNR ρ of the IR-PLDC should be optimized by 'maximizing' the square of the EXIT chart matching error function given by

$$J(\lambda_1, \ldots, \lambda_{P_{\rm in}}) = \int_0^1 e(\rho)^2 \, {\rm d}i, \qquad (7.54)$$

where the error function is given by

$$\mathbf{e}(\rho) = \underbrace{\begin{bmatrix} \Gamma_1(I_{A,1}), & \Gamma_2(I_{A,1}), & \cdots & \Gamma_P(I_{A,1}) \\ \Gamma_1(I_{A,2}), & \Gamma_2(I_{A,2}), & \cdots & \Gamma_P(I_{A,2}) \\ \vdots & \vdots & \ddots & \vdots \\ \Gamma_1(I_{A,l}), & \Gamma_2(I_{A,l}), & \cdots & \Gamma_P(I_{A,l}) \end{bmatrix} \begin{bmatrix} \lambda_1 \\ \lambda_2 \\ \vdots \\ \lambda_{P_{\rm in}} \end{bmatrix}}_{\Gamma_{\rm irr}} - \underbrace{\begin{bmatrix} \Gamma_{\rm rsc}^{-1}(I_{A,1}) \\ \Gamma_{\rm rsc}^{-1}(I_{A,2}) \\ \vdots \\ \Gamma_{\rm rsc}^{-1}(I_{A,l}) \end{bmatrix}}_{\Gamma_{\rm rsc}}, \qquad (7.55)$$

subject to the constraints imposed by Equation (7.52), where $I_{A,l}$ denotes the lth *a priori* information input of the PLDC components.

The gradient search method of maximizing $J(\lambda_1, \ldots, \lambda_P)$ of Equation (7.54) is similar to the algorithm proposed in [195]. More explicitly, the algorithm starts by setting the inner code rate $R_{\rm in}$ to the minimum value. If the set of weighting coefficients maximizing the area expression of Equation (7.54) is generated using the gradient search method of [195] and an open EXIT tunnel exists, $R_{\rm in}$ is increased by a small amount. The algorithm terminates at the highest possible $R_{\rm in}$ value, where an open convergence tunnel may no longer be found.

The reason that our proposed inner IR-PLDC scheme is seeking the solutions 'maximizing' the area expression of $J(\lambda_1, \ldots, \lambda_P)$ is justified as follows. The benefit of employing

Table 7.6: Design comparison of irregular schemes using either IRCCs having $\boldsymbol{\gamma} = [\gamma_1, \ldots, \gamma_{P_{\text{out}}}]$ or IR-PLDCs having $\boldsymbol{\lambda} = [\lambda_1, \ldots, \lambda_{P_{\text{in}}}]$, where the weighting coefficient vectors $\boldsymbol{\gamma}$ and $\boldsymbol{\lambda}$ quantify the fraction of bits encoded by each component code.

	IRCC	IR-PLDC
Design objective	Maximizing the achievable rates	
Component codes generated by	Puncturing a mother code	Varying T and Q
Area and rate relationship	Unique	Non-unique
Interleaver length requirement	High	Modest
Error function of Equation (7.54)	Minimizing	Maximizing
Aggregate code rate	$\frac{1}{R_{\text{out}}} = \sum_{i=1}^{P_{\text{out}}} \gamma_i \frac{1}{R_{i,\text{ircc}}}$	$\frac{1}{R_{\text{in}}} = \sum_{i=1}^{P_{\text{in}}} \lambda_i \frac{1}{R_{i,\text{ldc}}}$

irregular inner or outer codes for an iteratively detected scheme is to maximize the achievable rates. When using IRCCs as an outer code, minimizing the EXIT tunnel area corresponds to maximizing the achievable rate, owing to the area property discussed in Section 7.4.1. In other words, IRCCs are designed to find an outer EXIT curve that matches a given inner EXIT curve as closely as possible by maximizing the area under the EXIT curve. However, there is no one-to-one relationship between the inner aggregate rate R_{in} and the associated area under the EXIT curves. Consequently, the IR-PLDC scheme offers multiple area values under its EXIT curves for a given rate, as illustrated in Figure 7.37. A larger EXIT tunnel area potentially requires fewer outer iterations to achieve an infinitesimally low BER. Therefore, given an outer code, the design criterion for the inner IR-PLDC is to maximize the achievable rate, while having an infinitesimally low BER, which corresponds to maximizing the inner code rate R_{in} and simultaneously maximizing the EXIT tunnel area according to Equation (7.54), since the latter minimizes the number of iterations required. A detailed comparison of designing the proposed inner IR-PLDC and the IRCC of [195] as an outer code is provided in Table 7.6.

In the previous iteratively detected schemes in Sections 7.4.1 and 7.4.2, half-rate RSC(215) code is employed. It is natural to ask the question whether it is possible to employ an even simpler RSC code to reduce the decoding complexity, while operating near the MIMO channel's capacity. Hence, we designed our IR-PLDC system for a half-rate RSC(213) code, which has a decoding complexity of 217 ACS operations per LLR value compared with the 841 ACS operations imposed by the RSC(215) code. It is also worth mentioning that when plotting the EXIT chart, it is often assumed that the distribution of the LLRs is Gaussian, which is only sufficiently accurate when a high interleaver length is used in the schematic of Figure 7.32. In the context of the RSC(213) code, the Gaussian assumption of the EXIT chart becomes easier to satisfy in conjunction with a shorter interleaver, since the RSC(213) code imposes correlation over a factor of $\frac{3}{5}$ shorter segment of the encoded bit stream.

Figure 7.38 presents the EXIT chart and the corresponding decoding trajectory of the RSC(213)-coded IR-PLDC scheme of Figure 7.32 designed for operating at $\rho = 0$ dB, when using QPSK modulation in conjunction with a MMSE detector. The dotted lines are the EXIT curves of the $P_{\text{in}} = 11$ component codes of Table 7.5 and the solid line represents the EXIT curve of the inner IR-PLDC having the weighting coefficients given in Table G.1 of Appendix G. The resultant throughput of the system is $C(0\text{ dB}) = 1.1392$ (BPS Hz^{-1}),

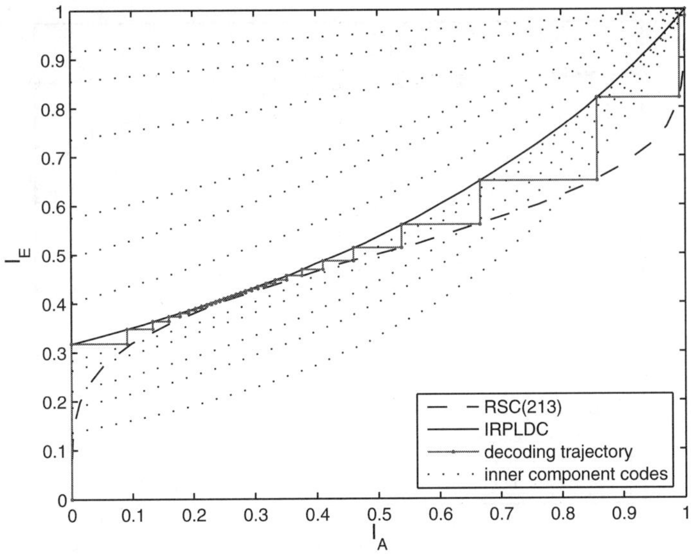

Figure 7.38: EXIT chart and the decoding trajectory of the RSC(213)-coded IR-PLDC scheme of Figure 7.32 recorded at $\rho = 0$ dB using QPSK modulation, when a MMSE detector was employed.

according to Equation (7.47). By maximizing R_{in} as well as maximizing the open EXIT tunnel area of Equation (7.54), the optimized EXIT curves of Figure 7.38 exhibit a significant tunnel area, where the decoding trajectory of Figure 7.38 shows that $k = 29$ outer iterations were required.

The corresponding BER of the RSC(213)-coded IR-PLDC system of Figure 7.32 designed for achieving an infinitesimally low BER at $\rho = 0$ dB using QPSK modulation is shown in Figure 7.39. There is a turbo cliff at $\rho = 0$ dB, when $k = 29$ outer iterations were carried out between the RSC(213) decoder and the IR-PLDC decoder. Note that the complexity required for achieving an infinitesimally low BER at $\rho = 0$ dB is quantified in terms of the number of ACS operations per LLR value. Given the number of outer iterations and the complexity of each PLDC component of Table 7.5 combined with the RSC(213) decoder, which requires 217 ACS operations, the total decoding complexity per LLR value was evaluated by considering the number of iterations as well as each component's complexity as follows:

$$\varpi(0 \text{ dB}) = 29 \cdot (0.146 \cdot 8562 + 0.034 \cdot 18126 + 0.82 \cdot 12\,546 + 217) = 3.5876 \times 10^5. \tag{7.56}$$

Naturally, the same design process can be extended to other SNR values. Figure 7.40 plots the maximum rates achieved by the proposed IR-PLDC scheme of Figure 7.32, when half-rate RSC(213) and RSC(215) codes were employed. Each point in Figure 7.40 was designed to achieve the maximum rate with the aid of specific weighting coefficients, which are listed in Tables G.1 and G.2 of Appendix G. The dotted lines of Figure 7.40 quantify the maximum achievable rates of the $P_{\text{in}} = 11$ component codes of Table 7.5. By simply adjusting the fraction of information fed into each component, as seen in Tables G.1 and G.2, the proposed system of Figure 7.32 employing irregular inner codes becomes capable of

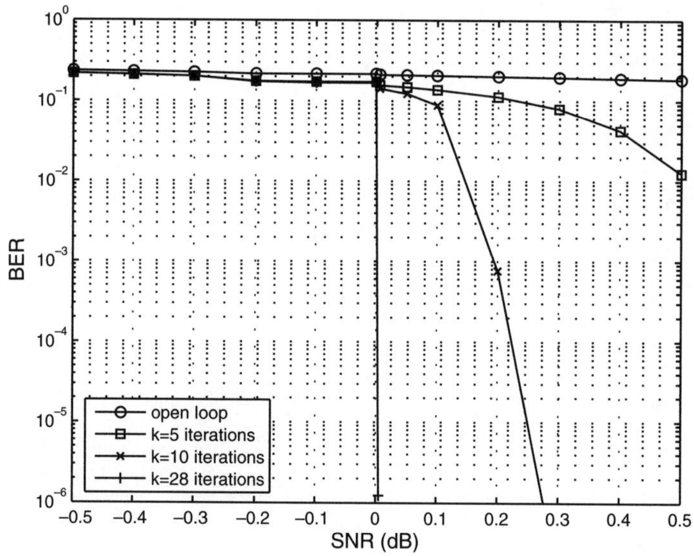

Figure 7.39: BER of the RSC(213)-coded IR-PLDC scheme of Figure 7.32 that is designed to achieve an infinitesimally low BER at $\rho = 0$ dB using QPSK modulation, when a MMSE detector was employed.

operating across a wide range of SNRs. Figure 7.40 clearly demonstrates the variation of the maximum rate, when encountering different SNR scenarios.

Also observe in Figure 7.40 that the IR-PLDC scheme employing the RSC(213) code is capable of achieving a higher maximum rate compared with the system employing the RSC(215) code, which is about 2.5 dB away from the MIMO channel's capacity. Compared with the EXIT curve of the RSC(215) code seen in Figure 7.17, the RSC(213) code's EXIT curve of Figure 7.38 has a lower I_E value for abscissa values of $I_A < 0.5$ and a higher I_E value in the rest of the abscissa range. The resultant shape of the outer EXIT curve forms a larger open EXIT tunnel area and hence enables the inner IR-PLDC code to provide a higher aggregate rate for the serial concatenated system. This implies that having a beneficial shape for the EXIT curves plays a more influential role in determining the achievable rates of the system than having a larger minimum distance, as it becomes explicit by comparing the performance of the RSC(213) and RSC(215) coded systems.

Tables G.1 and G.2 in Appendix G list the weighting coefficient vector $\boldsymbol{\lambda}$ for the IR-PLDC scheme of Figure 7.24 required for achieving maximum rates ranging from $\rho = -7$ dB to $\rho = 4$ dB, when the RSC(213) and RSC(215) codes were employed, respectively. The $P_{\text{in}} = 11$ component codes are listed from low-rate to high-rate components, where an entry of '0' implies that the specific PLDC component is inactivated during the transmission process. Observe further in Tables G.1 and G.2 that increasing the system's maximum rate upon increasing the SNR is achieved by appropriately adjusting the weighting coefficient vector $\boldsymbol{\lambda}$. However, when we have $\rho = 4$ dB in Table G.1, the maximum-rate PLDC component having a rate of $R_{3,\text{LDC}} = 2$ has already fully activated, which means that it has a weighting coefficient of $\lambda_3 = 1$. Therefore, no more rate increase is observed in Figure 7.40 when we have $\rho > 4$ dB. Naturally, further rate improvements can be achieved, if higher-rate PLDC components are employed.

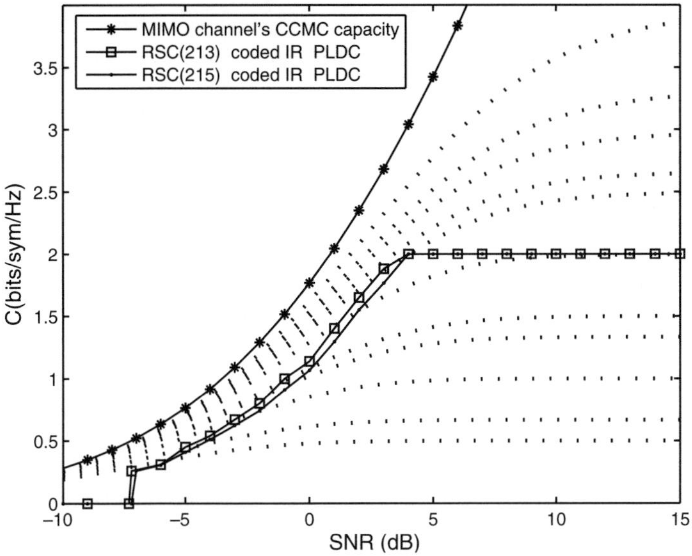

Figure 7.40: The maximum rates achieved by the IR-PLDC schemes of Figure 7.32 using RSC(213) and RSC(215) codes according to Tables G.1 and G.2, when QPSK modulation in conjunction with a MMSE detector were employed.

An alternative way of changing the shape of the EXIT curves is constituted by changing the precoder's memory size and/or the generator polynomials for the system of Figure 7.32. Previously, Figure 7.31 has indicated that changing the rate-one precoder's memory and generator polynomial does not change the area under the corresponding EXIT curves. Hence, Figure 7.41 portrays both the EXIT curves and the decoding trajectory of the IR-PLDC scheme of Figure 7.32 at $\rho = 0$ dB using the RSC(213) code, when the precoder's memory size was increased to three. Compared with Figure 7.38, the EXIT curves of Figure 7.41 are steeper. However, in this particular case, increasing the precoder's memory size does not benefit the system in terms of its maximum achievable rate, which is $C(0\ \text{dB}) = 0.7908$ (BPS Hz^{-1}), while $C(0\ \text{dB}) = 1.1392$ (BPS Hz^{-1}) was achieved for the IR-PLDC schemes having precoder memory size of one. Again, Figure 7.41 demonstrates that the shape of the inner and outer EXIT curves exhibits an influential role in the EXIT-chart-based design. The corresponding BER performance is shown in Figure 7.42.

7.5.1.3 Complexity-constrained RSC-coded IR-PLDCs

In Figures 7.38 and 7.41, we have characterized a number of IR-PLDC designs contrived for achieving the maximum rate at a certain SNR. Observe from Equation (7.56) that an excessive amount of complexity is required to decode each information bit. However, the mobile handsets normally cannot afford such high complexity. Therefore, in this section, we investigate the IR-PLDC designs of Figure 7.32 that are capable of achieving the maximum possible rate under a certain complexity constraint.

The total decoding complexity is constituted by two contributions. The first is the particular fraction of bits λ_i fed into each component code as well as each component's own complexity quantified in Table 7.5. Second, the number of outer iterations k required for reaching the $(1.0, 1.0)$ point on the EXIT chart constitutes a linear complexity factor. Observe

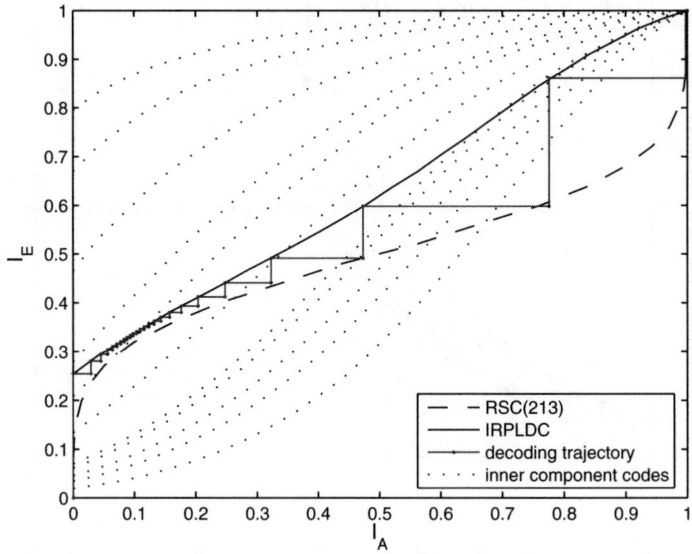

Figure 7.41: EXIT chart and the decoding trajectory of the RSC(213)-code IR-PLDC scheme of Figure 7.32 having a precoder memory of three recorded at $\rho = 0$ dB using QPSK modulation, when a MMSE detector was employed.

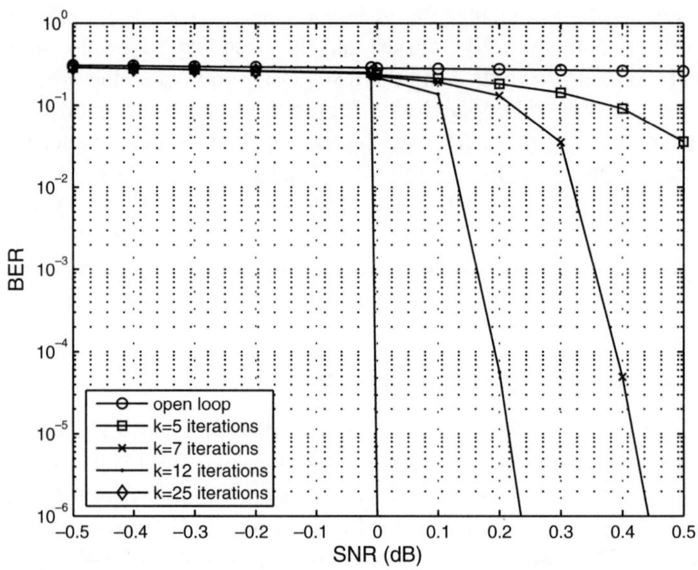

Figure 7.42: BER of the RSC(213)-coded IR-PLDC scheme of Figure 7.32 having a precoder memory of three that is designed to achieve an infinitesimally low BER at $\rho = 0$ dB using QPSK modulation, when a MMSE detector was employed.

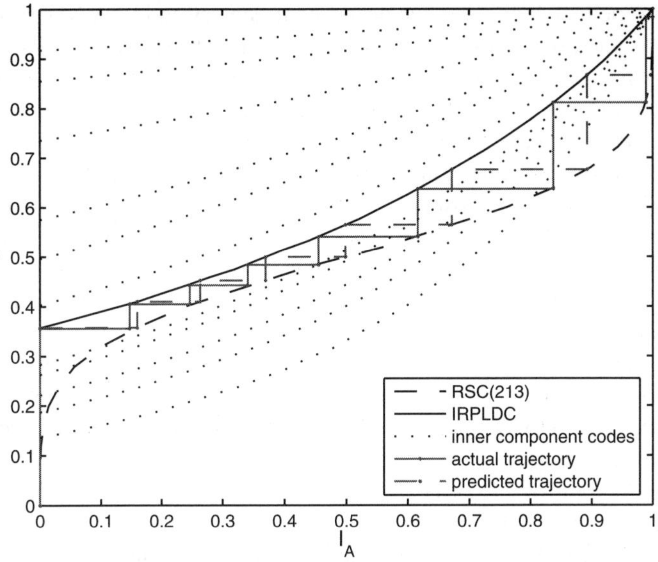

Figure 7.43: EXIT chart and the (predicted) decoding trajectory of the 'complexity-constrained' RSC(213)-coded IR-PLDC scheme of Figure 7.32 at $\rho = 0$ dB, when using QPSK modulation and a MMSE detector.

from Figure 7.38 that the IR-PLDC schemes designed for maximizing the achievable rate often result in a high decoding complexity, which is associated with a narrow EXIT tunnel. These high-complexity designs may be suitable for the BS. The flexibility of the IR-PLDCs also allows us to derive designs for mobile handsets, where the affordable complexity is more limited. We refer to this as a 'complexity-constrained' design.

For the above-mentioned 'maximum-rate' schemes, the design problem was formulated in Equation (7.54) and a gradient search [195] can be performed to find the most suitable weighting coefficient vector $\boldsymbol{\lambda}$. However, for the 'complexity-constrained' scheme, we are looking for the maximum achievable rate under a specific complexity constraint, which requires an extended search for the weighting coefficient vector $\boldsymbol{\lambda}$. More specifically, the complexity of an IR-PLDC scheme employing an RSC code at SNR ρ is restricted by

$$k \cdot (\varpi_1 \lambda_1 + \varpi_2 \lambda_2 + \cdots + \varpi_P \lambda_P + \varpi_{RSC}) \leq \varpi_{\text{target}}, \tag{7.57}$$

where ϖ_i is the complexity of the ith component code quantified in Table 7.5. The number of decoding iterations k required to achieve an infinitesimally low BER is predicted using a predicted decoding trajectory illustrated in Figure 7.43.

In the following design example, we optimize the IR-PLDC scheme of Figure 7.32 employing a half-rate RSC(213) code as our outer code at $\rho = 0$ dB. The complexity imposed at the receiver is restricted by $\varpi_{213,c}(0\text{ dB}) \leq \frac{1}{10}\varpi_{213}(0\text{ dB}) = 0.35876 \times 10^5$ ACS operations per LLR value.

Figure 7.43 shows the EXIT curves of our 'complexity-constrained' IR-PLDC scheme as well as the predicted decoding trajectory using interpolation and the actual decoding trajectory recorded using simulation. Firstly, a wider convergence tunnel is observed in Figure 7.43 compared with that of Figure 7.38, hence the number of decoding iterations can be effectively reduced. Furthermore, there is a difference between the predicted decoding trajectory generated using interpolation and the actual trajectory recorded using simulation.

The accuracy of the match between the decoding trajectory and the EXIT chart is dependent on the validity of the assumption of the LLRs' Gaussian distribution, which requires a large interleaver length. If the assumption has a limited accuracy, the actual decoding trajectory may exhibit an 'overshoot' problem. In this design example, the first interleaver Π_1 of Figure 7.32 is set to have a length of 10^6 bits and the length of the second interleaver Π_2 is equal to the corresponding number of bits fed into each component code. Note in Figure 7.43 that although no substantial 'overshoot' is visible, the resultant predicted and the actually encountered decoding trajectories exhibit an obvious difference after a few iterations. However, we may state that the number of decoding iterations predicted, k, using interpolation is sufficiently accurate. The results of our further investigations included here suggest that unless encountering narrow EXIT tunnels, where more than $k = 15$ iterations are necessary, the predicted number of iterations using EXIT-chart-based interpolation is typically quite accurate. On the other hand, when aiming for low-complexity designs, the 'wide EXIT tunnel assumption' is usually satisfied and hence the aforementioned inaccurate prediction is not encountered.

Furthermore, a rate of $C(0\text{ dB}) = 1.10929$ (BPS Hz^{-1}) is achieved in Figure 7.43, where the rate loss recorded is insignificant compared with the maximum achievable rate of $C(0\text{ dB}) = 1.1392$ (BPS Hz^{-1}) recorded for the previous 'maximum-rate' design at $\rho = 0$ dB. As a benefit of this 'complexity-constrained' design, only about 10% of the decoding complexity of the 'maximum-rate' design is imposed. The actual decoding complexity is

$$\begin{aligned}\varpi_{213,c}(0\text{ dB}) &= 7 \cdot (0.83 \cdot 4086 + 0.17 \cdot 5974 + 217)\\ &= 0.3237 \times 10^5 < 0.35876 \times 10^5,\end{aligned} \tag{7.58}$$

where the weighting coefficient vector is given by $\boldsymbol{\lambda} = [0, 0.83, 0, 0.17, 0, 0, 0, 0, 0, 0, 0]$.

Figure 7.44 portrays the BER performance of the RSC(213)-coded IR-PLDC scheme designed for satisfying the complexity constraint of Equation (7.57), when using QPSK modulation in conjunction with a MMSE detector. Recall that this system was designed for maintaining an infinitesimally low BER at $\rho = 0$ dB and $k = 7$ outer iterations were required by the system to achieve an infinitesimally low BER. In contrast to the previous 'maximum-rate' design of Figure 7.39, where a sharp turbo-cliff BER was observed, the BER of the complexity-constrained schemes dropped more gradually, owing to having a wider EXIT tunnel.

In order to elaborate a little further, Figure 7.45 quantifies the maximum achievable rate of the 'complexity-constrained' RSC-coded IR-PLDC scheme of Figure 7.32 having various complexity constraints. The minimum complexity required to enable the system to operate at $\rho = -2, -1$ and 0 dB is about 3576 ACS arithmetic operations, where a throughput of 0.5 was achieved in Figure 7.45. At the minimum decoding complexity point, the inner IR-PLDC only activates the minimum-complexity PLDC component of PLDC(2221) with a weighting coefficient $\lambda_0 = 1$. As a result of investing more complexity, a higher rate becomes achievable in conjunction with more complex PLDC components, while operating at a certain SNR ρ. Note that in the vicinity of the cliff region, the system becomes capable of approaching the performance of the maximum-rate design at a significant complexity reduction, as evidenced in Figures 7.43 and 7.44.

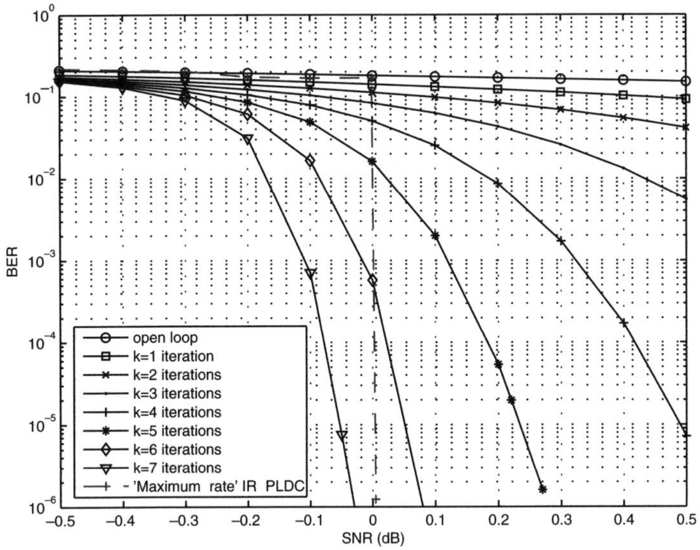

Figure 7.44: BER of the 'complexity-constrained' RSC(213)-coded IR-PLDC scheme of Figure 7.32 designed for achieving an infinitesimally low BER at $\rho = 0$ dB, when using QPSK modulation and a MMSE detector.

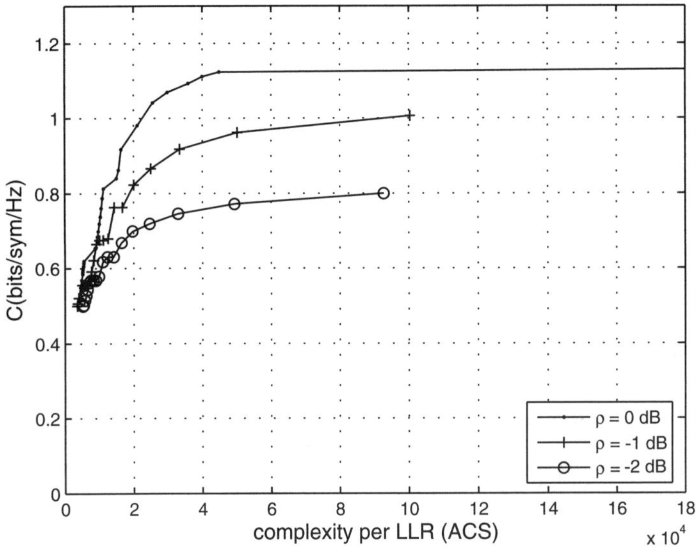

Figure 7.45: Maximum achievable rates against the required decoding complexity of the RSC(213)-coded IR-PLDC schemes of Figure 7.32 at $\rho = -2, -1, 0$ dB, when using QPSK modulation and a MMSE detector.

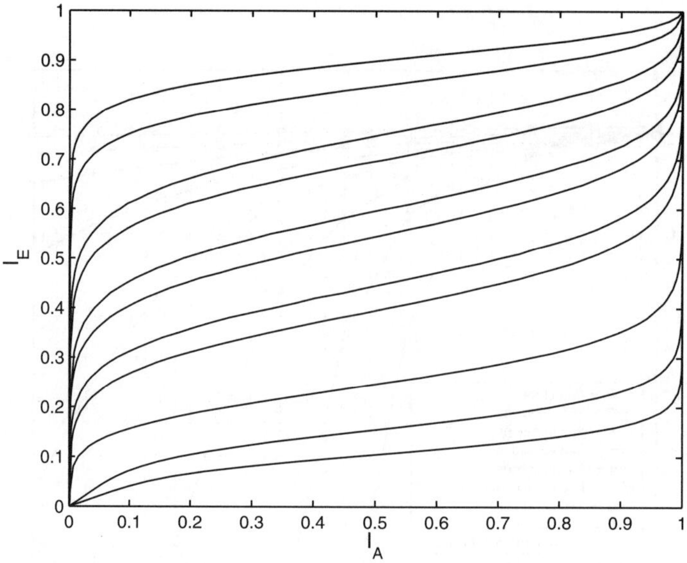

Figure 7.46: EXIT charts of the IRCC scheme having $P_{\text{out}} = 11$ component codes.

7.5.2 IR-PLDCs versus IRCCs

Since both irregular inner schemes, such as IR-PLDCs, and irregular outer schemes, i.e. IRCCs, have the ability of providing flexible EXIT curves, it is natural to investigate further in order to explore the difference between these two approaches.

In this section, we provide a detailed comparison of the IRCCs and IR-PLDCs in terms of the maximum achievable rates and the working SNR region. We adopt the RSC(213)-coded IR-PLDC scheme having a total number of $P = P_{\text{in}} + P_{\text{out}} = 11 + 1 = 12$ components, which has been extensively demonstrated in Section 7.5.1.2. Since the IR-PLDC component codes have a maximum rate of $R_{3,\text{LDC}} = 2$, the resultant maximum rate of the RSC(213)-coded IR-PLDC scheme of Figure 7.32 is limited to 2 (BPS Hz^{-1}) using QPSK modulation, as seen in Figure 7.40.

For a fair comparison, we also set the maximum rate of the IRCC-coded PLDC scheme to 2 (BPS Hz^{-1}), when QPSK modulation is employed. Furthermore, an IRCC scheme constituted by a set of $P_{\text{out}} = 11$ component codes was constructed in [176] from a systematic half-rate memory-four mother code defined by the octally represented generator polynomials of $(23, 35)_8$. Hence, we have a total number of $P = P_{\text{in}} + P_{\text{out}} = 1 + 11 = 12$ components. The EXIT chart characteristic of the resultant $P_{\text{out}} = 11$ IRCC component codes is shown in Figure 7.46, where the rates of the codes are $R_{i,\text{IRCC}} = [0.1, 0.15, 0.25, 0.4, 0.45, 0.55, 0.6, 0.7, 0.75, 0.85, 0.9]$, respectively. Each component encodes a specific fraction of the incoming bit stream, as quantified by the weighting coefficient vector $\gamma = [\gamma_1, \ldots, \gamma_{P_{\text{out}}}]$. Hence, the weighting coefficient vector γ is optimized with the aid of the iterative algorithm in [195], so that the EXIT curve of the resultant IRCC closely matches that of the inner code.

Figure 7.47 portrays the maximum achievable rates achieved by the IRCC-coded schemes using PLDC(2224) or PLDC(2221) as the inner codes according to Tables G.3 and G.4 in Appendix G, when employing QPSK modulation as well as a MMSE detector. The maximum achievable rates of the corresponding RSC(213)-coded IR-PLDC scheme

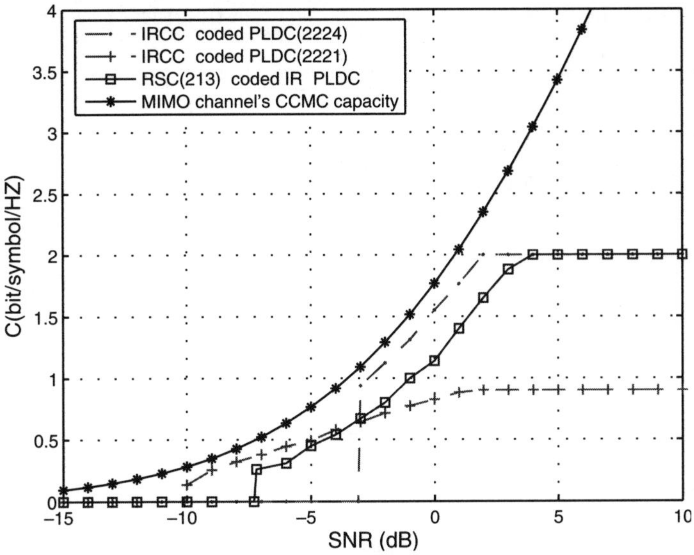

Figure 7.47: Maximum achievable rates plotted according to Tables G.3 and G.4 for the IRCC-coded schemes using PLDC(2224) and PLDC(2221) as the inner codes, when QPSK modulation in conjunction with a MMSE detector were employed.

configured according to Table G.1 is also plotted as the benchmark. When a rate-two PLDC(2224) code is employed, the resultant IRCC-coded scheme becomes capable of operating about 0.9 dB away from the MIMO channel's capacity of Equation (7.10), but no open EXIT tunnel is formed until we have $\rho \geq -3$ dB, which implies that it fails to function adequately in the low-SNR region. In contrast, when we use the half-rate PLDC(2221) arrangement as our inner code, the system performs adequately for SNRs in excess of $\rho = -10$ dB, as seen in Figure 7.47. However, this scheme suffers from a rate loss in the high SNR region, because the inner PLDC(2221) code fails to operate near the MIMO capacity. In contrast, the RSC(213)-coded IR-PLDC scheme operates about 2.5 dB away from the MIMO channel's capacity for SNRs spanning from $\rho = -7$ dB to $\rho = 4$ dB.

In conclusion, the IRCC-coded schemes are capable of operating about 0.9 dB away from the MIMO channel's capacity in either the high-SNR or the low-SNR region, but fail to operate across a number of SNRs. On the other hand, although the IR-PLDC aided scheme operates about 2.5 dB away from the MIMO channel's capacity, it can work in a wider SNR region.

7.5.3 IRCC-coded IR-PLDC Scheme

It is desirable to have a flexible system that is capable of working close to the MIMO channel's capacity and at the same time operating across a wide SNR region. We have demonstrated in Figure 7.47 that either IRCCs or IR-PLDCs are only capable of satisfying one of these design objectives. In this section, we demonstrate that by serially concatenating the IRCCs [195] and the proposed IR-PLDCs, such ambitious design objectives can be fulfilled at the same time.

Figure 7.48 portrays the schematic of the proposed IRCC-coded IR-PLDC scheme. The receiver is assumed to have the knowledge of the weighting coefficient vectors $\boldsymbol{\gamma}$ and $\boldsymbol{\lambda}$, as

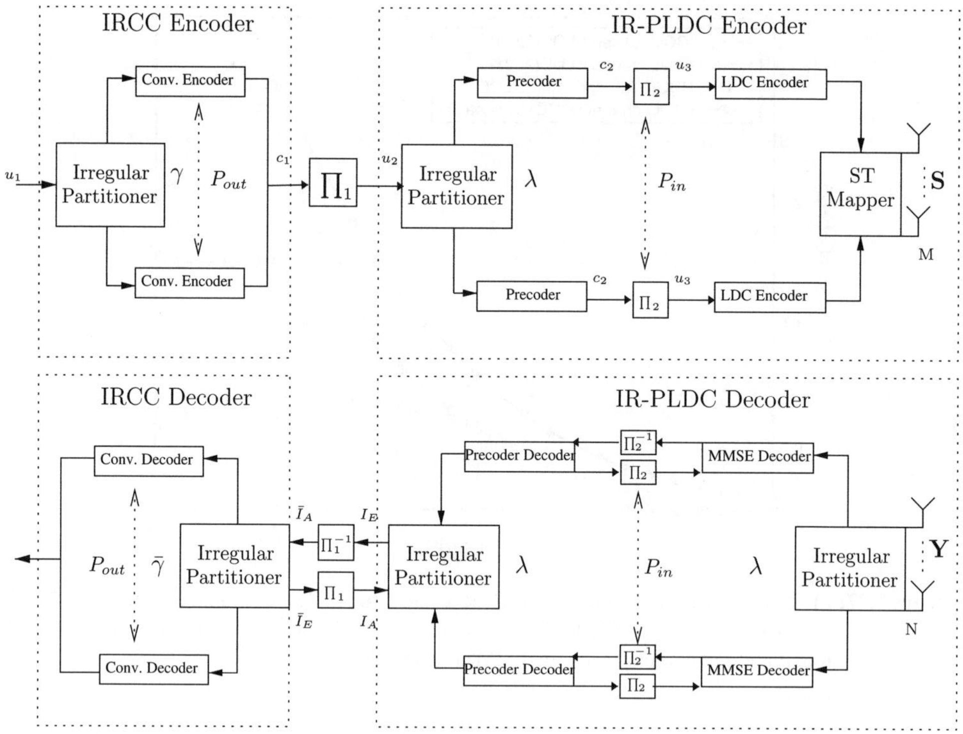

Figure 7.48: Schematic of the serial concatenated IRCC-coded IR-PLDC using iterative decoding.

well as that of the set of DCMs. The 'irregular partitioner' of Figure 7.48 feeds the required fraction of bits into the relevant component codes, according to the corresponding weighting coefficient vectors. For the sake of having a fair comparison with the schemes illustrated in Section 7.5.2, in the following design example, we consider a MIMO system equipped with $M = 2$ transmit as well as $N = 2$ receive antennas and a total number of $P = 12$ component codes are used, where we have $P_{\text{out}} = 6$ IRCC component codes and $P_{\text{in}} = 6$ IR-PLDC component codes. More specifically, the IRCC component codes of Figure 7.46 having a rate of $R_{i,\text{IRCC}} = [0.1, 0.25, 0.4, 0.55, 0.7, 0.9]$ and the IR-PLDC component codes having a rate of $R_{i,\text{LDC}} = [0.33, 0.5, 0.67, 1, 1.5, 2]$ in Table 7.5 were employed. Note that the IRCC encoder's 'irregular partitioner' of Figure 7.48 is based on the weighting coefficient vector $\gamma = [\gamma_1, \ldots, \gamma_{P_{\text{out}}}]$, whereas the 'irregular partitioner' of IRCC's decoder is determined by another vector $\bar{\gamma} = [\bar{\gamma}_1, \ldots, \bar{\gamma}_{P_{\text{out}}}]$. That is because γ quantifies the fraction of incoming information bits, while $\bar{\gamma}$ quantifies the fraction of incoming LLRs and they are related by

$$\bar{\gamma}_i = \frac{\gamma_i \times R_{\text{out}}}{R_{i,\text{IRCC}}}, \quad i = 1, \ldots, P_{\text{out}}. \tag{7.59}$$

We run an exhaustive search operation for all of the possible combinations of $\boldsymbol{\gamma}$ and $\boldsymbol{\lambda}$ under the following constraints.

- An open convergence tunnel must exist between the inner and outer EXIT curves in order to achieve an infinitesimally low BER, provided that the aid of the decoding trajectory arrives at the top right-hand corner of the EXIT chart.

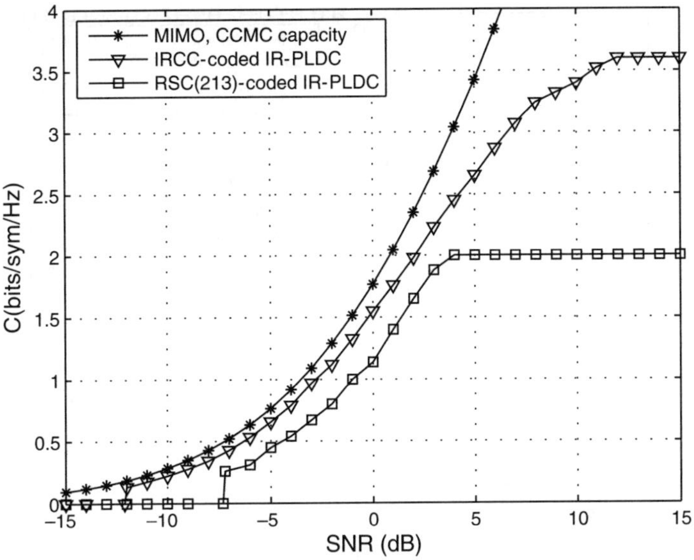

Figure 7.49: Maximum achievable rates plotted according to Tables G.5 and G.6 for the IRCC-coded IR-PLDC schemes of Figure 7.48, when QPSK modulation in conjunction with a MMSE detector were employed.

- The throughput $C(\rho) = \log_2(L) \cdot R_{\text{in}} \cdot R_{\text{out}}$ has to be maximized.
- The resultant open EXIT tunnel area has to be maximized, for the sake of minimizing the number of iterations required.
- We have $\gamma_1 + \gamma_2 + \cdots + \gamma_{P_{\text{out}}} = 1$ and $\lambda_1 + \lambda_2 + \cdots + \lambda_{P_{\text{in}}} = 1$.

The exhaustive search is set to have a step size of 0.05.

In Figure 7.49, we plotted the maximum rates achieved by the IRCC-coded IR-PLDC scheme of Figure 7.48 according to Tables G.5 and G.6, when QPSK modulation in conjunction with a MMSE detector were employed. The corresponding maximum achievable rates of the RSC(213)-coded IR-PLDC scheme of Figure 7.32 were also presented as the benchmark, which also employed $P = 12$ component codes. Observe in Figure 7.49 that employing an irregular design at both the inner and the outer codes enables the system to achieve an infinitesimally low BER for SNR starting from as low as $\rho = -12$ dB. Furthermore, the double-sided EXIT matching-based scheme is capable of operating about 0.9 dB from the MIMO channel's CCMC capacity quantified in Equation (7.10). Note that when we have $\rho > 2$ dB, the IRCC-coded IR-PLDC scheme of Figure 7.48 begins to gradually deviate from the MIMO capacity, owing to lack of high-rate IR-PLDC components. Naturally, the discrepancy observed in the high-SNR region can be eliminated, when adopting high-rate PLDCs as the component codes.

In order to illustrate this phenomenon a little further, Figure 7.50 characterizes the effect of the inner throughput $\log_2(L) \cdot R_{\text{in}}$ and the outer rate R_{out} of the IRCC-coded IR-PLDC scheme of Figure 7.48 designed for achieving the maximum rate according to Tables G.5 and G.6, when QPSK modulation and a MMSE detector were employed. The operating SNR range may be divided into two zones around $\rho = 3$ dB. Observe in Figure 7.50 that in the low-SNR region, namely in Zone 1, the proposed scheme achieved a near-capacity throughput

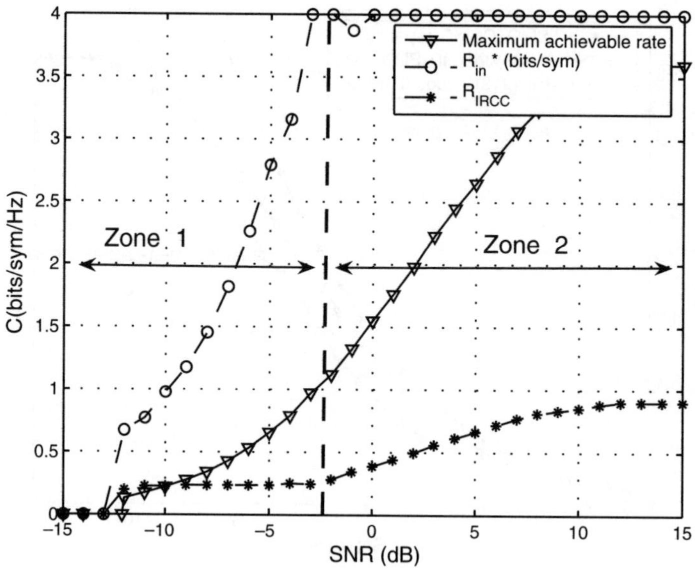

Figure 7.50: Inner throughput $\log_2(L) \cdot R_{\text{in}}$ and outer rate R_{out} for the IRCC-coded IR-PLDC scheme of Figure 7.48 to achieve maximum rate according to Tables G.5 and G.6, when QPSK modulation and a MMSE detector were employed.

owing to the flexibility provided by the IR-PLDCs. In contrast, when we have high SNRs, i.e. in Zone 2, the maximum rate is achieved by increasing the rate of the IRCC scheme. This observation further justifies the results presented in Figure 7.47.

As far as the decoding complexity is concerned, the original IRCC scheme of [176] imposed a potentially excessive complexity in the low-rate region, owing to the additional generator polynomials. However, flexible rates of the IRCC-coded IR-PLDC scheme of Figure 7.48 in the low-SNR region are achieved by adjusting the IR-PLDC's rate, where the associated complexity is low, as seen in Figure 7.50. In Zone 2, since high-rate IRCC components are activated, the overall decoding complexity is still manageable, despite the fact that the IR-PLDC scheme's complexity is increased.

The weighting coefficient vectors $\boldsymbol{\gamma}$ and $\boldsymbol{\lambda}$ of the IRCC and IR-PLDC schemes optimized for achieving the maximum aggregate rates are listed in Tables G.5 and G.6 of Appendix G. Observe in Table G.6 that typically only two components of the inner IR-PLDC are activated, while a number of IRCC components are required for the sake of minimizing the open EXIT tunnel area according to Table G.5. For example, when we have $\rho = 6$ dB, only $P_{\text{out}} + P_{\text{in}} = 2 + 2 = 4$ out of the total of $P = 12$ components were activated. As a result of having a limited range of EXIT curve shapes, the superimposed EXIT curves cannot be matched accurately; thus we observe that the system is operating about 1.6 dB away from the MIMO capacity at $\rho = 6$ dB. In contrast, in the low-SNR region, i.e. when we have $\rho = -4$ dB, $P_{\text{out}} + P_{\text{in}} = 5 + 2 = 7$ component codes were activated. Hence, a more accurate EXIT curve matching becomes possible. The resultant system operates 0.9 dB away from the MIMO capacity, as seen in Figure 7.49.

In Figure 7.51 we portray both the EXIT curves and the decoding trajectory of the IRCC-coded IR-PLDC scheme of Figure 7.48 recorded at $\rho = -4$ dB using QPSK modulation, when a MMSE detector was employed. The corresponding weighting coefficients are given

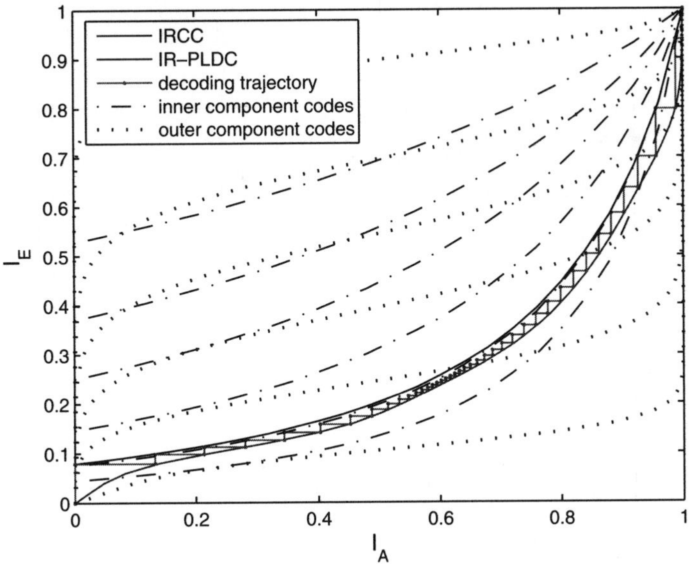

Figure 7.51: EXIT chart and the decoding trajectory of the IRCC-coded IR-PLDC scheme of Figure 7.48 at $\rho = -4$ dB according to Tables G.5 and G.6, when using QPSK modulation in conjunction with a MMSE detector.

in Tables G.5 and G.6. Observe in Figure 7.51 that an extremely narrow EXIT tunnel was formed, which required $k = 39$ outer iterations to reach the $(I_A, I_E) = (1.0, 1.0)$ point. The associated BER curve is plotted in Figure 7.52.

7.5.4 Summary

In order to pursue the design objective of operating in the vicinity of the MIMO channel's capacity across a wide range of SNRs, irregular system designs are necessary. In Sections 7.5.1–7.5.3, we have demonstrated the design process of irregular systems using IR-PLDCs and/or IRCCs. The factors affecting the design of near-capacity schemes using the irregular approach are summarized in Figure 7.53, where the decoding complexity and the EXIT matching accuracy of inner and outer EXIT curves are two main design factors. More specifically, the matching of the inner and outer codes' EXIT curves quantifies the maximum achievable rate, which is dependent on the affordable complexity. The specific shape of the EXIT curves is also related to the decoding complexity. For example, if the inner and outer EXIT curves are matched while having a high gradient, a large number of iterations are required to achieve an infinitesimally low BER. On the other hand, if the matched pair of EXIT curves are near-horizonal, a low number of iterations is required. The relationship between 'EXIT matching' and 'Complexity' is represented by the dotted line in Figure 7.53.

Irregular schemes are suitable for accurate EXIT curve matching, since they are capable of providing EXIT curves with flexible characteristics. The design of irregular outer codes can be accomplished, for example, by using IRCCs [176, 195] or IR-VLCs [307], where irregularity is achieved by employing various code rates. The area property quantified in Equation (7.50) suggests that maximizing the achievable rate corresponds to minimizing the open EXIT tunnel area. In contrast, the irregularity of the inner code can be created in various ways, as seen in Figure 7.53. The IR-PLDC approach proposed in this chapter creates a

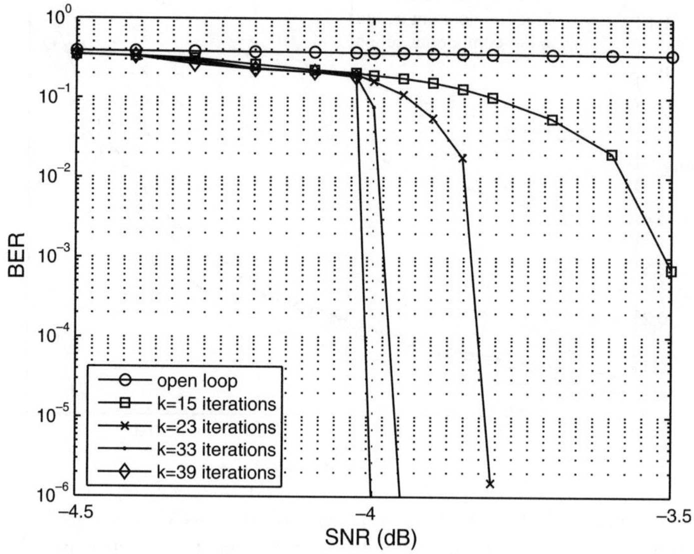

Figure 7.52: BER of the IRCC-coded IR-PLDC scheme of Figure 7.48 that is designed to achieve an infinitesimally low BER at $\rho = -4$ dB according to Tables G.5 and G.6, when using QPSK modulation in conjunction with a MMSE detector.

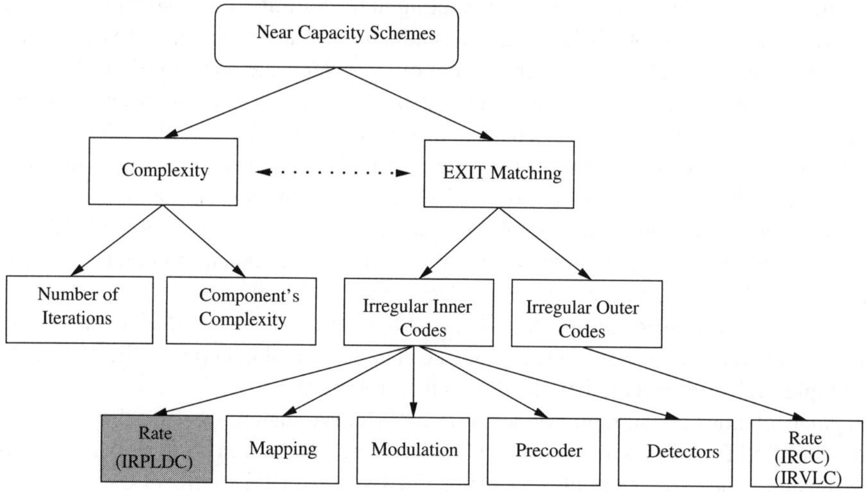

Figure 7.53: Factors affecting the design of near-capacity schemes using the irregular principle.

diverse set of EXIT curves using different rates, as exemplified in Figures 7.33 and 7.34. There are two main differences between the irregular inner and outer schemes. First, the inner code's EXIT curves are affected by the SNR encountered. In other words, there is a different set of inner EXIT curves for each SNR. This property further explains the reason that IRCC schemes may facilitate operation closer to the MIMO channel's capacity than IR-PLDCs within a certain SNR region, as evidenced in Figure 7.47. Second, the IR-PLDC scheme has a non-unique relationship between its aggregate rate $R_{\rm in}$ of Equation (7.53) and the area under the EXIT curves, which was demonstrated in Figure 7.37. Hence, our optimization objective is to maximize the achievable rate as well as maximize the open EXIT tunnel area. However, the employment of the IR-PLDCs potentially facilitates the resultant scheme to operate across a wide SNR range, as seen in Figure 7.49.

Figure 7.53 also lists other techniques of creating a diverse set of EXIT curves in the context of irregular inner codes. For example, changing the precoder's memories and/or the generator polynomials is capable of effectively changing the shape of the EXIT curves without affecting the area under them, as illustrated in Figure 7.31. Another possible approach of changing the EXIT curves' shape, while maintaining the same area, is to employ different mapping schemes for each component [46, 305]. Obviously, employing different modulation schemes for each component code can also adjust the effective throughput of the system. Hence, a diverse set of EXIT curve shapes can be generated. When the decoding complexity is taken into account, the inner code's irregularity can be complemented by using various detectors, such as ML, MMSE, SIC and Parallel Interference Cancelation (PIC) detectors. Finally, it is important to point out that the set of techniques seen in Figure 7.53 can be employed jointly to create an even more diverse inner irregularity.

7.6 Conclusion

In this chapter, after illustrating the design of STBCs, we have demonstrated that the family of LDCs constitutes a general framework, which accommodates the entire set of STBCs having different design objectives. Furthermore, we have proposed a novel method of optimizing the LDCs based on their DCMC capacity, as seen in Section 7.2.4. More particularly, in Section 7.3 we demonstrated the linkage between the existing STBC schemes and LDCs by characterizing their mathematical representations and the design philosophy under LDC's general framework. In Sections 7.4.1 we investigated the performance of two-stage serial concatenated LDCs, with the aid of EXIT charts. The employment of precoders for the sake of achieving an infinitesimally low BER was investigated in Section 7.4.2, while approaching the maximum achievable rate.

Motivated by the flexibility of the LDC structure combined with its near-capacity performance potential, in Section 7.5.1.2 we proposed a novel IR-PLDC scheme that is capable of approaching the maximum attainable rate across a wide range of SNRs, as demonstrated by Figures 7.38–7.42, when combined with a simple outer channel coder. In Figure 7.40, we showed that the proposed RSC(213)-coded IR-PLDC scheme is capable of operating about 2.5 dB from the MIMO channel's capacity. In the situation where the affordable decoding complexity is limited, in Section 7.5.1.3 we proposed a 'complexity-constrained' IR-PLDC scheme, which was designed with the aid of EXIT charts and maximizing the attainable rate under a specific complexity constraint. After a detailed examination of the advantages and drawbacks of irregular inner and outer encoding schemes in Section 7.5.2, in Section 7.5.3 we proposed an IRCC-coded IR-PLDC scheme, which is capable of operating about 0.9 dB from the MIMO channel's capacity for a wide range of SNRs, while achieving an infinitesimally low BER.

Chapter 8

Differential Space-Time Block Codes: A Universal Approach

8.1 Introduction and Outline

The primary focus of the codes discussed in Chapter 7 has been cases where only the receiver has knowledge of the CIR. In practice, knowledge of the CIR is typically acquired using a channel sounding sequence. However, an excessive number of training symbols may be required, especially when numerous antennas are involved. Hence, precious transmit power as well as valuable bandwidth is wasted. For example, a (4×4)-antenna-aided system requires the estimation of 16 channels, which imposes a high complexity in comparison to the idealized coherently detected system benefiting from perfect CIR estimation. Moreover, since the total transmit energy is shared by the multiple antennas, the energy available for training symbols is reduced compared with single-antenna-aided systems, which typically results in an increased channel estimation error. Furthermore, when the mobile station travels at a high speed, the channel's complex envelope varies dramatically; thus accurate MIMO channel estimation becomes a challenging task. Therefore, differentially encoded low-complexity schemes, dispensing with pilot-based channel estimation and invoking non-coherent detection, have been proposed in the literature.

The evolution of DSTBCs can be divided into two categories. More specifically, the first family is the Differential Orthogonal Space-Time Block Codes (DOSTBCs), which facilitates low complexity. Similar to their coherent counterparts [11, 12, 25], the class of DOSTBCs inherits certain restrictions owing to the orthogonal structure. The other family of DSTBCs is referred to as Differential Unitary Space-Time Modulation (DUSTM) in the literature, and was proposed independently by Hochwald *et al.* [63, 308] and Hughes [55]. This particular branch of DSTBCs has the advantage of supporting an arbitrary number of transmit and receive antennas and is capable of operating at high rates. However, the number of space-time constellation matrices to be designed grows exponentially with the effective throughput, which renders the design of such a large set of space-time matrices challenging. Nevertheless, a number of efforts have been reported in the literature to simplify the design problem by imposing different constraints [309–311].

Near-Capacity Multi-Functional MIMO Systems L. Hanzo, O. R. Alamri, M. El-Hajjar and N. Wu

In this chapter, we propose a unified structure for describing all of the DSTBCs found in the open literature and demonstrate that both DOSTBCs and DUSTMs are subsumed by this general framework. Furthermore, we extend the philosophy of LDCs into the differential encoding domain and propose a universal solution, namely the family of DLDCs based on the Cayley transform [10]. More explicitly, the novel contributions of this chapter are listed as follows [32, 185].

- Characterize the fundamental link between STBCs and DSTBCs and propose a unified structure to describe DSTBCs.
- Unify the family of DOSTBCs as well as the class of DUSTMs using the proposed general structure and explicitly characterize their different design perspectives.
- Propose the family of DLDCs based on the Cayley transform, which is capable of simultaneously achieving both high rate and full diversity.
- Propose turbo-detected channel-coded DSTBCs that are capable of achieving an infinitesimally low BER, while operating at low SNRs.

The outline of this chapter is as follows. Section 8.2 presents a universal system model for all of the DSTBC schemes found in the literature, which is generalized by extending the single-antenna-aided DPSK modulation scheme. Hence, the fundamental challenge of designing DSTBCs is outlined. In Section 8.3, we examine the family of DOSTBCs and explicitly show how DOSTBCs may be designed, where the orthogonality imposed may limit their application. Hence, Section 8.4 proposes a novel class of DSTBCs, namely DLDCs based on the Cayley transform. In Section 8.5, we amalgamate SP modulation with our DOSTBC design and propose a turbo-detected SP-aided DOSTBC scheme, aiming at operating at as low SNRs as possible at a given rate. Facilitated by the design flexibility provided by the IRCC-coded IR-PLDCs of Chapter 7, in Section 8.6 we propose the novel IRCC-coded IR-PDLDCs, which operate at high rates across a wide range of SNRs. We focus our attention on the specific restrictions introduced by the differential encoding structure when designing the irregular system having multiple component codes. Finally, our concluding remarks are provided in Section 8.7.

8.2 System Model

Before presenting the detailed design of DSTBCs, we briefly review the concept of differential coding schemes designed for single-transmit-antenna scenarios. More precisely, we present the details of the DPSK modulation scheme [248] and capture the main design concepts behind it. Later in this section, we generalize the single-antenna model to the MIMO environment and illustrate the specific constraints imposed by the differential encoding structure.

8.2.1 DPSK System Model for Single Antennas

Figure 8.1 portrays a schematic of the conventional DPSK modulation [248] designed for the single-antenna-aided systems. After the information bits have been mapped to a L-PSK symbol x_n, the nth transmitted symbol s_n generated using differential encoding is given by:

$$s_n = \begin{cases} x_n & (n = 1), \\ s_{n-1} \cdot x_n & (n > 1). \end{cases} \tag{8.1}$$

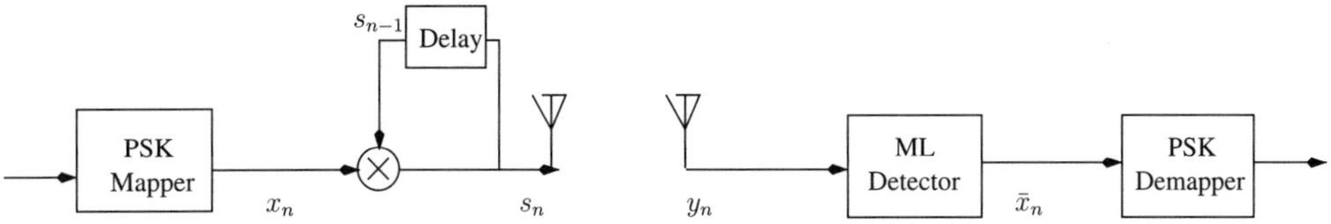

Figure 8.1: Schematic of the single-antenna-aided DPSK scheme.

The first symbol is an arbitrary reference symbol, which does not contain any information. Each differentially encoded symbol s_n is then transmitted over $T = 1$ time slot. At the receiver side, the corresponding received signal y_n seen in Figure 8.1 becomes

$$y_n = h_{1,1} s_n + v_n, \tag{8.2}$$

where $h_{1,1}$ represents the CIR between the single transmit and receive antenna pair and v_n denotes the zero-mean complex-valued Gaussian random variable with variance σ_0 determined by SNR ρ.

Since the transmitted signals depend on each other, the ML detector of Figure 8.1 is capable of detecting the information-bearing symbols x_n from successive received signals y_n. More explicitly, by combining Equations (8.1) and (8.2) we may have

$$\begin{aligned} y_n &= h_{1,1} s_{n-1} x_n + v_n \\ &= (y_{n-1} - v_{n-1}) x_n + v_n \\ &= y_{n-1} x_n + \underbrace{v_n - v_{n-1} x_n} \\ &= y_{n-1} x_n + \qquad \hat{v}_n. \end{aligned} \tag{8.3}$$

Observe in Equation (8.3) that x_n is related to the received signal y_n by the previous received signal y_{n-1}; thus the estimated symbol $\bar{x}_n$ of Figure 8.1 can be obtained using ML estimation and it is given by

$$\bar{x}_n = \arg\left\{ \min_{x_n \in L-PSK} (\|y_n - y_{n-1} x_n\|^2) \right\}. \tag{8.4}$$

Note that the equivalent noise $\hat{v}_n$ of Equation (8.3) has a variance of $2\sigma_0$, which is responsible for the well-known 3 dB SNR loss compared with its coherently detected counterpart having perfect CIR information.

The properties of the single-antenna-aided DPSK scheme are summarized as follows.

- The information is differentially encoded between successive transmission symbols; thus the information can be recovered without knowledge of the CIR, provided that it does not change substantially between them.

- A 3 dB performance penalty is expected compared with its coherently detected counterpart, owing to the doubled noise variance encountered during the differential detection.

- The transmitted symbol s_n generated by differential encoding using Equation (8.1) remains an L-PSK symbol.

Following a similar design philosophy, the differentially encoded structure can be generalized to MIMO applications.

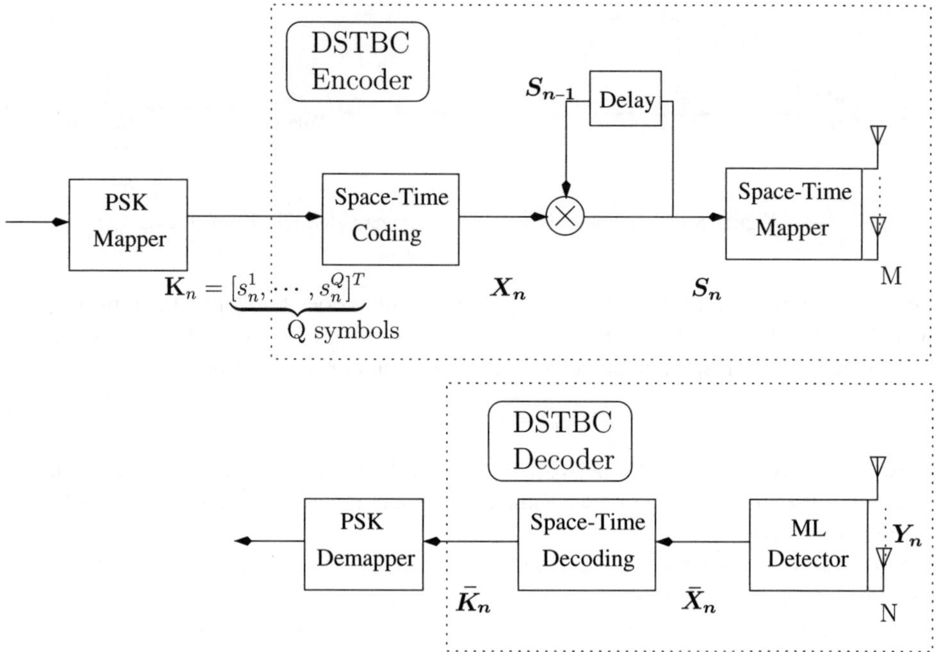

Figure 8.2: Schematic of a MIMO system equipped with M transmit and N receive antennas that employs DSTBCs, while transmitting Q symbols over T time slots using a differentially encoded space-time matrix $\mathbf{S}_n$.

8.2.2 DSTBC System Model for Multiple Antennas

We consider a MIMO wireless communication system equipped with M transmit as well as N receive antennas, which transmits a signal matrix $\mathbf{S}_n$ containing Q symbols during T time slots, as seen in Figure 8.2. During the transmission, the nth block of information bits is mapped to a symbol vector $\mathbf{K}_n = [s_n^1, \ldots, s_n^Q]^{\mathrm{T}}$ containing Q symbols drawn from an L-PSK constellation. Then $\mathbf{K}_n$ is further coded by the 'space-time coding' block of Figure 8.2, where the resultant space-time coded matrix $\mathbf{X}_n$ spans M spatial and T temporal slots. Note that the mapping between $\mathbf{K}_n$ and $\mathbf{X}_n$ is unique. In order to avoid channel estimation, the information has to be encoded into consecutive transmission matrices and the differentially encoded transmission matrix $\mathbf{S}_n$ is given by

$$\mathbf{S}_n = \begin{cases} \mathbf{U}_n & (n = 1), \\ \mathbf{S}_{n-1} \cdot \mathbf{X}_n & (n > 1), \end{cases} \tag{8.5}$$

where $\mathbf{U}_n$ is the reference matrix containing dummy information. The matrix multiplication of Equation (8.5) requires $\mathbf{X}_n$ to be a unitary matrix. Hence, we have $T = M$ and $\mathbf{U}_n \in \zeta^{M\times M}$, $\mathbf{X}_n \in \zeta^{M\times M}$ and $\mathbf{S}_n \in \zeta^{M\times M}$.

The nth received signal matrix $\mathbf{Y}_n$ having a size of $(N \times M)$ elements becomes

$$\mathbf{Y}_n = \mathbf{H}_n\mathbf{S}_n + \mathbf{V}_n, \tag{8.6}$$

where $\mathbf{V}_n \in \zeta^{N\times M}$ is assumed to be having independent samples of a zero-mean complex-valued Gaussian random process with variance σ_0 determined by SNR ρ. More importantly,

$\mathbf{H}_n \in \zeta^{N\times M}$ represents the Rayleigh fading coefficients. We assume that the scattering imposed by the propagation environment is sufficiently rich for each channel coefficient to be modeled as an i.i.d. zero-mean complex-valued Gaussian random variable having a common variance of 0.5 per real-valued dimension. Each channel of a transmit–receive antenna pair is assumed to be constant over T channel uses, which is then faded before the transmission of another block. The temporal correlation within each channel is governed by the normalized Doppler frequency f_d.

If the MIMO channel fades slowly and we assume that the channel envelope is constant over two consecutive transmission blocks, after combining Equations (8.5) and (8.6), we obtain

$$\begin{aligned}\mathbf{Y}_n &= \mathbf{H}_n\mathbf{S}_{n-1}\mathbf{X}_n + \mathbf{V}_n\\ &= (\mathbf{Y}_{n-1} - \mathbf{V}_{n-1})\mathbf{X}_n + \mathbf{V}_n\\ &= \mathbf{Y}_{n-1}\mathbf{X}_n + \underbrace{\mathbf{V}_n - \mathbf{V}_{n-1}\mathbf{X}_n}\\ &= \mathbf{Y}_{n-1}\mathbf{X}_n + \qquad \hat{\mathbf{V}}_n.\end{aligned} \tag{8.7}$$

Note that the CIR matrix $\mathbf{H}_n$ does not appear in the above equation. This implies that as long as the channel is approximately constant for $2T$ channel uses, the differential transmission scheme permits decoding without knowing the CIR.

Furthermore, the differential encoding of Equation (8.5) restricts the set of matrices $\mathbf{X}_n$ to be unitary, otherwise the product $\mathbf{S}_n = \mathbf{X}_n\mathbf{X}_{n-1}\ldots\mathbf{X}_1$ can go to zero, infinity, or both in different spatial and temporal directions. Moreover, when $\mathbf{X}_n$ is unitary, the additive noise term

$$\hat{\mathbf{V}}_n = \mathbf{V}_n - \mathbf{V}_{n-1}\mathbf{X}_n \tag{8.8}$$

is statistically independent of $\mathbf{X}_n$ having a variance of $2\sigma_0 \cdot \mathbf{I}$, which imposes the well-known 3 dB penalty. Since the additive noise term $\hat{\mathbf{V}}_n$ has independent complex Gaussian entries, the ML detector of Figure 8.2 stacks two consecutive received matrices $\mathbf{Y}_{n-1}$ and $\mathbf{Y}_n$, and the information-bearing matrix $\mathbf{X}_n$ can be estimated using

$$\bar{\mathbf{X}}_n = \arg\Big\{\min_{\mathbf{X}_n\in F}(\|\mathbf{Y}_n - \mathbf{Y}_{n-1}\mathbf{X}_n\|^2)\Big\}, \tag{8.9}$$

given all of the probabilities of $\mathbf{X}_n$. In fact, since each information vector $\mathbf{K}_n$ contains Q symbols selected from the L-PSK constellation, the total number of information vectors is

$$F = L^Q. \tag{8.10}$$

Then, the 'space-time decoding' block of Figure 8.2 can recover the corresponding symbol vector $\mathbf{K}_n$ by exploring the one-to-one relationship between $\mathbf{K}_n$ and $\mathbf{X}_n$.

The employment of differential 'modulation' schemes, such as DPSK, enables the single-antenna-aided system to recover the transmitted information without the knowledge of the CIR. However, when the extra spatial dimension becomes available, owing to the employment of multiple antennas at both the transmitter and the receiver, the information can be differentially encoded using the previous symbols as reference in both the spatial and temporal dimensions, instead of using the classic differentially encoded modulation schemes. Hence, the symbols transmitted in the DSTBC system of Figure 8.2 are drawn from an L-PSK constellation, rather than from a 'differential' L-PSK.

We now continue by offering a few remarks concerning the general DSTBC model of Figure 8.2.

- Compared with the single-antenna-aided DPSK scheme, the DSTBC scheme of Figure 8.2 employs space-time block coding to explore all of the available spatial and temporal slots.
- In order to carry out the differential encoding of Equation (8.5), the set of space-time coded matrices $\mathbf{X}_n$ has to be unitary.
- The design of DSTBC is separated from the design of modulation, as seen in Figure 8.2. Hence, various modulation schemes can be employed.

8.2.3 Link between STBCs and DSTBCs

Compared with the general STBC framework of Figure 7.1, the DSTBC structure portrayed in Figure 8.2 introduces a differential encoding unit, in order to forgo the burden of channel estimation. However, the differential encoding structure imposes a unitary constraint on the resultant space-time coded matrices. In other words, the challenge of designing DSTBCs can be described as that of designing a family of STBCs where all of the space-time matrices are unitary. From this point of view, it is straightforward to verify that DSTBCs inherit all of the properties of STBCs. Most importantly, the rank criterion and determinant criterion detailed in Section 7.3.1 can be used to maximize the achievable spatial diversity gain and coding gain.

For the coherently detected STBCs, we assumed that the receiver evaluated the CIR with the aid of training sequences and that the channel coefficients are statistically independent under these conditions. We have shown in Equation (7.11) that for large SNRs ρ the coherent detection-based MIMO capacity C_{MIMO} is approximately

$$C_{\text{MIMO}} \approx \min(M, N) \cdot \log_2\left[\frac{\rho}{M}\right] \quad (\text{BPS Hz}^{-1}), \tag{8.11}$$

which implies that the capacity gain of the coherent-detection-aided multiple-antenna-assisted channel is $\min(M, N)$ (BPS Hz^{-1}) for every 3 dB increase in the SNR. In order to address the scenario where the receiver has no *a priori* channel knowledge apart from the AWGN variance, Zheng and Tse [312] found the capacity of this non-coherent channel scenario to be:

$$C_{\text{MIMO}}^{\text{non}} \approx \bar{M}\left(1 - \frac{\bar{M}}{\Im}\right) \cdot \log_2\left[\frac{\rho}{M}\right] \quad (\text{BPS Hz}^{-1}), \tag{8.12}$$

where $\bar{M} = \min(M, N, \Im/2)$ and $\Im$ is the number of signaling intervals over which the channel may be deemed to be static. In [313], a detailed MI analysis was carried out between the coherent USTMs [54] and the non-coherent DUSTMs [308] under the assumption of $\Im = 2T$.

Equation (8.12) suggests that the capacity of the non-coherent MIMO channel is affected by its coherence time $\Im$. More explicitly, the non-coherent capacity of Equation (8.12) approaches the coherent capacity of Equation (8.11) upon increasing $\Im$ and $\Im > 2M$. Note that this model assumes that the channel is constant for $\Im$ signaling intervals and changes independently before the next. However, in our model presented in Section 8.2.2, we consider more realistic scenarios where the complex-valued CIR taps remain constant for $\Im = M$ channel uses and the temporal correlation is governed by the normalized Doppler frequency f_d. Since there is no closed-form representation of the CCMC capacity and DCMC capacity of the non-coherent channel, it is difficult to directly optimize DSTBCs from the capacity maximization perspective. Therefore, in this chapter we employ the conventional rank criterion and determinant criterion to optimize DSTBCs, rather than adopting a capacity-based approach in order to operate under various channel conditions.

8.3 DOSTBCs

In this section, we demonstrate the philosophy of DSTBCs based on an orthogonal design, namely the family of DOSTBCs. Similar to the OSTBCs detailed in Sections 7.3.2–7.3.5, the DOSTBCs' orthogonality facilitates low-complexity ML detection at the receivers. Inevitably, the orthogonal structure imposes various restrictions on the design of the space-time coded matrices $\mathbf{X}_n$ of Figure 8.2.

8.3.1 Differential Alamouti Codes

We commence our discussion from the well-known Alamouti scheme [11], where $M = 2$ transmit antennas are employed. To simplify our discussion, the case of using $N = 1$ receive antenna is considered, although employing multiple receive antennas is straightforward. For example, given two arbitrary L-PSK symbols t^1 and t^2, a $\mathbf{G}_2$ (see [11]) space-time matrix can be written as

$$\mathbf{G}_2(t^1, t^2) = \begin{pmatrix} t^1 & -t^{2^*} \\ t^2 & t^{1^*} \end{pmatrix}, \tag{8.13}$$

where each column represents a spatial dimension and each row denotes a temporal dimension. It is easy to verify that $\mathbf{G}_2$ is a unitary matrix.

The differential encoding method of [62] used for generating the nth $(n > 1)$ transmission matrix $\mathbf{S}_n$ of Figure 8.2 using the incoming L-PSK symbol vector $\mathbf{K}_n = [s_n^1, s_n^2]^{\mathrm{T}}$ can be written as

$$\begin{pmatrix} t_n^1 & -t_n^{2^*} \\ t_n^2 & t_n^{1^*} \end{pmatrix} = \begin{pmatrix} t_{n-1}^1 & -t_{n-1}^{2^*} \\ t_{n-1}^2 & t_{n-1}^{1^*} \end{pmatrix} \cdot \begin{pmatrix} s_n^1 & -s_n^{2^*} \\ s_n^2 & s_n^{1^*} \end{pmatrix}, \tag{8.14}$$

which can be represented as

$$\underbrace{\mathbf{G}_2(t_n^1, t_n^2)}_{\mathbf{S}_n} = \underbrace{\mathbf{G}_2(t_{n-1}^1, t_{n-1}^2)}_{\mathbf{S}_{n-1}} \cdot \underbrace{\mathbf{G}_2(s_n^1, s_n^2)}_{\mathbf{X}_n}. \tag{8.15}$$

The reference matrix $\mathbf{U}_n$ $(n = 1)$ transmitted during the first signaling block is also a $\mathbf{G}_2$ matrix obeying Equation (8.13), which contains two arbitrary symbols. When comparing Equation (8.15) with the general DSTBC encoding scheme of Equation (8.5), we observe that the challenge of designing a set of unitary space-time matrices $\mathbf{X}_n$ is tackled by employing the well-known $\mathbf{G}_2$ matrices. Thus, low-complexity ML detection becomes feasible.

The received signal matrix $\mathbf{Y}_n$ of Figure 8.2 can be written as $\mathbf{Y}_n = [y_n^1, y_n^2]$, where y_n^i denotes the received signal during the ith time slot of the nth block. Consequently, the orthogonal structure enables the receiver to decouple the symbol streams into independent symbols using linear combination, which is expressed as [62]

$$\begin{pmatrix} \bar{s}_n^1 \\ \bar{s}_n^2 \end{pmatrix} = \begin{pmatrix} y_{n-1}^{1^*} & y_{n-1}^2 \\ y_{n-1}^{2^*} & -y_{n-1}^1 \end{pmatrix} \cdot \begin{pmatrix} y_n^1 \\ y_n^{2^*} \end{pmatrix}. \tag{8.16}$$

Hence, the decoding statistics $\bar{s}_n^1$ and $\bar{s}_n^2$ may be expressed as

$$\bar{s}_n^1 = (|h_{1,1}|^2 + |h_{1,2}|^2)(|t_{n-1}^1|^2 + |t_{n-1}^2|^2)s_n^1 + \hat{v}_n^1, \tag{8.17}$$

$$\bar{s}_n^2 = (|h_{1,1}|^2 + |h_{1,2}|^2)(|t_{n-1}^1|^2 + |t_{n-1}^2|^2)s_n^2 + \hat{v}_n^2, \tag{8.18}$$

where $\hat{v}_n^i$ represents the ith combined noise component encountered during the nth received signal block. The employment of an L-PSK constellation together with the orthogonality constraint ensures that the term $(|t_{n-1}^1|^2 + |t_{n-1}^2|^2)$ remains a constant value. Hence, single-symbol ML decoding can be performed, provided that the channel's output power factor of $(|h_{1,1}|^2 + |h_{1,2}|^2)$ can be estimated. The estimation of the channel's output power requires a significantly lower complexity than that of the CIR itself. Observe that a transmit diversity order of $D_{tx} = 2$ has been achieved, since the chances are that even if $|h_{1,1}|^2$ is small, the independently faded $|h_{1,2}|^2$ value may not be.

If multiple receive antennas are employed, the well-known MRC technique [9] can be employed to attain a potential spatial receive diversity.

8.3.1.1 Using QAM Constellations

As more bits are transmitted per channel use with the aid of higher-order constellations, the SNR disadvantage of L-PSK over L-QAM increases [286]. In order to reduce the associated SNR penalty, DSTBCs based on Alamouti's STBC using QAM constellations were proposed [58], which increase the MED compared with that of the identical-throughput PSK constellation.

The differential encoding process of [58] used to generate the nth $(n > 1)$ transmission matrix $\mathbf{S}_n$ of Figure 8.2 employing the incoming L-QAM symbol vector $\mathbf{K}_n = [s_n^1, s_n^2]^{\mathrm{T}}$ can be rewritten as

$$\begin{pmatrix} t_n^1 & -t_n^{2^*} \\ t_n^2 & t_n^{1^*} \end{pmatrix} = \frac{1}{\theta} \begin{pmatrix} t_{n-1}^1 & -t_{n-1}^{2^*} \\ t_{n-1}^2 & t_{n-1}^{1^*} \end{pmatrix} \cdot \begin{pmatrix} s_n^1 & -s_n^{2^*} \\ s_n^2 & s_n^{1^*} \end{pmatrix}. \tag{8.19}$$

The difference between Equations (8.15) and (8.19) is the presence of the normalization factor given by $\theta = \sqrt{|t_{n-1}^1|^2 + |t_{n-1}^2|^2}$. Since QAM constellations do not maintain a constant modulus, the discrete amplitude of the symbols may render the peak power of the transmitted signals after differential encoding to become infinity or zero. Therefore, the introduction of the normalization factor θ is necessary. Furthermore, Equation (8.19) can be expressed as follows:

$$\underbrace{\mathbf{G}_2(t_n^1, t_n^2)}_{\mathbf{S}_n} = \frac{1}{\theta} \cdot \underbrace{\mathbf{G}_2(t_{n-1}^1, t_{n-1}^2)}_{\mathbf{S}_{n-1}} \cdot \underbrace{\mathbf{G}_2(s_n^1, s_n^2)}_{\mathbf{X}_n}. \tag{8.20}$$

Again, the received signal matrix $\mathbf{Y}_n$ is related to the transmission matrix $\mathbf{S}_n$ using Equation (8.6), where we have $\mathbf{Y}_n = [y_n^1, y_n^2]$. The receiver employs the linear combination process of Equation (8.16) to derive the following decision statistics:

$$\bar{s}_n^1 = (|h_{1,1}|^2 + |h_{1,2}|^2) \cdot \theta \cdot s_n^1 + \hat{v}_n^1, \tag{8.21}$$

$$\bar{s}_n^2 = (|h_{1,1}|^2 + |h_{1,2}|^2) \cdot \theta \cdot s_n^2 + \hat{v}_n^2, \tag{8.22}$$

which imply that in addition to estimating the channel's output power$(|h_{1,1}|^2 + |h_{1,2}|^2)$, the normalization factor θ also has to be estimated. According to [58], the normalization factor θ can be estimated by exploiting the knowledge of the received signal's auto-correlation function given by

$$\theta \approx \sqrt{\mathbf{Y}_n \mathbf{Y}_n^H / (|h_{1,1}|^2 + |h_{1,2}|^2)}. \tag{8.23}$$

Hence, again simple ML estimation can be performed to recover the transmitted QAM symbols.

8.3.2 DOSTBCs for Four Transmit Antennas

Retaining the property of low-complexity single-stream ML detection is desirable, especially for systems employing a high number of transmit antennas. There are a number of solutions in the literature [56, 59], which are capable of achieving full spatial diversity. However, these reported schemes only work for real-valued constellations, such as BPSK. Let us now assume that we have $M = 4$ transmit and $N = 1$ receive antennas. The well-known $\mathbf{G}_4(t^1, t^2, t^3, t^4)$ space-time matrix of [25, 56] is defined as

$$\mathbf{G}_4(t^1, t^2, t^3, t^4) = \begin{pmatrix} t^1 & t^2 & t^3 & t^4 \\ t^2 & -t^1 & -t^4 & t^3 \\ t^3 & t^4 & -t^1 & -t^2 \\ t^4 & -t^3 & t^2 & -t^1 \end{pmatrix}, \tag{8.24}$$

where we only retain the first four out of the eight time slots of the original $\mathbf{G}_4$ matrix, owing to the unitary constraint required by the differential encoding process.

Following the two-antenna-based philosophy of Equation (8.15), given the symbol vector $\mathbf{K}_n = [s_n^1, s_n^2, s_n^3, s_n^4]^{\mathrm{T}}$, the differential encoding of the original work [59] can be rewritten as

$$\underbrace{\mathbf{G}_4(t_n^1, t_n^2, t_n^3, t_n^4)}_{\mathbf{S}_n} = \underbrace{\mathbf{G}_4(t_{n-1}^1, t_{n-1}^2, t_{n-1}^3, t_{n-1}^4)}_{\mathbf{S}_{n-1}} \cdot \underbrace{\mathbf{G}_4(s_n^1, s_n^2, s_n^3, s_n^4)}_{\mathbf{X}_n}. \tag{8.25}$$

Again, the transmission matrix $\mathbf{S}_n$ and the set of space-time coded matrices $\mathbf{X}_n$ all accrue from the unitary matrix $\mathbf{G}_4$ of Equation (8.24). The reference matrix $\mathbf{U}_n = \mathbf{G}_4(t_n^1, t_n^2, t_n^3, t_n^4)$ transmits four arbitrary BPSK symbols during the first signaling block of channel use.

After receiving the nth signal matrix $\mathbf{Y}_n = [y_n^1, y_n^2, y_n^3, y_n^4]$, the orthogonality of the $\mathbf{G}_4$ scheme enables the receiver to separate the transmitted MIMO symbol streams into single decodable symbols after the following linear combination:

$$\begin{pmatrix} \bar{s}_n^1 \\ \bar{s}_n^2 \\ \bar{s}_n^3 \\ \bar{s}_n^4 \end{pmatrix} = \begin{pmatrix} y_{n-1}^1 & y_{n-1}^2 & y_{n-1}^3 & y_{n-1}^4 \\ -y_{n-1}^2 & y_{n-1}^1 & y_{n-1}^4 & -y_{n-1}^3 \\ -y_{n-1}^3 & -y_{n-1}^4 & y_{n-1}^1 & y_{n-1}^2 \\ -y_{n-1}^4 & y_{n-1}^3 & -y_{n-1}^2 & y_{n-1}^1 \end{pmatrix} \cdot \begin{pmatrix} y_n^1 \\ y_n^2 \\ y_n^3 \\ y_n^4 \end{pmatrix}. \tag{8.26}$$

Hence, the ith $(i = 1, 2, 3, 4)$ decision statistics can be derived by

$$\tilde{s}_n^i = (|h_{1,1}|^2 + |h_{1,2}|^2 + |h_{1,3}|^2 + |h_{1,4}|^2) s_n^i + \hat{v}_n^i, \tag{8.27}$$

where a transmit diversity order of $D_{tx} = 4$ is attained.

Again, note that the original $\mathbf{G}_4$ matrix in [25] occupies eight time slots, but the $\mathbf{G}_4$ matrix adopted in Equation (8.24) only preserves the first four rows in the interest of satisfying the unitary constraint of Equation (8.5). The absence of the conjugate complex operation in $\mathbf{G}_4$ of Equation (8.24) prevents the transmission of complex-valued symbols, since the presence of the conjugation is essential for decoupling the complex-valued transmitted symbols. As an alternative, the simple sub-optimum method of [59] is adopted, which merges the two real-valued dimensions constituted by two transmit antennas into the real and imaginary parts and then both of the resultant complex-valued symbols of $s_n^1 + is_n^2$ and $s_n^3 + is_n^4$ are transmitted by the four antennas. For each antenna, PAM is employed.

8.3.3 DOSTBCs based on QOSTBCs

In order to support high-rate communications, while benefiting from simple ML detection, the family of Differential Quasi-Orthogonal Space-Time Block Codes (DQOSTBCs) was proposed in [60, 61].

In order to support complex-valued modulated constellations in the context of four-antenna-aided DOSTBCs in Section 8.3.2, the family of DQOSTBCs relax the orthogonality of the space-time matrix. Hence, the transmitted MIMO symbol streams can only be decoupled into a combination of a number of symbols, rather than into individual symbols, where each symbol combination contains two symbols. More explicitly, a (4×4)-element QOSTBC matrix [14] is defined as follows:

$$\mathbf{G}_{Q4}(t^1, t^2, t^3, t^4) = \begin{pmatrix} t^1 & -t^{2^*} & t^3 & -t^{4^*} \\ t^2 & t^{1^*} & t^4 & t^{3^*} \\ -t^{3^*} & t^4 & t^{1^*} & -t^2 \\ -t^{4^*} & -t^3 & t^{2^*} & t^1 \end{pmatrix} = \begin{pmatrix} \mathbf{G}_2(t^1, t^2) & \mathbf{G}_2(t^3, t^4) \\ -\mathbf{G}_2^*(t^3, t^4) & \mathbf{G}_2^*(t^1, t^2) \end{pmatrix}. \tag{8.28}$$

Observe that $\mathbf{G}_{Q4}$ of Equation (8.28) is constructed using $\mathbf{G}_2$ matrices. Compared with the $\mathbf{G}_4$ matrix of Equation (8.24) employed for real-valued DOSTBCs, $\mathbf{G}_{Q4}$ of Equation (8.28) introduces the conjugate operation, which potentially enables the employment of complex-valued constellations. In other words, $\mathbf{G}_{Q4}$ of Equation (8.28) can be considered as the 'complex-valued' version of $\mathbf{G}_4$, which is capable of supporting a high throughput.

Given a symbol vector $\mathbf{K}_n = [s_n^1, s_n^2, s_n^3, s_n^4]^{\mathrm{T}}$ and assuming that $s_n^1, s_n^2, s_n^3, s_n^4$ are drawn from the same L-PSK constellation, the differential encoding process [61] of generating the nth transmission matrix $\mathbf{S}_n$ can be expressed as

$$\underbrace{\mathbf{G}_{Q4}(t_n^1, t_n^2, t_n^3, t_n^4)}_{\mathbf{S}_n} = \underbrace{\mathbf{G}_{Q4}(t_{n-1}^1, t_{n-1}^2, t_{n-1}^3, t_{n-1}^4)}_{\mathbf{S}_{n-1}} \cdot \underbrace{\mathbf{G}_{Q4}(s_n^1, s_n^2, s_n^3, s_n^4)}_{\mathbf{X}_n}. \tag{8.29}$$

It is necessary to point out that although $\mathbf{G}_{Q4}$ of Equation (8.28) itself is not a unitary matrix, it is constructed from the unitary matrix $\mathbf{G}_2$ of Equation (8.13). Hence, the multiplicative operation of Equation (8.29) is still valid.

Given the received signal matrix $\mathbf{Y}_n = [y_n^1, y_n^2, y_n^3, y_n^4]$, the receiver performs the following linear operations:

$$\begin{pmatrix} \bar{s}_n^1 \\ \bar{s}_n^2 \\ \bar{s}_n^3 \\ \bar{s}_n^4 \end{pmatrix} = \begin{pmatrix} y_{n-1}^1 & y_{n-1}^2 & y_{n-1}^3 & y_{n-1}^4 \\ -y_{n-1}^{2^*} & -y_{n-1}^{1^*} & y_{n-1}^{4^*} & -y_{n-1}^{3^*} \\ y_{n-1}^{3^*} & y_{n-1}^{4^*} & -y_{n-1}^{1^*} & -y_{n-1}^{2^*} \\ y_{n-1}^4 & -y_{n-1}^3 & -y_{n-1}^2 & y_{n-1}^1 \end{pmatrix} \begin{pmatrix} y_n^1 \\ y_n^{2^*} \\ y_n^{3^*} \\ y_n^4 \end{pmatrix}. \tag{8.30}$$

If we define

$$\phi_1 = |s_n^1|^2 + |s_n^2|^2 + |s_n^3|^2 + |s_n^4|^2, \tag{8.31}$$

$$\phi_2 = 2\,Real\{s_n^1 s_n^{4^*} - s_n^2 s_n^{3^*}\}, \tag{8.32}$$

$$\phi_3 = |h_{1,1}|^2 + |h_{2,1}|^2 + |h_{3,1}|^2 + |h_{4,1}|^2, \tag{8.33}$$

$$\phi_4 = 2\,Real\{h_{1,1}h_{4,1}^* - h_{2,1}h_{3,1}^*\}, \tag{8.34}$$

as well as

$$\psi_1 = \phi_1\phi_3 + \phi_2\phi_4, \tag{8.35}$$

$$\psi_2 = \phi_1\phi_4 + \phi_2\phi_3, \tag{8.36}$$

the resultant decision statistics after the linear combination of Equation (8.30) become

$$\bar{s}_n^1 = \psi_1 s_n^1 + \psi_2 s_n^4 + \hat{v}_n^1, \tag{8.37}$$
$$\bar{s}_n^2 = \psi_1 s_n^2 - \psi_2 s_n^3 + \hat{v}_n^2, \tag{8.38}$$
$$\bar{s}_n^3 = \psi_1 s_n^3 - \psi_2 s_n^2 + \hat{v}_n^3, \tag{8.39}$$
$$\bar{s}_n^4 = \psi_1 s_n^4 + \psi_2 s_n^1 + \hat{v}_n^4, \tag{8.40}$$

where $\hat{v}_n^i$ denotes the corresponding combined noise. Observe that the transmit symbols have been separated into two streams, namely $[s_n^1, s_n^4]$ and $[s_n^2, s_n^3]$. Hence, ML decoding can be employed to jointly decode each symbol group. In other words, DOSTBCs based on the quasi-orthogonal matrix can be considered as an alternative design, which achieves a higher rate than the DOSTBCs based on the orthogonal structure with a higher decoding complexity.

Note that the factor ϕ_1 of Equation (8.31) is a constant value, given that an L-PSK constellation is employed, while ϕ_3 of Equation (8.33) contributes the fourth spatial diversity order. However, the existence of ϕ_2 and ϕ_4 as the result of the relaxed orthogonality imposes interference on the decision statistics of Equations (8.37), (8.38), (8.39), (8.40), which potentially degrades the attainable diversity advantage. In fact, DQOSTBCs can only achieve second-order spatial transmit diversity [60] for a MIMO system employing $M = 4$ transmit antennas.

Recall that, as mentioned in Section 7.3.3, QOSTBCs are capable of achieving full diversity by introducing a constellation rotation [298, 299]. Similarly, this technique can also be applied to the family of DQOSTBCs. For example, the symbol pair $[s_n^1, s_n^4]$ is chosen from the conventional L-PSK constellation, whereas the symbols $[s_n^2, s_n^3]$ are chosen from the L-PSK constellation rotated by $\pi/4$ degrees. However, the diversity gain improvement accrues from the modulation diversity [299], rather than from the spatial diversity addressed in this chapter.

8.3.4 DOSTBCs based on LSTBCs and SSD-STBCs

The family of coherently detected LSTBCs [30] detailed in Section 7.3.4 and based on the so-called dispersion structure of Equation (7.32) are another candidate to design the set of space-time coded matrices $\mathbf{X}_n$ of Figure 8.2. We refer to these as the Differential Linear Space-Time Block Codes (DLSTBCs). The orthogonality of the set of dispersion matrices guarantees that the L-PSK symbols transmitted in parallel can be decoupled into individual single-stream symbols.

More explicitly, given the nth symbol vector $\mathbf{K}_n = [s_n^1, s_n^2, \ldots, s_n^Q]^{\mathrm{T}}$ containing Q single-stream symbols, the corresponding space-time coded matrix $\mathbf{X}_n$ may be expressed as

$$\mathbf{X}_n = \sum_{q=1}^{Q} (\alpha_n^q \mathbf{A}_q + j\beta_n^q \mathbf{B}_q), \tag{8.41}$$

where $s_n^q = \alpha_n^q + j\beta_n^q$. The dispersion matrices $\mathbf{A}_q$ and $\mathbf{B}_q$ have a size of $(M \times M)$ elements. According to Section 7.3.4, in order for the dispersion matrices to facilitate the separation of the transmission symbols, the set of matrices has to satisfy:

$$\mathbf{A}_{i_1}\mathbf{A}_i^H = \mathbf{I}, \quad \mathbf{B}_{i_1}\mathbf{B}_i^H = \mathbf{I} \quad (i = i_1), \tag{8.42}$$
$$\mathbf{A}_{i_1}\mathbf{A}_i^H = -\mathbf{A}_i\mathbf{A}_{i_1}^H, \quad \mathbf{B}_{i_1}\mathbf{B}_i^H = -\mathbf{B}_i\mathbf{B}_{i_1}^H \quad (i \neq i_1), \tag{8.43}$$
$$\mathbf{A}_{i_1}\mathbf{B}_i^H = \mathbf{B}_i\mathbf{A}_{i_1}^H \quad (1 < i, i_1 < Q). \tag{8.44}$$

Table 8.1: System parameters for the DOSTBC scheme of Figure 8.2.

Modulation	L-PSK or L-QAM
Mapping	Gray mapping
Number of transmit antennas (M)	2
Number of receive antennas (N)	1
Normalized Doppler frequency	$f_d = 10^{-2}$ or 10^{-3}
Detector	ML of Equation (8.9)

It is straightforward to check that under the above orthogonality constraints $\mathbf{X}_n$ is always a unitary matrix. Hence, the differential encoding of Equation (8.5) can be performed. The reference matrix is given by $\mathbf{U}_n = \mathbf{I}$, where $\mathbf{I}$ denotes an identity matrix having a size of $(M \times M)$.

Again, the orthogonality of the dispersion matrices allows us to avoid the joint detection of Q symbols. Instead, each transmitted symbol can be estimated individually by calculating [314]

$$\bar{s}_n^q = \arg\left\{ \min_{s_n^q \in L-PSK} (Real\{\mathrm{tr}(\mathbf{Y}_n^H \mathbf{Y}^{n-1} \mathbf{A}_q)\}\alpha_n^q + Real\{\mathrm{tr}(\mathbf{Y}_n^H \mathbf{Y}^{n-1} j\mathbf{B}_q)\}\beta_n^q)\right\}. \tag{8.45}$$

The SSD-STBCs detailed in Section 7.3.5 obey the same dispersion structure as that of LSTBCs. Furthermore, these two classes of STBCs share the same set of constraints on the dispersion matrices expressed as in Equations (8.42), (8.43) and (8.44). The only difference is that SSD-STBCs require $i \neq i_1$ in Equation (8.44). It is plausible that the resultant space-time coded matrices $\mathbf{X}_n$ remain unitary, therefore the standard differential encoding/decoding processes are still applicable. The resultant schemes are often referred to as Differential Single-Symbol-Decodable Space-Time Block Codes (DSSD-STBC) [315] in the literature.

8.3.5 Performance Results

In this section, we present our BER simulation results for a number of DOSTBC schemes outlined in Sections 8.3.1 to 8.3.4. All of the system parameters are listed in Table 8.1, unless otherwise specified.

Figure 8.3 portrays the BER performance of a family of DOSTBCs having $M = 2$ transmit and $N = 1$ receive antennas employing various modulation schemes, when communicating over i.i.d. Rayleigh fading channels having a normalized Doppler frequency of $f_d = 10^{-3}$. The orthogonal generator matrices of Equations (8.13) and (8.19) were used. In order to gauge the achievable BER performance, the coherently detected Alamouti's scheme [11] using BPSK modulation is plotted as the benchmark. Observe in Figure 8.3 that the DOSTBC scheme employing BPSK modulation suffers a 3 dB SNR loss compared with its coherent counterpart, owing to the doubled noise variance discussed in Section 8.2. Naturally, DOSTBCs employing high-order modulation schemes require more transmit power in order to maintain a certain BER performance. Observe in Figure 8.3 further that all of the curves exhibit similar shapes, which implies that all of them are capable of achieving a full transmit diversity order of $D_{tx} = 2$. The SNR advantage of L-QAM over the corresponding L-PSK constellation renders the QAM scheme our preferred choice in the

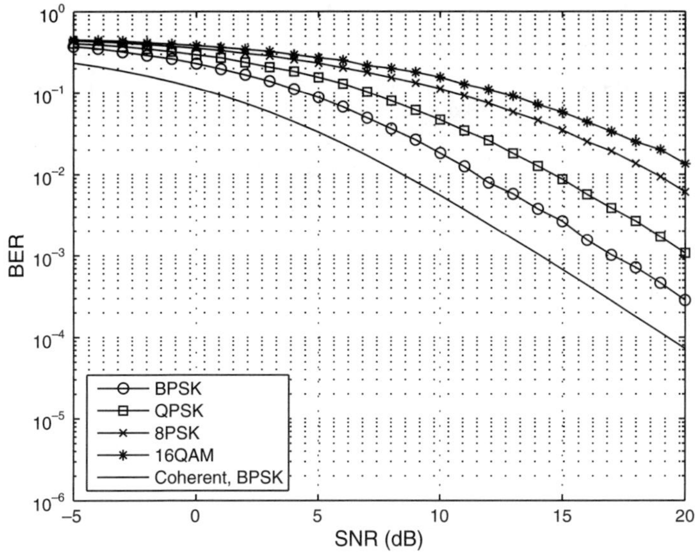

Figure 8.3: BER of a family of DOSTBCs of Figure 8.2 using the generator matrices of Equations (8.13) and (8.19), when employing various modulation schemes and communicating over i.i.d. Rayleigh fading channels having $f_d = 10^{-3}$. All of the system parameters are summarized in Table 8.1.

case of high-rate communications, which is a benefit of having larger distances among the QAM constellation points at a given average power.

Figure 8.4 characterizes the family of DOSTBCs using the generator matrices of Equations (8.13) and (8.24) employing $M = 1, 2, 3, 4$ transmit and $N = 1$ receive antennas, when communicating over Rayleigh fading channels having a normalized Doppler frequency of $f_d = 10^{-3}$. The corresponding differential encoding/decoding processes are described in Sections 8.3.1–8.3.4. Observe that a diversity order of $D = 1, 2, 3$ and 4 is achieved, respectively. Unsurprisingly, less transmit power is required for DOSTBCs having a higher-order diversity to maintain a specific BER level, which is a benefit of the higher complexity associated with the increased number of antennas.

In Figure 8.5, we characterize a class of BPSK-modulated DOSTBCs using the generator matrix of Equation (8.13) employing $M = 2$ transmit and $N = 1, 2, 3, 4$ receive antennas, for transmission over Rayleigh fading channels encountering a normalized Doppler frequency of $f_d = 10^{-3}$. Observe that the increased number of receive antennas substantially boosts the total attainable spatial diversity, since a diversity order of $D = 2, 4, 6, 8$ is achieved.

The effect of temporal channel correlation is quantified in Figure 8.6, where we employ a BPSK-modulated DOSTBC scheme having $M = 2$ transmit and $N = 1$ receive antennas. The grade of temporal channel correlation is governed by the normalized Doppler frequencies f_d. Ideally, the differential decoding procedure requires two consecutive transmission blocks to experience the same fading envelope. For example, when each block occupies $T = 2$ time slots, the fading has to be constant for four time slots, while for the four-transmit-antenna-aided DOSTBC requiring $T = 4$ time slots, a constant fading envelope has to be experienced for eight time slots. However, when we encounter a rapidly fading environment, this requirement may not be satisfied. Figure 8.6 explicitly plots the BER degradation imposed by rapid fading. When the channels exhibit relatively slow fading, i.e. $f_d = 10^{-3}$,

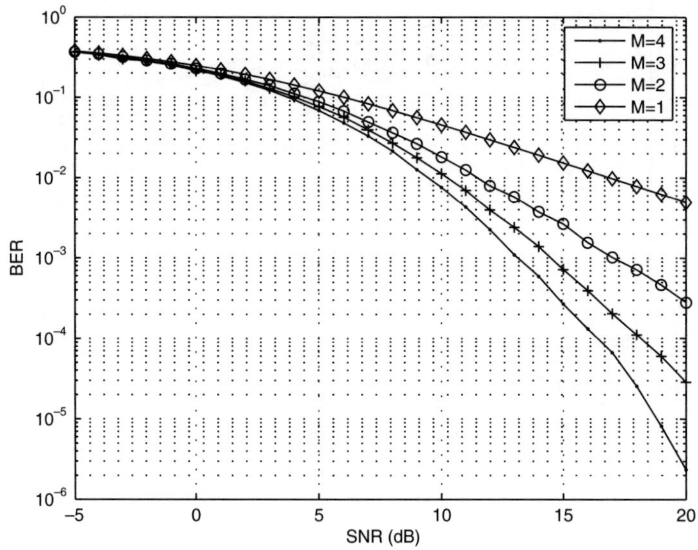

Figure 8.4: BER of a family of BPSK-modulated DOSTBCs of Figure 8.2 using the generator matrices of Equations (8.13) and (8.24) having $M = 1, 2, 3, 4$ and $N = 1$ antennas, when communicating over i.i.d. Rayleigh fading channels having $f_d = 10^{-3}$. All of the system parameters are summarized in Table 8.1.

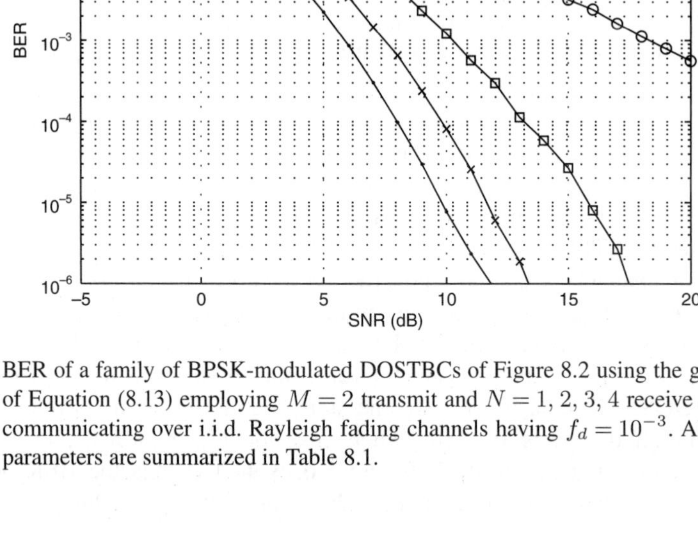

Figure 8.5: BER of a family of BPSK-modulated DOSTBCs of Figure 8.2 using the generator matrix of Equation (8.13) employing $M = 2$ transmit and $N = 1, 2, 3, 4$ receive antennas, when communicating over i.i.d. Rayleigh fading channels having $f_d = 10^{-3}$. All of the system parameters are summarized in Table 8.1.

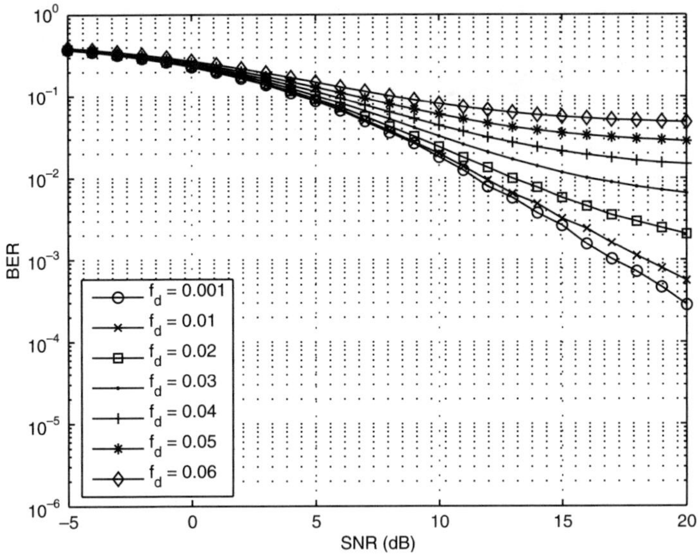

Figure 8.6: BER of a BPSK-modulated DOSTBC scheme of Figure 8.2 using the generator matrix of Equation (8.13), when communicating over i.i.d. Rayleigh fading channels having various normalized Doppler frequencies f_d. All of the system parameters are summarized in Table 8.1.

the best attainable BER performance is recorded. On the other hand, when the normalized fading rate reaches $f_d = 10^{-2}$, a significant BER degradation has already been observed. As for very rapid fading, such as $f_d = 0.04$, a BER floor above 10^{-2} has occurred. This implies that MIMO channels having a high coherence time $\Im$ potentially improve the performance of DOSTBCs in comparison with rapid fading. Indeed, the non-coherent MIMO channel's capacity of Equation (8.12) approaches the coherent MIMO channel's capacity expressed in Equation (8.11), when the value of $\Im$ is increased.

All of the above results are based on the assumption that all of the channels are faded independently. However, this assumption may not be readily satisfied, owing to the lack of sufficient antenna spacing at the mobile handsets. Hence, Figure 8.7 investigates the attainable BER performance of the BPSK-modulated DOSTBC scheme having $M = 2$ and $N = 1$ antennas, when communicating over spatially correlated Rayleigh fading channels encountering a normalized Doppler frequency of $f_d = 10^{-3}$. The spatial correlation model proposed in [316] was adopted, where the inter-antenna correlation is determined by the correlation coefficient δ. Furthermore, $\delta = 1$ implies that the two antennas' signals are perfectly correlated, while $\delta = 0$ represents the channels that are independently faded. Observe in Figure 8.7 that when we have $\delta = 0$, the best BER performance is observed. When the channel's fading correlation becomes higher, the associated BER degradation becomes more significant. When the spatial channels become identical as characterized by $\delta = 1$, the associated BER performance becomes identical to a single-antenna-aided-scheme using the differential BPSK constellation.

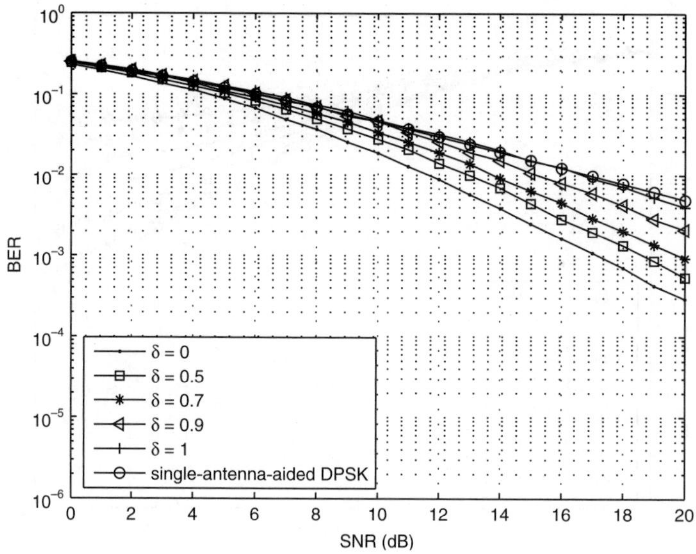

Figure 8.7: The effect of spatial correlation for a BPSK-modulated DOSTBC scheme of Figure 8.2 using the generator matrix of Equation (8.13), when communicating over Rayleigh fading channels having a normalized Doppler frequency of $f_d = 10^{-3}$. All of the system parameters are summarized in Table 8.1.

8.3.6 Summary

At this stage, it is beneficial to summarize the advantages and drawbacks of DOSTBCs detailed in Sections 8.3.1–8.3.4, before we move on to the more general structure of DLDCs. More explicitly, the benefits and constraints are summarized as follows.

- The number of time slots T used per transmission block has to be equal to M, a constraint which is imposed by the differential encoding structure of Equation (8.5).

- The number of symbols Q transmitted per space-time block is determined by the degree of orthogonality possessed by the space-time matrix $\mathbf{S}_n$.

- The number of transmit antennas M used in Figure 8.2 is limited and this constraint was imposed by the orthogonal design.

- The number of receive antennas N in Figure 8.2 is unrestricted.

- The power-related expressions have to be estimated, as exemplified by the channel's output power in Section 8.3.1.1 and the normalization factors considered in Section 8.3.3.

Note that apart from the constraint of $T = M$, all of the other constraints are imposed by the orthogonal structure, such as Equations (8.13), (8.24) and (8.28). Based on the general LDC framework detailed in Chapter 7, in the following section we propose a family of DLDCs which do not impose any constraints on the parameters Q and M of Figure 8.2 and avoid the estimation of the power-related factors in Sections 8.3.1.1 and 8.3.3.

8.4 DLDCs

The family of DLDCs proposed in this section inherits the design flexibility of the LDCs by adopting a non-orthogonal structure for the STC matrix. More particularly, in analogy to its non-differential counterpart, a DLDC($MNTQ$) scheme describes a differential coding scheme suitable for an $(M \times N)$-element MIMO configuration, while having a symbol rate of $R_{\text{DLDC}} = Q/T$.

8.4.1 Evolution to a Linear Structure

One of the early attempts in the literature to address the challenge of designing a set of unitary matrices $\mathbf{X}_n$ for the scheme of Figure 8.2 is constituted by the family of DUSTM schemes proposed simultaneously in [55, 63, 308]. A distinctive feature of DUSTM is that no orthogonality is imposed on the space-time coded matrix. The DUSTM schemes proposed in [55] and [308] were shown to have a good BER performance [208] when the constellation of matrices used for transmission form a group under the matrix multiplication of Equation (8.5). However, the philosophy of the DUSTMs is to directly design $F = L^Q$ space-time matrices $\mathbf{X}_n$ by maximizing the rank criterion of Equation (7.26) and the determinant criterion of Equation (7.27). Hence, this design problem becomes intractable, since the objective function may be non-convex and the associated search space increases exponentially with both the constellation size and the number of antennas. Although substantial efforts have been made [208, 317] to simplify the design of differential space-time constellation matrices, the optimization of high-order DUSTM schemes remains a challenging problem.

In Chapter 7, we have demonstrated that the family of LDCs transforms the design of $F = L^Q$ space-time matrices to the design of a single DCM χ, as seen in Equation (7.16). The DCM fully and uniquely specifies the characteristics of a particular LDC scheme. Similarly, by introducing the linear dispersion structure of Equation (8.46) into the design of differential space-time transmission schemes, the resultant DLDCs can also be fully described by a single DCM χ. Therefore, the problem of designing F space-time matrices for describing the DUSTMs can be simplified to that of designing a single matrix χ for the DLDCs.

Furthermore, Figure 8.8 illustrates the classification of various DSTBCs and highlights that the design of STBCs and DSTBCs is linked by the unitary constraint of Equation (8.5). Taking into account the discussions of Section 7.3, it is straightforward to show that DLDCs subsume the family of DOSTBCs, which impose different degrees of orthogonality on the space-time matrix. With the aid of the non-orthogonal structure of Equation (8.46), DLDCs become capable of supporting arbitrary $(M \times N)$ element MIMO antenna configurations and of supporting any Q value. More explicitly, given a symbol vector $\mathbf{K}_n = [s_n^1, \ldots, s_n^Q]^{\text{T}}$, the linear dispersion structure of the space-time coded matrix $\tilde{\mathbf{X}}_n \in \zeta^{M\times M}$ can be written as

$$\tilde{\mathbf{X}}_n = \sum_{q=1}^{Q} \mathbf{A}_q s_n^q, \tag{8.46}$$

which is a weighted superposition of the corresponding dispersion matrices $\mathbf{A}_q \in \zeta^{M\times M}$. The resultant linearly combined matrix $\tilde{\mathbf{X}}_n$ has to be unitary. All of the $\mathbf{A}_q$ matrices can be specified using the DCM χ defined by

$$\chi = [vec(\mathbf{A}_1), vec(\mathbf{A}_2), \ldots, vec(\mathbf{A}_Q)]. \tag{8.47}$$

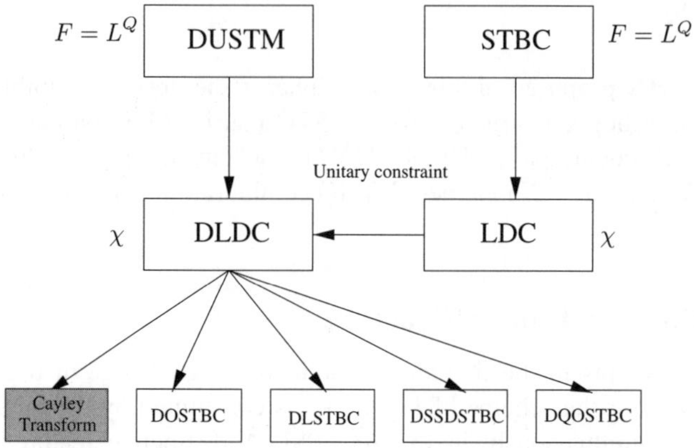

Figure 8.8: Classification of DSTBC techniques. The corresponding coherently detected schemes were classified in Figure 7.16.

However, the construction of a unitary matrix $\tilde{\mathbf{X}}_n$ obeying the linear combination structure of Equation (8.46) is challenging, because even if each individual dispersion matrix $\mathbf{A}_q$ is a unitary matrix, there is no guarantee that their weighted sum $\tilde{\mathbf{X}}_n$ will automatically be a unitary matrix.

8.4.2 Differential LDCs based on the Cayley Transform

The challenge of constructing a set of unitary matrices for differential encoding was addressed in [10,318], where the Cayley transform was employed. In Figure 8.9, we show the schematic of the DLDC scheme based on the Cayley transform. Compared with the general DSTBC framework of Figure 8.2, the Cayley transform [10, 319] of Figure 8.9 provides an efficient way of projecting the linearly structured matrix $\tilde{\mathbf{X}}_n$ into a unique unitary matrix $\mathbf{X}_n$, which potentially facilitates the differential encoding. Hence, the DLDCs generated using the Cayley transform remain a subset of the family of DLDCs, as seen in Figure 8.8.

8.4.2.1 The Cayley Transform

The philosophy behind DLDCs based on the Cayley transform is to design a set of linear-structured Hermitian space-time matrices $\tilde{\mathbf{X}}_n$ obeying Equation (8.46) and to uniquely map the resultant Hermitian matrices to the unitary matrices $\mathbf{X}_n$ of Figure 8.9. Designing Hermitian space-time matrices is attractive, since they are linear and hence the weighted sum of a number of Hermitian matrices also remains a Hermitian matrix, provided that the weighting coefficients are real valued.

The Cayley transform creates a unique mapping between a Hermitian matrix and a unitary matrix. More explicitly, the Cayley transform of the complex-valued $(M \times M)$-element matrix $\tilde{\mathbf{X}}_n$ of Equation (8.46) is defined as [10]

$$\mathbf{X}_n = (\mathbf{I} - j\tilde{\mathbf{X}}_n)(\mathbf{I} + j\tilde{\mathbf{X}}_n)^{-1}, \tag{8.48}$$

where $\mathbf{I}$ is the $(M \times M)$-element identity matrix and $j\tilde{\mathbf{X}}_n$ is assumed to have no eigenvalues at -1 in order to ensure that the $(\mathbf{I} + j\tilde{\mathbf{X}}_n)^{-1}$ item of Equation (8.48) exists. Note that $(\mathbf{I} - j\tilde{\mathbf{X}}_n)$, $(\mathbf{I} + j\tilde{\mathbf{X}}_n)$, $(\mathbf{I} - j\tilde{\mathbf{X}}_n)^{-1}$ and $(\mathbf{I} + j\tilde{\mathbf{X}}_n)^{-1}$ all commute. Hence, there are other

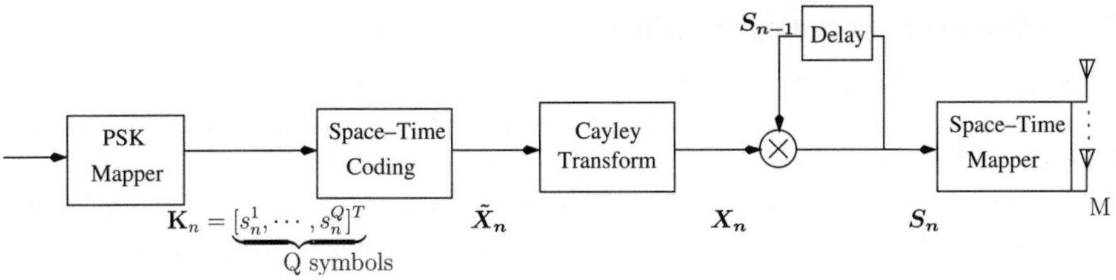

Figure 8.9: Schematic of a DLDC($MNTQ$) scheme equipped with M transmit antennas that employs the Cayley transform, while transmitting Q symbols over T time slots using $\mathbf{S}_n$.

equivalent ways to write this transform. The resultant matrix $\mathbf{X}_n$ is unitary, because

$$\begin{aligned}\mathbf{X}_n\mathbf{X}_n^H &= (\mathbf{I} - j\tilde{\mathbf{X}}_n)(\mathbf{I} + j\tilde{\mathbf{X}}_n)^{-1}[(\mathbf{I} - j\tilde{\mathbf{X}}_n)(\mathbf{I} + j\tilde{\mathbf{X}}_n)^{-1}]^H \\ &= (\mathbf{I} - j\tilde{\mathbf{X}}_n)(\mathbf{I} + j\tilde{\mathbf{X}}_n)^{-1}(\mathbf{I} - j\tilde{\mathbf{X}}_n)^{-1}(\mathbf{I} + j\tilde{\mathbf{X}}_n) \\ &= \mathbf{I},\end{aligned} \tag{8.49}$$

where we exploit the fact that $\tilde{\mathbf{X}}_n$ is Hermitian. The employment of the Cayley transform appears promising, because it is a unique one-to-one transform, which can be inverted to yield

$$j\tilde{\mathbf{X}}_n = (\mathbf{I} - \mathbf{X}_n)(\mathbf{I} + \mathbf{X}_n)^{-1}, \tag{8.50}$$

provided that the inverse exists or, equivalently, assuming that $j\tilde{\mathbf{X}}_n$ has no eigenvalue at -1. Some important properties of the Cayley transform are listed as follows [10].

- **Unique mapping.** A matrix with no eigenvalues at -1, such as $j\tilde{\mathbf{X}}_n$ of Equation (8.48), is unitary if and only if its Cayley transform is a skewed-Hermitian matrix. (A square matrix $\mathbf{A}$ is said to be a skewed-Hermitian matrix if it satisfies the relation $\mathbf{A}^H = -\mathbf{A}$. A skewed-Hermitian matrix can be obtained by $j\mathbf{A}$, provided that $\mathbf{A}$ is a Hermitian matrix.)

- **Eigenvalues.** The matrix $j\tilde{\mathbf{X}}_n$ and its Cayley transform $\mathbf{X}_n$ commute. Hence, they have the same eigenvectors, while their eigenvalues denoted by μ_i and ∇_i obey

$$\mu_i = \frac{1 - \nabla_i}{1 + \nabla_i}. \tag{8.51}$$

- **Diversity.** A set of unitary matrices $\mathbf{X}_n$ is referred to as fully diverse, which hence satisfies the rank criterion of Equation (7.26), if and only if its Cayley transform $j\tilde{\mathbf{X}}_n$ is also fully diverse. This allows the resultant DLDC schemes to achieve full diversity.

For the proofs of these properties, please refer to [10].

Hence, the Cayley transform is capable of preserving the characteristics of the Hermitian space-time matrices, when projecting them to unitary matrices.

It is necessary to point out a constraint imposed by designing in the Hermitian space. More precisely, real-valued modulated symbols s_q have to be employed in Equation (8.46) to ensure that the Cayley transformed Hermitian space-time matrix does result in a linear unitary matrix.

8.4.2.2 Differential Encoding/Decoding

With the aid of the Cayley transform introduced in Section 8.4.2.1, the differential encoding/decoding process can be further refined. More explicitly, when we have a real-valued modulated symbol vector $\mathbf{K}_n = [s_1, \ldots, s_Q]^{\mathrm{T}}$, the Hermitian space-time coded matrix $\tilde{\mathbf{X}}_n$ of Equation (8.46) is projected to a unitary matrix using the Cayley transform of Equation (8.48). Thus, the resultant unitary matrix $\mathbf{X}_n$ becomes suitable for the differential encoding process of Equation (8.5).

Given the received signal matrix $\mathbf{Y}_n$ and assuming that the channel's envelope remains constant over two consecutive transmission blocks, the resultant differential relationship can be expressed as

$$\begin{aligned}\mathbf{Y}_n &= \mathbf{H}_n\mathbf{S}_{n-1}\mathbf{X}_n + \mathbf{V}_n\\ &= (\mathbf{Y}_{n-1} - \mathbf{V}_{n-1})\mathbf{X}_n + \mathbf{V}_n\\ &= \mathbf{Y}_{n-1}\mathbf{X}_n + \mathbf{V}_n - \mathbf{V}_{n-1}\mathbf{X}_n.\end{aligned} \tag{8.52}$$

Then, conventional ML estimation can be performed as

$$\bar{\mathbf{X}}_n = \arg\Big\{\min_{\mathbf{X}_n \in F}(\|\mathbf{Y}_n - \mathbf{Y}_{n-1}\mathbf{X}_n\|^2)\Big\}. \tag{8.53}$$

The above detection method requires the receiver to carry out the Cayley transform in order to generate the candidate unitary matrix $\mathbf{X}_n$, which may impose an excessive decoding complexity at the receiver.

However, we can avoid the calculation of the Cayley transform of Equation (8.48) upon multiplying both sides of Equation (8.52) by $(\mathbf{I} + j\tilde{\mathbf{X}}_n)$, yielding

$$\begin{aligned}\mathbf{Y}_n(\mathbf{I} + j\tilde{\mathbf{X}}_n) &= \mathbf{Y}_{n-1}(\mathbf{I} - j\tilde{\mathbf{X}}_n) + \mathbf{V}_n(\mathbf{I} + j\tilde{\mathbf{X}}_n) - \mathbf{V}_{n-1}(\mathbf{I} - j\tilde{\mathbf{X}}_n)\\ \underbrace{\mathbf{Y}_n - \mathbf{Y}_{n-1}}_{\hat{\mathbf{Y}}_n} &= \underbrace{-(\mathbf{Y}_n + \mathbf{Y}_{n-1})i}_{\hat{\mathbf{H}}_n}\tilde{\mathbf{X}}_n + \underbrace{-\mathbf{V}_n(\mathbf{I} + j\tilde{\mathbf{X}}_n) - \mathbf{V}_{n-1}(\mathbf{I} - j\tilde{\mathbf{X}}_n)}_{\hat{\mathbf{V}}_n},\end{aligned} \tag{8.54}$$

The equivalent received signal matrix $\hat{\mathbf{Y}}_n$ and the equivalent CIR matrix $\hat{\mathbf{H}}_n$ in Equation (8.54) are both determined by the pair of consecutive received signal matrices $\mathbf{Y}_n$ and $\mathbf{Y}_{n-1}$. The corresponding noise matrix $\hat{\mathbf{V}}_n$ has independent columns with a covariance of

$$\hat{N}_0 = 2\sigma_0(\mathbf{I} + \tilde{\mathbf{X}}_n^2). \tag{8.55}$$

Furthermore, the linearity of the Hermitian matrix $\tilde{\mathbf{X}}_n$ of Equation (8.46) enables us to apply the vertical stacking operation $vec()$ to Equation (8.54), which results in

$$\bar{\mathbf{Y}}_n = \bar{\mathbf{H}}_n\chi\mathbf{K}_n + \bar{\mathbf{V}}_n, \tag{8.56}$$

where $\bar{\mathbf{Y}}_n \in \zeta^{NT\times 1}$, $\bar{\mathbf{H}}_n \in \zeta^{NT\times MT}$, $\chi \in \zeta^{MT\times Q}$, $\mathbf{K}_n \in \zeta^{Q\times 1}$, $\bar{\mathbf{V}}_n \in \zeta^{NT\times 1}$ and $\bar{\mathbf{H}}_n$ in Equation (8.56) is given by

$$\bar{\mathbf{H}}_n = \mathbf{I} \otimes \hat{\mathbf{H}}_n. \tag{8.57}$$

Thus, the simplified differential ML detection becomes

$$\bar{\mathbf{K}}_n = \arg\Big\{\min_{\mathbf{K}_n \in F}(\|\bar{\mathbf{Y}}_n - \bar{\mathbf{H}}_n\chi\mathbf{K}_n\|)^2\Big\}. \tag{8.58}$$

This is commonly referred to as the 'linearized ML detector' in the literature [10]. Since the resultant noise variance of Equation (8.55) becomes $\hat{N}_0 > 2\sigma_0\mathbf{I}$, the 'linearized ML detector' suffers from a further performance loss compared with the conventional differential ML detection of Equation (8.53).

We have argued in Section 8.2.3 that the main challenge of designing DSTBCs is to construct a set of space-time matrices obeying the unitary constraint of Equation (8.5). From this point of view, the problem of designing DSTBCs is related to that of STBCs. Therefore, it is straightforward to show that the LDCs' diversity property formulated in Theorem 7.3 is directly applicable to the DLDCs, where the maximum achievable diversity is $N \cdot \min(M, T)$. Since the differential encoding process restricts us to $T = M$, the maximum achievable spatial diversity order of the DLDCs becomes $D = NM$. The optimization of the DSTBCs from the capacity maximization perspective is challenging, because the capacity also depends on the temporal correlation of the channel. Therefore, the conventional rank criterion of Equation (7.26) and the determinant criterion of Equation (7.27) detailed in Section 7.3.1 are adopted in order to optimize the DLDCs based on the Cayley transform. More explicitly, we randomly generate a number of Hermitian dispersion matrices Q, where each matrix element obeys the Gaussian distribution. Hence, the achievable diversity order and coding gain of this particular DLDC$(MNTQ)$ scheme, which was portrayed in Figure 8.9, are examined after the Cayley transform. The specific code having the highest diversity order as well as coding gain is selected for the particular DLDC$(MNTQ)$ scheme. Throughout our investigations, we found that typically 100 000 random dispersion matrix search steps are required to generate a DLDC having a good BER performance.

8.4.2.3 Examples of DLDCs based on the Cayley Transform

Let us consider a simple system equipped with $M = 2$ transmit as well as $N = 2$ receive antennas and having an effective throughput of $C = 1$ (BPS Hz^{-1}). Hence, a DLDC(2222) scheme obeying Figure 8.9 and transmitting $Q = 2$ BPSK-modulated symbols during $T = 2$ channel uses, is required to achieve the target throughput. According to the optimization method outlined in Section 8.4.2.2, a particular choice of two Hermitian dispersion matrices are

$$\mathbf{A}_1 = \begin{pmatrix} -0.580 & -0.262 + j0.308 \\ -0.262 - j0.308 & 0.5799 \end{pmatrix}, \tag{8.59}$$

and

$$\mathbf{A}_2 = \begin{pmatrix} 0.404 & -0.366 + j0.450 \\ -0.366 - j0.450 & -0.404 \end{pmatrix}. \tag{8.60}$$

Therefore, the nth space-time coded matrices $\tilde{\mathbf{X}}_n$ of Figure 8.9 becomes

$$\tilde{\mathbf{X}}_n = \mathbf{A}_1 s_n^1 + \mathbf{A}_2 s_n^2, \tag{8.61}$$

where we have $s_n^1, s_n^2 \in [-1, 1]$. Hence, the Cayley transform of Equation (8.48) is employed to project the weighted Hermitian matrix $\tilde{\mathbf{X}}_n$ into a unitary matrix. The resultant four

Table 8.2: System parameters for the DLDCs of Figure 8.9 based on the Cayley transform.

Modulation	2PAM
Mapping	Gray mapping
Number of transmitters (M)	2
Number of receivers (N)	2
Normalized Doppler frequency	$f_d = 10^{-2}$
Detector	ML of Equation (8.53)

candidate unitary space-time coded matrices $\mathbf{X}_n^i$ $(i = 1, 2, 3, 4)$ of Figure 8.9 are

$$\begin{aligned}
\mathbf{X}_n^1 &= \begin{pmatrix} 0 - j0.176 & -0.758 - j0.628 \\ 0.758 - j0.629 & 0 + j0.176 \end{pmatrix}, \\
\mathbf{X}_n^2 &= \begin{pmatrix} 0 + j0.176 & 0.758 + j0.628 \\ -0.758 + j0.629 & 0 - j0.176 \end{pmatrix}, \\
\mathbf{X}_n^3 &= \begin{pmatrix} 0 - j0.984 & 0.142 + j0.104 \\ -0.142 + j0.103 & 0 + j0.984 \end{pmatrix}, \\
\mathbf{X}_n^4 &= \begin{pmatrix} 0 + j0.984 & -0.142 - j0.104 \\ 0.142 - j0.103 & 0 - j0.984 \end{pmatrix}.
\end{aligned} \tag{8.62}$$

For comparison, we may consider the set of space-time coded matrices $\mathbf{X}_n^i$ $(i = 1, 2, 3, 4)$ based on orthogonal designs, as illustrated in Equation (8.15) of Section 8.3.1, which are given by

$$\begin{aligned}
\mathbf{X}_1 &= \frac{1}{\sqrt{2}}\begin{pmatrix} 1 & -1 \\ 1 & 1 \end{pmatrix}, \quad \mathbf{X}_2 = \frac{1}{\sqrt{2}}\begin{pmatrix} 1 & 1 \\ -1 & 1 \end{pmatrix}, \\
\mathbf{X}_3 &= \frac{1}{\sqrt{2}}\begin{pmatrix} -1 & 1 \\ 1 & 1 \end{pmatrix}, \quad \mathbf{X}_4 = \frac{1}{\sqrt{2}}\begin{pmatrix} -1 & -1 \\ -1 & -1 \end{pmatrix}.
\end{aligned}$$

In Figure 8.10, we plot the BER performance of both the DLDC(2222) scheme based on the Cayley transform using Equation (8.62) and that of the DOSTBC scheme of Equation (8.63), when communicating over i.i.d. Rayleigh fading channels having $f_d = 10^{-2}$. Figure 8.10 shows that at $\text{BER} = 10^{-5}$ the DLDC's performance gain over the identical-throughput DOSTBC scheme is approximately 0.9 dB. This example shows that there are good codes within the DLDC structure based on the Cayley transform at low rates. Even more importantly, the power of the DLDCs lies in its flexibility to accommodate diverse antenna configurations as well as in its ability to achieve high rates, which is further demonstrated in Section 8.4.3.

8.4.3 Performance Results

In this section, we present our simulation results for a group of DLDCs based on the Cayley transform, which are associated with various parameter combinations $(MNTQ)$. All of the system parameters are listed in Table 8.2, unless otherwise specified.

Figure 8.11 characterizes the BER performance of a family of DLDCs having the parameters of $M = 2$, $N = 2$, $T = 2$ and $Q = 1, 2, 3, 4, 5$ using a ML detector, when

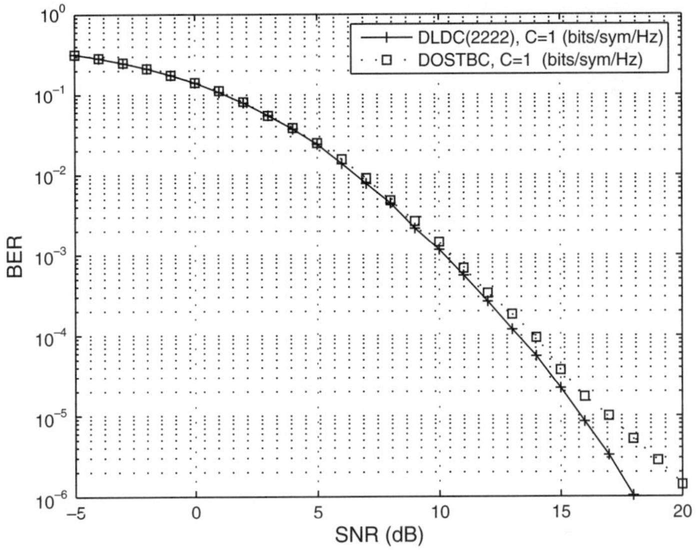

Figure 8.10: BER comparison of the DLDC(2222) of Figure 8.9 and the DOSTBC of [62] using BPSK modulation as well as a ML detector, when transmitting over i.i.d. Rayleigh fading channels having $f_d = 10^{-2}$.

transmitting over i.i.d. Rayleigh fading channels having a normalized Doppler frequency of $f_d = 10^{-2}$. The coherently detected LDC(2221) scheme of Figure 7.5 is plotted as the benchmark. Observe in Figure 8.11 that the DLDC(2221) scheme suffers from an approximately 3 dB performance loss compared with its coherently detected counterpart, where perfect CSI was assumed. We also observe in Figure 8.11 that all of the curves exhibit similar shapes, which implies that the maximum attainable diversity order of $D = 4$ has been achieved.

Figure 8.12 portrays the BER performance for a family of DLDCs having $M = 3$, $N = 2$, $T = 3$ and $Q = 1, 2, 3, 4, 5, 6$ using the differential ML detector of Equation (8.9), when transmitting over i.i.d. Rayleigh fading channels having a normalized Doppler frequency of $f_d = 10^{-2}$. This family of DLDCs is capable of achieving the maximum diversity order of $D = 6$. Similarly, the BER performance of the coherently detected LDC(3231) optimized using the method in Chapter 7 is given as a benchmark, where the usual 3 dB performance difference is observed, as a result of requiring no CSI.

Figure 8.13 further demonstrates the DLDCs' capability of achieving full spatial diversity upon employing $M = 2$ transmit and $N = 1, 2, 3$ receive antennas, when transmitting over i.i.d. Rayleigh fading channels encountering a normalized Doppler frequency of $f_d = 10^{-2}$. Upon increasing N, the total spatial diversity order achieved becomes $D = 2, 4, 6$, respectively.

A distinctive feature of the differentially encoded schemes is that the attainable BER performance is affected by the rate of fluctuation experienced by the MIMO channels, since the differential decoding of Equation (8.7) requires the channel's envelope to be constant over two consecutive transmission blocks. Figure 8.14 quantifies the SNR required for the DLDCs having $M = 2$, $N = 2$, $T = 2$ and $Q = 1, 2, 3, 4$ in order to maintain $\text{BER} = 10^{-4}$, when transmitting over i.i.d. Rayleigh fading channels encountering different normalized Doppler frequencies f_d. As expected, a higher transmission power is required for a DLDC scheme

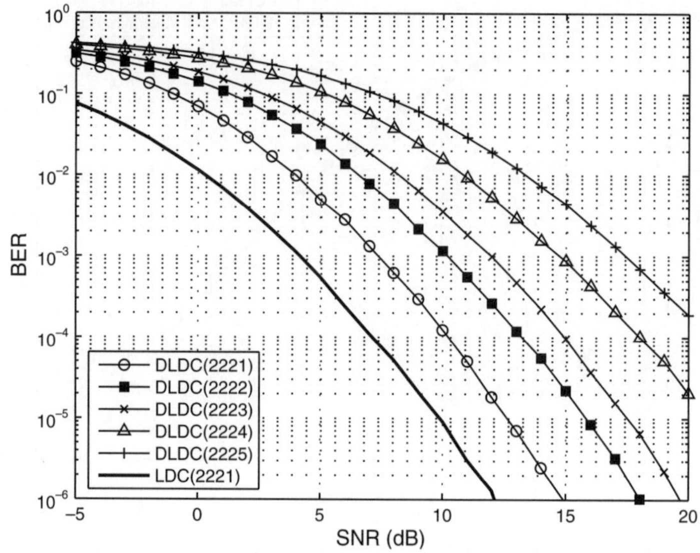

Figure 8.11: BER comparison of a family of DLDCs of Figure 8.9 having $M = 2$, $N = 2$, $T = 2$ and $Q = 1, 2, 3, 4, 5$, when transmitting over i.i.d. Rayleigh fading channels having $f_d = 10^{-2}$. All of the system parameters are summarized in Table 8.2.

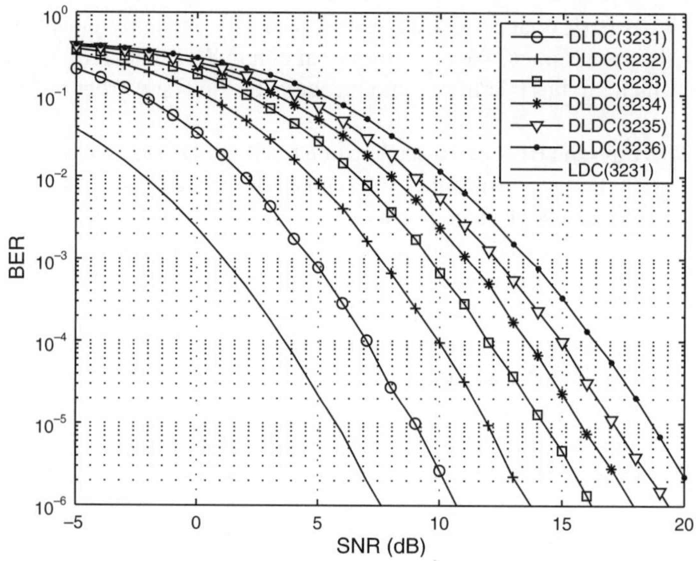

Figure 8.12: BER comparison of a family of DLDCs of Figure 8.9 having $M = 3$, $N = 2$, $T = 3$ and $Q = 1, 2, 3, 4, 5, 6$, when transmitting over i.i.d. Rayleigh fading channels having a normalized Doppler frequency of $f_d = 10^{-2}$. All of the system parameters are summarized in Table 8.2.

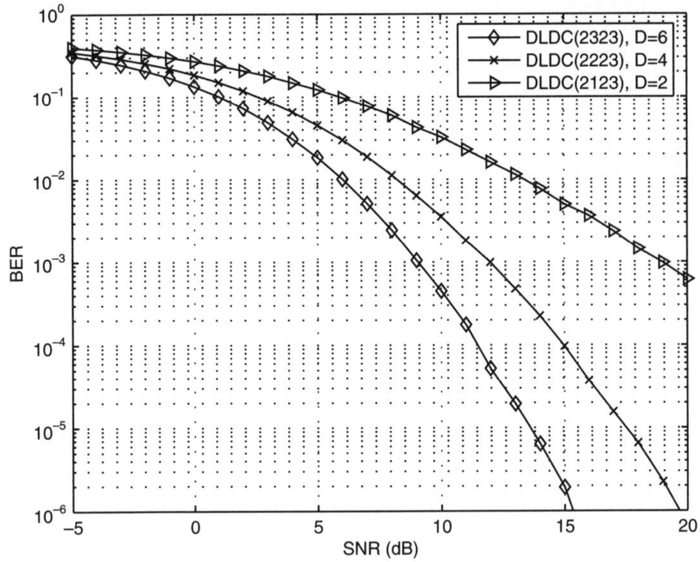

Figure 8.13: BER comparison of a family of DLDCs of Figure 8.9 having $M = 2$, $N = 1, 2, 3$, $T = 2$ and $Q = 3$, when transmitting over i.i.d. Rayleigh fading channels having $f_d = 10^{-2}$. All of the system parameters are summarized in Table 8.2.

to maintain a specific BER level, when the channel envelope fluctuates more rapidly. More specifically, an extra SNR of 11.4 dB is necessary for the DLDC(2221) scheme of Figure 8.9 to maintain $\text{BER} = 10^{-4}$, when f_d increases from 10^{-6} to 0.09. When we have $f_d > 0.09$, an error floor higher than $\text{BER} = 10^{-4}$ is observed. Also observe in Figure 8.14 that the achievable robustness against a high rate of channels' envelope fluctuation is reduced, when the throughput of the DLDC schemes increases by transmitting more symbols per space-time block. For example, an error floor higher than $\text{BER} = 10^{-4}$ occurred for the DLDC(2224) scheme having $f_d = 0.02$, whereas the DLDC(2221) arrangement is capable of maintaining the BER of 10^{-4} until the $f_d = 0.09$ value is exceeded.

The coherently detected LDCs detailed in Chapter 7 assume that the receiver has perfect CSI. Under this assumption, the differentially encoded schemes normally exhibit a 3 dB SNR performance loss, as shown in Figures 8.11 and 8.12. However, this assumption is unrealistic, owing to the inevitable channel estimation errors. Figure 8.15 explicitly quantifies the BER comparison of a LDC(3231) scheme having different degrees of channel estimation errors and the corresponding DLDC(3231) arrangement. We assume that the channel estimation errors obey the Gaussian distribution and the degree of the CSI estimation errors is governed by ω (dB) with respect to the received signal power. Hence, the perfect CSI scenario corresponds to $\omega = -\infty$. Observe in Figure 8.15 that the BER performance gradually degrades upon increasing ω. Also observe that there exists a crossover point between the BER curve of the DLDC(3231) scheme and that of its coherently detected counterpart LDC(3231) having different ω values. For example, when we have $\omega = -7$ dB, the crossover occurs at $\rho = 7$ dB. This implies that the DLDC(3231) scheme is capable of outperforming its coherently detected counterpart for SNR values in excess of $\rho > 7$ dB without the typically high complexity of MIMO channel estimation.

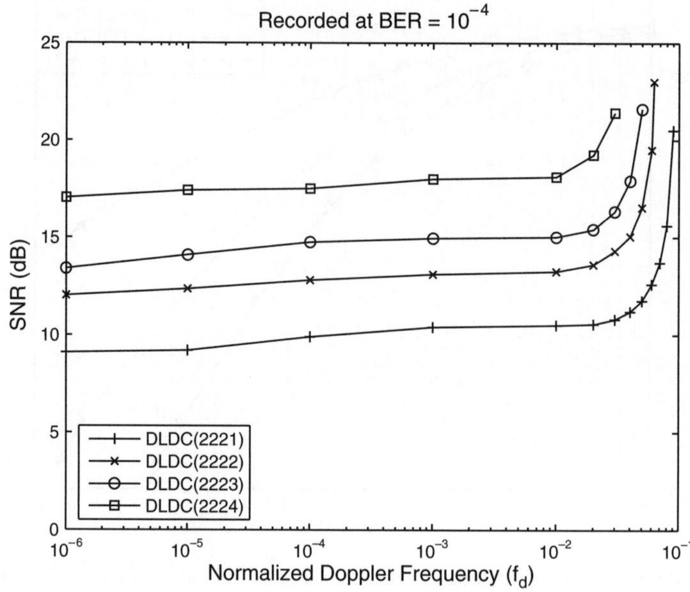

Figure 8.14: SNR required for a family of DLDCs of Figure 8.9 having $M = 2$, $N = 2$, $T = 2$ and $Q = 1, 2, 3, 4$ to maintain $\mathrm{BER} = 10^{-4}$, when transmitting over i.i.d. Rayleigh fading channels encountering various normalized Doppler frequency of f_d. All of the system parameters are summarized in Table 8.2.

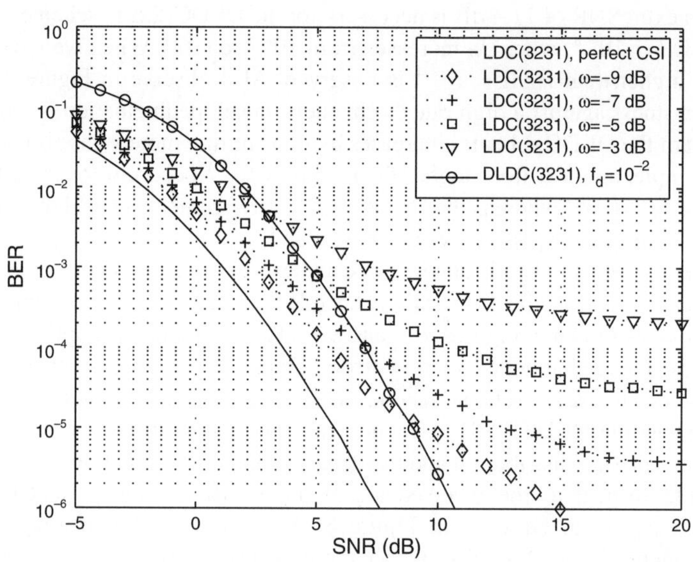

Figure 8.15: BER comparison of a LDC(3231) scheme of Figure 7.3 having imperfect CSI governed by ω and a non-coherent DLDC(3231) arrangement of Figure 8.9, when transmitting over i.i.d. Rayleigh-fading channels having a normalized Doppler frequency of $f_d = 10^{-2}$. All of the system parameters are summarized in Table 8.2.

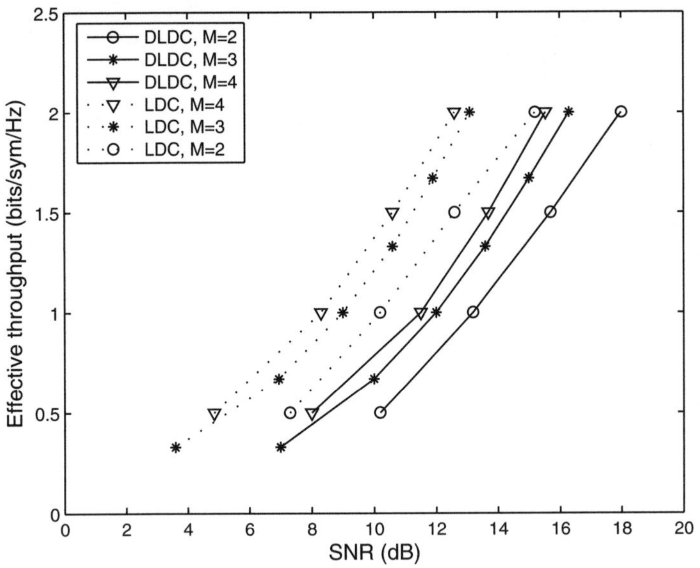

Figure 8.16: Effective throughput comparison of a family of DLDC schemes of Figure 8.9 employing $M = 2, 3, 4$ transmit and $N = 2$ receive antennas and the corresponding coherently detected LDC counterparts, when communicating over i.i.d. Rayleigh-fading channels having $f_d = 10^{-2}$. All of the system parameters are summarized in Table 8.2.

8.4.4 Summary

In this section, we summarize the performance trends of the DLDCs presented in Sections 8.3 and 8.4.3 in terms of their effective throughput and their coding gain, where the latter was defined as the SNR difference, expressed in decibels, at a BER of 10^{-4} between various DLDCs and single-antenna-aided DPSK systems having the identical effective throughput.

In Figure 8.16, we characterize the effective throughput of a family of DLDCs employing $M = 2, 3, 4$ transmit and $N = 2$ receive antennas, when transmitting over i.i.d. Rayleigh-fading channels having $f_d = 10^{-2}$. The increase of effective throughput was achieved by transmitting more symbols (Q) per space-time block, while 2PAM modulation was employed for all of the schemes. The effective throughput achieved by their coherently detected LDC counterparts was also shown in Figure 8.16. Observe that an approximately 3 dB SNR gap exists between the corresponding DLDC and LDC schemes.

Furthermore, in Figure 8.17 we characterize the effective throughput achieved by a family of 2PAM modulated DLDCs employing $M = 2$ transmit and $N = 2, 3, 4$ receive antennas, when communicating over i.i.d. Rayleigh-fading channels having a normalized Doppler frequency of $f_d = 10^{-2}$. Naturally, a certain throughput can be achieved using less power, when increasing the receive diversity order from $D_{rx} = 2$ to $D_{rx} = 4$. By comparing Figures 8.16 and 8.17, it becomes clear that for a given total diversity order receive diversity provides a more substantial improvement than the corresponding transmit diversity scheme, because the former improves the SNR by gleaning a high received signal power, whereas the transmit diversity shares the same total power over several antennas.

The coding gains of a family of DLDCs having an effective throughput of 1 or 2 (BPS Hz^{-1}) are listed in Table 8.3. The highest coding gains were recorded for the DLDC schemes employing 2PAM. Compared with the coding gains achieved by their LDC

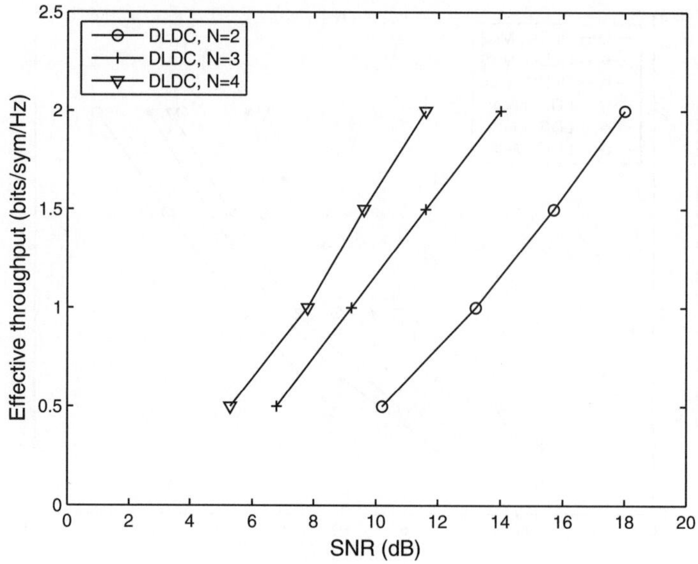

Figure 8.17: Effective throughput of a class of DLDC schemes of Figure 8.9 employing $N = 2, 3, 4$ receive and $M = 2$ transmit antennas, when communicating over i.i.d. Rayleigh-fading channels having $f_d = 10^{-2}$. All of the system parameters are summarized in Table 8.2.

Table 8.3: Coding gains of a family of DLDCs of Figure 8.9 having an effective throughput of 1 or 2 (BPS Hz^{-1}).

DLDC	Throughput	Modulation	Coding gain (dB)
DLDC(2221)	1	4PAM	18.1
DLDC(2222)	1	2PAM	23.3
DLDC(2221)	2	16PAM	11.4
DLDC(2222)	2	4PAM	18.7
DLDC(2224)	2	2PAM	21.5

counterparts listed in Tables 7.1 and 7.2, the DLDCs are capable of attaining similar coding gain over the single-antenna-aided systems, which was found to be approximately 23 dB.

8.5 RSC-coded Precoder-aided DOSTBCs

One crucial issue in the context of iteratively detected schemes [160] is the design of constellation mapping, which was studied in [320, 321]. All of the results indicate that the specific arrangement of the bits to the modulated signal constellation has substantial effect on the iterative system's performance. In [305], SP modulation was proposed for improving the performance of DOSTBCs [62], where the symbols transmitted in two time slots were jointly designed. As a result, the size of the new SP-modulated constellation became $L = 16$, if originally QPSK symbols were transmitted.

When the channel is classified as non-recursive, implying that it has a finite-duration CIR, then the achievable iteration gain of the receiver is limited, since there is a limited interleaver gain [322]. However, the channel can be rendered to appear recursive to the receiver, thus resulting in a useful interleaver gain by invoking a recursive inner encoder [168], namely a unity-rate precoder. There are other methods of rendering the impulse response duration infinite. For example, the differential encoding process of DOSTBC schemes potentially renders the CIR duration infinite, but unfortunately the conventional DOSTBC decoding strategy of Equations (8.17) and (8.18) is incapable of exploiting the recursion, hence rendering the DSTBC system more or less a diversity combiner. Another method of introducing recursion is the employment of differential constellations, such as DPSK. In [323], it has been argued that turbo-detected DPSK has the potential of outperforming coherent PSK, although this requires a number of iterative detection steps, which increases the detector's complexity. Therefore, unity-rate precoder remains the most flexible method of controlling the spreading of the extrinsic information using different feedback polynomials and various precoder memory sizes with the aid of EXIT chart analysis [169].

Based on the above discussions, in this section a RSC-coded SP-aided DOSTBC design was proposed. The novelty and rationale can be summarized as follows.

- The proposed scheme is capable of achieving both spatial and temporal diversity, without the high complexity of channel estimation in coherent detection.
- The employment of SP modulation allows us to intelligently amalgamate inner DOSTBC scheme and outer RSC code, where we jointly rather than separately design the space-time signals of two time slots.
- Various rate-one precoder designs are investigated. The spread of extrinsic information across the transmitted bit stream is governed by the precoder's memory. We demonstrate that the shape of the EXIT curve, which determines the convergence threshold, is controlled by the precoder's memory.
- We further improve the performance of the proposed serial concatenated receiver relying on classic convolutional codes by invoking IRCCs [176, 195] as an outer code. We jointly consider the flexibility provided by the specifically tailored EXIT characteristics of the IRCCs and the precoder. As a result, we demonstrate that an even lower SNR convergence threshold is attainable in comparison with the system using regular convolutional codes.

8.5.1 DOSTBC Design with SP Modulation

It was shown in [43] that the diversity product quantifying the coding advantage of a STBC scheme is determined by the MED of the vectors $\mathbf{K}_n = [s_n^1, s_n^2, \ldots, s_n^Q]^{\mathrm{T}}$. Therefore, it was proposed in [43] to use SP schemes that have the best known MED in the $2Q$-dimensional real-valued Euclidean space $\Re^{2Q}$.

In this section, DOSTBC schemes obeying the structure of Figure 8.2 and employing two transmit antennas are considered, where the symbols are differentially encoded using the generator matrix $\mathbf{G}_2$ of Equation (8.13). It has been demonstrated in Section 8.3.1 that when the DSTBC design is based on the orthogonal design, the received signals after the linear combination will be scaled versions of the transmitted symbols s_n^1 and s_n^2 as seen in Equations (8.17) and (8.18), which are corrupted by complex AWGN. This observation implies that the diversity product of DOSTBC scheme is also determined by the MED of all legitimate vectors $[s_n^1, s_n^2]$.

For the sake of generalizing our treatment, let us assume that there are L legitimate twin-time-slot vectors $[s_n^1, s_n^2]$, where L represents the number of sphere-packed modulated symbols. The encoder, then, has to choose the modulated symbol associated with each block of bits from these L legitimate symbols. In contrast to the independent DOSTBC signaling pulses s_n^1 and s_n^2, our aim is to design s_n^1 and s_n^2 jointly, so that they have the best MED from all other $(L-1)$ legitimate SP symbols, since this minimizes the system's SP symbol error probability.

Let $(a_{l,1}, a_{l,2}, a_{l,3}, a_{l,4})$, $l=0,\ldots,L-1$, be SP phasor points selected from the four-dimensional real-valued Euclidean space $\Re^4$, where each of the four elements $a_{l,1}$, $a_{l,2}$, $a_{l,3}$, $a_{l,4}$ gives one coordinate of the twin-time-slot complex-valued phasor points. Hence, s_n^1 and s_n^2 can be written as

$$\{s_n^1, s_n^2\} = \mathrm{SP}(a_{l,1}, a_{l,2}, a_{l,3}, a_{l,4}) = \{a_{l,1} + ia_{l,2}, a_{l,3} + ia_{l,4}\}. \tag{8.63}$$

In the four-dimensional real-valued Euclidean space $\Re^4$, the lattice $\bar{\Re}^4$ is defined as a SP having the best MED from all other $(L-1)$ legitimate SP constellation points in $\Re^4$ (see [221]). More specifically, $\bar{\Re}^4$ may be defined as a lattice that consists of all legitimate SP constellation points having integer coordinates $[a_1\ a_2\ a_3\ a_4]$ uniquely and unambiguously describing the legitimate combinations of the DOSTBC modulated symbols s_n^1 and s_n^2 of each time slot, but subjected to the SP constraint that the sum of $[a_1\ a_2\ a_3\ a_4]$ equals an integer constant. Let us assume that $\Re^\dagger\{[a_{l,1}, a_{l,2}, a_{l,3}, a_{l,4}] \in \Re^4 : 0 \le l \le L-1\}$ constitutes a set of L legitimate constellation points selected from the lattice $\bar{\Re}^4$ having a total energy of $\tilde{E} \triangleq \sum_{l=0}^{L-1}(|a_{l,1}|^2 + |a_{l,2}|^2 + |a_{l,3}|^2 + |a_{l,4}|^2)$, and introduce the set of complex constellation symbols, $\{SP^l : 0 \le l \le L-1\}$

$$\mathrm{SP}_l = \sqrt{\frac{2L}{\bar{E}}}(s_n^1, s_n^2), \quad l=0,1,\ldots,L-1, \tag{8.64}$$

whose diversity product is determined by the MED of the set of L legitimate constellation points in $\Re^\dagger$. For example, at the first layer of the SP lattice there are 24 constellation points according to all possible constellations of the values $[\pm1, \pm1, 0, 0]$ and the constellation set of $L=16$ points was found by computer search upon maximizing their Euclidean distance at a given total power $\bar{E}$.

In contrast to Alamouti's approach, our SP-aided DOSTBC allowed us to jointly design all of the space and time dimensions available. Hence, more extrinsic information can be obtained compared with conventional QPSK modulation. Furthermore, the best SP mapping that enables the creation of an open convergence tunnel can be chosen for the iteratively detected system. For the proposed scheme of Figure 8.18, Gray labeling is chosen, since there are no inner iterations between SP demapper and precoder. The exact choice of the $L=16$ SP constellation points is given in [46].

8.5.2 System Description

The block diagram of the three-stage scheme is shown in Figure 8.18, which employs two transmit and one receive antennas, although the concept may be readily applied to an arbitrary number of receive antennas.

First, a frame of information bits is encoded by a regular/irregular outer half-rate convolutional channel encoder. Then, the outer channel encoded bits of Figure 8.18 are

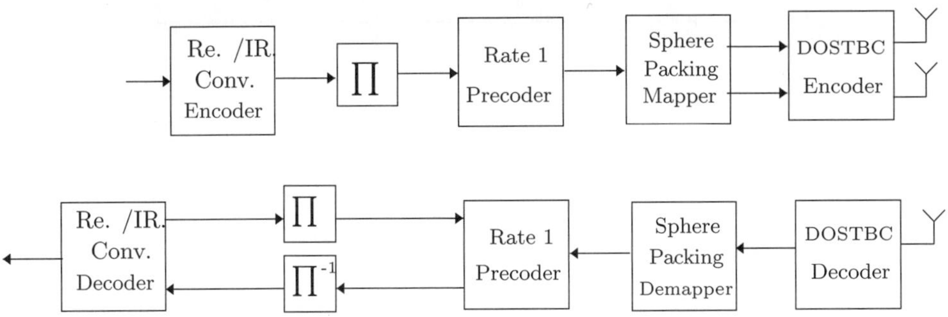

Figure 8.18: Schematic of the precoded SP-aided DOSTBC design using regular/irregular convolution codes.

permuted by a random interleaver. The interleaved bits are then encoded by a rate-one recursive precoder, which contributes towards achieving a high iteration gain. Initially a memory-one precoder is considered. Then, the SP mapper of Figure 8.18 maps each block of encoded bits to a legitimate constellation point $\mathrm{SP}^l_n \in \Re^\dagger$. The same mapper then represents the constellation point SP_l by two complex DOSTBC symbols s^1_n and s^2_n of the two time slots using Equations (8.63) and (8.64). Subsequently, the DOSTBC encoder of Figure 8.18 calculates the transmitted symbols using Equation (8.19), as detailed in Section 8.3.1. Furthermore, instead of directly transmitting the differentially encoded symbols, the proposed system may also be amalgamated with differential STS [13]. Since each spreading sequence of the STS scheme spans two symbol periods and transmitting two symbols, which is identical to the philosophy of DOSTBC, it can be argued that the employment of STS allows the system to accommodate two users without affecting the achievable single-user throughput. Similarly, multiple users may be supported with the aid of multiple antennas and longer spreading sequences.

The structure of the receiver is shown in Figure 8.18, where the DOSTBC decoder decouples the symbol stream according to Equations (8.21) and (8.22); then we have

$$\bar{s}^1_n = (|h_{1,1}|^2 + |h_{1,2}|^2) \cdot \theta \cdot s^1_n + \hat{v}^1_n, \tag{8.65}$$

$$\bar{s}^2_n = (|h_{1,1}|^2 + |h_{1,2}|^2) \cdot \theta \cdot s^2_n + \hat{v}^2_n, \tag{8.66}$$

where θ is given in Equation (8.23). Recall in Equations (8.63) and (8.64) that one SP constellation point is represented by s^1_n and s^2_n; thus we can derive

$$\overline{\mathrm{SP}}^l_n = (|h_{1,1}|^2 + |h_{1,2}|^2) \cdot \theta \cdot \mathrm{SP}^l_n + \tilde{v}^l_n. \tag{8.67}$$

Then, the decoded symbols of Equation (8.67) are passed to the SP demapper of Figure 8.18, where they are demapped to their LLR representation for each of the convolutional coded bits in a SP symbol. The rate-one precoder's decoder processes the *a priori* information fed back from the outer decoder and the output of the SP demapper of Figure 8.18. Then the extrinsic information is exchanged between the outer decoder and precoder's decoder for a consecutive number of iterations (k). Note that the SP demapper combined with the DOSTBC decoder is not involved in the iterative decoding process.

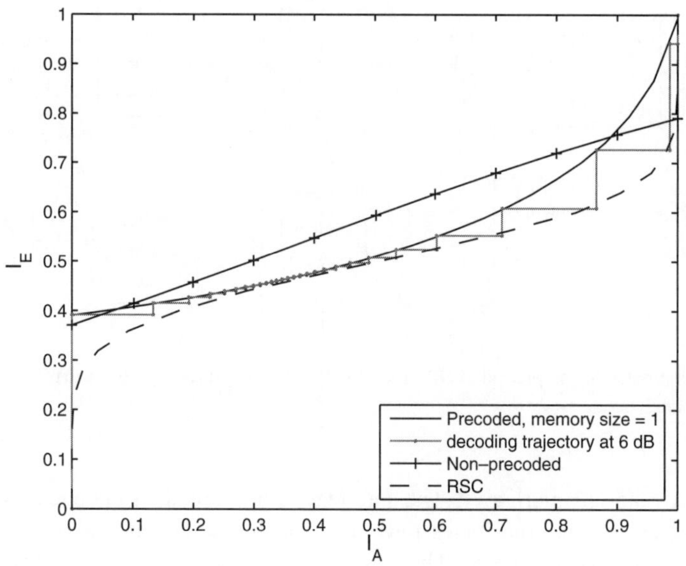

Figure 8.19: EXIT chart comparison of non-precoded and memory size one precoded system of Figure 8.18, when using the system parameters outlined in Table 8.4 and the actual decoding trajectory was recorded at $\rho = 6$ dB.

Table 8.4: System parameters for the precoded SP-aided DOSTBC system of Figure 8.18.

Modulation	SP ($L = 16$)
Mapping	Gray mapping
Number of transmit antennas	2
Number of receive antennas	1
Normalized Doppler frequency	$f_d = 0.01$
Regular outer channel code	RSC(2,1,5)
IRCC	Half-rate
Interleaver length	10^6 bits
Precoder rate	1
Precoder memory	1
Throughput	1 bit per channel use

8.5.3 EXIT Chart Analysis

The EXIT chart of the precoded system of Figure 8.18 is shown in Figure 8.19, when using a memory-one precoder together with RSC(215) code and the system parameters outlined in Table 8.4. This scheme is compared with a non-precoded system, when operating at SNR $\rho = 6$ dB. The non-precoded system also employs iterative decoding between the soft SP demapper and the outer channel decoder as described in [9,46], where the best AGM scheme that matches the outer code's EXIT curve was chosen as in [305].

Observe in Figure 8.19 that the precoded system reaches the point $I_E \approx 0.998$, where a convergence tunnel has taken shape, hence resulting in an infinitesimally low BER. In contrast, the non-precoded scheme reaches $I_E \approx 0.79$, as shown in Figure 8.19, hence only

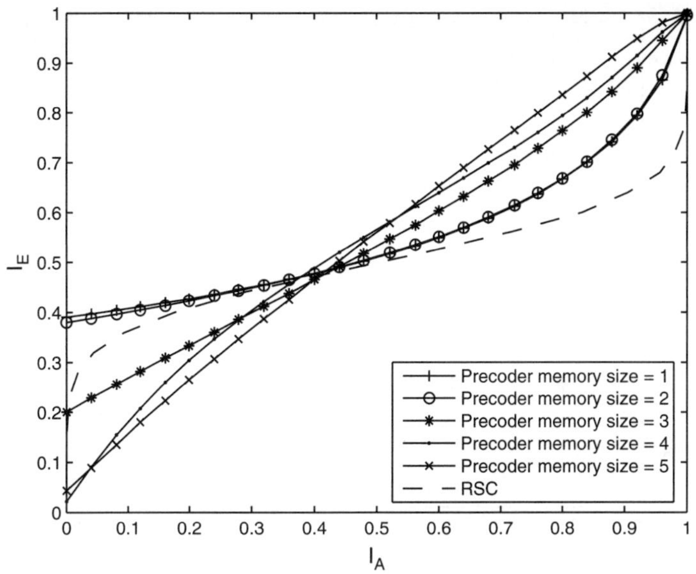

Figure 8.20: EXIT characteristics of precoded system of Figure 8.18 using Gray mapping for different precoder memory sizes, when employing the system parameters outlined in Table 8.4 and operating at $\rho = 6$ dB.

modest BER advantage can be achieved. In other words, it is expected that the precoded system is capable of reaching the typical turbo cliff, when considering the BER versus SNR performance, as shown in Section 8.5.4.

Figure 8.20 shows the EXIT curves of the inner decoder operating at $\rho = 6$ dB, when using various precoder memory sizes. The inner decoder is constituted by the precoder's decoder, SP demapper and DOSTBC decoder of Figure 8.18. The precoder's structure is shown in Figure 7.30. When the memory size of the precoder is increased, the EXIT curves of the inner decoder tend to start at lower I_E values, but rise more rapidly. For $\rho = 6$ dB, a memory-one precoded system has an open convergence tunnel, whereas a higher precoder memory tends to require a higher SNR value for an open convergence tunnel to take shape. However, once an open convergence tunnel has been formed, fewer iterations are required to reach the point of intersection by the systems employing precoders having a higher memory size. Hence, for this particular system employing an RSC(215) outer code, a precoder memory size of one is capable of achieving an infinitesimally low BER at the lowest possible SNR value. Note that the area under all of the EXIT curves recorded for various precoder memory sizes is constant, which indicates that the achievable channel capacity is not affected [270, 273] by changing the shape of the EXIT curves. Note that the same phenomenon has also been recorded for the coherently detected precoded STBCs of Figure 7.24, as justified in Figure 7.31.

The open EXIT tunnel area between the inner and outer decoder's curves in Figure 8.19 is not necessarily an unambiguous characteristic of the system, since the convergence tunnel is only narrow within a certain range. By shaping both the inner and outer EXIT curves, it is potentially possible to create an even narrower tunnel in conjunction with a lower SNR requirement. Since IRCCs exhibit flexible EXIT characteristics as extensively demonstrated in Section 7.5.2, we adopt an IRCC as our outer code, while keeping the overall code rate the

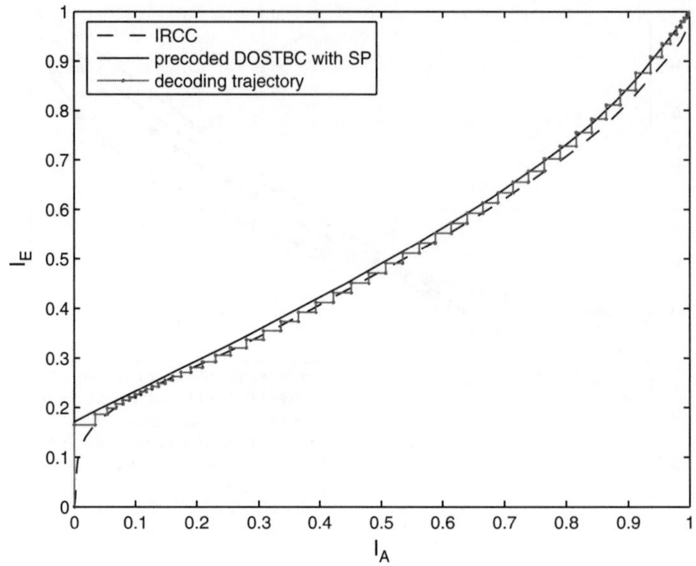

Figure 8.21: EXIT chart of the memory-size-three precoded system of Figure 8.18 using IRCC, when using the system parameters outlined in Table 8.4, and the actual decoding trajectory at $\rho = 5.5$ dB.

same as that of our previous RSC code. The IRCC scheme constituted by a set of $P_{\rm out} = 12$ component codes was employed, where the corresponding weighting coefficient vector is $\gamma = [\gamma_1, \ldots, \gamma_{P_{\rm out}}]$. The associated component rates of the IRCCs are $R_{i,\rm IRCC} = [0.1, 0.15, 0.25, 0.4, 0.45, 0.55, 0.6, 0.7, 0.75, 0.8, 0.85, 0.9]$, respectively. Hence, γ is optimized with the aid of the iterative algorithm of [195], so that the EXIT curve of the resultant IRCC closely matches that of the inner code. Since the inner code's EXIT curve benefits from the flexibility provided by the precoder's memory size, as shown in Figure 8.20, the IRCC is optimized for all of the different memory sizes featuring in Figure 8.20. Explicitly, our design objective is to find the lowest SNR value, where it is possible to form an open convergence tunnel.

Figure 8.21 illustrates both the EXIT curves of the IRCC optimized for the proposed system having a precoder memory of three as well as the bit-by-bit decoding trajectory at SNR $\rho = 5.5$ dB, where the optimized weighting coefficients are as follows: $\gamma = [0, 0.0206, 0.0199, 0.1567, 0, 0.0811, 0.1803, 0.0835, 0.2757, 0.0302, 0.0326, 0.1194]$. Observe that an extremely narrow EXIT tunnel is formed, which exhibits a 0.5 dB advantage over the RSC(215) coded scheme characterized in Figure 8.20.

Finally, the benefits of employing a SP scheme over the conventional QPSK modulation recorded for the iteratively detected schemes are characterized in Figure 8.22, where SP modulation employing $L = 16$ constellation points and various mapping schemes is compared with classic QPSK at $\rho = 6$ dB. Note that SP using Gray labeling is capable of generating higher extrinsic information than the various AGM schemes, since there is no information exchange between the SP demapper and the precoder's decoder [324]. Since the SP-aided DOSTBC facilitated the joint design of the space-time dimension available, more extrinsic information was obtained with the aid of the precoder compared with the conventional QPSK modulation. Therefore, the proposed SP-aided scheme using Gray

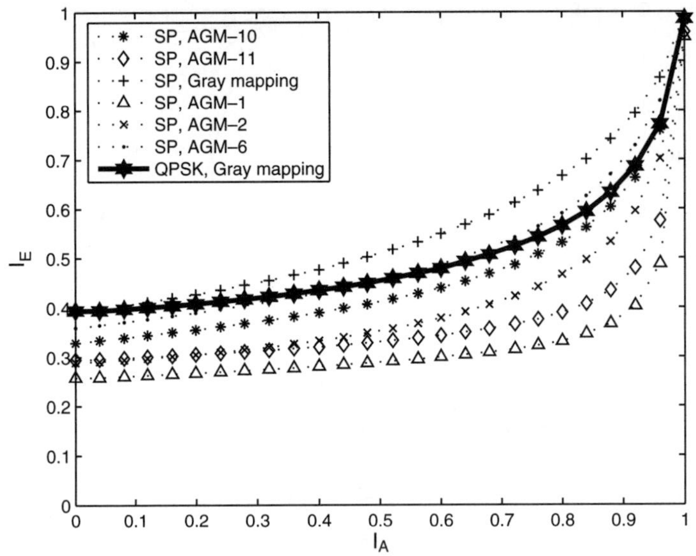

Figure 8.22: EXIT comparison of the system of Figure 8.18 employing SP using various mappings and employing QPSK with Gray mapping at $\rho = 6$ dB, when using the system parameters outlined in Table 8.4.

labeling exhibited a larger area under the EXIT curve compared with its QPSK counterpart, as seen in Figure 8.22.

8.5.4 Performance Results

We now proceed to characterize the achievable BER performance of the turbo-detected precoded SP-aided DOSTBC system of Figure 8.18. All of the simulation parameters are listed in Table 8.4, unless otherwise specified.

Figure 8.23 demonstrates the BER performance of the half-rate RSC/IRCC-coded SP-aided DOSTBC system. The precoding-aided RSC-coded system exhibits a turbo cliff at $\rho =$ 6 dB, where $k = 26$ iterations are required to reach the turbo-cliff region. This observation matches the prediction of the EXIT charts seen in Figure 8.19. Furthermore, the BER performance of the IRCC-coded scheme using a precoder memory of three is also shown in Figure 8.23. As predicted in Figure 8.21, the system achieved an infinitesimally low BER at $\rho = 5.5$ dB, where a 0.5 dB SNR gain is observed over the RSC-coded scheme having a precoder memory size of one. Naturally, a higher number of decoding iterations is needed in addition to the increased decoding complexity imposed by the precoder memory size of three.

Figure 8.24 quantifies the SNR ρ required for maintaining an open convergence tunnel against the normalized Doppler frequency f_d, for the proposed RSC-coded SP-aided scheme of Figure 8.18 using a memory-size-one precoder. At this SNR an infinitesimally low BER may be attained. The success of differential decoding depends on the channel's variation between two consecutive transmission blocks. When the normalized Doppler frequency obeys $f_d \leq 0.01$, the proposed system is robust against the CIR tap fluctuations and achieves an infinitesimally low BER around $\rho = 6$ dB. When the CIR fluctuates faster, a higher transmit power is required for an open convergence tunnel to form, as seen in Figure 8.24.

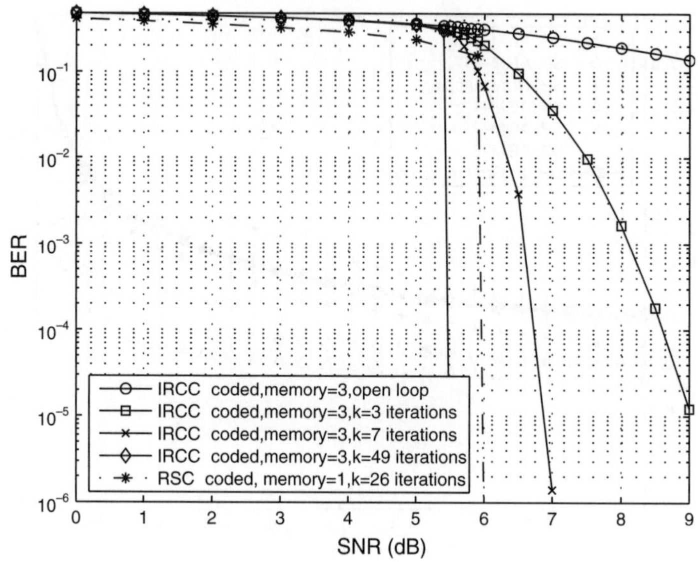

Figure 8.23: BER performance of a half-rate RSC/IRCC-coded SP-aided DOSTBC scheme of Figure 8.18 with unity-rate precoder having a memory size of one or three, when employing the system parameters outlined in Table 8.4.

Our proposed scheme successfully supports normalized Doppler frequencies up to $f_d = 0.06$, where $\rho = 12.2$ dB is necessary as evidenced by Figure 8.24. A normalized Doppler frequency higher than $f_d = 0.06$ will render the errors after differential decoding non-recoverable.

8.6 IRCC-coded Precoder-aided DLDCs

In this section, we propose an IRCC-aided IR-PDLDC scheme, which is capable of operating across a wide range of SNRs. The schematic of the proposed scheme is illustrated in Figure 8.25, which can again be viewed as a differentially encoded version of the IRCC-coded coherently detected IR-PLDC scheme detailed in Section 7.5.3. By replacing the coherently detected LDC coding blocks of Figure 7.48 with the non-coherent DLDC block, the resultant irregular scheme of Figure 8.25 becomes capable of supporting high rates, while dispensing with high complexity MIMO channel estimation. Therefore, similar design methods to those presented in Section 7.5.3 can be adopted. Hence, in this section we will focus on the proposed irregular system's features imposed by the non-coherent detection, rather than on the similarities to the features of the irregular coherent-detection-aided counterparts.

8.6.1 EXIT-chart-based IR-PDLDC Design

The family of DLDCs based on the Cayley transform of Equation (8.48) is employed as the inner space-time block, since they are capable of providing flexible rates and supporting arbitrary antenna configurations. We demonstrate how to generate an inner IR-PDLDC coding scheme containing $P_{\text{in}} = 6$ components for a MIMO configuration having $M = 3$ transmit and $N = 2$ receive antennas. Since $T = M$ is imposed by the differential encoding process, the resultant DLDCs have the potential of achieving the maximum attainable

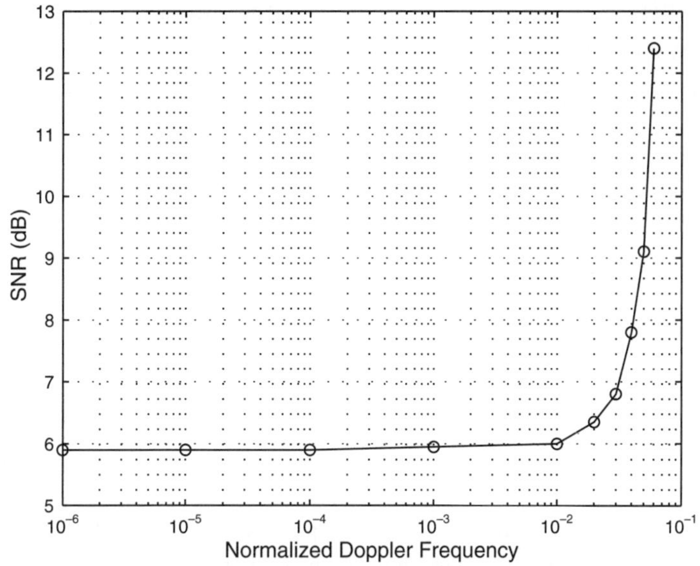

Figure 8.24: SNR required for the SP-aided RSC-coded DOSTBC of Figure 8.18 to achieve an infinitesimally low BER under different fading environments.

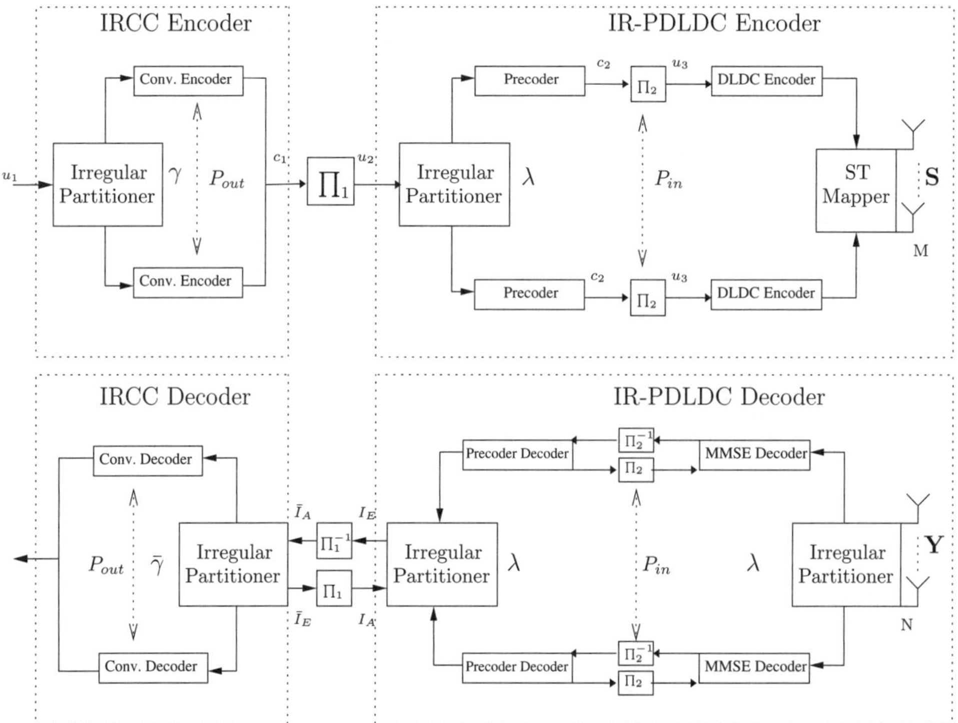

Figure 8.25: Schematic of the IRCC-coded IR-PDLDC using iterative decoding.

Table 8.5: The $P_{\text{in}} = 6$ component codes of the 2PAM modulated IR-PDLDC scheme of Figure 8.25 generated for a MIMO system having $M = 3$ and $N = 2$ antennas.

Index	M	N	T	Q	Rate	D	Inner iterations	Complexity
0	3	2	3	1	0.33	6	0	2399
1	⋮	⋮	⋮	2	0.67	6	1	6694
2	⋮	⋮	⋮	3	1	6	1	8590
3	⋮	⋮	⋮	4	1.33	6	1	10 486
4	⋮	⋮	⋮	5	1.67	6	1	12 382
5	⋮	⋮	⋮	6	2	6	1	14 278

spatial diversity order of $D = 6$, according to Theorem 7.3. By setting $Q = 1$, we have DLDC(3231), which can be optimized by maximizing the diversity gain using the rank criterion of Equation (7.26) and maximizing the coding gain using the determinant criterion of Equation (7.27). Consequently, we can generate different-rate DLDCs by changing the value of Q. Hence, by increasing the value of Q and maximizing both the diversity gain and the coding gain for each resultant DLDC$(MNTQ)$ component of the IR-PDLDC scheme, a set of meritorious DLDCs can be generated. Since Q single-stream transmitted symbols are jointly detected, low Q values are desirable for the sake of maintaining a low complexity. The resultant $P_{\text{in}} = 6$ component codes designed for our IR-PDLDC scheme are listed in Table 8.5. Hence, inner Precoded Differential Linear Dispersion Codes (PDLDCs) can be directly obtained by combining memory-one unity-rate precoders with DLDCs having the parameter combination $(MNTQ)$.

Figure 8.26 quantifies the maximum achievable rates for three 2PAM-modulated PDLDCs of Table 8.5 having $j = 0, 1$ inner iterations using an MMSE detector, when communicating over i.i.d. Rayleigh fading channels having $f_d = 10^{-2}$. For each set of comparisons, the DLDC's maximum achievable rate quantified after the MMSE detector of Figure 8.25 is plotted as the upper bound. Observe in Figure 8.26 that there exists a gap between the maximum achievable rate quantified both before and after the rate-one precoder when employing no inner iterations for different values of Q. However, the gap can be eliminated when employing $j = 1$ inner iterations. Therefore, the required number of inner iterations maximizing the achievable rate is listed in Table 8.5.

Figure 8.27 characterizes the maximum achievable rates for the PDLDCs of Table 8.5 communicating over Rayleigh fading channels having $f_d = 10^{-2}$, when using 2PAM modulation and a MMSE detector. Furthermore, the maximum achievable rates of the corresponding group of coherent PLDCs are also plotted under the same channel conditions as well as employing the same number of inner iterations. Observe in Figure 8.27 that PDLDCs require approximately 3 dB of extra power in comparison with the corresponding PLDCs to achieve an infinitesimally low BER when operating at a certain rate. However, the associated performance gap can be closed in the high-SNR region, where the noise variance becomes insignificant, compared with the signal's power.

Figure 8.28 characterizes the 2PAM-modulated PDLDC(3233) scheme's maximum achievable rates evaluated using Equation (7.47), when communicating over i.i.d. Rayleigh

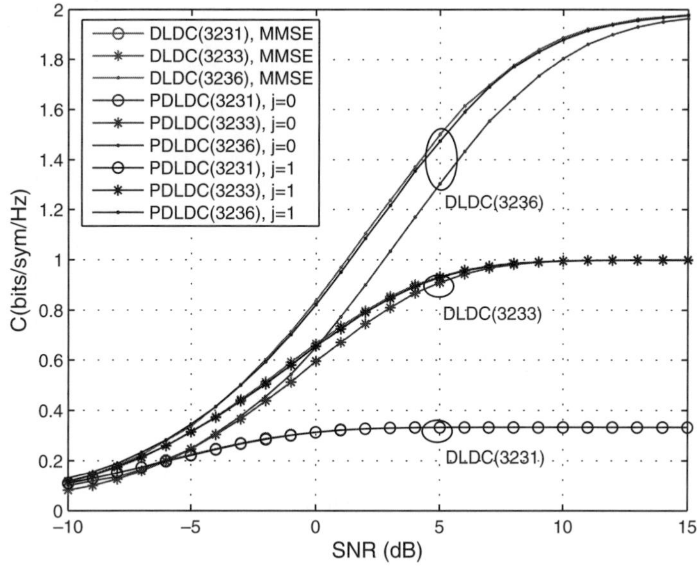

Figure 8.26: Maximum achievable rates for three 2PAM-modulated PDLDC schemes of Table 8.5 having $j = 0, 1$ inner iterations using a MMSE detector, when communicating over i.i.d. Rayleigh fading channels having $f_d = 10^{-2}$.

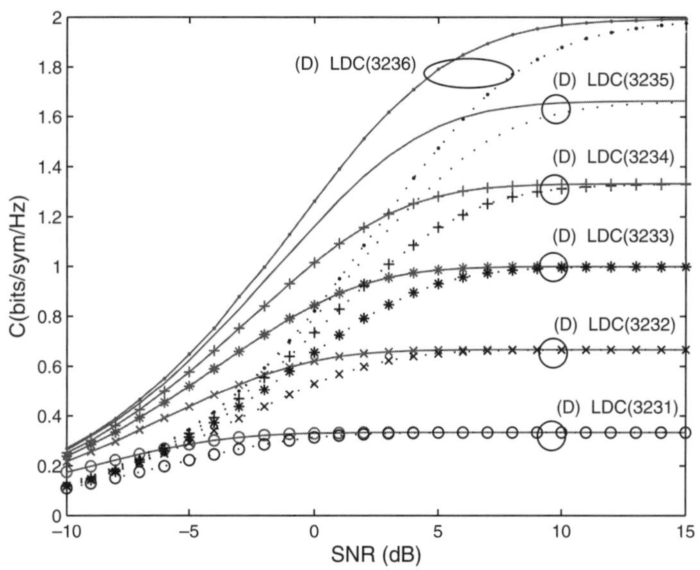

Figure 8.27: Maximum achievable rates for the 2PAM-modulated PDLDC schemes of Table 8.5 and their coherent counterparts PLDCs using a MMSE detector, when communicating over i.i.d. Rayleigh fading channels having $f_d = 10^{-2}$.

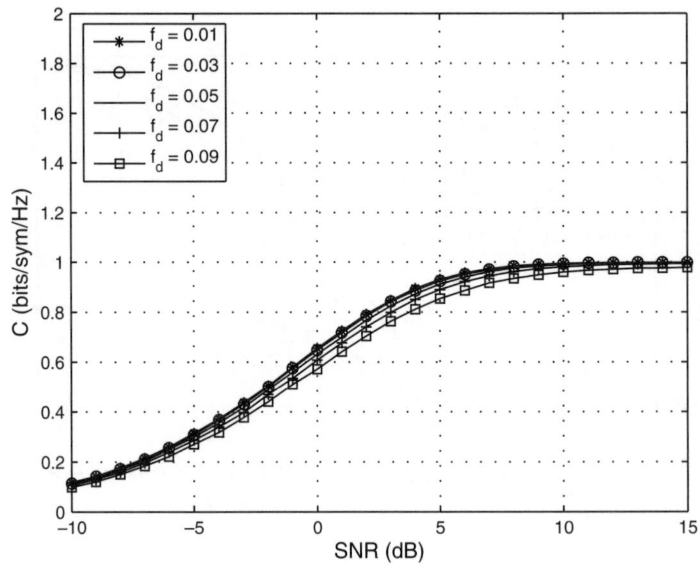

Figure 8.28: Maximum achievable rates of the 2PAM-modulated PDLDC(3233) scheme of Table 8.5 communicating over i.i.d. Rayleigh fading channels having different normalized Doppler frequencies f_d, when using a MMSE detector.

fading channels having different f_d values and employing a MMSE detector. Observe that the maximum achievable rate dropped gradually when the channel's fluctuation became faster and the associated rate loss becomes more apparent in the high-SNR region. This observation is consistent with those recorded for the non-coherent MIMO channel's capacity given in Equation (8.12), which suggests a capacity reduction when the corresponding coherence time $\Im$ is reduced. Also note that the maximum achievable rate increase is modest when we have $f_d < 0.03$.

Figure 8.29 characterizes the maximum achievable rates of the 2PAM-modulated PLDC(3233) having imperfect CSI governed by ω as well as that of the PDLDC(3233) scheme of Table 8.5, when encountering Rayleigh fading having $f_d = 10^{-2}$ and employing a MMSE detector. Observe in Figure 8.29 that at high SNRs the non-coherent PDLDC(3233) scheme is capable of achieving a higher throughput compared with its coherently detected counterparts, owing to the non-vanishing errors imposed by the imperfect CSI. More explicitly, when the channel estimation error power is $\omega = -7$ dB, the PDLDC(3233) scheme outperforms its coherent counterpart beyond the point of $\rho = 4$ dB. When ω reaches -3 dB, the coding advantage of the PDLDC(3233) scheme becomes promising for SNRs in excess of $\rho = -5$ dB.

8.6.2 Performance Results

In this section, we present some further numerical results for the scheme of Figure 8.25 designed for maximizing the throughput for SNRs in excess of certain thresholds, while maintaining an infinitesimally low BER. For all of the simulations, 2PAM modulation is employed and the first interleaver of Figure 8.25 is set to have a length of 10^6 bits. We employed a total number $P = 12$ component codes, where the IR-PDLDC scheme employs the $P_{\text{in}} = 6$ component codes of Table 8.5 and the IRCC contains $P_{\text{out}} = 6$ component codes

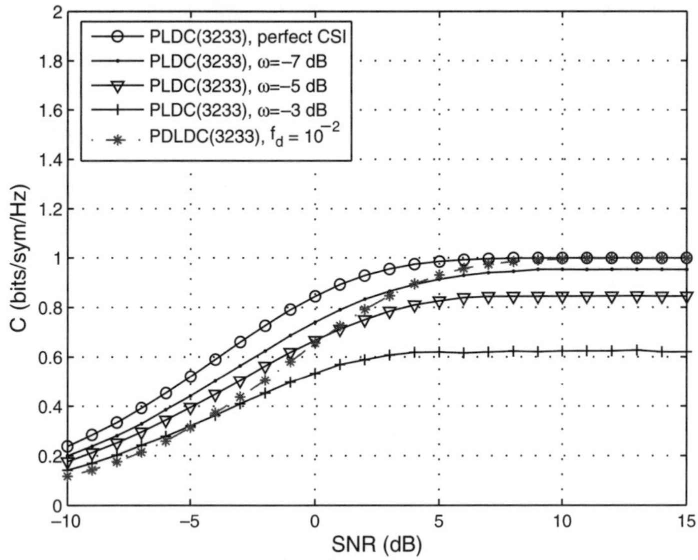

Figure 8.29: Maximum achievable rates of the 2PAM-modulated PDLDC(3233) scheme of Table 8.5 and the corresponding PLDC schemes having imperfect CSI governed by ω, when transmitting over i.i.d. Rayleigh-fading channels having $f_d = 10^{-2}$.

having a rate of $R_{i,\text{IRCC}} = [0.1, 0.25, 0.4, 0.55, 0.7, 0.9]$. The weighting coefficient vectors $\boldsymbol{\gamma}$ and $\boldsymbol{\lambda}$ are optimized using the method presented in Section 7.5.3, summarized in Tables G.7 and G.8.

Figure 8.30 presents the associated EXIT charts and the corresponding decoding trajectory of our IRCC-coded IR-PDLDC scheme of Figure 8.25 designed for operating at $\rho = -2$ dB using a MMSE detector, when communicating over i.i.d. Rayleigh fading channels having $f_d = 10^{-2}$. The dashed lines represent the EXIT curves of the IR-PDLDC's component codes of Table 8.5 and the dotted lines denote the EXIT curves for the set of IRCC components. The solid lines represent the aggregate EXIT curves of the IRCC and IR-PDLDC having the weighting coefficients listed in Tables G.7 and G.8. By simultaneously maximizing throughput and the EXIT tunnel area, the optimized EXIT curves of Figure 8.30 exhibit a narrow open tunnel, where the decoding trajectory shows that $k = 49$ outer iterations were required. The corresponding BER performance is shown in Figure 8.31.

Naturally, the same design process can be extended to a range of SNR values. Figure 8.32 plots the maximum rates achieved by the proposed IRCC-coded IR-PDLDC scheme of Figure 8.25 using a MMSE detector, when transmitting over i.i.d. Rayleigh fading channels having $f_d = 10^{-2}$. Each point of Figure 8.32 was designed to achieve the maximum throughput by adjusting the weighting coefficient vectors $\boldsymbol{\lambda}$ and $\boldsymbol{\gamma}$, listed in Tables G.7 and G.8 of Appendix G. The maximum rates achieved using the corresponding set of coherently detected IR-PLDCs of Figure 7.48 having perfect CSI are plotted as the benchmark. Similar to the observations inferred from Figure 8.27, the SNR penalty owing to differential encoding is approximately 3 dB. The proposed scheme is capable of operating in the SNR range in excess of $\rho = -12$ dB, while maintaining an infinitesimally low BER. However, the rate increase is limited to 1.8 (BPS Hz^{-1}) for SNRs beyond $\rho = 10$ dB. Clearly, a further rate increase can be achieved, when higher-rate DLDC components are employed.

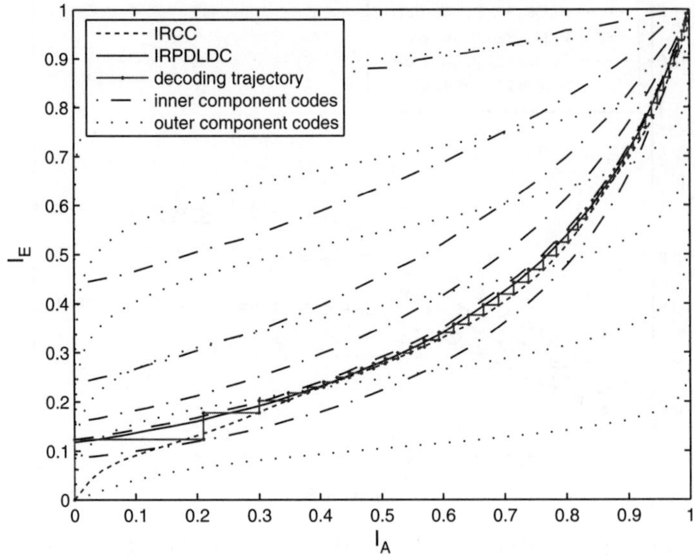

Figure 8.30: EXIT chart and the decoding trajectory of the IRCC-coded IR-PDLDC scheme of Figure 8.25 recorded at $\rho = -2$ dB using a MMSE detector and 2PAM modulation, when communicating over i.i.d. Rayleigh fading channels having $f_d = 10^{-2}$.

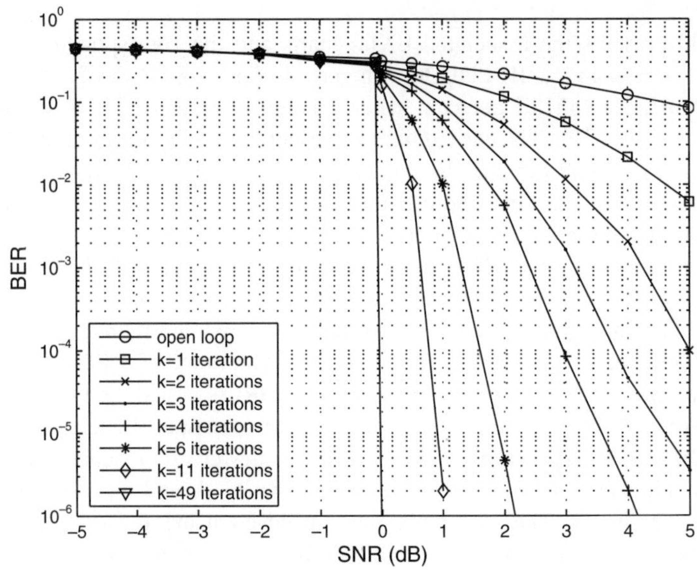

Figure 8.31: BER of the IRCC-coded IR-PDLDC scheme of Figure 8.25 designed for achieving an infinitesimally low BER at $\rho = -2$ dB using a MMSE detector and 2PAM modulation, when communicating over i.i.d. Rayleigh fading channels having $f_d = 10^{-2}$.

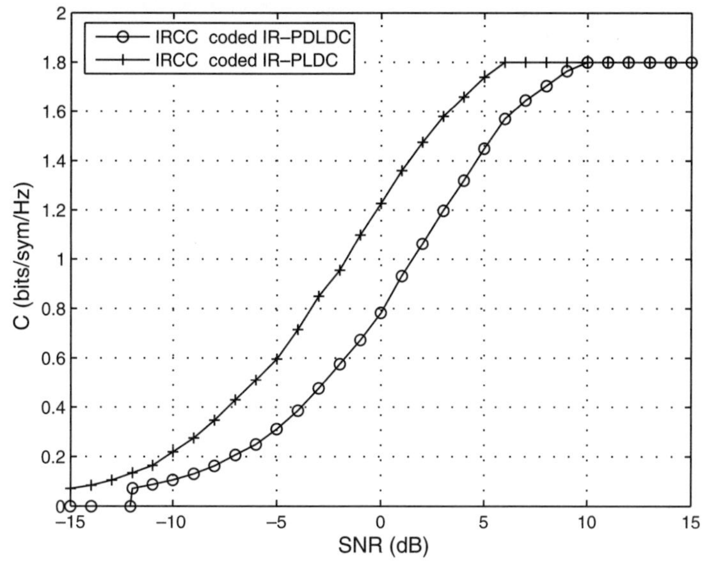

Figure 8.32: The maximum rates achieved by the IRCC-coded IR-P(D)LDC schemes of Figures 7.48 and 8.25 using a MMSE detector and 2PAM modulation according to Tables G.7, G.8, G.9 and G.10, when communicating over i.i.d. Rayleigh fading channels having $f_d = 10^{-2}$.

8.7 Conclusion

In this chapter, first a universal structure has been established in order to describe all of the existing DSTBCs found in the open literature. In Section 8.2, the challenge of designing DSTBCs obeying this general structure has been stated with the aim of designing a set of unitary space-time coded matrices. In Section 8.3, we have outlined the philosophy of the class of DSTBCs based on various orthogonal designs, where different degrees of orthogonality have been imposed on the associated space-time matrices. Furthermore, in Section 8.4, we proposed the family of non-orthogonal DLDCs based on the Cayley transform, which is suitable for operating at high rates and for arbitrary antenna configurations. We also demonstrated that the general framework of DLDCs subsumes the existing DOSTBC schemes presented in Figure 8.8.

In Section 8.5, the turbo principle was invoked for the proposed RSC-coded SP-aided DOSTBC scheme. The resultant scheme is capable of operating at a SNR of $\rho = 5.5$ dB at a rate of 1 (BPS Hz^{-1}) with the aid of IRCCs. In Section 8.6, we further investigated the performance of an irregular scheme, where the irregularity was imposed both at the outer code using the IRCC and at the inner code employing an IR-PDLDC. The amalgamated irregular differential scheme had a 3 dB SNR loss, when compared with its coherently detected irregular counterpart, although this loss was reduced when realistic imperfect CSI was used. More specifically, in Figures 8.15 and 8.29 we provided both a BER and a maximum achievable rate comparison between the DLDCs and the corresponding LDCs using realistic imperfect channel estimation. The results implied that DLDC schemes are more reliable for high-rate communications at high SNRs, when the channel estimation error is non-negligible. As a benefit of the DLDCs, the typically high-complexity MIMO channel estimation can be eliminated.

8.7 Conclusion

Chapter 9

Cooperative Space-Time Block Codes

9.1 Introduction and Outline

The space-time block coding techniques detailed in Chapters 7 and 8 provide promising solutions in the context of co-located MIMO systems requiring reliable wireless communications at high rates. However, it may not always be practical to accommodate multiple antennas at the MSs, owing to cost, size and other hardware limitations. A further limitation of co-located MIMO elements is that even at relatively large element separations their elements may not benefit from independent fading, when subjected to shadow-fading imposed for example by large-bodied vehicles or other shadowing local paraphernalia. As a remedy, High Speed Downlink Packet Access (HSDPA)-style adaptive modulations [325] as well as the concept of cooperative MIMOs [326] have been proposed for cellular systems as an attempt to attain a better communication efficiency beyond that permitted by a single node's local resources. More specifically, a group of mobile nodes, known as relays, 'share' their antennas with other users to create a virtual antenna array to provide spatial diversity gain. In order to deepen our discussion, Figure 9.1 illustrates the relay concept as well as potential applications of various cooperative MIMO systems in a cellular network. For example, when a user is behind a building or underground, as seen in Figure 9.1, direct communication with the BS becomes unreliable, owing to severe shadow fading and path loss. In order to maintain reliable wireless communications, a group of users in each others' vicinity may form a virtual antenna array in order to reliably forward the data between the source node and the BS.

A ground-breaking paper of Cover and El Gamal [116] proposed several cooperation strategies for the relays and extensively investigated the information-theoretic properties based on the assumption of encountering Line-of-Sight (LOS) Gaussian channels. More recent studies have focused on the more realistic assumptions of encountering fading channels [113, 114] and applied the relay-aided cooperation concept to wireless sensor networks [327]. It has been shown by several researchers [114,328] that considerable benefits can be achieved as a result of relay-aided cooperation, including the reduction of the outage probability and a substantial diversity gain as well as throughput improvement.

Near-Capacity Multi-Functional MIMO Systems L. Hanzo, O. R. Alamri, M. El-Hajjar and N. Wu

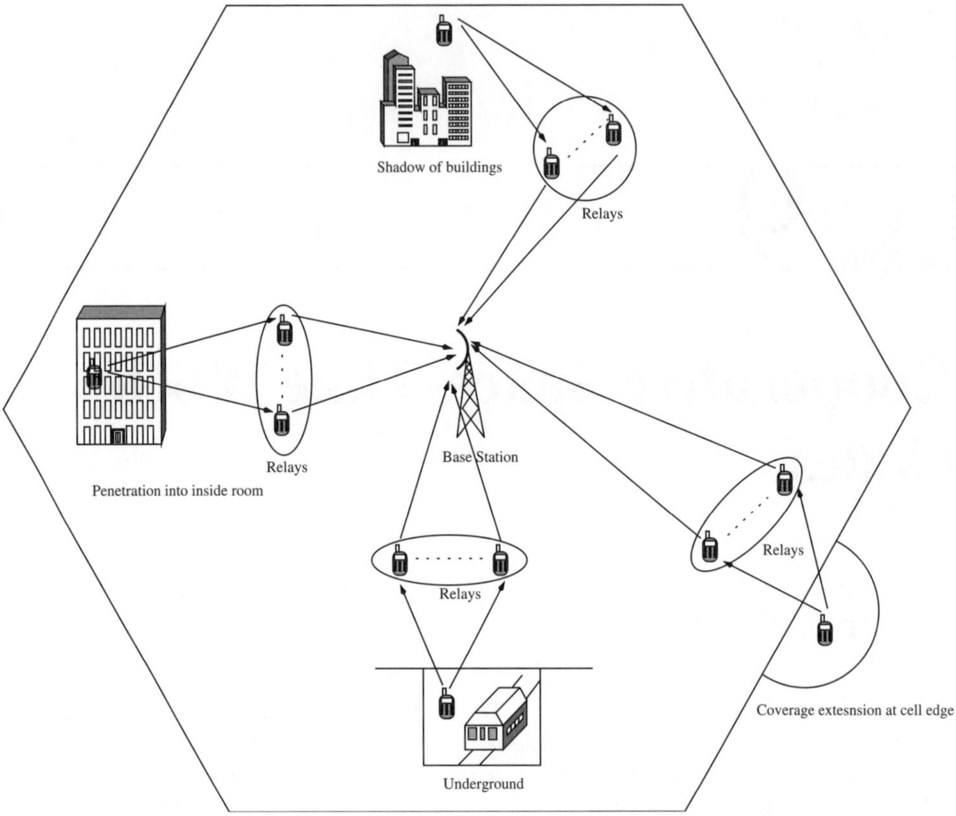

Figure 9.1: Cooperative MIMO systems in a cellular network.

Generally speaking, there are three types of cooperation [116]: Amplify-and-Forward (AF), Decode-and-Forward (DF) and Compress-and-Forward (CF). When using the AF cooperation scheme, the relay nodes simply amplify the received signal waveforms, but they amplify the signal and noise jointly and hence are unable to improve the SNR. On the other hand, in the DF strategy, the signals received at the relays are decoded and possibly re-encoded using different Forward Error Correction (FEC) codes, before being forwarded to the destination. Finally, the CF arrangement is also referred to as an observe-forward or quantize-forward technique by some researchers [328]. In its original form [116], the relay compresses, estimates or quantizes its observations without decoding the information.

Owing to the philosophical similarities between the cooperative MIMO and the co-located MIMO systems, numerous space-time block coding techniques have been 'transplanted' into relay-aided schemes in order to achieve cooperative diversity, based on either AF strategies [24, 125] or DF arrangements [113, 114]. It was Laneman and Wornell [24] who first proposed to employ OSTBCs for cooperative MIMO systems, where each relay transmits according to a different column of the OSTBC matrix. As argued in Chapter 7, the family of LDCs [10] becomes a more powerful design alternative to provide cooperative diversity gain for various relay-aided systems [329–331], owing to its remarkable design flexibility guaranteed by its linear structure. Furthermore, the authors of [128, 332, 333] exploited the diversity–multiplexing gain trade-off as a means of evaluating the fundamental limitations of different cooperation strategies. In a similar manner, the same trade-off was capitalized on by Zheng and Tse in the context of the co-located MIMO systems [23].

In an effort to introduce channel coding schemes into cooperative MIMO systems for the sake of attaining near error-free transmission, the authors of [123, 126] developed the so-called 'coded-cooperative' schemes, where each relay contributes extrinsic information in a manner similar to the 'parallel-concatenated' component of classic turbo codes [9]. More explicitly, after receiving the channel-coded signals broadcast by the source node, the relays decode and interleave the information, before it is re-encoded by another channel encoder. Hence, the BS becomes capable of exploiting the extrinsic information gleaned from various interleaved replicas of the transmitted information by employing a conventional turbo detector [9]. In [134, 137], the coded-cooperative-aided schemes have been further improved by forwarding the soft estimates of the transmitted bits from the relays, rather than forwarding the hard decoded bits. However, this family of coded-cooperative schemes inevitably imposes both a high complexity and a high delay at the relays. Furthermore, it increases the power consumption of the relays despite transmitting no source data from themselves.

In order to attain cooperative diversity as well as to eliminate high-complexity operations at the relays, in this chapter we propose a novel family of twin-layer CLDCs based on the AF protocol. Since the AF strategy typically relies on twin-phase transmissions, namely the broadcast interval and the cooperation interval, the proposed CLDC schemes also have a twin-layer structure, which allows us to explore each transmission interval's specific characteristics as discussed in more detail in Section 9.2.3. Furthermore, the novel class of twin-layer CLDCs inherits the LDCs' flexible yet powerful linear structure, where similar properties can be observed. We also propose a novel serial concatenated coded-cooperative system, and highlight its benefits in contrast to the coded-cooperative schemes of [123, 126] based on parallel concatenated philosophies. More explicitly, the novel features of the schemes proposed in this chapter are listed as follows.

- The proposed twin-layer CLDC schemes may be readily integrated with next-generation systems, such as the Third Generation Partnership Project's Long Term Evolution (3GPP-LTE) proposals.
- The twin-layer CLDC schemes advocated are capable of supporting an arbitrary number of relays equipped with single or multiple antennas, while adopting various modulation schemes.
- The power consumption as well as the complexity imposed by cooperation at the relays are insignificant, since only simple linear combination operations are performed.
- We propose a novel IRCC-aided IR-PCLDC using iterative decoding, which is designed based on EXIT charts [158].
- The proposed IRCC-coded IR-PCLDC scheme becomes capable of providing flexible rates, which may be harmonized with the SNR encountered.

The rest of this chapter is structured in two parts. In the first part, we commence by presenting a general system model for the CLDCs based on the AF protocol in Section 9.2.1 and outline the associated assumptions in Section 9.2.2, followed by the mathematical characterization of the twin-layer CLDC scheme. Hence, the linkage between LDCs designed for co-located MIMO systems and CLDCs designed for cooperative MIMO systems is outlined in Section 9.2.4. In the second part of our investigations, Section 9.3 introduces the

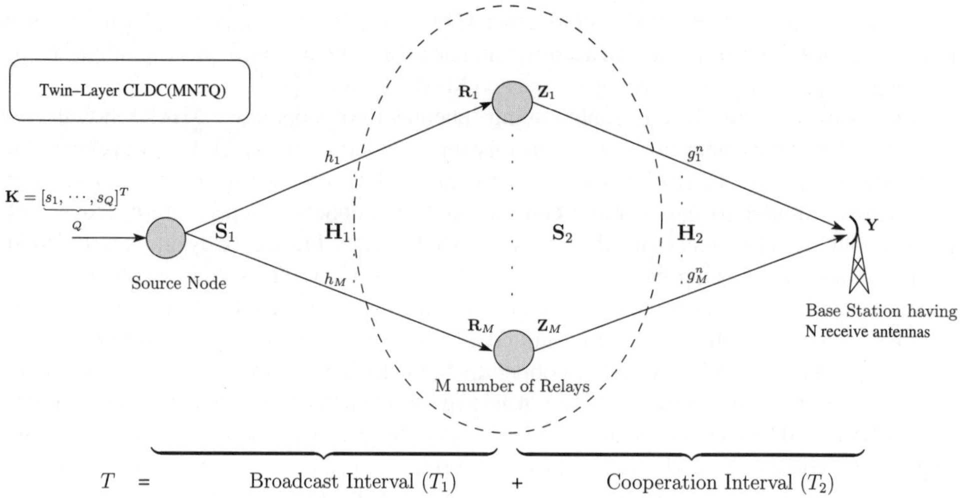

Figure 9.2: Schematic of the cooperation-aided UL system employing the twin-layer CLDCs.

irregular design philosophy in the context of cooperative systems and proposes a novel IRCC-coded IR-PCLDC scheme in order to maintain high rates across a wide SNR range, while maintaining an infinitesimally low BER. We focus our attention on the irregular system's distinctive features in the context of the cooperative MIMO systems. The corresponding irregular co-located MIMO system detailed in Section 7.5.3 is used as the benchmark. Finally, we summarize the findings of this chapter and provide our concluding remarks in Section 9.4.

9.2 Twin-layer CLDCs

9.2.1 System Model

In this section, we commence our discourse with the detailed description of the proposed twin-layer CLDCs based on the AF cooperation protocol, noting that this philosophy may be readily extended to DF and CF strategies. An UL scenario in a cellular network is considered, as exemplified in Figure 9.1, where a source node communicates with the BS having N receive antennas with the aid of M independent relays. For simplicity, each cooperating node is assumed to be equipped with a single transmit antenna, although multiple-antenna-aided nodes may also be incorporated into the proposed CLDCs.

Figure 9.2 portrays the schematic of the cooperation-aided UL system based on the proposed twin-layer CLDCs. As seen in Figure 9.2, each transmission block consists of two intervals, namely the broadcast interval and the cooperation interval. During the broadcast interval T_1, the source broadcasts an UL information-bearing vector $\mathbf{K} = [s_1, \ldots, s_Q]^{\mathrm{T}}$ containing Q L-PSK symbols to the relay nodes using the CLDC's first-layer dispersion matrix, which is described in detail in Section 9.2.3. During the cooperation interval T_2, the relays form a virtual antenna array and cooperatively transmit a space-time codeword $\mathbf{S}_2$ of Figure 9.2 to the BS based on the CLDC's second-layer dispersion matrices. Hence, the twin-layer CLDC scheme of Figure 9.2 based on the AF strategy can be fully specified by the parameters $(MNTQ)$, provided that the total number of channel uses T obeys $T = T_1 + T_2$. Throughout our investigations, a side-by-side comparison of the twin-layer cooperative

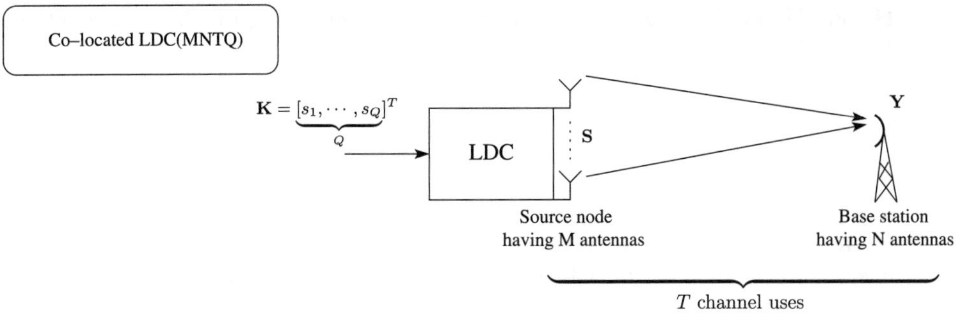

Figure 9.3: Schematic of a co-located UL MIMO system using LDCs.

CLDC($MNTQ$) scheme and the corresponding co-located LDC($MNTQ$) arrangement of Figure 9.3 is provided.

9.2.2 System Assumptions

In this section, we outline our assumptions in order to make the twin-layer CLDC scheme of Figure 9.2 as practical as possible. The assumptions and their rationale are summarized as follows.

- All of the relays of Figure 9.2 are assumed to transmit synchronously. Time-synchronous transmissions can be accomplished when the relative delays between the relays are significantly shorter than the symbol duration. Solutions designed for asynchronous cooperations have been proposed by a number of authors [334, 335], but this issue is beyond the scope of our discussions.

- All of the nodes of Figure 9.2 are assumed to have a single antenna and hence operate in half-duplex mode, i.e. at any point of time, a node can either transmit or receive. This constraint is imposed to prevent the high-power transmit signal from contaminating the low-power received signal, for example by the non-linear distortion-induced out-of-bound emissions routinely encountered at the transmitter. This may even potentially mislead the Automatic Gain Control (AGC) of the receiver into believing that a high-level signal was received, which would block the useful received signal arriving at a low power.

- All of the relays of Figure 9.2 transmit and receive on the same frequency as the source node, in order to avoid wasting or occupying additional bandwidth.

- No communication is permitted between the relays, in an effort to minimize the total network traffic. The relays may use the same unallocated time slot for their reception and transmission.

- Since the simple AF strategy is adopted, only linear combination operations are performed at the relays before retransmitting the signals dispersed to the cooperating MIMO elements to the BS, as is explicitly shown in the context of Equation (9.13).

- The BS seen in Figure 9.2 is equipped with N receive antennas. Moreover, the BS is assumed to have perfect knowledge of the CSI for all of the wireless links, i.e. it knows

$\mathbf{H}_1$ and $\mathbf{H}_2$ of Figure 9.2 in order to facilitate coherent detection. More particularly, $\mathbf{H}_1$ and $\mathbf{H}_2$ are given by

$$\mathbf{H}_1 = [h_1, \ldots, h_M]^{\mathrm{T}}, \quad \mathbf{H}_2 = \begin{pmatrix} g_1^1 & \cdots & g_m^1 \\ \vdots & \ddots & \vdots \\ g_1^N & \cdots & g_m^N \end{pmatrix}, \tag{9.1}$$

where h_m denotes the CIR between the source and the mth relay of the broadcast interval and g_m^n denotes the CIR between the mth relay and the nth receive antenna of the BS during the cooperation interval, and we have $m = (1, \ldots, M)$ and $n = (1, \ldots, N)$.

- All of the channel matrices $\mathbf{H}_1$ and $\mathbf{H}_2$ of Figure 9.2 are assumed to be representing quasi-static Rayleigh fading, i.e. the channel gains remain constant during T time slots and change independently at the beginning of the next. Furthermore, the channel gains are assumed to be spatially independent, while having a unit variance. Therefore, the resultant virtual antenna array elements can be considered to be subjected to perfectly independent fading. In contrast, the co-located MIMO elements of Figure 9.3 typically suffer from a certain degree of spatial correlation, owing to their insufficient spatial separation.

- We confine the total number of channel uses of the twin-layer CLDC scheme of Figure 9.2 to T, where $T = T_1 + T_2$. Hence, by appropriately adjusting the parameters T_1 or T_2, different degrees of freedom can be provided for the broadcast interval as well as for the cooperation interval.

- At any given time, the total transmit power of the twin-layer CLDC scheme of Figure 9.2 is normalized to unity. More explicitly, the source transmits at unit power during the broadcast interval, but remains silent during the cooperation interval. On the other hand, after listening for T_1 time slots, each relay transmits a signal vector with the power of $1/M$ in order to exploit the next unallocated T_2 time slots. Since M relays are involved in the cooperation, the total power remains unity during the cooperation interval.

9.2.3 Mathematical Representations

Based on the assumptions outlined in Section 9.2.2, we are now ready to provide a more insightful description of the twin-layer CLDC model of Figure 9.2. When a user intends to transmit data, the BS first functions as a control unit and selects a group of nearby users in order to assist the source node's UL transmission. More specifically, the BS assigns a dispersion matrix χ_1 obeying Equation (9.4) to the source node in order to enable its UL transmission during the broadcast interval. Each relay also receives a different dispersion matrix $\mathbf{B}_m$ obeying Equation (9.13) from the BS in order to attain diversity during the cooperation interval. These control information transmissions are necessary for initializing the cooperation-aided UL transmission of the source node.

Since the distance between the relays is typically far lower than the distance to the source, we could reasonably assume that the SNR between the source and the relays becomes

$$\rho_{SR} = \frac{1}{\sigma_{SR}}, \tag{9.2}$$

where σ_{SR} denotes the corresponding noise variance at the relays. During the cooperation interval, each relay transmits at a power of $1/M$, which implies that the total transmit power is evenly distributed across the M relays of Figure 9.2. Hence, the SNR at the BS can be written as

$$\rho_{RB} = \frac{1}{\sigma_{RB}}, \tag{9.3}$$

where σ_{RB} is the noise variance at the BS.

In the first stage of the cooperation-aided communication, namely the broadcast interval occupying T_1 time slots, the source node encodes each L-PSK modulated information vector $\mathbf{K} = [s_1, \ldots, s_Q]^{\mathrm{T}}$ to a transmission vector $\mathbf{S}_1 \in \zeta^{T_1 \times 1}$ based on

$$\underbrace{\begin{pmatrix} t^1 \\ \vdots \\ t^{T_1} \end{pmatrix}}_{\mathbf{S}_1} = \underbrace{\begin{pmatrix} x^{1,1} & \cdots & x^{1,Q} \\ \vdots & \ddots & \vdots \\ x^{T_1,1} & \cdots & x^{T_1,Q} \end{pmatrix}}_{\chi_1} \underbrace{\begin{pmatrix} s_1 \\ \vdots \\ s_Q \end{pmatrix}}_{\mathbf{K},} \tag{9.4}$$

where the first-layer dispersion matrix $\chi_1 \in \zeta^{T_1 \times Q}$ is responsible for dispersing the information vector $\mathbf{K}$ to all the T_1 temporal dimensions. According to the power constraint, the transmission vector is normalized to $E\{\mathrm{tr}(\mathbf{S}_1^H \mathbf{S}_1)\} = T_1$, which requires the dispersion matrix χ_1 to satisfy

$$\mathrm{tr}(\chi_1^H \chi_1) = T_1. \tag{9.5}$$

At the mth relay, the corresponding receive vector $\mathbf{R}_m = [r_m^1, \ldots, r_m^{T_1}]^{\mathrm{T}}$ becomes

$$\mathbf{R}_m = h_m \mathbf{S}_1 + \mathbf{V}_m. \tag{9.6}$$

By stacking the received signals from M relays, we arrive at

$$\underbrace{\begin{pmatrix} \mathbf{R}_1 \\ \vdots \\ \mathbf{R}_M \end{pmatrix}}_{\mathbf{R}} = \underbrace{\begin{pmatrix} h_1 \mathbf{I} \\ \vdots \\ h_M \mathbf{I} \end{pmatrix}}_{\bar{\mathbf{H}}_1} \mathbf{S}_1 + \underbrace{\begin{pmatrix} \mathbf{V}_1 \\ \vdots \\ \mathbf{V}_M \end{pmatrix}}_{\bar{\mathbf{V}}_1,} \tag{9.7}$$

where $\mathbf{R} = [\mathbf{R}_1, \ldots, \mathbf{R}_M]^{\mathrm{T}}$ and $\bar{\mathbf{V}}_1 \in \zeta^{MT_1 \times 1}$ denotes the combined noise vector having a variance of σ_{SR}. The equivalent channel matrix $\bar{\mathbf{H}}_1 \in \zeta^{MT_1 \times T_1}$ of Equation (9.7) can be represented by

$$\bar{\mathbf{H}}_1 = \mathbf{H}_1 \otimes \mathbf{I}, \tag{9.8}$$

where $\mathbf{I}$ denotes an identity matrix having a size of $(T_1 \times T_1)$. Furthermore, by combining Equations (9.4) and (9.7), we have

$$\mathbf{R} = \bar{\mathbf{H}}_1 \chi_1 \mathbf{K} + \bar{\mathbf{V}}_1. \tag{9.9}$$

During the cooperation interval of Figure 9.2, the relays cooperatively construct a space-time codeword based on the pre-assigned dispersion matrices $\mathbf{B}_m$ of Equation (9.13) using the received signal vectors $\mathbf{R}_m$ of Equation (9.6). First, the energy of the received signals has to be normalized during the broadcast interval. From Equation (9.6), we have

$$E\{\mathbf{R}_m^H \mathbf{R}_m\} = (1 + \sigma_{SR}) T_1. \tag{9.10}$$

Thus, before performing the second-layer dispersion operations, we multiply $\mathbf{R}_m$ of Equation (9.6) by the normalization factor

$$\theta = \sqrt{\frac{T_1}{E\{\mathbf{R}_m^H \mathbf{R}_m\}}} = \sqrt{\frac{1}{1+\sigma_{SR}}}, \tag{9.11}$$

so that we arrive at

$$\theta^2 E\{\mathbf{R}_m^H \mathbf{R}_m\} = T_1. \tag{9.12}$$

At the mth relay, the second-layer dispersion matrices $\mathbf{B}_m \in \zeta^{T_2 \times T_1}$ are employed to disperse the normalized received vector $\theta \mathbf{R}_m$ to the T_2 available time slots. Hence, the mth relay's transmitted vector is given by

$$\underbrace{\begin{pmatrix} z_m^1 \\ \vdots \\ z_m^{T_2} \end{pmatrix}}_{\mathbf{Z}_m} \underbrace{=}_{=} \theta \underbrace{\begin{pmatrix} b_m^{1,1} & \cdots & b_m^{1,T_1} \\ \vdots & \ddots & \vdots \\ b_m^{T_2,1} & \cdots & b_m^{T_2,T_1} \end{pmatrix}}_{\theta \cdot \mathbf{B}_m} \underbrace{\begin{pmatrix} r_m^1 \\ \vdots \\ r_m^{T_1} \end{pmatrix}}_{\mathbf{R}_m}, \tag{9.13}$$

which should satisfy the power constraint of $E\{\mathrm{tr}(\mathbf{Z}_m^H \mathbf{Z}_m)\} = T_2/M$ and z_m^i $(i = 1, \ldots, T_2)$ denotes the mth relay's transmit signal of the ith slot. Thus, the set of second-layer matrices $\mathbf{B}_m$ should obey

$$\mathrm{tr}(\mathbf{B}_m^H \mathbf{B}_m) = \frac{T_2}{M}. \tag{9.14}$$

Therefore, the 'virtual' space-time codeword $\mathbf{S}_2 \in \zeta^{M \times T_2}$ of Figure 9.2 can be formed by concatenating the dispersed vectors from all of the relays, which is given by

$$\mathbf{S}_2 = \begin{pmatrix} \mathbf{Z}_1^{\mathrm{T}} \\ \vdots \\ \mathbf{Z}_M^{\mathrm{T}} \end{pmatrix} = \theta \begin{pmatrix} (\mathbf{B}_1 \mathbf{R}_1)^{\mathrm{T}} \\ \vdots \\ (\mathbf{B}_M \mathbf{R}_M)^{\mathrm{T}} \end{pmatrix}. \tag{9.15}$$

Observe in Equation (9.15) that each relay contributes one row of the second-layer space-time codeword $\mathbf{S}_2$. Also note that each relay's received signal $\mathbf{R}_m$ formulated in Equation (9.6) may experience quite different channel conditions, owing to experiencing i.i.d. Rayleigh fading. Furthermore, the process of forming the cooperative space-time codeword $\mathbf{S}_2$ for the cooperation interval is visualized in Figure 9.4. More explicitly, the mth relay disperses the corresponding normalized received signal vector $\theta \mathbf{R}_m = \theta [r_m^1, \ldots, r_m^{T_1}]^{\mathrm{T}}$ to the T_2 temporal slots using the assigned dispersion matrix $\mathbf{B}_m$ obeying Equation (9.13), as seen in Figure 9.4. The resultant signal vector $\mathbf{Z}_m$ contributes one row of the space-time codeword $\mathbf{S}_2$ of Equation (9.15).

At the BS, the received signal matrix $\mathbf{Y} \in \zeta^{N \times M}$ of Figure 9.2 becomes

$$\mathbf{Y} = \mathbf{H}_2 \mathbf{S}_2 + \mathbf{V}_2. \tag{9.16}$$

Define the $row()$ operation as the vertical stacking of the rows of an arbitrary matrix. Subjecting both sides of Equation (9.16) to the $row()$ operation gives the equivalent system matrix:

$$\bar{\mathbf{Y}} = \bar{\mathbf{H}}_2 \mathbf{Z} + \bar{\mathbf{V}}_2, \tag{9.17}$$

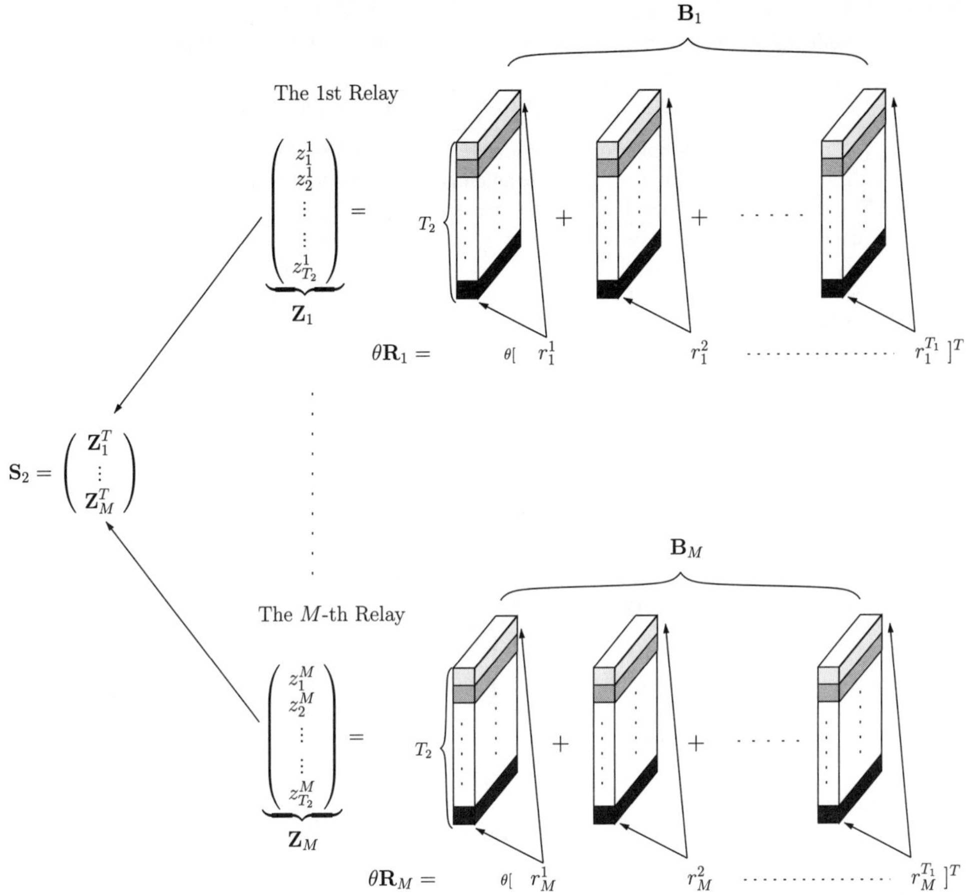

Figure 9.4: The cooperative space-time codeword $\mathbf{S}_2$ formulated based on Equations (9.13) and (9.15).

where $\bar{\mathbf{Y}} \in \zeta^{NT_2 \times 1}$ and $\bar{\mathbf{V}}_2 \in \zeta^{NT_2 \times 1}$ denotes the noise vector having a variance of σ_{RB}. The equivalent channel matrix $\bar{\mathbf{H}}_2 \in \zeta^{NT_2 \times MT_2}$ of Equation (9.17) is given by

$$\bar{\mathbf{H}}_2 = \mathbf{H}_2 \otimes \mathbf{I}, \tag{9.18}$$

and the equivalent transmission matrix $\mathbf{Z} \in \zeta^{MT_2 \times 1}$ of Equation (9.17) becomes

$$\underbrace{\begin{pmatrix} \mathbf{Z}_1 \\ \vdots \\ \vdots \\ \mathbf{Z}_M \end{pmatrix}}_{\mathbf{Z}} \begin{matrix} = \\ \\ = \end{matrix} \underbrace{\begin{pmatrix} \mathbf{B}_1 & \mathbf{0} & \cdots & \mathbf{0} \\ \mathbf{0} & \mathbf{B}_m & \cdots & \mathbf{0} \\ \mathbf{0} & \mathbf{0} & \ddots & \mathbf{0} \\ \mathbf{0} & \cdots & \cdots & \mathbf{B}_M \end{pmatrix}}_{\chi_2} \cdot \theta\mathbf{R}, \quad \chi_2 \cdot \theta\mathbf{R}, \tag{9.19}$$

where $\mathbf{0} \in \zeta^{T_2 \times T_1}$ denotes a zero matrix. The equivalent second-layer DCM $\chi_2 \in \zeta^{MT_2 \times MT_1}$ defined in Equation (9.19) fully characterizes the transmissions during the cooperation interval.

Table 9.1: Comparison of the LDCs of Figure 9.3 and the CLDCs of Figure 9.2.

	LDCs of Figure 9.3	CLDCs of Figure 9.2
M	Number of transmit antennas	Number of relays
N	Number of receive antennas	Number of receive antennas
T	Time slots per LDC block	$T = T_1 + T_2$
Q	Number of symbols	Number of symbols
R	$R_{\text{LDC}} = Q/T$	$R_{\text{CLDC}} = Q/T$
$\mathbf{K}$	Perfect knowledge	Obtained using Equation (9.6)
DCM	χ	$\{\chi_1, \chi_2\}$
Equivalent channel matrix	$\bar{\mathbf{H}}\chi$ of Equation (7.15)	$\bar{\mathbf{H}}_2\chi_2\bar{\mathbf{H}}_1\chi_1$ of Equation (9.20)

Further combining Equations (9.9), (9.17) and (9.19) we arrive at

$$\begin{aligned}\bar{\mathbf{Y}} &= \bar{\mathbf{H}}_2\mathbf{Z} + \bar{\mathbf{V}}_2 \\ &= \theta\bar{\mathbf{H}}_2\chi_2\mathbf{R} + \bar{\mathbf{V}}_2 \\ &= \theta\bar{\mathbf{H}}_2\chi_2(\bar{\mathbf{H}}_1\chi_1\mathbf{K} + \bar{\mathbf{V}}_1) + \bar{\mathbf{V}}_2 \\ &= \theta\,\underbrace{\bar{\mathbf{H}}_2\chi_2\bar{\mathbf{H}}_1\chi_1}\,\mathbf{K} + \underbrace{\theta\bar{\mathbf{H}}_2\chi_2\bar{\mathbf{V}}_1 + \bar{\mathbf{V}}_2} \\ &= \qquad \bar{\mathbf{H}}\mathbf{K} \qquad + \qquad \bar{\mathbf{V}}.\end{aligned} \tag{9.20}$$

The combined noise $\bar{\mathbf{V}}$ becomes colored with a covariance of $\sigma_0 = (\mathbf{I}_{T_2}\sigma_{RB} + \theta^2\bar{\mathbf{H}}_2\chi_2\chi_2^H\bar{\mathbf{H}}_2^H\sigma_{SR})$, owing to the noise amplification and recombination experienced at the relays. The equivalent channel matrix $\bar{\mathbf{H}} \in \zeta^{NT_2 \times Q}$ of Equation (9.20) is known to the BS, since the BS has the knowledge of the CIR matrices $\mathbf{H}_1$ as well as $\mathbf{H}_2$ of Figure 9.2 and that of the DCM pair $\{\chi_1, \chi_2\}$. Most importantly, Equation (9.20) demonstrates that the twin-layer CLDC structure originally formulated in Equations (9.4) and (9.13) can be merged into a 'single' equivalent channel matrix.

In order to perform ML detection, we have to whiten the noise, which is achieved by multiplying by ${\sigma_0}^{-1/2}$ on both sides of Equation (9.20). Hence, the ML estimation of the transmitted symbol vector $\mathbf{K} = [s_1, \ldots, s_Q]^{\mathrm{T}}$ can be written as

$$\bar{\mathbf{K}} = \arg\{\min(\|{\sigma_0}^{-1/2}\mathbf{Y} - {\sigma_0}^{-1/2}\bar{\mathbf{H}}\mathbf{K}_f\|^2)\}, \tag{9.21}$$

where $\mathbf{K}_f$ denotes all of the possible combinations of the Q transmitted symbols.

9.2.4 Link between CLDCs and LDCs

This section aims at characterizing the fundamental link between the LDCs designed for the co-located MIMO systems of Figure 9.3 and the CLDCs contrived for the cooperative MIMO systems of Figure 9.2. The similarities as well as the differences between the LDCs and the CLDCs are summarized in Table 9.1 and a range of remarks is offered as follows.

- The LDCs of Figure 9.3 employ M transmit antennas, whereas the CLDCs of Figure 9.2 form an identical-size virtual antenna array with the aid of M relays selected by the BS.

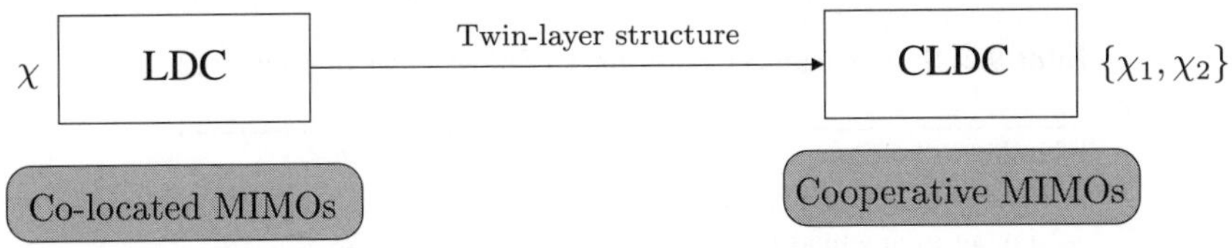

Figure 9.5: Link between LDCs designed for co-located MIMO systems and twin-layer CLDCs designed for cooperative MIMO systems.

- In both scenarios, the UL receiver of the BS is assumed to have N receive antennas.
- The LDCs of Figure 9.3 occupy a total of T channel slots per space-time block. For the sake of a fair comparison, the twin-layer CLDCs also span $T = T_1 + T_2$ time slots per block. More explicitly, the broadcast interval having T_1 slots is employed for the relays to receive the source information, whereas T_2 time slots are used for the cooperation phase.
- The symbol rate of both the LDCs and CLDCs is defined as $R = Q/T$, where Q is the number of L-PSK symbols transmitted per space-time block.
- When constructing the symbol vector $\mathbf{K} = [s_1, \ldots, s_Q]^{\mathrm{T}}$, each transmit antenna element of the LDCs of Figure 9.3 has direct access to the Q L-PSK symbols. In contrast, the CLDCs of Figure 9.2 only have access to the potentially channel-contaminated version of $\mathbf{K}$ received during the broadcast interval using Equation (9.6). We characterize the CLDCs' BER performance in Figure 9.7, when the relays have differently contaminated information vectors $\mathbf{K}$.
- LDCs can be fully characterized by a single DCM χ, as extensively demonstrated in Chapter 7. In contrast, the CLDCs are characterized by a DCM pair $\{\chi_1, \chi_2\}$, which was given in Equations (9.4) and (9.19), respectively, owing to the twin-layer structure shown in Figure 9.2. The twin-layer structure portrays the fundamental design difference between LDCs and CLDCs, and this relationship is shown explicitly in Figure 9.5.
- The LDCs' equivalent channel model of Equation (7.15) implies transmitting the symbol vector $\mathbf{K}$ through the channel characterized by the equivalent channel matrix $\bar{\mathbf{H}}\chi$. Similarly, Equation (9.20) indicates that the CLDCs transmit the symbol vector $\mathbf{K}$ through the channel characterized by the equivalent channel matrix $\bar{\mathbf{H}}_2\chi_2\bar{\mathbf{H}}_1\chi_1$. More explicitly, the DCM χ_1 of Equation (9.4) together with the CIR matrix $\bar{\mathbf{H}}_1$ of Equation (9.8) characterizes the transmission during the broadcast interval, while the cooperation interval is characterized by the DCM χ_2 of Equation (9.19) as well as by $\bar{\mathbf{H}}_2$ of Equation (9.18).

Observe furthermore in Equations (7.15) and (9.20) that both LDCs and CLDCs obey a similar equivalent system structure, where the equivalent CIR matrix is determined by the channel CIR experienced and the specific DCM employed. Recall from Section 7.2.4 that we optimized the LDCs' DCM χ based on the DCMC capacity of Equation (7.23) using the corresponding equivalent channel matrix of Equation (7.15). Similarly, for a CLDC scheme, we are looking for the particular pair of DCMs $\{\chi_1, \chi_2\}$ that maximizes the DCMC capacity of Equation (7.23). Accordingly, all the DCM pairs $\{\chi_1, \chi_2\}$ used in this chapter are listed in Appendix F.

Table 9.2: System parameters for the CLDC schemes of Figure 9.2.

Number of relays	M
Number of antenna per relay	1
Number of antennas at BS	N
Total channel uses	$T = T_1 + T_2$
Number of symbols per CLDC block	Q
SNR (dB)	$\rho_{SR} = \rho_{RB}$
Modulation	BPSK
Mapping	Gray mapping
Detector	ML of Equation (9.21)

9.2.5 Performance Results

In this section, we present our simulation results for a number of CLDC schemes obeying the structure of Figure 9.2, which are associated with parameters $(MNTQ)$. In most of our simulations, we set the SNR ρ_{SR} experienced at the relays to the SNR ρ_{RB} recorded at the BS and the resultant BER performance is plotted against ρ_{RB}. However, the scenarios where $\rho_{SR} \neq \rho_{RB}$ are also characterized in Figure 9.9. All of the system parameters are listed in Table 9.2, unless otherwise stated.

Figure 9.6 characterizes the BER of a family of CLDCs having $M = 4$, $N = 1$, $T = 8$ and $Q = 3$ as well as using different T_1 and T_2 values, when transmitting over uncorrelated i.i.d. Rayleigh fading channels. Since we set the total number of channel uses per CLDC block to a relatively high value of $T = 8$, we are able to vary the resource allocation by employing different T_1 and T_2 values during the broadcast and cooperation intervals. Observe in Figure 9.6 that the best achievable BER performance was recorded when we have $T_1 = 2$ and $T_2 = 6$. This phenomenon is related to the achievable cooperative diversity gain of the CLDC schemes of Figure 9.2.

More explicitly, observe in Figure 9.2 that during the broadcast interval, only first-order diversity can be achieved, because each cooperating node is equipped with a single antenna. During the cooperation interval, the space-time transmission matrix $\mathbf{S}_2$ of Equation (9.15) is formed across the relays, which potentially facilitates a spatial diversity order of $N \cdot \min(M, T_2)$ according to Theorem 7.3. Observe in Figure 9.6 that when we have $T_2 \geq (M = 4)$, the CLDCs achieve a cooperative diversity order of $D \approx 4$. The achievable performance is not exactly identical to that of fourth-order diversity, since the relays only have access to the noisy version of the transmitted information. When we have $(T_2 = 1, 2, 3) < M$, $D \approx 1, 2, 3$ can be observed from Figure 9.6, respectively. Again, observe in Figure 9.6 that the CLDC(4183) scheme having $T_1 = 2$ and $T_2 = 6$ exhibits the best BER performance, since it guaranteed the maximum achievable cooperative diversity order provided by the cooperation interval as well as protected the signals received by the relays during the broadcast interval. Figure 9.6 also plots the BER of the LDC(4183) scheme of Figure 9.3 designed for the co-located MIMO system as our benchmark. Observe in Figure 9.6 that a SNR gap of 8.3 dB is recorded at BER $= 10^{-4}$ between the BER curves of the LDC(4183) scheme and the CLDC(4183) arrangement having $T_1 = 2$ and $T_2 = 6$. The non-cooperative LDC scheme outperformed its cooperative counterpart, which is predominantly owing to the imperfect reception of the source data by the relays during the broadcast phase. A further reason is that only a reduced number of slots is available for increasing the diversity gain

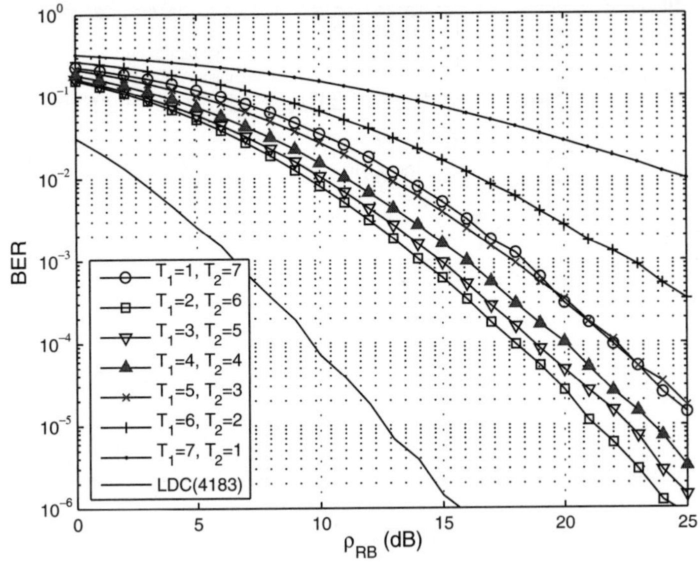

Figure 9.6: BER comparison of a group of CLDCs obeying the architecture of Figure 9.2 having $M = 4$, $N = 1$, $T = 8$ and $Q = 3$ while using different T_1 and T_2 values, when transmitting over uncorrelated i.i.d. Rayleigh fading channels. All of the system parameters are summarized in Table 9.2.

during the cooperation phase owing to dedicating $T_1 = 2$ slots to the broadcast phase, which imposes further SNR penalties.

In order to further explain this SNR gap in more quantitative terms, in Figure 9.7 we plot the BER performance of the CLDC(4183) scheme having $T_1 = 2$ and $T_2 = 6$, when transmitting over perfect broadcast channels or AWGN as well as over noisy Rayleigh-faded source-to-relay channels. The channels between the relays and the BS are assumed to be uncorrelated i.i.d. Rayleigh fading channels. When we have perfect source-to-relay links, namely the relays have access to the perfect source information, the CLDC(4183) scheme suffers about 1.8 dB SNR penalty in comparison with the LDC(4183) benchmark, since only $T_2 = 6$ time slots are employed for the cooperative transmission scenario compared with $T = 8$ slots available for the co-located MIMO systems. When the source-to-relay links are LOS AWGN channels, the associated SNR penalty in comparison with the LDC(4183) benchmark increases only modestly to 3.3 dB. Finally, when each relay's received signals $\mathbf{R}_m$ formulated in Equation (9.6) suffers further, owing to experiencing independent Rayleigh fading, the SNR gap widens more dramatically to 8.3 dB. It is worth noting at this stage that owing to the crucial impact of the broadcast phase integrity on the overall BER performance, it is anticipated that the best combination of T_1 and T_2 has the potential of improving the overall BER performance.

The BER performance gap between the CLDCs and the corresponding LDCs illustrated in Figure 9.6 was investigated under small-scale Rayleigh fading conditions. However, when the transmitted signals are subjected to large-scale fading effects, i.e. shadow fading, the cooperative MIMO system has the potential to outperform its co-located MIMO system counterpart. More particularly, Figure 9.8 characterizes the BER performance of the LDC(4183) scheme following the schematic of Figure 9.3, when communicating over

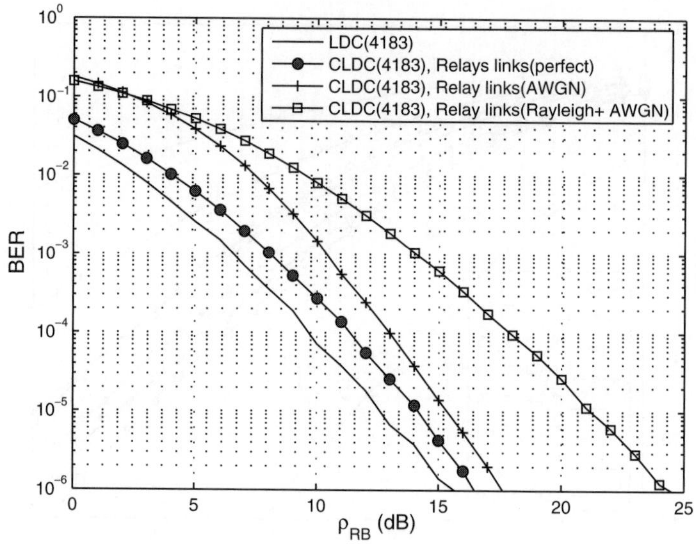

Figure 9.7: BER of the CLDC(4183) scheme of Figure 9.2 having $T_1 = 2$ and $T_2 = 6$. The source-to-relay channels were assumed to be either perfect, or a LOS AWGN or worse-case Rayleigh channels. The relay-to-BS channels remain uncorrelated i.i.d. Rayleigh fading channels. All of the system parameters are summarized in Table 9.2.

uncorrelated i.i.d. Rayleigh faded channels contaminated by large-scale shadow fading. The shadow fading effect is modeled to have log-normal distribution [220], which can be written as

$$h_{\text{slow}} = 10^{\bar{h}_s/10}, \tag{9.22}$$

where $\bar{h}_s$ is a random Gaussian variable with zero mean and standard deviation Ω (dB).

Observe in Figure 9.8 that when the shadow fading is insignificant, i.e. we have $\Omega = 0$ dB, the LDC(4183) scheme maintains the diversity order of $D \approx 4$, although suffering approximately 1.4 dB SNR penalty compared with the scenario in the absence of shadowing. When we have $\Omega = 6$ dB, the achievable diversity order is reduced to $D = 1$, since all of the communication channels tend to experience the same shadow fading. In contrast, the cooperative MIMO systems can eliminate the shadow fading effect by appropriately choosing a group of relays to assist their transmissions, as illustrated in Figure 9.1. Hence, Figure 9.8 demonstrates that the CLDC(4183) scheme is capable of offering more reliable transmissions than its co-located counterpart, when large-scale fading effects dominate the BER performance.

Figure 9.9 shows the BER performance of the CLDC(4183) scheme of Figure 9.2 having $T_1 = 2$ and $T_2 = 6$, when the source-to-relay Rayleigh fading channels have an SNR of ρ_{SR}, while the relay-to-BS Rayleigh fading channels have a SNR of ρ_{RB}. We use the BER curve of the CLDC(4183) scheme having $\rho_{SR} = \rho_{RB}$ as the benchmark, which was previously recorded in Figure 9.6. When we have $\rho_{SR} = \infty$, namely the source-to-relay links are Rayleigh fading channels without AWGN, the resultant BER curve exhibits 1.2 dB SNR gain over the benchmark. When we gradually decrease the value of ρ_{SR}, an error floor begins to emerge, owing to the noise of the broadcast interval formulated in Equation (9.20). Again, Figure 9.9 provides evidence of the importance of the received signals' integrity at relays.

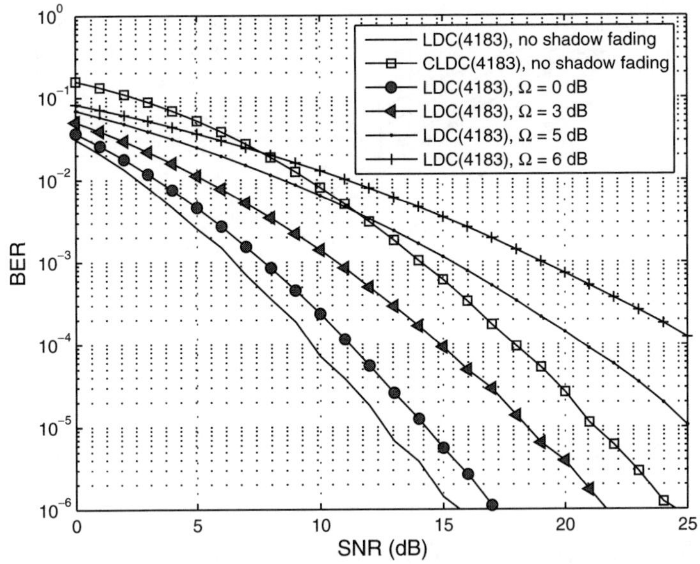

Figure 9.8: BER of the LDC(4183) scheme following the schematic of Figure 9.3, when communicating over uncorrelated i.i.d. Rayleigh faded channels contaminated by large-scale shadow fading governed by Ω.

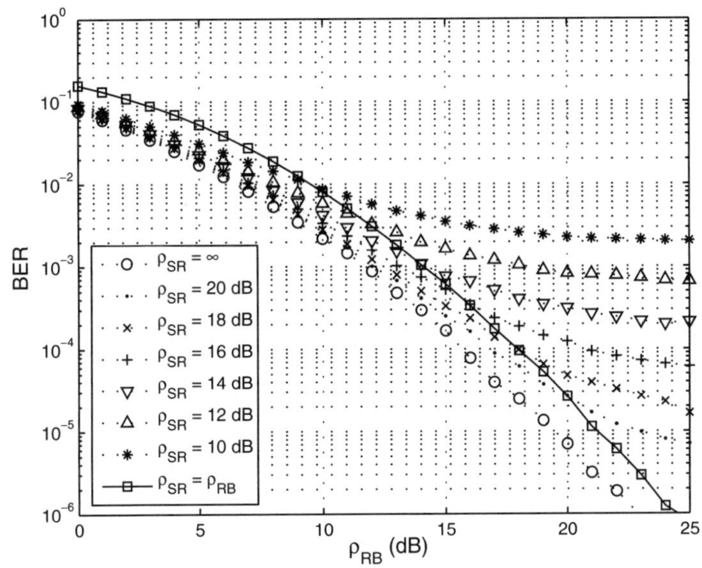

Figure 9.9: BER of the CLDC(4183) scheme of Figure 9.2 having $T_1 = 2$ and $T_2 = 6$, when communicating over uncorrelated i.i.d. Rayleigh faded channels at a SNR of ρ_{SR} and ρ_{RB}, respectively. All of the system parameters are summarized in Table 9.2.

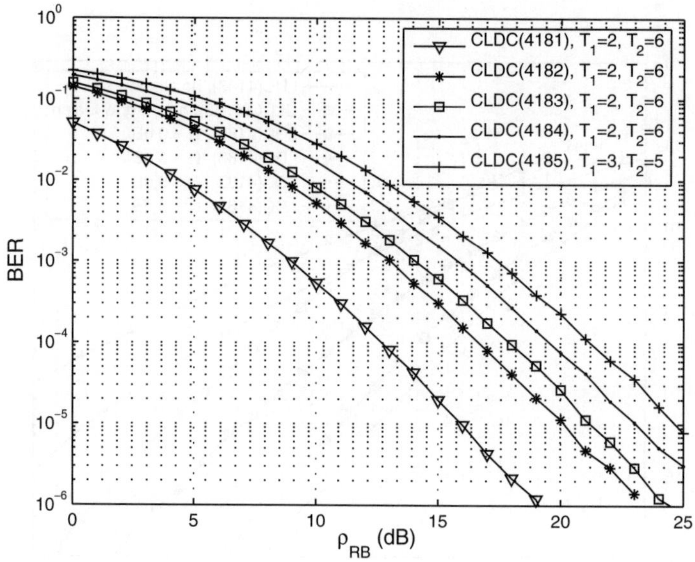

Figure 9.10: BER comparison of a group of CLDCs obeying the structure of Figure 9.2 having $M = 4$, $N = 1$, $T = 8$ and $Q = 1, 2, 3, 4, 5$, when transmitting over uncorrelated i.i.d. Rayleigh fading channels. All of the system parameters are summarized in Table 9.2.

Table 9.3: The best combination of T_1 and T_2 values for a group of CLDCs obeying the structure of Figure 9.2 having $M = 4$, $N = 1$, $T = 8$ and $Q = 1, 2, 3, 4, 5$.

	T	T_1	T_2	Diversity
CLDC(4181)	8	2	6	≈ 4
CLDC(4182)	8	2	6	≈ 4
CLDC(4183)	8	2	6	≈ 4
CLDC(4184)	8	2	6	≈ 4
CLDC(4185)	8	3	5	≈ 4

Figure 9.10 quantifies the BER performance of a group of CLDCs having $M = 4$, $N = 1$, $T = 8$ and $Q = 1, 2, 3, 4, 5$, when transmitting over uncorrelated i.i.d. Rayleigh fading channels. Previously, Figure 9.6 has demonstrated that the achievable BER performance is seriously affected by the integrity of the source-to-relay channels, depending on the specific choice of T_1 and T_2 values. Similar to Figure 9.6, we are capable of appropriately configuring a specific CLDC($418Q$) scheme by characterizing all of the possible T_1 and T_2 combinations and then choosing the particular configuration exhibiting the best achievable BER performance. A group of CLDCs configured using this method is listed in Table 9.3, while their BER performance is shown in Figure 9.10. Furthermore, observe in Table 9.3 that all of the CLDCs maintain $T_2 > M$, which potentially enables the CLDCs to achieve the maximum cooperative diversity order. When the value of Q is gradually increased, more time slots are necessary for the broadcast interval T_1 in order to increase the level of protection provided for the symbols received at the relays.

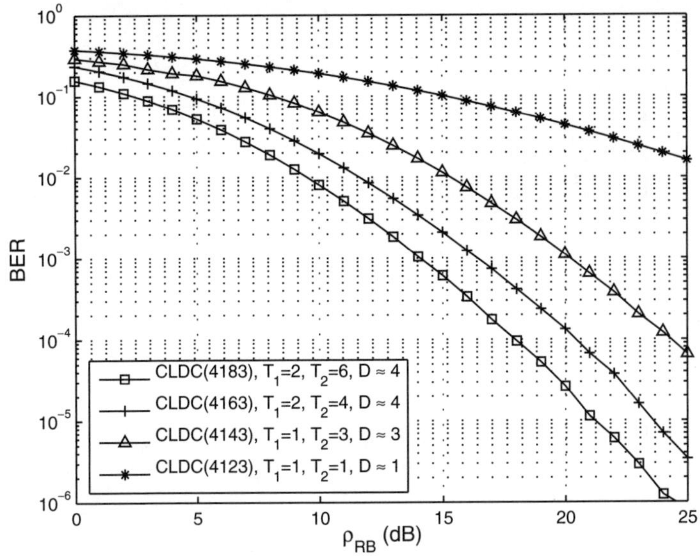

Figure 9.11: BER comparison of a group of CLDCs obeying the structure of Figure 9.2 and using $M = 4$, $N = 1$, $T = 2, 4, 6, 8$ and $Q = 3$, when transmitting over uncorrelated i.i.d. Rayleigh fading channels. All of the system parameters are summarized in Table 9.2.

Figure 9.11 characterizes a group of CLDCs obeying the structure of Figure 9.2 having $M = 4, N = 1, T = 2, 4, 6, 8$ and $Q = 3$, when transmitting over uncorrelated i.i.d. Rayleigh fading channels. Observe in Figure 9.11 that when the total number of slots available decreased from $T = 8$ to 4, the CLDC(4143) scheme having $T_1 = 1$ and $T_2 = 3$ exhibits a maximum diversity order of $D \approx 3$, since the cooperative diversity order is bounded by $N \cdot \min(M, T_2)$. When we have $T_1 = 1$ and $T_2 = 1$ corresponding to $T = 2$, the resultant CLDC(4123) scheme can only attain a cooperative diversity order of $D = 1$, despite the fact that $M = 4$ relays were employed to assist the cooperation-aided transmission. Again, Figure 9.11 demonstrates that the maximum achievable cooperative diversity order of a CLDC($MNTQ$) scheme is determined by both the total number of time slots T available and the number of relays M.

Figure 9.12 characterizes a group of CLDCs obeying the structure of Figure 9.2 having $M = 4, N = 1, 2, 3, T = 8$ and $Q = 3$ using $T_1 = 2$ as well as $T_2 = 6$, when communicating over uncorrelated i.i.d. Rayleigh fading channels. As expected, when the BS employs more receive antennas, the maximum achievable diversity order can be significantly improved as a benefit of the well-known receive diversity gain. In fact, for this family of CLDCs, Figure 9.12 suggests that a cooperative diversity order of $D \approx 4, 8, 12$ can be achieved, when increasing the value of N from 1 to 3.

Figure 9.13 attempts to characterize CLDCs' ability to provide cooperative diversity for the configuration of $M = 2, 3, 4,\ N = 1,\ T = 8$ and $Q = 3$, when transmitting over uncorrelated i.i.d. Rayleigh fading channels. Observe in Figure 9.13 that a diversity order of $D \approx 2, 3, 4$ can be achieved with the aid of $M = 2, 3, 4$ relays, respectively, provided that maintaining $T_2 > M$ is guaranteed. Recall from Figure 9.6 that the CLDC(4183) scheme of Table 9.3 constitutes the best configuration when we have $T_1 = 2$ and $T_2 = 6$. On the other hand, observe in Figure 9.13 that the CLDC(2183) scheme achieves the best performance when we have $T_1 = 3$ and $T_2 = 5$.

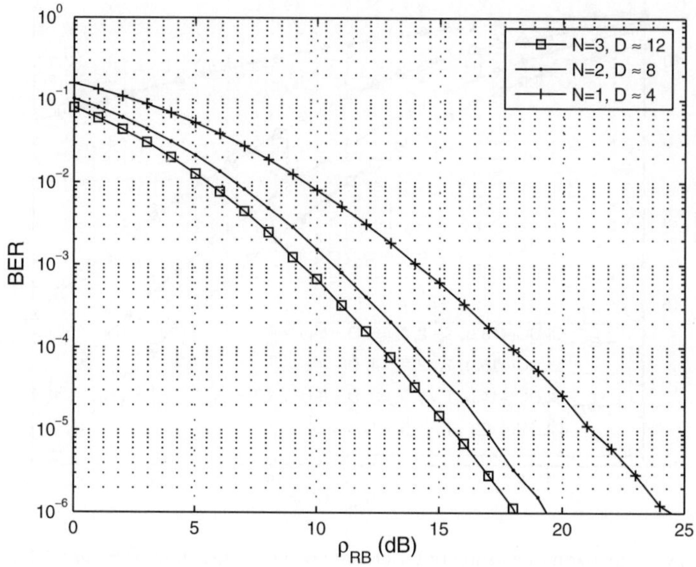

Figure 9.12: BER comparison of a group of CLDCs following the schematic of Figure 9.2 having $M = 4$, $N = 1, 2, 3$, $T = 8$ and $Q = 3$ using $T_1 = 2$ as well as $T_2 = 6$, when transmitting over uncorrelated i.i.d. Rayleigh fading channels. All of the system parameters are summarized in Table 9.2.

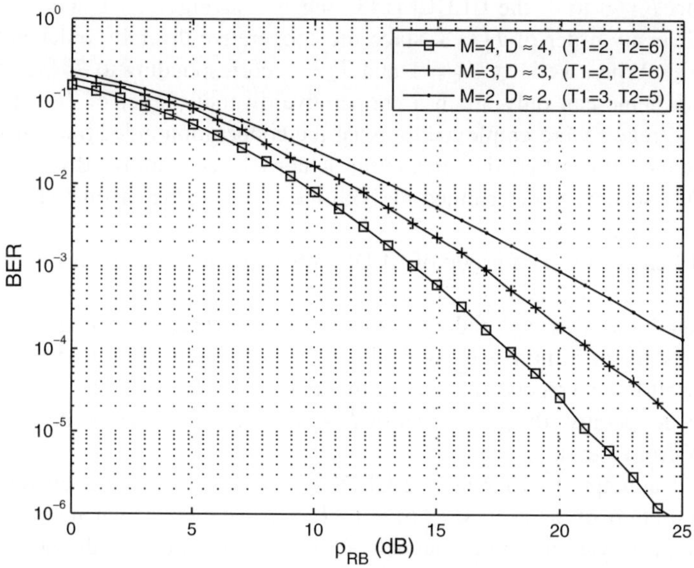

Figure 9.13: BER comparison of a group of CLDCs obeying the structure of Figure 9.2 having $M = 2, 3, 4$, $N = 1$, $T = 8$ and $Q = 3$, when transmitting over uncorrelated i.i.d. Rayleigh fading channels. All of the system parameters are summarized in Table 9.2.

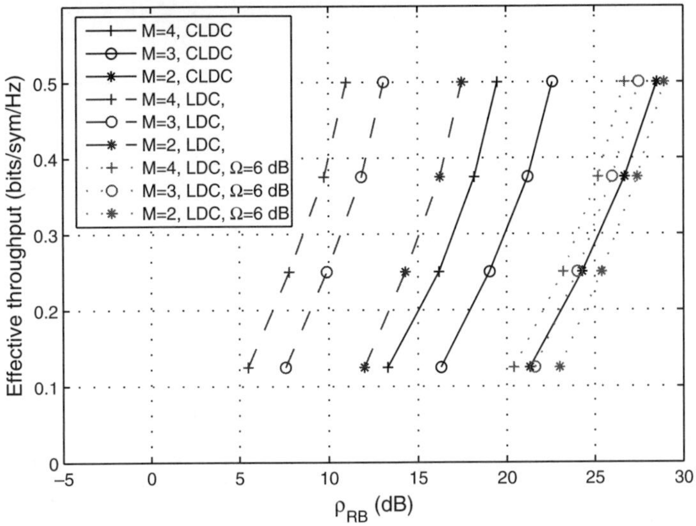

Figure 9.14: Effective throughput of a family of CLDCs obeying the structure of Figure 9.2 having $M = 2, 3, 4$, $N = 1$, $T = 8$ and $Q = 1, 2, 3, 4$ and the corresponding LDCs with/without shadowing, when transmitting over uncorrelated i.i.d. Rayleigh fading channels. All of the system parameters are summarized in Table 9.2.

In Figure 9.14, we characterize the effective throughput of a family of CLDCs employing the parameters of $M = 2, 3, 4$, $N = 1$, $T = 8$ and $Q = 1, 2, 3, 4$ recorded at $\text{BER} = 10^{-4}$, when transmitting over uncorrelated i.i.d. Rayleigh fading channels, where the throughput is calculated as $C = \log_2(L) \cdot R_{\text{CLDC}}$. The increase of the effective throughput was achieved by transmitting more symbols Q per CLDC block and BPSK modulation was employed by all of the schemes. The effective throughput achieved by the corresponding group of LDCs with/without shadowing is also shown as the benchmark. Similarly, the effective throughput achieved by a group of CLDCs employing different numbers of receive antennas $N = 1, 2, 3$, while having the parameters of $M = 4$, $T = 8$ and $Q = 1, 2, 3, 4$ is quantified in Figure 9.15. Observe in Figures 9.14 and 9.15 that CLDCs outperform the corresponding LDCs, when encountering large-scale shadowing governed by $\Omega = 6$ dB.

Finally, Table 9.4 lists the coding gains of a family of CLDCs obeying the structure of Figure 9.2 and the corresponding LDCs having the structure of Figure 9.3, when having an effective throughput of 0.5 (BPS Hz^{-1}) as well as using BPSK modulation. The coding gain was defined as the SNR difference, at a BER of 10^{-4}, between various LDCs/CLDCs and a single-antenna-aided system having the same effective throughput. Observe in Table 9.4 that the achieved coding gains gradually increase with the number of relays, and the family of LDCs having co-located MIMO elements suffers a substantial SNR penalty in the presence of shadowing.

9.3 IRCC-coded Precoder-aided CLDCs

The 'coded-cooperative' schemes proposed in [126, 134] were based on the DF cooperation strategy, which requires the relays to perform either hard or soft decoding. However, the DF-based coded-cooperative schemes have two impediments. First, the power consumption of the relays used for the decoding as well as for transmitting the re-encoded data may be quite

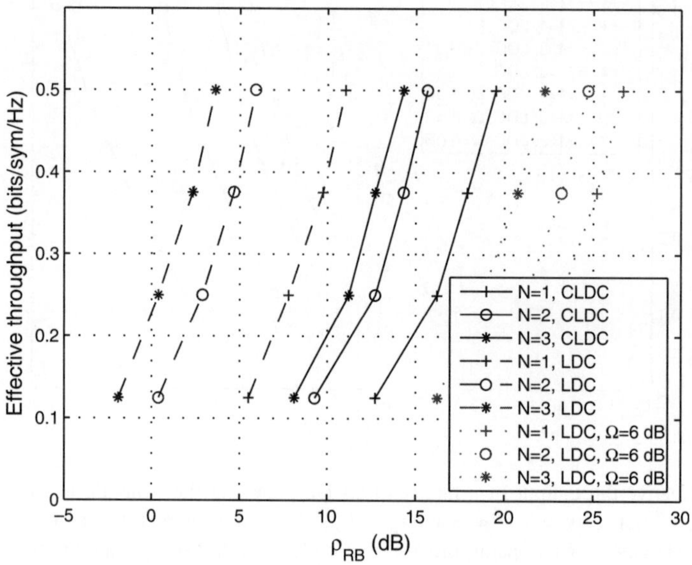

Figure 9.15: Effective throughput of a family of CLDCs obeying the structure of Figure 9.2 having $M = 4$, $N = 1, 2, 3$, $T = 8$ and $Q = 1, 2, 3, 4$ and the corresponding LDCs with/without shadowing, when transmitting over uncorrelated i.i.d. Rayleigh fading channels. All of the system parameters are summarized in Table 9.2.

Table 9.4: Coding gain comparison of a family of CLDCs obeying the structure of Figure 9.2 and the corresponding LDCs having the structure of Figure 9.3, which are extracted from Figures 9.14 and 9.15, when having an effective throughput of 0.5 (BPS Hz^{-1}) as well as using BPSK modulation.

CLDC				Coding	LDC				Coding gain	Coding gain
M	N	T	Q	gain (dB)	M	N	T	Q	no shadowing (dB)	$\Omega = 6$ dB (dB)
2	1	8	4	2.7	2	1	8	4	13.5	2.1
3	⋮	⋮	⋮	8.4	3	⋮	⋮	⋮	17.9	3.5
4	⋮	⋮	⋮	11.5	4	⋮	⋮	⋮	20.0	4.3
4	2	8	4	15.3	4	2	8	4	25.1	8.5
⋮	3	⋮	⋮	16.7	⋮	3	⋮	⋮	27.5	11.0

significant. Second, the resultant overall delay at the final destination may become excessive, which renders the support for delay-sensitive real-time interactive applications, such as video telephony, infeasible.

In this section, we propose a novel IRCC-aided IR-PCLDC scheme based on the AF protocol [116]. The proposed scheme becomes capable of operating at reduced SNRs as a benefit of adopting the sophisticated irregular near-capacity code-design principles of Section 7.5. More explicitly, we have demonstrated in Section 7.5.2 that the irregular design principles are applicable to both the inner and outer code. However, if only the inner code employs the irregular design, the resultant scheme may fail to closely approach the achievable capacity. Hence, we argued in the context of Figures 7.48 and 8.25 that irregularity should be imposed on both the inner code and the outer code, in order to achieve the best possible performance at an acceptable complexity.

The proposed scheme only requires the relays to perform low-complexity, low-delay linear combining according to Equation (9.13), rather than DF operations. Figure 9.16 portrays the schematic of the proposed IRCC-coded IR-PCLDC scheme using iterative decoding, where memory-one unit-rate precoders are employed. Compared with the IRCC-coded IR-PLDC scheme of Figure 7.48 designed for co-located MIMO systems, the schematic of Figure 9.16 employs a group of CLDCs for the relay-aided system's twin-phase transmissions. Owing to the striking similarity observed in Figures 9.16 and 7.48, our design methodology will be similar to that described in Section 7.5.3, which is appropriately adopted for the IRCC-coded IR-PCLDCs employed in a cooperative MIMO system. Hence, in this section we focus our attention on the proposed irregular system's features imposed by the twin-layer structure, rather than on the similarities with respect to the irregular co-located MIMO systems.

9.3.1 EXIT-chart-based IR-PCLDC Design

In this section, we describe the BPSK-modulated IRCC-coded IR-PCLDC scheme of Figure 9.16 having $P = 10$ component codes, where the IRCC employs $P_{\text{out}} = 6$ outer and the IR-PCLDC has $P_{\text{in}} = 4$ inner component codes, respectively. More explicitly, an IRCC scheme having a component rate of $R_{i,\text{IRCC}} = [0.1, 0.25, 0.4, 0.55, 0.7, 0.9]$ is adopted, where the corresponding EXIT characteristics have been shown in Figure 7.46. The BS is assumed to have $M = 4$ relays to assist the source node and it employs $N = 1$ receive antennas, as seen in Figure 9.2. All of the system parameters are listed in Table 9.5. As usual in irregular code design, we have to determine the specific fraction of the input bits to be encoded by each of the component codes, so that the corresponding inner and outer codes' EXIT curves match each other as closely as possible. This design philosophy was detailed in the context of Figure 7.49 of Section 7.5.2.

Table 9.6 lists the group of CLDCs used as the component codes of the IR-PCLDC scheme obeying the structure of Figure 9.16. The number of inner iterations (j) listed in Table 9.6 refers to the iterations carried out between the unity-rate precoder's decoder and the MMSE detector of Figure 9.16. Accordingly, the Precoded Cooperative Linear Dispersion Codes (PCLDCs) of Figure 9.16 are constituted by memory-one unity-rate precoders and CLDCs having parameters $(MNTQ)$. Furthermore, Table 9.7 constructs a group of LDCs having identical parameters to the CLDCs of Table 9.6, which are employed as the component codes of the IRCC-coded IR-PLDC scheme of Figure 7.48. Hence, we are able to carry out a fair comparison between the irregular system of Figure 9.16 invoked in cooperative MIMO systems and the corresponding irregular system of Figure 7.48 designed for co-located MIMO systems.

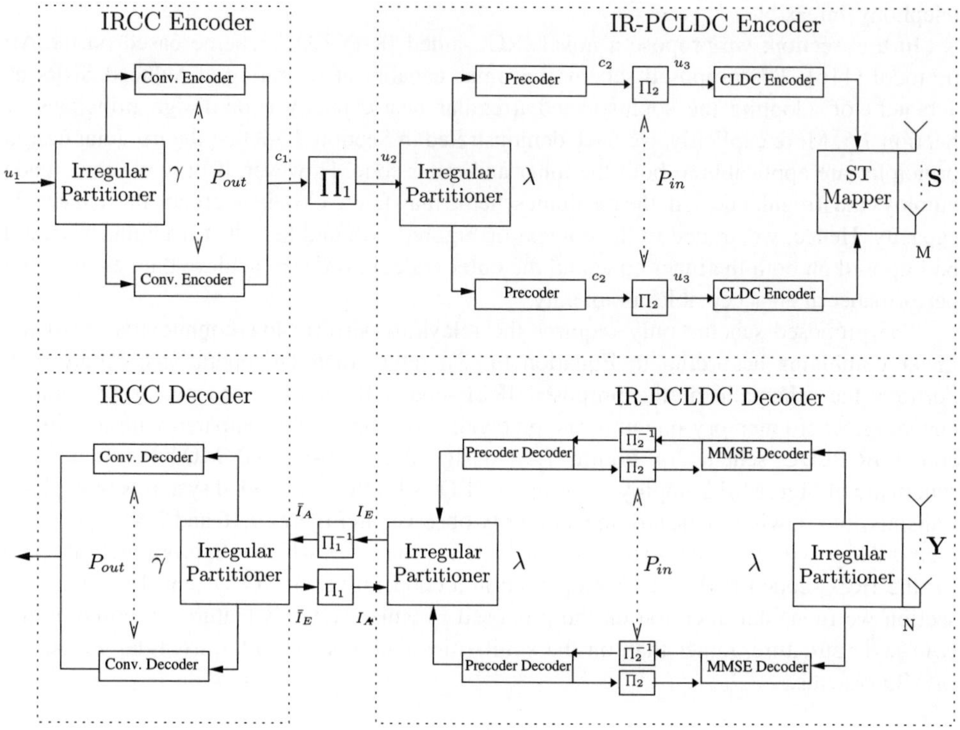

Figure 9.16: Schematic of the IRCC-coded IR-PCLDC scheme using iterative decoding.

Table 9.5: System parameters of the IRCC-coded IR-PCLDC scheme of Figure 9.16.

Number of relays	$M = 4$
Number of antennas per relay	1
Number of antennas at the BS	$N = 1$
Total number of slots	$T_1 + T_2 = 8$
Number of symbols per CLDC block	Q
SNR at the relays (dB)	$\rho_{SR} = 20$
SNR at the BS (dB)	ρ_{RB}
Modulation	BPSK
Mapping	Gray mapping
Detector	MMSE

Table 9.6: The $P_{\text{in}} = 4$ component codes of the BPSK-modulated IR-PCLDC scheme of Figure 9.16 generated for the cooperative MIMO system of Figure 9.2 having $M = 4$ and $N = 1$.

Index	M	N	T	T_1	T_2	Q	Rate	D	Inner iterations (j)
0	4	1	8	2	6	1	0.125	≈ 4	0
1	$\vdots$	$\vdots$	8	2	6	2	0.25	≈ 4	1
2	$\vdots$	$\vdots$	8	2	6	3	0.375	≈ 4	1
3	$\vdots$	$\vdots$	8	2	6	4	0.5	≈ 4	2

Table 9.7: The $P_{\text{in}} = 4$ component codes of the BPSK-modulated IR-PLDC scheme of Figure 7.48 generated for the co-located MIMO system of Figure 9.3 having $M = 4$ and $N = 1$ antennas.

Index	M	N	T	Q	Rate	D	Inner iterations (j)
0	4	1	8	1	0.125	4	0
1	$\vdots$	$\vdots$	$\vdots$	2	0.25	4	1
2	$\vdots$	$\vdots$	$\vdots$	3	0.375	4	1
3	$\vdots$	$\vdots$	$\vdots$	4	0.5	4	2

Figure 9.17 characterizes the EXIT charts of the BPSK-modulated PCLDC(4183) scheme of Table 9.6 obeying the structure of Figure 9.2, when communicating over i.i.d. Rayleigh fading channels. Assume that the inner code's EXIT curves can be perfectly matched by the EXIT curves of an outer code at any SNR ρ_{RB}, then the maximum achievable rate of a serial concatenated scheme can be approximated by evaluating the area under the EXIT curves. Given the rate of the inner block R_{in}, the maximum achievable rate is expressed as

$$C(\rho_{RB}) = \log_2(L) \cdot R_{\text{in}} \cdot R_{\text{out}}, \tag{9.23}$$

where R_{out} is approximated by the area under the inner code's EXIT curve [270] and L-PSK modulation is used. Note that in Sections 7.4.2 and 8.6.1, we have explicitly shown that there may be a potential maximum achievable rate gap, when it is measured before and after the unit-rate precoder. The associated gap can be closed by employing inner iterations (j), which has been shown in Figures 7.27 and 8.26. Similarly, there may be a potential gap between the maximum achievable rates recorded for CLDCs and PCLDCs based on each component code's EXIT characteristic and the associated gap can be eliminated by increasing the value of j, which is discussed in more detail in the following.

Figure 9.18 plots the maximum achievable rates of the CLDCs of Table 9.6 and the corresponding PCLDCs having different numbers of inner iterations j, when communicating over i.i.d. Rayleigh fading channels. The maximum achievable rates are measured based on the EXIT charts using Equation (9.23). When we use the PCLDC(4181) scheme, the employment of the precoder does not decrease the maximum achievable rates compared with the CLDC(4181) scheme even without inner iterations, i.e. at $j = 0$. When more symbols

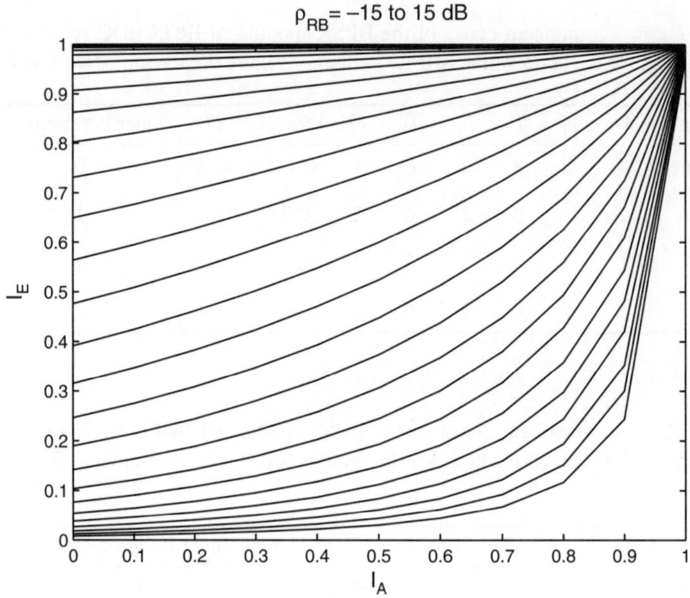

Figure 9.17: EXIT charts of the PCLDC(4183) scheme of Table 9.6 obeying the structure of Figure 9.2, when communicating over i.i.d. Rayleigh fading channels. All of the system parameters are summarized in Table 9.5.

have to be transmitted per CLDC block, i.e. when Q is increased, more inner iterations are necessary for the PCLDCs to approach the maximum achievable rates of the corresponding CLDCs. In particular, we found from Figure 9.18 that $j = 1$ is adequate for both the PCLDC(4182) and the PCLDC(4183) schemes and $j = 2$ is suitable for the PCLDC(4183) scheme. Again, the number of inner iterations necessary to close the maximum achievable rate gap is listed in Table 9.6.

Figure 9.19 characterizes the maximum achievable rates of the PCLDCs of Table 9.6 and the PLDC schemes of Table 9.7, when communicating over i.i.d. Rayleigh fading channels. Observe in Figure 9.19 that the family of PCLDCs summarized in Table 9.6 suffers from a substantial maximum rate loss compared with the class of its PLDC-aided counterparts. Recall that the CLDC's equivalent system model of Equation (9.20) suggests that the combined noise term $\bar{\mathbf{V}}$ consists of the noise variance σ_{SR} of the broadcast phase and that of the cooperation interval, namely σ_{RB}. In contrast, the PLDC scheme using identical parameters is only affected by the noise encountered at the BS. Again, it is the linear combination and amplification operations of Equation (9.13) carried out at the relays that impose the additional noise encountered during the broadcast interval. Naturally, the degradation imposed during the broadcast phase as characterized by σ_{SR} can be eliminated by adopting DF-related cooperation strategies, which may be able to ensure that the relays will have access to the perfect source information.

In order to elaborate a little further, Figure 9.20 characterizes the maximum achievable rates of the PCLDC(4183) schemes of Table 9.6, when transmitting over perfect broadcast channels, over noisy AWGN or Rayleigh-faded source-to-relay channels having a SNR of ρ_{SR}. The relay-to-BS channels are modeled as uncorrelated i.i.d. Rayleigh fading channels. More explicitly, even when the source-to-relay channels are perfect, which implies that the

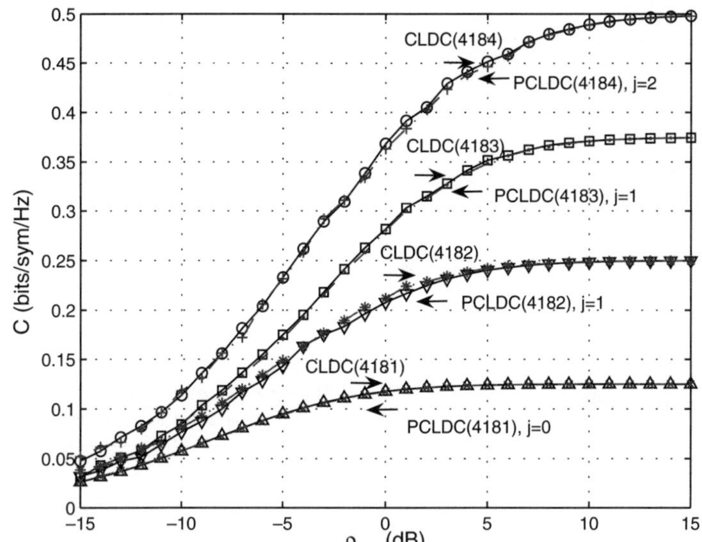

Figure 9.18: Maximum achievable rates of the CLDC and PCLDC schemes of Table 9.6 having different numbers of inner iterations j using a MMSE detector, when communicating over i.i.d. Rayleigh fading channels. All of the system parameters are summarized in Table 9.5.

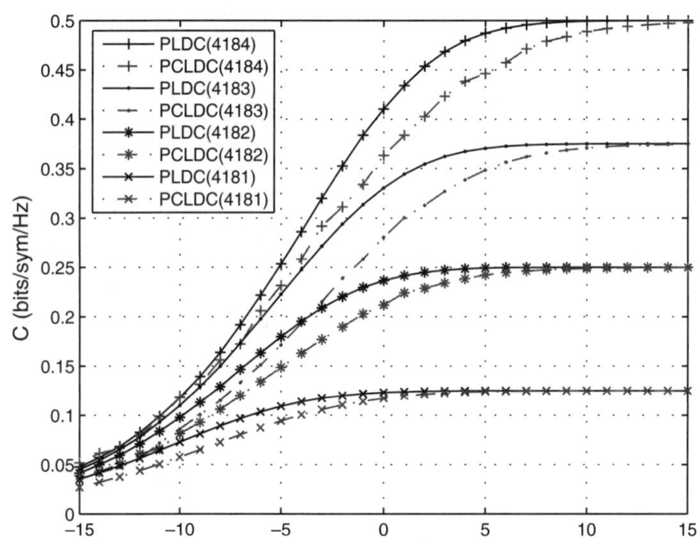

Figure 9.19: Maximum achievable rates of the PCLDC schemes of Table 9.6 and PLDC schemes of Table 9.7 using a MMSE detector, when communicating over i.i.d. Rayleigh fading channels. All of the system parameters are summarized in Table 9.5.

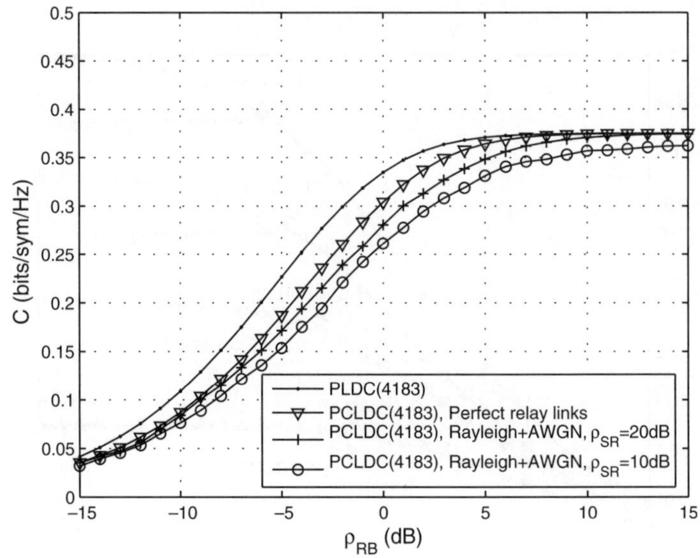

Figure 9.20: Maximum achievable rates of the PCLDC(4183) scheme of Table 9.6, when communicating over perfect channels, noisy media or Rayleigh-faded source-to-relay channels having a SNR of ρ_{SR}. All of the system parameters are summarized in Table 9.5.

relays have perfect source information, there exists a maximum achievable rate gap between the PCLDC(4183) schemes of Table 9.6 and the PLDC(4183) arrangements of Table 9.7. This is because the PCLDC(4183) scheme has to employ $T_1 = 2$ time slots to broadcast the source information to the relays. In other words, only $T_2 = 6$ slots or channel uses are available for the effective transmission of the data, whereas the PLDC(4183) scheme is capable of exploring all the dimensions provided by the $T = 8$ time slots.

Observe in Figure 9.20 that when we have more realistic Rayleigh-faded source-to-relay channels having a finite SNR of $\rho_{SR} = 20$ dB, a further maximum rate loss can be observed. Since each source-to-relay channel is independently faded and additionally corrupted by AWGN, each relay's received information inevitably becomes different, which implies that each relay will disperse potentially quite different information to the T_2 time slots, when relaying based on Equation (9.13). When the SNR of the source-to-relay links is reduced further to $\rho_{SR} = 10$ dB, the associated maximum achievable rate suffers an additional loss compared with that recorded for $\rho_{SR} = 20$ dB. Observe in Figure 9.20 that the associated rate reduction persists even in the high-SNR region, owing to the inherent noise imposed during the broadcast interval.

9.3.2 Performance Results

For all of the simulations, the first interleaver of Figure 9.16 is set to a length of 10^6 bits and all the simulation parameters are listed in Table 9.5. Again, we construct an identical IRCC-coded IR-PLDC scheme using Table 9.7 as the benchmark to quantify the achievable performance.

Our design objective is to maximize the effective throughput across the widest possible SNR range, which can be achieved by adaptively adjusting the weighting coefficient vectors

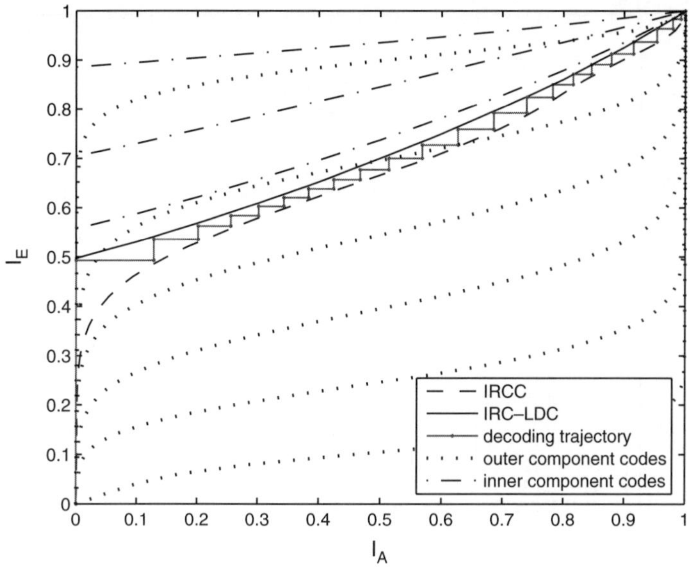

Figure 9.21: EXIT charts and the simulation-based decoding trajectory of the IRCC-coded IR-PCLDC scheme of Figure 9.16 recorded at $\rho_{RB} = 0$ dB, when communicating over i.i.d. Rayleigh fading channels. All of the system parameters are summarized in Table 9.5.

$\boldsymbol{\gamma} = [\gamma_1, \ldots, \gamma_{P_{\text{out}}}]$ of the IRCC scheme of Figure 7.46 and $\boldsymbol{\lambda} = [\lambda_1, \ldots, \lambda_{P_{\text{in}}}]$ of the IR-PCLDC scheme, respectively. More explicitly, an exhaustive search operation is carried out for all of the possible combinations of $\boldsymbol{\gamma}$ and $\boldsymbol{\lambda}$ in order to maximize the throughput $C(\rho) = \log_2(L) \cdot R_{\text{in}} \cdot R_{\text{out}}$ of Equation (9.23) under the following constraints:

- $\gamma_1 + \gamma_2 + \cdots + \gamma_{P_{\text{out}}} = 1$;
- $\lambda_1 + \lambda_2 + \cdots + \lambda_{P_{\text{in}}} = 1$;
- an open convergence tunnel must exist in the EXIT chart, which enables the irregular system to achieve an infinitesimally low BER using iterative decoding;
- the open EXIT tunnel area is maximized, for the sake of minimizing the number of outer iterations required.

Figure 9.21 presents the associated EXIT charts and the corresponding decoding trajectory of our IRCC-coded IR-PCLDC scheme designed for operating at $\rho_{RB} = 0$ dB, when communicating over i.i.d. Rayleigh fading channels. The dashed lines represent the EXIT curves of the IR-PCLDC's component codes of Table 9.6, while the dotted lines denote the EXIT curves for the set of IRCC components. The solid lines represent the aggregate EXIT curves of the IRCC scheme having the weighting coefficient vector of $\boldsymbol{\gamma} = [0, 0, 0, 0.2374, 0.5036, 0.2590]$ and the IR-PCLDC scheme having the weighting coefficient vector of $\boldsymbol{\lambda} = [0, 0, 0, 1]$. By simultaneously maximizing the achievable rate and the open EXIT tunnel area at $\rho_{RB} = 0$ dB, the proposed scheme becomes capable of achieving $C(0\ \text{dB}) = 0.3475$ (BPS Hz^{-1}) and the optimized EXIT curves of Figure 9.21 exhibit a narrow but still open tunnel, where the decoding trajectory shows that $k = 23$ outer iterations were required. The corresponding BER performance is portrayed in Figure 9.22.

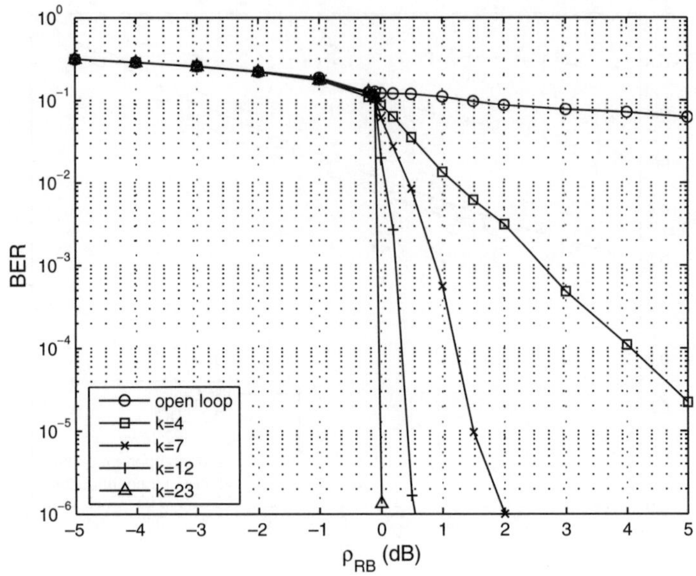

Figure 9.22: BER of the IRCC-coded IR-PCLDC scheme of Figure 9.16 designed for achieving an infinitesimally low BER at $\rho_{RB} = 0$ dB, when communicating over i.i.d. Rayleigh fading channels. All of the system parameters are summarized in Table 9.5.

By repeating the same design process as that used for $\rho_{RB} = 0$ dB in Figures 9.21 and 9.22, we are able to design the IRCC-coded IR-PCLDC scheme of Figure 9.16 for other operating SNR values. Figure 9.23 portrays the maximum achievable rates attained by the IRCC-coded IR-PCLDC scheme of Figure 9.16. Each component of Figure 9.16 was designed to achieve the maximum effective throughput using specific weighting coefficient vectors $\boldsymbol{\lambda}$ and $\boldsymbol{\gamma}$ listed in Tables G.11 and G.12. Furthermore, we also plot the maximum rates achieved by the IRCC-coded IR-PLDC scheme of Figure 7.48 designed for co-located MIMO systems according to Tables G.13 and G.14. Observe in Figure 9.23 that there is an effective throughput discrepancy between the irregular system designed for the co-located MIMO system of Figure 9.3 and the corresponding irregular scheme designed for the cooperative MIMO arrangement of Figure 9.2. This phenomenon is also evidenced in Figure 9.19 and has been further explained in Figure 9.20. Finally, the LDC(4184) scheme's CCMC capacity obtained using Equation (7.9) is also provided in order to give us an idea as to how far we are operating from the capacity limit.

9.4 Conclusion

In this chapter, we have extended the concept of LDCs designed for co-located MIMO systems in Chapter 7 to cooperative MIMO systems. Our key findings may be summarized as follows.

- In Sections 9.2.1–9.2.3, we established a general framework for the cooperative MIMO systems of Figure 9.2.
- Correspondingly, in Section 9.2.4, we have emphasized that the fundamental difference between co-located MIMO systems and cooperative MIMO systems is that the latter

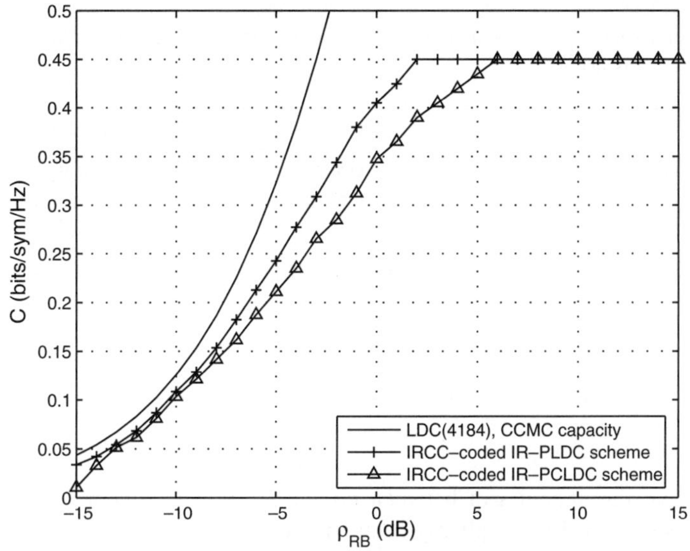

Figure 9.23: The maximum rates achieved by the IRCC-coded IR-PCLDC scheme of Figure 9.16 according to Tables G.11 and G.12 and the corresponding IRCC-coded IR-PLDC scheme of Figure 7.48 using Tables G.13 and G.14. All of the system parameters are summarized in Table 9.5.

relies on a twin-phase transmission regime, which provides us with a high grade of freedom in terms of appropriately sharing the total number of slots between the broadcast and cooperation phases.

- In Section 9.2.5, we demonstrated the achievable BER performance of CLDC schemes having various parameter combinations. The effective throughput values achieved by the uncoded CLDCs have been summarized in Figures 9.14 and 9.15 and the achievable coding gains were listed in Table 9.4.

- In Section 9.3 we imposed irregularity on the relay-aided system based on the AF cooperation protocol, and demonstrated that the resultant IRCC-coded IR-PCLDC scheme of Figure 9.16 becomes capable of operating at a throughput that is close to the corresponding IR-PLDC benchmark as seen in Figure 9.23, while maintaining an infinitesimally low BER. The associated weighting coefficient vectors $\boldsymbol{\lambda}$ and γ are listed in Tables G.11 and G.12, respectively.

- Finally, we summarized the characteristics of both co-located MIMO systems using LDCs and those of cooperative MIMO schemes using CLDCs in Table 9.8. Both of the schemes are capable of supporting arbitrary $(MNTQ)$ parameter combinations as well as employing arbitrary modulation schemes. Furthermore, the achievable total diversity order is jointly determined by the minimum spatial and temporal dimensions. For uncoded systems, the LDCs outperform the CLDCs when encountering a small-scale fading environment. In contrast, the CLDCs tend to function more reliably than the LDCs in the context of large-scale fading, i.e. shadowing. The corresponding trends have been further explained in Figures 9.7 and 9.8. As far as near-capacity coded systems are concerned, which are implemented using IR-PLDCs/IR-PCLDCs, again,

Table 9.8: Characteristics of Co-located MIMO systems using LDCs and Cooperative MIMO arrangements using CLDCs.

		Co-located MIMO	Cooperative MIMO
Similarity	$(MNTQ)$	Arbitrary	Arbitrary
	Modulation	Arbitrary	Arbitrary
	Diversity	$N \cdot \min(M, T)$	$\approx N \cdot \min(M, T_2)$
Uncoded	Small-scale fading	Better than CLDCs	Trends in Figure 9.7
	Large-scale fading	Trends in Figure 9.8	Better than LDCs
Coded	Small-scale fading	Better than CLDCs	Explained in Figure 9.20
	Throughput	Compared in Figure 9.23	

the effective throughput gap between co-located MIMO schemes and cooperative MIMO systems has been explicitly detailed in Figure 9.20.

Part III

Differential Turbo Detection of Multi-functional MIMO-aided Multi-user and Cooperative Systems

List of Symbols in Part III

General Notation

- The superscript * is used to indicate complex conjugation. Therefore, a^* represents the complex conjugate of the variable a.
- The superscript ^{T} is used to indicate the matrix transpose operation. Therefore, $\mathbf{a}^{\mathrm{T}}$ represents the transpose of the matrix $\mathbf{a}$.
- The superscript $^{\mathcal{H}}$ is used to indicate the complex conjugate transpose operation. Therefore, $\mathbf{a}^{\mathcal{H}}$ represents the complex conjugate transpose of the matrix $\mathbf{a}$.
- The notation $\hat{x}$ represents the estimate of x.

Special Symbols

$a_{l,i}$	The ith coordinate of the lth SP symbol $\mathbf{s}^l$.
$\mathcal{A}$	The area under a curve.
$\mathbf{A}$	The *a priori* probability matrix of a non-binary decoder.
$\mathbf{A}_{\mathrm{ldpc}}$	The *a priori* probability matrix of the non-binary LDPC decoder.
$\mathbf{A}_{\mathrm{urc}}$	The *a priori* probability matrix of the symbol-based unity-rate decoder.
B	The number of binary bits corresponding to a constellation symbol.
$\mathbf{b}$	A block of B binary bits.
b_i	The binary bit at position i in $\mathbf{b}$.
c	Outer channel coded bits.
$\mathbf{C}$	The space-time signal matrix.
c_t^i	The complex symbol transmitted by transmit antenna i at time slot t.
$\mathbb{C}^n$	The n-dimensional complex space.
$C_{\mathrm{DCMC}}^{\mathrm{STBC\text{-}SP}}$	The DCMC capacity of STBC-SP schemes.
D	The dimension of a D-dimensional signal set.
$\mathcal{D}$	The depth of the random interleaver.
$\mathbf{D}_{\mathrm{ldpc}}$	The APP matrix of the non-binary LDPC decoder.
$\mathbf{D}_{\mathrm{urc}}$	The APP matrix of the symbol-based unity-rate decoder.
$E[k]$	The expected value of k.
E_b	Bit energy.

E_s	Symbol energy.
E_{total}	The total energy of a constellation set.
f_D	The normalized Doppler frequency.
G	The feedforward generator polynomial of RSC codes.
G_r	The feedback generator polynomial of RSC codes.
$\mathbf{G}_{2^k}$	An orthogonal design of size $(2^k \times 2^k)$.
h_i	The CIR from transmit antenna i for single-receive-antenna systems.
$h_{i,j}$	The CIR from transmit antenna i to receive antenna j.
I_0	The bit-wise unconditional MI.
I_{3S}	The number of three-stage iterations.
I_A	The MI associated with the *a priori* information.
I_{A_D}	The MI associated with the *a priori* LLR values $L_{D,a}$ of the outer channel decoder.
I_{A_M}	The MI associated with the *a priori* LLR values $L_{M,a}$ of the SP demapper.
I_E	The MI associated with the extrinsic information.
I_{E_D}	The MI associated with the extrinsic LLR values $L_{D,e}$ of the outer channel decoder.
I_{E_M}	The MI associated with the extrinsic LLR values $L_{M,e}$ of the SP demapper.
I_{ext}	The number of external joint iterations.
I_{int}	The number of LDPC internal iterations.
$\mathbf{I}_n$	The identity matrix of size $(n \times n)$.
$\mathcal{K}$	The constraint length of RSC codes.
K	The rank of a matrix.
K_{ldpc}	LDPC output block length.
L	The size of the legitimate modulation constellation S.
$L_{D,a}$	The *a priori* LLR values of the outer channel decoder.
$L_{D,e}$	The extrinsic LLR values of the outer channel decoder.
$L_{D,i,p}$	The LLR values of the original uncoded systematic information bits.
$L_{D,p}$	The *a posteriori* LLR values of the outer channel decoder.
$L_{M,a}$	The *a priori* LLR values of the SP demapper.
$L_{M,e}$	The extrinsic LLR values of the SP demapper.
$L_{M,p}$	The *a posteriori* LLR values of the SP demapper.
N_t	Number of transmit antennas.
N_r	Number of receive antennas.
n_A	The zero-mean Gaussian random variable used for modeling the *a priori* information input.
n_f	A normalization factor.
P	Number of subcodes in a family of subcodes (e.g. IRCCs).
$\mathbf{Q}$	The soft-metric probability matrix produced by the symbol-based SP demodulator.
R	Coding rate.
$\mathbf{R}$	The channel correlation matrix.
r_i	The received SP symbol at time instant t.
$\mathbb{R}^n$	The n-dimensional real-valued Euclidean space.
$\mathbf{s}$	A SP symbol.
S	The legitimate constellation set.
S_0^k	The subset of the legitimate constellation set S that contains all symbols having $b_k = 0$.

S_1^k	The subset of the legitimate constellation set S that contains all symbols having $b_k = 1$.
$\mathbf{s}^l$	The lth SP legitimate symbol.
T	The number of time slots needed for transmitting a specific number of symbols.
T_s	Signaling period.
T_{sp}	The transfer function from SP to complex signals.
T_{sym}	Symbol period.
$\mathbf{v}$	Non-binary LDPC-encoded integer symbols.
$\mathbf{w}$	A four-dimensional Gaussian random variable.
W	Bandwidth.
y_t^j	The received complex signal at receive antenna j at time slot t.
z_t^j	The complex AWGN at receive antenna j at time instant t.
α_i	Weight coefficient of the ith subcode.
ζ_R	The diversity product for time-correlated fading channels having a correlation matrix $\mathbf{R}$.
ζ_{rapid}	The diversity product for rapid fading channels.
ζ_{static}	The diversity product for quasi-static fading channels.
η	Bandwidth efficiency.
θ_i	The phase shift of the channel impulse response h_i.
π	Interleaver.
π^{-1}	Deinterleaver.
ρ	The SNR.
σ_n^2	The complex noise's variance.

Chapter 10

Differential Space-Time Spreading

10.1 Introduction

An effective and practical way of counteracting the effect of wireless channels is to provide diversity, which can be achieved by employing STC [4,11,12,49,244,336–341]. STC employs multiple transmit antennas, where coding is performed in both the spatial and temporal domains in order to introduce correlation between signals transmitted from the multiple antennas in different time slots. The spatial-temporal correlation is imposed in order to exploit the fact that the individual MIMO links are likely to experience independent fading and hence to mitigate the effects of transmission errors at the receiver. STC can achieve a substantial transmit diversity and power gain over its spatially uncoded counterpart without bandwidth expansion. There are numerous well-established coding structures, including STBCs [11,12], STTCs [337] and Layered Space-Time (LST) codes [69]. A central issue in designing all of these schemes is the exploitation of multipath effects in order to achieve diversity performance gains [245].

In practice, the CSI of each link between each transmit and each receive antenna pair has to be estimated at the coherent receiver either blindly or using training symbols. In such a coherent system, it is assumed that the channel does not change dramatically during a transmitted frame of data [49]. However, channel estimation invoked for all of the transmit and receive antennas substantially increases both the cost and complexity of the receiver. Furthermore, when the CSI fluctuates dramatically from burst to burst, an increased number of training symbols has to be transmitted, potentially resulting in an undesirably high transmission overhead and wastage of transmission power.

Alternatively, it is beneficial to develop low-complexity techniques that do not require any channel information at the receiver. For a single transmit antenna, it is well known that differential schemes, such as DPSK [248], can be demodulated without the use of channel estimates. Differential schemes have been used widely in practical cellular mobile communication systems [6, 245, 248]. It is natural to consider extensions of differential schemes to MIMO systems. A detection algorithm designed for Alamouti's scheme [11] was proposed in [52], where the channel encountered at time instant t was estimated using the pair of symbols detected at time instant $t-1$. The algorithm, nonetheless, has to estimate the channel during the very first time instant using training symbols and hence is not truly differential. Tarokh and Jafarkhani [53, 62] proposed a differential encoding and decoding technique for Alamouti's scheme [11] using real-valued phasor constellations and hence the transmitted signal can be demodulated either with or without CSI at the receiver. The resultant differential-decoding-aided non-coherent receiver performs within 3 dB from the coherent receiver assuming perfect knowledge of the CIR at the receiver. The complex constellation was also restricted to PSK schemes, which was extended to QAM constellations in [58]. This extension, however, requires knowledge of the channel power in order to appropriately normalize the received signal. The channel power in [58] was estimated blindly using the received differentially encoded signals without invoking any channel estimation techniques or transmitting any pilot symbols. The proposed DSTBC was then extended to multiple antennas [56] using a real-valued phasor constellation. Afterwards, the authors of [58, 59] developed a differential space-time block coding scheme that supports non-constant modulus constellations combined with four transmit antennas.

The novelty and rationale of this chapter can be summarized as follows.

1. DSTS is a MIMO-aided scheme, which is advocated for the sake of achieving a high transmit diversity gain. This facilitates low-complexity differential detection, rather than using a more complex receiver employing both channel estimation [342] for all MIMO links and coherent detection. Moreover, the system benefits from the multi-user support capability of the STS scheme. Furthermore, the high diversity order of the system results in a Gaussian-like channel error distribution as a function of time, i.e. the bit index, which improves the attainable system performance.

2. In addition, the system is combined with multi-dimensional SP modulation [43, 46], which is capable of maximizing the coding advantage of the transmission scheme by jointly designing and detecting the sphere-packed DSTS symbols.

3. We quantify the capacity of the DSTS-SP scheme for transmission over both Rayleigh and Gaussian channels.

The rest of the chapter is organized as follows. First, differential encoding designed for a single transmit antenna is described briefly in Section 10.2. Then, the encoding and decoding processes of the DSTS scheme employing two transmit antennas are carried out in Section 10.3 for conventional as well as for SP modulation. Section 10.3.6 presents the capacity analysis of the SP-modulation-aided DSTS (DSTS-SP) scheme employing two transmit antennas and a variable number of receive antennas, while in Section 10.3.7 we present our comparative study of the various twin-antenna-aided DSTS schemes. In Section 10.4 we demonstrate how a four-antenna-aided DSTS scheme can be combined with conventional real- and complex-valued as well as a novel SP-modulation scheme and quantify the attainable capacity of the four-antenna-aided DSTS-SP scheme. Our conclusions are presented in Section 10.5, followed by a chapter summary in Section 10.6.

10.2 DPSK

Before presenting the details of DSTS designed for multiple transmit antennas, we review the concept of differential encoding/decoding for a single transmit antenna. To be more precise, the following section presents the details of DPSK modulation and captures the main ideas behind it [248].

In DPSK modulation, the demodulator does not have to perform channel estimation. The two consecutive transmitted symbols depend on each other and the demodulator detects the transmitted data symbol by observing two successive symbols. It is assumed that the

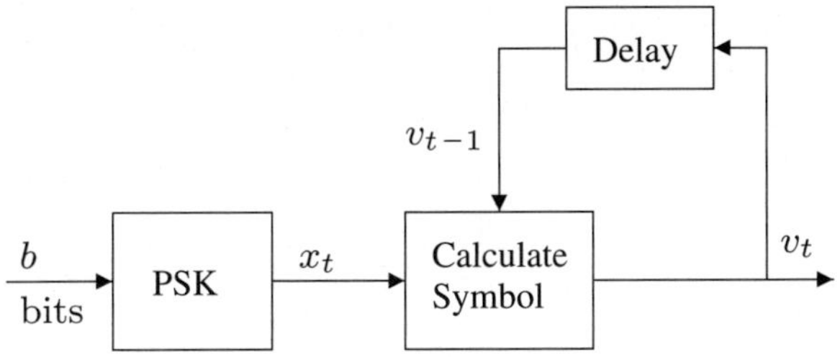

Figure 10.1: Transmitter block diagram for DPSK modulation.

channel has a phase response that is approximately constant for two symbol periods and this is approximately valid for the case of slow fading channels. The information is essentially transmitted by first providing a single dummy reference symbol, followed by differentially phase-modulated symbols [245].

Let us assume that we transmit the modulated symbol v_t at time instant t, when the CIR between the transmitter and receiver is h_t, and that the noise sample is n_t with a variance of σ_n^2. Therefore, the received signal at time instant t is

$$r_t = h_t \cdot v_t + n_t. \tag{10.1}$$

For DPSK modulation, the transmitted symbol v_t at time instant t is obtained from $v_t = x_t \cdot v_{t-1}$, as shown in Figure 10.1, where x_t is a non-differentially PSK-modulated symbol and v_{t-1} is the symbol transmitted at time instant $t-1$. To detect the signal transmitted at time instant t, the receiver computes $r_t \cdot r_{t-1}^*$, where * represents the complex conjugate operation. Then the receiver finds the legitimate symbol of the QPSK constellation closest to $r_t \cdot r_{t-1}^*$ as the estimates of the transmitted symbol [49]. To further augment the rationale behind the above scheme and taking into consideration the assumption that the channel encountered is a slow fading one, i.e. that we have $h_t \approx h_{t-1} = h$, we arrive at

$$\begin{aligned} r_t \cdot r_{t-1}^* &= (h_t \cdot v_t + n_t) \cdot (h_{t-1} \cdot v_{t-1} + n_{t-1})^* \\ &= |h|^2 \cdot v_t \cdot v_{t-1}^* + h \cdot v_t \cdot n_{t-1}^* + n_t \cdot h^* \cdot v_{t-1}^* + n_t \cdot n_{t-1}^* \\ &= |h|^2 \cdot x_t \cdot |v_{t-1}|^2 + N \\ &= |h|^2 \cdot x_t + N, \end{aligned} \tag{10.2}$$

where $|v_{t-1}|^2 = 1$, N is a Gaussian noise process having a variance of $\sigma_N^2 \approx 2 \cdot h \cdot \sigma_n^2$ and the path gain h is assumed to be constant during the modulation instants of $t-1$ and t. Therefore, the optimal estimate of x_t is given by

$$\tilde{x}_t = \arg\min_{x_t} |r_t \cdot r_{t-1}^* - |h|^2 \cdot x_t|^2. \tag{10.3}$$

Based on Equation (10.3), it becomes clear that the decoded output does not depend on either earlier demodulation decisions or the CSI; rather, it depends only on the received symbols of two consecutive symbol periods. The noise power experienced by the receiver of the differential decoding scheme is $\sigma_N^2 \approx 2 \cdot h \cdot \sigma_n^2$, which is about twice that of the coherent scheme. Therefore, for the same transmission power, the received SNR of the differential detection scheme is approximately half of that of the coherent detection scheme using perfect channel knowledge at the receiver. This translates to a 3 dB SNR difference for the performance of these two systems, i.e. the coherently detected system using perfect channel knowledge at the receiver outperforms the differentially detected scheme by 3 dB.

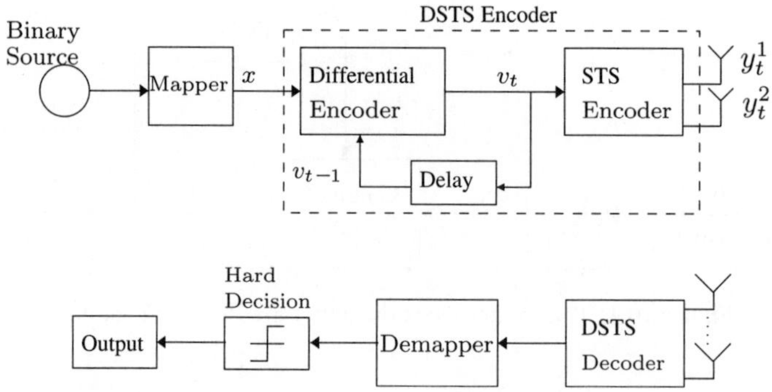

Figure 10.2: The twin-antenna-aided DSTS system block diagram.

10.3 DSTS Design using Two Transmit Antennas

As is widely recognized, coherent detection schemes require CSI, which is acquired by transmitting training symbols. However, high-accuracy MIMO channel estimation imposes a high complexity on the receiver. This renders differential encoding and decoding an attractive design alternative, despite the associated E_b/N_0 loss.

The transmitted and received DSTS symbols are encoded and decoded based on the differential relationship among subsequent symbols as illustrated in Section 10.2 for classic DPSK. For the sake of simplicity, in what follows we consider having a single receive antenna, although the extension to systems having more than one receive antenna is straightforward.

10.3.1 Twin-antenna-aided DSTS Encoding using Conventional Modulation

According to Figure 10.2, it becomes clear that the DSTS encoder can be divided into two main stages. The differential encoding takes place before STS and the differentially encoded symbols are then spread as exemplified in simple graphical terms in Figure 10.3 (see [51]), where two symbols are transmitted using two transmit antennas within two time slots.

The DSTS encoding algorithm operates as follows. At time instant $t = 0$, the arbitrary dummy reference symbols v_0^1 and v_0^2 are passed to the STS encoder for transmission from antennas one and two, respectively. The dummy symbols v_0^1 and v_0^2 usually carry no information. At time instants $t \geq 1$, a block of $2B$ bits arrive at the mapper, where each set of B bits is mapped to a symbol x_t^k, $k = 1, 2$, selected from a 2^B-ary constellation. Assume that v_t^1 and v_t^2 are the differentially encoded symbols, then differential encoding of Figure 10.2 is carried out as follows:

$$v_t^1 = \frac{(x_t^1 \cdot v_{t-1}^1 + x_t^2 \cdot v_{t-1}^{2*})}{\sqrt{(|v_{t-1}^1|^2 + |v_{t-1}^2|^2)}}, \tag{10.4}$$

$$v_t^2 = \frac{(x_t^1 \cdot v_{t-1}^2 - x_t^2 \cdot v_{t-1}^{1*})}{\sqrt{(|v_{t-1}^1|^2 + |v_{t-1}^2|^2)}}, \tag{10.5}$$

where the superscript * represents the complex conjugate operation.

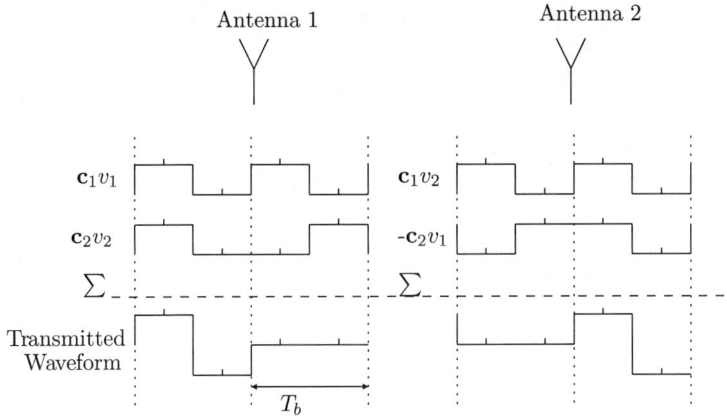

Figure 10.3: Illustration of STS using two transmit antennas transmitting two bits within $2T_b$ duration. $v_1 = v_2 = 1$ were assumed and $\bar{\mathbf{c}}_1 = [+1+1-1-1+1+1-1-1]$ and $\bar{\mathbf{c}}_2 = [+1+1-1-1-1-1+1+1]$.

The differentially encoded symbols are then spread with the aid of the spreading codes $\bar{\mathbf{c}}_1$ and $\bar{\mathbf{c}}_2$ to both transmit antennas, where $\bar{\mathbf{c}}_1$ and $\bar{\mathbf{c}}_2$ are generated from the same user-specific spreading code $\bar{\mathbf{c}}$ by ensuring that the two spreading codes $\bar{\mathbf{c}}_1$ and $\bar{\mathbf{c}}_2$ become orthogonal using the simple code-concatenation rule of Walsh–Hadamard codes, yielding longer codes and hence a proportionately reduced per antenna throughput according to

$$\bar{\mathbf{c}}_1 = [\bar{\mathbf{c}} \quad \bar{\mathbf{c}}], \tag{10.6}$$

$$\bar{\mathbf{c}}_2 = [\bar{\mathbf{c}} \quad -\bar{\mathbf{c}}]. \tag{10.7}$$

The differentially encoded data is then divided into two half-rate substreams and the two consecutive symbols are then spread to both transmit antennas using the mapping of

$$\mathbf{y}_t^1 = \frac{1}{\sqrt{2}}(\bar{\mathbf{c}}_1 \cdot v_t^1 + \bar{\mathbf{c}}_2 \cdot v_t^{2*}), \tag{10.8}$$

$$\mathbf{y}_t^2 = \frac{1}{\sqrt{2}}(\bar{\mathbf{c}}_1 \cdot v_t^2 - \bar{\mathbf{c}}_2 \cdot v_t^{1*}), \tag{10.9}$$

which is exemplified in simple graphical terms in Figure 10.3.

10.3.2 Receiver and Maximum Likelihood Decoding

We assume that the channel is modeled as a temporally correlated narrowband Rayleigh fading channel, where the channel coefficients are spatially independent, associated with a normalized Doppler frequency of $f_D = f_d T_s = 0.01$, where f_d is the Doppler frequency and T_s is the symbol duration. The complex AWGN of $n = n_I + jn_Q$ contaminates the received signal, where n_I and n_Q are two independent zero-mean Gaussian random variables having a variance of $\sigma_n^2 = \sigma_{n_I}^2 = \sigma_{n_Q}^2 = N_0/2$ per dimension, with $N_0/2$ representing the double-sided noise power spectral density expressed in W Hz^{-1}.

The received signal at the output of the single receiver antenna can be represented as

$$\mathbf{r}_t = h_1 \cdot \mathbf{y}_t^1 + h_2 \cdot \mathbf{y}_t^2 + \mathbf{n}_t, \tag{10.10}$$

where h_1 and h_2 denote the narrowband complex-valued CIRs corresponding to the first and second transmit antennas, respectively, where it is assumed that the channel coefficients remain unchanged for two consecutive transmitted vectors $\mathbf{y}^k$, while $\mathbf{n}_t$ is a complex-valued Gaussian random variable with a covariance matrix of $\sigma_n^2 \cdot \mathbf{I}_{SF}$, with SF representing the SF of the spreading codes $\bar{\mathbf{c}}_1$ and $\bar{\mathbf{c}}_2$ and $\mathbf{I}_{SF}$ is the identity matrix of size $SF \times SF$.

The received signal $\mathbf{r}_t$ is then correlated with $\bar{\mathbf{c}}_1$ and $\bar{\mathbf{c}}_2$ according to the following operations:

$$d_t^1 = \bar{\mathbf{c}}_1^\dagger \cdot \mathbf{r}_t = \frac{1}{\sqrt{2}} \cdot h_1 \cdot v_t^1 + \frac{1}{\sqrt{2}} \cdot h_2 \cdot v_t^2 + \bar{\mathbf{c}}_1^\dagger \cdot \mathbf{n}_t, \tag{10.11}$$

$$d_t^2 = \bar{\mathbf{c}}_2^\dagger \cdot \mathbf{r}_t = \frac{1}{\sqrt{2}} \cdot h_1 \cdot v_t^{2*} - \frac{1}{\sqrt{2}} \cdot h_2 \cdot v_t^{1*} + \bar{\mathbf{c}}_2^\dagger \cdot \mathbf{n}_t, \tag{10.12}$$

where the superscript † represents the Hermitian or the conjugate transpose operation.

Following the received signal's correlation with $\bar{\mathbf{c}}_1$ and $\bar{\mathbf{c}}_2$, we arrive at two data symbols that are then differentially decoded by using the received data of two consecutive time slots as follows:

$$\begin{aligned} \tilde{x}_t^1 &= d_t^1 \cdot d_{t-1}^{1*} + d_t^{2*} \cdot d_{t-1}^2 \\ &= \tfrac{1}{2} \cdot (|h_1|^2 + |h_2|^2) \cdot \sqrt{|v_{t-1}^1|^2 + |v_{t-1}^2|^2} \cdot x_t^1 + N_1, \end{aligned} \tag{10.13}$$

$$\begin{aligned} \tilde{x}_t^2 &= d_t^1 \cdot d_{t-1}^{2*} - d_t^{2*} \cdot d_{t-1}^1 \\ &= \tfrac{1}{2} \cdot (|h_1|^2 + |h_2|^2) \cdot \sqrt{|v_{t-1}^1|^2 + |v_{t-1}^2|^2} \cdot x_t^2 + N_2, \end{aligned} \tag{10.14}$$

where N_1 and N_2 are zero-mean complex-valued Gaussian random variables having variances of $\sigma_N^2 = \sigma_{N_1}^2 = \sigma_{N_2}^2 \approx 2 \cdot \chi_{2N_t}^2 \cdot \sigma_n^2$, with $N_t = 2$ being the number of transmit antennas and $\chi_{2N_t}^2 = \frac{1}{2} \cdot (|h_1|^2 + |h_2|^2) \cdot \sqrt{|v_{t-1}^1|^2 + |v_{t-1}^2|^2}$ representing a chi-squared distributed random variable having $2N_t = 4$ degrees of freedom.

We can observe from Equations (10.13) and (10.14) that the proposed method guarantees achieving a diversity gain, since the two transmit antennas' signals are independently faded according to the values of h_1 and h_2, while using a low-complexity decoding algorithm. Moreover, since $\bar{\mathbf{c}}_1$ and $\bar{\mathbf{c}}_2$ are derived by appropriately concatenating the user-specific code $\bar{\mathbf{c}}$, no extra spreading codes are required for carrying out the STS operation and the two symbols of the two transmit antennas are transmitted in two time slots.

The previous decoding operation has been carried out for the case of constant modulus constellations such as PSK. Non-constant modulus constellations [248] can also be transmitted using the proposed DSTS scheme. According to Equations (10.13) and (10.14), the DSTS decoded signal has a multiplicative factor of

$$\chi_{2N_t}^2 = \tfrac{1}{2} \cdot (|h_1|^2 + |h_2|^2) \cdot (\sqrt{|v_{t-1}^1|^2 + |v_{t-1}^2|^2})$$

as compared with the original transmitted symbols regardless of the effect of noise. Therefore, in order to obtain the original transmitted multilevel constellation symbols, the multiplicative factor must be compensated for by estimating the channel's output power $(|h_1|^2 + |h_2|^2)$ as well as the power of the previously transmitted symbols $(\sqrt{|v_{t-1}^1|^2 + |v_{t-1}^2|^2})$. The power of the previously transmitted symbols can be estimated from that of the symbols received during the previous time slot. Furthermore, to estimate the channel's output power, the following simple computation can be carried out:

$$d^1_{t-1} \cdot d^{1*}_{t-1} + d^2_{t-1} \cdot d^{2*}_{t-1} = \tfrac{1}{2} \cdot (|h_1|^2 + |h_2|^2) \cdot (|v^1_{t-1}|^2 + |v^2_{t-1}|^2) + w, \qquad (10.15)$$

where w is a zero-mean complex-valued Gaussian random variable having a variance of $\sigma_w^2 \approx 2 \cdot \chi^2_{2N_t} \cdot \sigma_n^2$. Therefore, using the estimate of the signal power of the previous transmitted symbols, i.e. $(|v^1_{t-1}|^2 + |v^2_{t-1}|^2)$, as well as the result of Equation (10.15), the channel's power transfer function of $(|h_1|^2 + |h_2|^2)$ can be calculated. Then the received signal can be normalized by the channel's power transfer function and the transmitted signal power estimates, before the demodulation process takes place.

The above encoding/decoding operations have been carried out for the case of a single receive antenna, but these arguments may be readily extended to an arbitrary number of receive antennas, where the resultant signals are appropriately combined, before passing them to the differential detector.

Therefore, to further augment the rationale behind the above arguments, we contrast the decoding processes of the coherent and differentially encoded STS schemes, while employing non-constant modulus constellations.

1. The coherently decoded signal can be represented as $\tilde{x}_t = \frac{1}{2} \cdot (|h_1|^2 + |h_2|^2) \cdot x_t + N_{\text{coh}}$, while the differentially decoded signal is given by Equation (10.13) as $\tilde{x}_t = \frac{1}{2} \cdot (|h_1|^2 + |h_2|^2) \cdot \sqrt{|v^1_{t-1}|^2 + |v^2_{t-1}|^2} \cdot x_t + N_{\text{diff}}$.

2. The coherent decoder has to estimate the CIR and hence it has full knowledge of h_1 and h_2, which is required for normalizing the decoded signal and estimating the transmitted symbol x_t. In contrast, the differential-encoding-aided receiver does not employ channel estimation, although it is required to estimate $(|h_1|^2 + |h_2|^2)$ and $(\sqrt{|v^1_{t-1}|^2 + |v^2_{t-1}|^2})$ for demodulating the non-constant modulus transmitted symbol x.

3. To elaborate a little further, the differential decoder does not employ any channel estimation for quantifying $(|h_1|^2 + |h_2|^2)$. First, the differential decoder estimates $(\sqrt{|v^1_{t-1}|^2 + |v^2_{t-1}|^2})$ from the symbols received during the previous time slot and then it employs Equation (10.15) for estimating the channel's output power by using the received signal only.

10.3.3 Design using SP Modulation[1]

The design concept of maximizing the diversity product[2] [54, 208] was generalized in [43] in order to account for the effects of the temporal correlation exhibited by the fading channel. In order to maximize the achievable coding advantage for DSTS signals that attain full diversity, we construct a class of DSTS signals in conjunction with SP modulation [46, 221], which is referred to as DSTS-SP.

According to Equations (10.13) and (10.14), the DSTS-decoded symbols $\tilde{x}^1$ and $\tilde{x}^2$ represent scaled versions of the transmitted symbols x^1 and x^2 corrupted by the complex-valued AWGN. Assuming that there are L legitimate SP-modulated signals, which the DSTS

[1]This section is based on the design of the SP modulation of [343, 344].

[2]The diversity product or coding advantage is defined as the estimated SNR gain over an uncoded system having the same diversity order as the coded system [12].

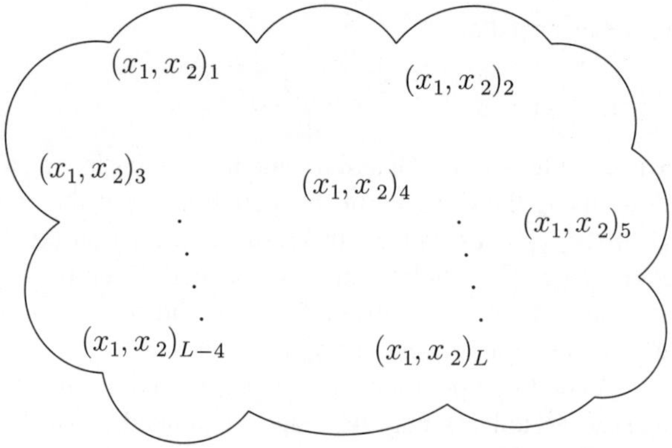

Figure 10.4: The L legitimate two-dimensional complex vectors.

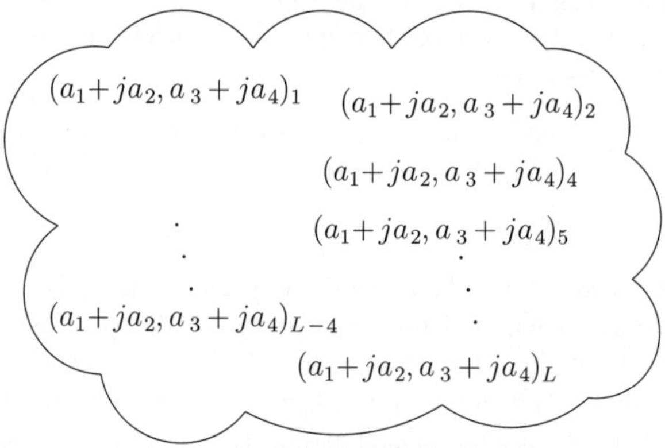

Figure 10.5: The L legitimate two-dimensional complex vectors represented by their real and imaginary components.

encoder can choose from, this observation implies that the diversity product is determined by the MED of the L two-dimensional complex-valued vectors $(x_1, x_2)_l \in \mathbb{C}^2, l = 0, \ldots, L-1$. Therefore, in order to maximize the diversity product, the L two-dimensional complex vectors must be designed by ensuring that they have the best MED in the two-dimensional complex-valued space $\mathbb{C}^2$, as illustrated in Figure 10.4. If each of the L two-dimensional complex vectors is expressed using its real and imaginary components, then we have

$$(x_1, x_2)_l \quad \Longleftrightarrow \quad (a_1 + ja_2, a_3 + ja_4)_l. \tag{10.16}$$

Hence, each of these complex vectors can be represented as shown in Figure 10.5. It may be observed directly from Figure 10.5 that the design problem can be readily transformed from the two-dimensional complex-valued space $\mathbb{C}^2$ to the four-dimensional real-valued Euclidean space $\mathbb{R}^4$, as portrayed in Figure 10.6. It was proposed in [43] to use SP for combining the individual antennas' signals into a joint design, since SP-modulated symbols have the best known MED in the $2(k+1)$-dimensional real-valued Euclidean space $\mathbb{R}^{2(k+1)}$ (see [221]), which directly maximizes the diversity product.

To summarize, according to Section 10.3.1, x^1 and x^2 represent independent conventional PSK/QAM modulated symbols transmitted from the first and second transmit antennas and no effort is made to jointly design a signal constellation for the various combinations of x^1

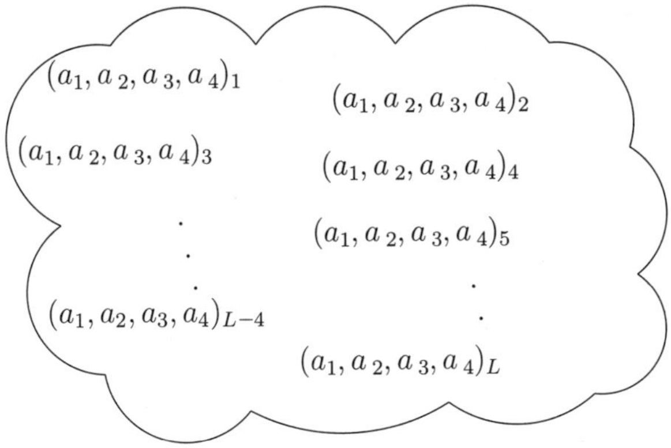

Figure 10.6: The L legitimate four-dimensional real-valued vectors.

and x^2. In contrast, in the case of SP, these symbols are designed jointly, in order to further increase the attainable Euclidean distance and, hence, the resultant diversity product or coding advantage, as suggested previously. Assuming that there are L legitimate vectors $(x^{l,1}, x^{l,2})$, $l = 0, 1, \ldots, L-1$, where L represents the number of sphere-packed modulated symbols, the transmitter then has to choose the modulated signal from these L legitimate symbols to be transmitted over the two DSTS antennas, where the twin-antenna-aided DSTS-SP bandwidth efficiency is given by $(\log_2 L/2)$ bits per channel use.

In contrast to the independent transmitted signal design of Section 10.3.1, the aim is to design $x^{l,1}$ and $x^{l,2}$ jointly, so that they have the best MED from all other $(L-1)$ legitimate SP symbols, since this minimizes the system's SP symbol error probability. Let $(a^{l,1}, a^{l,2}, a^{l,3}, a^{l,4})$, $l = 0, 1, \ldots, L-1$, be legitimate phasor points of the four-dimensional real-valued Euclidean space $\mathbb{R}^4$; hence the two time slots' complex-valued phasor points $x^{l,1}$ and $x^{l,2}$ may be written as

$$\begin{aligned}\{x^{l,1}, x^{l,2}\} &= T_{\text{sp}}(a^{l,1}, a^{l,2}, a^{l,3}, a^{l,4}) \\ &= \{a^{l,1} + ja^{l,2}, a^{l,3} + ja^{l,4}\},\end{aligned} \tag{10.17}$$

where the SP function T_{sp} represents the mapping of the SP symbols $(a^{l,1}, a^{l,2}, a^{l,3}, a^{l,4})$ to the complex-valued symbols $x^{l,1}$ and $x^{l,2}$, $l = 0, \ldots, L-1$.

In the four-dimensional real-valued Euclidean space $\mathbb{R}^4$, the lattice D_4 is defined as a SP having the best MED from all other $(L-1)$ legitimate constellation points in $\mathbb{R}^4$ (see [221]). More specifically, D_4 may be defined as a lattice that consists of all legitimate sphere-packed constellation points having integer coordinates $(a^{l,1}, a^{l,2}, a^{l,3}, a^{l,4})$, $l = 0, \ldots, L-1$, uniquely and unambiguously describing the L legitimate combinations of the two time slots' modulated DSTS symbols, but subjected to the SP constraint of [221]:

$$a^{l,1} + a^{l,2} + a^{l,3} + a^{l,4} = \kappa_l, \quad l = 0, \ldots, L-1, \tag{10.18}$$

where κ_l may assume any even integer value. Alternatively, D_4 may be defined as the integer span of the vectors $\mathbf{u}_1$, $\mathbf{u}_2$, $\mathbf{u}_3$ and $\mathbf{u}_4$ that form the rows of the following generator matrix [221]

$$\begin{bmatrix}\mathbf{u}_1 \\ \mathbf{u}_2 \\ \mathbf{u}_3 \\ \mathbf{u}_4\end{bmatrix} \triangleq \begin{bmatrix}2 & 0 & 0 & 0 \\ 1 & 1 & 0 & 0 \\ 1 & 0 & 1 & 0 \\ 1 & 0 & 0 & 1\end{bmatrix}. \tag{10.19}$$

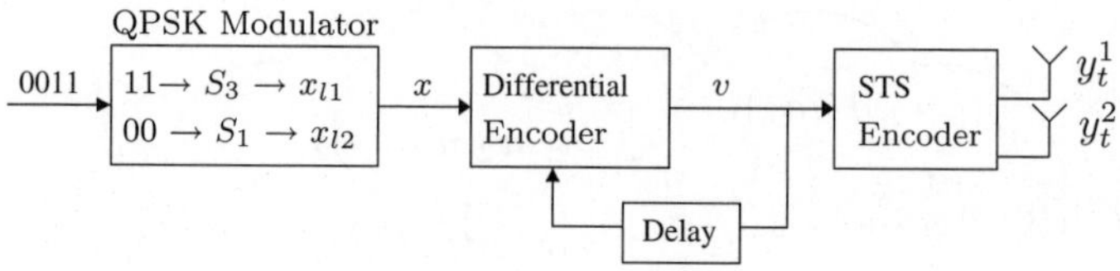

Figure 10.7: Transmission of two QPSK symbols using two-antenna-aided DSTS system.

We may infer from the above definition in Equation (10.19) that D_4 contains the centers $(2, 0, 0, 0)$, $(1, 1, 0, 0)$, $(1, 0, 1, 0)$ and $(1, 0, 0, 1)$. It also contains all linear combinations of these points [221].

Assuming that $\mathbf{S} = \{\mathbf{s}^l = [a^{l,1}, a^{l,2}, a^{l,3}, a^{l,4}] \in \mathbb{R}^4 : 0 \leq l \leq L-1\}$ constitutes a set of L legitimate constellation points from the lattice D_4 having a total energy of

$$E_{\text{total}} \triangleq \sum_{l=0}^{L-1} (|a^{l,1}|^2 + |a^{l,2}|^2 + |a^{l,3}|^2 + |a^{l,4}|^2), \tag{10.20}$$

and upon introducing the notation

$$\begin{aligned} \mathbf{C}_l &= \sqrt{\frac{2L}{E_{\text{total}}}}(x^{l,1}, x^{l,2}) \\ &= \sqrt{\frac{2L}{E_{\text{total}}}}(a^{l,1} + ja^{l,2}, a^{l,3} + ja^{l,4}), \quad l = 0, 1, \ldots, L-1, \end{aligned} \tag{10.21}$$

we have a set of complex SP constellation symbols, $\{\mathbf{C}_l\colon 0 \leq l \leq L-1\}$, whose diversity product is determined by the MED of the set of L legitimate SP constellation points in $\mathbf{S}$.

The following example illustrates how SP modulation is combined with the twin-antenna-aided DSTS scheme as compared with the conventionally modulated DSTS system.

Example 10.3.1. Assume that there are $L = 16$ different legitimate signals $(x^{l,1}, x^{l,2})$, $l = 0, 1, \ldots, 15$, that can be transmitted by the DSTS encoder. We will compare two modulation schemes, namely conventional QPSK and SP modulation.

- **Conventional QPSK modulation.** There are four legitimate complex-valued QPSK symbols (S_0, S_1, S_2, S_3) that can be used to represent independently any of the $x^{l,1}$ and $x^{l,2}$ symbols, $l = 0, 1, \ldots, 15$. The transmission scheme using two consecutive time slots is portrayed in Figure 10.7.

- **SP modulation.** We need $L = 16$ SP phasor points $(a^{l,1}, a^{l,2}, a^{l,3}, a^{l,4})$ from the lattice D_4 in order to jointly represent each pair of signals $(x^{l,1}, x^{l,2})$, $l = 0, 1, \ldots, 15$ according to Equation (10.21) as depicted in Figure 10.8.

10.3.4 SP Constellation Construction

Since the signal constructed from the SP scheme[3] of Equation (10.21) is multiplied by a factor that is inversely proportional to $\sqrt{E_{\text{total}}}$, namely by $\sqrt{2L/E_{\text{total}}}$, it is desirable to choose a specific subset of L SP constellation points from the entire set of legitimate SP

[3]This section is based on the design of the SP modulation of [343, 344].

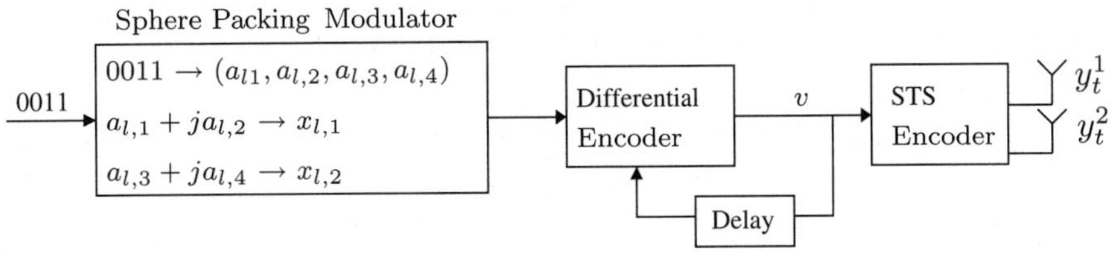

Figure 10.8: Transmission of sphere-packed symbols using two-antenna-aided DSTS system.

constellation points hosted by D_4, which results in the minimum total energy E_{total}, while maintaining a certain MED amongst the SP symbols. Viewing this design trade-off from a different perspective, if more than L SP points satisfy the minimum total energy constraint, an exhaustive computer search is carried out for determining the optimum choice of the L SP constellation points out of all possible points, which possess the highest MED, hence minimizing the SP-symbol error probability. The legitimate constellation points hosted by D_4 are categorized into layers or shells based on their Euclidean norms or energy (i.e. the distance from the origin) as seen in Table 10.1. For example, it was shown in [221] that the first layer consists of 24 legitimate constellation points hosted by D_4 having an identical minimum energy of $E_{\text{total}} = 2$. In simple terms, it may be readily verified that the SP symbol centered at $(0, 0, 0, 0)$ has 24 minimum-distance or closest-neighbor SP symbols around it, centered at the points $(\pm 1, \pm 1, 0, 0)$, where any choice of signs and any ordering of the coordinates is legitimate [221].

Table 10.1 provides a summary of the constellation points hosted by the first ten SP layers in the four-dimensional lattice D_4. In order to generate the full list of SP constellation points for a specific layer, we have to apply all legitimate permutations and signs for the corresponding constellation points given in Table 10.1.

10.3.5 Bandwidth Efficiency of the Twin-antenna-aided DSTS System

In the two-antenna-aided DSTS encoder, the data is serial-to-parallel converted into two substreams. The new bit duration of each parallel substream, or equivalently the symbol duration, becomes $T_s = 2T_b$ as illustrated in [13, 51] and as exemplified in simple graphical terms in Figure 10.3. According to Section 10.3.1, the DSTS transmitter using two transmit antennas transmits two real- or complex-valued conventional modulated symbols in two time slots. Therefore, the two-transmit-antenna DSTS code rate is $2/2 = 1$ and then according to the number of BPS B, the system's bandwidth efficiency becomes equal to B bits per channel use. For example, in the case of QPSK we have $B = 2$ BPS, which results in an effective system bandwidth efficiency of 2 bits per channel use. Table 10.2 presents the bandwidth efficiency of the twin-antenna-aided DSTS system for different conventional modulated constellation sizes.

On the other hand, for a two-transmit-antenna system using SP modulation, one SP symbol is transmitted in two time slots. Therefore, the DSTS-SP code rate is $1/2$ and then according to the number of BPS B_{sp}, the SP system's bandwidth efficiency becomes equal to $B_{\text{sp}}/2$ bits per channel use. For example, in the case of SP with $L = 16$ constellation, we have $B_{\text{sp}} = 4$ BPS, which results in an effective system's bandwidth efficiency of 2 bits per channel use, which is identical to the system's bandwidth efficiency of the QPSK-modulated twin-antenna-aided DSTS system. Table 10.3 presents the bandwidth efficiency of the twin-antenna-aided DSTS-SP system for different constellation sizes.

Table 10.1: The first ten layers of the lattice D_4.

Layer	Constellation points				Norm	Number of combinations
0	0	0	0	0	0	1
1	±1	±1	0	0	2	24
2	±2	0	0	0	4	8
	±1	±1	±1	±1	4	16
3	±2	±1	±1	0	6	96
4	±2	±2	0	0	8	24
5	±2	±2	±1	±1	10	96
	±3	±1	0	0	10	48
6	±3	±1	±1	±1	12	64
	±2	±2	±2	0	12	32
7	±3	±2	±1	0	14	192
8	±2	±2	±2	±2	16	16
	±4	0	0	0	16	8
9	±4	±1	±1	0	18	96
	±3	±2	±2	±1	18	192
	±3	±3	0	0	18	24
10	±4	±2	0	0	20	48
	±3	±3	±1	±1	20	96

Table 10.2: Bandwidth efficiency of twin-antenna-aided DSTS systems for different conventional modulation signal sets.

Modulation	BPS	Bandwidth efficiency (bits per channel use)
BPSK	1	1
QPSK	2	2
8-PSK	3	3
16-QAM	4	4
64-QAM	6	6

10.3.6 Capacity of the Two-antenna-aided DSTS-SP Scheme

The capacity of a single-input-single-output AWGN channel was quantified by Shannon in 1948 [272]. Since then, substantial research efforts have been invested in finding channel codes that would produce an arbitrarily low probability of error. Shannon's channel capacity was only defined for a CCMC [6], where the channel input is a continuous-amplitude discrete-time Gaussian-distributed signal and the capacity is only restricted by either the signaling energy or the bandwidth [293]. In contrast, in the context of discrete-amplitude QAM [248] and PSK [6] signals, we encounter a DCMC [6].

With the advent of MIMO systems, the MIMO channel capacity is of immediate interest. Thus, the channel capacity of MIMO systems was found for CCMC in [3, 245, 251, 345] and then Ng *et al.* [293] developed the DCMC channel capacity of the STBC-aided MIMO system combined with multi-dimensional signal sets. In this section, we present the CCMC and DCMC capacities of the twin-antenna-aided DSTS system using multi-dimensional

Table 10.3: Bandwidth efficiency of twin-antenna-aided DSTS-SP systems for different SP signal set sizes L.

L	BPS	Bandwidth efficiency (bits per channel use)
4	2	1
8	3	1.5
16	4	2
32	5	2.5
64	6	3
128	7	3.5
256	8	4
512	9	4.5
1024	10	5
2048	11	5.5
4096	12	6

SP modulation and employing N_r receive antennas. The same analysis can be performed to generate the capacity of the DSTS system employing two-dimensional conventional modulation schemes.

The complex-valued channel output symbols received during two consecutive DSTS time slots at receive antenna r, $r \in [1, \ldots, N_r]$, are DSTS decoded in order to extract the estimates $\tilde{x}^1$ and $\tilde{x}^2$ of the most likely transmitted symbols x^1 and x^2 resulting in

$$\begin{aligned}\tilde{x}^1_{t,r} &= \tfrac{1}{2} \cdot (|h_{1,r}|^2 + |h_{2,r}|^2) \cdot \sqrt{|v^1_{t-1}|^2 + |v^2_{t-1}|^2} \cdot x^1_t + N_{1,r} \\ &= \chi^2_{2N_t,r} \cdot x^1_t + N_{1,r},\end{aligned} \tag{10.22}$$

$$\begin{aligned}\tilde{x}^2_{t,r} &= \tfrac{1}{2} \cdot (|h_{1,r}|^2 + |h_{2,r}|^2) \cdot \sqrt{|v^1_{t-1}|^2 + |v^2_{t-1}|^2} \cdot x^2_t + N_{2,r} \\ &= \chi^2_{2N_t,r} \cdot x^2_t + N_{2,r},\end{aligned} \tag{10.23}$$

where $h_{1,r}$ and $h_{2,r}$ denote the narrowband complex-valued CIRs corresponding to the first and second transmit antennas and the rth receive antenna, respectively. Here $N_{1,r}$ and $N_{2,r}$ are zero-mean complex-valued Gaussian random variables having variances of $\sigma^2_N = \sigma^2_{N_{1,r}} = \sigma^2_{N_{2,r}} \approx 2 \cdot \chi^2_{2N_t,r} \cdot \sigma^2_n$ and $\chi^2_{2N_t,r} = \frac{1}{2} \cdot (|h_{1,r}|^2 + |h_{2,r}|^2) \cdot \sqrt{|v^1_{t-1}|^2 + |v^2_{t-1}|^2}$ represents the chi-squared distributed random variable having $2N_t = 4$ degrees of freedom.

The received sphere-packed symbol $\tilde{\mathbf{s}}_r$ at receive antenna r is then constructed from the estimates $\tilde{x}^1_{t,r}$ and $\tilde{x}^2_{t,r}$ using the inverse function of T_{sp} introduced in Equation (10.17) as

$$\tilde{\mathbf{s}}_r = T^{-1}_{\text{sp}}(\tilde{x}^1_{t,r}, \tilde{x}^2_{t,r}), \tag{10.24}$$

where we have $\tilde{\mathbf{s}} = [\tilde{a}^1\ \tilde{a}^2\ \tilde{a}^3\ \tilde{a}^4] \in \mathbb{R}^4$. Therefore, the received sphere-packed symbol $\tilde{\mathbf{s}}_r$ at receive antenna r can be written as

$$\tilde{\mathbf{s}}_r = \chi^2_{2N_t,r} \cdot \mathbf{s}^l + \acute{\mathbf{N}}_r, \tag{10.25}$$

where we have $\mathbf{s}^l = [a^{l,1}\ a^{l,2}\ a^{l,3}\ a^{l,4}] \in S,\ 0 \le l \le L-1$. Furthermore, $\acute{\mathbf{N}}_r = [\acute{N}_{1,r}\ \acute{N}_{2,r}\ \acute{N}_{3,r}\ \acute{N}_{4,r}] \in \mathbb{R}^4$ is a four-dimensional real-valued Gaussian random variable having a covariance matrix of $\sigma^2_{\acute{N}} \cdot \mathbf{I}_D = \sigma^2_N \cdot \mathbf{I}_D = 2 \cdot \chi^2_{2N_t,r} \cdot \sigma^2_n \cdot \mathbf{I}_D = 2 \cdot \chi^2_{2N_t,r} \cdot (N_0/2) \cdot \mathbf{I}_D$, where we have $D = 4$, since the SP symbol constellation $\mathbf{S}$ is four-dimensional.

Therefore, the received sphere-packed symbol $\tilde{\mathbf{s}}$ can be written as

$$\tilde{\mathbf{s}} = \sum_{r=1}^{N_r} \chi^2_{2N_t,r} \cdot \mathbf{s}^l + \acute{\mathbf{N}}. \tag{10.26}$$

The conditional probability of receiving a four-dimensional signal $\tilde{\mathbf{s}}$, given that a four-dimensional L-ary signal $\mathbf{s}^l \in \mathbf{S}$, $l \in [0, \ldots, L-1]$, was transmitted over the Rayleigh channel of Equation (10.26), is given by the following PDF [6]:

$$\begin{aligned} p(\tilde{\mathbf{s}}|\mathbf{s}^l) &= \frac{1}{\prod_{d=1}^{D=4} \sqrt{\pi N_0 \sum_{r=1}^{N_r} \chi^2_{2N_t,r}[d]}} \cdot \exp\left(\sum_{d=1}^{D=4} \sum_{r=1}^{N_r} \frac{-(\tilde{\mathbf{s}}_r[d] - \chi^2_{2N_t,r}[d] \cdot \mathbf{s}^l[d])^2}{\chi^2_{2N_t,r}[d] \cdot N_0} \right) \\ &= \frac{1}{\prod_{d=1}^{D=4} \sqrt{\pi N_0 \sum_{r=1}^{N_r} \chi^2_{2N_t,r}[d]}} \cdot \exp\left(\sum_{d=1}^{D=4} \sum_{r=1}^{N_r} \frac{-(\tilde{a}^d_r - \chi^2_{2N_t,r}[d] \cdot a^{l,d})^2}{\chi^2_{2N_t,r}[d] \cdot N_0} \right). \end{aligned} \tag{10.27}$$

The channel capacity for the DSTS MIMO system employing D-dimensional L-ary SP signaling over the DCMC [6] can be derived from that of the discrete memoryless channel [249] as

$$\begin{aligned} C_{\text{DCMC}} = \max_{p(\mathbf{s}^0),\ldots,p(\mathbf{s}^{L-1})} \sum_{l=0}^{L-1} \underbrace{\int_{-\infty}^{\infty} \cdots \int_{-\infty}^{\infty}}_{D\text{-fold}} p(\tilde{\mathbf{s}}|\mathbf{s}^l) \cdot p(\mathbf{s}^l) \\ \cdot \log_2\left(\frac{p(\tilde{\mathbf{s}}|\mathbf{s}^l)}{\sum_{k=0}^{L-1} p(\tilde{\mathbf{s}}|\mathbf{s}^k) \cdot p(\mathbf{s}^k)} \right) d\tilde{\mathbf{s}} \quad \text{(BPS)}, \end{aligned} \tag{10.28}$$

where $p(\mathbf{s}^l)$ is the probability of occurrence for the transmitted SP symbol $\mathbf{s}^l$ and $p(\tilde{\mathbf{s}}|\mathbf{s}^l)$ is expressed in Equation (10.27). The right-hand side of Equation (10.28) is maximized when the transmitted SP symbols are equiprobably distributed, i.e. when we have $p(\mathbf{s}^l) = 1/L$, $l = 0, \ldots, L-1$, which leads to achieving the full capacity [249]. The right-hand side of Equation (10.28) may be further simplified as follows [293]:

$$\begin{aligned} \log_2\left(\frac{p(\tilde{\mathbf{s}}|\mathbf{s}^l)}{\sum_{k=0}^{L-1} p(\tilde{\mathbf{s}}|\mathbf{s}^k) \cdot p(\mathbf{s}^k)} \right) &= -\log_2\left(\frac{1}{L} \sum_{k=0}^{L-1} \frac{p(\tilde{\mathbf{s}}|\mathbf{s}^k)}{p(\tilde{\mathbf{s}}|\mathbf{s}^l)} \right) \\ &= \log_2(L) - \log_2 \sum_{k=0}^{L-1} \exp(\Psi_{l,k}), \end{aligned} \tag{10.29}$$

where $\Psi_{l,k}$ is expressed as [293]

$$\begin{aligned} \Psi_{l,k} &= \sum_{d=1}^{D=4} \sum_{r=1}^{N_r} \left(\frac{-(\tilde{a}^d_r - \chi^2_{2N_t,r}[d] \cdot a^{k,d})^2}{\chi^2_{2N_t,r}[d] \cdot N_0} + \frac{(\tilde{a}^d_r - \chi^2_{2N_t,r}[d] \cdot a^{l,d})^2}{\chi^2_{2N_t,r}[d] \cdot N_0} \right) \\ &= \sum_{d=1}^{D=4} \sum_{r=1}^{N_r} \left(\frac{-(\chi^2_{2N_t,r}[d] \cdot (a^{l,d} - a^{k,d}) + \acute{N}_{d,r})^2 + (\acute{N}_{d,r})^2}{\chi^2_{2N_t,r}[d] \cdot N_0} \right). \end{aligned} \tag{10.30}$$

Now, substituting Equation (10.29) into Equation (10.28) and observing that we have $p(\mathbf{s}^l) = 1/L$, yields [293]

$$C_{\text{DCMC}} = \frac{\log_2(L)}{L} \sum_{l=0}^{L-1} \underbrace{\int_{-\infty}^{\infty} \cdots \int_{-\infty}^{\infty}}_{D\text{-fold}} p(\tilde{\mathbf{s}}|\mathbf{s}^l)\, d\tilde{\mathbf{s}}$$

$$- \frac{1}{L} \sum_{l=0}^{L-1} \underbrace{\int_{-\infty}^{\infty} \cdots \int_{-\infty}^{\infty}}_{D\text{-fold}} p(\tilde{\mathbf{s}}|\mathbf{s}^l) \log_2 \sum_{k=0}^{L-1} \exp(\Psi_{l,k})\, d\tilde{\mathbf{s}}$$

$$= \log_2(L) - \frac{1}{L} \sum_{l=0}^{L-1} E\left[\log_2 \sum_{k=0}^{L-1} \exp(\Psi_{l,k}) \,\middle|\, \mathbf{s}^l\right] \quad \text{(BPS)}, \tag{10.31}$$

where $E[A \mid B]$ is the expectation of A conditioned on B. The expectation in Equation (10.31) can be estimated using a sufficiently high number of h and $\hat{\mathbf{N}}_r$ realizations with the aid of Monte Carlo simulations for $r = 1, \ldots, N_r$.

The bandwidth efficiency η_{DCMC} of the DCMC is computed by normalizing the DCMC capacity C_{DCMC} with respect to the product of W and T, where W is the bandwidth and T is the signaling period of the finite-energy signaling waveform. Furthermore, it was reported in [346, pp. 348–351], that the constellation dimension D is given by $D = 2WT$. Explicitly, the bandwidth efficiency as a function of the capacity is given by [250, 293, 346]

$$\eta_{\text{DCMC}}(\text{SNR}) = \frac{C_{\text{DCMC}}(\text{SNR})}{WT} = \frac{C_{\text{DCMC}}(\text{SNR})}{D/2} \quad (\text{bit s}^{-1}\ \text{Hz}^{-1}). \tag{10.32}$$

For a multiple-antenna-aided transmitter using N_r coherent detectors provided with perfect knowledge of the channel coefficients at the receiver, the CCMC's [6] bandwidth efficiency can be formulated as follows [251]:

$$\eta_{\text{CCMC}}^{\text{coherent}}(\text{SNR}) = E\left[\log_2\left(1 + \sum_{r=1}^{N_r} \chi^2_{2N_t,r} \frac{SNR}{N_t}\right)\right] \quad (\text{bit s}^{-1}\ \text{Hz}^{-1}), \tag{10.33}$$

where the expectation $E[\cdot]$ is taken over $\chi^2_{2N_t,r}$.

Figures 10.9, 10.10 and 10.11 show the DCMC capacity evaluated from Equation (10.31) for the four-dimensional SP-modulation-assisted DSTS as well as STS schemes for $L = 4$, 16 and 64, when employing $N_t = 2$ transmit antennas as well as $N_r = 1$, 2 and 4 receive antennas, respectively. The CCMC [6] capacity of the MIMO scheme is also plotted for comparison in Figures 10.9, 10.10 and 10.11 based on [251].

Figures 10.12, 10.13 and 10.14 quantify and compare the achievable bandwidth efficiency of various SP-modulated DSTS schemes with those of their identical-throughput conventionally modulated DSTS counterparts. The specific modulation type employed by the various schemes is outlined in Table 10.4. The figures explicitly illustrate that a higher bandwidth efficiency may be attained when employing SP modulation in conjunction with DSTS schemes having $N_t = 2$ transmit antennas as compared with an equivalent system employing conventional PSK modulation.

10.3.7 Performance of the Two-antenna-aided DSTS System

In this section, the two-antenna-aided DSTS schemes of Sections 10.3.1 and 10.3.3 are considered. Simulation results are provided for systems having different bandwidth efficiencies

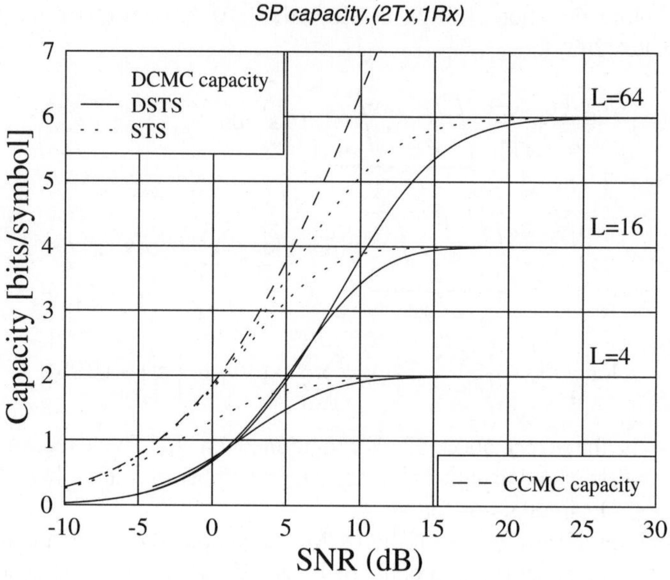

Figure 10.9: Capacity comparison of coherent and differential STS-SP-based schemes evaluated from Equation (10.31) and using $L = 4$, 16 and 64, when employing $N_t = 2$ transmit and $N_r = 1$ receive antennas for communicating over a correlated Rayleigh fading channel having a normalized Doppler frequency of $f_D = 0.01$.

Table 10.4: Conventional and SP modulation employed for different bandwidth efficiency rates.

Bandwidth efficiency	Conventional modulation	SP modulation
1	BPSK	$L = 4$
2	QPSK	$L = 16$
3	8-PSK	$L = 64$
4	16-PSK	$L = 256$

in conjunction with appropriate conventional and SP modulation schemes, as outlined in Table 10.4. Observe that two consecutive time slots are required for transmitting a single SP symbol when using the two-antenna-aided system. In contrast, two conventionally modulated symbols are transmitted during the same time period. Therefore, the bandwidth efficiency of the SP modulation scheme has to be twice as high as that of the conventional modulation scheme in order to compensate for the potential rate loss and to produce systems having an identical overall bandwidth efficiency. This explains the specific choices of L in Table 10.4. Our results are presented in terms of the BER and SP-SER performance curves for various systems employing $N_r = 1, 2, 3$ and 4 receive antennas for communication over a temporally correlated narrowband Rayleigh fading channel having a normalized Doppler frequency of $f_D = 0.01$.

Figures 10.15 and 10.16 compare the BER performance of the differentially encoded and the coherently detected STS while using BPSK-modulated signals, two transmit antennas and a SF of four in conjunction with one and four receive antennas, respectively. The coherent

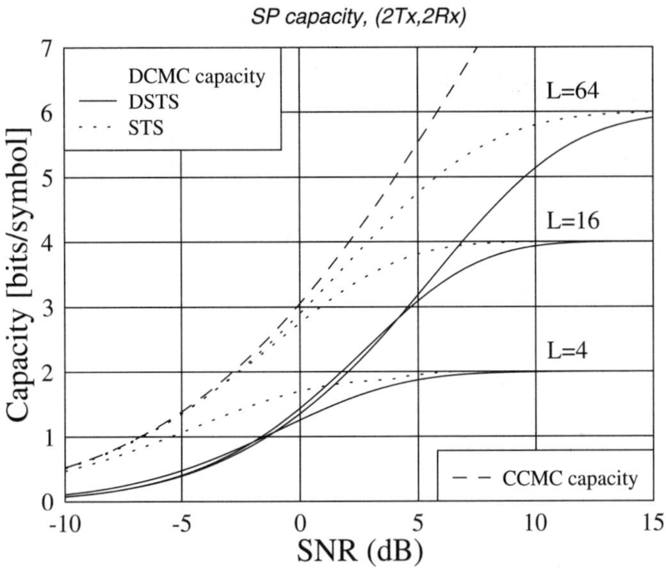

Figure 10.10: Capacity comparison of coherent and differential STS-SP-based schemes evaluated from Equation (10.31) and using $L = 4$, 16 and 64, when employing $N_t = 2$ transmit and $N_r = 2$ receive antennas for communicating over a correlated Rayleigh fading channel having a normalized Doppler frequency of $f_D = 0.01$.

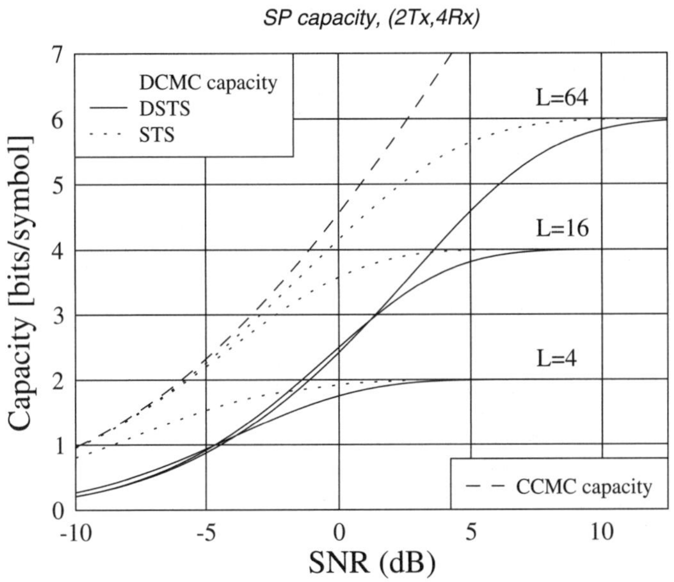

Figure 10.11: Capacity comparison of coherent and differential STS-SP-based schemes evaluated from Equation (10.31) and using $L = 4$, 16 and 64, when employing $N_t = 2$ transmit and $N_r = 4$ receive antennas for communicating over a correlated Rayleigh fading channel having a normalized Doppler frequency of $f_D = 0.01$.

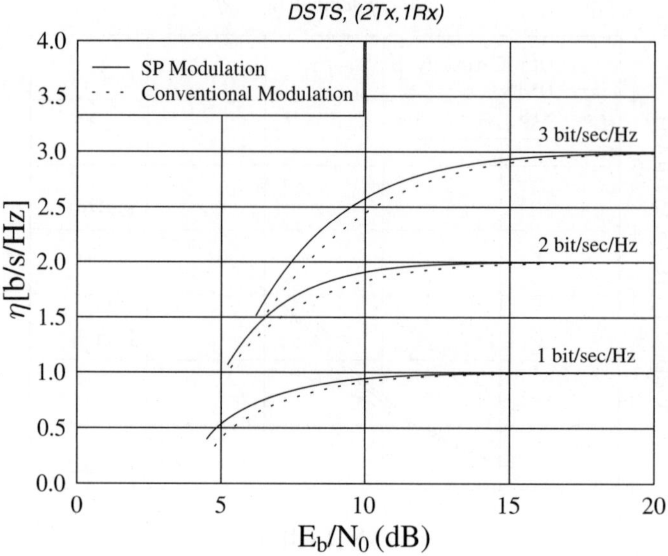

Figure 10.12: Bandwidth efficiency comparison of SP- and conventional-modulation-aided DSTS schemes evaluated from Equation (10.32), when employing $N_t = 2$ transmit and $N_r = 1$ receive antennas for communicating over a correlated Rayleigh fading channel having a normalized Doppler frequency of $f_D = 0.01$.

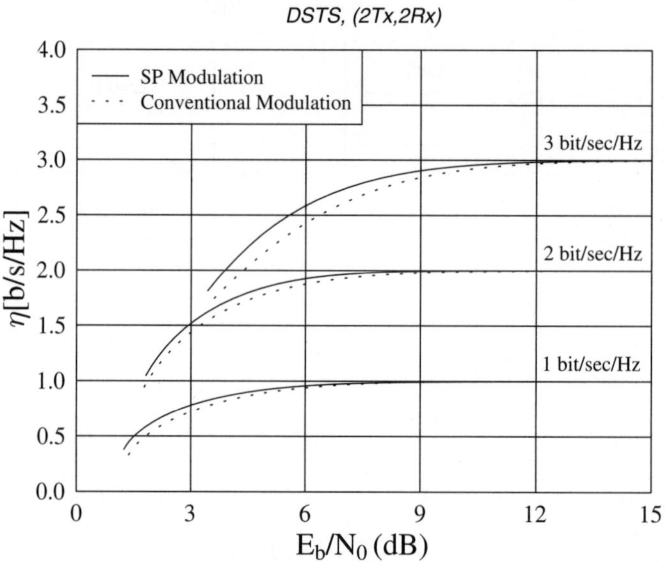

Figure 10.13: Bandwidth efficiency comparison of SP- and conventional-modulation-aided DSTS schemes evaluated from Equation (10.32), when employing $N_t = 2$ transmit and $N_r = 2$ receive antennas for communicating over a correlated Rayleigh fading channel having a normalized Doppler frequency of $f_D = 0.01$.

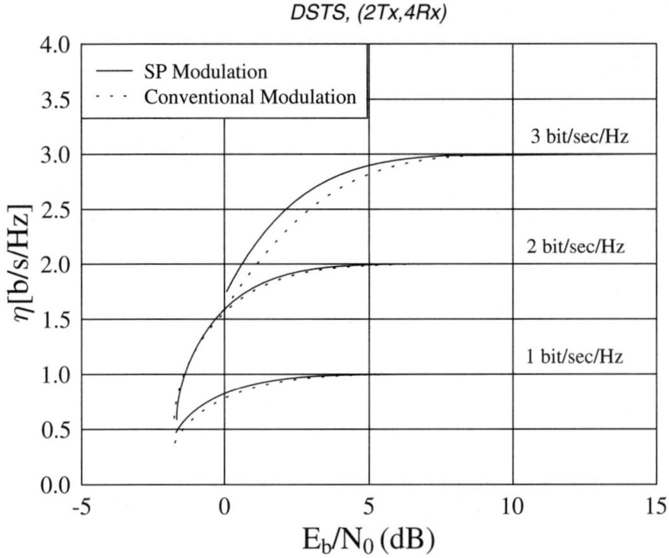

Figure 10.14: Bandwidth efficiency comparison of SP- and conventional-modulation-aided DSTS schemes evaluated from Equation (10.32), when employing $N_t = 2$ transmit and $N_r = 4$ receive antennas for communicating over a correlated Rayleigh fading channel having a normalized Doppler frequency of $f_D = 0.01$.

STS results are generated for the idealized scenario, where perfect channel knowledge is assumed at the receiver. As shown in Figures 10.15 and 10.16, the error doubling induced by the differential decoding results in a 3 dB performance loss as compared with coherent detection-aided STS benefiting from perfect channel knowledge at the receiver. Again, this is mainly due to the fact that according to Equation (10.13) differential decoding results in doubling the noise power as compared with the coherently detected signals. However, the differential encoding/decoding process eliminates the complexity of channel estimation required in coherent detection-aided schemes, which is directly proportional to the product of the number of transmit and receive antennas and also depends on the characteristics of the channel. In addition, it becomes clear from Figures 10.15 and 10.16 that the BER performance of both the DSTS and STS schemes is independent of the number of users in the system, which is a benefit of the fact that the spreading codes used are orthogonal and the channel is frequency-flat faded.

Furthermore, in order to study the effect of channel estimation error on the performance of the coherently detected STS signals, we contaminate the channel information at the receiver side with noise. We add AWGN to the channel information at the receiver side to model the effect of errors that may occur due to the channel estimation. Although the channel estimation error typically does not obey a Gaussian distribution, this simplified investigation gives us an insight concerning the effects of channel estimation errors on the system performance degradation of coherent systems. Figure 10.17 compares the BER performance of the DSTS and the STS scheme, while using two transmit antennas, one receive antenna, BPSK modulation, a SF of four and a single user. As discussed previously, coherent systems assuming perfect channel knowledge at the receiver outperform their differentially encoded, non-coherently detected counterparts by about 3 dB. However, when we add noise to the CSI used by the coherent STS scheme, we see that the performance degrades, as shown in

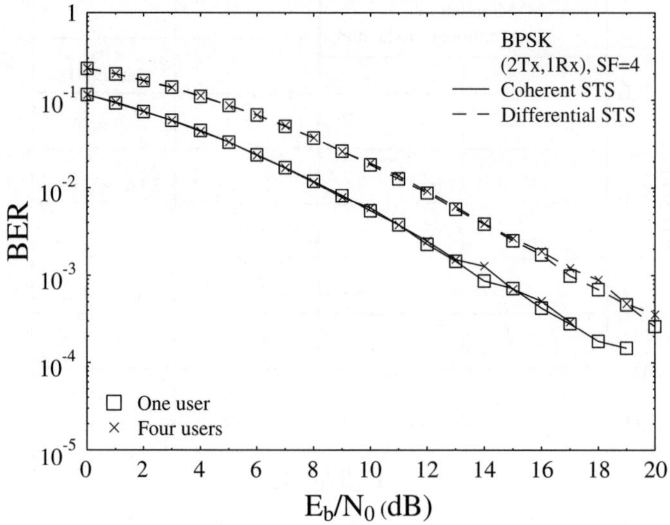

Figure 10.15: Comparison of the BER performance of coherent and differential STS, while using a BPSK-modulated signal, two transmit antennas, one receive antenna, a SF of four and a variable number of users.

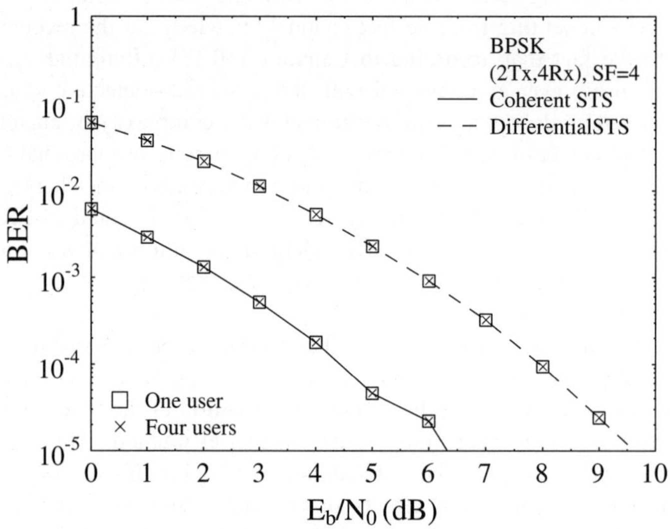

Figure 10.16: Comparison of the BER performance of coherent and differential STS, while using a BPSK-modulated signal, two transmit antennas, four receive antennas, a SF of four and a variable number of users.

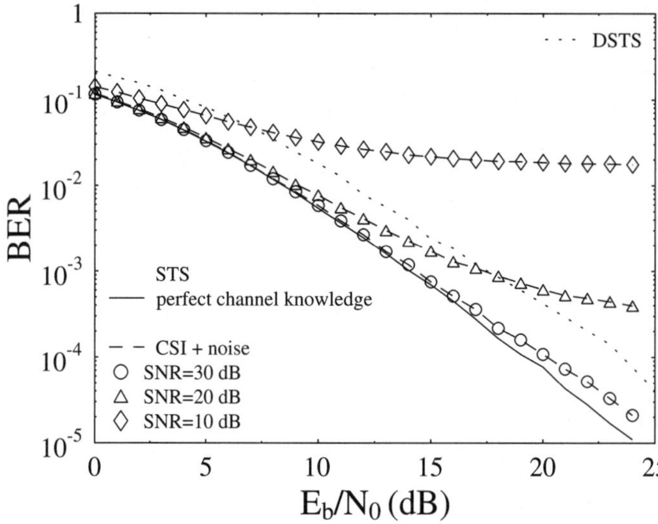

Figure 10.17: Comparison of the BER performance of coherent and differentially encoded non-coherent STS while using a BPSK-modulated signal, two transmit antennas, one receive antenna and a SF of four for supporting a single user. The CSI in the coherent STS is contaminated with AWGN in order to compare the performance when there is a channel estimation error.

Figure 10.17. More quantitatively, Figure 10.17 shows that when the power of the channel estimation noise added to the CSI is increased and hence the corresponding CSI SNR is 20 dB or less, the performance of the coherent STS scheme tends to exhibit an error floor and its BER curve crosses the BER curve of the DSTS scheme. Beyond this cross-over point the DSTS outperforms the STS. Therefore, the DSTS constitutes a convenient and low-complexity design alternative to the coherent STS scheme, since the DSTS scheme eliminates the complexity of channel estimation and also results in a better performance when the channel estimation error is high.

On the other hand, Figure 10.18 compares the BER performance of the DSTS scheme using BPSK, QPSK and 16-QAM signals. Moreover, a comparison between the BER performance of the differentially and coherently detected STS when using SP modulation is provided in Figure 10.19. As for the conventional modulation schemes, the differentially decoded system performs within 3 dB of the coherently detected system using perfect channel knowledge at the receiver.

Furthermore, in order to understand the effects of varying the Doppler frequency on the performance of the DSTS system, Figure 10.20 shows the attainable BER performance of the QPSK-modulated system using two transmit antennas, one receive antenna and a SF of four, while communicating over a temporally correlated narrowband Rayleigh fading channel and varying the Doppler frequency. As shown in the figure, when the normalized Doppler frequency increases from $f_D = 0.0001$ to 0.01, the system's performance remains similar. However, as the Doppler frequency increases beyond 0.01, the achievable BER performance substantially degrades. This is predominantly due to the fact that increasing the Doppler frequency makes the channel fast fading and thus the differential decoding scheme, which relies on the fact that the subsequent symbols experience similar fading, performs poorly.

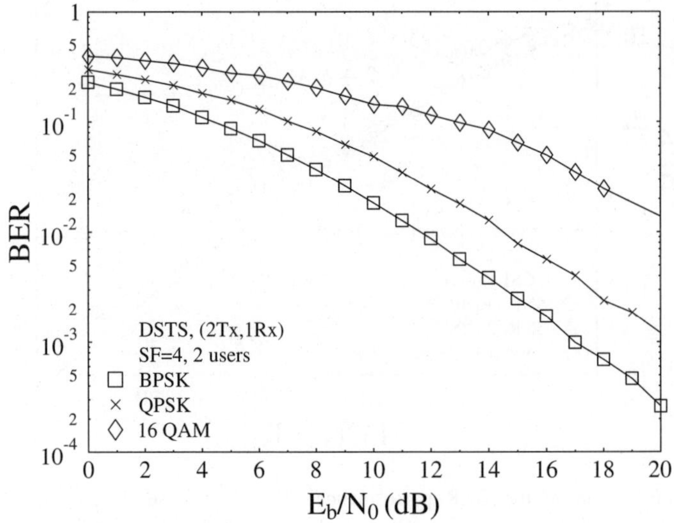

Figure 10.18: Comparison of the BER performance of DSTS while using PSK and QAM modulated signals, two transmit antennas, one receive antenna, a SF of four and two users.

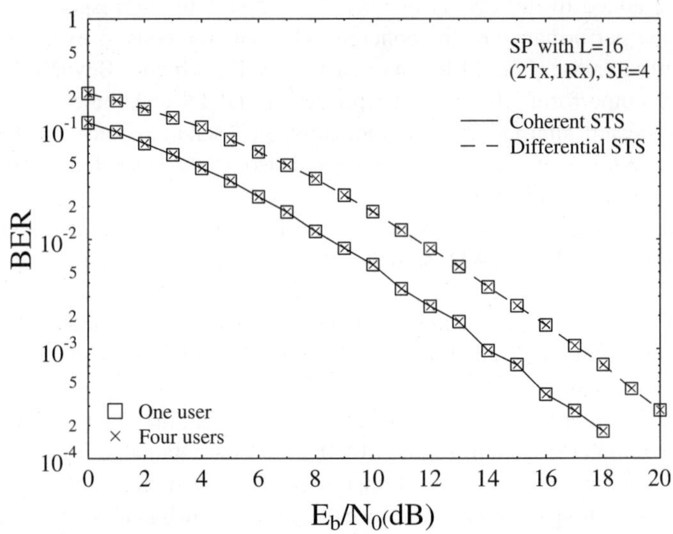

Figure 10.19: Comparison of the BER performance of coherent and differential STS, while employing SP-modulated signal with $L = 16$, two transmit antennas, one receive antenna, a SF of four and a variable number of users.

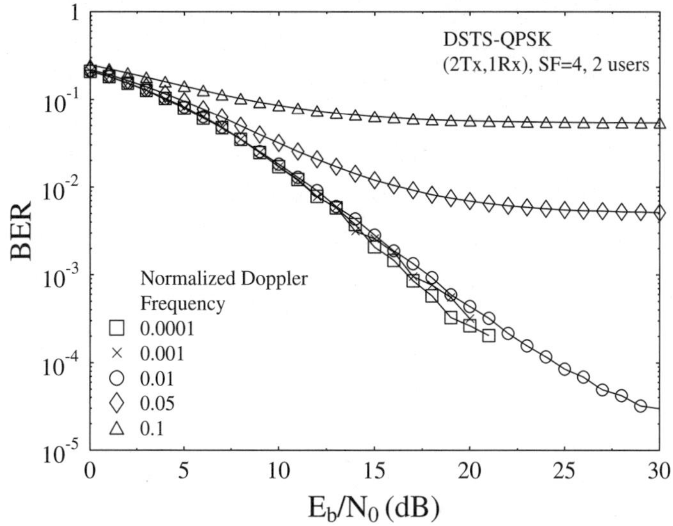

Figure 10.20: Comparison of the BER performance of differential STS while employing SP-modulated signal with $L = 16$, two transmit antennas, one receive antenna, a SF of four and two users for a different normalized Doppler frequency values.

Figure 10.21 shows the SP-SER performance curves of the DSTS scheme in conjunction with different conventional as well as SP modulations at various bandwidth efficiency values, as outlined in Table 10.4. All systems employ two transmit antennas for communication over a correlated Rayleigh fading channel associated with $f_D = 0.01$. Moreover, the system uses a SF of four and here accommodates two users. It is evident from Figure 10.21 that for a particular bandwidth efficiency, the two curves corresponding to the conventional modulation and to the SP modulation schemes have the same asymptotic slope (i.e. diversity order). This observation is based on the fact that the DSTS scheme is capable of achieving full diversity, similar to Alamouti's STBC scheme [11]. Accordingly, it is not expected that the asymptotic slope of the performance curves would improve by merely employing new modulation schemes without introducing another level of channel coding. The resultant BER performance curves are shown in Figure 10.22. Note that SP modulation attains a better SP-SER performance than the conventionally modulated DSTS scheme and this is expected, since SP was specifically designed for improving the DSTS SP-SER as compared with conventional DSTS schemes. However, observe in Figure 10.22 that the BER performances of SP modulation and conventional modulation are identical for systems having a bandwidth efficiency of 2 bits per channel use because it can be shown [46] that at this throughput they constitute specific manifestations of each other. In contrast, the DSTS-SP BER performance recorded for 3 bits per channel use is marginally worse than that of the conventionally modulated DSTS schemes, as shown in Figure 10.22, which demonstrates that the marginal advantage of conventional modulation over SP modulation diminishes at high SNRs.

Figures 10.23–10.28 illustrate the beneficial effect of increasing the number of receive antennas from two to four, respectively. Observe in Figures 10.24, 10.26 and 10.28 that the BER performance of SP modulation improves in comparison with that of conventional modulation upon increasing the number of receive antennas, especially for schemes having

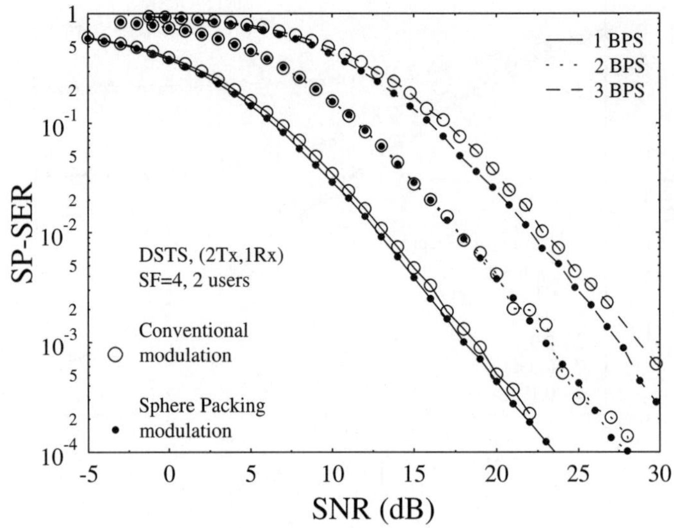

Figure 10.21: Performance comparison of the SP-SER of DSTS in combination with conventional modulation and SP modulation for different bandwidth efficiency rates as outlined in Table 10.4 while employing two transmit antennas, one receive antenna, a SF of four, two users and communicating over a correlated Rayleigh fading channel associated with $f_D = 0.01$.

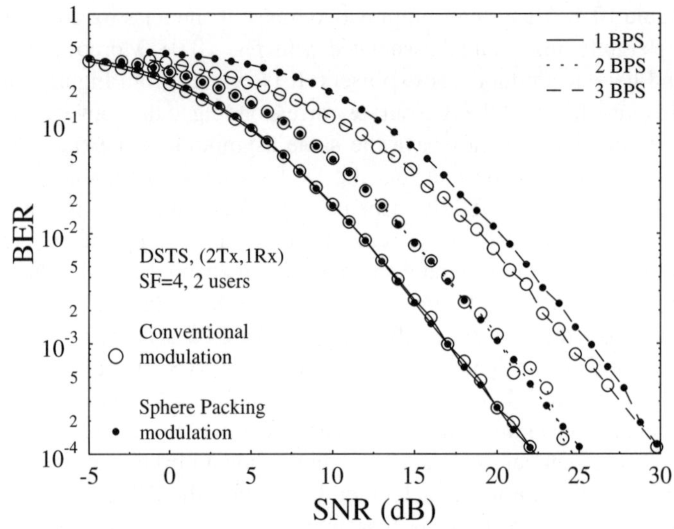

Figure 10.22: Performance comparison of the BER of DSTS in combination with conventional modulation and SP modulation for different bandwidth efficiency rates as outlined in Table 10.4 while employing two transmit antennas, one receive antenna, a SF of four, two users and communicating over a correlated Rayleigh fading channel associated with $f_D = 0.01$.

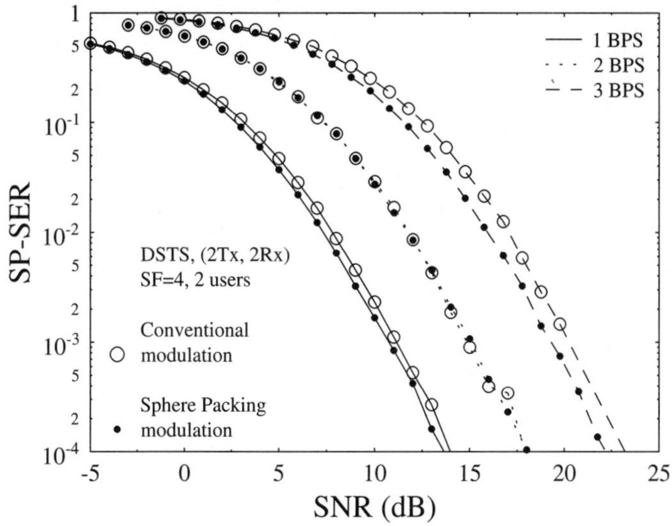

Figure 10.23: Performance comparison of the SP-SER of DSTS in combination with conventional modulation and SP modulation for different bandwidth efficiency rates as outlined in Table 10.4 while employing two transmit antennas, two receive antennas, a SF of four, two users and communicating over a correlated Rayleigh fading channel associated with $f_D = 0.01$.

bandwidth efficiencies of 1 and 3 bits per channel use. Observe, however, in Figures 10.21–10.28 that both the BER and SP-SER performance curves of QPSK modulation as well as those of the identical-throughput SP modulation having $L = 16$ are identical. Again, this phenomenon is due to the fact that QPSK modulation is a special case of the SP modulation constellation when combined with DSTS. More specifically, let us consider the DSTS signal when $x_{l,1}$ and $x_{l,2}$ are chosen independently from the QPSK modulation constellation; then the 16 legitimate combined signals produced will be identical to the 16 legitimate signals constructed using Equation (10.21), where $(a_{l,1}, a_{l,2}, a_{l,3}, a_{l,4})$, $l = 0, \ldots, 15$, correspond to the $L = 16$ SP constellation points hosted by D_4.

Finally, the attainable coding gains of SP modulation over conventional modulation are summarized in Table 10.5 for the schemes characterized in Figures 10.21–10.28 at a SP-SER of 10^{-4}, when communicating over a correlated Rayleigh fading channel associated with $f_D = 0.01$.

10.4 DSTS Design Using Four Transmit Antennas

In the following section, we present the design of the DSTS system using four transmit antennas that can be implemented together with real- and complex-valued phasor constellations as well as with SP modulation.

10.4.1 Design using Real-valued Constellations

A high-level block diagram of the four-antenna-aided DSTS scheme in shown in Figure 10.29, where the DSTS encoder is divided into two main stages. The differential encoding takes place, after which the differentially encoded symbol matrices are space-time

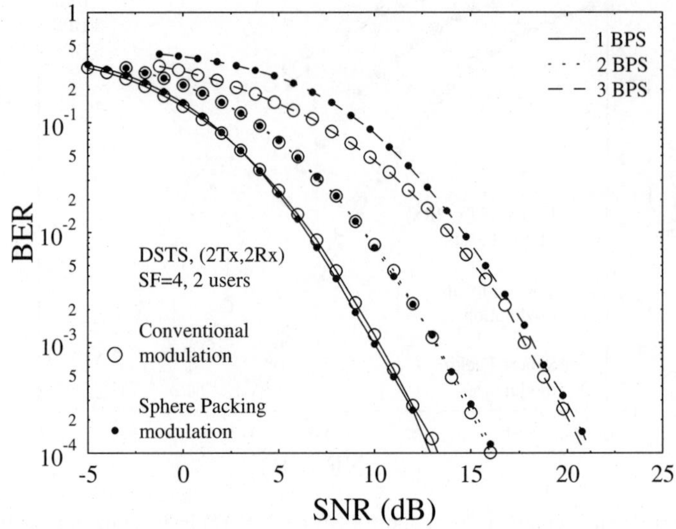

Figure 10.24: Performance comparison of the BER of DSTS in combination with conventional modulation and SP modulation for different bandwidth efficiency rates as outlined in Table 10.4 while employing two transmit antennas, two receive antennas, a SF of four, two users and communicating over a correlated Rayleigh fading channel associated with $f_D = 0.01$.

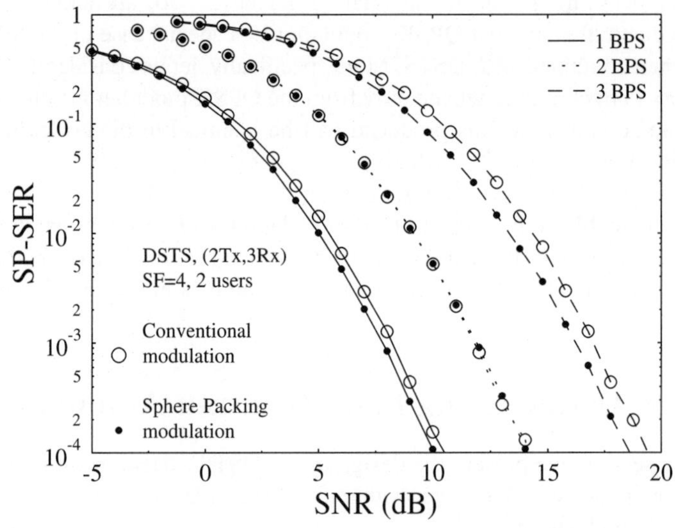

Figure 10.25: Performance comparison of the SP-SER of DSTS in combination with conventional modulation and SP modulation for different bandwidth efficiency rates as outlined in Table 10.4 while employing two transmit antennas, three receive antennas, a SF of four, two users and communicating over a correlated Rayleigh fading channel exhibiting $f_D = 0.01$.

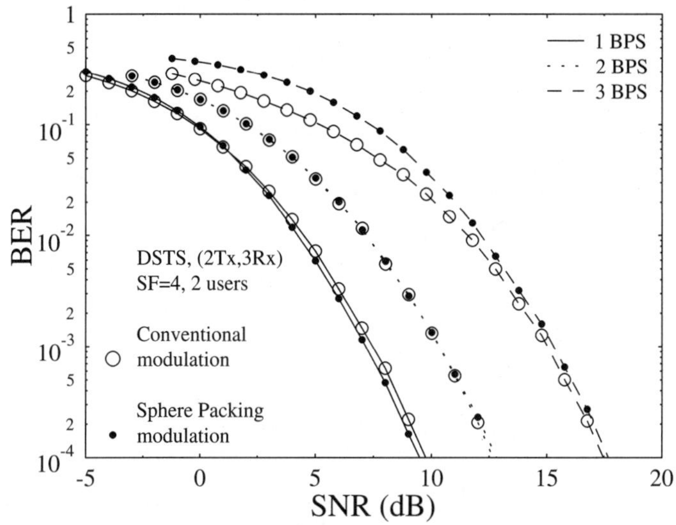

Figure 10.26: Performance comparison of the BER of DSTS in combination with conventional modulation and SP modulation for different bandwidth efficiency rates as outlined in Table 10.4 while employing two transmit antennas, three receive antennas, a SF of four, two users and communicating over a correlated Rayleigh fading channel exhibiting $f_D = 0.01$.

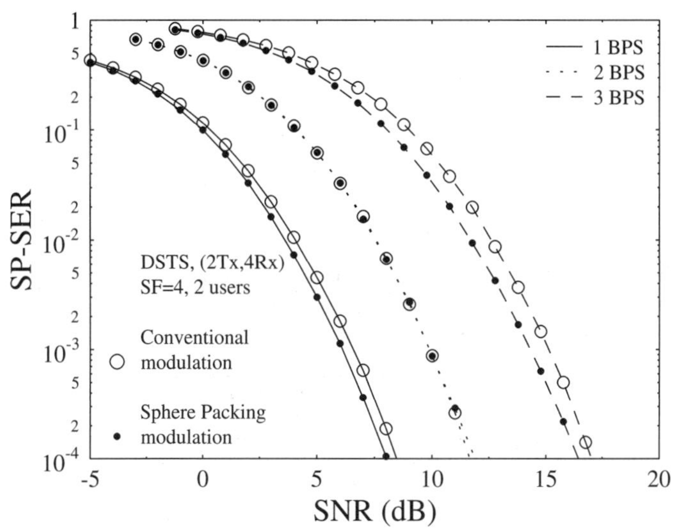

Figure 10.27: Performance comparison of the SP-SER of DSTS in combination with conventional modulation and SP modulation for different bandwidth efficiency rates as outlined in Table 10.4 while employing two transmit antennas, four receive antennas, a SF of four, two users and communicating over a correlated Rayleigh fading channel exhibiting $f_D = 0.01$.

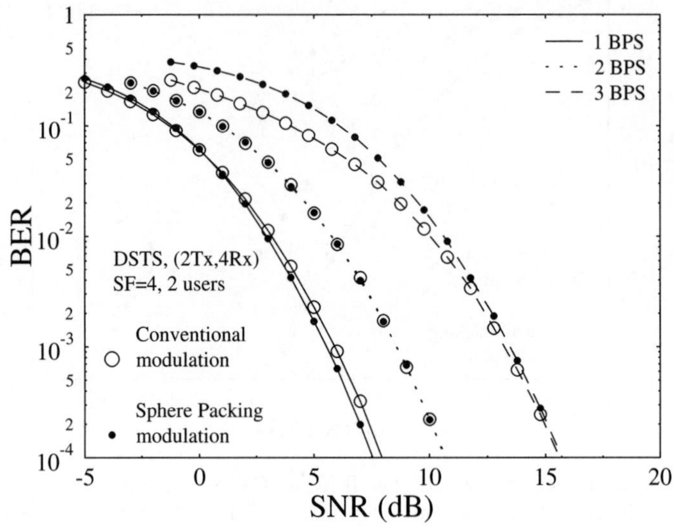

Figure 10.28: Performance comparison of the BER of DSTS in combination with conventional modulation and SP modulation for different bandwidth efficiency rates as outlined in Table 10.4 while employing two transmit antennas, four receive antennas, a SF of four, two users and communicating over a correlated Rayleigh fading channel exhibiting $f_D = 0.01$.

Table 10.5: Coding gains of SP modulation over conventional modulation at SP-SER of 10^{-4} for the schemes of Figures 10.21, 10.23, 10.25 and 10.27, when communicating over a correlated Rayleigh fading channel associated with $f_D = 0.01$.

Number of receive antennas	Coding gains (dB)		
	1 BPS	2 BPS	3 BPS
1	0.20	0.0	0.70
2	0.25	0.0	0.90
3	0.30	0.0	0.95
4	0.30	0.0	1.00

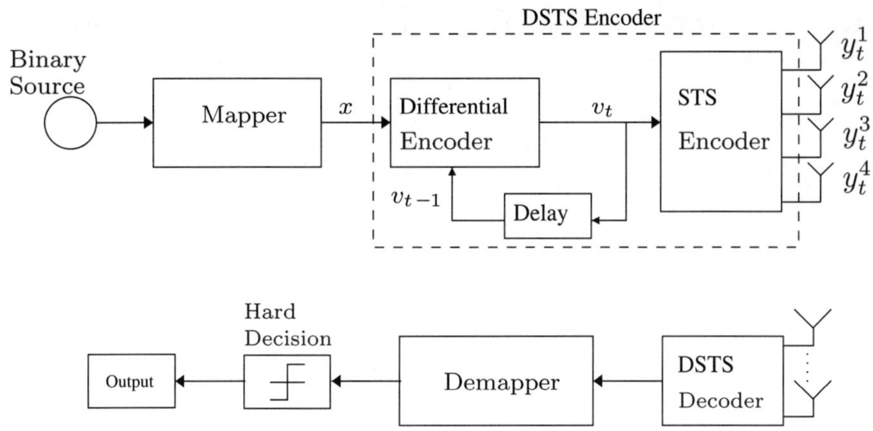

Figure 10.29: The four-transmit-antenna DSTS system block diagram.

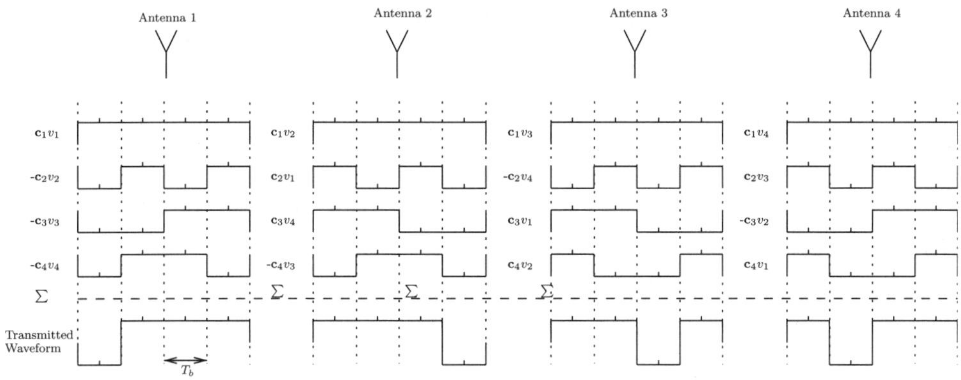

Figure 10.30: Illustration of STS using four transmit antennas transmitting four bits within $4T_b$ duration. Here $v_1 = v_2 = v_3 = v_4 = 1$ were assumed and $\bar{\mathbf{c}}_1 = [+1+1+1+1+1+1+1+1]$, $\bar{\mathbf{c}}_2 = [+1+1-1-1+1+1-1-1]$, $\bar{\mathbf{c}}_3 = [+1+1+1+1-1-1-1-1]$ and $\bar{\mathbf{c}}_4 = [+1+1-1-1-1-1+1+1]$.

spread. Moreover, the basic principle of the four-antenna-aided STS is exemplified in simple graphical terms in Figure 10.30, where an eight-chip orthogonal spreading code was used for spreading each bit of duration T_b to an interval of $T_s = 4T_b$.

The DSTS encoding and decoding algorithms operate as follows. At time instant $t = 0$, the arbitrary dummy reference real-valued symbols v_0^1, v_0^2, v_0^3 and v_0^4 are transmitted from antennas one, two, three and four, respectively. At time instants $t \geq 1$, a block of $4B$ bits arrives at the modulator of Figure 10.29, where each set of B bits is mapped to a real-valued modulated symbol x_t^k, $k = 1, 2, 3, 4$, selected from a 2^B-ary constellation.

Assuming v_t^k to be the symbol transmitted from antenna k, $k = 1, 2, 3, 4$, at time instant t, differential encoding is carried out as follows:

$$\mathbf{V_t} = \begin{pmatrix} v_t^1 \\ v_t^2 \\ v_t^3 \\ v_t^4 \end{pmatrix} = x_t^1 \cdot \begin{pmatrix} v_{t-1}^1 \\ v_{t-1}^2 \\ v_{t-1}^3 \\ v_{t-1}^4 \end{pmatrix} + x_t^2 \cdot \begin{pmatrix} v_{t-1}^2 \\ -v_{t-1}^1 \\ v_{t-1}^4 \\ -v_{t-1}^3 \end{pmatrix} + x_t^3 \cdot \begin{pmatrix} v_{t-1}^3 \\ -v_{t-1}^4 \\ -v_{t-1}^1 \\ v_{t-1}^2 \end{pmatrix} + x_t^4 \cdot \begin{pmatrix} v_{t-1}^4 \\ v_{t-1}^3 \\ -v_{t-1}^2 \\ -v_{t-1}^1 \end{pmatrix}. \tag{10.34}$$

The vector $\mathbf{V_t}$ of Equation (10.34) is normalized by the magnitude of the previously computed vector $\mathbf{V_{t-1}}$ before transmission in order to limit the peak power and hence the out-of-band power emissions.

The differentially encoded symbols are then spread with the aid of the spreading codes $\bar{\mathbf{c}}_1$, $\bar{\mathbf{c}}_2$, $\bar{\mathbf{c}}_3$ and $\bar{\mathbf{c}}_4$, which are generated from the same user-specific spreading code $\bar{\mathbf{c}}$ by ensuring that they are orthogonal using the simple code-concatenation rule of Walsh–Hadamard codes, yielding longer codes and hence a proportionately reduced per-antenna throughput according to:

$$\bar{\mathbf{c}}_1^{\mathrm{T}} = [\bar{\mathbf{c}} \quad \bar{\mathbf{c}} \quad \bar{\mathbf{c}} \quad \bar{\mathbf{c}}], \tag{10.35}$$

$$\bar{\mathbf{c}}_2^{\mathrm{T}} = [\bar{\mathbf{c}} \quad -\bar{\mathbf{c}} \quad \bar{\mathbf{c}} \quad -\bar{\mathbf{c}}], \tag{10.36}$$

$$\bar{\mathbf{c}}_3^{\mathrm{T}} = [\bar{\mathbf{c}} \quad \bar{\mathbf{c}} \quad -\bar{\mathbf{c}} \quad -\bar{\mathbf{c}}], \tag{10.37}$$

$$\bar{\mathbf{c}}_4^{\mathrm{T}} = [\bar{\mathbf{c}} \quad -\bar{\mathbf{c}} \quad -\bar{\mathbf{c}} \quad \bar{\mathbf{c}}]. \tag{10.38}$$

The differentially encoded data is then divided into four quarter-rate substreams and the four consecutive symbols are then spread to the four transmit antennas as shown in Figures 10.29 and 10.30 using the mapping of

$$\mathbf{y}_t^1 = \frac{1}{\sqrt{4}}(\bar{\mathbf{c}}_1 \cdot v_t^1 - \bar{\mathbf{c}}_2 \cdot v_t^2 - \bar{\mathbf{c}}_3 \cdot v_t^3 - \bar{\mathbf{c}}_4 \cdot v_t^4), \tag{10.39}$$

$$\mathbf{y}_t^2 = \frac{1}{\sqrt{4}}(\bar{\mathbf{c}}_1 \cdot v_t^2 + \bar{\mathbf{c}}_2 \cdot v_t^1 + \bar{\mathbf{c}}_3 \cdot v_t^4 - \bar{\mathbf{c}}_4 \cdot v_t^3), \tag{10.40}$$

$$\mathbf{y}_t^3 = \frac{1}{\sqrt{4}}(\bar{\mathbf{c}}_1 \cdot v_t^3 - \bar{\mathbf{c}}_2 \cdot v_t^4 + \bar{\mathbf{c}}_3 \cdot v_t^1 + \bar{\mathbf{c}}_4 \cdot v_t^2), \tag{10.41}$$

$$\mathbf{y}_t^4 = \frac{1}{\sqrt{4}}(\bar{\mathbf{c}}_1 \cdot v_t^4 + \bar{\mathbf{c}}_2 \cdot v_t^3 - \bar{\mathbf{c}}_3 \cdot v_t^2 + \bar{\mathbf{c}}_4 \cdot v_t^1). \tag{10.42}$$

Assuming the channel to be temporally correlated narrowband Rayleigh fading, the received signal at the output of the single receive antenna can be represented as

$$\mathbf{r}_t = h_1 \cdot \mathbf{y}_t^1 + h_2 \cdot \mathbf{y}_t^2 + h_3 \cdot \mathbf{y}_t^3 + h_4 \cdot \mathbf{y}_t^4 + \mathbf{n}_t, \tag{10.43}$$

where h_1, h_2, h_3 and h_4 denote the narrowband complex-valued CIRs corresponding to the four transmit antennas, while $\mathbf{n}_t$ is a complex-valued Gaussian random variable having a covariance matrix of $\sigma_n^2 \cdot \mathbf{I}_{SF}$, where SF is the SF of the per-antenna spreading code $\bar{\mathbf{c}}_k$, $k = 1, 2, 3, 4$.

The received signal $\mathbf{r}_t$ is then correlated with $\bar{\mathbf{c}}_1$, $\bar{\mathbf{c}}_2$, $\bar{\mathbf{c}}_3$ and $\bar{\mathbf{c}}_4$ according to the following operation: $d_t^k = \bar{\mathbf{c}}_k^{\dagger} \cdot \mathbf{r}_t$, $k \in [1, \dots, 4]$. After the correlation operation we arrive at four data symbols represented by

$$d_t^1 = \frac{1}{\sqrt{4}} \cdot h_1 \cdot v_t^1 + \frac{1}{\sqrt{4}} \cdot h_2 \cdot v_t^2 + \frac{1}{\sqrt{4}} \cdot h_3 \cdot v_t^3 + \frac{1}{\sqrt{4}} \cdot h_4 \cdot v_t^4 + \bar{\mathbf{c}}_1^\dagger \cdot \mathbf{n}_t, \quad (10.44)$$

$$d_t^2 = -\frac{1}{\sqrt{4}} \cdot h_1 \cdot v_t^2 + \frac{1}{\sqrt{4}} \cdot h_2 \cdot v_t^1 - \frac{1}{\sqrt{4}} \cdot h_3 \cdot v_t^4 + \frac{1}{\sqrt{4}} \cdot h_4 \cdot v_t^3 + \bar{\mathbf{c}}_2^\dagger \cdot \mathbf{n}_t, \quad (10.45)$$

$$d_t^3 = -\frac{1}{\sqrt{4}} \cdot h_1 \cdot v_t^3 + \frac{1}{\sqrt{4}} \cdot h_2 \cdot v_t^4 + \frac{1}{\sqrt{4}} \cdot h_3 \cdot v_t^1 - \frac{1}{\sqrt{4}} \cdot h_4 \cdot v_t^2 + \bar{\mathbf{c}}_3^\dagger \cdot \mathbf{n}_t, \quad (10.46)$$

$$d_t^4 = -\frac{1}{\sqrt{4}} \cdot h_1 \cdot v_t^4 - \frac{1}{\sqrt{4}} \cdot h_2 \cdot v_t^3 + \frac{1}{\sqrt{4}} \cdot h_3 \cdot v_t^2 + \frac{1}{\sqrt{4}} \cdot h_4 \cdot v_t^1 + \bar{\mathbf{c}}_4^\dagger \cdot \mathbf{n}_t. \quad (10.47)$$

To derive the decoder equations of the DSTS receiver, the received signals in Equations (10.44)–(10.47) are rearranged in vectorial form as follows:

$$\mathbf{R}_{t-1}^1 = (d_{t-1}^1, d_{t-1}^2, d_{t-1}^3, d_{t-1}^4), \quad (10.48)$$

$$\mathbf{R}_{t-1}^2 = (-d_{t-1}^2, d_{t-1}^1, d_{t-1}^4, -d_{t-1}^3), \quad (10.49)$$

$$\mathbf{R}_{t-1}^3 = (-d_{t-1}^3, -d_{t-1}^4, d_{t-1}^1, d_{t-1}^2), \quad (10.50)$$

$$\mathbf{R}_{t-1}^4 = (-d_{t-1}^4, d_{t-1}^3, -d_{t-1}^2, d_{t-1}^1), \quad (10.51)$$

$$\mathbf{R}_t = (d_t^1, d_t^2, d_t^3, d_t^4). \quad (10.52)$$

To decode the transmitted symbols x_t^k, $k = 1, 2, 3, 4$, the decoder uses Equations (10.48)–(10.52) and computes

$$\begin{aligned}\tilde{x}_t^k = \mathrm{Re}\{\mathbf{R}_t \cdot \mathbf{R}_{t-1}^k\} &= \frac{1}{4} \cdot \sum_{i=1}^{4} |h_i|^2 \cdot \sqrt{\sum_{j=1}^{4} |v_{t-1}^j|^2} \cdot x_t^k + N_k \\ &= \chi_{2N_t}^2 \cdot x_t^k + N_k, \end{aligned} \quad (10.53)$$

where $\mathrm{Re}\{\cdot\}$ denotes the real part of a complex number, $\chi_{2N_t}^2$ represents a chi-squared random variable having $2N_t = 8$ degrees of freedom and N_k denotes the noise term having a variance of $\chi_{2N_t}^2 \cdot N_0/2$. The receiver estimates x_t^k based on Equation (10.53) by employing a ML decoder.

According to Equation (10.53), the receiver only has to estimate $\sum_{i=1}^4 |h_i|^2$ in order to decode non-constant modulus real-valued constellations, such as PAM. In other words, the receiver does not have to estimate the individual CIR tap values of h_i, $i = 1, 2, 3, 4$; only $\sum_{i=1}^4 |h_i|^2$ and $\sqrt{\sum_{j=1}^4 |v_{t-1}^j|^2}$ have to be estimated in order to recover PAM modulated information from the received signal of Equation (10.53). A simple channel power estimator may be derived by computing the autocorrelation of the received signal as follows [58]:

$$E\{d_t^i \cdot d_t^i\} = \sum_{i=1}^{4} |h_i|^2 + \sigma_n^2. \quad (10.54)$$

The power of the previously transmitted symbols $\sqrt{\sum_{j=1}^4 |v_{t-1}^j|^2}$ can be estimated from the previous output of the decoder [58].

We can observe from Equation (10.53) that the proposed method guarantees achieving a full diversity gain, while using a low-complexity decoding algorithm. Since $\bar{\mathbf{c}}_k$, $k = 1, 2, 3, 4$, are derived by appropriately concatenating the user-specific code $\bar{\mathbf{c}}$, no extra spreading codes are required for carrying out the STS operation and the four symbols of the four transmitters are transmitted in four time slots.

10.4.2 Design using Complex-valued Constellations

QAM signals represented by complex-valued constellations can also be transmitted using the proposed four-antenna-aided DSTS scheme. Accordingly, we assume that at time instant t, a block of $4B$ bits arrives at the encoder, where each $2B$ bits are modulated using an M-ary complex-valued constellation, so that we have $2B = \log_2 M$. The modulator outputs the two complex symbols x_{tc}^1 and x_{tc}^2, conveying the original $4B$ bits, where the postscript c is used to denote complex symbols. Now x_{tc}^1 and x_{tc}^2 are mapped to x_t^k, $k = 1, 2, 3, 4$, defined in the previous section as follows

$$(x_t^1, x_t^2, x_t^3, x_t^4) = (\text{Re}\{x_{tc}^1\}, \text{Im}\{x_{tc}^1\}, \text{Re}\{x_{tc}^2\}, \text{Im}\{x_{tc}^2\}). \quad (10.55)$$

Similarly to real-valued constellations, $\tilde{x}_t^k$, $k = 1, 2, 3, 4$, can be estimated using Equation (10.53), which is then used to recover the original complex symbols $\tilde{x}_{tc}^1$ and $\tilde{x}_{tc}^2$ as follows:

$$(\tilde{x}_{tc}^1, \tilde{x}_{tc}^2) = (\tilde{x}_t^1 + j\tilde{x}_t^2, \tilde{x}_t^3 + j\tilde{x}_t^4). \quad (10.56)$$

10.4.3 Design using SP Modulation

According to Equation (10.53), the decoded signals represent scaled versions of x_t^1, x_t^2, x_t^3 and x_t^4 corrupted by the complex-valued AWGN. This observation implies that the diversity product of the four-antenna-aided DSTS scheme is determined by the MED of all legitimate vectors $(x_t^1, x_t^2, x_t^3, x_t^4)$. The idea is to jointly design the legitimate four-component vectors $(x_t^1, x_t^2, x_t^3, x_t^4)$ so that they are represented by a single phasor point selected from a SP constellation corresponding to a four-dimensional real-valued lattice having the best known MED in the four-dimensional real-valued space $\mathbb{R}^4$. For the sake of generalizing our treatment, let us assume that there are L legitimate vectors $(x^{l,1}, x^{l,2}, x^{l,3}, x^{l,4})$, $l = 0, 1, \ldots, L-1$, where L represents the number of four-component sphere-packed modulated symbols. The transmitter, then, has to choose the modulated signal from these L legitimate symbols, which have to be differentially space-time spread and transmitted from the four transmit antennas. The bandwidth efficiency of the four-antenna-aided DSTS-SP system is $(\log_2 L)/4$ bits per channel use.

In contrast to the independent transmitted signal design of Section 10.4.1, the aim is to design $x^{l,1}$, $x^{l,2}$, $x^{l,3}$, $x^{l,4}$ jointly, so that they have the best MED from all other $(L-1)$ legitimate SP symbols, since this minimizes the system's SP symbol error probability. Let $(a^{l,1}, a^{l,2}, a^{l,3}, a^{l,4})$, $l = 0, 1, \ldots, L-1$, be legitimate phasor points of the four-dimensional real-valued Euclidean space $\mathbb{R}^4$. Hence, $x^{l,1}$, $x^{l,2}$, $x^{l,3}$, $x^{l,4}$ may be written as

$$\begin{aligned}\{x^{l,1}, x^{l,2}, x^{l,3}, x^{l,4}\} &= T_{\text{sp}}(a^{l,1}, a^{l,2}, a^{l,3}, a^{l,4}) \\ &= \{a^{l,1}, a^{l,2}, a^{l,3}, a^{l,4}\}. \end{aligned} \quad (10.57)$$

Assuming that $\mathbf{S} = \{\mathbf{s}^l = [a^{l,1}, a^{l,2}, a^{l,3}, a^{l,4}] \in \mathbb{R}^4 : 0 \le l \le L-1\}$ constitutes a set of L legitimate constellation points from the lattice D_4 having a total energy of

$$E_{\text{total}} \triangleq \sum_{l=0}^{L-1} (|a^{l,1}|^2 + |a^{l,2}|^2 + |a^{l,3}|^2 + |a^{l,4}|^2), \quad (10.58)$$

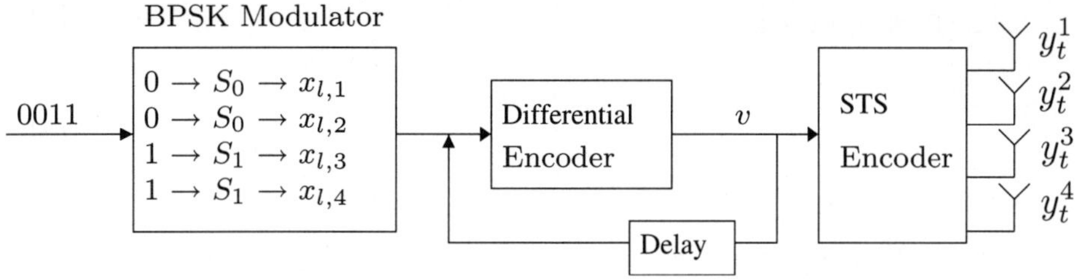

Figure 10.31: Transmission of four BPSK symbols using a four-antenna-aided DSTS scheme.

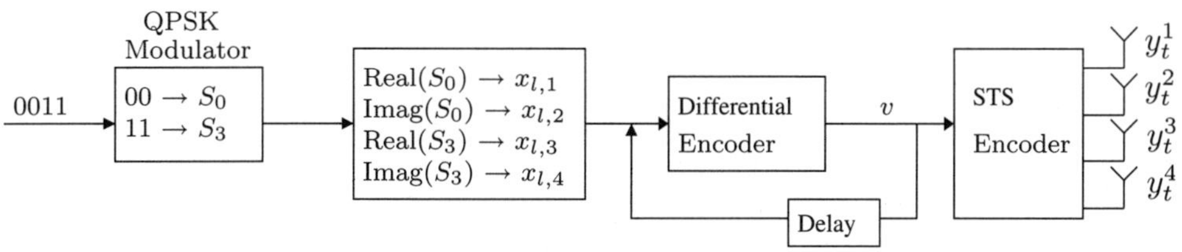

Figure 10.32: Transmission of two QPSK symbols using a four-antenna-aided DSTS scheme.

and upon introducing the notation

$$\begin{aligned}\mathbf{C}_l &= \sqrt{\frac{2L}{E_{\text{total}}}}(x^{l,1}, x^{l,2}, x^{l,3}, x^{l,4}) \\ &= \sqrt{\frac{2L}{E_{\text{total}}}}(a^{l,1}, a^{l,2}, a^{l,3}, a^{l,4}), \quad l = 0, 1, \ldots, L-1,\end{aligned} \tag{10.59}$$

we have a set of constellation symbols, $\{\mathbf{C}_l : 0 \le l \le L-1\}$, leading to the design of DSTS signals whose diversity product is determined by the MED of the set of L legitimate constellation points in $\mathbf{S}$.

The following example illustrates how SP modulation is implemented in combination with the four-antenna-aided DSTS scheme as compared with the conventionally modulated DSTS scheme.

Example 10.4.1. Assume that there are $L = 16$ different legitimate symbols $(x^{l,1}, x^{l,2}, x^{l,3}, x^{l,4})$, $l = 0, 1, \ldots, 15$, that can be used by the encoder. We compare three modulation schemes, namely conventional BPSK, conventional QPSK and SP modulation.

- **Conventional BPSK modulation.** There are two real-valued legitimate symbols (S_0, S_1) that can be used to independently represent any of the $x^{l,1}$, $x^{l,2}$, $x^{l,3}$ and $x^{l,4}$, $l = 0, 1, \ldots, 15$ symbols. The transmission scheme processing the signals in four consecutive time slots is outlined in Figure 10.31.

- **Conventional QPSK modulation.** There are four complex-valued legitimate symbols (S_0, S_1, S_2, S_3) that can be used to independently represent any of the $x^{l,1}$, $x^{l,2}$, $x^{l,3}$ and $x^{l,4}$, $l = 0, 1, \ldots, 15$ symbols. The transmission scheme processing the signals of four consecutive time slots is highlighted in Figure 10.32.

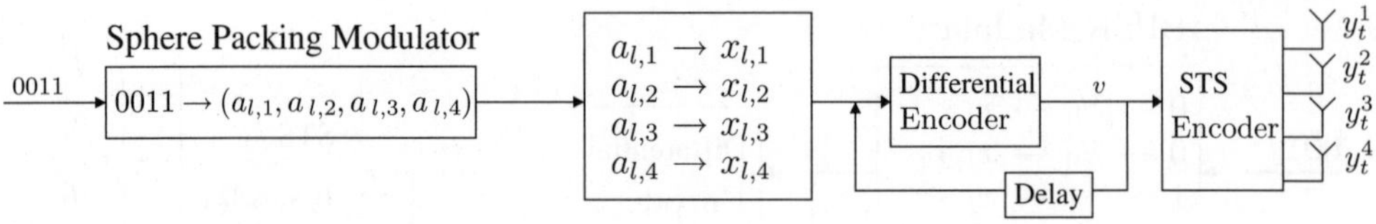

Figure 10.33: Transmission of a SP symbol using a four-antenna-aided DSTS scheme.

Table 10.6: Bandwidth efficiency of four-antenna-aided DSTS-SP systems for different SP signal set sizes L.

L	BPS	Bandwidth efficiency (bits per channel use)
4	2	0.5
8	3	0.75
16	4	1
32	5	1.25
64	6	1.5
128	7	1.75
256	8	2
512	9	2.25
1024	10	2.5
2048	11	2.75
4096	12	3

- **SP modulation.** We need $L = 16$ SP phasor points $(a^{l,1}, a^{l,2}, a^{l,3}, a^{l,4})$ from the lattice D_4 in order to jointly represent each signal $(x^{l,1}, x^{l,2}, x^{l,3}, x^{l,4})$, $l = 0, 1, \ldots, 15$ according to Equation (10.59), as depicted in Figure 10.33.

10.4.4 Bandwidth Efficiency of the Four-antenna-aided DSTS Scheme

In the four-antenna-aided DSTS encoder, the data is serial-to-parallel converted into four substreams. The new bit duration of each parallel substream or, equivalently, the symbol duration becomes $T_s = 4T_b$ as illustrated in Figure 10.30. According to Section 10.4.3, the DSTS transmitter using four transmit antennas transmits one SP symbol in four time slots. Therefore, the DSTS-SP code rate becomes $1/4$ and then according to the number of BPS B_{sp}, the DSTS-SP system's bandwidth efficiency becomes $B_{\text{sp}}/4$. For example, in the case of SP using $L = 16$, we have $B_{\text{sp}} = 4$ BPS, which results in an effective bandwidth efficiency of 1 bit per channel use. Table 10.6 presents the bandwidth efficiency of the four-antenna-aided DSTS system for different SP-modulated constellation sizes.

The effective bandwidth efficiency for the DSTS-SP system is different from that of conventionally modulated DSTS schemes, which can also be categorized as real or complex valued. For the case of real-valued modulation constellations, the DSTS system using four transmit antennas transmits four symbols in four time slots, which gives an effective bandwidth efficiency of B bits per channel use. However, for complex-valued constellations, such as QPSK for example, the DSTS scheme transmits two complex-valued symbols, each

Table 10.7: Bandwidth efficiency of four-antenna-aided DSTS systems for different conventional modulation signal sets.

Modulation	BPS	Bandwidth efficiency (bits per channel use)
BPSK	1	1
QPSK	2	1
8-PSK	3	1.5
16-QAM	4	2
64-QAM	6	3

conveying one bit per in-phase plus one bit per quadrature component in four time slots. This also results in an effective system bandwidth efficiency of 1 bit per channel use, which is identical to that of the four-antenna-aided DSTS-SP system in conjunction with $L = 16$. Table 10.7 presents the bandwidth efficiency of the four-antenna-aided DSTS system for different conventionally modulated constellation sizes.

10.4.5 Capacity of the Four-antenna-aided DSTS-SP Scheme

According to Equation (10.53), the DSTS-SP decoded signal can be modeled as

$$\tilde{\mathbf{s}}_r = \chi^2_{2N_t,r} \cdot \mathbf{s}^l + \acute{\mathbf{N}}_r, \tag{10.60}$$

where we have

$$\mathbf{s}^l = [a^{l,1}\ a^{l,2}\ a^{l,3}\ a^{l,4}] \in S, \quad 0 \le l \le L-1,$$

$$\chi^2_{2N_t,r} = \frac{1}{4} \cdot \sum_{i=1}^{4} |h_{ir}|^2 \cdot \sqrt{\sum_{j=1}^{4} |v_t^j|^2}.$$

Therefore, the capacity of the four-antenna-aided DSTS-SP scheme can be derived using the same method as that in Section 10.3.6 to arrive at Equation (10.31).

Figures 10.34, 10.35 and 10.36 show the DCMC capacity evaluated from Equation (10.31) for the four-dimensional SP-modulation-assisted DSTS as well as STS schemes for $L = 4$, 16 and 64, when employing $N_t = 4$ transmit antennas as well as $N_r = 1$, 2 and 4 receive antennas, respectively. The CCMC [6] capacity of the MIMO scheme was also plotted for comparison in Figures 10.34, 10.35 and 10.36, based on [251].

Figure 10.37 compares the achievable bandwidth efficiency of various SP-modulated DSTS schemes, while employing $N_t = 4$ transmit antennas and $N_r = 1,\ 2$ and 4 receive antennas. The figures explicitly illustrate that a higher bandwidth efficiency may be attained when employing SP modulation in conjunction with DSTS schemes having $N_t = 4$ transmit antennas, as the number N_r of receive antennas increases.

10.4.6 Performance of the Four-antenna-aided DSTS Scheme

In this section, the four-antenna-aided DSTS scheme is considered. Simulation results are provided for systems having different bandwidth efficiencies in conjunction with appropriate real- and complex-valued conventional as well as SP modulation, when communicating over

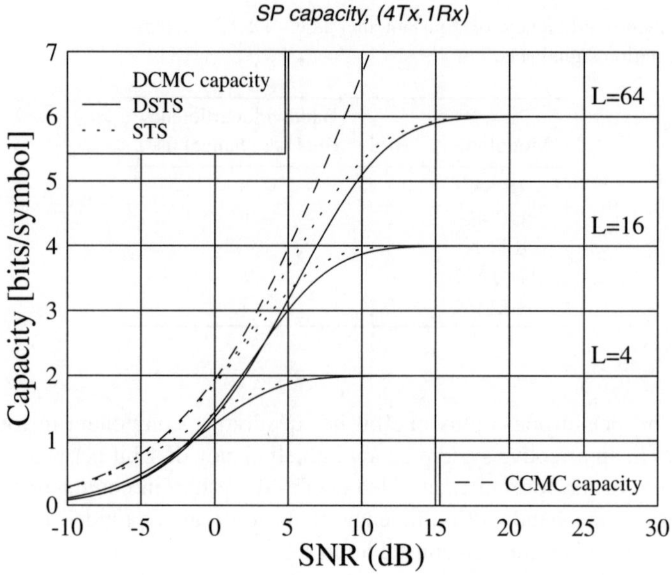

Figure 10.34: Capacity comparison of coherent and differential STS-SP-based schemes using $L = 4$, 16 and 64, when employing $N_t = 4$ transmit and $N_r = 1$ receive antennas for communicating over a correlated Rayleigh fading channel having a normalized Doppler frequency of $f_D = 0.01$.

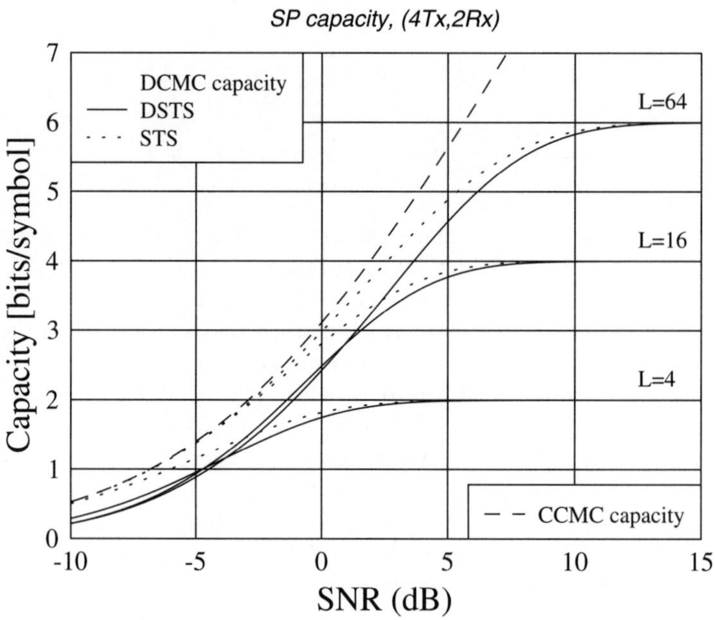

Figure 10.35: Capacity comparison of coherent and differential STS-SP-based schemes using $L = 4$, 16 and 64, when employing $N_t = 4$ transmit and $N_r = 2$ receive antennas for communicating over a correlated Rayleigh fading channel having a normalized Doppler frequency of $f_D = 0.01$.

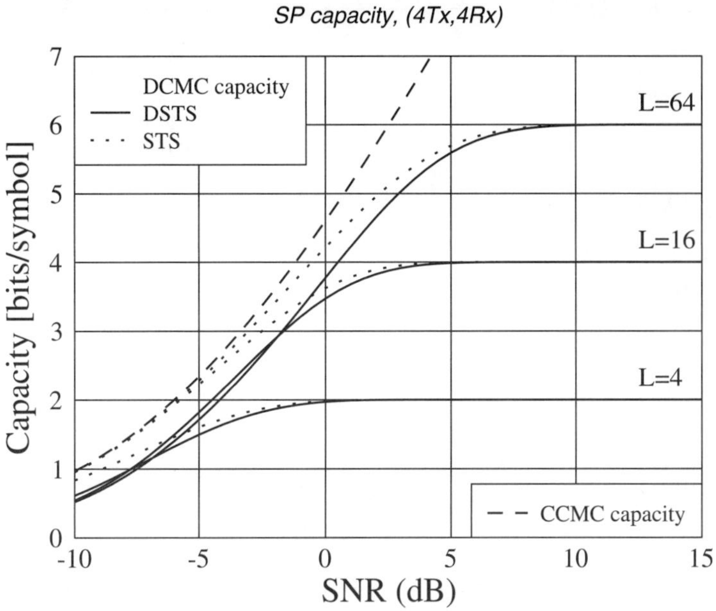

Figure 10.36: Capacity comparison of coherent and differential STS-SP-based schemes using $L = 4$, 16 and 64, when employing $N_t = 4$ transmit and $N_r = 4$ receive antennas for communicating over a correlated Rayleigh fading channel having a normalized Doppler frequency of $f_D = 0.01$.

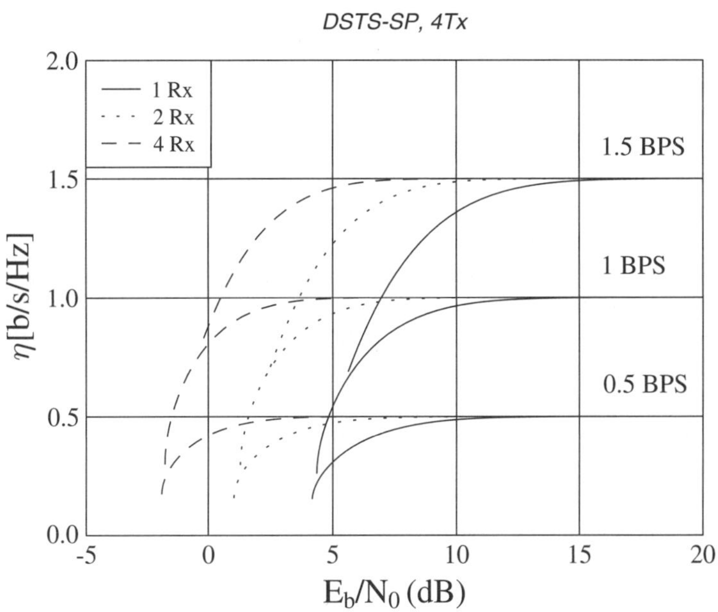

Figure 10.37: Bandwidth efficiency of DSTS-based schemes employing SP modulation scheme for $L = 4$, 16 and 64, when employing $N_t = 4$ transmit and $N_r = 1$, 2 and 4 receive antennas for communicating over a correlated Rayleigh fading channel having a normalized Doppler frequency of $f_D = 0.01$.

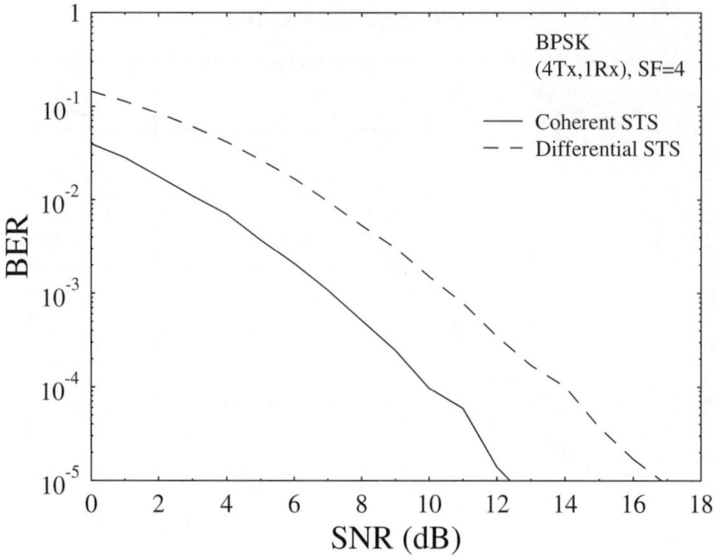

Figure 10.38: Comparison of the BER performance of coherent and differential STS, while using a BPSK-modulated signal, four transmit antennas, one receive antenna and a SF of four and supporting two users for communicating over a correlated Rayleigh fading channel having a normalized Doppler frequency of $f_D = 0.01$.

a correlated narrowband Rayleigh fading channel having a normalized Doppler frequency of $f_D = 0.01$.

Figure 10.38 compares the BER performance of the differentially encoded and of the coherently detected STS scheme, while using BPSK, four transmit antennas, one receive antenna, a SF of four and supporting two users. Again, coherent detection requires CSI at the receiver for decoding the received signal; however, in this case we assumed that the CIR is perfectly known at the receiver side. As shown in Figure 10.38, differential encoding results in a 3 dB performance loss compared with coherent detection, when assuming perfect channel knowledge. This is mainly due to the fact that differential decoding results in doubling the noise power compared with that recorded for coherently detected signals. However, using the differential encoding results in eliminating the complexity of channel estimation required by coherent detection schemes, where in this case the receiver is expected to estimate $4 \times N_r$ channel links. Furthermore, Figures 10.39 and 10.40 compare the BER performance of the DSTS encoded BPSK- and QPSK-modulated signals, while using two and four transmit antennas, one receive antenna, a SF of four and supporting two users. It transpires from the figures that increasing the diversity order of the system by increasing the number of transmit antennas improves the attainable system performance. For the case of BPSK modulation, the system performance improves by almost $E_b/N_0 = 10$ dB at a BER of 10^{-5}. Figure 10.41 compares the BER performance of the DSTS system using BPSK, QPSK, 8-PSK as well as 16-QAM signals.

Figure 10.42 shows the SP-SER performance curves of the DSTS scheme in conjunction with different conventional as well as SP modulation schemes at various bandwidth efficiency values, as outlined in Table 10.4. However, the figure does not show a comparison between the BPSK-modulated and the SP-modulated signals using $L = 4$, owing to the fact that

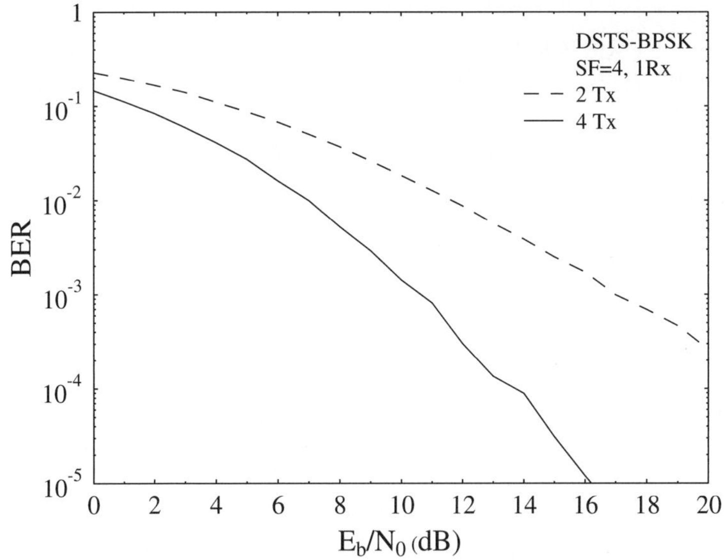

Figure 10.39: Comparison of the BER performance of the DSTS-assisted BPSK-modulated signal while using two and four transmit antennas, one receive antenna and a SF of four and supporting two users for communicating over a correlated Rayleigh fading channel having a normalized Doppler frequency of $f_D = 0.01$.

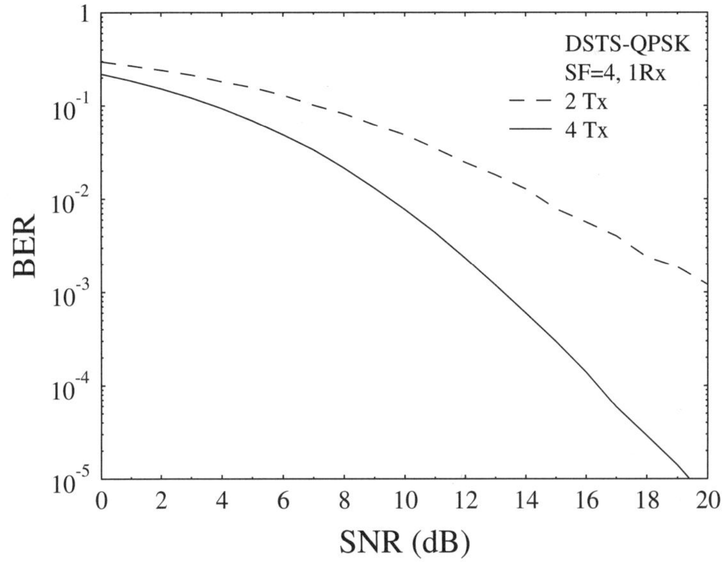

Figure 10.40: Comparison of the BER performance of the DSTS-aided QPSK-modulated signal while using two and four transmit antennas, one receive antenna and a SF of four and supporting two users for communicating over a correlated Rayleigh fading channel having a normalized Doppler frequency of $f_D = 0.01$.

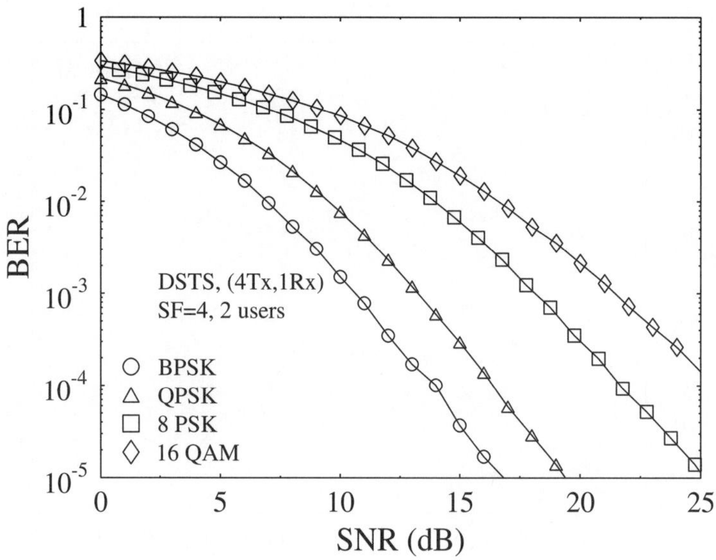

Figure 10.41: Comparison of the BER performance of DSTS while using BPSK, QPSK, 8-PSK and 16-QAM modulated signals, four transmit antennas, one receive antenna and a SF of four and supporting two users for communicating over a correlated Rayleigh fading channel having a normalized Doppler frequency of $f_D = 0.01$.

the two systems have different bandwidth efficiencies, when combined with four-antenna-aided DSTS. All systems employ four transmit antennas for communication over a correlated narrowband Rayleigh fading channel associated with $f_D = 0.01$. Moreover, the system uses a SF of four, while supporting two users. Figure 10.42 suggests that the SP-SER performance of DSTS schemes may be improved by employing SP modulation. The resultant BER performance curves are shown in Figure 10.43. The BER performances of SP modulation and conventional modulation are identical for systems having a bandwidth efficiency of 1 BPS, but as suggested in Figure 10.43 they are different for the higher bandwidth efficiency schemes. As depicted in Figure 10.43, the BER performance of the conventional-modulation-based four-antenna-aided DSTS scheme is better than that of the SP-aided system. This is due to the fact that SP modulation was designed specifically for improving the SP-SER, rather than the BER, and this explains the advantage of SP modulation in Figure 10.42.

To elaborate a little further, Figures 10.44–10.49 illustrate the beneficial effect of increasing the number of receive antennas from two to four respectively. Observe in Figures 10.45, 10.47 and 10.49 that as expected, the BER performance of SP modulation improves in comparison to that of conventional modulation when increasing the number of receive antennas, especially for schemes having a bandwidth efficiency of 1.5 bit per channel use. Observe, however, in Figures 10.43–10.48 that both the BER and SP-SER performance curves of QPSK modulation as well as those of the identical-throughput SP modulation having $L = 16$ are identical. This phenomenon is due to the fact that QPSK modulation is a special case of the SP modulation as discussed in Section 10.3.7.

Finally, the attainable coding gains of SP modulation over conventional modulation are summarized in Table 10.8 for the schemes characterized in Figures 10.42–10.49 at a SP-SER of 10^{-4}, when communicating over a correlated Rayleigh fading channel associated with $f_D = 0.01$.

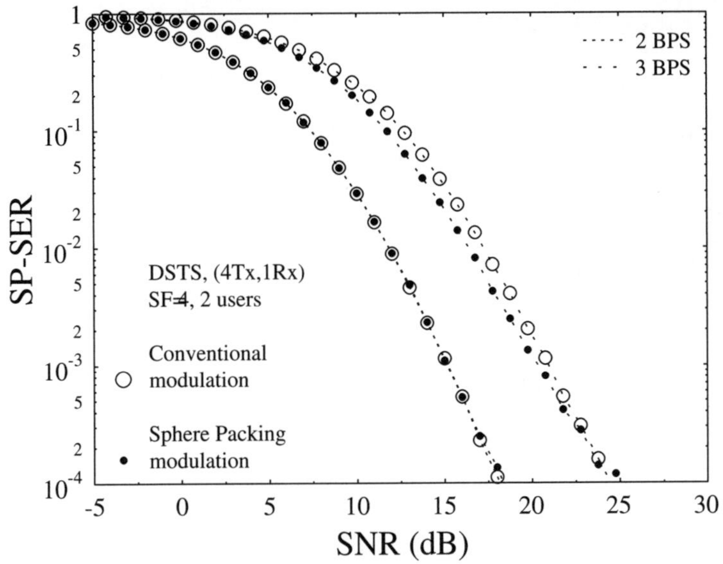

Figure 10.42: Performance comparison of the SP-SER of DSTS in combination with conventional modulation and SP modulation for different bandwidth efficiency values as outlined in Table 10.4 while employing four transmit antennas, one receive antenna and a SF of four and supporting two users, when communicating over a correlated Rayleigh fading channel associated with $f_D = 0.01$.

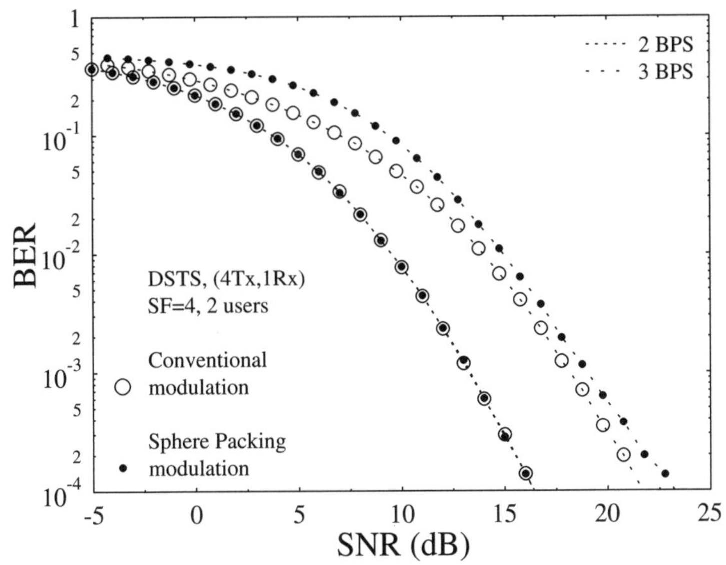

Figure 10.43: Performance comparison of the BER of DSTS in combination with conventional modulation and SP modulation for different bandwidth efficiency values as outlined in Table 10.4 while employing four transmit antennas, one receive antenna and a SF of four and supporting two users, when communicating over a correlated Rayleigh fading channel associated with $f_D = 0.01$.

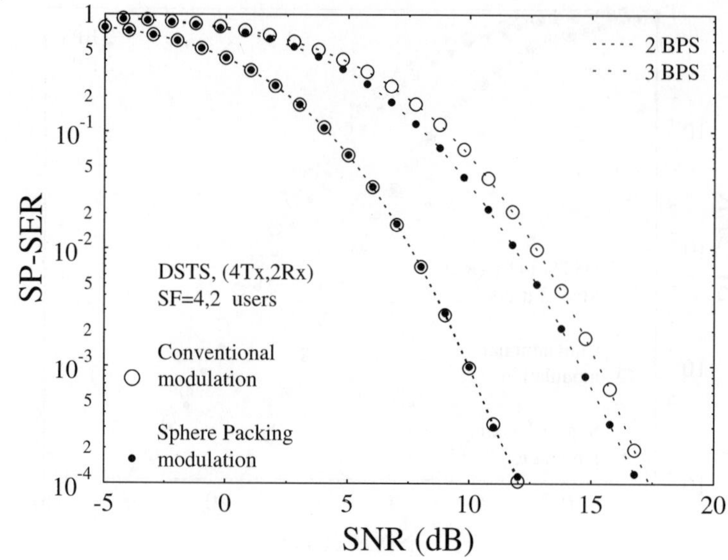

Figure 10.44: Performance comparison of the SP-SER of DSTS in combination with conventional modulation and SP modulation for different bandwidth efficiency values as outlined in Table 10.4 while employing four transmit antennas, two receive antennas and a SF of four and supporting two users, when communicating over a correlated Rayleigh fading channel associated with $f_D = 0.01$.

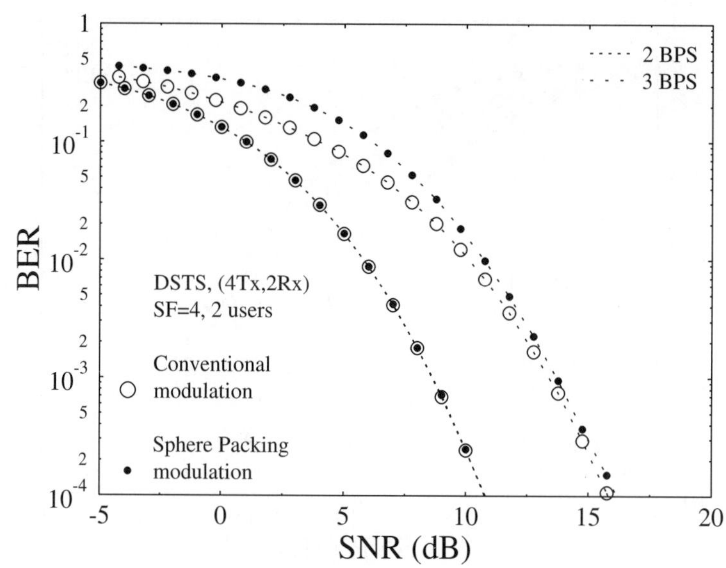

Figure 10.45: Performance comparison of the BER of DSTS in combination with conventional modulation and SP modulation for different bandwidth efficiency values as outlined in Table 10.4 while employing four transmit antennas, two receive antennas and a SF of four and supporting two users, when communicating over a correlated Rayleigh fading channel associated with $f_D = 0.01$.

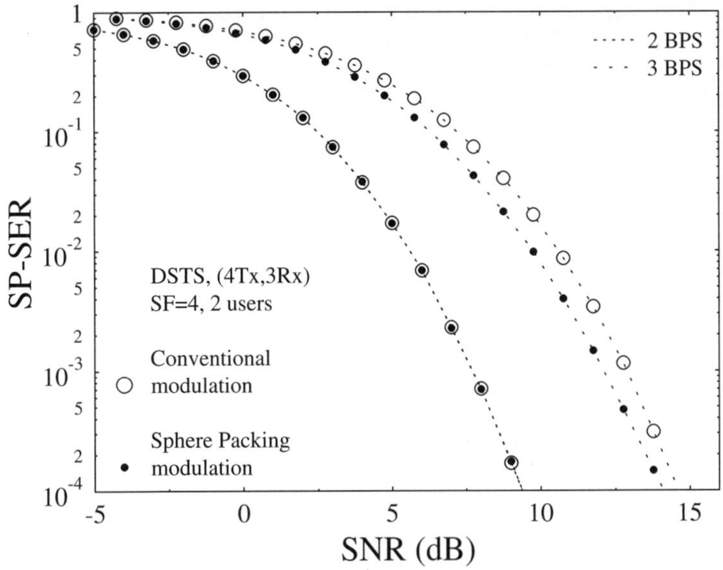

Figure 10.46: Performance comparison of the SP-SER of DSTS in combination with conventional modulation and SP modulation for different bandwidth efficiency values as outlined in Table 10.4 while employing four transmit antennas, three receive antennas and a SF of four and supporting two users, when communicating over a correlated Rayleigh fading channel associated with $f_D = 0.01$.

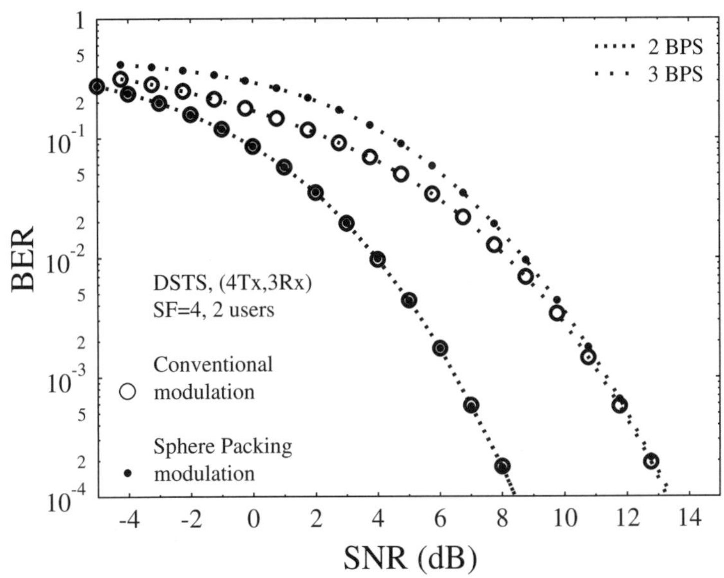

Figure 10.47: Performance comparison of the BER of DSTS in combination with conventional modulation and SP modulation for different bandwidth efficiency values as outlined in Table 10.4 while employing four transmit antennas, three receive antennas and a SF factor of four and supporting two users, when communicating over a correlated Rayleigh fading channel associated with $f_D = 0.01$.

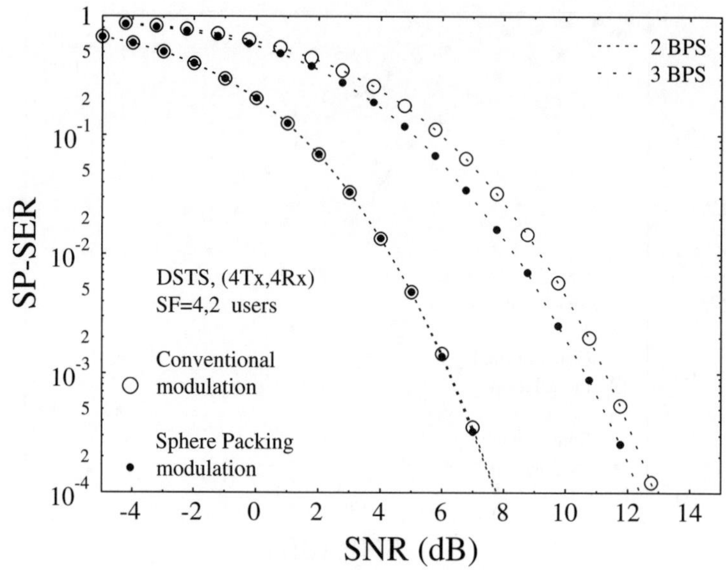

Figure 10.48: Performance comparison of the SP-SER of DSTS in combination with conventional modulation and SP modulation for different bandwidth efficiency values as outlined in Table 10.4 while employing four transmit antennas, four receive antennas and a SF of four and supporting two users, when communicating over a correlated Rayleigh fading channel associated with $f_D = 0.01$.

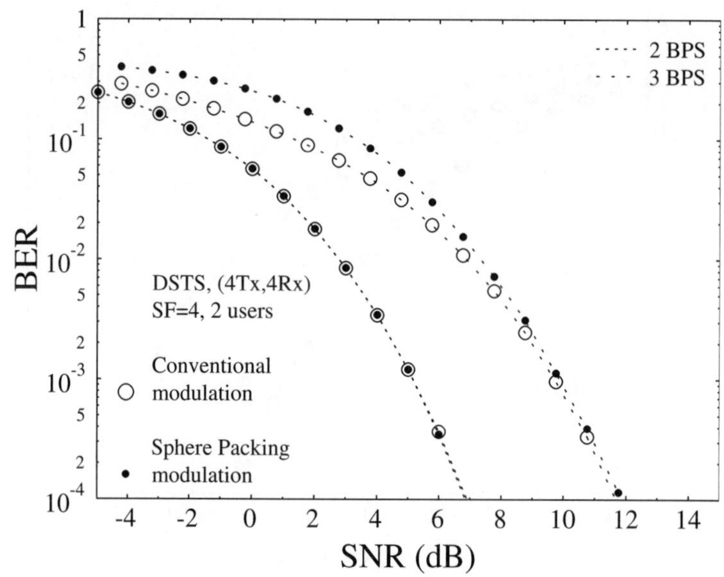

Figure 10.49: Performance comparison of the BER of DSTS in combination with conventional modulation and SP modulation for different bandwidth efficiency values as outlined in Table 10.4 while employing four transmit antennas, four receive antennas and a SF of four and supporting two users, when communicating over a correlated Rayleigh fading channel associated with $f_D = 0.01$.

Table 10.8: Coding gains of SP modulation over conventional modulation at SP-SER of 10^{-4} for the schemes of Figures 10.42, 10.44, 10.46 and 10.48, when communicating over a correlated Rayleigh fading channel associated with $f_D = 0.01$.

Number of receive antennas	Coding gains (dB)	
	2 BPS	3 BPS
1	0.0	0.00
2	0.0	0.40
3	0.0	0.80
4	0.0	0.85

10.5 Chapter Conclusions

In this chapter, we have introduced the concept of DSTS employing two and four transmit antennas and demonstrated that the systems can be combined with conventional real- and complex-valued constellations. Furthermore, in order to maximize the diversity product, we combined the DSTS with SP modulation, which has the best known MED in the $2(k+1)$-dimensional real-valued Euclidean space $\mathbb{R}^{2(k+1)}$ (see [221]). The capacity analysis provided in Sections 10.3.6 and 10.4.5 demonstrated that the DSTS-SP design is capable of attaining potential performance improvements over a conventionally modulated DSTS design. The simulation results presented in Sections 10.3.7 and 10.4.6 demonstrated that the DSTS system is capable of providing a full diversity gain, while employing two and four transmit antennas. Tables 10.5 and 10.8 summarize the coding gains of the SP-modulation-aided DSTS schemes over conventional modulated DSTS schemes at a SP-SER of 10^{-4}, when communicating over a correlated narrowband Rayleigh fading channel and employing two and four transmit antennas, respectively.

10.6 Chapter Summary

This chapter first reviewed the concept of differential encoding in Section 10.2. It was shown that differential encoding requires no CSI at the receiver and thus eliminates the complexity of channel estimation at the expense of a 3 dB performance loss compared with the coherently detected system assuming perfect channel knowledge at the receiver. In Section 10.3, we outlined the encoding and decoding processes of the DSTS scheme, when combined with conventional modulation, such as PSK and QAM. In Section 10.3.3, the philosophy of DSTS using SP modulation was introduced based on the fact that the diversity product of the DSTS design is improved by maximizing the MED of the DSTS symbols, which is motivated by the fact that SP has the best known MED in the real-valued space. Section 10.3.4 discussed the problem of constructing a SP constellation having a particular size L. The constellation points were first chosen based on the minimum energy criterion. Then, an exhaustive computer search was conducted for all of the SP symbols having the lowest possible energy, in order to find the specific set of L points having the best MED from all of the other constellation points satisfying the minimum energy criterion. The capacity of DSTS-SP schemes employing $N_t = 2$ transmit antennas was derived in Section 10.3.6 followed by the performance characterization of a twin-antenna-aided DSTS scheme in Section 10.3.7

demonstrating that the DSTS scheme is capable of providing full diversity. In addition to that, the results demonstrated that DSTS-SP schemes are capable of outperforming DSTS schemes that employ conventional modulation (PSK, QAM), when comparing the SP-SER performance.

The four-antenna-aided DSTS design was characterized in Section 10.4, where it was demonstrated how the DSTS scheme can be combined with conventional real- and complex-valued constellations as well as with SP modulation. It was also demonstrated that the four-dimensional SP modulation scheme is constructed differently in the case of two transmit antennas than when employing four transmit antennas. The capacity analysis of the four-antenna-aided DSTS-SP scheme was also derived for different bandwidth efficiency systems, while employing a variable number of receive antennas, in Section 10.4.5. Finally, Section 10.4.6 presents the simulation results obtained for the four-antenna-aided DSTS scheme when combined with conventional as well as SP modulations.

Tables 10.5 and 10.8 summarize the coding gains of SP-modulation-aided DSTS schemes over conventional modulated DSTS schemes at a SP-SER of 10^{-4}, when communicating over a correlated narrowband Rayleigh fading channel and employing two and four transmit antennas, respectively.

In the next chapter, we demonstrate that further performance improvement can be attained by the concatenation of these schemes with channel codes and performing iterative detection by exchanging extrinsic information between the different component decoders/demapper at the receiver side. The convergence behavior of the iteratively detected system is studied using EXIT charts.

Chapter 11

Iterative Detection of Channel-coded DSTS Schemes

11.1 Introduction

In Chapter 10, a DSTS scheme has been proposed for transmission over temporally correlated narrowband Rayleigh fading channels using conventional PSK and QAM modulations as well as a SP modulation scheme. The DSTS arrangement is a MIMO scheme that is capable of attaining a full diversity gain as well as exploiting the combined advantages of differential encoding and multi-user support capability of STS [13]. The performance of the DSTS scheme can be enhanced by combining it with iterative-detection-aided schemes, where iterative decoding is carried out by exchanging extrinsic information between the different constituent decoders and demappers.

The turbo principle of [146] was extended to multiple parallel concatenated codes in [147], to serially concatenated codes in [148] and to multiple serially concatenated codes in [149]. In [158], the employment of the turbo principle was considered for iterative soft demapping in the context of multilevel modulation schemes combined with channel decoding, where a soft symbol-to-bit demapper was used between the multilevel demodulator and the binary channel decoder. The iterative soft demapping principle of [158] was extended to SP-aided STBC schemes in [46], where the SP demapper of [43] was modified in [46] for the sake of accepting the *a priori* information passed to it from the channel decoder as extrinsic information.

Furthermore, it was shown in [166] that a recursive inner code is needed in order to maximize the interleaver gain and to avoid the formation of a BER floor when employing iterative decoding. In [168], unity-rate inner codes were employed for designing low-complexity iterative-detection-aided schemes suitable for bandwidth- and power-limited systems having stringent BER requirements.

Recently, studying the convergence behavior of iterative decoding has attracted considerable research attention [46,167,169,172,173,175,181,255,256,258,347]. In [169], ten Brink proposed the employment of the EXIT characteristics between a concatenated decoder's output and input for describing the flow of extrinsic information through the soft-in soft-out constituent decoders.

Motivated by the performance improvements reported in the previous chapter and in [43, 46, 348], the novelty and rationale of this chapter can be summarized as follows.

1. The bit-to-SP-symbol mapper is designed using an EXIT-chart-based procedure, which allows us to achieve diverse design objectives. For example, we can design a system having the lowest possible turbo-cliff SNR, but tolerating the formation of an error floor. Alternatively, we can design a system having a low error floor, but exhibiting a slightly higher turbo-cliff SNR.

2. A unity-rate precoder is introduced, which is capable of completely eliminating the system's error floor as well as operating at the lowest possible turbo-cliff SNR without significantly increasing the associated complexity or interleaver delay.

3. We propose a novel technique for computing the maximum achievable rate of the system using EXIT charts and we show that the achievable rate obtained using EXIT charts closely matches the capacity limits computed in Section 10.3.6.

4. As a benefit of the proposed solution, it will be demonstrated in Section 11.2.4 that the iteratively detected twin-antenna-aided DSTS-SP scheme is capable of providing an E_b/N_0 gain of at least 14.9 dB at a BER of 10^{-5} over the equivalent bandwidth efficiency uncoded DSTS-SP scheme. Furthermore, the AGM-1-based iteratively detected twin-antenna-aided DSTS-SP scheme is capable of performing within 2.3 dB of the maximum achievable rate limit obtained using EXIT charts at $\text{BER} = 10^{-5}$.

5. A URC is amalgamated with the iteratively detected four-antenna-aided DSTS-SP system in order to eliminate the error floor and to operate as close as possible to the system's capacity. Explicitly, the system employing no URC precoding in conjunction with AGM-1 attains a coding gain of 12 dB at a BER of 10^{-5} and performs within 1.82 dB of the maximum achievable rate limit. In contrast, the URC-precoded system outperforms its non-precoded counterpart and operates within 0.92 dB of the maximum achievable rate limit obtained using EXIT charts.

The rest of the chapter is organized as follows. An overview of the iterative-detection-based DSTS scheme is presented in Section 11.2 for systems employing two and four transmit antennas as well as conventional and SP modulation schemes. In Section 11.2.2 we introduce the EXIT chart analysis technique and illustrate how to generate the EXIT charts for visualizing the interaction of the inner as well as outer codes. Our performance results and discussions of the iteratively detected DSTS scheme are presented in Section 11.2.4. In Section 11.3 unity-rate inner code is combined with the iterative-detection-based system in order to eliminate the error floor and perform as close as possible to the system's capacity. Our conclusions are presented in Section 11.4, followed by the chapter's summary in Section 11.5.

11.2 Iterative Detection of RSC-coded DSTS Schemes

A block diagram of the iterative-detection-aided DSTS system is shown in Figure 11.1, where the transmitted source bit stream $\mathbf{u}$ is convolutionally encoded by a half-rate RSC code and then interleaved by a random bit interleaver Π. After bit interleaving, the conventional mapper first maps B channel-coded bits $\mathbf{b} = b_0, \ldots, b_{B-1} \in \{0, 1\}$ to a conventionally modulated PSK or QAM symbol x. In contrast, the SP mapper maps B_{sp} (the notation B_{sp} is used for the SP modulation to differentiate it from that for the conventional modulation, as illustrated

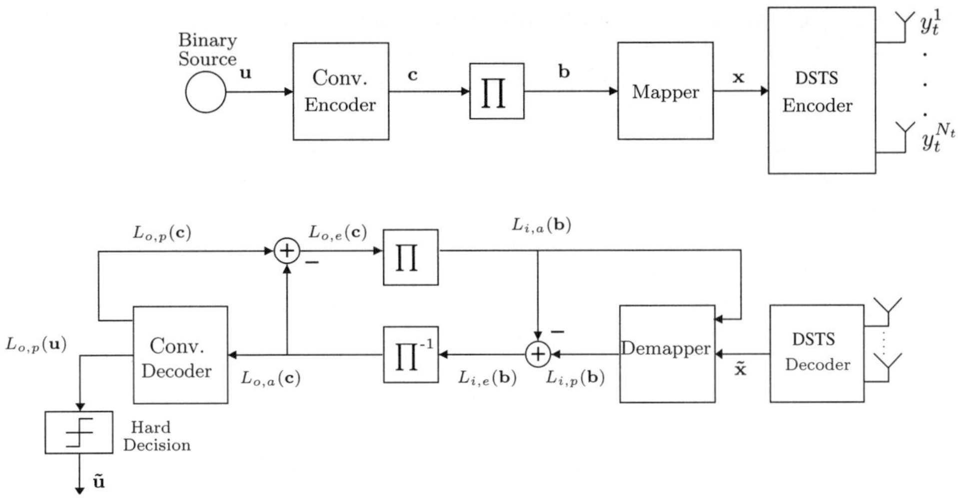

Figure 11.1: Iteratively-detected DSTS system block diagram.

in Chapter 10) channel-coded bits $\mathbf{b} = b_0, \ldots, b_{B_{\text{sp}}-1} \in \{0, 1\}$ to a SP symbol $\mathbf{s}^l \in \mathbf{S}$, $l = 0, 1, \ldots, L-1$ as described in Chapter 10, such that we have $\mathbf{s}^l = \text{map}_{\text{sp}}(\mathbf{b})$, where $B_{\text{sp}} = \log_2 L$ and L represents the number of modulated symbols in the sphere-packed signaling alphabet, as described in Chapter 10. Subsequently, we have a set of symbols that can be transmitted using DSTS.

In this chapter, we consider transmission over a temporally correlated narrowband Rayleigh fading channel, associated with a normalized Doppler frequency of $f_D = f_d T_s = 0.01$, where f_d is the Doppler frequency and T_s is the symbol duration, while the spatial channel coefficients are independent. The complex AWGN of $n = n_I + jn_Q$ contaminates the received signal, where n_I and n_Q are two independent zero-mean Gaussian random variables having a variance of $\sigma_n^2 = \sigma_{n_I}^2 = \sigma_{n_Q}^2 = N_0/2$ per dimension, with $N_0/2$ representing the double-sided noise power spectral density expressed in W Hz^{-1}.

In the receiver, the soft-in soft-out RSC decoder iteratively exchanges extrinsic information with the soft demapper, as shown in Figure 11.1. The RSC decoder invokes the Bahl–Cocke–Jelinek–Raviv (BCJR) algorithm [164] on the basis of bit-based trellis [349]. All BCJR calculations are performed in the logarithmic probability domain and using a lookup table for correcting the Jacobian approximation in the Log-MAP algorithm [9, 165].

The extrinsic soft information, represented in the form of LLRs [261], is iteratively exchanged between the demapper and the RSC decoder for the sake of assisting each other's operation, as detailed in [148]. In Figure 11.1, $L(\cdot)$ denotes the LLRs of the bits concerned, where the subscript i indicates the inner demapper, while o corresponds to outer RSC decoding. In addition, the subscripts a, p and e denote the dedicated role of the LLRs with a, p and e indicating *a priori*, *a posteriori* and extrinsic information, respectively.

As shown in Figure 11.1, the received and DSTS-decoded complex-valued symbols $\tilde{\mathbf{x}}$ are demapped to their LLR representation for each of the channel-coded bits per symbol. The *a priori* LLR values $L_{i,a}(\mathbf{b})$ of the demapper are subtracted from the *a posteriori* LLR values $L_{i,p}(\mathbf{b})$ for the sake of generating the extrinsic LLR values $L_{i,e}(\mathbf{b})$, and then the LLRs $L_{i,e}(\mathbf{b})$ are deinterleaved by a soft-bit deinterleaver, as seen in Figure 11.1. Next, the soft bits $L_{o,a}(\mathbf{c})$ are passed to the RSC decoder in order to compute the *a posteriori* LLR values $L_{o,p}(\mathbf{c})$ provided by the Log-MAP algorithm [165] for all the channel-coded bits. During the last iteration, only the LLR values $L_{o,p}(\mathbf{u})$ of the original uncoded systematic information bits are required, which are passed to the hard decision decoder of Figure 11.1 in order to determine the estimated transmitted source bits. As seen in Figure 11.1, the extrinsic information $L_{o,e}(\mathbf{c})$ is generated by subtracting the *a priori* information from the *a posteriori* information according to $(L_{o,p}(\mathbf{c}) - L_{o,a}(\mathbf{c}))$, which is then fed back to the demapper as the *a priori* information $L_{i,a}(\mathbf{b})$ after appropriately reordering them using the interleaver of Figure 11.1. The demapper of Figure 11.1 exploits the *a priori* information for the sake of providing improved *a posteriori* LLR values, which are then passed to the channel decoder and then back to the demapper for further iterations.

11.2.1 Iterative Demapping

For the sake of simplicity, we consider a system having a single receive antenna, although its extension to several receive antennas is feasible. As discussed in Section 10.3.6, the received channel output symbols are first DSTS decoded and diversity combined in order to extract the estimates $\tilde{\mathbf{x}}$ of the most likely transmitted symbols $\mathbf{x}$:

$$\tilde{x}_t = \chi^2_{2N_t} \cdot x_t + N, \tag{11.1}$$

where $\chi^2_{2N_t}$ represents a chi-squared distributed random variable having $2N_t$ degrees of freedom, as defined in Section 10.3.2 and Section 10.4.1. Furthermore, N_t is the number of transmit antennas and N is a zero-mean complex-valued Gaussian random variable having a variance of $\sigma_N^2 \approx 2 \cdot \chi^2_{2N_t} \cdot \sigma_n^2 = 2 \cdot \chi^2_{2N_t} \cdot N_0/2$.

11.2.1.1 Conventional Modulation

The decoded symbol $\tilde{x}$ can be written as

$$\tilde{x} = \chi^2_{2N_t} \cdot x + N. \tag{11.2}$$

According to Equation(11.2), the conditional PDF $p(\tilde{x}|x)$ of receiving a symbol $\tilde{x}$, given that symbol x was transmitted is given by

$$p(\tilde{x}|x) = \frac{1}{\prod_{d=1}^{D} \sqrt{\pi \cdot N_0 \cdot \chi^2_{2N_t}[d]}} \cdot \exp\left(\sum_{d=1}^{D} \frac{-(\tilde{x}[d] - \chi^2_{2N_t}[d] \cdot x[d])^2}{\chi^2_{2N_t}[d] \cdot N_0}\right), \tag{11.3}$$

where D represents the dimension of the symbol constellation used. Hence, we have $D = 1$ for real-valued constellations and $D = 2$ for complex-valued constellations.

The received symbol $\tilde{x}$ carries B channel-coded and interleaved bits $\mathbf{b} = b_0, \ldots, b_{B-1} \in \{0, 1\}$. The LLR value of bit b_k for $k = 0, \ldots, B-1$ can be written as [348]

$$L(b_k|\tilde{x}) = L_a(b_k) + \ln \frac{\sum_{x \in S_{1k}} p(\tilde{x}|x) \cdot \exp(\sum_{j=0, j\neq k}^{B-1} b_j L_a(b_j))}{\sum_{x \in S_{0k}} p(\tilde{x}|x) \cdot \exp(\sum_{j=0, j\neq k}^{B-1} b_j L_a(b_j))}, \tag{11.4}$$

where S_{1k} and S_{0k} are subsets of the symbol constellation $\mathbf{S}$, so that $\mathbf{S}_{1k} \triangleq \{x \in \mathbf{S} : b_k = 1\}$ and, likewise, $\mathbf{S}_{0k} \triangleq \{x \in \mathbf{S} : b_k = 0\}$. In other words, S_{ik} represents all symbols of the set S, where we have $b_k \in \{0, 1\}$, $k = 0, \ldots, B-1$. Using Equation (11.3), we can write Equation (11.4) as

$$\begin{aligned} L(b_k|\tilde{x}) &= L_a(b_k) + \ln\Bigg\{\sum_{x \in S_{1k}} \exp\Bigg[\Bigg(\sum_{d=1}^{D} \frac{-(\tilde{x}[d] - \chi^2_{2N_t}[d] \cdot x[d])^2}{\chi^2_{2N_t}[d] \cdot N_0}\Bigg) + \sum_{j=0, j\neq k}^{B-1} b_j L_a(b_j)\Bigg]\Bigg\} \\ &\quad \times \Bigg\{\sum_{x \in S_{0k}} \exp\Bigg[\Bigg(\sum_{d=1}^{D} \frac{-(\tilde{x}[d] - \chi^2_{2N_t}[d] \cdot x[d])^2}{\chi^2_{2N_t}[d] \cdot N_0}\Bigg) + \sum_{j=0, j\neq k}^{B-1} b_j L_a(b_j)\Bigg]\Bigg\}^{-1} \\ &= L_{i,a} + L_{i,e}. \end{aligned} \tag{11.5}$$

11.2.1.2 SP Modulation

As detailed in Chapter 10, the mapping of the SP symbols to the DSTS scheme's antennas is different for two and four transmit antennas. A received sphere-packed symbol $\tilde{\mathbf{s}}$ is constructed from the estimates $\tilde{x}$ as

$$\tilde{\mathbf{s}} = T_{\text{sp}}^{-1}(\tilde{x}^1, \tilde{x}^2), \tag{11.6}$$

for the case of two transmit antennas according to Equation (10.24). In contrast, for the case of four-antenna-aided DSTS-SP, $\tilde{\mathbf{s}}$ is constructed from the estimates $\tilde{x}$ as

$$\tilde{\mathbf{s}} = T_{\text{sp}}^{-1}(\tilde{x}^1, \tilde{x}^2, \tilde{x}^3, \tilde{x}^3), \tag{11.7}$$

where $\tilde{\mathbf{s}} = \{[\tilde{a}^1, \tilde{a}^2, \tilde{a}^3, \tilde{a}^4] \in \mathbb{R}^4\}$. However, for both two- and four-antenna-aided DSTS schemes, the received sphere-packed symbol $\tilde{\mathbf{s}}$ can be written as

$$\tilde{\mathbf{s}} = \chi^2_{2N_t} \cdot \mathbf{s}^l + \mathbf{N}, \tag{11.8}$$

where we have $\mathbf{s}^l \in \mathbf{S}$, $0 \leq l \leq L-1$, and $\mathbf{N}$ is a four-dimensional Gaussian random variable having a covariance matrix of $\sigma_N^2 \cdot \mathbf{I}_4 \approx 2 \cdot \chi^2_{2N_t} \cdot \sigma_n^2 \cdot \mathbf{I}_4$, since the SP symbol constellation $\mathbf{S}$ is four-dimensional.

The conditional probability of receiving a four-dimensional signal $\tilde{\mathbf{s}}$, given that a four-dimensional L-ary signal $\mathbf{s}^l \in \mathbf{S}$, $l \in [0, \ldots, L-1]$, was transmitted over the Rayleigh channel of Equation (11.8) is given by Equation (10.27) and repeated here for convenience

$$p(\tilde{\mathbf{s}}|\mathbf{s}^l) = \frac{1}{\prod_{d=1}^{D=4} \sqrt{\pi \cdot N_0 \cdot \chi^2_{2N_t}[d]}} \cdot \exp\Bigg(\sum_{d=1}^{D=4} \frac{-(\tilde{a}_r^d - \chi^2_{2N_t}[d] \cdot a^{l,d})^2}{\chi^2_{2N_t}[d] \cdot N_0}\Bigg), \tag{11.9}$$

where $D = 4$ is used, since a four-dimensional SP symbol constellation is employed.

The SP symbol $\tilde{\mathbf{s}}$ carries B_{sp} channel-coded bits $\mathbf{b} = b_0, \ldots, b_{B_{\text{sp}}-1} \in \{0, 1\}$. The LLR value of bit b_k for $k = 0, \ldots, B_{\text{sp}} - 1$ can be written as [348]

$$L(b_k|\tilde{\mathbf{s}}) = L_a(b_k) + \ln \frac{\sum_{\mathbf{s}^l \in \mathbf{S}_{1k}} p(\tilde{\mathbf{s}}|\mathbf{s}^l) \cdot \exp(\sum_{j=0, j\neq k}^{B_{\text{sp}}-1} b_j L_a(b_j))}{\sum_{\mathbf{s}^l \in \mathbf{S}_{0k}} p(\tilde{\mathbf{s}}|\mathbf{s}^l) \cdot \exp(\sum_{j=0, j\neq k}^{B_{\text{sp}}-1} b_j L_a(b_j))}, \tag{11.10}$$

where $\mathbf{S}_{1k}$ and $\mathbf{S}_{0k}$ are subsets of the symbol constellation $\mathbf{S}$, so that $\mathbf{S}_{1k} \triangleq \{\mathbf{s}^l \in \mathbf{S} : b_k = 1\}$ and, likewise, $\mathbf{S}_{0k} \triangleq \{\mathbf{s}^l \in \mathbf{S} : b_k = 0\}$. In other words, $\mathbf{S}_{ik}$ represents all symbols of the set

$\mathbf{S}$, where we have $b_k \in \{0, 1\}$, $k = 0, \ldots, B_{\text{sp}} - 1$. Using Equation (11.9), we can write Equation (11.10) as

$$
\begin{aligned}
&L(b_k|\tilde{\mathbf{s}}) \\
&= L_a(b_k) + \ln\Bigg\{\sum_{\mathbf{s}^l \in \mathbf{S}_{1k}} \exp\Bigg[\Bigg(\sum_{d=1}^{D=4} \frac{-(\tilde{a}_r^d - \chi_{2N_t}^2[d] \cdot a^{l,d})^2}{\chi_{2N_t}^2[d] \cdot N_0}\Bigg) + \sum_{j=0, j\neq k}^{B_{\text{sp}}-1} b_j L_a(b_j)\Bigg]\Bigg\} \\
&\quad \times \Bigg\{\sum_{\mathbf{s}^l \in \mathbf{S}_{0k}} \exp\Bigg[\Bigg(\sum_{d=1}^{D=4} \frac{-(\tilde{a}_r^d - \chi_{2N_t}^2[d] \cdot a^{l,d})^2}{\chi_{2N_t}^2[d] \cdot N_0}\Bigg) + \sum_{j=0, j\neq k}^{B_{\text{sp}}-1} b_j L_a(b_j)\Bigg]\Bigg\}^{-1} \\
&= L_{i,a} + L_{i,e}.
\end{aligned} \tag{11.11}
$$

11.2.2 EXIT Chart Analysis

The concept of EXIT charts was proposed in [169, 172] for predicting the convergence behavior of iterative decoders, where the evolution of the input/output mutual information exchange between the inner and outer decoders in consecutive iterations was examined. The application of EXIT charts is based on two main assumptions, which are realistic when using long interleavers, namely that the *a priori* LLR values are fairly uncorrelated and that the PDF of the *a priori* LLR values is Gaussian distributed.

The following analysis is presented for the SP modulation case and the same analysis can be extended for the conventional modulation with the difference of mapping the constellation points to the transmit antennas as discussed in Chapter 10.

11.2.2.1 Transfer Characteristics of the Demapper

As seen in Figure 11.1, the inputs of the demapper are the DSTS-decoded and diversity-combined data stream $\tilde{\mathbf{x}}$ and the *a priori* information $L_{i,a}(\mathbf{b})$ generated by the outer channel decoder. The demapper outputs the *a posteriori* LLR $L_{i,p}(\mathbf{b})$, then subtracts the *a priori* LLR and hence produces the extrinsic LLR $L_{i,e}(\mathbf{b})$. Based on the previously mentioned two assumptions, the *a priori* input $L_{i,a}(\mathbf{b})$ can be modeled by applying an independent zero-mean Gaussian random variable having a variance of σ_a^2.

In conjunction with the outer channel coded and interleaved bits $b \in \{0, 1\}$ of Figure 11.1 or equivalently $d \in \{-1, +1\}$, the *a priori* input $L_{i,a}(\mathbf{b})$ can be written as [169]

$$L_{i,a} = \frac{\sigma_a^2}{2} \cdot d + n_a, \tag{11.12}$$

since $L_{i,a}$ is a LLR value obeying the Gaussian distribution [261]. Accordingly, the conditional PDF of the *a priori* input $L_{i,a}(\mathbf{b})$ is

$$p_a(\zeta|d) = \frac{1}{\sqrt{2\pi}\sigma_a} \cdot \exp\bigg(-\frac{(\zeta - (\sigma_a^2/2) \cdot d)^2}{2\sigma_a^2}\bigg). \tag{11.13}$$

The MI $I_{i,a}(\mathbf{b}) = I(\mathbf{b}; L_{i,a}(\mathbf{b}))$ or, equivalently, $I_{i,a}(\mathbf{b}) = I(\mathbf{d}; L_{i,a}(\mathbf{b}))$, $0 \leq I_{i,a} \leq 1$, between the outer coded and interleaved bit stream $\mathbf{b}$ and the *a priori* LLR values $L_{i,a}(\mathbf{b})$ is used to quantify the information content of the *a priori* knowledge [262]:

$$I_{i,a}(\mathbf{b}) = \frac{1}{2} \cdot \sum_{d=-1,+1} \int_{-\infty}^{+\infty} p_a(\zeta|d) \cdot \log_2 \frac{2 \cdot p_a(\zeta|d)}{p_a(\zeta|d=-1) + p_a(\zeta|d=+1)} \, \mathrm{d}\zeta. \tag{11.14}$$

Using Equation (11.13), Equation (11.14) can be expressed as

$$I_{i,a}(\sigma_a) = 1 - \frac{1}{2\sigma_a^2} \int_{-\infty}^{+\infty} \exp\left(-\frac{(\zeta - \sigma_a^2/2)^2}{2\sigma_a^2}\right) \cdot \log_2[1 + e^{-\zeta}] \, d\zeta. \tag{11.15}$$

It was shown in [174] that the mutual information between the equiprobable bits $\mathbf{d}$ and their respective LLRs L for symmetric and consistent[1] L-values always simplifies to

$$\begin{aligned} I(d; L) &= 1 - \int_{-\infty}^{+\infty} p(L|d = +1) \cdot \log_2[1 + e^{-L}] \, dL \\ &= 1 - E_{d=+1}\{\log_2[1 + e^{-L}]\}. \end{aligned} \tag{11.16}$$

In order to quantify the information content of the extrinsic LLR values $L_{i,e}(\mathbf{b})$ at the output of the demapper, the mutual information $I_{i,e}(\mathbf{b}) = I(\mathbf{b}; L_{i,e}(\mathbf{b}))$ can be used, which is computed as in (11.14) using the PDF p_e of the extrinsic output.

Considering $I_{i,e}(\mathbf{b})$ as a function of both $I_{i,a}(\mathbf{b})$ and the E_b/N_0 value encountered, the demapper's extrinsic information transfer characteristic is defined as [169, 172]:

$$I_{i,e}(\mathbf{b}) = T_i(I_{i,a}(\mathbf{b}), E_b/N_0). \tag{11.17}$$

Figure 11.2 illustrates how the EXIT characteristic $I_{i,e}(\mathbf{b})$ is calculated for a specific $(I_{i,a}(\mathbf{b}),\ E_b/N_0)$ input combination. First, the wireless channel noise variance σ_n is computed according to the specific E_b/N_0 value considered. Then, a specific value of $I_{i,a}(\mathbf{b})$ is selected to compute σ_a, where the EXIT curve has to be evaluated using $\sigma_a = J^{-1}(I_{i,a})$. Afterwards, Equation (11.12) is used to generate $L_{i,a}(\mathbf{b})$, as shown in Figure 11.2, which is applied as the *a priori* LLR input of the demapper. Finally, the MI of $I_{i,e}(\mathbf{b}) = I(\mathbf{b}; L_{i,e}(\mathbf{b}))$, $0 \le I_{i,e}(\mathbf{b}) \le 1$, between the outer coded and interleaved bit stream $\mathbf{b}$ and the LLR values $L_{i,e}(\mathbf{b})$ is calculated with the aid of the PDF p_e of the extrinsic output $L_{i,e}(\mathbf{b})$. This requires the determination of the distribution p_e by means of Monte Carlo simulations. However, according to [181] the MI can be estimated using a sufficiently large number of samples even for non-Gaussian or unknown distributions, which may be expressed as [181]:

$$\begin{aligned} I(\mathbf{b}; L_{i,e}(\mathbf{b})) &= 1 - E_{d=+1}\{\log_2[1 + \exp(-L_{i,e}(\mathbf{b}))]\} \\ &\approx 1 - \frac{1}{B_{\text{sp}}} \sum_{i=1}^{B_{\text{sp}}} \log_2[1 + \exp(-\mathbf{b}(i) \cdot L_{i,e}(\mathbf{b}(i)))]. \end{aligned} \tag{11.18}$$

Figure 11.3 shows the EXIT characteristics of the SP demapper in conjunction with $L = 16$ and various SP mapping schemes, while using the twin-antenna-aided DSTS scheme for transmission over temporally correlated Rayleigh fading channels. As seen in Figure 11.3, Gray mapping does not provide any iteration gain upon increasing the mutual information at the input of the demapper. However, using a variety of different AGM schemes [46, 348] results in different EXIT characteristics, as illustrated by the different slopes seen in Figure 11.3. Any mapping that is different from the classic Gray mapping may be referred to as AGM. The nine different AGM mapping schemes characterized in Figure 11.3 are specifically selected from all of the possible mapping schemes for $L = 16$ in order to demonstrate the different EXIT characteristics associated with different bit-to-symbol mappings. We tested the performance of all legitimate AGM schemes in order to find the best performer. The Gray mapping and AGM schemes considered in this chapter are listed in Appendix H.

[1]The LLR values are symmetric if their PDF is symmetric $p(-\zeta/d = +1) = p(\zeta/d = -1)$. In addition, all LLR values with symmetric distributions satisfy the consistency condition $p(-\zeta/d) = e^{-d\zeta} p(\zeta/d)$ (see [174]).

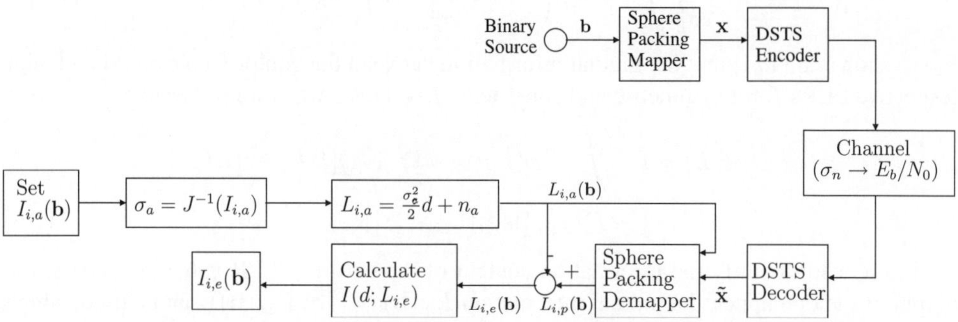

Figure 11.2: Demapper's transfer characteristics evaluation procedure block diagram.

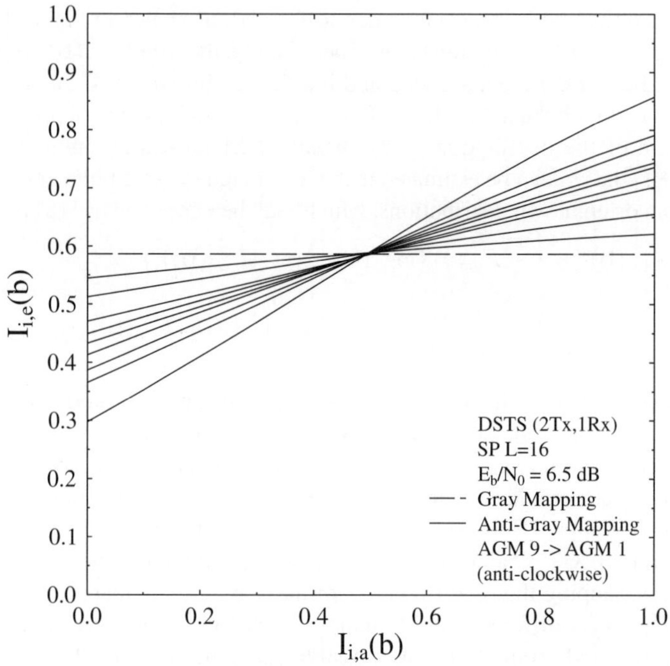

Figure 11.3: SP demapper EXIT characteristics for different bits-to-SP-symbol mappings at $E_b/N_0 =$ 6.5 dB for $L = 16$, while using the twin-antenna-aided DSTS scheme for transmission over temporally correlated Rayleigh fading channels.

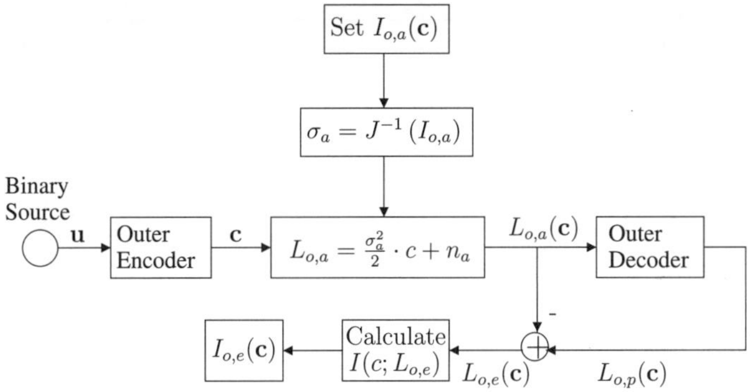

Figure 11.4: Evaluation of the outer channel decoder transfer characteristics.

11.2.2.2 Transfer Characteristics of the Outer Decoder

The relationship between the outer channel decoder LLR input $L_{o,a}(\mathbf{c})$ and extrinsic output $L_{o,e}(\mathbf{c})$ can be described by the extrinsic transfer characteristic of the outer channel decoder. According to Figure 11.1, the input of the outer channel decoder consists only of the *a priori* input $L_{o,a}(\mathbf{c})$ provided by the SP demapper after appropriately reordering the corresponding extrinsic LLR $L_{i,e}(\mathbf{b})$. Therefore, the EXIT characteristics of the outer channel decoder are independent of the E_b/N_0 value and hence $I_{o,e}(\mathbf{c})$ may be written as

$$I_{o,e}(\mathbf{c}) = T_o(I_{o,a}(\mathbf{c})), \tag{11.19}$$

where $I_{o,a}(\mathbf{c}) = I(\mathbf{c};\, L_{o,a}(\mathbf{c}))$, $0 \leq I_{o,a} \leq 1$, is the MI between the outer channel coded bit stream $\mathbf{c}$ and the *a priori* LLR values $L_{o,a}(\mathbf{c})$ and similarly $I_{o,e}(\mathbf{c}) = I(\mathbf{c};\, L_{o,e}(\mathbf{c}))$, $0 \leq I_{o,e} \leq 1$, is the MI between the outer channel coded bit stream $\mathbf{c}$ and the extrinsic LLR values $L_{o,e}(\mathbf{c})$. The computational model of evaluating the EXIT characteristics of the outer channel decoder is shown in Figure 11.4. As seen in the figure, the procedure is similar to that of the SP demapper shown in Figure 11.2, except that its value is independent of the E_b/N_0 value. Again, $I_{o,e} = I(\mathbf{c};\, L_{o,e}(\mathbf{c}))$ can be computed either by evaluating the histogram approximation of p_e [169, 172] and then applying Equation (11.14) or, more conveniently, by the time averaging method [181] of Equation (11.18) as

$$\begin{aligned} I(\mathbf{c};\, L_{o,e}(\mathbf{c})) &= 1 - E\{\log_2[1 + \exp(-L_{o,e})]\} \\ &\approx 1 - \frac{1}{B_{\mathrm{sp}}} \sum_{i=1}^{B_{\mathrm{sp}}} \log_2[1 + \exp(-\mathbf{c}(i) \cdot L_{o,e}(\mathbf{c}_i))]. \end{aligned} \tag{11.20}$$

The EXIT characteristics of various half-rate RSC codes having different constraint lengths K are shown in Figure 11.5. The generator polynomials employed are given in the figure's legend in octal form, where G_r is the feedback polynomial and G is the feed-forward polynomial. Figure 11.5 demonstrates that for $I_{o,a} > 0.5$, the set of RSC codes having higher constraint lengths converge faster upon increasing $I_{o,a}$ than the RSC codes having smaller constraint lengths. Furthermore, we note that the extrinsic characteristics of the RSC codes having constraint lengths of $K = 4$ and $K = 5$ are close to each other and that they depend on the generator polynomial used.

Moreover, the EXIT characteristics of several half-rate RSC codes having a constraint length of $K = 3$ and variable generator polynomials are shown in Figure 11.6. All of the

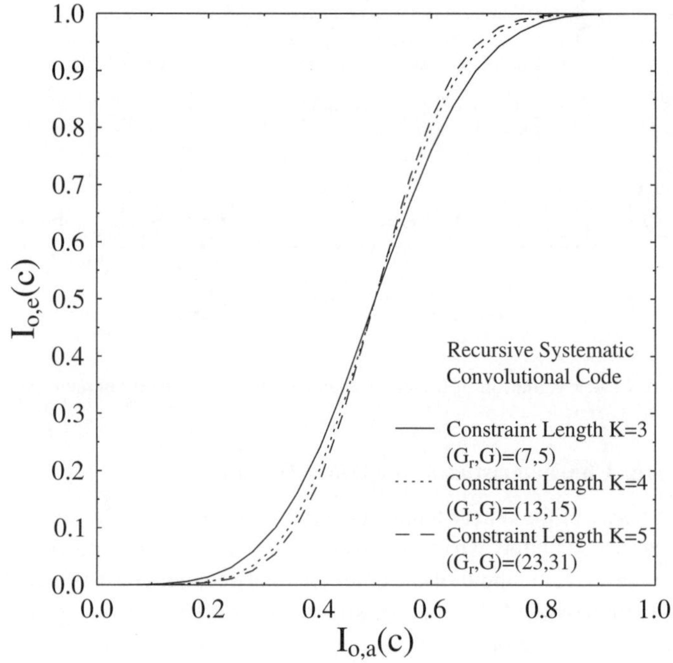

Figure 11.5: EXIT characteristics of half-rate RSC codes having different constraint lengths.

possible generator polynomials having a constraint length of $K = 3$ are employed. Explicitly, we have $(G_r, G) = (4, 4)$ to $(G_r, G) = (7, 7)$, where the generator polynomial is presented in octal form. According to Figure 11.6, the code having a generator polynomial $(G_r, G) = (5, 7)$ converges faster than the other codes.

11.2.2.3 EXIT Chart

The exchange of extrinsic information in the system of Figure 11.1 can be visualized by plotting the extrinsic information characteristics of the inner demapper and the outer RSC decoder in an EXIT chart [169, 172]. The outer RSC decoder's extrinsic output information $I_{o,e}(\mathbf{c})$ becomes the demapper's *a priori* input information $I_{i,a}(\mathbf{b})$, which is represented on the x-axis of the EXIT chart. Similarly, on the y-axis we plot the demapper's extrinsic output information $I_{i,e}(\mathbf{b})$, which becomes the outer RSC decoder's *a priori* input information $I_{o,a}(\mathbf{c})$. The EXIT curves presented in this section correspond to the system employing a half-rate RSC code having constraint length $K = 3$, denoted as RSC(2,1,3), in conjunction with an octal generator polynomial $(G_r, G) = (7, 5)$.

Figure 11.7 shows the EXIT chart of an iteratively detected RSC-coded DSTS-SP scheme employing two transmit antennas and AGM-1 of Figure 11.3 when communicating over a temporally correlated Rayleigh fading channel having a normalized Doppler frequency $f_D = 0.01$. Ideally, in order for the exchange of extrinsic information between the SP demapper and the outer RSC decoder to converge at a specific E_b/N_0 value, the EXIT curve of the SP demapper at the E_b/N_0 value of interest and the extrinsic transfer characteristics curve of the outer RSC decoder should only intersect at the $(1.0, 1.0)$ point. If this condition is satisfied, then a so-called convergence tunnel [169, 172] appears on the EXIT chart. The

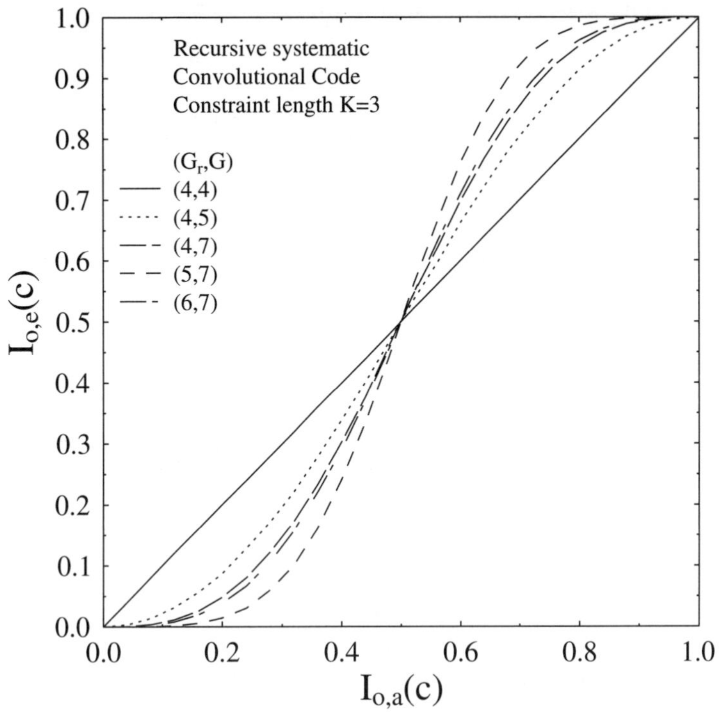

Figure 11.6: EXIT characteristics of half-rate RSC codes with constraint length $K = 3$ and different generator polynomials.

narrower the tunnel, the more iterations are required for reaching the $(1.0, 1.0)$ point and the closer the performance is to the channel capacity. If, however, the two extrinsic transfer characteristics intersect at a point close to the line at $I_{o,e}(\mathbf{c}) = 1.0$ rather than at the $(1.0, 1.0)$ point, then a moderately low BER may be still achieved, although it will remain higher than in the schemes where the intersection is at the $(1.0, 1.0)$ point. These types of tunnels are referred to here as semi-convergent tunnels. Observe in Figure 11.7 that a semi-convergent tunnel exists at $E_b/N_0 = 7.0$ dB. This implies that according to the predictions of the EXIT chart seen in Figure 11.7, the iterative decoding process is expected to converge at an E_b/N_0 value between 6.5 and 7.0 dB. These EXIT-chart-based convergence predictions are usually verified by the actual iterative decoding trajectory, as discussed in Section 11.2.4.

After analyzing the EXIT chart of the DSTS-SP scheme employing two transmit antennas and the optimum[2] SP AGM-1 of Figure 11.3, it is worth investigating the effect of employing different constellations. Figure 11.8 shows the EXIT chart of the system employing AGM-3 of Figure 11.3. As seen in the figure, a semi-convergent tunnel exists at an E_b/N_0 value of 6.0 dB, which is lower than that recorded for the optimum AGM mapping AGM-1. However, the intersection between the EXIT curve of the SP demapper employing AGM-3 at $E_b/N_0 =$ 6.0 dB and that of the outer RSC decoder is almost at $I_{i,e}(\mathbf{b}) = 0.75$, while the SP AGM-1 demapper's extrinsic transfer characteristic and the outer RSC decoder's extrinsic transfer curve intersect at $I_{i,e}(\mathbf{b}) = 0.85$ for $E_b/N_0 = 7.0$ dB. Therefore, the iterative detection of

[2]The optimum SP AGM is selected from Figure 11.3 as the mapping that results in the highest intersection point between the EXIT curve of the SP demapper and the EXIT curve of the outer RSC decoder at a specific E_b/N_0 value.

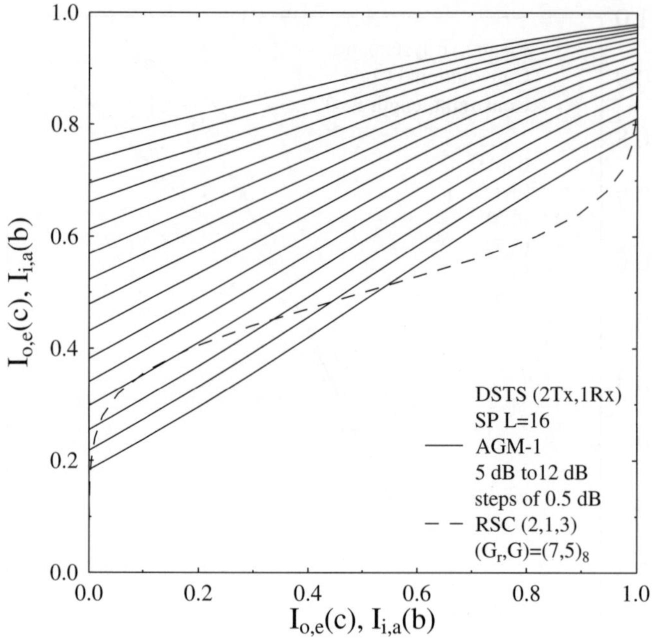

Figure 11.7: EXIT chart of an iteratively detected RSC-coded DSTS-SP scheme employing two transmit antennas and AGM-1 of Figure 11.3 in combination with the outer RSC(2,1,3) code, while communicating over a temporally correlated Rayleigh fading channel exhibiting $f_D = 0.01$.

the DSTS-SP system employing AGM-3 converges at a lower E_b/N_0 than that of the AGM-1-aided system, although the AGM-1-based system converges to a lower BER value. Their difference becomes more explicit as we move towards higher index AGMs such as AGM-8 shown in Figure 11.9. In Figure 11.9 a semi-convergent tunnel exists at $E_b/N_0 = 5.0$ dB. However, the SP AGM-8 demapper's EXIT curve and the outer RSC decoder's EXIT curve intersect near $I_{i,e}(\mathbf{b}) = 0.4$ at $E_b/N_0 = 5.0$ dB, i.e. the BER performance dramatically improves at $E_b/N_0 = 5.0$ dB, but fails to decay to an infinitesimally low BER value, as shown in Section 11.2.4.

Figure 11.10 depicts the EXIT chart of the iteratively detected RSC-coded DSTS-QPSK scheme employing two transmit antennas and AGM in combination with outer RSC code having constraint lengths $K = 3$ when communicating over a correlated Rayleigh fading channel having $f_D = 0.01$. Observe in Figure 11.10 that a semi-convergent tunnel exists at $E_b/N_0 = 5.0$ dB; however, the extrinsic transfer characteristic curve of the QPSK demapper and that of the outer RSC decoder intersect at almost $I_{i,e}(\mathbf{b}) = 0.55$ at $E_b/N_0 = 5.0$ dB, i.e. the performance curve converges at $E_b/N_0 = 5.0$ dB but not to a very low BER value compared with the DSTS-SP system employing AGM-1.

Finally, the EXIT chart of the iteratively detected RSC-coded DSTS-SP scheme employing four transmit antennas and AGM-1 of Figure 11.3 in combination with outer RSC code having constraint lengths $K = 3$ is shown in Figure 11.11. Observe in the figure that a semi-convergent tunnel exists at $E_b/N_0 = 7.0$ dB, which is similar to the behavior of the system employing two transmit antennas as characterized in Figure 11.7. However, note that the DSTS-SP schemes employing two and four transmit antennas have different bandwidth

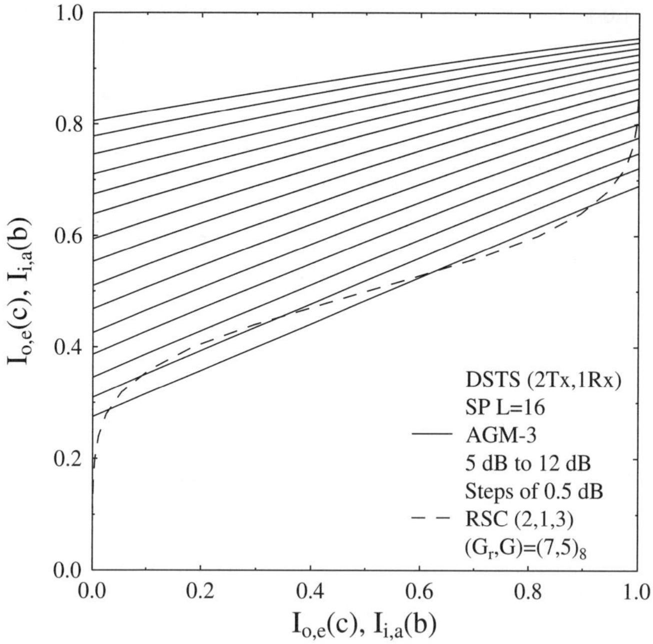

Figure 11.8: EXIT chart of an iteratively detected RSC-coded DSTS-SP scheme employing two transmit antennas and AGM-3 of Figure 11.3 in combination with the outer RSC(2,1,3) code, while communicating over a temporally correlated Rayleigh fading channel associated with $f_D = 0.01$.

efficiencies. Therefore, we plot in Figure 11.12 the EXIT curves of the 1 bit per channel use bandwidth efficiency two-antenna-aided DSTS-SP scheme in conjunction with $L = 16$ and AGM-1 together with the 0.5 bits per channel use two-antenna-aided DSTS-SP scheme employing $L = 4$ and AGM as well as with the 0.5 bits per channel use four-antenna-aided DSTS-SP scheme employing $L = 16$ and AGM-1. Figure 11.12 shows the EXIT curves of the demapper for $E_b/N_0 = 6.5$ and 7.0 dB. As shown in the figure, the four-antenna-aided system has a higher intersection point between the SP AGM-1 demapper extrinsic transfer characteristics curve and the outer RSC decoder extrinsic transfer characteristics curve compared with the equivalent-bandwidth-efficiency system employing two transmit antennas as well as with the higher-bandwidth-efficiency system employing two transmit antennas.

11.2.3 Maximum Achievable Bandwidth Efficiency

In Section 10.3.6 we derived the capacity and bandwidth efficiency of the DSTS-SP scheme. In this section, a procedure for computing an upper limit on the maximum achievable bandwidth efficiency of the system is proposed. It was argued in [176,270] that the maximum achievable bandwidth efficiency of the system is equal to the area under the EXIT curve of the inner code provided that the bit stream $\mathbf{b}$ has independently and uniformly distributed bits, the channel is an erasure channel, the inner code is unity-rate and the MAP algorithm is used for decoding. Assuming that the area under the EXIT curve of the inner decoder, i.e. the SP demapper in this case, is represented by $\mathcal{A}_i$, the maximum achievable rate for the outer

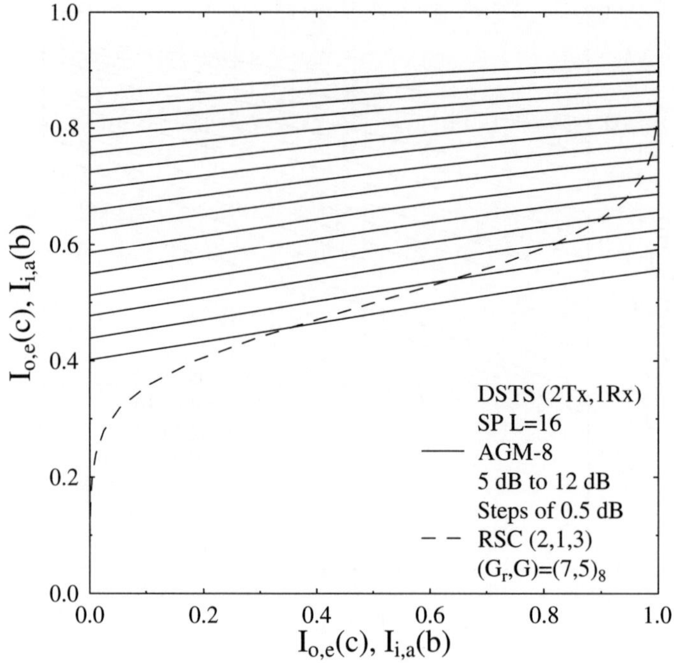

Figure 11.9: EXIT chart of an iteratively detected RSC-coded DSTS-SP scheme employing two transmit antennas and AGM-8 of Figure 11.3 in combination with the outer RSC(2,1,3) code, while communicating over a temporally correlated Rayleigh fading channel associated with $f_D = 0.01$.

code is given by $R_{\max} = \mathcal{A}_i(E_b/N_0)$ (see [176]) at a specific E_b/N_0 value. In other words, if $\mathcal{A}_i$ is calculated for different E_b/N_0 values, the maximum achievable bandwidth efficiency may be formulated as a function of the E_b/N_0 value as follows:

$$\begin{aligned}\eta_{\max}(\underline{E_b/N_0}) &= B_{\text{sp}} \cdot R_{\text{DSTS-SP}} \cdot R_{\max} \\ &\approx B_{\text{sp}} \cdot R_{\text{DSTS-SP}} \cdot \mathcal{A}_i(E_b/N_0) \quad (\text{BPS Hz}^{-1}), \qquad (11.21)\end{aligned}$$

where $B_{\text{sp}} = \log_2(L)$ is the number of bits per SP symbol, $R_{\text{DSTS-SP}} = \frac{1}{2}$ for the $N_t = 2$ transmit antennas case and $R_{\text{DSTS-SP}} = \frac{1}{4}$ for the $N_t = 4$ transmit antennas case. In addition, $\underline{E_b/N_0}$ and E_b/N_0 are related as follows:

$$\underline{E_b/N_0} = E_b/N_0 + 10\log\left(\frac{R_o}{\mathcal{A}_i(E_b/N_0)}\right) \quad (\text{dB}), \qquad (11.22)$$

where R_o is the original outer code rate used for generating the EXIT curve of the inner decoder/demapper corresponding to the different $\mathcal{A}_i$ values. The simple procedure shown in Algorithm 11.1 may be used to calculate the maximum achievable bandwidth efficiency of Equation (11.21) for $E_b/N_0 \in [\rho_{\min}, \rho_{\max}]$, assuming that $R_{\text{arbitrary}}$ is an arbitrary rate and ϵ is a small constant.

Observe that $\rho_{\min}$ and $\rho_{\max}$ are adjusted accordingly in order to produce the desired range of the resultant $\underline{E_b/N_0}$ values. Furthermore, the output of Algorithm 11.1 is independent of the specific choice of R_o, since Equation (11.22) would always adjust the $\underline{E_b/N_0}$ values, regardless of R_o.

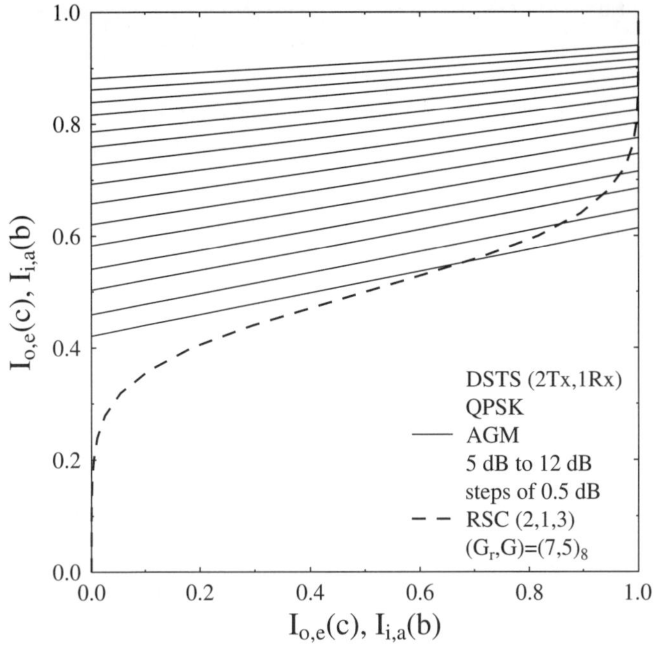

Figure 11.10: EXIT chart of an iteratively detected RSC-coded DSTS-QPSK scheme employing two transmit antennas and AGM in combination with the outer RSC(2,1,3) code, while communicating over a temporally correlated Rayleigh fading channel associated with $f_D = 0.01$.

The MIMO channel capacity curves of the four-dimensional SP-modulation-assisted DSTS scheme in conjunction with $L = 16$ are shown in Figures 11.13 and 11.14 for two and four transmit antennas, respectively. These two figures portray the DCMC bandwidth efficiency curve as well as the maximum achievable rate of the system derived from the EXIT curves according to Algorithm 11.1. Observe in Figures 11.13 and 11.14 that the maximum achievable rate of the system derived from the EXIT curves is quite close to the DCMC bandwidth efficiency. Note that the maximum achievable rate obtained from the EXIT charts and the bandwidth efficiency limit calculated using Equation (10.32) were only proven to be equal for the family of binary erasure channels [270]. Nonetheless, similar experimentally verified trends have been observed for both AWGN and ISI channels [174, 176], when APP-based decoders are used for all decoder blocks [270]. However, our DSTS decoder employs a very simple decoding algorithm that utilizes only two or four consecutively received symbols, despite the fact that all of the symbols are interdependent. Therefore, the decoder employed is suboptimum and if a trellis-based DSTS decoder, such as the MAP algorithm [164], is employed, then the maximum achievable rate obtained from the EXIT chart might match the capacity limit computed. Nevertheless, the complexity of the MAP algorithm is high in return for a modest gain of 0.35 dB observed in Figure 11.13 and the 0.2 dB gain seen in Figure 11.14.

At a bandwidth efficiency of $\eta = 1$ bit s^{-1} Hz^{-1}, the DCMC capacity limit of the two-antenna-aided DSTS-SP scheme is $E_b/N_0 \approx 4.85$ dB. Furthermore, at a bandwidth efficiency of $\eta = 1$ bit s^{-1} Hz^{-1}, the maximum achievable rate of the same scheme derived from the

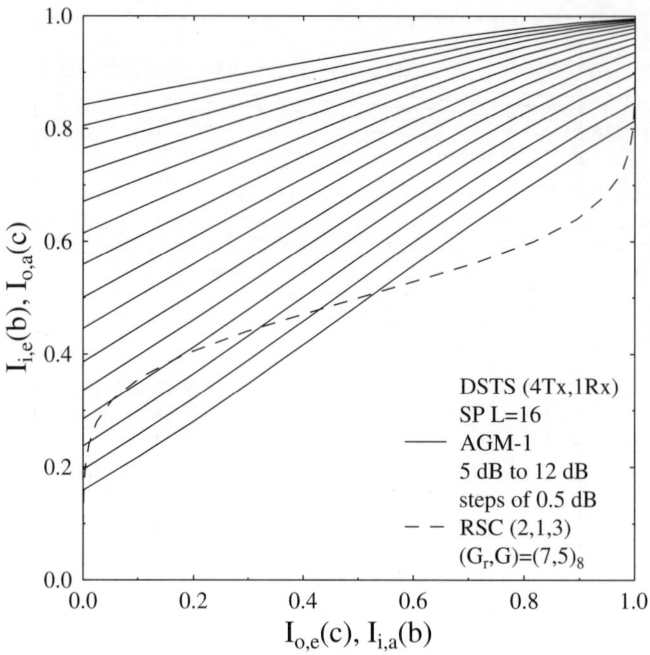

Figure 11.11: EXIT chart of an iteratively detected RSC-coded DSTS-SP scheme employing four transmit antennas and AGM-1 of Figure 11.3 in combination with the outer RSC(2,1,3) code, while communicating over a temporally correlated Rayleigh fading channel associated with $f_D = 0.01$.

EXIT curves is $E_b/N_0 = 5.2$ dB. On the other hand, the DCMC capacity limit of the four-antenna-aided DSTS-SP scheme is $E_b/N_0 \approx 4.95$ dB at a bandwidth efficiency of 0.5 bit s^{-1} Hz^{-1}, while the maximum achievable rate limit derived from the EXIT curves at the same bandwidth efficiency is $E_b/N_0 = 5.18$ dB.

11.2.4 Results and Discussion

In this section, we consider a DSTS system employing two and four transmit antennas and a single receive antenna in order to demonstrate the performance improvements achieved by the proposed iteratively detected SP-aided system. All simulation parameters are listed in Table 11.1.

As mentioned in Section 11.2.2.3, the EXIT-chart-based convergence predictions can be verified by the actual iterative decoding trajectory. Figure 11.15 records the trajectory of the iteratively detected RSC-coded DSTS-SP scheme in conjunction with two transmit antennas and AGM-1 of Figure 11.3 in combination with the system parameters outlined in Table 11.1, while operating at $E_b/N_0 = 7.0$ dB and employing an interleaver depth of $D_{\text{int}} = 1\,000\,000$ bits. The steps seen in the figure represent the actual EXIT between the demapper and the outer RSC channel decoder. Since a long interleaver is employed, the assumptions outlined at the beginning of Section 11.2.2 are justified and hence the EXIT-chart-based convergence prediction becomes accurate.

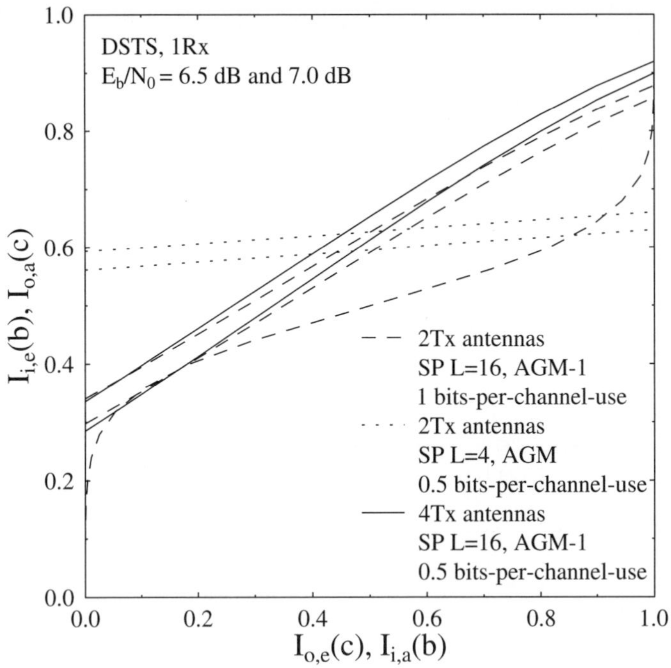

Figure 11.12: Comparison of the EXIT charts of an iteratively detected RSC-coded DSTS-SP scheme employing two and four transmit antennas in conjunction with $L = 16$ and AGM-1 together with the two-antenna-aided DSTS-SP scheme employing $L = 4$ and AGM.

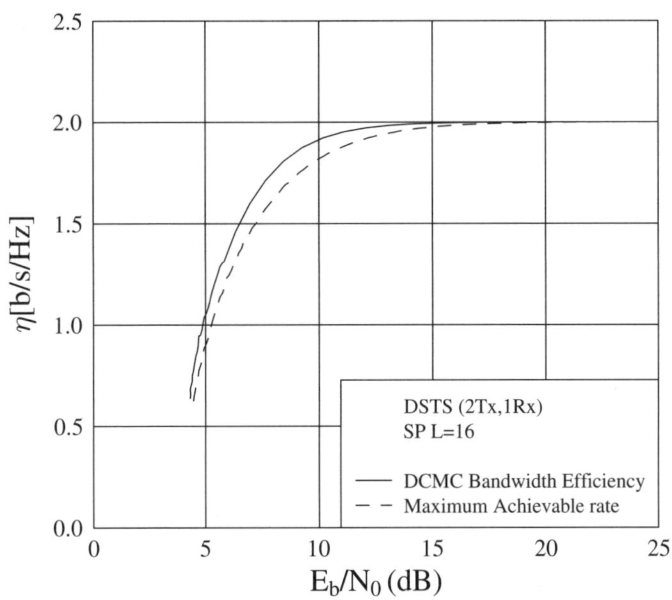

Figure 11.13: Comparison of the DCMC bandwidth efficiency and the maximum achievable rate obtained using EXIT charts of the two-antenna-aided DSTS-SP in conjunction with $L = 16$.

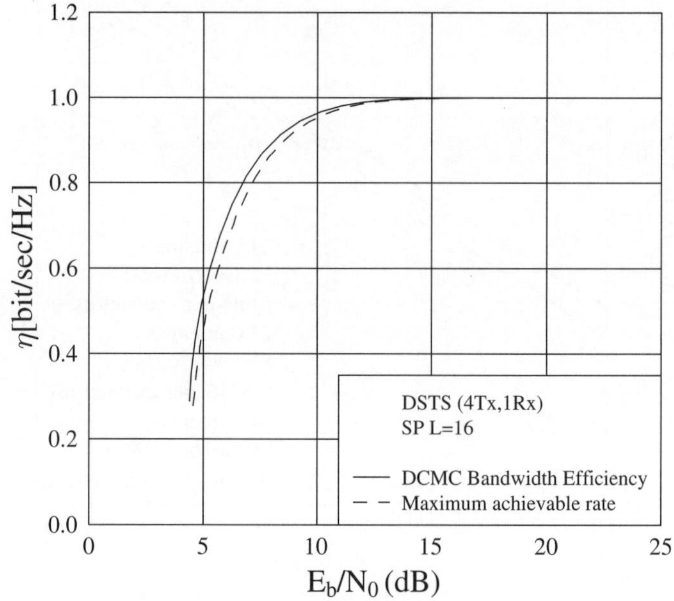

Figure 11.14: Comparison of the DCMC bandwidth efficiency and the maximum achievable rate obtained using EXIT charts of the four-antenna-aided DSTS-SP in conjunction with $L = 16$.

Table 11.1: Iteratively-detected RSC-coded DSTS-SP system parameters.

SP modulation	$L = 16$
Number of transmit antennas	2 and 4
Number of receive antennas	1
Channel	Correlated Rayleigh fading
Normalized Doppler frequency	0.01
Outer channel code	RSC(2,1,3)
Generator polynomial	$(G_r, G) = (7, 5)_8$
Spreading code	Walsh–Hadamard code
Spreading factor	8
Number of users	4

Step 1: Let $R_o = R_{\text{arbitrary}}$.

Step 2: Let $E_b/N_0 = \rho_{\min}$ (dB).

Step 3: Calculate N_0.

Step 4: Let $I_{i,a}(\mathbf{b}) = 0$.

Step 5: Activate the SP demapper.

Step 6: Save $I_{i,e}(\mathbf{b}) = T_i(I_{i,a}(\mathbf{b}), E_b/N_0)$.

Step 7: Let $I_{i,a}(\mathbf{b}) = I_{i,a}(\mathbf{b}) + \epsilon$.
If $I_{i,a}(\mathbf{b}) \leq 1.0$, go to **Step 5**.

Step 8: Calculate $\mathcal{A}_i(E_b/N_0) = \int_0^1 T_i(i, E_b/N_0)\ di$.

Step 9: Calculate $\underline{E_b/N_0}$ using Equation (11.22).

Step 10: Save $\eta_{\max}(\underline{E_b/N_0})$ of Equation (11.21).

Step 11: Let $E_b/N_0 = E_b/N_0 + \epsilon$.
If $E_b/N_0 \leq \rho_{\max}$ (dB), go to **Step 3**.

Step 12: Output $\eta_{\max}(\underline{E_b/N_0})$ from **Step 10**.

Algorithm 11.1: Maximum achievable bandwidth efficiency using EXIT charts.

Moreover, Figures 11.16–11.18 record the trajectory of the iteratively detected RSC-coded DSTS-SP scheme in conjunction with two transmit antennas and AGM-1 of Figure 11.3 in combination with the system parameters outlined in Table 11.1, while operating at $E_b/N_0 = 7.0$ dB associated with interleaver depths of $D_{\text{int}} = 100\,000$, 10 000 and 1000 bits, respectively. The decoding trajectory in Figure 11.16 employs an interleaver depth of $D_{\text{int}} = 100\,000$ bits; as seen in the figure, the decoding trajectory is different from that observed in Figure 11.15. In other words, the system employing a shorter interleaver requires more iterations to reach the highest intersection point between the EXIT curves of the demapper and the outer RSC code, in addition to the fact that as $I_{i,a}(\mathbf{b})$ increases, the trajectory does not match the EXIT curve of the RSC decoder; this might be due to the fact that in conjunction with an interleaver depth of $D_{\text{int}} = 100\,000$ bits, the LLR distribution is no longer Gaussian for high $I_{i,a}(\mathbf{b})$ values.

On the other hand, the decoding trajectories shown in Figures 11.17 and 11.18 are different from the EXIT chart prediction because shorter interleavers are used and thus the assumptions at the beginning of Section 11.2.2 are not valid. The BER performance of the iteratively detected RSC-coded DSTS-SP scheme in conjunction with two transmit antennas and AGM-1 in combination with the system parameters outlined in Table 11.1 is shown in Figure 11.19 when using $I = 10$ iterations and varying the interleaver depth. As seen in the figure, upon increasing the interleaver depth from $D_{\text{int}} = 1000$ to 10 000 bits, the system's performance dramatically improves. Upon further increasing the interleaver depth from $D_{\text{int}} = 10\,000$ to 100 000 bits, the attainable performance improves, but not as much as when increasing it from 1000 to 10 000 bits. Furthermore, increasing the interleaver depth beyond

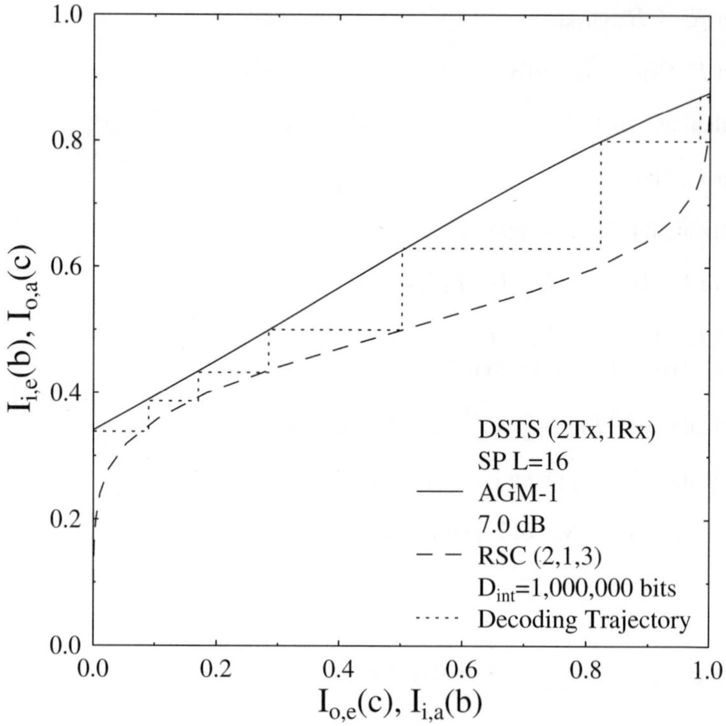

Figure 11.15: Decoding trajectory of the iteratively detected half-rate RSC-coded DSTS-SP scheme in conjunction with two transmit antennas and AGM-1 of Figure 11.3 employing the system parameters outlined in Table 11.1, while operating at $E_b/N_0 = 7.0$ dB with an interleaver depth of $D_{\text{int}} = 1\,000\,000$ bits.

$D_{\text{int}} = 200\,000$ bits does not significantly improve the achievable system performance. In addition, observe that a turbo cliff appears at $E_b/N_0 = 8.5$ dB upon increasing the interleaver depth to $D_{\text{int}} = 10\,000$ bits, while a turbo cliff occurs at $E_b/N_0 = 7.5$ dB when using an interleaver depth of $D_{\text{int}} = 100\,000$ bits. In addition, the system employing an interleaver depth of $D_{\text{int}} \geq 200\,000$ bits converges at $E_b/N_0 = 7.0$ dB, as predicted by the EXIT curve of Figure 11.7. Furthermore, as the interleaver depth increases, the system's performance approaches the capacity limit, as shown in Figure 11.19. However, owing to the error floor observed for the BER curves in Figure 11.19, as the E_b/N_0 increases, the system's BER curve diverges from the capacity limit. Explicitly, the system performs within 2.3 dB from the maximum achievable rate limit at $\text{BER} = 10^{-5}$ and within 3.3 dB from the same limit at $\text{BER} = 10^{-6}$.

Figure 11.20 plots the BER performance of the iteratively detected RSC-coded DSTS-SP scheme employing AGM-1 versus the number of iterations while using different interleaver depths ranging from $D_{\text{int}} = 1000$ to $800\,000$ bits in combination with the system parameters outlined in Table 11.1, while operating at $E_b/N_0 = 7.0$ dB. The plot investigates the BER performance versus the complexity of the system quantified in terms of the number of iterations. As shown in the figure, when using short interleavers, increasing the number of iterations results in no significant BER performance improvement, which is the case for the interleavers with depths of $D_{\text{int}} = 1000$ bits and $10\,000$ bits. However, as the interleaver

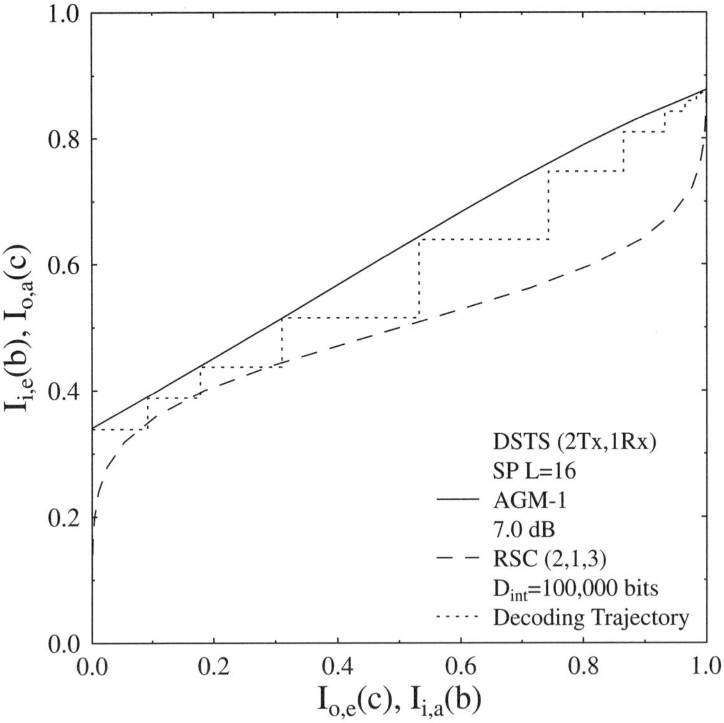

Figure 11.16: Decoding trajectory of the iteratively detected half-rate RSC-coded DSTS-SP scheme in conjunction with two transmit antennas and AGM-1 of Figure 11.3 employing the system parameters outlined in Table 11.1, while operating at $E_b/N_0 = 7.0$ dB with an interleaver depth of $D_{\text{int}} = 100\,000$ bits.

becomes longer, the achievable system performance improves upon increasing the number of iterations. Moreover, as the interleaver depth increases, the system requires fewer iterations to saturate, as shown in Figure 11.20. For example, for the case of an interleaver depth of $D_{\text{int}} = 800\,000$ bits, it is shown in Figure 11.20 that after $I = 7$ iterations, there is no more improvement in the attainable system performance, while the system employing $D_{\text{int}} = 400\,000$ bits requires one more iteration before the system's performance saturates according to Figure 11.20.

Figures 11.21 and 11.22 record the trajectories of the iteratively detected RSC-coded DSTS-SP schemes in conjunction with two transmit antennas and the system parameters outlined in Table 11.1 while employing AGM-3 and AGM-8, respectively. Figure 11.21 records the trajectory of the system employing AGM-3 while operating at $E_b/N_0 = 6.0$ dB, while Figure 11.22 records the trajectory of the system employing AGM-8 while operating at $E_b/N_0 = 5.5$ dB, when considering an interleaver depth of $D_{\text{int}} = 1\,000\,000$ bits. The BER performance of the iteratively detected half-rate RSC-coded DSTS-SP scheme recorded in conjunction with two transmit antennas and different Gray mapping and AGM mapping schemes, while using the system parameters outlined in Table 11.1, is shown in Figure 11.23, when applying $I = 10$ iterations. Figure 11.23 also plots the performance curves for the equivalent bandwidth efficiency of 1 bit per channel use employing uncoded DSTS in conjunction with SP $L = 4$ and BPSK. Observe in Figure 11.23 that the Gray-mapping- and

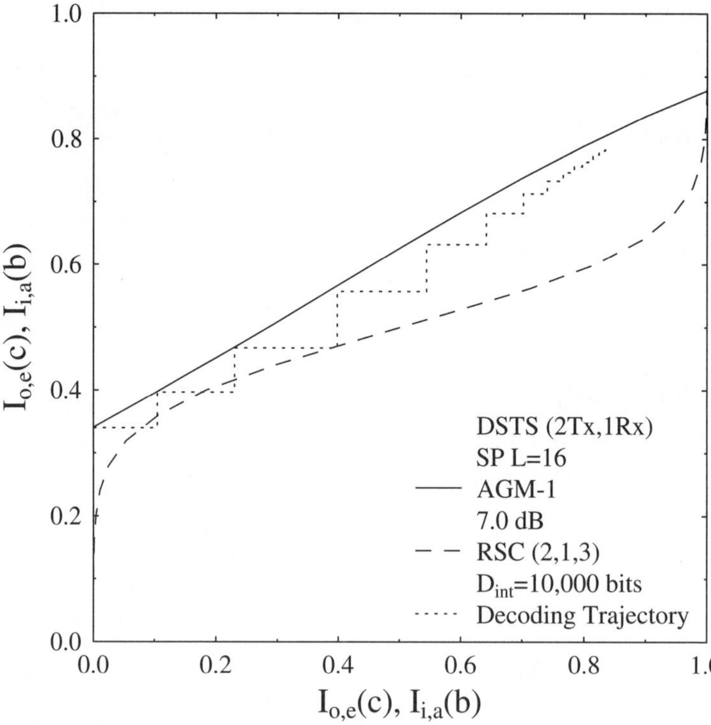

Figure 11.17: Decoding trajectory of the iteratively detected half-rate RSC-coded DSTS-SP scheme in conjunction with two transmit antennas and AGM-1 of Figure 11.3 employing the system parameters outlined in Table 11.1, while operating at $E_b/N_0 = 7.0$ dB with an interleaver depth of $D_{\text{int}} = 10\,000$ bits.

AGM-8-based systems have a similar performance and this can be justified by referring to the EXIT chart of Figure 11.3, where the EXIT curves of the Gray-mapping- and AGM-8-based systems have similar slopes. More explicitly, the system employing AGM-8 outperforms that employing Gray mapping by an E_b/N_0 of 0.6 dB at $\text{BER} = 10^{-6}$, while the system employing AGM-3 outperforms that employing Gray mapping by an E_b/N_0 of 2.7 dB at $\text{BER} = 10^{-6}$. In addition, the AGM-1-aided systems outperform the Gray-mapping-based system by an E_b/N_0 of 4.7 dB at $\text{BER} = 10^{-6}$.

According to the EXIT chart predictions of Section 11.2.2.3, the system employing AGM does not reach the $I_{i,e}(\mathbf{b}) = 1.0$ point and thus must have an error floor, which clearly appears in Figure 11.23. Moreover, it is clear from the figure that the AGM-3-aided system converges at a lower E_b/N_0 value as compared with the AGM-1-based system, which implies that at lower E_b/N_0 values it is better to use AGM-3 rather than AGM-1. However, as the E_b/N_0 value increases, the AGM-1-based system starts to outperform that employing AGM-3; therefore whether to use AGM-1 or AGM-3 depends on the application or on the range of E_b/N_0 values of interest for the specific application considered. Furthermore, the performance results of Figure 11.23 match with the EXIT chart predictions of Section 11.2.2. In addition to that, it is obvious from Figure 11.23 that at low E_b/N_0 values, the system employing AGM-3 approaches the system capacity more closely than the AGM-1-based system. However, as the E_b/N_0 value increases, the AGM-3-based system exhibits an error

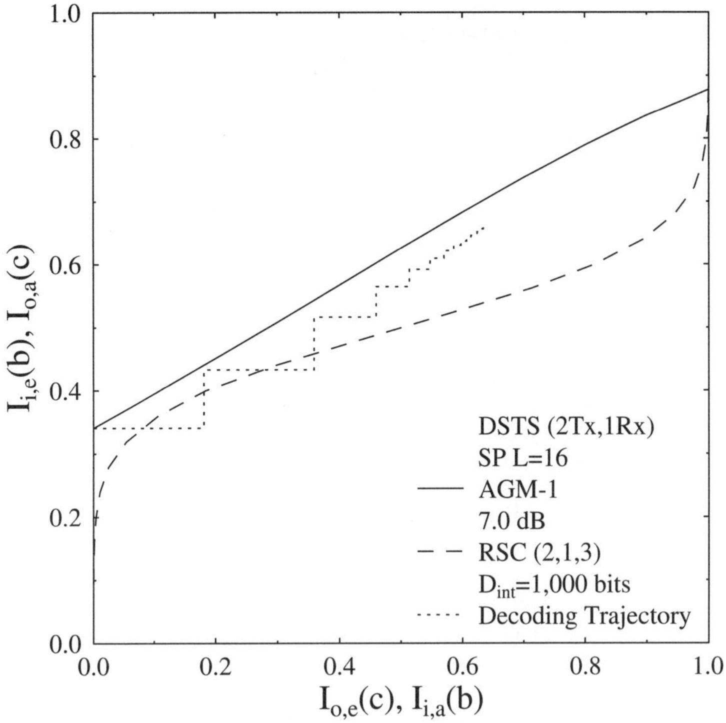

Figure 11.18: Decoding trajectory of the iteratively detected half-rate RSC-coded DSTS-SP scheme in conjunction with two transmit antennas and AGM-1 of Figure 11.3 employing the system parameters outlined in Table 11.1, while operating at $E_b/N_0 = 7.0$ dB with an interleaver depth of $D_{\text{int}} = 1000$ bits.

floor and thus moves away from the capacity limit at lower BER values. On the other hand, the AGM-1-based system's performance is closer to the capacity limit than that of the AGM-3-, AGM-8-or Gray-mapping-based systems' performance at low BER values, although there is an error floor.

Figure 11.24 compares the attainable performance of the proposed RSC-coded DSTS-SP scheme employing both AGM-1 and Gray mapping of the bits to the SP symbol, which are also contrasted to that of an identical bandwidth efficiency 1 bit per channel use uncoded DSTS-SP scheme using $L = 4$ and a conventional DSTS-BPSK design transmitting two independent BPSK symbols over the two antennas, when communicating over a correlated Rayleigh fading channel and employing the system parameters of Table 11.1. In Figure 11.24, an interleaver depth of $D_{\text{int}} = 1\,000\,000$ bits was employed and a normalized Doppler frequency of $f_D = 0.01$ was used. Observe in the figure that the two Gray-mapping-based DSTS-SP BER curves are exactly the same when $I = 0$ as well as $I = 10$ iterations were employed, which is evident from the flat curve of the Gray mapping in Figure 11.3. In contrast, AGM-1 achieved a substantial performance improvement in conjunction with iterative demapping and decoding. Explicitly, Figure 11.24 demonstrates that a coding advantage of about 22.5 dB was achieved at a BER of 10^{-6} after $I = 10$ iterations by the convolutional-coded AGM-based DSTS-SP system over both the uncoded DSTS-SP and the DSTS-BPSK schemes for transmission over the correlated Rayleigh fading channel

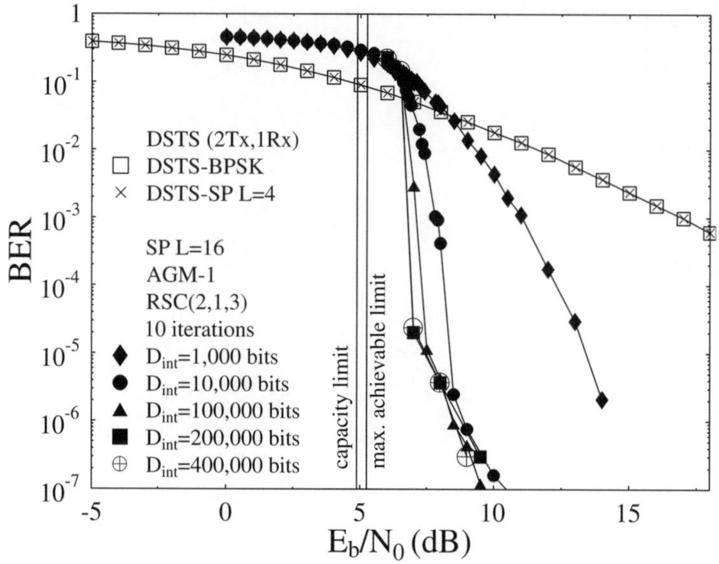

Figure 11.19: Performance comparison of the AGM-1 based RSC-coded two-transmit-antennas DSTS-SP schemes in conjunction with $L = 16$ against an identical bandwidth efficiency of 1 bit per channel use uncoded DSTS-SP scheme using $L = 4$ and against the conventional DSTS-BPSK scheme, when employing the system parameters outlined in Table 11.1 and using different interleaver depths after $I = 10$ iterations.

considered. In addition, a coding advantage of approximately 4.7 dB was attained over the RSC-coded Gray-mapping-based DSTS-SP scheme. Finally, the AGM-1-based system performs within 2.3 dB from the maximum achievable rate limit at BER of 10^{-5} and within 3.3 dB from the same limit at BER of 10^{-6}.

The coding gain of the iteratively detected DSTS-SP systems employing AGM-1, AGM-3 and AGM-8 is monitored in Figure 11.25 against the number of iterations employed at BER of 10^{-5} for an interleaver depth of $D_{\text{int}} = 1\,000\,000$ bits. The coding gain is measured versus the performance of the uncoded equivalent bandwidth efficiency system employing DSTS-SP in conjunction with $L = 4$ and Gray mapping. Figure 11.25 shows that increasing the number of iterations tends to attain a gradually eroding coding gain. Furthermore, it is observed that the coding gain of the AGM-1-based system is higher than that employing AGM-3 at BER $=10^{-5}$ and this is verified by the performance curves of Figure 11.23. Moreover, note that when no iterations are employed, the AGM-8-based system has a higher coding gain than the other two systems.

Figure 11.26 depicts the coding gain of the iteratively detected DSTS-SP systems employing AGM-1 versus the number of iterations employed at BER of 10^{-5} for different interleaver depths D_{int}. It becomes explicit from the figure that as the interleaver depth increases, the coding gain increases and the system performance approaches the maximum achievable rate limit. Moreover, increasing the interleaver depth beyond $D_{\text{int}} = 100\,000$ bits results in a modest improvement in the system's performance.

A comparison between the performance of the SP-aided and that of the equivalent bandwidth efficiency conventionally modulated iteratively detected two-antenna-aided DSTS

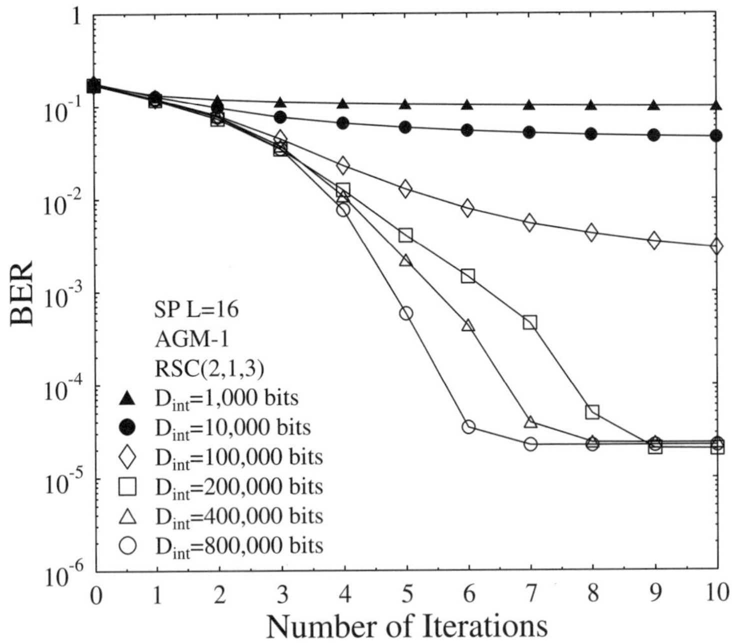

Figure 11.20: Comparison of the BER performance versus the number of iterations for the iteratively detected half-rate RSC-coded DSTS-SP scheme in conjunction with two transmit antennas and AGM-1 of Figure 11.3, while employing the system parameters outlined in Table 11.1 for different interleaver depths recorded at $E_b/N_0 = 7.0$ dB.

system is shown in Figure 11.27. The proposed iteratively detected two-antenna-aided DSTS-SP scheme provides an improved performance over an equivalent-throughput DSTS scheme dispensing with SP modulation, as evidenced in Figure 11.27, demonstrating that the AGM-1-aided DSTS-SP scheme using $L = 16$ exhibits an E_b/N_0 gain of around 3.4 dB at a BER of 10^{-6} over the identical bandwidth efficiency 1 bit per channel use DSTS-QPSK scheme.

Figure 11.28 records the trajectory of the iteratively detected RSC-coded DSTS-SP scheme in conjunction with four transmit antennas and AGM-1 in combination with the system parameters of Table 11.1 while operating at $E_b/N_0 = 7.0$ dB with interleaver depth of $D_{\text{int}} = 100\,000$ bits. The decoding trajectory shown in Figures 11.28 matches with the EXIT chart prediction of Figure 11.11.

Figure 11.29 compares the attainable performance of the proposed RSC-coded four-antenna-aided DSTS-SP scheme employing both AGM-1 and Gray mapping of the bits to the SP symbol, which are also contrasted with that of an identical bandwidth efficiency 0.5 bits per channel use uncoded DSTS-SP scheme using $L = 4$, when communicating over a correlated Rayleigh fading channel. In Figure 11.29, an interleaver depth of $D_{\text{int}} = 1\,000\,000$ bits was employed in conjunction with the system parameters of Table 11.1. Observe in the figure that the two Gray-mapping-based DSTS-SP BER curves are exactly the same, regardless of whether no iterations or $I = 10$ iterations were employed, similarly to the two transmit antennas case and as exemplified in Figure 11.3. In contrast, the AGM-based system achieves a substantial performance improvement in conjunction with iterative demapping and decoding. Explicitly, Figure 11.29 demonstrates that a coding advantage of about 16.7 dB was achieved at a BER of 10^{-6} after $I = 10$ iterations by the RSC-coded AGM-1-based DSTS-SP

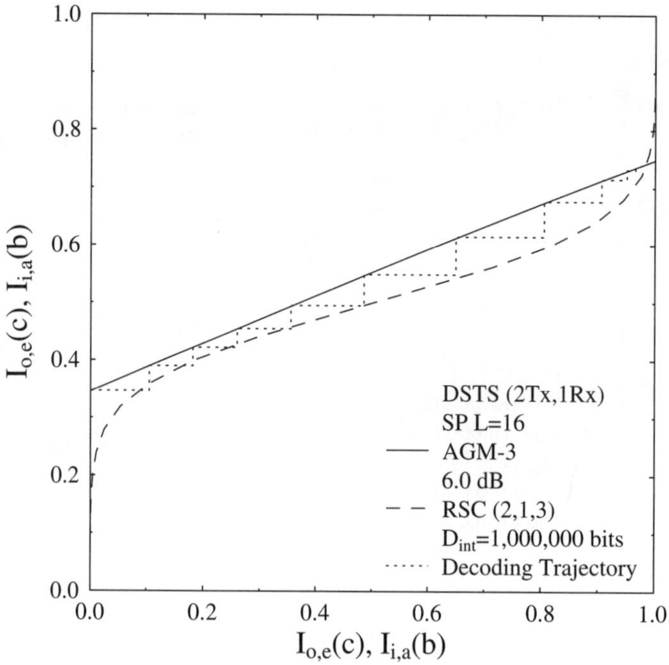

Figure 11.21: Decoding trajectory of the iteratively detected half-rate RSC-coded DSTS-SP scheme in conjunction with two transmit antennas and AGM-3 of Figure 11.3 employing the system parameters outlined in Table 11.1, while operating at $E_b/N_0 = 6.0$ dB with an interleaver depth of $D_{\rm int} = 1\,000\,000$ bits.

system over the uncoded DSTS-SP scheme. In addition, a coding advantage of approximately 3 dB was attained over the 0.5 BPS throughput RSC-coded GM-based DSTS-SP scheme. Finally, after $I = 10$ iterations, the AGM-1-based four-antenna-aided system performs within 1.82 dB from the maximum achievable rate limit at BER of 10^{-5} and within 2.12 dB from the same limit at BER of 10^{-6}.

Furthermore, Figure 11.30 provides a performance comparison of the iteratively detected RSC-coded four-antenna-aided DSTS scheme in conjunction with SP $L = 16$ employing AGM-1, AGM-3 and AGM-8 as well as the identical-bandwidth-efficiency QPSK-aided scheme. As evidenced in Figure 11.30, the AGM-1-aided system outperforms the QPSK-aided system by 3.2 dB at BER of 10^{-6} and the AGM-3-assisted system outperforms the QPSK-aided system by 1.3 dB at the same BER. However, the QPSK-aided scheme outperforms its identical-bandwidth-efficiency AGM-8-aided counterpart by almost 0.6 dB at BER of 10^{-6}.

Finally, Figure 11.31 compares the performance of the AGM-1-based RSC-coded DSTS-SP scheme in conjunction with $L = 16$, when employing four transmit antennas, and that of the AGM-based system employing two transmit antennas as well as SP $L = 4$, while using an interleaver depth of $D_{\rm int} = 1\,000\,000$ bits, $I = 10$ iterations and the system parameters outlined in Table 11.1. The four-antenna-aided system outperforms its two-antenna-aided counterpart by approximately 3.2 dB at a BER of 10^{-6}.

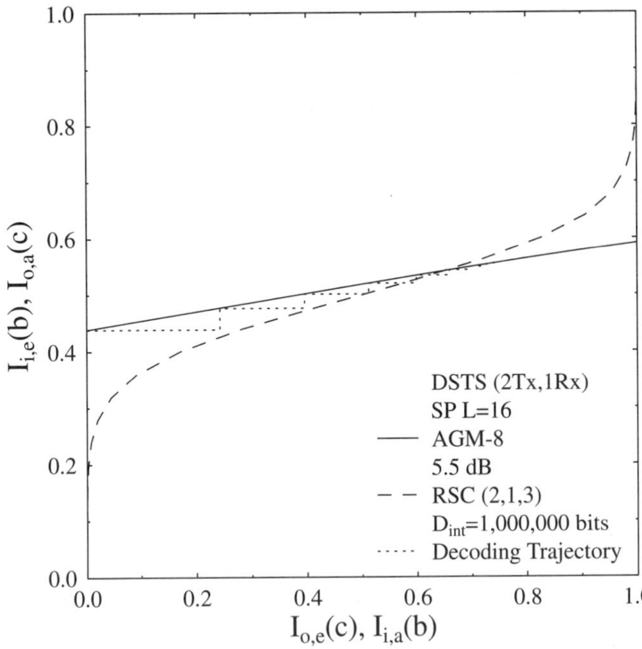

Figure 11.22: Decoding trajectory of the iteratively detected half-rate RSC-coded DSTS-SP scheme in conjunction with two transmit antennas and AGM-8 of Figure 11.3 employing the system parameters outlined in Table 11.1, while operating at $E_b/N_0 = 5.5$ dB with an interleaver depth of $D_{\text{int}} = 1\,000\,000$ bits.

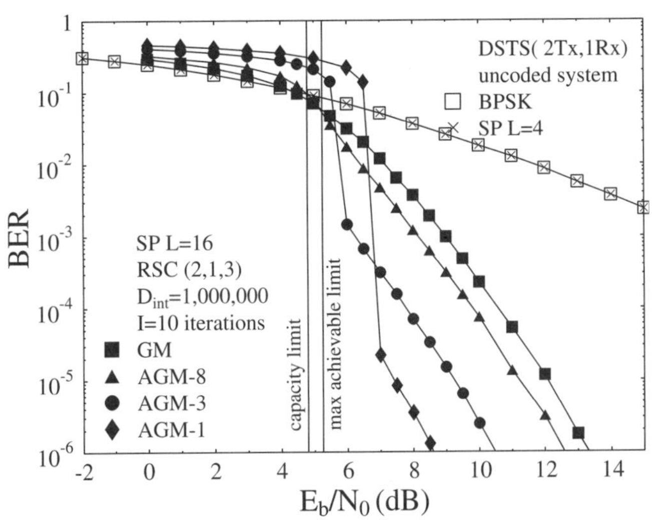

Figure 11.23: Performance comparison of different AGM- and Gray-mapping-based iteratively detected RSC-coded two-transmit-antennas DSTS-SP schemes in conjunction with $L = 16$ against an identical bandwidth efficiency of 1 bit per channel use uncoded DSTS-SP scheme using $L = 4$ and against the conventional DSTS-BPSK scheme, when employing the system parameters outlined in Table 11.1 with an interleaver depth of $D_{\text{int}} = 1\,000\,000$ bits after $I = 10$ iterations.

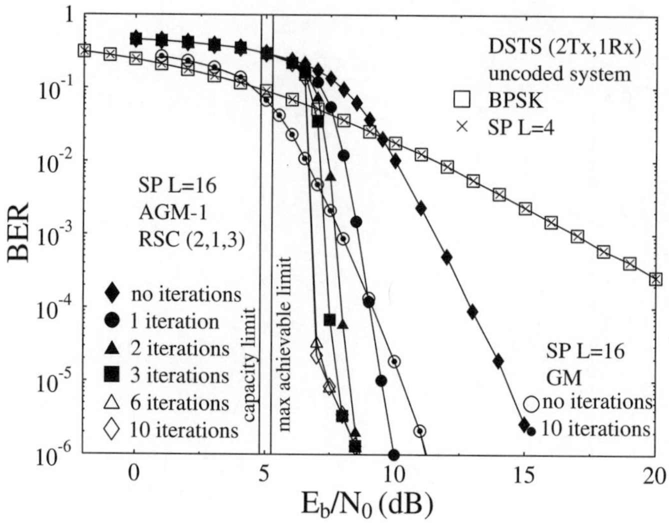

Figure 11.24: Performance comparison of AGM-1- and Gray-mapping-based RSC-coded two-transmit-antennas DSTS-SP schemes in conjunction with $L = 16$ against an identical bandwidth efficiency of 1 bit per channel use uncoded DSTS-SP scheme using $L = 4$ and against a conventional DSTS-BPSK scheme, when employing the system parameters outlined in Table 11.1 and using an interleaver depth of $D_{\text{int}} = 1\,000\,000$ bits for a variable number of iterations I.

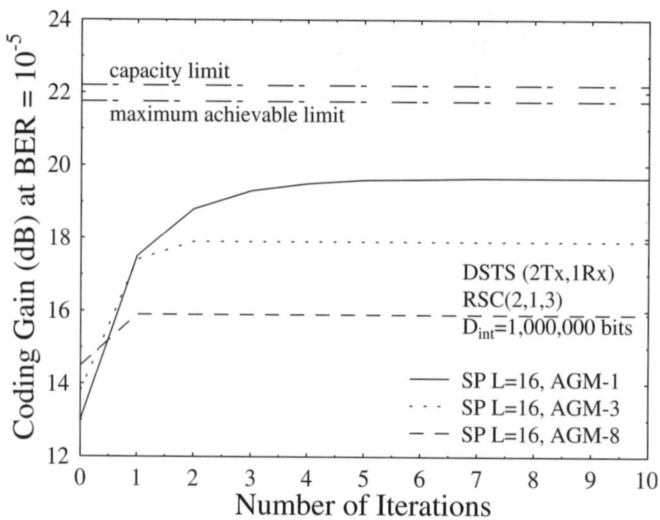

Figure 11.25: Coding gain of the iteratively detected half-rate RSC-coded two-transmit-antennas DSTS-SP scheme against the number of iterations employed at a BER of 10^{-5} when employing AGM-1, AGM-3 and AGM-8 with an interleaver depth of $D_{\text{int}} = 1\,000\,000$ bits.

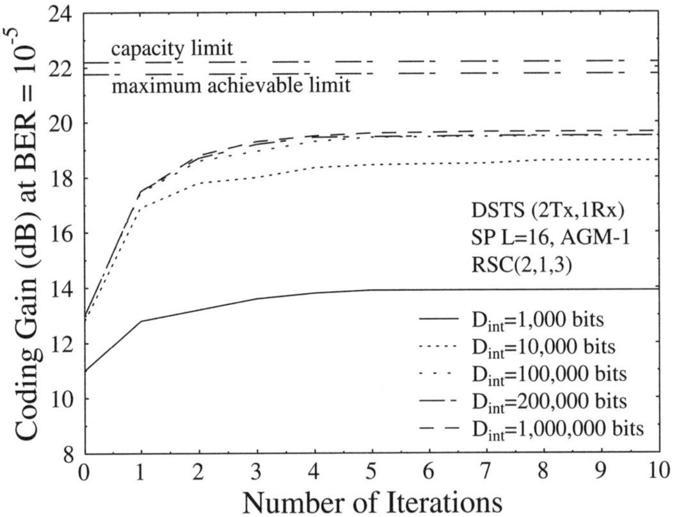

Figure 11.26: Coding gain of the iteratively detected half-rate RSC-coded two-transmit-antennas DSTS-SP scheme against the number of iterations employed at a BER of 10^{-5} when employing AGM-1 and varying the interleaver depth D_{int} between 1000 and 1 000 000 bits.

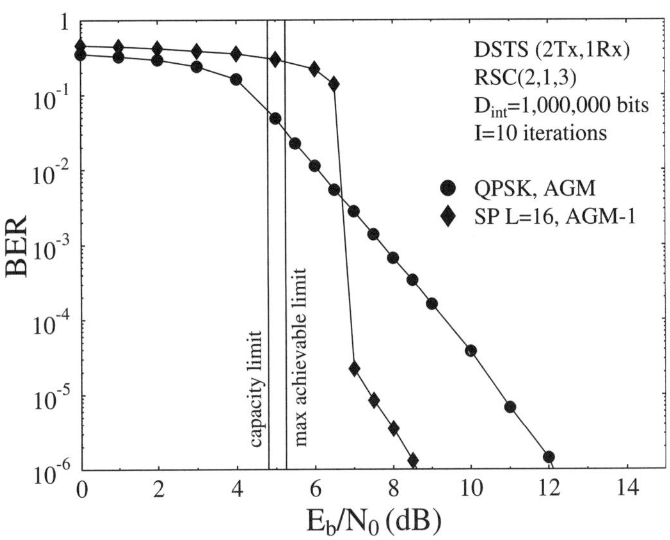

Figure 11.27: Performance comparison of an AGM-1-based iteratively detected RSC-coded two-antenna-aided DSTS-SP scheme in conjunction with $L = 16$ and the equivalent bandwidth efficiency of 1 bit per channel use AGM-based iteratively detected RSC-coded DSTS-QPSK scheme, while using an interleaver depth of $D_{\text{int}} = 1\,000\,000$ bits for $I = 10$ iterations and using the system parameters outlined in Table 11.1.

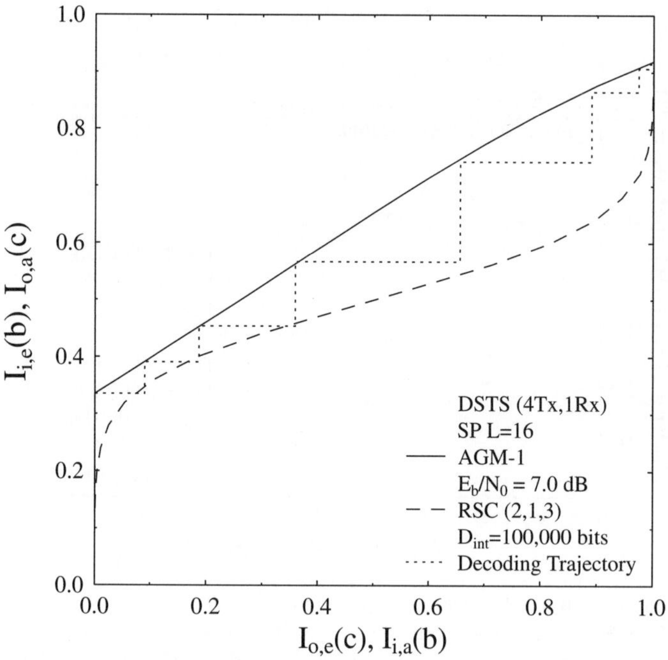

Figure 11.28: Decoding trajectory of the iteratively detected half-rate RSC-coded DSTS-SP scheme in conjunction with four transmit antennas and AGM-1 of Figure 11.3 employing the system parameters outlined in Table 11.1, while operating at $E_b/N_0 = 7.0$ dB with an interleaver depth of $D_{\text{int}} = 100\,000$ bits.

11.2.5 Application: Soft-bit-assisted Iterative AMR-WB Source Decoding and Iterative Detection of Channel-coded DSTS-SP System

The classic Shannonian source and channel coding separation theorem [272] has limited applicability in the context of finite-complexity, finite-delay lossy speech [350] and video [351] codecs, where the different encoded bits exhibit different error sensitivity. These arguments are particularly valid when the limited-complexity limited-delay source encoders fail to remove all of the redundancy from the correlated speech or video source signal. Fortunately, this residual redundancy may be beneficially exploited for error protection by intelligently exchanging soft information amongst the various receiver components.

These powerful iterative decoding principles may be further enhanced by exploiting the innovative concept of soft speech bits, which was developed by Fingscheidt and Vary [352, 353], culminating in the formulation of Iterative Source and Channel Decoding (ISCD) [354]. More explicitly, in ISCD the source and channel decoders iteratively exchange extrinsic information for the sake of improving the overall system performance. As a further development, in [158] the turbo principle [9] was employed for iterative soft demapping in multilevel modulation [248] schemes combined with channel coding which resulted in an enhanced BER performance. Thus, ISCD may be beneficially combined with iterative soft demapping in the context of multilevel modulation and amalgamated with a number of other sophisticated wireless transceiver components. In the resultant multi-stage scheme,

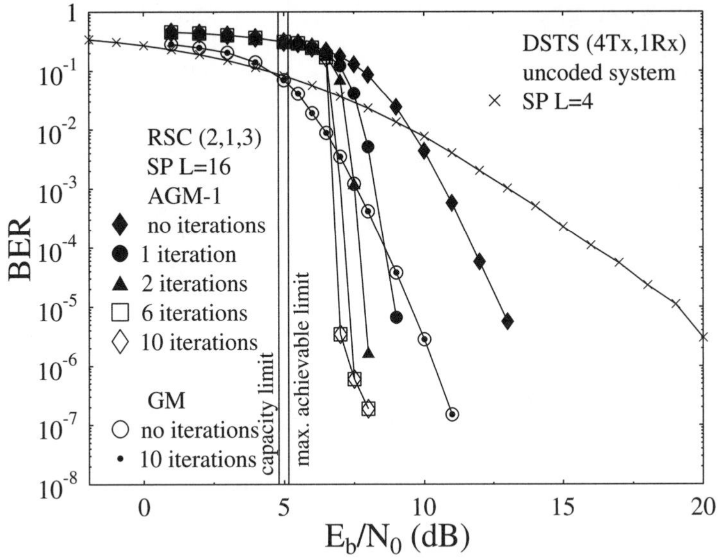

Figure 11.29: Performance comparison of AGM-1-and Gray-mapping-based RSC-coded four-antenna-aided DSTS-SP scheme in conjunction with $L = 16$ against an identical bandwidth efficiency of 0.5 bits per channel use uncoded DSTS-SP $L = 4$ scheme when employing the system parameters outlined in Table 11.1 and using an interleaver depth of $D_{\text{int}} =$ 1 000 000 bits for a variable number of iterations I.

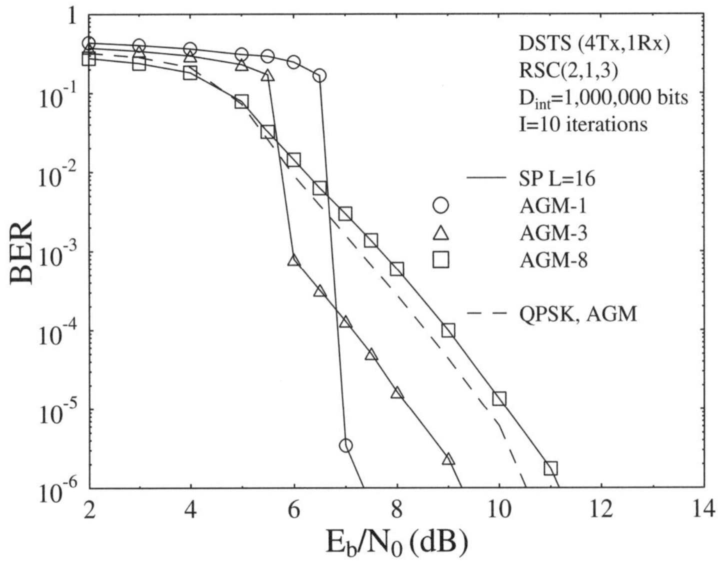

Figure 11.30: Performance comparison of different AGM-based RSC-coded four-antenna-aided DSTS-SP schemes in conjunction with $L = 16$ and an AGM-based RSC-coded DSTS-QPSK scheme, while using an interleaver depth of $D_{\text{int}} = 1\,000\,000$ bits for $I = 10$ iterations and using the system parameters outlined in Table 11.1.

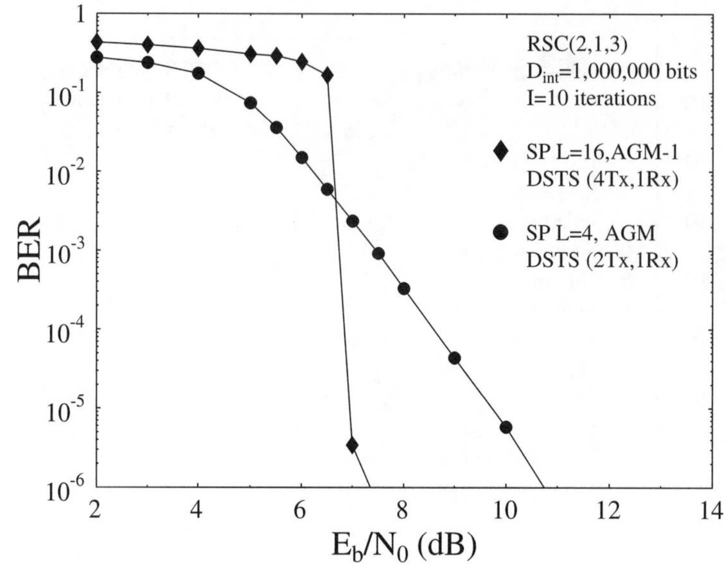

Figure 11.31: Performance comparison of an AGM-1-based RSC-coded four-antenna-aided DSTS-SP scheme in conjunction with $L = 16$ against an equivalent bandwidth efficiency two-antenna-aided DSTS-SP scheme employing $L = 4$, while using an interleaver depth of $D_{\text{int}} = 1\ 000\ 000$ bits after $I = 10$ iterations and using the system parameters outlined in Table 11.1.

extrinsic information is exchanged amongst three receiver components, namely the demapper, the channel decoder and the soft-input soft-output source decoder in the spirit of [355].

Explicitly, we propose and investigate the jointly optimized ISCD scheme of Figure 11.32 invoking the AMR-WB speech codec [356], which is protected by a RSC code. The resultant bit stream is transmitted using DSTS amalgamated with SP modulation [197] over a temporally correlated narrowband Rayleigh fading channel. An efficient iterative turbo-detection scheme is utilized for exchanging extrinsic information between the constituent decoders. Figure 11.32 shows the schematic of the proposed arrangement, referred to as DSTS-SP RSC AMR-WB, where the extrinsic information gleaned is exchanged amongst all three constituent decoders, namely the SP demapper, the RSC decoder and the AMR-WB decoder.

The AMR-WB speech encoder produces a frame of speech coded parameters, namely $\{\mathbf{v}_{1\mathcal{T}}, \mathbf{v}_{2\mathcal{T}}, \ldots, \mathbf{v}_{\kappa\mathcal{T}}, \ldots, \mathbf{v}_{36,\mathcal{T}}\}$, where $\mathbf{v}_{\kappa,\mathcal{T}}$ denotes an encoded parameter, with $\kappa \in [1, \ldots, K_\kappa]$ denoting the index of each parameter in the encoded speech frame and $K_\kappa = 36$, whilst τ denotes the time index referring to the current encoded frame index. Then, $\mathbf{v}_{\kappa,\mathcal{T}}$ is quantized and mapped to the bit sequence $\mathbf{c}_{1,\kappa\mathcal{T}} = [c(1)_{1,\kappa\mathcal{T}}\ c(2)_{1,\kappa\mathcal{T}}, \ldots, c(M)_{1,\kappa\mathcal{T}}]$, where M is the total number of bits assigned to the κth parameter. Then, the outer interleaver Π_{out} permutes the bits of the sequence $\mathbf{c}_1$ yielding $\mathbf{u}$ of Figure 11.32. Afterwards, the interleaved bit stream $\mathbf{u}$ is RSC encoded to produce the bit stream $\mathbf{c}_2$, which is then interleaved by the interleaver Π_{in} of Figure 11.32. After bit interleaving, the SP mapper maps blocks of B_{sp} channel-coded bits $\mathbf{b} = b_0, \ldots, b_{B_{\text{sp}}-1} \in \{0, 1\}$ to the L legitimate four-dimensional SP-modulated symbols $\mathbf{s}^l \in \mathbf{S}$. The SP-modulated symbols $\mathbf{x}$ are then transmitted using the DSTS scheme of Section 10.3.3.

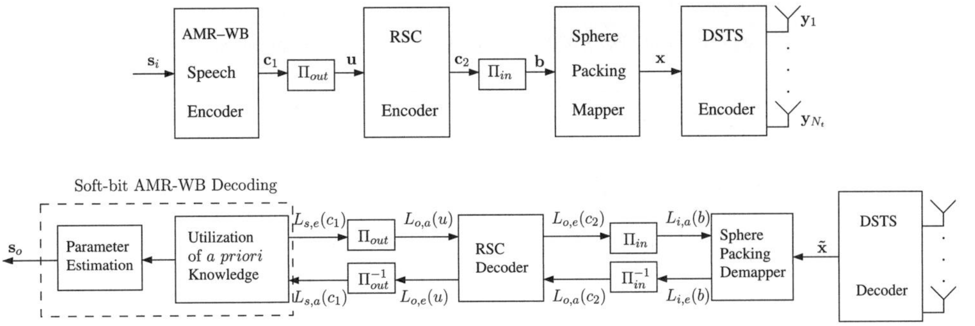

Figure 11.32: Block diagram of the DSTS-SP RSC AMR-WB scheme.

At the receiver side, as shown in Figure 11.32, the received complex-valued symbols are first decoded by the DSTS decoder in order to produce the received SP soft symbols $\tilde{\mathbf{x}}$. Then, iterative demapping/decoding is carried out between the SP demapper, the RSC decoder and the soft-input soft-output AMR-WB speech decoder, where extrinsic information is exchanged between the three constituent demapper/decoders. In the speech decoder, the residual redundancy[3] is exploited as *a priori* information in computing the extrinsic LLR values and estimating the speech parameters. During the last iteration, speech parameter estimation is carried out in the AMR-WB speech decoder in order to generate the transmitted source data estimate $\mathbf{s}_o$.

In the following, we characterize the attainable performance of the proposed DSTS-SP RSC AMR-WB scheme using both the BER and the Segmental Signal-to-Noise Ratio (SegSNR) [350] evaluated at the speech decoder's output as a function of the channel SNR. All simulation parameters are listed in Table 11.2. In our simulations we employed one inner iteration between the SP demapper and the RSC decoder followed by one outer iteration between the RSC decoder and the AMR-WB decoder. The system performance is compared versus a benchmark scheme where no outer iterations are carried out between the AMR-WB decoder and the RSC decoder.

Figure 11.33 depicts the BER performance of the DSTS-SP RSC AMR-WB scheme and that of its corresponding DSTS-SP RSC benchmark counterpart. It can be seen from Figure 11.33 that the DSTS-SP RSC AMR-WB scheme outperforms the DSTS-SP RSC benchmark scheme by about 1 dB at $\text{BER} = 10^{-4}$ after $I_{\text{sys}} = 4$ system iterations, where we define a system iteration cycle as having an inner iteration followed by a single outer iteration, which is referred to as I_{sys}. The AMR-WB-decoded scheme has a lower BER at its speech-decoded output than its benchmark dispensing with speech decoding, because the extrinsic information exchange between the AMR-WB decoder and the RSC decoder has the potential of improving the attainable BER. Note that the turbo effect of the BER seen in the figures in Section 11.2.4 is absent in Figure 11.33. This is due to the fact that the system employs a short interleaver depth of $D_{\text{int}} = 317$ bits, which corresponds to a delay of 20 ms in speech transmission. This result is similar to that using an interleaver depth of $D_{\text{int}} = 1000$ bits in Figure 11.19.

Figure 11.34 depicts the speech SegSNR performance of the proposed DSTS-SP RSC AMR-WB scheme together with that of the benchmark scheme. In this context the residual

[3]For a detailed discussion about the residual redundancy, please refer to [199].

Table 11.2: DSTS-SP RSC AMR-WB system parameters.

Source coding	AMR-WB
Bit rates (kbit/s)	15.85
Speech frame length (ms)	20
Sampling rate (kHz)	16
Channel coding	RSC code
Code rate	1/2
Code memory	$K = 7$
Code generator (G_r, G)	$(217, 110)_8$
Modulation scheme	SP ($L = 16$)
MIMO scheme	DSTS
Number of transmitters, N_t	2
Number of receivers, N_r	1
Spreading code	Walsh–Hadamard code
Spreading factor	8
Number of users	4
Channel	Correlated Rayleigh fading
Normalized Doppler frequency	0.01
System bandwidth efficiency	1 bit per channel use

redundancy inherent in the encoded source is exploited twice, first while computing the extrinsic information and secondly during the Markov-model-based parameter estimation [357]. It can be seen from Figure 11.34 that the exploitation of the residual redundancy inherent in the encoded source during the decoding process benefiting from zero-order Markov-model-based parameter estimation performs approximately 0.5 dB better in terms of the required channel E_b/N_0 value than its corresponding hard-speech-decoding-based counterpart when allowing a SegSNR degradation of 1.0 dB in comparison to the maximum attainable SegSNR maintained over perfectly error-free channels. In addition, iteratively exchanging the soft information amongst the three receiver components of the amalgamated DSTS-SP RSC AMR-WB scheme has resulted in a further E_b/N_0 gain of about 2.6 dB after $I_{\text{sys}} = 4$ system iterations, when tolerating a SegSNR degradation of 1 dB.

11.3 Iterative Detection of RSC-coded and Unity-rate Precoded Four-antenna-aided DSTS-SP System

As mentioned in Section 11.1, it was shown in [166] that a recursive inner code is needed in order to maximize the interleaver gain and to avoid the formation of a BER floor when employing iterative decoding. In [168], unity-rate inner codes were employed for designing low-complexity iterative-detection-aided schemes suitable for bandwidth- and power-limited systems having stringent BER requirements. In this section we consider an iteratively

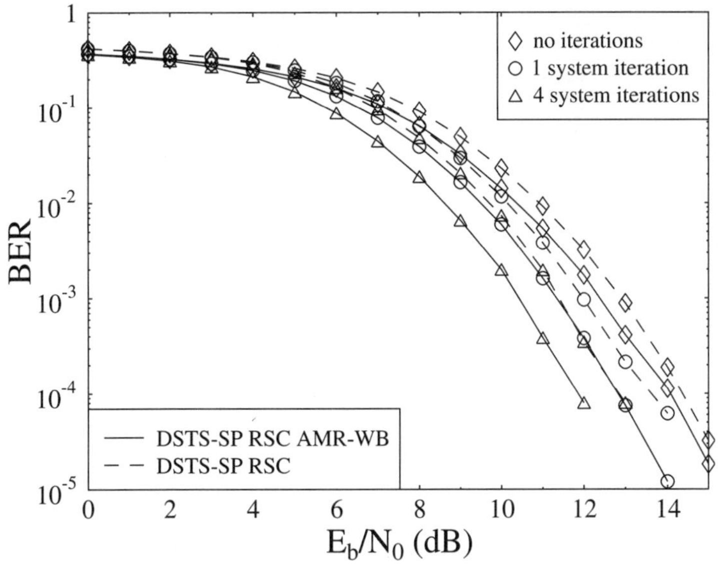

Figure 11.33: BER performance of the jointly optimized DSTS-SP RSC AMR-WB scheme of Figure 11.32 employing the system parameters of Table 11.2, when communicating over a temporally correlated narrowband Rayleigh fading channel.

detected RSC-coded and unity-rate precoded DSTS-SP scheme, where iterative detection is carried out between the outer RSC decoder and the inner URC decoder.

11.3.1 System Overview

The schematic of the proposed DSTS system is shown in Figure 11.35, where the transmitted source bits are convolutionally encoded and then interleaved by a random bit interleaver. A half-rate memory-two RSC code was employed having a generator polynomial with octal representation of $(G_r, G) = (7, 5)_8$. After channel interleaving the symbols are precoded by a URC encoder. The SP mapper of Figure 11.35 maps B_{sp} channel-coded and precoded bits $\mathbf{b} = b_0, \dots, b_{B_{\text{sp}}-1} \in \{0, 1\}$ to a SP symbol $\mathbf{s}^l \in S$, $l = 0, 1, \dots, L-1$, so that we have $\mathbf{s}^l = \text{map}_{\text{sp}}(\mathbf{b})$, where $B_{\text{sp}} = \log_2 L$ and L represents the number of modulated symbols in the sphere-packed signaling alphabet. Subsequently, each of the four components of a SP symbol is transmitted using DSTS via four transmit antennas, as detailed in Section 10.4.3.

In the following, we consider transmission over a temporally correlated narrowband Rayleigh fading channel associated with a normalized Doppler frequency of $f_D = f_d T_s = 0.01$. The complex AWGN of $n = n_I + jn_Q$ contaminates the received signal, where n_I and n_Q are two independent zero-mean Gaussian random variables having a variance of $\sigma^2_{n_I} = \sigma^2_{n_Q} = N_0/2$ per dimension and N_0 represents the double-sided noise power bandwidth density expressed in W Hz^{-1}.

As shown in Figure 11.35, the received complex-valued symbols are first decoded by the DSTS decoder to produce a received SP symbol $\tilde{\mathbf{x}}$, which is fed into the SP demapper. The output of the demapper represents the LLR metric $L_M(\mathbf{b})$ passed from the SP demapper to the URC decoder. As seen in Figure 11.35, the URC decoder processes the information forwarded by the demapper in conjunction with the *a priori* information $L_{i,a}(\mathbf{u}_2)$ passed from the RSC decoder, in order to generate the APP. The *a priori* LLR values of the URC decoder

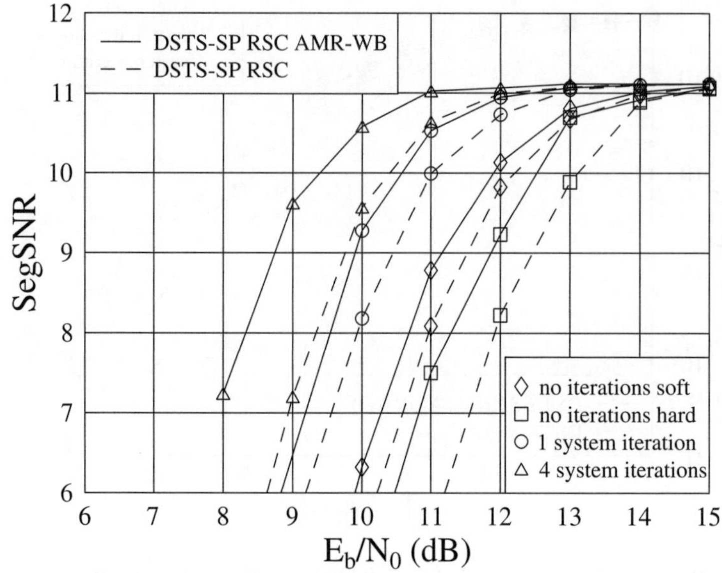

Figure 11.34: Average SegSNR performance of the jointly optimized DSTS-SP RSC AMR-WB scheme of Figure 11.32 employing the system parameters of Table 11.2 in comparison with the DSTS-SP RSC benchmark scheme, when communicating over a temporally correlated narrowband Rayleigh fading channel.

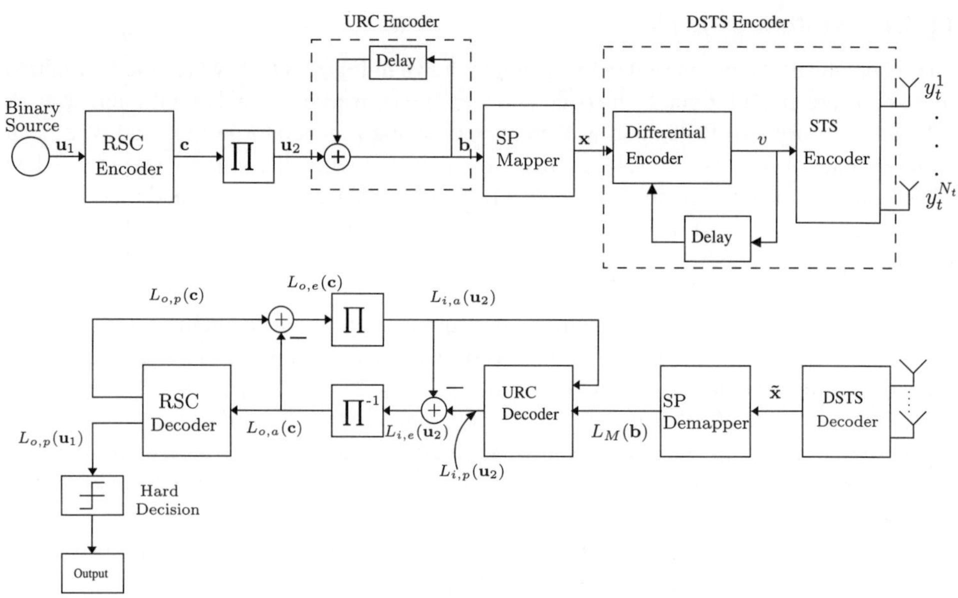

Figure 11.35: The iteratively detected RSC-coded and URC-precoded DSTS-SP system block diagram.

Table 11.3: RSC-coded and URC-precoded DSTS-SP system parameters.

SP modulation	$L = 16$
Number of transmit antennas	4
Number of receive antennas	1
Channel	Temporally correlated Rayleigh fading
Normalized Doppler frequency	0.01
Outer channel code	RSC (2, 1, 3)
Generator	$(G_r, G) = (7, 5)_8$
Precoder	URC
Generator	$G = (G_r, G) = (3, 2)_8$
Spreading code	Walsh–Hadamard code
Spreading factor	8
Number of users	4

are subtracted from the *a posteriori* LLR values for the sake of generating the extrinsic LLR values $L_{i,e}(\mathbf{u}_2)$ and then the LLRs $L_{i,e}(\mathbf{u}_2)$ are deinterleaved by a soft-bit deinterleaver, as seen in Figure 11.35. Next, the soft bits $L_{o,a}(\mathbf{c})$ are passed to the RSC decoder of Figure 11.35 in order to compute the *a posteriori* LLR values $L_{o,p}(\mathbf{c})$ for all of the channel-coded bits $\mathbf{c}$. During the last iteration, only the LLR values $L_{o,p}(\mathbf{u}_1)$ of the original uncoded systematic information bits are required, which are passed to the hard decision decoder of Figure 11.35 in order to determine the estimated transmitted source bits. As seen in Figure 11.35, the extrinsic information $L_{o,e}(\mathbf{c})$, is generated by subtracting the *a priori* information from the *a posteriori* information according to $(L_{o,p}(\mathbf{c}) - L_{o,a}(\mathbf{c}))$, which is then fed back to the URC decoder as the *a priori* information $L_{i,a}(\mathbf{u}_2)$ after appropriately reordering them using the interleaver of Figure 11.35. The URC decoder of Figure 11.35 exploits the *a priori* information for the sake of providing improved *a posteriori* LLR values, which are then passed to the half-rate RSC decoder and then back to the URC decoder for further iterations.

11.3.2 Results and Discussion

In this section, we consider a DSTS-SP scheme using four transmit antennas and a single receive antenna in order to demonstrate the performance improvements achieved by the proposed system of Figure 11.35. All simulation parameters are listed in Table 11.3.

Figure 11.36 depicts the EXIT chart for the iterative-detection-aided channel-coded DSTS-SP system employing $L = 16$ and Gray mapping in conjunction with the half-rate RSC outer code, URC inner code and the system parameters outlined in Table 11.3 for different E_b/N_0 values. The Gray mapping was used in this case because no iterations are invoked between the SP demapper and the decoders and thus, in this case, it is better to use Gray mapping that results in a higher initial mutual information and hence a higher starting point in the EXIT curve. Ideally, in order for the exchange of extrinsic information between the URC decoder and the RSC decoder to converge at a specific E_b/N_0 value, the EXIT curve of the URC decoder and that of the outer RSC decoder should only intersect at a point near the $I_{o,e}(\mathbf{c}) = 1.0$ line. If this condition is satisfied, then a so-called convergence tunnel [169,172] appears in the EXIT chart. It is plausible that the narrower the tunnel, the more iterations are required for reaching the $I_{o,e}(\mathbf{c}) = 1.0$ line. Observe from the figure that a convergence tunnel is formed at an E_b/N_0 of 6.5 dB. This implies that, according to the predictions of the

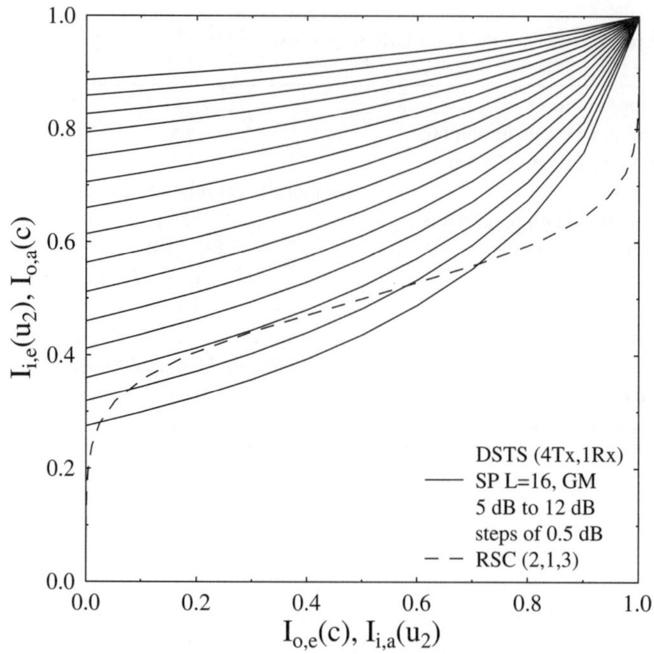

Figure 11.36: EXIT chart of a RSC-coded and URC-precoded DSTS-SP scheme employing Gray mapping in conjunction with $L = 16$, while using an interleaver depth of $D_{\text{int}} = 1\,000\,000$ bits and the system parameters outlined in Table 11.3.

EXIT chart seen in Figure 11.36, the iterative decoding process is expected to converge at $E_b/N_0 \in [6.0, 6.5]$ dB. The EXIT-chart-based convergence predictions can be verified by the actual iterative decoding trajectory of Figure 11.37, where the trajectory at $E_b/N_0 = 6.5$ dB is recorded while using an interleaver depth of $D_{\text{int}} = 1\,000\,000$ bits. The steps seen in the figure represent the actual EXIT between the URC decoder and the outer RSC channel decoder. Since a long interleaver is employed, the assumptions outlined at the beginning of Section 11.2.2 are justified and hence the EXIT-chart-based convergence prediction becomes accurate.

Furthermore, a comparison between the convergence behavior of the precoded and the non-precoded systems has been shown in Figure 11.38 for $E_b/N_0 = 6.0$ and 6.5 dB as well as an interleaver depth of $D_{\text{int}} = 1\,000\,000$ bits. The non-precoded system corresponds to the iterative-detection-aided system of Section 11.2. Observe in Figure 11.38 that the precoded system's EXIT curve emerges from a higher point than that of both the AGM-1- and the AGM-3-aided non-precoded systems. On the other hand, the precoded system's EXIT curve reaches the $(1.0, 1.0)$ point for all E_b/N_0, as compared with $I_{o,e}(\mathbf{c}) = 0.9$ for the AGM-1-based system and $I_{o,e}(\mathbf{c}) = 0.81$ for the AGM-3-based system for $E_b/N_0 = 6.5$ dB. Furthermore, note that the precoded system has a convergence tunnel at an E_b/N_0 value, which is only slightly higher than 6.0 dB as compared with having a convergence tunnel at $E_b/N_0 > 6.5$ dB for the AGM-1-aided non-precoded system and an $E_b/N_0 = 6.0$ dB for the AGM-3-aided non-precoded system. Hence, the precoded system converges at an E_b/N_0 value lower than that of the AGM-1-aided non-precoded system, which is close to that of the AGM-3-based non-precoded system. However, the precoded system reaches the point

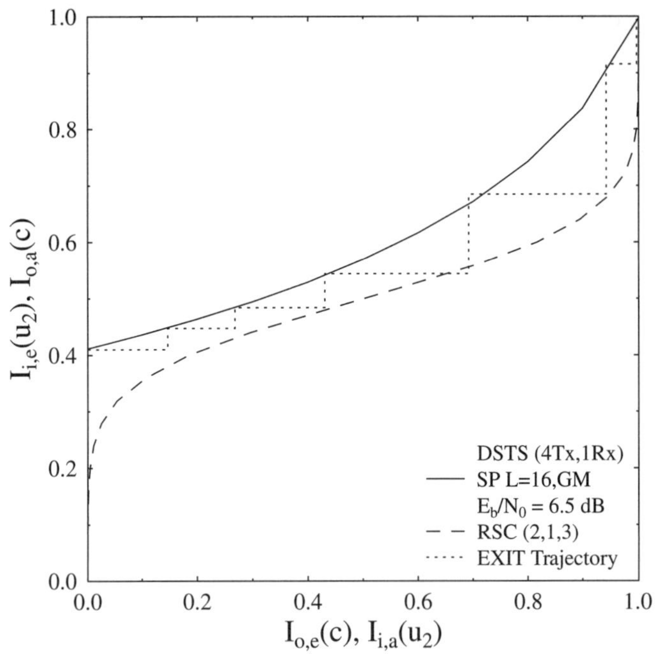

Figure 11.37: Decoding trajectory of the iteratively detected RSC-coded and URC-precoded DSTS-SP scheme employing Gray mapping in conjunction with $L = 16$ and the system parameters outlined in Table 11.3 while operating at $E_b/N_0 = 6.5$ dB.

of $(1.0, 1.0)$ in the EXIT curve resulting in an infinitesimally low BER. In contrast, the non-precoded systems do not reach the $(1.0, 1.0)$ point, as shown in Figure 11.38, which results in achieving only a modest BER performance associated with an error floor as E_b/N_0 increases.

Figure 11.39 compares the attainable performance of the RSC-coded four-antenna-aided DSTS-SP scheme employing Gray mapping of the bits to the SP symbol while using the URC precoder with the non-precoded system, and these are also contrasted with that of an identical bandwidth efficiency 0.5 bits per channel use uncoded DSTS-SP scheme using $L = 4$, when communicating over a temporally correlated Rayleigh fading channel and employing the system parameters of Table 11.3. In Figure 11.39, an interleaver depth of $D_{\text{int}} = 1\,000\,000$ bits was employed. Observe in the figure that the two Gray-mapping-based non-precoded DSTS-SP BER curves are exactly the same, regardless of whether no iterations or $I = 10$ decoding iterations were employed; this is similar to the situation discussed in Section 11.2.4. In contrast, the precoded DSTS-SP system employing Gray mapping achieved a substantial performance improvement in conjunction with iterative demapping and decoding. That is due to the fact that no iteration was employed between the demapper and the decoder, while the iterations were employed between the RSC decoder and the URC decoder. Explicitly, the figure demonstrates that a coding advantage of about 4.2 dB was achieved at a BER of 10^{-6} after $I = 10$ iterations by the RSC-coded and URC-precoded Gray-mapping-based DSTS-SP system over the non-precoded RSC-coded and Gray-mapping-based DSTS-SP scheme.

In addition, Figure 11.40 depicts our performance comparison between two iteratively detected DSTS schemes, namely that of the non-precoded system employing AGM-1, AGM-3 and AGM-8 constellation mapping, as well as the performance of the precoded systems in conjunction with Gray mapping, while employing an interleaver depth of $D_{\text{int}} = 1\,000\,000$

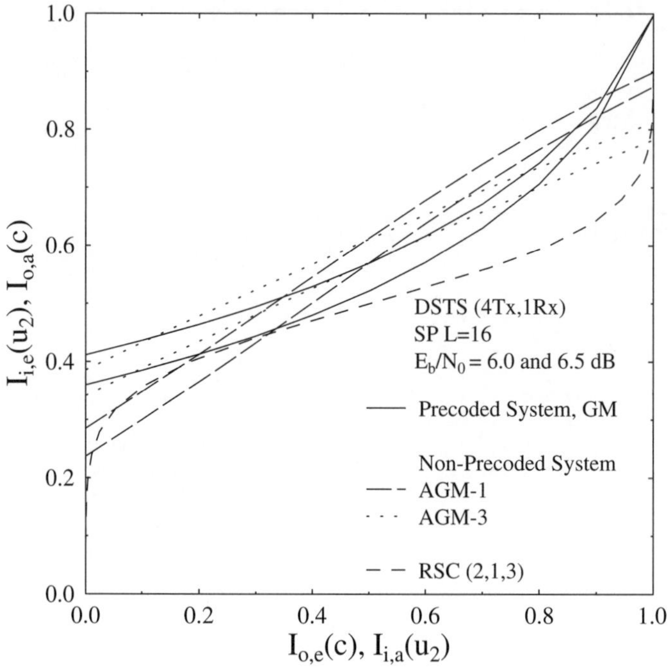

Figure 11.38: Comparison of the convergence behavior of the precoded and non-precoded DSTS-SP systems employing Gray mapping and AGM in conjunction with $L = 16$, based on their EXIT characteristics while using an interleaver depth of $D_{\mathrm{int}} = 1\,000\,000$ bits and the system parameters outlined in Table 11.3 and operating at E_b/N_0 of 6 and 6.5 dB.

bits, $I = 10$ iterations and using the system parameters of Table 11.3. The results of Figure 11.40 demonstrate that the non-precoded system employing AGM-3 has approached the system capacity quite closely. However, as the E_b/N_0 value increases we note in Figure 11.40 that the BER performance reaches a point where further BER improvements require a more substantial E_b/N_0 increase and this is justified by the EXIT chart predictions of Figure 11.38. Moreover, the non-precoded system employing AGM-1 converges at $E_b/N_0 = 7.0$ dB to a lower BER than the AGM-3-based system. Similarly to the AGM-3-based result, the BER performance of the AGM-1-aided non-precoded system converges to a low BER value at E_b/N_0 of 7.0 dB, after which the system's BER performance exhibits an error floor, as shown in Figure 11.40. Furthermore, Figure 11.40 demonstrates that the proposed precoded system converges at $E_b/N_0 = 6.1$ dB and exhibits an infinitesimally low BER, as suggested by the EXIT chart of Figure 11.37. More explicitly, the URC precoded DSTS-SP scheme of Figure 11.40 using $L = 16$ and Gray mapping exhibits an E_b/N_0 gain of 1.2 dB at a BER of 10^{-6} over the AGM-1-based non-precoded system and 3.1 dB at a BER of 10^{-6} over the AGM-3-based non-precoded system. Finally, the iteratively detected RSC-coded and URC-precoded DSTS-SP system employing $L = 16$ in conjunction with Gray mapping performs within 0.92 dB from the maximum achievable rate limit and within 1.3 dB from the system capacity limit at BER $= 10^{-6}$. In contrast, the non-precoded system employing AGM-1 performs within 2.12 dB from the maximum achievable rate limit at the same BER and the non-precoded system employing AGM-3 performs within 4.02 dB from the same limit at the same BER.

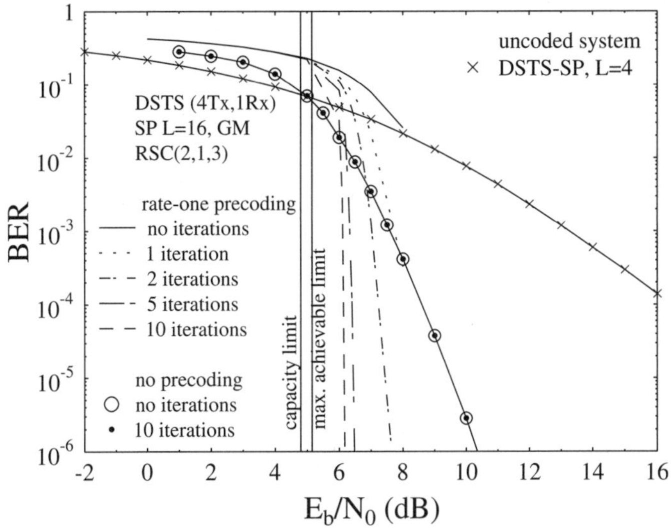

Figure 11.39: Performance comparison of URC-precoded and non-precoded Gray-mapping-based RSC-coded four-antenna-aided DSTS-SP schemes in conjunction with $L = 16$ against an identical bandwidth efficiency 0.5 bits per channel use uncoded DSTS-SP in conjunction with $L = 4$ when employing the system parameters outlined in Table 11.3 and using an interleaver depth of $D_{\text{int}} = 1\,000\,000$ bits for a variable number of iterations.

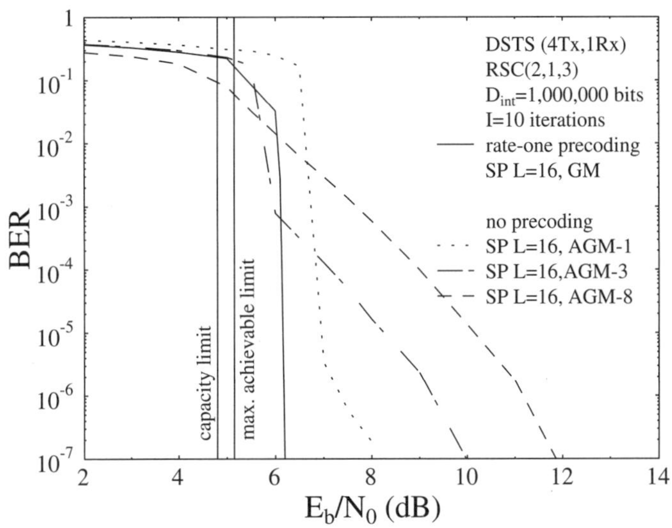

Figure 11.40: Performance comparison of different AGM-based half-rate RSC-coded and Gray-mapping-based half-rate RSC-coded and URC-precoded four-antenna-aided DSTS-SP schemes in conjunction with $L = 16$ while using an interleaver depth of $D_{\text{int}} = 1\,000\,000$ bits, $I = 10$ iterations and the system parameters outlined in Table 11.3.

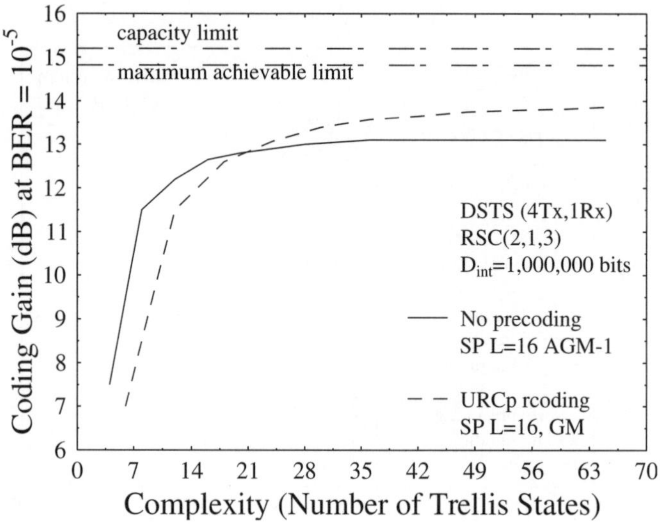

Figure 11.41: Coding gain comparison of the non-precoded and the URC-precoded iteratively detected RSC-coded four-antenna-aided DSTS-SP schemes plotted versus the number of trellis states at a BER of 10^{-5} when employing an interleaver depth of $D_{\text{int}} = 1\,000\,000$ bits.

Finally, Figure 11.41 compares the coding gain achieved for the iteratively detected RSC-coded four-antenna-aided DSTS-SP system when no precoding is employed and that when URC precoding is used. The figure plots the coding gain versus the number of trellis states, rather than versus the number of iterations as in Figure 11.25, since the two systems compared in Figure 11.41 have different numbers of trellis states, while that in Figure 11.25 employed the same trellis structure. Observe in Figure 11.41 that the non-precoded system has a lower complexity than the precoded system at a distance of 1.82 dB from the maximum achievable rate limit, where the non-precoded system approaches an infinitesimally low BER. The precoded system is capable of performing equally well in BER terms, while operating about 1 dB closer to the maximum achievable rate limit than the non-precoded system. However, this is achieved at the cost of almost doubling the complexity, as seen in Figure 11.41.

11.3.3 Application: Iteratively Detected Irregular Variable-length Coded and Unity-rate Precoded DSTS-SP Schemes

The schematic of the iteratively detected irregular variable-length coded (VLC) and unity-rate precoded DSTS-SP system is shown in Figure 11.42, where the VLC-encoded bits $\mathbf{c}_1$ are interleaved by a random bit interleaver and then the interleaved bit stream $\mathbf{u}_2$ is encoded by a URC encoder. After URC encoding, the DSTS-SP modulator maps B_{sp} coded bits $\mathbf{b} = b_0, \ldots, b_{B_{\text{sp}}-1} \in \{0, 1\}$ to a SP symbol $\mathbf{x}$ as discussed in Section 10.4.3. Subsequently, we have a set of SP symbols that can be transmitted with the aid of DSTS within two time slots using two transmit antennas. The schemes considered in this section differ from those in Section 11.3.1 in their choice of the outer source codec. Specifically, we consider an IR-VLC codec and an equivalent-rate regular VLC-based benchmark scheme. We refer to these two schemes as the IR-VLC and VLC DSTS-SP arrangements, as appropriate.

The schemes considered are designed for facilitating the near-capacity transmission of source symbol sequences over a correlated narrowband Rayleigh fading channel. We consider

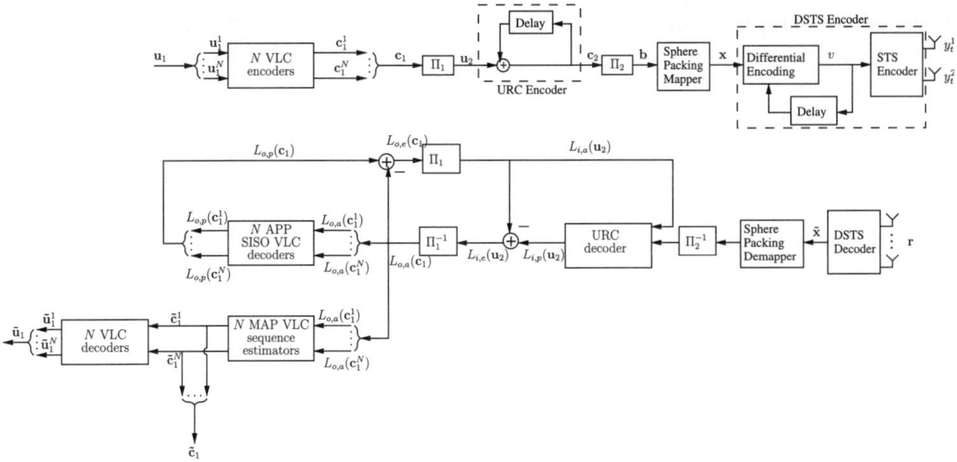

Figure 11.42: Schematic of the IR-VLC and VLC DSTS-SP schemes. In the IR-VLC DSTS-SP scheme we have $N = 15$ different irregularly encoded protection classes, whilst $N = 1$ in the VLC DSTS-SP scheme.

$K = 16$-ary source symbol values that have the probabilities of occurrence resulting from the Lloyd-Max (LM) quantization [358] of independent Gaussian distributed source samples. More explicitly, we consider the 4-bit LM quantization of a Gaussian source. Note that these occurrence probabilities vary by more than an order of magnitude between 0.0082 and 0.1019. These probabilities correspond to entropy or average information values between 3.29 and 6.93 bits, motivating the application of VLC and giving an overall source entropy of $E = 3.77$ bits per VLC symbol.

In the transmitter shown in Figure 11.42, the source symbol frame $\mathbf{u}_1$ comprises $J =$ 15 000 4-bit source symbols having the $K = 16$-ary values $\{u_{1,j}\}_{j=1}^{J} \in [1 \ldots K]$. These 4-bit source symbols are decomposed into N different protection classes $\{\mathbf{u}_1^n\}_{n=1}^{N}$, where we opted for $N = 15$ in the case of the IR-VLC DSTS-SP scheme and $N = 1$ in the case of the VLC DSTS-SP scheme. The number of symbols in the source symbol frame $\mathbf{u}_1$ that are decomposed into the source symbol frame component $\mathbf{u}_1^n$ is specified as J^n, where we have $J^1 = J$ in the case of the VLC DSTS-SP scheme. In contrast, in the case of the IR-VLC DSTS-SP scheme, the specific values of $\{J^n\}_{n=1}^{N}$ may be chosen in order to shape the EXIT curve of the IR-VLC codec, so that it does not cross the EXIT curve of the precoder.

Each of the N source symbol frame components $\{\mathbf{u}_1^n\}_{n=1}^{N}$ is VLC-encoded using the corresponding codebook from the set of N VLC codebooks $\{\mathbf{VLC}^n\}_{n=1}^{N}$, having a range of coding rates $\{R^n\}_{n=1}^{N} \in [0, 1]$. The specific source symbols having the value of $k \in [1 \ldots K]$ and encoded by the specific VLC codebook $\mathbf{VLC}^n$ are represented by the codeword $\mathbf{VLC}^{n,k}$, which has a length of $I^{n,k}$ bits. The J^n VLC codewords that represent the J^n source symbols in the source symbol frame component $\mathbf{u}_1^n$ are concatenated to provide the transmission frame component $\mathbf{c}_1^n = \{\mathbf{VLC}^{n,u_{1,j^n}^n}\}_{j^n=1}^{J^n}$.

Owing to the variable length of the VLC codewords, the number of bits comprised by each transmission frame component $\mathbf{c}_1^n$ will typically vary from frame to frame. In order to facilitate the VLC decoding of each transmission frame component $\mathbf{c}_1^n$, it is necessary to explicitly convey its length $I^n = \sum_{j^n=1}^{J^n} I^{n,u_{1,j^n}^n}$ to the receiver with the aid of side information. Furthermore, this highly error-sensitive side information must be reliably protected against transmission errors. This may be achieved using a low-rate block code or repetition code, for example. For the sake of avoiding obfuscating details, this is not explicitly shown in Figure 11.42.

In the transmitter of Figure 11.42, the N transmission frame components $\{\mathbf{c}_1^n\}_{n=1}^N$ are concatenated. As shown in Figure 11.42, the resultant transmission frame $\mathbf{c}_1$ has a length of $\sum_{n=1}^N I^n$ bits. Following interleaving Π_1, the transmission frame $\mathbf{u}_2$ is precoded [168] by the URC and then interleaved again before being SP modulated for transmission using DSTS.

In the receiver, the URC decoder and the VLC decoder iteratively exchange extrinsic information, as shown in Figure 11.42. In parallel to the formation of the bit-based transmission frame $\mathbf{c}_1$ from N components, the *a priori* LLRs $L_{o,a}(\mathbf{c}_1)$ are decomposed into N components, as shown in Figure 11.42. Each of the N VLC decoding processes is provided with the *a priori* LLR sub-frame $L_{o,a}(\mathbf{c}_1^n)$ and in response it generates the *a posteriori* LLR sub-frame $L_{o,p}(\mathbf{c}_1^n)$, $n \in [1 \ldots N]$. These *a posteriori* LLR sub-frames are concatenated in order to provide the *a posteriori* LLR frame $L_{o,p}(\mathbf{c}_1)$, as shown in Figure 11.42.

During the final decoding iteration, N bit-based MAP VLC sequence estimation processes are invoked instead of soft-in soft-out VLC decoding, as shown in Figure 11.42. In this case, each transmission frame component $\mathbf{c}_1^n$ is estimated from the corresponding *a priori* LLR frame component $L_{o,a}(\mathbf{c}_1^n)$. The resultant transmission frame component estimates $\tilde{\mathbf{c}}_1^n$ may be concatenated to provide the transmission frame estimate $\tilde{\mathbf{c}}_1$. In addition, the transmission frame component estimates $\tilde{\mathbf{c}}_1^n$ may be VLC decoded to provide the source symbol frame component estimates $\tilde{\mathbf{u}}_1^n$.

11.3.3.1 IR-VLC Design using EXIT Chart Analysis

The IR-VLC DSTS-SP scheme employs $N = 15$ component VLC codebooks $\{\mathbf{VLC}^n\}_{n=1}^N$ having approximately equally spaced coding rates in the range $[0.26, 0.95]$. In each case, we employ a Variable Length Error Correcting (VLEC) codebook [359] that is tailored to the source symbol values' probabilities of occurrence and having the maximum minimum free distance that can be achieved at the particular coding rate considered. In contrast, in the VLC DSTS-SP scheme, we employ just $N = 1$ VLC codebook, which is identical to the VLC codebook $\mathbf{VLC}^{10}$ of the IR-VLC DSTS-SP scheme, having a coding rate of $R = 0.5$, as shown in Figure 11.43. Note that this coding rate results in an average interleaver length of $J \cdot E/R = 113\,100$ bits and a bandwidth efficiency of 1 bit per channel use, if we ignore the negligible overhead of conveying the side information and assume ideal Nyquist filtering having a zero excess bandwidth. We note furthermore that for the proposed DSTS-SP system, this bandwidth efficiency is associated with an E_b/N_0 maximum achievable rate bound of 5.2 dB as described in Section 11.2.3.

Figure 11.43 shows the EXIT curves that characterize the VLC decoding of the VLC codebooks together with the precoder's EXIT curves recorded for E_b/N_0 values of 5.5 and 6.0 dB. Figure 11.43 also shows the EXIT curve of the IR-VLC scheme. This is obtained as the appropriately weighted superposition of the $N = 15$ unequal protection component VLC codebooks' EXIT curves, where the weight applied to the EXIT curve of the component VLC codebook $\mathbf{VLC^n}$ is related to the number of source symbols that are employed for encoding J^n (see [174]). Using the approach of [174], the values of $\{J^n\}_{n=1}^N$ given in Figure 11.43 were designed for ensuring that the IR-VLC coding rate matches that of our regular VLC scheme, namely VLC^{10}, and so that the IR-VLC EXIT curve does not cross the precoder's EXIT curve at an E_b/N_0 value of 5.5 dB. We note that only four out of the $N = 15$ VLC components were chosen by the proposed EXIT-chart matching procedure for encoding a non-zero number of source symbols. As shown in Figure 11.43, the presence of the resultant

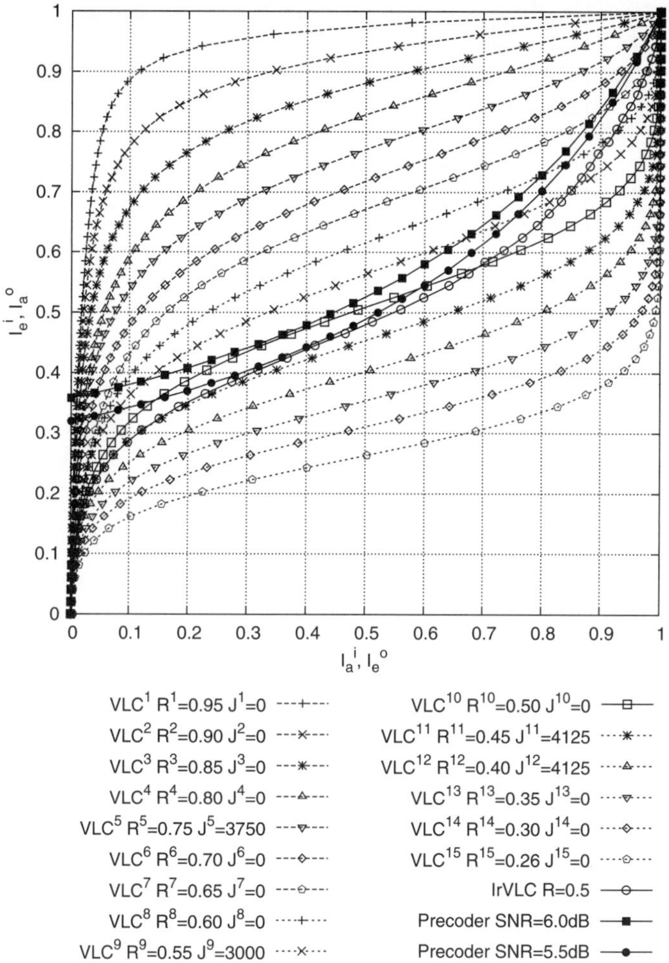

Figure 11.43: VLC EXIT curves and URC-precoded DSTS-SP EXIT curves.

open EXIT chart tunnel implies that an infinitesimally low SER may be achieved by the IR-VLC DSTS-SP scheme for E_b/N_0 values in excess of 5.5 dB, which is just 0.3 dB from the maximum achievable rate bound of 5.2 dB. In contrast, no open EXIT chart tunnel is maintained for E_b/N_0 values below 6.0 dB in the case of the VLC DSTS-SP benchmark scheme. This value of E_b/N_0 is 0.8 dB from the DSTS-SP capacity bound, representing a discrepancy that is 2.67 times that of the IR-VLC DSTS-SP scheme.

11.3.3.2 Performance Results

We consider a SP modulation scheme associated with $L = 16$ in conjunction with Gray mapping for assigning the source bits to the SP symbols, while employing a twin-antenna-aided DSTS system and a single receiver antenna in order to demonstrate the performance improvements achieved by the proposed system.

Figure 11.44 shows the EXIT curve of the IR-VLC scheme employed as well as the EXIT curves of the precoded DSTS-SP system together with the decoding trajectories at E_b/N_0

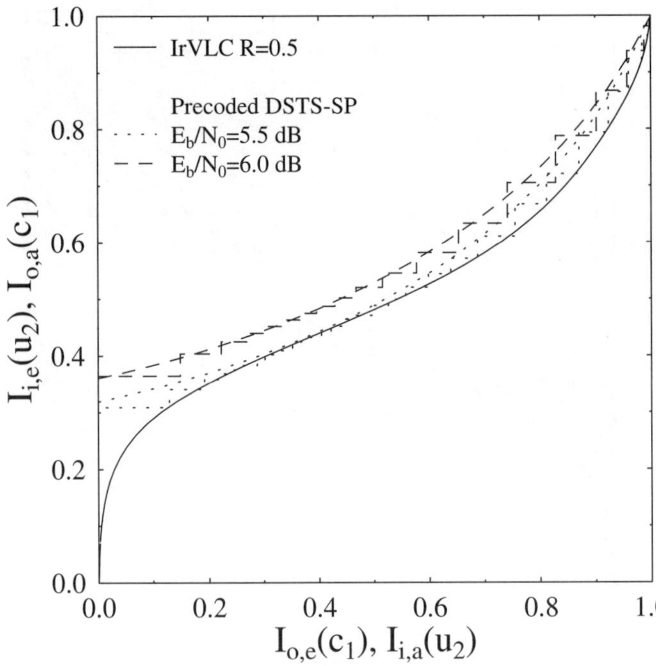

Figure 11.44: Decoding trajectory of the iteratively detected IR-VLC DSTS-SP scheme operating at $E_b/N_0 = 5.5$ and 6.0 dB.

values of both 5.5 and 6.0 dB. The system communicates over a correlated narrowband Rayleigh fading channel associated with a normalized Doppler frequency of $f_D = 0.01$ and employs a random interleaver having an average depth of $D_{\text{int}} = 113\,100$ bits. The decoding trajectory recorded for $E_b/N_0 = 6.0$ dB and shown in Figure 11.44 is the one we obtained from simulations, where the system did not converge below an E_b/N_0 value of 6.0 dB, although the EXIT curve of Figure 11.43 predicted an open tunnel at E_b/N_0 of 5.5 dB. The EXIT-chart predictions are accurate if we employ a specially designed interleaver that is capable of removing the correlation of the data imposed by both the correlated channel employed and the differential encoding that introduces more correlation to the data. Figure 11.44 also shows the decoding trajectory at $E_b/N_0 = 5.5$ dB. However, this trajectory was generated by simulating the effect of a random interleaver capable of eliminating the correlation, i.e. by generating uncorrelated LLRs at the input of the precoder's decoder. Therefore, the EXIT curve predictions can be fulfilled if we succeed in designing an interleaver having a reasonable length that can be used for eliminating the correlation imposed by the DSTS scheme and by the channel employed. The performance of the IR-VLC-aided system did not match with the EXIT-chart prediction of Figure 11.43, while the performance of the system employing the RSC as an outer code did match with the EXIT-chart prediction of the system. This might be due to the fact that the IR-VLC scheme is more sensitive to the correlation exhibited by the data than is the RSC code.

Figure 11.45 compares the attainable performance of the IR-VLC-aided and of the VLC-aided DSTS-SP systems, when communicating over a correlated narrowband Rayleigh fading channel with a normalized Doppler frequency of $f_D = 0.01$. The EXIT analysis of Figure 11.43 predicted a difference of 0.5 dB between the performance of the two systems.

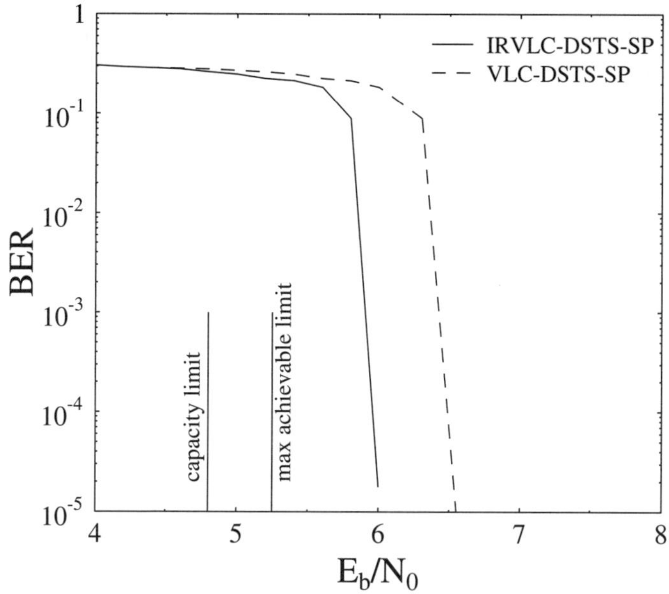

Figure 11.45: Performance comparison of the IR-VLC and VLC DSTS-SP systems while employing an average interleaver length of 113 100 bits and 40 iterations.

In Figure 11.45 we present the corresponding BER curves having the same E_b/N_0 difference as the EXIT curve prediction, but with a shift of 0.5 dB from the prediction. In other words, as mentioned in the previous paragraph, owing to the employment of an interleaver that is incapable of eliminating the effect of correlation, the IR-VLC-aided system converges at $E_b/N_0 = 6.0$ dB and the VLC-aided system at $E_b/N_0 = 6.5$ dB. The BER curves presented in Figure 11.44 were recorded after 40 decoding iterations between the VLC decoder and the precoder's decoder, as shown in Figure 11.42.

11.4 Chapter Conclusions

In this chapter, we have proposed a novel system that exploits the advantages of both iterative detection [348] and the DSTS schemes employing two and four transmit antennas [197]. The proposed DSTS scheme benefits from a substantial diversity gain without the need for any CSI. Moreover, our investigations demonstrated that significant performance improvements may be achieved when the AGM DSTS-SP is combined with outer channel decoding and iterative detection exchanging extrinsic information between the decoder and the demapper, as compared with the Gray-mapping-based systems. Subsequently, EXIT charts were used to search for bit-to-symbol mapping schemes that converge at lower E_b/N_0 values. Several DSTS-SP mapping schemes covering a wide range of extrinsic transfer characteristics were investigated. When using an appropriate bits-to-symbol mapping scheme and ten detection iterations, gains of about 19.5 dB were obtained by the convolutional-coded twin-antenna-aided DSTS-SP schemes over the identical-throughput uncoded DSTS-SP benchmark scheme discussed in Chapter 10. Furthermore, the AGM-1-based iteratively detected twin-antenna-aided DSTS-SP scheme is capable of performing within 2.3 dB of the maximum achievable rate limit obtained using EXIT charts at $\text{BER} = 10^{-5}$.

Table 11.4: Iteratively detected RSC-coded DSTS system coding gain and distance from maximum achievable rate limit at BER = 10^{-5} (GM: Gray mapping).

		Coding gain (dB)	Distance from maximum achievable rate limit (dB)
DSTS (2Tx,1Rx) No URC precoding	SP $L = 16$, GM	14.9	6.9
	SP $L = 16$, AGM-1	19.5	2.3
	SP $L = 16$, AGM-3	17.75	4.05
	SP $L = 16$, GM-8	15.9	5.9
	QPSK, AGM	16.1	5.7
DSTS (4Tx,1Rx) No URC precoding	SP $L = 16$, GM	9.5	4.32
	SP $L = 16$, AGM-1	12	1.82
	SP $L = 16$, AGM-3	10.9	2.92
	SP $L = 16$, GM-8	8.9	4.92
	QPSK, AGM	9.2	4.62
DSTS (4Tx,1Rx) URC precoding	SP $L = 16$, GM	12.9	0.92

Table 11.5: Iteratively detected RSC-coded DSTS system coding gain and distance from maximum achievable rate limit at BER = 10^{-6} (GM: Gray mapping).

		Coding gain (dB)	Distance from maximum achievable rate limit (dB)
DSTS (2Tx,1Rx) No URC precoding	SP $L = 16$, GM	17.8	8
	SP $L = 16$, AGM-1	22.5	3.3
	SP $L = 16$, AGM-3	20.5	5.3
	SP $L = 16$, GM-8	18.4	7.4
	QPSK, AGM	19	6.8
DSTS (4Tx,1Rx) No URC precoding	SP $L = 16$, GM	13.7	5.12
	SP $L = 16$, AGM-1	16.7	2.12
	SP $L = 16$, AGM-3	14.8	4.02
	SP $L = 16$, GM-8	12.9	5.92
	QPSK, AGM	13.5	5.32
DSTS (4Tx,1Rx) URC precoding	SP $L = 16$, GM	17.9	0.92

In addition, the chapter characterized the benefits of precoding, when concatenated with the outer channel code, suggesting that an E_b/N_0 gain of at least 1.2 dB can be obtained over the uncoded system at a BER of 10^{-6}, depending on the mapping scheme used. Explicitly, the four-antenna-aided DSTS-SP system employing no URC precoding attains a coding gain of 12 dB at a BER of 10^{-5} and performs within 1.82 dB of the maximum achievable rate limit. In contrast, the URC-precoded system outperforms its non-precoded counterpart and operates within 0.92 dB of the maximum achievable rate limit obtained using EXIT charts.

11.5 Chapter Summary

In this chapter, two realizations of a novel iterative-detection-aided DSTS-SP scheme have been presented, namely an iteratively detected RSC-coded DSTS-SP scheme as well as an iteratively detected RSC-coded and URC-precoded DSTS-SP arrangement. The iteratively detected RSC-coded DSTS-SP scheme was described in Section 11.2. In Section 11.2.1, we showed how the DSTS-SP demapper was modified for exploiting the *a priori* knowledge provided by the channel decoder, which is essential for the employment of iterative detection.

The concept of the EXIT chart was introduced in Section 11.2.2 as a tool designed for studying iterative-detection-aided schemes. We proposed nine different AGM schemes in Figure 11.3 that were specifically selected from all of the possible mapping schemes for $L = 16$, in order to create the different extrinsic information transfer characteristics associated with different bit-to-SP-symbol mapping schemes. The Gray mapping as well as the various AGM mapping schemes considered in this chapter are detailed in Appendix H. In Section 11.2.2.2, we explained the procedure of computing the EXIT characteristics of an outer decoder in a serially concatenated scheme. Then, we proposed a novel technique for computing the maximum achievable bandwidth efficiency of the system based on the EXIT charts in Section 11.2.3, followed by a discussion of the system's performance. Section 11.2.5 presented an application of the iteratively detected RSC-coded DSTS-SP system, where an AMR-WB source codec was employed by the system in order to demonstrate the attainable performance improvements.

In Section 11.3 we proposed an iteratively detected RSC-coded and URC-precoded DSTS-SP scheme that is capable of eliminating the error floor exhibited by the previous system, which hence performed closer to the system's achievable capacity. In Section 11.3.1 we presented an overview of the system operation, followed by a discussion of the results in Section 11.3.2. In Section 11.3.3 we presented an application of the proposed URC-precoded DSTS-SP system, while employing IR-VLCs as our outer code for the sake of achieving a near-capacity performance.

Finally, Tables 11.4 and 11.5 present the coding gains as well as the distance from the maximum achievable rate limit for the iteratively detected RSC-coded DSTS system, while employing SP as well as QPSK modulation schemes. The tables present the results for both the two- and four-antenna-aided DSTS schemes, when both systems optionally employ URC precoding.

In the next chapter we consider the design of adaptive iteratively detected DSTS-SP schemes designed for attaining the highest possible system throughput, while maintaining a specific quality of service exemplified in terms of the attainable BER.

11.5 Chapter summary

Chapter 12

Adaptive DSTS-assisted Iteratively Detected SP Modulation

12.1 Introduction

Mobile radio signals are subject to propagation path loss as well as small-scale fading and large-scale fading. Owing to the nature of the fading channel, transmission errors occur in bursts, when the channel exhibits deep fades or when there is a sudden surge of multiple access interference or ISI [51]. In mobile communication systems, power control techniques [360] are used to mitigate the effects of path loss and slow fading. However, in order to counteract the problem of fast fading and co-channel interference, agile power control algorithms are required [51, 361]. On the other hand, adaptive-rate transmission [362, 363] can be used to overcome these problems due to the time-variant fluctuations of the channel's quality or for mitigating the effects of shadow fading, when, for example, all of the transmit diversity antennas experience correlated fading, as exemplified by the effects of large-bodied vehicles. In adaptive-rate transmission the information rate is varied according to the channel's quality rather than according to the users' requirements.

In recent years, the concept of intelligent multi-mode multimedia transceivers has emerged in the context of wireless systems [362]. The fundamental limitation of wireless systems is constituted by the TD and FD channel fading, that is exemplified in terms of the SINR fluctuations experienced by wireless modems [362, 364]. Furthermore, the continued increase in demand for all types of wireless services including voice, data and multimedia increases the need for higher data rates. Therefore, no fixed mode transceiver may be expected to provide an attractive performance at a reasonable complexity and interleaver delay. Hence, advanced adaptive MIMO techniques, coded modulation as well as adaptive modulation and coding arrangements have to be invoked, which are capable of near-instantaneous HSDPA-style reconfiguration.

Adaptive modulation and coding techniques that track the time-varying characteristics of wireless channels can be used to significantly increase the data rate, reliability and spectrum efficiency of wireless communication systems. In recent years various adaptive coding and modulation arrangements have been proposed [51, 365, 366]. A whole range of

Near-Capacity Multi-Functional MIMO Systems L. Hanzo, O. R. Alamri, M. El-Hajjar and N. Wu

different transmission parameters can be adapted, including the transmission power [367], the system bandwidth [368], the modulation scheme [365, 369] and the SF of DS-CDMA systems [370] in addition to the code rate or interleaver length of channel-coded systems. The fundamental goal of near-instantaneous adaptation is to ensure that the most efficient mode is used in the face of rapidly-fluctuating time-variant channel conditions based on appropriate activation criteria. As a benefit, near-instantaneous adaptive systems are capable of achieving a higher effective throughput compared with their non-adaptive counterparts. When the channel quality is low, a lower information rate is chosen in order to reduce the number of errors. Conversely, when the channel quality is high, a higher information rate is used to increase the throughput of the system [362].

In [371], various multi-rate systems were compared, including variable SF-based, multi-code and adaptive-modulation-aided schemes. According to the multi-code philosophy, the SF is kept constant for all users, but multiple spreading codes are assigned to users having higher bit-rate requirements. Multiple data rates can also be supported by a variable-SF scheme, where the chip rate is kept fixed but the data rate is varied by varying the SF of different users. Hence, the lower the SF, the higher the data rate. Moreover, in [371] multilevel modulation schemes were investigated, where higher-rate users were assigned higher-order modulation modes transmitting several BPS. However, it was concluded that the performance experienced by users requiring higher rates was significantly worse than that experienced by those requiring lower rates. Furthermore, the employment of L-ary orthogonal modulation supporting variable rate transmission was investigated in [372].

The transmission rate of each user can be adapted according to the channel quality in order to mitigate the effects of channel quality fluctuations. The performance of two different methods of combating the channel variations was analyzed in [373]. More specifically, these two methods were based on either the adaptation of the transmission power in order to compensate for the channel quality variations or the switching of the information rate in order to suit the prevalent channel conditions. In [374], the authors investigated an adaptive packet transmission-based CDMA scheme, where the transmission rate is modified by varying both the channel code rate and the processing gain of the CDMA user, employing the carrier-to-interference-plus-noise ratio as the channel quality parameter.

In this chapter, we also propose to vary the information rate in accordance with the channel quality. The instantaneous channel quality can be estimated at the receiver and the chosen information rate can then be communicated to the transmitter via explicit signaling in a closed-loop scheme. Conversely, in an open-loop scheme, by assuming reciprocity in the UL and the downlink (DL) channel of Time Division Duplex (TDD) systems, the information rate for the DL transmission is chosen according to the channel quality estimate related to the UL and vice versa. Explicitly, in this chapter we investigate a novel adaptive DSTS-SP-aided technique for supporting a wide range of bit rates, where the transmission of data from the four antennas is adapted by activating two different transmission schemes according to the near-instantaneous channel SNR conditions. Moreover, the transmission bit rate is adjusted with the aid of a Variable Spreading Factor (VSF). More explicitly, given a fixed bandwidth and a fixed chip rate, the system enables a user to benefit from a higher bit rate while using a lower SF when the instantaneous SNR is sufficiently high. Furthermore, an iteratively detected variable-rate RSC code is employed for further enhancing the system's attainable BER performance, where the code rate may be increased for the sake of increasing the achievable system throughput as the channel quality improves. As a further benefit, the proposed system exploits the implementation advantages of low-complexity differential encoding/decoding, although this is achieved at the cost of the typical SNR degradation of differential encoding. Therefore, the achievable integrity and bit-rate enhancements of the system are determined by the following factors.

- The specific transmission configuration used for transmitting data from the four antennas.
- The SF used.
- The RSC encoder's code rate.

This chapter is organized as follows. In Section 12.2, a brief system overview is presented. In Section 12.3 the adaptive DSTS-SP scheme varying the transmission configuration of the four antennas is detailed, followed by Section 12.4, which justifies the purpose of adapting the SF used. The iteratively detected variable code rate DSTS-SP system is discussed in Section 12.5. In Section 12.6, we quantify the performance of the proposed system, and finally we conclude with our findings in Section 12.7.

12.2 System Overview

The adaptive DSTS system considered in this chapter employs four transmit antennas and a single receive antenna. A block diagram of the proposed system is shown in Figure 12.1, where four-dimensional SP modulation and real-valued orthogonal spreading were employed. A suitable transmission mode is selected according to the near-instantaneous channel conditions, which is quantified by the SNR in our case. A low-complexity technique of determining the channel conditions to be expected at the receiver in the context of TDD is that of exploiting the correlation between the fading envelope of the UL and the DL, since the UL and the DL slots are transmitted at the same frequency and hence are likely to fade coincidentally, unless frequency-selective fading is encountered. Therefore, when transmitting a frame, transmitter A estimates the SNR of receiver B at the other end of the link based on the SNR estimate of receiver A and selects the most appropriate transmission mode accordingly. The proposed adaptive transceiver assumes the availability of a reliable modem-mode signaling link between the transmitter and the receiver, such as the control channel of the HSDPA system.

As shown in Figure 12.1, the transmitted source bits have two different paths to follow. In the upper path shown in the figure, the transmitted source bits are convolutionally encoded and then interleaved by a random bit interleaver. A variable-rate RSC code is employed, where the code rate is varied between $R = \frac{1}{6}$ and $\frac{1}{2}$ depending on the near-instantaneous channel conditions. Moreover, when the channel SNR is sufficiently high for the target system performance to be met without channel coding, the transmitted source bits are not channel coded at all. After deciding on whether to invoke channel coding or not, the SP mapper maps B_{sp} bits $\mathbf{b} = b_0, \ldots, b_{B_{sp}-1} \in \{0, 1\}$ to a SP symbol $\mathbf{s}^l \in S$, $l = 0, 1, \ldots, L-1$, so that we have $\mathbf{s}^l = \text{map}_{\text{sp}}(\mathbf{b})$, where $B_{\text{sp}} = \log_2 L$ and L represents the number of modulated symbols in the SP signaling alphabet, as described in Chapter 10. Subsequently, each of the four components of a SP symbol is then transmitted using DSTS via four transmit antennas in two or four consecutive time slots, depending on the channel conditions, as detailed in Section 12.3. Furthermore, a VSF is employed by each user so that the system's effective bit rate can be enhanced along with any improvement in the channel conditions. Therefore, creating signaling modes that enable reliable communication even in poor channel conditions renders the system robust. In contrast, under good channel conditions the spectrally efficient modes are activated, in order to increase the effective throughput.

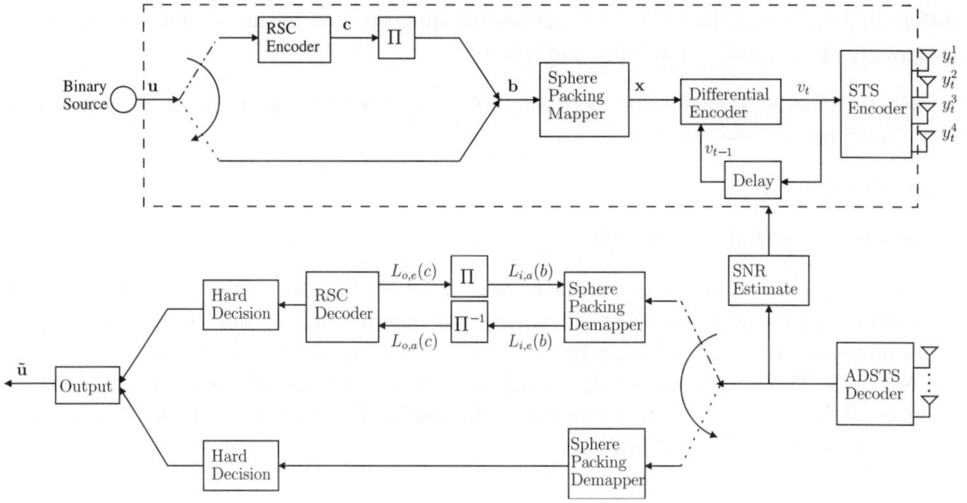

Figure 12.1: The proposed adaptive DSTS-assisted iteratively detected SP-aided system model.

In this chapter, we consider transmissions over a correlated narrowband Rayleigh fading channel, associated with a normalized Doppler frequency of $f_D = f_d T_s = 0.01$, where f_d is the Doppler frequency and T_s is the symbol duration. The complex AWGN of $n = n_I + jn_Q$ contaminates the received signal, where n_I and n_Q are two independent zero-mean Gaussian random variables having a variance of $\sigma_n^2 = \sigma_{n_I}^2 = \sigma_{n_Q}^2 = N_0/2$ per dimension with $N_0/2$ representing the noise power spectral density expressed in W Hz^{-1}.

At the receiver side of Figure 12.1, the DSTS decoder decodes the received signal according to the received modem-mode side information or in other words according to the information fed back from the receiver to the transmitter before the transmission of the specific frame. More explicitly, the receiver has to decide whether the data was channel coded and what code rate was used for encoding. If no channel coding was employed, then the received signal will be directly demodulated by the SP demapper of Figure 12.1. In contrast, if channel coding was employed, then the decoder has to decide which code-rate was employed. As shown in Figure 12.1, the received complex-valued symbols are demapped to their LLR representation for each of the B_{sp} channel-coded bits per SP symbol. The *a priori* LLR values of the demodulator are subtracted from the *a posteriori* LLR values for the sake of generating the extrinsic LLR values $L_{i,e}(\mathbf{b})$ and then the extrinsic LLRs are deinterleaved by a soft-bit deinterleaver, again as seen in Figure 12.1. Next, the soft bits $L_{o,a}(\mathbf{c})$ are passed to the convolutional decoder in order to compute the *a posteriori* LLR values for all of the channel-coded bits. During the last iteration, only the LLR values of the original uncoded systematic information bits are required, which are passed to the hard decision decoder of Figure 12.1 in order to determine the transmitted source bits.

12.3 Adaptive DSTS-assisted SP Modulation

The main philosophy of the proposed adaptive DSTS-SP scheme is to maximize the system's throughput while maintaining the required QoS, namely the target BER performance. The scheme considered in this chapter consists of $N_t = 4$ transmit antennas and $N_r = 1$ receive

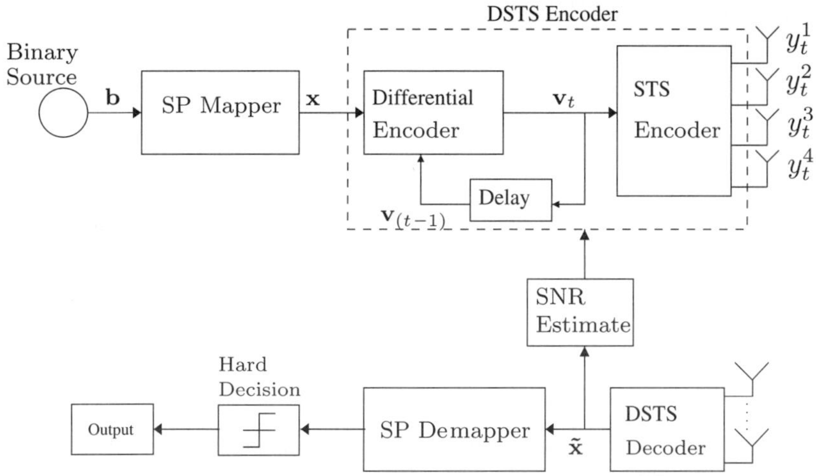

Figure 12.2: The proposed adaptive DSTS-assisted SP-aided system model.

antenna, although its extension to $N_r > 1$ antennas is straightforward. The transmitter schematic of the kth user and the receiver schematic of the reference user are shown in Figure 12.2, where SP modulation and real-valued spreading are employed.

The proposed adaptive system switches between two DSTS schemes depending on the channel's SNR. When a low channel quality is encountered and thus the short-term BER is higher than the required target BER, a high-diversity four-antenna-aided scheme is employed by the transceiver. However, as the channel quality improves and the system's BER performance becomes better than the target BER, then a lower-diversity higher-throughput twin-layer four-antenna-aided DSTS-SP configuration is used.

In the four-antenna-aided DSTS encoder, the data is serial-to-parallel converted to four substreams. The new bit duration of each parallel substream, which is referred to as the symbol duration, becomes $T_s = 4T_b$ as illustrated in Section 10.4 and in [13, 51]. The four-antenna-aided DSTS transmitter conveys one SP symbol in four time slots. Therefore, the DSTS-SP signaling rate becomes $\frac{1}{4}$ and the effective throughput quantified in terms of the symbol rate is related to the BPS B_{sp} as $B_{\text{sp}}/4$. However, this fixed-mode four-antenna-aided DSTS-SP system is unable to maximize the throughput as a function of the channel SNR quality. For example, at a high SNR value, the system provides a lower BER than the target BER value, which imposes a low effective throughput. Thus, an intelligent high-efficiency DSTS system must be capable of monitoring the near-instantaneous channel quality and of adapting the DSTS scheme's mode of operation. When the channel SNR encountered is low and hence the resultant BER is higher than the target BER, a low-throughput DSTS transmitter mode is activated, which exhibits a high diversity gain. In contrast, when the channel quality is high, and hence the resultant BER is lower than the target BER, then a high-throughput DSTS-assisted transmitter mode having a lower transmit diversity gain is activated.

12.3.1 Single-layer Four-antenna-aided DSTS-SP System

The high-diversity four-antenna-aided DSTS-SP mode of operation acts as described in Section 10.4.3. At time instant $t = 0$, the arbitrary reference symbols v_0^1, v_0^2, v_0^3 and v_0^4

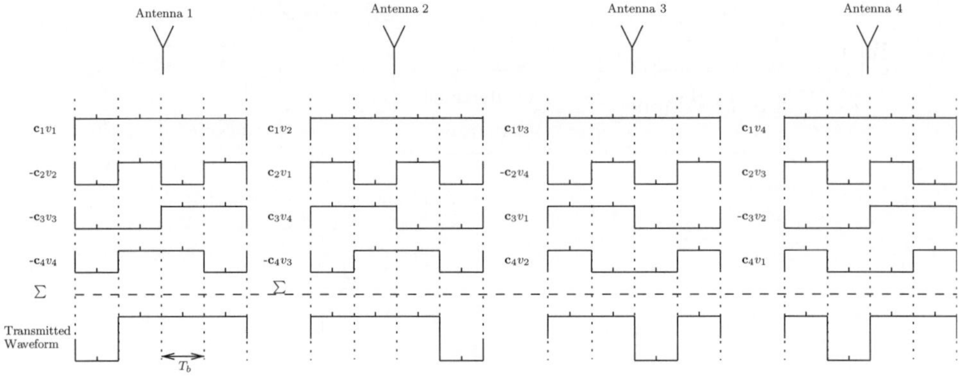

Figure 12.3: Illustration of STS using four transmit antennas transmitting four bits within $4T_b$ duration. Here $v_1 = v_2 = v_3 = v_4 = 1$ were assumed and $\bar{\mathbf{c}}_1 = [+1+1+1+1+1+1+1+1]$, $\bar{\mathbf{c}}_2 = [+1+1-1-1+1+1-1-1]$, $\bar{\mathbf{c}}_3 = [+1+1+1+1-1-1-1-1]$ and $\bar{\mathbf{c}}_4 = [+1+1-1-1-1-1+1+1]$.

are transmitted from the four antennas. At time instants $t \geq 1$, a block of B_{sp} bits arrives at the SP mapper of Figure 12.2, where the B_{sp} bits are mapped to a real-valued four-dimensional SP symbol selected from the set $\mathbf{S} = \{\mathbf{s}^l = [a_{l,1}, a_{l,2}, a_{l,3}, a_{l,4}] \in \mathbb{R}^4 : 0 \leq l \leq L-1\}$, where L is the number of legitimate SP constellation points having a total energy of $E \triangleq \sum_{l=0}^{L-1}(|a_{l,1}|^2 + |a_{l,2}|^2 + |a_{l,3}|^2 + |a_{l,4}|^2)$.

The differentially encoded symbols are then spread with the aid of the spreading codes $\bar{\mathbf{c}}_1$, $\bar{\mathbf{c}}_2$, $\bar{\mathbf{c}}_3$ and $\bar{\mathbf{c}}_4$, which are generated from the same user-specific spreading code $\bar{\mathbf{c}}$ by ensuring that they are orthogonal using the simple code-concatenation rule of Walsh–Hadamard codes, yielding longer codes and hence a proportionately reduced per-antenna throughput.

The differentially encoded data is then divided into four quarter-rate substreams and the four consecutive differentially encoded symbols are then spread to the four transmit antennas using the mapping of

$$\mathbf{y}_t^1 = \frac{1}{\sqrt{4}}(\bar{\mathbf{c}}_1 \cdot v_t^1 - \bar{\mathbf{c}}_2 \cdot v_t^2 - \bar{\mathbf{c}}_3 \cdot v_t^3 - \bar{\mathbf{c}}_4 \cdot v_t^4), \tag{12.1}$$

$$\mathbf{y}_t^2 = \frac{1}{\sqrt{4}}(\bar{\mathbf{c}}_1 \cdot v_t^2 + \bar{\mathbf{c}}_2 \cdot v_t^1 + \bar{\mathbf{c}}_3 \cdot v_t^4 - \bar{\mathbf{c}}_4 \cdot v_t^3), \tag{12.2}$$

$$\mathbf{y}_t^3 = \frac{1}{\sqrt{4}}(\bar{\mathbf{c}}_1 \cdot v_t^3 - \bar{\mathbf{c}}_2 \cdot v_t^4 + \bar{\mathbf{c}}_3 \cdot v_t^1 + \bar{\mathbf{c}}_4 \cdot v_t^2), \tag{12.3}$$

$$\mathbf{y}_t^4 = \frac{1}{\sqrt{4}}(\bar{\mathbf{c}}_1 \cdot v_t^4 + \bar{\mathbf{c}}_2 \cdot v_t^3 - \bar{\mathbf{c}}_3 \cdot v_t^2 + \bar{\mathbf{c}}_4 \cdot v_t^1), \tag{12.4}$$

where $\bar{\mathbf{c}}_1$, $\bar{\mathbf{c}}_2$, $\bar{\mathbf{c}}_3$ and $\bar{\mathbf{c}}_4$ are four STS-related orthogonal codes having a period of $4T_b$. The transmitted signal's waveforms corresponding to the four transmission antennas are shown in Figure 12.3, which shows that four real-valued symbols are jointly transmitted within a time duration of $4T_b$ with the aid of four transmit antennas. Hence, this scheme transmits one SP symbol during the interval of $4T_b$ and hence the effective transmission rate becomes $R_b = 1 \times 1/(4T_b) = 1/(4T_b)$.

12.3.2 Twin-layer Four-antenna-aided DSTS-SP System

Again, in its most robust but lowest-throughput mode, the above scheme transmits one SP symbol in $4T_b$ duration and hence the effective transmission rate becomes $1/(4T_b)$. In contrast, when the channel quality improves, the transmitter divides the four antennas into two groups of two antennas each for the sake of increasing the effective throughput by creating two independent second-order DSTS-aided subchannels. In this case, at time instants $t \geq 1$, a block of $2B_{\text{sp}}$ bits arrives at the SP mapper of Figure 12.2, where each of the B_{sp} bits is mapped to a four-dimensional SP symbol selected from the set $\mathbf{S} = \{\mathbf{s}^l = [a_{l,1}, a_{l,2}, a_{l,3}, a_{l,4}] \in \mathbb{R}^4 : 0 \leq l \leq L-1\}$. The four components of the four-dimensional SP symbol are combined according to Equation (10.21), where we have $\{x_t^1, x_t^2\} = \{a_{l,1} + ja_{l,2}, a_{l,3} + ja_{l,4}\}$ and $\{x_t^3, x_t^4\} = \{a_{l,5} + ja_{l,6}, a_{l,7} + ja_{l,8}\}$.

In this mode, the differential encoding is carried out as follows:

$$\begin{aligned}
v_t^1 &= \frac{(x_t^1 \cdot v_{t-1}^1 + x_t^2 \cdot v_{t-1}^{2*})}{\sqrt{|v_{t-1}^1|^2 + |v_{t-1}^2|^2}}, \\
v_t^2 &= \frac{(x_t^1 \cdot v_{t-1}^2 - x_t^2 \cdot v_{t-1}^{1*})}{\sqrt{|v_{t-1}^1|^2 + |v_{t-1}^2|^2}}, \\
v_t^3 &= \frac{(x_t^3 \cdot v_{t-1}^3 + x_t^4 \cdot v_{t-1}^{4*})}{\sqrt{|v_{t-1}^3|^2 + |v_{t-1}^4|^2}}, \\
v_t^4 &= \frac{(x_t^3 \cdot v_{t-1}^4 - x_t^4 \cdot v_{t-1}^{3*})}{\sqrt{|v_{t-1}^3|^2 + |v_{t-1}^4|^2}}.
\end{aligned} \tag{12.5}$$

The differentially encoded symbols are then spread with the aid of the spreading codes $\bar{\mathbf{c}}_1$, $\bar{\mathbf{c}}_2$, $\bar{\mathbf{c}}_3$ and $\bar{\mathbf{c}}_4$ to the transmit antennas. In this case, each user is assigned two spreading codes $\bar{\mathbf{c}}$, $\bar{\mathbf{c}'}$. The spreading codes used in this second-order diversity scenario are generated as follows:

$$\begin{aligned}
\bar{\mathbf{c}}_1^{\mathrm{T}} &= [\bar{\mathbf{c}} \quad \bar{\mathbf{c}}], \\
\bar{\mathbf{c}}_2^{\mathrm{T}} &= [\bar{\mathbf{c}} \quad -\bar{\mathbf{c}}], \\
\bar{\mathbf{c}}_3^{\mathrm{T}} &= [\bar{\mathbf{c}'} \quad \bar{\mathbf{c}'}], \\
\bar{\mathbf{c}}_4^{\mathrm{T}} &= [\bar{\mathbf{c}'} \quad -\bar{\mathbf{c}'}],
\end{aligned} \tag{12.6}$$

where $\bar{\mathbf{c}}_1$, $\bar{\mathbf{c}}_2$, $\bar{\mathbf{c}}_3$ and $\bar{\mathbf{c}}_4$ are the four DSTS-related orthogonal codes having a period of $2T_b$. The DSTS operation and the resultant transmitted waveforms from the four antennas are shown in Figure 12.4. Figure 12.4 shows that the first and second antennas transmit jointly one SP symbol within a time duration of $2T_b$, while the third and fourth antennas jointly transmit another SP symbol within the same $2T_b$ duration. However, the operation of the first and second antennas is independent of that of the third and fourth antennas.

The differentially encoded data is then spread to the transmit antennas and transmitted using the mapping of

$$\mathbf{y}_t^1 = \frac{1}{\sqrt{4}}(\bar{\mathbf{c}}_1 \cdot v_t^1 + \bar{\mathbf{c}}_2 \cdot v_t^{2*}), \tag{12.7}$$

$$\mathbf{y}_t^2 = \frac{1}{\sqrt{4}}(\bar{\mathbf{c}}_1 \cdot v_t^2 - \bar{\mathbf{c}}_2 \cdot v_t^{1*}), \tag{12.8}$$

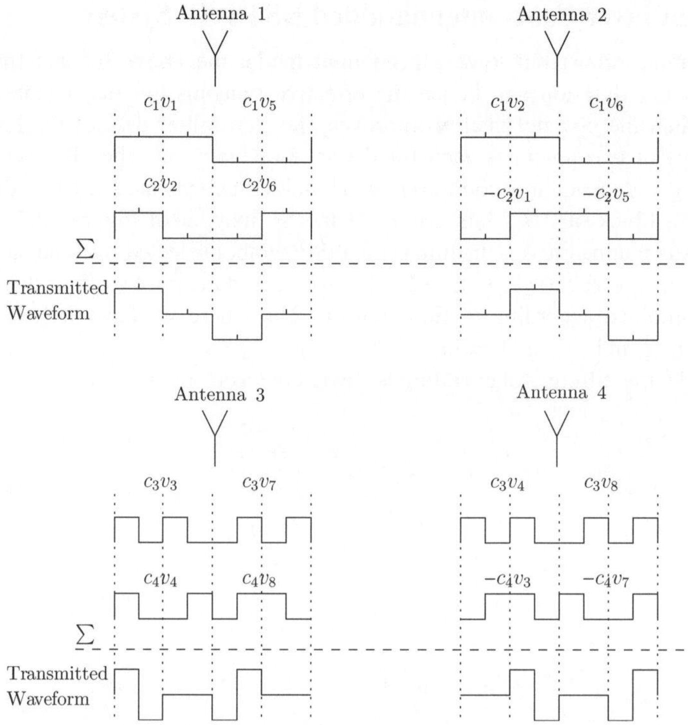

Figure 12.4: Illustration of STS using twin-layer four transmit antennas transmitting four bits within $2T_b$ duration. Here $v_1 = v_2 = v_3 = v_4 = 1$ and $v_5 = v_6 = v_7 = v_8 = -1$ were assumed and $\bar{\mathbf{c}} = [+1 + 1]$ and $\bar{\mathbf{c}'} = [+1 - 1]$.

$$\mathbf{y}_t^3 = \frac{1}{\sqrt{4}}(\bar{\mathbf{c}}_3 \cdot v_t^3 + \bar{\mathbf{c}}_4 \cdot v_t^{4*}), \tag{12.9}$$

$$\mathbf{y}_t^4 = \frac{1}{\sqrt{4}}(\bar{\mathbf{c}}_3 \cdot v_t^4 - \bar{\mathbf{c}}_4 \cdot v_t^{3*}). \tag{12.10}$$

The signal at the output of the single receiver antenna can be represented as

$$\mathbf{r}_t = h_1 \cdot \mathbf{y}_t^1 + h_2 \cdot \mathbf{y}_t^2 + h_3 \cdot \mathbf{y}_t^3 + h_4 \cdot \mathbf{y}_t^4 + \mathbf{n}_t. \tag{12.11}$$

The received signal $\mathbf{r}_t$ is then correlated with $\bar{\mathbf{c}}_1$, $\bar{\mathbf{c}}_2$, $\bar{\mathbf{c}}_3$ and $\bar{\mathbf{c}}_4$ according to Section 10.3.2. Differential decoding is carried out by using the received data of two consecutive time slots in a similar fashion to Equation (10.13) in order to arrive at $\frac{1}{2} \cdot \sum_{i=1}^{2} |h_i|^2 \cdot \sqrt{\sum_{j=1}^{2} |v_t^j|^2} \cdot x_t^k + N_k$.

Based on the above, the twin-layer four-antenna-aided DSTS scheme transmits two SP symbols in $2T_b$ duration. Specifically, the first and second antennas transmit one SP symbol in the same fashion as in Section 10.3 with the aid of $\bar{\mathbf{c}}_1$ and $\bar{\mathbf{c}}_2$ in two time slots, while the third and fourth transmit antennas transmit another SP symbol in the same time period with the aid of $\bar{\mathbf{c}}_3$ and $\bar{\mathbf{c}}_4$. Thus, this scheme transmits two SP symbols in $2T_b$ duration resulting in an effective transmission rate of $2 \times 1/2T_b = 1/T_b$, which is effectively four times that of the single-layer four-antenna-aided DSTS-SP system. In other words, the twin-layer scheme transmits four SP symbols in $4T_b$ time duration compared with a single SP symbol in the single-layer DSTS scheme of Section 12.3.1.

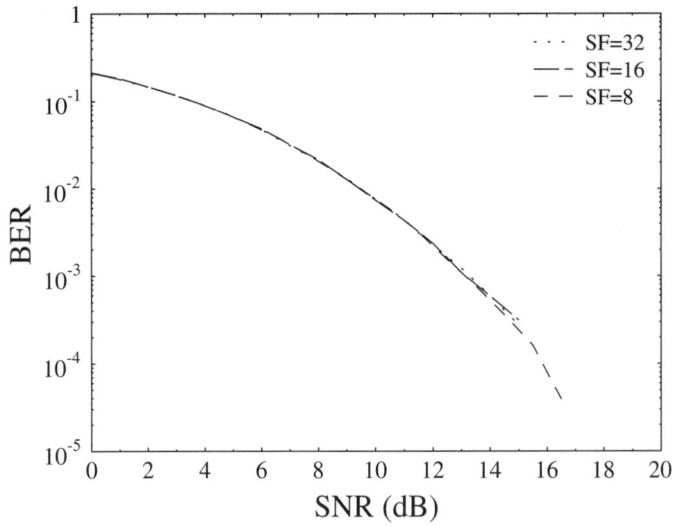

Figure 12.5: BER versus SNR performance of the DSTS-SP scheme in conjunction with $L = 16$ and three different SFs.

12.4 VSF-based Adaptive Rate DSTS

In this section we discuss the employment of VSF codes in adaptive-rate DSTS-SP systems, where the chip rate is kept constant and hence so is the bandwidth, while the effective bit rate is varied by varying the SF over the course of transmission. For example, when stipulating a constant chip rate, the number of bits transmitted using a SF of 4 is half of that when employing SF $= 2$.

When the channel quality is high, a low SF can be used in order to increase the throughput and, conversely, when the channel conditions are hostile, a high SF is employed for maintaining the target BER performance [51]. To elaborate further, Figure 12.5 shows the BER performance of the DSTS-SP system versus the SNR in conjunction with different SFs. Figure 12.5 illustrates that using orthogonal VSFs does not affect the BER versus $\text{SNR} = E_s/N_0$ performance, where E_s is the spread symbol's energy. Hence, no performance gain is attained when comparing the adaptive-rate scheme with a fixed-rate scheme [51]. However, Figure 12.6 demonstrates that plotting the BER curves versus the chip SNR (CSNR) defined as $E_s/N_0/\text{SF}$ results in SF-dependent BER performance. Therefore, it can be concluded that upon accommodating the channel quality fluctuations using VSFs, the spread symbol's SNR is varied accordingly for the sake of maintaining a constant CSNR according to the expression $\text{SNR} = \text{CSNR} \times \text{SF}$ (see [51]). For example, when the channel quality is high and hence a low SF is used, the transmitter power, which is proportional to the SNR at a given fixed N_0 value, is also reduced. Therefore, the CSNR is always maintained at the specific value associated with the highest SF. In other words, the VSF regime accommodates the channel quality variations by adapting the SF according to the near-instantaneous channel quality without increasing the transmitted power above that associated with the highest SF [51]. When the SF is decreased, the SNR is proportionately decreased.

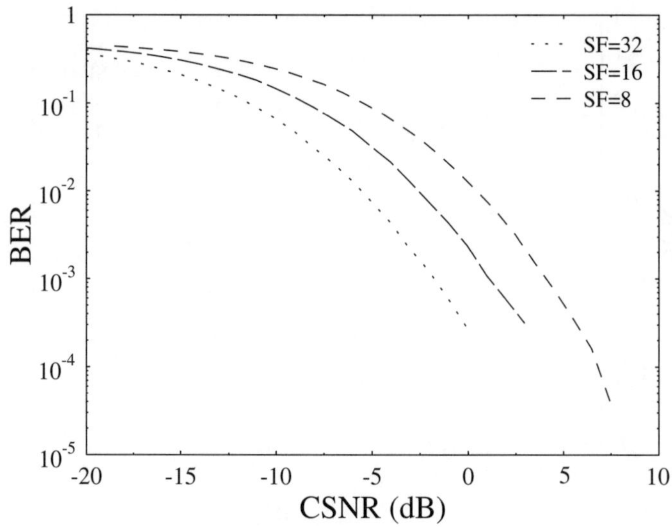

Figure 12.6: BER versus CSNR performance of the DSTS-SP scheme in conjunction with $L = 16$ and three different SFs, where SNR = CSNR × SF.

12.5 Variable-code-rate Iteratively Detected DSTS-SP System

As already discussed in Chapter 10, the detected DSTS signals can be represented by Equation (10.22), where a received SP symbol $\tilde{\mathbf{s}}$ is constructed from the estimates $\tilde{a}_1$, $\tilde{a}_2$, $\tilde{a}_3$ and $\tilde{a}_4$. In Chapter 11 we presented a detailed account of how iterative detection is carried out. The SP symbol $\tilde{\mathbf{s}}$ carries B_{sp} channel-coded bits $\mathbf{b} = b_0, \ldots, b_{B_{\text{sp}}-1} \in \{0, 1\}$, and the corresponding LLRs of the bits can be computed in a similar manner to that discussed in Chapter 11.

Naturally, the code rate of the RSC code employed affects both the system's performance and the throughput. More specifically, as the RSC code rate increases, the system's effective throughput increases at the expense of degrading the system's performance. Therefore, when the channel SNR is low and hence the resultant BER is higher than the target BER, a powerful low-throughput low-rate RSC code is activated. In contrast, when the channel quality is high and hence the resultant BER is lower than the target BER, a higher-throughput scheme corresponding to a higher-rate RSC code is activated.

12.6 Results and Discussion

We consider SP modulation associated with $L = 16$ and DSTS employing four transmit antennas and a single receive antenna in order to demonstrate the performance improvements achieved by the proposed adaptive system. All simulation parameters are listed in Table 12.1, where BE represents the bandwidth efficiency of the system in BPS Hz^{-1}. The target BER of the system is selected to be 10^{-3}.

Figure 12.7 plots the BER as well as the bandwidth efficiency performance of the proposed adaptive iteratively detected DSTS-SP system. Owing to employing time-variant modes, the performance results are plotted versus SNR on the x-axis. The SNR can be

Table 12.1: Proposed adaptive system parameters.

SP Modulation	$L = 16$
Number of transmit antennas	4
Number of receive antennas	1
Channel	Correlated Rayleigh fading with $f_D = 0.01$
Mode 1	Four Tx antennas DSTS-SP system SF $= 32$, RSC code rate $= \frac{1}{4}$, BE $= 0.25$
Mode 2	Four Tx antennas DSTS-SP system SF $= 32$, RSC code rate $= \frac{1}{2}$, BE $= 0.5$
Mode 3	Two Groups of two Tx antennas DSTS-SP system, SF $= 16$, RSC code rate $= \frac{1}{6}$, BE $= 4/3$
Mode 4	Two Groups of two Tx antennas DSTS-SP system, SF $= 16$, RSC code rate $= \frac{1}{4}$, BE $= 2$
Mode 5	Two Groups of two Tx antennas DSTS-SP system, SF $= 16$, RSC code rate $= \frac{1}{2}$, BE $= 4$
Mode 6	Two Groups of two Tx antennas DSTS-SP system, SF $= 8$, No channel coding, BE $= 16$

converted to a E_b/N_0 value upon dividing the SNR by the bandwidth efficiency of the system. The performance of the adaptive system is evaluated by analyzing the BER and the bandwidth efficiency expressed in bit s^{-1} Hz^{-1}. The BER curve of the adaptive DSTS-SP system, which can be read by referring to the y-axis on the left of the figure, is plotted along with those of the non-adaptive modes. The system employs an interleaver depth ranging between 48 000 and 16 000 bits depending on the code rate of the RSC code employed. Moreover, six iterative detection iterations are employed in conjunction with the system parameters outlined in Table 12.1. The BER performance reaches the target BER around SNR $= 5$ dB and it does not exceed the target BER for higher SNRs, while switching between the different transmission modes.

The y-axis on the right-hand side of Figure 12.7 plots the achievable effective bandwidth efficiency of the proposed adaptive iteratively detected DSTS-SP system, while employing an interleaver depth ranging from 48 000 to 16 000 bits depending on the transmission mode employed. Six iterative detection iterations and the system parameters outlined in Table 12.1 were invoked. Depending on the channel quality quantified in terms of the channel SNR, the transmitter activates one of the transmission modes outlined in Table 12.1. The bandwidth efficiency of the system varies from 0.25 bit s^{-1} Hz^{-1} for the minimum-throughput mode to 16 bit s^{-1} Hz^{-1} for the highest-throughput mode. For example, if we calculate the bandwidth efficiency of mode 1, employing the $L = 16$ SP modulation scheme using SF $= 32$ and a RSC code rate of $\frac{1}{4}$ yields $4 \cdot \frac{1}{4} \cdot \frac{1}{4} = 0.25$ bit s^{-1} Hz^{-1}, since each SP symbol is transmitted in four time slots. Similarly, for the highest-throughput mode using the $L = 16$ SP modulation scheme, SF $= 8$ and no channel coding, we arrive at $2 \cdot 4 \cdot 4 \cdot \frac{1}{2} = 16$ bit s^{-1} Hz^{-1}, since two SP symbols are transmitted in two time slots.

Finally, Figure 12.8 portrays the mode selection probability of the proposed adaptive iteratively detected DSTS-SP system. It is clear from the figure that as the average SNR increases, the higher-throughput systems are employed more often.

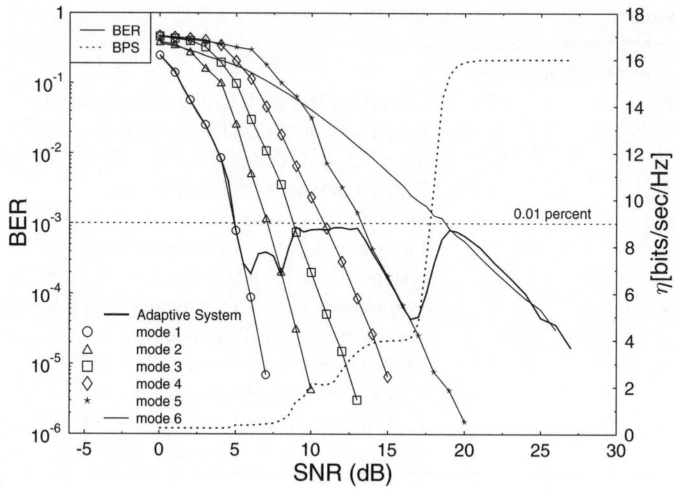

Figure 12.7: BER and throughput performance of the proposed iteratively detected adaptive DSTS-SP system for a target BER of 10^{-3}.

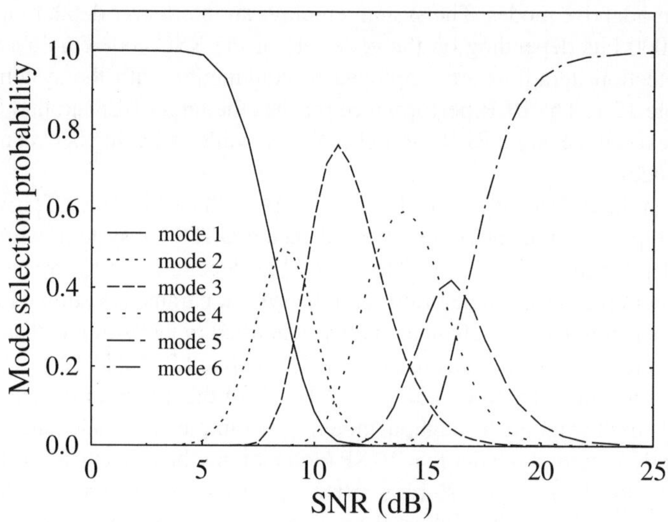

Figure 12.8: The mode selection probability of the proposed adaptive DSTS-SP system at a target BER of 10^{-3}.

12.7 Chapter Conclusion and Summary

In this chapter we proposed a novel adaptive DSTS system that exploits the advantages of differential encoding and iterative demapping as well as SP modulation, while adapting the system parameters for the sake of achieving the highest possible bandwidth efficiency as well as maintaining a given target BER. The proposed adaptive DSTS-SP scheme benefits from a substantial diversity gain, while using four transmit antennas without the need for pilot-assisted channel envelope estimation and coherent detection. The proposed scheme reaches the target BER of 10^{-3} at a SNR of about 5 dB and maintains it for SNRs in excess of this value, while increasing the effective throughput. The system's bandwidth efficiency varies from 0.25 bit s^{-1} Hz^{-1} to 16 bit s^{-1} Hz^{-1}.

Furthermore, in the presence of large-scale shadow fading, MIMO systems perform poorly, since the diversity gain that is based on the fact that the spatial channels fade independently decreases in the presence of shadow fading. Therefore, the adaptive MIMO scheme presented constitutes a feasible design alternative in the presence of shadow fading.

Chapter 13

Layered Steered Space-Time Codes

13.1 Introduction

Time-varying multipath fading imposes a fundamental limitation on wireless transmissions, which can be counteracted by employing MIMO schemes [9]. More explicitly, information theoretic studies of [3, 251] have revealed that a MIMO system attains a higher capacity than a single-input single-output system. In [15], Wolniansky *et al.* proposed the multi-layer MIMO structure known as the V-BLAST scheme, whose transceiver is capable of providing a tremendous increase of a specific user's effective bandwidth efficiency without the need for any increase in the transmitted power or in the system's bandwidth.

On the other hand, STBCs [11, 12] constitute a powerful transmit diversity scheme, which uses low-complexity linear processing at the receiver and is capable of achieving the maximum possible diversity gain. Since the V-BLAST structure is capable of achieving the maximum multiplexing gain, while the STBC scheme attains the maximum antenna diversity gain, it was proposed in [17] to combine the benefits of these two techniques for the sake of providing both antenna diversity and bandwidth efficiency gains. This hybrid scheme was improved in [18] by optimizing the decoding order of the different antenna layers. Furthermore, beamforming [16] constitutes an effective technique of reducing the multiple-access interference, where the antenna gain is increased in the direction of the desired user whilst reducing the gain towards the interfering users. In order to achieve additional performance gains, beamforming has also been combined with STBC to attain a higher SNR gain [20].

The concept of combining orthogonal transmit diversity designs with the principle of SP was introduced by Su *et al.* [43] in order to maximize the achievable coding advantage, where it was demonstrated that the proposed SP-aided STBC scheme was capable of outperforming the conventional orthogonal-design-based STBC schemes of [11, 12]. A further advance was proposed in [46], where the SP demapper of [43] was modified for the sake of accepting the *a priori* information passed to it from the channel decoder as extrinsic information.

In [158], the employment of the iterative detection principle [146] was considered for iterative soft demapping in the context of multilevel modulation schemes combined with channel decoding, where a soft symbol-to-bit demapper was used between the multilevel demodulator and the binary channel decoder. It was also demonstrated in [166] that a recursive inner code is needed in order to maximize the interleaver gain and to avoid the formation of a BER floor, when employing iterative decoding. In [168], unity-rate inner codes were employed for designing low-complexity iteratively detected schemes suitable for bandwidth- and power-limited systems having stringent BER requirements.

Recently, studying the convergence behavior of iterative decoding has attracted considerable attention. In [169], ten Brink proposed the employment of the so-called EXIT characteristics between a concatenated decoder's output and input for describing the flow of extrinsic information through the soft-in soft-out constituent decoders. The concept of EXIT chart analysis has been extended to three-stage concatenated systems in [171, 174, 178].

The novelty and rationale of the proposed system can be summarized as follows.

1. We amalgamate the merits of V-BLAST, STC and beamforming for the sake of achieving a multiplexing gain, a diversity gain and a beamforming gain. The resultant scheme is referred to here as a LSSTC. In addition, the system is combined with multi-dimensional SP modulation, which is capable of maximizing the coding advantage of the transmission scheme by jointly designing and detecting the sphere-packed space-time symbols.

2. We quantify the capacity of the LSSTC-SP scheme for transmission over both Rayleigh and Gaussian channels. Furthermore, we propose a novel technique for quantifying the maximum achievable rate of the system using EXIT charts.

3. We propose three near-capacity iteratively detected LSSTC-SP receiver structures, where iterative detection is carried out between an outer code's Decoder I, an intermediate code's Decoder II and a LSSTC-SP demapper. The three proposed schemes differ in the number of inner iterations employed between Decoder II and the SP demapper, as well as in the choice of the outer code, which is either a regular RSC code or an IRCC [174, 176]. On the other hand, the intermediate code employed is a URC, which is capable of completely eliminating the system's error floor as well as of operating at the lowest possible turbo-cliff SNR without significantly increasing either the associated complexity or the interleaver delay. Furthermore, a comparison between the three iteratively-detected schemes reveals that a carefully designed two-stage[1] iterative detection scheme is capable of operating sufficiently close to capacity at a lower complexity, when compared with a three-stage[2] system employing RSC or a two-stage system employing IRCC as an outer code.

The rest of the chapter is organized as follows. In Section 13.2 we present the encoding and decoding algorithms of the novel LSSTC scheme and demonstrate how the scheme can be combined with conventional modulation as well as with multi-dimensional SP modulation. In Section 13.3 we quantify the capacity of the LSSTC scheme, followed by a discussion about the iterative detection schemes invoked and the two-dimensional as well as three-dimensional EXIT charts employed in Section 13.4. The procedure of computing the maximum achievable rate of the LSSTC-SP scheme using EXIT charts is detailed in Section 13.4.3. The attainable

[1] A two-stage system employs iterations between the outer code's decoder and the intermediate code's decoder, but no iterations are employed between the intermediate code's decoder and the SP demapper.

[2] A three-stage system employs iterations between the SP demapper and the intermediate code's decoder, which we refer to as inner iterations, as well as between the outer code's decoder and the intermediate code's decoder, which are referred to as outer iterations.

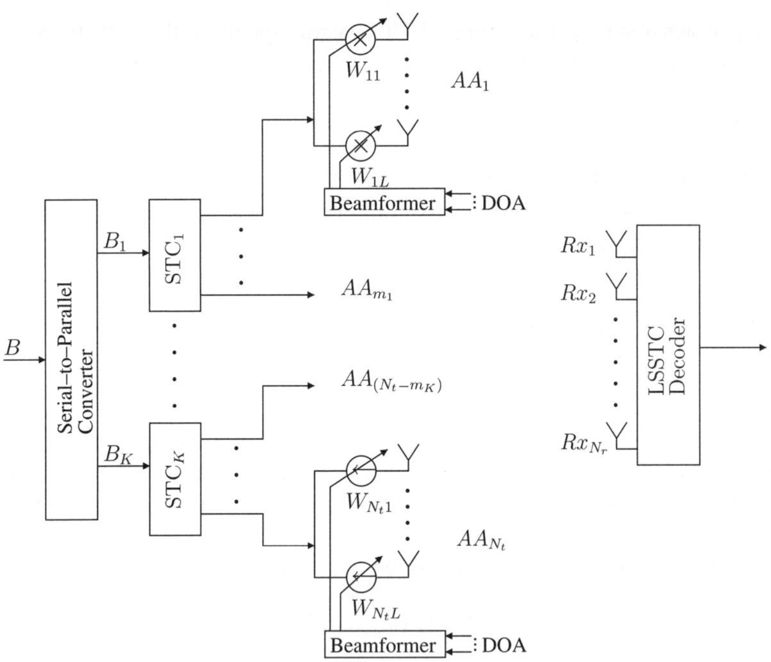

Figure 13.1: LSSTC system block diagram.

performance of the proposed schemes is studied comparatively in Section 13.5, followed by our conclusions in Section 13.6. In Section 13.7 we provide a chapter summary discussing both the main contributions and the organization of this chapter.

13.2 LSSTCs

13.2.1 LSSTC using Conventional Modulation

A block diagram of the proposed LSSTC scheme is illustrated in Figure 13.1. The antenna architecture employed in Figure 13.1 has N_t transmit AAs spaced sufficiently far apart in order to experience independent fading and hence to achieve transmit diversity. The L_{AA} elements of each of the AAs are spaced at a distance of $d = \lambda/2$ for the sake of achieving a beamforming gain, where λ represents the carrier's wavelength. Furthermore, the receiver is equipped with $N_r \geqslant N_t$ antennas. According to Figure 13.1, a block of B input information symbols is serial-to-parallel converted to K groups of symbol streams of length $B_1, B_2, \ldots, B_K$, where $B_1 + B_2 + \cdots + B_K = B$. Each group of B_k symbols, $k \in [1, K]$, is then encoded by a component space-time code STC_k associated with m_k transmit AAs, where $m_1 + m_2 + \cdots + m_K = N_t$.

In this contribution, we consider transmissions over a temporally correlated narrowband Rayleigh fading channel associated with a normalized Doppler frequency of $f_D = f_d T_s = 0.01$, with f_d being the Doppler frequency and T_s the symbol duration, while the spatial channel coefficients are independent. The complex AWGN of $n = n_I + jn_Q$ contaminates the received signal, where n_I and n_Q are two independent zero-mean Gaussian random variables having a variance of $N_0/2$ per dimension.

The L_{AA}-dimensional spatial-temporal CIR vector spanning the mth transmitter AA, $m \in [1, \ldots, N_t]$, and the nth receiver antenna, $n \in [1, \ldots, N_r]$, can be expressed as

$$\mathbf{h}_{nm}(t) = \mathbf{a}_{nm}(t)\delta(t - \tau_k) = [a_{nm,0}(t), \ldots, a_{nm,(L_{AA}-1)}(t)]^{\mathrm{T}}\delta(t - \tau_k), \tag{13.1}$$

where τ_k is the signal's delay and $a_{nm,l}(t)$ is the CIR with respect to the mnth link and the lth element of the mth AA. Based on the assumption that the array elements are separated by half a wavelength, we have

$$\mathbf{a}_{nm}(t) = \alpha_{nm}(t) \cdot \mathbf{d}_{nm}, \tag{13.2}$$

where $\alpha_{nm}(t)$ is a Rayleigh faded envelope,

$$\mathbf{d}_{nm} = [1, \exp(j[\pi \sin(\psi_{nm})]), \ldots, \exp(j[(L_{AA} - 1)\pi \sin(\psi_{nm})])]^{\mathrm{T}} \tag{13.3}$$

and ψ_{nm} is the nmth link's DOA. As for the AA specific DOA, we consider a scenario where the distance between the transmitter and the receiver is significantly higher than that between the AAs and thus we can assume that the signals arrive at the different AAs in parallel, i.e. the DOA at the different AAs is the same.

The received baseband data matrix $\mathbf{Y}$ can be expressed as

$$\mathbf{Y} = \mathbf{HWX} + \mathbf{N}, \tag{13.4}$$

where $\mathbf{X}$ represents the transmitted symbols matrix, $\mathbf{H}$ is an $(N_r \times N_t)$ matrix whose entries of $\mathbf{h}_{nm}$ are defined in Equation (13.1) and $\mathbf{N}$ denotes the AWGN matrix whose entries have a variance of $N_0/2$ per dimension. Furthermore, $\mathbf{W}$ is a diagonal AA weight matrix, whose diagonal entry $\mathbf{w}_{mn}$ is the L_{AA}-dimensional weight vector for the mth beamformer AA and the nth receive antenna. In this scenario, the MRC-criterion-based transmit beamformer, which constitutes an effective solution to maximizing the antenna gain, is the optimum beamformer. Let

$$\mathbf{w}_{mn} = \mathbf{d}_{nm}^{\dagger}, \tag{13.5}$$

where the superscript † represents the Hermitian of the matrix. Then the received signal can be expressed as

$$\mathbf{Y} = L\widehat{\mathbf{H}}\mathbf{X} + \mathbf{N}, \tag{13.6}$$

where $\widehat{\mathbf{H}}$ is an $(N_r \times N_t)$ matrix, whose entries are α_{nm}. Moreover, $\mathbf{Y}$ can be written as

$$\mathbf{Y} = L\sum_{k=1}^{K} \widehat{\mathbf{H}}_k\mathbf{x}_k + \mathbf{N}, \tag{13.7}$$

where $\mathbf{x}_k$ represents the component STC used at layer k, with $k \in [1, \ldots, K]$.

The most beneficial decoding order of the STC layers is determined on the basis of detecting the highest-power layer first for the sake of a high correct detection probability. For simplicity, let us consider the case of $K = 2$ STBC layers and that layer 1 is detected first, which allows us to eliminate the interference caused by the signal of layer 2. However, the proposed concept is applicable to arbitrary STCs and to an arbitrary number of layers K. For this reason, the decoder of layer 1 has to compute a matrix $\mathbf{Q}$, so that we have $\mathbf{Q} \cdot \widehat{\mathbf{H}}_2 = 0$. Therefore, the decoder computes an orthonormal basis for the left null space of $\widehat{\mathbf{H}}_2$ and assigns the vectors of the basis to the rows of $\mathbf{Q}$. Multiplying $\mathbf{Q}$ by $\mathbf{Y}$ suppresses the interference of layer 2 originally imposed on layer 1 and generates a signal which can be decoded using ML STBC detection. Then, the decoder subtracts the remodulated contribution

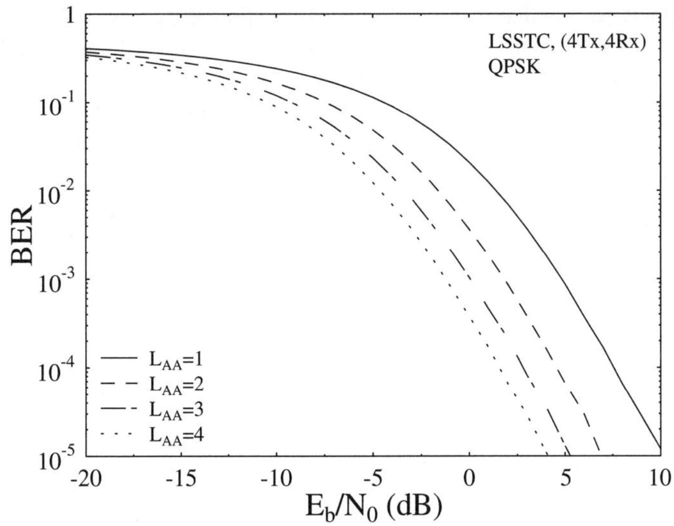

Figure 13.2: BER performance of a QPSK-modulated $N_t \times N_r = 4 \times 4$ LSSTC system for a variable number L_{AA} of elements per AA.

of the decoded symbols of layer 1 from the composite twin-layer received signal $\mathbf{Y}$. Finally, the decoder applies direct STBC decoding to the second layer, since the interference imposed by the first layer has been eliminated. This group-interference cancelation procedure can be generalized to any N_t and K values.

We consider a system employing $N_t \times N_r = 4 \times 4$ antennas and $K = 2$ layers in order to demonstrate the performance improvements achieved by a DL scheme, where a BS employing $N_t = 4$ transmit antennas is communicating with a laptop receiver employing $N_r = 4$ back-plane antennas. The system employs QPSK modulation and considers transmission over a temporally correlated Rayleigh fading channel. In addition, we assume that the CSI is perfectly known at the receiver. We also assume that the transmitter has full knowledge of the DOA without any estimation or estimation errors. Figure 13.2 shows the effect of increasing the DL BS beamforming gain achieved upon increasing the number of beam-steering elements L_{AA} in the AA, while maintaining the same total number of AAs. As shown in Figure 13.2, when the number of beam-steering elements L_{AA} increases, the achievable BER performance substantially improves.

13.2.2 LSSTC using SP Modulation

According to the previous discussion, it becomes clear that the decoded signal represents a scaled version of the transmitted signal corrupted by noise. This observation implies that the diversity product of the LSSTC scheme is determined by the MED of all legitimate transmitted vectors. Hence, in order to maximize the achievable coding advantage, it was proposed in [43] to use SP schemes that maximize the MED of the transmitted signal vectors. Our idea is to jointly design the legitimate m_k-component complex-valued vectors $(x^1, x^2, \ldots, x^{m_k})$ transmitted from layer k, $k \in [0, \ldots, K]$, so that they are represented by a single phasor point selected from a SP constellation corresponding to a $2m_k$-dimensional real-valued lattice having the best known MED in the $2m_k$-dimensional real-valued space $\mathbb{R}^{2m_k}$.

In what follows we assume that each layer is constituted by a twin-AA STBC scheme, i.e. we have $m_k = 2$, for all $k \in [0, K]$, which means that the SP design required is $2m_k = 4$-dimensional. To summarize, according to the previous discussion, x^1 and x^2 represent independent conventional PSK-modulated symbols transmitted from the first and second transmit AA and no effort is made to jointly design a signal constellation for the various combinations of x^1 and x^2. In contrast, in the case of SP, these symbols are designed jointly in order to further increase the attainable coding advantage. Assuming that there are L legitimate vectors $(x^{l,1}, x^{l,2})$, $l = 0, 1, \ldots, L-1$, where L represents the number of sphere-packed modulated symbols, the transmitter then has to choose the modulated signal from these L legitimate symbols to be transmitted over the two AAs in layer $k \in [1, \ldots, K]$.

In the four-dimensional real-valued Euclidean space $\mathbb{R}^4$, the lattice D_4 is defined as a SP constellation having the best MED from all other $(L-1)$ legitimate four-component constellation points in $\mathbb{R}^4$ (see [221]). More specifically, D_4 may be defined as a lattice that consists of all legitimate sphere-packed constellation points having integer coordinates $[a^{l,1}\ a^{l,2}\ a^{l,3}\ a^{l,4}]$ subjected to the SP constraint of

$$a^{l,1} + a^{l,2} + a^{l,3} + a^{l,4} = \kappa, \tag{13.8}$$

where κ is an even integer [221]. Assuming that $\mathbf{S} = \{\mathbf{s}^l = [a^{l,1}, a^{l,2}, a^{l,3}, a^{l,4}] \in \mathbb{R}^4 : 0 \leq l \leq L-1\}$ constitutes a set of L legitimate constellation points from the lattice D_4 having a total energy of

$$E_{\text{total}} \triangleq \sum_{l=0}^{L-1} (|a^{l,1}|^2 + |a^{l,2}|^2 + |a^{l,3}|^2 + |a^{l,4}|^2), \tag{13.9}$$

upon introducing the notation

$$\mathbf{C}_l = \{x^{l,1}, x^{l,2}\} = T_{\text{sp}}(a^{l,1}, a^{l,2}, a^{l,3}, a^{l,4}) = \{a^{l,1} + ja^{l,2}, a^{l,3} + ja^{l,4}\}, \tag{13.10}$$

we have a set of constellation symbols, $\{\mathbf{C}_l\colon 0 \leq l \leq L-1\}$, leading to the design of LSSTC signals, whose diversity product is determined by the MED of the set of L legitimate constellation points in $\mathbf{S}$.

Figure 13.3 depicts the BER performance of the SP-modulated LSSTC scheme in conjunction with Gray mapping and $L = 16$, while employing $N_t = 4$, $N_r = 4$ and a variable number L_{AA} of elements per AA. The results in Figure 13.3 correspond to the scenario where the channel is assumed to be perfectly known at the receiver. Figure 13.3 shows the effect of increasing the DL BS beamforming gain achieved by increasing the number of beam-steering elements L_{AA} in the AA for the system employing $N_t \times N_r = 4 \times 4$ antennas and $K = 2$ layers, while maintaining the same total number of AAs. As shown in Figure 13.3, when the number of beam-steering elements L_{AA} increases, the achievable BER performance substantially improves.

13.3 Capacity of LSSTCs

Upon using the decoding order of $(1, 2, \ldots, K)$, group k will have a diversity order of $m_k \times (N_r - N_t + m_1 + m_2 + \cdots + m_k) = m_k \times N_{rk}$. Thus, the LSSTC decoded signal of layer k, assuming perfect interference cancelation, can be described as

$$\tilde{x}_k = L \sum_{r=1}^{N_{rk}} \sum_{t=1}^{m_k} \alpha_{rt} \cdot x_k + \Delta = \sum_{r=1}^{N_{rk}} \chi^2_{2m_k r} \cdot x_k + \Delta_k, \tag{13.11}$$

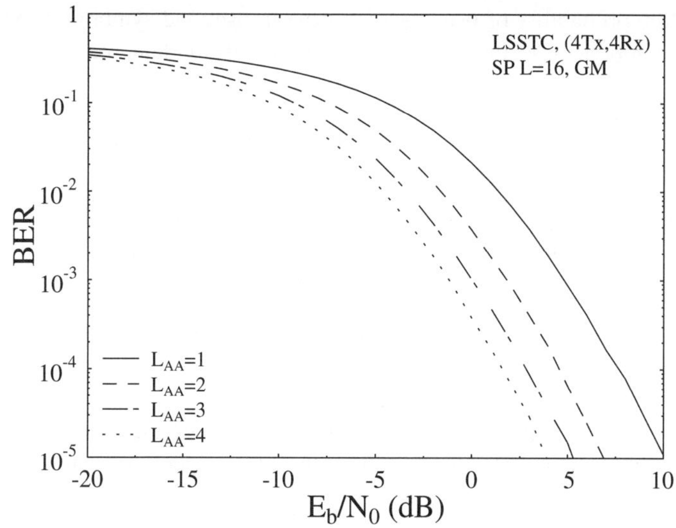

Figure 13.3: BER performance of the SP-modulated $N_t \times N_r = 4 \times 4$ LSSTC system in conjunction with $L = 16$ for a variable number L_{AA} of elements per AA.

where $\chi^2_{2m_k r} = L \sum_{t=1}^{m_k} \alpha_{rt}$ represents a chi-squared distributed random variable having $2m_k$ degrees of freedom and Δ_k is the AWGN after decoding, which has a variance of $\chi^2_{2m_k r} \cdot N_0/2$ per dimension.

The received sphere-packed symbol $\tilde{\mathbf{s}}_k$ at layer k is then constructed from the estimates $\tilde{x}_k^1$ and $\tilde{x}_k^2$ using the inverse function of T_{sp} introduced in Equation (13.10) as $\tilde{\mathbf{s}}_k = T_{\text{sp}}^{-1}(\tilde{x}_k^1, \tilde{x}_k^2)$, where we have $\tilde{\mathbf{s}} = [\tilde{a}^1\ \tilde{a}^2\ \tilde{a}^3\ \tilde{a}^4] \in \mathbb{R}^4$. Therefore, the received sphere-packed symbol $\tilde{\mathbf{s}}_k$ can be written as

$$\tilde{\mathbf{s}}_k = \sum_{r=1}^{N_{rk}} \chi^2_{2m_k r} \mathbf{s}_k + \acute{\Delta}_k, \tag{13.12}$$

where we have $\acute{\Delta}_k = [\acute{\Delta}_{k1}\ \acute{\Delta}_{k2}\ \acute{\Delta}_{k3}\ \acute{\Delta}_{k4}] \in \mathbb{R}^4$, which is a four-dimensional real-valued Gaussian random variable having a covariance matrix of

$$\sigma^2_{\acute{\Delta}_k} \cdot \mathbf{I}_D = \sigma^2_{\Delta_k} \cdot \mathbf{I}_D = \chi^2_{2m_{k,r}} \cdot N_0/2 \cdot \mathbf{I}_D, \tag{13.13}$$

where we have $D = 4$, since the SP symbol constellation $\mathbf{S}$ is four-dimensional.

Let $\tilde{\mathcal{S}} = (\tilde{\mathbf{s}}_1, \tilde{\mathbf{s}}_2, \ldots, \tilde{\mathbf{s}}_K)^{\text{T}}$, $\mathcal{S} = (\mathbf{s}_1, \mathbf{s}_2, \ldots, \mathbf{s}_K)^{\text{T}}$ and note that in conjunction with K groups there are $M = (L)^K$ possible SP phasor combinations, where SP in conjunction with L is used for transmission. Thus, the achievable capacity of the MIMO system proposed for transmission over the DCMC can be derived from that of the discrete memoryless channel as [249, 293]:

$$C_{\text{DCMC}} = \max_{p(\mathcal{S}^1),\ldots,p(\mathcal{S}^M)} \sum_{i=1}^{M} \underbrace{\int_{-\infty}^{\infty} \cdots \int_{-\infty}^{\infty}}_{D\text{-fold}} p(\tilde{\mathcal{S}}|\mathcal{S}_i) \cdot p(\mathcal{S}_i) \cdot \log_2\left(\frac{p(\tilde{\mathcal{S}}|\mathcal{S}_i)}{\sum_{v=1}^{M} p(\tilde{\mathcal{S}}|\mathcal{S}_v) \cdot p(\mathcal{S}_v)}\right) \cdot d\tilde{\mathcal{S}} \quad \text{(BPS)}, \tag{13.14}$$

where $p(\mathcal{S}_i)$ is the probability of occurrence for the transmitted symbol vector $\mathcal{S}_i$. Furthermore, since the components of the vectors $\mathcal{S}$ and $\tilde{\mathcal{S}}$ are independent, we have

$$p(\tilde{\mathcal{S}}|\mathcal{S}) = \prod_{k=1}^{K} p(\tilde{\mathbf{s}}_k|\mathbf{s}_k). \tag{13.15}$$

Furthermore, according to Equation (13.12), the conditional probability $p(\tilde{\mathbf{s}}_k|\mathbf{s}_k)$ of receiving a four-dimensional signal $\tilde{\mathbf{s}}_k$ at layer k is given by the following PDF:

$$p(\tilde{\mathbf{s}}_k|\mathbf{s}_k) = \frac{1}{\prod_{d=1}^{D=4} \sqrt{\pi N_0 \sum_{r=1}^{N_{rk}} \chi^2_{2m_kr}[d]}} \cdot \exp\left(\sum_{d=1}^{D=4}\sum_{r=1}^{N_{rk}} \frac{-\left(\tilde{\mathbf{s}}_k[d] - \chi^2_{2m_kr}[d]\cdot \mathbf{s}_k[d]\right)^2}{\chi^2_{2m_kr}[d]\cdot N_0}\right). \tag{13.16}$$

Moreover, C_{DCMC} in Equation (13.14) is maximized when the transmitted symbols are equiprobably distributed, i.e. for $p(\mathcal{S}_i) = 1/M$ (see [293]). Hence, we can write

$$\begin{aligned}
\log_2\left(\frac{p(\tilde{\mathcal{S}}|\mathcal{S}_i)}{\sum_{v=1}^{M} p(\tilde{\mathcal{S}}|\mathcal{S}_v)\cdot p(\mathcal{S}_v)}\right) &= \log_2\left(\frac{p(\tilde{\mathcal{S}}|\mathcal{S}_i)}{1/M\cdot\sum_{v=1}^{M} p(\tilde{\mathcal{S}}|\mathcal{S}_v)}\right)\\
&= \log_2(M) - \log_2\left(\sum_{v=1}^{M}\frac{p(\tilde{\mathcal{S}}|\mathcal{S}_v)}{p(\tilde{\mathcal{S}}|\mathcal{S}_i)}\right) = \log_2(M) - \log_2\left(\sum_{v=1}^{M}\prod_{k=1}^{K}\frac{p(\tilde{\mathbf{s}}_k|\mathbf{s}_{kv})}{p(\tilde{\mathbf{s}}_k|\mathbf{s}_{ki})}\right)\\
&= \log_2(M) - \log_2\left(\sum_{v=1}^{M}\prod_{k=1}^{K}\exp(\Psi_{k,vi})\right),
\end{aligned} \tag{13.17}$$

where the term $\Psi_{k,vi}$ is given by

$$\begin{aligned}
\Psi_{k,vi} &= \sum_{d=1}^{D=4}\sum_{r=1}^{N_{rk}}\left(-\frac{(\tilde{\mathbf{s}}_k[d] - \chi^2_{2m_kr}[d]\mathbf{s}_{k,v}[d])^2}{\chi^2_{2m_kr}[d]\cdot N_0} + \frac{(\tilde{\mathbf{s}}_k[d] - \chi^2_{2m_kr}[d]\mathbf{s}_{k,i}[d])^2}{\chi^2_{2m_kr}[d]\cdot N_0}\right)\\
&= \sum_{d=1}^{D=4}\sum_{r=1}^{N_{rk}}\left(-\frac{(\chi^2_{2m_kr}[d](\mathbf{s}_{k,i}[d] - \mathbf{s}_{k,v}[d]) + \acute{\Delta}_i[d])^2}{\chi^2_{2m_kr}[d]\cdot N_0} + \frac{(\acute{\Delta}_i[d])^2}{\chi^2_{2m_kr}[d]\cdot N_0}\right).
\end{aligned} \tag{13.18}$$

Finally, Equation (13.14) can be simplified to

$$C_{\mathrm{DCMC}} = \log_2(M) - \frac{1}{M}\sum_{i=1}^{M}\prod_{k=1}^{K} E\left[\log_2\left(\sum_{v=1}^{M}\prod_{k=1}^{K}\exp(\Psi_{k,vi})\right)\,\middle|\, \mathbf{s}_i\right] \quad \text{(BPS)}, \tag{13.19}$$

where $E[A|B]$ is the expectation of A conditioned on B.

On the other hand, the CCMC capacity of the proposed LSSTC scheme can be expressed as [249, 293]

$$C_{\mathrm{CCMC}} = \sum_{k=1}^{K} E\left[\frac{D}{2}\log_2\left(1 + \sum_{r=1}^{N_{rk}} \chi^2_{2m_k,r}\frac{\mathrm{SNR}}{N_t}\right)\right] \quad \text{(BPS)}. \tag{13.20}$$

Finally, the bandwidth efficiency is related to the capacity according to [293]

$$\eta = \frac{C}{D/2} \quad (\text{bit s}^{-1}\ \text{Hz}^{-1}). \tag{13.21}$$

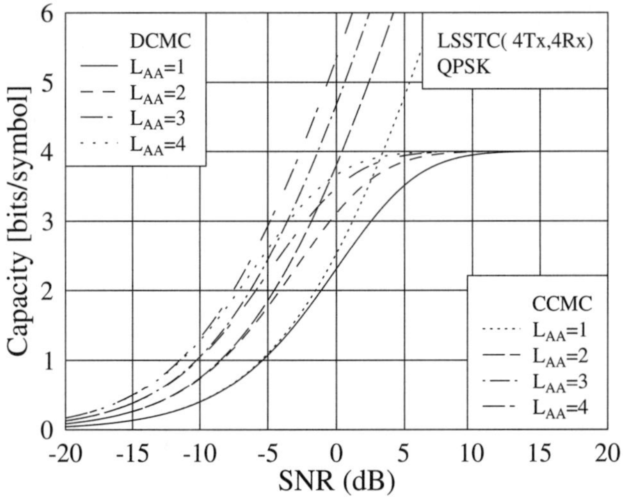

Figure 13.4: Capacity of a QPSK-modulated $N_t \times N_r = 4 \times 4$ LSSTC scheme for variable L_{AA} values.

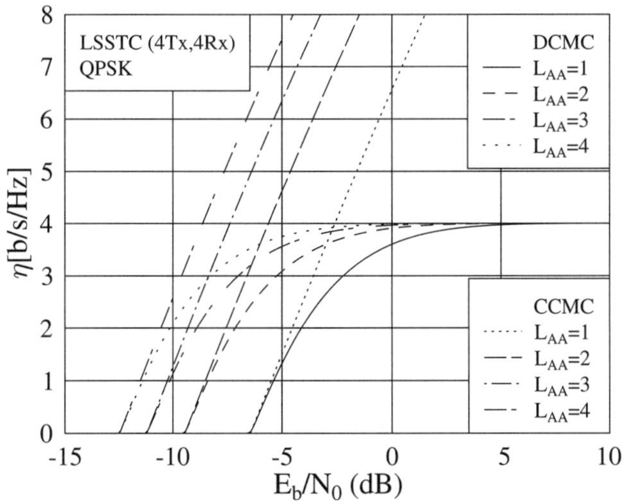

Figure 13.5: Bandwidth efficiency of a QPSK-modulated $N_t \times N_r = 4 \times 4$ LSSTC scheme for variable L_{AA} values.

Figure 13.4 shows the DCMC capacity evaluated from Equation (13.19) for the QPSK-assisted LSSTC, when employing $N_t = 4$ transmit antennas and $N_r = 4$ receive antennas as well as $L_{AA} = 1$, 2, 3 and 4 elements per transmit AA. The CCMC [6] capacity of the same multi-functional MIMO scheme is also plotted for comparison in Figure 13.4 based on Equation (13.20). Furthermore, Figure 13.5 compares the achievable bandwidth efficiency of the QPSK-modulated LSSTC scheme in conjunction with $N_t \times N_r = 4 \times 4$ while varying the number of elements L_{AA} per AA. Figure 13.5 shows that as the number of elements per AA increases the achievable bandwidth efficiency improves. Figure 13.5 also compares the achievable CCMC bandwidth efficiency for various L_{AA} values.

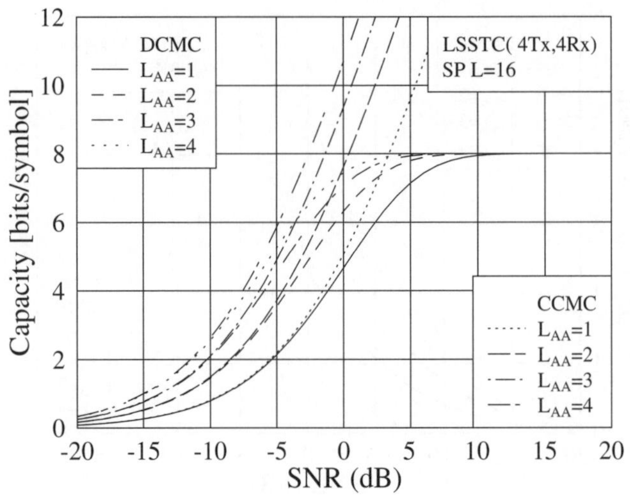

Figure 13.6: Capacity of the SP-modulated $N_t \times N_r = 4 \times 4$ LSSTC scheme in conjunction with $L = 16$ for variable L_{AA} values.

On the other hand, Figure 13.6 shows the DCMC capacity evaluated from Equation (13.19) for the LSSTC-SP scheme in conjunction with $L = 16$, when employing $N_t = 4$ transmit antennas, $N_r = 4$ receive antennas and a variable number of elements per AA L_{AA}. The CCMC [6] capacity of the same multi-functional MIMO scheme was also plotted for comparison in Figure 13.6 based on Equation (13.20). Furthermore, Figure 13.7 compares the achievable bandwidth efficiency of the LSSTC-SP scheme in conjunction with $L = 16$, $N_t \times N_r = 4 \times 4$, while varying the number of elements L_{AA} per AA. Figure 13.7 demonstrates that as the number of elements per AA increases, the achievable bandwidth efficiency improves. Figure 13.7 also compares the achievable CCMC bandwidth efficiency for various L_{AA} values.

Finally, Figure 13.8 compares the attainable bandwidth efficiency of the LSSTC multi-functional MIMO scheme, when QPSK and SP in conjunction with $L = 16$ are employed. As seen in Figure 13.8, when $L_{AA} = 1$ was employed, the SP-assisted system had a slightly higher bandwidth efficiency than the QPSK-aided system. On the other hand, Figure 13.8 demonstrates that increasing the number of elements per AA to $L_{AA} = 4$ results in improving the bandwidth efficiency of the SP-aided system compared with that of its QPSK-assisted counterpart.

13.4 Iterative Detection and EXIT Chart Analysis

The block diagram of the LSSTC-aided iteratively detected SP modulation is shown in Figure 13.9. The transmitted source bits $\mathbf{u}_1$ are encoded by the outer channel Encoder I having a rate of R_I. The outer channel encoded bits $\mathbf{c}_1$ are then interleaved by a random bit interleaver Π_1, where the randomly permuted bits $\mathbf{u}_2$ are fed through the URC Encoder II. The encoded bits $\mathbf{c}_2$ at the output of the URC encoder are interleaved by a second random bit interleaver Π_2, producing the permuted bit stream $\mathbf{b}$. After bit interleaving, the SP mapper maps blocks of B_{sp} channel-coded bits $\mathbf{b} = b_0, \ldots, b_{B_{\text{sp}}-1} \in \{0, 1\}$ to the L legitimate

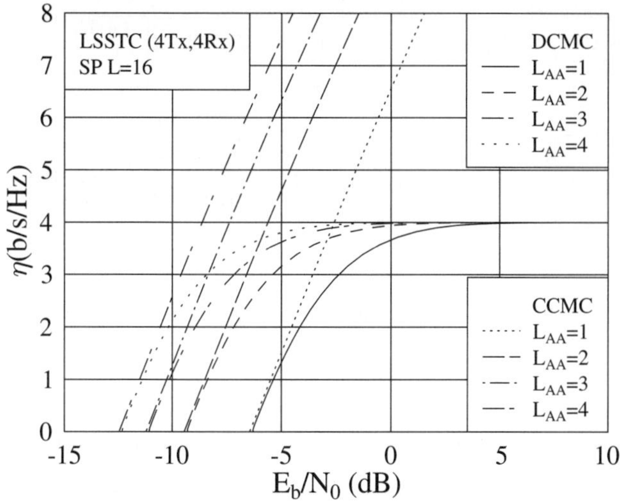

Figure 13.7: Bandwidth efficiency of the SP-modulated $N_t \times N_r = 4 \times 4$ LSSTC scheme in conjunction with $L = 16$ for variable L_{AA} values.

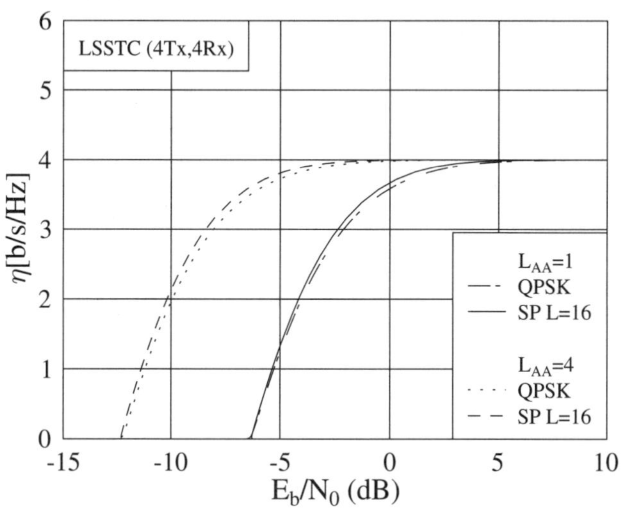

Figure 13.8: Comparison of the bandwidth efficiency of the QPSK- and the SP-aided schemes in conjunction with the $L = 16$, $N_t \times N_r = 4 \times 4$ based LSSTC scheme for $L_{AA} = 1$ and $L_{AA} = 4$.

four-dimensional SP-modulated symbols $\mathbf{s}^l \in \mathbf{S}$. The SP-modulated symbols $\mathbf{s}$ are then transmitted using the LSSTC transmitter of Section 13.2.

At the receiver side, as shown in Figure 13.9, the received complex-valued symbols are first decoded by the LSSTC-SP decoder in order to produce the received SP soft symbols $\tilde{\mathbf{s}}$. Then, iterative demapping/decoding is carried out between the SP demapper, the soft-in soft-out URC Decoder II and the soft-in soft-out Decoder I, where extrinsic information can be exchanged between the three constituent demapper/decoder modules.

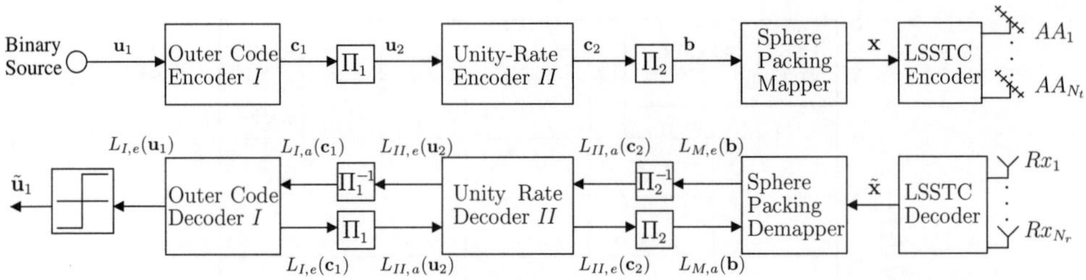

Figure 13.9: Block diagram of the LSSTC employing SP modulation in conjunction with a unity-rate precoder and an outer code.

Table 13.1: LSSTC-aided iterative detection system parameters.

SP modulation	$L = 16$
Number of transmitter AAs N_t	4
Number of elements per AA L_{AA}	4
Number of receiver antennas N_r	4
Channel	Correlated Rayleigh fading
Normalized Doppler frequency	0.01
Encoder I	
System 1	RSC(2,1,3), $(G_r, G) = (7, 5)_8$
System 2	Half-rate IRCC
System 3	RSC(2,1,2), $(G_r, G) = (3, 2)_8$
Encoder II	Unity-rate code
Generator	$(G_r, G) = (3, 2)_8$
Interleaver depth D_{int}	180 000 bits

More specifically, $L_{\cdot,a}(\cdot)$ in Figure 13.9 represents the *a priori* information, expressed in terms of the LLRs of the corresponding bits, whereas $L_{\cdot,e}(\cdot)$ represents the extrinsic LLRs of the corresponding bits. The iterative process is performed for a number of consecutive iterations. During the last iteration, only the LLR values $L_{I,p}(\mathbf{u_1})$ of the original data information bits $\mathbf{u}_1$ are required, which are passed to a hard decision decoder in order to determine the estimated transmitted source bits $\tilde{\mathbf{u}}_1$, as shown in Figure 13.9.

In this chapter we present three iterative-detection-aided SP-assisted LSSTC schemes. The first and second systems, referred to as System 1 and System 2, respectively, employ no iterations between the URC Decoder II and the SP demapper of Figure 13.9, i.e. no inner iterations. However, the two systems differ in the choice of the outer Encoder I, namely while System 1 employs a regular RSC code, System 2 uses an IRCC [174, 176]. Finally, System 3 invokes three-stage iterative detection exchanging extrinsic information between the SP demapper, the URC Decoder II and the outer RSC Decoder I. In what follows, all of the results presented characterize a LSSTC-SP scheme using $(N_t \times N_r) = (4 \times 4)$ and $L_{AA} = 4$ elements per AA in conjunction with the system parameters outlined in Table 13.1.

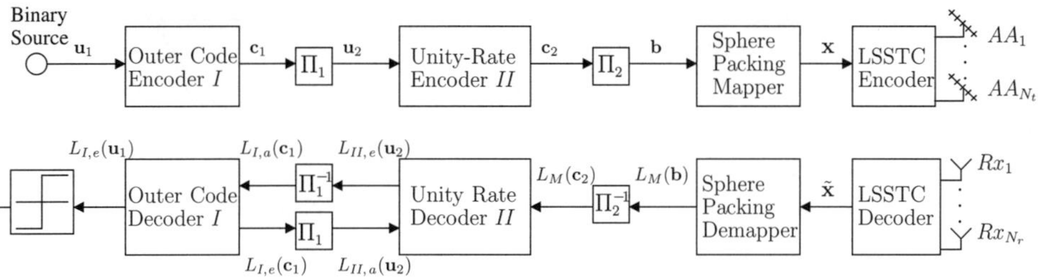

Figure 13.10: Block diagram of the LSSTC employing SP modulation in conjunction with a unity-rate precoder and an outer code. The receiver employs iterative detection between the outer code's decoder and the precoder's decoder.

13.4.1 Two-stage Iterative Detection Scheme

In this section System 1 and System 2 are described, where the exchange of extrinsic information is carried out between the outer Decoder I and the URC Decoder II only, i.e. no iterations are carried out between the URC Decoder II and the SP demapper. As seen in Figure 13.10, the URC Decoder II processes the information forwarded by the SP demapper in conjunction with the *a priori* information $L_{II,a}(\mathbf{u}_2)$ in order to generate the *a posteriori* probability. The *a priori* LLR values of the URC decoder are subtracted from the *a posteriori* LLR values for the sake of generating the extrinsic LLR values $L_{II,e}(\mathbf{u}_2)$, as seen in Figure 13.10. Next, the soft bits $L_{I,a}(\mathbf{c}_1)$ are passed to the outer Decoder I of Figure 13.10 in order to compute the *a posteriori* LLR values for all of the channel-coded bits. As seen in Figure 13.10, the extrinsic information $L_{I,e}(\mathbf{c}_1)$ is then fed back to the URC Decoder II as the *a priori* information $L_{II,a}(\mathbf{u}_2)$ after appropriately reordering its bits using the interleaver Π_1 of Figure 13.10. The soft-in soft-out Decoder II of Figure 13.10 exploits the *a priori* information for the sake of providing improved extrinsic LLR values, which are then passed to the outer Decoder I and then back to the Decoder II for further iterations.

13.4.1.1 Two-dimensional EXIT Charts

As discussed in Chapter 11, the main objective of employing EXIT charts [169, 172] is to predict the convergence behavior of the iterative decoding process by examining the evolution of the input/output MI exchange between the constituent decoders in consecutive iterations.

Let $I_{\cdot,a}(x)$, $0 \leq I_{\cdot,a}(x) \leq 1$, denote the MI between the *a priori* LLRs $L_{\cdot,a}(x)$ and the corresponding bits x and let $I_{\cdot,e}(x)$, $0 \leq I_{\cdot,e}(x) \leq 1$, denote the MI between the extrinsic LLRs $L_{\cdot,e}(x)$ and the corresponding bits x.

In the two-stage iterative detector of System 1 and System 2 of Figure 13.10, the EXIT characteristics of Decoders I and II can be described by the following two EXIT functions [169, 178]:

$$I_{I,e}(\mathbf{c}_1) = T_{I,c_1}[I_{I,a}(\mathbf{c}_1)] \tag{13.22}$$

$$I_{II,e}(\mathbf{u}_2) = T_{II,u_2}[I_{II,a}(\mathbf{u}_2),\, E_b/N_0]. \tag{13.23}$$

Figure 13.11 shows the EXIT chart of System 1 employing an iteratively detected RSC-coded LSSTC-SP system in conjunction with $L = 16$ and Gray mapping, where iterations are carried out between the outer half-rate RSC code and the inner URC decoders, while no iterations are invoked between the URC decoder and the SP demapper. The system employs

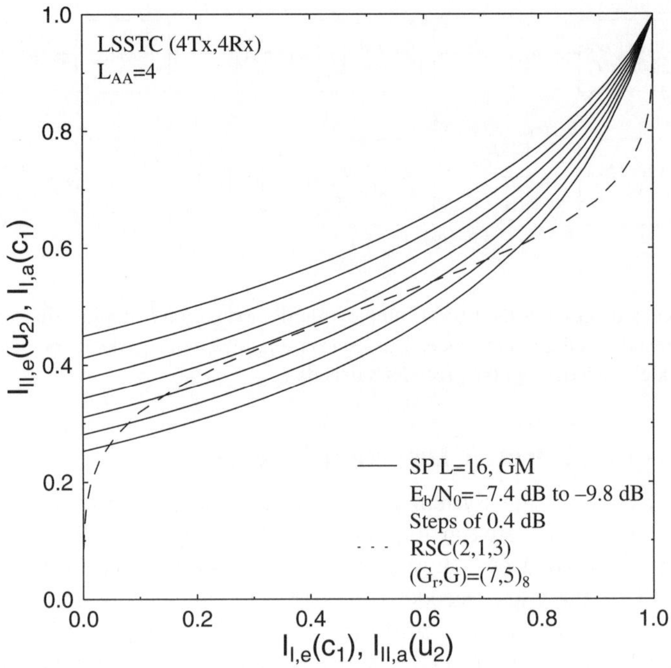

Figure 13.11: EXIT chart of a RSC-coded and URC-precoded LSSTC-SP System 1 employing Gray mapping in conjunction with $L = 16$ and the system parameters outlined in Table 13.1.

a half-rate memory-two RSC code, denoted as RSC(2,1,3), in conjunction with an octal generator polynomial of $(G_r, G) = (7, 5)_8$, where G_r is the feedback polynomial. Encoder II is a simple URC scheme, described by the pair of octal generator polynomials $(G_r, G) = (3, 2)_8$. Furthermore, the EXIT chart of Figure 13.11 was generated for the LSSTC-SP system employing $(N_t, N_r) = (4, 4)$ using $L_{AA} = 4$ elements per AA in conjunction with the system parameters of Table 13.1. Gray mapping was used in this case, because no iterations are invoked between the SP demapper and the decoders, hence it is better to use Gray mapping that results in a higher initial MI and hence a higher starting point for the EXIT curve. Observe from Figure 13.11 that an open convergence tunnel is formed around $E_b/N_0 = -8.5$ dB. This implies that according to the predictions of the EXIT chart seen in Figure 13.11, the iterative decoding process is expected to converge at $E_b/N_0 = -8.5$ dB. The EXIT-chart-based convergence predictions can be verified by the Monte Carlo simulation-based iterative decoding trajectory of Figure 13.12, where the trajectory was recorded at $E_b/N_0 = -8.5$ dB, while using an interleaver depth of $D_{\text{int}} = 180\,000$ bits and the rest of the system parameters outlined in Table 13.1. The steps seen in the figure represent the actual extrinsic information exchange between the URC's decoder and the outer RSC channel decoder.

13.4.1.2 EXIT Tunnel-area Minimization for Near-capacity Operation using IRCCs

It is a well-understood property of the conventional two-dimensional EXIT charts that a narrow but marginally open EXIT tunnel represents a near-capacity performance [176]. Therefore, we invoke IRCCs for the sake of appropriately shaping the EXIT curves by minimizing the area within the EXIT tunnel using the procedure of [174, 176].

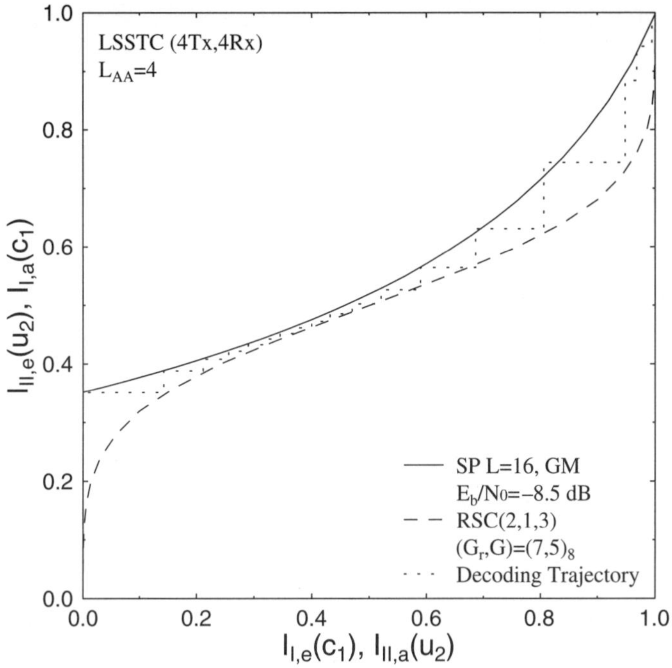

Figure 13.12: Decoding trajectory of the iteratively detected RSC-coded and URC-precoded LSSTC-SP System 1 employing Gray mapping in conjunction with $L = 16$ and the system parameters outlined in Table 13.1 while operating at $E_b/N_0 = -8.5$ dB.

Let $\mathcal{A}_I$ and $\bar{\mathcal{A}}_I$ be the areas under the EXIT curve of Decoder I and its inverse, respectively. Similarly, the area $\mathcal{A}_{II}$ is defined as that under the EXIT curve of the URC Decoder II. It was observed in [174, 273] that for the APP-based outer Decoder I, the area $\bar{\mathcal{A}}_I$ may be approximated by $\bar{\mathcal{A}}_I \approx R_I$, where the equality $\bar{\mathcal{A}}_I = R_I$ was later shown in [270] for the family of BECs. The area property of $\bar{\mathcal{A}}_I \approx R_I$ implies that the lowest SNR convergence threshold occurs when we have $\mathcal{A}_{II} = R_I + \epsilon$, where ϵ is an infinitesimally small number, provided that the following convergence constraints hold [176]:

$$T_{II,u_2}(0) > 0, \quad T_{II,u_2}(1) = 1, \quad T_{II,u_2}(i) > T_{I,c_1}^{-1}(i), \quad \text{for all } i \in [0, 1). \tag{13.24}$$

Observe in Figure 13.12, however, that there is a wide tunnel between the EXIT curve $T_{II,u_2}(i)$ and the EXIT curve $T_{I,c_1}^{-1}(i)$ of the outer half-rate RSC code at $E_b/N_0 = -8.5$ dB, especially when $i < 0.2$ and $i > 0.6$. This implies that the BER curve is farther from the achievable capacity than necessary. More quantitatively, the area under the EXIT curve $T_{II,u_2}(i)$ is $\mathcal{A}_{II} \approx 0.5625$ at $E_b/N_0 = -8.5$ dB, which is larger than the outer code rate of $R_I = 0.50$. Therefore, according to Figure 13.12 and to the area property of $\bar{\mathcal{A}}_I \approx R_I$, a lower E_b/N_0 convergence threshold may be attained, provided that the constraints outlined in Equation (5.7) are satisfied. In other words, the EXIT curve $T_{I,c_1}^{-1}(i)$ of the outer code should match the EXIT curve $T_{II,u_2}(i)$ of Figure 13.12 more closely. Hence we invoke IRCCs as outer codes that exhibit flexible EXIT characteristics, which can be optimized for more closely matching the EXIT curve $T_{II,u_2}(i)$ of Figure 13.12, converting the near-capacity code optimization to a simple curve-fitting problem.

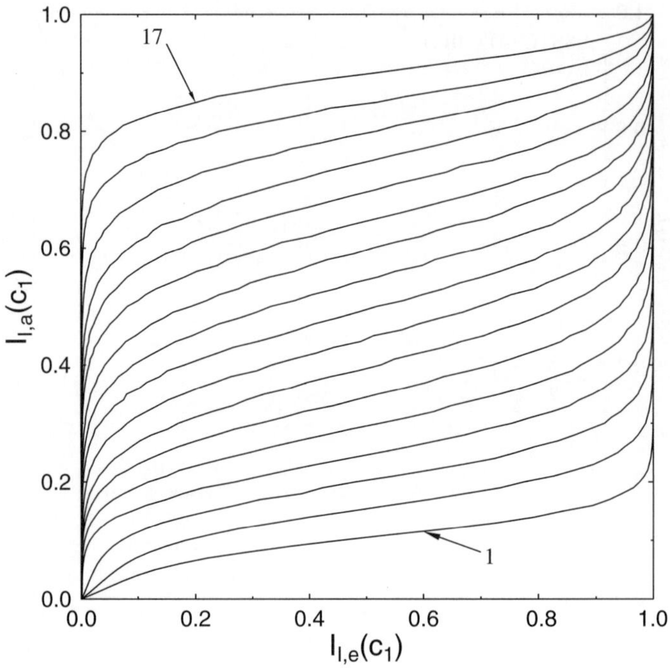

Figure 13.13: EXIT functions of the 17 subcodes used in the IRCC.

An IRCC scheme constituted by a set of $P = 17$ subcodes was constructed in [176] from a systematic half-rate memory-four mother code defined by the octally represented generator polynomials $(G_r, G) = (31, 27)_8$. Each of the $P = 17$ subcodes has a different code rate R_I^i, for all $i \in [1, 17]$, where puncturing was employed to obtain the rates of $R_I^i > 0.5$ and the code rates of $R_I^i < 0.5$ were created by adding more generators and by puncturing. The two additional generators employed in [176] are defined by the octally represented polynomials of $G_1 = (35)_8$ and $G_2 = (35)_8$, where the resultant $P = 17$ subcodes have coding rates spanning the range of $[0.1, 0.9]$. Each of the $P = 17$ subcodes encodes a specific fraction of the uncoded bits determined by the weighting coefficient α_i, $i = 1, \ldots, P$. Assuming an overall average code rate of $R_I = 0.5$, the following conditions must be satisfied:

$$\sum_{i=1}^{P} \alpha_i = 1, \quad R_I = \sum_{i=1}^{P} \alpha_i R_I^i \quad \text{and} \quad \alpha_i \in [0, 1], \quad \text{for all } i. \tag{13.25}$$

The EXIT function $T_{I,c_1}(I_{I,a}(c_1))$ corresponding to the IRCC may be constructed from the EXIT functions of the $P = 17$ subcodes, $T_{I,c_1}^i(I_{I,a}(c_1))$, $i = 1, \ldots, P$. More specifically, the EXIT function $T_{I,c_1}(I_{I,a}(c_1))$ of the IRCC is the weighted superposition of the $P = 17$ EXIT functions $T_{I,c_1}^i(I_{I,a}(c_1))$, $i = 1, \ldots, P$, as follows [176]:

$$T_{I,c_1}(I_{I,a}(c_1)) = \sum_{i=1}^{P} \alpha_i T_{I,c_1}^i(I_{I,a}(c_1)). \tag{13.26}$$

Figure 13.13 shows the EXIT functions of the $P = 17$ subcodes used in [176]. Hence, the coefficients α_i are optimized with the aid of the iterative algorithm of [174], so that the EXIT

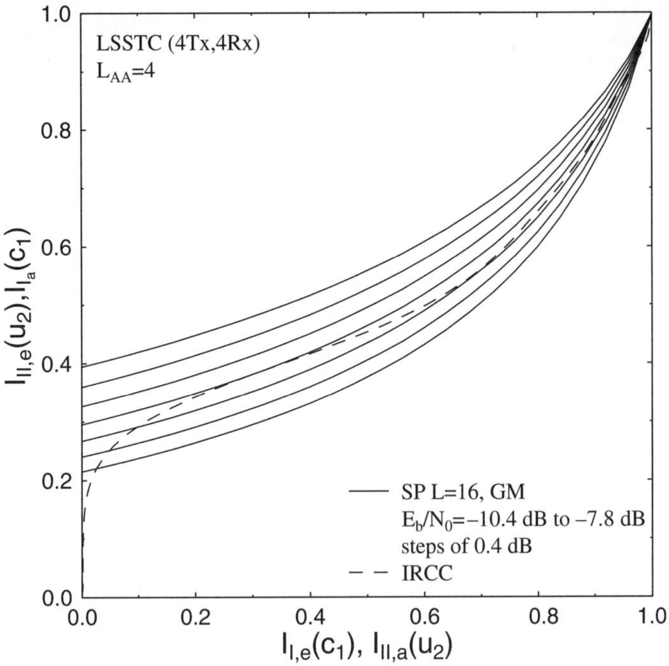

Figure 13.14: EXIT chart of an IRCC-coded and URC-precoded LSSTC-SP System 2 employing Gray mapping in conjunction with $L = 16$ and the system parameters outlined in Table 13.1.

curve of the resultant IRCC closely matches the EXIT curve $T_{II,u_2}(i)$ at the specific E_b/N_0 value, where we have $\mathcal{A}_{II} \approx 0.50$. It is observed that $\mathcal{A}_{II} \approx 0.501$ at $E_b/N_0 = -9.4$ dB. However, observe in Figure 13.14 that we have an open tunnel at $E_b/N_0 = -9.2$ dB, indicating that this E_b/N_0 value is close to the lowest attainable convergence threshold, when employing a half-rate outer code. Figure 13.14 also shows the EXIT curve of the resultant IRCC, where the optimized weighting coefficients are as follows:

$$[\alpha_1, \alpha_2, \ldots, \alpha_{16}, \alpha_{17}]$$
$$= [0, 0, 0, 0, 0, 0, 0.571927, 0.167114, 0, 0, 0, 0.0194083, 0.170044, 0, 0, 0, 0.0715566]. \quad (13.27)$$

Figure 13.14 shows the EXIT chart of System 2 employing an iterative-detection-aided IRCC-coded LSSTC-SP system using Gray mapping, where the iterations are carried out between the outer half-rate IRCC code and the inner URC decoders, while no iterations are invoked between the URC decoder and the SP demapper. The system parameters of Table 13.1 are used for producing the EXIT curves of Figure 13.14. Observe from Figure 13.14 that an open convergence tunnel is formed around $E_b/N_0 = -9.2$ dB. This implies that according to the predictions of the EXIT chart of Figure 13.14, the iterative decoding process is expected to converge at $E_b/N_0 = -9.2$ dB. However, observe in Figure 13.14 that the open tunnel at $E_b/N_0 = -9.2$ dB is quite narrow and thus it requires a large number of iterations to converge at $E_b/N_0 = -9.2$ dB. This issue is discussed further in the complexity analysis in Section 13.5.

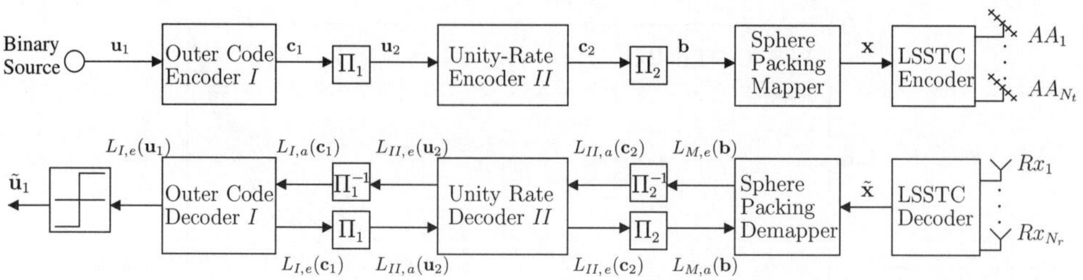

Figure 13.15: Block diagram of the LSSTC employing SP modulation in conjunction with a unity-rate precoder and an outer code. The receiver employs iterative detection between the three constituent decoders, namely between the outer code decoder, the precoder's decoder and the SP demapper.

13.4.2 Three-stage Iterative Detection Scheme

As shown in Figure 13.15, the received complex-valued symbols are first decoded by the LSSTC decoder in order to produce the received SP soft symbols $\tilde{\mathbf{s}}$, where each SP symbol represents a block of B_{sp} coded bits as described in Section 13.2.2. Then, iterative demapping/decoding is carried out between the SP demapper, soft-in soft-out URC Decoder II and soft-in soft-out outer Decoder I, where extrinsic information is exchanged between the three constituent demapper/decoders. The iterative process is performed for a number of consecutive iterations. During the last iteration, only the LLR values $L_{I,p}(\mathbf{u}_1)$ of the original uncoded systematic information bits $\mathbf{u}_1$ are required, which are passed to a hard decision decoder in order to determine the estimated transmitted source bits $\tilde{\mathbf{u}}_1$ as shown in Figure 13.15.

13.4.2.1 Three-dimensional EXIT Charts

In this section System 3 is described, where iterative detection is employed by exchanging extrinsic information between the SP demapper, the URC Decoder II and the outer Decoder I. As seen from Figure 13.15, the input of Decoder II is constituted by the *a priori* input $L_{II,a}(\mathbf{c}_2)$ and the *a priori* input $L_{II,a}(\mathbf{u}_2)$ after appropriately ordering the data provided by the SP demapper and Decoder I, respectively. Therefore, the EXIT characteristics of Decoder II can be described by the following two EXIT functions [169, 178]:

$$I_{II,e}(\mathbf{c}_2) = T_{II,c_2}[I_{II,a}(\mathbf{u}_2), I_{II,a}(\mathbf{c}_2)], \tag{13.28}$$

$$I_{II,e}(\mathbf{u}_2) = T_{II,u_2}[I_{II,a}(\mathbf{u}_2), I_{II,a}(\mathbf{c}_2)], \tag{13.29}$$

which are illustrated by the three-dimensional surfaces drawn in dotted lines in Figures 13.16 and 13.17, respectively. On the other hand, the EXIT characteristics of the SP demapper as well as those of Decoder I are each dependent on a single *a priori* input, namely on $L_{M,a}(\mathbf{b})$ and $L_{I,a}(\mathbf{c}_1)$, respectively, both of which are provided by the URC Decoder II after appropriately ordering the bits, as seen in Figure 13.15. The EXIT characteristics of the SP demapper are also dependent on the E_b/N_0 value. Consequently, the corresponding EXIT functions for the SP demapper and Decoder I, respectively, may be written as

$$I_{M,e}(\mathbf{b}) = T_{M,b}[I_{M,a}(\mathbf{b}), E_b/N_0], \tag{13.30}$$

$$I_{I,e}(\mathbf{c}_1) = T_{I,c_1}[I_{I,a}(\mathbf{c}_1)], \tag{13.31}$$

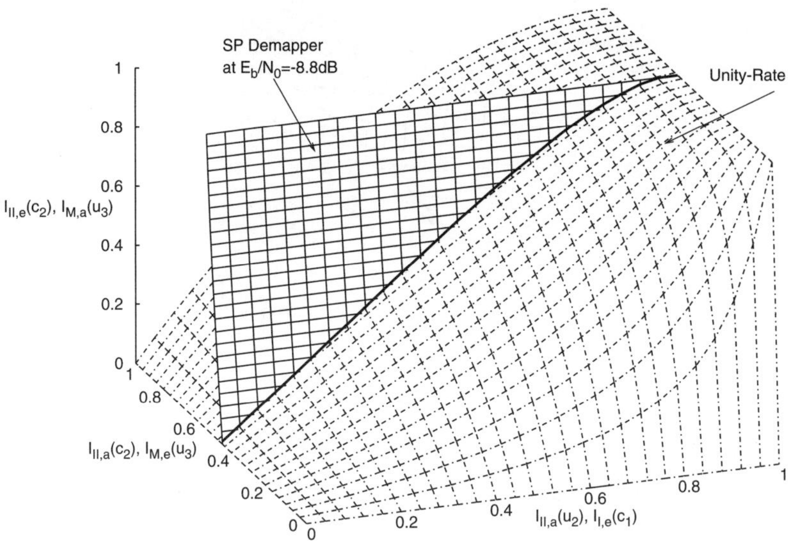

Figure 13.16: Three-dimensional EXIT chart of the URC Decoder II and the SP demapper at $E_b/N_0 = -8.8$ dB.

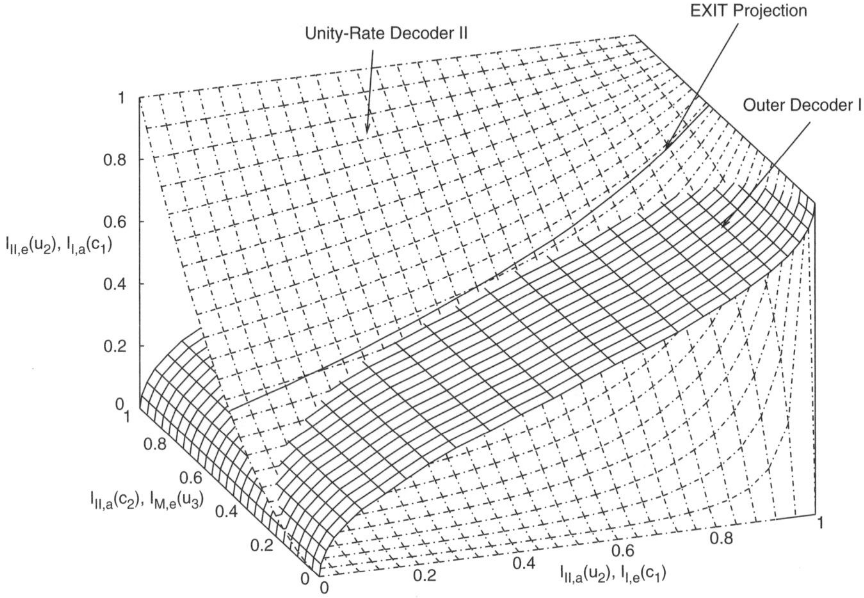

Figure 13.17: Three-dimensional EXIT chart of the URC Decoder II and the RSC Decoder I with projection from Figure 13.16.

which are illustrated by the three-dimensional surfaces drawn in solid lines in Figures 13.16 and 13.17, respectively.

Equations (13.28)–(13.31) can be represented with the aid of two three-dimensional EXIT charts. More specifically, the three-dimensional EXIT chart of Figure 13.16 is used to plot Equation (13.28) and Equation (13.30), which describe the EXIT relation between the

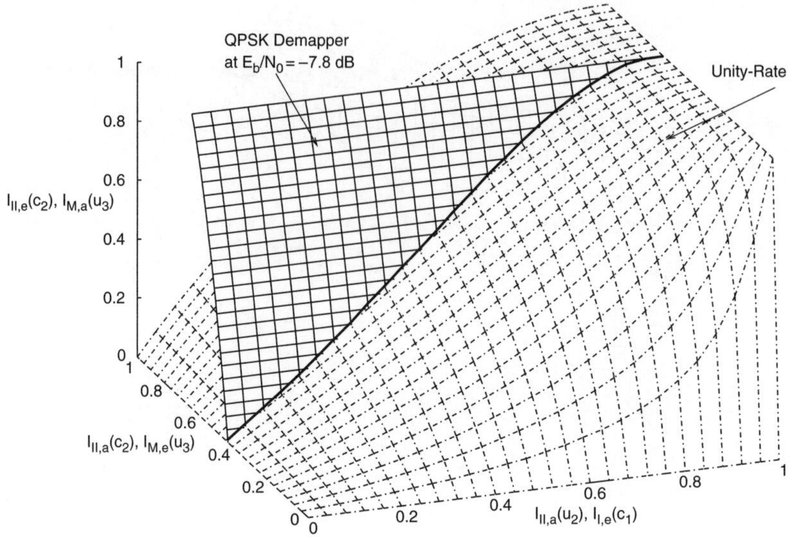

Figure 13.18: Three-dimensional EXIT chart of the URC Decoder II and the QPSK demapper at $E_b/N_0 = -7.8$ dB.

SP demapper and Decoder II. Similarly, the three-dimensional EXIT chart of Figure 13.17 can be used to describe the EXIT relation between Decoder II and Decoder I by plotting Equation (13.29) and Equation (13.31). Furthermore, for the sake of comparison we plot the three-dimensional EXIT curves of the QPSK-modulated LSSTC scheme employing a three-stage iterative-detection-aided receiver and the same system parameters as the SP system of Figures 13.16 and 13.17. Figure 13.18 describes the EXIT relation between the QPSK demapper and the URC Decoder II, while Figure 13.19 describes the EXIT relation between Decoder II and Decoder I together with the EXIT projection from Figure 13.18.

13.4.2.2 Two-dimensional EXIT Chart Projection

As observed in Figures 13.16 and 13.17, it is cumbersome to interpret the three-dimensional EXIT charts. Hence, in this section we derive their unique and unambiguous two-dimensional representations, which may be interpreted more readily.

The intersection of the surfaces in Figure 13.16, shown as a thick solid line, portrays the achievable performance when exchanging MI between the SP demapper and the URC Decoder II for different fixed values of $I_{II,a}(\mathbf{u}_2)$ spanning the range of $[0, 1]$. Each $[I_{II,a}(\mathbf{u}_2),\ I_{II,a}(\mathbf{c}_2),\ I_{II,e}(\mathbf{c}_2)]$ point belonging to the intersection line of Figure 13.16 uniquely specifies a three-dimensional point $[I_{II,a}(\mathbf{u}_2),\ I_{II,a}(\mathbf{c}_2),\ I_{II,e}(\mathbf{u}_2)]$ in Figure 13.17, according to the EXIT function of Equation (13.29). Therefore, the line corresponding to the $[I_{II,a}(\mathbf{u}_2), I_{II,a}(\mathbf{c}_2), I_{II,e}(\mathbf{c}_2)]$ points along the thick line of Figure 13.16 is projected to the solid line shown in Figure 13.17. The two-dimensional projection of the solid line in Figure 13.17 at $I_{II,a}(\mathbf{c}_2) = 0$ onto the plane spanned by the lines $[I_{II,a}(\mathbf{u}_2), I_{II,e}(\mathbf{u}_2)]$ and $[I_{I,e}(\mathbf{c}_1), I_{I,a}(\mathbf{c}_1)]$ is shown in Figure 13.20 at $E_b/N_0 = -8.8$ dB for all possible SP AGM schemes of Appendix H. This projected EXIT curve may be written as

$$I_{II,e}(\mathbf{u}_2) = T^{p}_{II,u_2}[I_{II,a}(\mathbf{u}_2),\ E_b/N_0]. \tag{13.32}$$

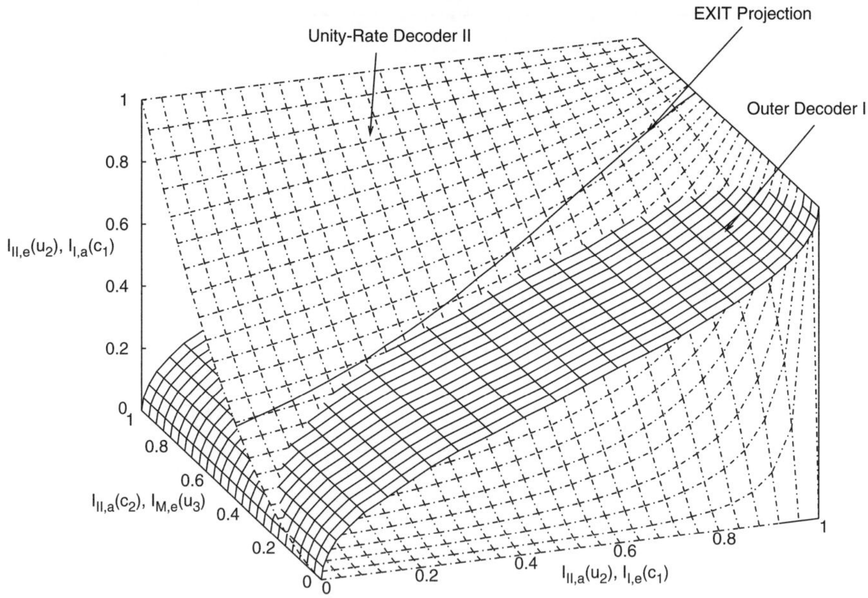

Figure 13.19: Three-dimensional EXIT chart of the URC Decoder II and the RSC Decoder I with projection from Figure 13.18.

Observe in Figure 13.20 the variety of curves that result from using different mapping schemes in the three-stage iterative-detection-aided system. Figure 13.20 shows the two-dimensional EXIT projection at $E_b/N_0 = -8.8$ dB. As seen in the figure, an open tunnel exists at $E_b/N_0 = -8.8$ dB for the system employing AGM-6, while an open tunnel exists for the other mapping schemes at different E_b/N_0 values higher than -8.8 dB. Therefore, according to the EXIT chart prediction, the system employing AGM-6 exhibits an open tunnel at $E_b/N_0 = -8.8$ dB and thus it is expected that the system employing AGM-6 exhibits an infinitesimally low BER at E_b/N_0 of -8.8 dB.

The intersection of the surfaces in Figure 13.16, shown as a thick solid line, portrays the best achievable performance, when exchanging MI between the SP demapper and the URC Decoder II for different fixed values of $I_{II,a}(\mathbf{u}_2)$ spanning the range of [0, 1]. The best achievable performance is that corresponding to the AGM-6-assisted system, as portrayed in Figure 13.20. Therefore, the line corresponding to the $[I_{II,a}(\mathbf{u}_2), I_{II,a}(\mathbf{c}_2), I_{II,e}(\mathbf{c}_2)]$ points along the thick line of Figure 13.16 is projected to the solid line shown in Figure 13.17, while the two-dimensional projection of the solid line in Figure 13.17 at $I_{II,a}(\mathbf{c}_2) = 0$ onto the plane spanned by the lines $[I_{II,a}(\mathbf{u}_2), I_{II,e}(\mathbf{u}_2)]$ and $[I_{I,e}(\mathbf{c}_1), I_{I,a}(\mathbf{c}_1)]$ is shown in Figure 13.21 at $E_b/N_0 = -8.8$ dB. Figure 13.21 shows the two-dimensional-projected EXIT curve of the combined SP demapper and the URC Decoder II at $E_b/N_0 = -8.8$ dB, when employing the best possible AGM scheme, namely AGM-6. Figure 13.21 records the two-dimensional projected EXIT curves for a variable number of inner iterations between the SP demapper and Decoder II. As observed from Figure 13.21, when no inner iterations are employed, the system becomes essentially a two-stage arrangement employing AGM instead of Gray mapping for System 1 and System 2. According to Figure 13.21, when no inner iterations are carried out, the system requires $E_b/N_0 > -8.8$ dB for maintaining an open tunnel. However, observe that when 1, 2 and 20 inner iterations are carried out, the open

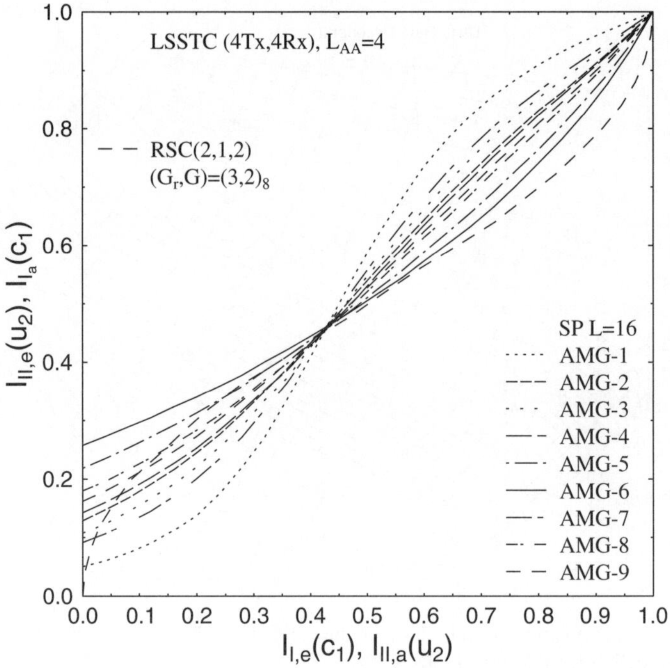

Figure 13.20: Two-dimensional projection of the EXIT charts of the three-stage RSC-coded LSSTC-SP System 3, when employing all possible AGM-aided SP in conjunction with $L = 16$, while using an interleaver length of $D_{\text{int}} = 180\,000$ bits and the system parameters outlined in Table 13.1 at $E_b/N_0 = -8.8$ dB.

EXIT tunnel is formed at $E_b/N_0 = -8.8$ dB. Therefore, in our further investigations we use a single inner iteration that produces the same result and imposes the lowest complexity. This implies that according to the predictions of the two-dimensional EXIT chart seen in Figure 13.21, the iterative decoding process is expected to converge to the $(1.0, 1.0)$ point and hence an infinitesimally low BER may be attained beyond $E_b/N_0 = -8.8$ dB. This expectation is confirmed by the decoding trajectory of Figure 13.22, which was recorded for an interleaver depth of $D_{\text{int}} = 180\,000$ bits in conjunction with the system parameters outlined in Table 13.1.

Figure 13.23 shows the two-dimensional-projected EXIT curve of the combined QPSK demapper and the URC Decoder II at $E_b/N_0 = -7.8$ dB, when employing the AGM scheme [263]. Figure 13.23 records the two-dimensional projected EXIT curves for a single inner iteration between the QPSK demapper and Decoder II. Observe in Figure 13.23 that no open tunnel exists between the EXIT projection and the outer code's EXIT curve at an E_b/N_0 below -7.8 dB. Therefore, a comparison between Figures 13.21 and 13.23 shows the flexibility that the multi-dimensional modulation scheme has in the choice of the mapping schemes compared with the two-dimensional QPSK scheme. Hence, according to the predictions of the two-dimensional EXIT chart seen in Figure 13.23, the iterative decoding process is expected to converge to the $(1.0, 1.0)$ point and hence an infinitesimally low BER may be attained beyond $E_b/N_0 = -7.8$ dB.

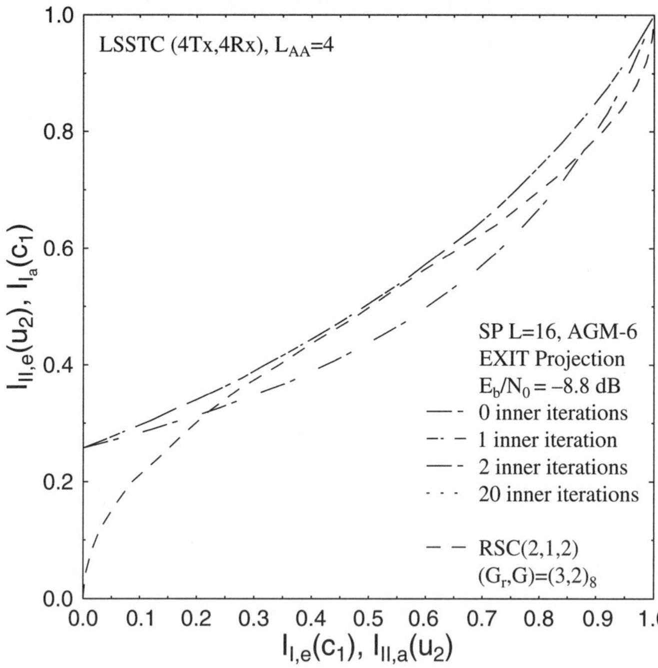

Figure 13.21: Two-dimensional projection of the EXIT charts of the three-stage RSC-coded LSSTC-SP scheme, when employing the best possible AGM-6-aided SP in conjunction with $L = 16$, while using an interleaver depth of $D_{\text{int}} = 180\ 000$ bits and the system parameters outlined in Table 13.1.

13.4.3 Maximum Achievable Bandwidth Efficiency

The MIMO channel's bandwidth efficiency curves recorded for the four-dimensional SP-modulation-assisted LSSTC scheme in conjunction with $(N_t \times N_r) = (4 \times 4)$ and $L_{AA} = 4$ elements per AA are shown in Figure 13.24, portraying both the DCMC and CCMC bandwidth efficiency curves as well as the maximum achievable rate of the system derived from the EXIT curves according to the algorithm of Section 11.2.3. Observe the discrepancy between the two bandwidth efficiency curves shown in Figure 13.24 that are calculated using Equation (13.21) and Equation (11.21), which is due to the fact that Equation (13.21) was computed for the case where perfect interference cancelation is considered at the receiver. Therefore, Equation (13.21) constitutes an upper bound on the system's bandwidth efficiency, while Equation (11.21) constitutes a tighter bound on the maximum achievable bandwidth efficiency of the system considered in this chapter.

Similarly, we plot in Figure 13.25 the MIMO channel's bandwidth efficiency curves recorded for the QPSK-modulation-assisted LSSTC scheme in conjunction with $(N_t \times N_r) = (4 \times 4)$ and $L_{AA} = 4$ elements per AA. Figure 13.25 portrays both the DCMC and CCMC bandwidth efficiency curves as well as the maximum achievable rate of the system derived from the EXIT curves according to the algorithm of Section 11.2.3.

Figure 13.24 shows that at a bandwidth efficiency of $\eta = 2$ bit s^{-1} Hz^{-1}, which is the bandwidth efficiency of the system employing the parameters of Table 13.1, the DCMC bandwidth efficiency limit seen for the LSSTC-SP scheme in Figure 13.25 is about $E_b/N_0 =$

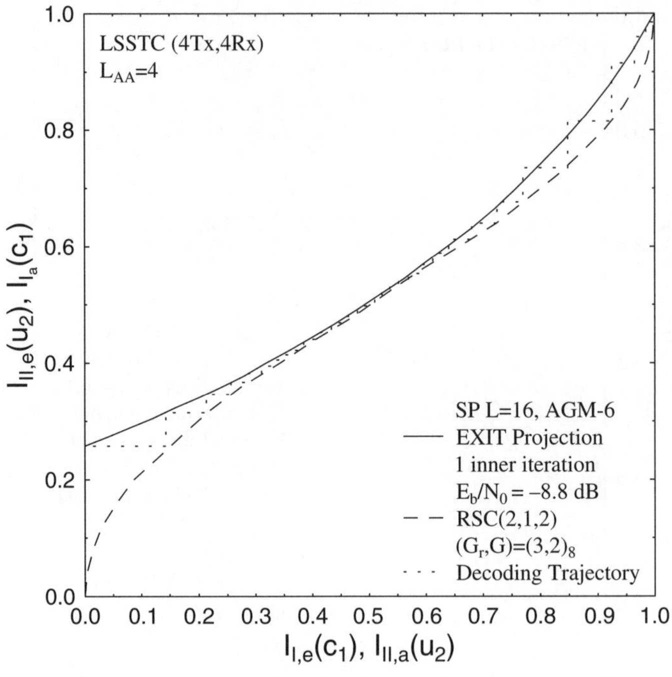

Figure 13.22: Decoding trajectory of the iteratively detected three-stage RSC-coded and URC-precoded LSSTC-SP System 3 employing Gray mapping in conjunction with $L = 16$ and the system parameters outlined in Table 13.1, while operating at $E_b/N_0 = -8.8$ dB.

-10.15 dB, while the maximum achievable rate limit obtained using EXIT charts is at $E_b/N_0 = -9.45$ dB.

Figure 13.25 shows that at a bandwidth efficiency of $\eta = 2$ bit s^{-1} Hz^{-1} the DCMC bandwidth efficiency limit seen for the LSSTC-SP scheme in Figure 13.25 is about $E_b/N_0 = -10$ dB, while the maximum achievable rate limit obtained using EXIT charts is at $E_b/N_0 = -9.34$ dB.

13.5 Results and Discussion

In this section, we consider a LSSTC system associated with the system parameters outlined in Table 13.1 in order to demonstrate the performance improvements achieved by the proposed system. We employ SP in conjunction with $L = 16$ and QPSK modulation and hence the overall bandwidth efficiency of the system is 2 bit s^{-1} Hz^{-1}. The results presented in this section correspond to the scenario where the channel is perfectly known at the receiver. We also assume that the transmitter has full knowledge of the DOA without any estimation or estimation errors.

Figure 13.26 compares the performance of the proposed System 1 employing the LSSTC-SP scheme in conjunction with $L = 16$ and Gray mapping together with the system parameters of Table 13.1 for different numbers of iterations against that of an uncoded LSSTC-SP scheme using $L = 4$, which has an identical bandwidth efficiency of 2 bit s^{-1} Hz^{-1}. Figure 13.26 shows the performance of the iteratively detected RSC-coded LSSTC-SP scheme, when employing an interleaver depth of $D_{\text{int}} = 180\,000$ bits and while

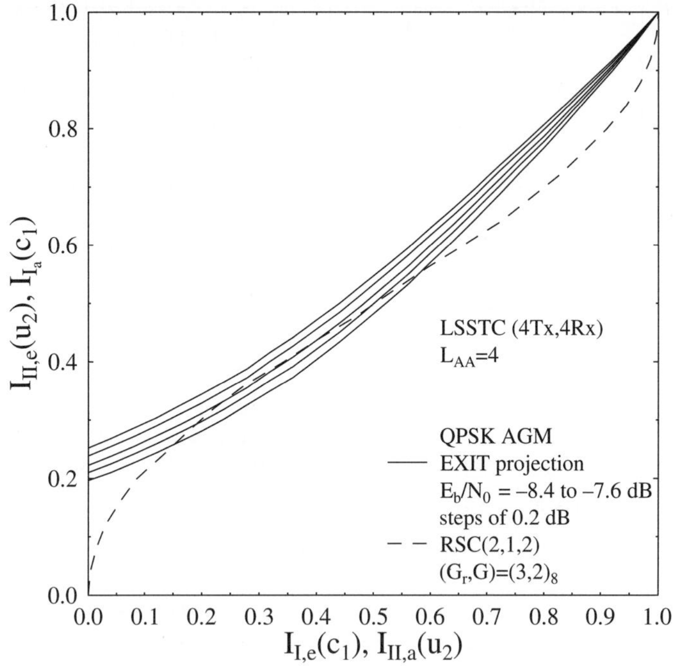

Figure 13.23: Two-dimensional projection of the EXIT charts of the three-stage RSC-coded LSSTC-QPSK scheme, when employing the AGM-aided QPSK in conjunction with an interleaver depth of $D_{\text{int}} = 180\,000$ bits and the system parameters outlined in Table 13.1.

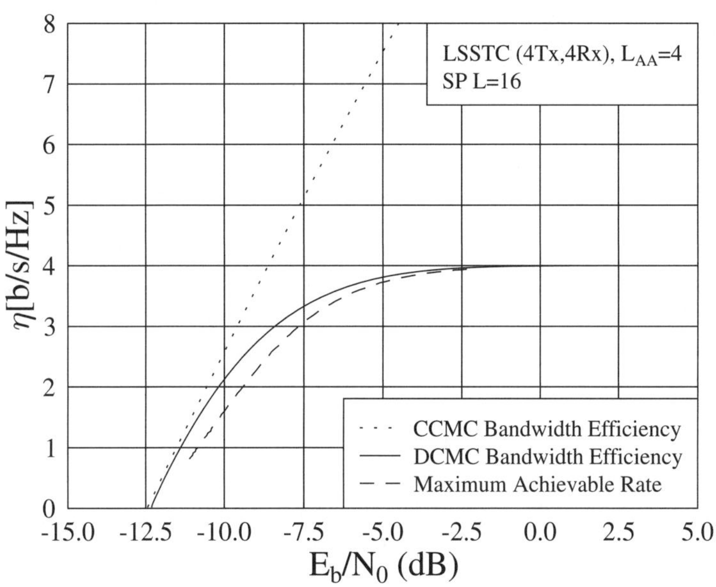

Figure 13.24: Bandwidth efficiency of the four-AAs-aided LSSTC-SP system in conjunction with $L = 16$ employing four elements per AA for both DCMC and CCMC together with the maximum achievable rate obtained using EXIT charts.

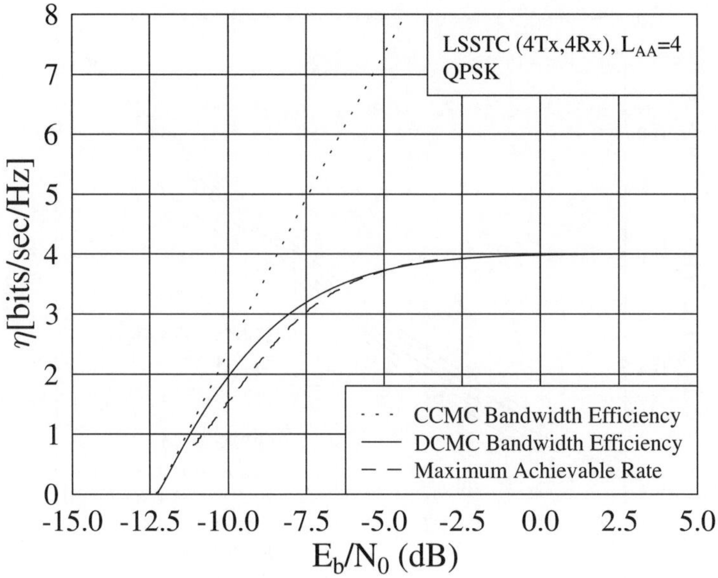

Figure 13.25: Bandwidth efficiency of the four-AAs-aided LSSTC-QPSK system in conjunction with four elements per AA for both DCMC and CCMC together with the maximum achievable rate obtained using EXIT charts.

communicating over a temporally correlated Rayleigh fading channel associated with a normalized Doppler frequency of $f_D = 0.01$. Explicitly, Figure 13.26 demonstrates that a coding advantage of about 16.5 dB was achieved at a BER of 10^{-6} after 17 iterations by System 1 over the equivalent-throughput uncoded LSSTC-SP scheme employing $L = 4$. Furthermore, Figure 13.26 demonstrates that the BER performance closely matches the EXIT-chart-based prediction of Figure 13.12, where the system approaches an infinitesimally low BER at $E_b/N_0 = -8.5$ dB after 17 iterations. Finally, according to Figure 13.26, System 1 performs within 0.9 dB from the maximum achievable rate limit of Figure 13.24 obtained using the EXIT chart and within 1.65 dB from the LSSTC-SP system's bandwidth efficiency limit.

On the other hand, Figure 13.27 compares the performance of the proposed System 2 employing the LSSTC-SP scheme in conjunction with $L = 16$ and Gray mapping, together with the system parameters of Table 13.1 for different numbers of iterations, where the system has a bandwidth efficiency of 2 bit s^{-1} Hz^{-1}. The system of Figure 13.27 employs the IRCC of Figure 13.14 as an outer code and iterative detection is carried out between the IRCC decoder and the unity rate code's decoder. Furthermore, the system employs an interleaver depth of $D_{\text{int}} = 180\,000$ bits and communicates over a temporally correlated Rayleigh fading channel associated with a normalized Doppler frequency of $f_D = 0.01$. Explicitly, Figure 13.27 demonstrates that the system approaches an infinitesimally low BER at $E_b/N_0 = -9.0$ dB after 100 iterations. However, according to the EXIT chart of Figure 13.14, it is predicted that the system's BER performance converges at $E_b/N_0 = -9.2$ dB. This is mainly due to the fact that the convergence tunnel in the EXIT chart of Figure 13.14 is quite narrow and the system requires more than 100 iterations for matching the EXIT chart prediction of maintaining an infinitesimally low BER at an E_b/N_0 value of -9.2 dB. Finally, according to Figure 13.27, System 2 performs within 0.4 dB of the

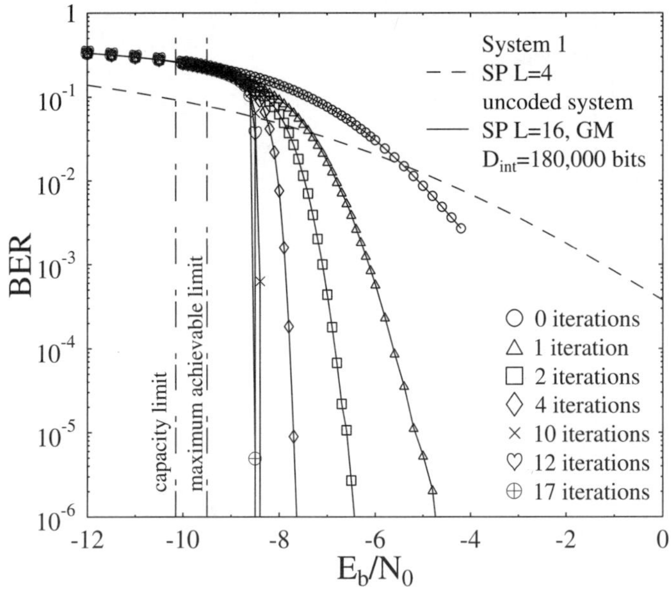

Figure 13.26: Performance comparison of the proposed LSSTC-SP-aided System 1 employing iterative detection between a half-rate RSC decoder and a URC decoder employing Gray-mapping-aided SP in conjunction with $L = 16$, while using an interleaver length of $D_{\text{int}} = 180\,000$ bits and the system parameters outlined in Table 13.1 for a variable number of iterations.

maximum achievable rate limit obtained using the EXIT chart of Figure 13.24 and within 1.1 dB from the LSSTC-SP system's bandwidth efficiency limit.

Figure 13.28 compares the attainable performance of the proposed System 3 employing the LSSTC-SP scheme in conjunction with $L = 16$ and AGM-6 and the system parameters of Table 13.1 recorded for different numbers of iterations. In Figure 13.28, an interleaver depth of $D_{\text{int}} = 180\,000$ bits was employed for communication over a temporally correlated Rayleigh fading channel associated with a normalized Doppler frequency of 0.01. Figure 13.28 demonstrates that the BER performance closely matches the EXIT-chart-based prediction of Figure 13.22, where the system approaches an infinitesimally low BER at $E_b/N_0 = -8.8$ dB after 46 iterations. Finally, according to Figure 13.28, System 3 performs within 0.6 dB of the maximum achievable rate limit obtained using the EXIT chart of Figure 13.24 and within 1.35 dB of the LSSTC-SP system's bandwidth efficiency limit.

Furthermore, Figure 13.29 compares the attainable performance of the iteratively detected RSC-coded LSSTC scheme employing QPSK modulation in conjunction with the system parameters of Table 13.1 for different numbers of iterations. Figure 13.29 shows the BER curve for the three-stage system, where the decoder employs iterative detection exchanging extrinsic information between the three constituent decoders/demapper, namely the QPSK demapper, the URC Decoder II and the RSC Decoder I. Figure 13.29 demonstrates that the BER performance closely matches the EXIT-chart-based prediction of Figure 13.23, where the system approaches an infinitesimally low BER at $E_b/N_0 = -7.8$ dB after 38 iterations. Finally, according to Figure 13.29, the proposed QPSK-modulated system performs within 1.54 dB of the maximum achievable rate limit obtained using the EXIT chart of Figure 13.24 and within 2.2 dB of the system's bandwidth efficiency limit.

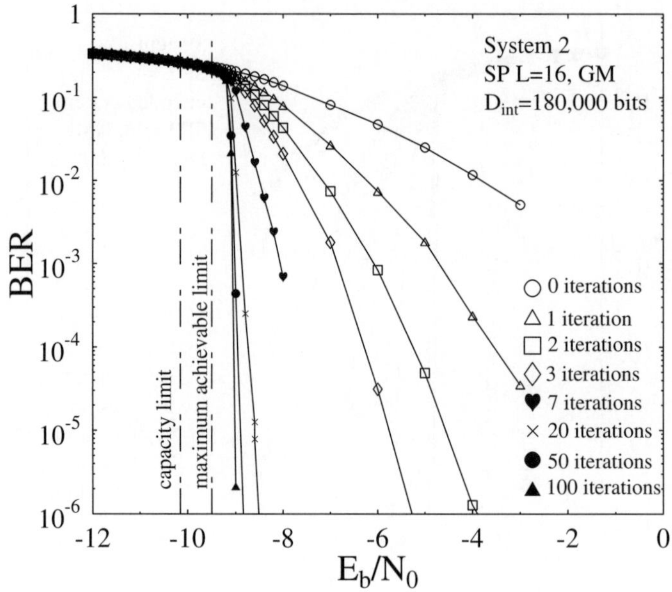

Figure 13.27: Performance comparison of the proposed LSSTC-SP aided System 2 employing iterative detection between a half-rate IRCC decoder and a URC decoder employing Gray-mapping-aided SP in conjunction with $L = 16$, while using an interleaver length of $D_{\text{int}} = 180\,000$ bits and the system parameters outlined in Table 13.1 for a variable number of iterations.

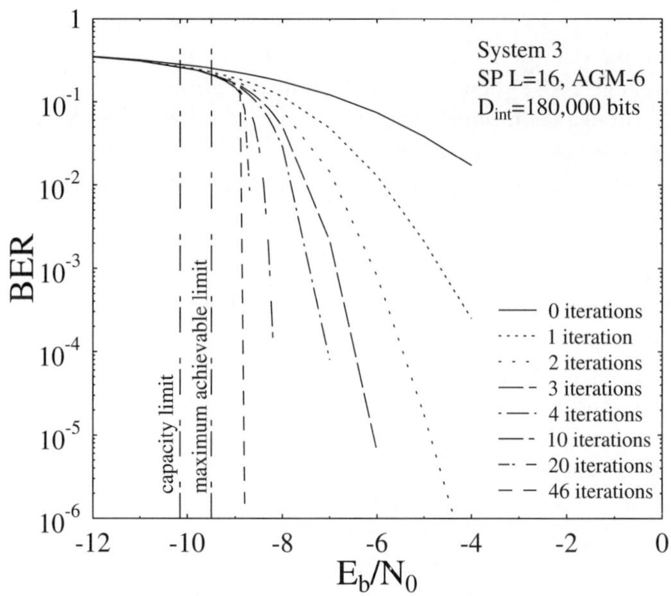

Figure 13.28: Performance comparison of the proposed LSSTC-SP-aided System 3 employing three-stage iterative detection between a half-rate RSC decoder and a URC decoder as well as a SP demapper employing AGM-aided SP in conjunction with $L = 16$, while using an interleaver depth of $D_{\text{int}} = 180\,000$ bits and the system parameters outlined in Table 13.1 for a variable number of iterations.

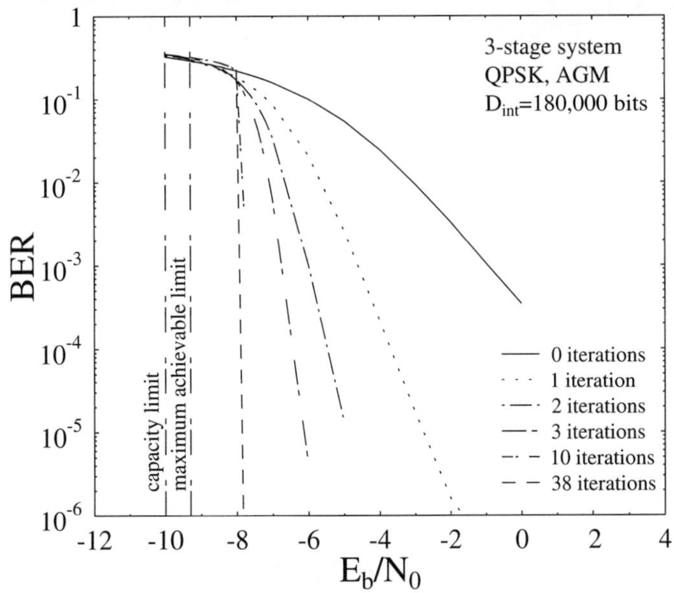

Figure 13.29: Performance comparison of the proposed LSSTC QPSK-aided System 3 employing iterative detection between a half-rate RSC decoder, a URC decoder and a QPSK demapper employing AGM-aided QPSK, while using an interleaver depth of $D_{\text{int}} =$ 180 000 bits and the system parameters outlined in Table 13.1 for a variable number of iterations.

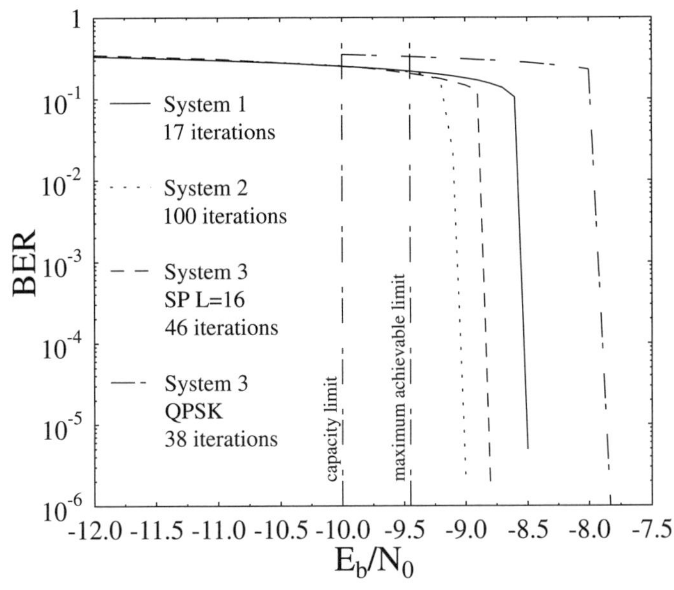

Figure 13.30: Performance comparison of the three proposed LSSTC SP-aided systems employing two-stage iteration between an outer code and URC decoders, as well as a three-stage iterative system between an outer RSC, intermediate URC decoders and a SP demapper. The BER performance of the three-stage system employing QPSK is also plotted.

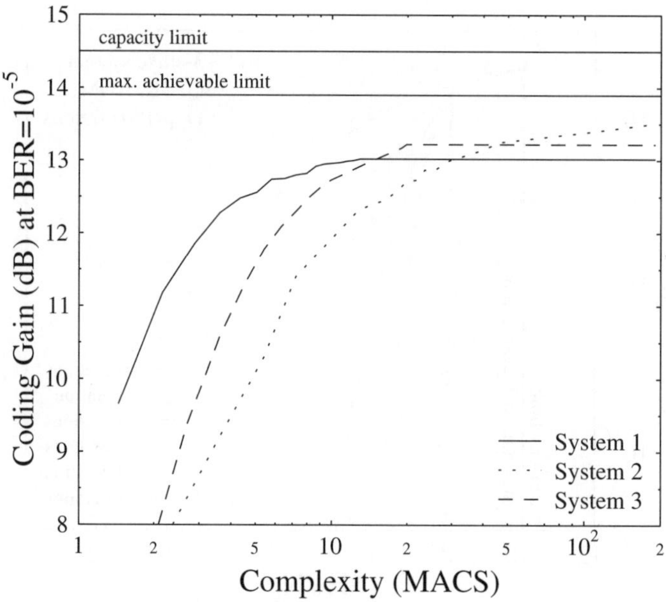

Figure 13.31: Comparison of the coding gain at a BER of 10^{-5} versus the complexity in terms of the number of trellis states of the three proposed LSSTC SP-aided systems while employing the system parameters in Table 13.1. (MACS: Million Add–Compare–Select operations.)

A comparison between the three proposed SP-modulated systems and the QPSK-modulated three-stage iteratively detected system is presented in Figure 13.30. Figure 13.30 compares the maximum achievable performance of the proposed iterative-detection-aided LSSTC schemes, while employing an interleaver depth of $D_{\text{int}} = 180\,000$ bits and the system parameters of Table 13.1. Observe that the performance of System 1 as well as that of the SP-modulated and QPSK-modulated System 3 matches the EXIT chart predictions of Figures 13.12, 13.22 and 13.23 after $I = 17$, 46 and 38 iterations, respectively. Explicitly, the SP-modulated System 3 converges at E_b/N_0 of -8.8 dB, as predicted by the EXIT chart of Figure 13.22, while the QPSK-modulated System 3 converges at E_b/N_0 of -7.8 dB, as predicted by the EXIT chart of Figure 13.23. Hence, the SP-modulated System 3 performs within 0.6 dB of the maximum achievable rate limit obtained using the EXIT charts, while the QPSK-modulated System 3 performs within 1.54 dB of the maximum achievable rate limit obtained using the EXIT charts. However, System 2 does not closely match the EXIT-chart-based convergence prediction at $E_b/N_0 = -9.2$ dB even after $I = 100$ iterations, when employing an interleaver depth of $D_{\text{int}} = 180\,000$ bits. This is due to the fact that at $E_b/N_0 = -9.2$ dB, the EXIT tunnel of Figure 13.14 is narrow and thus requires a large number of iterations, which is significantly higher than $I = 100$. Thus, using the system parameters outlined in Table 13.1 and $I = 100$ decoding iterations, System 2 converges at $E_b/N_0 = -9.0$ dB, which is 0.4 dB from the maximum achievable rate obtained using the EXIT chart. Finally, System 1 performs within 0.9 dB of the maximum achievable rate limit.

Finally, Figure 13.31 shows the coding gain achieved at a BER of 10^{-5} for each system versus the detection complexity expressed in terms of the number of trellis states. Figure 13.31 demonstrates that System 1 employing the two-stage iteratively detected system in conjunction with the RSC(2,1,3) code has the lowest complexity at a distance of 0.9 dB from the maximum achievable rate limit, where System 1 converges and hence approaches an infinitesimally low BER. System 3 is capable of performing equally well in BER terms, while operating 0.3 dB closer to the maximum achievable rate limit than System 1. However, this is achieved at the cost of almost doubling the complexity, as seen in Figure 13.31. On the other hand, in order to operate as close as $E_b/N_0 = 0.2$ dB from the maximum achievable rate limit, System 2 using IRCCs has to be employed in order to further reduce the EXIT tunnel's area. However, to match the EXIT chart predictions of Figure 13.14, the system requires a large number of iterations between Decoder I and Decoder II. When employing as many as 100 decoding iterations, System 2 becomes capable of performing within 0.4 dB of the maximum achievable rate limit, although this is achieved at the cost of a complexity that is 20 times that required for operating within 0.9 dB of the maximum achievable rate limit using System 1 and ten times that necessitated by operating within 0.6 dB of the maximum achievable rate limit, when employing System 3. Note that the coding gain seen in Figure 13.31 for System 2 does not saturate and, thus, provided that a higher number of iterations is affordable, a higher coding gain can be attained, as expected from the EXIT chart of Figure 13.14.

13.6 Chapter Conclusions

In this chapter, we proposed a novel multi-functional MIMO scheme that combines the benefits of STC and V-BLAST as well as beamforming. The system is also combined with multi-dimensional SP modulation facilitating the joint design of the AA's space-time signals and, hence, maximizing the coding advantage of the transmission scheme. We also quantified the capacity of the LSSTC-SP scheme and computed an upper limit on the achievable bandwidth efficiency of the system, which is based on EXIT charts. Furthermore, we proposed three near-capacity iteratively detected LSSTC-SP schemes, where iterative detection is carried out between an outer code decoder, an intermediate code decoder and an LSSTC-SP demapper. The three proposed schemes differ in the number of inner iterations employed between Decoder II and the SP demapper, as well as in the choice of the outer code, which is either a regular RSC code or an IRCC. On the other hand, the intermediate code employed is a URC, which is capable of completely eliminating the system's error floor as well as operating at the lowest possible turbo-cliff SNR without significantly increasing the associated complexity or interleaver delay. Explicitly, the system can operate within 0.9, 0.6 and 0.4 dB of the maximum achievable rate limit. However, to operate within 0.6 dB of the maximum achievable rate limit, the system imposes twice the complexity compared with a system operating within 0.9 dB of this limit. On the other hand, to operate as close as 0.4 dB from the maximum achievable rate limit, the system imposes 20 times higher complexity than that operating within 0.9 dB of the maximum achievable rate limit. The proposed design principles are applicable to an arbitrary number of antennas and diverse antenna configurations as well as modem schemes.

13.7 Chapter Summary

In this chapter, we have proposed a multi-functional MIMO scheme that combines the benefits of V-BLAST codes, of space-time codes and of beamforming. Thus, the proposed system benefits from the multiplexing gain of V-BLAST, from the diversity gain of space-time codes and from the SNR gain of the beamformer. This multi-functional MIMO scheme is referred to as a LSSTC. To further enhance the attainable system performance and to maximize the coding advantage of the proposed transmission scheme, the system was also combined with multi-dimensional SP modulation.

In Section 13.3 we quantified the capacity of the proposed multi-functional MIMO scheme and presented the capacity limits for a system employing $N_t = 4$ transmit antennas, $N_r = 4$ receive antennas and a variable number L_{AA} of elements per AA. Furthermore, in Section 13.4.3 we derived an upper bound of the achievable bandwidth efficiency for the system based on the EXIT charts obtained for the iteratively detected system.

To further enhance the achievable system performance, the proposed MIMO scheme was serially concatenated with an outer code combined with a URC, where three different receiver structures were presented by varying the iterative detection configuration of the constituent decoders/demapper as outlined in Table 13.1. In Section 13.4.1 we provided a brief description of the iteratively detected two-stage RSC-coded LSSTC-SP scheme, where extrinsic information was exchanged between the outer RSC decoder and the inner URC decoder, while no iterations were carried out between the URC decoder and the SP demapper. The schematic of the system is shown in Figure 13.10. The convergence behavior of the iterative-detection-aided system was analyzed using EXIT charts in Section 13.4.1.1. In Section 13.4.1.2, we employed the powerful technique of EXIT tunnel-area minimization for near-capacity operation. More specifically, we exploited the well-understood properties of conventional two-dimensional EXIT charts that a narrow but nonetheless open EXIT tunnel represents a near-capacity performance. Consequently, we invoked IRCCs for the sake of appropriately shaping the EXIT curves by minimizing the area within the EXIT tunnel using the procedure of [174, 176].

In Section 13.4.2 we presented a three-stage iteratively detected RSC-coded LSSTC scheme, where extrinsic information was exchanged between the three constituent demapper/decoders, namely the outer RSC decoder, the inner URC decoder and the demapper. Three-dimensional EXIT charts were presented in Section 13.4.2.1, followed by Section 13.4.2.2 where the simplified two-dimensional projections of the three-dimensional EXIT charts were provided. In Figure 13.20 we portrayed the wide choice of the SP mapping schemes available and demonstrated that the system employing AGM-6 provided the best achievable performance amongst all possible mapping schemes. In contrast, QPSK modulation has only a single AGM scheme, which was characterized in Figure 13.23.

In Section 13.5 we discussed our performance results and characterized the three proposed iteratively detected LSSTC schemes, while employing the system parameters outlined in Table 13.1. Explicitly, the SP-aided system can operate within 0.9, 0.6 and 0.4 dB of the maximum achievable rate limit. However, when operating within 0.6 dB of the maximum achievable rate limit, the system imposes twice the complexity compared with a system operating within 0.9 dB of this limit. On the other hand, to operate as close as 0.4 dB from the maximum achievable rate limit, the system imposes a 20 times higher complexity as that operating within 0.9 dB of the same limit. The proposed design principles are applicable to an arbitrary number of antennas and diverse antenna configurations as well as modem schemes. In contrast, the QPSK-modulated iteratively detected three-stage system is capable of operating within 1.54 dB of the maximum achievable rate limit and thus the SP-modulated system outperforms its QPSK-aided counterpart by about 1 dB at a BER of 10^{-6}.

Chapter 14

DL LSSTS-aided Generalized MC DS-CDMA

14.1 Introduction

In Chapter 13 we presented a multifunctional MIMO scheme that combines the benefits of space-time codes and of the V-BLAST scheme as well as of beamforming. In other words, the LSSTC of Chapter 13 benefits from a spatial diversity gain and a multiplexing gain as well as beamforming gain. The correct decoding of the LSSTC of Chapter 13 requires that the number of receive antennas is higher than or equal to the number of transmit antennas. This condition makes the LSSTC scheme less practical, especially when we consider the DL scheme of a BS communicating with a MS. The LSSTC scheme can, however, be conveniently applied for communicating between two BSs or between a BS and a laptop. To make the multi-functional MIMO presented in Chapter 13 more practical, we present in this chapter a multi-functional MIMO scheme employing four DL transmit and two receive antennas, which constitutes a rank-deficient scheme, where the channel matrix is non-invertible. However, we use linear decoding to decode the twin-antenna-based received signal. The proposed multi-functional MIMO combines the benefits of STS, V-BLAST and generalized MC DS-CDMA as well as beamforming. Therefore, the proposed scheme benefits from a spatial diversity gain, a frequency diversity gain, a multiplexing gain and beamforming gain.

Alamouti [11] discovered a witty transmit diversity scheme using two transmit antennas, which was generalized by Tarokh *et al.* [12, 25] to an arbitrary number of transmit antennas, defining the concept of STBCs. Inspired by the philosophy of STBCs, Hochwald *et al.* [13] proposed the transmit diversity concept known as STS for the DL of WCDMA [51] that is capable of achieving the highest possible transmit diversity gain.

In [19] the authors presented a transmission scheme referred to as Double Space-Time Transmit Diversity (D-STTD), which consists of two STBC layers at a transmitter that is equipped with four transmit antennas, while the receiver is equipped with two antennas. The decoding of D-STTD presented in [19] is based on a linear decoding scheme presented in [105], where the authors presented a broad overview of STC and signal processing designed for high-data-rate wireless communications. In [105] a two-user scheme was

Near-Capacity Multi-Functional MIMO Systems L. Hanzo, O. R. Alamri, M. El-Hajjar and N. Wu

presented, where each user is equipped with a two-antenna-aided STBC block transmitting at the same carrier frequency and in the same time slot. A two-antenna-aided receiver was implemented for the sake of decoding the users' data, while eliminating the interference imposed by the users on each others' data. A zero-forcing decoder designed for the D-STTD was presented in [102] for the sake of reducing the decoding complexity. Finally, [96, 106] present further results that compare the performance of STBC versus D-STTD and extend the applicability of the scheme to more than two STBC layers.

On the other hand, beamforming [16] constitutes an effective technique of reducing the multiple-access interference, where the antenna gain is increased in the direction of the desired user whilst reducing the gain towards the interfering users. Several attempts have been made to design hybrid MIMO schemes combining STBC with beamforming [20, 98–100, 107, 375]. In order to achieve additional performance gains, beamforming has also been combined with STBC to attain a higher SNR gain [20]. In [375] eigen-beamforming was combined with STBC while allocating the transmitted power equally between the different antenna elements. On the other hand, ideal beamforming was combined with STBCs in [107] for the sake of demonstrating the performance gains attained by such a combination.

In addition, MC CDMA [51, 376] is based on a combination of code division and multicarrier or OFDM techniques [376]. In [22], a generalized MC DS-CDMA scheme was proposed that includes the subclasses of both multitone [377] and orthogonal MC DS-CDMA [378] as special cases. In addition, in [379–381] STS has been combined with beamforming and with generalized MC DS-CDMA for the sake of combining the benefits of spatial diversity, frequency diversity and beamforming gain.

Furthermore, in [158], the employment of the iterative decoding principle [146] was considered for iterative soft demapping in the context of multilevel modulation schemes combined with channel decoding. It was also demonstrated in [166] that a recursive inner code is needed in order to maximize the interleaver gain and to avoid the formation of a BER floor, when employing iterative decoding. In [168], unity-rate inner codes were employed for designing low-complexity iterative-detection-aided schemes suitable for bandwidth- and power-limited systems having stringent BER requirements. In [169], ten Brink proposed the employment of the so-called EXIT characteristics between a concatenated decoder's input and output for describing the flow of extrinsic information through the soft-in soft-out constituent decoders.

In a nutshell, in this chapter we propose a system that combines the benefits of STS, V-BLAST and beamforming as well as generalized MC DS-CDMA. The proposed system is referred to as LSSTS-aided generalized MC DS-CDMA. The system is characterized by the spatial diversity gain of the STS, the multiplexing gain of the V-BLAST and the frequency diversity gain of the generalized MC DS-CDMA as well as beamforming gain. In the generalized MC DS-CDMA scheme considered in this chapter, the subcarrier frequencies are arranged in a way that guarantees that the same STS signal is spread to and, hence, transmitted by the specific V subcarriers having the maximum possible frequency separation, so that they experience independent fading and achieve the maximum attainable frequency diversity. Therefore, the novelty and rationale of the proposed system can be summarized as follows.

1. We amalgamate the merits of V-BLAST, STC, beamforming and generalized MC DS-CDMA for the sake of achieving a multiplexing gain and a spatial and frequency diversity gain as well as beamforming gain. We propose a transmission scheme equipped with four transmit and two receive antennas and employ a low-complexity linear receiver to decode the received signal.

2. We demonstrate that the number of users supported is substantially increased by invoking combined spreading in both the TD and the FD. We also use a user-grouping technique for minimizing the Multi-User Interface (MUI) imposed, when employing TD and FD spreading in the LSSTS-aided generalized MC DS-CDMA DL scheme.

3. We propose three iteratively detected LSSTS schemes, where iterative detection is carried out by exchanging extrinsic information between two serially concatenated channel codes. We use EXIT charts to analyze the convergence behavior of the proposed iterative-detection-aided schemes and propose a novel LLR post-processing technique for improving the iteratively detected systems' performance. The three iterative-detection-aided schemes differ in the way the channel coding is implemented in the different STS layers. The overall code-rate of Systems 1-3 is identical.

   (a) In the first scheme, referred to as System 1, a single outer and a single inner channel code are used to encode the bits transmitted.

   (b) In the second scheme, namely System 2, a single outer code is implemented, whose output is split into two substreams each of which is encoded using a separate inner code.

   (c) In contrast, in the third proposed scheme, referred to as System 3, the input data bit stream is first split into two different substreams, where a pair of different outer as well as inner codes are implemented in the different substreams.

   We show that the three systems exhibit a similar complexity quantified in terms of the total number of trellis states encountered, which determines the number of ACS arithmetic operations. Similarly, we demonstrate that provided that we employ sufficiently long interleavers, the three systems attain a similar BER performance. In contrast, when shorter interleavers are employed, System 1 performs better than System 2, which in turn performs better than System 3. This is due to the fact that the interleaver depths of Systems 2 and 3 are lower than that of System 1 since the bit stream is split into two substreams in Systems 2 and 3, which constrains the interleaver to be shorter and hence the correlation in the extrinsic information becomes higher, which eventually decays the BER performance.

The rest of the chapter is organized as follows. In Section 14.2 we present the encoding and decoding algorithms of the LSSTS-aided generalized MC DS-CDMA scheme and demonstrate how the scheme benefits from the diversity gain, the multiplexing gain and the beamforming gain. In Section 14.3 we present how the TD and FD spreading can be combined in order to increase the number of users supported by the system and we introduce the user-grouping technique for reducing the MUI. Iterative detection of the proposed system is discussed in Section 14.4, where we introduce the LLR post-processing technique followed by a comparison of the attainable performance of the proposed schemes in Section 14.5. We present our conclusions in Section 14.6 followed by a brief chapter summary discussing both the main contributions and the organization of this chapter in Section 14.7.

14.2 LSSTS-aided Generalized MC DS-CDMA

In this section we describe the proposed LSSTS-aided generalized MC DS-CDMA scheme designed for achieving spatial diversity gain, frequency diversity gain and multiplexing gain as well as beamforming gain. The antenna architecture employed in Figure 14.1 for the proposed scheme is equipped with $N_t = 4$ transmit AAs spaced sufficiently far apart in order to experience independent fading. The L_{AA} elements of each of the AAs are spaced at a distance of half the wavelength for the sake of achieving beamforming. Furthermore, the receiver is equipped with $N_r = 2$ antennas. The system can support K users transmitting at the same time and using the same carrier frequencies, while they can be differentiated by the user-specific spreading code $\bar{\mathbf{c}}_k$, where $k \in [1, K]$. In addition, in the generalized MC DS-CDMA considered, the subcarrier frequencies are arranged in a way that guarantees that the same STS signal is spread to and hence transmitted by the specific V subcarriers having the maximum possible frequency separation, so that they experience independent fading and achieve the maximum attainable frequency diversity.

14.2.1 Transmitter Model

The system considered employs the generalized MC DS-CDMA scheme of [22] using UV subcarriers. The transmitter schematic of the kth user is shown in Figure 14.1, where a block of UN_t data symbols $\mathbf{x}$ is S/P converted to U parallel subblocks. Afterwards, each set of N_t symbols is S/P converted to $G = 2$ groups, where each group is encoded using the $N_{tg} = 2$ antenna-aided STS procedure of [13], where the transmitted signal is spread to N_{tg} transmit antennas with the aid of the orthogonal spreading codes of $\{\bar{\mathbf{c}}_{k,1}, \bar{\mathbf{c}}_{k,2}, \ldots, \bar{\mathbf{c}}_{k,N_{tg}}\}$, $k = 1, 2, \ldots, K$.

The spreading codes $\bar{\mathbf{c}}_{k,1}$ and $\bar{\mathbf{c}}_{k,2}$ are generated from the same user-specific spreading code $\bar{\mathbf{c}}_k$ by ensuring that the two spreading codes $\bar{\mathbf{c}}_{k,1}$ and $\bar{\mathbf{c}}_{k,2}$ become orthogonal using the simple code-concatenation rule of Walsh–Hadamard codes, yielding longer codes and hence a proportionately reduced per-antenna throughput according to

$$\bar{\mathbf{c}}_{k,1} = [\bar{\mathbf{c}}_k \quad \bar{\mathbf{c}}_k], \tag{14.1}$$

$$\bar{\mathbf{c}}_{k,2} = [\bar{\mathbf{c}}_k \quad -\bar{\mathbf{c}}_k]. \tag{14.2}$$

The discrete symbol duration of the orthogonal STS codes is $N_{tg}N_e$, where N_e represents the kth user's TD spreading factor. Each of the U subblocks is then divided into two half-rate substreams and the two consecutive symbols in each substream are then spread to both transmit antennas using the mapping of

$$\mathbf{s}_{k,u1} = (\bar{\mathbf{c}}_{k,1} \cdot x_{k,u1} + \bar{\mathbf{c}}_{k,2} \cdot x^*_{k,u2}), \tag{14.3}$$

$$\mathbf{s}_{k,u2} = (\bar{\mathbf{c}}_{k,1} \cdot x_{k,u2} - \bar{\mathbf{c}}_{k,2} \cdot x^*_{k,u1}), \tag{14.4}$$

$$\mathbf{s}_{k,u3} = (\bar{\mathbf{c}}_{k,1} \cdot x_{k,u3} + \bar{\mathbf{c}}_{k,2} \cdot x^*_{k,u4}), \tag{14.5}$$

$$\mathbf{s}_{k,u4} = (\bar{\mathbf{c}}_{k,1} \cdot x_{k,u4} - \bar{\mathbf{c}}_{k,2} \cdot x^*_{k,u3}), \tag{14.6}$$

which is exemplified in simple graphical terms in Figure 10.3.

The UN_t outputs of the UG STS blocks modulate a group of subcarrier frequencies $\{f_{u,1}, f_{u,2}, \ldots, f_{u,V}\}$. Since each of the U subblocks is spread to and hence conveyed with the aid of V subcarriers, a total of UV subcarriers are required in the MC DS-CDMA system considered. The UV subcarrier signals are superimposed on each other in order to form the complex modulated signal. The subcarrier frequencies are arranged in a way that guarantees that the same STS signal is spread to and hence transmitted by the specific V subcarriers having the maximum possible frequency separation, so that they experience independent fading and achieve the maximum attainable frequency diversity. Finally, according to the kth user's channel information, the UVN_t signals of the kth user are weighted by the transmit

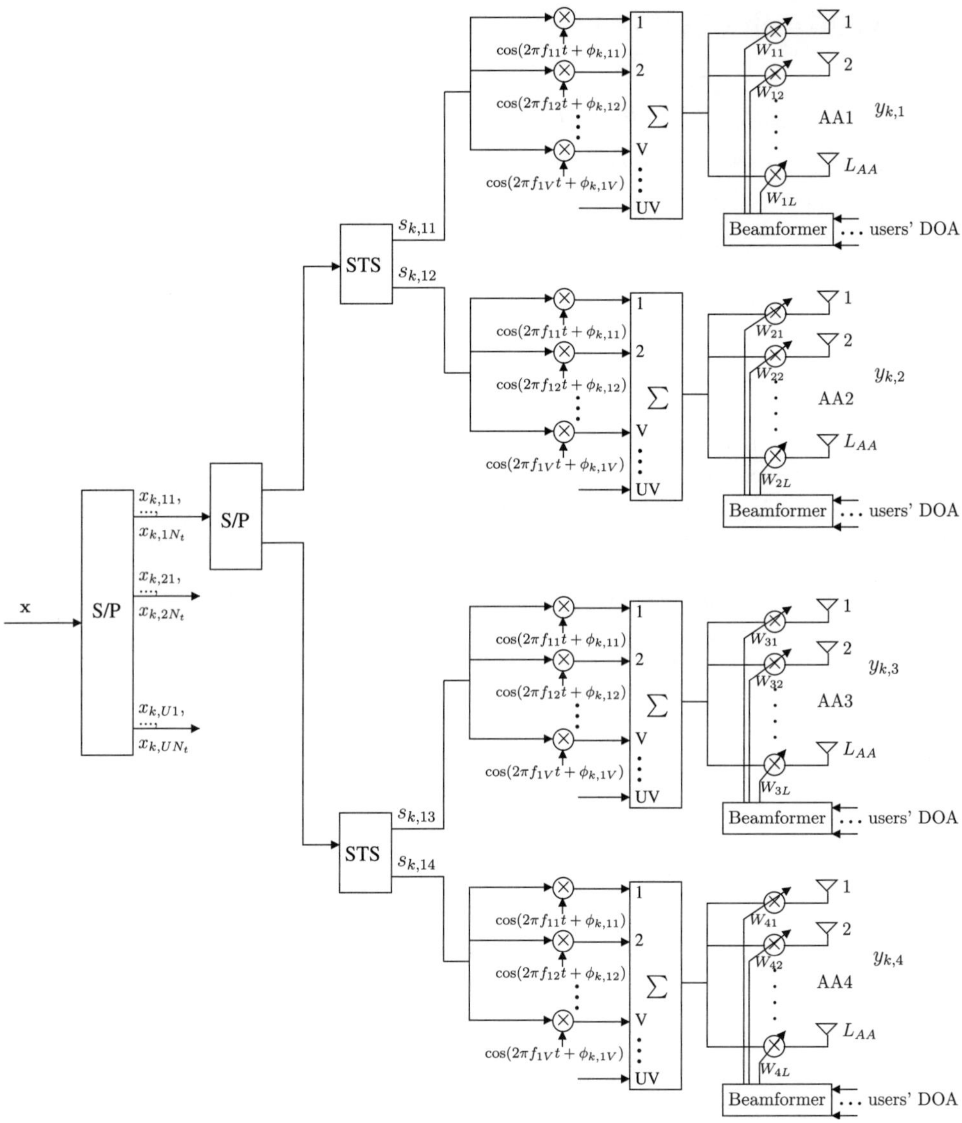

Figure 14.1: The k_{th} user's LSSTS-aided generalized MC DS-CDMA transmitter system model.

weight vector $\mathbf{w}_{uv,n}^{(k)}$ determined for the uvth subcarrier of the kth user, which is generated for the nth AA.

The k_{th} user's transmitted signal can be written as follows:

$$\begin{aligned}\mathbf{y}_{k,1}(t) &= \mathbf{y}_{k,1}(\tau + aN_tT_s)\\ &= \sum_{u=1}^{U}\sum_{v=1}^{V}\sqrt{\frac{2P_k}{VL_{AA}}\frac{1}{N_tN_{tg}}}\mathbf{w}_{uv,1}^{k}\mathbf{s}_{k,u1}\cos(2\pi f_{uv}\tau + \phi_{k,uv}),\end{aligned} \tag{14.7}$$

$$\begin{aligned}\mathbf{y}_{k,2}(t) &= \mathbf{y}_{k,2}(\tau + aN_tT_s)\\ &= \sum_{u=1}^{U}\sum_{v=1}^{V}\sqrt{\frac{2P_k}{VL_{AA}}\frac{1}{N_tN_{tg}}}\mathbf{w}^k_{uv,2}\mathbf{s}_{k,u2}\cos(2\pi f_{uv}\tau + \phi_{k,uv}),\end{aligned} \tag{14.8}$$

$$\begin{aligned}\mathbf{y}_{k,3}(t) &= \mathbf{y}_{k,3}(\tau + aN_tT_s)\\ &= \sum_{u=1}^{U}\sum_{v=1}^{V}\sqrt{\frac{2P_k}{VL_{AA}}\frac{1}{N_tN_{tg}}}\mathbf{w}^k_{uv,3}\mathbf{s}_{k,u3}\cos(2\pi f_{uv}\tau + \phi_{k,uv}),\end{aligned} \tag{14.9}$$

$$\begin{aligned}\mathbf{y}_{k,4}(t) &= \mathbf{y}_{k,4}(\tau + aN_tT_s)\\ &= \sum_{u=1}^{U}\sum_{v=1}^{V}\sqrt{\frac{2P_k}{VL_{AA}}\frac{1}{N_tN_{tg}}}\mathbf{w}^k_{uv,4}\mathbf{s}_{k,u4}\cos(2\pi f_{uv}\tau + \phi_{k,uv}),\end{aligned} \tag{14.10}$$

where T_s is the symbol duration, $a = 0, 1, \ldots, 0 \leq \tau < N_tT_s$, P_k/V represents the transmitted power of each subcarrier, the factor L_{AA} in the denominator is due to beamforming and the factor N_tN_{tg} in the denominator suggests that the STS scheme using N_t transmit antennas and N_{tg} orthogonal spreading codes distributes its power proportionally in space and time.

The bandwidth efficiency of the proposed system can be formulated as follows. Assuming that the system employs a modulation scheme transmitting B BPS, then the bandwidth efficiency of the LSSTS-aided generalized MC DS-CDMA is given by $2UB$ bits per channel use.

14.2.2 Receiver Model

Let us assume that there are N_t transmit AAs at the BS, which are located sufficiently far apart from each other, having an antenna spacing of 10λ, where λ represents the carrier's wavelength. The channel impulse response vector $h_{uv,nm}(t)$ spanning the nth transmit antenna array, $n \in [1, N_t]$, and the mth receive antenna, $m \in [1, N_r]$, while employing the uvth subcarrier can be expressed as

$$\begin{aligned}\mathbf{h}^k_{uv,nm}(t) &= [h^k_{uv,nm0}(t), h^k_{uv,nm1}(t), \ldots, h^k_{uv,nm(L_{AA}-1)}(t)]^{\mathrm{T}}\\ &= \mathbf{a}^k_{uv,nm}(t)\delta(t-\tau_k)\\ &= [a^k_{uv,nm0}(t), a^k_{uv,nm1}(t), \ldots, a^k_{uv,nm(L_{AA}-1)}(t)]\delta(t-\tau_k),\end{aligned} \tag{14.11}$$

where τ_k is the signal's delay and $a_{uv,nml}(t)$ is the CIR with respect to the nmth link, uvth subcarrier and the lth element of the nth AA. Based on the assumption that the array elements are separated by half a wavelength, we can simplify $\mathbf{a}^k_{uv,nm}(t)$ according to

$$\begin{aligned}\mathbf{a}^k_{uv,nm}(t) &= \alpha^k_{uv,nm}(t)\cdot\mathbf{d}^k_{nm}\\ &= \alpha^k_{uv,nm}(t)[1, \exp(j[\pi\sin(\psi_{nm})]), \ldots, \exp(j[(L_{AA}-1)\pi\sin(\psi_{nm})])]^{\mathrm{T}},\end{aligned} \tag{14.12}$$

where $\alpha_{nm}(t)$ is a Rayleigh faded envelope,

$$\mathbf{d}^k_{nm} = [1, \exp(j[\pi\sin(\psi_{nm})]), \ldots, \exp(j[(L_{AA}-1)\pi\sin(\psi_{nm})])]^{\mathrm{T}}, \tag{14.13}$$

and ψ_{nm} is the nmth link's DOA. As for the AA-specific DOA, we consider a scenario where the distance between the transmitter and the receiver is significantly higher than that between the AAs and thus we can assume that the signals arrive at the different AAs in parallel, i.e. the DOA at the different AAs is the same.

Assuming that the K users' data expressed in the form of Equations (14.7)–(14.10) are transmitted synchronously over a dispersive Rayleigh fading channel characterized by the CIR of Equation (14.11), the complex-valued received signal of user 1 over the two receive antennas can be expressed as

$$\mathbf{z}_1^1 = \sum_{k=1}^{K}\sum_{u=1}^{U}\sum_{v=1}^{V}(\mathbf{h}_{uv,11}\mathbf{y}_{k,uv1} + \mathbf{h}_{uu,21}\mathbf{y}_{k,uv2} + \mathbf{h}_{uv,31}\mathbf{y}_{k,uv3} + \mathbf{h}_{uv,41}\mathbf{y}_{k,uv4}) + \mathbf{n}_1, \tag{14.14}$$

$$\mathbf{z}_2^1 = \sum_{k=1}^{K}\sum_{u=1}^{U}\sum_{v=1}^{V}(\mathbf{h}_{uv,12}\mathbf{y}_{k,uv1} + \mathbf{h}_{uv,22}\mathbf{y}_{k,uv2} + \mathbf{h}_{uv,32}\mathbf{y}_{k,uv3} + \mathbf{h}_{uv,42}\mathbf{y}_{k,uv4}) + \mathbf{n}_2. \tag{14.15}$$

In Equations (14.7)–(14.10), $\mathbf{w}_{uv,nm}^k$ represents the weight vector of the desired user derived from the nmth antenna link and the uvth subcarrier, which is generated by the MRC beamformer [362] with the aid of CSI. Let $\mathbf{w}_{uv,nm}^1 = \mathbf{d}_{nm}^{1\dagger}$, then the $k = 1$st user's received signal for the uvth subcarrier can be simplified to

$$\mathbf{z}_{uv,1}^1 = L_{AA}[\alpha_{uv,11}\mathbf{y}_{1,uv2} + \alpha_{uv,21}\mathbf{y}_{1,uv2} + \alpha_{uv,31}\mathbf{y}_{1,uv3} + \alpha_{uv,41}\mathbf{y}_{1,uv4}] + \mathbf{n}_{uv,1}, \tag{14.16}$$

$$\mathbf{z}_{uv,2}^1 = L_{AA}[\alpha_{uv,12}\mathbf{y}_{1,uv2} + \alpha_{uv,22}\mathbf{y}_{1,uv2} + \alpha_{uv,32}\mathbf{y}_{1,uv3} + \alpha_{uv,42}\mathbf{y}_{1,uv4}] + \mathbf{n}_{uv,2}. \tag{14.17}$$

The two received signals $\mathbf{z}_{uv,1}$ and $\mathbf{z}_{uv,2}$, corresponding to the first and second receive antennas, respectively, are then correlated with $\bar{\mathbf{c}}_{1,1}$ and $\bar{\mathbf{c}}_{1,2}$ of Equations (14.1) and (14.2) according to the following operations:

$$\begin{aligned} r_{uv,11}^1 &= \bar{\mathbf{c}}_{1,1}^\dagger \cdot \mathbf{z}_{uv,1}^1 \\ &= \sqrt{\frac{2P_1}{VL_{AA}}\frac{1}{N_t N_{tg}}} L_{AA}[\alpha_{uv,11}x_{1,u1} + \alpha_{uv,21}x_{1,u2} + \alpha_{uv,31}x_{1,u3} + \alpha_{uv,41}x_{1,u4}] \\ &\quad + \bar{\mathbf{c}}_{1,1}^\dagger \cdot \mathbf{n}_{uv,1}, \end{aligned} \tag{14.18}$$

$$\begin{aligned} r_{uv,12}^1 &= \bar{\mathbf{c}}_{1,2}^\dagger \cdot \mathbf{z}_{uv,1}^1 \\ &= \sqrt{\frac{2P_1}{VL_{AA}}\frac{1}{N_t N_{tg}}} L_{AA}[-\alpha_{uv,11}x_{1,u2}^* + \alpha_{uv,21}x_{1,u1}^* - \alpha_{uv,31}x_{1,u4}^* + \alpha_{uv,41}x_{1,u3}^*] \\ &\quad + \bar{\mathbf{c}}_{1,2}^\dagger \cdot \mathbf{n}_{uv,1}, \end{aligned} \tag{14.19}$$

$$\begin{aligned} r_{uv,21}^1 &= \bar{\mathbf{c}}_{1,1}^\dagger \cdot \mathbf{z}_{uv,2}^1 \\ &= \sqrt{\frac{2P_1}{VL_{AA}}\frac{1}{N_t N_{tg}}} L_{AA}[\alpha_{uv,12}x_{1,u1} + \alpha_{uv,22}x_{1,u2} + \alpha_{uv,32}x_{1,u3} + \alpha_{uv,42}x_{1,u4}] \\ &\quad + \bar{\mathbf{c}}_{1,1}^\dagger \cdot \mathbf{n}_{uv,2}, \end{aligned} \tag{14.20}$$

$$
\begin{aligned}
r^1_{uv,22} &= \bar{\mathbf{c}}^{\dagger}_{1,2} \cdot \mathbf{z}^1_{uv,2} \\
&= \sqrt{\frac{2P_1}{VL_{AA}} \frac{1}{N_t N_{tg}}} L_{AA}[-\alpha_{uv,12} x^*_{1,u2} + \alpha_{uv,22} x^*_{1,u1} - \alpha_{uv,32} x^*_{1,u4} + \alpha_{uv,42} x^*_{1,u3}] \\
&\quad + \bar{\mathbf{c}}^{\dagger}_{1,2} \cdot \mathbf{n}_{uv,2}.
\end{aligned}
\tag{14.21}
$$

The received and despread signals of Equations (14.18)–(14.21) can be written in a matrix form as follows:

$$
\begin{aligned}
\mathbf{r}^1_{uv} &= \begin{bmatrix} r^1_{uv,11} \\ r^{1*}_{uv,12} \\ r^1_{uv,21} \\ r^{1*}_{uv,22} \end{bmatrix} \\
&= \sqrt{\frac{2P_1 L_{AA}}{V} \frac{1}{N_t N_{tg}}} \begin{bmatrix} \alpha_{uv,11} & \alpha_{uv,21} & \alpha_{uv,31} & \alpha_{uv,41} \\ \alpha^*_{uv,21} & -\alpha^*_{uv,11} & \alpha^*_{uv,41} & -\alpha^*_{uv,31} \\ \alpha_{uv,12} & \alpha_{uv,22} & \alpha_{uv,32} & \alpha_{uv,42} \\ \alpha^*_{uv,22} & -\alpha^*_{uv,12} & \alpha^*_{uv,42} & -\alpha^*_{uv,32} \end{bmatrix} \cdot \begin{bmatrix} x_{1,u1} \\ x_{1,u2} \\ x_{1,u3} \\ x_{1,u4} \end{bmatrix} \\
&\quad + \begin{bmatrix} \bar{\mathbf{c}}^{\dagger}_{1,1} \cdot \mathbf{n}_1 \\ (\bar{\mathbf{c}}^{\dagger}_{1,2} \cdot \mathbf{n}_1)^* \\ \bar{\mathbf{c}}^{\dagger}_{1,1} \cdot \mathbf{n}_2 \\ (\bar{\mathbf{c}}^{\dagger}_{1,2} \cdot \mathbf{n}_2)^* \end{bmatrix}.
\end{aligned}
\tag{14.22}
$$

Therefore, the received and despread signal matrix $\mathbf{r}$ can be written as[1]

$$
\mathbf{r} = \mathbf{H} \cdot \mathbf{X} + \mathbf{N}. \tag{14.23}
$$

The channel matrix $\mathbf{H}$ can be represented as

$$
\mathbf{H} = \begin{bmatrix} \mathbf{H}_1 & \mathbf{H}_2 \\ \mathbf{G}_1 & \mathbf{G}_2 \end{bmatrix}, \tag{14.24}
$$

where

$$
\mathbf{H}_1 = \begin{bmatrix} \alpha_{uv,11} & \alpha_{uv,21} \\ \alpha^*_{uv,21} & -\alpha^*_{uv,11} \end{bmatrix}, \quad \mathbf{H}_2 = \begin{bmatrix} \alpha_{uv,31} & \alpha_{uv,41} \\ \alpha^*_{uv,41} & -\alpha^*_{uv,31} \end{bmatrix},
$$

$$
\mathbf{G}_1 = \begin{bmatrix} \alpha_{uv,12} & \alpha_{uv,22} \\ \alpha^*_{uv,22} & -\alpha^*_{uv,12} \end{bmatrix} \quad \text{and} \quad \mathbf{G}_2 = \begin{bmatrix} \alpha_{uv,32} & \alpha_{uv,42} \\ \alpha^*_{uv,42} & -\alpha^*_{uv,32} \end{bmatrix}.
$$

In addition, the transmitted symbol matrix can be written as

$$
\mathbf{X} = \begin{bmatrix} \mathbf{X}_1 \\ \mathbf{X}_2 \end{bmatrix}, \quad \text{where } \mathbf{X}_1 = \begin{bmatrix} x_{1,u1} \\ x^*_{1,u2} \end{bmatrix} \quad \text{and} \quad \mathbf{X}_2 = \begin{bmatrix} x_{1,u3} \\ x^*_{1,u4} \end{bmatrix}.
$$

Hence, the received and despread signal can be represented as

$$
\mathbf{r} = \begin{bmatrix} \mathbf{r}_1 \\ \mathbf{r}_2 \end{bmatrix} = \begin{bmatrix} \mathbf{H}_1 & \mathbf{H}_2 \\ \mathbf{G}_1 & \mathbf{G}_2 \end{bmatrix} \cdot \begin{bmatrix} \mathbf{X}_1 \\ \mathbf{X}_2 \end{bmatrix} + \begin{bmatrix} \mathbf{N}_1 \\ \mathbf{N}_2 \end{bmatrix}. \tag{14.25}
$$

[1] In the following analysis we remove the subscript uv for simplicity of notation.

The decoding is carried out in two steps: first the interference cancelation is performed according to [19, 105] followed by the STS decoding procedure of [13]. The interference cancelation employed completely eliminates the interference of the two layers on each other as follows. The received and despread signal matrix $\mathbf{r}$ is multiplied by a matrix $\mathbf{Q}$ yielding

$$\mathbf{Q}\cdot\mathbf{r}=\begin{bmatrix}\tilde{\mathbf{r}}_1\\ \tilde{\mathbf{r}}_2\end{bmatrix}=\begin{bmatrix}\widetilde{\mathbf{H}} & \mathbf{0}\\ 0 & \widetilde{\mathbf{G}}\end{bmatrix}\cdot\begin{bmatrix}\mathbf{X}_1\\ \mathbf{X}_2\end{bmatrix}+\begin{bmatrix}\widetilde{\mathbf{N}}_1\\ \widetilde{\mathbf{N}}_2\end{bmatrix}. \tag{14.26}$$

According to Equation (14.26) the modified received signal $\tilde{\mathbf{r}}_1$ depends only on signals transmitted from the first STS layer and the modified received signal $\tilde{\mathbf{r}}_2$ depends only on signals transmitted from the second STS layer. It was shown in [105] that a solution for $\mathbf{Q}$ is given by

$$\mathbf{Q}=\begin{bmatrix}\mathbf{I}_2 & -\mathbf{G}_1\mathbf{G}_2^{-1}\\ -\mathbf{H}_2\mathbf{H}_1^{-1} & \mathbf{I}_2\end{bmatrix}, \tag{14.27}$$

where $\mathbf{I}_2$ is the identity matrix of dimension 2×2.

Hence, $\widetilde{\mathbf{H}}$ and $\widetilde{\mathbf{G}}$ can be expressed as

$$\widetilde{\mathbf{H}}=\mathbf{H}_1-\mathbf{G}_1\mathbf{G}_2^{-1}\mathbf{H}_2, \tag{14.28}$$

$$\widetilde{\mathbf{G}}=\mathbf{G}_2-\mathbf{H}_2\mathbf{H}_1^{-1}\mathbf{G}_1. \tag{14.29}$$

An important observation is that the matrices $\widetilde{\mathbf{H}}$ and $\widetilde{\mathbf{G}}$ have the same structure as that of the channel matrix $\mathbf{H}_1$. Hence, the above process will transform the decoding of the LSSTS signal into two separate problems that can be solved by the simple decoding process of STS [13].

Finally, after combining the $k=1$st user's identical replicas of the same signal transmitted by spreading over V subcarriers, the decision variables corresponding to the symbols transmitted in the uth subblock can be expressed as

$$\widetilde{x}_{1,u}=\sum_{v=1}^{V}\widetilde{x}_{1,uv}. \tag{14.30}$$

The decoded signal can be expressed as

$$\widetilde{x}=\sqrt{\frac{2P_1L_{AA}}{V}\frac{1}{N_tN_{tg}}}\sum_{v=1}^{V}(|\widetilde{\alpha}_{uv,1}|^2+|\widetilde{\alpha}_{uv,2}|^2)x+\eta. \tag{14.31}$$

Therefore, according to Equation (14.31) the decoded signal has a diversity order of $2V$. More explicitly, second-order spatial diversity is attained from the STS operation and a diversity order of V is achieved as a benefit of spreading by the generalized MC DS-CDMA scheme, where the subcarrier frequencies are arranged in a way that guarantees that the same STS signal is spread to and, hence, transmitted by the specific V subcarriers having the maximum possible frequency separation, so that they experience fading as independently as possible.

We consider a system employing BPSK modulation, L_{AA} elements per AA, V subcarriers and a TD spreading factor of $N_e=32$ for the sake of demonstrating the performance improvements achieved by the proposed system. The transmitter is equipped with $N_t=4$ AAs, while the receiver has $N_r=2$ antennas. We assume the availability of perfect channel

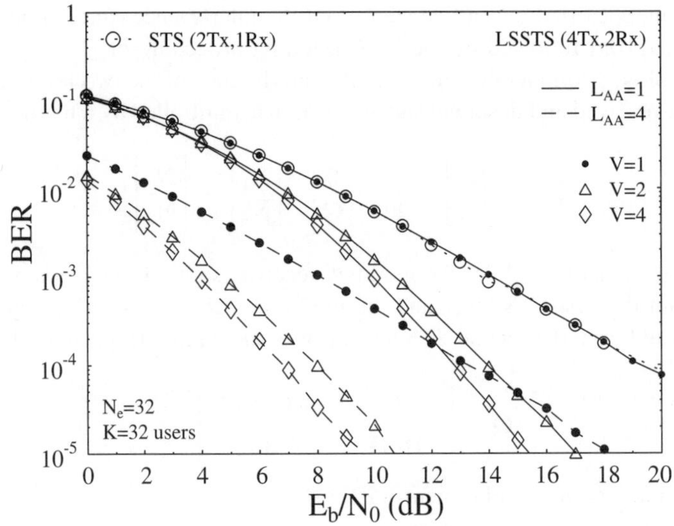

Figure 14.2: BER performance of the proposed system in Figure 14.1 employing $N_t = 4$ AAs and $N_r = 2$ antennas in conjunction with a varying number L_{AA} of elements per AA as well as a varying number of subcarriers V, while employing $K = 32$ users and a spreading factor $N_e = 32$. The per-user throughput is 2 bits per channel use.

knowledge both at the receiver and at the beamformer. The resultant per-user throughput is 2 bits per channel use. Figure 14.2 portrays the benefits of the different components employed in the system, namely the MC DS-CDMA, the beamforming, the STS and the V-BLAST components. When a single carrier is employed, i.e. we have $V = 1$, and $L_{AA} = 1$ element per AA, the system's performance is identical to that of the STS scheme of [13]. Therefore, the system has a diversity order of two, while the bandwidth efficiency of the proposed system is twice that of the STS scheme of [13]. In addition, Figure 14.2 shows the beamforming gain achieved upon increasing the number of beam-steering elements L_{AA} in the AA, while maintaining the same total number of AAs. As shown in the figure, when the number of beam-steering elements L_{AA} increases, the achievable BER performance substantially improves. Furthermore, to increase the achievable diversity order, the system employs $V > 1$ subcarriers, as shown in Figure 14.2. Hence, the proposed system has a diversity order of $2V$ owing to the employment of LSSTS-aided generalized MC DS-CDMA and the throughput becomes twice that of a system employing only a single STS block, which is a benefit of the V-BLAST structure. Figure 14.2 quantifies the advantages of increasing both L_{AA} and V in the proposed system, where increasing L_{AA} increases the SNR gain of the system while increasing V improves the attainable diversity order.

14.3 Increasing the Number of Users by Employing TD and FD Spreading

In the previous section, the DS spreading used by the generalized MC DS-CDMA system was carried out in the TD only based on orthogonal Walsh–Hadamard codes. It was proposed in [382, 383] to employ spreading in the FD for the MC CDMA schemes for the sake of exploiting the attainable diversity gain in the FD. In the generalized MC DS-CDMA scheme considered, the transmitted data stream can be spread in both the TD and the FD in order to support more users or to achieve the maximum attainable frequency diversity gain [384]. When FD spreading is employed, the FD spreading is applied after STS TD spreading in Figure 14.1 by multiplying the data symbols of the V subcarriers by the V chip values of a spreading code invoked for spreading the data in the FD across the V subcarriers. Hence, the SF of the FD spreading code is equal to the number of subcarriers V. The resultant bandwidth in this case is identical to that when the TD-only spreading is considered. Therefore, in this case the system benefits from TD as well as FD spreading, which allows for increasing the number of users, as described in Section 14.3.1.

At the receiver side, the received signal is first despread in the TD and then despread by the FD spreading code of length V. Furthermore, the number of users supported by employing generalized MC DS-CDMA using both TD and FD spreading is equal to $N_e \cdot V$. In other words, the N_e users spread in the TD will have a unique spreading code in the FD and the users having a different FD spreading code can share the same TD spreading code. Hence, the complexity of implementing separate TD and FD Multi-User Detectors (MUDs) for short spreading codes is expected to be significantly lower than that of a single TD MUD designed for long codes, as exemplified by comparing a 64-chip TD-only scheme to that using 8-chip TD and 8-chip FD spreading. The total number of users supported becomes $V \cdot \mathcal{K}_{\max} = V \cdot N_e$, which is V times the number of users supported by the scheme employing TD-only spreading.

14.3.1 Transmitter Model

The transmitter schematic of the proposed system employing TD and FD spreading is shown in Figure 14.1, which has been presented as the system model in Section 14.2 in conjunction with TD-only spreading. The transmitter model is the same as in Figure 14.1, except that the V-depth FD repetition scheme in Figure 14.1 is replaced by the V-depth FD spreading arrangement of Figure 14.3. Let us assume that the kth user's FD spreading code can be represented as $\bar{\mathbf{c}}'_k = \{c'_k[1], c'_k[2], \ldots, c'_k[V]\}$. In addition, the TD spreading code is denoted by $\bar{\mathbf{c}}_k$ as in Section 14.2.1.

As shown in Figure 14.1 and as discussed in Section 14.2, the data stream of duration T_b is S/P converted to U parallel substreams. Afterwards, each set of N_t symbols is S/P converted to $G = 2$ groups, where each group is encoded using the $N_{tg} = 2$ antenna-aided STS procedure using the spreading code $\bar{\mathbf{c}}_k$, as described in Equations (14.3)–(14.6). Afterwards, when employing TD and FD spreading, instead of using data repetition over V subcarriers as shown in Figure 14.1, the U subblocks generated after the TD spreading are now further spread across the FD using the previously introduced FD spreading codes $\bar{\mathbf{c}}'$, as depicted in Figure 14.3. When employing TD-only spreading, the number of users $\mathcal{K}_{\max}$ that can be supported by the system is equal to the spreading factor of the TD spreading code used, i.e. $\mathcal{K}_{\max} = N_e$. On the other hand, the total number of orthogonal codes that can be used for the FD spreading is equal to V. This implies that if V users share the same TD spreading code, they can be distinguished by their FD spreading codes. Hence, employing TD and FD spreading increases the number of users to $V \cdot \mathcal{K}_{\max}$, as compared with $\mathcal{K}_{\max}$ for the system employing TD-only spreading.

The TD and FD orthogonal spreading codes can be assigned as follows. If the number of users is still less than $\mathcal{K}_{\max}$, the users will be assigned different TD spreading codes, while sharing the same FD spreading code. The resultant scheme in this case is equivalent to that described in Section 14.2. When the number of users is in the range of $v \cdot \mathcal{K}_{\max} \leq K \leq (v+1) \cdot \mathcal{K}_{\max}$, where $v = 1, 2, \ldots, V-1$, then the same TD orthogonal spreading code

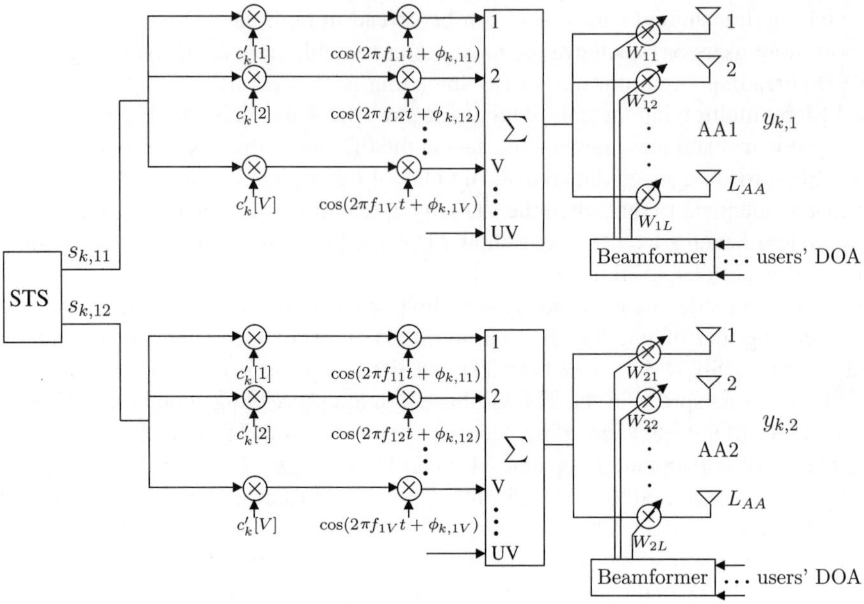

Figure 14.3: The first STS block of Figure 14.1 employing TD and FD spreading.

will be assigned to v users, where these users sharing the same TD code will be assigned different FD orthogonal spreading codes. Hence, the users sharing the same TD spreading code can be distinguished by their corresponding FD spreading code.

Since the subcarrier signals are transmitted over independently fading channels, the orthogonality of the FD spreading codes cannot be retained in frequency-selective fading channels. Hence, MUI is inevitably introduced, which degrades the attainable BER performance, when the number of users sharing the same TD spreading code increases.

When employing TD and FD spreading, the kth user's transmitted signals can be expressed as

$$\begin{aligned}\mathbf{y}_{k,n}(t) &= \mathbf{y}_{k,n}(\tau + aN_tT_s)\\ &= \sum_{u=1}^{U}\sum_{v=1}^{V}\sqrt{\frac{2P_k}{VL_{AA}}\frac{1}{N_tN_{tg}}}\mathbf{w}^k_{uv,n}\mathbf{s}_{k,un}c'_k[v]\cos(2\pi f_{uv}\tau + \phi_{k,uv}),\end{aligned} \tag{14.32}$$

where $n \in [1, N_t]$ represents the number of the transmit antenna, T_s is the symbol duration, $a = 0, 1, \ldots, 0 \le \tau < N_tT_s$, P_k/V represents the transmitted power of each subcarrier, the factor L_{AA} in the denominator is due to beamforming and the factor N_tN_{tg} in the denominator suggests that the STS scheme using N_t transmit antennas and N_{tg} orthogonal spreading codes distributes its power proportionally in space and time.

14.3.2 Receiver Model

Let us assume that there are $1 \le K' \le V$ users sharing the same TD spreading code but they are distinguished by their FD spreading codes. Then, when the $K'\mathcal{K}_{\max}$ users' signals are transmitted over frequency selective fading channels, the complex-valued received signal of user 1 can be expressed as

$$\begin{aligned}\mathbf{z}_m^1 &= \sum_{k=1}^{K'\mathcal{K}_{\max}} \sum_{u=1}^{U}\sum_{v=1}^{V}\sum_{n=1}^{N_t} \mathbf{h}_{uv,nm}\mathbf{y}_{k,uvn} + \mathbf{n}_m \\ &= \sqrt{\frac{2P_k}{VL_{AA}}\frac{1}{N_tN_{tg}}} \sum_{k=1}^{K'\mathcal{K}_{\max}} \sum_{u=1}^{U}\sum_{v=1}^{V}\sum_{n=1}^{N_t} \mathbf{a}_{uv,nm}\mathbf{w}_{uv,n}^1\mathbf{s}_{k,un}c_k'[v]\cos(2\pi f_{uv}\tau + \phi_{k,uv}) \\ &\quad + \mathbf{n}_m. \end{aligned} \tag{14.33}$$

Since orthogonal multicarrier signals and orthogonal TD STS spreading codes are used for the synchronous DL transmission over per-subcarrier Rayleigh flat fading, there is no interference between the users employing different TD spreading codes or using different subcarrier signals. The receiver in this case performs two main operations. The first operation consists mainly of multicarrier demodulation followed by STS decoding, which is similar to the decoding process of Section 14.2.2. The first part of the decoding operation provides V outputs corresponding to the V subcarriers conveying the same data. The second operation in this case corresponds to FD despreading of the V subcarrier outputs by the FD orthogonal spreading code.

Following the multicarrier demodulation and the TD despreading operations of Equation (14.33), the $k = 1$st user's data mapped to the uvth subcarrier can be expressed as

$$\begin{aligned} r_{uv,mg}^1 &= \bar{\mathbf{c}}_{1,g}^{\dagger} \cdot \mathbf{z}_{uv,m}^1 \\ &= \sum_{k=1}^{K'} \sqrt{\frac{2P_1}{VL_{AA}}\frac{1}{N_tN_{tg}}} [\mathbf{a}_{uv,1m}\mathbf{w}_{uv,1m}^1 e_{k,u1} + \mathbf{a}_{uv,2m}\mathbf{w}_{uv,2m}^1 e_{k,u2} \\ &\quad + \mathbf{a}_{uv,3m}\mathbf{w}_{uv,3m}^1 e_{k,u3} + \mathbf{a}_{uv,4m}\mathbf{w}_{uv,4m}^1 e_{k,u4}] \cdot c_k'[v] + \bar{\mathbf{c}}_{1,g}^{\dagger} \cdot \mathbf{n}_{uv,m}, \end{aligned} \tag{14.34}$$

where $e_{k,ui}$ may assume the values of $x_{k,u1}$, $x_{k,u2}$, $x_{k,u3}$, $x_{k,u4}$ or their conjugates as compared with Equations (14.18)–(14.21).

In Equation (14.34), $\mathbf{w}_{uv,nm}^k$ represents the weight vector of the desired user derived from the nmth antenna link and the uvth subcarrier, which is generated by the MRC beamformer [362] with the aid of CSI. Let $\mathbf{w}_{uv,nm}^1 = \mathbf{d}_{nm}^{1\dagger}$, where $\mathbf{d}_{nm}^1$ is defined in Equation (14.13), then Equation (14.34) can be simplified to

$$\begin{aligned} r_{uv,mg}^1 = &\sqrt{\frac{2P_1}{VL_{AA}}\frac{1}{N_tN_{tg}}} \\ &\times \Bigg[L_{AA}(\alpha_{uv,1m}e_{k,u1} + \alpha_{uv,2m}e_{k,u2} + \alpha_{uv,3m}e_{k,u3} + \alpha_{uv,4m}e_{k,u4}) \cdot c_k'[v] \\ &+ \sum_{k=2}^{K'}(\alpha_{uv,1m}\mathbf{d}_{1m}^k\mathbf{d}_{1m}^{1\dagger}e_{k,u1} + \alpha_{uv,2m}\mathbf{d}_{2m}^k\mathbf{d}_{2m}^{1\dagger}e_{k,u2} \\ &+ \alpha_{uv,3m}\mathbf{d}_{3m}^k\mathbf{d}_{3m}^{1\dagger}e_{k,u3} + \alpha_{uv,4m}\mathbf{d}_{4m}^k\mathbf{d}_{4m}^{1\dagger}e_{k,u4}) \cdot c_k'[v]\Bigg] + \bar{\mathbf{c}}_{1,g}^{\dagger} \cdot \mathbf{n}_{uv,m}. \end{aligned} \tag{14.35}$$

By employing the decoding scheme of Section 14.2.2 and after despreading the V decision variables with the aid of the V-chip FD spreading code $\overline{\mathbf{c}}'$, we arrive at

$$\widetilde{x} = \sqrt{\frac{2P_1}{VL_{AA}} \frac{1}{N_t N_{tg}}} \left[L_{AA} \sum_{v=1}^{V} (|\widetilde{\alpha}_{uv,1}|^2 + |\widetilde{\alpha}_{uv,2}|^2)x + \overline{\mathbf{c}}_1' \mathbf{i} \right] + \eta, \qquad (14.36)$$

where $\mathbf{i}$ represents a V-dimensional interference vector and η represents the noise term after STS demodulation and FD despreading.

We observe from Equation (14.36) that MUI is inevitably introduced, since the orthogonality of the FD spreading codes cannot be retained over frequency-selective fading channels. Observe that the desired user's signal is not interfered with by the signals of the users employing different orthogonal TD spreading codes, when assuming synchronous DL transmission as well as flat fading of the individual subcarriers. The users sharing the same TD spreading code and employing different FD spreading codes interfere with each other. Therefore, the MUI can be reduced, if we carefully select the $(K' - 1)$ potentially interfering users, namely those which have the lowest FD interference coefficient with respect to the desired user, from the entire set of all of the $K'\mathcal{K}_{\max}$ users [379, 381]. More explicitly, when selecting the specific users for the sake of sharing the same TD spreading code with the desired user, it must be ensured that their FD interference remains low and, hence, they remain distinguishable. The user grouping technique is discussed in the following section.

14.3.3 User Grouping Technique

As mentioned in the previous section, some MUI is inevitably imposed, when communicating over frequency-selective fading channels. However, the MUI can be reduced if the $(K' - 1)$ users are carefully grouped, so that the specific users having the lowest FD interference coefficient with respect to the desired user share the same TD spreading code with the desired user. More explicitly, the users sharing the same TD spreading code are carefully selected from the entire set of all of the $K'\mathcal{K}_{\max}$ users. The user grouping algorithm [379, 381] used is suboptimal, yet its performance improvements justify its employment, as we show in Figure 14.5.

The above-mentioned interference coefficient can be defined as $\rho_{1k} = \mathbf{d}^k \mathbf{d}^{1\dagger}$, which can be evaluated before transmission ensues, based on the assumption that the users' DOAs are perfectly known at the beamformer [379, 381]. The user grouping technique can be represented by the block diagram of Figure 14.4 and operates as follows.

In the absence of any prior knowledge, the initial value of ρ_{th} is set to zero, where ρ_{th} represents a threshold interference coefficient. The users having a FD interference coefficient of $\rho_{k_1 k_2} < \rho_{\text{th}}$ are deemed to be the users interfering with each other. Furthermore, when the k_1th user shares the same TD spreading code with the k_2th user, they belong to the same TD group and are differentiated by their FD spreading codes. The algorithm aims to ensure that the users sharing the same TD sequence have the lowest possible FD interference coefficient. The selection procedure will continue until all of the users have been grouped. However, if the threshold value ρ_{th} was set too low, some users cannot be allocated to any of the TD user groups owing to imposing a FD interference coefficient lower than ρ_{th}. In this scenario, ρ_{th} is increased by a given step size of $0 < \mu < L_{AA}$. Based on the increased threshold value, another user allocation attempt is initiated. The process continues until all of the users are grouped. Following this user-grouping procedure, the effect of the interfering signals imposed on the desired user's signal becomes less pronounced. Therefore,

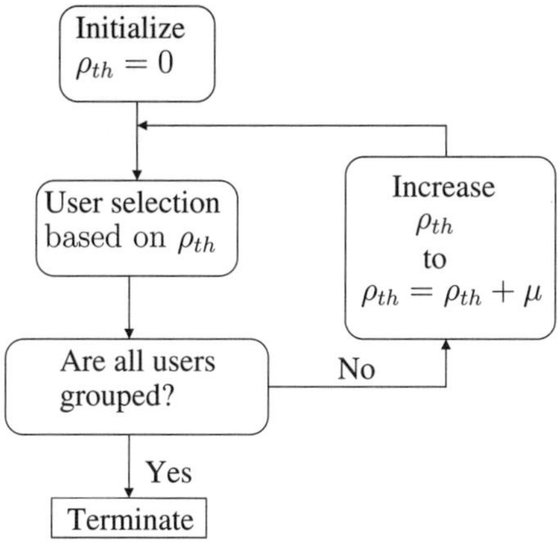

Figure 14.4: Block diagram of the user grouping technique.

the achievable BER performance is improved. In addition, when a new user joins or leaves the communication system, then the user grouping has to be updated [379, 381].

In Figure 14.5 we plot the BER performance of the proposed DL LSSTS-aided generalized MC DS-CDMA system using $N_t = 4$ transmit AAs, $N_r = 2$ receive antennas, $V = 4$ subcarriers, $L_{AA} = 4$ elements per AA and a TD spreading factor $N_e = 32$. The system also employs BPSK modulation for $K = 1$, 32 and 64 users. We assume having perfect channel knowledge at both the receiver and the beamformer. The resultant per-user throughput is 2 bits per channel use. Figure 14.5 shows that the performance of the system supporting $K = 32$ users is identical to that of the system serving a single user, since no interference is encountered by the $K = 32$ users employing different orthogonal 32-chip Walsh codes as their TD direct sequence (DS) spreading in the synchronous DL. Let us now consider TD and FD spreading, which is employed for the sake of supporting $K = 64$ users. Consequently, MUI is inevitably introduced among the users sharing the same TD spreading code. This becomes clear in Figure 14.5 for the case of $K = 64$ users when no user grouping was employed, since the performance of the system supporting $K = 64$ users is significantly worse than that supporting a single user or even $K = 32$ users. However, when user grouping is employed by the LSSTS system for the sake of reducing the MUI imposed, the performance of the system supporting $K = 64$ users substantially improves. Consequently, as a benefit of the user grouping technique, the BER performance of the 64-user system is only slightly inferior in comparison with that serving a single user.

14.4 Iterative Detection and EXIT Chart Analysis

In this section we design an iteratively detected receiver for the proposed system using iterative detection of serially concatenated RSC codes and URCs combined with the QPSK-assisted LSSTS-aided generalized MC DS-CDMA scheme. We present three different transceiver structures referred to as Systems 1, 2 and 3, as shown in Figures 14.6, 14.7 and 14.8, respectively.

In the structure of System 1 seen in Figure 14.6 the transmitted source bits $\mathbf{u}_1$ are encoded by the outer RSC code's Encoder I having a rate of $R_I = \frac{1}{2}$. The outer channel encoded bits $\mathbf{c}_1$ are then interleaved by a random bit interleaver Π_1, where the randomly permuted bits $\mathbf{u}_2$

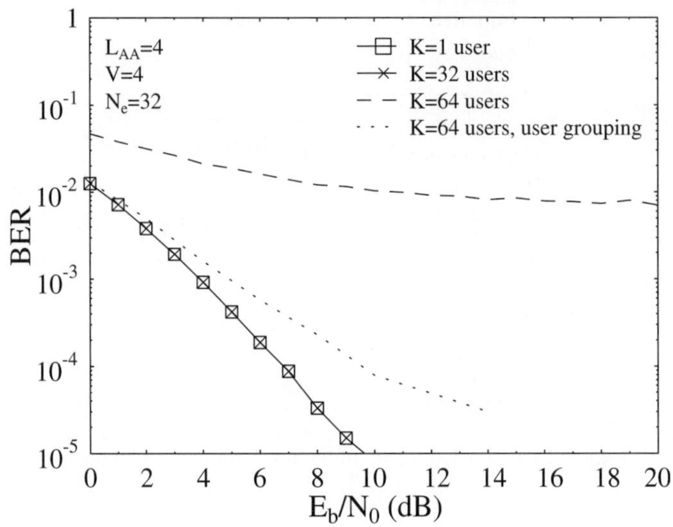

Figure 14.5: BER performance of the proposed system in conjunction with a varying number of users, where both TD and FD spreading as well as user grouping were employed to improve the achievable system performance, while suppressing the MUI. The per-user throughput is 2 bits per channel use.

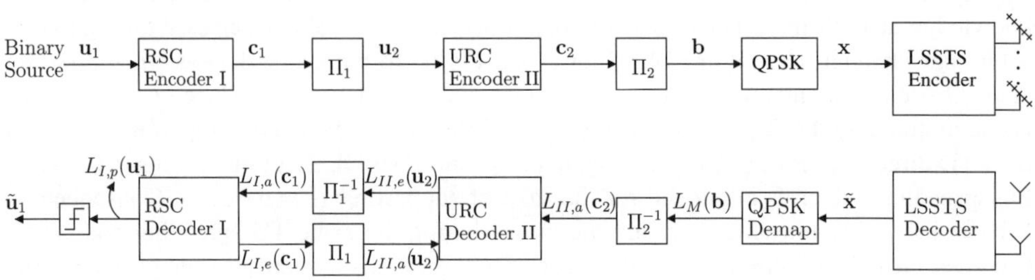

Figure 14.6: Block diagram of the proposed DL System 1 employing QPSK modulation in series with a unity-rate precoder and an outer RSC code.

are fed through the URC Encoder II. The encoded bits $\mathbf{c}_2$ at the output of the URC encoder are interleaved by a second random bit interleaver Π_2, producing the permuted bit stream $\mathbf{b}$. The interleaver Π_2 is used in order to mitigate the correlation in the soft data sequence $L_M(\mathbf{b})$. After bit interleaving, the QPSK modulator maps blocks of B channel-coded bits to their legitimate symbols, which are then transmitted using the transmitter structure of Figure 14.1.

At the receiver side, the soft-in soft-out RSC decoder iteratively exchanges extrinsic information with the URC decoder, as shown in Figure 14.6. The extrinsic soft information, represented in the form of LLRs [261], is iteratively exchanged between the URC and the RSC decoders for the sake of assisting each other's operation, as detailed in [148]. In Figure 14.6, $L(\cdot)$ denotes the LLRs of the bits concerned, where the subscript I indicates the RSC decoder, while II corresponds to the inner URC decoder. In addition, the subscripts a, p and e denote the dedicated role of the LLRs, with a, p and e indicating *a priori*, *a posteriori*

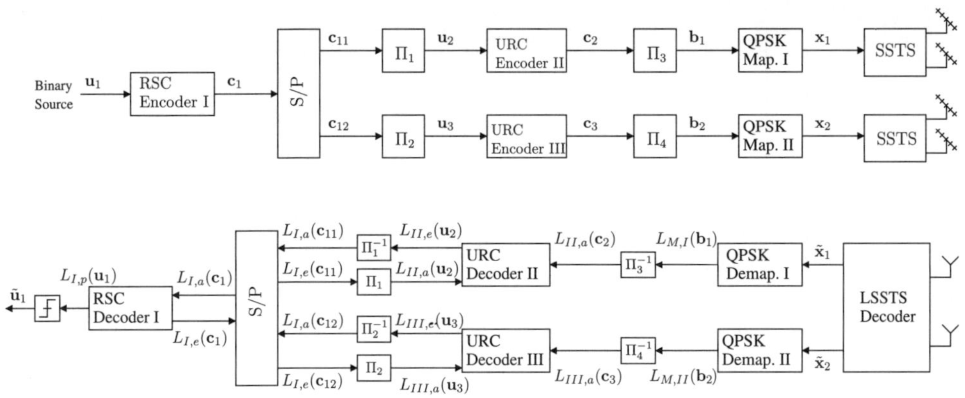

Figure 14.7: Block diagram of the proposed DL System 2 employing an RSC code in series with two parallel branches corresponding to a URC encoder in series with a QPSK mapper transmitting through two steered STS blocks.

and extrinsic information, respectively. Furthermore, the LLR $L_M(\mathbf{b})$ denotes the soft output of the QPSK demapper.

As shown in Figure 14.6, the received and decoded complex-valued symbol stream $\tilde{\mathbf{x}}$ is then fed into the QPSK demapper. The output of the demapper represents the LLR metric $L_M(\mathbf{b})$ passed from the QPSK demapper to the URC decoder. As seen in Figure 14.6, the URC decoder processes the information forwarded by the demapper in conjunction with the *a priori* information in order to generate the APP. The *a priori* LLR values of the URC decoder are subtracted from the *a posteriori* LLR values for the sake of generating the extrinsic LLR values $L_{II,e}(\mathbf{u}_2)$ and then the LLRs $L_{II,e}(\mathbf{u}_2)$ are deinterleaved by a soft-bit deinterleaver, as seen in Figure 14.6. Next, the soft bits $L_{I,a}(\mathbf{c}_1)$ are passed to the RSC decoder of Figure 14.6 in order to compute the *a posteriori* LLR values $L_{II,p}(\mathbf{c}_1)$ provided by the Log-MAP algorithm [165] for all the channel-coded bits $\mathbf{c}_1$. During the last iteration, only the LLR values $L_{I,p}(\mathbf{u}_1)$ of the original uncoded systematic information bits are required, which are passed to the hard-decision decoder of Figure 14.6 in order to determine the estimated transmitted source bits. As seen in Figure 14.6, the extrinsic information $L_{I,e}(\mathbf{c}_1)$ is fed back to the URC decoder as the *a priori* information $L_{I,a}(\mathbf{u}_2)$ after appropriately reordering it using the interleaver of Figure 14.6. The URC decoder exploits the *a priori* information for the sake of providing improved *a posteriori* LLR values, which are then passed to the half-rate RSC decoder and then back to the URC decoder for further iterations.

In the structure of Figure 14.7, denoted as System 2, the transmitted source bits $\mathbf{u}_1$ are encoded by the outer RSC code's Encoder I having a rate of $R_I = \frac{1}{2}$. The outer channel encoded bits $\mathbf{c}_1$ are then S/P converted to two parallel streams $\mathbf{c}_{11}$ and $\mathbf{c}_{12}$. Each bit stream is then interleaved by a random bit interleaver, where the interleaved bits in each stream are encoded by a corresponding URC encoder. Note that the URC encoders in each stream are identical. The URC encoded bits in each stream are interleaved by random bit interleavers and then mapped to QPSK symbols and transmitted using steered STS. Similarly to the URC encoders, the QPSK modulators I and II are identical. The two STS blocks transmit different data at the same time and employ the same subcarriers for the generalized MC DS-CDMA as described in Section 14.2.

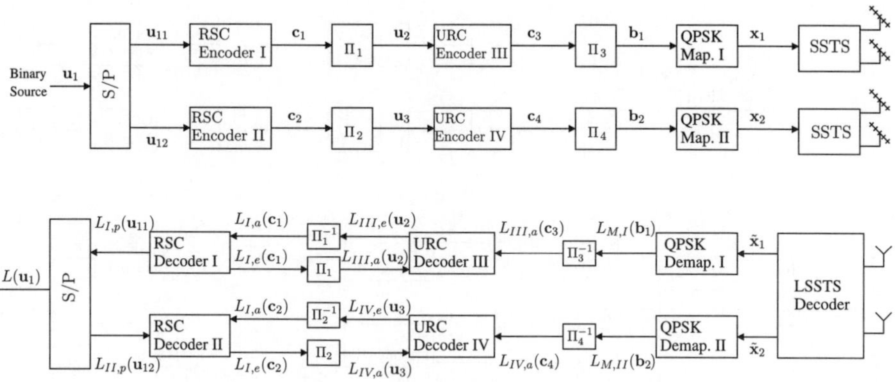

Figure 14.8: Block diagram of the proposed DL System 3 employing two parallel branches of RSC encoder in series with a URC encoder and transmitting through a QPSK-aided steered STS.

At the receiver side, the decoding process of Section 14.2.2 is employed, where the decoded symbols are passed to their corresponding branch as shown in Figure 14.7. In each branch, the decoded symbols are passed to the QPSK demapper to produce the corresponding LLR $L_{M,i}(\mathbf{b}_i)$, $i = 1, 2$, values. The demapper's soft output is deinterleaved by a soft bit deinterleaver and passed to the URC decoders as *a priori* information. The URC decoder utilizes the LLR information passed to it from the demapper as well as the RSC decoder to produce the extrinsic LLR values $L_{i,e}(\mathbf{u}_i)$, $i = 2, 3$. The extrinsic output of the URC decoders is deinterleaved and then Parallel-to-Serial (P/S) converted to be passed to the RSC decoder as *a priori* information. The RSC decoder utilizes the information passed from the URC decoders to produce the extrinsic LLR $L_{I,e}(\mathbf{c}_1)$. The extrinsic output of the RSC decoder is then S/P converted to be passed to the URC decoders of each branch, which in turn exploit the *a priori* information for the sake of providing improved *a posteriori* LLR values, which are then passed to the half-rate RSC decoder and then back to the URC decoders for further iterations.

Finally, in the structure of Figure 14.8, referred to as System 3, the transmitted source bits $\mathbf{u}_1$ are first S/P converted to two parallel substreams $\mathbf{u}_{11}$ and $\mathbf{u}_{12}$. Each substream is encoded by the outer RSC code's encoder having a rate of $R_I = \frac{1}{2}$. The outer channel encoded bits $\mathbf{c}$ of each bit stream are then interleaved by a random bit interleaver, where the interleaved bits in each stream are then encoded by a corresponding URC encoder. The URC encoded bits in each stream are then interleaved by random bit interleavers to be finally mapped to QPSK symbols and transmitted using steered STS. Note that the RSC encoders I and II are identical, the URC encoders III and IV are identical as well as the QPSK modulators I and II are identical.

At the receiver side of System 3 seen in Figure 14.8, the decoding process of Section 14.2.2 is employed, where the decoded symbols are passed to their corresponding branch, as shown in Figure 14.8. Each branch then applies iterative detection exchanging extrinsic information between the corresponding RSC and URC decoders, as shown in Figure 14.8. The output $L_{I,p}(\mathbf{u}_{11})$ and $L_{II,p}(\mathbf{u}_{12})$ of the RSC decoders is then P/S converted to produce a single stream $L(\mathbf{u}_1)$, which is passed to the hard-decision decoder of Figure 14.8.

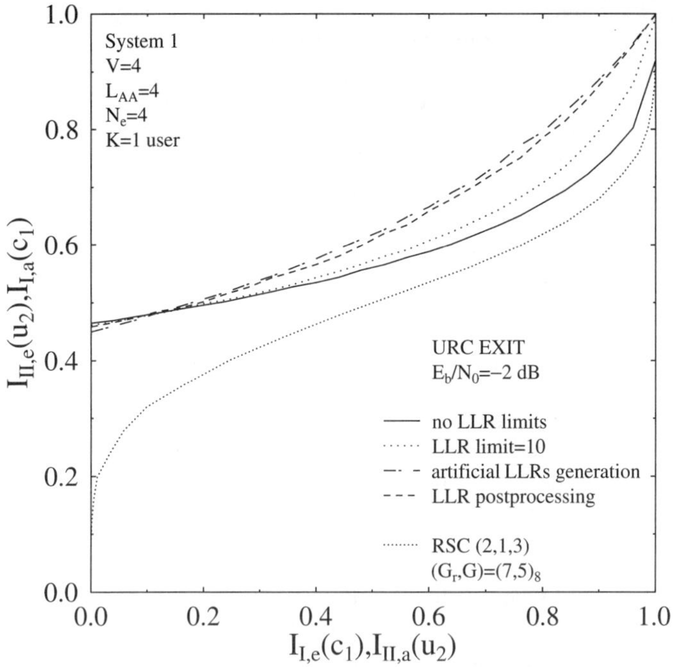

Figure 14.9: EXIT chart of a RSC-coded and URC-precoded proposed System 1 of Figure 14.6 employing Gray-mapping-aided QPSK in conjunction with $N_t = 4$, $N_r = 2$, $V = 4$, $L_{AA} = 4$, $K = 1$ user, $E_b/N_0 = -2$ dB and the remaining system parameters outlined in Table 14.1.

14.4.1 EXIT Charts and LLR Post-processing

As discussed in Chapter 11, the main objective of employing EXIT charts [169, 172] is to predict the convergence behavior of the iterative decoding process by examining the evolution of the input/output MI exchange between the constituent decoders in consecutive iterations. Again, the application of EXIT charts is based on two main assumptions, which are realistic when using high interleaver depths, namely that the *a priori* LLR values are uncorrelated and that they satisfy the consistency condition.

Let $I_{\cdot,a}(x)$, $0 \leq I_{\cdot,a}(x) \leq 1$, denote the MI between the *a priori* LLRs $L_{\cdot,a}(x)$ as well as the corresponding bits x and let $I_{\cdot,e}(x)$, $0 \leq I_{\cdot,e}(x) \leq 1$, denote the MI between the extrinsic LLRs $L_{\cdot,e}(x)$ and the corresponding bits x.

Figure 14.9 shows the EXIT chart of System 1 depicted in Figure 14.6 employing an iteratively detected RSC-coded and URC-precoded LSSTS system in conjunction with Gray-mapping-aided QPSK modulation, where iterations are carried out between the outer half-rate RSC code and the inner URC decoders, while no iterations are invoked between the URC decoder and the QPSK demapper. The system employs a half-rate memory-two RSC code, denoted as RSC(2,1,3), in conjunction with an octal generator polynomial of $(G_r, G) = (7, 5)_8$, where G_r is the feedback polynomial and G is the feedforward polynomial. Encoder II is a URC encoder, described by the pair of octal generator polynomials $(G_r, G) = (3, 2)_8$. Furthermore, the EXIT chart of Figure 14.9 was generated for the system employing $N_t = 4$ transmit AAs and $N_r = 2$ antennas, while using $L_{AA} = 4$ elements per AA in conjunction with $V = 4$ subcarriers and the system parameters outlined in Table 14.1.

Table 14.1: System parameters.

Modulation scheme	QPSK
Mapping	Gray mapping
Number of transmitter AAs N_t	4
Number of elements per AA	L_{AA}
Number of receiver antennas N_r	2
Number of subcarriers V	4
TD SF N_e	4
FD SF	V
Number of users	K
Outer encoder	RSC(2,1,3),
Generator	$(G_r, G) = (7, 5)_8$
Inner encoder	URC
Generator	$(G_r, G) = (3, 2)_8$
Interleaver depth	D_{int} bits

Observe in Figure 14.9 that there are several EXIT curves for the URC decoder for the same E_b/N_0 value. Let us first consider the dark line marked by the legend 'no LLR limits'. This EXIT curve corresponds to the URC decoder of Figure 14.6, which has a recursive encoder at the transmitter, and hence it is expected that the EXIT curve of the URC decoder will reach the $(1.0, 1.0)$ point of perfect convergence in the EXIT curve while using sufficiently long interleavers, as discussed in Section 11.2.4. However, the EXIT curve of the proposed System 1 characterized in Figure 14.9 shows that the EXIT curves of the URC decoder do not reach the $(1.0, 1.0)$ point. As a first step in circumventing this problem, we attempt to limit the maximum and minimum of the LLR values $L_M(\mathbf{b})$ for the sake of avoiding the problem of numerical overflow in the computer's memory. Limiting the LLR values allowed the URC EXIT curve to reach the $(1.0, 1.0)$ point, as shown in Figure 14.9 by the dotted line associated with the legend 'LLR limit $= 10$'. On the other hand, for the sake of testing the accuracy of the URC EXIT curve, while imposing a limit on the LLR values, we generated artificial Gaussian distributed and uncorrelated LLRs $L_M(\mathbf{b})$ that satisfy the consistency condition. The resultant EXIT curve in this case is represented by the dotted line having the legend 'artificial LLRs generation'. The artificial LLRs are generated assuming the transmitted bits are known at the receiver and it is used as a benchmark for testing the accuracy of our results. As shown in Figure 14.9, the curves corresponding to the case where the LLR's dynamic range is limited and where the artificial LLRs are generated are quite different. Therefore, limiting the LLR values does not solve the problem.

In order to understand this problem, let us return to the basics of the LLR and MI. The soft information pertaining to bits is typically represented using the LLRs within the receiver. Here, the particular LLR $L_M(b_i)$ in the frame $L_M(\mathbf{b})$ that pertains to the bit b_i from the frame $\mathbf{b}$ is specified according to

$$L_M(b_i) = \log\left(\frac{P(b_i = 1)}{P(b_i = 0)}\right), \tag{14.37}$$

where $P(b_i) \in [0, 1]$ is the probability that the bit b_i has the logical value zero or one within the transmitter. Note that the logarithmic domain is employed since it provides symmetry resulting in LLRs having a positive or negative sign, when there is a higher confidence in a

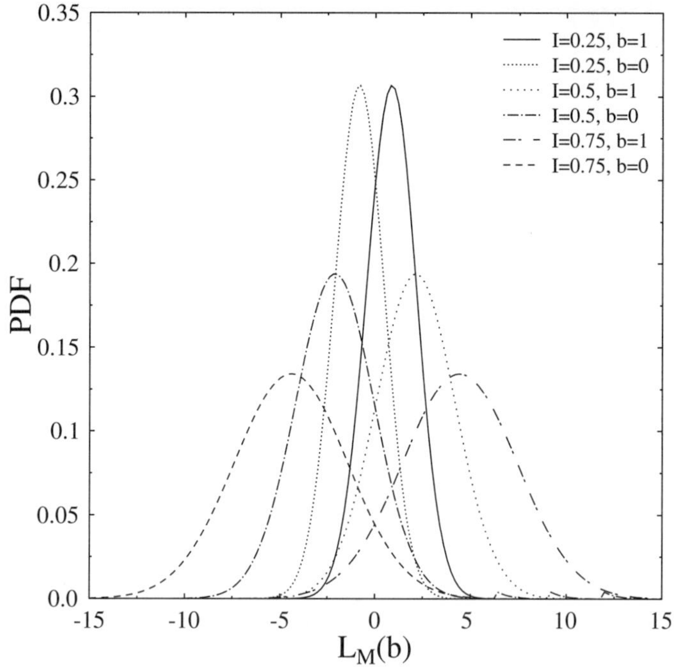

Figure 14.10: Gaussian distributed LLRs $L_M(b)$ corresponding to logical zero- and one-valued bits $\mathbf{b}$ for various MI I.

logical one- or a logical zero-valued bit, respectively. Furthermore, the level of this confidence increases with the LLR's magnitude.

On the other hand, as described in [385], the MI between a LLR frame $L(\mathbf{b})$ and the corresponding bit stream $\mathbf{b}$ depends on the distribution of the LLR values. More specifically, if the distribution of the LLR values that correspond to logical zero-valued bits is equal to that of the LLR values pertaining to logical one-valued bits, then the MI will be zero. In this case, the LLR values are unreliable and the hard decision based on the LLR values will result in an error rate of 50%.

As the reliability of the LLR values increases, the LLR distributions corresponding to the bits one and zero will move apart, as shown in Figure 14.10. As the MI I increases, the two LLR distributions will move apart and will overlap only at the tails of the distributions, giving a higher MI value. This results in a high confidence in the hard decoding based on the LLR values and a reduced probability of bit errors at the receiver. In Figure 14.10 the LLRs can be seen to have Gaussian distributions, although other distributions may be encountered in practice, depending on the transmission channel employed.

Observe in Figure 14.10 that the LLR abscissa value along the x-axis can be calculated by computing $\log(P(b=1)/P(b=0))$, where $P(b)$ can be computed from the y-axis of the figure. This property has been tested for the LLR $L_M(\mathbf{b})$ in the iteratively detected DSTS system of Section 11.3 and for the proposed System 1 of Figure 14.6. For the DSTS system, we show that this property is true by plotting $\log(P(b=1)/P(b=0))$ versus $L_M(\mathbf{b})$ in Figure 14.11. As shown in Figure 14.11, the result is a diagonal line, which means that $L_M(b_i) = \log(P(b_i=1)/P(b_i=0))$ as expected.

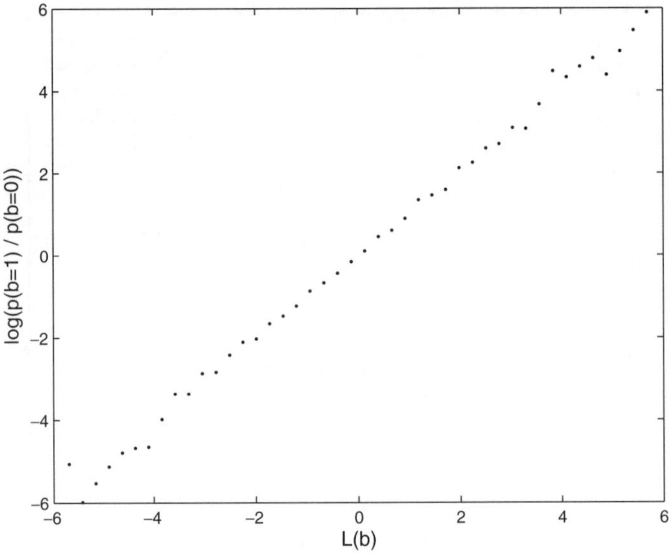

Figure 14.11: Logarithm of the probabilities in Figure 14.10 versus the corresponding QPSK demapper LLR output $L_M(\mathbf{b})$ for the DSTS system of Chapter 10.

The same experiment has been carried out for the proposed System 1 of Figure 14.6. We plotted $\log(P(b=1)/P(b=0))$ versus $L_M(\mathbf{b})$ and the result is shown in Figure 14.12. Theoretically, the result should be identical to Figure 14.11, where we have a diagonal line. However, we observe in Figure 14.12 that the result is a non-linear function. The reason for this behavior is the fact that the input $\tilde{\mathbf{x}}$ of the QPSK demapper is not Gaussian distributed, although we calculate the LLR values $L_M(\mathbf{b})$ assuming that the input data stream $\tilde{\mathbf{x}}$ is Gaussian distributed. A trivial solution to this problem is to try to find the probability distribution of the LSSTS decoder's output $\tilde{\mathbf{x}}$ and compute the LLRs in the QPSK demapper using the correct PDF. However, it is not straightforward to find a mathematical formula to model the PDF of $\tilde{\mathbf{x}}$. On the other hand, it is possible to compute the LLRs based on the histogram of the received and decoded data $\tilde{\mathbf{x}}$. However, computing the histogram for every received frame is a complex and time-consuming process.

By considering the specific relationship between the probabilities and LLR values seen in Figure 14.12 we can find a relationship that can transform the result of Figure 14.12 into that seen in Figure 14.11. The y-axis of Figure 14.12 is the correct LLR value and hence we have to transform the x-axis LLR value into its corresponding y-axis value. An empirical transformation of the $L_M(\mathbf{b})$ LLR values has been computed for the sake of correcting the relationship between the LLR values and their corresponding probabilities. The transformation is applied to the LLR values at the output of the QPSK demapper and hence it is referred to as LLR post-processing. The empirical transformation can be expressed as

$$\text{LLR}_{\text{out}} = \frac{L_M(\mathbf{b})}{1.25(\log_2(V)+1) - 0.75\lceil K/N_e - 1\rceil}, \tag{14.38}$$

where LLR_{out} represents the LLR passed from the QPSK demapper to the deinterleaver Π_2 of Figure 14.6 after the transformation of the LLRs. This transformation is referred to as LLR post-processing, since it is applied at the receiver side after computing the LLRs in the QPSK demapper.

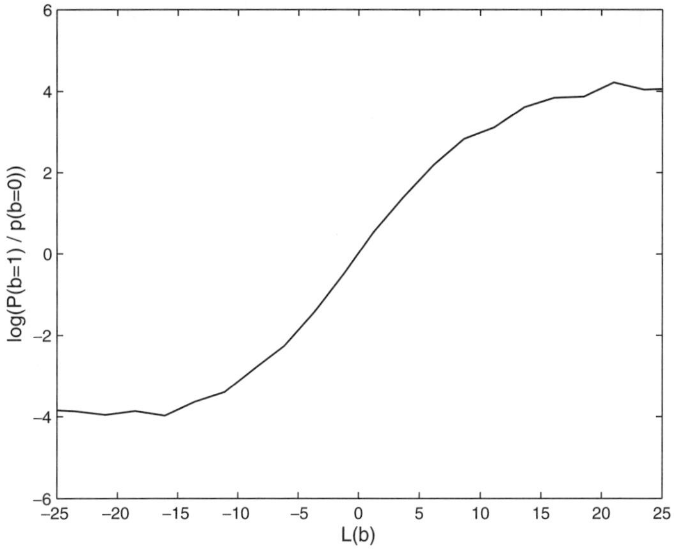

Figure 14.12: Logarithm of the probabilities in Figure 14.10 versus the corresponding QPSK demapper LLR output $L_M(\mathbf{b})$ for the proposed System 1 shown in Figure 14.6.

Figure 14.9 also shows the EXIT curve of the inner URC decoder after the LLR post-processing technique was employed. As shown in Figure 14.9, the EXIT curve of the system where the post-processing is employed is similar to that where the artificial LLRs were considered. Hence, the proposed LLR post-processing employing the LLR post-processing technique solves the problem of the non-Gaussian decoded data passed from the MIMO decoder to the QPSK demapper and at the same time eliminates the complexity of the histogram estimation for every received frame.

Observe from Figure 14.13 that an open EXIT chart convergence tunnel is formed around $E_b/N_0 = -3$ dB for System 1 employing $V = 4$ subcarriers, $L_{AA} = 4$ elements per AA, $N_e = 4$ and supporting $K = 1$ user, while using the remaining system parameters outlined in Table 14.1. This implies that according to the predictions of the EXIT chart seen in Figure 14.13, the iterative decoding process is expected to converge at an E_b/N_0 of at least -3 dB. The EXIT-chart-based convergence predictions can be verified by the Monte Carlo simulation-based iterative decoding trajectory of Figure 14.14, where the trajectory was recorded at $E_b/N_0 = -2.8$ dB, while using an interleaver depth of $D_{\text{int}} = 160\,000$ bits, $V = 4$ subcarriers, $L_{AA} = 4$ elements per AA, $N_e = 4$ and $K = 1$ user in conjunction with the system parameters outlined in Table 14.1. The steps seen in Figure 14.14 represent the actual extrinsic information exchange between the URC's decoder and the outer RSC channel decoder.

On the other hand, increasing the number of users beyond the TD spreading factor N_e and employing both TD and FD spreading combined with the user grouping technique of Section 14.3.3 degrades the system's performance, as shown in Figure 14.5. The TD and FD spreading as well as the user grouping technique have been applied to System 1 of Figure 14.6. Iterative detection has been carried out by exchanging extrinsic information between the RSC decoder and the URC decoder at the receiver side in addition to the LLR post-processing technique, which was applied for the sake of correcting the LLR output of the QPSK demapper. The resultant EXIT chart is shown in Figure 14.15, where a comparison

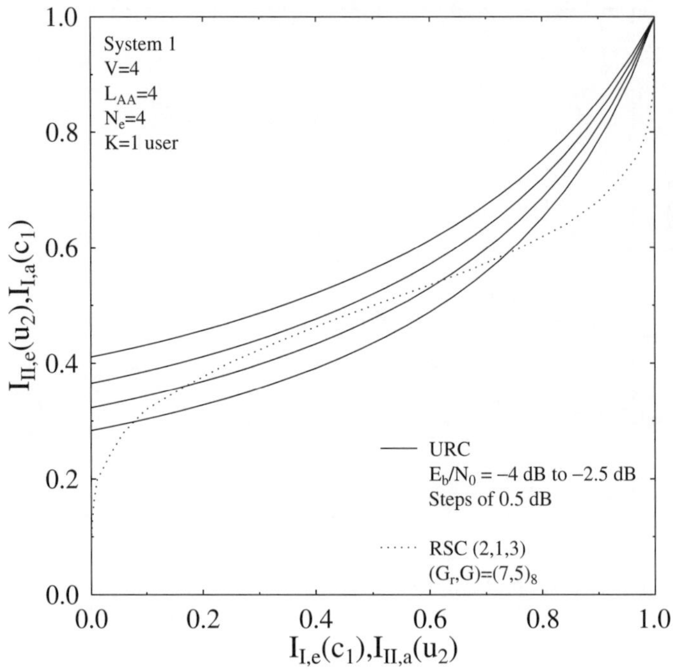

Figure 14.13: EXIT chart of the RSC-coded and URC-precoded System 1 of Figure 14.6 employing an interleaver depth of $D_{\rm int} = 160\,000$ bits, $V = 4$ subcarriers, $L_{AA} = 4$ elements per AA, $N_e = 4$ and $K = 1$ user in conjunction with the system parameters outlined in Table 14.1.

between the EXIT curves of the URC decoder is offered for both $K = 1$ and 8 users. Figure 14.15 portrays the EXIT chart of a system employing $N_t = 4$ transmit AAs, $N_r = 2$ receive antennas, $V = 4$ subcarriers, $L_{AA} = 4$ elements per AA, $N_e = 4$ and the system parameters outlined in Table 14.1. As shown in Figure 14.15, it is expected that the single-user system outperforms the overloaded eight-user system by E_b/N_0 of about 0.2 dB.

Figure 14.16 shows the EXIT chart of the iteratively detected System 2 of Figure 14.7. The EXIT curve of the inner URC decoder was recorded for one of the two substreams seen in Figure 14.7. The EXIT curve of the two substreams is identical, since the interference cancellation scheme of Section 14.2.2 completely eliminates the interference imposed by one of the STS layers on the other. The EXIT chart of Figure 14.16 was recorded for the system employing $V = 4$ subcarriers, $L_{AA} = 4$ elements per AA, $N_e = 4$, $K = 1$ user and the remaining system parameters outlined in Table 14.1. The EXIT-chart-based convergence predictions can be verified by the Monte Carlo simulation-based iterative decoding trajectory of Figure 14.16, where the trajectory was recorded at $E_b/N_0 = -2.8$ dB, while using an interleaver depth of $D_{\rm int} = 160\,000$ bits. The steps seen in Figure 14.16 represent the actual extrinsic information exchange between the URCs' decoders and the outer RSC channel decoder.

On the other hand, Figure 14.17 shows the EXIT chart of the iteratively detected System 3 of Figure 14.8. The EXIT curves of the inner URC decoder as well as the outer RSC code's decoder were recorded for one of the two substreams seen in Figure 14.8. Again, the EXIT curve of the two substreams is identical since the interference cancelation scheme

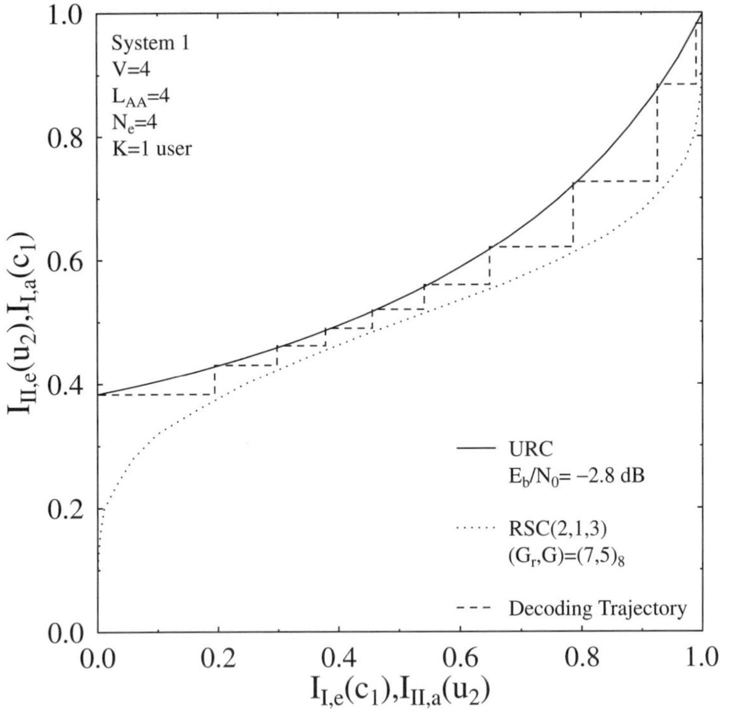

Figure 14.14: Decoding trajectory of the iteratively detected RSC-coded and URC-precoded System 1 seen in Figure 14.6 employing an interleaver depth of $D_{\rm int} = 160\,000$ bits, $V = 4$ subcarriers, $L_{AA} = 4$ elements per AA, $N_e = 4$ and $K = 1$ user, while operating at $E_b/N_0 = -2.8$ dB in conjunction with the system parameters outlined in Table 14.1.

of Section 14.2.2 completely cancels the interference imposed by one of the STS layers on the other. The EXIT chart of Figure 14.17 was recorded for the system employing $V = 4$ subcarriers, $L_{AA} = 4$ elements per AA, $N_e = 4$, $K = 1$ user and the system parameters outlined in Table 14.1. The EXIT-chart-based convergence predictions can be verified by the Monte Carlo simulation-based iterative decoding trajectory of Figure 14.17, where the trajectory was recorded at $E_b/N_0 = -2.8$ dB, while using an interleaver depth of $D_{\rm int} = 160\,000$ bits.

As observed in Figures 14.14, 14.16 and 14.17, the three systems, namely Systems 1, 2 and 3, may be expected to have a similar BER performance based on these EXIT chart predictions. This is due to the fact that the interference cancelation operation of the proposed system outlined in Section 14.2.2 completely eliminates any interference imposed by one of the STS layers on the other layer. Therefore, the iteratively detected Systems 1–3 have similar BER performances according to our EXIT chart prediction provided that we employ sufficiently long interleavers, which are capable of eliminating the correlation of the extrinsic information.

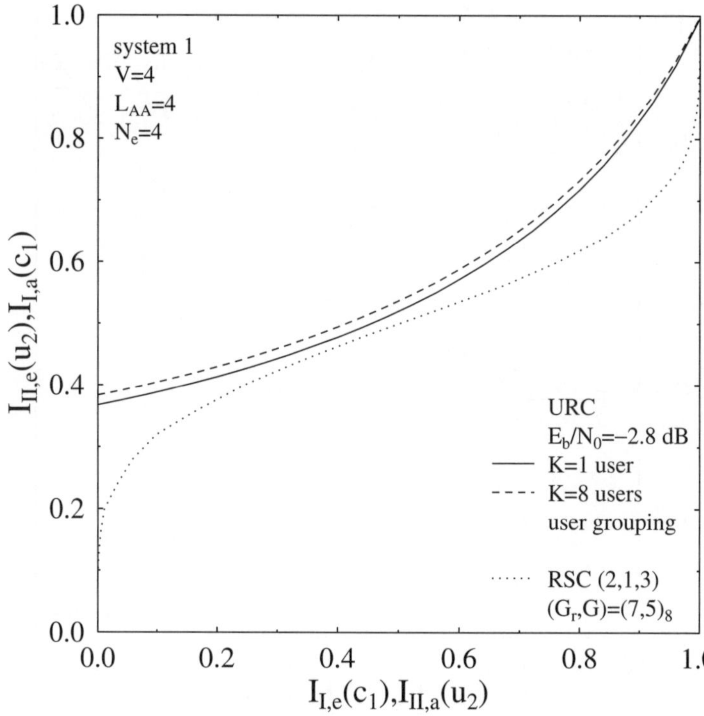

Figure 14.15: EXIT chart of a RSC-coded and URC-precoded System 1 of Figure 14.6 employing an interleaver depth of $D_{\text{int}} = 160\,000$ bits, $V = 4$ subcarriers, $L_{AA} = 4$ elements per AA, $N_e = 4$ and $K = 1$ or 8 users in conjunction with the system parameters outlined in Table 14.1.

14.5 Results and Discussion

In this section, we consider a LSSTS system associated with $N_t = 4$ transmit AAs, $N_r = 2$ receive antennas, $V = 4$ subcarriers, $L_{AA} = 4$ elements per AA, $N_e = 4$ and the system parameters outlined in Table 14.1, in order to demonstrate the performance improvements achieved by the proposed systems, namely Systems 1, 2 and 3. We employ Gray-mapping-aided QPSK modulation. In addition, perfect channel knowledge is assumed at the receiver as well as at the transmit beamformer.

Figure 14.18 compares the BER performance of the proposed System 1 supporting $K = 1$ user in conjunction with Gray-mapping-aided QPSK for different numbers of iterations. Figure 14.18 portrays the performance of the iteratively detected RSC-coded and URC-precoded System 1, when employing an interleaver depth of $D_{\text{int}} = 160\,000$ bits. Figure 14.18 demonstrates that the BER performance closely matches the EXIT-chart-based prediction of Figure 14.14, where the system approaches an infinitesimally low BER at $E_b/N_0 = -2.8$ dB after ten iterations. On the other hand, Figure 14.19 compares the BER performance of the iterative-detection-aided System 1, while employing $V = 4$ subcarriers, $L_{AA} = 4$ elements per AA, $N_e = 4$ and ten decoding iterations in conjunction with Gray-mapping-aided QPSK for $K = 1$ and 8 users. The system of Figure 14.19 employs TD and FD spreading for the sake of increasing the number of users to $K = 8$, which is twice the TD spreading factor, i.e. twice the number of users supported by employing TD-only spreading,

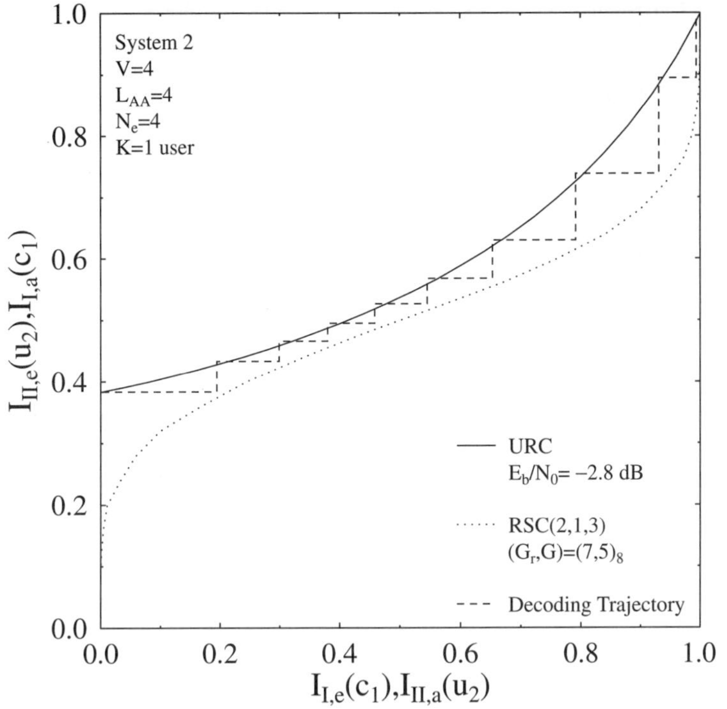

Figure 14.16: Decoding trajectory of the iteratively detected System 2 of Figure 14.7 employing an interleaver depth of $D_{\text{int}} = 160\,000$ bits, $V = 4$ subcarriers, $L_{AA} = 4$ elements per AA, $N_e = 4$ and $K = 1$ user, while operating at $E_b/N_0 = -2.8$ dB in conjunction with the system parameters outlined in Table 14.1. The inner decoder's EXIT curve was generated for one branch of the system seen in Figure 14.7.

in addition to applying the user grouping technique of Section 14.3.3. Furthermore, we employ the LLR post-processing technique of Equation (14.38) for the sake of correcting the LLR output of the QPSK demapper. According to the EXIT chart predictions seen in Figure 14.15, the system employing $K = 1$ and 8 users has an E_b/N_0 requirement difference of about 0.2 dB. According to Figure 14.19, when the system employs an interleaver depth of $D_{\text{int}} = 160\,000$ bits and ten decoding iterations, the system supporting a single user outperforms that supporting $K = 8$ users by about 0.45 dB at a BER of 10^{-5}.

As discussed previously, according to the EXIT chart prediction of Figures 14.14, 14.16 and 14.17 the BER performance of Systems 1–3 is identical, when employing a long interleaver as well as a sufficient number of decoding iterations. Observe from Figures 14.6, 14.7 and 14.8 that the interleaver depth for Systems 2 and 3 is half of that for System 1 when considering the same frame length for the input data bit stream $\mathbf{u}_1$. This is due to the fact that the input bit streams in Systems 2 and 3 are split into two parallel equal-length bit streams. In what follows we refer to the interleaver Π_1 depth of System 1 in Figure 14.6 as D_{int} and it is equal to the RSC code rate R times the input bit stream frame length. Hence the interleaver Π_1 depth of Systems 2 and 3 in Figures 14.7 and 14.8 is $D_{\text{int}}/2$. Figure 14.20 compares the BER performance of the proposed System 2 employing $V = 4$ subcarriers, $L_{AA} = 4$ elements per AA, $N_e = 4$ and supporting $K = 1$ user in conjunction with Gray-mapping-aided QPSK for different numbers of iterations. The figure shows the performance of the iteratively detected

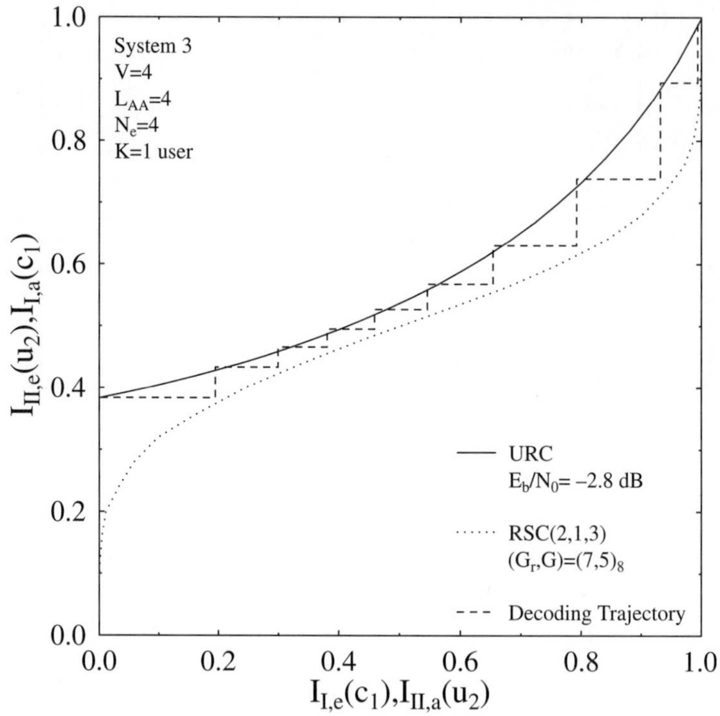

Figure 14.17: Decoding trajectory of the iteratively detected System 3 of Figure 14.8 employing an interleaver depth of $D_{\text{int}} = 160\,000$ bits, $V = 4$ subcarriers, $L_{AA} = 4$ elements per AA, $N_e = 4$ and $K = 1$ user, while operating at $E_b/N_0 = -2.8$ dB in conjunction with the system parameters outlined in Table 14.1. The inner and outer EXIT curve was generated for one branch of the system seen in Figure 14.8.

RSC-coded and URC precoded System 2, when employing a bit sequence $\mathbf{u}_1$ having a length of 80 000 bits, i.e. $D_{\text{int}} = 160\,000$ bits. This means that the interleaver Π_1 of Figure 14.7 has a depth of 80 000 bits. Figure 14.20 demonstrates that the BER performance closely matches the EXIT-chart-based predictions of Figure 14.16, where the system approaches a BER below 10^{-5} at E_b/N_0 in excess of -2.8 dB after ten iterations. Similarly, we plot in Figure 14.21 the BER performance comparison of the proposed System 3 employing $V = 4$ subcarriers, $L_{AA} = 4$ elements per AA, $N_e = 4$ and $K = 1$ user in conjunction with Gray-mapping-aided QPSK for different numbers of iterations and employing an interleaver Π_1 of Figure 14.8 having depth of $80\,000$ bits, i.e. $D_{\text{int}} = 160\,000$ bits. Figure 14.21 demonstrates that the BER performance closely matches the EXIT-chart-based predictions of Figure 14.17, where the system performance approaches a BER below 10^{-5} at E_b/N_0 in excess of -2.8 dB after ten iterations.

To comment briefly on the associated complexity, for an interleaver depth of D_{int} and after I iterations, System 1 encounters $(I+1)(4D_{\text{int}} - 16)$ trellis states, while System 2 invokes $(I+1)(4D_{\text{int}} - 17)$ trellis states. On the other hand, System 3 has $(I+1)(4D_{\text{int}} - 32)$ trellis states. Hence, for the system parameters of Table 14.1 employed in the above investigations, i.e. for $D_{\text{int}} = 160\,000$ bits and $I = 10$ iterations, System 1 encounters a total of $7\,039\,824$ trellis states, System 2 employs $7\,039\,813$ trellis states and finally System 3 requires $7\,039\,648$ trellis states. Therefore, we may conclude that the three systems also

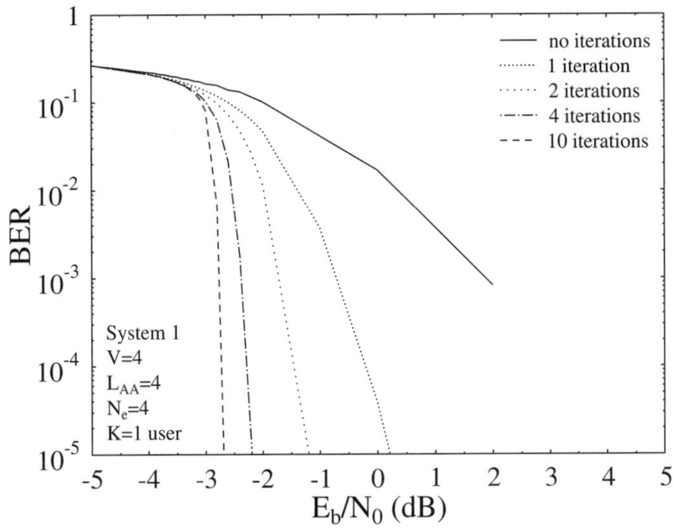

Figure 14.18: Performance comparison of the GM-based RSC-coded and URC-precoded System 1 of Figure 14.6 in conjunction with $V = 4$ subcarriers, $L_{AA} = 4$ elements per AA, $N_e = 4$ and $K = 1$ user and the system parameters outlined in Table 14.1, when using an interleaver depth of $D_{\text{int}} = 160\,000$ bits for a variable number of iterations.

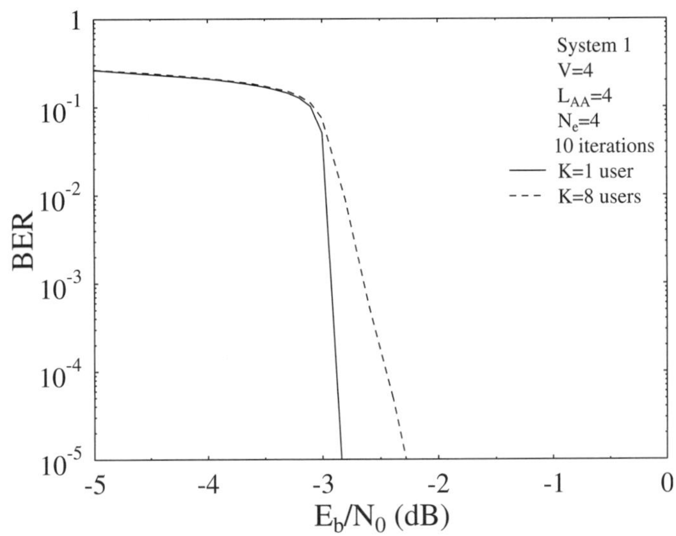

Figure 14.19: Performance comparison of the Gray-mapping-based RSC-coded and URC-precoded System 1 of Figure 14.6 in conjunction with $V = 4$ subcarriers, $L_{AA} = 4$ elements per AA, $N_e = 4$ and ten iterations and the system parameters outlined in Table 14.1, when using an interleaver depth of $D_{\text{int}} = 160\,000$ bits for $K = 1$ and 8 users.

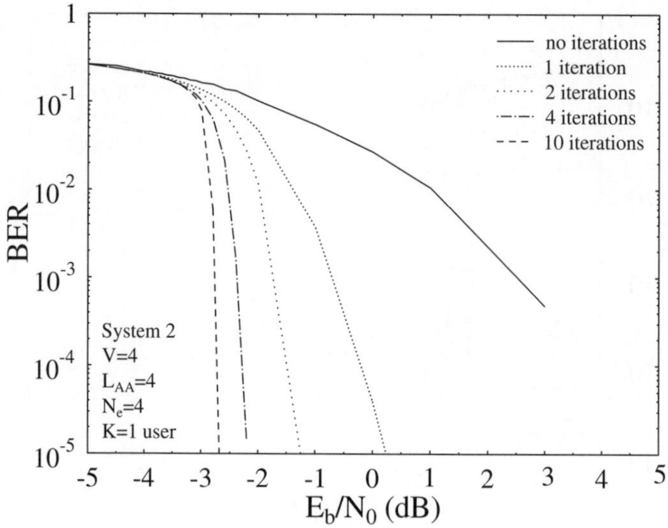

Figure 14.20: Performance comparison of the Gray-mapping-based RSC-coded and URC-precoded System 2 of Figure 14.7 in conjunction with $V = 4$ subcarriers, $L_{AA} = 4$ elements per AA, $N_e = 4$, $K = 1$ user and the system parameters outlined in Table 14.1, when using an interleaver depth of $D_{\text{int}} = 160\,000$ bits for a variable number of iterations.

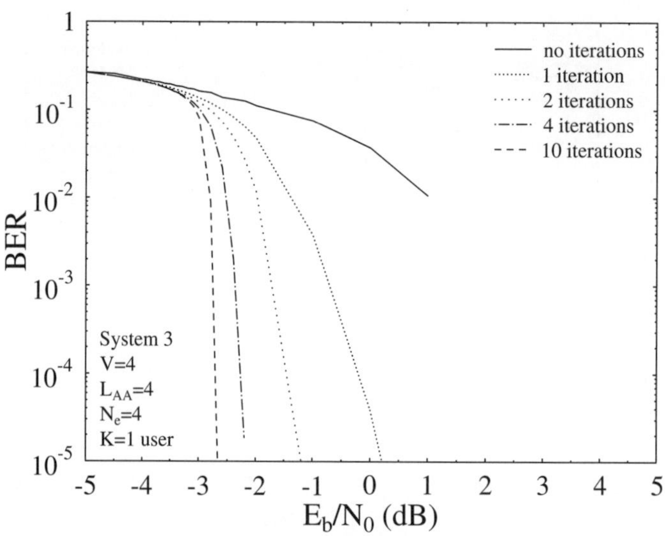

Figure 14.21: Performance comparison of the Gray-mapping-based RSC-coded and URC-precoded System 3 of Figure 14.8 in conjunction with $V = 4$ subcarriers, $L_{AA} = 4$ elements per AA, $N_e = 4$ and $K = 1$ user and the system parameters outlined in Table 14.1, when using an interleaver depth of $D_{\text{int}} = 160\,000$ bits for a variable number of iterations.

have a similar complexity, although System 3 is the least complex in terms of the number of trellis states used throughout the iterative decoding process, which determines the number of ACS arithmetic operations. Note that System 2 employs one more mapper and URC encoder and decoder blocks than System 1, while System 3 employs one more mapper, one URC encoder/decoder and an extra RSC encoder/decoder in comparison with System 1.

We have shown that while employing a high interleaver depth $D_{\text{int}} = 160\,000$ bits, Systems 1–3 have both a similar performance and a similar complexity, although Systems 2 and 3 use more encoder and decoder components than System 1. On the other hand, splitting the bit stream into two parallel substreams in Systems 2 and 3 implies that their interleaver depth becomes lower than that of System 1, while considering the same number of input bits. Hence, in what follows we study the effect of the interleaver depth on the achievable performance of the three systems.

Figure 14.22 shows the BER performance comparison of Systems 1–3 while employing different interleaver depths D_{int} varying between 80 000 bits and 2000 bits. Figure 14.22 compares the achievable BER performance of the proposed systems employing $V = 4$ subcarriers, $L_{AA} = 4$ elements per AA, $N_e = 4$ and supporting $K = 1$ user in conjunction with Gray-mapping-aided QPSK after $I = 10$ decoding iterations. Observe in Figures 14.18–14.21 and in Figures 14.22(a) and (b) that when long interleavers are employed, i.e. when the interleaver depth is $D_{\text{int}} = 160\,000$, 80 000 bits or 40 000 bits, the three systems have a similar BER performance. However, when we employ shorter interleavers, we can observe from Figure 14.22 that the BER performance of System 3 becomes inferior to that of System 2, which in turn performs worse than System 1. This is due to the fact that the interleaver depth of Systems 2 and 3 is half of that of System 1, which implies that the extrinsic information remains more correlated and hence prevents the decoding trajectory from reaching the $(1.0, 1.0)$ point of perfect convergence, as discussed in Section 11.2.4. Observe also in Figure 14.22 that the performance of System 3 is lower than that of System 2 and this is due to the fact that the trellis of the RSC decoder in System 2 is longer than that of System 3 and hence a high LLR value will improve the attainable performance right across the entire trellis in System 2, while it will only benefit one of the two RSC constituent trellises in System 3.

Therefore, we may conclude that although Systems 1–3 have a similar overall complexity in terms of the number of their trellis states encountered, the BER performance of the three systems remains similar only when an interleaver depth of $D_{\text{int}} \geqslant 40\,000$ bits is employed. In contrast, if we employ shorter interleavers, we can observe from Figure 14.22 that the BER performance of System 1 degrades to a lesser extent in comparison to that of System 2, which in turn performs better than System 3, as portrayed in Figure 14.22.

14.6 Chapter Conclusions

In this chapter, we proposed a novel multi-functional DL MIMO scheme that combines the benefits of STC, V-BLAST and generalized MC DS-CDMA as well as beamforming. The system proposed in this chapter differs from that of Chapter 13 in terms of the decoding procedure, which allows the receiver of this chapter to have fewer receive antennas than the number of transmit antenna arrays. Hence, the system proposed in this chapter can be applied where relatively small handsets are used at the receiver. The proposed system employing $N_t = 4$ transmit antenna arrays has a per-user throughput that is twice that of a system employing only a single STS block, which was the case in [13]. On the other hand, employing generalized MC DS-CDMA and assuming that the subcarrier frequencies are arranged in a

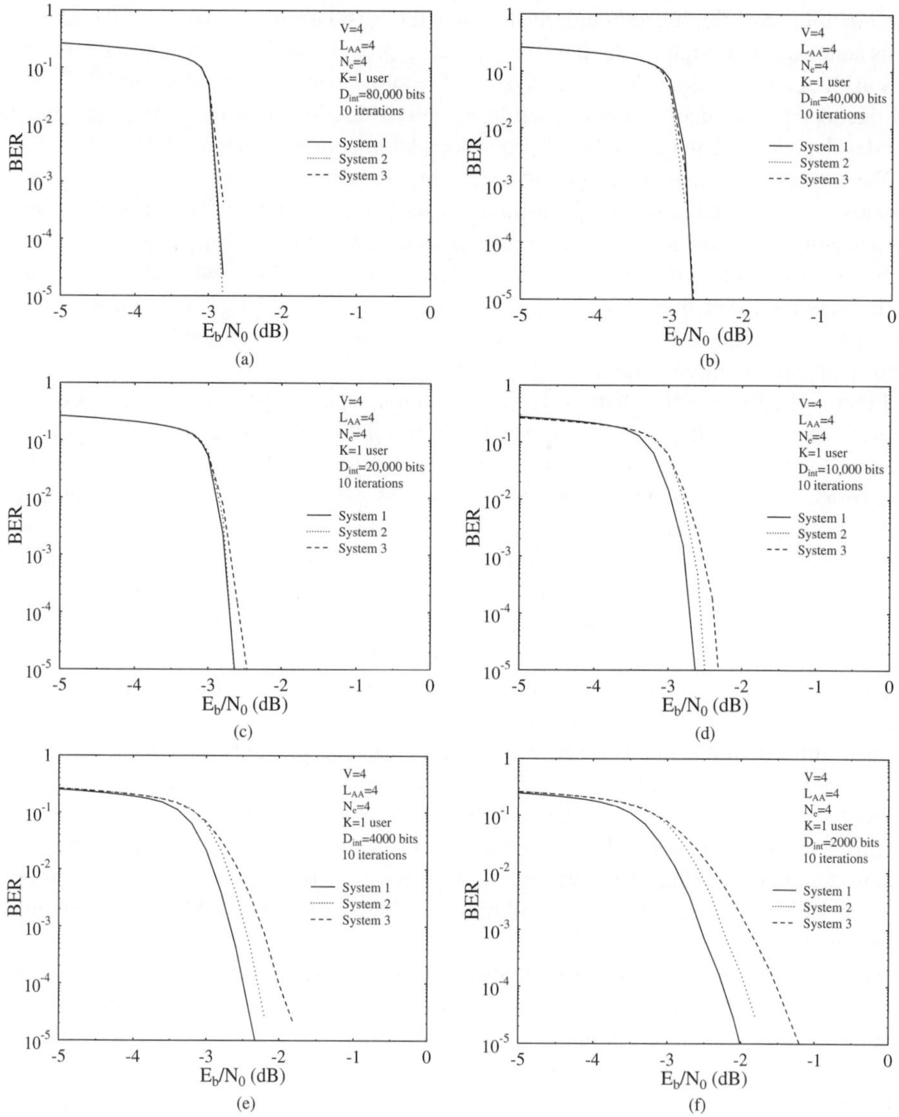

Figure 14.22: Performance comparison of Gray-mapping-based RSC-coded and URC-precoded Systems 1, 2 and 3 in conjunction with $V = 4$ subcarriers, $L_{AA} = 4$ elements per AA, $N_e = 4$ and $K = 1$ user and the system parameters outlined in Table 14.1, when varying the interleaver depth for ten decoding iterations.

way that guarantees that the same STS signal is spread to and, hence, transmitted by the specific V subcarriers having the maximum possible frequency separation, the diversity order of the system employing V subcarriers increases V-fold compared with that employing a single subcarrier. In addition, in order to increase the number of users so that the system can support more than N_e users, where N_e is the TD spreading factor, TD and FD spreading was employed. We also employed a user-grouping technique for the sake of minimizing the multi-user interference imposed on each other by the users sharing the same TD spreading code. In order to improve the performance of the proposed system, we presented the three iterative-detection-aided structures of Figures 14.6, 14.7 and 14.8.

The three iterative-detection-aided systems ultimately resulted in a similar BER performance and also had a similar complexity provided that we employ an interleaver depth of $D_{\text{int}} \geqslant 40\,000$ bits. Explicitly, after ten decoding iterations and employing an interleaver depth of $D_{\text{int}} = 160\,000$ bits, the three systems attained a BER below 10^{-5} at E_b/N_0 values in excess of -2.8 dB. On the other hand, System 1 of Figure 14.6 had $7\,039\,824$ trellis states, System 2 of Figure 14.7 employed $7\,039\,813$ trellis states and, finally, System 3 of Figure 14.8 required $7\,039\,648$ trellis states, when considering $D_{\text{int}} = 160\,000$ bits and $I = 10$ decoding iterations. On the other hand, if we employ interleavers having a depth shorter than $40\,000$ bits, we can observe from Figure 14.22 that System 1 performs better than Systems 2 and 3, since the interleaver depth of Systems 2 and 3 is half that employed in System 1. Finally, the single-user iteratively detected System 1 employing $N_e = 4$ and $V = 4$ outperformed the $K = 8$ users system by an E_b/N_0 of about 0.45 dB at a BER of 10^{-5}.

14.7 Chapter Summary

In this chapter, we have proposed a multi-functional multi-user MIMO scheme that combined the benefits of V-BLAST, of space-time codes, of generalized MC DS-CDMA and of beamforming. Thus, the proposed system benefits from the multiplexing gain of V-BLAST, from the spatial diversity gain of space-time codes, from the frequency diversity gain of the generalized MC DS-CDMA and from the SNR gain of the beamformer. This multi-functional MIMO scheme was referred to as LSSTS-aided generalized MC DS-CDMA.

In Section 14.2 we introduced the proposed MIMO scheme, where we illustrated how the transmitter of the different users was constructed. Then, we outlined the decoding process that takes place at the receiver side, where two-stage decoding was employed. First, the interference cancelation was carried out, where the interference imposed by one of the STS blocks on the other was perfectly canceled by the interference cancelation technique employed. Afterwards, STS decoding was carried out in the same way as proposed in [13]. In order to increase the number of users supported by the system, FD spreading was applied in the generalized MC DS-CDMA in addition to the TD spreading of the STS. In this case, the users can share the same TD spreading code and then they are distinguished by their FD spreading code. This results in the users sharing the same TD spreading code imposing multi-user interference on each other. Hence, we employed a user grouping technique that minimizes the FD interference coefficient for the users in the same TD group. The user grouping technique was described in Section 14.3.3.

To further enhance the achievable system performance, the proposed MIMO scheme was serially concatenated with an outer code combined with a URC, where three different iteratively detected systems were presented, referred to as Systems 1, 2 and 3. System 1, shown in Figure 14.6, employed the serial concatenation of a RSC encoder and a URC encoder with the proposed QPSK-modulated LSSTS scheme. At the receiver side, iterative detection was carried out between the RSC decoder and the URC decoder. In addition, in the structure of Figure 14.7, denoted as System 2, the transmitted source bits were encoded by the outer RSC code's encoder. The outer channel encoded bits were then S/P converted to two parallel streams. Each bit stream was then encoded by a corresponding URC encoder followed by a QPSK modulator in each substream. The data in each stream was then transmitted using SSTS. At the receiver side of System 2, iterative detection was carried out between the RSC decoder and the two URC decoders, where the LLRs were S/P converted from the RSC decoder to the URC decoders and they were then P/S converted, when passed from the URC decoders to the RSC decoder. Finally, in the structure of Figure 14.8, denoted as System 3, the transmitted source bits were first S/P converted to two parallel substreams, where each substream was encoded by the outer RSC code's encoder followed by the URC encoder. The URC encoded bits in each stream were then mapped to QPSK symbols and transmitted using SSTS. At the receiver side of System 3, seen in Figure 14.8, the decoding process of Section 14.2.2 was employed, where the decoded symbols were passed to their corresponding branch as shown in Figure 14.8. Each branch then applied iterative detection, exchanging extrinsic information between the corresponding RSC and URC decoders, as shown in Figure 14.8. The extrinsic output of the RSC decoders was then P/S converted to produce a LLR stream, which was passed to the hard-decision decoder of Figure 14.8.

We used EXIT charts in order to study the convergence behavior of the proposed systems and in Section 14.4.1 we proposed a LLR post-processing scheme for the soft output of the QPSK demapper, in order to improve the achievable system performance. In Section 14.5 we discussed our performance results and characterized the three proposed iteratively detected schemes, while employing $N_t = 4$ transmit AAs, $N_r = 2$ receive antennas, L_{AA} elements per AA, V subcarriers and K users. We demonstrated that the three proposed systems attain a BER lower than 10^{-5} at E_b/N_0 in excess of -2.8 dB, while employing $D_{\text{int}} = 160\,000$ bits and $I = 10$ iterations. Finally, for the system employing $N_e = 4$ and $V = 4$, the single-user iteratively detected System 1 outperformed the eight-user system by an E_b/N_0 of about 0.45 dB at a BER of 10^{-5}.

Chapter 15

Distributed Turbo Coding

15.1 Introduction

Wireless channels suffer from multipath propagation of signals, which results in a time-variant received signal strength, when the transmitter or receiver are moving. During severe fading of a specific propagation path, the received signal cannot be correctly decoded, unless some less attenuated multipath versions of it are available at the decoder side. This can be arranged by introducing transmit diversity, for example. In conventional MIMO systems constituted by co-located MIMO elements, transmit diversity is generated by transmitting different versions of the signal from different antennas located at the same BS or MS. Transmit diversity results in a significantly improved BER performance, when the different transmit antennas are positioned sufficiently far apart to ensure that the paths from each antenna to the destination experience independent fading.

The antenna spacing in co-located MIMOs is assumed to be sufficiently large so that the assumption of statistical independence of the different paths from the different antennas is justified. However, satisfying the assumption of a sufficient high antenna spacing may be impractical for pocket-sized wireless devices, which are typically limited in size and hardware complexity to a single transmit antenna. On the other hand, spatial fading correlations caused by insufficient antenna spacing at the transmitter or receiver of a MIMO system result in a degraded capacity as well as BER performance for MIMO systems, as shown in Figure 15.1 for a twin-antenna-aided STBC system [11]. Spatial correlation is typically introduced as a result of large-scale shadow fading that affects the transmission links between the different transmit and receive antennas [5]. Figure 15.1 compares the BER performance of a single-transmit and single-receive antenna system with that of a twin-antenna-aided STBC system affected by the large-scale shadow fading. As shown in Figure 15.1, the performance of MIMO systems degrades as the variance of the shadow fading increases and the single-antenna-aided system succeeds in outperforming the MIMO system when the shadow-fading variance is higher than 5 dB.

Hence, we can surmise that transmit diversity methods are not readily applicable to compact wireless communicators owing to the size as well as complexity constraints that limit the use of multiple transmit antennas. For example, in wireless mobile systems the size of the mobile unit is a limiting factor in incorporating several antennas that are sufficiently far apart for attaining statistically uncorrelated fading between the different transmit and receive

Near-Capacity Multi-Functional MIMO Systems L. Hanzo, O. R. Alamri, M. El-Hajjar and N. Wu

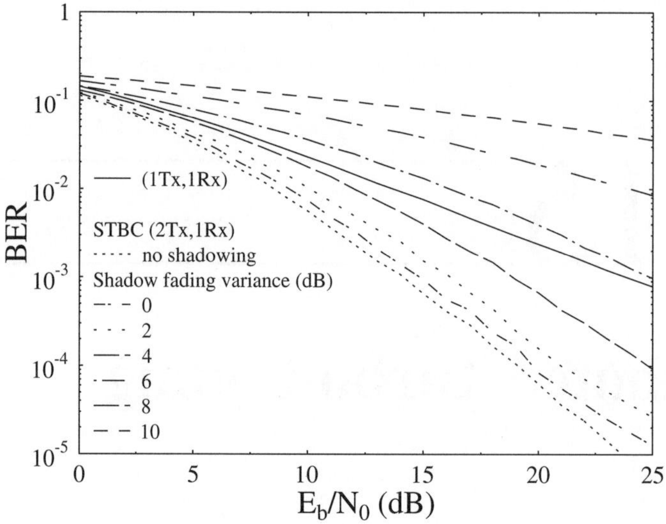

Figure 15.1: Effect of large-scale shadow fading on the performance of a twin-antenna-aided STBC system.

antennas. Recently, cooperative communication techniques [113, 114, 386] were proposed for eliminating correlation amongst the diversity paths by cooperatively activating the single antennas of several MSs, hence effectively creating a distributed MIMO scheme. In other words, single-antenna-aided users support each other by 'sharing their antennas' and thus generate a virtual multi-antenna environment [121]. Again, in cooperative communications it is possible to guarantee that the cooperating users are sufficiently far apart in order to attain independent fading. Since the signals transmitted from different users undergo independent fading, spatial diversity can be achieved by the concerted action of the cooperating partners' antennas.

In this chapter we design a DTC scheme, where two users cooperate in a two-phase cooperation scheme. During the first phase of cooperation, each user sends their own data to the other user, followed by the second phase where both users transmit their own data as well as the data of the other user after interleaving and channel coding. The two users employ multi-dimensional SP modulation and then transmit their data simultaneously. The receiver applies interference cancelation for the sake of eliminating the interference imposed by one user's data on the other. We employ SP modulation for the sake of attaining further iteration gains, as described in Chapter 11. The data transmitted from the second user's transmitter is an interleaved version of the bit stream transmitted from the first user's transmitter. Hence, at the receiver side, the interference canceler outputs two data streams corresponding to the data transmitted from the first and the second users' transmitters, respectively. Afterwards, iterative detection is employed between the SP demapper and the channel decoder in each data stream (or decoding branch) as well as between the two channel decoders of the two branches, hence forming a distributed turbo code. In addition, we study the effects of different Inter-User Channel (IUC) characteristics on the attainable BER performance of the proposed DTC system, where we compare the attainable BER performance, when the IUC is perfect, Gaussian, Ricean and Rayleigh faded.

The rest of the chapter is organized as follows. In Section 15.2 we present an overview of cooperative communications, followed by the description of the proposed DTC system in Section 15.3. In Section 15.4 we characterize the DTC-aided system with the aid of simulation results, while considering different IUC characteristics. We conclude in Section 15.5 and present a brief chapter summary in Section 15.6.

15.2 Background of Cooperative Communications

In this section we present a brief literature overview of cooperative communications and introduce the basic ideas behind the concept of user cooperation.

The basic idea behind cooperative communications can be traced back to the idea of the relay channel, which was introduced in 1971 by Van der Meulen [115]. Cover and El Gamal characterized the relay channel from an information-theoretic point of view in [116]. The relay model comprises three components: a source transmitting data, a destination receiving the data from the source and a relay receiving the data from the source and then transmitting it to the destination. Cooperative communications may be interpreted as a generalization of the relay channel, where the source and the relay transmit their own data as well as the other's data, which results in the destination receiving multiple copies of the same data from the source as well as from the relays, hence benefiting from a spatial diversity gain and eventually from an improved BER performance for the two users. Cooperative techniques benefit from the broadcast nature of wireless signals, where the signal transmitted from a specific user to a specific destination can be 'overheard' by neighboring users. These users can process the signal they overhear without compromising the security of the data and then transmit the processed data to the destination.

In [118] Sendonaris *et al.* generalized the relay model to multiple nodes that transmit their own data as well as serve as relays for each other. The scheme proposed in [118] was referred to as 'user cooperation diversity', where the authors examined the achievable rate regions and outage probabilities for this particular scheme. In [113, 114] Sendonaris *et al.* presented a simple user-cooperation methodology based on a DF signaling scheme using CDMA. The orthogonality of the different spreading codes of the different users makes it possible for the intended receiver to distinguish between the information transmitted from different cooperating users. Laneman *et al.* [119] reported the gains achieved in terms of an improved data rate and reduced sensitivity to channel variations, where it was concluded that cooperation effectively mimics a multi-antenna scenario with the aid of single-antenna terminals.

Cooperative communications has been shown to offer significant performance gains in terms of various performance metrics including diversity gains [24, 124, 144] as well as multiplexing gains [128]. In the following sections we review the mainstream cooperative methods used for signaling the data between the different users and the destination.

15.2.1 Amplify-and-Forward

Each user receives a noise-contaminated version of the other users' signals. In the case of an AF signaling strategy [119], the relay simply amplifies the noisy version of the signal without improving its SNR and retransmits it to the destination. The destination then combines the information sent by the source and the relay and makes a final decision on the transmitted bits [387]. Although the relay amplifies the noise in addition to amplifying the desired signal, the destination still observes two independently faded versions of the signal, thus benefiting from a diversity gain as compared with non-cooperative schemes.

AF was fist proposed and analyzed by Laneman *et al.* in [119], where the authors have shown that in the case of two-user cooperation, AF is capable of achieving a diversity order of two. In the scheme of [119], it was assumed that the destination has perfect knowledge of the IUC so that optimal decoding can be performed, where the IUC knowledge was assumed to be acquired by exchanging this IUC information between the nodes with the aid of side-information signaling or by blind estimation [387]. More elaborate AF algorithms and more general linear relaying schemes have been considered in [125, 388].

15.2.2 Decode-and-Forward

In the DF signaling scheme the relay decodes the partners' signals and then re-encodes the detected bits before their retransmission [387]. The destination combines the signal received from the source and the relay, hence creating spatial diversity. A witty low-complexity DF signaling strategy can be found in [113, 114], where two users were paired to cooperate with each other using CDMA. Each signaling period is divided into three time slots, where in the first and second time slots each user transmits their own data. Each user's data is broadcast and hence can be detected by the other user. In the second time slot, each user detects the other user's data. In the third time slot, both users transmit a linear combination of their own second time slot data and the partners' second time slot data, each multiplied by the appropriate spreading code. The allocation of transmitted power for the three time slots is determined by the average power constraint of each user. Explicitly, when the IUC has a high SNR, more power can be allocated to cooperation. Otherwise, less power is assigned for cooperation. The advantage of this signaling regime is its appealing simplicity and adaptability to channel variations. In addition, it is required that the destination has acquired knowledge of the CIR of the IUC for the sake of optimal decoding [387].

In order to avoid the problem of error propagation, selection DF was proposed by Laneman *et al.* in [124], where the relay detects the signal from the source and only forwards the signal when the instantaneous SNR for the IUC is below a certain threshold. Otherwise, the source continues its transmission to the destination in the form of repetition or more powerful codes [387]. If the measured SNR falls below the threshold, the relay transmits what it receives from the source using either AF or DF, in order to attain a diversity gain. Another signaling strategy is referred to as incremental relaying [124], which can be viewed as an extension to incremental redundancy or Hybrid Automatic Repeat Request (HARQ). In this case, the relay retransmits in case the destination provides a negative acknowledgement in an attempt to attain a diversity gain.

15.2.3 Coded Cooperation

Coded cooperation [120, 132] combines the concept of cooperative communications with channel coding. Coded cooperation maintains the same information rate, code rate, bandwidth and transmit power as a comparable non-cooperative system. The basic idea is that each user attempts to transmit incremental redundancy for their partner. Whenever the IUC is not favorable, the users automatically revert to a non-cooperative mode [387]. The key to the efficiency of coded cooperation is that all of this is managed automatically with the aid of sophisticated code design, with no feedback between the users [387].

Each user encodes their data bits using a Cyclic Redundancy Check (CRC) followed by a specific code from a family of Rate Compatible Punctured Convolutional (RCPC) codes [120]. Each user's encoded codeword is divided into two segments containing N_1 and N_2 bits, where $N_1 + N_2 = N$ and N is the total codeword length of the encoded sequence. In the first time slot, each user transmits their own first set of N_1 bits, where the encoded codeword is punctured to N_1 bits and hence the N_1 bits transmitted constitute a valid or legitimate codeword. The remaining N_2 bits are the punctured bits. Each user then attempts to decode the transmission of the other user. If this attempt is deemed to be successful based on the CRC code, the user computes and transmits the N_2 bits of the other user in the second time slot. Otherwise, the user transmits their own N_2 bits. Thus, each user always transmits a total of $N = N_1 + N_2$ bits per source block over the two time slots [120, 387].

The users act independently in the second time slot, with no knowledge of whether their own first frame was decoded correctly. As a result, there are four possible cases for the transmission of the second frame: both users cooperate, neither user cooperates, user 1 cooperates and user 2 does not or vice versa. It was suggested in [126] that the destination successively decodes according to all possibilities and checks the CRC code's success for each case. If the CRC fails for all possibilities, then the destination will select the specific codeword with the lowest path metric from the Viterbi algorithm.

In addition, it was proposed in [24] to use distributed space-time codes for the relay channel, demonstrating its benefits from an information-theoretic point of view. In [126] Janani *et al.* proposed space-time cooperation in addition to implementing turbo coding by exchanging extrinsic information between the data received from the source and the relay. Furthermore, a method designed for achieving cooperative diversity using rate compatible punctured codes was proposed in [120, 132]. In [122, 123], it was proposed to employ distributed turbo codes by exchanging extrinsic information between the data received from the source and that received from the relay, where the relay applies interleaving for the data received from the source and then uses an appropriate channel code before retransmission.

15.3 DTC

In this section we propose a specific DTC scheme based on the system architecture shown in Figure 15.2, where the users u_1 and u_2 cooperate in a two-phase cooperation scenario. The difference between our proposed system and those presented in [120, 122, 123, 126, 132] is that the two users in our design transmit their own data in addition to the other user's data, while the systems in [120, 122, 123, 126, 132] have a single active user and a relay transmitting the data of the active user only. In addition, in our proposed system we study the effect of the IUC characteristics on the attainable performance of the proposed DTC system. As shown in Figure 15.2 two users cooperate in order to communicate with a BS. In this way the users achieve a diversity gain in the case where the two users transmit in a space-time coded manner, or attain a multiplexing gain, if they transmit in a BLAST-like scheme. In the proposed scheme, the two users cooperate in two phases, where they exchange their data in the first cooperation phase and then they both transmit simultaneously their own data as well as the data of the other user in the second phase of cooperation.

The proposed system operates in a half-duplex mode, where none of the transceivers can transmit and receive at the same time. The transceivers operate in a TDD mode, where different transmitters transmit in different time slots. As shown in Figure 15.2, in time slot t_1 user 1 transmits their data to user 2. Similarly, in time slot t_2 user 2 shares their data with user 1. Hence, the first two time slots, i.e. time slots t_1 and t_2, comprise the first phase of cooperation. In the second phase of cooperation, i.e. in time slot t_3, the two users transmit their data simultaneously to the BS after appropriate interleaving and channel coding.

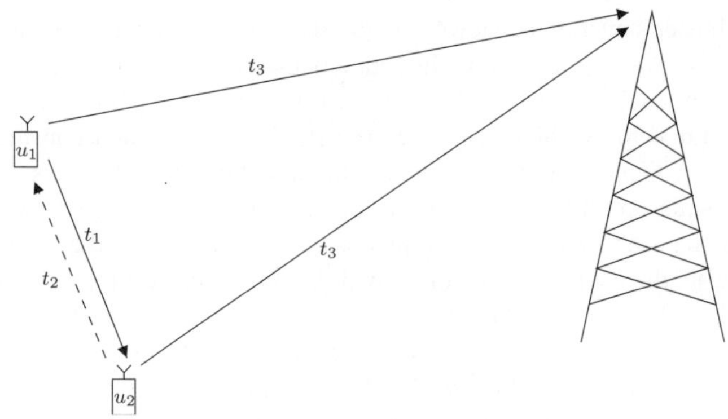

Figure 15.2: DTC system model.

A more detailed block diagram of the proposed scheme is shown in Figure 15.3. In time slot t_1, user 1 transmits their data bit stream $\mathbf{a}_1$ to user 2, where the received and decoded estimate of $\mathbf{a}_1$ is denoted by $\widetilde{\mathbf{a}}_1$. Hence, the DF signaling strategy is used in the proposed DTC scheme. Similarly, user 2 transmits their data bit stream $\mathbf{a}_2$ to user 1, where the received and decoded bit stream is denoted by $\widetilde{\mathbf{a}}_2$. There are several scenarios for the inter-user communication, i.e. for the phase-one cooperation. The simplest is phase-one cooperation, where each user transmits the source bits directly without any channel coding and the other user applies hard-decision decoding of the received signal. Another possible scenario is where each user encodes their data using a channel code of a specific code rate and then the receiver decodes the received stream by passing it through the modulated symbol-to-channel-coded-bits demapper as well as the channel decoder and may employ iterative detection by exchanging extrinsic information between the demapper and the channel decoder.

After the inter-user communication phase has been concluded during time slots t_1 and t_2, the two users now both have their own data as well as an estimate of the data of the other user. Both users concatenate their own data with the estimate of the other user's data, as shown in Figure 15.3 and detailed as follows. The first user appends the estimate of the second user's data $\widetilde{\mathbf{a}}_2$ with their own data $\mathbf{a}_1$ and then encodes the resultant bit stream $\mathbf{b}_1$ by a RSC code. The channel coded bit stream $\mathbf{c}_1$ is interleaved by the random bit interleaver Π_1 of Figure 15.3 and then the interleaved bits $\mathbf{b}_4$ are modulated by the SP Mapper I of Figure 15.3. Similarly, user 2 appends their own data $\mathbf{a}_2$ to the estimate of the user 1 data $\widetilde{\mathbf{a}}_1$ and then interleaves the resultant bit frame $\mathbf{b}_2$ by the bit interleaver Π_2 of Figure 15.3. The interleaved bit stream $\mathbf{b}_3$ is channel coded by a RSC code and then the encoded bit stream $\mathbf{c}_2$ is interleaved. Finally, the interleaved bit stream $\mathbf{b}_5$ is modulated by the SP Mapper II of Figure 15.3.

The two users simultaneously transmit their SP modulated symbol streams $\mathbf{x}_1$ and $\mathbf{x}_2$ assuming that there is perfect synchronization between the two users' MSs. The two users' MSs normalize their transmit power so that the transmit power in the two phases of cooperation is equivalent to the case when there is no cooperation.

At the BS, low-complexity ZF Interference Cancelation (IC) is applied, as described in [15]. The IC decoder outputs the two streams of decoded data $\widetilde{\mathbf{x}}_1$ and $\widetilde{\mathbf{x}}_2$ corresponding to the data transmitted by user 1 and user 2, respectively, as shown in Figure 15.3. After the IC stage, estimates of the transmitted data streams $\widetilde{\mathbf{x}}_1$ and $\widetilde{\mathbf{x}}_2$ are passed to the SP demappers of Figure 15.3. The SP demappers I and II of Figure 15.3 utilize the data received from the IC together with the *a priori* information $L_{MI,a}(\mathbf{b}_4)$ and $L_{MII,a}(\mathbf{b}_5)$ passed to them from the RSC decoders for the sake of providing the improved extrinsic information $L_{MI,e}(\mathbf{b}_4)$

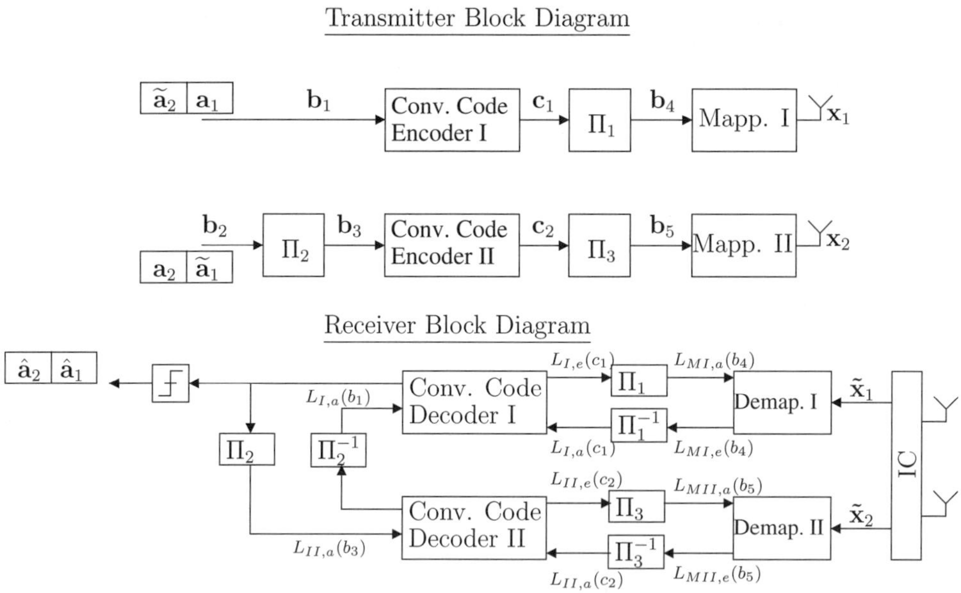

Figure 15.3: Block diagram of the two users' transmitters and the BS receiver.

and $L_{MII,e}(\mathbf{b}_5)$, which is then passed to the RSC decoders as the *a priori* information $L_{I,a}(\mathbf{c}_1)$ and $L_{II,a}(\mathbf{c}_2)$. Afterwards, iterative detection is carried out by exchanging extrinsic information between the two RSC decoders of the two branches. The two RSC decoders employ iterative detection for the sake of providing improved extrinsic information to the SP demappers of Figure 15.3. The iteration employed between the two RSC decoders is similar in its concept to that employed in classic turbo codes [9, 389], which motivates the terminology of distributed turbo coding.

Figure 15.4 shows the EXIT chart of the proposed DTC system. In Figure 15.4 we plot the EXIT curve of the inner SP demappers in conjunction with $L = 16$ and AGM-1 listed in Appendix H. The EXIT curves of the inner SP demappers seen in Figure 15.4 are shown for E_b/N_0 between 0 and 6 dB in steps of 0.5 dB. Figure 15.4 also shows the inverted EXIT curve of the outer DTC. The outer codes applied are half-rate memory-two RSC codes in conjunction with an octal generator polynomial of $(G_r, G) = (7, 5)_8$, where G is the feedforward polynomial and G_r is the feedback polynomial. We plot two curves corresponding to the outer DTC in Figure 15.4, where the first, represented by the long dashed line evolving mostly above the dashed curve, corresponds to two iterations between the two outer RSC codes' decoders, while the other curve marked by the dashed line corresponds to six iterations between the two RSC codes' decoders. In this case, in the first phase of cooperation, each user modulates the source bits and transmits the QPSK-modulated symbols to the other user without incorporating any channel coding. In addition, the EXIT curves are plotted for the system where the IUC is considered to be perfect. As shown in Figure 15.4, an open convergence tunnel is formed around $E_b/N_0 = 3.0$ dB. This implies that according to the predictions of the EXIT chart seen in Figure 15.4, the iterative decoding process is expected to converge for $E_b/N_0 > 3.0$ dB. On the other hand, observe in Figure 15.4 that the point of intersection between the EXIT curves of the inner and the outer codes approaches the

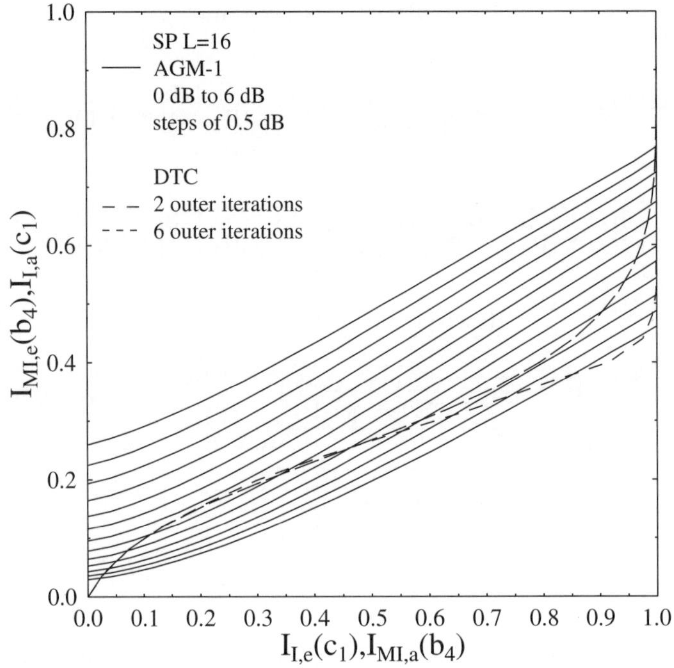

Figure 15.4: EXIT chart of the DTC system, where cooperation is employed between two users employing half-rate RSC codes while UL communication is performed over a narrowband Rayleigh fading channel.

1.0 mutual information points of the x-axis, as the number of outer iterations or distributed turbo coding iterations increases from $I_{\text{out}} = 2$ to 6. Hence, it is expected that employing $I_{\text{out}} = 6$ outer iterations between the RSC decoders will result in a better performance than that of the system employing two decoding iterations between the outer RSC codes.

On the other hand, Figure 15.5 shows the EXIT curve of the benchmark scheme, where two users cooperate by transmitting their own data simultaneously without transmitting the other user's data. In other words, each user encodes their own bit stream by an outer RSC code and then the encoded bit stream is interleaved by a random bit interleaver. Afterwards, the interleaved bit stream is modulated by a SP mapper and then transmitted from the user's single-antenna-aided MS. At the receiver side, IC is employed in order to eliminate the interference imposed by each of the users' data on the other user's data. Iterative detection is then carried out between the outer RSC decoder and the inner SP demapper in each branch and no iterations are employed between the two outer RSC codes' decoders, since the data in the two branches of Figure 15.3 in this case is different. Figure 15.5 shows the EXIT chart of the benchmark scheme in conjunction with an inner SP demapper employing $L = 16$ and AGM-1 of Appendix H for E_b/N_0 between 0 and 6 dB in steps of 0.5 dB. The outer code in this case is a half-rate memory-two RSC code in conjunction with an octal generator polynomial of $(G_r, G) = (7, 5)_8$. As shown in Figure 15.5, an open convergence tunnel is formed around $E_b/N_0 = 4.0$ dB. This implies that according to the predictions of the EXIT chart seen in Figure 15.5, the iterative decoding process is expected to converge at $E_b/N_0 > 4.0$ dB. In addition, a comparison between Figures 15.4 and 15.5 shows that the

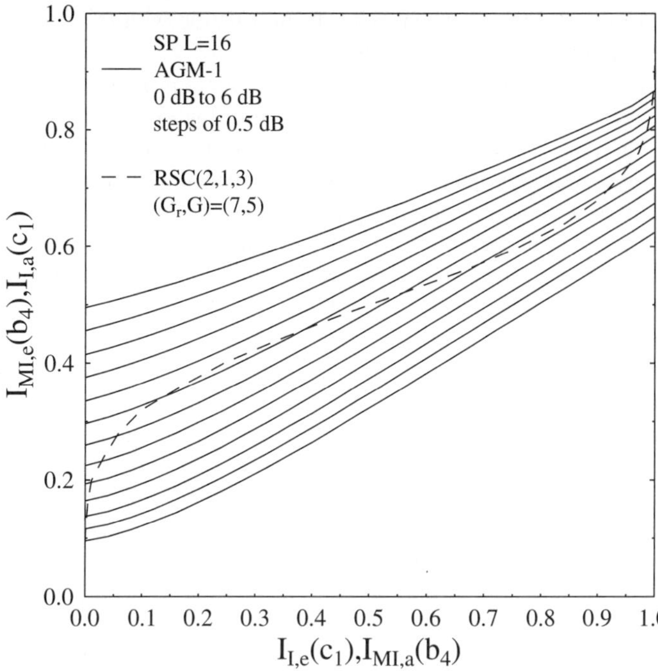

Figure 15.5: EXIT chart of the benchmark scheme where iterative detection is carried out between the outer RSC decoder and the inner SP demapper of each user without employing any iteration between the RSC decoders of the two users. In this case, each user transmits their own data only and, hence, there is no inter-user communication involved.

benchmark scheme exhibits a higher error floor as compared with the error floor at low BER values for the proposed DTC scheme.

15.4 Results and Discussion

In this section, we consider the achievable performance of the proposed DTC system, where two users cooperate in order to simultaneously transmit their data and then apply the ZF IC at the receiver for the sake of eliminating the interference imposed by each of the users' data on the other user's data. Each user's MS is equipped with a single antenna and each applies a half-rate memory-two RSC code for coding the bits, which are then mapped to symbols using a SP mapper in conjunction with $L = 16$ and AGM-1. In this section, we consider UL transmissions over a narrowband uncorrelated Rayleigh fading channel, where coherent detection is applied at the receiver side. It is also assumed that the receiver has perfect knowledge of the UL channel impulse response.

The proposed DTC scheme corresponds to a cooperation between two users, where each user is equipped with a single-antenna terminal. The system operates in a TDD mode, where the two users exchange their data in two time slots and then they transmit their joint data to the BS. Each user encodes their bit stream by an outer RSC code and then maps the encoded bits to SP symbols using a SP mapper. At the BS, ZF IC is applied, where the decoder outputs two data streams corresponding to the data received from user 1 and user 2. The decoded symbols are demapped to soft LLR values by the SP demappers of Figure 15.3.

The SP demapper of each branch exchanges soft information with the RSC decoder in the same branch after appropriate interleaving and deinterleaving. Afterwards, the RSC decoders of the two branches exchange extrinsic information in order to provide improved information for the SP demappers, which in turn pass extrinsic information to the RSC decoders. As shown in Figure 15.4, in order to arrive at as low an error floor as possible, it is required to employ $I_{\text{out}} = 6$ outer iterations between the two RSC decoders. Hence, in what follows, we refer to an iteration as a system iteration I_{sys}, where the SP demappers exchange extrinsic information with the RSC decoders once in an inner iteration, followed by $I_{\text{out}} = 6$ outer decoding iterations between the two parallel RSC decoders, before passing the improved extrinsic information back to the SP demappers.

The benchmark scheme for our proposed system is also a cooperative scheme exchanging extrinsic information between single-antenna-aided users. However, in the benchmark scheme, the two users do not exchange their information; rather, each user transmits their own data after channel coding, interleaving and then SP mapping. The two users transmit their data simultaneously and then ZF IC is applied at the receiver, where each user's data is passed to a separate branch of iterative detection between the SP demapper and the outer RSC decoder. In this case, no iterative detection is carried out by exchanging extrinsic information between the two RSC decoders, since the two branches have different data.

The bandwidth efficiency of the proposed DTC scheme can be analyzed as follows. During the first time slot t_1, user 1 transmits their data to user 2; this is followed by time slot t_2, where user 2 transmits their data to user 1. Afterwards, during the second phase of cooperation, each user has to transmit the data of the two users, which requires two time slots. Hence, a total of four time slots are required for the transmission of the two users' data to the BS. In contrast, the benchmark scheme requires only a single time slot for the transmission of the same amount of data to the BS by the two users. Hence, the proposed scheme has a factor of four lower throughput compared with the benchmark, but as a benefit, the proposed scheme attains a better BER performance according to the EXIT chart predictions of Figures 15.4 and 15.5, which is demonstrated in Figure 15.7 below.

Figure 15.6 shows the BER performance of a V-BLAST system having co-located MIMO elements, where the transmitter has two antennas and the receiver is also equipped with two antennas, while communicating over narrowband Rayleigh fading channels also affected by large-scale shadow fading. The source bits are encoded by a half-rate RSC code and then interleaved by a random bit interleaver, where the interleaved bits are mapped to SP symbols before transmission. At the receiver side, iterative detection is carried out by exchanging extrinsic information between the outer RSC code and the SP demapper. Figure 15.6 shows the effect of shadow fading on the attainable performance of a MIMO scheme having co-located elements, where the shadow fading imposes correlation on the impulse response of the channels between the two transmit antennas and the receive antennas. In Figure 15.6 the RSC code used is a half-rate memory-two code with an octal generator polynomial of $(G_r, G) = (5, 7)_8$ and the BER curves correspond to $I_{\text{inner}} = 3$ decoding iterations between the SP demapper and the RSC decoder. As shown in Figure 15.6, when the shadow fading standard deviation increases, the attainable BER performance degrades. Therefore, distributed MIMO or cooperative communications represent an intelligent way of retaining the gains of the MIMO systems having independently fading co-located elements without being affected by the shadow fading, while relying on single-antenna-aided MSs.

Figure 15.7 compares the attainable BER performance of the proposed DTC scheme and of the benchmark scheme, while considering UL transmissions over a narrowband Rayleigh fading channel. In this case, the IUC is considered to be perfect, i.e. the data exchanged

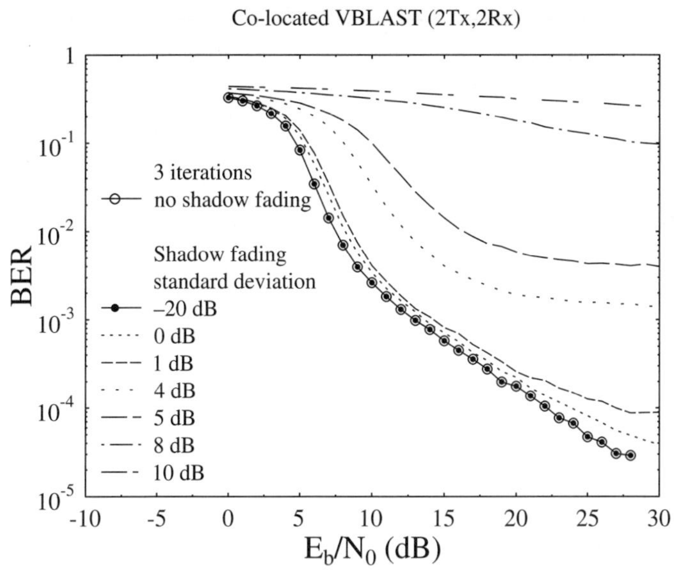

Figure 15.6: BER performance of a V-BLAST scheme employing two transmit and two receive antennas in conjunction with iterative detection between an outer RSC decoder and a SP demapper at the receiver side. The effect of shadow fading on the attainable performance is shown.

between the two users is recovered perfectly without any errors. In Figure 15.7 an interleaver depth of $D_{\text{int}} = 80\,000$ bits was employed and SP modulation in conjunction with $L = 16$ and AGM-1 is used. Observe in Figure 15.7 that the benchmark scheme suffers from a fairly pronounced error floor and that no iteration gain may be obtained after employing more than one inner iteration between the SP demapper and the RSC decoder in each branch. This result confirms the EXIT chart predictions of Figure 15.5, where the EXIT curves of the inner and outer codes intersect for $E_b/N_0 = 6.0$ dB at a point different from the 1.0 point on the x-axis indicating the lack of convergence to an infinitesimally low BER. On the other hand, observe in Figure 15.7 that the attainable BER performance of the proposed DTC-aided system exhibits no error floor. This can be explained from the EXIT chart of Figure 15.4, where the EXIT curves of the inner and outer codes intersect at the 1.0 point on the x-axis, when there is an open tunnel. The 1.0 point on the x-axis corresponds to the case where the outer decoder has perfect extrinsic information available at its output. This means that for the E_b/N_0 values where there is an open tunnel, the EXIT curves of the inner and outer codes intersect at the 1.0 point of perfect convergence, i.e. at the point where the outer decoder has perfect extrinsic information at its output, which results in an infinitesimally low BER. Again, for the proposed DTC scheme, a system iteration I_{sys} is constituted by a single inner iteration between the SP demapper and the RSC decoder of each branch, followed by $I_{\text{outer}} = 6$ iterations between the two RSC decoders in the two branches. Explicitly, the proposed DTC system outperforms its benchmark scheme by E_b/N_0 of about 25 dB at a BER of 10^{-5} after employing $I_{\text{sys}} = 5$ system iterations.

Figure 15.8 shows the attainable BER performance of the proposed DTC scheme, while considering UL transmission over a narrowband Rayleigh fading channel, where the IUC considered is a LOS AWGN channel, i.e. Gaussian noise is added to the received data without any amplitude or phase attenuation. In Figure 15.8 an interleaver depth of $D_{\text{int}} = 80\,000$ bits

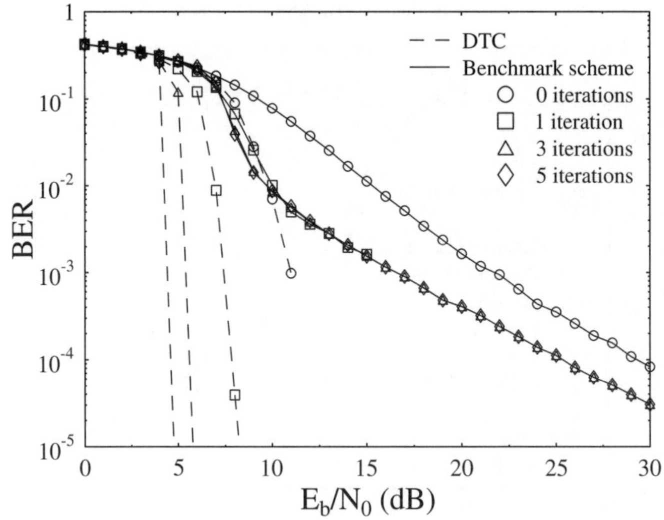

Figure 15.7: BER performance comparison of the proposed DTC system and the benchmark scheme in conjunction with an interleaver depth of $D_{\text{int}} = 80\,000$ bits for a variable number of system iterations I_{sys}, where two users are cooperating and communication is over a narrowband UL Rayleigh fading channel. The IUC in this case is considered to be perfect.

was employed and SP modulation in conjunction with $L = 16$ and AGM-1 was used. Figure 15.8 also shows the BER performance of the DTC system, when a perfect IUC is considered. The BER curves seen in Figure 15.8 correspond to the system where $I_{\text{sys}} = 4$ system iterations are employed. The notation $\text{SNR} + 10$ dB in Figure 15.8 means that the IUC SNR is 10 dB higher than the UL channel SNR, where SNR represents the UL channel SNR. Observe in Figure 15.8 that the attainable BER performance of the DTC system, when the IUC is constituted by an AWGN channel having a 10 dB higher SNR than that of the UL channel, is equivalent to that of the system considering a perfect IUC. However, when the IUC SNR is 5 dB higher than the UL channel SNR, an error floor is formed in Figure 15.8. This is due to the fact that the iterative detection at the receiver side assumes that the data in the two RSC decoders is identical, while the inter-user communication induces errors in the data available at the users' terminals, and hence results in a discrepancy in the data transmitted from the two users' terminals, which results in the error floor of Figure 15.8.

Figure 15.9 shows the attainable BER performance of the proposed DTC scheme, while considering UL transmission over a narrowband Rayleigh fading channel, when the IUC considered is a Ricean channel. In Figure 15.9 an interleaver depth of $D_{\text{int}} = 80\,000$ bits was employed and SP modulation in conjunction with $L = 16$ and AGM-1 is used. Figure 15.9 also shows the BER performance of the DTC-aided system, when a perfect IUC is considered. The BER curves in Figure 15.9 correspond to the system where $I_{\text{sys}} = 4$ system iterations are employed. Observe in Figure 15.9 that the attainable BER performance of the DTC-assisted system is equivalent to that of the system considering a perfect IUC, when the Ricean K-factor is 10 dB and the IUC SNR is 30 dB higher than the UL channel SNR. However, when the IUC SNR becomes less than 20 dB higher than the UL channel SNR, the BER performance degrades, where an error floor is formed. As the Ricean K-factor increases to 13 dB, no error floor is formed in the BER curve even for an IUC SNR of 10 dB higher than the UL channel SNR. On the other hand, in Figure 15.10 a Rayleigh faded IUC is considered

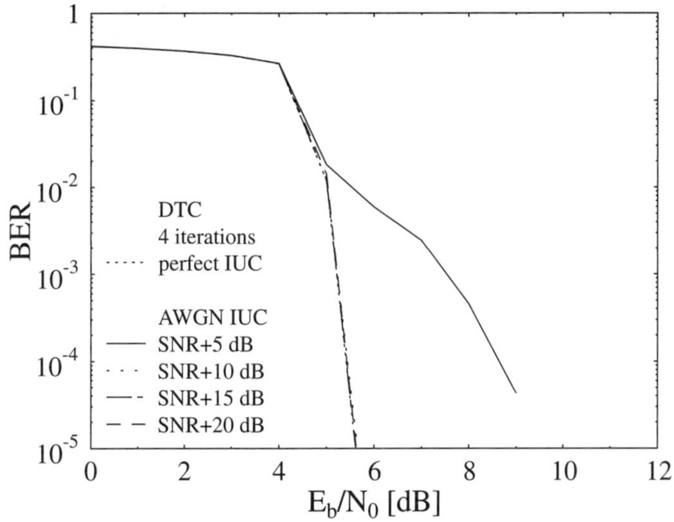

Figure 15.8: Comparison of the attainable BER performance of the proposed DTC scheme, while considering a AWGN IUC for variable IUC SNR values. The figure corresponds to $I_{\text{sys}} = 4$ iterations in conjunction with an interleaver depth of $D_{\text{int}} = 80\,000$ bits.

for the DTC-aided system. Observe in Figure 15.10 that an error floor is formed for all IUC SNR values, even when the IUC SNR goes as high as 50 dB above the UL channel SNR. This is due to the fact that transmission over Rayleigh fading channels imposes errors in the data received by the two users from each other and hence this means that there will be a difference between the data transmitted from the two users, which affects the turbo detection process at the receiver side.

Figures 15.7–15.10 show the BER performance of the proposed DTC-aided system, while considering different IUC characteristics. Explicitly, Figure 15.8 shows the BER performance of the UL transmission, while the IUC is considered to be AWGN with variable IUC SNR values. Figure 15.9 shows the same BER performance, while the IUC is Ricean with a variable Ricean K-factor and a variable IUC SNR. In addition, in Figure 15.10 we show the UL BER performance of the DTC scheme, while considering a Rayleigh IUC with variable IUC SNR values. Observe in the figures that an error floor is formed at a specific IUC SNR value, depending on the different IUC characteristics. To understand this phenomenon further, in Figure 15.11 we plot the BER performance of a single-input single-output QPSK-modulated system, while considering inter-user transmissions over AWGN, Rayleigh and Ricean channels. Observe in Figures 15.8–15.10 that an error floor is formed for the UL transmission of the proposed DTC scheme, when transmission over the IUC induces errors in the data exchanged between the two users. It can be shown that for the proposed DTC scheme to attain a reasonable BER performance with no error floor, the inter-user BER should be less than 10^{-6}.

The previous results outlined in Figures 15.7–15.10 show that the proposed DTC scheme attains a good BER performance without any error floor, provided that the inter-user communications can maintain a BER below 10^{-6}. Observe in the BER results of Figure 15.10 that an error floor in formed for inter-user transmission over a Rayleigh IUC for all IUC SNR values. All of the previous results correspond to the system where no channel coding was employed in the inter-user communication, i.e. each user maps the source bits to QPSK

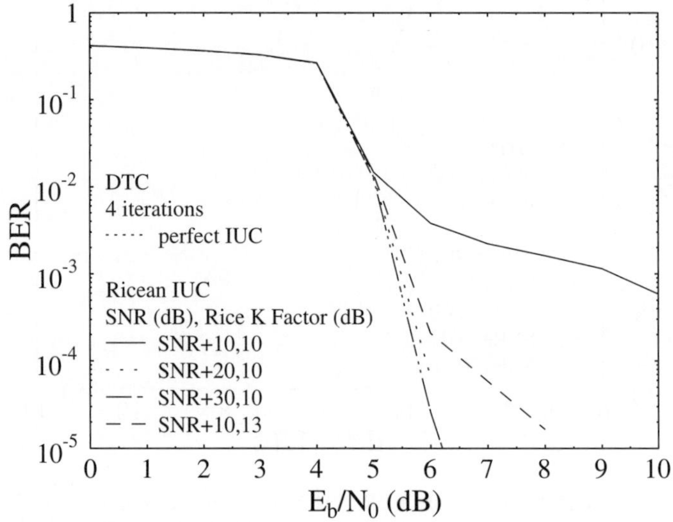

Figure 15.9: Comparison of the attainable BER performance of the proposed DTC scheme, while considering a Ricean IUC for a variable IUC SNR having different Ricean K-factors. The figure corresponds to $I_{\rm sys} = 4$ iterations in conjunction with an interleaver depth of $D_{\rm int} = 80\,000$ bits.

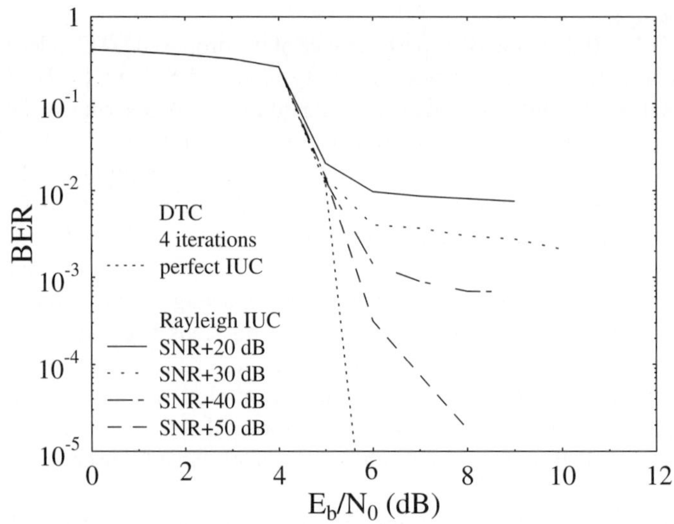

Figure 15.10: Comparison of the attainable BER performance of the proposed DTC scheme, while considering a Rayleigh IUC for variable IUC SNR values. The figure corresponds to $I_{\rm sys} = 4$ iterations in conjunction with an interleaver depth of $D_{\rm int} = 80\,000$ bits.

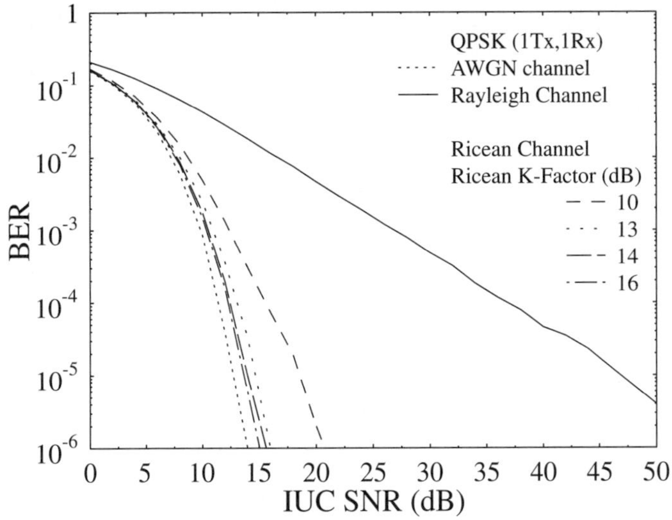

Figure 15.11: Attainable BER performance of a single-input single-output QPSK-modulated system, while considering inter-user transmission over an AWGN channel, a Rayleigh channel and a Ricean channel with a variable Ricean K-factor.

symbols and transmits them to the other user. In order to improve the performance of the inter-user communication, especially when the IUC is Rayleigh faded, each user can benefit from the RSC code it is equipped with and then encode the data bits, before mapping the encoded bits to QPSK symbols for transmission. This improves the BER performance of the IUC and allows the attainable BER performance of the DTC UL scheme to improve, as shown in Figure 15.12 when compared with Figure 15.10. In Figure 15.12 the two users employ a half-rate RSC code for the exchange of their data. Observe in Figure 15.12 that the DTC UL scheme attains a BER performance with no error floor, when the IUC SNR is 15 dB higher than the UL channel SNR, while for the uncoded inter-user communications scenario, an error floor was formed even when the inter-user Rayleigh channel had a SNR 50 dB higher than the UL channel SNR.

15.5 Chapter Conclusions

In this chapter, we have proposed a DTC scheme, where two users cooperate in order to improve their attainable BER performance. Each user's single-antenna-aided transmitter is constituted by a RSC code, an interleaver and a SP mapper. In the first phase of cooperation, the two users exchange their data by transmission over the inter-user channel. Each user then has their own data as well as the other user's data. In the second phase of cooperation, each user concatenates their own data and the other user's data. The first user applies RSC channel coding to the resultant data bit stream and then the channel coded bits are interleaved by a random bit interleaver. The interleaved bits are then mapped to 4-bit symbols by a SP mapper, where the symbols are transmitted from a single antenna. The second user applies the same procedure, but interleaves the data bit stream before channel encoding. The receiver applies the ZF IC algorithm of [15] and outputs two data streams that are passed to two branches, where each branch has a SP demapper, a deinterleaver and a RSC decoder. Iterative detection is carried out by exchanging extrinsic information between the SP demappers and the RSC

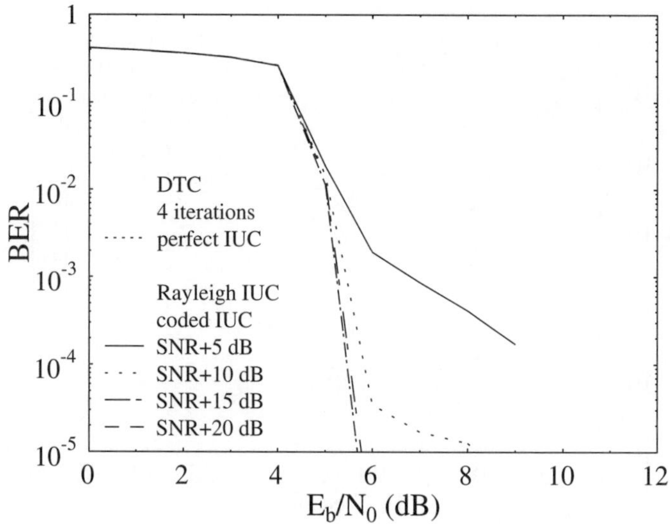

Figure 15.12: Comparison of the attainable BER performance of the proposed DTC scheme, while considering a Rayleigh IUC for variable IUC SNR values as well as employing RSC channel coding for inter-user communications. The figure corresponds to $I_{\text{sys}} = 4$ iterations in conjunction with an interleaver depth of $D_{\text{int}} = 80\,000$ bits.

decoders in each branch as well as between the two RSC decoders of the two branches, thus forming a turbo code.

The proposed DTC scheme was benchmarked against another cooperative scheme, where the two users do not exchange their data. In the benchmark scheme, each user encodes their own data bits by a RSC encoder and interleaves the coded bits, and then the interleaved bits are mapped to SP symbols for transmission from a single antenna. In the benchmark scheme, the iterative detection is carried out by exchanging extrinsic information between the SP demapper and the RSC decoder of each branch without employing any iterations between the RSC decoders of the two branches.

The proposed DTC scheme attained a throughput that is a factor of four lower than that of the benchmark scheme. However, the proposed scheme attained an E_b/N_0 gain of about 25 dB at BER of 10^{-5} when compared with the benchmark scheme, while considering perfect IUC and employing $I_{\text{sys}} = 5$ system iterations in conjunction with an interleaver depth of $D_{\text{int}} = 80\,000$ bits, where a system iteration corresponds to a single inner iteration between the SP demapper and the RSC decoder of each branch, followed by $I_{\text{outer}} = 6$ iterations between the RSC decoders of the two branches.

In addition, a study of the inter-user channel characteristics' effects on the performance of the proposed UL DTC scheme was conducted. First, in phase one of the cooperation, each user transmits their own data to the other user by mapping the source bits to QPSK symbols and then transmitting them through a single antenna. When the users exchange their data over an AWGN channel with IUC SNR being 10 dB higher than the UL channel SNR, the attainable BER performance of the DTC scheme is similar to that when the inter-user channel is assumed to be perfect. Furthermore, while considering a Rayleigh faded inter-user channel, an error floor is formed for the attainable BER performance of the proposed scheme even for IUC SNR as high as 50 dB higher than the UL channel SNR. It was shown in Figure 15.11 that in order to eliminate the error floor in the BER performance of the proposed scheme, the

BER performance of the inter-user channel is required to be less than 10^{-6}. Hence, in order to improve the attainable system performance, it is suggested to use channel coding, while the users are exchanging their data during the first phase of cooperation.

15.6 Chapter Summary

In this chapter, we have proposed a novel cooperative communication scheme referred to as DTC. In the proposed scheme, two users are cooperating, where each user's transmitter constitutes of a RSC code and then an interleaver followed by a SP mapper. The two users exchange their data in the first phase of cooperation followed by the second phase, which is the UL transmission of the combined data from the two users' antennas, protected by channel coding and interleaving. The second user interleaves the data bits before channel coding, so that at the receiver side iterative detection can be applied by exchanging extrinsic information between the two channel decoders after appropriately arranging the bits in the required order by interleaving and deinterleaving.

In Section 15.2 we provided an overview of cooperative communications, elaborating on the major cooperative signaling schemes, including AF, DF and coded cooperation. In Section 15.3 we presented our DTC scheme, where we proposed a two-phase cooperation scheme. In the first phase, the two users exchange their data, while in the second phase they simultaneously transmit their data to the BS. In Section 15.4 we presented our performance results and compared the proposed scheme with a benchmark scheme employing no exchange of the users' data, i.e. each user only transmits their own data. In Section 15.4 we also studied the effect of varying the IUC characteristics on the performance of the UL DTC scheme, where we showed that the IUC BER should not be higher than 10^{-6} in order to avoid the formation of an error floor in the UL. In Section 15.4 it was shown that an error floor is formed when the IUC considered is Rayleigh-faded even if the IUC SNR is as high as 50 dB higher than the UL channel SNR, as shown in Figure 15.10. In order to improve the system performance, it was shown in Figure 15.12 that the attainable BER performance can be improved when the users exchange their data over a Rayleigh channel, if the users employ channel coding in the first phase of cooperation.

In Section 15.4, we have shown that the data exchanged between the two users in the first phase of cooperation requires the IUC BER to be lower than 10^{-6} in order to eliminate any error floor in the UL performance of the DTC-aided system. We tackled this problem by using channel coding in the first phase of cooperation, where each user's data is channel coded and then interleaved before mapping the bits to QPSK symbols for transmission. Observe in Figure 15.12 that using channel coding in the first phase of cooperation does not entirely eliminate the error floor for IUC SNR being 10 dB higher than the UL channel SNR. Therefore, another solution has to be devised in order to improve the attainable performance, when there are errors in the data exchanged between the users in the first phase of cooperation.

In the proposed scheme, DF signaling was used in the first phase of cooperation, where a hard decision was performed on the received data. In [129], it was argued that the DF signaling loses the benefit of soft information. Hence, in [129] the employment of soft DF signaling was proposed, where all operations are performed in a soft-input soft-output fashion. It was shown in [129] that the soft DF outperforms the hard-decoded DF and the AF signaling schemes. In [137] soft DF was used, where the soft information was quantized, encoded and transmitted using superimposed modulation to the destination. The scheme proposed in [137] is a practical method devised for encoding the soft information without the need for reducing the system's throughput or increasing the system's bandwidth.

In [129, 134, 137] soft information has been used in diverse systems where there are errors in the first phase of cooperation and it was shown that soft DF attains a better performance than hard DF. Therefore, a potential improvement of our proposed system is to employ soft DF, where the MAP decoder used at the BS has to be modified in order to decode the data received from the two users.

Chapter 16

Conclusions and Future Research

In this chapter, a summary of the book is provided in Section 16.1, while a few future research ideas are presented in Section 16.2.

16.1 Summary and Conclusions

16.1.1 Chapter 1: Problem Formulation, Objectives and Benefits

In Chapter 1, we provided a general overview of the various subjects related to the systems considered in this book. More specifically, in Section 1.1, we discussed the key characteristics of the mobile radio channel in contrast to the Gaussian channel. In addition, we introduced the three most commonly used diversity techniques, namely temporal diversity, frequency diversity and antenna or space diversity. We then classified the entire family of MIMO systems in Section 1.2 and discussed the diversity versus multiplexing trade-offs in the context of multi-functional MIMO systems.

The various aspects of coherent versus non-coherent detection techniques designed for STBCs using both co-located and distributed or cooperative antenna elements of single-antenna-aided mobile handsets were introduced in Section 1.3. In Section 1.4, we presented a historical overview of STC and MIMO systems, along with the related state-of-the-art current research related to all aspects of the book.

The philosophy of concatenated schemes and iterative decoding was also discussed in Section 1.4 where Tables 1.12 and 1.13 summarized the main contributions in this burgeoning field. In Section 1.4 we also provided an overview of recent techniques devised for studying the convergence behavior of iterative decoding. Chapter 1 also presented a general overview of how the book is organized, where the outline of the book was described in Section 1.6.1. Finally, the novel contributions of the book were summarized in Section 1.6.2.

16.1.2 Chapter 2: Space-Time Block Code Design using Sphere Packing

In Chapter 2, we considered the theory and design of uncoded STBC-SP schemes. We first summarized the design criteria of space-time coded communication systems in Section 2.2, where the channel model along with the design criteria invoked both for quasi-static and for rapid fading channels are described in Sections 2.2.1 and 2.2.2, respectively. In Section 2.3, we emphasized the design criteria relevant for time-correlated fading channels, where both the pairwise error probability and the corresponding design criterion were presented in Section 2.3.2. The concept of diversity product, which was introduced in [54, 208] and generalized in [43] in order to account for the effect of the temporal correlation was discussed in Section 2.3.3.1. Furthermore, both lower and upper bounds on the generalized diversity product discussed in Section 2.3.3.1 were provided in Section 2.3.3.2.

In Section 2.4, orthogonal space-time designs combined with SP modulation were considered for space-time signals, where the motivation behind the adoption of SP modulation in conjunction with orthogonal design was discussed in Section 2.4.2. The signal design derived for $N_t = 2$ transmit antennas was provided in Section 2.4.3 in order to further illustrate the concept of combining an orthogonal space-time design with SP. Section 2.4.4 discussed the problem of constructing a SP constellation having a particular size L. Constellation points were first chosen based on the minimum energy criterion. Then, an exhaustive computer search was conducted to find the set of L points having the best MED from the entire set of constellation points satisfying the minimum energy criterion. The capacity of STBC-SP schemes employing $N_t = 2$ transmit antennas was derived in Section 2.4.5, demonstrating that STBC-SP schemes exhibited a higher capacity than conventionally modulated STBC schemes. Finally, the performance of STBC-SP schemes was presented in Section 2.5, demonstrating that STBC-SP schemes are capable of outperforming STBC schemes that employ conventional modulation (i.e. PSK, QAM). The coding gains achieved by the SP-assisted STBC over conventional modulated STBC at a SP-SER and a BER of 10^{-4} recorded for the schemes characterized in Figures 2.14–2.25 are summarized in Tables 2.6 and 2.7.

16.1.3 Chapter 3: Turbo Detection of Channel-coded STBC-SP Schemes

In Chapter 3, we demonstrated that the performance of STBC-SP systems can be further improved by concatenating SP-aided modulation with channel coding and performing demapping as well as channel decoding iteratively. The SP demapper of [43] was further developed for the sake of accepting the *a priori* information passed to it from the channel decoder as extrinsic information.

In this chapter, two realizations of a novel bit-based iterative-detection-aided STBC-SP scheme were presented, namely a RSC-coded turbo-detected STBC-SP scheme and a binary LDPC-coded turbo-detected STBC-SP arrangement. Our system overview was provided in Section 3.2. The LDPC-coded scheme of Figure 3.2 did not require channel interleaving, since the LDPC parity check matrix is randomly constructed, where each of the parity check equations is checking several random bit positions in a codeword, which has a similar effect to that of the channel interleaver. In Section 3.3, we showed how the STBC-SP demapper was modified for exploiting the *a priori* knowledge provided by the channel decoder, which is essential for the employment of iterative demapping and decoding.

EXIT chart analysis was invoked in Section 3.4 in order to study and design the turbo-detected schemes proposed in Section 3.2. Measuring the demapper's EXIT characteristics was explained in Section 3.4.1 and in Figure 3.4. We proposed ten different AGM schemes in Figure 3.5 that are specifically selected from all of the possible mapping schemes for $L = 16$ in order to demonstrate the different extrinsic information transfer characteristics associated with different bit-to-symbol mapping schemes. Observe that the slope of the EXIT curves corresponding to the different AGM schemes in Figure 3.5 increases gradually in fine steps. This characteristic is essential for the sake of designing near-capacity turbo-detected systems.

Both the Gray mapping and the various AGM mapping schemes considered in this book are detailed in Appendix A. In Section 3.4.2, we discussed how the EXIT characteristics of an outer decoder in a serially concatenated scheme may be calculated. Figure 3.6 summarizes the calculation process.

The performance of the turbo-detected bit-based STBC-SP schemes was presented in Section 3.5. Firstly, we considered the performance of the RSC-coded turbo-detected STBC-SP scheme in Section 3.5.1. The relation between the achievable BER and the mutual information at the input as well as at the output of the outer RSC decoder was discussed in Section 3.5.1.1. The predictions of our EXIT chart analysis outlined in Section 3.4.3 were verified by generating the actual decoding trajectories in Section 3.5.1.2. The effect of interleaver depth was also addressed in Section 3.5.1.2, since matching the predictions of the EXIT chart analysis is only guaranteed when employing large interleaver depths. The BER performance of the proposed RSC-coded STBC-SP scheme was compared in Section 3.5.1.3 with that of an uncoded STBC-SP scheme [43] and with that of an RSC-coded conventionally modulated STBC scheme. Secondly, we considered the performance of the LDPC-coded turbo-detected STBC-SP scheme in Section 3.5.2. The relation between the achievable BER and the mutual information at the input of the outer LDPC decoder was discussed in Section 3.5.2.1 and Figure 3.22. The effect of the LDPC output block length K_{ldpc} on the achievable performance was investigated in Section 3.5.2.2, while the effect of internal LDPC iterations and joint external iterations was studied in Section 3.5.2.3.

When using an appropriate bits-to-symbol mapping scheme and ten turbo-detection iterations, gains of about 20.4 and 21.2 dB were obtained by the convolutional-coded and LDPC-coded STBC-SP schemes, respectively, over the identical-throughput 1 BPS uncoded STBC-SP benchmark scheme [43]. Table 16.1 summarizes the coding gains of AGM-9-based RSC-coded STBC-SP schemes in conjunction with $L = 16$ and RSC-coded QPSK-modulated STBC schemes against an identical-throughput 1 BPS uncoded STBC scheme at BER of 10^{-3} and 10^{-5}, when employing the system parameters outlined in Table 3.3 and using different interleaver depths after ten external joint iterations. In order to highlight the advantage of employing SP modulation over conventional modulation schemes, Figure 16.1 shows coding gain comparisons of AGM-9-based RSC-coded STBC-SP schemes in conjunction with $L = 16$ and the conventional RSC-coded QPSK-modulated STBC schemes of Table 16.1.

16.1.4 Chapter 4: Turbo Detection of Channel-coded DSTBC-SP Schemes

In Chapters 2 and 3, we assumed that the CSI was perfectly known at the receiver. This, however, requires sophisticated channel estimation techniques, which imposes excess cost and complexity. In Chapter 4, we considered the design of novel SP-modulated differential STBC schemes that require no channel estimation, where we described in Section 4.2.1 how DSTBC schemes are constructed using SP modulation. The performance of uncoded DSTBC-SP schemes was considered in Section 4.2.2, where we compared the performance of different DSTBC-SP schemes against equivalent conventional DSTBC schemes under various channel conditions. Simulation results were provided for systems having different BPS rates in conjunction with appropriate conventional and SP modulation schemes, as outlined in Table 4.1. In Section 4.2.2.1, the channel was assumed to be constant over the transmission period of one frame, which was referred to as a block-fading Rayleigh channel. The attainable coding gains of SP modulation over conventional modulation at a SP-SER of 10^{-4} were summarized in Table 4.2, while the BER performance was illustrated

Table 16.1: Coding gains of AGM-9-based RSC-coded STBC-SP schemes in conjunction with $L = 16$ and RSC-coded QPSK-modulated STBC schemes against an identical-throughput 1 BPS uncoded STBC scheme at BER of 10^{-3} and 10^{-5}, when employing the system parameters outlined in Table 3.3 and using different interleaver depths after ten external joint iterations.

	E_b/N_0 (dB)		Gain (dB)	
	BER $= 10^{-3}$	BER $= 10^{-5}$	BER $= 10^{-3}$	BER $= 10^{-5}$
Uncoded STBC, BPSK	14.1	24.2	0	0
Uncoded STBC-SP, $L = 4$	14.00	24.00	0.1	0.2
RSC-coded, STBC-SP, $L = 16$				
RSC (2,1,5), AGM-9				
Interleaver depth $= 10^3$ bits	3.8	4.6	10.3	19.6
Interleaver depth $= 10^4$ bits	2.9	4.0	11.2	20.2
Interleaver depth $= 10^5$ bits	2.6	4.0	11.5	20.2
Interleaver depth $= 10^6$ bits	2.5	3.8	11.6	20.4
RSC-coded, STBC, QPSK				
RSC (2,1,5), set-partitioning				
Interleaver depth $= 10^3$ bits	3.5	6.0	10.6	18.2
Interleaver depth $= 10^4$ bits	3.2	5.8	10.9	18.4
Interleaver depth $= 10^5$ bits	3.2	5.7	10.9	18.5
Interleaver depth $= 10^6$ bits	3.1	5.7	11.0	18.5

in Figures 4.2–4.9. In Section 4.2.2.2, the channel was assumed to be constant over the transmission period of one SP symbol (i.e. two consecutive time slots). This type of channel was referred to in Section 2.5 as a SPSI channel. The channel was also assumed to be correlated and to have a normalized Doppler frequency of $f_D = 0.01$. Table 4.3 summarized the coding gains of SP modulation over conventional modulation at a SP-SER of 10^{-4}. The BER performance of DSTBC-SP schemes, when communicating over the SPSI-correlated Rayleigh fading channel considered, was characterized in Figures 4.10–4.17.

In Section 4.3, we proposed novel bit-based RSC-coded turbo-detected DSTBC-SP schemes. The system's architecture was outlined in Section 4.3.1, where the schematic of the proposed arrangement was provided in Figure 4.18. The EXIT chart analysis of Section 3.4 was employed in Section 4.3.2 in order to design and analyze the convergence behavior of the proposed turbo-detected RSC-coded DSTBC-SP schemes. Figures 4.20 and 4.21 illustrated two AGM-based DSTBC-SP schemes in conjunction with outer RSC codes having constraint lengths of $\mathcal{K} = 3$ and 5, which are specifically designed for low BER floors. Appendix C provides the complete list of EXIT charts for the RSC-coded turbo-detected DSTBC-SP scheme of Figure 4.18, when employing the mapping schemes of Figure 4.19 in combination with various outer RSC codes.

In Section 4.3.3, we investigated the performance of the proposed RSC-coded DSTBC-SP schemes, when employing the simulation parameters listed in Table 4.5. The actual decoding trajectories of the proposed AGM-10-based scheme of Figure 4.20 were provided in Figures 4.22–4.25, when using various interleaver depths. Similarly, the actual decoding trajectories of the proposed AGM-9-based scheme of Figure 4.21 were provided in Figures 4.29–4.32, when using various interleaver depths. Finally, the BER performance of both schemes was demonstrated in Figures 4.27, 4.28 and 4.34–4.37.

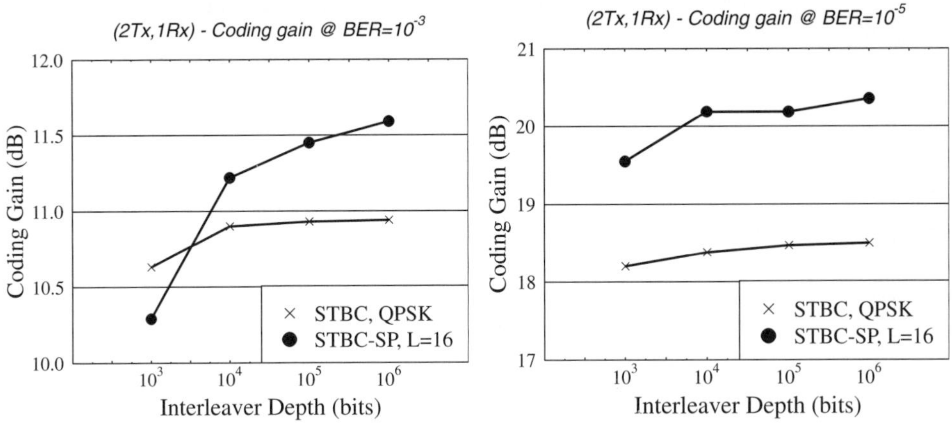

Figure 16.1: Coding gains comparison of AGM-9-based RSC-coded STBC-SP schemes in conjunction with $L = 16$ and RSC-coded QPSK-modulated STBC schemes against an identical-throughput 1 BPS uncoded STBC scheme at BER of 10^{-3} and 10^{-5}, when employing the system parameters outlined in Table 3.3 and using different interleaver depths after ten external joint iterations.

Several DSTBC-SP mapping schemes covering a wide range of extrinsic transfer characteristics were investigated. When using an appropriate bit-to-symbol mapping scheme and ten turbo-detection iterations, E_b/N_0 gains of about 23.8 and 3.2 dB were obtained by the RSC-coded DSTBC-SP scheme over the identical-throughput 1 BPS uncoded DSTBC-SP benchmark scheme and over a turbo-detected system based on the DSTBC scheme of [58, 62]. Table 16.2 summarizes the coding gains of AGM-10-based RSC-coded DSTBC-SP schemes in conjunction with $L = 16$ and RSC-coded QPSK-modulated DSTBC schemes against an identical-throughput 1 BPS uncoded DSTBC scheme at BERs of 10^{-3} and 10^{-5}, when employing the system parameters outlined in Table 4.5 and using different interleaver depths after ten external joint iterations. Figure 16.2 presents coding gain comparisons of AGM-10-based RSC-coded DSTBC-SP schemes in conjunction with $L = 16$ and the conventional RSC-coded QPSK-modulated DSTBC schemes of Table 16.2. In addition, Table 16.3 summarizes the coding gains of AGM-9-based RSC-coded DSTBC-SP schemes in conjunction with $L = 16$ and RSC-coded QPSK-modulated DSTBC schemes against an identical-throughput 1 BPS uncoded DSTBC scheme at BERs of 10^{-3} and 10^{-5}, when employing the system parameters outlined in Table 4.5 and using different interleaver depths after ten external joint iterations. Coding gain comparisons of AGM-9-based RSC-coded DSTBC-SP schemes in conjunction with $L = 16$ and the classic RSC-coded QPSK-modulated DSTBC schemes of Table 16.3 are provided in Figure 16.3.

16.1.5 Chapter 5: Three-stage Turbo-detected STBC-SP Schemes

Conventional two-stage turbo-detected schemes introduced in Chapters 3 and 4 suffered from a BER floor, preventing them from achieving infinitesimally low BER values, since the inner coding stage is of a non-recursive nature. In Chapter 5, we circumvented this deficiency by proposing a three-stage turbo-detected STBC-SP scheme, where a rate-one recursive inner precoder is employed to avoid having a BER floor. Section 5.2 provided a brief description of the proposed three-stage system, where the schematic of the entire

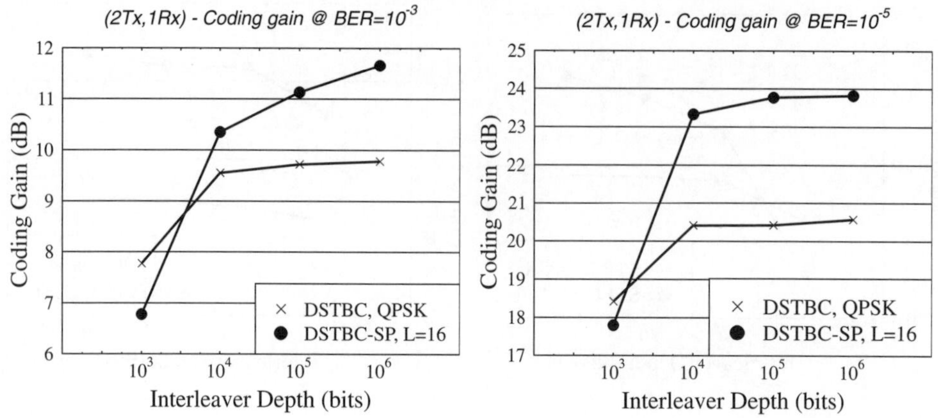

Figure 16.2: Coding gains comparison of AGM-10-based RSC-coded DSTBC-SP schemes in conjunction with $L = 16$ and RSC-coded QPSK-modulated DSTBC schemes against an identical-throughput 1 BPS uncoded DSTBC scheme at BER of 10^{-3} and 10^{-5}, when employing the system parameters outlined in Table 4.5 and using different interleaver depths after ten external joint iterations.

Table 16.2: Coding gains of AGM-10-based RSC-coded DSTBC-SP schemes in conjunction with $L = 16$ and RSC-coded QPSK-modulated DSTBC schemes against an identical-throughput 1 BPS uncoded DSTBC scheme at BER of 10^{-3} and 10^{-5}, when employing the system parameters outlined in Table 4.5 and using different interleaver depths after ten external joint iterations.

	E_b/N_0 (dB)		Gain (dB)	
	BER $= 10^{-3}$	BER $= 10^{-5}$	BER $= 10^{-3}$	BER $= 10^{-5}$
Uncoded DSTBC, BPSK	17.4	31.1	0	0
Uncoded DSTBC-SP, $L = 4$	17.3	31.0	0.1	0.1
RSC-coded, DSTBC-SP, $L = 16$				
RSC (2,1,3), AGM-10				
Interleaver depth $= 10^3$ bits	10.6	13.3	6.8	17.8
Interleaver depth $= 10^4$ bits	7.0	7.8	10.4	23.3
Interleaver depth $= 10^5$ bits	6.2	7.3	11.2	23.8
Interleaver depth $= 10^6$ bits	5.7	7.3	11.7	23.8
RSC-coded, DSTBC, QPSK				
RSC (2,1,3), set-partitioning				
Interleaver depth $= 10^3$ bits	9.6	12.7	7.8	18.4
Interleaver depth $= 10^4$ bits	7.8	10.7	9.6	20.4
Interleaver depth $= 10^5$ bits	7.6	10.7	9.8	20.4
Interleaver depth $= 10^6$ bits	7.6	10.5	9.8	20.6

Table 16.3: Coding gains of AGM-9-based RSC-coded DSTBC-SP schemes in conjunction with $L = 16$ and RSC-coded QPSK-modulated DSTBC schemes against an identical-throughput 1 BPS uncoded DSTBC scheme at BER of 10^{-3} and 10^{-5}, when employing the system parameters outlined in Table 4.5 and using different interleaver depths after ten external joint iterations.

	E_b/N_0 (dB)		Gain (dB)	
	BER $= 10^{-3}$	BER $= 10^{-5}$	BER $= 10^{-3}$	BER $= 10^{-5}$
Uncoded DSTBC, BPSK	17.4	31.1	0	0
Uncoded DSTBC-SP, $L = 4$	17.3	31.0	0.1	0.1
RSC-coded, DSTBC-SP, $L = 16$				
RSC (2,1,5), AGM-9				
Interleaver depth $= 10^3$ bits	10.3	13.8	7.1	17.3
Interleaver depth $= 10^4$ bits	6.8	7.8	10.6	23.3
Interleaver depth $= 10^5$ bits	5.8	7.4	11.6	23.7
Interleaver depth $= 10^6$ bits	5.7	7.2	11.7	23.9
RSC-coded, DSTBC, QPSK				
RSC (2,1,5), set-partitioning				
Interleaver depth $= 10^3$ bits	9.3	12.2	8.1	18.9
Interleaver depth $= 10^4$ bits	6.9	9.4	10.5	21.7
Interleaver depth $= 10^5$ bits	6.7	9.2	10.7	21.9
Interleaver depth $= 10^6$ bits	6.6	8.9	10.8	22.2

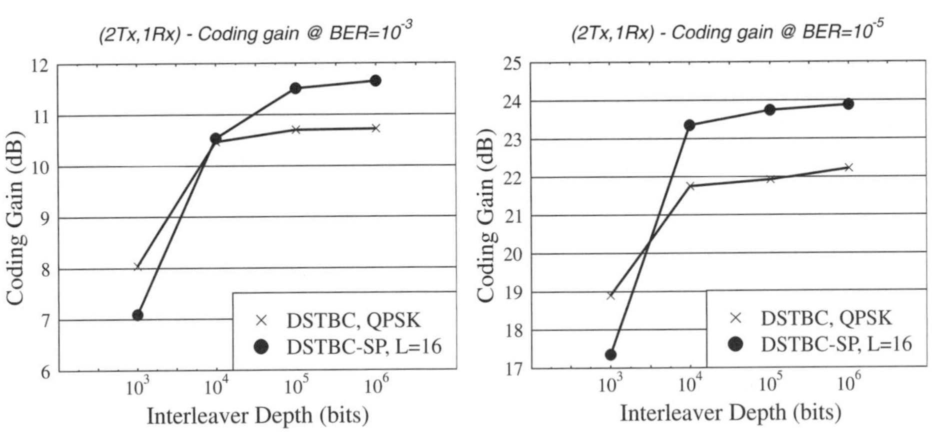

Figure 16.3: Coding gains comparison of AGM-9-based RSC-coded DSTBC-SP schemes in conjunction with $L = 16$ and RSC-coded QPSK-modulated DSTBC schemes against an identical-throughput 1 BPS uncoded DSTBC scheme at BER of 10^{-3} and 10^{-5}, when employing the system parameters outlined in Table 4.5 and using different interleaver depths after ten external joint iterations.

system was shown in Figure 5.1. We considered three different types of channel codes for encoder I, namely a repeater, a RSC code and an IRCC. The resultant schemes are outlined in Table 5.1. Our three-dimensional EXIT chart analysis was presented in Section 5.3.2, where its simplified two-dimensional projections were provided in Section 5.3.3. In Section 5.3.4, we employed the powerful technique of EXIT tunnel-area minimization for near-capacity operation. More specifically, we exploited the well-understood properties of conventional two-dimensional EXIT charts that a narrow but nonetheless open EXIT tunnel represents a near-capacity performance. Consequently, we invoked IRCCs for the sake of appropriately shaping the EXIT curves by minimizing the area within the EXIT tunnel using the procedure of [175, 176].

In Section 5.4, an upper bound on the maximum achievable rate was calculated based on the EXIT chart analysis. More explicitly, a procedure was proposed for calculating a tighter upper bound of the maximum achievable bandwidth efficiency of STBC-SP schemes based on the area property of $\bar{A}_I \approx R_I$ of the EXIT charts discussed in Section 5.3.4. The proposed procedure was applied in Section 5.4 for calculating the maximum achievable bandwidth efficiency of the three-stage turbo-coded STBC-SP scheme associated with the SP signal constellation size of $L = 16$ considered in this chapter. The design procedure was summarized in Algorithm 5.1.

The performance of the three-stage turbo-detected STBC-SP schemes was demonstrated and characterized in Section 5.5, where all simulation parameters were outlined in Section 5.5.1 and Table 5.2. In Section 5.5.2, we considered the performance of the three-stage RA-coded STBC-SP scheme, where the actual decoding trajectories, BER performance and the effect of interleaver depth on the achievable performance were discussed in Sections 5.5.2.1, 5.5.2.2 and 5.5.3.3, respectively. We observed from the BER curves seen in Figure 5.13 that the performance of the RA-coded STBC-SP scheme was limited by a BER floor, despite the employment of a recursive inner precoder. This observation is attributed to the convergence tunnel's bottleneck seen in Figure 5.8. The performance of the three-stage RSC-coded as well as IRCC-coded STBC-SP schemes were characterized in Sections 5.5.3 and 5.5.4, respectively, where the proposed IRCC-coded three-stage scheme operated within about 1.0 dB of the capacity limit and within 0.5 dB of the maximum achievable bandwidth efficiency limit. The effect of interleaver depth on the attainable performance of both the RSC-coded and the IRCC-coded schemes was investigated in Sections 5.5.3.3 and 5.5.4.3, respectively. The E_b/N_0 distance to capacity was summarized in Figures 5.19 and 5.24 for the three-stage RSC-coded as well as for the IRCC-coded STBC-SP schemes, respectively, when using the system parameters outlined in Table 5.2 and employing various interleaver depths as well as different numbers of three-stage iterations. Finally, in Section 5.5.5, the performance of both the three-stage RSC-coded and IRCC-coded STBC-SP schemes was compared, when employing various interleaver depths, while using different numbers of three-stage iterations. More specifically, Figures 5.25–5.27 compared the achievable coding gain of the three-stage RSC-coded and IRCC-coded STBC-SP schemes against Alamouti's identical-throughput 1 BPS conventional $\mathbf{G}_2$-BPSK scheme at a BER of both 10^{-3} and 10^{-5} in combination with the system parameters outlined in Table 5.2, when employing different numbers of three-stage iterations and the interleaver depths of $\mathcal{D} = 10^4$, 10^5 and 10^6 bits, respectively.

Figure 16.4 compares the achievable E_b/N_0 distance to capacity for the three-stage RSC-coded as well as for the IRCC-coded STBC-SP schemes of Table 5.1 in combination with the system parameters outlined in Table 5.2, when employing various interleaver depths and different numbers of three-stage iterations. Observe in Figure 16.4 that regardless of

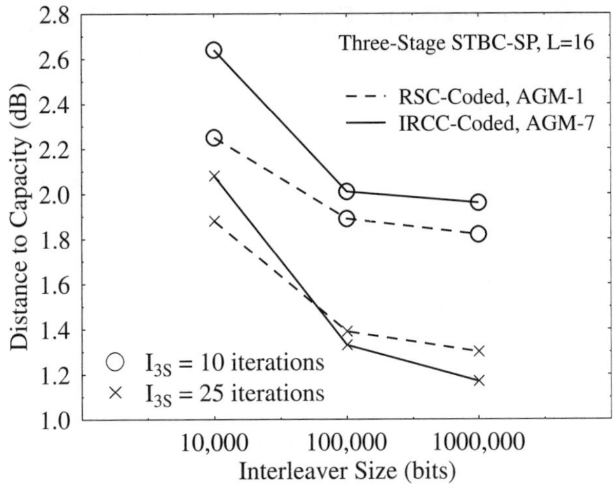

Figure 16.4: Achievable distance to capacity for the three-stage RSC-coded as well as for the IRCC-coded STBC-SP schemes in combination with the system parameters outlined in Table 5.2, when employing various interleaver depths and different numbers of three-stage iterations.

the interleaver depth, the RSC-coded scheme always outperforms the IRCC-coded scheme, when using a low number of three-stage[1] iterations. In contrast, upon increasing the number of three-stage iterations and the interleaver depth, the IRCC-coded scheme approaches the capacity limit faster than the RSC-coded scheme. Figure 16.4 also demonstrates that the AGM-1-based RSC-coded scheme is capable of providing attractive performance improvements compared with the IRCC-coded scheme. This indicates that the AGM-1 scheme and the RSC outer code constitute a meritorious system, which was facilitated by employing the best of the ten different AGM schemes of Figure 3.5.

16.1.6 Chapter 6: Symbol-based Channel-coded STBC-SP Schemes

In all previous chapters, iterative decoding was employed at the bit level. In contrast, in Chapter 6, we explored a range of further design options and proposed a purely symbol-based scheme, where symbol-based turbo detection was carried out by exchanging extrinsic information between an outer non-binary LDPC code and a rate-one non-binary inner precoder. The motivation behind the development of this symbol-based scheme is that a reduced transmit power may be required, when symbol-based rather than bit-based iterative decoding is employed [183].

The system's architecture was presented in Section 6.2. Section 6.2.1 provided a detailed description of the proposed symbol-based and turbo-detected scheme, where a non-binary LDPC code was combined with symbol-based interleaving. The equivalent bit-based scheme was described in Section 6.2.2, where a binary LDPC code was combined with a bit-based interleaver. Symbol-based iterative decoding was discussed in Section 6.3, where it was demonstrated how the *a priori* information $\mathbf{A}$ is removed from the decoded APP matrix $\mathbf{D}$ with the aid of symbol-based element-wise division for the sake of generating the extrinsic probability matrix $\mathbf{E}$.

[1]A three-stage iteration is referred to as a system iteration in Table 5.2.

Section 6.4 provided our non-binary EXIT chart analysis. More specifically, in Section 6.4.1 we demonstrated how non-binary EXIT charts can be generated without generating an L-dimensional histogram [184] since the complexity of this operation may become higher than conducting full-scale BER or SER simulations, when the number of BPS is high. In Section 6.4.2, we addressed the problem of generating the *a priori* symbol probabilities, when the binary bits within each non-binary symbol are assumed be either independent or not. More specifically, the binary bits within each non-binary symbol were assumed to be independent of each other, when a binary LDPC code was employed, which is equivalent to bit interleaving, as alluded to in Section 6.2. Generating the *a priori* symbol probabilities for this particular assumption was presented in Section 6.4.2.1. In contrast, the binary bits of each non-binary symbol are no longer independent when employing a non-binary LDPC code and performing symbol-based decoding. Accordingly, a detailed procedure was described in Section 6.4.2.2 for creating the *a priori* symbol probabilities, when the binary bits of each non-binary symbol may no longer be assumed to be independent. The results of our non-binary EXIT chart analysis were provided in Section 6.4.3, where the novel non-binary EXIT charts were used for studying the convergence of the proposed symbol-based schemes in Section 6.4.3.1. On the other hand, in Section 6.4.3.2 non-binary EXIT charts were used for exploring the convergence of the bit-based LDPC-coded STBC-SP schemes. The EXIT charts of both the symbol-based and bit-based schemes were compared in Section 6.4.3.3. It was explicitly demonstrated in Figures 6.11–6.13 and Table 6.2 that the symbol-based schemes required a lower transmit power and a lower number of decoding iterations to achieve a performance comparable with that of their bit-based counterparts.

The performance of the symbol-based and bit-based LDPC-coded STBC-SP schemes was investigated in Section 6.5, when employing the system parameters outlined in Section 6.5.1 and Table 6.1. First, the actual decoding trajectories were presented in Section 6.5.2, where the mismatch seen in Figures 6.14 and 6.15 between the actual trajectories and the EXIT curves was a consequence of employing a finite interleaver depth of $K_{\text{ldpc}} = 6000$ bits. The attainable BER performance of both the symbol-based and the bit-based schemes was demonstrated in Section 6.5.3, where Figure 6.16 compares the achievable performance of all symbol-based schemes outlined in Table 6.1 against that of their bit-based counterparts. Furthermore, Table 6.3 lists the approximate SNR values where a turbo cliff occurs, based on the BER curves seen in Figure 6.16. The effect of employing various interleaver depths or, equivalently, LDPC output block lengths on the achievable performance was considered in Figures 6.17–6.20 of Section 6.5.4. Table 16.4 summarizes the SNR values required for achieving a BER of 10^{-5} by both the symbol-based and the bit-based schemes after $I_{\text{ext}} = 5$ joint external iterations and $I_{\text{int}} = 3$ internal LDPC iterations, when employing the system parameters outlined in Table 6.1 and using various LDPC output block lengths.

16.1.7 Chapter 7: Linear Dispersion Codes: An EXIT Chart Perspective

In Chapter 7 a novel LDC structure was introduced in order to design STBCs that are capable of fully exploiting the degrees of freedom provided by both the spatial and temporal dimensions. This simple yet powerful architecture provides a unified framework, which not only subsumes all of the major STBC representatives found in the open literature but reveals further insightful STBC design principles. Secondly, we proposed novel 'irregular' coding as well as transceiver schemes and their design methodologies, in the pursuit of near-capacity operation across a wide range of SNRs. Three particular applications of linear-structure-based irregular schemes were investigated, namely the IR-PLDCs of co-located MIMO systems of

Table 16.4: The approximate SNR values required for achieving a BER of 10^{-5} by both the symbol-based and the bit-based schemes after $I_{\text{ext}} = 5$ joint external iterations and $I_{\text{int}} = 3$ internal LDPC iterations, when employing the system parameters outlined in Table 6.1 and using various LDPC output block lengths.

		SNR to achieve BER $= 10^{-5}$ (dB)	
	K_{ldpc} (bits)	Symbol-based scheme	Bit-based scheme
Scheme 1	1488	1.8	2.5
	3000	1.3	1.9
	6000	1.1	1.6
	12 000	0.9	1.4
Scheme 2	1488	3.7	4.3
	3000	3.1	3.9
	6000	2.8	3.4
	12 000	2.6	3.2
Scheme 3	1488	5.9	6.5
	3000	5.3	5.7
	6000	5.0	5.4
	12 000	4.7	5.3

Chapter 7, the IR-PDLDCs of co-located MIMO schemes outlined in Chapter 8 and the IR-PCLDCs of cooperative MIMO arrangements discussed in Chapter 9. These are detailed in the following sections, together with a discussion of future research ideas.

Chapter 7 exemplified the application of the LDCs designed for coherently detected co-located MIMO systems. This application was motivated by the fundamental challenge of effectively exploiting the additional spatial dimensions offered by the multiple antennas employed at both the transmitter and the receiver. The concept of LDCs introduced in Section 7.2 is radical, since their inherent linearity simplifies the challenge of designing a potentially excessive number of space-time codewords. As an explicit benefit, it allows us to design a single DCM χ defined in Equation (7.16), regardless of the number of antennas employed as well as of the modulation schemes used. Two LDC models were investigated in Sections 7.2.2 and 7.2.3, which are suitable for describing OSTBCs and non-orthogonal STBCs, respectively. Furthermore, Section 7.2.4 proposed to optimize the DCM χ from a capacity maximization perspective, which was achieved by maximizing the LDCs' DCMC capacity using Equation (7.23). Section 7.2.5 characterized the BER performance as well as the associated achievable capacity of a family of LDCs, which are summarized in Table 16.5.

Section 7.3.1 examined the existing STBC design philosophies found in the open literature, including the rank criterion [52] and the determinant criterion [52] as well as the diversity–rate trade-offs and the diversity–multiplexing gain trade-offs. Furthermore, Section 7.3.1 summarized the STBC's beneficial property of TSS. As explicitly shown in Table 7.3, the grade of TSS serves as an indicator of the specific trade-off struck in terms of the achievable rate, diversity, complexity, orthogonality and design flexibility. If full TSS is guaranteed, the performance of the associated STBC is bounded by the rate–diversity trade-off. In contrast, if partial TSS or non-TSS is observed, the resultant STBC arrangement could simultaneously achieve both a high throughput and full diversity using a non-orthogonal

Table 16.5: BER and capacity characteristics of the family of LDC($MNTQ$) schemes of Figure 7.3, when employing QPSK modulation as well as the ML detector of Equation (7.18).

LDC($MNTQ$)	BER	Capacity
Changing the value of Q	Figure 7.5	Figures 7.6 and 7.7
Changing the value of T	Figure 7.8	Figure 7.9
Changing the value of M or N	Figure 7.12	Figure 7.13
Changing the DCM χ	Figure 7.10	Figure 7.11

structure. In other words, the LDCs' ability to encompass the entire spectrum of TSS by simply applying the corresponding DCM χ allows the family of LDCs to meet various design objectives. Even more appealingly, the choice of χ is non-unique, since the solution to Equation (7.23) is non-unique. In Sections 7.3.2–7.3.7, we explicitly characterized the mathematical representation as well as the design philosophy of STBCs characterized in the literature and those of LDCs. In fact, the LDCs subsume all of the known STBC schemes by imposing various restrictions on the set of the dispersion matrices.

As far as serial-concatenated channel-coded STBC schemes are concerned, Section 7.4 analyzed various iteratively detected RSC-coded LDCs with/without a unity-rate precoder using EXIT charts [158, 169]. For the non-precoded scheme of Figure 7.17, we demonstrated in Figures 7.18 and 7.22 that the corresponding scheme was unable to operate near the capacity, since the inner code's EXIT curves became near-horizontal and hence the resultant BER performance improved only modestly. The employment of the unity-rate precoder potentially allows the associated scheme of Figure 7.24 to achieve an infinitesimally low BER, since the inner and outer EXIT curves reach the $(1.0, 1.0)$ point of the EXIT chart, as exemplified in Figures 7.26 and 7.29. However, the precoded scheme of Figure 7.24 still operates far from the MIMO channel's capacity, since the open EXIT tunnel area remains relatively high. Table 7.4 explicitly characterized the coding gains achieved by the precoded/non-precoded schemes discussed in Section 7.4.

In order to achieve near-capacity operation associated with an infinitesimally low BER, Section 7.5 employed the sophisticated irregular design principle for serial-concatenated systems. More explicitly, we proposed the novel RSC-coded IR-PLDC scheme of Figure 7.32, where the irregularity was imposed on the inner code. The maximum achievable rates of Figure 7.40 showed that this scheme becomes capable of operating as close as 2.5 dB from the MIMO channel's capacity for SNRs in excess of a certain threshold. The open EXIT tunnel area was maximized at the target SNR, under the condition that the system's effective throughput was maximized. Maximizing the effective throughput in order to operate near the attainable capacity is our prime concern. On the other hand, the open EXIT tunnel area also has to be maximized in order to reduce the associated decoding complexity, which was justified in Figures 7.36 and 7.37. Section 7.5.1.3 demonstrated that the system of Figure 7.32 is capable of maintaining a moderate complexity, if we allow the system to operate further away from the achievable capacity.

In Section 7.5.2, the irregular design principle was applied solely to the outer code. The resultant IRCC-coded LDC scheme operated approximately 0.9 dB from the MIMO channel's capacity within a limited SNR region, as seen in Figure 7.47. This was because the irregular outer code is capable of shaping the EXIT curve more flexibly than the irregular inner code, provided that an open EXIT tunnel existed. In contrast, the irregular inner IR-PLDC scheme of Figure 7.32 was capable of creating open EXIT tunnels based on the SNR encountered by employing different-rate LDC component codes. Figure 7.47 clearly demonstrated the above-mentioned observations by comparing the maximum achievable rates of the IRCC-coded PLDCs and RSC-coded IR-PLDCs, when employing the same number of component codes. Accordingly, Section 7.5.3 investigated the scenario where the irregularity was imposed on both the inner and outer codes. The resultant IRCC-coded IR-PLDC scheme of Figure 7.48 became capable of operating within approximately 0.9 dB of the capacity across a wide SNR region, as demonstrated in Figures 7.49 and 7.50.

Section 7.5.4 summarized the potential methods of constructing a near-capacity scheme using the irregular principle. More explicitly, the irregularity of the outer code can be created by employing different rate component codes [176, 195]. In contrast, there are diverse ways of creating a set of EXIT curves having distinguished characteristics for the inner code, such as using different code rates, as well as diverse mapping and modulation schemes, as portrayed in Figure 7.53. Inevitably, near-capacity operation imposes a high decoding complexity as well as a high interleaver delay, since high-complexity component codes have to be employed and the associated number of iterations to achieve an infinitesimally low BER is also high. In conclusion, the proposed irregular scheme of Figure 7.48 can operate near the attainable capacity across diverse SNRs at a potentially high complexity and high delay.

16.1.8 Chapter 8: Differential Space-Time Block Codes: A Universal Approach

Chapter 8 considered the application of linear-dispersion-style system structures and the irregular transceiver design philosophy in the context of non-coherently detected DSTBCs. The motivation behind DSTBCs stated in Section 8.1 is to dispense with the burden of high-complexity MIMO channel estimation required by the coherently detected STBCs of Chapter 7, while still achieving a beneficial diversity gain.

Section 8.2 characterized the DSTBC's structure of Figure 8.2 from two different perspectives. First, we examined the principles that facilitate the employment of differential encoding/decoding for a single-antenna-aided scheme and then demonstrated that the same design philosophy can be applied to multiple-antenna-aided systems. Secondly, the schematic of Figure 8.2 was portrayed from a specific perspective, demonstrating that DSTBCs can be considered as STBCs combined with a differential encoder. This perspective allowed us to bridge the design of STBCs and DSTBCs, since the challenge of designing DSTBCs can be described as that of designing a family of STBCs, where all of the space-time matrices are unitary.

Hence, in Section 8.3 we were able to investigate various DOSTBC schemes found in the open literature using the general framework of Figure 8.2. Since DOSTBCs are based on various orthogonal matrices, the unitary constraint of Equation (8.5) is satisfied automatically. The orthogonality of the resultant designs also enables the receiver to perform low-complexity ML detection, since the symbol streams can be decoupled into separately decoded groups. Unfortunately, the orthogonal design principle imposes numerous constraints on the DOSTBCs. For example, the number of antennas that may be employed is limited and the number of symbols transmitted per space-time block is also restricted. As expected, the BER performance of the family of DOSTBCs typically suffers a 3 dB SNR penalty as the result of the doubled noise variance of the differential detection formulated in Equation (8.9) when compared with the corresponding LDCs of Chapter 7, although the 3 dB difference is reduced when realistic CIR estimation is used at the receiver.

Table 16.6: BER summary of the family of DLDC($MNTQ$) schemes of Figure 8.2 based on the Cayley transform, when employing 2PAM modulation and the ML detector of Equation (8.53).

DLDC($MNTQ$)	BER
Changing the value of Q	Figure 8.11
Changing the value of T or M	Figure 8.12
Changing the value of N	Figure 8.13
The effect of fading	Figure 8.14
Compared with LDCs with perfect CSI	Figures 8.11 and 8.12
Compared with LDCs with imperfect CSI	Figure 8.15

In order to eliminate the constraints imposed by the orthogonal structure, Section 8.4 proposed the class of DLDCs based on the Cayley transform [10]. In other words, DLDCs offer an alternative method of designing a set of unitary space-time matrices, instead of employing the set of orthogonal matrices detailed in Section 8.3. More explicitly, the linear dispersion structure of Equation (8.46) can be employed in the Hermitian space, so that the weighted sum of a number of Hermitian matrices remains also a Hermitian matrix, provided that the weighting coefficients are real valued. Furthermore, the Cayley transform [10] detailed in Section 8.4.2 provides an efficient way of projecting the weighted Hermitian matrix to a unique unitary matrix, which facilitates the differential encoding of Equation (8.5). Recall from Section 7.3 that we classified the various STBC techniques in Figure 7.16 and demonstrated that LDCs subsume all of the members of the entire family. Similarly, it is straightforward to derive the family of DOSTBCs from the DLDCs by simply imposing different degrees of orthogonality on the dispersion matrices, which was portrayed in Figure 8.8.

Table 16.6 summarizes the BER performances presented in Section 8.4.3 for the family of DLDCs based on the Cayley transform. The DLDCs' ability to accommodate flexible antenna configurations as well as facilitating various parameter combinations was demonstrated in Figures 8.11–8.13. Note that the unitary matrix constraint shown in Section 8.2.3 and the restriction of employing real-valued modulation are imposed by the differential encoding process and by the Cayley transform, respectively. In addition, Figure 8.14 investigated the DLDCs' achievable performance under different fading conditions, since the success of differential decoding relies on the degree of the channels' stationarity. As expected, a substantial performance degradation was observed in Figure 8.14, when the channels' fluctuation became severe. Figures 8.11 and 8.12 verified the 3 dB SNR difference between the DSTBCs based on the Cayley transform and the corresponding LDCs having perfect CSI. However, when more realistic channel estimation scenarios were considered, the DLDCs became capable of outperforming their coherently detected LDC counterparts, as explicitly shown in Figure 8.15. These findings remained valid even when the channels were experiencing moderately rapid Rayleigh fading having $f_d = 10^{-2}$. The achievable coding gain of the DLDCs based on the Cayley transform was recorded in Table 8.3. Similar coding advantages have been observed also in Tables 7.1 and 7.2, when employing the corresponding LDCs.

In Section 8.5, we proposed a novel RSC-coded SP-aided DOSTBC scheme employing a unity-rate precoder and using iterative decoding, as an attempt to jointly design the space-time block coding and the modulation scheme. The philosophy behind SP is that the symbols transmitted during a space-time codeword should be designed jointly, rather than chosen separately from the modulation constellations. Hence, the SP modulation potentially requires all of the transmitted symbols to experience identical fading, which can be satisfied by the DSTBC schemes of Section 8.3, which were based on orthogonal designs. In contrast, the DSTBCs of Section 8.4 were based on the non-orthogonal structure, and the resultant DLDCs were less applicable to SP modulation. The joint STBC-SP design is particularly attractive, since the SP modulation is capable of gleaning more extrinsic information than that of the conventional L-PSK modulation schemes, which allows the inner code to create a larger open tunnel in the EXIT chart, as seen in Figure 8.22. Accordingly, this coding advantage comes at the price of increasing the decoding complexity.

Furthermore, Section 8.5 investigated the achievable BER performance of the regular/irregular-coded precoded DOSTBC schemes using SP modulation at a throughput of 1 (BPS Hz^{-1}). Since the irregular outer scheme provides a higher grade of flexibility in shaping the outer code's EXIT curve, the resultant IRCC-coded DOSTBC scheme becomes capable of achieving an infinitesimally low BER at $\rho = 5.5$ dB in comparison to the SNR of $\rho = 6$ dB required by the regular-coding-aided scheme, as seen in Figure 8.23.

Section 8.6 further extended the irregular principle in the context of non-coherently detected systems, where the irregularity was imposed on both the inner code and the outer code. The design method of the resultant IRCC-coded IR-PDLDC scheme is similar to that of its coherently detected counterpart detailed in Section 7.5.3. Hence, we focused our attention on the distinctive features imposed by the differential encoding structure of Section 8.4.2. First, we investigated the maximum achievable rates of the IR-PDLDCs of Figure 8.25 and that of the IR-PLDCs of Figure 7.48 associated with perfect CSI. Figure 8.27 explicitly demonstrated the well-known 3 dB SNR difference between the coherently detected and non-coherently detected schemes, as a result of the doubled noise variance of Equation (8.7). Furthermore, Figure 8.29 recorded the maximum achievable rates of the IR-PLDCs of Figure 7.48 in the presence of channel estimation errors, as governed by the value of ω. Observe in Figure 8.29 that the IR-PDLDC scheme of Figure 8.25 was capable of achieving a higher throughput than its coherently detected IR-PLDC counterpart in the high-SNR region, when the channel estimation was imperfect. The maximum achievable rates of the IR-PDLDC scheme of Figure 8.25 under various fading conditions were also investigated. As expected, Figure 8.28 suggested that the maximum achievable rates decreased gradually when the channel conditions fluctuated rapidly. Finally, the effective throughput achieved by the IRCC-coded IR-PDLDC scheme was shown in Figure 8.32. Again, a 3 dB SNR difference was observed, when compared with the corresponding IRCC-coded IR-PLDC benchmark.

16.1.9 Chapter 9: Cooperative Space-Time Block Codes

The application of the linear dispersion structure as well as of the irregular design philosophy of cooperative MIMO systems was demonstrated in Chapter 9. The employment of relay-aided transmissions was motivated by the realistic propagation scenario, where the wireless links between the source and the destination suffer from severe shadowing effects or from path loss. Hence, a group of relays can be employed to invoke cooperation transmissions in order to overcome these large-scale fading effects, which may dramatically erode the benefits of co-located MIMO elements, as exemplified in Figure 9.1. In general, there are three types of cooperation strategies, namely DF, AF and CF. However, Chapter 9 focused on the AF arrangement, which requires low-complexity relays.

Accordingly, Section 9.2.1 proposed a novel twin-layer CLDC scheme, which was specifically tailored to the two-phase transmission regime employed by the AF relay protocol.

Table 16.7: BER performance summary of the family of CLDC$(MNTQ)$ schemes of Figure 9.2, when employing BPSK modulation as well as the ML detector of Equation (9.21).

CLDC$(MNTQ)$	BER
Changing the value of Q	Figure 9.10
Changing the value of T_1 or T_2	Figure 9.6
Changing the value of T	Figure 9.11
Changing the value of M	Figure 9.13
Changing the value of N	Figure 9.12
Changing the value of ρ_{SR}	Figure 9.9
Compared with LDCs without shadowing	Figure 9.7
Compared with LDCs with shadowing	Figure 9.8

Furthermore, in Section 9.2.2 we outlined the assumptions stipulated, in order to make the twin-layer CLDC scheme of Figure 9.2 applicable to the 3GPP-LTE system. The mathematical model of the twin-layer CLDC scheme presented in Section 9.2.3 revealed that the CLDCs have a similar system structure to that of the LDCs of Chapter 7. Recall that the LDC's equivalent channel matrix of Equation (7.15) is characterized by the DCM χ. In contrast, the CLDC's equivalent channel matrix of Equation (9.20) is determined by a pair of DCMs (χ_1, χ_2) in order to describe their twin-phase transmission. More specifically, Section 9.2.4 listed the similarities and differences between the LDCs obeying the structure of co-located MIMO systems outlined in Figure 9.3 and the CLDCs having the structure of cooperative MIMO systems as seen in Figure 9.2.

Table 16.7 summarizes the achievable BER performance quantified in Section 9.2.5 for a family of CLDCs obeying the structure of Figure 9.2 as well as having various $(MNTQ)$ parameter combinations. Similar to the LDCs of Chapter 7 and to the DLDCs of Chapter 8, the flexible linear dispersion matrix-based structure formulated in Equations (9.4) and (9.13) allows the CLDCs to support arbitrary $(MNTQ)$ parameter combinations, as listed in Table 16.7. In particular, we exploited the CLDC's special characteristics inherited from the twin-phase transmission of the cooperative MIMO systems. More explicitly, Figure 9.6 demonstrated that the number of time slots used for the broadcast/cooperation interval has a dominant impact on the achievable BER performance, since it determines the symbol's integrity received at the relays and the achievable diversity gains at the destination. Furthermore, when the source-to-relay links experienced different SNRs ρ_{SR}, the corresponding CLDC's performance was characterized in Figure 9.9.

Table 16.7 also compares the performance of the CLDCs and the LDCs having identical parameters, when the channels were subjected both to small-scale Rayleigh fading and to large-scale shadowing. In the context of Rayleigh fading environments, the LDCs of co-located MIMO schemes outperformed the corresponding CLDCs of cooperative MIMO systems, since the noise introduced at the relays dominates the attainable performance, which was analyzed in Figure 9.7. However, in the presence of large-scale shadowing, where all of the LDCs' wireless channels tend to fade together, the CLDCs exhibited a significant advantage, as seen in Figure 9.8. The effective throughput of a class of CLDCs as well as of the corresponding group of LDCs operating with/without shadowing was characterized in Figures 9.14 and 9.15. The corresponding coding gains were extracted from these figures, and are listed in Table 9.4.

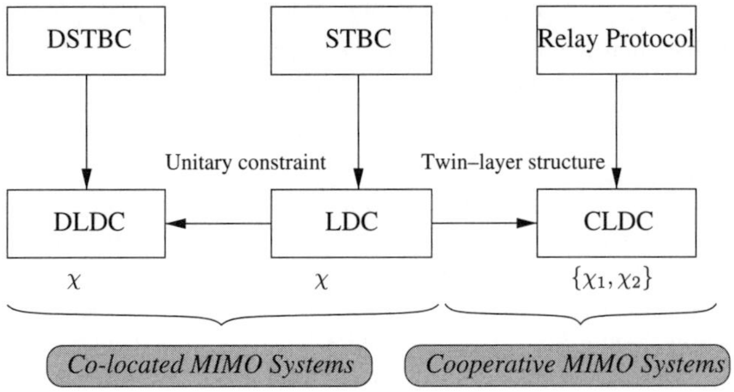

Figure 16.5: Link between the LDCs/DLDCs designed for co-located MIMO systems and the twin-layer CLDCs designed for cooperative MIMO systems.

Section 9.3 demonstrated that the irregular design principle applied to the co-located MIMO systems of Chapters 7 and 8 can also be applied to the cooperative MIMO systems. Hence, the resultant IRCC-coded IR-PCLDC scheme of Figure 9.16 based on the AF cooperation protocol became capable of achieving a flexible effective throughput according to the SNR encountered. Throughout the investigations, an IRCC-coded IR-PLDC scheme of Figure 7.48 having identical parameters was constructed as the benchmark, when communicating over small-scale Rayleigh fading channels. Figure 9.19 showed that there was a maximum achievable rate gap between the IR-PCLDCs and the corresponding IR-PLDCs, since the relays only have access to imperfect source information, as highlighted in Figure 9.20. The effective throughput of the IRCC-coded IR-PCLDC scheme of Figure 9.16 was characterized in Figure 9.23, where a similar rate gap was recorded compared with the co-located MIMO benchmark.

16.1.9.1 Linking LDCs, DLDCs and CLDCs

Although each individual link between the LDCs and the DLDCs/CLDCs has been characterized in Figures 8.8 and 9.5, respectively, Figure 16.5 portrays the fundamental design guidelines of the LDC, DLDC and CLDC techniques investigated in this book. As shown in Chapter 7, the LDCs characterized by a single DCM χ provide a unified solution to the design of STBCs used in co-located MIMO systems. The linear dispersion matrix-based design philosophy can be projected to the differential encoding domain with the aid of the unitary constraint of Section 8.2.3, as seen in Figure 16.5. The resultant DLDCs of Chapter 8 can also be characterized by the DCM χ of Equation (8.56). Furthermore, in the context of the relay-aided networks, the concept of the linear dispersion structure remains applicable, provided that the DCM pair $\{\chi_1, \chi_2\}$ of Equation (9.20) is available to the CLDCs in order to characterize the twin-phase transmission regime required by the cooperative MIMO systems of Figure 9.2.

Table 16.8 summarizes the appropriate application scenarios and the characteristics of the LDCs of Chapter 7, those of the DLDCs of Chapter 8 and those of the CLDCs of Chapter 9. Since co-located MIMO systems have access to all of the spatial and temporal dimensions at the transmitter, each L-PSK symbol is dispersed to all of the slots and the aggregate space-time codeword is the weighted sum of all of the dispersed symbols. Hence, we refer to this operation as 'symbol-based' linear dispersion, which has been visualized in Figure 7.4. In contrast, for a cooperative MIMO system, each relay functions as a virtual element of a

Table 16.8: The suitable application scenarios and the characteristics of the LDCs of Chapter 7, the DLDCs of Chapter 8 and the CLDCs of Chapter 9.

	LDC	DLDC	CLDC
Application	Co-located MIMO	Co-located MIMO	Cooperative MIMO
Suitable fading	Small-scale	Small-scale	Large-scale
CSI requirement	Perfect CSI	Non-CSI	Perfect CSI
Dispersion based on	Equation (7.12)	Equation (8.46)	Equations (9.4) and (9.13)
Visualized dispersion	Figure 7.4	Figure 7.4	Figure 9.4
DCM listed in	Appendix D	Appendix E	Appendix F
Schematic of the uncoded system	Figure 7.3	Figure 8.2	Figure 9.2
Uncoded BER	Table 16.5	Table 16.6	Table 16.7
Coding gain	Tables 7.1 and 7.2	Table 8.3	Table 9.4
Uncoded throughput	Figures 7.14 and 7.15	Figures 8.16 and 8.17	Figures 9.14 and 9.15
Schematic of the irregular system	Figure 7.48	Figure 8.25	Figure 9.16
Coded throughput	Figure 7.49	Figure 8.32	Figure 9.23
Weighting coefficient	Appendix G	Appendix G	Appendix G

MIMO antenna array, where only T_2 temporal dimensions are accessible. Hence, each relay disperses all of the received signals to the temporal dimensions and each relay contributes one row of the aggregate space-time codeword formulated in Equation (9.15). Therefore, we refer to it as 'relay-based' linear dispersion, where the dispersion process was shown graphically in Figure 9.4.

Finally, we constructed a set of comparisons between the family of LDCs, DLDCs and CLDCs in order to evaluate their advantages as well as their limitations, when communicating in small-scale or large-scale fading scenarios as well as their when having perfect or imperfect CSI at the receiver. All of the simulation parameters are listed in Table 16.9. Observe in Table 16.9 that we set $M = T$ and assume that the channels were subjected to Rayleigh fading having $f_d = 10^{-2}$ in order to enable adequate operation of the DLDCs. Hence, the group of LDCs and DLDCs have the potential to achieve the full diversity order of $D = N \cdot \min(M, T)$ in comparison to the reduced maximum diversity order $D \approx N \cdot \min(M, T_2)$ of the CLDCs.

- For Comparison A of Table 16.9, Figure 16.6 characterizes the achievable throughput of the group of LDCs, DLDCs and CLDCs recorded at $\text{BER} = 10^{-4}$, when the wireless channels were subjected to small-scale Rayleigh fading and perfect CSI was available at the receiver. Observe in Figure 16.6 that the class of LDCs is capable of operating at the lowest SNR at a certain throughput. The group of DLDCs suffers from a 3 dB SNR penalty in comparison with that of the LDCs, since no CSI was exploited. The family of CLDCs operates at SNRs further away from that of the LDCs, owing to the reduced achievable diversity order as well as for the reasons detailed in Section 9.2.5.

- For Comparison B of Table 16.9, Figure 16.7 characterizes the effective throughput of the LDCs, the DLDCs and the CLDCs of Table 16.9 recorded at $\text{BER} = 10^{-4}$, when the wireless channels were subjected to small-scale Rayleigh fading and the receiver

Table 16.9: Comparison of the LDCs of Figure 7.3, the DLDCs of Figure 8.2 and the CLDCs of Figure 9.2, when communicating over small-scale/large-scale fading channels and having perfect/imperfect CSI at the receiver.

	LDC	DLDC	CLDC
M	3	3	3
N	2	2	2
T	3	3	$T_1 = 1, T_2 = 2$
Q	1,2,3	1,2,3	1,2,3
Modulation	BPSK	BPSK	BPSK
Mapping	Gray mapping	Gray mapping	Gray mapping
ML detector	Equation (7.18)	Equation (8.9)	Equation (9.21)
Doppler frequency	$f_d = 10^{-2}$	$f_d = 10^{-2}$	$f_d = 10^{-2}$
Diversity	$D = 6$	$D = 6$	$D \approx 4$
Comparison A	Small-scale Rayleigh fading, perfect CSI in Figure 16.6		
Comparison B	Small-scale Rayleigh fading, imperfect CSI in Figure 16.7		
Comparison C	Large-scale shadowing, perfect CSI in Figure 16.8		

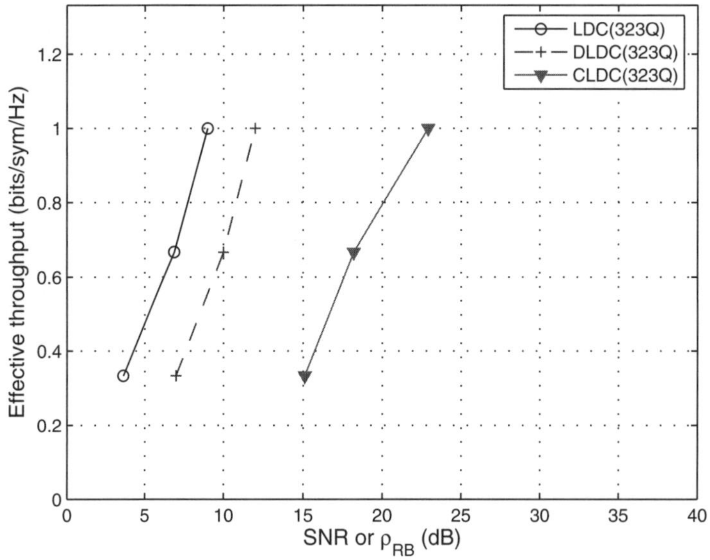

Figure 16.6: Throughput comparison for the LDCs of Figure 7.3, the DLDCs of Figure 8.2 and the CLDCs of Figure 9.2 recorded at $\mathrm{BER} = 10^{-4}$, when communicating over small-scale Rayleigh fading channels and assuming that the perfect CSI was known by the receiver. All of the system parameters are summarized in Table 16.9.

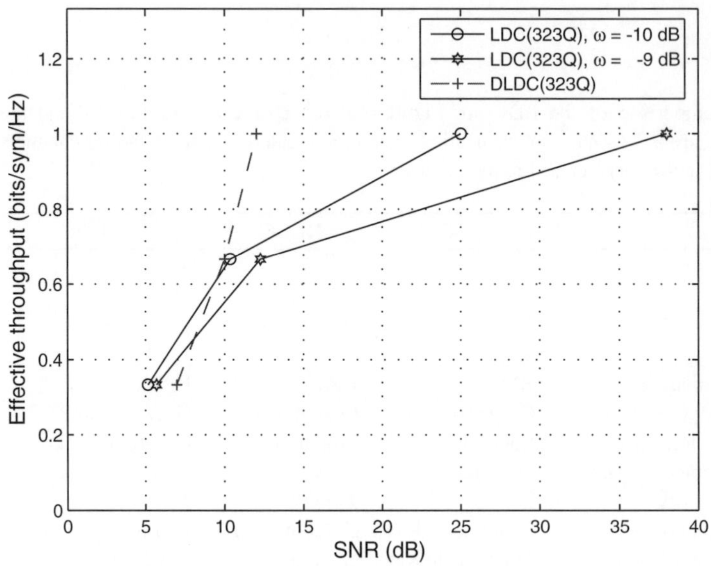

Figure 16.7: Throughput comparison for the LDCs of Figure 7.3, the DLDCs of Figure 8.2 and the CLDCs of Figure 9.2 recorded at $\text{BER} = 10^{-4}$, when communicating over small-scale Rayleigh fading channels and having imperfect CSI governed by ω (dB). All of the system parameters are summarized in Table 16.9.

has imperfect CSI governed by the parameter ω (dB). Observe in Figure 16.7 that the family of DLDCs demonstrated a significant advantage over the LDCs at a high throughput, even when the channel estimation errors were as low as $\omega = -10$ dB. Since the group of CLDCs has an error floor higher than $\text{BER} = 10^{-4}$, the associated throughput curve was omitted from Figure 16.7. This phenomenon suggested that the CLDCs are more sensitive to the channel estimation errors than the LDCs. This is because the CLDCs require the knowledge of $M(N+1)$ channels in comparison to MN links necessitated by the LDCs designed for co-located MIMO systems.

- Finally in Comparison C of Table 16.9, Figure 16.8 characterizes the throughput of the LDCs, of the DLDCs and of the CLDCs of Table 16.9 recorded at $\text{BER} = 10^{-4}$, when the communication channels were subjected to shadowing and the receiver had access to perfect CSI. Observe in Figure 16.8 that the family of CLDCs designed for cooperative MIMO systems has the best ability to combat the effect of large-scale shadowing with the aid of relays. Compared with the small-scale Rayleigh fading performance curves of Figure 16.6, the SNR required for the group of LDCs in Table 16.9 to maintain a BER of 10^{-4} increased by about 17 dB, even though the receiver had access to perfect CSI.

Note that the above-mentioned observations based on Figures 16.6, 16.7 and 16.8 are consistent with the results recorded in Table 16.8. Again, our investigations indicate that the LDCs obeying the structure of Figure 7.3 are ideal for small-scale fading environments, when near-perfect CSI is available. On the other hand, the DLDCs having the structure of Figure 8.2 constitute the most appropriate solution, when the CSI is unavailable or the channel estimation imposes severe errors. When large-scale shadowing dominates the achievable

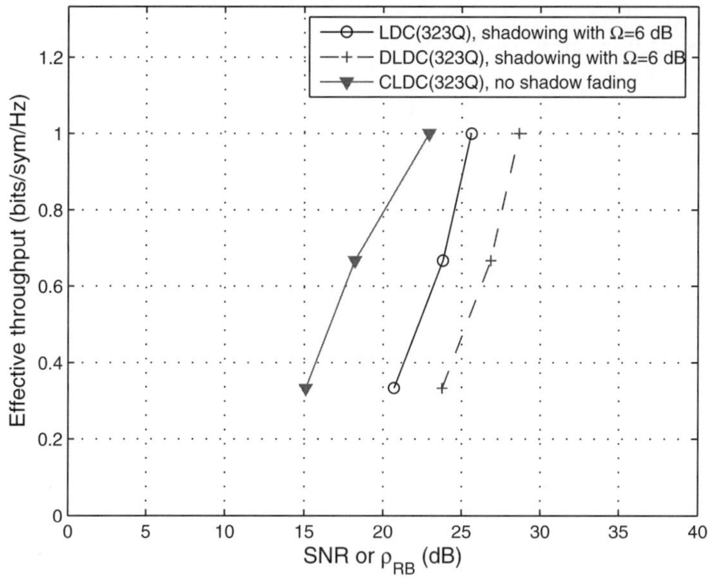

Figure 16.8: Throughput comparison for the LDCs of Figure 7.3, the DLDCs of Figure 8.2 and the CLDCs of Figure 9.2 recorded at $\text{BER} = 10^{-4}$, when the channels were subjected to large-scale shadowing and assuming that the perfect CSI was known at the receiver. All of the system parameters are summarized in Table 16.9.

performance, the family of CLDCs obeying the structure of Figure 9.2 remains capable of maintaining reliable wireless communications.

16.1.10 Chapter 10: Differential Space-Time Spreading

DSTS employing either two or four transmit antennas was proposed in Chapter 10 as a non-coherent MIMO scheme that eliminates the potentially high complexity of MIMO channel estimation at the expense of a 3 dB performance loss compared with the corresponding coherently detected system using perfect channel knowledge at the receiver. DSTS is capable of providing transmit diversity gains, while at the same time supporting multiple users employing different spreading codes. DSTS was designed to work with real- as well as complex-valued conventional modulation schemes including BPSK, QPSK and 16-QAM. In addition, DSTS was combined with multi-dimensional SP modulation in order to attain a higher coding gain, since SP has the best known MED in the $2(k + 1)$-dimensional real-valued Euclidean space $\mathbb{R}^{2(k+1)}$ (see [221]).

In Section 10.3, we outlined both the encoding and decoding algorithms of the twin-antenna-aided DSTS scheme, when combined with conventional modulation. Afterwards, in Section 10.3.3, the philosophy of DSTS using SP modulation, referred to as DSTS-SP, was introduced based on the fact that the diversity product of the DSTS design is improved by maximizing the MED of the DSTS symbols [343]. This was motivated by the fact that SP has the best known MED in the real-valued space. In addition, the capacity of the DSTS scheme employing $N_t = 2$ transmit antennas was derived in Section 10.3.6, where it was shown that the DSTS-SP scheme attains a higher bandwidth efficiency than that of the conventional DSTS scheme dispensing with SP. The performance characterization of a twin-antenna-aided DSTS scheme was provided in Section 10.3.7, demonstrating that the DSTS scheme

Table 16.10: Coding gains of SP modulation over conventional modulation at SP-SER of 10^{-4} for the schemes of Figures 10.21, 10.23, 10.25 and 10.27, when employing twin-antenna-aided DSTS and communicating over a correlated Rayleigh fading channel associated with $f_D = 0.01$.

Number of receive antennas	Coding gain (dB) 1 BPS	2 BPS	3 BPS
1	0.20	0.0	0.70
2	0.25	0.0	0.90
3	0.30	0.0	0.95
4	0.30	0.0	1.00

is capable of providing both a full diversity gain and a multi-user capability. In addition to that, the results demonstrated that DSTS-SP schemes are capable of outperforming DSTS schemes that employ conventional modulation, as shown in Table 16.10, which summarizes the coding gains of DSTS-SP over conventional modulated DSTS schemes at a SP-SER of 10^{-4}, when communicating over a correlated narrowband Rayleigh fading channel. Observe in Table 16.10 that there is no gain for the DSTS-SP system over the conventional DSTS system for the 2 BPS case. This is due to the fact that the QPSK modulation is a special case of the SP modulation, as discussed in Section 10.3.7.

The four-antenna-aided DSTS design was characterized in Section 10.4, where it was demonstrated that the DSTS scheme can be combined with both conventional real- and complex-valued modulated constellations as well as with SP modulation. It was also shown that the four-dimensional SP modulation scheme is constructed differently in the case of two transmit antennas than when employing four transmit antennas. The capacity analysis of the four-antenna-aided DSTS-SP scheme was also derived for systems having a different bandwidth efficiency, while employing a variable number of receive antennas in Section 10.4.5. Finally, in Section 10.4.6 we presented the simulation results obtained for the four-antenna-aided DSTS scheme, when combined with both conventional and SP modulation.

16.1.11 Chapter 11: Iterative Detection of Channel-coded DSTS Schemes

Further performance improvements can be attained by the DSTS system, when combined with channel coding and employing iterative detection at the receiver by exchanging extrinsic information between the constituent decoders/demappers. In Chapter 11, two realizations of a novel iterative-detection-aided DSTS-SP scheme were presented, namely an iteratively detected RSC-coded DSTS-SP scheme as well as an iteratively detected RSC-coded and URC-based precoded DSTS-SP arrangement. The iteratively detected RSC-coded DSTS-SP scheme was described in Section 11.2. In Section 11.2.1 we showed how the DSTS-SP demapper was modified for exploiting the *a priori* knowledge provided by the channel decoder, which is essential for the employment of iterative detection.

The concept of EXIT charts was introduced in Section 11.2.2 as a semi-analytical tool for analyzing the convergence behavior of iterative-detection-aided schemes. We have used several different AGM schemes, whose EXIT curves are shown in Figure 16.9, which were

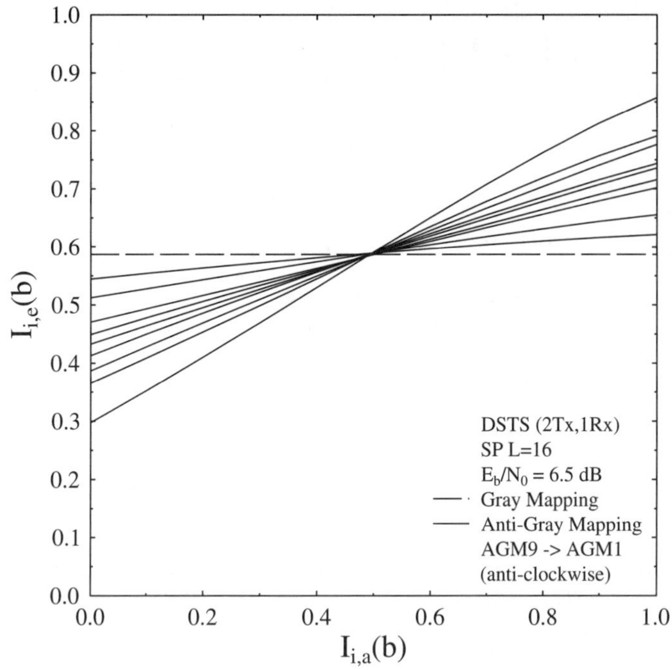

Figure 16.9: SP demapper EXIT characteristics for different bits to SP symbol mappings at $E_b/N_0 = 6.5$ dB for $L = 16$, while considering transmissions using the twin-antenna-aided DSTS scheme.

specifically selected from all of the possible mapping schemes for $L = 16$. Both the Gray mapping and the various AGM schemes considered in Figure 16.9 are detailed in Appendix H. We have analyzed different iteratively detected schemes using EXIT charts and shown how EXIT charts allow us to satisfy diverse design objectives. For example, we can design a system having the lowest possible turbo-cliff SNR, but tolerating the formation of an error floor. Alternatively, we can design a system having no error floor, but having a slightly higher turbo-cliff SNR. To elaborate further, the BER performance of the iteratively detected half-rate RSC-coded DSTS-SP scheme recorded in conjunction with two transmit antennas and different Gray mapping and AGM schemes is shown in Figure 16.10, when applying $I = 10$ iterations. Observe in Figure 16.10 that the Gray mapping- and AGM-8 based-systems have a similar performance and this can be justified by referring to the EXIT chart of Figure 16.9, where the EXIT curves of the Gray-mapping- and AGM-8-based systems have similar slopes. On the other hand, observe in Figure 16.10 that the AGM-3-based system exhibits a turbo cliff at an E_b/N_0 value lower than the AGM-1-based system, while the AGM-1-aided system attains a lower error floor than its AGM-3 counterpart.

On the other hand, in order to show the effects of the interleaver depth on the performance of iteratively detected schemes, Figure 16.11 compares the BER performance of the iteratively detected RSC-coded DSTS-SP scheme employing AGM-1 versus the number of iterations, while using different interleaver depths ranging from $D_{\text{int}} = 1\,000$ to 800 000 bits and operating at $E_b/N_0 = 7.0$ dB. The plot investigates the BER performance versus the complexity of the system quantified in terms of the number of iterations. As shown in the figure, when using short interleavers, increasing the number of iterations results in no

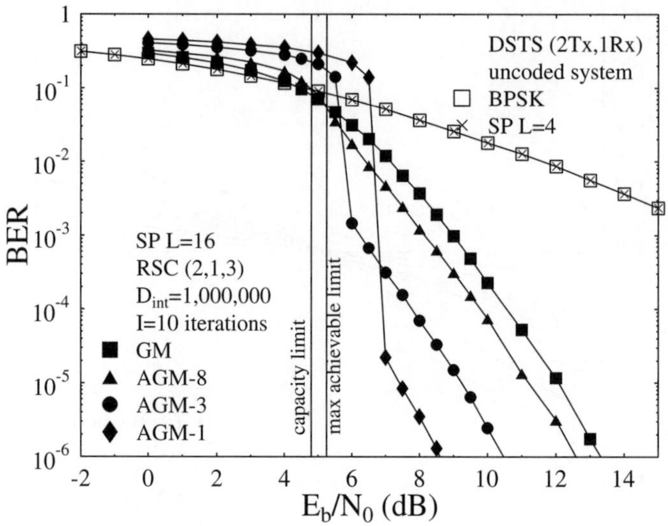

Figure 16.10: Performance comparison of different AGM- and Gray-mapping-based iteratively detected RSC-coded two transmit antennas DSTS-SP schemes in conjunction with $L = 16$ against an identical spectral efficiency of 1 bit per channel use uncoded DSTS-SP scheme using $L = 4$ and against the conventional DSTS-BPSK scheme, when employing an interleaver depth of $D_{\text{int}} = 1\,000\,000$ bits after $I = 10$ iterations.

significant BER performance improvement, which is the case for the interleavers with depths of $D_{\text{int}} = 1\,000$ and $10\,000$ bits. However, as the interleaver becomes longer, i.e. as the correlation of the extrinsic LLRs is reduced, the achievable system performance improves upon increasing the number of iterations. Moreover, as the interleaver depth increases, the system requires fewer iterations to achieve its best attainable performance, as shown in Figure 16.11. For example, for the case of an interleaver depth of $D_{\text{int}} = 800\,000$ bits, it is shown in Figure 16.11 that after $I = 7$ iterations, there is no more improvement in the attainable system performance, while the system employing $D_{\text{int}} = 400\,000$ bits requires one more iteration before the system's performance saturates.

We also proposed a novel technique for computing the maximum achievable bandwidth efficiency of the system based on EXIT charts in Section 11.2.3, where it was shown that the maximum achievable bandwidth efficiency based on EXIT charts closely matches with the analytical calculation carried out in Chapter 10 for the bandwidth efficiency of the DSTS-SP system.

In order to eliminate the error floor exhibited by the iteratively detected system discussed previously, in Section 11.3 we proposed an iteratively detected RSC-coded and URC-precoded DSTS-SP scheme, which performed closer to the system's achievable rate. This is due to the fact that the URC has a recursive encoder and, hence, when applying a sufficiently long interleaver, the EXIT chart of the URC decoder is capable of reaching the point of perfect convergence at $(1.0, 1.0)$. Figure 16.12 compares the EXIT chart of the iteratively detected four-antenna-aided DSTS-SP scheme, while applying the non-precoded AGM-1 and AGM-3 SP as well as the precoded Gray-mapping-aided SP modulation schemes. As shown in Figure 16.12, the non-precoded scheme's inner decoder EXIT curves do not reach the point of perfect convergence at $(1.0, 1.0)$ and hence an error floor will be formed, where there is

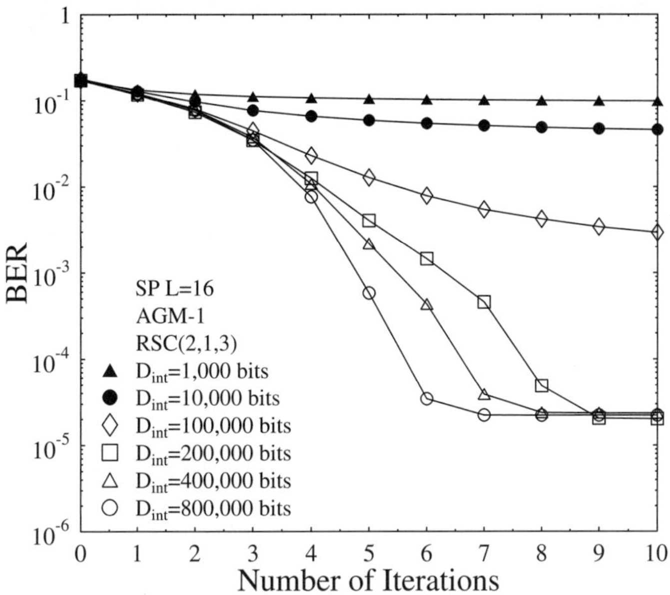

Figure 16.11: Comparison of the BER performance versus the number of iterations for the iteratively detected half-rate RSC-coded DSTS-SP scheme in conjunction with two transmit antennas and AGM-1 of Figure 16.9, while employing different interleaver depths recorded at E_b/N_0 of 7.0 dB.

no error floor in the BER of the precoded system, since the EXIT curve of the URC decoder reaches the point of perfect convergence at $(1.0, 1.0)$.

Explicitly, when using an appropriate bits-to-symbol mapping scheme and ten turbo-detection iterations, gains of about 19.5 dB were obtained by the RSC-coded twin-antenna-aided DSTS-SP schemes over the identical-throughput uncoded DSTS-SP benchmark scheme described in Chapter 10. Furthermore, the AGM-1-based iteratively detected twin-antenna-aided DSTS-SP scheme is capable of performing within 2.3 dB from the maximum achievable rate limit obtained using EXIT charts at $\text{BER} = 10^{-5}$, when employing an interleaver depth of $D_{\text{int}} = 1\,000\,000$ bits. In addition, Chapter 11 characterized the benefits of precoding, when concatenated with the outer channel code, suggesting that an E_b/N_0 gain of at least 1.2 dB can be obtained over the non-precoded system at a BER of 10^{-5}, depending on the mapping scheme used. Explicitly, the four-antenna-aided DSTS-SP system employing no URC precoding attains a coding gain of 12 dB at a BER of 10^{-5} and performs within 1.82 dB from the maximum achievable rate limit, when employing an interleaver depth of $D_{\text{int}} = 1\,000\,000$ bits. In contrast, the URC-aided precoded system outperforms its non-precoded counterpart and operates within 0.92 dB of the maximum achievable rate limit obtained using EXIT charts, when employing an interleaver depth of $D_{\text{int}} = 1\,000\,000$ bits.

Finally, Tables 16.11 and 16.12 present the coding gains as well as the distance from the maximum achievable rate limit for the iteratively detected RSC-coded DSTS system, while employing SP as well as QPSK modulation schemes. The coding gain is measured against the performance of the identical-throughput uncoded DSTS-SP system. The tables present the results for both the two- and four-antenna-aided DSTS schemes, when both systems

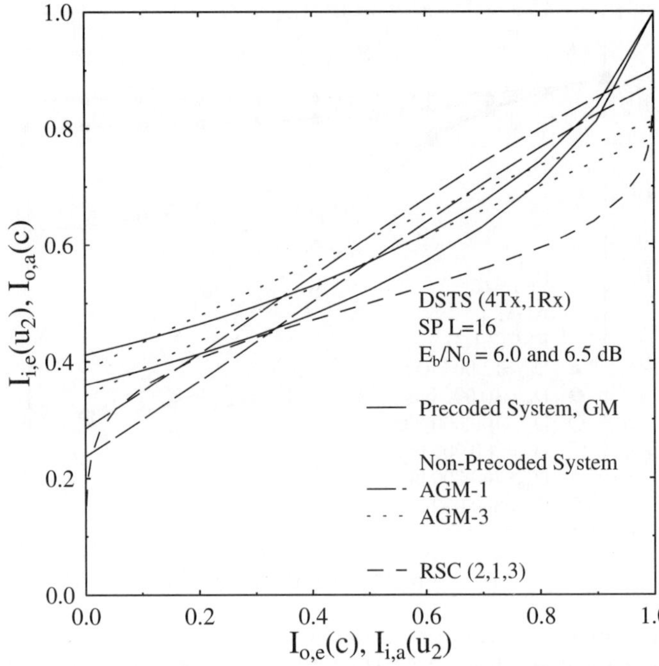

Figure 16.12: Comparison of the convergence behavior of the precoded and non-precoded DSTS-SP systems employing Gray mapping and AGM in conjunction with $L = 16$, based on their EXIT characteristics while operating at E_b/N_0 of 6 and 6.5 dB.

optionally employ URC precoding at $\text{BER} = 10^{-5}$ and 10^{-6}, while employing an interleaver depth of $D_{\text{int}} = 1\,000\,000$ bits.

16.1.12 Chapter 12: Adaptive DSTS-assisted Iteratively Detected SP Modulation

On the other hand, in order to maximize the throughput of the DSTS-SP scheme, while maintaining a certain target QoS, an adaptive DSTS-SP system was proposed in Chapter 12. The proposed adaptive DSTS-SP system exploits the advantages of differential encoding and iterative decoding as well as SP modulation, while adapting the system parameters for the sake of achieving the highest possible spectral efficiency, as well as maintaining a given target BER. The proposed adaptive DSTS-SP scheme benefits from a substantial diversity gain, while using four transmit antennas without the need for pilot-assisted channel estimation and coherent detection. The proposed scheme reaches the target BER of 10^{-3} at a SNR of about 5 dB and maintains it for SNRs in excess of this value, while increasing the effective throughput. The system's bandwidth efficiency varies from 0.25 to 16 bit s^{-1} Hz^{-1}.

The achievable integrity and bit rate enhancements of the adaptive DSTS-SP system are determined by the following factors: the specific transmission configuration used for transmitting data from the four antennas, the spreading factor used and the RSC encoder's code rate.

Table 16.11: Iteratively detected RSC-coded DSTS system coding gain and distance from maximum achievable rate limit at $\text{BER} = 10^{-5}$, when employing an interleaver depth of $D_{\text{int}} =$ 1 000 000 bits.

		Coding gain (dB)	Distance from maximum achievable rate limit (dB)
DSTS (2Tx,1Rx) No URC precoding	SP $L = 16$, Gray mapping	14.9	6.9
	SP $L = 16$, AGM-1	19.5	2.3
	SP $L = 16$, AGM-3	17.75	4.05
	SP $L = 16$, AGM-8	15.9	5.9
	QPSK, AGM	16.1	5.7
DSTS (4Tx,1Rx) No URC precoding	SP $L = 16$, Gray mapping	9.5	4.32
	SP $L = 16$, AGM-1	12	1.82
	SP $L = 16$, AGM-3	10.9	2.92
	SP $L = 16$, AGM-8	8.9	4.92
	QPSK, AGM	9.2	4.62
DSTS (4Tx,1Rx) URC precoding	SP $L = 16$, Gray mapping	12.9	0.92

Table 16.12: Iteratively detected RSC-coded DSTS system coding gain and distance from maximum achievable rate limit at $\text{BER} = 10^{-6}$, when employing an interleaver depth of $D_{\text{int}} =$ 1 000 000 bits.

		Coding gain (dB)	Distance from maximum achievable rate limit (dB)
DSTS (2Tx,1Rx) No URC precoding	SP $L = 16$, Gray mapping	17.8	8
	SP $L = 16$, AGM-1	22.5	3.3
	SP $L = 16$, AGM-3	20.5	5.3
	SP $L = 16$, AGM-8	18.4	7.4
	QPSK, AGM	19	6.8
DSTS (4Tx,1Rx) No URC precoding	SP $L = 16$, Gray mapping	13.7	5.12
	SP $L = 16$, AGM-1	16.7	2.12
	SP $L = 16$, AGM-3	14.8	4.02
	SP $L = 16$, AGM-8	12.9	5.92
	QPSK, AGM	13.5	5.32
DSTS (4Tx,1Rx) URC precoding	SP $L = 16$, Gray mapping	17.9	0.92

16.1.13 Chapter 13: Layered Steered Space-Time Codes

In Chapter 13, we proposed a multi-functional MIMO scheme that combines the benefits of the V-BLAST scheme and of space-time codes as well as of beamforming. Thus, the proposed system benefits from the multiplexing gain of the V-BLAST, from the diversity gain of the space-time codes and from the SNR gain of beamforming. The multi-functional MIMO scheme was referred to as a LSSTC. To further enhance the attainable system performance

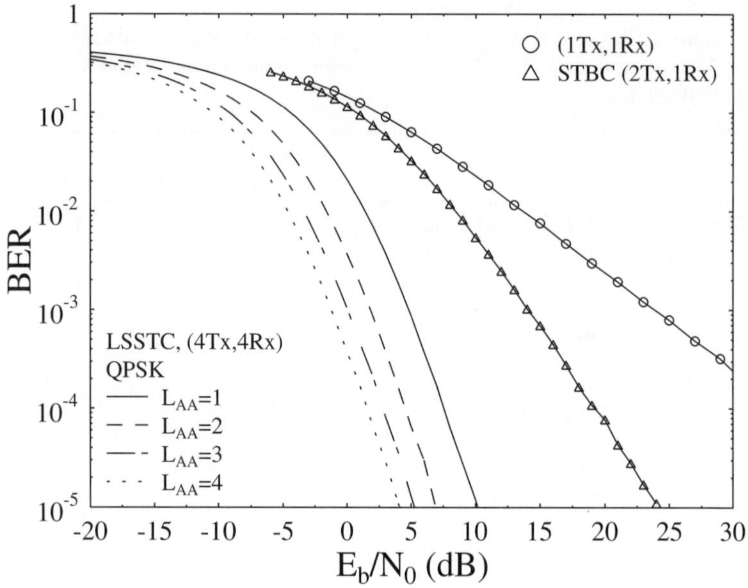

Figure 16.13: Comparison of the attainable BER performance of a QPSK-modulated $N_t \times N_r = 4 \times 4$ LSSTC system for variable L_{AA} values, with that of the twin-antenna-aided STBC scheme and the single-transmit- and single-receive-antenna-aided scheme.

and to maximize the coding advantage of the proposed transmission scheme, the system was combined with multi-dimensional SP modulation.

Figure 16.13 compares the attainable BER performances of a SISO system, of a twin-antenna-aided STBC system and of a four-transmit- and four-receive-antenna-aided LSSTC system. The multiplexing gain of the LSSTC scheme is exemplified by the fact that the LSSTC system used in Figure 16.13 has a throughput that is twice that of a system employing a twin-antenna-aided STBC scheme. In addition, observe in Figure 16.13 that the LSSTC scheme attains a diversity gain that is exemplified in terms of the gain attained by the LSSTC scheme over a single-antenna-aided system as well as over the twin-antenna-aided STBC scheme. On the other hand, notice that an increased SNR gain is obtained by the LSSTC scheme, when more beamforming elements per AA are used. Therefore, Figure 16.13 shows the diversity gain as well as the beamforming gain of the proposed LSSTC scheme and the multiplexing gain is exemplified by the fact that the LSSTC's throughput is higher than that of a twin-antenna-aided STBC scheme.

In Section 13.3 we quantified the capacity of the proposed LSSTC scheme and presented the capacity limits for a system employing $N_t = 4$ transmit AAs, $N_r = 4$ receive antennas and a variable number L_{AA} of elements per AA. Furthermore, in Section 13.4.3 we quantified an upper bound for the achievable bandwidth efficiency of the system based on the EXIT charts obtained for the iteratively detected system. It was shown that there is a discrepancy between the maximum achievable rate limit obtained using the EXIT chart and the analytical bandwidth efficiency. This is due to the fact that the capacity of the LSSTC scheme was analyzed for the case where perfect interference cancelation was assumed, while the proposed system employed a low-complexity, but error-prone ZF IC scheme.

In order to further enhance the achievable system performance, the proposed LSSTC scheme was serially concatenated with both an outer code and a URC, where three different receiver structures were created by varying the iterative detection configuration of the constituent decoders/demapper. As a benchmark scheme, we proposed a two-stage iteratively detected RSC-coded LSSTC-SP scheme, where extrinsic information was exchanged between the outer RSC decoder and the inner URC decoder, while no iterations were carried out between the URC decoder and the Gray-mapping-based SP demapper. This system was referred to as System 1. The convergence behavior of the iterative-detection-aided system was analyzed using EXIT charts.

In Section 13.4.1.2, we employed the powerful technique of EXIT tunnel-area minimization for near-capacity operation. More specifically, we exploited the well-understood properties of EXIT charts that a narrow but nonetheless open EXIT tunnel represents a near-capacity performance. Consequently, we invoked IRCCs for the sake of appropriately shaping the EXIT curves by minimizing the area within the EXIT tunnel using the procedure of [174, 176]. The IRCC-aided system was referred to as System 2.

In Section 13.4.2 we presented a three-stage iteratively detected RSC-coded LSSTC scheme, where extrinsic information was exchanged between the three constituent decoders/demapper, namely the outer RSC decoder, the inner URC decoder and the SP demapper. The three-stage system was referred to as System 3.

Explicitly, the SP-aided LSSTC system, employing $N_t = 4$ transmit antennas, $N_r = 4$ receive antennas and $L_{AA} = 4$ elements per AA, is capable of operating within 0.9, 0.6 and 0.4 dB of the maximum achievable rate limit, as shown in Figure 16.14. However, to operate within 0.6 dB of the maximum achievable rate limit, the system imposes twice the complexity compared with a system operating within 0.9 dB of this limit. On the other hand, to operate as close as 0.4 dB from the maximum achievable rate limit, the system imposes a 20 times higher complexity than that for operating within 0.9 dB of the maximum achievable rate limit. In contrast, the QPSK-modulated three-stage iteratively detected system is capable of operating within 1.54 dB of the maximum achievable rate limit and thus the SP-modulated system outperforms its QPSK-aided counterpart by about 1 dB at a BER of 10^{-6}. The proposed design principles are applicable to an arbitrary number of antennas and diverse antenna configurations as well as to various modem schemes. The complexity of the proposed schemes is compared in Figure 16.15, where the coding gain attained by the different schemes at a BER of 10^{-5} has been plotted versus the corresponding complexity expressed in terms of the number of trellis states, which is directly proportional to the number of ACS operations.

Finally, Table 16.13 presents the coding gains as well as the distance from the maximum achievable rate limit for the proposed iteratively detected systems, namely the SP-aided Systems 1–3 and the QPSK-aided System 3. The table presents the results for the LSSTC system at $\text{BER} = 10^{-6}$, while employing an interleaver depth of $D_{\text{int}} = 180\,000$ bits.

The LSSTC scheme of Chapter 13 is characterized by a diversity gain and a multiplexing gain as well as a beamforming gain. However, a drawback of the design is that it requires the number of receive antennas to be at least as high as the number of transmit AAs. Therefore, for a system employing four transmit AAs, the receiver requires four antennas for correct decoding, which implies that the LSSTC scheme cannot be used in a DL transmission from a BS to a pocket-sized mobile phone due to the size limitation of implementing four antennas. The LSSTC scheme can, however, be conveniently applied for communicating between two BSs or between a BS and a laptop.

16.1.14 Chapter 14: DL LSSTS-aided Generalized MC DS-CDMA

In order to make the LSSTC scheme more practical, in Chapter 14 we presented a multi-functional MIMO scheme employing four DL transmit and two receive antennas. The

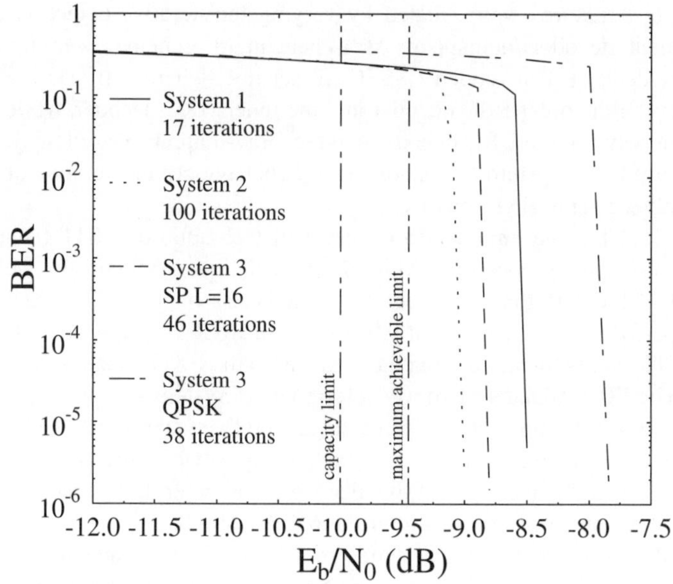

Figure 16.14: Performance comparison of the three proposed LSSTC-SP aided systems employing two-stage iteration between an outer code and a URC decoder, as well as that of the three-stage iterative information exchange between an outer RSC decoder, an intermediate URC decoder and a SP demapper.

Table 16.13: Iteratively detected LSSTC system coding gain and distance from maximum achievable rate limit at BER $= 10^{-6}$ in conjunction with $N_t = 4$ transmit AAs, $N_r = 4$ receive antennas, $L_{AA} = 4$ elements per AA and $D_{\text{int}} = 180\,000$ bits.

LSSTC (4Tx,4Rx) $L_{AA} = 4$	Coding gain (dB)	Distance from maximum achievable rate limit (dB)
System 1	14.5	0.9
System 2	15.0	0.4
SP-aided System 3	14.8	0.6
QPSK-aided System 3	13.1	1.6

proposed multi-functional MIMO scheme of Chapter 14 combines the benefits of STS, V-BLAST and generalized MC DS-CDMA as well as beamforming. The proposed scheme of Chapter 14 is referred to as LSSTS. The LSSTS scheme benefits from a spatial diversity gain, a frequency diversity gain and a multiplexing gain as well as a beamforming gain.

Figure 16.16 compares the attainable BER performance of LSSTS-assisted generalized MC DS-CDMA with that of the LSSTC scheme. The figure also shows the BER performance of the twin-antenna-aided STBC and the SISO benchmark systems. The LSSTS scheme employs four transmit AAs and two receive antennas, while the LSSTC scheme of Figure 16.16 employs four transmit AAs and four receive antennas. Observe in Figure 16.16 that the BER performance of the LSSTS scheme is identical to that of the STBC scheme when a single subcarrier is used. This means that the LSSTS scheme attains a spatial diversity gain of two,

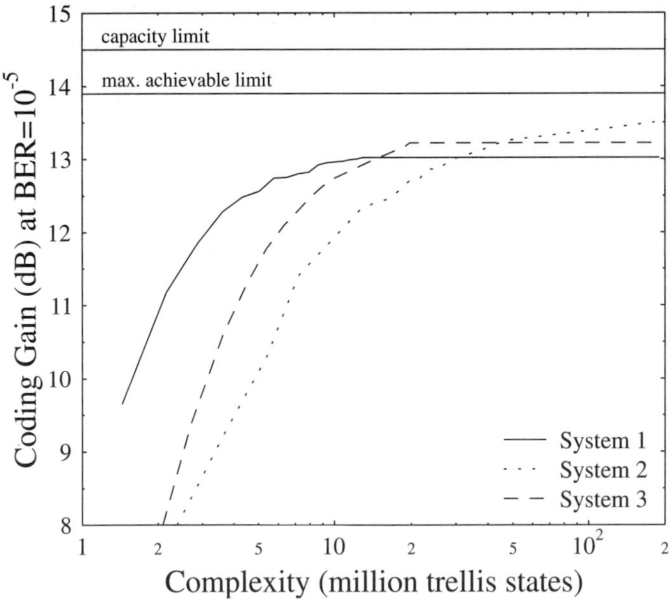

Figure 16.15: Comparison of the coding gain at a BER of 10^{-5} versus the complexity in million trellis states of the three proposed LSSTC-SP-aided systems.

while attaining a multiplexing gain that is twice that of a twin-antenna-aided STBC scheme. In addition, increasing the number of subcarriers V improves the attainable BER performance as shown in Figure 16.16 for the LSSTS scheme. Hence, the LSSTS scheme is also capable of attaining frequency diversity gain, when the subcarrier frequencies are arranged in a way that guarantees that the same STS signal is spread to and, hence, transmitted by the specific V subcarriers having the maximum possible frequency separation, so that they experience independent fading. On the other hand, comparing the BER performance of the LSSTS scheme employing $V = 1$ subcarrier with that of the LSSTC scheme shows that the LSSTC scheme attains a better BER performance. This is due to the fact that the LSSTC scheme employs more receive antennas than the LSSTS scheme and, hence, the LSSTC scheme is capable of attaining a higher spatial diversity gain.

In Section 14.3 we demonstrated that the number of users supported by the LSSTS scheme can be substantially increased by invoking combined spreading in both the TD and the FD. We also used a novel user-grouping technique for minimizing the multi-user interference imposed, when employing both TD and FD spreading in the LSSTS-aided generalized MC DS-CDMA DL scheme.

Furthermore, in order to further improve the attainable performance of the LSSTS-assisted generalized MC DS-CDMA, we proposed three iteratively detected LSSTS schemes, where iterative detection was carried out by exchanging extrinsic information between two serially concatenated channel codes. We used EXIT charts to analyze the convergence behavior of the proposed iterative-detection-aided schemes and proposed a novel LLR post-processing technique for improving the iteratively detected systems' performance.

In order to elaborate a little further on the LLR post-processing technique, observe in Figure 16.17 that there are several EXIT curves for the URC decoder at the same E_b/N_0 value. Let us first consider the dark line marked by the legend 'no LLR limits'. This EXIT

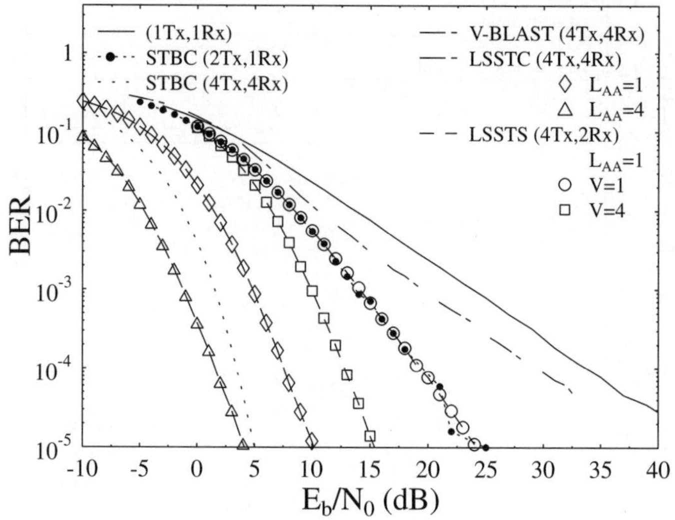

Figure 16.16: Comparison of the attainable BER performance of a QPSK-modulated LSSTS-assisted generalized MC DS-CDMA and LSSTC systems with that of the twin-antenna-aided STBC scheme and the single-transmit- and single-receive-antenna-aided scheme.

curve corresponds to a URC decoder, which has a recursive encoder at the transmitter, and hence it is expected that the EXIT curve of the URC decoder will indeed reach the $(1.0, 1.0)$ point of perfect convergence in the EXIT curve, as discussed in [196]. However, Figure 16.17 also shows that the EXIT curve of the URC decoder does not reach the $(1.0, 1.0)$ point. Limiting the maximum and minimum of the LLR values allowed the URC EXIT curve to reach the $(1.0, 1.0)$ point, as shown in Figure 16.17 by the dotted line associated with the legend 'LLR limit $= 10$'.

On the other hand, for the sake of testing the accuracy of the URC EXIT curve, while imposing a limit on the LLR values, we generated artificial Gaussian distributed and uncorrelated LLRs, where the resultant EXIT curve is represented by the dotted line having the legend 'artificial LLR generation'. As shown in Figure 16.17, the curves corresponding to the case where the LLRs' dynamic range is limited and where the artificial LLRs are generated are quite different. Therefore, limiting the LLR values did not solve the problem. The reason for this behavior is the fact that the output of the LSSTS decoder that is passed to the QPSK demapper is not Gaussian distributed, although the LLR values in the demapper are calculated assuming Gaussian distribution. Therefore, in order to eliminate the complexity of computing an analytical formula for the PDF of the LSSTS decoded data or computing the LLRs based on the histogram of the data, we devised the LLR post-processing technique of Section 14.4.1 as a transformation for the output LLR of the demapper. Figure 16.17 shows that the system employing the LLR post-processing technique attains a similar EXIT curve to the case where artificial LLRs were generated.

The three iterative-detection-aided LSSTS schemes differed in the way the channel coding was implemented in the different STS layers, while the overall code-rate of the three systems was kept identical.

1. In the first scheme, referred to as System 1, a single outer and a single inner channel code was used to encode the bits transmitted.

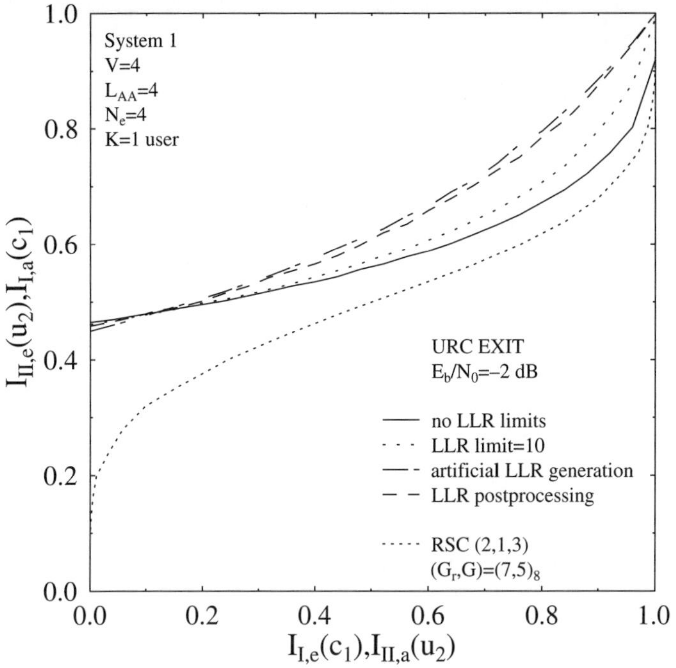

Figure 16.17: EXIT chart of a RSC-coded and URC-precoded proposed System 1 of Figure 14.6 employing Gray-mapping-aided QPSK in conjunction with $N_t = 4$, $N_r = 2$, $V = 4$, $L_{AA} = 4$, $K = 1$ user and $E_b/N_0 = -2$ dB.

2. In the second scheme, namely System 2, a single outer code was implemented, whose output was split into two substreams, each of which was encoded using a separate inner code.

3. In contrast, in the third proposed scheme, referred to as System 3, the input data bit stream was first split into two different substreams, where a pair of different outer as well as inner codes were implemented in the different substreams.

It was shown in Chapter 14 that the three systems exhibit a similar complexity quantified in terms of the total number of trellis states encountered, which determines the number of ACS arithmetic operations. Similarly, we demonstrated that, provided that we employed sufficiently long interleavers, the three systems attained a similar BER performance. In contrast, when shorter interleavers were employed, System 1 performed better than System 2, which, in turn, performed better than System 3. This is due to the fact that the interleaver depth of Systems 2 and 3 is lower than that of System 1, since the bit stream is split into two substreams in Systems 2 and 3, which constrains the interleaver to be shorter and hence the correlation in the extrinsic information becomes higher, which eventually degrades the BER performance.

16.1.15 Chapter 15: Distributed Turbo Coding

The MIMO schemes presented in Sections 16.1.10 and 16.1.13 are considered to be co-located MIMO systems, i.e. the multiple antennas at the transmitter and receiver are

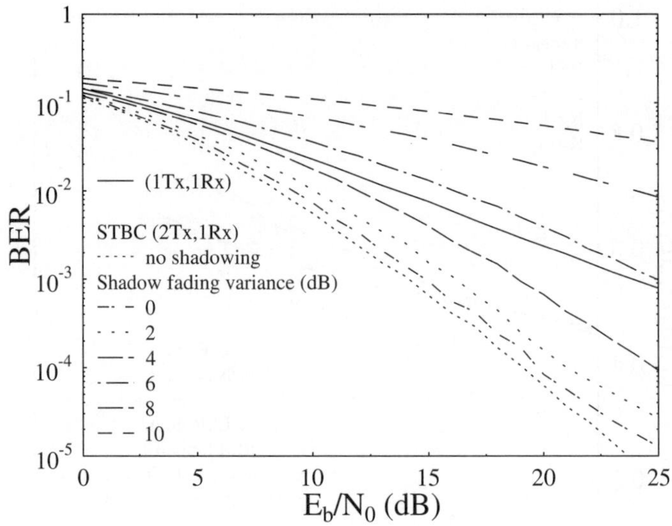

Figure 16.18: Effect of large-scale shadow fading on the performance of STBC systems.

connected physically to the same station. In addition, in order to attain a diversity gain, it is required that the CIR between the different transmit antennas and the receive antenna be uncorrelated or statistically independent. This is possible if the antenna spacing is sufficiently large so that the assumption of statistical independence of the different paths from the different antennas is justified. However, the assumption of sufficient antenna spacing may be impractical for pocket-sized wireless devices, which are typically limited in size and hardware complexity to a single transmit antenna.

On the other hand, spatial fading correlations caused by insufficient antenna spacing at the transmitter or receiver of a MIMO system results in degradation in the capacity as well as the BER performance of MIMO systems as shown in Figure 16.18 for a twin-antenna STBC system [11]. Correlation is likely to be introduced as a result of large-scale shadow fading that affects the transmission links between the different transmit and receive antennas [5]. Figure 16.18 compares the BER performance of a single-transmit- and single-receive-antenna-aided system with that of a twin-antenna-aided STBC system affected by the large-scale shadow fading. As shown in Figure 16.18, the performance of MIMO systems degrades as the shadow fading effects increase and the single-antenna-aided system performs better than a MIMO system when the shadow fading variance is higher than 5 dB.

Cooperative communications were introduced recently for attaining spatial diversity, where it is typically possible to guarantee the spatial separation of the transmit antennas. This is due to the fact that the different antennas belong to different MSs, which are assumed to be far enough apart to attain statistically independent fading from the different antennas. Therefore, since the signals transmitted from different users undergo independent fading, spatial diversity can be achieved with the aid of the cooperating partners' antennas.

In Chapter 15 we proposed a DTC scheme, where two users cooperate by appropriately sending their own data coupled with the other user's data after interleaving and channel coding. Cooperation in the proposed DTC scheme is carried out in two phases. During the first phase of cooperation, the two users exchange their data in two time slots. Hence, after the first phase of cooperation the two users have their own data as well as the data of the

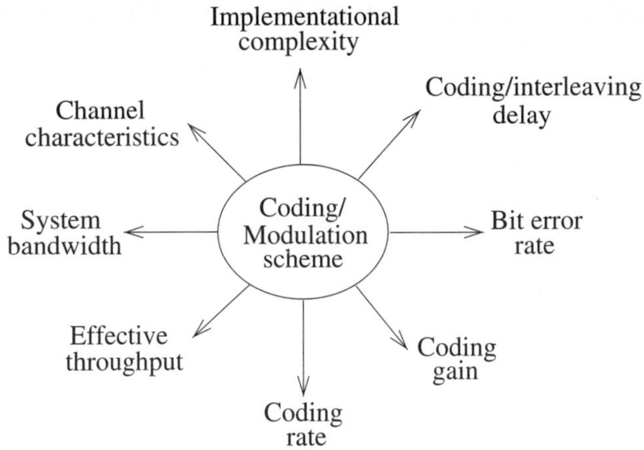

Figure 16.19: Factors affecting the design of channel coding and modulation schemes.

other user. Then, during the second phase of cooperation, each user employs channel coding and interleaving before mapping the bits to multi-dimensional SP symbols that are then simultaneously transmitted from the two users' antennas. At the receiver side, interference cancelation is applied to the received data, where two branches of decoded data are output from the interference canceler. Afterwards, iterative detection is carried out between the demapper and the channel code's decoder in each branch as well as between the channel codes' decoders in the two branches.

The proposed DTC is compared against a benchmark scheme where no exchange of data is carried out between the two users. In contrast, each user transmits their own data after channel coding and interleaving. At the receiver side, interference cancellation is performed on the received signal and then iterative detection is carried out between the demapper and the channel code's decoder in each branch of the decoded signal.

The DTC scheme has a four times lower throughput than the benchmark scheme. However, the DTC scheme is capable of attaining an E_b/N_0 gain of more than 25 dB at a BER of 10^{-5} over the benchmark scheme due to the fact that the benchmark scheme has an error floor, while the DTC scheme does not.

In addition, in Section 15.4 we studied the effect of errors induced in the data exchanged between the different users in the first phase of cooperation on the performance of the UL transmission in the second phase of cooperation. We considered transmission over AWGN, Rayleigh and Ricean inter-user channels and studied the effects of transmission over the different channels on the performance of the DTC scheme. It was shown in Section 15.4 that an error rate higher than 10^{-6} in the inter-user communication results in an error floor in the attainable BER performance of the cooperative UL transmission. Therefore, it was proposed to consider soft data relaying in the DTC scheme and this is discussed further in terms of our future research ideas outlined in the next section.

16.2 Future Research Ideas

In this section, a few future research ideas are presented. Future research has to carefully consider the stylized design trade-offs summarized in Figure 16.19, which illustrates the factors affecting the design of channel coding and modulation schemes [9].

16.2.1 Generalized Turbo-detected SP-assisted Orthogonal Design

In this book we have demonstrated that potential performance improvements may be attained when performing iterative decoding and demapping between outer channel codes and an inner demapper defined by a multi-dimensional lattice, such as the D_4 lattice. The techniques and procedures proposed in this book may be readily extended to SP demappers characterized by higher-dimensional lattices, where the larger constellation size in the higher-dimensional space renders the mapping design more flexible. Multi-dimensional constellations had been shown to be beneficial in the design of trellis-coded modulation as early as 1987 [390–393]. During the preparation of this book, multi-dimensional labeling has been used for QPSK-based bit-interleaved coded modulation employing iterative decoding for transmission over a single antenna [394, 395]. Further improvements on multi-dimensional constellation labeling were proposed in [396–399]. More recently, multi-dimensional constellation labeling has been proposed for bit-interleaved space-time-coded modulation using iterative decoding [400], where the labeling of two 16-QAM symbols was designed jointly and was optimized using the so-called Reactive Tabu Search (RTS) of [401].

The orthogonal transmit diversity of Equation (2.27), which is reproduced here for convenience [207]

$$\mathbf{G}_{2^k}(x_1, x_2, \ldots, x_{k+1}) = \begin{bmatrix} \mathbf{G}_{2^{k-1}}(x_1, x_2, \ldots, x_k) & x_{k+1}\mathbf{I}_{2^{k-1}} \\ -x_{k+1}^{*}\mathbf{I}_{2^{k-1}} & \mathbf{G}_{2^{k-1}}^{\mathcal{H}}(x_1, x_2, \ldots, x_k) \end{bmatrix}, \quad (16.1)$$

for $k = 1, 2, \ldots$, may be designed with the aid of the SP lattice that has the best known MED in the $2(k+1)$-dimensional real-valued Euclidean space $\mathbb{R}^{2(k+1)}$ (see [43]). Table 16.14 outlines all known SP lattices having up to $D = 24$ dimensions and exhibiting the best MED [221] that correspond to the orthogonal transmit diversity of Equation (16.1). As a further extension, the techniques proposed in this book and in [394–400] may be implemented for optimizing the constellation labeling of the specific SP lattices seen in Table 16.14. In the spirit of Figure 16.19, this implies designing high-rate schemes that achieve infinitesimally low BERs and are capable of providing near-capacity performance.

16.2.2 Precoder Design for Short Interleaver Depths

In Chapter 5, rate-one precoders were successfully employed for improving the iteration gain of three-stage iterative detection schemes. However, only rate-one precoders having a single shift register were considered in Chapter 5. Our future work will include the employment of rate-one precoders having several shift register stages, which will provide us with a higher degree of freedom in terms of the associated generator polynomials and the constraint length of the precoders. In addition, precoders having arbitrary coding rates may be employed, while maintaining the overall system's throughput for the sake of further increasing the design space. The three-dimensional EXIT charts of Section 5.3 will be used for the sake of finding optimum combinations of generator polynomials, the constraint length and the coding rate of the precoders. The ultimate aim of employing novel precoding techniques is that of improving the system's attainable performance, while maximizing its throughput and minimizing its delay, which are critical design trade-offs, as seen in Figure 16.19. For example, systems optimized for convergence thresholds close to their information-theoretic limits using the EXIT chart analysis only approach these predictions when employing high interleaver depths, which imposes impractically long delays in certain applications. However, recently a novel design procedure has been proposed in [176, 402] for creating systems exhibiting beneficial

Table 16.14: SP lattices with the best known MED [221] corresponding to the orthogonal transmit diversity of Equation (16.1).

k	Dimension $2(k+1)$	Lattice	Number of constellation points in first layer
1	4	D_4	24
2	6	E_6	72
3	8	E_8	240
4	10	P_{10c}	372
5	12	K_{12}	756
6	14	Λ_{14}	1422
7	16	Λ_{16}	4320
8	18	Λ_{18}	7398
9	20	Λ_{20}	17 400
10	22	Λ_{22}	49 896
11	24	Λ_{24}	196 560

decoding convergence after a fixed number of iterations. The design procedure is based on the observation that EXIT chart predictions are usually accurately satisfied for the first few iterations, regardless of the depth of the interleaver employed. This indicates that the design procedure of [176, 402] would produce systems having high performance, while employing low interleaver depths. The multi-dimensional constellation labeling proposed in Section 16.2.1 and the design procedure of [176,402] may be jointly invoked for the sake of designing systems achieving an infinitesimally low BER, while imposing a practical delay.

16.2.3 Improving the Coding Gain of V-BLAST Schemes

STBCs [11, 12, 25] and STTCs [8] constitute specific examples of MIMO transmit diversity schemes, where the signals transmitted from different antennas are jointly designed in space and time for the sake of minimizing the attainable error rate, while sacrificing the achievable multiplexing gain. There is another class of MIMO schemes that focuses on maximizing the data rate with the aid of spatial multiplexing, where the individual antennas' signals are transmitted independently. An example of this class is the spatial multiplexing scheme proposed by Foschini [69], which is referred to as the V-BLAST scheme. In contrast to space-time codes, these schemes maximize the achievable throughput by sacrificing the attainable diversity gain. Our future work considers improving the coding gain of V-BLAST schemes, while maintaining their full multiplexing gain. The idea is to jointly design the complex symbols that are transmitted from all antennas at a particular time instant, so that they are represented by a single phasor point selected from a SP constellation corresponding to an n-dimensional real-valued lattice having the best known MED in the n-dimensional real-valued space $\mathbb{R}^n$, where n is twice the number of transmit antennas, i.e. $n = 2 \cdot N_t$. For example, systems having $N_t = 4$ transmit antennas may be designed using the lattice E_8, which is defined as the SP having the best MED in $\mathbb{R}^8$ (see [221]) and shown in Table 16.14. The optimum detection method for the proposed scheme is ML decoding where all legitimate phasor points of the n-dimensional lattice are tentatively tested, in order to find the specific phasor point that minimizes the Euclidean norm with respect to the received signal. However, the size of the signal space becomes excessive for a large number of transmit antennas and for a high BPS throughput rendering exhaustive search impractical, if not impossible. However, detection methods based on ZF techniques, which invoke matrix inversion for the sake of finding the best estimate, may be employed. ZF techniques exhibit a low complexity, but they are less powerful when compared with ML detection. A detection method that provides a better compromise between complexity and accuracy was proposed in [69] and is known as nulling and canceling combined with optimum ordering. A recently proposed detection method that has attracted considerable attention is sphere decoding [67, 243], which attains similar performance to that of the excessive-complexity optimum ML detection at a significantly lower complexity. A reduced-complexity near-optimum detection method was proposed in [403]. Careful design of the legitimate constellation points from the n-dimensional real-valued lattice is expected to further decrease the decoding complexity imposed.

16.2.4 Adaptive Closed-loop Co-located MIMO Systems

In this book we demonstrated that substantial performance improvements may be attained by co-located MIMO elements using a group of LDCs, when only the receiver has the knowledge of CSI. On the other hand, when the transmitter has access to full or partial CSI, we have a closed-loop MIMO system. The CSI available at the transmitter has been shown to be beneficial for optimizing STBC schemes in terms of their achievable link capacity and throughput [404]. Closed-loop MIMO systems can be implemented by employing a dedicated CSI feedback channel using FDD or sharing the same wireless channels in TDD systems.

Hence, we propose the adaptive MIMO scheme portrayed in Figure 16.20 for further study in order to exploit the CSI information at the transmitter, which jointly considers the design of modulation, STC and the decoding complexity. The main idea is to dynamically adjust the system's throughput based on the near-real-time channel condition as well as the affordable complexity at the receiver. More explicitly, the channel estimator of Figure 16.20 provides the CSI required by a sphere decoder [67, 405, 406] and determines the number of CSI bits feedback to the transmitter. For example, a single bit can be used to indicate for the transmitter of Figure 16.20 whether to increase or decrease the throughput by adjusting the modulation order or by changing the value of Q of the LDCs. Alternatively, if the receiver's affordable complexity is limited, for example by its battery consumption, a single bit can be transmitted to inform the transmitter of Figure 16.20 to reduce the throughput, to employ antenna selection [407] or even to decrease the value of T. Naturally, more precise control of the transmitter is attainable based on the near-instantaneous channel quality and on the affordable complexity at the receiver, when more feedback bits are provided by the channel estimator of Figure 16.20.

16.2.5 Improved Performance Cooperative MIMO Systems

The CLDCs proposed in Chapter 9 were designed for cooperative MIMO systems based on the AF cooperation protocol [116]. We offer the following suggestions in order to address some of the unresolved design issues as well as to improve the system's achievable performance.

- Observe in Figure 9.7 that the system's performance is gravely affected by the integrity of the signals received at the relays. Unfortunately, the source-to-relay links have to operate without diversity assistance, as seen in Figure 9.2. Hence, we expect the code design of the broadcast interval to be improved by increasing the achievable detection

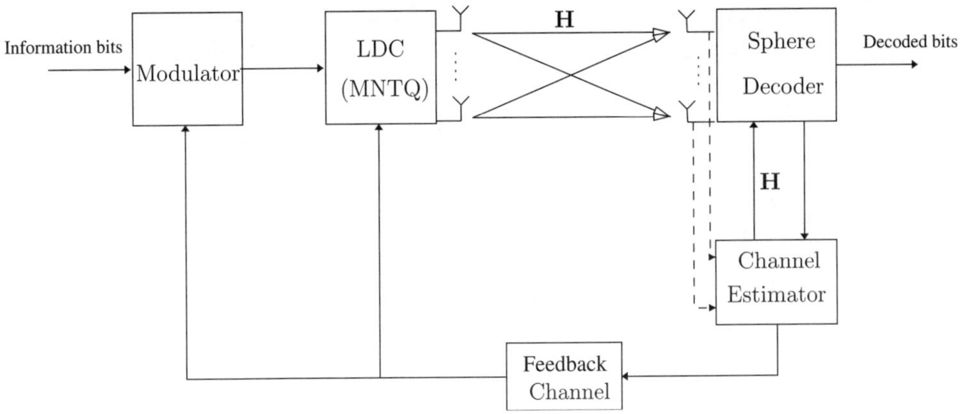

Figure 16.20: Schematic of the adaptive co-located MIMO system using LDCs.

reliability at the relays, which can be achieved, for example, by Luby Transform (LT) codes [408–410].

- Figure 16.7 demonstrated that the CLDCs of Figure 9.2 are sensitive to the channel estimation errors, since both the source-to-relay and the relay-to-BS channels have to be estimated. Therefore, differential encoding schemes [411, 412] are desirable to eliminate the impediments imposed by the channel estimation process.

- The DF cooperation strategy can be employed instead of the AF protocol investigated in Chapter 9 in order to eliminate the ambiguity imposed by the relays. However, if the relays make an erroneous decision concerning the source information, the errors may be propagated further at the BS. Hence, the DF protocol is suitable for the scenarios where the source-to-relay links are highly reliable and the BS has the knowledge of which relay is involved in the cooperation.

- Finally, owing to the distributed nature of the cooperative MIMO systems, it is straightforward to employ the multi-functional MIMO concept originally designed for the co-located MIMO systems [130]. For example, we can configure a group of relays to form a beam [413] or configure another group of relays to employ STC [329], according to their specific geographic distributions.

16.2.6 Differential Multi-functional MIMO

As discussed in Chapter 10, the channel estimation complexity increases with the product of the number of transmit and receive antennas. In addition, channel estimation errors degrade the performance of the MIMO systems when coherent detection is employed. A solution for eliminating the complexity of MIMO channel estimation is to employ non-coherent detection dispensing with channel estimation.

The multi-functional MIMOs presented in Chapters 13 and 14 use coherent detection, while assuming perfect channel knowledge at the receiver. However, channel estimation, which is a complex process, includes channel estimation errors that degrade the attainable BER performance of the system. Figure 16.21 compares the attainable BER performance of the LSSTC scheme of Chapter 13 both when considering perfect channel knowledge and

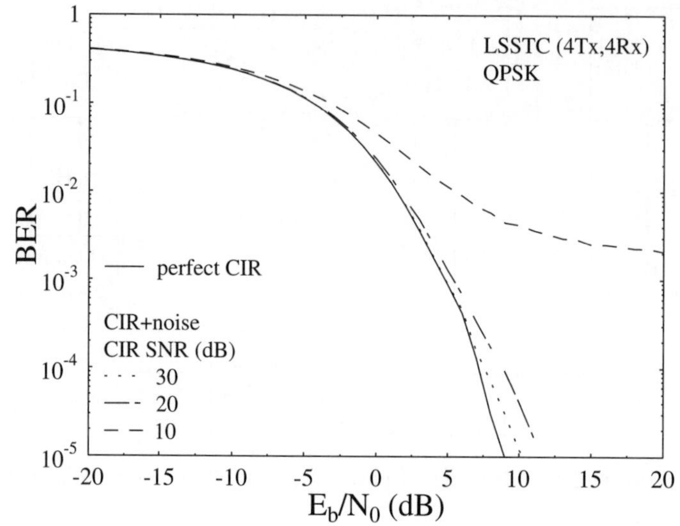

Figure 16.21: Effect of channel estimation error on the performance of the LSSTC scheme of Chapter 13. Gaussian noise was added to the CIR at the receiver for the sake of inducing errors in the channel knowledge.

when modeling the channel estimation error as Gaussian noise imposed on the channel impulse response at the receiver side. Observe that as the noise variance increases, i.e. as the channel estimation error increases, the BER performance degrades and an error floor is formed. Therefore, an attractive way of eliminating the potentially high-complexity channel estimation as well as the performance degradation due to channel estimation errors is to design non-coherent receivers that do not require any channel knowledge. Differential schemes, such as DSTS, employ non-coherent detectors at the expense of 3 dB performance degradation compared with the coherent scheme using perfect channel knowledge.

Therefore, as a further extension to the multi-functional MIMO schemes of Chapters 13 and 14, we can design differential multi-functional MIMO schemes that do not require any channel knowledge. In addition, it is expected that the differential scheme will have a 3 dB performance degradation when compared with the coherent scheme assuming perfect channel knowledge at the receiver. However, when channel estimation is employed, differential detection eliminates the complexity of channel estimation and we may even attain a better BER performance than that of the coherent scheme, when the channel estimation is not reliable.

16.2.7 Multi-functional Cooperative Communication Systems

MIMO systems require more than one transmit antenna, but satisfying this need may be impractical for pocket-sized wireless devices, which are typically limited in both size and hardware complexity to a single transmit antenna. Furthermore, as most wireless systems support multiple users, user cooperation [113, 114, 386] can be employed, where users support each other by 'sharing their antennas' and thus generate a virtual multiple antenna environment [121]. Since the signals transmitted from different users undergo independent fading, spatial diversity can be achieved with the aid of the cooperating partners' antennas.

Several cooperative communication schemes have been proposed in the literature [24,113, 114,119,121,124,128,144,386,414], where cooperative communications have been shown to offer significant performance gains in terms of various performance metrics, including diversity gains as well as multiplexing gains. Hence, a potential research idea is to investigate the design of cooperative communication schemes that are characterized by diversity gain and multiplexing gain as well as beamforming gain. In other words, we propose to design multi-functional cooperative communication schemes. An UL scheme can be implemented, where each MS may be equipped with a single antenna or a single antenna array. Users can be combined in a way so that the nearest users can transmit in a STC manner, where the channels from the different MSs to the UL receiver are assumed to be statistically independent. In addition, different groups of users employing STC may transmit their data at the same time and using the same carrier frequency, like V-BLAST, in order to increase the attainable throughput of the system.

On the other hand, the user cooperation is usually implemented as two-phase cooperation. The first phase corresponds to the phase where the users exchange their data so that they can assist each other in the second phase of cooperation, where the users communicate with the BS. Hence, the total throughput of the system depends on the number of time slots as well as on the way the different users exchange their data. In addition, the performance of the system depends on the signaling scheme used by the different users for relaying the data of the other users. Therefore, the multi-functional cooperative communication scheme can be designed and studied in the context of completely different signaling schemes, evaluating their effect on the attainable BER performance as well as on the attainable throughput.

On the other hand, instead of considering a cooperative UL scheme, where the different users communicate with a BS, it is possible to consider an *ad-hoc* network, where the different users cooperate with other users. In this case, the receiving users can also cooperate in order to reliably decode their received signals and hence each receiving user can decode their own data. The receivers have to exchange their received signals as well as the CIRs of the channels between the transmitting users and each receiving user. This can be carried out in a single time slot for each receiving user using CDMA spreading, where each receiving user transmits at the same time its received signal with the CIRs from the transmitting users.

16.2.8 Soft Relaying and Power Optimization in DTC

In Chapter 15 we proposed a DTC scheme that combines the concepts of cooperative communications and turbo coding. In the proposed scheme, we considered equal power allocation for the two phases of cooperation as well as for the two users. However, in a practical scenario, the two cooperating users must be closer to each other than to the BS. Hence, power can be allocated more efficiently so that less power can be allocated for the first phase of cooperation and more power can be allocated to the second phase, while keeping the total transmit power in the two phases of cooperation constant. In addition, the transmit power can be optimally shared between the two users so that the user with the better channel conditions can transmit more power.

Half-rate channel coding has been used in the first phase of communication in the DTC scheme of Chapter 15 in order to improve the attainable BER performance of the UL transmission. The results presented in Chapter 15 were considered for the system, where the half-rate codes were employed in the two phases of cooperation. This means that the system is wasting the available bandwidth. Hence, it is possible to distribute the code rate between the two phases of cooperation so that the system utilizes the available bandwidth as efficiently as possible, while providing a good BER performance. In addition, iterative detection can be carried out during the first phase of cooperation, where the number of iterations can be adapted depending on the IUC SNR. In other words, when the IUC SNR is high enough, no iterations may be applied and then as the IUC SNR decreases, the number of decoding iterations can be increased in order to maintain a good BER performance in the first phase of cooperation, which eventually affects the performance of the transmission in the second phase of cooperation.

On the other hand, soft relaying has been proposed as a method for combining the main advantages of both AF and DF signaling strategies. In [129], soft DF has been shown to outperform the DF and AF signaling, where it was argued that the DF signaling loses soft information and, hence, all operations were performed in a LLR domain. A more detailed study on the soft DF was carried out in [137], where it was shown how the soft information can be quantized, encoded and then modulated using superimposed modulation in order to maintain the system's throughput and bandwidth constant. Soft information relaying has also been used in [134] in order to pass soft information from the relay to the BS using BPSK modulation. Furthermore, in [130, 135] distributed source coding techniques have been adopted for employment in wireless cooperative communications in order to improve the inter-user performance. Therefore, based on the performance improvements reported in the literature while using soft information relaying and based on the fact that the performance of the DTC scheme of Chapter 15 is highly dependent on the inter-user channel characteristics, it is of potential research interest to investigate the effect of using soft information relaying. In other words, in the DTC scheme the two users transmit soft estimates of the other users' data instead of performing hard decoding and losing the advantage of soft information.

16.3 Closing Remarks

Throughout this book we have endeavored to depict the range of contradictory system design trade-offs associated with the conception of MIMO systems. Our intention was to present the material in an unbiased fashion and sufficiently richly illustrated in terms of the associated design trade-offs so that readers will be able to find recipes and examples for solving their own particular wireless communications problems. In this rapidly evolving field it is a challenge to complete a timely, yet self-contained treatise, since new advances are being discovered at an accelerating pace, which the authors would like to report on. Our sincere hope is that you, our valued readers, have found the book a useful source of information, but above all a catalyst for further research.

Appendix A

Gray Mapping and AGM Schemes for SP Modulation of Size $L = 16$

In this appendix, Gray mapping and the ten different AGM schemes introduced in Chapter 3 for STBC-SP signals of size $L = 16$ are described in detail. More specifically, for all mapping schemes, constellation points of the lattice D_4 are given for each integer index $l = 0, 1, \ldots, 15$. Observe that all mapping schemes use the same 16 constellation points, which were optimized in Section 2.4.4 and Example 2.4.2. The normalization factor of these constellation points is $\sqrt{2L/E_{\text{total}}} = 1$ as described in Equation (2.63). The constellation points corresponding to each mapping scheme are given in Tables A.1–A.11.

Table A.1: Gray mapping.

	Points from D_4					Points from D_4			
Integer index	a_1	a_2	a_3	a_4	Integer index	a_1	a_2	a_3	a_4
0	−1	−1	0	0	8	−1	0	0	−1
1	0	−1	−1	0	9	0	0	−1	−1
2	0	−1	+1	0	10	0	0	+1	−1
3	+1	−1	0	0	11	+1	0	0	−1
4	−1	0	0	+1	12	−1	+1	0	0
5	0	0	−1	+1	13	0	+1	−1	0
6	0	0	+1	+1	14	0	+1	+1	0
7	+1	0	0	+1	15	+1	+1	0	0

Table A.2: AGM-1.

	Points from D_4					Points from D_4			
Integer index	a_1	a_2	a_3	a_4	Integer index	a_1	a_2	a_3	a_4
0	+1	−1	0	0	8	−1	0	0	−1
1	0	−1	−1	0	9	0	0	−1	−1
2	0	−1	+1	0	10	0	0	+1	−1
3	−1	−1	0	0	11	+1	0	0	−1
4	−1	0	0	+1	12	−1	+1	0	0
5	0	0	−1	+1	13	0	+1	−1	0
6	0	0	+1	+1	14	0	+1	+1	0
7	+1	0	0	+1	15	+1	+1	0	0

Table A.3: AGM-2.

	Points from D_4					Points from D_4			
Integer index	a_1	a_2	a_3	a_4	Integer index	a_1	a_2	a_3	a_4
0	+1	+1	0	0	8	−1	0	0	−1
1	0	−1	−1	0	9	0	0	−1	−1
2	0	−1	+1	0	10	0	0	+1	−1
3	+1	−1	0	0	11	+1	0	0	−1
4	−1	0	0	+1	12	−1	+1	0	0
5	0	0	−1	+1	13	0	+1	−1	0
6	0	0	+1	+1	14	0	+1	+1	0
7	+1	0	0	+1	15	−1	−1	0	0

Table A.4: AGM-3.

	Points from D_4					Points from D_4			
Integer index	a_1	a_2	a_3	a_4	Integer index	a_1	a_2	a_3	a_4
0	+1	0	0	−1	8	0	−1	−1	0
1	+1	0	0	+1	9	0	0	−1	+1
2	−1	0	0	−1	10	−1	−1	0	0
3	0	0	+1	+1	11	−1	0	0	+1
4	0	−1	+1	0	12	0	0	−1	−1
5	+1	−1	0	0	13	0	+1	−1	0
6	0	0	+1	−1	14	+1	+1	0	0
7	0	+1	+1	0	15	−1	+1	0	0

Table A.5: AGM-4.

	Points from D_4					Points from D_4			
Integer index	a_1	a_2	a_3	a_4	Integer index	a_1	a_2	a_3	a_4
0	0	0	−1	+1	8	+1	−1	0	0
1	0	0	−1	−1	9	0	0	+1	−1
2	−1	+1	0	0	10	−1	0	0	+1
3	−1	0	0	−1	11	−1	−1	0	0
4	+1	0	0	+1	12	0	−1	−1	0
5	+1	+1	0	0	13	+1	0	0	−1
6	0	+1	−1	0	14	0	−1	+1	0
7	0	+1	+1	0	15	0	0	+1	+1

Table A.6: AGM-5.

	Points from D_4					Points from D_4			
Integer index	a_1	a_2	a_3	a_4	Integer index	a_1	a_2	a_3	a_4
0	0	+1	+1	0	8	−1	0	0	+1
1	0	0	+1	−1	9	−1	−1	0	0
2	−1	0	0	−1	10	0	0	−1	+1
3	−1	+1	0	0	11	0	−1	−1	0
4	0	−1	+1	0	12	+1	−1	0	0
5	+1	+1	0	0	13	0	0	−1	−1
6	0	0	+1	+1	14	+1	0	0	+1
7	+1	0	0	−1	15	0	+1	−1	0

Table A.7: AGM-6.

	Points from D_4					Points from D_4			
Integer index	a_1	a_2	a_3	a_4	Integer index	a_1	a_2	a_3	a_4
0	−1	0	0	+1	8	0	0	+1	+1
1	0	−1	−1	0	9	−1	0	0	−1
2	0	0	+1	−1	10	0	−1	+1	0
3	0	+1	−1	0	11	0	+1	+1	0
4	0	0	−1	+1	12	−1	−1	0	0
5	+1	−1	0	0	13	−1	+1	0	0
6	+1	+1	0	0	14	+1	0	0	+1
7	+1	0	0	−1	15	0	0	−1	−1

Table A.8: AGM-7.

	Points from D_4					Points from D_4			
Integer index	a_1	a_2	a_3	a_4	Integer index	a_1	a_2	a_3	a_4
0	−1	−1	0	0	8	0	0	−1	−1
1	−1	+1	0	0	9	0	0	−1	+1
2	+1	−1	0	0	10	0	0	+1	−1
3	+1	+1	0	0	11	0	0	+1	+1
4	0	−1	−1	0	12	−1	0	0	−1
5	0	−1	+1	0	13	−1	0	0	+1
6	0	+1	−1	0	14	+1	0	0	−1
7	0	+1	+1	0	15	+1	0	0	+1

Table A.9: AGM-8.

	Points from D_4					Points from D_4			
Integer index	a_1	a_2	a_3	a_4	Integer index	a_1	a_2	a_3	a_4
0	0	−1	−1	0	8	0	+1	+1	0
1	−1	−1	0	0	9	+1	+1	0	0
2	−1	0	0	−1	10	+1	0	0	+1
3	0	0	−1	−1	11	0	0	+1	+1
4	0	−1	+1	0	12	0	+1	−1	0
5	−1	+1	0	0	13	+1	−1	0	0
6	−1	0	0	+1	14	+1	0	0	−1
7	0	0	−1	+1	15	0	0	+1	−1

Table A.10: AGM-9.

	Points from D_4					Points from D_4			
Integer index	a_1	a_2	a_3	a_4	Integer index	a_1	a_2	a_3	a_4
0	+1	+1	0	0	8	0	+1	−1	0
1	+1	0	0	−1	9	0	0	+1	+1
2	+1	0	0	+1	10	0	0	+1	−1
3	−1	+1	0	0	11	0	−1	−1	0
4	0	+1	+1	0	12	+1	−1	0	0
5	0	0	−1	+1	13	−1	0	0	−1
6	0	0	−1	−1	14	−1	0	0	+1
7	0	−1	+1	0	15	−1	−1	0	0

Table A.11: AGM-10.

	Points from D_4					Points from D_4			
Integer index	a_1	a_2	a_3	a_4	Integer index	a_1	a_2	a_3	a_4
0	−1	−1	0	0	8	0	+1	−1	0
1	+1	0	0	−1	9	0	0	+1	+1
2	+1	0	0	+1	10	0	0	+1	−1
3	−1	+1	0	0	11	0	−1	−1	0
4	0	+1	+1	0	12	+1	−1	0	0
5	0	0	−1	+1	13	−1	0	0	−1
6	0	0	−1	−1	14	−1	0	0	+1
7	0	−1	+1	0	15	+1	+1	0	0

Appendix B

EXIT Charts of Various Bit-based Turbo-detected STBC-SP Schemes

In this appendix, we present the EXIT charts of various bit-based turbo-detected STBC-SP schemes employing all of the mapping schemes described in Appendix A in combination with RSC codes and binary LDPC codes, when communicating over a SPSI-correlated Rayleigh fading channel having a normalized Doppler frequency of $f_D = 0.1$.

B.1 EXIT Charts of RSC-coded STBC-SP Schemes

We present EXIT charts of the RSC-coded STBC-SP Scheme shown in Figure 3.1, when using a single receive antenna and employing half-rate RSC codes having the parameters outlined in Table B.1, where G_r is the recursive polynomial. Furthermore, I_{A_M} and I_{E_M} were defined in Section 3.4.1, while I_{A_D} and I_{E_D} were defined in Section 3.4.2.

Table B.1: Parameters of the RSC codes.

		Generator polynomials (in octals)	
Code number	Constraint length	G_r	G
RSC-1	3	05	07
RSC-2	4	15	17
RSC-3	5	35	23
RSC-4	6	53	75
RSC-5	7	133	171
RSC-6	8	247	371
RSC-7	9	561	753

Near-Capacity Multi-Functional MIMO Systems L. Hanzo, O. R. Alamri, M. El-Hajjar and N. Wu

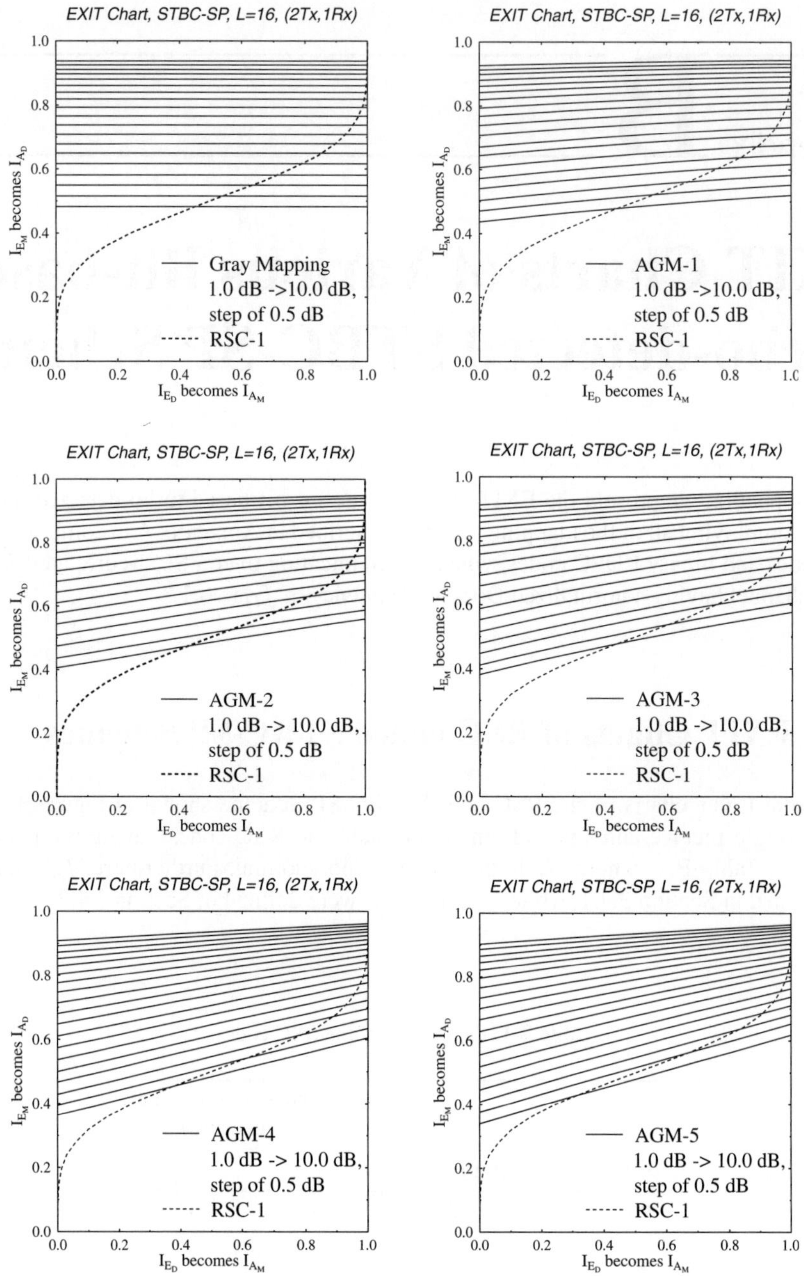

Figure B.1: RSC-1 coded STBC-SP schemes.

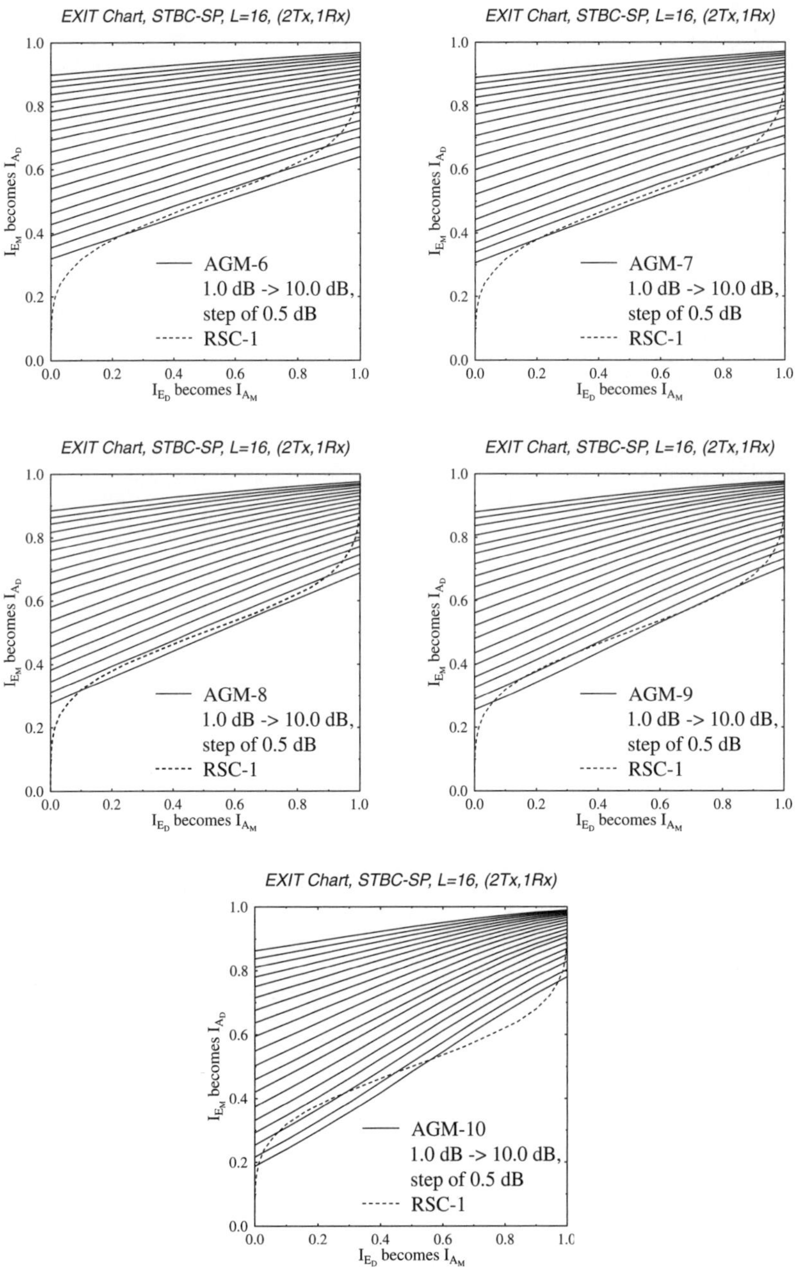

Figure B.1: Continued.

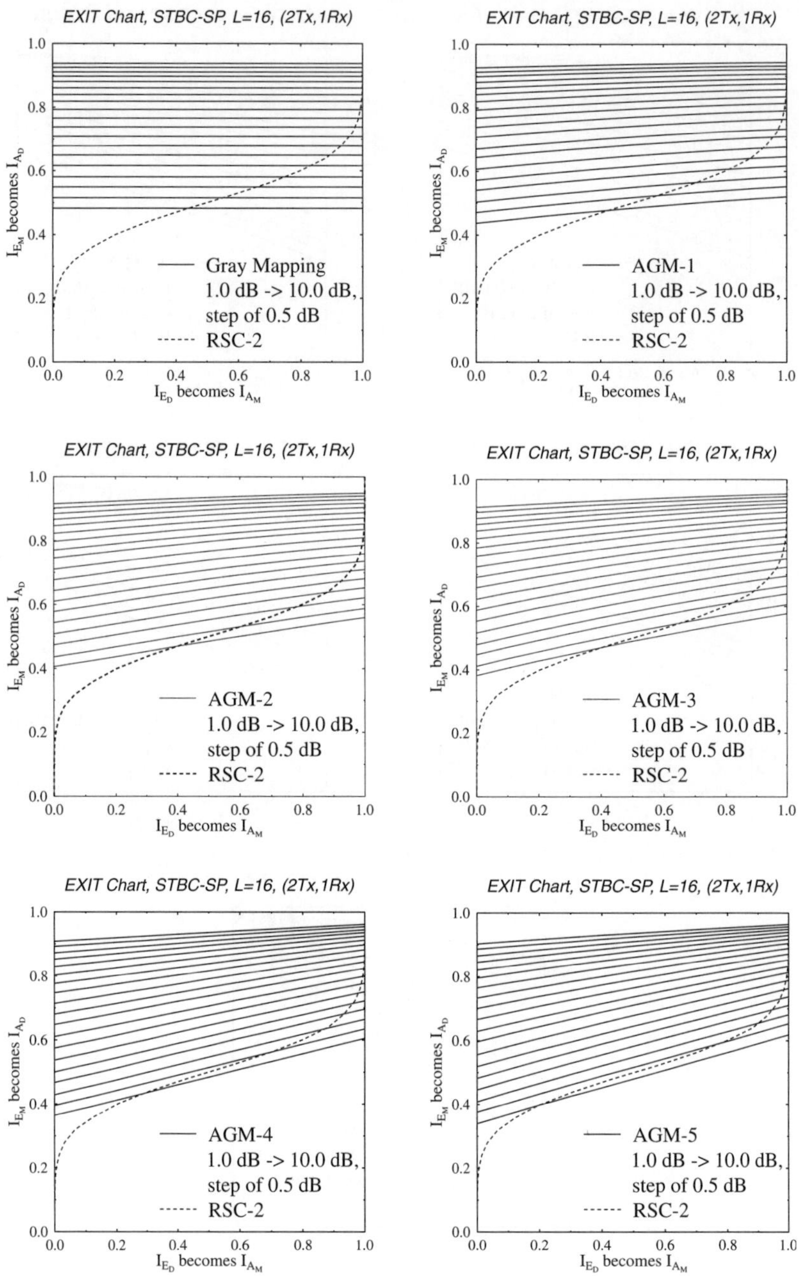

Figure B.2: RSC-2 coded STBC-SP schemes.

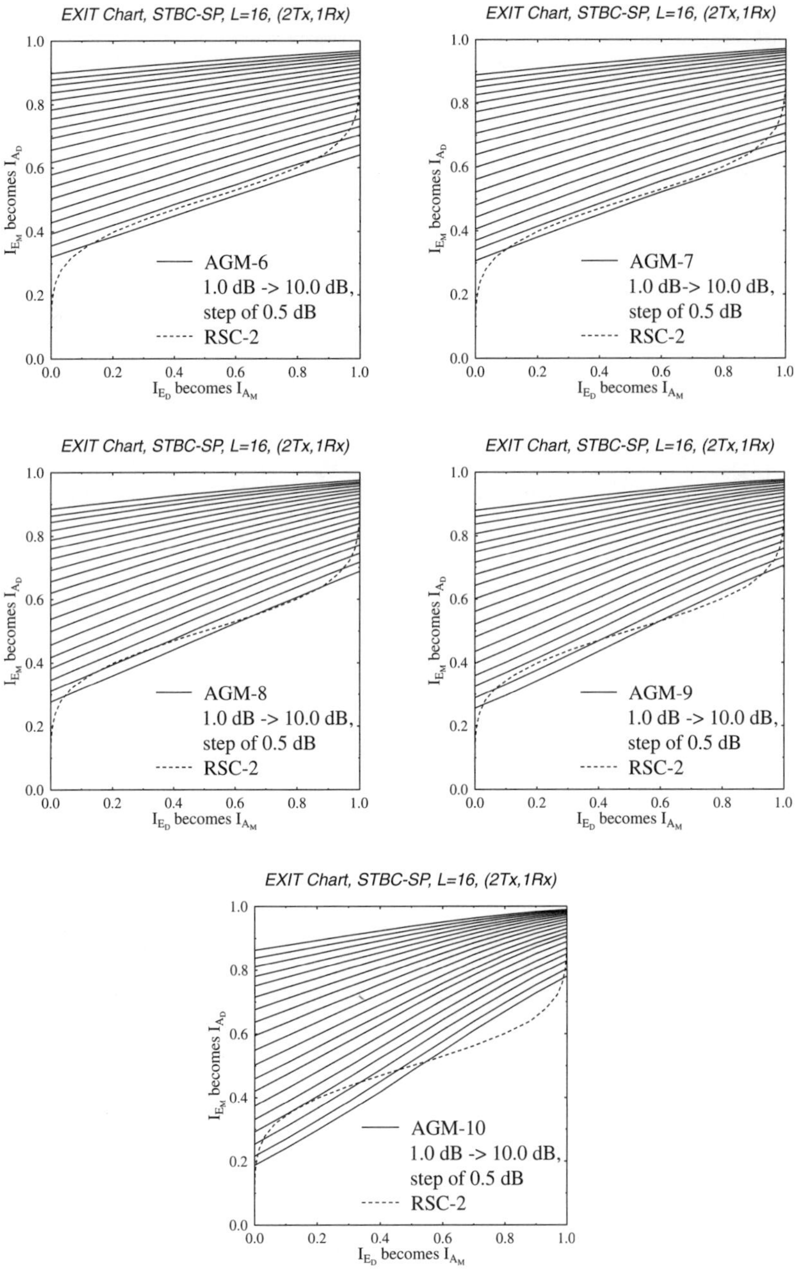

Figure B.2: Continued.

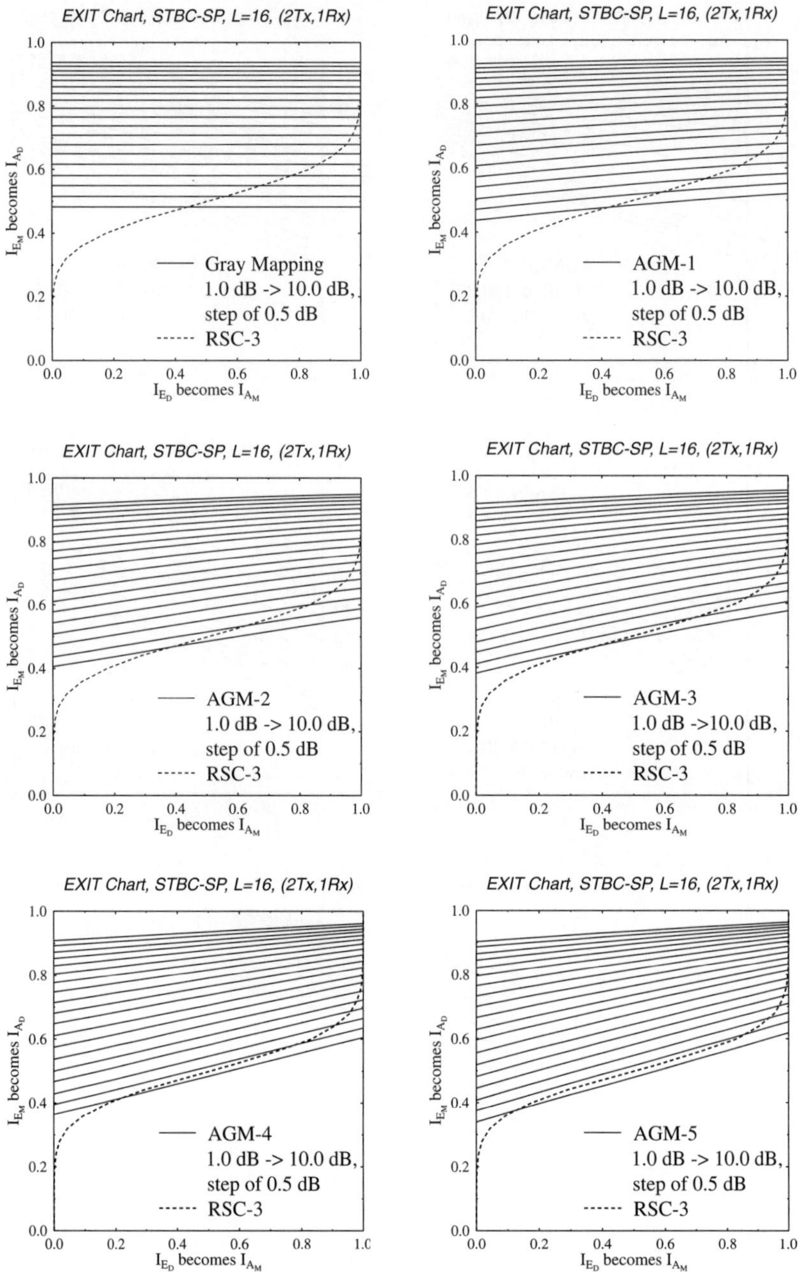

Figure B.3: RSC-3 coded STBC-SP schemes.

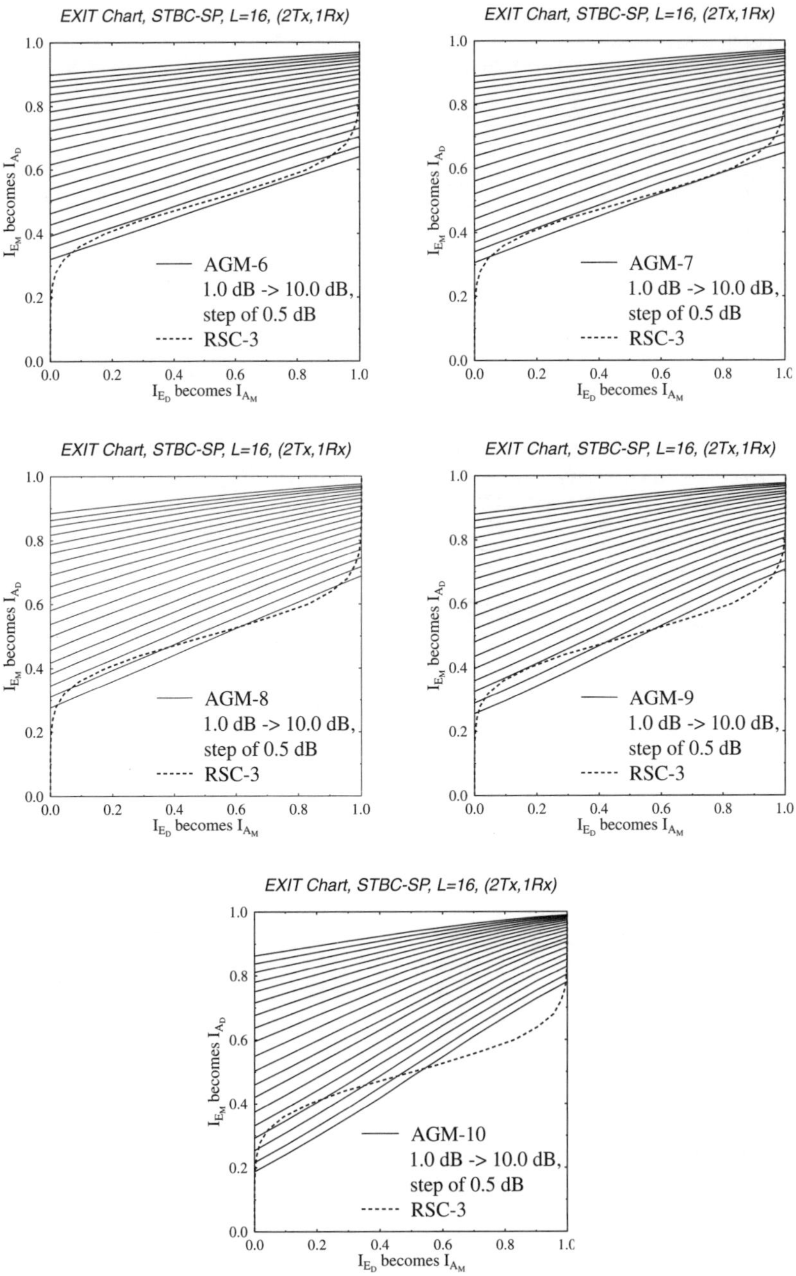

Figure B.3: Continued.

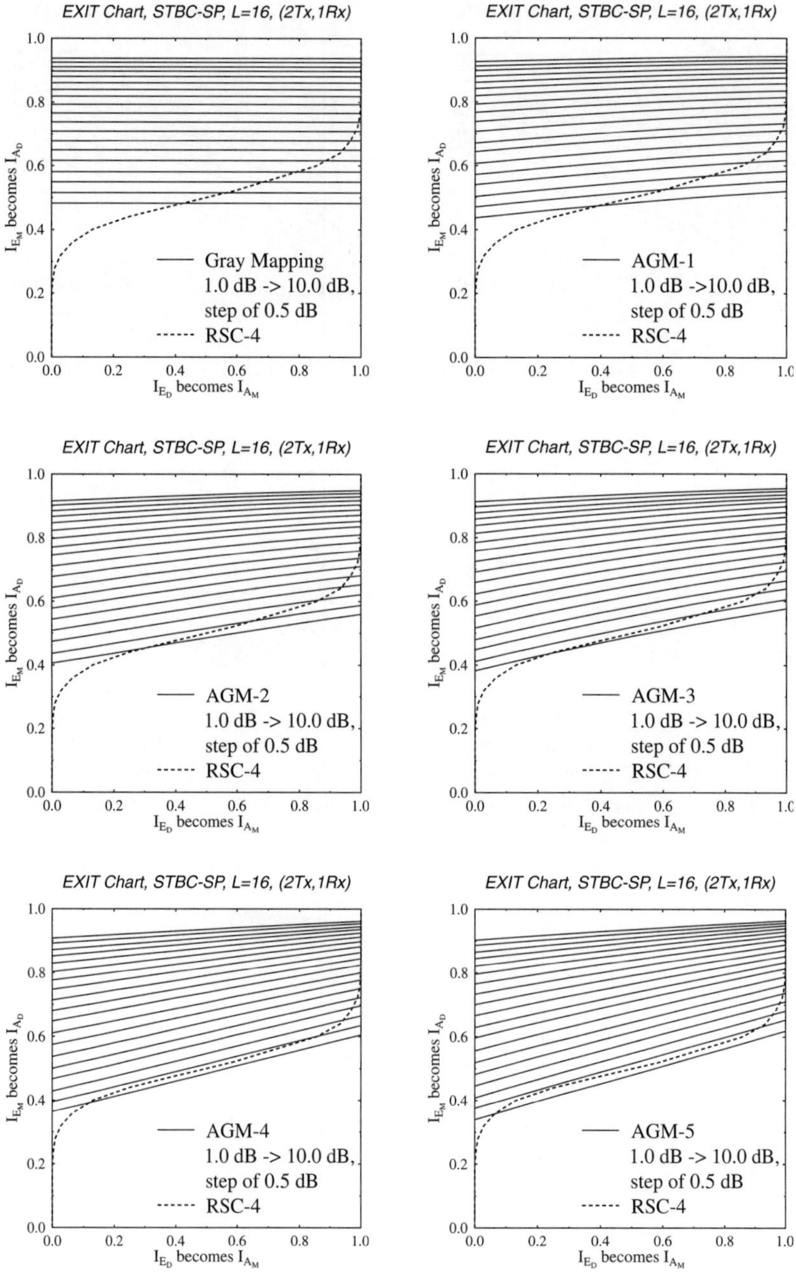

Figure B.4: RSC-4 coded STBC-SP schemes.

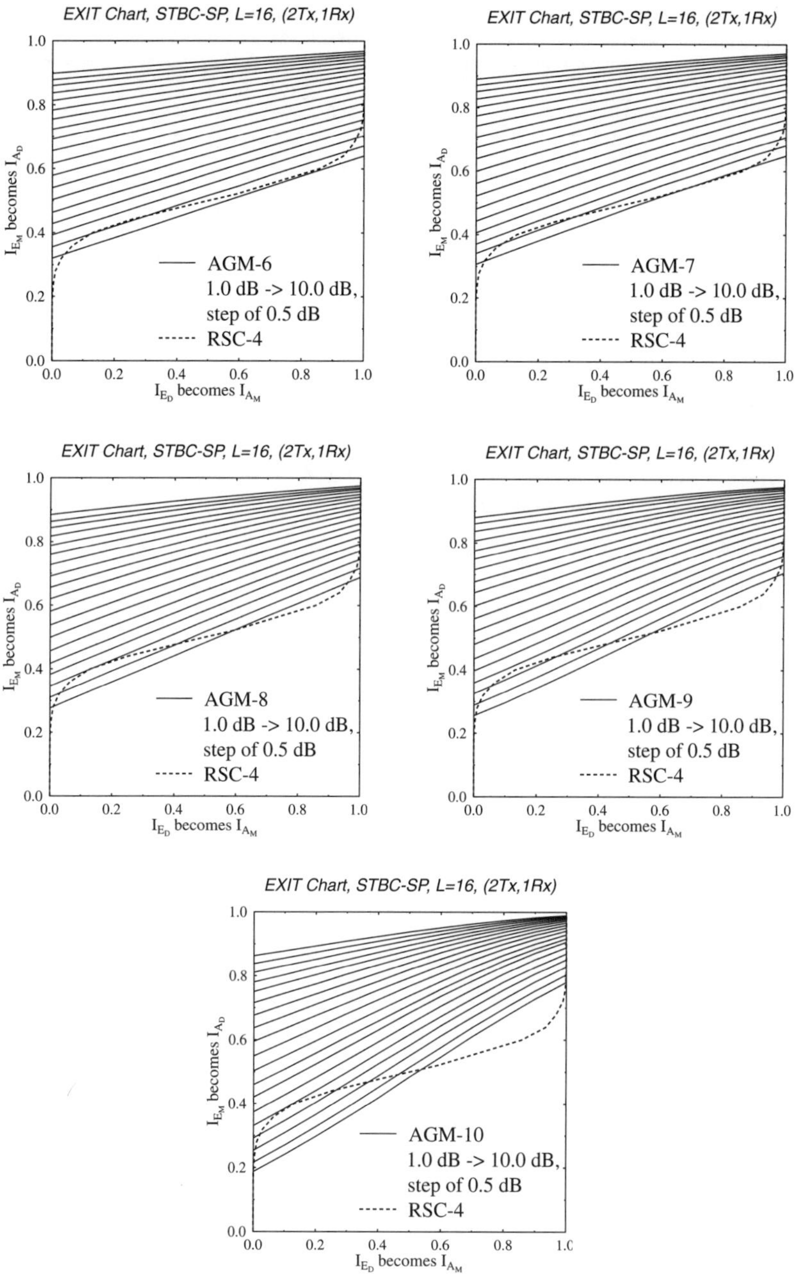

Figure B.4: Continued.

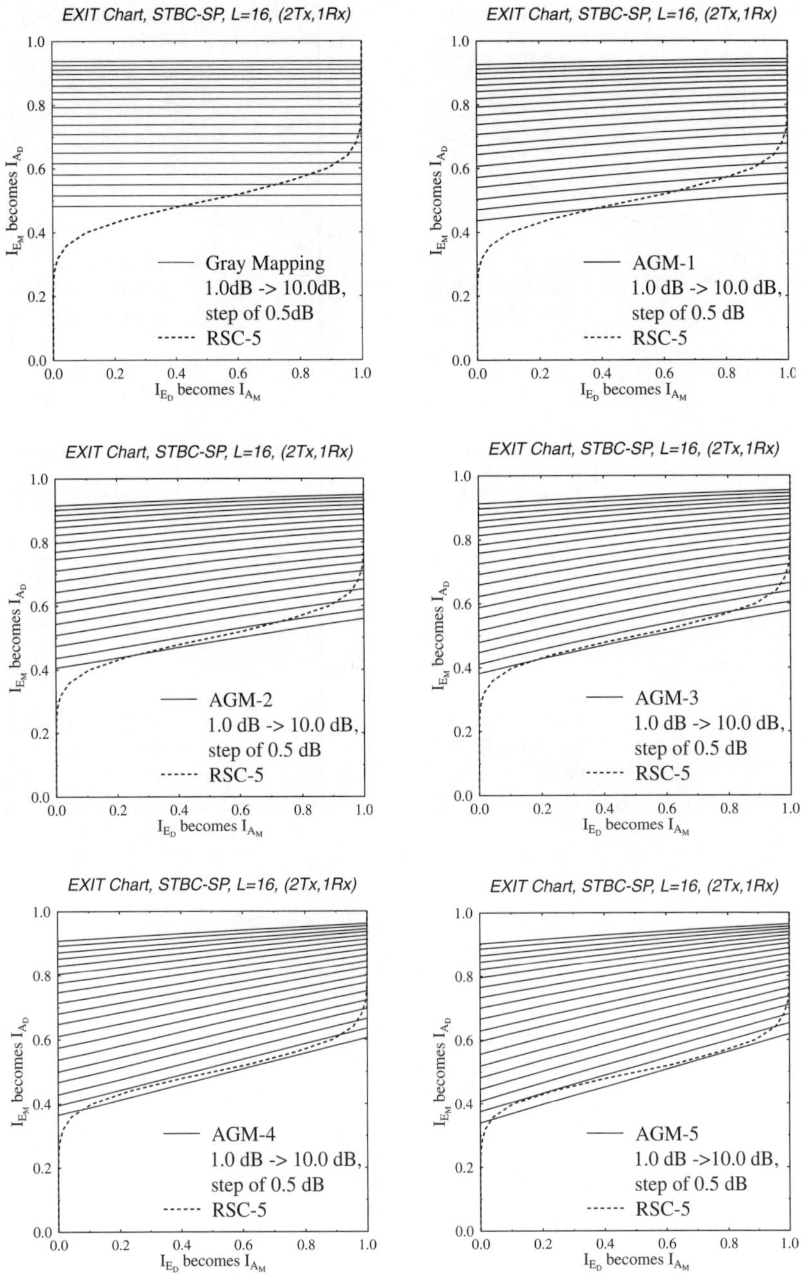

Figure B.5: RSC-5 coded STBC-SP schemes.

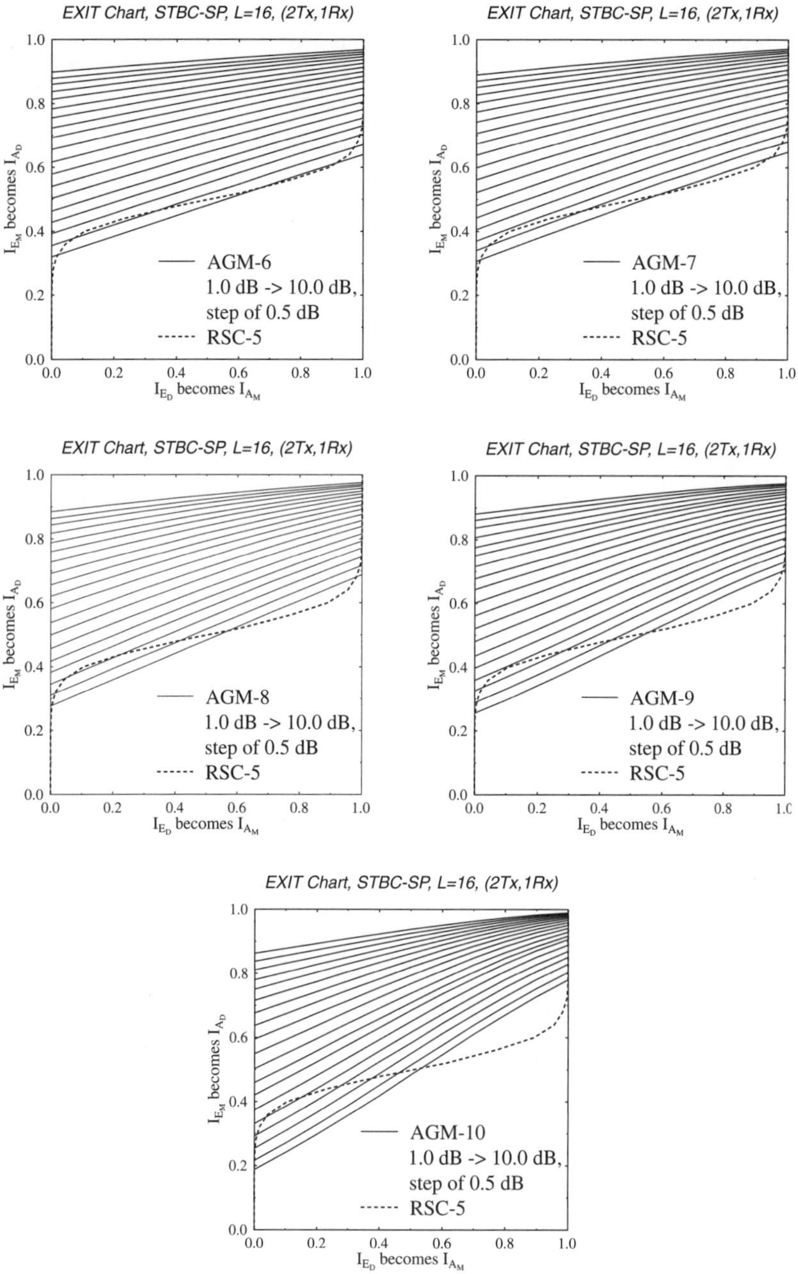

Figure B.5: Continued.

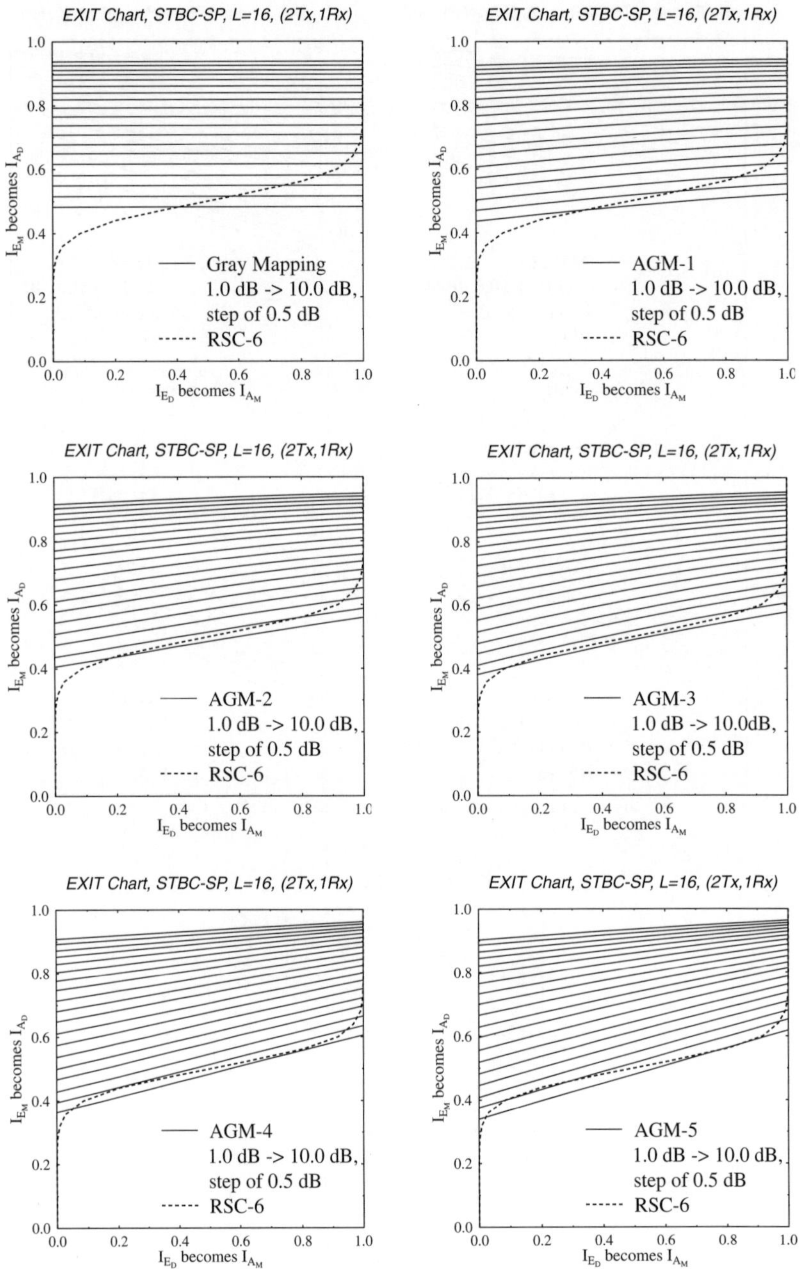

Figure B.6: RSC-6 coded STBC-SP schemes.

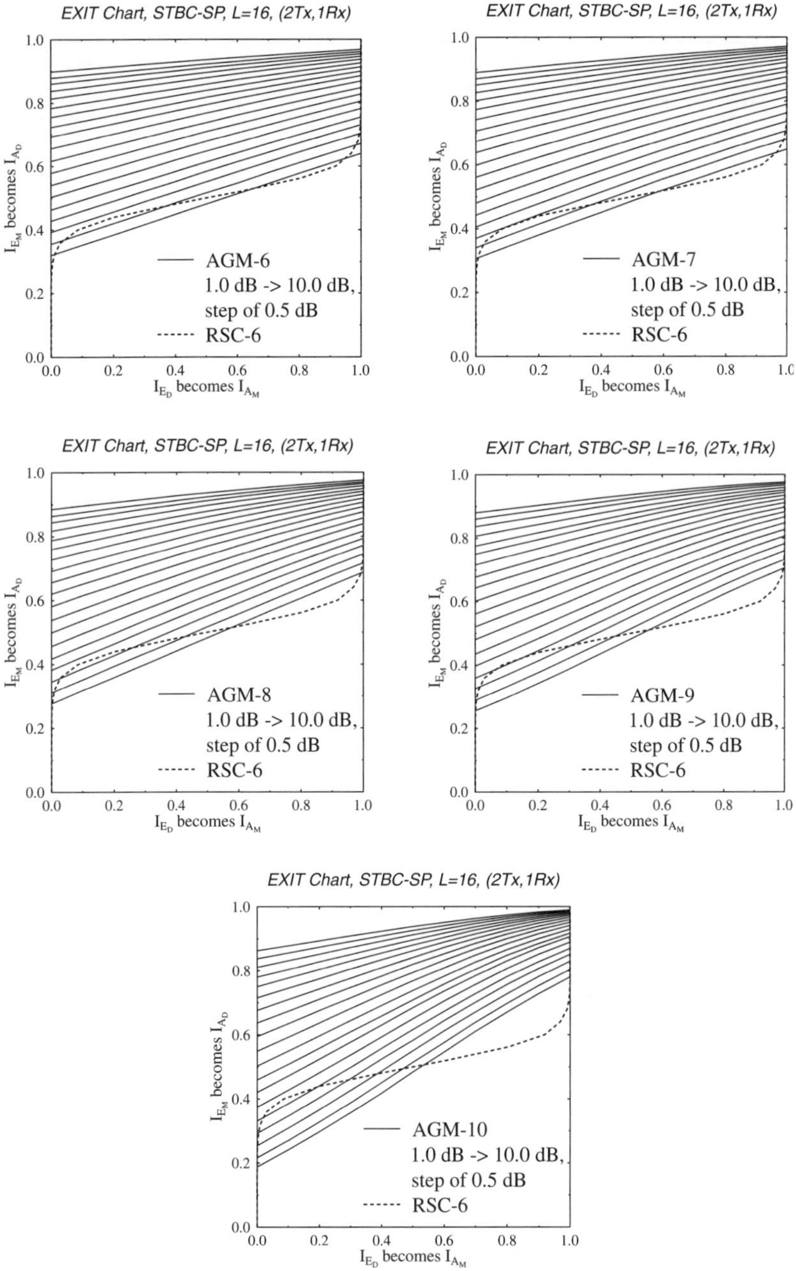

Figure B.6: Continued.

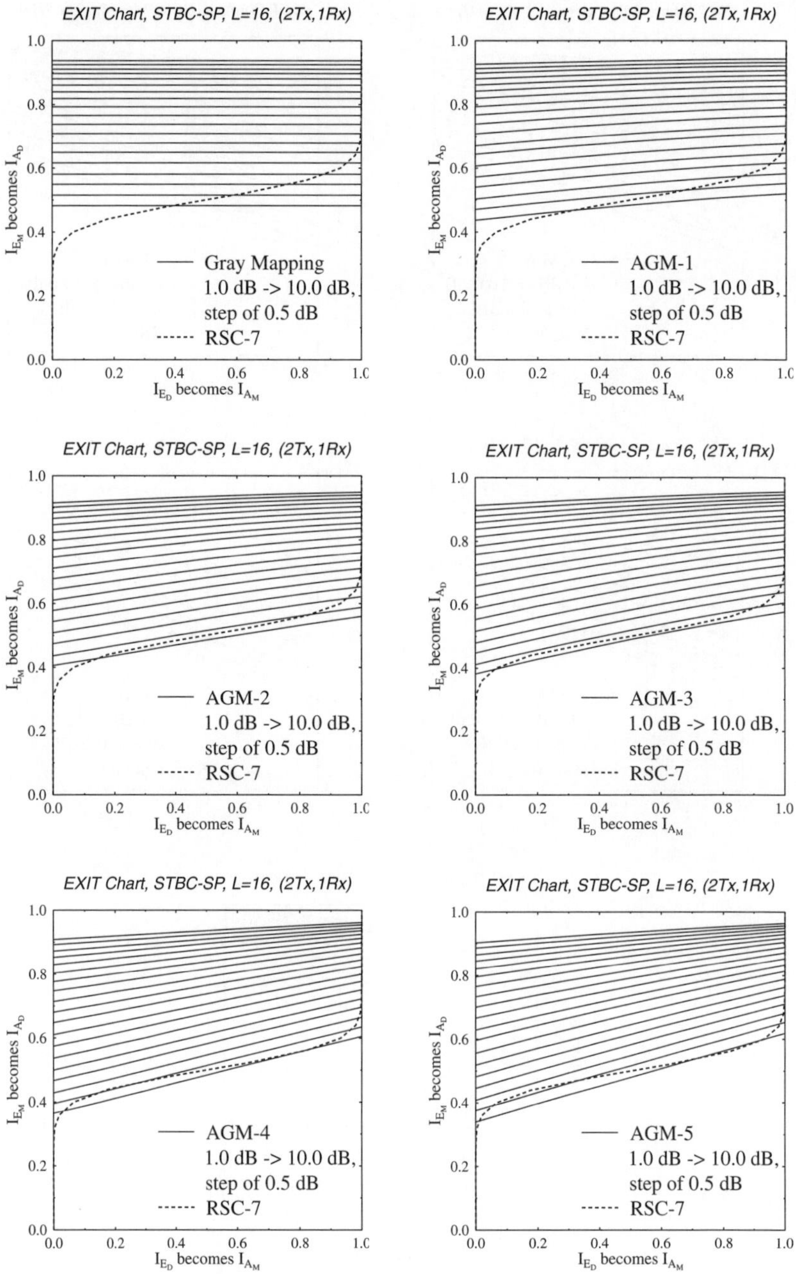

Figure B.7: RSC-7 coded STBC-SP schemes.

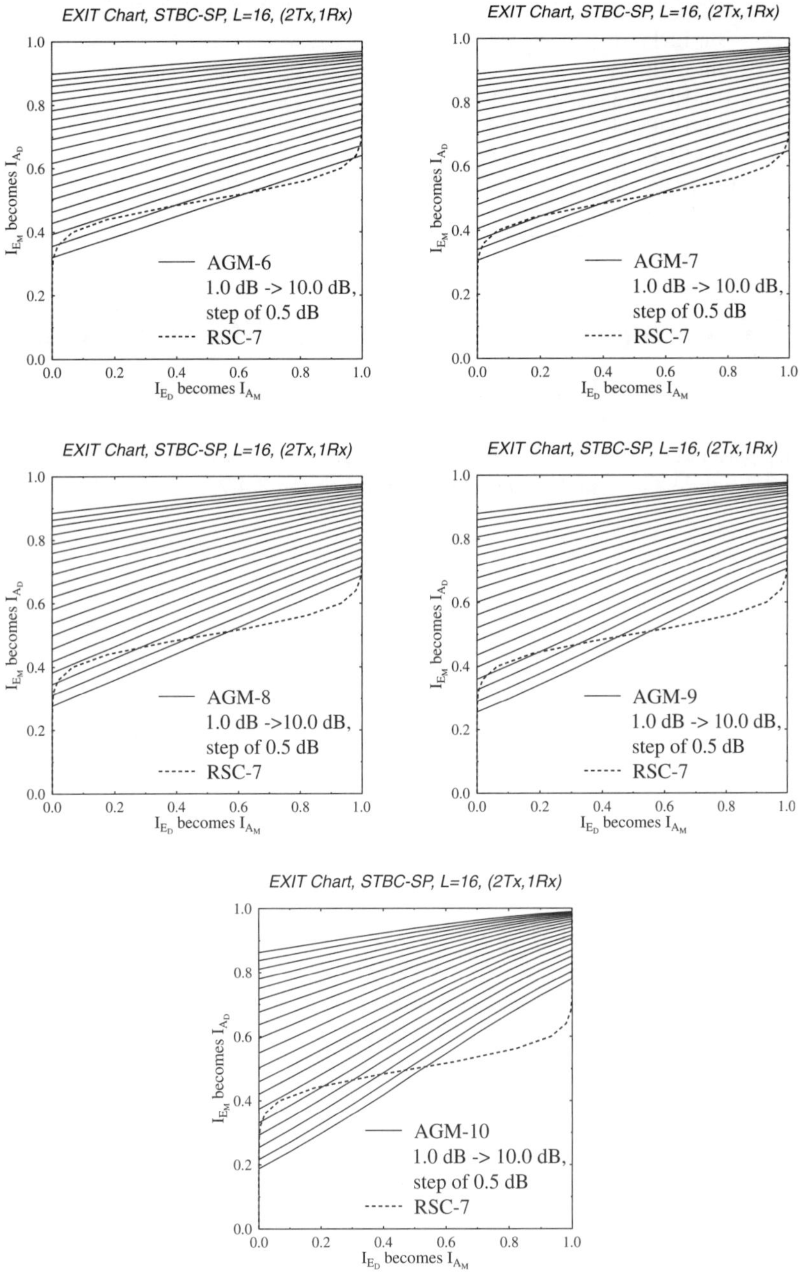

Figure B.7: Continued.

Table B.2: Parameters of the binary LDPC codes.

Code number	Column weight	Number of internal LDPC iterations
Bin-LDPC-1	2.5	1
Bin-LDPC-2	2.5	2
Bin-LDPC-3	2.5	3
Bin-LDPC-4	2.5	4
Bin-LDPC-5	2.5	5
Bin-LDPC-10	2.5	10
Bin-LDPC-20	2.5	20

B.2 EXIT Charts of LDPC-coded STBC-SP Schemes

This section provides the EXIT charts for the binary LDPC-coded STBC-SP schemes of Figure 3.2, when using a single receive antenna and employing half-rate binary LDPC codes having the parameters outlined in Table B.2. Furthermore, I_{A_M} and I_{E_M} were defined in Section 3.4.1, while I_{A_D} and I_{E_D} were defined in Section 3.4.2.

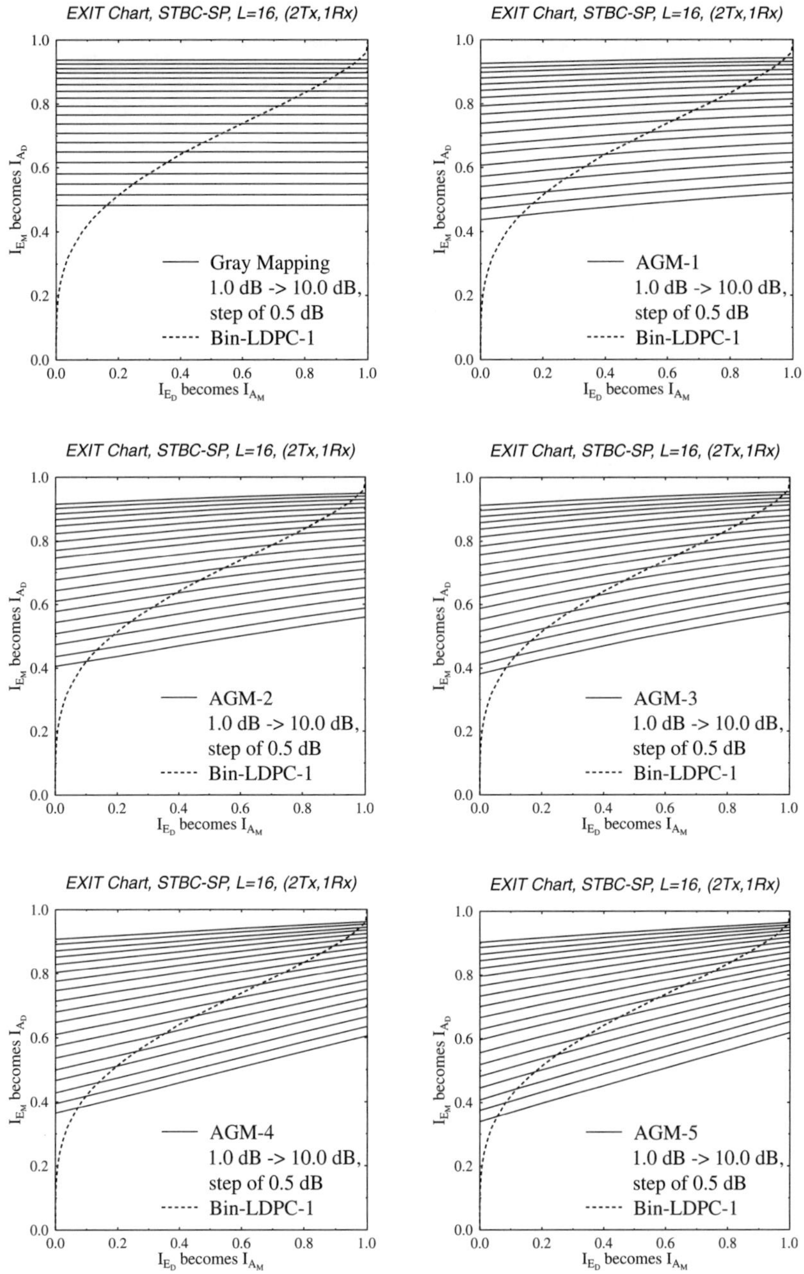

Figure B.8: Bin-LDPC-1 coded STBC-SP schemes.

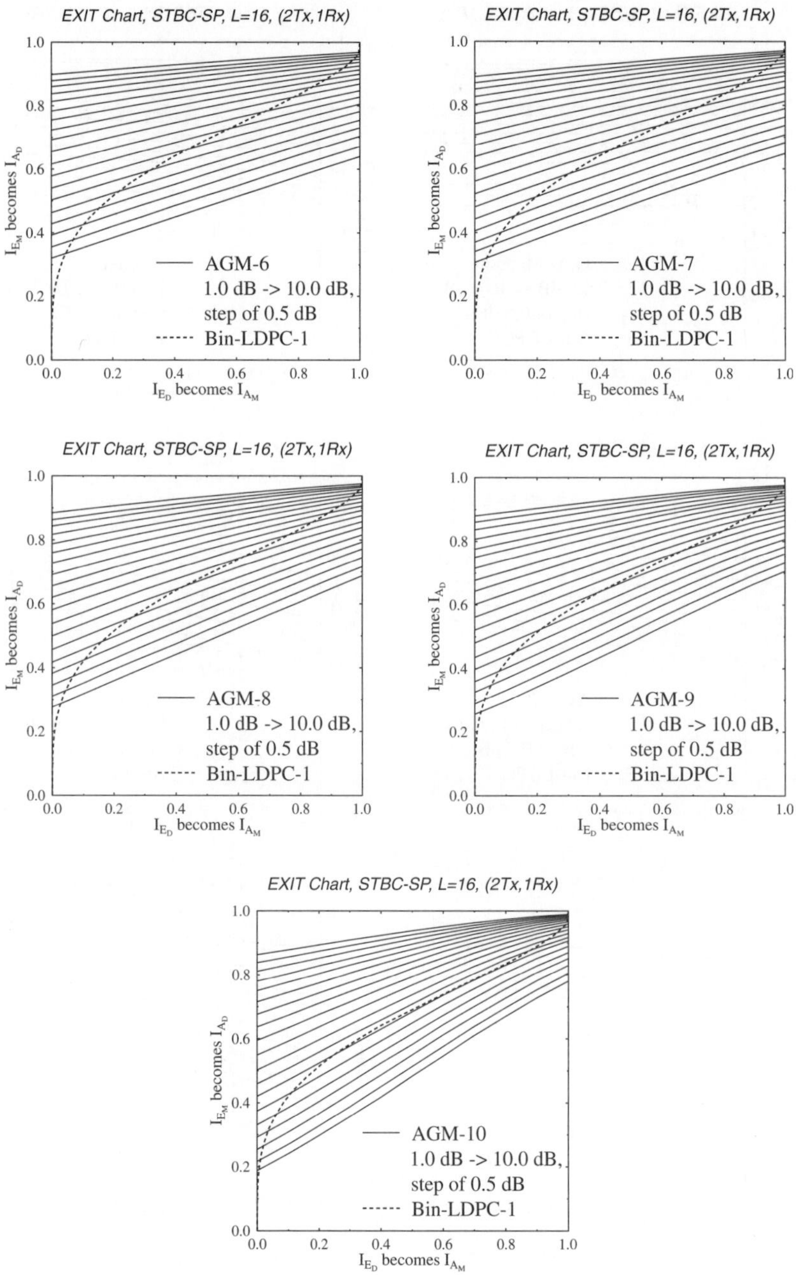

Figure B.8: Continued.

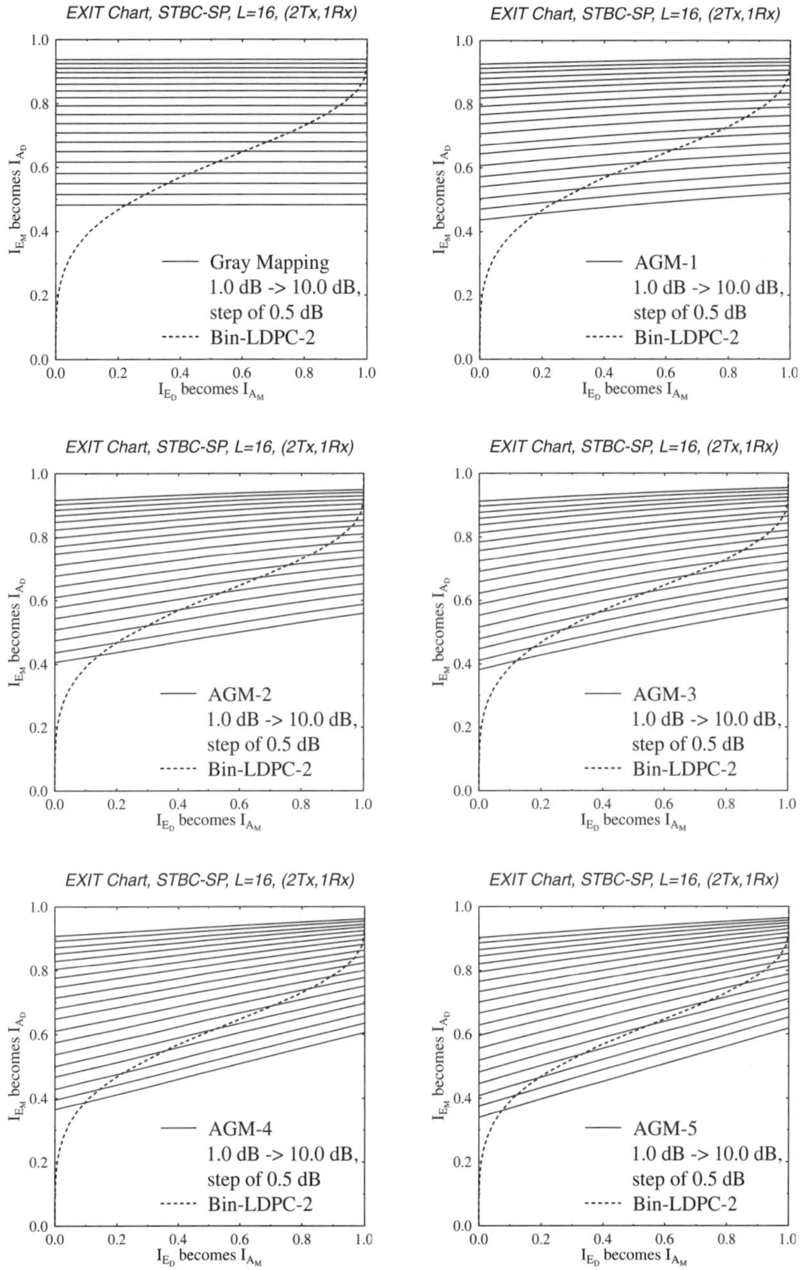

Figure B.9: Bin-LDPC-2 coded STBC-SP schemes.

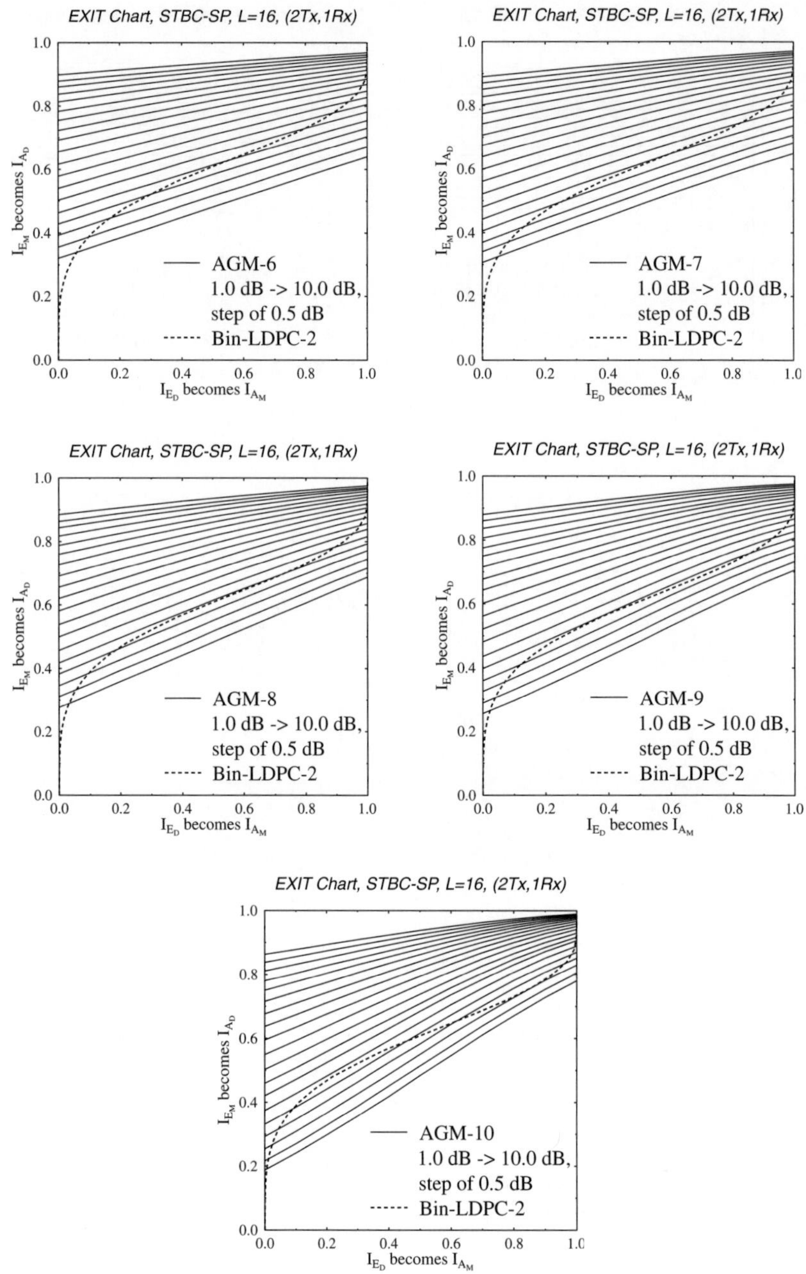

Figure B.9: Continued.

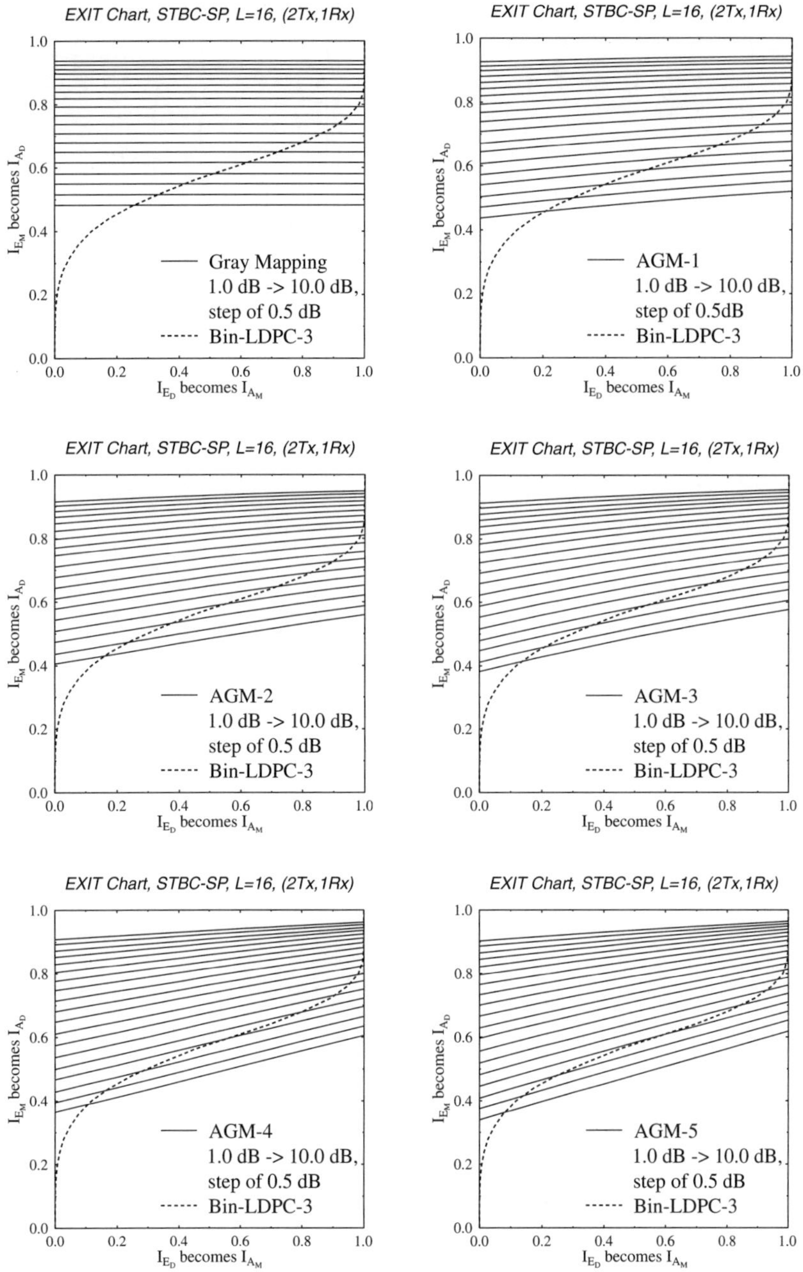

Figure B.10: Bin-LDPC-3 coded STBC-SP schemes.

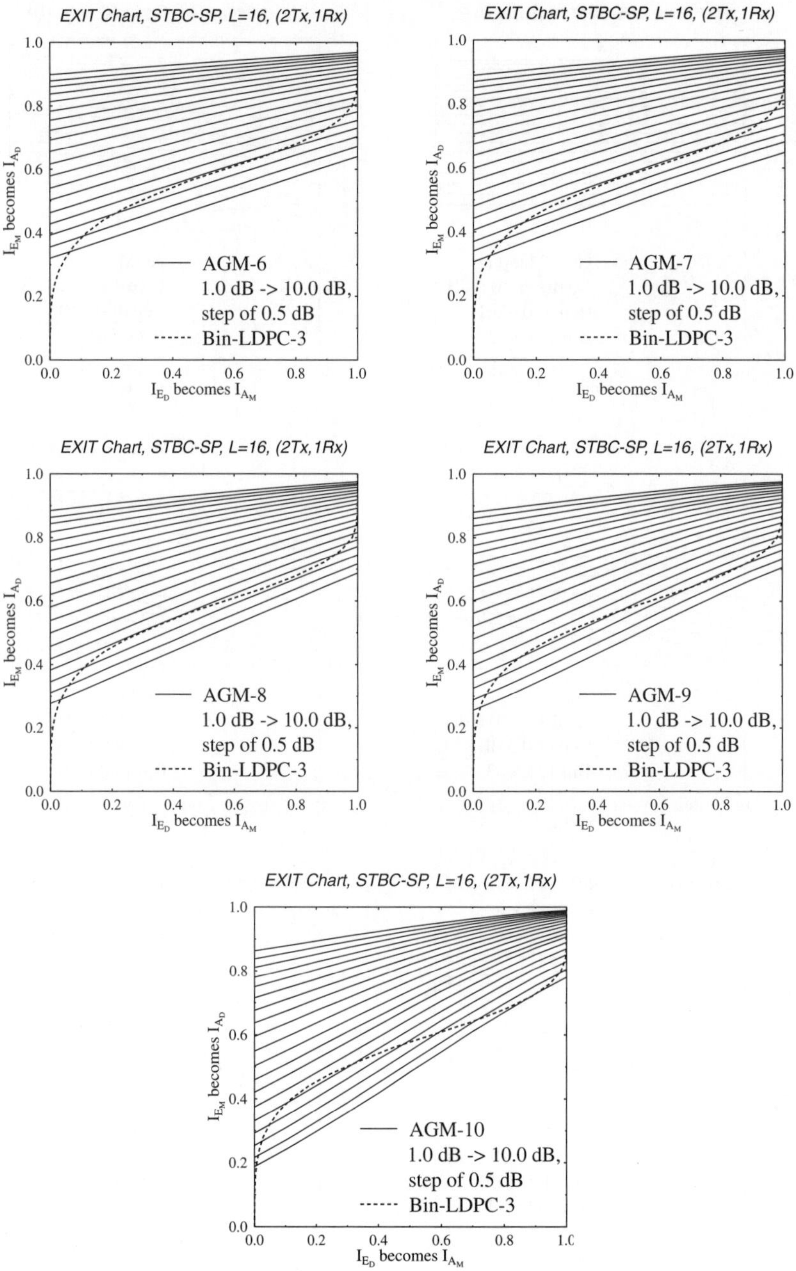

Figure B.10: Continued.

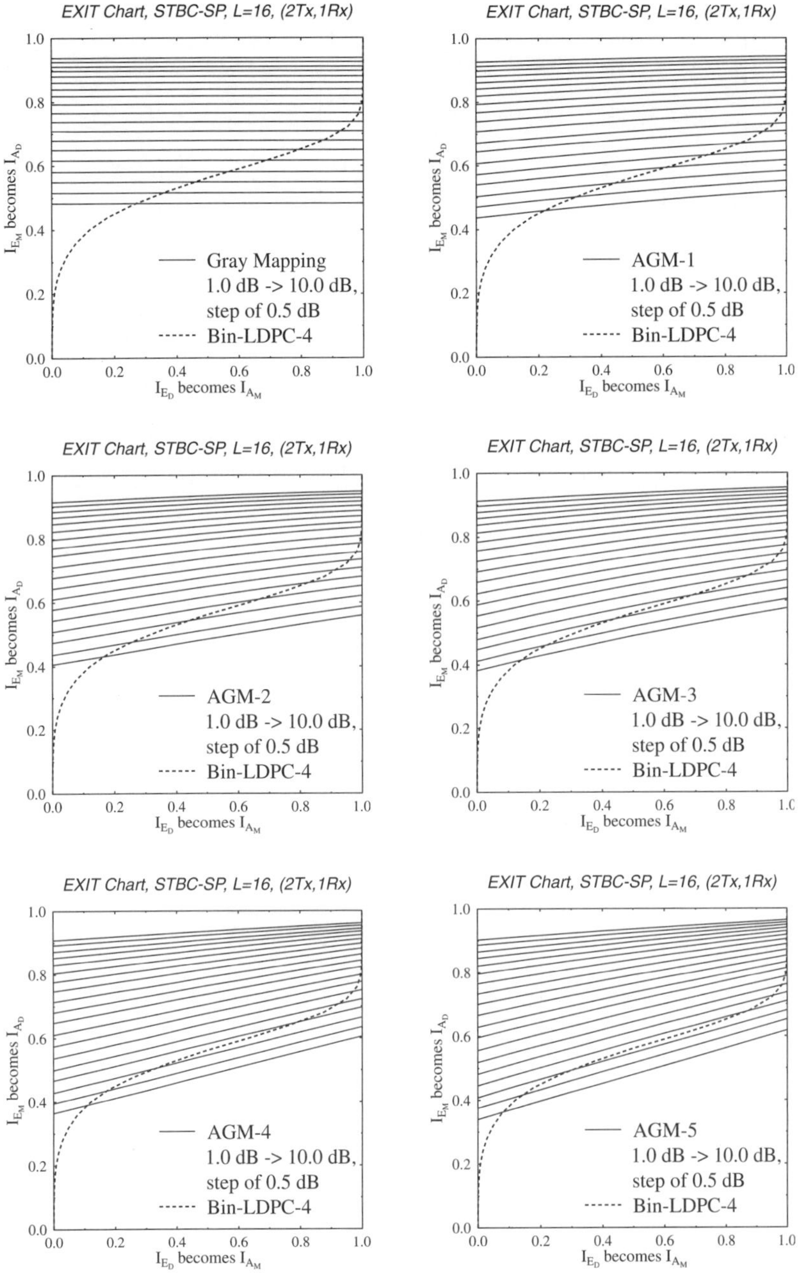

Figure B.11: Bin-LDPC-4 coded STBC-SP schemes.

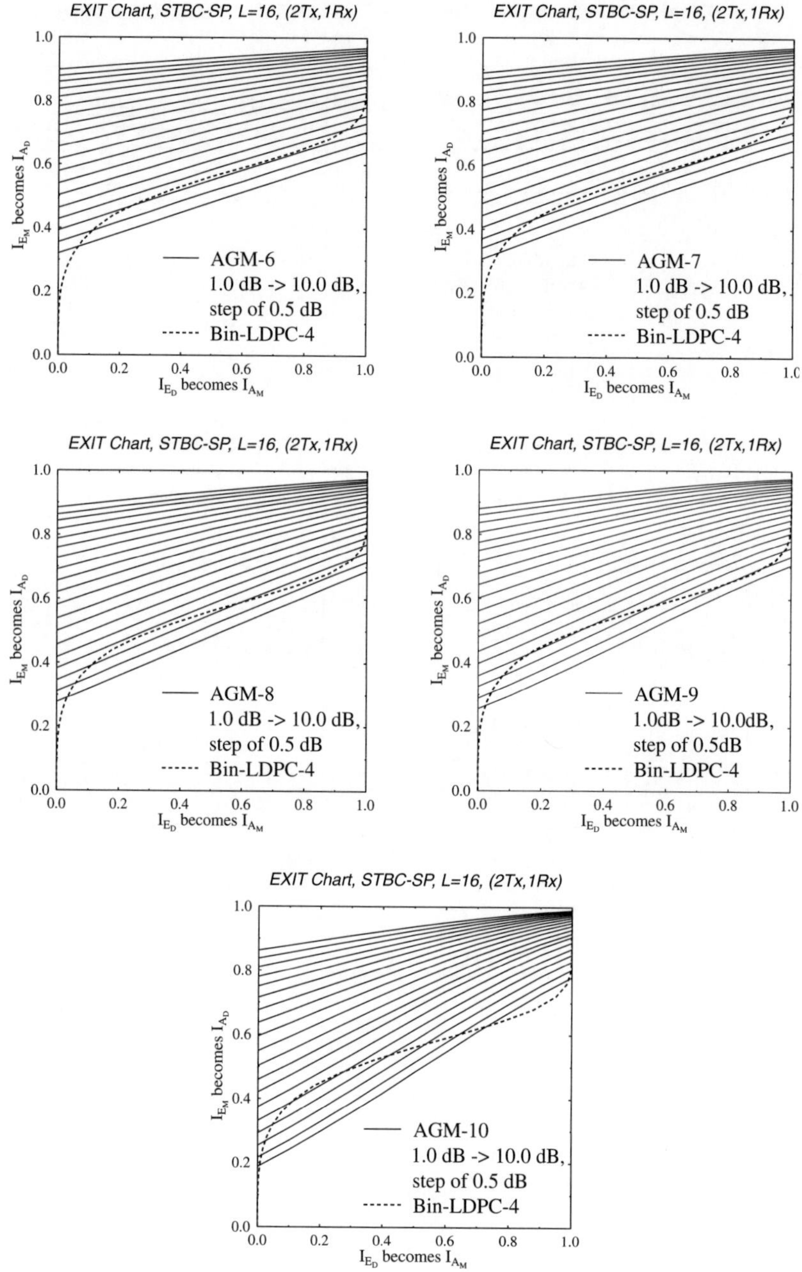

Figure B.11: Continued.

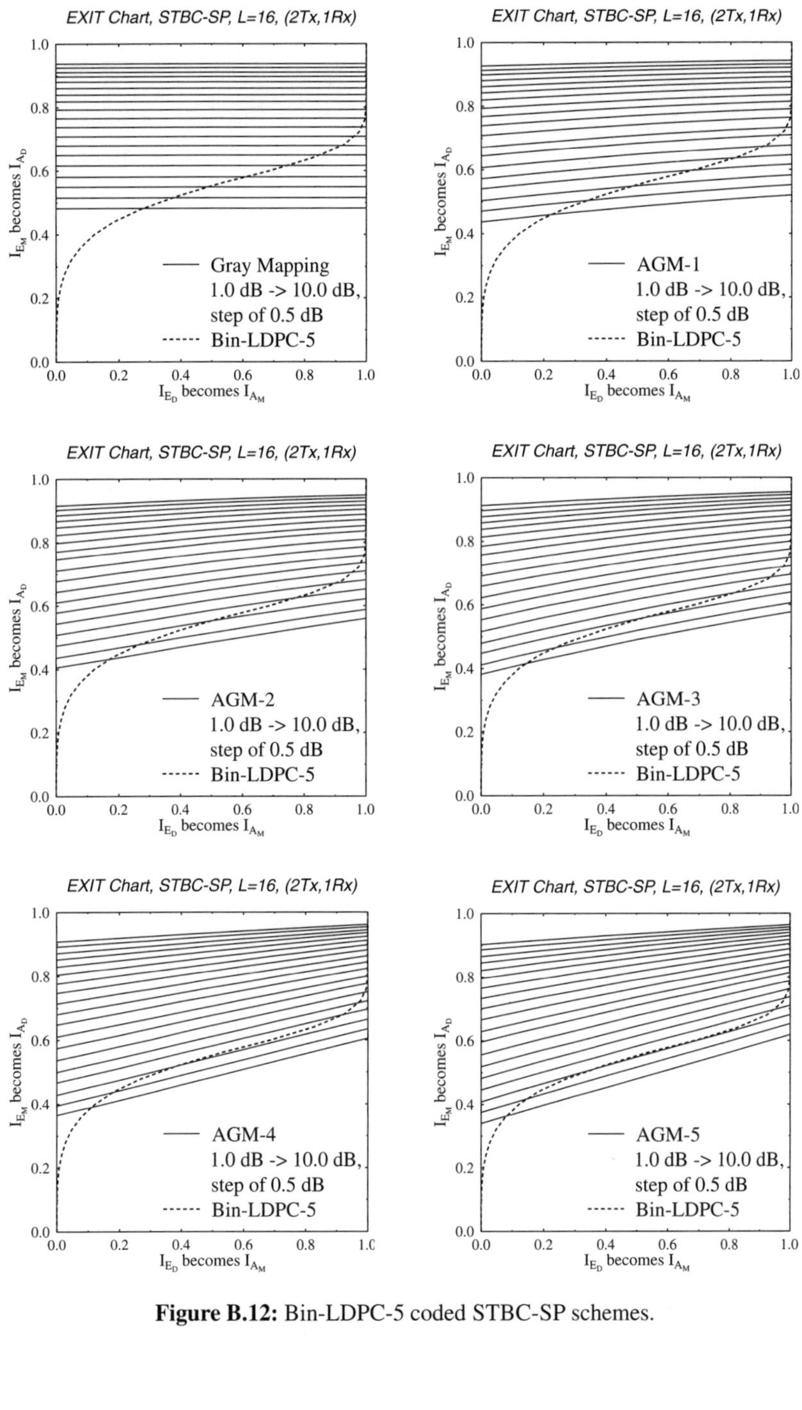

Figure B.12: Bin-LDPC-5 coded STBC-SP schemes.

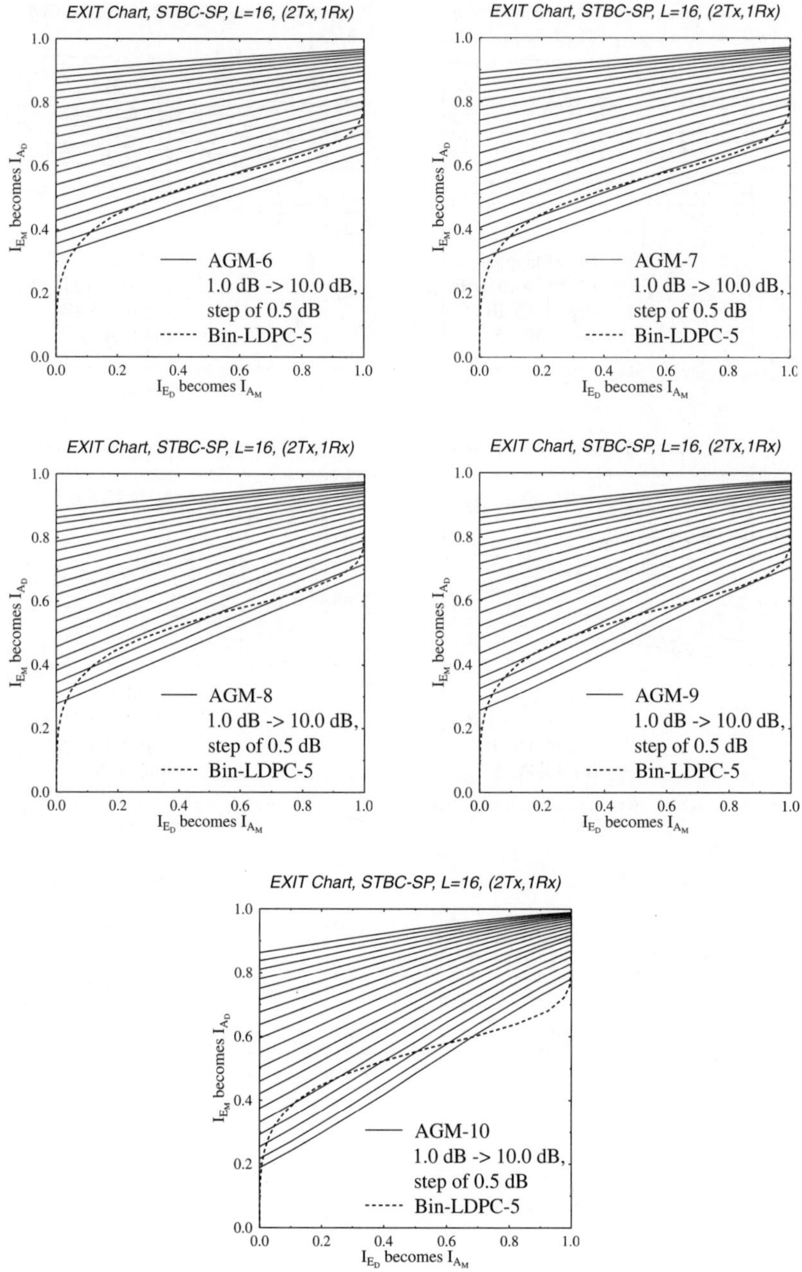

Figure B.12: Continued.

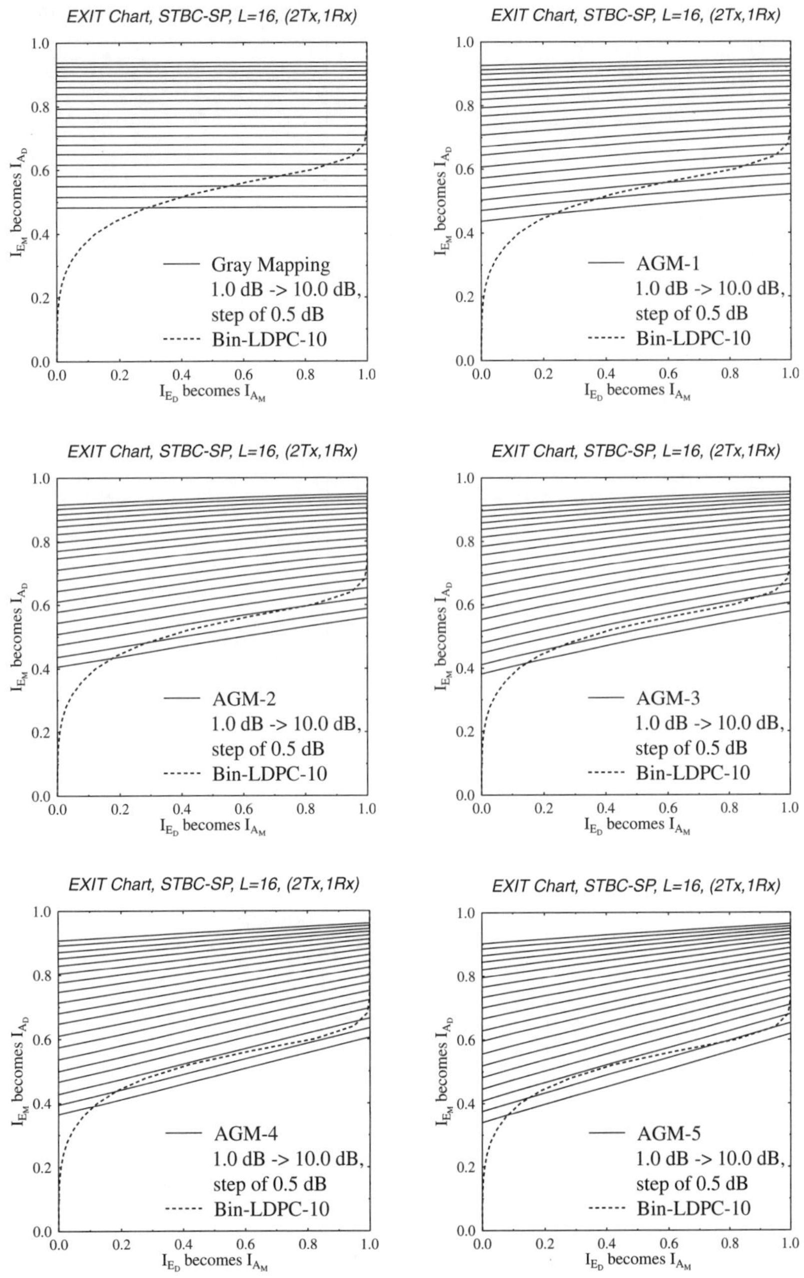

Figure B.13: Bin-LDPC-10 coded STBC-SP schemes.

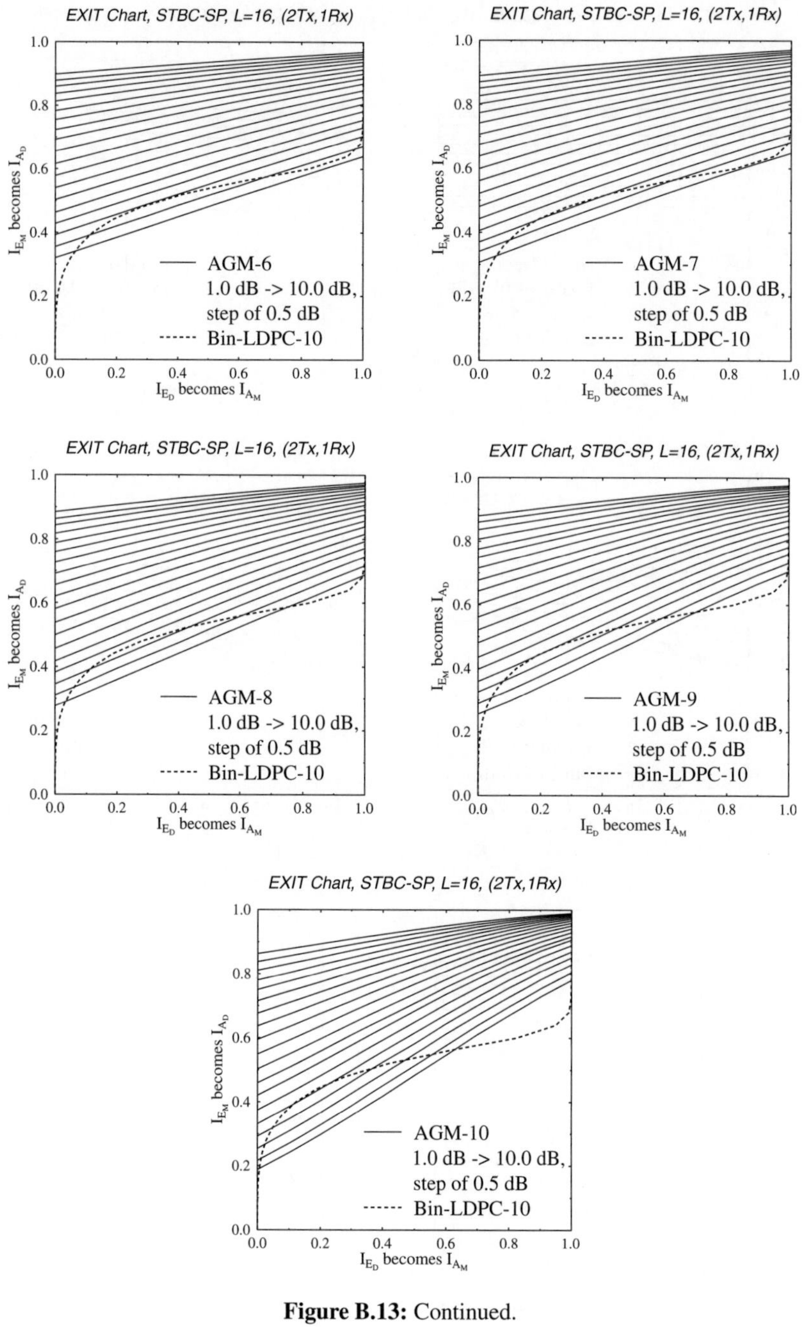

Figure B.13: Continued.

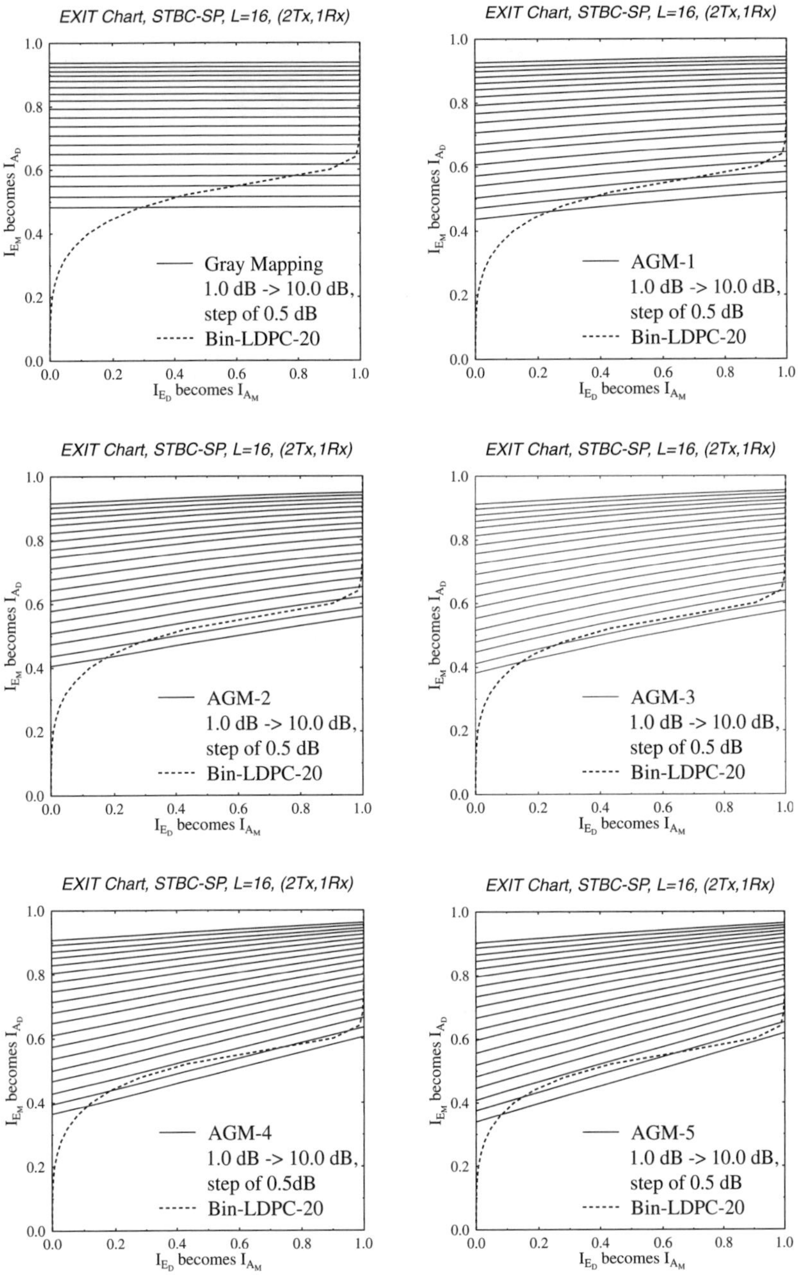

Figure B.14: Bin-LDPC-20 coded STBC-SP schemes.

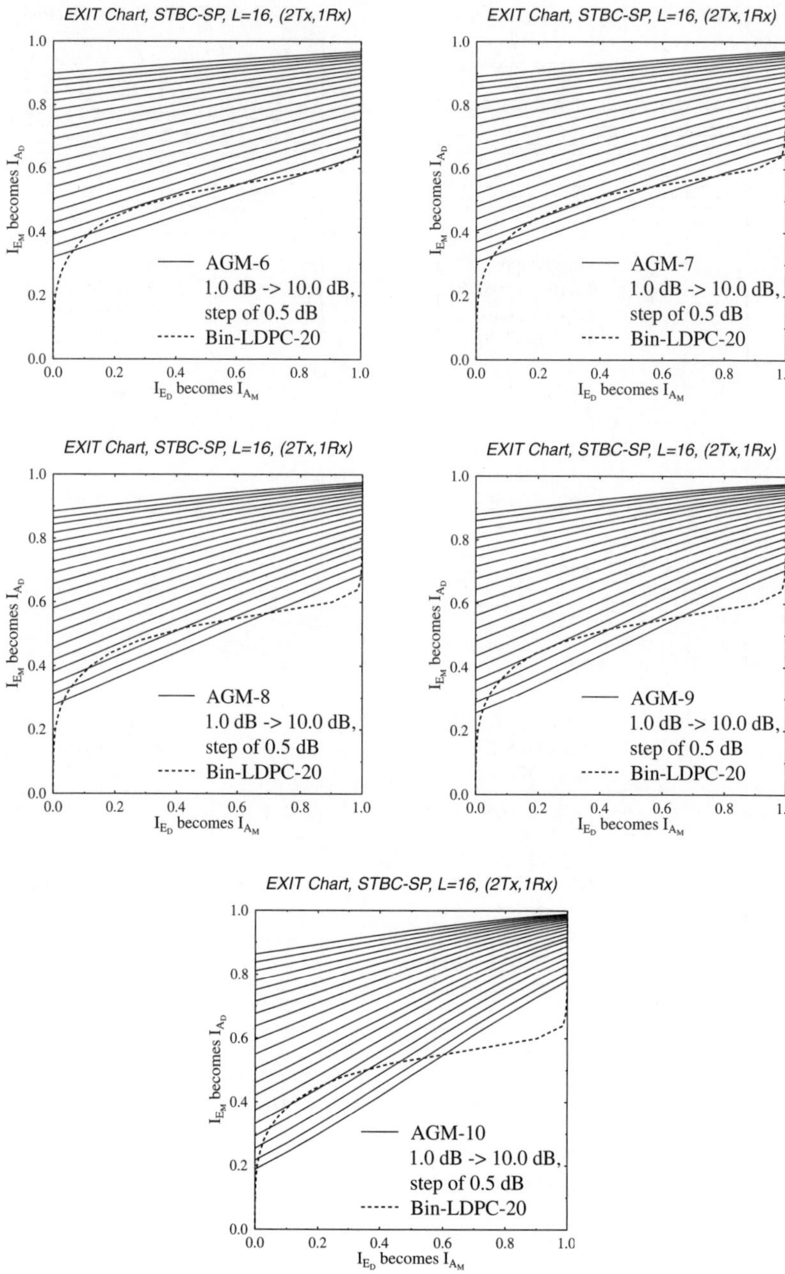

Figure B.14: Continued.

Appendix C

EXIT Charts of Various Bit-based Turbo-detected DSTBC-SP Schemes

In this appendix, we present the EXIT charts of various bit-based turbo-detected DSTBC-SP schemes employing all of the mapping schemes described in Appendix A in combination with RSC codes when communicating over a SPSI-correlated Rayleigh fading channel having a normalized Doppler frequency of $f_D = 0.01$. We present EXIT charts of the RSC-coded DSTBC-SP Scheme shown in Figure 4.18, when using a single receive antenna and employing half-rate RSC codes having the parameters outlined in Table C.1, where G_r is the recursive polynomial. Furthermore, I_{A_M} and I_{E_M} were defined in Section 3.4.1, while I_{A_D} and I_{E_D} were defined in Section 3.4.2.

Table C.1: Parameters of the RSC codes.

		Generator polynomials (in octals)	
Code number	Constraint length	G_r	G
RSC-1	3	05	07
RSC-2	4	15	17
RSC-3	5	35	23
RSC-4	6	53	75
RSC-5	7	133	171
RSC-6	8	247	371
RSC-7	9	561	753

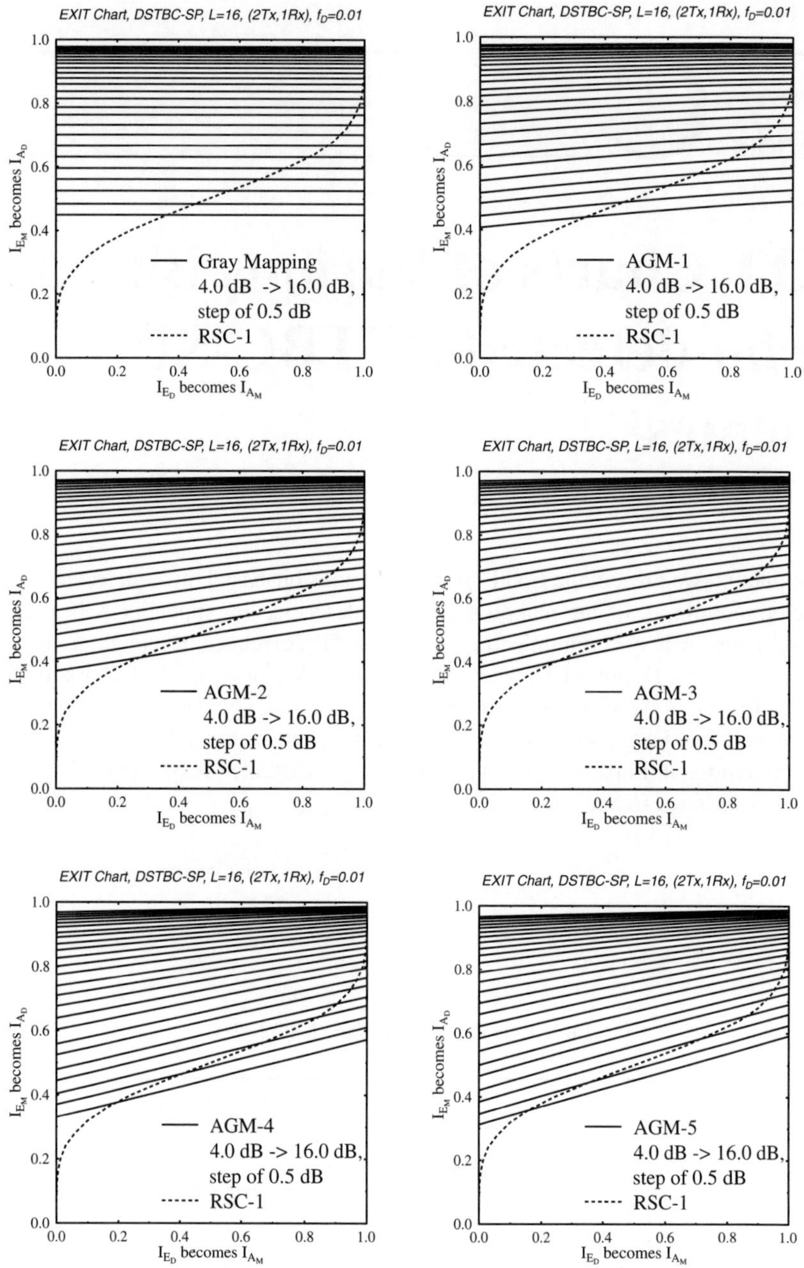

Figure C.1: RSC-1 coded DSTBC-SP schemes.

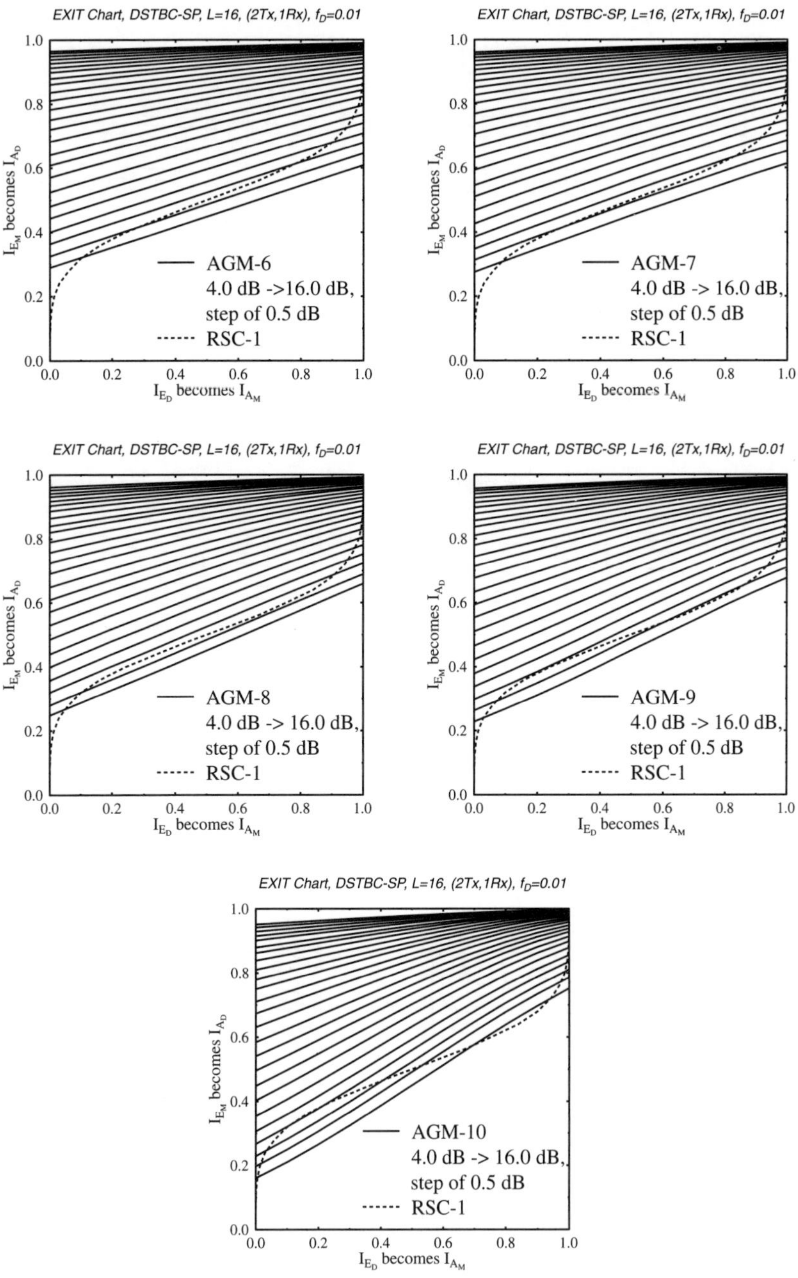

Figure C.1: Continued.

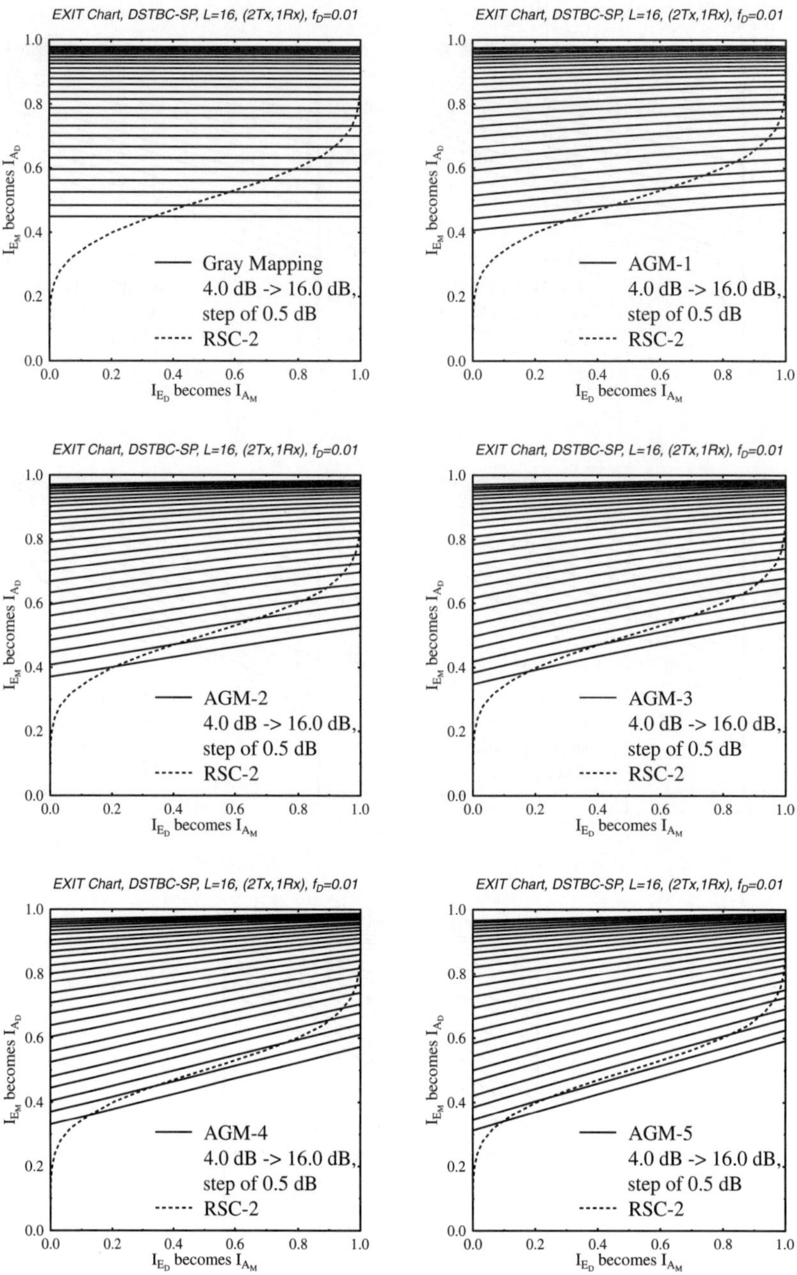

Figure C.2: RSC-2 coded DSTBC-SP schemes.

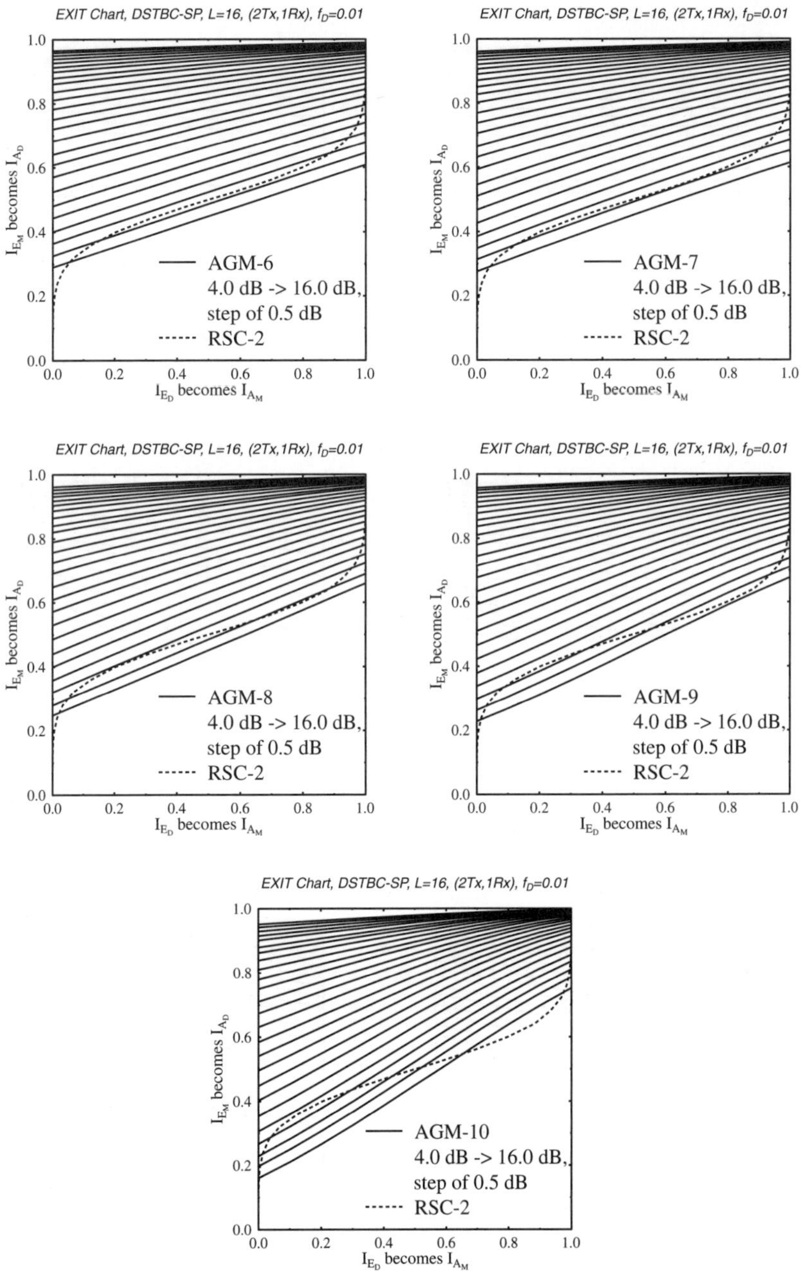

Figure C.2: Continued.

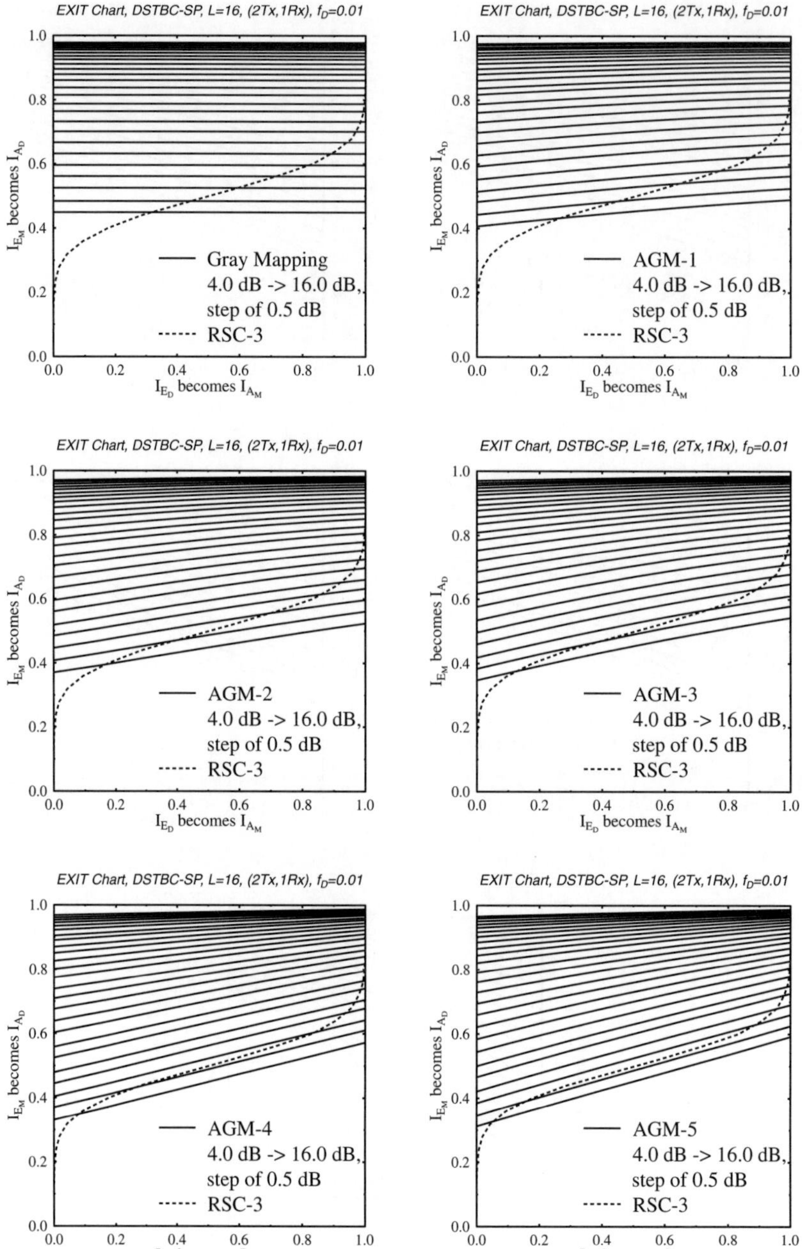

Figure C.3: RSC-3 coded DSTBC-SP schemes.

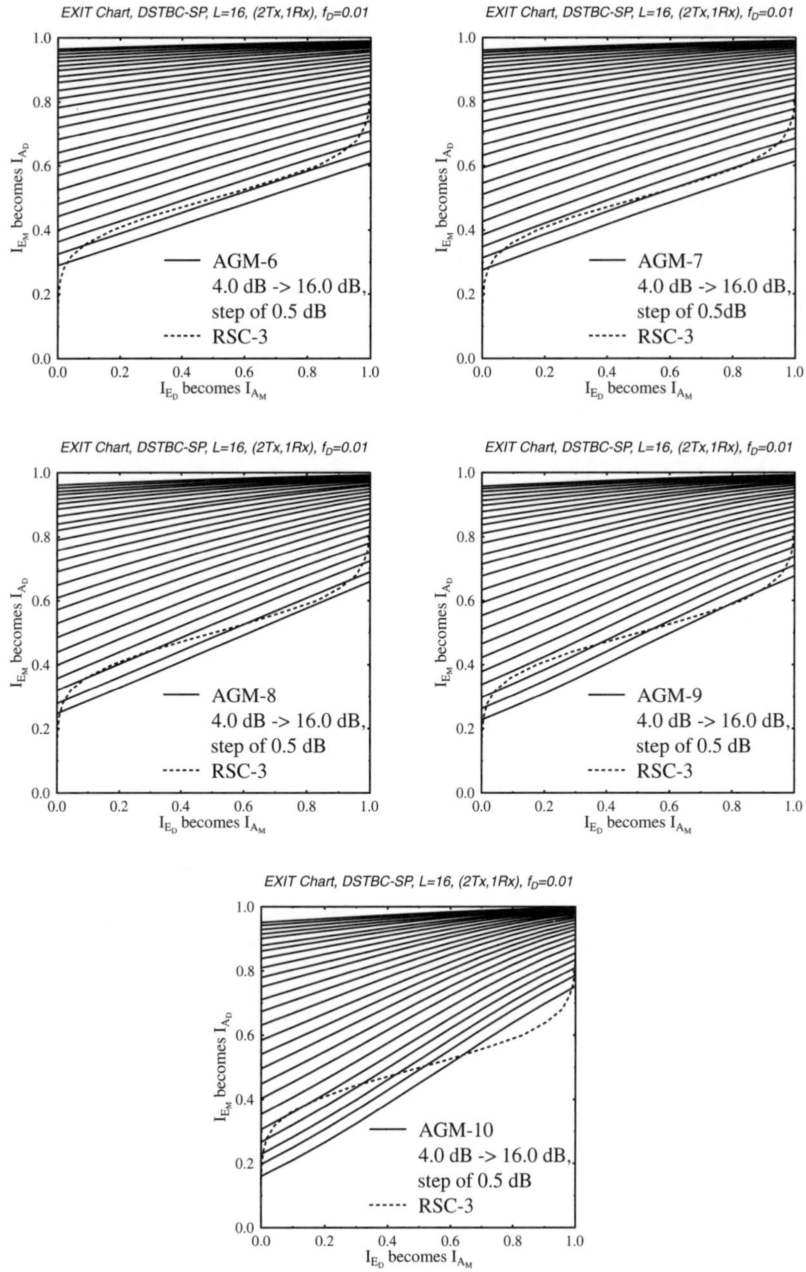

Figure C.3: Continued.

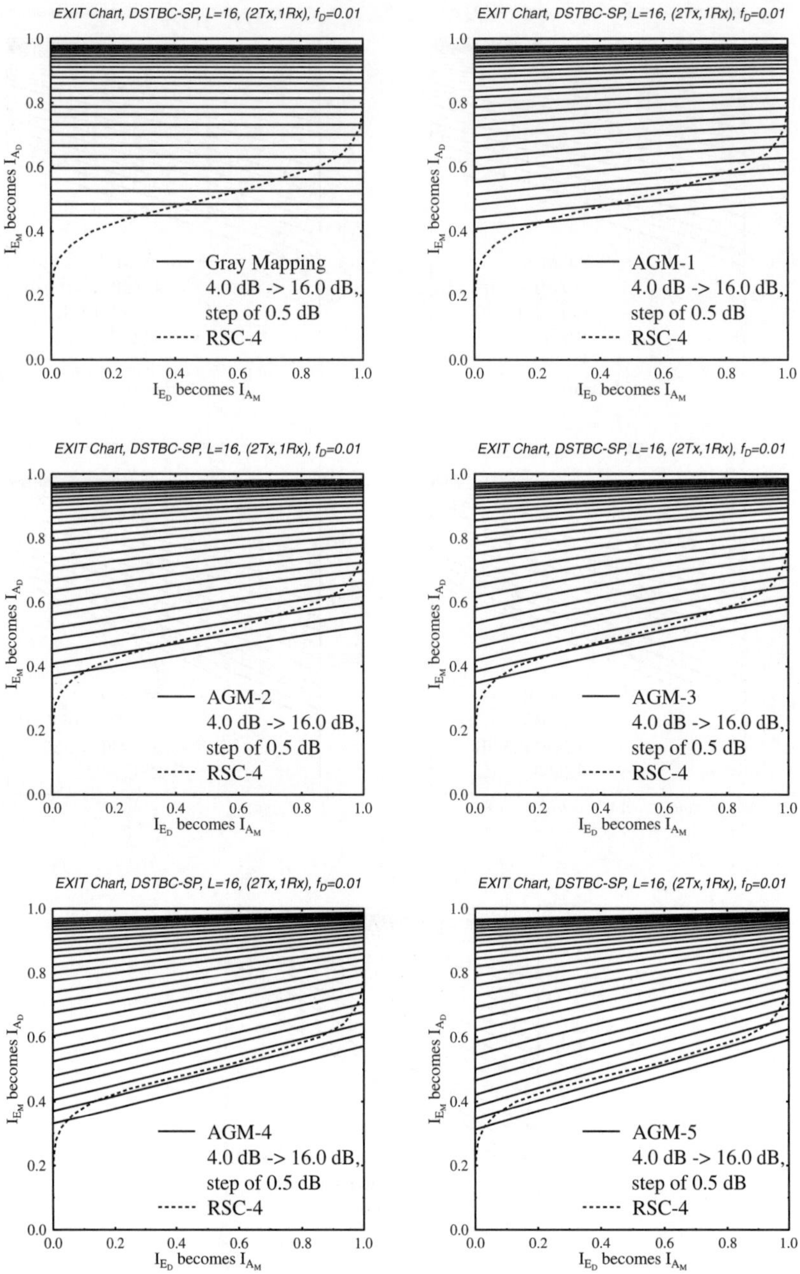

Figure C.4: RSC-4 coded DSTBC-SP schemes.

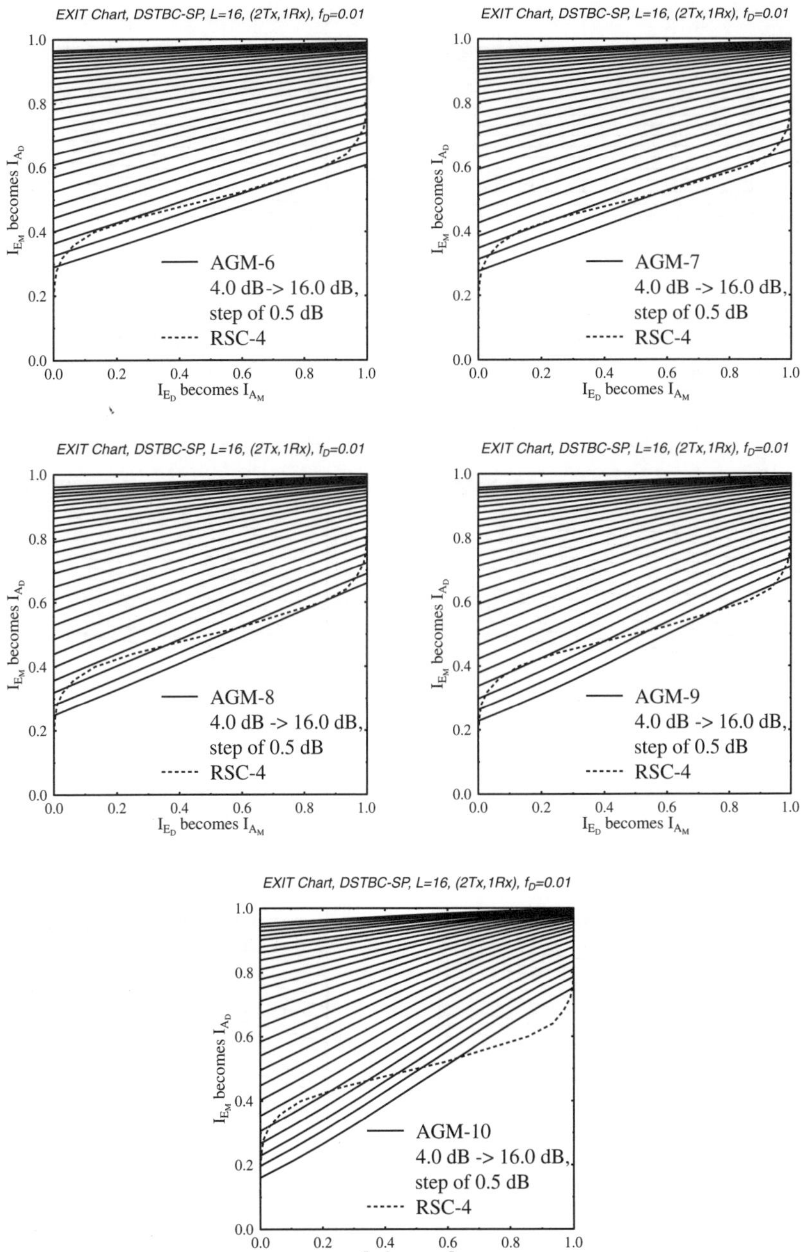

Figure C.4: Continued.

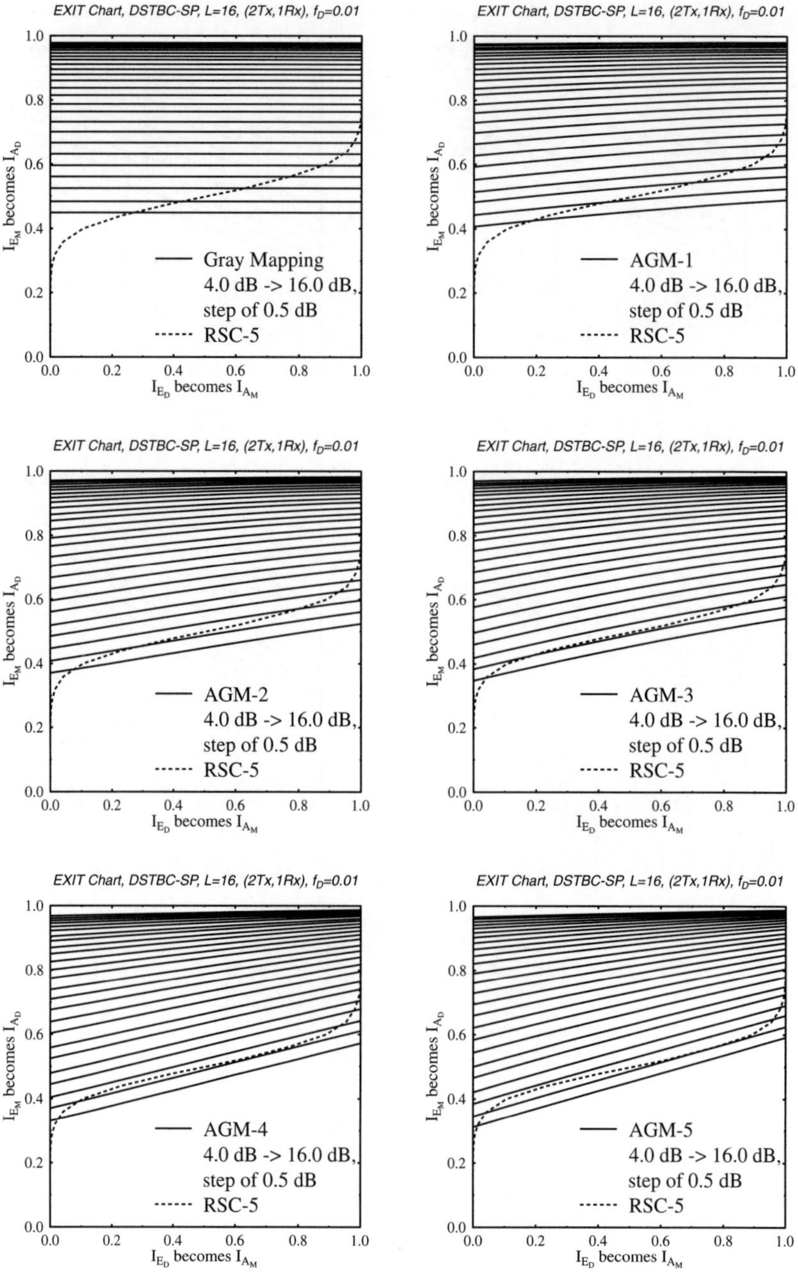

Figure C.5: RSC-5 coded DSTBC-SP schemes.

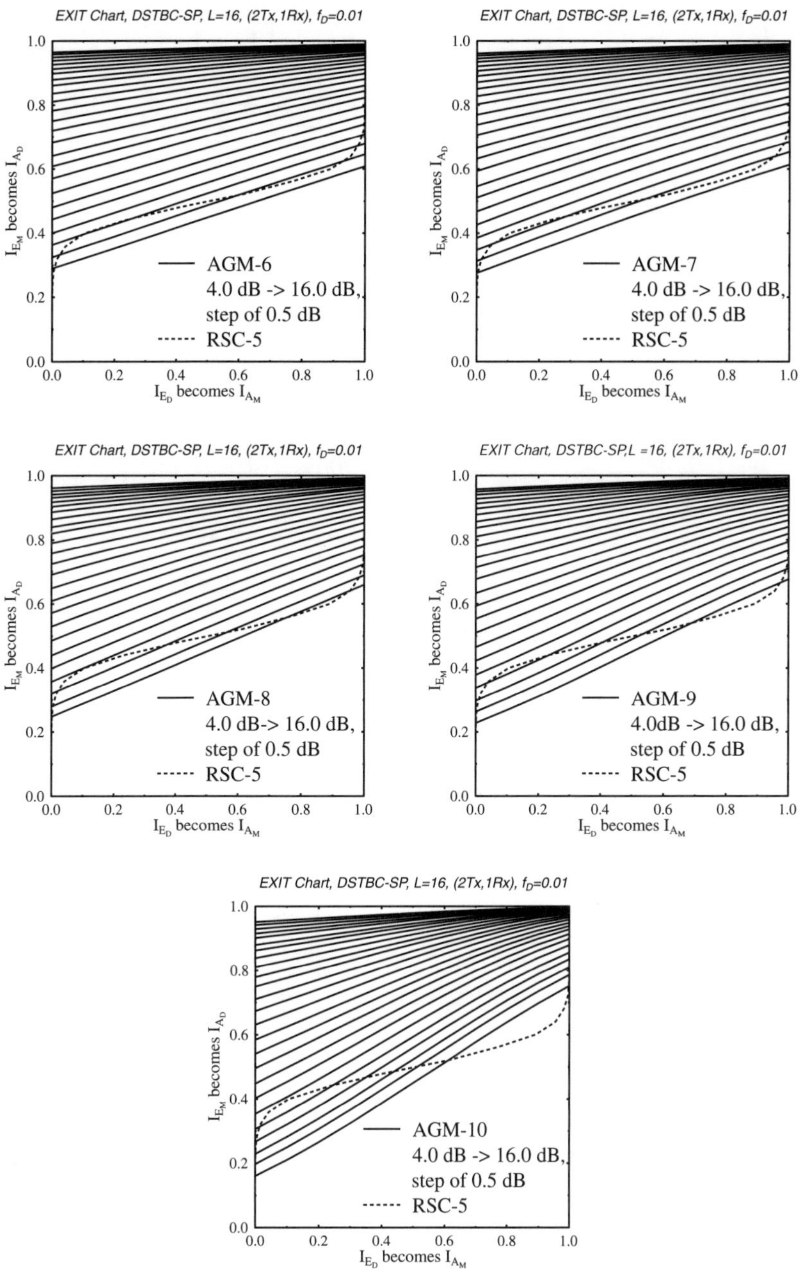

Figure C.5: Continued.

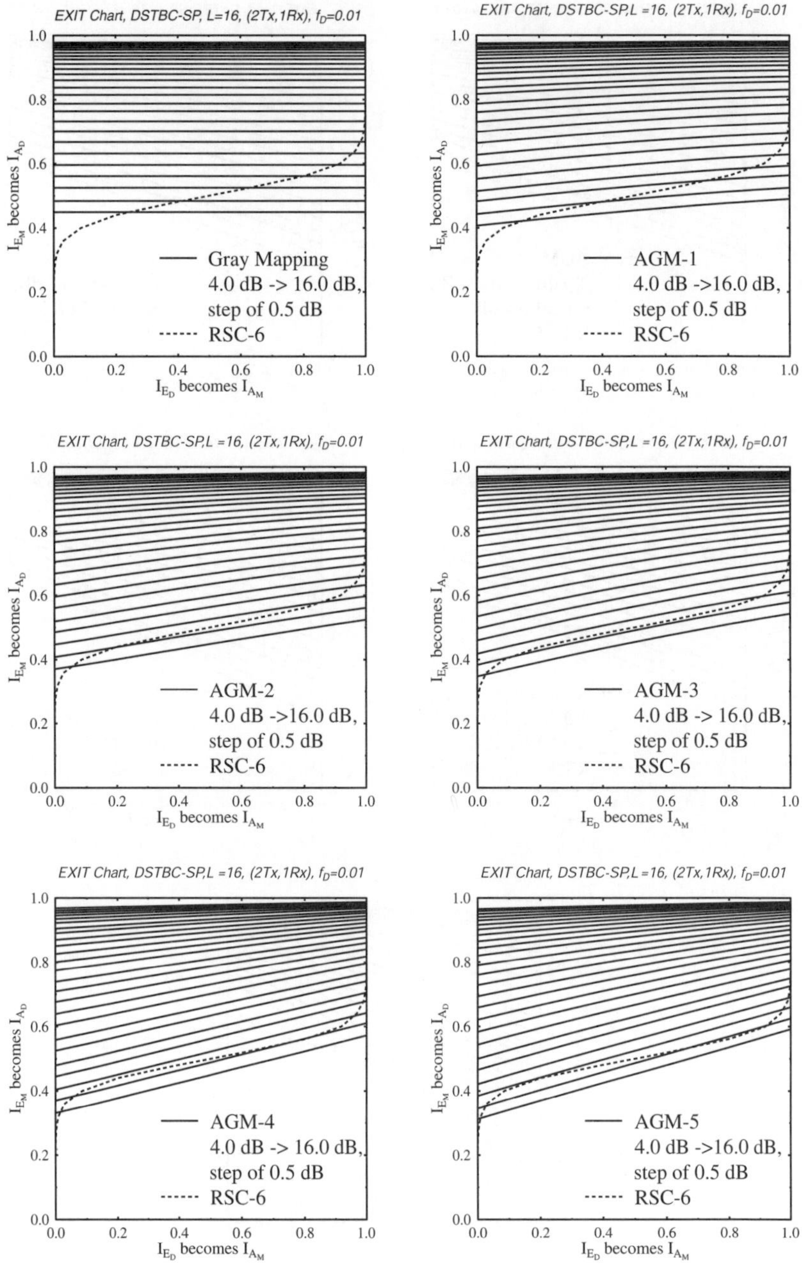

Figure C.6: RSC-6 coded DSTBC-SP schemes.

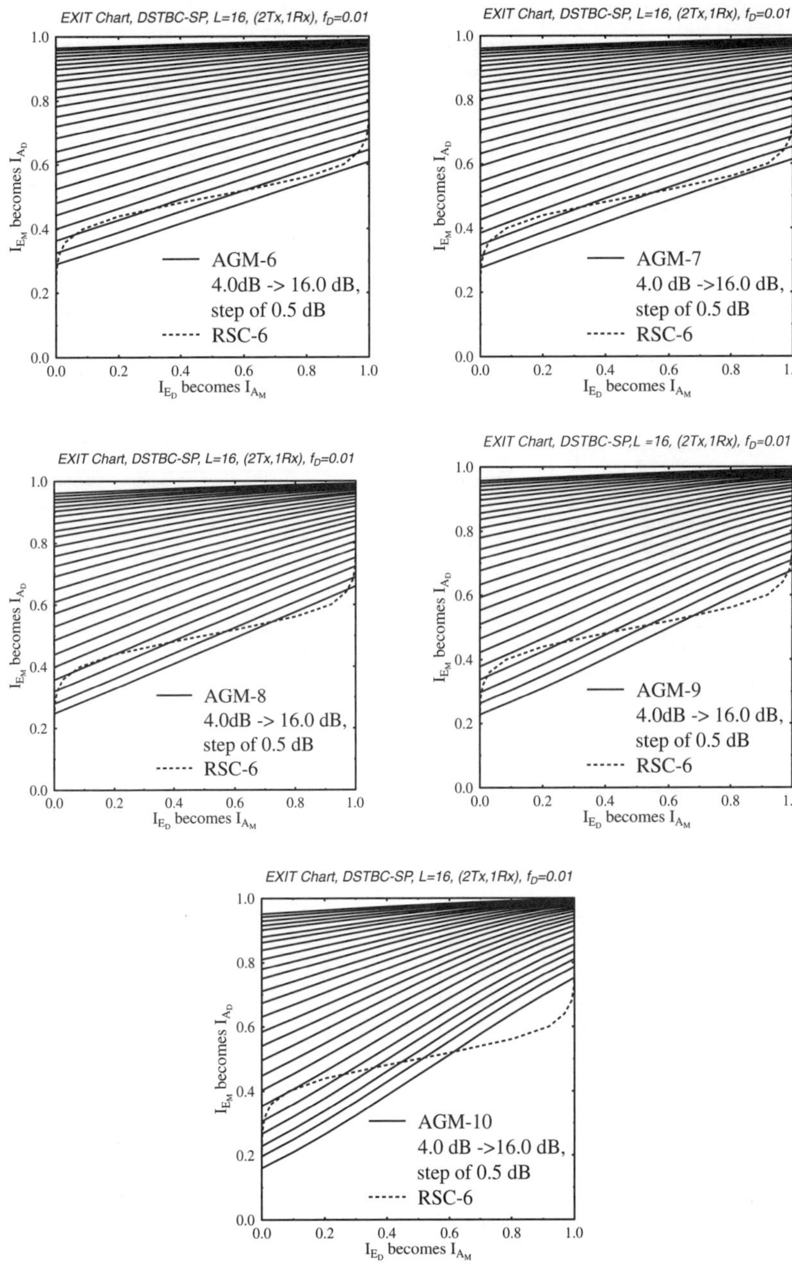

Figure C.6: Continued.

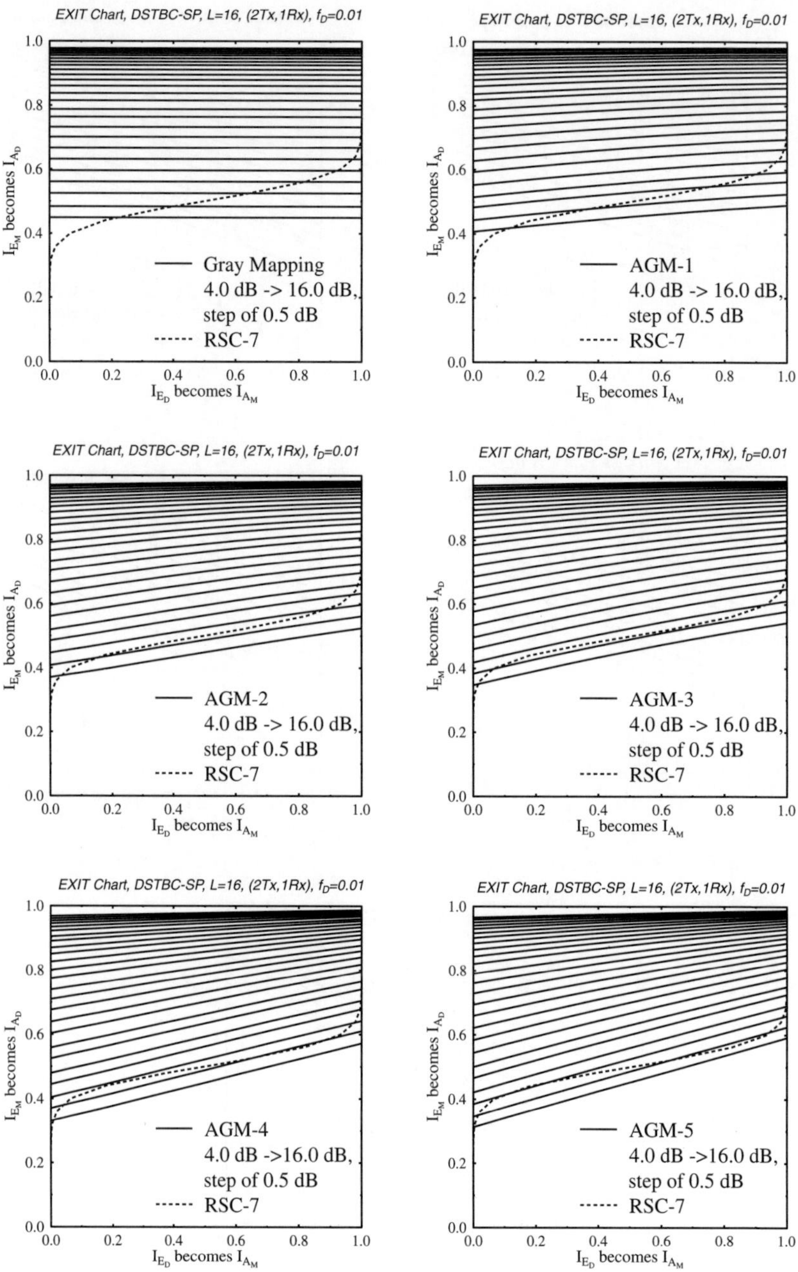

Figure C.7: RSC-7 coded DSTBC-SP schemes.

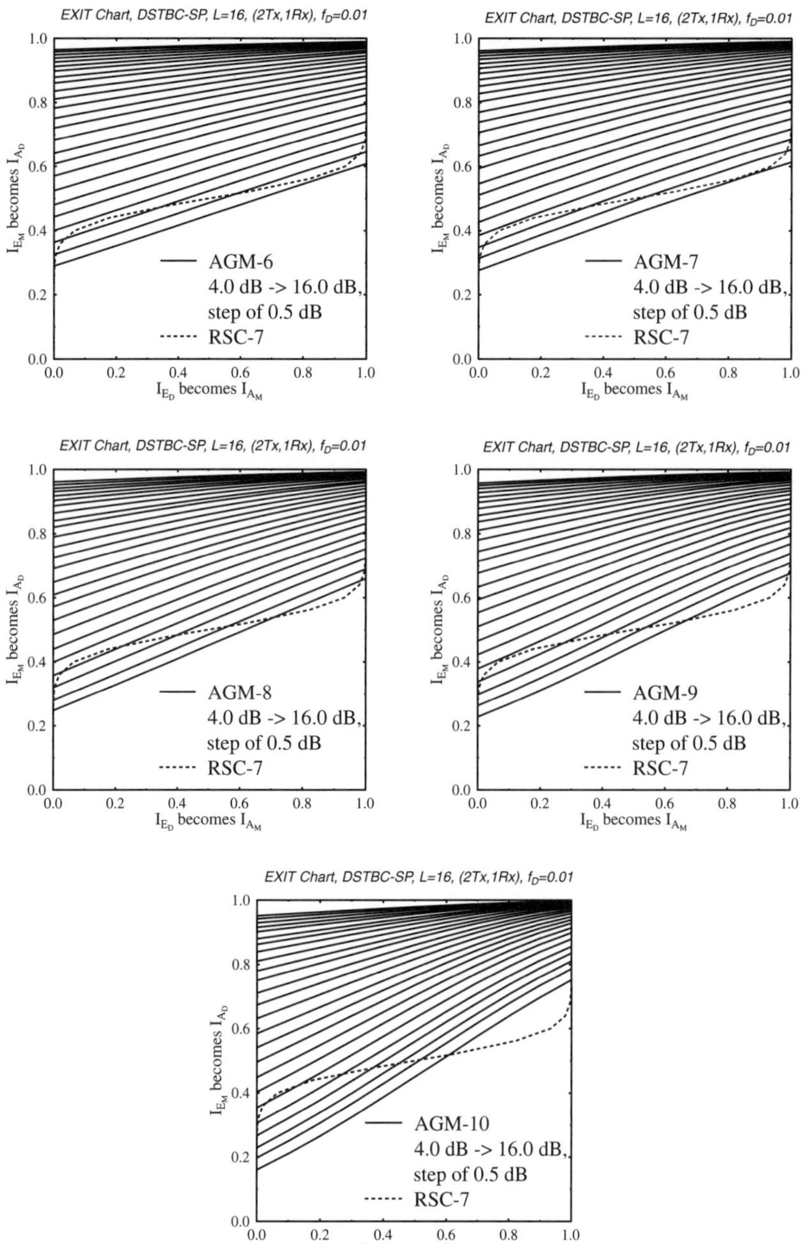

Figure C.7: Continued.

Appendix D

LDCs' χ for QPSK Modulation

$$\chi_{2212} = \begin{pmatrix} -0.405864 + j0.401752 & 0.266975 - j0.320303 \\ -0.373855 - j0.184668 & -0.533834 - j0.20286 \end{pmatrix}$$

$$\chi_{2221} = \begin{pmatrix} -0.92995 - j0.166142 \\ -0.0493851 - j0.269278 \\ 0.0470708 - j0.056298 \\ 0.0470708 - j0.056298 \end{pmatrix}$$

$$\chi_{2222} = \begin{pmatrix} -0.37759 - j0.481206 & -0.0903934 + j0.4462 \\ 0.283719 - j0.242103 & -0.515433 + j0.18240 \\ -0.0725764 + j0.3777 & -0.443492 + j0.2987 \\ 0.571661 + j0.1095 & 0.336265 + j0.30786 \end{pmatrix}$$

$$\chi_{2223} = \begin{pmatrix} -0.246026 - j0.27331 & 0.513074 + j0.18723 & -0.05485 + j0.0888 \\ -0.0256093 - j0.3472 & -0.323245 - j0.091157 & -0.50436 + j0.3351 \\ 0.261717 + j0.34385 & 0.238277 - j0.1696 & -0.20259 + j0.49595 \\ 0.119714 - j0.45733 & 0.36222 - j0.19698 & 0.04141 + j0.01799 \end{pmatrix}$$

$$\chi_{2224} = \begin{pmatrix} -0.02667 - j0.3869 & 0.3181 - j0.1179 & -0.1357 - j0.3378 & 0.2356 + j0.2157 \\ 0.02776 - j0.2499 & 0.3341 + j0.0361 & -0.3109 + j0.1713 & -0.2236 + j0.3845 \\ 0.18745 - j0.11353 & 0.1395 + j0.4161 & -0.3828 + j0.0743 & 0.03857 - j0.325 \\ -0.48671 + j0.0384 & 0.0176 + j0.2812 & -0.01802 - j0.298 & -0.3046 - j0.0097 \end{pmatrix}$$

$$\chi_{2225} = \left(\begin{matrix} 0.0995 - j0.1970 & -0.1595 + j0.2814 & -0.2239 - j0.2483 & -0.0999 + j0.4198 \\ -0.1929 - j0.1008 & -0.2018 - j0.2046 & 0.25229 - j0.1654 & 0.1984 - j0.09254 \\ 0.2651 + j0.08348 & 0.1991 - j0.2574 & -0.1421 + j0.0309 & -0.2946 + j0.211 \\ -0.3677 + j0.0415 & -0.2348 - j0.2526 & -0.3051 + j0.2967 & -0.2156 - j0.0704 \end{matrix} \right.$$
$$\left. \begin{matrix} 0.1881 - j0.115 \\ 0.0275 - j0.4799 \\ -0.1925 - j0.3569 \\ 0.1069 + j0.0135 \end{matrix} \right)$$

Near-Capacity Multi-Functional MIMO Systems L. Hanzo, O. R. Alamri, M. El-Hajjar and N. Wu

$$\chi_{2226} = \begin{pmatrix} -0.0118453 + j0.176595 & -0.166926 + j0.18957 & -0.21592 - j0.278066 & 0.05431 + j0.29694 & -0.35861 - j0.18914i & -0.0552 - j0.14966 \\ 0.271779 + j0.155392 & -0.243656 - j0.32982 & 0.08774 + j0.089246 & -0.16584 + j0.26809 & -0.00187 + j0.22265 & -0.26267 + j0.01512 \\ -0.025031 - j0.40389 & 0.13664 + j0.237754 & 0.0384657 - j0.0742 & -0.34909 + j0.11268 & -0.01572 + j0.13145 & -0.26573 - j0.17709 \\ 0.0695421 + j0.142181 & 0.215694 + j0.04254 & 0.363479 - j0.360688 & 0.13103 + j0.08859 & -0.00548 + j0.31543 & 0.17058 - j0.10373 \end{pmatrix}$$

$$\chi_{2231} = \begin{pmatrix} -0.180116 - j0.437363 \\ 0.0607185 - j1.08843 \\ 0.569829 - j0.398904 \\ -0.578465 + j0.03283 \\ 0.754848 - j0.265569 \\ 0.355719 - j0.039016 \end{pmatrix}$$

$$\chi_{2232} = \begin{pmatrix} -0.135059 + j0.67473 & 0.613496 + j0.16956 \\ -0.439605 + j0.294438 & -0.648091 - j0.135779 \\ -0.22639 - j0.0018267 & -0.499265 - j0.121784 \\ 0.44635 + j0.440015 & -0.183282 - j0.203973 \\ -0.42659 + j0.0988857 & 0.24671 - j0.464298 \\ 0.327112 + j0.0607891 & -0.196041 - j0.0474298 \end{pmatrix}$$

$$\chi_{2233} = \begin{pmatrix} -0.197896 - j0.397516 & 0.447064 - j0.556957 & -0.233644 - j0.268266 \\ -0.169231 - j0.569433 & -0.289421 + j0.173695 & -0.0991308 + j0.0559252 \\ 0.406543 + j0.205873 & -0.221729 + j0.0415185 & -0.26662 + j0.10425 \\ -0.253797 - j0.115446 & -0.446467 + j0.257081 & 0.119519 - j0.58575 \\ -0.160325 + j0.34845 & 0.165754 - j0.115912 & 0.530789 - j0.137715 \\ 0.0136372 - j0.131207 & -0.0164136 + j0.136043 & 0.101885 - j0.33176 \end{pmatrix}$$

$$\chi_{2234} = \begin{pmatrix} -0.38169 + j0.20106 & -0.1807 + j0.3158 & -0.2639 - j0.1252 & -0.111 + j0.0384 \\ -0.20817 - j0.26138 & -0.2739 - j0.04583 & 0.14763 + j0.3791 & 0.4402 - j0.1636 \\ 0.32969 + j0.07748 & 0.0083 + j0.2443 & -0.1939 - j0.01956 & 0.3213 - j0.5498 \\ 0.0757 - j0.51343 & -0.1736 + j0.3061 & -0.1701 - j0.2706 & -0.0333 + j0.087 \\ 0.1787 - j0.1289 & 0.06474 + j0.326 & -0.2501 + j0.1633 & 0.0441 + j0.0789 \\ -0.1390 - j0.01618 & 0.2131 - j0.4481 & -0.3875 - j0.3456 & 0.1384 - j0.2716 \end{pmatrix}$$

$$\chi_{2235} = \begin{pmatrix} -0.1654 - j0.0065 & -0.1782 - j0.2422 & 0.5564 - j0.1301 & -0.0665 - j0.2825 & 0.0651 - j0.0545 \\ 0.2076 - j0.4481 & 0.1119 - j0.0566 & 0.0511 - j0.1929 & 0.0484 - j0.1940 & 0.0041 - j0.3703 \\ -0.1103 - j0.1275 & -0.0387 + j0.1548 & -0.1579 + j0.0478 & -0.2659 - j0.034 & 0.5917 - j0.1534 \\ 0.3900 - j0.1748 & -0.0481 - j0.3926 & -0.1776 - j0.12803 & -0.0648 + j0.1935 & -0.0529 + j0.0267 \\ 0.1113 + j0.2462 & -0.3847 + j0.0538 & -0.1801 - j0.0061 & 0.3683 - j0.2535 & -0.0440 - j0.2628 \\ 0.2079 - j0.0364 & -0.3899 + j0.0948 & 0.2835 + j0.2138 & -0.2471 + j0.3183 & -0.0813 - j0.0314 \end{pmatrix}$$

$$\chi_{2241} = \begin{pmatrix} -0.221114 - j0.494598 \\ 0.0980142 + j0.226399 \\ -0.746841 + j0.225963 \\ 0.373998 + j0.912255 \\ 0.536852 + j0.584814 \\ -0.263779 + j0.21009 \\ 0.691621 - j0.683595 \\ 0.39136 + j0.47113 \end{pmatrix}$$

$$\chi_{2243} = \begin{pmatrix} -0.453242 + j0.374656 & -0.126301 - j0.282369 & -0.0967778 + j0.139826 \\ 0.273518 + j0.0266982 & -0.467402 - j0.2064 & -0.172853 + j0.411392 \\ -0.420225 + j0.141066 & -0.22024 + j0.25243 & -0.048435 + j0.275407 \\ -0.321455 + j0.242619 & 0.365645 - j0.23303 & 0.0147895 + j0.105005 \\ -0.241918 + j0.0266229 & 0.264479 + j0.165391 & -0.25357 + j0.77101 \\ -0.117268 + j0.602912 & -0.0038704 + j0.523916 & -0.0801905 - j0.248962 \\ -0.212304 + j0.171448 & -0.0686492 - j0.465512 & -0.094808 - j0.327579 \\ 0.181526 + j0.0970578 & 0.263082 - j0.118006 & -0.359276 + j0.208113 \end{pmatrix}$$

$$\chi_{2245} = \begin{pmatrix} -0.0494 - j0.0683 & 0.1747 + j0.1919 & 0.155907 - j0.07955 & -0.0702 - j0.1889 & 0.2557 - j0.2321 \\ -0.2073 + j0.2180 & -0.1061 + j0.1024 & 0.150372 - j0.14881 & 0.3176 + j0.1941 & -0.0581 - j0.2098 \\ 0.1671 + j0.0559 & -0.2076 + j0.3527 & -0.374529 + j0.01133 & 0.0323 + j0.0361 & -0.3035 - j0.1819 \\ -0.1159 + j0.3092 & -0.2626 + j0.1096 & -0.0287448 + j0.49359 & 0.0487 + j0.338 & 0.1980 - j0.1296 \\ 0.0581 - j0.1775 & -0.0212 + j0.491 & 0.170009 - j0.240328 & -0.4753 + j0.3518 & -0.0558 + j0.0209 \\ 0.2598 + j0.0239 & 0.2756 - j0.2007 & -0.138142 + j0.14911 & -0.2003 + j0.0388 & 0.0203 - j0.5863 \\ -0.0128 - j0.4441 & 0.0879 - j0.0186 & 0.193049 + j0.292598 & 0.1077 - j0.0086 & 0.1329 - j0.1430 \\ -0.4211 + j0.2911 & 0.2578 + j0.1749 & 0.00639021 + j0.2981 & -0.2370 - j0.2067 & -0.2577 - j0.0005 \end{pmatrix}$$

$$\chi_{2247} = \begin{pmatrix} -0.12986 - j0.23838 & -0.1156 - j0.0763 & -0.0329 - j0.19681 & -0.2163 - j0.25286 & 0.0846305 - j0.180206 & -0.250361 + j0.228041 & -0.273297 - j0.0244441 \\ 0.10971 - j0.4252 & -0.34448 + j0.011 & -0.0039 - j0.0972 & 0.37495 + j0.0361 & -0.219413 + j0.0949425 & 0.0424736 + j0.078488 & 0.0242912 + j0.129786 \\ 0.2896 + j0.0998 & -0.1052 - j0.0735 & -0.3172 - j0.2053 & -0.09719 - j0.2700 & 0.0758912 + j0.0340755 & 0.229038 - j0.1981 & 0.0925592 + j0.317003 \\ -0.04043 - j0.1204 & -0.0581 - j0.0706 & 0.251 - j0.02953 & -0.2822 + j0.18163 & -0.0318609 + j0.517846 & 0.104264 + j0.0231903 & 0.0330019 + j0.0672416 \\ 0.07016 - j0.0673 & -0.0223 - j0.4401 & 0.262 - j0.2702 & -0.111 + j0.1022 & -0.150191 - j0.307028 & 0.157935 - j0.169761 & 0.020939 - j0.0485428 \\ -0.11586 + j0.052 & -0.0511 - j0.1605 & -0.0376 - j0.2777 & -0.027 + j0.08845 & -0.0483534 - j0.106107 & -0.197735 + j0.244426 & 0.505831 + j0.0244521 \\ -0.10619 + j0.3817 & -0.2639 + j0.07324 & 0.156 + j0.074 & 0.1380 - j0.1307 & -0.22209 + j0.053245 & -0.270766 - j0.157822 & 0.104142 + j0.0777363 \\ 0.111 - j0.0041 & 0.1363 + j0.30387 & -0.093 + j0.2368 & 0.2099 + j0.110 & 0.126008 - j0.00882476 & -0.13067 + j0.27887 & -0.0235858 + j0.295523 \end{pmatrix}$$

$$\chi_{2324} = \begin{pmatrix} -0.01049 + j0.349 & -0.3895 + j0.00368 & -0.2532 + j0.10835 & 0.2535 + j0.2935 \\ 0.10824 - j0.4266 & 0.0331 + j0.2476 & -0.42283 - j0.0105 & -0.064 + j0.247 \\ -0.03331 + j0.272 & 0.3437 + j0.00559 & -0.42152 - j0.207 & 0.12481 - j0.2643 \\ -0.2982 + j0.1408 & -0.1849 + j0.365 & -0.0832 + j0.1318 & -0.3952 - j0.2074 \end{pmatrix}$$

$$\chi_{3224} = \begin{pmatrix} -0.2425 - j0.1850 & -0.2895 + j0.0503 & -0.3121 - j0.08697 & -0.14691 - j0.27664 \\ 0.05667 - j0.3875 & -0.1549 - j0.0539 & -0.0144 + j0.3689 & -0.0822 + j0.2394 \\ 0.04526 - j0.0398 & -0.0516 - j0.2582 & -0.2578 + j0.0461 & 0.03451 + j0.2839 \\ -0.1681 - j0.2590 & 0.16140 + j0.1316 & -0.0037 - j0.31776 & -0.146 + j0.3754 \\ -0.0712 + j0.2685 & -0.2593 - j0.3806 & 0.2515 - j0.03864 & -0.2128 + j0.1755 \\ -0.0902 + j0.2629 & -0.1638 + j0.187 & -0.1553 - j0.0118 & -0.105 + j0.0808 \end{pmatrix}$$

$$\chi_{3324} = \begin{pmatrix} -0.0562 - j0.361 & 0.0629 - j0.2637 & 0.146287 - j0.1437 & 0.05466 + j0.3514 \\ 0.2767 - j0.122 & -0.214 - j0.3854 & -0.07529 + j0.183 & 0.2074 - j0.1908 \\ -0.147 + j0.185 & 0.058 - j0.3382 & -0.2978 - j0.0407 & -0.316 - j0.07860 \\ 0.1720 + j0.168 & -0.143 - j0.088 & 0.28631 - j0.3333 & -0.166 + j0.1585 \\ 0.2228 - j0.1502 & -0.098 - j0.005 & -0.1803 - j0.2818 & -0.244 - j0.2426 \\ 0.281 + j0.0958 & 0.0990 + j0.257 & -0.1367 + j0.0672 & 0.1262 + j0.0167 \end{pmatrix}$$

Appendix E

DLDCs' χ for 2PAM Modulation

$$\chi_{3231} = \begin{pmatrix} -0.370502 + j0 \\ 0.167623 + j0.701519 \\ -0.255219 - j0.524332 \\ 0.167623 - j0.701519 \\ -0.173728 - j0 \\ -0.659946 + j0.162501 \\ -0.255219 + j0.524332 \\ -0.659946 - j0.162501 \\ -0.433707 - j0 \end{pmatrix}$$

$$\chi_{3232} = \begin{pmatrix} -0.0549 + j0 & 0.1476 - j0 \\ -0.26993 + j0.676 & -0.3309 - j0.3760 \\ -0.1187 - j0.0593 & 0.0868 - j0.4113 \\ -0.2699 - j0.676 & -0.3309 + j0.376 \\ -0.039712 - j0 & 0.0493 + j0 \\ 0.09487 - j0.1667 & -0.3406 - j0.2528 \\ -0.1187 + j0.0593 & 0.0868 + j0.4113 \\ 0.0949 + j0.1667 & -0.3406 + j0.2528 \\ 0.5718 + j0 & 0.5105 + j0 \end{pmatrix}$$

$$\chi_{3233} = \begin{pmatrix} -0.2498 + j0 & 0.0754 - j0 & 0.3885 - j0 \\ 0.2403 + j0.01 & -0.1796 - j0.3199 & 0.11496 - j0.1835 \\ -0.0634 + j0.4061 & 0.2034 - j0.2178 & -0.1827 + j0.3217 \\ 0.2403 - j0.01 & -0.1796 + j0.3199 & 0.11496 + j0.1835 \\ -0.385619 - j0 & -0.4054 + j0 & 0.2958 + j0 \\ -0.3044 + j0.2739 & 0.0656 + j0.3614 & 0.388977 + j0.20423 \\ -0.0634 - j0.4061 & 0.2034 + j0.2178 & -0.1827 - j0.3217 \\ -0.3044 - j0.2739 & 0.0656 - j0.3614 & 0.3889 - j0.2043 \\ 0.00019 + j0 & -0.3367 + j0 & -0.0898 - j0 \end{pmatrix}$$

Near-Capacity Multi-Functional MIMO Systems L. Hanzo, O. R. Alamri, M. El-Hajjar and N. Wu

$$\chi_{3234} =$$
$$\begin{pmatrix} -0.1158 + j0 & -0.1032 + j0 & -0.3165 + j0 & -0.530825 + j0 \\ -0.0123 + j0.3571 & -0.0046 + j0.1659 & -0.1119 + j0.0726 & -0.0613 - j0.1763 \\ 0.4215 + j0.0889 & -0.1255 + j0.2798 & 0.0538 + j0.0852 & -0.1628 + j0.2166 \\ -0.0123 - j0.3571 & -0.0046 - j0.1659 & -0.1119 - j0.0726 & -0.0613 + j0.1763 \\ -0.227013 - j0 & -0.3148 + j0 & 0.452988 - j0 & -0.2747 - j0 \\ -0.0655 - j0.1575 & 0.2654 + j0.3468 & -0.0263 + j0.1942 & -0.0749 - j0.279 \\ 0.4215 - j0.089 & -0.1255 - j0.2798 & 0.0538 - j0.085 & -0.1627 - j0.21661 \\ -0.0656 + j0.1575 & 0.265 - j0.347 & -0.0263 - j0.1942 & -0.0748 + j0.279 \\ 0.01569 + j0 & -0.1254 - j0 & 0.5585 + j0 & 0.0956 - j0 \end{pmatrix}$$

$$\chi_{3235} =$$
$$\begin{pmatrix} -0.2309 + j0 & 0.086 + j0 & -0.0066 - j0 & 0.4227 - j0 \\ -0.185 - j0.0977 & -0.011 - j0.0934 & 0.1107 + j0.317 & -0.2262 + j0.0991 \\ 0.1390 + j0.230 & -0.3061 + j0.3981 & 0.0569 + j0.1810 & 0.1893 + j0.0475 \\ -0.1853 + j0.0977 & -0.011 + j0.093 & 0.110 - j0.317 & -0.2262 + j0.0991 \\ -0.3457 - j0 & 0.1243 + j0 & -0.4062 - j0 & 0.0413 + j0 \\ -0.0037 - j0.1011 & 0.0217 + j0.1582 & -0.033 - j0.0552 & 0.3123 + j0.1146 \\ 0.1390 + j0.230 & -0.3061 + j0.3981 & 0.0569 + j0.1810 & 0.1893 + j0.0475 \\ -0.0037 - j0.1011 & 0.0217 + j0.1582 & -0.033 - j0.0552 & 0.3123 + j0.1146 \\ 0.4168 + j0 & -0.0622 - 0 & -0.3593 + j0 & -0.0039 - j0 \end{pmatrix}$$
$$\begin{pmatrix} 0.0989 - j0 \\ 0.3942 - j0.0362 \\ 0.0065 - j0.0236 \\ 0.3942 - j0.0362 \\ -0.2193 + j0 \\ 0.1821 + j0.1746 \\ 0.0065 - j0.0236 \\ 0.1821 + j0.1746 \\ 0.3163 + j0 \end{pmatrix}$$

$$\chi_{3236} =$$
$$\begin{pmatrix} -0.2857 + j0 & 0.0504628 + j0 & -0.342589 + j0 & 0.0572714 + j0 \\ 0.06459 - j0.1708 & -0.0988 - j0.2411 & 0.1939 + j0.064 & -0.0722 + j0.0244 \\ -0.03706 + j0.0443 & -0.3769 + j0.0992 & 0.0474 + j0.102 & -0.01643 - j0.3795 \\ 0.06459 - j0.1708 & -0.0988 - j0.2411 & 0.1939 + j0.064 & -0.0722 + j0.0244 \\ 0.220344 + j0 & 0.0480425 - j0 & 0.00109 - j0 & 0.2958 - j0 \\ -0.1052 - j0.1247 & 0.0599 - j0.05052 & 0.0105 - j0.274 & 0.1271 - j0.1936 \\ -0.03706 + j0.0443 & -0.3769 + j0.0992 & 0.0474 + j0.102 & -0.01643 - j0.3795 \\ -0.1052 - j0.1247 & 0.0599 - j0.05052 & 0.0105 - j0.274 & 0.1271 - j0.1936 \\ 0.4931 + j0 & -0.2078 - j0 & -0.35099 - j0 & -0.041168 + j0 \end{pmatrix}$$
$$\begin{pmatrix} 0.143171 - j0 & -0.109684 + j0 \\ 0.0774 - j0.387 & -0.395 - j0.02217 \\ 0.2569 - j0.047 & 0.15752 + j0.0851 \\ 0.0774 - j0.387 & -0.395 - j0.02217 \\ -0.1186 - j0 & 0.1071 - j0 \\ 0.0513 - j0.0326 & -0.1857 - j0.0782 \\ 0.2569 - j0.047 & 0.15752 + j0.0851 \\ 0.0513 - j0.0326 & -0.1857 - j0.0782 \\ 0.0999088 - j0 & -0.13375 + j0 \end{pmatrix}$$

$$\chi_{2221} =$$
$$\begin{pmatrix} -0.404626 + j0 \\ 0.820031 + j0.404755 \\ 0.820031 - j0.404755 \\ 0.404626 + j0 \end{pmatrix}$$

$$\chi_{2222} =$$
$$\begin{pmatrix} -0.5802 + j0 & 0.404212 - j0 \\ -0.2624 - j0.3077 & -0.366 - j0.45 \\ -0.2624 + j0.3077 & -0.366 + j0.45 \\ 0.579896 + j0 & -0.40439 + j0 \end{pmatrix}$$

$$\chi_{2223} = \begin{pmatrix} -0.328116 + j0 & -0.469847 - j0 & 0.0222762 + j0 \\ 0.0913 + j0.46418 & -0.0892 - j0.3194 & -0.5631 + j0.1259 \\ 0.0913 - j0.46418 & -0.0892 + j0.3194 & -0.5631 - j0.1259 \\ 0.333784 + j0 & 0.475327 + j0 & -0.0202509 - j0 \end{pmatrix}$$

$$\chi_{2224} = \begin{pmatrix} -0.372 + j0 & 0.4744 + j0 & 0.0984 - j0 & -0.3559 + j0 \\ -0.4041 - j0.091 & -0.1984 - j0.2643 & -0.1903 + j0.3864 & 0.1056 - j0.1499 \\ -0.4041 + j0.091 & -0.1984 + j0.2643 & -0.1903 - j0.3864 & 0.1056 + j0.1499 \\ 0.1348 + j0 & -0.2377 - j0 & -0.3454 + j0 & -0.5532 + j0 \end{pmatrix}$$

$$\chi_{2225} = \begin{pmatrix} -0.1415 + j0 & -0.4042 - j0 & 0.4097 - j0 & -0.2627 + j0 \\ 0.0316 + j0.0217 & -0.1391 + j0.3455 & -0.3227 + j0.1897 & -0.3517 - j0.3026 \\ 0.0316 - j0.0217 & -0.1391 - j0.3455 & -0.3227 - j0.1897 & -0.3517 + j0.3026 \\ -0.6517 + j0 & 0.0119 - j0 & -0.0283 - j0 & -0.0047 + j0 \end{pmatrix.$$
$$\begin{matrix} 0.2824 + j0 \\ -0.0422 - j0.0512 \\ -0.0422 + j0.0512 \\ -0.27265 - j0 \end{matrix}\Bigg)$$

Appendix F

CLDCs' χ_1 and χ_2 for BPSK Modulation

Recall from Section 9.2.3 that we defined the DCM $\chi_1 \in \zeta^{T_1 \times Q}$ for the broadcast interval and an analogous matrix $\chi_2 \in \zeta^{MT_2 \times MT_1}$ for characterizing the cooperation interval, which is given by

$$\chi_2 = \begin{pmatrix} \mathbf{B}_1 & \mathbf{0} & \cdots & \mathbf{0} \\ \mathbf{0} & \mathbf{B}_m & \cdots & \mathbf{0} \\ \mathbf{0} & \mathbf{0} & \ddots & \mathbf{0} \\ \mathbf{0} & \cdots & \cdots & \mathbf{B}_M \end{pmatrix}, \tag{F.1}$$

where $\mathbf{B}_m \in \zeta^{T_2 \times T_1}$ and $\mathbf{0} \in \zeta^{T_2 \times T_1}$ denotes a zero matrix.

Since the size of the DCM χ_2 is often quite large, we define another matrix $\bar{\chi}_2$ having a size of $(T_1 T_2 \times M)$ in order to present the dispersion matrices in a more compact form, which is given as follows:

$$\bar{\chi}_2 = [vec(\mathbf{B}_1), vec(\mathbf{B}_2), \ldots, vec(\mathbf{B}_M)]. \tag{F.2}$$

$$\chi_1(4181, T_1 = 2, T_2 = 6) = \begin{pmatrix} -0.8054 + j0.9896 \\ 0.4786 + j0.3782 \end{pmatrix}$$

$$\bar{\chi}_2(4181, T_1 = 2, T_2 = 6) = \begin{pmatrix} -0.2925 + j0.3594 & -0.1667 - j0.1819 & -0.392 - j0.0055i & -0.4077 + j0.0371 \\ 0.1738 + j0.1374 & -0.2843 - j0.2997 & -0.0778 + j0.225 & -0.0721 - j0.0102 \\ 0.2104 - j0.0556 & 0.0098 + j0.005 & -0.1882 + j0.3942 & 0.0884 - j0.3008 \\ 0.1120 + j0.5111 & 0.1562 - j0.1450 & 0.4430 - j0.1178 & 0.3282 + j0.3345 \\ 0.0622 - j0.3963 & 0.1381 - j0.1562 & 0.2122 + j0.1585 & 0.0442 - j0.3906 \\ 0.0388 - j0.0508 & -0.3664 + j0.5434 & i0.0146 - j0.2605 & 0.2845 + j0.1540 \\ 0.3884 - j0.0272 & -0.2456 - j0.3130 & 0.1882 + j0.0160 & 0.2412 - j0.2033 \\ -0.2449 + j0.5105 & -0.0337 + j0.2887 & -0.3098 + j0.1832 & 0.0047 + j0.2537 \\ -0.0747 - j0.01823 & 0.3680 - j0.1673 & -0.0550 + j0.1429 & 0.1947 - j0.2236 \\ 0.0187 + j0.1511 & -0.0563 + j0.5262 & 0.004 + j0.3746 & 0.4934 - j0.0909 \\ -0.0406 - j0.057 & -0.1773 - j0.1177 & 0.1378 - j0.1408 & 0.0713 + j0.4279 \\ 0.4732 + j0.1411 & -0.1312 + j0.03579 & -0.6160 + j0.0524 & 0.0756 + j0.2294 \end{pmatrix}$$

Near-Capacity Multi-Functional MIMO Systems L. Hanzo, O. R. Alamri, M. El-Hajjar and N. Wu

$$\chi_1(4182, T_1=2, T_2=6) = \begin{pmatrix} -0.4838-j0.5383 & 0.0551-j0.6878 \\ 0.6889-j0.0394 & -0.5271-j0.4959 \end{pmatrix}$$

$$\bar{\chi}_2(4182, T_1=2, T_2=6) = \begin{pmatrix} -0.3545-j0.3944 & -0.0045+j0.6001 & -0.2679+j0.1594 & -0.0344+j0.1286 \\ 0.5047-j0.0289 & -0.1781+j0.2782 & 0.3350-j0.2350 & -0.0116-j0.0919 \\ 0.0884-j0.0396 & -0.0583-j0.1296 & -0.1513-j0.0147 & -0.0293+j0.5856 \\ -0.1015+j0.0157 & -0.1118+j0.2598 & -0.0291-j0.3369 & 0.0869+j0.1181 \\ 0.0281-j0.0831 & -0.0895-j0.3184 & 0.1179+j0.3124 & -0.0375+j0.3506 \\ -0.3200+j0.2884 & 0.2563-j0.0737 & 0.0302+j0.4853 & 0.1793+j0.4494 \\ -0.2509-j0.2475 & 0.1173+j0.2456 & 0.0007-j0.0166 & 0.0695-j0.4795 \\ -0.0898-j0.1111 & -0.2637-j0.2359 & -0.0459+j0.4471 & -0.0289+j0.3445 \\ -0.3867-j0.1989 & 0.1538+j0.1944 & 0.3467-j0.2887 & 0.0147-j0.0058 \\ -0.0959-j0.2959 & 0.2651-j0.1216 & -0.3553+j0.1135 & -0.0917-j0.4221 \\ 0.3866+j0.3213 & 0.3797-j0.1122 & 0.0699+j0.3765 & 0.2363+j0.0601 \\ 0.1384-j0.2185 & -0.2572-j0.4259 & 0.1976+j0.1392 & -0.1166+j0.3687 \end{pmatrix}$$

$$\chi_1(4183, T_1=2, T_2=6) = \begin{pmatrix} -0.0128+j0.1408 & -0.1711-j0.6604 & -0.0316-j0.7167 \\ 0.6999-j0.1752 & -0.0823-j0.5151 & 0.1068+j0.4426 \end{pmatrix}$$

$$\bar{\chi}_2(4183, T_1=2, T_2=6) = \begin{pmatrix} -0.0085+j0.0936 & -0.2769-j0.3746 & -0.0056-j0.3750 & -0.3079-j0.1600 \\ -0.1138-j0.4392 & 0.2488-j0.0769 & 0.3535+j0.0142 & 0.1272-j0.1987 \\ -0.0210-j0.4766 & 0.1693-j0.2010 & -0.2999+j0.4355 & -0.4265-j0.0862 \\ 0.3327-j0.0510 & 0.0353-j0.4967 & -0.1331+j0.0459 & -0.3847+j0.0256 \\ -0.0484-j0.4290 & -0.0961-j0.1811 & -0.0457-j0.0633 & 0.2161-j0.2943 \\ 0.0827+j0.0342 & 0.1226-j0.3019 & -0.4175-j0.0662 & -0.2331+j0.2198 \\ -0.0546+j0.0046 & 0.4167-j0.0401 & 0.0030-j0.5396 & 0.2180+j0.0096 \\ -0.1339+j0.3918 & -0.1557-j0.1280 & -0.2019+j0.0647 & -0.2823-j0.0337 \\ -0.3404-j0.0024 & -0.0704-j0.1223 & -0.0780-j0.0001 & -0.4304-j0.1147 \\ 0.3968+j0.1526 & -0.2195-j0.2403 & 0.1937+j0.4835 & -0.3821-j0.1044 \\ 0.3512-j0.0938 & 0.3787-j0.4351 & 0.1464+j0.0522 & -0.2407+j0.1828 \\ -0.1449+j0.3546 & -0.2746+j0.0123 & 0.3319+j0.0466 & 0.3085-j0.2824 \end{pmatrix}$$

$$\chi_1(4184, T_1=2, T_2=6) = \begin{pmatrix} -0.3439-j0.1960 & -0.2670+j0.5943 & 0.0378-j0.2339 & -0.5589-j0.2244 \\ 0.4529+j0.3324 & -0.2748+j0.2500 & -0.2868+j0.6279 & -0.2585-j0.0544 \end{pmatrix}$$

$$\bar{\chi}_2(4184, T_1=2, T_2=6) = \begin{pmatrix} -0.2136-j0.1217 & -0.1473+j0.2378 & -0.5139+j0.2171 & -0.4358+j0.1618 \\ -0.1658+j0.3691 & 0.1116+j0.3585 & -0.0132+j0.1322 & 0.1501-j0.2588 \\ 0.0235-j0.1452 & 0.1192-j0.1431 & -0.0742-j0.0147 & -0.0530-j0.2310 \\ -0.3471-j0.1394 & -0.2079-j0.1048 & -0.0224-j0.2400 & -0.2920-j0.0522 \\ 0.4673+j0.3173 & -0.2050+j0.3654 & 0.1226+j0.3303 & 0.3573+j0.3383 \\ 0.0140-j0.2121 & 0.5133+j0.0535 & -0.4051+j0.2629 & -0.0502+j0.2359 \\ -0.1735+j0.4573 & -0.1149+j0.0692 & -0.1740+j0.4664 & -0.2181-j0.2347 \\ 0.1287+j0.0947 & -0.3074+j0.2104 & -0.1755+j0.2553 & -0.4391-j0.1735 \\ -0.2762+j0.2484 & -0.1262+j0.5668 & -0.2510-j0.2168 & -0.3286+j0.2278 \\ 0.2131-j0.3684 & -0.2299+j0.2989 & 0.1540+j0.1193 & -0.2135+j0.1627 \\ 0.1405+j0.0502 & 0.3103+j0.0011 & -0.1605-j0.3728 & -0.2639-j0.0263 \\ 0.3712+j0.0784 & 0.1318+j0.0128 & -0.0165-j0.3054 & -0.3214+j0.1372 \end{pmatrix}$$

$$\chi_1(4185, T_1=3, T_2=5) = \left(\begin{matrix} -0.0085+j0.5196 & -0.1005-j0.1172 & -0.0170-j0.1127 & -0.5283+j0.3177 \\ -0.0616+j0.2309 & -0.6130-j0.1501 & -0.2593+j0.6090 & -0.0591-j0.2384 \end{matrix} \right.$$

$$\left. \begin{matrix} 0.5568+j0.0553 \\ -0.1480-j0.1556 \end{matrix} \right)$$

$$\bar{\chi}_2(4185, T_1 = 3, T_2 = 5) =$$
$$\begin{pmatrix} -0.2598 + j0.1219 & -0.1623 + j0.1250 & -0.1791 - j0.0678 & -0.0977 - j0.2066 \\ -0.2342 - j0.0991 & -0.1540 - j0.0662 & 0.0046 + j0.2690 & -0.4254 + j0.1552 \\ 0.2079 + j0.4125 & 0.0601 - j0.0075 & -0.0954 + j0.2249 & 0.1165 - j0.1197 \\ -0.0114 - j0.2283 & 0.1254 - j0.1603 & 0.3015 - j0.2845 & -0.2112 + j0.1640 \\ 0.0598 - j0.0215 & -0.5079 + j0.2087 & -0.0836 - j0.2628 & 0.2362 - j0.0650 \\ 0.0140 - j0.2674 & -0.1948 - j0.0350 & 0.1646 + j0.3185 & 0.4048 + j0.1807 \\ -0.2124 + j0.0197 & -0.1559 + j0.1745 & -0.2042 + j0.0061 & -0.0278 - j0.0694 \\ -0.0334 - j0.1312 & -0.0408 + j0.3978 & -0.2833 - j0.2402 & -0.1846 + j0.0142 \\ -0.0684 - j0.3077 & 0.2902 - j0.1976 & -0.1323 - j0.2358 & 0.0374 + j0.1211 \\ -0.4062 - j0.1295 & 0.1988 - j0.0065 & -0.0100 - j0.1876 & 0.2910 - j0.2820 \\ 0.0248 + j0.2084 & 0.0558 + j0.0024 & 0.1346 + j0.1628 & 0.0678 - j0.1194 \\ -0.1086 - j0.2842 & 0.4243 + j0.1725 & -0.0048 + j0.0181 & 0.2355 + j0.0890 \\ -0.0351 - j0.2917 & 0.1414 + j0.0171 & -0.1205 + j0.1984 & -0.1622 + j0.2352 \\ 0.1008 - j0.2576 & 0.3358 + j0.1943 & 0.1748 - j0.2001 & -0.4138 + j0.2568 \\ 0.0755 + j0.3339 & -0.0599 + j0.1712 & 0.2218 + j0.4450 & -0.0264 + j0.1221 \end{pmatrix}$$

$$\chi_1(4163, T_1 = 2, T_2 = 4) =$$
$$\begin{pmatrix} -0.3303 - j0.3193 & -0.2146 - j0.4589 & -0.5325 + j0.4987 \\ 0.6794 - j0.2491 & -0.1937 + j0.4730 & -0.2186 + j0.4090 \end{pmatrix}$$

$$\bar{\chi}_2(4163, T_1 = 2, T_2 = 4) =$$
$$\begin{pmatrix} -0.2298 - j0.2221 & -0.2837 - j0.3488 & -0.3169 + j0.2518 & -0.0656 + j0.0739 \\ -0.1493 - j0.3193 & 0.2514 + j0.2502 & 0.0231 - j0.2635 & 0.5389 + j0.1393 \\ -0.3705 + j0.3470 & 0.0860 - j0.3256 & -0.4126 - j0.2860 & 0.1674 + j0.3785 \\ -0.1218 + j0.0339 & 0.0393 - j0.2390 & -0.1148 + j0.0319 & -0.0071 + j0.0954 \\ 0.1353 - j0.2457 & -0.1822 + j0.2684 & -0.3193 + j0.0761 & -0.1773 - j0.2913 \\ -0.0005 - j0.3831 & -0.0257 + j0.1558 & -0.3253 - j0.3234 & -0.1483 - j0.0600 \\ 0.3018 - j0.0231 & -0.1419 + j0.0838 & 0.3821 + j0.1380 & 0.0475 + j0.0984 \\ 0.2990 + j0.3059 & 0.5750 - j0.1099 & 0.0423 + j0.1229 & 0.1575 + j0.5669 \end{pmatrix}$$

$$\chi_1(4143, T_1 = 1, T_2 = 3) =$$
$$\begin{pmatrix} -0.6323 + j0.2230 & -0.1558 - j0.5720 & 0.4167 + j0.1593 \end{pmatrix}$$

$$\bar{\chi}_2(4143, T_1 = 1, T_2 = 3) =$$
$$\begin{pmatrix} -0.5476 + j0.1931 & -0.5760 + j0.2573 & -0.2554 - j0.2909 & -0.3169 + j0.3566 \\ -0.1349 - j0.4953 & 0.5261 - j0.0371 & -0.0629 + j0.7045 & -0.3277 - j0.3445 \\ 0.3609 + j0.1380 & -0.2679 + j0.0463 & 0.2362 + j0.2100 & -0.1683 + j0.5176 \end{pmatrix}$$

Appendix G

Weighting Coefficient Vectors $\boldsymbol{\lambda}$ and γ

This appendix presents all of the weighting coefficient vectors $\boldsymbol{\lambda}$ of the inner irregular scheme and γ of the outer irregular scheme used in Sections 7.5, 8.6 and 9.3 of this book. More explicitly:

- Table G.1 shows λ of the RSC(213)-coded IR-PLDC scheme of Figure 7.32;
- Table G.2 shows λ of the RSC(215)-coded IR-PLDC scheme of Figure 7.32;
- Table G.3 shows γ of the IRCC-coded PLDC(2224) scheme of Figure 7.24;
- Table G.4 shows γ of the IRCC-coded PLDC(2221) scheme of Figure 7.24;
- Table G.5 shows γ of the IRCC-coded IR-PLDC scheme of Figure 7.48;
- Table G.6 shows λ of the IRCC-coded IR-PLDC scheme of Figure 7.48;
- Table G.7 shows γ of the IRCC-coded IR-PDLDC scheme of Figure 8.25;
- Table G.8 shows λ of the IRCC-coded IR-PDLDC scheme of Figure 8.25;
- Table G.9 shows γ of the IRCC-coded IR-PLDC scheme of Figure 7.48;
- Table G.10 shows λ of the IRCC-coded IR-PLDC scheme of Figure 7.48;
- Table G.11 shows γ of the IRCC-coded IR-PCLDC scheme of Figure 9.16;
- Table G.12 shows λ of the IRCC-coded IR-PCLDC scheme of Figure 9.16;
- Table G.13 shows γ of the IRCC-coded IR-PLDC scheme of Figure 7.48;
- Table G.14 shows λ of the IRCC-coded IR-PLDC scheme of Figure 7.48.

Table G.1: The weighting coefficient vector $\boldsymbol{\lambda}$ for the QPSK-modulated IR-PLDC scheme of Figure 7.32 designed for achieving maximum rates, while using a RSC(213) code.

SNR (dB)	Max rate	R_{in}	(2241) $=0.25$	(2231) $=0.33$	(2221) $=0.5$	(2232) $=0.67$	(2243) $=0.75$	(2222) $=1.0$	(2245) $=1.25$	(2234) $=1.33$	(2223) $=1.5$	(2235) $=1.67$	(2224) $=2.0$
−7	0.26	0.26	0.447	0.553	0	0	0	0	0	0	0	0	0
−6	0.31	0.31	0	0.777	0.223	0	0	0	0	0	0	0	0
−5	0.45	0.45	0	0.222	0.778	0	0	0	0	0	0	0	0
−4	0.54	0.54	0	0	0.637	0.363	0	0	0	0	0	0	0
−3	0.67	0.67	0	0	0.0209	0.9539	0	0	0.0252	0	0	0	0
−2	0.80	0.80	0	0	0.25	0	0	0.75	0	0	0	0	0
−1	1.009	1.009	0	0	0	0	0	0.99	0	0	0	0	0.01
0	1.1392	1.1392	0	0	0	0.146	0.034	0	0	0.82	0	0	0
1	1.402	1.402	0	0	0	0	0	0.1402	0	0	0.8598	0	0
2	1.65	1.65	0	0	0	0	0	0	0	0	0.5314	0.1889	0.2797
3	1.88	1.88	0	0	0	0	0	0	0	0	0	0.08	0.92
4	2.0	2.0	0	0	0	0	0	0	0	0	0	0	1

Table G.2: The weighting coefficient vector $\boldsymbol{\lambda}$ for the QPSK-modulated IR-PLDC scheme of Figure 7.32 designed for achieving maximum rates, while using a RSC(215) code.

SNR (dB)	Max rate	R_{in}	(2241) $= 0.25$	(2231) $= 0.33$	(2221) $= 0.5$	(2232) $= 0.67$	(2243) $= 0.75$	(2222) $= 1.0$	(2245) $= 1.25$	(2234) $= 1.33$	(2223) $= 1.5$	(2235) $= 1.67$	(2224) $= 2.0$
−7	0.26	0.26	0.7729	0.2271	0	0	0	0	0	0	0	0	0
−6	0.306	0.306	0.069	0.8469	0.0841	0	0	0	0	0	0	0	0
−5	0.41	0.41	0	0.381	0.619	0	0	0	0	0	0	0	0
−4	0.51	0.51	0	0	0.847	0.153	0	0	0	0	0	0	0
−3	0.622	0.622	0	0	0.216	0.784	0	0	0	0	0	0	0
−2	0.74	0.74	0	0	0	0.8	0	0	0	0	0.2	0	0
−1	0.9315	0.9315	0	0	0	0.1472	0	0.8528	0	0	0	0	0
0	1.065	1.065	0	0	0	0.2511	0	0	0	0.7489	0	0	0
1	1.296	1.296	0	0	0	0	0	0.5335	0.0179	0	0	0	0.4486
2	1.55	1.55	0	0	0	0	0	0	0	0	0.8573	0	0.1427
3	1.765	1.765	0	0	0	0	0	0	0	0	0	0.693	0.307
4	2.0	2.0	0	0	0	0	0	0	0	0	0	0	1

Table G.3: The weighting coefficient vector γ of the IRCC for the IRCC-coded PLDC(2224) scheme of Figure 7.24 designed for achieving maximum rates, while using the QPSK modulation in conjunction with a MMSE detector.

SNR (dB)	Max rate	R_{out}	$R_{\mathrm{ircc},1} = 0.10$	$R_{\mathrm{ircc},2} = 0.15$	$R_{\mathrm{ircc},3} = 0.25$	$R_{\mathrm{ircc},4} = 0.40$	$R_{\mathrm{ircc},5} = 0.45$	$R_{\mathrm{ircc},6} = 0.55$	$R_{\mathrm{ircc},7} = 0.6$	$R_{\mathrm{ircc},8} = 0.7$	$R_{\mathrm{ircc},9} = 0.75$	$R_{\mathrm{ircc},10} = 0.85$	$R_{\mathrm{ircc},11} = 0.9$
−3	0.940	0.235	0.255	0	0.182	0	0.268	0	0	0.183	0	0	0.112
−2	1.124	0.281	0.075	0.198	0	0.161	0.108	0.175	0	0	0.208	0.032	0.043
−1	1.310	0.328	0	0.186	0.164	0.215	0	0.126	0.042	0.136	0.075	0	0.056
0	1.550	0.388	0.006	0.06	0.233	0.152	0	0.168	0.054	0.169	0	0	0.158
1	1.764	0.436	0	0	0.244	0.179	0	0.195	0	0.209	0	0.075	0.098
2	2.0	0.5	0	0	0.181	0.178	0	0.114	0.09	0.237	0	0.033	0.167

Table G.4: The weighting coefficient vector γ of the IRCC for the IRCC-coded PLDC(2221) scheme of Figure 7.24 designed for achieving maximum rates, while using the QPSK modulation in conjunction with a MMSE detector.

SNR (dB)	Max rate	R_{out}	$R_{\text{ircc},1} = 0.10$	$R_{\text{ircc},2} = 0.15$	$R_{\text{ircc},3} = 0.25$	$R_{\text{ircc},4} = 0.40$	$R_{\text{ircc},5} = 0.45$	$R_{\text{ircc},6} = 0.55$	$R_{\text{ircc},7} = 0.6$	$R_{\text{ircc},8} = 0.7$	$R_{\text{ircc},9} = 0.75$	$R_{\text{ircc},10} = 0.85$	$R_{\text{ircc},11} = 0.9$
−10	0.136	0.136	0.5705	0.0439	0.2553	0.1303	0	0	0	0	0	0	0
−9	0.256	0.256	0.0467	0.2843	0.1405	0.2517	0	0	0.1111	0.1657	0	0	0
−8	0.322	0.322	0.0098	0.1364	0.2722	0	0.2336	0.0749	0	0.1589	0	0	0.1142
−7	0.38	0.38	0	0.0062	0.3765	0.1320	0	0.2514	0	0	0.1664	0	0.0675
−6	0.445	0.445	0	0	0.2804	0.2494	0	0.0686	0.087	0.2008	0	0	0.1138
−5	0.492	0.492	0.0054	0	0	0.2395	0.1033	0.0599	0	0.2387	0.0643	0.0569	0.232
−4	0.581	0.581	0	0	0	0	0.2795	0.3589	0	0	0.2058	0.0038	0.1556
−3	0.642	0.642	0.0026	0	0	0	0	0.2984	0.2497	0.2158	0	0.0808	0.1527
−2	0.715	0.715	0	0	0	0	0	0	0.1801	0.655	0	0	0.1649
−1	0.772	0.772	0.0019	0	0	0	0	0	0	0	0.7381	0.0457	0.2143
0	0.824	0.824	0	0	0	0	0	0	0	0	0.41	0.176	0.414
1	0.881	0.881	0	0	0	0	0	0	0	0	0	0.3691	0.6309
2	0.9	0.9	0	0	0	0	0	0	0	0	0	0	1

Table G.5: The weighting coefficient vector γ of the IRCC for the IRCC-coded IR-PLDC scheme of Figure 7.48, when using QPSK modulation in conjunction with a MMSE detector.

SNR (dB)	Max rate	R_{out}	$R_{\text{ircc},1} = 0.10$	$R_{\text{ircc},3} = 0.25$	$R_{\text{ircc},4} = 0.40$	$R_{\text{ircc},6} = 0.55$	$R_{\text{ircc},8} = 0.70$	$R_{\text{ircc},11} = 0.90$
−12	0.1316	0.1975	0.3544	0.0633	0.3038	0.2785	0	0
−11	0.175	0.2275	0.2418	0.2747	0.0879	0.2453	0.1503	0
−10	0.222	0.2275	0.2418	0.2747	0.0879	0.2453	0.1503	0
−9	0.2762	0.235	0.234	0.2128	0.2553	0	0.2979	0
−8	0.3418	0.235	0.234	0.2128	0.2553	0	0.2979	0
−7	0.4272	0.235	0.234	0.2128	0.2553	0	0.2979	0
−6	0.532	0.235	0.234	0.2128	0.2553	0	0.2979	0
−5	0.6558	0.235	0.234	0.2128	0.2553	0	0.2979	0
−4	0.7894	0.25	0.2	0.25	0.16	0.11	0.28	0
−3	0.97	0.245	0.2245	0.2041	0.1633	0.1224	0.2857	0
−2	1.12	0.28	0.1429	0.2679	0.1429	0.1964	0.2499	0
−1	1.3258	0.3425	0.073	0.2555	0.1752	0.1606	0.2044	0.1313
0	1.55	0.3875	0.0387	0.2258	0.2065	0.1419	0.2710	0.1161
1	1.76	0.44	0	0.2273	0.2273	0.1250	0.3182	0.1022
2	1.98	0.495	0	0.101	0.3232	0.1111	0.2828	0.1819
3	2.23	0.5575	0	0.0224	0.3229	0.0987	0.3139	0.2421
4	2.45	0.6125	0	0	0.1959	0.2245	0.2857	0.2939
5	2.65	0.6623	0	0	0.0302	0.4567	0.1057	0.4074
6	2.87	0.7175	0	0	0	0.2683	0.2927	0.4390
7	3.07	0.7675	0	0	0	0.1075	0.3648	0.5277
8	3.24	0.81	0	0	0	0	0.3889	0.6111
9	3.32	0.83	0	0	0	0	0.2952	0.7048
10	3.4	0.85	0	0	0	0	0.2059	0.7941
11	3.52	0.88	0	0	0	0	0.0795	0.9205
12	3.6	0.89	0	0	0	0	0	1

Table G.6: The weighting coefficient vector $\boldsymbol{\lambda}$ of the IR-PLDC for the IRCC-coded IR-PLDC scheme of Figure 7.48, when using QPSK modulation in conjunction with a MMSE detector.

SNR (dB)	Max rate	R_{in}	(2231) $=0.33$	(2221) $=0.5$	(2232) $=0.67$	(2222) $=1.0$	(2223) $=1.5$	(2224) $=2.0$
−12	0.1316	0.337	1	0	0	0	0	0
−11	0.175	0.3846	0.6	0.4	0	0	0	0
−10	0.222	0.4878	0.1	0.8	0.1	0	0	0
−9	0.2762	0.5882	0	0.4	0.6	0	0	0
−8	0.3418	0.7273	0	0	0.75	0.25	0	0
−7	0.4272	0.9091	0	0	0.2	0.8	0	0
−6	0.532	1.1321	0	0	0	0.7	0.2	0.1
−5	0.6558	1.3954	0	0	0	0.3	0.4	0.3
−4	0.7894	1.5789	0	0	0	0	0.2	0.8
−3	0.97	2	0	0	0	0	0	1
−2	1.12	2	0	0	0	0	0	1
−1	1.3258	1.9355	0	0	0	0	0.1	0.9
0	1.55	2	0	0	0	0	0	1
1	1.76	2	0	0	0	0	0	1
2	1.98	2	0	0	0	0	0	1
3	2.23	2	0	0	0	0	0	1
4	2.45	2	0	0	0	0	0	1
5	2.65	2	0	0	0	0	0	1
6	2.87	2	0	0	0	0	0	1
7	3.07	2	0	0	0	0	0	1
8	3.24	2	0	0	0	0	0	1
9	3.32	2	0	0	0	0	0	1
10	3.4	2	0	0	0	0	0	1
11	3.52	2	0	0	0	0	0	1
12	3.6	2	0	0	0	0	0	1

Table G.7: The weighting coefficient vector γ of the IRCC for the IRCC-coded IR-PDLDC scheme of Figure 8.25 transmitting over Rayleigh fading channels having $f_d = 10^{-2}$, when using 2PAM or, equivalently, BPSK modulation in conjunction with a MMSE detector.

SNR (dB)	Max rate	R_{out}	$R_{\text{ircc},1} = 0.10$	$R_{\text{ircc},3} = 0.25$	$R_{\text{ircc},4} = 0.40$	$R_{\text{ircc},6} = 0.55$	$R_{\text{ircc},8} = 0.70$	$R_{\text{ircc},11} = 0.90$
−12	0.0708	0.2125	0.3059	0.1765	0.0941	0.2588	0.1647	0
−11	0.087	0.235	0.234	0.2128	0.2553	0	0.2979	0
−10	0.1058	0.2275	0.2637	0.1648	0.2637	0	0.3078	0
−9	0.1302	0.2275	0.2637	0.1648	0.2637	0	0.3078	0
−8	0.1621	0.2675	0.1682	0.2804	0.0748	0.3084	0	0.1682
−7	0.2064	0.29	0.1379	0.2586	0.1379	0.1897	0.1207	0.1552
−6	0.3113	0.2825	0.1592	0.2212	0.1416	0.1947	0.1239	0.1594
−5	0.3113	0.2825	0.1592	0.2212	0.1416	0.1947	0.1239	0.1594
−4	0.3857	0.2825	0.1592	0.2212	0.1416	0.1947	0.1239	0.1594
−3	0.4754	0.29	0.1379	0.2586	0.1379	0.1897	0.1207	0.1552
−2	0.5755	0.305	0.1311	0.2049	0.1967	0.0902	0.2295	0.1476
−1	0.6716	0.3425	0.0876	0.2189	0.1168	0.2409	0.2044	0.1314
0	0.7822	0.395	0.0506	0.1582	0.2531	0.0696	0.3544	0.1141
1	0.9320	0.48	0	0.1823	0.1667	0.1719	0.2917	0.1874
2	1.0627	0.5425	0	0.0691	0.2581	0.1014	0.3226	0.2488
3	1.1980	0.605	0	0	0.2645	0.0909	0.3471	0.2975
4	1.32	0.66	0	0	0.1515	0.0833	0.4243	0.3409
5	1.45	0.725	0	0	0	0.2276	0.3379	0.4345
6	1.57	0.785	0	0	0	0.0701	0.3567	0.5732
7	1.645	0.8225	0	0	0	0.0334	0.2553	0.7113
8	1.705	0.8525	0	0	0	0.0323	0.1232	0.8445
9	1.765	0.8825	0	0	0	0.0312	0	0.9688
10	1.8	0.9	0	0	0	0	0	1

Table G.8: The weighting coefficient vector $\boldsymbol{\lambda}$ of the IR-PDLDC for the IRCC-coded IR-PDLDC scheme of Figure 8.25 transmitting over Rayleigh fading channels having $f_d = 10^{-2}$, when using 2PAM or, equivalently, BPSK modulation in conjunction with a MMSE detector.

SNR (dB)	Max rate	R_{in}	(3231) $= 0.33$	(3232) $= 0.67$	(3233) $= 1.0$	(3234) $= 1.33$	(3235) $= 1.67$	(3236) $= 2.0$
−12	0.0708	0.3333	1	0	0	0	0	0
−11	0.0870	0.3703	0.8	0.2	0	0	0	0
−10	0.1058	0.4651	0.5	0.3	0.2	0	0	0
−9	0.1302	0.5722	0.25	0.6	0	0.05	0.1	0
−8	0.1621	0.6061	0.1	0.9	0	0	0	0
−7	0.2064	0.7117	0	0.9	0	0	0.05	0.05
−6	0.2489	0.8811	0	0.35	0.55	0	0.1	0
−5	0.3113	1.1019	0	0.15	0.3	0.35	0.2	0
−4	0.3857	1.3652	0	0.1	0.05	0.15	0.7	0
−3	0.4754	1.6393	0	0	0	0.1	0.85	0.05
−2	0.5755	1.887	0	0	0	0	0.3	0.7
−1	0.6716	1.9608	0	0	0	0	0.1	0.9
0	0.7822	1.9802	0	0	0	0	0.05	0.95
1	0.9320	1.9418	0	0	0	0	0.15	0.85
2	1.0627	1.9608	0	0	0	0	0.1	0.9
3	1.1980	1.9802	0	0	0	0	0.05	0.95
4	1.32	2.0	0	0	0	0	0	1
5	1.45	2.0	0	0	0	0	0	1
6	1.57	2.0	0	0	0	0	0	1
7	1.645	2.0	0	0	0	0	0	1
8	1.705	2.0	0	0	0	0	0	1
9	1.765	2.0	0	0	0	0	0	1
10	1.8	2.0	0	0	0	0	0	1

Table G.9: The weighting coefficient vector γ of the IRCC for the IRCC-coded IR-PLDC scheme of Figure 7.48 transmitting over Rayleigh fading channels having $f_d = 10^{-2}$, when using 2PAM or, equivalently, BPSK modulation in conjunction with a MMSE detector.

SNR (dB)	Max rate	R_{out}	$R_{\mathrm{ircc},1}$ $= 0.10$	$R_{\mathrm{ircc},3}$ $= 0.25$	$R_{\mathrm{ircc},4}$ $= 0.40$	$R_{\mathrm{ircc},6}$ $= 0.55$	$R_{\mathrm{ircc},8}$ $= 0.70$	$R_{\mathrm{ircc},11}$ $= 0.90$
−15	0.0675	0.1975	0.3544	0.1266	0.1013	0.4177	0	0
−14	0.0879	0.235	0.234	0.1228	0.2553	0.09	0.2979	0
−13	0.1104	0.2125	0.2823	0.2353	0.1882	0.1294	0.1648	0
−12	0.14	0.2275	0.2418	0.2747	0.0879	0.2418	0.1538	0
−11	0.1774	0.235	0.234	0.1228	0.2553	0.09	0.2979	0
−10	0.2186	0.235	0.234	0.1228	0.2553	0.09	0.2979	0
−9	0.2749	0.235	0.234	0.1228	0.2553	0.09	0.2979	0
−8	0.3469	0.235	0.234	0.1228	0.2553	0.09	0.2979	0
−7	0.4292	0.235	0.234	0.1228	0.2553	0.09	0.2979	0
−6	0.5099	0.2575	0.1748	0.2913	0.1553	0.1068	0.2718	0
−5	0.5958	0.3575	0.042	0.3147	0.1678	0.1538	0.1958	0.1259
−4	0.715	0.3575	0.042	0.3147	0.1678	0.1538	0.1958	0.1259
−3	0.85	0.425	0	0.2647	0.1882	0.1941	0.2471	0.1059
−2	0.955	0.4775	0	0.1047	0.377	0.0576	0.3665	0.0942
−1	1.099	0.555	0	0	0.3243	0.1982	0.3153	0.1622
0	1.223	0.6175	0	0	0.1296	0.3117	0.3401	0.2186
1	1.36	0.68	0	0	0	0.3235	0.4118	0.2647
2	1.475	0.745	0	0	0	0.0738	0.5638	0.3624
3	1.58	0.79	0	0	0	0	0.4873	0.5127
4	1.66	0.83	0	0	0	0	0.2952	0.7048
5	1.74	0.87	0	0	0	0	0.1207	0.8793
6	1.8	0.9	0	0	0	0	0	1

Table G.10: The weighting coefficient vector $\boldsymbol{\lambda}$ of the IR-PLDC for the IRCC-coded IR-PLDC scheme of Figure 7.48 transmitting over Rayleigh fading channels having $f_d = 10^{-2}$, when using 2PAM or, equivalently, BPSK modulation in conjunction with a MMSE detector.

SNR (dB)	Max rate	R_{in}	(3231) $= 0.33$	(3232) $= 0.67$	(3233) $= 1.0$	(3234) $= 1.33$	(3235) $= 1.67$	(3236) $= 2.0$
−15	0.0675	0.3419	0.95	0.05	0	0	0	0
−14	0.0879	0.3738	0.8	0.15	0.05	0	0	0
−13	0.1104	0.5195	0.45	0.05	0.5	0	0	0
−12	0.14	0.6154	0.1	0.85	0.05	0	0	0
−11	0.1774	0.7547	0	0.65	0.35	0	0	0
−10	0.2186	0.9302	0	0.15	0.85	0	0	0
−9	0.2749	1.1696	0	0	0.45	0.5	0.05	0
−8	0.3469	1.476	0	0	0.05	0.45	0.4	0.1
−7	0.4292	1.8265	0	0	0	0.05	0.35	0.6
−6	0.5099	1.9802	0	0	0	0	0.05	0.95
−5	0.5958	1.6667	0	0	0	0	1	0
−4	0.715	2	0	0	0	0	0	1
−3	0.85	2	0	0	0	0	0	1
−2	0.955	2	0	0	0	0	0	1
−1	1.099	2	0	0	0	0	0	1
0	1.223	2	0	0	0	0	0	1
1	1.36	2	0	0	0	0	0	1
2	1.475	2	0	0	0	0	0	1
3	1.58	2	0	0	0	0	0	1
4	1.66	2	0	0	0	0	0	1
5	1.74	2	0	0	0	0	0	1
6	1.8	2	0	0	0	0	0	1

Table G.11: The weighting coefficient vector γ of the IRCC for the IRCC-coded IR-PCLDC scheme of Figure 9.16 transmitting over i.i.d. Rayleigh fading channels, when using BPSK modulation in conjunction with a MMSE detector.

SNR (dB)	Max rate	R_{out}	$R_{\text{ircc},1} = 0.10$	$R_{\text{ircc},3} = 0.25$	$R_{\text{ircc},4} = 0.40$	$R_{\text{ircc},6} = 0.55$	$R_{\text{ircc},8} = 0.70$	$R_{\text{ircc},11} = 0.90$
−15	0.0171	0.13	0.7308	0	0	0	0.2692	0
−14	0.0329	0.1975	0.3797	0	0.3038	0.1392	0.1773	0
−13	0.0513	0.205	0.3415	0.0610	0.2927	0.1341	0.1707	0
−12	0.0671	0.235	0.234	0.2128	0.2553	0	0.2979	0
−11	0.0810	0.235	0.234	0.2128	0.2553	0	0.2979	0
−10	0.1034	0.2275	0.2637	0.1648	0.2637	0	0.3078	0
−9	0.1213	0.2425	0.2062	0.3093	0	0.3402	0.1443	0
−8	0.1488	0.2975	0.1176	0.2941	0.1345	0.1849	0.1176	0.1513
−7	0.1638	0.3275	0.0763	0.3053	0.1221	0.2519	0.1069	0.1375
−6	0.1975	0.395	0	0.3481	0.1013	0.3481	0.0886	0.1139
−5	0.2238	0.4475	0	0.1676	0.3575	0.0615	0.3128	0.1006
−4	0.235	0.47	0	0.1064	0.4255	0	0.3723	0.0958
−3	0.2813	0.5625	0	0.0222	0.2133	0.2933	0.3111	0.1601
−2	0.285	0.57	0	0	0.2807	0.1930	0.3684	0.1579
−1	0.3125	0.625	0	0	0.128	0.264	0.392	0.216
0	0.3475	0.695	0	0	0	0.2374	0.5036	0.2590
1	0.3763	0.7525	0	0	0	0.0365	0.6047	0.3588
2	0.39	0.78	0	0	0	0	0.5385	0.4615
3	0.405	0.81	0	0	0	0	0.3889	0.6111
4	0.42	0.84	0	0	0	0	0.25	0.75
5	0.435	0.87	0	0	0	0	0.1207	0.8793
6	0.45	0.9	0	0	0	0	0	1

Table G.12: The weighting coefficient vector $\boldsymbol{\lambda}$ of the IR-PCLDC for the IRCC-coded IR-PCLDC scheme of Figure 9.16 transmitting over i.i.d. Rayleigh fading channels, when using BPSK modulation in conjunction with a MMSE detector.

SNR (dB)	Max rate	R_{in}	(4181) $= 0.125$	(4182) $= 0.25$	(4183) $= 0.375$	(4184) $= 0.5$
−15	0.0171	0.01316	0.9	0.1	0	0
−14	0.0329	0.16667	0.5	0.5	0	0
−13	0.0513	0.25	0	1	0	0
−12	0.0671	0.2857	0	0.75	0	0.25
−11	0.0810	0.3448	0	0.45	0	0.55
−10	0.1034	0.4545	0	0.1	0.1	0.8
−9	0.1213	0.5	0	0	0	1
−8	0.1488	0.5	0	0	0	1
−7	0.1638	0.5	0	0	0	1
−6	0.1975	0.5	0	0	0	1
−5	0.2238	0.5	0	0	0	1
−4	0.235	0.5	0	0	0	1
−3	0.2813	0.5	0	0	0	1
−2	0.285	0.5	0	0	0	1
−1	0.3125	0.5	0	0	0	1
0	0.3475	0.5	0	0	0	1
1	0.3763	0.5	0	0	0	1
2	0.39	0.5	0	0	0	1
3	0.405	0.5	0	0	0	1
4	0.42	0.5	0	0	0	1
5	0.435	0.5	0	0	0	1
6	0.45	0.5	0	0	0	1

Table G.13: The weighting coefficient vector γ of the IRCC for the IRCC-coded IR-PLDC scheme of Figure 7.48 transmitting over Rayleigh fading channels, when using BPSK modulation in conjunction with a MMSE detector.

SNR (dB)	Max rate	R_{out}	$R_{\text{ircc},1}$ $= 0.10$	$R_{\text{ircc},3}$ $= 0.25$	$R_{\text{ircc},4}$ $= 0.40$	$R_{\text{ircc},6}$ $= 0.55$	$R_{\text{ircc},8}$ $= 0.70$	$R_{\text{ircc},11}$ $= 0.90$
−15	0.0338	0.2275	0.2418	0.2747	0.0879	0.2418	0.1538	0
−14	0.0422	0.2125	0.2824	0.2353	0.1882	0.1294	0.1647	0
−13	0.0542	0.235	0.2340	0.2128	0.2553	0	0.2979	0
−12	0.0684	0.235	0.2340	0.2128	0.2553	0	0.2979	0
−11	0.0870	0.235	0.2340	0.2128	0.2553	0	0.2979	0
−10	0.1085	0.235	0.2340	0.2128	0.2553	0	0.2979	0
−9	0.1288	0.2575	0.1748	0.2913	0.1553	0.1068	0.2718	0
−8	0.1537	0.3125	0.08	0.4	0	0.264	0.112	0.144
−7	0.1825	0.365	0.0273	0.3425	0.1096	0.3014	0.0959	0.1233
−6	0.2127	0.4325	0	0.2023	0.3237	0.1272	0.2428	0.1040
−5	0.2425	0.485	0	0.0773	0.3711	0.1701	0.2887	0.0928
−4	0.2775	0.555	0	0	0.3243	0.1982	0.3153	0.1622
−3	0.3088	0.6175	0	0.0202	0.0324	0.4453	0.2834	0.2187
−2	0.3438	0.6875	0	0	0	0.28	0.4582	0.2618
−1	0.38	0.76	0	0	0	0	0.6447	0.3553
0	0.395	0.79	0	0	0	0	0.4873	0.5127
1	0.415	0.83	0	0	0	0	0.2952	0.7048
2	0.45	0.9	0	0	0	0	0	1

Table G.14: The weighting coefficient vector $\boldsymbol{\lambda}$ of the IR-PLDC for the IRCC-coded IR-PLDC scheme of Figure 7.48 transmitting over Rayleigh fading channels, when using BPSK modulation in conjunction with a MMSE detector.

SNR (dB)	Max rate	R_{in}	(4181) $= 0.125$	(4182) $= 0.25$	(4183) $= 0.375$	(4184) $= 0.5$
−15	0.0338	0.1485	0.7	0.25	0.05	0
−14	0.0422	0.1987	0.4	0.3	0.05	0.25
−13	0.0542	0.2308	0.1	0.85	0.05	0
−12	0.0684	0.2913	0	0.65	0.2	0.15
−11	0.0870	0.3704	0	0.05	0.9	0.05
−10	0.1085	0.4615	0	0.05	0.1	0.85
−9	0.1288	0.5	0	0	0	1
−8	0.1537	0.4918	0	0	0.05	0.95
−7	0.1825	0.5	0	0	0	1
−6	0.2127	0.4918	0	0	0.05	0.95
−5	0.2425	0.5	0	0	0	1
−4	0.2775	0.5	0	0	0	1
−3	0.3088	0.5	0	0	0	1
−2	0.3438	0.5	0	0	0	1
−1	0.38	0.5	0	0	0	1
0	0.395	0.5	0	0	0	1
1	0.415	0.5	0	0	0	1
2	0.45	0.5	0	0	0	1

Appendix H

Gray Mapping and AGM Schemes for SP Modulation of Size $L = 16$

In this appendix, Gray mapping and the nine different AGM schemes introduced in Chapter 11 for DSTS-SP signals of size $L = 16$ are described in detail. More specifically, for all mapping schemes, constellation points of the lattice D_4 are given for each integer index $l = 0, 1, \ldots, 15$. Observe that all mapping schemes use the same 16 constellation points. The normalization factor of these constellation points is $\sqrt{2L/E_{\text{total}}} = 1$ as described in Equation (10.21). The constellation points corresponding to each mapping scheme are given in Tables H.1–H.10.

Table H.1: Gray mapping.

	Points from D_4					Points from D_4			
Integer index	a_1	a_2	a_3	a_4	Integer index	a_1	a_2	a_3	a_4
0	−1	−1	0	0	8	−1	0	0	−1
1	0	−1	−1	0	9	0	0	−1	−1
2	0	−1	+1	0	10	0	0	+1	−1
3	+1	−1	0	0	11	+1	0	0	−1
4	−1	0	0	+1	12	−1	+1	0	0
5	0	0	−1	+1	13	0	+1	−1	0
6	0	0	+1	+1	14	0	+1	+1	0
7	+1	0	0	+1	15	+1	+1	0	0

Near-Capacity Multi-Functional MIMO Systems L. Hanzo, O. R. Alamri, M. El-Hajjar and N. Wu

Table H.2: AGM-1.

	Points from D_4					Points from D_4			
Integer index	a_1	a_2	a_3	a_4	Integer index	a_1	a_2	a_3	a_4
0	+1	−1	0	0	8	−1	0	0	−1
1	0	−1	−1	0	9	0	0	−1	−1
2	0	−1	+1	0	10	0	0	+1	−1
3	−1	−1	0	0	11	+1	0	0	−1
4	−1	0	0	+1	12	−1	+1	0	0
5	0	0	−1	+1	13	0	+1	−1	0
6	0	0	+1	+1	14	0	+1	+1	0
7	+1	0	0	+1	15	+1	+1	0	0

Table H.3: AGM-2.

	Points from D_4					Points from D_4			
Integer index	a_1	a_2	a_3	a_4	Integer index	a_1	a_2	a_3	a_4
0	+1	+1	0	0	8	−1	0	0	−1
1	0	−1	−1	0	9	0	0	−1	−1
2	0	−1	+1	0	10	0	0	+1	−1
3	+1	−1	0	0	11	+1	0	0	−1
4	−1	0	0	+1	12	−1	+1	0	0
5	0	0	−1	+1	13	0	+1	−1	0
6	0	0	+1	+1	14	0	+1	+1	0
7	+1	0	0	+1	15	−1	−1	0	0

Table H.4: AGM-3.

	Points from D_4					Points from D_4			
Integer index	a_1	a_2	a_3	a_4	Integer index	a_1	a_2	a_3	a_4
0	+1	0	0	−1	8	0	−1	−1	0
1	+1	0	0	+1	9	0	0	−1	+1
2	−1	0	0	−1	10	−1	−1	0	0
3	0	0	+1	+1	11	−1	0	0	+1
4	0	−1	+1	0	12	0	0	−1	−1
5	+1	−1	0	0	13	0	+1	−1	0
6	0	0	+1	−1	14	+1	+1	0	0
7	0	+1	+1	0	15	−1	+1	0	0

Table H.5: AGM-4.

	Points from D_4					Points from D_4			
Integer index	a_1	a_2	a_3	a_4	Integer index	a_1	a_2	a_3	a_4
0	0	0	−1	+1	8	+1	−1	0	0
1	0	0	−1	−1	9	0	0	+1	−1
2	−1	+1	0	0	10	−1	0	0	+1
3	−1	0	0	−1	11	−1	−1	0	0
4	+1	0	0	+1	12	0	−1	−1	0
5	+1	+1	0	0	13	+1	0	0	−1
6	0	+1	−1	0	14	0	−1	+1	0
7	0	+1	+1	0	15	0	0	+1	+1

Table H.6: AGM-5.

	Points from D_4					Points from D_4			
Integer index	a_1	a_2	a_3	a_4	Integer index	a_1	a_2	a_3	a_4
0	0	+1	+1	0	8	−1	0	0	+1
1	0	0	+1	−1	9	−1	−1	0	0
2	−1	0	0	−1	10	0	0	−1	+1
3	−1	+1	0	0	11	0	−1	−1	0
4	0	−1	+1	0	12	+1	−1	0	0
5	+1	+1	0	0	13	0	0	−1	−1
6	0	0	+1	+1	14	+1	0	0	+1
7	+1	0	0	−1	15	0	+1	−1	0

Table H.7: AGM-6.

	Points from D_4					Points from D_4			
Integer index	a_1	a_2	a_3	a_4	Integer index	a_1	a_2	a_3	a_4
0	−1	0	0	+1	8	0	0	+1	+1
1	0	−1	−1	0	9	−1	0	0	−1
2	0	0	+1	−1	10	0	−1	+1	0
3	0	+1	−1	0	11	0	+1	+1	0
4	0	0	−1	+1	12	−1	−1	0	0
5	+1	−1	0	0	13	−1	+1	0	0
6	+1	+1	0	0	14	+1	0	0	+1
7	+1	0	0	−1	15	0	0	−1	−1

Table H.8: AGM-7.

	Points from D_4					Points from D_4			
Integer index	a_1	a_2	a_3	a_4	Integer index	a_1	a_2	a_3	a_4
0	−1	−1	0	0	8	0	0	−1	−1
1	−1	+1	0	0	9	0	0	−1	+1
2	+1	−1	0	0	10	0	0	+1	−1
3	+1	+1	0	0	11	0	0	+1	+1
4	0	−1	−1	0	12	−1	0	0	−1
5	0	−1	+1	0	13	−1	0	0	+1
6	0	+1	−1	0	14	+1	0	0	−1
7	0	+1	+1	0	15	+1	0	0	+1

Table H.9: AGM-8.

	Points from D_4					Points from D_4			
Integer index	a_1	a_2	a_3	a_4	Integer index	a_1	a_2	a_3	a_4
0	0	−1	−1	0	8	0	+1	+1	0
1	−1	−1	0	0	9	+1	+1	0	0
2	−1	0	0	−1	10	+1	0	0	+1
3	0	0	−1	−1	11	0	0	+1	+1
4	0	−1	+1	0	12	0	+1	−1	0
5	−1	+1	0	0	13	+1	−1	0	0
6	−1	0	0	+1	14	+1	0	0	−1
7	0	0	−1	+1	15	0	0	+1	−1

Table H.10: AGM-9.

	Points from D_4					Points from D_4			
Integer index	a_1	a_2	a_3	a_4	Integer index	a_1	a_2	a_3	a_4
0	+1	+1	0	0	8	0	+1	−1	0
1	+1	0	0	−1	9	0	0	+1	+1
2	+1	0	0	+1	10	0	0	+1	−1
3	−1	+1	0	0	11	0	−1	−1	0
4	0	+1	+1	0	12	+1	−1	0	0
5	0	0	−1	+1	13	−1	0	0	−1
6	0	0	−1	−1	14	−1	0	0	+1
7	0	−1	+1	0	15	−1	−1	0	0

Glossary

16-QAM	16-level Quadrature Amplitude Modulation
3G	Third Generation
3GPP-LTE	Third Generation Partnership Project's Long Term Evolution
8-PSK	Eight-level Phase Shift Keying
AA	Antenna Array
ACS	Add–Compare–Select
AF	Amplify-and-Forward
AGC	Automatic Gain Control
AGM	Anti-Gray Mapping
AMR-WB	Adaptive Multi-rate Wideband
APP	*A Posteriori* Probability
AWGN	Additive White Gaussian Noise
BEC	Binary Erasure Channel
BER	Bit error ratio, the number of the bits received incorrectly
BICM	Bit-Interleaved Coded Modulation
BLAST	Bell Labs Layered Space-Time
BLER	Block Error Ratio
BPS	Bits per Symbol
BPSK	Binary Phase Shift Keying
BS	Base Station

CCMC	Continuous-input Continuous-output Memoryless Channel
CDMA	Code Division Multiple Access
CF	Compress-and-Forward
CIR	Channel Impulse Response
CLDC	Cooperative Linear Dispersion Code
CRC	Cyclic Redundancy Check
CSI	Channel State Information
CSNR	Chip Signal-to-Noise Ratio
D_4	The lattice corresponding to the SP having the best MED in the four-dimensional real-valued Euclidean space R^4
D-BLAST	Diagonal Bell Labs Layered Space-Time
DCM	Dispersion Character Matrix
DCMC	Discrete-input Continuous-output Memoryless Channel
DF	Decode-and-Forward
DL	Downlink
DLDC	Differential Linear Dispersion Code
DLSTBC	Differential Linear Space-Time Block Codes
DMC	Discrete Memoryless Channel
DOA	Direction of Arrival
DOSTBC	Differential Orthogonal Space-Time Block Codes
DPSK	Differential Phase Shift Keying
DS	Direct Sequence
DS-CDMA	Direct Sequence Code Division Multiple Access
DSSD-STBC	Differential Single-Symbol-Decodable Space-Time Block Codes
D-STTD	Double Space-Time Transmit Diversity
DSTS	Differential Space-Time Spreading
DSTS-SP	SP-modulation-aided DSTS
DSTBC	Differential Space-Time Block Code
DSTBC-SP	SP-modulation-aided DSTBC

D-STTD	Double Space-Time Transmit Diversity
DTC	Distributed Turbo Coding
DUSTM	Differential Unitary Space-Time Modulation
E_b/N_0	Ratio of bit energy to noise power spectral density
EGC	Equal Gain Combining
EXIT	Extrinsic Information Transfer
FD	Frequency Domain
FEC	Forward Error Correction
FER	Frame Error Rate
FFT	Fast Fourier Transform
GF	Galois Field
GSIC	Group Successive Interference Cancellation
HARQ	Hybrid Automatic Repeat Request
HSDPA	High Speed Downlink Packet Access
IC	Interference Cancelation
i.i.d.	Independent and Identically Distributed
IIR	Infinite Impulse Response
IRCC	Irregular Convolutional Code
IR-PCLDC	Irregular Precoded Cooperative Linear Dispersion Codes
IR-PLDC	Irregular Precoded Linear Dispersion Code
IR-VLC	Irregular Variable Length Code
ISCD	Iterative Source and Channel Decoding
ISI	Inter-Symbol Interference
IUC	Inter-User Channel
LDC	Linear Dispersion Code
LDPC	Low Density Parity Check
LLR	Log-Likelihood Ratio
LM	Lloyd-Max

LOS	Line-of-Sight
LSSTC	Layered Steered Space-Time Code
LSSTS	Layered Steered Space-Time Spreading
LST	Layered Space-Time
LSTBC	Linear Space-Time Block Code
LT	Luby Transform
MACS	Million Add-Compare-Select operations
MAP	Maximum *A Posteriori*
MC DS-CDMA	Multicarrier Direct Sequence Code Division Multiple Access
MED	Minimum Euclidean Distance
MI	Mutual Information
MIMO	Multiple-Input Multiple-Output
ML	Maximum Likelihood
MMSE	Minimum Mean Square Error
MRC	Maximum Ratio Combining
MS	Mobile Station
MUD	Multi-User Detectors
MUI	Multi-User Interface
OFDM	Orthogonal Frequency-Division Multiplexing
OSTBC	Orthogonal Space-Time Block Code
P/S	Parallel-to-Serial
PAM	Pulse-Amplitude Modulation
PCLDC	Precoded Cooperative Linear Dispersion Code
PDF	Probability Density Function
PDLDC	Precoded Differential Linear Dispersion Code
PIC	Parallel Interference Cancelation
PLDC	Precoded Linear Dispersion Code
PSEP	Pairwise Symbol Error Probability
PSK	Phase Shift Keying

QAM	Quadrature Amplitude Modulation
QoS	Quality of Service
QOSTBC	Quasi-Orthogonal Space-Time Block Code
QPSK	Quadrature Phase Shift Keying
RA	Repeat-Accumulate
RCPC	Rate Compatible Punctured Convolutional
RSC	Recursive Systematic Convolutional
RTS	Reactive Tabu Search
S/P	Serial-to-Parallel
SC	Selection Combining
SCC	Serial Concatenated Code
SDM	Spatial Division Multiplexing
SDMA	Space Division Multiple Access
SegSNR	Segmental Signal-to-Noise Ratio
SER	Symbol Error Ratio
SF	Spreading Factor
SIC	Successive Interface Cancellation
SINR	Signal-to-Interference-plus-Noise Ratio
SISO	Single-Input Single-Output
SNR	Signal-to-Noise Ratio, noise energy compared to the signal energy
SP	Sphere Packing
SP-SER	Sphere Packing Symbol Error Ratio
SPSI	Sphere Packing Symbol Invariant
SSD-STBC	Single-Symbol-Decodable Space-Time Block Code
ST-SER	Space-Time Symbol Error Rate
STBC	Space-Time Block Code
STBC-SP	Space-Time Block Code using Sphere Packing modulation
STC	Space-Time Coding

STP	Space-Time Processing
STS	Space-Time Spreading
STTC	Space-Time Trellis Coding
TASTBC	Threaded Algebraic Space-Time Block Code
TCM	Trellis Coded Modulation
TD	Time Domain
TDD	Time Division Duplex
TSS	Transmit Symbol Separability
TVLT	Time Variant Linear Transformation
UL	Uplink
URC	Unity Rate Code
USTM	Unitary Space-Time Modulation
VAA	Virtual Antenna Array
V-BLAST	Vertical Bell Laboratories Layered Space-Time
VLC	Variable Length Code
VLEC	Variable Length Error Correcting
VSF	Variable Spreading Factor
WLAN	Wireless Local Area Network
WCDMA	Wideband Code Division Multiple Access
ZF	Zero Forcing

Bibliography

[1] Gesbert, D., Shafi, M., Shiu, D.-S., Smith P.J. and Naguib, A. (2003) From theory to practice: an overview of MIMO space-time coded wireless systems. *IEEE Journal on Selected Areas in Communications*, **21**, 281–302.

[2] FCC (2006) Third Generation Wireless Systems. www.fcc.gov/3G/#sec2 (September 2006).

[3] Foschini, G.J. and Gans, M. (1998) On limits of wireless communication in a fading environment when using multiple antennas. *Wireless Personal Communications*, **6**(3), 311–335.

[4] Hochwald, B.M., Caire, G., Hassibi, B. and Marzetta, T.L. (2003) The academic and industrial embrace of space-time methods. *IEEE Transactions on Information Theory*, **49**, 2329–2331.

[5] Rappaport, T.S. (1996) *Wireless Communications: Principles and Practice*, Prentice-Hall, Englewood Cliffs, NJ.

[6] Proakis, J.G. (2001) *Digital communications*, McGraw-Hill, New York.

[7] Pottie, G.J. (1995) System design choices in personal communications. *IEEE Personal Communications*, **2**, 50–67.

[8] Tarokh, V., Seshardi, N. and Calderbank, A. (1998) Space-time codes for high data rate wireless communication: performance criterion and code construction. *IEEE Transactions on Information Theory*, **44**, 744–765.

[9] Hanzo, L., Liew, T.H. and Yeap, B.L. (2002) *Turbo Coding, Turbo Equalisation and Space Time Coding for Transmission over Fading Channels*, John Wiley & Sons, Ltd, Chichester/IEEE Press, Piscataway, NJ.

[10] Hassibi, B. and Hochwald, B.M. (2002) High-rate codes that are linear in space and time. *IEEE Transactions on Information Theory*, **48**, 1804–1824.

[11] Alamouti, S.M. (1998) A simple transmit diversity technique for wireless communications. *IEEE Journal on Selected Areas in Communications*, **16**(8), 1451–1458.

[12] Tarokh, V., Jafarkhani, H. and Calderbank, A.R. (1999) Space-time block codes from orthogonal designs. *IEEE Transactions on Information Theory*, **45**(5), 1456–1467.

[13] Hochwald, B., Marzetta, T.L. and Papadias, C.B. (2001) A transmitter diversity scheme for wideband CDMA systems based on space-time spreading. *IEEE Journal on Selected Areas in Communications*, **19**(1), 48–60.

[14] Jafarkhani, H. (2001) A quasi-orthogonal space-time block code. *IEEE Transactions on Communications*, **49**(1), 1–4.

[15] Wolniansky, P.W., Foschini, G.J., Golden, G.D. and Valenzuela, R.A. (1998) *V-BLAST: An Architecture for Realizing Very High Data Rates over the Rich-scattering Wireless Channel*. Proceedings of the International Symposium on Signals, Systems and Electronics, September, Pisa, pp. 295–300.

Near-Capacity Multi-Functional MIMO Systems L. Hanzo, O. R. Alamri, M. El-Hajjar and N. Wu

[16] Blogh, J. and Hanzo, L. (2002) *Third-generation systems and intelligent wireless networking: smart antennas and adaptive modulation*, John Wiley & Sons, Ltd, Chichester/IEEE Press, Piscataway, NJ.

[17] Tarokh, V., Naguib, A., Seshadri, N. and Calderbank, A.R. (1999) Combined array processing and space-time coding. *IEEE Transactions on Information Theory*, **45**(4), 1121–1128.

[18] Tao, M. and Cheng, R.S. (2004) Generalized layered space-time codes for high data rate wireless communications. *IEEE Transactions on Wireless Communications*, **3**(4), 1067–1075.

[19] Onggosanusi, E.N., Dabak, A.G. and Schmidl, T.A. (2002) *High rate space-time block coded scheme: performance and improvement in correlated fading channels*. Proceedings of the IEEE Wireless Communications and Networking Conference, March, vol. 1, pp. 194–199.

[20] Jongren, G., Skoglund, M. and Ottersten, B. (2002) Combining beamforming and orthogonal space-time block coding. *IEEE Transactions on Information Theory*, **48**, 611–627.

[21] El-Hajjar, M., Alamri, O., Wang, J., Zummo, S. and Hanzo, L. (2008) Layered steered space-time codes using multi-dimensional sphere-packing modulation. *IEEE Transactions on Wireless Communications*, **7**(3), 943–952.

[22] Yang, L.-L. and Hanzo, L. (2002) Performance of generalized multicarrier DS-CDMA over Nakagami-m fading channels. *IEEE Transactions on Communications*, **50**, 956–966.

[23] Zheng, L.Z. and Tse, D. (2003) Diversity and multiplexing: a fundamental tradeoff in multiple-antenna channels. *IEEE Transactions on Information Theory*, **49**, 1073–1096.

[24] Laneman, J.N. and Wornell, G.W. (2003) Distributed space-time-coded protocols for exploiting cooperative diversity in wireless networks. *IEEE Transactions on Information Theory*, **49**(10), 2415–2425.

[25] Tarokh, V., Jafarkhani, H. and Calderbank, A.R. (1999) Space-time block coding for wireless communications: performance results. *IEEE Journal on Selected Areas in Communications*, **17**(3), 451–460.

[26] Lu, H.F. and Kumar, P.V. (2003) Rate-diversity tradeoff of space-time codes with fixed alphabet and optimal constructions for PSK modulation. *IEEE Transactions on Information Theory*, **49**, 2747–2751.

[27] Heath, R.W.J. and Paulraj, A. (2002) Linear dispersion codes for MIMO systems based on frame theory. *IEEE Transactions on Signal Processing*, **50**, 2429–2441.

[28] Chen, H.J., Haimovich, A. and Sjogren, J.A. (2007) Turbo space-time codes with time varying linear transformations. *IEEE Transactions on Wireless Communications*, **6**, 486–493.

[29] Gamal, H.E. and Damen, M.O. (2003) Universal space-time coding. *IEEE Transactions on Information Theory*, **49**, 1097–1119.

[30] Ganesan, G. and Stoica, P. (2001) *Differential modulation using space-time block codes*. Proceedings of the Conference Record of the 35th Asilomar Conference on Signals, Systems and Computers, vol. 1, Pacific Grove, CA, pp. 236–240.

[31] Wu, N. and Hanzo, L. (2008) *Near-capacity irregular precoded linear dispersion codes*. Proceedings of the IEEE International Conference on Communications, May, Beijing, China, pp. 4501–4505.

[32] Wu, N. and Hanzo, L. (2008) *Irregular precoder-aided differential linear dispersion codes*. Proceedings of the Vehicular Technology Conference, May, (Marina Bay, Singapore), pp. 285–289.

[33] Brennan, D.G. (1959) Linear diversity combining techniques. *Proceedings of the IRE*, **47**, 1075–1102.

[34] Wittneben, A. (1991) *Base station modulation diversity for digital simulcast*. Proceedings of the 41st IEEE Vehicular Technology Conference: Gateway to the Future Technology in Motion, May, St. Louis, MO, pp. 848–853.

[35] Wittneben, A. (1993) *A new bandwidth efficient transmit antenna modulation diversity scheme for linear digital modulation*, Proceedings of the IEEE International Conference on Communications, vol. 3, May, Geneva, pp. 1630–1634.

[36] Seshadri, N. and Winters, J.H. (1993) *Two signaling schemes for improving the error performance of frequency-division-duplex transmission systems using transmitter antenna diversity*. Proceedings of the IEEE Vehicular Technology Conference, May, Secaucus, NJ, pp. 508–511.

[37] Winters, J.H. (1994) *The diversity gain of transmit diversity in wireless systems with Rayleigh fading*. Proceedings of the IEEE International Conference on Communications, May, New Orleans, LA, pp. 1121–1125.

[38] Eng, T., Kong, N. and Milstein, L.B. (1996) Comparison of diversity combining techniques for Rayleigh-fading channels. *IEEE Transactions on Communications*, **44**, 1117–1129.

[39] Guey, J.-C., Fitz, M.P., Bell, M.R. and Kuo, W.-Y. (1999) Signal design for transmitter diversity wireless communication systems over Rayleigh fading channels. *IEEE Transactions on Communications*, **47**, 527–537.

[40] Stoica, P. and Ganesan, G. (2002) *Space-time block codes: trained, blind and semi-blind detection*. Proceedings of the IEEE International Conference on Acoustics, Speech and Signal Processing, vol. 2, Orlando, FL, pp. 1609–1612.

[41] Brennan, D.G. (2003) Linear diversity combining techniques. *Proceedings of the IEEE*, **91**, 331–356.

[42] Wang, H. and Xia, X.-G. (2003) Upper bounds of rates of complex orthogonal space-time block codes. *IEEE Transactions on Information Theory*, **49**, 2788–2796.

[43] Su, W., Safar, Z. and Liu, K.J.R. (2003) *Space-time signal design for time-correlated Rayleigh fading channels*. Proceedings of the IEEE International Conference on Communications (ICC), vol. 5, May, Anchorage, Alaska, pp. 3175–3179.

[44] Zhang, H. and Gulliver, T.A (2005) Capacity and error probability analysis for orthogonal space-time block codes over fading channels. *IEEE Transactions on Wireless Communications*, **4**, 808–819.

[45] Liew, T.-H. and Hanzo, L. (2006) Space-time trellis and space-time block coding versus adaptive modulation and coding aided OFDM for wideband channels. *IEEE Transactions on Vehicular Technology*, **55**, 173–187.

[46] Alamri, O., Ng, S.X., Guo, F., Zummo, S. and Hanzo, L. (2007) *Sphere-packing modulation aided space-time coding using non-binary LDPC-coded iterative-detection*. Proceedings of the IEEE Wireless Communications and Networking Conference (WCNC), Hong Kong, March.

[47] Luo, P. and Leib, H. (2008) Class of full-rank space-time codes combining orthogonal designs with delay diversity. *IEEE Transactions on Vehicular Technology*, **57**, 260–272.

[48] Winters, J.H. (1998) The diversity gain of transmit diversity in wireless systems with Rayleigh fading. *IEEE Transactions on Vehicular Technology*, **47**, 119–123.

[49] Jafarkhani, H. (2005) *Space-Time Coding: Theory and Practice*, Cambridge University Press, Cambridge.

[50] Tarokh, V., Naguib, A., Seshadri, N. and Calderbank, A.R. (1999) Space-time codes for high data rate wireless communication: performance criteria in the presence of channel estimation errors, mobility, and multiple paths. *IEEE Transactions on Communications*, **47**, 199–207.

[51] Hanzo, L., Yang, L.-L., Kuan, E.-L. and Yen, K. (2003) *Single and multi-carrier DS-CDMA: Multi-user detection, space-time spreading, synchronisation, networking and standards*, John Wiley & Sons, Ltd, Chichester/IEEE Press, Piscataway, NJ.

[52] Tarokh, V., Alamouti, S.M. and Poon, P. (1998) *New detection schemes for transmit diversity with no channel estimation*. Proceedings of the IEEE International Conference on Universal Personal Communications, vol. 2, October, Florence, Italy, pp. 917–920.

[53] Tarokh, V. and Jafarkhani, H. (1999) *A differential detection scheme for transmit diversity*. Proceedings of the IEEE Wireless Communications and Networking Conference, vol. 3, September, pp. 1043–1047.

[54] Hochwald, B.M. and Sweldens, W. (2000) Differential unitary space-time modulation. *IEEE Transactions on Communications*, **48**(12), 2041–2052.

[55] Hughes, B.L. (2000) Differential space-time modulation. *IEEE Transactions on Information Theory*, **46**(7), 2567–2578.

[56] Jafarkhani, H. and Tarokh, V. (2001) Multiple transmit antenna differential detection from generalized orthogonal designs. *IEEE Transactions on Information Theory*, **47**(6), 2626–2631.

[57] Schober, R. and Lampe, L. (2002) Noncoherent receivers for differential space-time modulation. *IEEE Transactions on Communications*, **50**, 768–777.

[58] Hwang, C.-S., Nam, S.H., Chung, J. and Tarokh, V. (2003) Differential space time block codes using nonconstant modulus constellations. *IEEE Transactions on Signal Processing*, **51**(11), 2955–2964.

[59] Nam, S.H., Hwang, C.-S., Chung, J. and Tarokh, V. (2004) *Differential space time block codes using QAM for four transmit antennas*. Proceedings of the IEEE International Conference on Communications, vol. 2, pp. 952–956.

[60] Zhu, Y. and Jafarkhani, H. (2005) Differential modulation based on quasi-orthogonal codes. *IEEE Transactions on Wireless Communications*, **4**(6), 3005–3017.

[61] Song, L.-Y. and Burr, A.G. (2007) General differential modulation scheme for quasi-orthogonal space-time block codes with partial or full transmit diversity. *IET Communications*, **1**, 256–266.

[62] Tarokh, V. and Jafarkhani, H. (2000) A differential detection scheme for transmit diversity. *IEEE Journal on Selected Areas in Communications*, **18**(7), 1169–1174.

[63] Hochwald, B.M. and Marzetta, T.L. (2000) Unitary space-time modulation for multiple-antenna communications in Rayleigh flat fading. *IEEE Transactions on Information Theory*, **46**, 543–564.

[64] Biglieri, E., Taricco, G. and Tulino, A. (2002) Performance of space-time codes for a large number of antennas. *IEEE Transactions on Information Theory*, **48**, 1794–1803.

[65] Paulraj, A.J., Gore, D.A., Nabar, R.U. and Bolcskei, H. (2004) An overview of MIMO communications – a key to gigabit wireless. *Proceedings of the IEEE*, **92**, 198–218.

[66] Viterbo, E. and Boutros, J. (1999) A universal lattice code decoder for fading channels. *IEEE Transactions on Information Theory*, **45**, 1639–1642.

[67] Damen, O., Chkeif, A. and Belfiore, J.-C. (2000) Lattice code decoder for space-time codes. *IEEE Communications Letters*, **4**, 161–163.

[68] Agrell, E., Eriksson, T., Vardy, A. and Zeger, K. (2002) Closest point search in lattices. *IEEE Transactions on Information Theory*, **48**, 2201–2214.

[69] Foschini, G.J. (1996) Layered space-time architecture for wireless communication in a fading environment when using multiple antennas. *Bell Laboratories Technical Journal*, **1**, 41–59.

[70] Golden, G.D., Foschini, C.J., Valenzuela, R.A. and Wolniansky, P.W. (1999) Detection algorithm and initial laboratory results using V-BLAST space-time communication architecture. *Electronics Letters*, **35**, 14–16.

[71] Benjebbour, A., Murata, H., and Yoshida, S. (2001) *Comparison of ordered successive receivers for space-time transmission*. Proceedings of the IEEE Vehicular Technology Conference, vol. 4, October, Atlantic City, NJ, pp. 2053–2057.

[72] Sellathurai, M. and Haykin, S. (2002) Turbo-BLAST for wireless communications: theory and experiments. *IEEE Transactions on Signal Processing*, **50**, 2538–2546.

[73] Wubben, D., Bohnke, R., Kuhn, V. and Kammeyer, K.-D. (2003) MMSE extension of *V-BLAST based on sorted QR decomposition*. Proceedings of the IEEE Vehicular Technology Conference, vol. 1, October, pp. 508–512.

[74] Zhu, H., Lei, Z. and Chin, F.P.S. (2004) An improved square-root algorithm for BLAST. *IEEE Signal Processing Letters*, **11**, 772–775.

[75] Huang, Y., Zhang, J. and Djuric, P.M. (2005) Bayesian detection for BLAST. *IEEE Transactions on Signal Processing*, **53**, 1086–1096.

[76] Verdu, S. (2003) *Multiuser Detection*, Cambridge University Press, Cambridge.

[77] Hassibi, B. (2000) *An efficient square-root algorithm for BLAST*. Proceedings of the IEEE International Conference on Acoustics, Speech and Signal Processing, vol. 2, Istanbul, June.

[78] Wubben, D., Bohnke, R., Rinas, J., Kuhn, V. and Kammeyer, K.D. (2001) Efficient algorithm for decoding layered space-time codes. *Electronics Letters*, **37**, 1348–1350.

[79] Zha, W. and Blostein, S.D. (2002) *Modified decorrelating decision-feedback detection of BLAST space-time system*. Proceedings of the IEEE International Conference on Communications, vol. 1, April/May, New York, pp. 335–339.

[80] Xiaofeng, T., Zhuizhuan, Y., Haiyan, Q., Ping, Z., Haas, H. and Costa, E. (2002) *New sub-optimal detection algorithm of layered space-time code*. Proceedings of the IEEE Vehicular Technology Conference, vol. 4, May, pp. 1791–1794.

[81] Choi, J. (2004) A bi-directional zero-forcing BLAST receiver. *IEEE Transactions on Signal Processing*, **52**, 2670–2673.

[82] Choi, W.-J., Cheong, K.-W. and Cioffi, J.M. (2000) *Iterative soft interference cancellation for multiple antenna systems*. Proceedings of the IEEE Wireless Communications and Networking Conference, vol. 1, September, Chicago, IL, pp. 304–309.

[83] Baro, S., Bauch, G., Pavlic, A. and Semmler, A. (2000) *Improving BLAST performance using space-time block codes and turbo decoding*. Proceedings of the IEEE Global Telecommunications Conference (GLOBECOM), vol. 2, San Francisco, CA, pp. 1067–1071.

[84] Jing, X., Wang, H., Ming, C. and Cheng, S. (2003) *A novel BLAST detection algorithm based on instantaneous error ordering*. Proceedings of the IEEE International Conference on Communications, vol. 5, May, pp. 3056–3060.

[85] Shen, C., Zhuang, H., Dai, L. and Zhou, S. (2003) Detection algorithm improving V-BLAST performance over error propagation. *Electronics Letters*, **39**, 1007–1008.

[86] Yao, H. and Wornell, G.W. (2002) *Lattice-reduction-aided detectors for MIMO communication systems*. Proceedings of the IEEE Global Telecommunications Conference, vol. 1, November, pp. 424–428.

[87] Wubben, D., Bohnke, R., Kuhn, V. and Kammeyer, K.-D. (2004) *Near-maximum-likelihood detection of MIMO systems using MMSE-based lattice reduction*. Proceedings of the IEEE International Conference on Communications, vol. 2, June, pp. 798–802.

[88] Elkhazin, A., Plataniotis, K. and Pasupathy, S. (2004) *Group MAP BLAST detector*. Proceedings of the IEEE International Conference on Acoustics, Speech and Signal Processing, vol. 4, May, pp. 785–788.

[89] Sellathurai, M. and Haykin, S. (2003) T-BLAST for wireless communications: first experimental results. *IEEE Transactions on Vehicular Technology*, **52**, 530–535.

[90] Compton, R.T. (1988) *Adaptive Antennas: Concepts and Performance*, Prentice-Hall, Englewood Cliffs, NJ.

[91] Godara, L.C. (1997) Applications of antenna arrays to mobile communications I: Performance improvement, feasibility and system considerations. *Proceedings of the IEEE*, **85**, 1031–1060.

[92] Godara, L.C. (1997) Application of antenna arrays to mobile communications II: Beam-forming and direction-of-arrival considerations. *Proceedings of the IEEE*, **85**, 1195–1245.

[93] Chryssomallis, M. (2000) Smart antennas. *IEEE Antennas and Propagation Magazine*, **42**, 129–136.

[94] Naguib, A.F., Seshadri, N. and Calderbank, A.R. (1998) *Applications of space-time block codes and interference suppression for high capacity and high data rate wireless systems*. Proceedings of the Conference Record of the 32nd Asilomar Conference on Signals, Systems and Computers, vol. 2, Pacific Grove, CA, pp. 1803–1810.

[95] Huang, H. and Viswanathan, H. (2000) *Multiple antennas and multiuser detection in high data rate CDMA systems*. Proceedings of the IEEE 51st Vehicular Technology Conference, vol. 1, May, Tokyo, pp. 556–560.

[96] Stamoulis, A. Al-Dhahir, N. and Calderbank, A. (2001) *Further results on interference cancellation and space-time block codes*. Proceedings of the Conference Record of the 35th Asilomar Conference on Signals, Systems and Computers, vol. 1, November, Pacific Grove, CA, pp. 257–261.

[97] Huang, H., Viswanathan, H. and Foschini, G.J. (2002) Multiple antennas in cellular CDMA systems: transmission, detection and spectral efficiency. *IEEE Transactions on Wireless Communications*, **1**, 383–392.

[98] Soni, R.A., Buehrer, R.M. and Benning, R.D. (2002) Intelligent antenna system for CDMA2000. *IEEE Signal Processing Magazine*, **19**, 54–67.

[99] Liu, J. and Gunawan, E. (2003) Combining ideal beamforming and Alamouti space-time block codes. *Electronics Letters*, **39**, 1258–1259.

[100] Zhu, F. and Lim, M.S. (2004) Combined beamforming with space-time block coding using double antenna array group. *Electronics Letters*, **40**, 811–813.

[101] Zhao, L. and Dubey, V.K. (2005) Detection schemes for space-time block code and spatial multiplexing combined system. *IEEE Communications Letters*, **9**, 49–51.

[102] Lee, H., Powers, E.J. and Kang, J. (2005) *Low-complexity ZF detector for D-STTD systems in time-selective fading channels*. Proceedings of the IEEE 62nd Vehicular Technology Conference, vol. 3, September, pp. 2043–2047.

[103] Sellathurai, M., Ratnarajah, T. and Guinand, P. (2007) Multirate layered space-time coding and successive interference cancellation receivers in quasi-static fading channels. *IEEE Transactions on Wireless Communications*, **6**, 4524–4533.

[104] Ekbatani, S. and Jafarkhani, H. (2008) Combining beamforming and space-time coding using quantized feedback. *IEEE Transactions on Wireless Communications*, **7**, 898–908.

[105] Naguib, A.F., Seshadri, N. and Calderbank, A.R. (2000) Increasing data rate over wireless channels. *IEEE Signal Processing Magazine*, **17**, 76–92.

[106] Al-Dhahir, N., Fragouli, C., Stamoulis, A., Younis, W. and Calderbank, R. (2002) Space-time processing for broadband wireless access. *IEEE Communications Magazine*, **40**, 136–142.

[107] Meng, W., Gu, L. and Li, C. (2005) *The combined beamforming and space-time block coding technique for downlink transmission*. Proceedings of the International Conference on Wireless Networks, Communications and Mobile Computing, vol. 1, June, pp. 481–486.

[108] Li, X., Chindapol, A. and Ritcey, J.A. (2002) Bit-interleaved coded modulation with iterative decoding and 8 PSK signaling. *IEEE Transactions on Communications*, **50**, 1250–1257.

[109] Negi, R., Tehrani, A.M. and Cioffi, J.M. (2002) Adaptive antennas for space-time codes in outdoor channels. *IEEE Transactions on Communications*, **50**, 1918–1925.

[110] Nabar, R., Bolcskei, H. and Paulraj, A.J. (2001) *Transmit optimization for spatial multiplexing in the presence of spatial fading correlation*. Proceedings of the IEEE Global Telecommunications Conference (GLOBECOM), vol. 1, November, San Antonio, TX, pp. 131–135.

[111] Hong, Z., Liu, K., Heath, R.W.Jr. and Sayeed, A.M. (2003) Spatial multiplexing in correlated fading via the virtual channel representation. *IEEE Journal on Selected Areas in Communications*, **21**, 856–866.

[112] Kim, H. and Chun, J. (2004) *MIMO structure which combines the spatial multiplexing and beamforming*. Proceedings of the IEEE Vehicular Technology Conference, vol. 1, May, pp. 108–112.

[113] Sendonaris, A., Erkip, E. and Aazhang, B. (2003) User cooperation diversity Part I: System description. *IEEE Transactions on Communications*, **51**(11), 1927–1938.

[114] Sendonaris, A., Erkip, E. and Aazhang, B. (2003) User cooperation diversity Part II: Implementation aspects and performance analysis. *IEEE Transactions on Communications*, **51**(11), 1939–1948.

[115] Van der Meulen, E.C. (1971) Three-terminal communication channels. *Advanced Applied Probability*, **3**(1), 120–154.

[116] Cover, T. and El Gamal, A. (1979) Capacity theorems for the relay channel. *IEEE Transactions on Information Theory*, **25**(5), 572–584.

[117] Willems, F. (1983) The discrete memoryless multiple access channel with partially cooperating encoders. *IEEE Transactions on Information Theory*, **29**(3), 441–445.

[118] Sendonaris, A., Erkip, E. and Aazhang, B. (1998) *Increasing uplink capacity via user cooperation diversity*. Proceedings of the IEEE International Symposium on Information Theory, Cambridge, MA, August.

[119] Laneman, J.N., Wornell, G.W. and Tse, D.N.C. (2001) *An efficient protocol for realizing cooperative diversity in wireless networks*. Proceedings of the IEEE International Symposium on Information Theory, Washington, DC.

[120] Hunter, T.E. and Nosratinia, A. (2002) *Cooperation diversity through coding*. Proceedings of the IEEE International Symposium on Information Theory.

[121] Dohler, M., Lefranc, E. and Aghvami, H. (2002) *Space-time block codes for virtual antenna arrays*. Proceedings of the 13th IEEE International Symposium on Personal, Indoor and Mobile Radio Communications, vol. 1, September, pp. 414–417.

[122] Valenti, M.C. and Zhao, B. (2003) *Distributed turbo codes: towards the capacity of the relay channel*. Proceedings of the IEEE Vehicular Technology Conference VTC-Fall, vol. 1, October, pp. 322–326.

[123] Zhao, B. and Valenti, M.C. (2003) Distributed turbo coded diversity for relay channel. *Electronics Letters*, **39**, 786–787.

[124] Laneman, J.N., Tse, D.N. and Wornell, G.W. (2004) Cooperative diversity in wireless networks: Efficient protocols and outage behavior. *IEEE Transactions on Information Theory*, **50**, 3062–3080.

[125] Nabar, R.U., Bolcskei, H. and Kneubuhler, F.W. (2004) Fading relay channels: performance limits and space-time signal design. *IEEE Journal on Selected Areas in Communications*, **22**, 1099–1109.

[126] Janani, M., Hedayat, A., Hunter, T. and Nosratinia, A. (2004) Coded cooperation in wireless communications: space-time transmission and iterative decoding. *IEEE Transactions on Signal Processing*, **52**, 362–371.

[127] Stefanov, A. and Erkip, E. (2004) Cooperative coding for wireless networks. *IEEE Transactions on Communications*, **52**, 1470–1476.

[128] Azarian, K., El Gamal, H. and Schniter, P. (2005) On the achievable diversity–multiplexing tradeoff in half-duplex cooperative channels. *IEEE Transactions on Information Theory*, **51**, 4152–4172.

[129] Sneessens, H. and Vandendorpe, L. (2005) *Soft decode and forward improves cooperative communications*. Proceedings of the 6th IEE International Conference on 3G and Beyond, November, Washington, DC, pp. 1–4.

[130] Hu, R. and Li, J. (2005) *Exploiting Slepian-Wolf codes in wireless user cooperation*. Proceedings of the IEEE 6th Workshop on Signal Processing Advances in Wireless Communications, June, pp. 275–279.

[131] Yu, M. and Li, J. (2005) *Is amplify-and-forward practically better than decode-and-forward or vice versa?* Proceedings of the IEEE International Conference on Acoustics, Speech and Signal Processing, March, vol. 3.

[132] Hunter, T.E. and Nosratinia, A. (2006) Diversity through coded cooperation. *IEEE Transactions on Wireless Communications*, **5**, 283–289.

[133] Hunter, T.E., Sanayei, S. and Nosratinia, A. (2006) Outage analysis of coded cooperation. *IEEE Transactions on Information Theory*, **52**, 375–391.

[134] Li, Y., Vucetic, B., Wong, T.F. and Dohler, M. (2006) Distributed turbo coding with soft information relaying in multihop relay networks. *IEEE Journal on Selected Areas in Communications*, **24**, 2040–2050.

[135] Hu, R. and Li, J. (2006) *Practical compress–forward in user cooperation: Wyner-Ziv Cooperation*. Proceedings of the IEEE International Symposium on Information Theory, July, Seattle, WA, pp. 489–493.

[136] Host-Madsen, A. (2006) Capacity bounds for cooperative diversity. *IEEE Transactions on Information Theory*, **52**, 1522–1544.

[137] Bui, T. and Yuan, J. (2007) *A decode and forward cooperation scheme with soft relaying in wireless communication*. Proceedings of the IEEE 8th Workshop on Signal Processing Advances in Wireless Communications (SPAWC), June, Helsinki, pp. 1–5.

[138] Khormuji, M.N. and Larsson, E.G. (2007) *Improving collaborative transmit diversity by using constellation rearrangement*. Proceedings of the IEEE Wireless Communications and Networking Conference, March, Kowloon, pp. 803–807.

[139] Bao, X. and Li, J. (2007) Efficient message relaying for wireless user cooperation: Decode-Amplify-Forward (DAF) and Hybrid DAF and coded-cooperation. *IEEE Transactions on Wireless Communications*, **6**, 3975–3984.

[140] Xiao, L., Fuja, T., Kliewer, J. and Costello, D. (2007) A network coding approach to cooperative diversity. *IEEE Transactions on Information Theory*, **53**, 3714–3722.

[141] Yue, G., Wang, X., Yang, Z. and Host-Madsen, A. (2008) Coding schemes for user cooperation in low-power regimes. *IEEE Transactions on Signal Processing*, **56**, 2035–2049.

[142] Zhang, W., Li, Y., Xia, X.-G., Ching, P.C. and Ben Letaief, K. (2008) Distributed space-frequency coding for cooperative diversity in broadband wireless Ad Hoc networks. *IEEE Transactions on Wireless Communications*, **7**, 995–1003.

[143] Wang, T. and Giannakis, G.B. (2008) Complex field network coding for multiuser cooperative communications. *IEEE Journal on Selected Areas in Communications*, **26**, 561–571.

[144] Mitran, P., Ochiai, H. and Tarokh, V. (2005) Space-time diversity enhancements using collaborative communications. *IEEE Transactions on Information Theory*, **51**, 2041–2057.

[145] Forney, G. (1966) *Concatenated Codes*, MIT Press, Cambridge, MA.

[146] Berrou, C., Glavieux, A. and Thitimajshima, P. (1993) *Near Shannon limit error-correcting coding and decoding: Turbo-codes. 1*. Proceedings of the IEEE International Conference on Communications, vol. 2, May, Geneva, Switzerland, pp. 1064–1070.

[147] Divsalar, D. and Pollara, F. (1995) Multiple turbo codes for deep-space communications. Telecommunications and Data Acquisition Progress Report 42–121. Jet Propulsion Laboratory, Pasadena, CA, May.

[148] Benedetto, S. and Montorsi, G. (1996) Serial concatenation of block and convolutional codes. *IEE Electronics Letters*, **32**, 887–888.

[149] Benedetto, S., Divsalar, D., Montorsi, G. and Pollara, F. (1998) Analysis, design, and iterative decoding of double serially concatenated codes with interleavers. *IEEE Journal on Selected Areas in Communications*, **16**, 231–244.

[150] Raphaeli, D. and Zarai, Y. (1997) *Combined turbo equalization and turbo decoding*. Proceedings of the IEEE Global Telecommunications Conference (GLOBECOM), vol. 2, November, Phoenix, AZ, pp. 639–643.

[151] Raphaeli, D. and Zarai, Y. (1998) Combined turbo equalization and turbo decoding. *IEEE Communications Letters*, **2**, 107–109.

[152] Toegel, M., Pusch, W. and Weinrichter, H. (2000) *Combined serially concatenated codes and turbo-equalization*. Proceedings of the 2nd International Symposium on Turbo Codes, September, Brest, France, pp. 375–378.

[153] Ramamurthy, R. and Ryan, W.E. (2001) Convolutional double accumulate codes (or double turbo DPSK). *IEEE Communications Letters*, **5**, 157–159.

[154] Douillard, C., Jezequel, M., Berrou, C., Picart, A., Didier, P. and Glavieux, A. (1995) Iterative correction of intersymbol interference: turbo equalization. *European Transactions on Telecommunications*, **6**, 507–511.

[155] Benedetto, S., Divsalar, D., Montorsi, G. and Pollara, F. (1997) A soft-input soft-output APP module for iterative decoding of concatenated codes. *IEEE Communications Letters*, **1**, 22–24.

[156] Caire, G., Taricco, G. and Biglieri, E. (1997) *Bit-interleaved coded modulation*. Proceedings of the IEEE International Symposium on Information Theory (ISIT), June–July, Ulm, Germany, p. 96.

[157] Caire, G., Taricco, G. and Biglieri, E. (1998) Bit-interleaved coded modulation. *IEEE Transactions on Information Theory*, **44**, 927–946.

[158] ten Brink, S., Speidel, J. and Yan, R.-H. (1998) *Iterative demapping and decoding for multilevel modulation*. Proceedings of the IEEE Global Telecommunications Conference (GLOBCOM), vol. 1, November, Sydney, NSW, pp. 579–584.

[159] Li, X., and Ritcey, J.A. (1997) Bit-interleaved coded modulation with iterative decoding. *IEEE Communications Letters*, **1**, 169–171.

[160] Li, X. and Ritcey, J.A. (1998) Bit-interleaved coded modulation with iterative decoding using soft feedback. *IEE Electronics Letters*, **34**, 942–943.

[161] Li, X. and Ritcey, J.A. (1999) Trellis-coded modulation with bit interleaving and iterative decoding. *IEEE Journal on Selected Areas in Communications*, **17**, 715–724.

[162] Wang, X. and Poor, H.V. (1999) Iterative (turbo) soft interference cancellation and decoding for coded CDMA. *IEEE Transactions on Communications*, **47**, 1046–1061.

[163] Sezgin, A., Wuebben, D. and Kuehn, V. (2003) *Analysis of mapping strategies for turbo-coded space-time block codes*. Proceedings of the IEEE Information Theory Workshop, March/April, France, Paris, pp. 103–106.

[164] Bahl, L., Cocke, J., Jelinek, F. and Raviv, J. (1974) Optimal decoding of linear codes for minimizing symbol error rate. *IEEE Transactions on Information Theory*, **20**(2), 284–287.

[165] Robertson, P., Villebrun, E. and Hoeher, P. (1995) *A comparison of optimal and sub-optimal MAP decoding algorithms operating in the Log domain*. Proceedings of the International Conference on Communications, June, Seattle, WA, pp. 1009–1013.

[166] Benedetto, S., Divsalar, D., Montorsi, G. and Pollara, F. (1998) Serial concatenation of interleaved codes: performance analysis, design, and iterative decoding. *IEEE Transactions on Information Theory*, **44**, 909–926.

[167] Divsalar, D., Dolinar, S. and Pollara, F. (2000) *Low complexity turbo-like codes*. Proceedings of the 2nd International Symposium on Turbo Codes and Related Topics, September, Brest, France, pp. 73–80.

[168] Divsalar, D., Dolinar, S. and Pollara, F. (2000) *Serial turbo trellis coded modulation with rate-1 inner code*. Proceedings of the IEEE International Symposium on Information Theory (ISIT), June, Sorrento, Italy, p. 194.

[169] ten Brink, S. (2000) Designing iterative decoding schemes with the extrinsic information transfer chart. *AEÜ International Journal of Electronics and Communications*, **54**, 389–398.

[170] Lee, I. (2001) The effect of a precoder on serially concatenated coding systems with an ISI channel. *IEEE Transactions on Communications*, **49**, 1168–1175.

[171] ten Brink, S. (2001) Code characteristic matching for iterative decoding of serially concatenated codes. *Annals of Telecommunications*, **56**, 394–408.

[172] ten Brink, S. (2001) Convergence behavior of iteratively decoded parallel concatenated codes. *IEEE Transactions on Communications*, **49**(10), 1727–1737.

[173] El Gamal, H. and Hammons, A.R. (2001) Analyzing the turbo decoder using the Gaussian approximation. *IEEE Journal on Selected Areas in Communications*, **47**, 671–686.

[174] Tüchler, M. (2002) *Convergence prediction for iterative decoding of threefold concatenated systems*. Proceedings of the IEEE Global Telecommunications Conference (GLOBECOM), vol. 2, November, Taipei, Taiwan, pp. 1358–1362.

[175] Tüchler, M., ten Brink, S. and Hagenauer, J. (2002) *Measures for tracing convergence of iterative decoding algorithms*. Proceedings of the 4th International ITG Conference on Source and Channel Coding, January, Berlin, Germany, pp. 53–60.

[176] Tüchler, M. (2004) Design of serially concatenated systems depending on the block length. *IEEE Transactions on Communications*, **52**, 209–218.

[177] Lifang, Li., Divsalar, D. and Dolinar, S. (2005) Iterative demodulation, demapping and decoding of coded non-square QAM, in *IEEE Transactions on Communications*, **53**, 16–19.

[178] Brännström, F., Rasmussen, L.K. and Grant, A. (2005) Convergence analysis and optimal scheduling for multiple concatenated codes. *IEEE Transactions on Information Theory*, **51**, 3354–3364.

[179] Maunder, R.G., Wang, J., Ng, S.X., Yang, L.-L. and Hanzo, L. (2008) On the performance and complexity of irregular variable length codes for near-capacity joint source and channel coding. *IEEE Transactions on Wireless Communications*, **7**, 1338–1347.

[180] Narayanan, K.R. (2001) Effect of precoding on the convergence of turbo equalization for partial response channels. *IEEE Journal on Selected Areas in Communications*, **19**, 686–698.

[181] Hagenauer, J. (2004) *The EXIT chart – Introduction to extrinsic information transfer in iterative processing*. Proceedings of the European Signal Processing Conference, September, Vienna, Austria, pp. 1541–1548.

[182] Kliewer, J., Ng, S.X. and Hanzo, L. (2006) Efficient computation of EXIT functions for non-binary iterative decoding. *IEEE Transactions on Communications*, **54**, 2133–2136.

[183] Scanavino, B., Montorsi, G. and Benedetto, S. (2001) *Convergence properties of iterative decoders working at bit and symbol level*. Proceedings of the IEEE Global Telecommunications Conference (GLOBECOM), vol. 2, November, San Antonio, TX, pp. 1037–1041.

[184] Kliewer, J., Ng, S.X. and Hanzo, L. (2006) *On the computation of EXIT characteristics for symbol-based iterative decoding*. Proceedings of the 4th International Symposium on Turbo Codes in connection with 6th International ITG-Conference on Source and Channel Coding, Munich, Germany, April.

[185] Wu, N., Alamri, O., Ng, S.X. and Hanzo, L. (2008) Precoded sphere packing aided bit-interleaved differential space-time coded modulation using iterative decoding. *IEEE Transactions on Vehicular Technology*, **57**, 1311–1316.

[186] Alamri, O., Yeap, B.L. and Hanzo, L. (2004) *Turbo detection of channel-coded space-time signals using sphere packing modulation*. Proceedings of the IEEE Vehicular Technology Conference (VTC), vol. 4, September, Los Angeles, CA, pp. 2498–2502.

[187] Alamri, O., Yeap, B.L. and Hanzo, L. (2005) *Turbo detection of channel-coded space-time signals using sphere packing modulation*. Proceedings of the 3rd International Workshop on Signal Processing for Wireless Communications (SPWC), June, King's College, London, UK.

[188] Alamri, O., Yeap, B.L. and Hanzo, L. (2007) A turbo detection and sphere packing modulation aided space-time coding scheme. *IEEE Transactions on Vehicular Technology*, **56**(2), 575–582.

[189] Alamri, O., Nan, Wu. and Hanzo, L. (2006) *A differential turbo detection aided sphere packing modulated space-time coding scheme*. Proceedings of IEEE Vehicular Technology Conference (VTC), May, Melbourne, Australia.

[190] Alamri, O., Wang, J., Ng, S.X., Yang, L.L. and Hanzo, L. (2006) Near-capacity three-stage turbo detection of irregular convolutional coded joint sphere-packing modulation and space-time coding. *IEEE Transactions on Communications*, to appear.

[191] Alamri, O., Wang, J., Ng, S.X., Yang, L.L. and Hanzo, L. (2007) *Near-capacity transceiver design using EXIT-curve fitting: three-stage turbo detection of irregular convolutional coded joint sphere-packing modulation and space-time coding*. Proceedings of the IEEE International Conference on Communications (ICC), June, Glasgow, Scotland.

[192] Alamri, O., Guo, F., Jiang, M. and Hanzo, L. (2005) *Turbo detection of symbol-based non-binary LDPC-coded space-time signals using sphere packing modulation*. Proceedings of the IEEE Vehicular Technology Conference (VTC), vol. 1, September, Dallas, TX, pp. 540–544.

[193] Alamri, O., Ng, S.X., Guo F. and Hanzo, L. (2006) Non-binary LDPC-coded sphere-packed transmit diversity. *IEEE Transactions on Vehicular Technology*, to appear.

[194] Alamri, O., Ng, S.X., Guo, F. and Hanzo, L. (2006) *A purely symbol-based precoded and LDPC-coded iterative-detection assisted sphere-packing modulated space-time coding scheme*. Proceedings of the IEEE Wireless Communications and Networking Conference (WCNC), vol. 3, April, Las Vegas, NV, pp. 1201–1206.

[195] Tüchler, M. and Hagenauer, J. (2002) *EXIT charts of irregular codes*. Proceedings of the 36th Annual Conference on Information Science and Systems (CISS), March, Princeton, NJ, USA.

[196] El-Hajjar, M., Alamri, O., Ng, S.X. and Hanzo, L. (2008) Turbo detection of precoded sphere packing modulation using four transmit antennas for differential space-time spreading. *IEEE Transactions on Wireless Communications*, **7**, 943–952.

[197] El-Hajjar, M., Alamri, O. and Hanzo, L. (2006) *Differential space-time spreading using four transmit antennas and iteratively detected sphere packing modulation*. Proceedings of the IEEE 9th International Symposium on Spread Spectrum Techniques and Applications, August, Manaus-Amazon, pp. 322–326.

[198] Othman, N., El-Hajjar, M., Alamri, O. and Hanzo, L. (2009) Iterative AMR-WB source and channel-decoding of differential space-time spreading assisted sphere packing modulation. *IEEE Transactions on Vehicular Technology*, **58**(1), 484–490.

[199] Othman, N., El-Hajjar, M., Alamri, O. and Hanzo, L. (2007) *Soft-bit assisted iterative AMR-WB source-decoding and turbo-detection of channel-coded differential space-time spreading using sphere packing modulation.* Proceedings of the IEEE Vehicular Technology Conference (VTC), April, pp. 2010–2014.

[200] Othman, N.S., El-Hajjar, M., Pham, A.Q., Alamri, O., Ng, S.X. and Hanzo, L. (2008) *Over-complete source-mapping aided AMR-WB MIMO transceiver using three-stage iterative detection.* Proceedings of the IEEE International Conference on Communications, May, Beijing, China, pp. 751–755.

[201] El-Hajjar, M., Alamri, O. and Hanzo, L. (2007) *Adaptive differential space-time-spreading-assisted turbo-detected sphere packing modulation.* Proceedings of the IEEE Wireless Communications and Networking Conference, March, pp. 645–649.

[202] El-Hajjar, M. and Hanzo, L. (2007) Layered steered space-time codes and their capacity. *Electronics Letters*, **43**, 680–682.

[203] El-Hajjar, M., Alamri, O. and Hanzo, L. (2007) *Layered steered space-time codes using iterative detection.* Proceedings of the IEEE Workshop on Signal Processing Systems, October, Shanghai, China, pp. 35–39.

[204] El-Hajjar, M., Alamri, O., Maunder, R.G. and Hanzo, L. (2008) Layered steered space-time spreading using generalised MC DS-CDMA. *IEEE Transactions on Vehicular Technology*, submitted.

[205] Ganesan, G. and Stoica, P. (2001) Space-time block codes: a maximum SNR approach. *IEEE Transactions on Information Theory*, **47**, 1650–1656.

[206] Tirkkonen, O. and Hottinen, A. (2002) Square-matrix embeddable space-time block codes for complex signal constellations. *IEEE Transactions on Information Theory*, **48**, 384–395.

[207] Su, W. and Xia, X.G. (2003) On space-time block codes from complex orthogonal designs. *Wireless Personal Communications,* **25**, 1–26.

[208] Shokrollahi, A., Hassibi, B., Hochwald, B.M. and Sweldens, W. (2001) Representation theory for high-rate multiple-antenna code design. *IEEE Transactions on Information theory*, **47**, 2335–2367.

[209] Hassibi, B. and Hochwald, B.M. (2002) Cayley differential unitary space-time codes. *IEEE Transactions on Information Theory*, **48**, 1485–1503.

[210] Liang, X.-B. and Xia, X.-G. (2002) Unitary signal constellations for differential space-time modulation with two transmit antennas: parametric codes, optimal designs, and bounds. *IEEE Transactions on Information Theory*, **48**, 2291–2322.

[211] Grimm, J., Fitz, M. and Krogmeier, J. (1998) *Further results on space-time coding for Rayleigh fading*. Proceedings of the 36th Allerton Conference on Communications, Control and Computing, Illinois, USA, pp. 391–400.

[212] Hammons Jr, A.R. and El Gamal, H. (2000) On the theory of space-time codes for PSK modulation. *IEEE Transactions on Information Theory*, **46**, 524–542.

[213] Safar, Z. and Liu, K.J. (2002) Systematic design of space-time trellis codes for diversity and coding advantages. *EURASIP Journal on Applied Signal Processing*, 2002:3, 221–235.

[214] Yan, Q. and Blum, R. (2001) *Robust space-time block coding for rapid fading channels.* Proceedings of the IEEE Global Telecommunications Conference (GLOBECOM), vol. 1, November, San Antonio, TX, pp. 460–464.

[215] Zummo, S. and Al-Semari, S. (2000) *Space-time coded QPSK for rapid fading channels.* Proceedings of the IEEE International Symposium on Personal, Indoor and Mobile Radio Communications (PIMRC), vol. 1, London, UK, pp. 504–508.

[216] Firmanto, W., Vucetic, B. and Yuan, J. (2001) Space-time TCM with improved performance on fast fading channels. *IEEE Communications Letters*, **5**, 154–156.

[217] Golub, G.H. and Van-Loan, C.F. (1996) *Matrix Computations*, Johns Hopkins University Press, Baltimore, MD.

[218] Fitz, M., Grimm, J. and Siwamogsatham, S. (1999) *A new view of performance analysis techniques in correlated Rayleigh fading*. Proceedings of the IEEE Wireless Communications and Networking Conference (WCNC), vol. 1, September, New Orleans, LA, pp. 139–144.

[219] Siwamogsatham, S., Fitz, M.P. and Grimm, J. (2002) A new view of performance analysis of transmit diversity schemes in correlated Rayleigh fading. *IEEE Transactions on Information Theory*, **48**, 950–956.

[220] Jakes, W.C. (1974) *Microwave Mobile Communications*, John Wiley & Sons, Ltd, New York.

[221] Conway, J.H. and Sloane, N.J. (1999) *Sphere Packings, Lattices and Groups*, Springer, Berlin.

[222] Hales, T.C. (2000) Cannonballs and honeycombs. *Notices of the American Mathematical Society*, **47**, 440–449.

[223] Hales, T.C. (2005) A proof of the Kepler conjecture. *Annals of Mathematics*, **162**, 1065–1185.

[224] Smith, H.J.S. (1867) On the orders and genera of quadratic forms containing more than three indeterminates. *Royal Society of London Proceedings Series I*, **16**, 197–208.

[225] Korkine, A. and Zolotare, G. (1873) Sur les formes quadratiques. *Mathematische Annalen* **6**, 366–389.

[226] Levenshtein, V.I. (1979) On bounds for packing in n-dimensional Euclidean space. *Soviet Mathematics Doklady*, **20**, 417–421.

[227] Odlyzko, A.M. and Sloane, N.J.A. (1979) New bounds on the number of unit spheres that can touch a unit sphere in n dimensions. *Journal of Combinatorial Theory* A**26**, 210–214.

[228] Hilbert, D. (1901) Mathematische Probleme. *Archiv der Mathematik und Physik*, **3**, 44–63.

[229] Milnor, J. (1976) Hilbert's problem 18: on crystallographic groups, fundamental domains and on sphere packing. *Proceedings Symposium Pure Mathematics*, **28**, 491–506.

[230] Baranovskii, E.P. (1971) Packings, coverings, partitionings and certain other distributions in spaces of constant curvature. *Progress in Mathematics*, **9**, 209–253.

[231] Coxeter, H.S.M. (1968) *Twelve Geometric Essays*, Southern Illinois University Press, Carbondsale, IL.

[232] Toth, G. (1964) *Regular Figures*, Pergamon, Oxford.

[233] Hahn, T. (2002) *International tables for crystallography*, Kluwer, Dortrech.

[234] Hammer, J. (1977) *Unsolved Problems Concerning Lattice Points*, Pitman, San Francisco, CA.

[235] Schwarzenberger, R.L.E. (1980) *N-Dimensional Crystallography*, Pitman, San Francisco, CA.

[236] Cassels, J.W.S. (1971) *An Introduction to the Geometry of Numbers*, Springer, Berlin.

[237] Hancock, H. (1964) *Development of the Minkowski Geometry of Numbers*, Dover, New York.

[238] Keller, O.H. (1954) *Geometrie der Zahlen*, Springer, Berlin.

[239] Aird, T.J. and Rice, J.R. (1977) Systematic Search in High Dimensional Sets. *SIAM Journal on Numerical Analysis*, **14**, 296–312.

[240] Sobol, I.M. (1979) On the Systematic Search in a Hypercube. *SIAM Journal on Numerical Analysis*, **16**, 790–793.

[241] Renka, R.J. (1984) Interpolation on the Surface of a Sphere. *ACM Transactions on Mathematical Software*, **10**, 437–439.

[242] Dobkin, D. and Lipton, R.J. (1976) Multidimensional Searching Problems. *SIAM Journal on Computing*, **5**, 181–186.

[243] Fincke, U. and Pohst, M. (1985) Improved methods for calculating vectors of short length in a lattice, including a complexity analysis. *Mathematics of Computation*, **44**, 463–471.

[244] Hochwald, B.M., Caire, G., Hassibi, B. and Marzetta, T.L. Eds. (2003) Special issue on space-time transmission, reception, coding and signal processing. *IEEE Transactions on Information Theory*, **49**, 2329–2806.

[245] Vucetic, B. and Yuan, J. (2003) *Space-Time Coding*, John Wiley & Sons, Ltd, Chichester/IEEE Press, Piscataway, NJ.

[246] Ng, S.X. and Hanzo, L. (2006) On the MIMO channel capacity of multidimensional signal sets. *IEEE Transactions on Vehicular Technology*, **55**, **2**, 528–536.

[247] Ng, S.X. and Hanzo, L. (2002) Space-time IQ-interleaved TCM and TTCM for AWGN and Rayleigh fading channels. *IEE Electronics Letters*, **38**, 1553–1555.

[248] Hanzo, L., Ng, S.X., Keller, T. and Webb, W.T. (2004) *Quadrature Amplitude Modulation: From Basics to Adaptive Trellis-Coded, Turbo-Equalised and Space-Time Coded OFDM, CDMA and MC-CDMA System*, John Wiley & Sons, Ltd, Chichester/IEEE Press, Piscataway, NJ, 2nd edn.

[249] Gallager, R. (1968) *Information Theory and Reliable Communication*, John Wiley & Sons, Ltd, New York.

[250] McIllree, P.E. (1995) Channel capacity calculations for *M*-ary *N*-dimensional signal sets. MS Thesis. School of Electronic Engineering, University of South Australia, Adelaide, Australia.

[251] Telatar, E. (1999) Capacity of multi-antenna Gaussian channels. *European Transactions on Telecommunication*, **10**, 585–595.

[252] Goff, S.L., Glavieux, A. and Berrou, C. (1994) *Turbo-codes and high spectral efficiency modulation*. Proceedings of the IEEE International Conference on Communications (ICC), May, New Orleans, LA, pp. 645–649.

[253] Mittelholzer, T., Lin, X. and Massey, J. (1997) *Multilevel turbo coding for M-ary quadrature and amplitude modulation*. Proceedings of the International Symposium on Turbo Codes and Related Topics, September, Brest, France, pp. 127–134.

[254] Robertson, P. (1997) *An overview of bandwidth efficient turbo coding schemes*. Proceedings of the International Symposium on Turbo Codes and Related Topics, September, Brest, France, pp. 103–110.

[255] Richardson, T.J. and Urbanke, R. (2001) The capacity of low-density parity-check codes under message-passing decoding. *IEEE Transactions on Information Theory*, **47**, 599–618.

[256] Richardson, T.J., Shokrollahi, A. and Urbanke, R. (2001) Design of capacity-approaching low-density parity-check codes. *IEEE Transactions on Information Theory*, **47**, 619–637.

[257] Chung, S.Y., Forney, G.D., Richardson, T.J. and Urbanke, R. (2001) On the design of low-density parity-check codes within 0.0045 dB of the Shannon limit. *IEEE Communication Letter*, **5**, 58–60.

[258] Peleg, M., Sason, I., Shamai, S. and Elia, A. (1999) On interleaved differentially encoded convolutional codes. *IEEE Transactions on Information Theory*, **45**, 2572–2582.

[259] Tüchler, M., ten Brink, S. and Hagenauer, J. (2002) *Measures for tracing convergence of iterative decoding algorithms*. Proceedings of the 4th International ITG Conference on Source and Channel Coding, January, Berlin, Germany, pp. 53–60.

[260] Leon-Garcia, A. (1994) *Probability and Random Processes for Electrical Engineering*, Addison-Wesley, New York, 2nd edn.

[261] Hagenauer, J., Offer, E. and Papke, L. (1996) Iterative decoding of binary block and convolutional codes. *IEEE Transactions on Information Theory*, **42**(2), 429–445.

[262] Cover, T.M. and Thomas, J.A. (1991) *Elements of Information Theory*. John Wiley & Sons, Ltd, Chichester/IEEE Press, Piscataway, NJ.

[263] Ungerboeck, G. (1982) Channel coding with multilevel/phase signals. *IEEE Transactions on Information Theory*, **28**(1), 55–67.

[264] Hwang, C.-S., Nam, S.H., Chung, J. and Tarokh, V. (2003) *Differential space time block codes using QAM constellations*. Proceedings of the IEEE International Symposium on Personal, Indoor and Mobile Radio Communications, (PIMRC), vol. 2, September, Beijing, China, pp. 1693–1697.

[265] Breiling, M. and Hanzo, L. (2000) The super-trellis structure of turbo codes. *IEEE Transactions on Information Theory*, **46**, 2212–2228.

[266] Divsalar, D., Dolinar, S. and Pollara, F. (2000) *Serial concatenated trellis coded modulation with rate-1 inner code*. Proceedings of the IEEE Global Telecommunications Conference (GLOBECOM), vol. 2, November-December, San Francisco, CA, pp. 777–782.

[267] Tullberg, H.M. and Siegel, P.H. (2001) *Serial concatenated trellis coded modulation with inner rate-1 accumulate code*. Proceedings of the IEEE Global Telecommunications Conference (GLOBECOM), vol. 2, November, San Antonio, TX, pp. 936–940.

[268] ten Brink, S. (2001) *Convergence of multi-dimensional iterative decoding schemes*. Proceedings of the 35th Asilomar Conference on Signals, Systems, and Computers, vol. 1, November, Pacific Grove, CA, pp. 270–274.

[269] Brännström, F., Rasmussen, L.K. and Grant, A. (2003) *Optimal scheduling for multiple serially concatenated codes*. Proceedings of the 3rd International Symposium on Turbo Codes and Related Topics, September, Brest, France, pp. 383–386.

[270] Ashikhmin, A., Kramer, G. and ten Brink, S. (2004) Extrinsic information transfer functions: model and erasure channel properties. *IEEE Transactions on Information Theory*, **50**, 2657–2673.

[271] Divsalar, D., Jin, H. and McEliece, R.J. (1998) *Coding theorems for turbo-like codes*. Proceedings of the 36th Allerton Conference on Communications, Control, and Computing, September, Illinois, USA, pp. 201–210.

[272] Shannon, C.E. (1948) A mathematical theory of communication. *Bell Systems Technical Journal*, **27**, 379–423.

[273] Ashikhmin, A., Kramer, G. and ten Brink, S. (2002) *Extrinsic information transfer functions: A model and two properties*. Proceedings of the 36th Annual Conference on Information Sciences and Systems (CISS), March, Princeton, NJ, USA.

[274] Gallager, R. (1962) Low density parity check codes. *IEEE Transactions on Information Theory*, **8**, 21–28.

[275] Richardson, T. and Urbanke, R. (2003) The renaissance of Gallager's low-density parity-check codes. *IEEE Communications Magazine*, **41**, 126–131.

[276] MacKay, D.J. and Neal, R.M. (1996) Near Shannon limit performance of low density parity check codes. *IEE Electronics Letters*, **32**.

[277] Davey, M.C. and MacKay, D.J. (1998) Low density parity check codes over GF(q). *IEEE Communications Letters*, **2**, 165–167.

[278] Grant, A. (2001) *Convergence of non-binary iterative decoding*. Proceedings of the IEEE Global Telecommunications Conference (GLOBECOM), vol. 2, November, San Antonio, TX, pp. 1058–1062.

[279] Land, I., Hoeher, P. and Gligorevic, S. (2004) *Computation of symbol-wise mutual information in transmission systems with logAPP decoders and application to EXIT charts*. Proceedings of the International ITG Conference on Source and Channel Coding (SCC), January, Erlangen, Germany, pp. 195–202.

[280] Chen, H. and Haimovich, A. (2004) EXIT charts for turbo trellis-coded modulation. *IEEE Communications Letters*, **8**, 668–670.

[281] Ng, S.X., Chung, J.Y. and Hanzo, L. (2005) Turbo-detected unequal protection MPEG-4 wireless video telephony using multi level coding, trellis coded modulation and space-time trellis coding. *IEE Proceedings on Communications*, **152**, 1116–1124.

[282] Ng, S.X., Wang, J., Tao, M., Yang, L.-L. and Hanzo, L. (2007) Iteratively decoded variable length space-time coded modulation: code construction and convergence analysis. *IEEE Transactions on Wireless Communications*, **6**(5), 1953–1963.

[283] Damen, M.O., Gamal, H.E. and Beaulieu, N.C. (2003) Linear threaded algebraic space-time constellations. *IEEE Transactions on Information Theory*, **49**, 2372–2388.

[284] Ma, X.L. and Giannakis, G.B. (2003) Full-diversity full-rate complex-field space-time coding. *IEEE Transactions on Signal Processing*, **51**(11), 2917–2930.

[285] Gohary, R.H. and Davidson, T.N. (2005) Design of linear dispersion codes: asymptotic guidelines and their implementation. *IEEE Transactions on Wireless Communications*, **4**, 2892–2906.

[286] Proakis, J.G. (1989) *Digital Communications*, McGraw-Hill, New York, 2nd ed.

[287] Daubechies, I. (1992) *Ten Lectures on Wavelets*, SAIM, Philadelphia, PA.

[288] Belfiore, J.C., Rekaya, G. and Viterbo, E. (2005) The golden code: a 2×2 full-rate space-time code with nonvanishing determinants. *IEEE Transactions on Information Theory*, **51**, 1432–1436.

[289] Hong, Y., Viterbo, E. and Belfiore, J.C. (2007) Golden space-time trellis coded modulation. *IEEE Transactions on Information Theory*, **53**, 1689–1705.

[290] Wu, N. and Hanzo, L. (2008) Near-capacity irregular convolutional-coding aided irregular precoded linear dispersion codes. *IEEE Transactions on Vehicular Technology*.

[291] Damen, M.O., Gamal, H.E. and Caire, G. (2003) On maximum-likelihood detection and the search for the closest lattice point. *IEEE Transactions on Information Theory*, **49**, 2389–2402.

[292] Wang, J.B., Wang, X.D. and Madihian, M. (2005) On the optimum design of space-time linear-dispersion codes. *IEEE Transactions on Wireless Communications*, **4**, 2928–2938.

[293] Ng, S.X., Jörg Kliewer, Alamri, O. and Hanzo, L. (2006) *On the design of turbo trellis coded modulation schemes using symbol-based EXIT charts*. Proceedings of the IEEE Vehicular Technology Conference (VTC), September, Montreal, Canada.

[294] Horn, R.A. and Johnson, C.R. (1985) *Matrix Analysis*, Cambridge University Press, Cambridge.

[295] Lu, H.F. and Kumar, P.V. (2005) A unified construction of space-time codes with optimal rate-diversity tradeoff. *IEEE Transactions on Information Theory*, **51**, 1709–1730.

[296] Damen, M.O. and Beaulieu, N.C. (2003) On diagonal algebraic space-time block codes. *IEEE Transactions on Communications*, **51**, 911–919.

[297] Sharma, N. and Papadias, C.B. (2003) Improved quasi-orthogonal codes through constellation rotation. *IEEE Transactions on Communications*, **51**(3), 332–335.

[298] Su, W.F. and Xia, X.G. (2002) *Quasi-orthogonal space-time block codes with full diversity*. Proceedings of the IEEE Global Telecommunications Conference, vol. 2, November, pp. 1098–1102.

[299] Su, W.F. and Xia, X.G. (2004) Signal constellations for quasi-orthogonal space-time block codes with full diversity. *IEEE Transactions on Information Theory*, **50**(10), 2331–2347.

[300] Ganesan, G. and Stoica, P. (2000) *Space-time diversity using orthogonal and amicable orthogonal designs*. Proceedings of the IEEE International Conference on Acoustics, Speech, and Signal Processing, vol. 5, June, Istanbul, pp. 2561–2564.

[301] Sandhu, S. and Paulraj, A. (2000) Space-time block codes: a capacity perspective. *IEEE Communications Letters*, **4**, **12**, 384–386.

[302] Sandhu, S. and Paulraj, A. (2001) *Unified design of linear space-time block codes*. Proceedings of the IEEE Global Telecommunications Conference, vol. 2, November, San Antonio, TX, pp. 1073–1077.

[303] Yuen, C., Guan, Y.L. and Tjhung, T.T. (2006) *Algebraic relationship between amicable orthogonal designs and quasi-orthogonal STBC with minimum decoding complexity*. Proceedings of the IEEE International Conference on Communications, vol. 11, June, Istanbul, pp. 4882–4887.

[304] Yuen, C., Guan, Y.L. and Tjhung, T.T. (2005) Quasi-orthogonal STBC with minimum decoding complexity. *IEEE Transactions on Wireless Communications*, **4**, 2089–2094.

[305] Alamri, O., Wu, N. and Hanzo, L. (2006) *A differential turbo detection aided sphere packing modulated space-time coding scheme*. Proceedings of the Vehicular Technology Conference, vol. 5, May, Melbourne, pp. 2474–2478.

[306] Wang, J., Ng, S.X., Wolfgang, A., Yang, L.-L., Chen, S. and Hanzo, L. (2006) *Near-capacity three-stage MMSE turbo equalisation using irregular convolutional codes*. Proceedings of the International Symposium on Turbo Codes, April, Munich, Germany.

[307] Maunder, R.G., Wang, J., Ng, S.X., Yang, L.L. and Hanzo, L. (2007) Iteratively decoded irregular variable length coding and trellis coded modulation, in *IEEE Workshop on Signal Processing Systems*, (Shanghai), October, pp. 222–227.

[308] Hochwald, B.M. and Sweldens, W. (2000) Differential unitary space-time modulation. *IEEE Transactions on Communications*, **48**(12), 2041–2052.

[309] Cheng, S., Nallanathan, A. and Kam, P.Y. (2004) A new class of signal constellations for differential unitary space-time modulation (DUSTM). *IEEE Communications Letters*, **8**, 1–3.

[310] Soh, T.P., Ng, C.S. and Kam, P.Y. (2005) Improved signal constellations for differential unitary space-time modulations with more than two transmit antennas. *IEEE Communications Letters*, **9**, 7–9.

[311] Himsoon, T., Su, W.F. and Liu, K.J.R. (2005) Differential unitary space-time signal design using matrix rotation structure. *IEEE Signal Processing Letters*, **12**, 45–48.

[312] Zheng, L. and Tse, D.N.C. (2002) Communication on the Grassmann manifold: a geometric approach to the noncoherent multiple-antenna channel. *IEEE Transactions on Information Theory*, **48**(2), 359–383.

[313] Moustakas, A.L., Simon, S.H. and Marzetta, T.L. (2006) Capacity of differential versus nondifferential unitary space-time modulation for MIMO channels. *IEEE Transactions on Information Theory*, **52**, 3622–3634.

[314] Ganesan, G. and Stoica, P. (2002) Differential modulation using space-time block codes. *IEEE Signal Processing Letters*, **9**(2), 57–60.

[315] Yuen, C., Guan, Y.L. and Tjhung, T.T. (2006) Single-symbol-decodable differential space-time modulation based on QO-STBC. *IEEE Transactions on Wireless Communications*, **5**(12), 3329–3334.

[316] Ertel, R. and Reed, J.H. (1998) Generation of two equal power correlated Rayleigh fading envelopes. *IEEE Communications Letters*, **2**(10), 276–278.

[317] Hajiaghayi, M. and Tellambura, C. (2007) Unitary signal constellations for differential space-time modulation. *IEEE Communications Letters*, **11**, 25–27.

[318] Oggier, F. and Hassibi, B. (2007) Algebraic Cayley differential space-time codes. *IEEE Transactions on Information Theory*, **53**, 1911–1919.

[319] Wang, J.B., Wang, X.D. and Madihian, M. (2005) Design of minimum error-rate Cayley differential unitary space-time codes. *IEEE Journal on Selected Areas in Communications*, **23**, 1779–1787.

[320] Chindapol, A. and Ritcey, J.A. (2001) Design, analysis, and performance evaluation for BICM-ID with square QAM constellations in Rayleigh fading channels. *IEEE Journal on Selected Areas in Communications*, **19**(5), 944–957.

[321] Li, Y.B. and Xia, X.G. (2005) Constellation mapping for space-time matrix modulation with iterative demodulation/decoding. *IEEE Transactions on Communications*, **53**(5), 764–768.

[322] Narayanan, K.R. and Stüber, G.L. (1999) A serial concatenation approach to iterative demodulation and decoding. *IEEE Transactions on Communications*, **47**(7), 956–961.

[323] Hoeher, P. and Lodge, J. (1999) Turbo DPSK: iterative differential PSK demodulation and channel decoding. *IEEE Transactions on Communications*, **47**(6), pp. 837–843.

[324] Huang, Y. and Ritcey, J.A. (2005) Improved 16-QAM constellation labeling for BI-STCM-ID with the Alamouti scheme. *IEEE Communications Letters*, **9**, 157–159.

[325] Hanzo, L., Blogh, J. and Ni, S. (2008) *3G, HSPA and FDD versus TDD Networking: Smart Antennas and Adaptive Modulation*, John Wiley & Sons, Ltd, Chichester/IEEE Press, Piscataway, NJ, 2nd ed.

[326] Fan, Y. and Thompson, J. (2007) MIMO configurations for relay channels: Theory and practice. *IEEE Transactions on Wireless Communications*, **6**, 1774–1786.

[327] Dohler, M., Li, Y., Vucetic, B., Aghvami, A.H., Arndt, M. and Barthel, D. (2006) Performance analysis of distributed space-time block-encoded sensor networks. *IEEE Transactions on Vehicular Technology*, **55**, 1776–1789.

[328] Kramer, G., Gastpar, M. and Gupta, P. (2005) Cooperative strategies and capacity theorems for relay networks. *IEEE Transactions on Information Theory*, **51**, 3037–3063.

[329] Liang, K.C., Wang, X.D. and Berenguer, I. (2007) Minimum error-rate linear dispersion codes for cooperative relays. *IEEE Transactions on Vehicular Technology*, **56**, 2143–2157.

[330] Jing, Y.D. and Hassibi, B. (2006) Distributed space-time coding in wireless relay networks. *IEEE Transactions on Wireless Communications*, **5**(12), 3524–3536.

[331] Yiu, S., Schober, R. and Lampe, L. (2006) Distributed space-time block coding. *IEEE Transactions on Communications*, **54**, 1195–1206.

[332] Stauffer, E., Oyman, O., Narasimhan, R. and Paulraj, A. (2007) Finite-SNR diversity-multiplexing tradeoffs in fading relay channels. *IEEE Journal on Selected Areas in Communications*, **25**, 245–257.

[333] Yang, S. and Belfiore, J.C. (2007) Optimal space-time codes for the MIMO amplify-and-forward cooperative channel. *IEEE Transactions on Information Theory*, **53**, 647–663.

[334] Wei, S. (2007) Diversity–multiplexing tradeoff of asynchronous cooperative diversity in wireless networks. *IEEE Transactions on Information Theory*, **53**, 4150–4172.

[335] Shang, Y. and Xia, X.G. (2006) Shift-full-rank matrices and applications in space-time trellis codes for relay networks with asynchronous cooperative diversity. *IEEE Transactions on Information Theory*, **52**, 3153–3167.

[336] Liu, J., Li, J. and Larsson, E.G. (2002) Differential space-time block code modulation for DS-CDMA systems. *EURASIP Journal on Applied Signal Processing*, **3**, 289–296.

[337] Seshadri, N., Tarokh, V. and Calderbank, A.R. (1997) *Space-time codes for wireless communication: code construction*. Proceedings of the IEEE Vehicular Technology Conference, vol. 2, Phoenix, AZ, pp. 637–641.

[338] Foschini, G.J., Chizhik, D., Gans, M.J., Papadias, C. and Valenzuela, R.A. (2003) Analysis and performance of some basic space-time architectures. *IEEE Journal on Selected Areas in Communications*, **21**(3), 303–320.

[339] Ganesan, G. and Stoica, P. (2001) Utilizing space-time diversity for wireless communications. *Wireless Personal Communications*, **18**, 149–163.

[340] Ganesan, G. and Stoica, P. (2001) Space-time diversity using orthogonal and amicable orthogonal designs. *Wireless Personal Communications*, **18**, 165–178.

[341] Ganesan, G., Stoica, P. and Larsson, E. (2004) Orthogonal space-time block codes with feedback. *Wireless Personal Communications*, **28**, 287–312.

[342] Hanzo, L., Münster, M., Choi, B.J. and Keller, T. (2003) *OFDM and MC-CDMA for broadband multi-user communications, WLANs and broadcasting*, John Wiley & Sons, Ltd, Chichester/IEEE Press, Piscataway, NJ.

[343] Alamri, O. (2006) *Turbo detection of sphere packing modulation aided space-time coding schemes*. PhD thesis, School of Electronics and Computer Science, University of Southampton, September 2006.

[344] Alamri, O., Yeap, B.L. and Hanzo, L. (2004) *Turbo detection of channel-coded space-time signals using sphere packing modulation*. Prodeedings of the IEEE Vehicular Technology Conference, vol. 4, September 2004, Los Angeles, CA, pp. 2498–2502.

[345] Shin, H. and Lee, J.H. (2003) Capacity of multiple-antenna fading channels: spatial fading correlation, double scattering and keyhole. *IEEE Transactions on Information Theory*, **49**(10), 2636–2647.

[346] Wozencraft, J.M. and Jacobs, I.M. (1965) *Principles of communication engineering*, John Wiley & Sons, Ltd, Chichester/IEEE Press, Piscataway, NJ, USA.

[347] Sae-Young Chung, Richardson, T.J. and Urbanke, R.L. (2001) Analysis of sum-product decoding of low-density parity-check codes using a Gaussian approximation. *IEEE Transactions on Information Theory*, **47**, 657–670.

[348] ten Brink, S., Speidel, J. and Han, R.-H. (1998) Iterative demapping for QPSK modulation. *Electronics Letters*, **34**(15), 1459–1460.

[349] Balakirsky, V.B. (1997) *Joint source-channel coding with variable length codes*. Proceedings of the IEEE International Symposium on Information Theory, June, Ulm, Germany, p. 419.

[350] Hanzo, L., Somerville, F.C. and Woodard, J.P. (2001) *Voice compression and communications: Principles and applications for fixed wireless channels*, John Wiley & Sons, Ltd, Chichester/IEEE Press, Piscataway, NJ.

[351] Hanzo, L., Cherriman, P.J. and Streit, J. (2001) *Wireless video communications: second to third generation and beyond*, John Wiley & Sons, Ltd, Chichester/IEEE Press, Piscataway, NJ.

[352] Fingscheidt, T. and Vary, P. (1997) *Speech decoding with error concealment using residual source redundancy*. Proceedings of the IEEE Workshop on Speech Coding for Telecommunication, September, pp. 91–92.

[353] Fingscheidt, T. and Vary, P. (2001) Softbit speech decoding: A new approach to error concealment. *IEEE Transactions on Speech and Audio Processing*, **9**, 240–251.

[354] Adrat, M., Vary, P. and Spittka, J. (2001) *Iterative source-channel decoder using extrinsic information from softbit-source decoding*. Proceedings of the IEEE International Conference on Acoustics, Speech and Signal Processing, May, pp. 2653–2656.

[355] Clevorn, T., Brauers, J., Adrat, M. and Vary, P. (2005) Turbo decodulation: Iterative combined demodulation and source-channel decoding. *IEEE Communications Letters*, **9**, 820–822.

[356] Bessette, B., Salami, R., Lefebvre, R., Jelinek, M., Rotola-Pukkila, J., Vainio, J., Mikkola, H. and Jarvinen, K. (2002) The adaptive multirate wideband speech codec (AMR-WB). *IEEE Transactions on Speech and Audio Processing*, **10**, 620–636.

[357] Hindelang, T., Fingscheidt, T., Seshadri, N. and Cox, R.V. (2000) *Combined source/channel (de-)coding: can* a priori *information be used twice?* Proceedings of the IEEE International Conference on Communications, June, pp. 1208–1212.

[358] Lloyd, S. (1982) Least squares quantization in PCM. *IEEE Transactions on Information Theory*, **28**, 129–137.

[359] Buttigieg, V. and Farrell, P.G. (2000) Variable-length error-correcting codes. *IEE Proceedings – Communications*, **147**, 211–215.

[360] Dimitriou, N., Tafazolli, R. and Sfikas, G. (2000) Quality of service for multimedia CDMA. *IEEE Communications Magazine*, **38**, 88–94.

[361] Gilhousen, K.S., Jacobs, I.M., Padovani, R., Viterbi, A.J., Weaver, L.A.Jr. and Wheatley, C.E. (1991) On the capacity of a cellular CDMA system. *IEEE Transactions on Vehicular Technology*, **40**, 303–312.

[362] Hanzo, L., Wong, C.H. and Yee, M.S. (2002) *Adaptive wireless transceivers: turbo-coded, turbo-equalized and space-time coded TDMA, CDMA and OFDM systems*, John Wiley & Sons, Ltd, Chichester/IEEE Press, Piscataway, NJ.

[363] Bello, P.A. (1998) Sample size required in error-rate measurement on fading channels. *Proceedings of the IEEE*, **86**, 1435–1441.

[364] Yang, L.-L. and Hanzo, L. (2005) Adaptive space-time-spreading-assisted wideband CDMA systems communicating over dispersive Nakagami-m fading channels. *EURASIP Journal on Wireless Communications and Networking*, **2**, 216–230.

[365] Webb, W.T. and Steele, R. (1995) Variable rate QAM for mobile radio. *IEEE Transactions on Communications*, **43**(7), 2223–2230.

[366] Goldsmith, A.J. and Chua, S.-G. (1997) Variable-rate variable-power MQAM for fading channels. *IEEE Transactions on Communications*, **45**, 1218–1230.

[367] Hayes, J. (1968) Adaptive feedback communications. *IEEE Transactions on Communications*, **16**, 29–34.

[368] Cavers, J. (1972) Variable-rate transmission for Rayleigh fading channels. *IEEE Transactions on Communications*, **20**(1), 15–22.

[369] Choi, B.-J. and Hanzo, L. (2003) Optimum mode-switching-assisted constant-power single- and multi-carrier adaptive modulation. *IEEE Transactions on Vehicular Technology*, **52**(3), 536–560.

[370] Yang, L.-L. and Hanzo, L. (2004) Adaptive rate DS-CDMA systems using variable spreading factors. *IEEE Transactions on Vehicular Technology*, **53**(1), 72–81.

[371] Ottosson, T. and Svensson, A. (1998) On schemes for multirate support in DS-CDMA systems. *Wireless Personal Communications,* **6**, 265–287.

[372] Ojanpera, T. and Prasad, R. (1998) *Wideband CDMA for third generation mobile communications*, Artech House, Norwood, MA.

[373] Kim, S.W. (1999) Adaptive rate and power DS-CDMA communications in fading channels. *IEEE Communications Letters*, **3**, 85–87.

[374] Abeta, S., Sampei, S. and Morinaga, N. (1996) *Channel activation with adaptive coding rate and processing gain control for cellular DS-CDMA systems*. Proceedings of the IEEE 46th Vehicular Technology Conference, vol. 2, April/May, Atlanta, GA, pp. 1115–1119.

[375] Linnan, L., Jingming, K. and Zesong, F. (2006) *Combining Eigen-beamforming and space-time block coding with equal power allocation*. Proceedings of the International Conference on Communication Technology (ICCT), November, Guilin, China, pp. 1–4.

[376] Hanzo, L. and Keller, T. (2006) *OFDM and MC-CDMA: A Primer*, John Wiley & Sons, Ltd, Chichester/IEEE Press, Piscataway, NJ.

[377] Vandendorpe, L. (1995) Multitone spread spectrum multiple access communications system in a multipath Rician fading channel. *IEEE Transactions on Vehicular Technology*, **44**, 327–337.

[378] Sourour, E.A. and Nakagawa, M. (1996) Performance of orthogonal multicarrier CDMA in a multipath fading channel. *IEEE Transactions on Communications*, **44**, 356–367.

[379] El-Hajjar, M., Hu, B., Yang, L.-L. and Hanzo, L. (2007) Coherent and differential downlink space-time steering aided generalised multicarrier DS-CDMA. *IEEE Transactions on Wireless Communications*, **6**, 3857–3863.

[380] El-Hajjar, M., Tee, R.Y.S., Bin, H., Yang, L.-L. and Hanzo, L. (2007) *Downlink steered space-time spreading assisted generalised Multicarrier DS-CDMA using sphere packing aided multilevel coding*. Proceedings of the IEEE 66th Vehicular Technology Conference, September, Baltimore, MD, USA, pp. 472–476.

[381] Bin Hu, (2006) *Multicarrier DS-CDMA communication systems using smart antennas*. PhD thesis, School of Electronics and Computer Science, University of Southampton, April.

[382] Gui, X. and Ng, T.S. (1999) Performance of asynchronous orthogonal multicarrier CDMA system in frequency selective fading channel. *IEEE Transactions on Communications*, **47**, 1084–1091.

[383] Hara, S. and Prasad, R. (1997) Overview of multicarrier CDMA, *IEEE Communications Magazine*, vol. 35 126–133.

[384] Yang, L.-L. and Hanzo, L. (2005) Performance of broadband multicarrier DS-CDMA using space-time spreading-assisted transmit diversity. *IEEE Transactions on Wireless Communications*, **4**(3), 885–894.

[385] ten Brink, S. (1999) Convergence of iterative decoding. *Electronics Letters*, **35**(13), 1117–1119.

[386] Scutari, G. and Barbarossa, S. (2005) Distributed space-time coding for regenerative relay networks. *IEEE Transactions on Wireless Communications*, **4**, 2387–2399.

[387] Nosratinia, A., Hunter, T.E. and Hedayat, A. (2004) Cooperative communication in wireless networks. *IEEE Communications Magazine*, **42**(10), 74–80.

[388] Jing, Y. and Hassibi, B. (2004) *Wireless networks, diversity and space-time codes*. Proceedings of the IEEE Information Theory Workshop, October, 463–468.

[389] Berrou, C. and Glavieux, A. (1996) Near optimum error correcting coding and decoding: turbo-codes. *IEEE Transactions on Communications*, **44**, 1261–1271.

[390] Wei, L.-F. (1987) Trellis-coded modulation with multidimensional constellations. *IEEE Transactions on Information Theory*, **33**, 483–501.

[391] Wei, L.-F. (1989) Rotationally invariant trellis-coded modulations with multidimensional M-PSK. *IEEE Journal on Selected Areas in Communications*, **7**, 1281–1295.

[392] Deng, R.H. and Costello-Jr, D.J. (1989) High rate concatenated coding systems using multidimensional bandwidth-efficient trellis inner codes. *IEEE Transactions on Communications*, **37**, 1091–1096.

[393] Pietrobon, S.S., Deng, R.H., Lafanechere, A., Ungerboeck, G. and Costello-Jr, D.J. (1990) Trellis-coded multidimensional phase modulation. *IEEE Transactions on Information Theory*, **36**, 63–89.

[394] Tran, N.H. and Nguyen, H.H. (2004) *Improving the performance of QPSK BICM-ID by mapping on the hypercube*. Proceedings of the IEEE Vehicular Technology Conference (VTC), vol. 2, September, Los Angeles, CA, pp. 1299–1303.

[395] Tran, N.H. and Nguyen, H.H. (2006) Design and performance of BICM-ID systems with hypercube constellations. *IEEE Transactions on Wireless Communications*, **5**, 1169–1179.

[396] Bäro, S. (2004) *Turbo-detection in MIMO systems: bit labeling and pre-coding*. Proceedings of the 5th International ITG Conference on Source and Channel Coding, January, Erlangen, Germany.

[397] Simoens, F., Wymeersch, H. and Moeneclaey, M. (2004) *Spatial mapping for MIMO systems*. Proceedings of the IEEE Information Theory Workshop, October, San Antonio, TX, pp. 187–192.

[398] Simoens, F., Wymeersch, H., Bruneel, H. and Moeneclaey, M. (2005) Multidimensional mapping for bit-interleaved coded modulation with BPSK/QPSK signaling. *IEEE Communications Letters*, **9**, 453–455.

[399] Wymeersch, H., Simoens, F., Bruneel, H. and Moeneclaey, M. (2005) *Multi-dimensional modulation for UWB communications*. Proceedings of the IEEE Workshop on Signal Processing Advances in Wireless Communications (SPAWC), June, New York, NY, pp. 42–46.

[400] Mohammed, A.S., Hidayat, W. and Bossert, M. (2006) *Multidimensional 16-QAM constellation labeling of BI-STCM-ID with the Alamouti scheme*. Proceedings of the IEEE Wireless Communications and Networking Conference (WCNC), April, Las Vegas, NV.

[401] Battiti, R. and Tecchiolli, G. (1994) Reactive search: toward self-tuning heuristics. *Operations Research Society of America (ORSA) Journal on Computing*, **6**(2), 126–140.

[402] Tüchler, M. (2003) *Design of serially concatenated systems for long or short block lengths*. Proceedings of the IEEE International Conference on Communications (ICC), vol. 4, May, Anchorage Alaska, USA, pp. 2948–2952.

[403] Akhtman, J., Wolfgang, A., Chen, S. and Hanzo, L. (2007) Reduced-complexity near-optimum detection for MIMO-OFDM: Parts I and II. *IEEE Transactions on Wireless Communications*, **6**(5), 1900–1909.

[404] Maaref, A. and Aissa, S. (2005) Capacity of space-time block codes in MIMO Rayleigh fading channels with adaptive transmission and estimation errors. *IEEE Transactions on Wireless Communications*, **4**, 2568–2578.

[405] Wong, K.K., Paulraj, A. and Murch, R.D. (2007) Efficient high-performance decoding for overloaded MIMO antenna systems. *IEEE Transactions on Wireless Communications*, **6**, 1833–1843.

[406] Zhang, L., Lei, H.P., Zhang, X. and Yang, D.C. (2007) *Efficient complex sphere decoding framework for linear dispersion space-time block codes*. Proceedings of the IEEE International Symposium on Personal, Indoor and Mobile Radio Communications, September, Greece, pp. 1–4.

[407] Deng, D., Zhao, M. and Zhu, J.K. (2006) *Transmit antenna selection for linear dispersion codes based on linear receiver*. Proceedings of the Vehicular Technology Conference, vol. 6, May, Melbourne, pp. 2927–2931.

[408] Luby, M. (2002) *LT codes*. Proceedings of the IEEE Symposium on Foundations of Computer Science, November, Vancouver, Canada, pp. 271–280.

[409] Wei, C.Y., Nguyen, T., Wu, N., Akhtman, J., Yang, L.L. and Hanzo, L. (2007) *Luby transform coding aided iterative detection for downlink SDMA systems*. Proceedings of the IEEE Workshop on Signal Processing Systems, October, Shanghai, China, pp. 397–402.

[410] Wu, N., Nguyen, T.D., Wei, C.Y., Yang, L.L. and Hanzo, L. (2008) *Integrated LT coding, bit interleaved differential space time coding and sphere packing modulation for the wireless internet*. Proceedings of the Vehicular Technology Conference, May, Marina Bay, Singapore, pp. 344–348.

[411] Kiran, T. and Rajan, B.S. (2007) Partially-coherent distributed space-time codes with differential encoder and decoder. *IEEE Journal on Selected Areas in Communications*, **25**, 426–433.

[412] Wang, G.Y., Zhang, Y.M. and Amin, M. (2006) Differential distributed space-time modulation for cooperative networks. *IEEE Transactions on Wireless Communications*, **5**, 3097–3108.

[413] Mudumbai, R., Barriac, G. and Madhow, U. (2007) On the feasibility of distributed beamforming in wireless networks. *IEEE Transactions on Wireless Communications*, **6**, 1754–1763.

[414] El-Hajjar, M., Zummo, S. and Hanzo, L. (2007) *Near-instantaneously adaptive cooperative schemes based on space-time block codes and V-BLAST*. Proceedings of the IEEE Vehicular Technology Conference (VTC2007-Spring), April, pp. 2200–2204.

Index

Author Index

Near-Capacity Multi-Functional MIMO Systems L. Hanzo, O. R. Alamri, M. El-Hajjar and N. Wu

O

P

R

S

T

U

V

W

X

Y

Z